Numerical Mathematics an

M. Feistauer • V. Dolejší • P. Knobloch • K. Najzar
Editors

Numerical Mathematics and Advanced Applications

Proceedings of ENUMATH 2003
the 5th European Conference
on Numerical Mathematics and
Advanced Applications

Prague, August 2003

Springer

Editors

Miloslav Feistauer
Vít Dolejší
Petr Knobloch
Karel Najzar

Charles University Prague
Faculty of Mathematics and Physics
Department of Numerical Mathematics
Sokolovská 83, 186 75 Praha 8

Czech Republic

email:
feist, dolejsi, knobloch, knaj@karlin.mff.cuni.cz

Library of Congress Control Number: 2004107015
Mathematics Subject Classification (2000): 65-XX, 74-XX, 76-XX, 78-XX

ISBN 978-3-642-62288-5 ISBN 978-3-642-18775-9 (eBook)
DOI 10.1007/978-3-642-18775-9

Originally published by Springer-Verlag Berlin Heidelberg New York in 2004
Softcover reprint of the hardcover 1st edition 2004

Cover Design: *design & production GmbH*, Heidelberg
Typesetting: Computer to film by author's data
Printed on acid-free paper 40/3142XT 5 4 3 2 1 0

Preface

These proceedings collect the major part of the lectures given at ENUMATH2003, the European Conference on Numerical Mathematics and Advanced Applications, held in Prague, Czech Republic, from 18 August to 22 August, 2003.

The importance of numerical and computational mathematics and scientific computing is permanently growing. There is an increasing number of different research areas, where numerical simulation is necessary. Let us mention fluid dynamics, continuum mechanics, electromagnetism, phase transition, cosmology, medicine, economics, finance, etc. The success of applications of numerical methods is conditioned by changing its basic instruments and looking for new appropriate techniques adapted to new problems as well as new computer architectures.

The ENUMATH conferences were established in order to provide a forum for discussion of current topics of numerical mathematics. They seek to convene leading experts and young scientists with special emphasis on contributions from Europe. Recent results and new trends are discussed in the analysis of numerical algorithms as well as in their applications to challenging scientific and industrial problems.

The first ENUMATH conference was organized in Paris in 1995, then the series continued by the conferences in Heidelberg 1997, Jyväskylä 1999 and Ischia Porto 2001. It was a great pleasure and honour for the Czech numerical community that it was decided at Ischia Porto to organize the ENUMATH2003 in Prague. It was the first time when this conference crossed the former Iron Courtain and was organized in a postsocialist country.

The ENUMATH2003 was organized by the Faculty of Mathematics and Physics of the Charles University in cooperation with the Department of Mathematics of the Institute of Chemical Technology in Prague. The Charles University, the oldest university in the Middle Europe, was founded in 1348. In the middle ages, mathematics was studied in Prague at the Artistic Faculty, later at the Philosophical Faculty and in the 20th century it belonged to the Faculty of Natural Sciences till 1952, when the Faculty of Mathematics and Physics was founded. As follows from historical sources, already in the 15th century the students of the Charles University had the opportunity to be trained in "Computational Mathematics". Křišťan from Prachatice, who is in the Czech history well–known as a Church reformer (he was a friend of the famous reformer Jan Hus), or a personal doctor of medicine of the Czech and German-Roman King Wenceslas IV, was a professor of astronomy and mathematics at the Charles University. He wrote lecture notes with the title "Algoritmus Prosaicus", where he describes among other approximate methods for the realization of various mathematical operations. The contemporary Czech school of numerical and applied mathematics was born much later in the second part of the 20th century. It is connected mainly with the names of Ivo Babuška, who can be considered as a father of the Czech numerical mathematics, Karel Rektorys, Milan Práger, Emil Vitásek, Miloš Zlámal, Alexander Ženíšek, Ivan Hlaváček, a famous specialist in PDE's Jindřich Nečas (who influenced a number of Czech numerical analysts) and Jan Polášek (who founded the Prague school of CFD).

These proceedings contain a selection of invited plenary lectures, papers presented in minisymposia and works communicated within the sessions. All contributions of these proceedings have been reviewed by members of the Scientific Committee. At this occasion we want to thank the members of the Program Committee (F. Brezzi, M. Feistauer, R. Glowinski, R. Jeltsch, Yu. Kuznetsov, J. Periaux, R. Rannacher) and the members of the Scientific Committee (O. Axelsson, C. Bernardi, C. Canuto, M. Griebel, R. Hoppe, G. Kobelkov, M. Křížek, P. Neittaanmäki, O. Pironneau, A. Quarteroni, C. Schwab, E. Süli, W. Wendland) for their scientific support. We are grateful to the plenary speakers R. Blaheta, A. Bermúdez, T. Gallouët, J. Haslinger, R. Hiptmair, T. Hughes, J. Rappaz, A. Russo, A. Tveito and V. Schulz for coming to Prague and richly contributing to the success of the conference.

We are very much obliged to A. Klíč, the head of the Department of Mathematics of the Institute of Chemical Technology, and our colleagues J. Felcman, J. Segethová and E. Plandorová from the local organizing committee for the cooperation in the organization of the conference. Finally, we are gratefully indebted to O. Ulrych for the TEX editing and the preparation of the camera-ready manuscript of the proceedings. Lastly, we thank all participants for coming and animating the meeting.

We believe that this volume will be an invaluable instrument for obtaining an overview of the latest and newest results and aspects of numerical mathematics and scientific computing and their applications.

M. Feistauer
V. Dolejší
P. Knobloch
K. Najzar
editors

Table of Contents

Part I

Plenary Lectures

Numerical Analysis of Finite Element Methods for Eddy Current Problems. Applications to Electrode Simulation

Alfredo Bermúdez[1], Rodolfo Rodríguez[2] and Pilar Salgado[3]

[1] Departamento de Matemática Aplicada, Universidade de Santiago de Compostela, Spain, *mabermud@usc.es*
[2] GI^{2}MA, Departamento de Ingeniería Matemática, Universidad de Concepción, Chile, *rodolfo@ing-mat.udec.cl*
[3] Departamento de Matemática Aplicada, Universidade de Santiago de Compostela, Spain, *mpilar@usc.es*

Summary. The objective of this work is to introduce and numerically solve a 3D-mathematical model for steady thermoelectrical behavior of electrodes in a metallurgical electric furnace. The mathematical model couples the time-harmonic eddy current model with the heat transfer equations in a bounded 3D-domain. An important part of the paper deals with the analysis and numerical solution of the eddy current model in a bounded domain.

1 Introduction

Silicon is produced industrially by reduction of silicon dioxide with carbon by a reaction which can be written in a simple way as follows:

$$\mathrm{Si\ O_2 + 2C = Si + 2CO}.$$

This reaction takes place in submerged arc furnaces which use three-phase alternating current. A simple sketch of the furnace can be seen in Figure 1. It consists of a cylindrical pot containing charge materials and three electrodes disposed conforming an equilateral triangle.

Electrodes are the main components of reduction furnaces and their purpose is to conduct the electric current which enters the electrode through the "contact clamps" (see Figure 1). The electric current goes down crossing the column length comprised between the contact clamps and the lower end of the column generating heat by Joule effect. At the tip of the electrode an electric arc is produced, reaching temperatures of about 2500 °C which are needed for the reduction chemical reactions to take place.

Classical electrodes extensively used in industry include *pure graphite*, *pre-baked* and *Søderberg* electrodes. The latter are the most used in ferro-silicon industry and they are composed by paste consisting of a carbon aggregate and a tar binding which are fed into a steel casing; the casing have steel

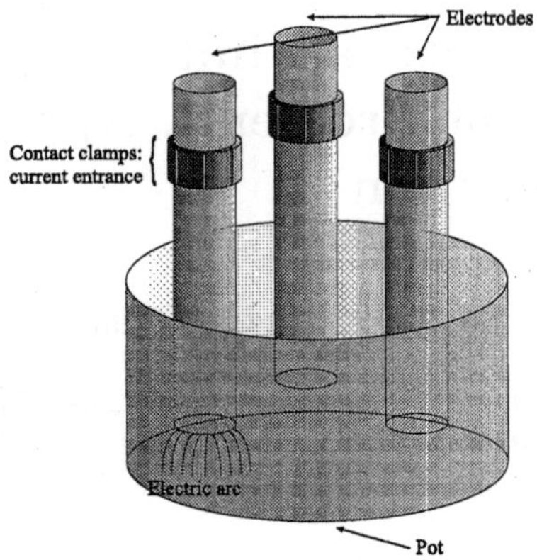

Fig. 1. A reduction furnace

fins attached to its inner part, which are placed radially in the cylinder. The great amount of heat generated by Joule effect is partially employed to bake the paste; this is a crucial process during which the initially soft/liquid non-conductive paste at the top of the electrode becomes a solid conductor. The advantages of Søderberg electrodes with respect to pure graphite or prebaked electrodes are that they are built in larger sizes and cost less. However, as the electrode is consumed, it has to be slipped and the steel casing moves with the carbon body so it melts and pollutes silicon. This is why they cannot be used to obtain *silicon metal* or silicon with metallurgical quality, which is used as alloying of other metals as aluminum. Thus prebaked electrodes have been for many years the only alternative for commercial silicon metal production.

In the early nineties, the Spanish company Ferroatlántica S.L. built a new compound electrode named ELSA ([14]) which serves for the production of silicon metal. It seems to be the solution for all silicon furnaces because its cost can be up to one third the price of a prebaked electrode.

ELSA electrode consists of a central column of baked carbonaceous material, graphite or similar, surrounded by a Søderberg-like paste (see Figure 2). There is a steel casing without fins that contains the paste until it is baked at the contact clamps zone. Two different slipping systems exist, one for the casing and another one for the central column; the combination of both systems is necessary so as to slip the casing as little as possible and also to carry out the correct extrusion of the carbon electrode. Then, unlike in the case of Søderberg electrodes, the casing is not consumed and it is possible to produce silicon with metallurgical quality. The result is that the furnace operation is similar to that of prebaked electrodes, but the compound electrode is less expensive. The disadvantage is that slipping velocity is not free as in prebaked electrodes, because the paste has to be baked before leaving the casing, so it is necessary a minimum period of time between slippages. Thus, baking of paste is a crucial point in the working of this type of electrodes.

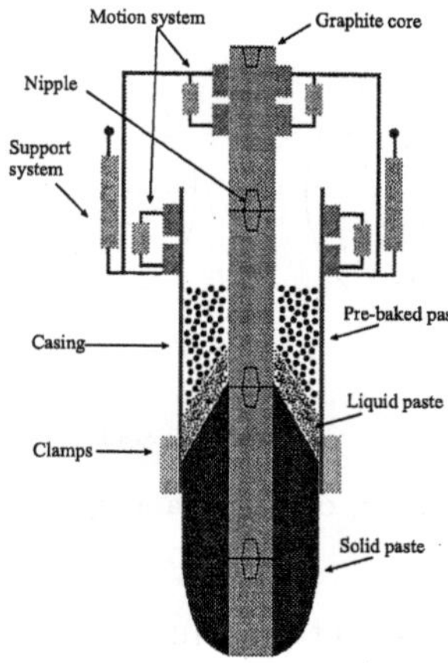

Fig. 2. Sketch of ELSA electrode

In general, the design and control parameters of electrodes are very complex and numerical simulation plays an important role at this point. Modeling the involved phenomena in a computer allows us to analyze the influence of changing a parameter without the need of expensive and difficult tests. Thus, during the last 20 years, an important number of mathematical models and computer programs have been developed in order to simulate the thermoelectrical behavior of classical electrodes (see for instance [15, 17, 18]). In particular, the mathematical models based on cylindrical symmetry have been the most extensively used. However, ELSA electrode works in a different manner from the classical electrodes. While classical electrode has only a constitutive material, compound electrode combines a good electric current conductor as graphite with a paste which becomes a good conductor only at high temperatures. Not only the core of graphite is important in the movement of the column but also in the distribution of current inside the electrode. Moreover, unlike Søderbeg electrodes, the non existence of fins gives a geometrical axisymmetry (see Figure 3).

This is why we first developed a finite element method based on cylindrical symmetry to compute the electric current and temperature distribution in a radial section of the electrode [5]. While the axisymmetric model has given valuable information on important electrode parameters, the assumption of cylindrical symmetry makes necessary to neglect the following facts:

- The electromagnetic effect caused on one electrode by the two others, that is the so called "proximity effect". This arises because the magnetic field generated by each electrode induces eddy currents in the two others.
- Thermal boundary conditions are not axisymmetric. Indeed, the temperature of the air around the electrode is greater on the surfaces oriented toward the furnace center.

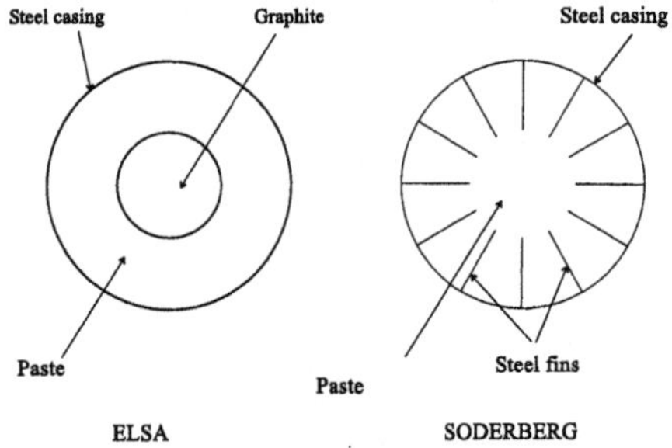

Fig. 3. Cross section of ELSA and Søderberg electrode

- The current entrance in the electrode through the contact clamps is not axisymmetric. The current is transferred to the contact clamps through copper bus tubes which in its turn are connected to three transformers with different phases. Then, in each electrode, half of the clamps receive current from one transformer while the other ones are connected to a second transformer.

These points can only be considered by using a pure three-dimensional model. Moreover, 3D-models are always needed to simulate Søderberg electrodes because the presence of fins breaks cylindrical symmetry (see Figure 3). Thus, we have developed a three dimensional thermoelectrical model which is enough general to model any kind of electrodes and even the complete furnace. In this paper, we describe two different mathematical models and analyze them from mathematical and numerical points of view.

The electromagnetic problem is obtained from the time-harmonic Maxwell equations assuming the frequency is low enough as to neglect the term involving the displacement current in Ampere's law. This is the so-called *eddy current* model. Because of many interesting applications in electrical engineering, numerical simulation of eddy current problems have led to a great number of publications in recent years (see for instance [1, 2, 3, 10, 11, 12, 13]). We notice that Maxwell equations concern the whole space, but we are interested in solving the problem in a bounded domain, so we have to define suitable boundary conditions and this need represents the main difficulty to study the problem in a bounded domain. Thus, we start introducing the eddy current problem in the whole furnace, including the electrodes and the air around, and defining natural and essential boundary conditions. In a second step we change this model, by introducing realistic boundary conditions, to compute the electromagnetic fields in only one electrode. Finally, we couple the electromagnetic model with a thermal one. Coupling between Maxwell and heat transfer equations is due to Joule effect which is the source term in the heat equation, and to the fact that thermoelectrical parameters depend on temperature.

The outline of the paper is as follows: In Section 2 we deal with the mathematical and numerical analysis of the electromagnetic problem in a bounded 3D domain which includes conductors and dielectrics. We introduce a weak formulation which involves the magnetic field in the conductor domain and a scalar magnetic potential in the dielectric one. This hybrid formulation is discretized by using Nédélec edge finite elements for the magnetic field and standard piecewise linear continuous elements for the magnetic potential. The resulting discrete problems are studied and error estimates are obtained under mild smoothness assumptions on the solution. Section 3 is devoted to propose and analyze a finite element method to solve the electromagnetic problem only in one electrode. We introduce a weak formulation of the problem in terms of the magnetic field and deal with boundary conditions directly related with the intensities which enter the domain. Lagrange multipliers are introduced to impose these "non standard" boundary conditions and the resulting mixed formulations are studied following classical techniques. In Section 4, we couple the electromagnetic problem with the thermal one and give a result concerning existence of solution. We end the paper by reporting, in Section 5, some numerical results obtained for ELSA and Søderberg electrodes.

2 The electromagnetic problem in the whole furnace

In order to consider all the facts which are neglected in the axisymmetric models, we start proposing a model to solve the eddy current problem in a bounded domain like the one presented in Figure 4, which includes not only *conductors* (the electrodes and wires supplying the electric current), but *dielectrics* as well (the air).

2.1 The eddy current problem

Eddy currents are usually modeled by the low-frequency harmonic Maxwell equations. Assuming alternating electric current of angular frequency ω, they are

$$\mathbf{curl}\,\mathbf{H} = \mathbf{J}, \tag{1}$$

$$\mathrm{i}\omega\mu\mathbf{H} + \mathbf{curl}\,\mathbf{E} = \mathbf{0}, \tag{2}$$

$$\operatorname{div}\mathbf{B} = 0, \tag{3}$$

$$\operatorname{div}\mathbf{D} = \rho, \tag{4}$$

with

$$\mathbf{B} = \mu\mathbf{H}, \quad \mathbf{D} = \epsilon\mathbf{E}, \quad \mathbf{J} = \sigma\mathbf{E}, \tag{5}$$

where $\mathbf{H}$, $\mathbf{J}$, $\mathbf{B}$, $\mathbf{E}$, and $\mathbf{D}$ are the complex amplitudes associated with the magnetic field, the current density, the magnetic induction, the electric field and the electric displacement, respectively; ρ is the electric charge density, μ is the magnetic permeability, ϵ is the electric permittivity and σ is the electric conductivity.

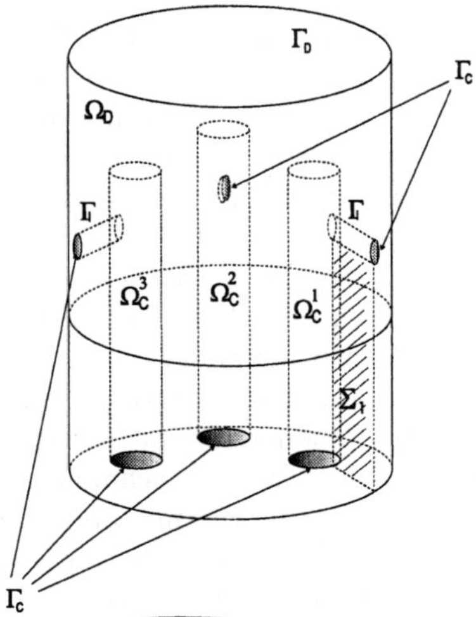

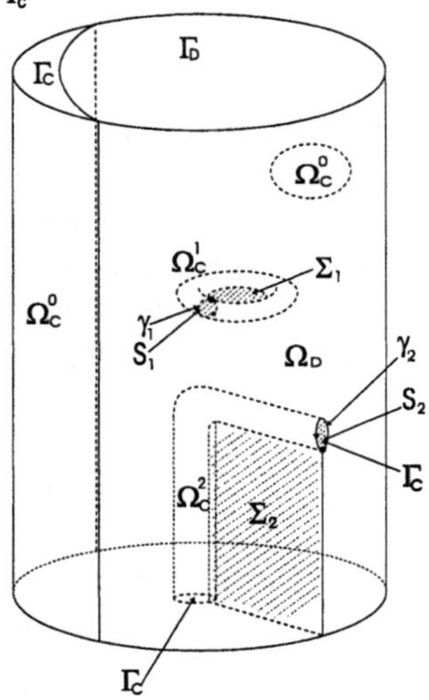

Fig. 4. Sketch of the furnace

Fig. 5. Sketch of a general domain

We will solve these equations in a bounded domain Ω, which consists of two parts, Ω_{C} and Ω_{D}, occupied by conductors and dielectrics, respectively (see Figure 5). The boundary of the domain Ω also splits into two parts: $\Gamma_{\mathrm{C}} := \partial\Omega_{\mathrm{C}} \cap \partial\Omega$ and $\Gamma_{\mathrm{D}} := \partial\Omega_{\mathrm{D}} \cap \partial\Omega$. Finally, we denote by $\Gamma_{\mathrm{I}} := \partial\Omega_{\mathrm{C}} \cap \partial\Omega_{\mathrm{D}}$, the interface between dielectric and conductors. The boundary conditions added to the eddy current model are

$$\mathbf{E} \times \mathbf{n} = \mathbf{g} \qquad \text{on } \Gamma_{\mathrm{C}}, \tag{6}$$

$$\mathbf{H} \times \mathbf{n} = \mathbf{f} \qquad \text{on } \Gamma_{\mathrm{D}}, \tag{7}$$

with $\mathbf{g}$ and $\mathbf{f}$ being given tangential vector fields (i.e., satisfying $\mathbf{g} \cdot \mathbf{n} = 0$ on Γ_C and $\mathbf{f} \cdot \mathbf{n} = 0$ on Γ_D) and $\mathbf{n}$ an outward unit normal vector to $\partial\Omega$.

We remark that (6) is the natural condition for the conducting part of the boundary, while (7) is imposed on the dielectric part and allows taking into account all of the electromagnetic effects outside the domain.

We will introduce and analyze a finite element method to solve this problem in domains of general topology. To attain this goal, we will consider a formulation introduced by Bossavit and Vérité [13], which involves the magnetic field in the conductor domain and a scalar magnetic potential in the dielectric one. Then, as a first step, we start analyzing a weak formulation of the problem in terms of the magnetic field.

2.2 A magnetic field formulation of the eddy current problem

Let us assume that Ω is simply connected, with a Lipschitz-continuous connected boundary. The subdomains Ω_C and Ω_D are also assumed to have Lipschitz-continuous boundaries, although not necessarily connected. Finally, the boundaries of Γ_C, Γ_D, and Γ_I are also assumed to have Lipschitz-continuous boundaries.

Let us consider the following closed subspaces of $\mathrm{H}(\mathbf{curl}, \Omega)$,

$$\mathcal{V} = \{\mathbf{G} \in \mathrm{H}(\mathbf{curl}, \Omega) : \ \mathbf{curl}\, \mathbf{G} = \mathbf{0} \text{ in } \Omega_D\},$$
$$\mathcal{V}^0 = \left\{\mathbf{G} \in \mathcal{V} : \ \mathbf{G} \times \mathbf{n} = \mathbf{0} \text{ in } \mathrm{H}_{00}^{-1/2}(\Gamma_D)^3\right\},$$

where $\mathrm{H}_{00}^{-1/2}(\Gamma_D)^3$ denotes the dual space of $\mathrm{H}_{00}^{1/2}(\Gamma_D)^3$ which, in its turn, is the space of functions defined on Γ_D that extended by $\mathbf{0}$ on $\partial\Omega \setminus \Gamma_D$ belong to $\mathrm{H}^{1/2}(\partial\Omega)^3$. We assume that $\mu, \epsilon, \sigma \in \mathrm{L}^\infty(\Omega)$, and that there exist constants, $\underline{\mu}$, $\underline{\epsilon}$, and $\underline{\sigma}$, such that

$$\mu(\mathbf{x}) \geq \underline{\mu} > 0, \qquad \epsilon(\mathbf{x}) \geq \underline{\epsilon} > 0, \quad \text{a.e. in } \Omega,$$
$$\sigma(\mathbf{x}) \geq \underline{\sigma} > 0, \quad \text{a.e. in } \Omega_C, \qquad \sigma(\mathbf{x}) = 0 \quad \text{in } \Omega_D.$$

We suppose that the boundary data $\mathbf{g}$ satisfies $\mathbf{g} \times \mathbf{n} \in \mathrm{H}_{00}^{1/2}(\Gamma_C)^3$. On the other hand, concerning the boundary data $\mathbf{f}$, we suppose there exists a field $\mathbf{H_f} \in \mathcal{V}$ such that

$$\mathbf{H_f} \times \mathbf{n} = \mathbf{f} \text{ in } \mathrm{H}_{00}^{-1/2}(\Gamma_D)^3.$$

Then, multiplying the equation (2) by a test function of the space $\mathcal{V}^0$, integrating in Ω, and using Green's formula, (1), (6), and (7), we obtain the following weak formulation in terms of the magnetic field $\mathbf{H}$.

Problem MP.- *To find* $\mathbf{H} \in \mathcal{V}$ *such that*

$$\mathbf{H} \times \mathbf{n} = \mathbf{f} \textit{ in } \mathrm{H}_{00}^{-1/2}(\Gamma_D)^3, \tag{8}$$

$$\mathrm{i}\omega \int_\Omega \mu \mathbf{H} \cdot \bar{\mathbf{G}} + \int_{\Omega_C} \frac{1}{\sigma}\, \mathbf{curl}\, \mathbf{H} \cdot \mathbf{curl}\, \bar{\mathbf{G}} = \langle \mathbf{g} \times \mathbf{n}, \mathbf{G} \times \mathbf{n} \rangle_{\Gamma_C} \quad \forall \mathbf{G} \in \mathcal{V}^0. \tag{9}$$

Theorem 1. *If there exists* $\mathbf{H_f} \in \boldsymbol{\mathcal{V}}$ *such that* $\mathbf{H_f} \times \mathbf{n} = \mathbf{f}$ *in* $\mathrm{H}_{00}^{-1/2}(\Gamma_{\mathrm{D}})^3$, *then problem* **MP** *has a unique solution.*

Once the magnetic field $\mathbf{H}$ is known, the current density $\mathbf{J}$ and the electric field $\mathbf{E}$ can be computed in conductors, namely, $\mathbf{J} = \mathbf{curl}\,\mathbf{H}$ and $\mathbf{E} = (\frac{1}{\sigma}\mathbf{J})|_{\Omega_{\mathrm{C}}}$. These are the magnitudes actually needed in most applications and satisfy the Maxwell equations (1)–(5) and boundary conditions (6)–(7) (see Theorem 3.2 in [7]).

2.3 Introducing a magnetic potential

In this section we show how problem **MP** can be transformed by replacing the magnetic field in the dielectric domain Ω_{D} by a (scalar) magnetic potential.

Let $\Omega_{\mathrm{C}} = \bigcup_{j=0}^{j=J} \Omega_{\mathrm{C}}^j$, with Ω_{C}^0 being the union of all the connected components of Ω_{C} such that $\Omega \setminus \Omega_{\mathrm{C}}^0$ is simply connected, and Ω_{C}^j, $j = 1, \ldots, J$, the remaining connected components of Ω_{C} (see Figure 5).

We assume that for each Ω_{C}^j, $j = 1, \ldots, J$, there exists an open "cut" surface $\Sigma_j \subset \Omega_{\mathrm{D}}$ such that $\partial\Sigma_j \subset \partial\Omega_{\mathrm{D}}$ and $\widetilde{\Omega}_{\mathrm{D}} := \Omega_{\mathrm{D}} \setminus \bigcup_{j=0}^{j=J} \overline{\Sigma}_j$ is pseudo-Lipschitz and simply connected (see Figure 5). We also assume that each of these surfaces Σ_j is connected, and $\overline{\Sigma}_j \cap \overline{\Sigma}_k = \emptyset$ for $j \neq k$ (see, for instance, [4]).

For any function $\widetilde{\Psi} \in \mathrm{H}^1(\widetilde{\Omega}_{\mathrm{D}})$, we denote by $[\![\widetilde{\Psi}]\!]_{\Sigma_j}$ the jump of $\widetilde{\Psi}$ through Σ_j. The gradient of $\widetilde{\Psi}$ in $\mathcal{D}'(\widetilde{\Omega}_{\mathrm{D}})$ can be extended to $\mathrm{L}^2(\Omega_{\mathrm{D}})^3$ and will be denoted by $\widetilde{\mathbf{grad}}\,\widetilde{\Psi}$.

Let Θ be the linear space of $\mathrm{H}^1(\widetilde{\Omega}_{\mathrm{D}})$ defined by

$$\Theta = \left\{\widetilde{\Psi} \in \mathrm{H}^1(\widetilde{\Omega}_{\mathrm{D}}) : \ [\![\widetilde{\Psi}]\!]_{\Sigma_j} = \text{constant}, \ j = 1, \ldots, J\right\}.$$

Then, for $\widetilde{\Psi} \in \mathrm{H}^1(\widetilde{\Omega}_{\mathrm{D}})$, we have that $\widetilde{\mathbf{grad}}\,\widetilde{\Psi} \in \mathrm{H}(\mathbf{curl}, \Omega_{\mathrm{D}})$ if and only if $\widetilde{\Psi} \in \Theta$, in which case $\mathbf{curl}\,(\widetilde{\mathbf{grad}}\,\widetilde{\Psi}) = \mathbf{0}$ (see Lemma 3.11 in [4]). Then, for all $\mathbf{G} \in \boldsymbol{\mathcal{V}}$ there exist $\widetilde{\Psi} \in \Theta$ such that $\mathbf{G}|_{\Omega_{\mathrm{D}}} = \widetilde{\mathbf{grad}}\,\widetilde{\Psi}$.

We introduce the following notation: for $\mathbf{G}_{\mathrm{C}} \in \mathrm{L}^2(\Omega_{\mathrm{C}})^3$ and $\mathbf{G}_{\mathrm{D}} \in \mathrm{L}^2(\Omega_{\mathrm{D}})^3$, we denote by $(\mathbf{G}_{\mathrm{C}}|\mathbf{G}_{\mathrm{D}})$ the field $\mathbf{G} \in \mathrm{L}^2(\Omega)^3$ defined a.e. by

$$\mathbf{G}(\mathbf{x}) := \begin{cases} \mathbf{G}_{\mathrm{C}}(\mathbf{x}) & \text{if } \mathbf{x} \in \Omega_{\mathrm{C}}, \\ \mathbf{G}_{\mathrm{D}}(\mathbf{x}) & \text{if } \mathbf{x} \in \Omega_{\mathrm{D}}. \end{cases}$$

Let us denote by $\boldsymbol{\mathcal{W}}$ the linear space given by

$$\boldsymbol{\mathcal{W}} := \left\{(\mathbf{G}, \widetilde{\Psi}) \in \mathrm{H}(\mathbf{curl}, \Omega_{\mathrm{C}}) \times (\Theta/\mathbb{C}) : \ (\mathbf{G}|\,\widetilde{\mathbf{grad}}\,\widetilde{\Psi}) \in \mathrm{H}(\mathbf{curl}, \Omega)\right\}.$$

Similarly, we define the closed subspace of $\boldsymbol{\mathcal{W}}$

$$\boldsymbol{\mathcal{W}}^0 := \left\{(\mathbf{G}, \widetilde{\Psi}) \in \boldsymbol{\mathcal{W}} : \ \widetilde{\mathbf{grad}}\,\widetilde{\Psi} \times \mathbf{n} = \mathbf{0} \text{ in } \mathrm{H}_{00}^{-1/2}(\Gamma_{\mathrm{D}})^3\right\}.$$

By using this notation we can define the following problem:
Problem HP.- *To find* $(\mathbf{H}, \widetilde{\Phi}) \in \mathcal{W}$ *such that*

$$\widetilde{\mathbf{grad}}\, \widetilde{\Phi} \times \mathbf{n} = \mathbf{f} \ \textit{in} \ \mathrm{H}_{00}^{-1/2}(\Gamma_{\mathrm{D}})^3,$$

$$\mathrm{i}\omega \int_{\Omega_{\mathrm{C}}} \mu \mathbf{H} \cdot \bar{\mathbf{G}} + \int_{\Omega_{\mathrm{C}}} \frac{1}{\sigma} \mathbf{curl}\, \mathbf{H} \cdot \mathbf{curl}\, \bar{\mathbf{G}} + \mathrm{i}\omega \int_{\Omega_{\mathrm{D}}} \mu\, \widetilde{\mathbf{grad}}\, \widetilde{\Phi} \cdot \widetilde{\mathbf{grad}}\, \bar{\widetilde{\Psi}} =$$
$$= \langle \mathbf{g} \times \mathbf{n}, \mathbf{G} \times \mathbf{n} \rangle_{\Gamma_{\mathrm{C}}} \qquad \forall (\mathbf{G}, \widetilde{\Psi}) \in \mathcal{W}^0.$$

This is the well known magnetic field/magnetic potential hybrid formulation of the eddy current problem introduced by Bossavit and Vérité [13]. The main advantage with respect to formulation (8)–(9) lies in the fact that a vector field is replaced by a scalar one in the dielectric domain.

Theorem 2. *Under the assumptions of Theorem 1, problem* **HP** *has a unique solution* $(\mathbf{H}, \widetilde{\Phi})$, *with* $(\mathbf{H} | \widetilde{\mathbf{grad}}\, \widetilde{\Phi})$ *being the unique solution of problem* **MP**.

2.4 Numerical solution

In this section we first introduce a discretization of problem **MP** and then we obtain a discrete version of problem **HP** equivalent to the previous one.

Let us assume Ω, Ω_{C}, and Ω_{D} are Lipschitz polyhedra, and consider a family of regular tetrahedral meshes $\{\mathcal{T}_h\}$ of Ω such that, for every mesh $\mathcal{T}_h$, each element $K \in \mathcal{T}_h$ is contained either in $\overline{\Omega}_{\mathrm{C}}$ or in $\overline{\Omega}_{\mathrm{D}}$ (h stands as usual for the corresponding mesh-size).

The magnetic field is discretized by using Nédélec edge finite elements (see [19]). In particular, $\mathbf{H}$ is approximated in each tetrahedron K by a polynomial vector field in the space

$$\mathcal{N}(K) := \left\{ \mathbf{G}_h \in \mathcal{P}_1(K)^3 : \ \mathbf{G}_h(\mathbf{x}) = \mathbf{a} \times \mathbf{x} + \mathbf{b}, \ \mathbf{a}, \mathbf{b} \in \mathbb{C}^3, \ \mathbf{x} \in K \right\}.$$

Then, fields in $\mathrm{H}(\mathbf{curl}, \Omega)$ will be approximated in the following finite dimensional space:

$$\mathcal{N}_h(\Omega) := \{ \mathbf{G}_h \in \mathrm{H}(\mathbf{curl}, \Omega) : \ \mathbf{G}_h|_K \in \mathcal{N}(K) \ \forall K \in \mathcal{T}_h \}.$$

In order to use these elements to discretize problem **MP**, we have to use an approximation $\mathbf{f}_{\mathrm{I}}$ of the boundary data $\mathbf{f}$ such that a discrete version of equation (8) can hold true. To attain this goal, we will use the two-dimensional Nédélec interpolant of $\mathbf{n} \times \mathbf{f}$ on the triangular mesh induced by $\mathcal{T}_h$ on the polyhedral surface Γ_{D}. This interpolant and several of its properties are described in detail in [7].

Then, in order to discretize problem **MP**, we introduce the following finite-dimensional spaces,

$$\mathcal{V}_h := \{\mathbf{G}_h \in \mathcal{N}_h(\Omega) : \ \mathbf{curl}\,\mathbf{G}_h = \mathbf{0} \ \text{ in } \Omega_{\mathrm{D}}\},$$
$$\mathcal{V}_h^0 := \{\mathbf{G}_h \in \mathcal{V}_h : \ \mathbf{G}_h \times \mathbf{n} = \mathbf{0} \ \text{ on } \Gamma_{\mathrm{D}}\},$$

and obtain the following discrete magnetic problem:

Problem DMP.- *Find* $\mathbf{H}_h \in \mathcal{V}_h$ *such that*

$$\mathbf{H}_h \times \mathbf{n} = \mathbf{f}_{\mathrm{I}} \quad \text{on } \Gamma_{\mathrm{D}},$$
$$\mathrm{i}\omega \int_{\Omega} \mu \mathbf{H}_h \cdot \bar{\mathbf{G}}_h + \int_{\Omega_{\mathrm{C}}} \frac{1}{\sigma} \mathbf{curl}\,\mathbf{H}_h \cdot \mathbf{curl}\,\bar{\mathbf{G}}_h = \int_{\Gamma_{\mathrm{C}}} \mathbf{g} \times \mathbf{n} \cdot \bar{\mathbf{G}}_h \times \mathbf{n} \qquad \forall \mathbf{G}_h \in \mathcal{V}_h^0.$$

Theorem 3. *Let us assume that the solution* $\mathbf{H}$ *of problem* **MP** *satisfies* $\mathbf{H}|_{\Omega_{\mathrm{C}}} \in \mathrm{H}^r(\mathbf{curl}, \Omega_{\mathrm{C}})$ *and* $\mathbf{H}|_{\Omega_{\mathrm{D}}} \in \mathrm{H}^r(\Omega_{\mathrm{D}})^3$, *with* $r \in (\frac{1}{2}, 1]$. *Then,* $\mathbf{f}_{\mathrm{I}}$ *is well defined by the 2D Nédélec interpolant of* $\mathbf{n} \times \mathbf{f}$, *problem* **DMP** *has a unique solution* $\mathbf{H}_h$ *and the following error estimate holds:*

$$\|\mathbf{H} - \mathbf{H}_h\|_{\mathrm{H}(\mathbf{curl},\Omega)} \le C h^r \left[\|\mathbf{H}\|_{\mathrm{H}^r(\mathbf{curl},\Omega_{\mathrm{C}})} + \|\mathbf{H}\|_{\mathrm{H}^r(\Omega_{\mathrm{D}})^3} \right].$$

However, problem **DMP** is actually just a "theoretical" method in that its solution requires to impose somehow the curl-free condition in the definition of $\mathcal{V}_h$ to trial and test functions. Then, we will handle this curl-free condition by introducing a discrete multiple-valued magnetic potential in the dielectric domain.

We assume that the cut surfaces Σ_j are polyhedral and that the meshes are compatible with them, in the sense that each $\overline{\Sigma}_j$ is union of faces of tetrahedra $K \in \mathcal{T}_h$, for each mesh $\mathcal{T}_h$. Therefore, $\mathcal{T}_h^{\Omega_{\mathrm{D}}} := \{K \in \mathcal{T}_h : \ K \subset \overline{\Omega}_{\mathrm{D}}\}$ can also be seen as a mesh of $\widetilde{\Omega}_{\mathrm{D}}$.

In order to introduce an approximation of the space Θ, let us denote

$$\mathcal{L}_h(\widetilde{\Omega}_{\mathrm{D}}) := \left\{ \widetilde{\Psi}_h \in \mathrm{H}^1(\widetilde{\Omega}_{\mathrm{D}}) : \ \widetilde{\Psi}_h|_K \in \mathcal{P}_1(K) \ \forall K \in \mathcal{T}_h^{\Omega_{\mathrm{D}}} \right\}.$$

Then, we consider the family of finite dimensional subspaces of Θ given by

$$\Theta_h := \{\widetilde{\Psi}_h \in \mathcal{L}_h(\widetilde{\Omega}_{\mathrm{D}}) : \ [\![\widetilde{\Psi}_h]\!]_{\Sigma_j} = \text{constant}, \ j = 1, \dots, J\}.$$

The following lemma shows that each curl-free vector field in $\mathcal{N}_h(\Omega_{\mathrm{D}})$ admits a multiple-valued potential in Θ_h (see [7]).

Lemma 1. *Let* $\mathbf{G}_h \in \mathrm{L}^2(\Omega_{\mathrm{D}})^3$. *Then* $\mathbf{G}_h \in \mathcal{N}_h(\Omega_{\mathrm{D}})$ *with* $\mathbf{curl}\,\mathbf{G}_h = \mathbf{0}$ *in* Ω_{D} *if and only if there exists* $\widetilde{\Psi}_h \in \Theta_h$ *such that* $\mathbf{G}_h = \widetilde{\mathbf{grad}}\,\widetilde{\Psi}_h$ *in* Ω_{D}. *Such* $\widetilde{\Psi}_h$ *is unique up to an additive constant.*

Let us introduce the following families of finite-dimensional approximations of $\mathcal{W}$ and $\mathcal{W}^0$, respectively:

$$\mathcal{W}_h := \left\{ (\mathbf{G}_h, \widetilde{\Psi}_h) \in \mathcal{N}_h(\Omega_{\mathrm{C}}) \times (\Theta_h/\mathbb{C}) : \ (\mathbf{G}_h | \, \widetilde{\mathbf{grad}} \, \widetilde{\Psi}_h) \in \mathrm{H}(\mathbf{curl}, \Omega) \right\},$$
$$\mathcal{W}_h^0 := \left\{ (\mathbf{G}_h, \widetilde{\Psi}_h) \in \mathcal{W}_h : \ \widetilde{\mathbf{grad}} \, \widetilde{\Psi}_h \times \mathbf{n} = \mathbf{0} \text{ on } \Gamma_{\mathrm{D}} \right\}.$$

Thus, we define the following discrete problem which is equivalent to problem **DMP**:

Problem DHP.- *To find* $(\mathbf{H}_h, \widetilde{\Phi}_h) \in \mathcal{W}_h$ *such that*

$$\widetilde{\mathbf{grad}} \, \widetilde{\Phi}_h \times \mathbf{n} = \mathbf{f}_{\mathrm{I}} \quad \text{on } \Gamma_{\mathrm{D}},$$
$$\mathrm{i}\omega \int_{\Omega_{\mathrm{C}}} \mu \mathbf{H}_h \cdot \bar{\mathbf{G}}_h + \int_{\Omega_{\mathrm{C}}} \frac{1}{\sigma} \mathbf{curl}\, \mathbf{H}_h \cdot \mathbf{curl}\, \bar{\mathbf{G}}_h + \mathrm{i}\omega \int_{\Omega_{\mathrm{D}}} \mu \, \widetilde{\mathbf{grad}} \, \widetilde{\Phi}_h \cdot \widetilde{\mathbf{grad}} \, \bar{\widetilde{\Psi}}_h$$
$$= \int_{\Gamma_{\mathrm{C}}} \mathbf{g} \times \mathbf{n} \cdot \bar{\mathbf{G}}_h \times \mathbf{n} \qquad \forall (\mathbf{G}_h, \widetilde{\Psi}_h) \in \mathcal{W}_h^0.$$

Theorem 4. *Let us assume that the solution* $(\mathbf{H}, \widetilde{\Phi})$ *of problem* **HP** *satisfies* $\mathbf{H} \in \mathrm{H}^r(\mathbf{curl}, \Omega_{\mathrm{C}})$ *and* $\widetilde{\mathbf{grad}} \, \widetilde{\Phi} \in \mathrm{H}^r(\Omega_{\mathrm{D}})^3$*, with* $r \in (\frac{1}{2}, 1]$*. Then, problem* **DHP** *is well posed, it has a unique solution* $(\mathbf{H}_h, \widetilde{\Phi}_h)$*, and*

$$\|\mathbf{H} - \mathbf{H}_h\|_{\mathrm{H}(\mathbf{curl}, \Omega_{\mathrm{C}})} + \| \widetilde{\mathbf{grad}} \, \widetilde{\Phi} - \widetilde{\mathbf{grad}} \, \widetilde{\Phi}_h \|_{\mathrm{L}^2(\Omega_{\mathrm{D}})^3}$$
$$\leq C h^r \left[\|\mathbf{H}\|_{\mathrm{H}^r(\mathbf{curl}, \Omega_{\mathrm{C}})} + \| \widetilde{\mathbf{grad}} \, \widetilde{\Phi} \|_{\mathrm{H}^r(\Omega_{\mathrm{D}})^3} \right].$$

Effective procedures to solve numerically the problem **DMP** are described in [7]. In particular, numerical techniques to impose the following constraints are studied:

1. $(\mathbf{G}_h | \, \widetilde{\mathbf{grad}} \, \widetilde{\Psi}_h) \in \mathrm{H}(\mathbf{curl}, \Omega)$, which arise in the definition of $\mathcal{W}_h$.
2. $[\![\widetilde{\Psi}_h]\!]_{\Sigma_j} =$ constant, which arise in the definition of Θ_h.
3. The boundary condition $\widetilde{\mathbf{grad}} \, \widetilde{\Phi}_h \times \mathbf{n} = \mathbf{f}_{\mathrm{I}}$ on Γ_{D}.

The first constraint is imposed by eliminating the degrees of freedom of $\mathbf{G}_h$ associated with the edges $\ell \in \Gamma_I$ in terms of those of $\widetilde{\Phi}_h$ corresponding to the vertices of the mesh on this interface.

The second constraint is handled by distinguishing the degrees of freedom of $\widetilde{\Psi}_h$ on one side of the surface Σ_j from those on the other side, and by eliminating ones of them in terms of the others and the current intensities through each conductor Ω_{C}^j.

The third constraint is imposed by means of a Lagrange multiplier, increasing in this way the number of unknowns but with the advantage that the computer implementation is quite straightforward.

We have developed a MATLAB code which implements the method described above. To validate the computer code and to test the performance and convergence properties of the method, we have solved a problem with known analytical solution (see [7] for further details).

3 The electromagnetic problem in one electrode

3.1 Statement of the problem

The model described in the previous section presents some drawbacks. First, it is highly complex and its numerical solution takes a lot of time. On the other hand, it is difficult to obtain the boundary data $\mathbf{f}$ from realistic data such as intensities or potentials, which usually are the only data we know. Then, we are going to propose an alternative approach which consists in solving the eddy current problem in one electrode which is a particular bounded conducting domain. We are going to analyze a weak formulation of this problem in terms of the magnetic field, considering realistic boundary conditions from the point of view of applications. In particular, following Bossavit [12], we will consider boundary conditions directly related with the input current intensities which enter the electrode. We will impose these boundary conditions by means of Lagrange multipliers and study the resulting mixed formulations.

Since we only consider the conducting domain, we will get an important saving in computer time when compared with the model of the whole furnace, and we will still be able to consider some important effects which are not taken into account by the axisymmetric models, although not the proximity effect.

We consider a bounded conducting domain Ω having a Lipschitz-continuous and connected boundary. However, it is not necessary that Ω be simply connected. Let $\partial\Omega$ be the boundary of the domain Ω which splits into two parts: $\partial\Omega = \bar{\Gamma}_{\mathrm{E}} \cup \bar{\Gamma}_{\mathrm{J}}$. The surface Γ_{E} corresponds to the tip of the electrode where the electric arc arises. In its turn, the rest of the electrode boundary splits as follows:

$$\bar{\Gamma}_{\mathrm{J}} = \bar{\Gamma}_{\mathrm{J}}^0 \cup \bar{\Gamma}_{\mathrm{J}}^1 \cup \cdots \cup \bar{\Gamma}_{\mathrm{J}}^N,$$

where Γ_{J}^n, $n = 1, \ldots, N$, are the parts of the boundary connected to the wires supplying electric current to the electrode, and $\Gamma_{\mathrm{J}}^0 = \Gamma_{\mathrm{J}} \setminus (\bar{\Gamma}_{\mathrm{J}}^1 \cup \cdots \cup \bar{\Gamma}_{\mathrm{J}}^N)$ is the remaining part (see Figure 6). We also assume $\bar{\Gamma}_{\mathrm{J}}^n \cap \bar{\Gamma}_{\mathrm{E}} = \emptyset$ and $\bar{\Gamma}_{\mathrm{J}}^n \cap \bar{\Gamma}_{\mathrm{J}}^m = \emptyset$, $m, n = 1, \ldots, N$, $m \neq n$.

Our goal is to solve the eddy current equations (1)–(5) subject to the following boundary conditions:

$$\mathbf{E} \times \mathbf{n} = \mathbf{0} \quad \text{on } \Gamma_{\mathrm{E}}, \tag{10}$$

$$\int_{\Gamma_{\mathrm{J}}^n} \mathbf{curl}\,\mathbf{H} \cdot \mathbf{n} = I_n, \quad n = 1, \ldots, N, \tag{11}$$

$$\mathbf{E} \times \mathbf{n} = \mathbf{0} \quad \text{on } \Gamma_{\mathrm{J}}^n, \quad n = 1, \ldots, N, \tag{12}$$

$$\mathbf{curl}\,\mathbf{H} \cdot \mathbf{n} = 0 \quad \text{on } \Gamma_{\mathrm{J}}^0, \tag{13}$$

$$\mu \mathbf{H} \cdot \mathbf{n} = 0 \quad \text{on } \partial\Omega, \tag{14}$$

where the only data I_n, $n = 1, \ldots, N$, are the current intensities through each wire.

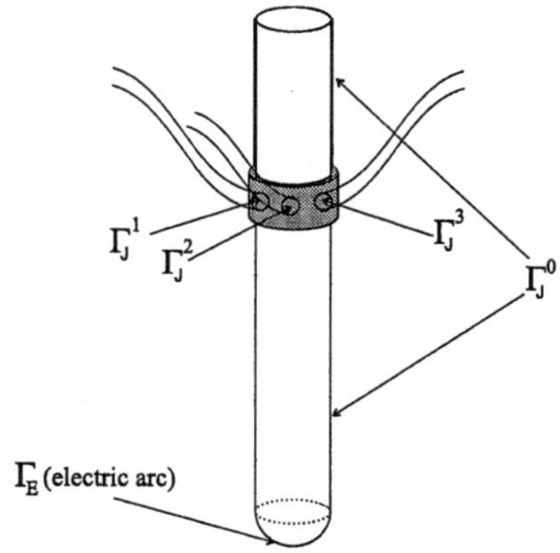

Fig. 6. Example of domain

Condition (10) is the natural one to model the free current exit on the electrode tip. Conditions (11) and (13) take into account the input intensities and the fact that there is no current flow through Γ_{J}^0, respectively. Conditions (12) and (14) have been proposed by Bossavit [12] in a more general setting. They will appear as natural boundary conditions of the weak formulation of our problem. The former implies the assumption that the electric current is normal to the surface on the current entrance, whereas the latter means that the magnetic field is tangential to the conductor surface. Of course, condition (14) is not always fulfilled, but it is a good approximation in our model problem.

Next, we analyze a weak formulation of this problem in terms of the magnetic field and propose a finite element method for its numerical solution.

3.2 Analysis of the weak formulation of the problem

To obtain a weak formulation of the eddy current problem (1)–(5) with boundary conditions (10)–(14) in terms of the magnetic field, we notice that the boundary condition (14) implies that the tangential component of $\mathbf{E}$ on the boundary of Ω is a gradient. In particular, we obtain that $\mathbf{E} \times \mathbf{n} = -\nabla\phi \times \mathbf{n}$ on $\partial\Omega$ for some scalar function ϕ with $\phi = 0$ on Γ_{E}, because of (10). Moreover, because of (12), $\phi|_{\Gamma_{\mathrm{J}}^n}$ must be constant. Then, multiplying the equation (2) by a test function $\mathbf{G}$ such that $\mathbf{curl}\,\mathbf{G} \cdot \mathbf{n} = 0$ on Γ_{J}^0 and $\int_{\Gamma_{\mathrm{J}}^n} \mathbf{curl}\,\mathbf{G} \cdot \mathbf{n} = 0$, $n = 1, \dots, N$, using Green's formula, and taking into account that $\mathbf{E} = \frac{1}{\sigma}\,\mathbf{curl}\,\mathbf{H}$, we obtain

$$\mathrm{i}\omega \int_\Omega \mu\mathbf{H} \cdot \bar{\mathbf{G}} + \int_\Omega \frac{1}{\sigma}\,\mathbf{curl}\,\mathbf{H} \cdot \mathbf{curl}\,\bar{\mathbf{G}} = 0.$$

Let $\boldsymbol{\mathcal{X}} := \mathrm{H}(\mathbf{curl}, \Omega)$ and $a\colon \boldsymbol{\mathcal{X}} \times \boldsymbol{\mathcal{X}} \longrightarrow \mathbb{C}$ be the sesquilinear continuous and elliptic form defined by

$$a(\mathbf{H},\mathbf{G}) := \mathrm{i}\omega \int_{\Omega} \mu \mathbf{H}\cdot\bar{\mathbf{G}} + \int_{\Omega} \frac{1}{\sigma}\,\mathbf{curl}\,\mathbf{H}\cdot\,\mathbf{curl}\,\bar{\mathbf{G}}.$$

Let $\mathcal{L}$ be the following closed subspace of $\mathrm{H}^{1/2}_{00}(\Gamma_{\mathrm{J}})$:

$$\mathcal{L} := \left\{\nu \in \mathrm{H}^{1/2}_{00}(\Gamma_{\mathrm{J}}) : \ \nu|_{\Gamma_{\mathrm{J}}^n} = \text{constant},\ n = 1,\dots,N\right\}.$$

Given $\mathbf{I} = (I_1,\dots,I_N) \in \mathbb{C}^N$, let us consider the closed linear manifold of $\mathcal{X}$,

$$\mathcal{W}(\mathbf{I}) := \left\{\mathbf{G} \in \mathcal{X} : \ \langle\,\mathbf{curl}\,\mathbf{G}\cdot\mathbf{n},\nu\rangle_{\Gamma_{\mathrm{J}}} = \sum_{n=1}^{N}\int_{\Gamma_{\mathrm{J}}^n} I_n\,\bar{\nu} \quad \forall\nu\in\mathcal{L}\right\},$$

and its associated subspace

$$\mathcal{W}(\mathbf{0}) = \left\{\mathbf{G} \in \mathcal{X} : \ \langle\,\mathbf{curl}\,\mathbf{G}\cdot\mathbf{n},\nu\rangle_{\Gamma_{\mathrm{J}}} = 0 \quad \forall\nu\in\mathcal{L}\right\}.$$

We introduce the following problem:
Problem PI.- *For any* $\mathbf{I}\in\mathbb{C}^N$, *find* $\mathbf{H}\in\mathcal{W}(\mathbf{I})$ *such that*

$$a(\mathbf{H},\mathbf{G}) = 0 \qquad \forall\mathbf{G}\in\mathcal{W}(\mathbf{0}).$$

Theorem 5. *Given* $\mathbf{I}\in\mathbb{C}^N$, *problem* **PI** *has a unique solution* $\mathbf{H}$.

To avoid dealing with functions that satisfy the constraints involved in $\mathcal{W}(\mathbf{I})$ and $\mathcal{W}(\mathbf{0})$, we consider a mixed formulation of the problem. It consists in handling the boundary conditions (11) and (13) in a weak sense by introducing a Lagrange multiplier defined on Γ_{J}.

Let b be the sesquilinear form defined in $\mathcal{X}\times\mathcal{L}$ by

$$b(\mathbf{G},\nu) := \langle\,\mathbf{curl}\,\mathbf{G}\cdot\mathbf{n},\nu\rangle_{\Gamma_{\mathrm{J}}}.$$

The mixed problem associated with problem **PI** is the following:
Problem MPI.- *Given* $\mathbf{I}\in\mathbb{C}^N$, *find* $\mathbf{H}\in\mathcal{X}$ *and* $\lambda\in\mathcal{L}$ *such that*

$$a(\mathbf{H},\mathbf{G}) + b(\bar{\mathbf{G}},\lambda) = 0 \qquad \forall\mathbf{G}\in\mathcal{X},$$

$$b(\mathbf{H},\nu) = \sum_{n=1}^{N}\int_{\Gamma_{\mathrm{J}}^n} I_n\,\bar{\nu} \qquad \forall\nu\in\mathcal{L}.$$

Theorem 6. *Given* $I\in\mathbb{C}^N$, *let* $\mathbf{H}\in\mathcal{X}$ *be the solution of problem* **PI**. *Then, there exists a unique* $\lambda\in\mathcal{L}$ *such that* $(\mathbf{H},\lambda)$ *is the only solution of problem* **MPI**. *Furthermore, the following estimate holds:*

$$\|\mathbf{H}\|_{\mathcal{X}} + \|\lambda\|_{\mathrm{H}^{1/2}_{00}(\Gamma_{\mathrm{J}})} \le C\,|\mathbf{I}|\,.$$

The proof is based on the classical Babuška-Brezzi theory. In particular we prove the inf-sup condition for the bilinear form b by using results concerning vector potentials in $\mathbb{R}^3$ (see [9]).

Theorem 3.5 in [9] shows that the solution of problem **MPI**, together with $\mathbf{E} = \frac{1}{\sigma}\,\mathbf{curl}\,\mathbf{H}$ and $\mathbf{J} = \mathbf{curl}\,\mathbf{H}$, satisfy the Maxwell equations (1)–(5) and the boundary conditions (10)–(14) in a suitable weak sense. Moreover, from that theorem, we also have that the Lagrange multiplier is an *electric surface potential* on Γ_{J}, namely,

$$\mathbf{n} \times (\mathbf{E} \times \mathbf{n}) = -\mathbf{n} \times (\nabla \bar{\lambda}^* \times \mathbf{n}) =: -\,\mathbf{grad}_{\Gamma} \bar{\lambda}^* \quad \text{on } \Gamma_{\mathrm{J}},$$

λ^* being a lifting of λ to Ω such that $\lambda^* \in \mathrm{H}^1(\Omega)$ and $\lambda^*|_{\Gamma_{\mathrm{E}}} = 0$.

3.3 Finite element discretization

In this section we introduce a discretization of the mixed problem **MPI** and study its convergence properties. To this goal, we assume that Ω is a Lipschitz polyhedron and that Γ_{J}^n are polyhedral surfaces for all $n = 0, \ldots, N$. Consequently, Γ_{E} is also a polyhedral surface. We also assume that σ is piecewise smooth (e.g., $\mathcal{C}^2$) on a polyhedral partition of Ω.

We consider a family of shape-regular tetrahedral meshes $\{\mathcal{T}_h\}$ of Ω. We assume that the meshes are compatible with the splitting of the boundary of the domain in the sense that, $\forall K \in \mathcal{T}_h$ with a face T lying on $\partial\Omega$,

- either $T \subset \bar{\Gamma}_{\mathrm{E}}$ or $T \subset \bar{\Gamma}_{\mathrm{J}}^n$ for some $n = 0, \ldots, N$;
- $\sigma|_T$ is smooth.

The magnetic field, which is a function of $\boldsymbol{\mathcal{X}} = \mathrm{H}(\mathbf{curl}, \Omega)$, is discretized by the lowest-order Nédélec edge finite elements described in Section 2, i.e. we define $\boldsymbol{\mathcal{X}}_h = \boldsymbol{\mathcal{N}}_h(\Omega)$ as an approximation of $\boldsymbol{\mathcal{X}}$.

Let $\mathcal{T}_h^{\Gamma_{\mathrm{J}}}$ be the triangular mesh induced by $\mathcal{T}_h$ on the polyhedral surface Γ_{J} and consider the following finite-dimensional space:

$$\mathcal{Q}_h^1(\Gamma_{\mathrm{J}}) := \left\{ q_h \in \mathrm{H}_0^1(\Gamma_{\mathrm{J}}) : \; q_h|_T \in \mathcal{P}_1(T) \; \forall T \in \mathcal{T}_h^{\Gamma_{\mathrm{J}}} \right\}.$$

The Lagrange multiplier will be approximated in the finite dimensional space

$$\mathcal{L}_h := \left\{ \nu_h \in \mathcal{Q}_h^1(\Gamma_{\mathrm{J}}) : \; \nu_h|_{\Gamma_{\mathrm{J}}^n} = \text{constant}, \; n = 1, \ldots, N \right\}.$$

We define the following discrete problem

Problem DMPI.- *Given $\mathbf{I} \in \mathbb{C}^N$, find $\mathbf{H}_h \in \boldsymbol{\mathcal{X}}_h$ and $\lambda_h \in \mathcal{L}_h$ such that*

$$a(\mathbf{H}_h, \mathbf{G}_h) + b(\bar{\mathbf{G}}_h, \lambda_h) = 0 \qquad \forall \mathbf{G}_h \in \boldsymbol{\mathcal{X}}_h,$$

$$b(\mathbf{H}_h, \nu_h) = \sum_{n=1}^{N} \int_{\Gamma_{\mathrm{J}}^n} I_n \, \bar{\nu}_h \qquad \forall \nu_h \in \mathcal{L}_h.$$

Theorem 7. *Given* $\mathbf{I} \in \mathbb{C}^N$*, problem* **DMPI** *attains a unique solution* $(\mathbf{H}_h, \lambda_h)$. *Furthermore, if the solution* $(\mathbf{H}, \lambda)$ *of problem* **MPI** *satisfies* $\mathbf{H} \in \mathrm{H}^r(\mathbf{curl}, \Omega)$ *with* $1/2 < r \leq 1$*, then the following error estimate holds true:*

$$\|\mathbf{H} - \mathbf{H}_h\|_{\mathcal{X}} \leq Ch^r \|\mathbf{H}\|_{\mathrm{H}^r(\mathbf{curl}, \Omega)}.$$

References

1. Alonso, A., Valli, A. (1997): A domain decomposition approach for heterogeneous time-harmonic Maxwell equations, Comput. Methods Appl. Mech. Engrg., **143**, 97–112.
2. Alonso, A., Valli, A. (1999): An optimal domain decomposition preconditioner for low-frequency time-harmonic Maxwell equations, Math. Comp., **68**, 607–631.
3. Ammari, H., Buffa, A., Nédélec, J.-C. (2000): A justification of eddy currents model for the Maxwell equations, SIAM J. Appl. Math., **60**, 1805–1823.
4. Amrouche, C., Bernardi, C., Dauge. M., Girault, V. (1998): Vector potentials in three-dimensional non-smooth domains, Math. Methods Appl. Sci., **21**, 823–864.
5. Bermúdez, A., Bullón, J., Pena, F., Salgado, P. (2003), A numerical method for transient simulation of metallurgical compound electrodes, Finite Elem. Anal. Des., **39**, 283–299.
6. Bermúdez, A., Muñoz, R. (1999): Existence of solution of a coupled problem arising in the thermoelectrical simulation of an electrode, Quart. of Appl. Math., **57**, 621–636.
7. Bermúdez, A., Rodríguez, R., Salgado, P. (2002): A finite element method with Lagrange multipliers for low-frequency harmonic Maxwell equations, SIAM J. Numer. Anal., **40**, 1823–1849.
8. Bermúdez, A., Rodríguez, R., Salgado, P. (2003): Numerical analysis of the electric field formulation of an eddy currents problem, C. R. Acad. Sci. Paris, Serie I, **337**, 359–364.
9. Bermúdez, A., Rodríguez, R., Salgado, P., Numerical treatment of realistic boundary conditions for the eddy current problem in an electrode via Lagrange multipliers, Math. Comp., (to appear).
10. Bossavit, A. (1991): The computation of eddy-currents in dimension 3 by using mixed finite elements and boundary elements in association, Math. Comput. Modelling, **15**, 33–42.
11. Bossavit, A. (1999): "Hybrid" electric-magnetic methods in eddy-current problems, Comput. Methods Appl. Mech. Engrg., **178**, 383–391.
12. Bossavit, A. (2000): Most general "non-local" boundary conditions for the Maxwell equation in a bounded region, COMPEL, **19**, 239–245.
13. Bossavit, A., Vérité, J.C. (1982): A mixed FEM-BIEM method to solve 3-D eddy current problems, IEEE Trans. Mag., **18**, 431–435.
14. Bullón, J., Gallego,V. (1994): The use of a compound electrode for the production of silicon metal. Electric Furnace Conference Proceedings, Vol. 52, Iron & Steel Society, Warrendale, PA, 371–374.
15. D'Ambrosio, P., Letizia, I. (1980): Temperature and internal stress distribution of carbon electrodes used in an electric arc furnace for the production of silicon metal. Proceedings of Carbon'80, Baden-Baden, 526–529.

16. Howison, S.D., Rodrigues, J.F., Shillor, M. (1993): Stationary solutions to the thermistor problem, J. Math. Anal. Appl., **174** 2, 573-588.
17. Innvær, R., Fidje, K., Sira, T. (1987): 3-dimensional calculations on smelting electrodes. Reprinted in MIC-Model. Identif. Control **8**, 103–115.
18. Innvær, R., Olsen, L. (1980): Practical use of mathematical models for Søderberg electrodes. In: Electric Furnace Conference Proceedings, vol. 38. Iron & Steel Society, Warrendale, PA, 40–47.
19. Nédélec, J.-C. (1980): Mixed finite elements in $\mathbb{R}^3$, Numer. Math., **35**, 315–341.
20. Salgado, P. (2002): Mathematical and numerical analysis of some electromagnetic problems. Application to the simulation of metallurgical electrodes. Ph. D. Thesis, Universidade de Santiago de Compostela, Spain.

Space Decomposition Preconditioners and Parallel Solvers

Radim Blaheta

Institute of Geonics AS CR
Studentská 1768, 708 00 Ostrava - Poruba, Czech Republic
blaheta@ugn.cas.cz

Summary. This paper uses the general framework of space decomposition – subspace correction for providing an overview of Schwarz-type preconditioners. The considered preconditioners are one-level and two-level Schwarz methods based on an overlapping domain decomposition, a two-level method with the coarse grid space created by aggregation and a new two-level method with interfaces in the coarse grid space. Beside the description and analysis, we discuss also some implementation details, the use of inexact subproblem solvers etc. The efficiency of preconditioners is illustrated by numerical examples.

1 Introduction

This paper concerns the numerical solution of large scale linear systems arising from the finite element (FE) solution of boundary value problems. Especially, the attention is payed to that numerical methods, which are suitable for implementation on high performance parallel computers.

This motivates the interest in space decomposition (SD) preconditioners described in Section 2. These preconditioners provide a general framework, which has many specific applications. This paper concentrates on Schwarz-type preconditioners, which are usually based on the overlapping domain decomposition (DD). This class itself involves many variants of the basic technique and some of them will be reported in this paper. There are also many other useful decompositions, which can be used for the construction of preconditioners. Let us mention two examples: the hierarchical decomposition of FE spaces (cf. [13, 20, 10]) or displacement decomposition for elasticity problems (cf. e.g. [4] and the references therein).

An overview of the standard one-level and two-level overlapping DD methods can be found in Sections 3 and 4. These methods are also called Schwarz methods according to the pioneering work of H.A. Schwarz from 1870, cf. [19]. But the real and rapid development of these methods, motivated by the interest in parallel computing, starts in the second half of 1980's.

In Section 5, we describe a less standard two-level Schwarz method with the auxiliary global problem created by aggregation of unknowns. This method is useful in many cases, when application of the standard two-level methods requires a lot of extra work involved in creating an auxiliary coarse grid and relating this grid to the original one.

Section 6 is devoted to the description of a new two-level method, which uses non-overlapping DD and interaction of subdomain problems only via a special coarse grid space with interfaces.

Some aspects of implementation of the preconditioners on parallel computers, the use of inexact subproblem solvers and use of nonsymmetric multiplicative or hybrid preconditioners are discussed in Section 7. The preconditioners, which are not symmetric positive definite, can be efficiently implemented within the generalized preconditioned conjugate gradient method (GPCG), cf. [4] and the references therein. In Section 8, we provide some numerical results illustrating the efficiency of the described methods and corresponding solvers.

In the final section, we summarize the results and mention some topics not covered in the paper.

2 Abstract SD preconditioners

Our aim is the solution of abstract discrete symmetric elliptic problem in the following form

$$\text{find } u \in V: \quad a(u,v) = l(v) \quad \forall v \in V, \tag{1}$$

where V is a finite dimensional subspace of a Hilbert space $\mathcal{V} = \mathcal{V}(\Omega)$ of functions defined in a domain $\Omega \subset R^d$ $(d = 2,3)$, a is a bounded symmetric positive definite (SPD) bilinear form on $\mathcal{V}$ and l is a bounded linear functional on $\mathcal{V}$.

To be more specific, we shall consider elliptic boundary value problems with the bilinear form

$$a(u,v) = \int_\Omega \sum_{i,j=1}^{d} k_{ij} \frac{\partial u}{\partial x_i} \frac{\partial v}{\partial x_j} dx \tag{2}$$

defined in the Sobolev space $H^1(\Omega)$ equipped with the seminorm $|\cdot|_{H^1(\Omega)}$ and the norm $\|\cdot\|_{H^1(\Omega)}$. We assume that $\mathcal{K} = (k_{ij})$ is a symmetric positive definite $d \times d$ matrix, which guarantees the existence of positive constants κ_1, κ_2, such that

$$\kappa_1 |v|^2_{H^1(\Omega)} \leq a(v,v) \leq \kappa_2 |v|^2_{H^1(\Omega)} \qquad \forall v \in H^1(\Omega). \tag{3}$$

Sometimes, we shall assume that $\mathcal{V} \subset H^1(\Omega)$ is such that

$$\kappa_0 \|v\|^2_{H^1(\Omega)} \leq a(v,v) \qquad \forall v \in \mathcal{V}. \tag{4}$$

Let us note that most of the presented results are valid also for other elliptic problems, e.g. problems of linear elasticity.

Using an inner product $\langle\, , \rangle$ in V, the problem (1) can be rewritten into the following operator form

$$\text{find } u \in V\colon \quad Au = b, \tag{5}$$

where $A\colon\ V \to V$ and $b \in V$ are determined by the identities

$$\langle Au, v\rangle = a(u,v), \qquad \langle b, v\rangle = l(v) \qquad \forall\, u, v \in V\,.$$

Further, we shall assume that there is a decomposition of the space V,

$$V = V_1 + \ldots + V_m, \tag{6}$$

where V_k ($k = 1, \ldots, m$) are subspaces of V, which are not necessarily linearly independent. For each subspace V_k, we introduce

- operator $A_k\colon\ V_k \to V_k$ defined by $\langle A_k u, v\rangle = a(u,v) \quad \forall\, u, v \in V_k$,
- prolongation operator $I_k\colon\ V_k \to V$ given by the inclusion $V_k \subset V$,
- restriction operator $R_k\colon\ V \to V_k$ given by orthogonal projection to V_k, i.e. $\forall\, v \in V\colon\ R_k v \in V_k$ and $\langle v - R_k v, w\rangle = 0\ \forall w \in V_k$.

Note that I_k and R_k are adjoint with respect to $\langle\,,\rangle$ and $A_k = R_k A I_k$.

The decomposition (6) allows to introduce *space decomposition (SD) preconditioners.* The SD preconditioner is an operator G, which can be used for a cheap computation of the pseudoresidual $g = Gr$ from a given residual r. The pseudoresidual should provide an approximation of the error $A^{-1}r$ or at least its direction. The computation of pseudoresiduals g is realized by the following *SD algorithm,*

$g = 0$
for $k = 1, \ldots, m$ **do**
  $g \leftarrow g + I_k A_k^{-1} R_k\, z_k$
end

This SD algorithm represents several types of preconditioners:

1. The **additive preconditioner** $G_A\colon\ r \to g$ arises if $z_k = r$. In this case,

$$G_A = \sum_{k=1}^{m} B_k, \qquad B_k = I_k A_k^{-1} R_k\,, \tag{7}$$

is symmetric and positive definite with respect to $\langle\,,\rangle$. Moreover,

$$G_A A = \sum_{k=1}^{m} P_k, \qquad P_k = I_k A_k^{-1} R_k A\,, \tag{8}$$

where P_k are projections $V \to V_k$, which are orthogonal with respect to the inner product induced by A, $\langle u, v\rangle_A = \langle Au, v\rangle$.

2. The **multiplicative preconditioner** $G_M: r \to g$ arises if $z_k = r - Ag$. In this case, it is easy to verify by induction that

$$G_M = [I - (I - P_m) \dots (I - P_1)]A^{-1} = A^{-1}[I - (I - AB_m) \dots (I - AB_1)] . \tag{9}$$

This preconditioner is not symmetric. To obtain a symmetric multiplicative operator, it is possible to continue in the subspace corrections in the reverse order. It gives

$$G_{SM} = [I - (I - P_1) \dots (I - P_m) \dots (I - P_1)]A^{-1} . \tag{10}$$

For this preconditioner, it is easy to show that $G_{SM}A$ is symmetric in $\langle\, ,\rangle_A$, which implies that G_{SM} is symmetric in $\langle\, ,\rangle$. Similarly, it is easy to show that $G_{SM}A$ and G_{SM} are positive semidefinite. The positive definiteness of G_{SM} is equivalent to the fact that $I - G_M A$ is convergent (see Theorem 2).

3. The **hybrid preconditioner** $G_H : r \to g$ arises if some residuals are updated and others are kept. For example, we can update only the residual after the first subspace correction, which gives

$$G_H = B_1 + \sum_{k=2}^{m} B_k(I - AB_1) = [P_1 + \sum_{k=2}^{m} P_k(I - P_1)]A^{-1} . \tag{11}$$

This preconditioner can be again symmetrized to the form

$$G_{SH} = [P_1 + (I - P_1) \sum_{k=2}^{m} P_k(I - P_1)]A^{-1} . \tag{12}$$

It shows that the preconditioned operator $G_{SH}A$ is decomposed,

$$G_{SH}A = P_1 + (I - P_1) \sum_{k=1}^{m} P_k(I - P_1) , \tag{13}$$

i.e. $G_{SH}A$ is equal to identity on range $R(P_1)$ and to the additively preconditioned operator $G_A A$ on range $R(I - P_1)$. Thus, we can expect that the hybrid preconditioner will be more efficient than the additive one, cf. also [17].

Further in this paper, we shall exploit the following two assumptions, which characterize the considered decomposition:

(A1) There exists a constant K_0 such that

$$\forall v \in V \;\; \exists v_k \in V_k : \;\; v = v_1 + \dots + v_m, \quad \sum_k \| v_k \|_A^2 \leq K_0 \| v \|_A^2 . \tag{14}$$

(A2) There exists a constant K_1 such that

$$\forall v \in V \;\; \forall v_k \in V_k: \;\; v = v_1 + \ldots + v_m: \quad \|v\|_A^2 \le K_1 \sum_k \|v_k\|_A^2. \qquad (15)$$

Note that $\|v\|_A = \sqrt{\langle v, v\rangle_A}$. A trivial upper bound is $K_1 = m$. But we are interested in m independent bounds for K_1, which can be found e.g. by considering the angles of subspaces V_k. If

$$\begin{aligned} \varepsilon_{kl} &= \cos(V_k, V_l) \\ &= \sup\{\langle v_k, v_l\rangle_A : \; v_k \in V_k, \; v_l \in V_l, \; \|v_k\|_A = 1, \; \|v_l\|_A = 1\}, \end{aligned}$$

$\mathcal{E} = (\varepsilon_{kl})$ and $\rho(\mathcal{E})$ is the spectral radius of $\mathcal{E}$, then

$$K_1 \le \rho(\mathcal{E}) \le \max_k \sum_l \varepsilon_{kl}. \qquad (16)$$

Sometimes, one of the subspaces, say V_1, plays an exceptional role. Then it may be useful to consider $\mathcal{E}_1 = (\varepsilon_{kl}, \; k, l \ne 1)$ and the estimate

$$K_1 \le 2(1 + \rho(\mathcal{E}_1)) \le 2 + 2 \max_{k \ne 1} \sum_{l \ne 1} \varepsilon_{kl}. \qquad (17)$$

Theorem 1. *Let $\lambda_{\min}(G_A A)$, $\lambda_{\max}(G_A A)$ and $\operatorname{cond}(G_A A)$ denote respectively the smallest and largest eigenvalue and the condition number of the additively preconditioned operator A. Then*

$$\lambda_{\min}(G_A A) \ge 1/K_0, \quad \lambda_{\max}(G_A A) \le K_1, \quad \operatorname{cond}(G_A A) \le K_0 K_1. \qquad (18)$$

Remark 1. The estimate of $\lambda_{\min}(G_A A)$ is well known as Lions' lemma, see [16]. The proof of Theorem 1 as well as some historical remarks can be found e.g. in books [13, 20, 18].

Theorem 2. *For the multiplicative preconditioner, we get*

$$\|I - G_{SM}A\|_A = \|I - G_M A\|_A^2 \le \left[1 - \frac{1}{K_0(1+K_1)^2}\right]^2, \qquad (19)$$

$$\operatorname{cond}(G_{SM}A) \le \left[1 - \|I - G_{SM}A\|_A\right]^{-1}. \qquad (20)$$

Remark 2. The proof of the estimate (19) is more technical, see [7]. The complete proof can be found e.g. in [20, 18]. The estimate (20) easily follows from (19).

3 Schwarz preconditioners based on overlapping DD

A suitable space decomposition (5) can be constructed via overlapping decomposition of the computational domain Ω. Now, we describe and analyse a typical model situation. We solve the boundary value problem (1) in $\Omega \subset R^d$ by the finite element method with linear triangular or tetrahedral elements. Let $\mathcal{T}_h$ be a regular finite element division of Ω, $h \sim \max\{\text{diam}(e),\, e \in \mathcal{T}_h\}$. Moreover, let us assume that there is a division of Ω into m non-overlapping subdomains $\Omega_1^0, \ldots, \Omega_m^0$. Let Ω_k^δ be subdomains of Ω aligned with the division $\mathcal{T}_h$ and such that $\Omega_k^\delta \supset \{x \in \Omega\colon \text{dist}(x, \Omega_k^0) \leq \delta/2\}$. Let us also denote $H_s \sim \max\{\text{diam}(\Omega_k^\delta),\, k = 1, \ldots, m\}$ and assume that each point of Ω belongs to at most m_c subdomains, see Fig. 1.

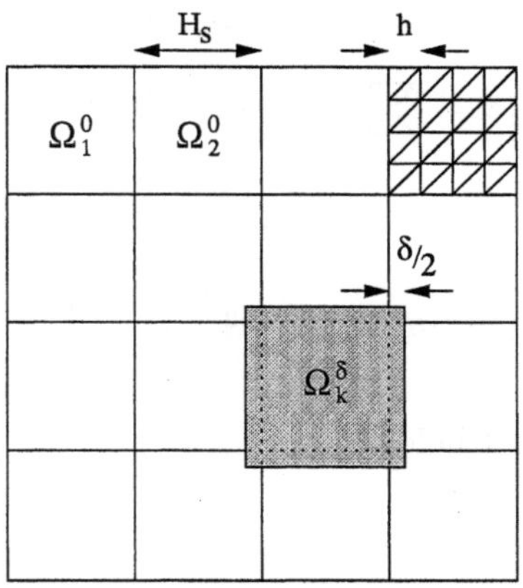

Fig. 1. Decomposition of Ω, subdomains Ω_k^0, Ω_k^δ, fine triangulations $\mathcal{T}_h$, $m_c = 4$

The overlapping domain decomposition

$$\Omega = \Omega_1^\delta \cup \ldots \cup \Omega_m^\delta \tag{21}$$

now induces a decomposition (6) of the finite element space $V = V_h$ into the subspaces $V_1, \ldots, V_m$,

$$V_k = \{v \in V\colon v = 0 \text{ in } \Omega \setminus \Omega_k^\delta\}. \tag{22}$$

Moreover, we get the following characterization of this decomposition.

Theorem 3. *Under the above assumptions, the decomposition* $V = V_1 + \ldots + V_m$ *with the local FE spaces* (22) *fulfils the assumptions* (A1), (A2) *with the constants*

$$K_0 = C(1 + \delta^{-2}), \qquad K_1 = m_c. \tag{23}$$

Remark 3. The proof of Theorem 3 can be found in the works of M. Dryja and O. Widlund, see the books [13, 20] and the references given there. We sketch the proof here, because it contains ideas important for understanding the further development of the method in the next sections.

Sketch of the proof.

- For the overlapping domain decomposition, there is a partition of the unity $\Theta_1, \dots, \Theta_m \in C^\infty(R^d)$, $0 \le \Theta_k \le 1$,

$$\Theta_k = 0 \text{ in } R^d \setminus \Omega_k^\delta, \quad \sum_{k=1}^m \Theta_k = 1, \quad \|\text{grad}\, \Theta_k\|_{L^\infty(\Omega)} = O(\delta^{-1}).$$

- Using this partition of unity and the piecewise linear interpolation $\Pi_h\colon\ C(\Omega) \to V_h$, we can decompose any element $v \in V_h$ as follows,

$$v = \sum_k v_k, \quad v_k = \Pi_h(\Theta_k v).$$

- Property (A1):

$$\sum_k \|v_k\|_A^2 \le \kappa_2 \sum_k |v_k|^2_{H^1(\Omega_k^\delta)} \le \kappa_2 \sum_k \sum_{T\subset\Omega_k^\delta,\ T\in\mathcal{T}_h} |\Pi_h(\Theta_k v)|^2_{H^1(T)}.$$

On the element level, it holds that

$$|\Pi_h(\Theta_k v)|^2_{H^1(T)} \le \mu \left[\delta^{-2}\|v\|^2_{L_2(T)} + |v|^2_{H^1(T)}\right],$$

see e.g. [10], Lemma 7.4.17 or [12]. Moreover, $T \subset \mathcal{T}_h$ can belong to at most m_c subdomains Ω_k^δ. Thus,

$$\begin{aligned}\sum_k \|v_k\|_A^2 &\le \kappa_2\, \mu\, m_c \sum_{T\in\mathcal{T}_h} \left[\delta^{-2}\|v\|^2_{L_2(T)} + |v|^2_{H^1(T)}\right] \\ &\le \kappa_2\, \mu\, m_c \left(\delta^{-2}+1\right) \|v\|^2_{H^1(\Omega)} \\ &\le \kappa_2\, \mu\, m_c\, \kappa_0^{-1} \left(\delta^{-2}+1\right) \|v\|^2_A .\end{aligned}$$

Note that κ_0, κ_1, κ_2 are constants from (3), (4).

- Property (A2):

$$\begin{aligned}\|\sum_k v_k\|_A^2 &= \sum_{ij} a(v_i, v_j) = \sum_{ij}\sum_{T\in\mathcal{T}_h} a_T(v_i, v_j) = \sum_{T\in\mathcal{T}_h}\sum_{ij} a_T(v_i, v_j) \\ &\le \sum_{T\in\mathcal{T}_h}\sum_{ij} \sqrt{a_T(v_i, v_i)}\sqrt{a_T(v_j, v_j)} \\ &= \sum_{T\in\mathcal{T}_h}\sum_{i} \sqrt{a_T(v_i, v_i)} \sum_j \sqrt{a_T(v_j, v_j)},\end{aligned}$$

where $a_T(v_i, v_i)$ denotes the restriction of the bilinear form a to T. Each $T \in \mathcal{T}_h$ belongs to at most m_c subdomains and therefore $\sum \sqrt{a_T(v_i, v_i)}$ has at most m_c nonzero terms. The Cauchy-Bunyakowski-Schwarz (C.B.S.) inequality thus gives

$$\sum_i \sqrt{a_T(v_i, v_i)} \leq \sqrt{m_c} \sqrt{\sum_i a_T(v_i, v_i)}\,.$$

Hence,

$$\| \sum_k v_k \|_A^2 \leq \sum_{T \in \mathcal{T}_h} m_c \sum_k a_T(v_k, v_k) = m_c \sum_k \| v_k \|_A^2 \,.$$

□

Normally, the overlap δ is kept proportional to the size of subdomain H_s, i.e. $\delta = \beta H_s$, where β is the proportionality constant. In this case, a combination of Theorems 1, 2, 3 says that $\mathrm{cond}(G_A A)$ and $\mathrm{cond}(G_{SM} A)$ deteriorate with the increasing number of subdomains ($H_s \to 0$, $\delta \to 0$). A remedy can be found in adding an auxiliary coarse grid FE space V_0 to the space decomposition with local FE spaces (22), see the next section.

4 Two–level Schwarz preconditioners

Let us consider the extended decomposition

$$V = V_0 + V_1 + \ldots + V_m\,, \tag{24}$$

where $V = V_h$ is the FE space corresponding to the FE division $\mathcal{T}_h$, $V_1, \ldots, V_m$ are the local FE spaces (22) and $V_0 = V_H$ is the FE space corresponding to a coarser FE division $\mathcal{T}_H$ of the domain Ω. We assume that $\mathcal{T}_h$ is a refinement of $\mathcal{T}_H$, which guarantees that $V_H \subset V_h$. Moreover, we assume that $H \leq H_s$.

In a more general case of non-nested grids, we could use a prolongation (interpolation) $I_H^h: \; V_H \to V_h$ ($I_H^h = \Pi_h$) and choose $V_0 = I_H^h V_H$, see [12]. The subproblem operator A_0 corresponding to V_0 is determined by restriction of the variational formulation to V_0. There is also the relation $A_0 = (I_H^h)^* A I_H^h$.

Theorem 4. *The decomposition* (24) *fulfils the assumptions* (A1), (A2) *with the constants*

$$K_0 = C(1 + \delta^{-2} H^2)\,, \qquad K_1 = 2(1 + m_c)\,. \tag{25}$$

Sketch of the proof.

- For the analysis, we shall use the fact there is a mapping $Q\colon \; V_h \to V_0$ and constants σ_1, σ_2 such that

$$|Qv|_{H^1(\Omega)} \leq \sigma_1 |v|_{H^1(\Omega)}\,, \tag{26}$$
$$\|v - Qv\|_{L_2(\Omega)} \leq \sigma_2 H |v|_{H^1(\Omega)}\,. \tag{27}$$

The above properties can be proved e.g. if Q is the L_2-orthogonal projection into V_0, see [8].

- For any $v \in V$, we take a decomposition $v = v_0 + v_1 + \ldots + v_m$, $v_0 = Qv$, $v_k = \Pi_h(\Theta_k(v - v_0))$ for $k = 1, \ldots, m$.
- For this decomposition,

$$\|v_0\|_A^2 \leq \kappa_2 |v_0|^2_{H^1(\Omega)} \leq \kappa_2 \sigma_1 |v|^2_{H^1(\Omega)} \leq \kappa_2 \sigma_1^2 \kappa_1^{-1} \|v\|_A^2\,.$$

- Further,

$$\begin{aligned}\sum_{k=1}^{m} \|v_k\|_A^2 &\leq \kappa_2 \sum_{k=1}^{m} |v_k|^2_{H^1(\Omega_k^\delta)} = \kappa_2 \sum_{k=1}^{m} \sum_{T \subset \Omega_k^\delta,\, T \in \mathcal{T}_h} |v_k|^2_{H^1(T)} \\ &\leq \kappa_2\, \mu\, m_c \left[\delta^{-2} \|v - v_0\|^2_{L_2(\Omega)} + |v - v_0|^2_{H^1(\Omega)}\right] \\ &\leq \kappa_2\, \mu\, m_c \left[\sigma_2^2\, \delta^{-2} H^2 |v|^2_{H^1(\Omega)} + 2|v|^2_{H^1(\Omega)} + 2|v_0|^2_{H^1(\Omega)}\right] \\ &\leq \kappa_2\, \mu\, m_c \left[2 + 2\sigma_1^2 + \sigma_2^2\, \delta^{-2} H^2\right] |v|^2_{H^1(\Omega)} \\ &\leq c \left[1 + \delta^{-2} H^2\right] \|v\|_A^2\end{aligned}$$

- The above estimates show that, $K_0 = C(1 + \delta^{-2}H^2)$.
- The estimate of K_1 follows easily from (17) and the estimate (23). □

Remark 4. The proof of Theorem 4 can be found again in the works of M. Dryja and O. Widlund, see also [13, 20, 10]. The estimates (25) can be strengthened. In [14], it is proved that $K_0 = C(1 + \delta^{-1}H)$. This estimate is sharp, cf. [9].

Remark 5. The constant C depends on κ_1, κ_2, which indicates some dependence on the anisotropy and jumps in coefficients of the bilinear form, cf. [2]. On the contrary to Theorem 3, the assumption (4) was not used here.

If $\mathcal{T}_h$ arises as a refinement of a coarser FE division $\mathcal{T}_H$, then the use of $V_0 = V_H$ is fully natural. In other cases, it may be impractical and costly to construct an extra division $\mathcal{T}_H$ together with the interpolation I_H^h and coarse grid operator $A_0 = A_H$. For these cases, it may be advantageous to use another construction of the auxiliary global space V_0 by aggregation. This construction will be described in the next section.

5 Two–level Schwarz preconditioners with aggregations

Let $V = V_h = \operatorname{span}\{\phi_1^h, \ldots, \phi_n^h\}$ and let the index set $\{1, \ldots, n\}$ be decomposed into groups $G_1, \ldots, G_N$. Then it is possible to define aggregated basis functions ψ_k and the space $V_0 \subset V$ as follows,

$$\psi_k = \sum_{i \in G_k} \phi_i^h , \qquad V_0 = \operatorname{span}\{\psi_1, \dots, \psi_N\} . \tag{28}$$

We shall assume that the aggregations are regular, i.e. there is a constant $\bar{\beta}$ such that each $\operatorname{supp}\psi_k$ contains a ball with diameter $\bar{\beta}H$, where

$$H \sim \max_k \operatorname{diam}(\operatorname{supp}\psi_k) \leq H_s .$$

As a consequence, there are positive constants C_1, C_2 such that $C_1 H^d \leq |\operatorname{supp}\psi_k| \leq C_2 H^d$. The space V_0 from (28) together with the local FE spaces $V_1, \dots, V_m$ from (22) create a new decomposition

$$V = V_0 + V_1 + \ldots + V_m . \tag{29}$$

Theorem 5. *The decomposition* (29) *fulfils the assumptions* (A1), (A2) *with the constants*

$$K_0 = C(1 + h^{-1}H + \delta^{-2}H^2) , \qquad K_1 = 2(1 + m_c) . \tag{30}$$

Sketch of the proof. We can follow the same procedure as in the proof of Theorem 4 but instead of L_2-projection to V_h, we shall consider an averaging operator $\bar{Q}$.

– Let us consider the mapping $\bar{Q}\colon V_h \to V_0$ defined as follows:

$$\text{for } v \in V_h , \quad \bar{Q}v = \sum_{k=1}^{N} \alpha_k \psi_k , \quad \alpha_k = \frac{1}{|\operatorname{supp}\psi_k|} \int_{\operatorname{supp}\psi_k} v(x)dx . \tag{31}$$

For this operator, it holds that

$$|\bar{Q}v|_{H^1(\Omega)} \leq \bar{\sigma}_1 \sqrt{\frac{H}{h}}\, |v|_{H^1(\Omega)} , \tag{32}$$

$$\|v - \bar{Q}v\|_{L_2(\Omega)} \leq \bar{\sigma}_2 H \,|v|_{H^1(\Omega)} , \tag{33}$$

see [11, 15] for the details.

– For any $v \in V$, we take the decomposition

$$v = v_0 + v_1 + \ldots + v_m , \quad v_0 = \bar{Q}v , \quad v_k = \Pi_h(\Theta_k(v - v_0)) , \quad k = 1, \dots, m .$$

– Then

$$\|v_0\|_A^2 \leq \kappa_2 |v_0|_{H^1(\Omega)}^2 \leq \kappa_2 \bar{\sigma}_1^2 \frac{H}{h} |v|_{H^1(\Omega)}^2 \leq \kappa_2 \bar{\sigma}_1^2 \kappa_1^{-1} \frac{H}{h} \|v\|_A^2 .$$

– Further

$$\sum_{k=1}^{m} \|v_k\|_A^2 \le \kappa_2 \sum_{k=1}^{m} |v_k|^2_{H^1(\Omega_k^\delta)} = \kappa_2 \sum_{k=1}^{m} \sum_{T\subset\Omega_k^\delta,\, T\in\mathcal{T}_h} |v_k|^2_{H^1(T)}$$
$$\le \kappa_2\, \mu\, m_c \left[\delta^{-2}\|v - v_0\|^2_{L_2(\Omega)} + |v - v_0|^2_{H^1(\Omega)}\right]$$
$$\le \kappa_2\, \mu\, m_c \left[\sigma_2^2\, \delta^{-2} H^2 + 2 + 2\bar{\sigma}_1^2 \frac{H}{h}\right] |v|^2_{H^1(\Omega)}$$
$$\le c\left[1 + h^{-1}H + \delta^{-2}H^2\right] \|v\|_A^2 \,.$$

– The above estimates give (30). The estimate of K_1 is the same as in Theorem 4. □

The space V_0 created by aggregation was firstly introduced in multigrid context, see e.g. [3] and the references therein. The properties of the basis functions $\{\psi_k\}$ can be improved by smoothing and the smoothed aggregations can be again used in two-level Schwarz preconditioners, see [11].

Note that the paper [15] describes the use of massive aggregation, which means aggregation of all degrees of freedom in each subdomain. According to our experience, less massive aggregation is more efficient and better balanced with the other subproblems for smaller numbers of subdomains.

For structured grids, the aggregations can be done by regular clustering. Some algorithms for creating aggregations for unstructured grids are described e.g. in [11].

6 Two-level Schwarz preconditioners with interfaces

The considered two-level Schwarz preconditioners involve two ways of interaction between the subproblems: via the overlap and via the coarse grid. In this section, we would like to show that the interaction via the coarse grid is sufficient if the coarse grid involves the interfaces between the subdomains.

Thus, let $\Omega_1^0, \ldots, \Omega_m^0$ be a non overlapping decomposition of the domain Ω and each Ω_k^0 be aligned with a fine FE division $\mathcal{T}_h$ of the domain Ω. Let

$$V_k = \{v \in V :\ v = 0 \text{ in } \Omega \setminus \Omega_k^0\}\,, \quad W = V_1 + \ldots + V_m\,. \tag{34}$$

Then $W \subset V$, but $W \neq V$. If there is a coarse grid space V_0 such that

$$V = V_0 + V_1 \ldots + V_m = V_0 + W\,, \tag{35}$$

then V_0 should contain all degrees of freedom corresponding to the interface $\Gamma = \bigcup \Gamma_{kl}$, where $\Gamma_{kl} = (\partial\Omega_k \cap \partial\Omega_l) \setminus \partial\Omega$.

There are several possibilities how to construct a proper space V_0. The first possibility is to use a coarse grid $\mathcal{T}_H$ providing enough nodes on the interface Γ, i.e. node$(\mathcal{T}_H) \cap \Gamma$ = node$(\mathcal{T}_h) \cap \Gamma$. The coarse grid $\mathcal{T}_H$ should be also

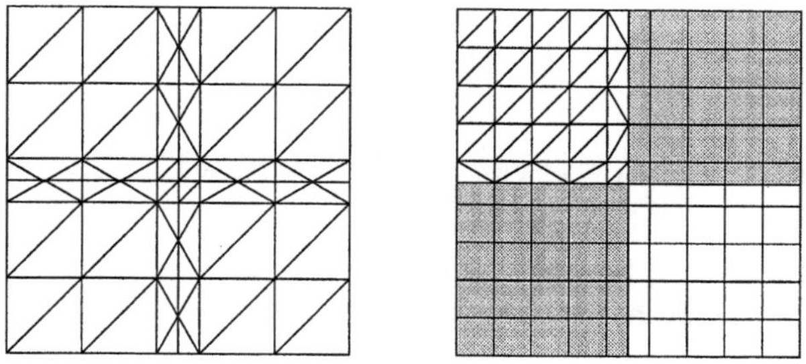

Fig. 2. A coarse grid with interfaces (left) and its refinement with the domain decomposition (right)

compatible with the subdomains Ω_k^0 and $\mathcal{T}_h$ should be a refinement of $\mathcal{T}_H$. Such situation can be seen in Fig. 2.

Another more flexible possibility how to create a proper space V_0 is the construction by aggregation. This construction can be similar as in the previous section, the only difference is that we restrict the aggregation to that degrees of freedom, which do not correspond to the nodes on the interfaces between the subdomains, see Fig. 3.

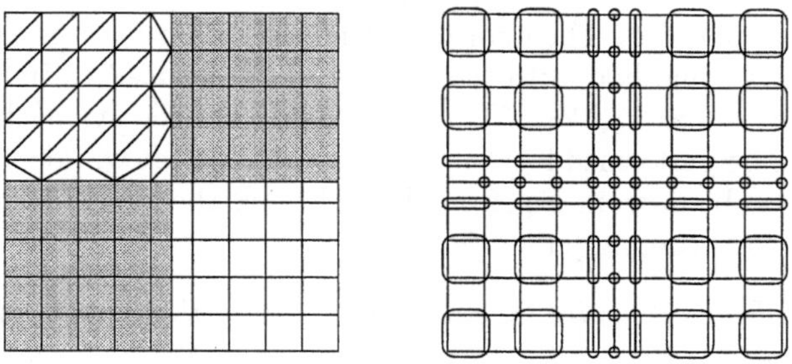

Fig. 3. Fine grid and domain decomposition (left) and aggregation out of the interfaces between the subdomains (right)

Theorem 6. *The decomposition* (35) *with subspaces* $V_1, \ldots, V_m$ *corresponding to non-overlapping subdomains and the coarse grid space* V_0 *with interfaces fulfils the assumptions* (A1), (A2) *with the constants*

$$K_0 = 1/(1-\gamma), \qquad K_1 = 2, \tag{36}$$

where $\gamma = \cos(V_0, W_0) < 1$ *for any* $W_0 \subset W$ *such that* $V = V_0 \oplus W_0$.

Proof. Any $v \in V$ can be written as $v = v_0 + w_0$, $v_0 \in V_0$, $w_0 \in W_0 \subset W$. Moreover, $w_0 = v_1 + \ldots + v_m$, where $v_k \in V_k$ for $k = 1, \ldots, m$. The subspaces $V_1, \ldots, V_m$ corresponding to the non-overlapping subdomains are $\langle\,,\rangle_A$ orthogonal, thus

$$\sum_{k=1}^{m} \|v_k\|_A^2 = \|v_1 + \ldots + v_m\|_A^2 = \|w_0\|_A^2,$$

$$\sum_{k=0}^{m} \|v_k\|_A^2 = \|v_0\|_A^2 + \|w_0\|_A^2 \le \frac{1}{1-\gamma}\|v_0 + w_0\|_A^2 = \frac{1}{1-\gamma}\|v\|_A^2 .$$

On the other hand, for any $v \in V$, $v = v_0 + v_1 + \ldots + v_m$, where $v_k \in V_k$ for $k = 0, 1, \ldots, m$, we have

$$\|v\|_A^2 \le 2\|v_0\|_A^2 + 2\|v_1 + \ldots + v_m\|_A^2 = 2\sum_{k=0}^{m} \|v_k\|_A^2 .$$

□

Note that the constant $\gamma < 1$ appeared also in the analysis of the hybrid preconditioner G_H. We simply get

$$\|I - G_H A\|_A = \| \left(I - \sum_{k=1}^{m} P_k\right) (I - P_0) \|_A = \gamma , \tag{37}$$

because $\sum_{k=1}^{m} P_k$ is now again $\langle\, ,\rangle_A$ orthogonal projection. Note also that $G_H = G_M$ in this case. More details can be found in [5].

As a consequence of Theorem 6, the investigation of convergence of the two-level Schwarz method with interfaces can be reduced to investigation of the constant $\gamma = \cos(V_0, W_0)$. This investigation of the strengthened C.B.S. inequality can be performed locally on macroelements. It will be illustrated for a simple 2D problem from Fig. 2 in the following theorem. More results will be given in a forthcoming paper [5].

Theorem 7. *Let us consider the problem* (1) *with orthotropic bilinear form* (2) *on the domain $\Omega \subset R^2$ and the situation similar to Fig. 2. It means that there is a coarse grid $\mathcal{T}_H$ with coarse rectangular triangles inside the subdomains and special rectangular macroelements and fine triangles along the boundaries of the subdomains, see Fig. 2, 4. There is also a refined triangulation $\mathcal{T}_h$, which consists from congruent rectangular isosceles triangles with two axiparallel sides. These triangles arise from $\mathcal{T}_H$ by division of each coarse triangle and each boundary macroelement into four congruent triangles, see Fig. 4.*

Let V_0 be the FE space corresponding to $\mathcal{T}_H$ and W be the union of subspaces V_k defined in (34). *Let W_0 contain the functions from W, which are zero in the nodes belonging to $\mathcal{T}_H$. The nonzero coefficients of the bilinear form (k_{11} and k_{22})) are assumed to be constant on the macroelements from $\mathcal{T}_H$.*

Then

$$\gamma = \cos(V_0, W_0) \le \max\left(\sqrt{\frac{1}{2}}, \sqrt{\frac{k_{11}}{k_{11}+k_{22}}}, \sqrt{\frac{k_{22}}{k_{11}+k_{22}}}\right). \tag{38}$$

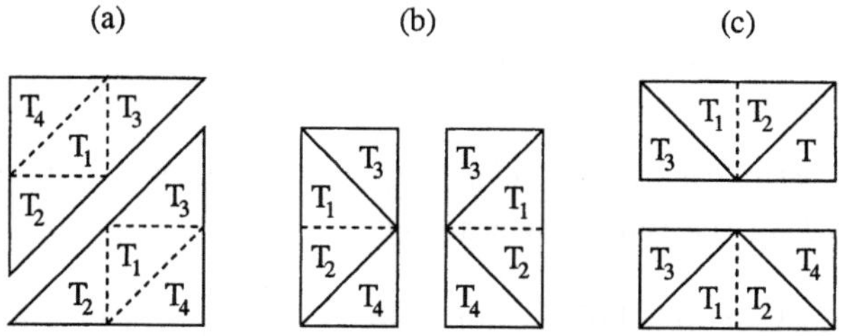

Fig. 4. The coarse triangles with refinement (a) and boundary macroelements with 3 triangles and their refinement into 4 triangles (b),(c)

Proof. The C.B.S. constant can be investigated locally on the inner and boundary macroelements. If

$$a_E(v,w) = \int_E \sum_i k_{ii} \frac{\partial v}{\partial x_i} \frac{\partial w}{\partial x_i} dx$$

and

$$a_E(v,w) \leq \gamma_E \sqrt{a_E(v,v)}\, \sqrt{a_E(w,w)}\,,$$

for each macroelement E and functions v and w, which are restrictions of $v \in V_0$ and $w \in W_0$ to the macroelement E, then

$$\gamma \leq \max_E \gamma_E .$$

For the inner macroelements (coarse triangles), it is possible to show that $\gamma_E \leq \sqrt{\frac{1}{2}}$, see e.g. [1] (the estimate is a simple generalization of Remark 3.1.).

Now consider an interface macroelement E of the type shown in Fig. 4(b). For $v \in V_0$, we get

$$\frac{\partial v}{\partial x_1}\,|_{T_1} = \frac{\partial v}{\partial x_1}\,|_{T_2}, \qquad \frac{\partial v}{\partial x_2}\,|_{T_1} = \frac{\partial v}{\partial x_2}\,|_{T_2} .$$

For $w \in W_0$, we get $\frac{\partial w}{\partial x_1} = \frac{\partial w}{\partial x_2} = 0$ in T_3 and T_4. Further,

$$\frac{\partial w}{\partial x_2}\,|_{T_1} = -\frac{\partial w}{\partial x_2}\,|_{T_2}\,, \qquad \frac{\partial w}{\partial x_1}\,|_{T_1} = \frac{\partial w}{\partial x_1}\,|_{T_2}\,, \qquad \frac{\partial w}{\partial x_1}\,|_{T_1} = \frac{\partial w}{\partial x_2}\,|_{T_1} .$$

Thus,

$$a_E(v,v) = \int_E k_{11} \left(\frac{\partial v}{\partial x_1}\right)^2 + k_{22} \left(\frac{\partial v}{\partial x_2}\right)^2 dx \geq 2 \int_{T_1} k_{11} \left(\frac{\partial v}{\partial x_1}\right)^2 dx,$$

$$a_E(w,w) = \int_E k_{11} \left(\frac{\partial w}{\partial x_1}\right)^2 + k_{22} \left(\frac{\partial w}{\partial x_2}\right)^2 dx = 2 \int_{T_1} (k_{11} + k_{22}) \left(\frac{\partial w}{\partial x_1}\right)^2 dx,$$

$$a_E(v,w) = \int\limits_{T_1 \cup T_2} \left(k_{11} \frac{\partial v}{\partial x_1} \frac{\partial w}{\partial x_1} + k_{22} \frac{\partial v}{\partial x_2} \frac{\partial w}{\partial x_2}\right) dx = 2 \int\limits_{T_1} k_{11} \frac{\partial v}{\partial x_1} \frac{\partial w}{\partial x_1} dx$$

$$\leq \sqrt{\frac{k_{11}}{k_{11}+k_{22}}}\sqrt{a_E(v,v)}\sqrt{a_E(w,w)} \quad \text{i.e. } \gamma_E = \sqrt{\frac{k_{11}}{k_{11}+k_{22}}}.$$

A similar estimate $\gamma_E = \sqrt{\frac{k_{22}}{k_{11}+k_{22}}}$ can be proved for the interface macroelement E of the type shown in Fig. 4(c). □

Note that for isotropic bilinear form, we get $\gamma = \sqrt{\frac{1}{2}}$, which indicates very good efficiency of the two-level Schwarz preconditioners with interfaces. The efficiency is not deteriorated by jumps in coefficients if these jumps do not occur within the macroelements. On the other hand, the efficiency deteriorates if the subdomain boundaries cut the main direction of a strong anisotropy. Therefore, for strongly anisotropic problems, it may be useful to use domain decompositions cutting only the weak direction of the anisotropy. The numerical (grid) anisotropy has of course the same effect as the physical anisotropy.

The idea of using Schwarz preconditioner with no overlap but a suitable coarse grid appeared also in the paper [2], but in a context with a different construction of the coarse grid problem.

7 Parallel implementation of the preconditioners

Let us consider the solution of a symmetric elliptic boundary value problem by the FE method. Then we have to assemble and solve linear algebraic system

$$Au = b, \qquad u, b \in R^n, \tag{39}$$

with a symmetric positive definite $n \times n$ matrix. This system is an algebraic representation of the problem (5). The system (39) is mostly solved by the preconditioned conjugate gradient (CG) method.

The solution of large-scale systems requires to use powerful parallel computers and the domain decomposition can be then exploited for parallel implementation of all main operations:

- assembling of the system,
- matrix–by–vector multiplication,
- computation of vector updates and scalar products,
- construction of preconditioners.

This parallel implementation uses decomposition of data (vectors, matrices) into blocks corresponding to the subdomains. The overlapping DD induces decomposition of data to overlapping blocks $v \equiv \{v_k^\delta\}$ etc. The matrix–by–vector multiplication can be implemented with both overlapping and non–overlapping blocks. The first case may be advantageous if the extended diagonal blocks $\{A_{kk}^\delta\}$, $k = 1, \ldots, m$, cover all nonzero elements of A.

Blocks of data are mapped to the processors of the parallel computer and the domain decomposition or more precisely the decomposition of the FE grid should be done with a special care to the computational load of processors. This is not difficult for structured grids. For a suitable decomposition of unstructured grids, it is possible to use several graph algorithms, see e.g. [18]. The decomposition should also respect anisotropies as mentioned in Section 6 and further in Section 8. For two-level preconditioners, it is additionally needed to balance the size of the auxiliary global problem.

The space decomposition preconditioners require to solve the subproblems of the type $A_k w_k = v_k$. Although these subproblems are smaller than the original problem, it may be still too expensive or inefficient to solve these systems exactly by direct solvers. Then the exact solution of the subproblems can be avoided by different means.

First, we can use an approximation $\bar{A}_k$ to A_k for computing approximate values of $A_k^{-1} v_k$. This approximation can be given by incomplete factorization of A_k or by applying a fixed number of iterations with some linear stationary iterative method as SSOR or multigrid. The use of approximate operators $\bar{A}_k$ is covered by the theory described e.g. in [20].

Second, we can use inner CG iterations for solving $A_k w_k = v_k$ approximately up to some lower relative inner accuracy $\|A_k w_k - v_k\| \leq \varepsilon_0 \|v_k\|$, where ε_0 is say 10^{-1}. This choice gives SD preconditioners, which are not more linear operators. We can speak about nonlinear preconditioners. These preconditioners can be implemented within the standard preconditioned CG method, but the resulting inner-outer iterative method can fail or lead to a slow convergence. The reason for such difficulties is the loss of orthogonality and sometimes also loss of linear independence of the search directions in CG method with nonlinear preconditioning.

A remedy is to store (some) search directions and orthogonalize the new one to the stored ones. Such flexible or generalized preconditioned CG method is described e.g. in [4] and the references therein. In the next section, we shall use the generalized preconditioned GPCG[s] method with orthogonalization of the search direction to s previous ones.

The GPCG[s] method also allows to use nonsymmetric multiplicative or nonsymmetric hybrid preconditioners. These nonsymmetric preconditioners are substantially cheaper than the symmetrized variants requiring at least one extra update of the residual and frequently leading to similar convergence rates as the symmetrized variants.

The convergence of GPCG[s] with nonlinear approximate additive SD preconditioners can be proved with the aid of the convergence theory, which can be found e.g. in [4]. The convergence of GPCG[s] with nonlinear approximate multiplicative or hybrid preconditioner $G = G_M, G_H$ can be proved if $\|I - \omega G A\|_A < 1$ for some $\omega \in (0, 1\rangle$, see again [4].

8 Numerical results

The efficiency of various SD preconditioners described in this paper can be compared by solving a simple model problem

$$k_{11}\frac{\partial^2 u}{\partial x_1^2} + k_{22}\frac{\partial^2 u}{\partial x_2^2} = f \quad \text{in } \Omega = \langle 0,2\rangle \times \langle 0,3\rangle\,,$$
$$u = 0 \quad \text{on } \partial\Omega\,,$$

discretized by linear triangular FE. We use the uniform grid with the mesh size $h = 1/30$, subdomains $\Omega_k = \langle 0,2\rangle \times \langle x_k, x_{k+1}\rangle$ and overlap $\delta = 2h$. The subproblems are solved exactly. The required numbers of iterations for the accuracy $\varepsilon = 10^{-3}$ and various additive (AP) and hybrid (HP) SD preconditioners can be seen in Table 1. The hybrid preconditioners are used in nonsymmetric form in combination with GPCG[1]. The coarse grid is either nested coarse triangular grid, aggregation with clustering 2×2 square macroelements or the same aggregation with interface. For each number of subdomains, the first column shows the numbers of iterations for $k_{11} = k_{22} = 1$, the second one for $k_{11} = 1,\ k_{22} = 1/100$ and the last one for $k_{11} = 1,\ k_{22} = 100$.

Table 1. The numbers of iterations for the relative accuracy $\varepsilon = 10^{-3}$. AP denotes additive preconditioner, HP denotes nonsymmetric hybrid preconditioner + GPCG[1].

		Number of subdomains														
Type	Coarse grid space	4			8			12			16			24		
AP	one level	16	4	26	22	4	38	23	5	51	31	5	61	37	6	78
AP	nested, $H = 3h$	7	7	16	8	7	20	8	7	23	7	7	26	8	7	31
HP	nested, $H = 3h$	6	6	13	6	7	16	6	7	20	6	7	22	6	6	27
AP	regular agg's $2h$	13	8	19	15	8	23	16	9	27	17	9	30	17	9	36
HP	regular agg's $2h$	10	7	15	11	7	19	11	7	22	11	8	25	11	8	29
AP	agg's $2h$ + int.	14	8	22	14	8	30	14	8	36	14	9	40	14	9	46
HP	agg's $2h$ + int.	8	4	14	7	5	16	8	5	19	8	5	22	8	5	25

For demonstration of efficiency of parallel solvers based on CG method and additive Schwarz preconditioners, we give in Table 2 the numbers of iterations and computer times from solving large-scale linear system with the dimension of about 4 million arising from FE discretization of 3D elasticity problem, which has a practical application in geomechanics. More details can be found e.g. in [6]. The adopted discretization uses linear tetrahedral FE.

The computations use one directional decomposition up to 8 subdomains and coarse grid problem created by regular aggregation of 2×2×2 or 5×5×5 hexahedral macroelements. The computations were performed on a small Linux cluster consisting of 8 PCs with AMD Athlon/1400 processors and 768 MB memories. The computers are interconnected via standard Fast Ethernet 100 Mbit/s network.

Table 2. Results from the solution of the large-scale 3D elasticity problem with the relative accuracy $\varepsilon = 10^{-4}$. $\#P$ denotes the number of exploited processors, m is the number of subdomains.

	No coarse grid			Aggregation 2×2×2			Aggregation 5×5×5		
$\#m$	#P	#It	T [s]	#P	#It	T [s]	#P	#It	T [s]
2	2	104	428	3	45	268	3	56	243
3	3	113	321	4	47	242	4	59	176
4	4	121	268	5	51	240	5	62	144
5	5	128	234	6	53	236	6	65	127
6	6	133	211	7	55	244	7	67	114
7	7	136	193	8	57	265	8	70	105

9 Concluding remarks

We used the space decomposition framework for providing an overview of the Schwarz-type domain decomposition preconditioners. We described a large variety of these preconditioners, still not including some newly developed variants.

Our experience shows that the Schwarz preconditioners can be used for development of efficient and scalable parallel solvers at least when working on smaller parallel computing systems. These methods are flexible enough for balancing the computational load of the processors, adopting inexact solvers etc. On the other hand, we showed that a special care should be devoted to physical or numerical anisotropy and similar difficulties.

The Schwarz technique can be also used for solving other problems as nonsymmetric problems of convection-diffusion type, parabolic and nonlinear problems and many others. The general SD framework is suitable also for developing preconditioners based on hierarchical or physical decompositions, see e.g. [13, 20, 21, 4].

Acknowledgement: Many thanks are due to P. Byczanski and J. Starý for preparing the numerical results. The work was supported by the grant S3086102 of the Academy of Sciences of the Czech Republic.

References

1. Axelsson, O., Blaheta, R. (2001): Two simple derivations of universal bounds for the C.B.S. inequality constant. Report 0133, Dept. Math., University Nijmegen 2001. To appear in Appl. Math.
2. Bjørstadt, P.E., Dryja, M., Vainikko, E. (1997): Additive Schwarz methods without subdomain overlap and with new coarse spaces. In: Glowinski, R. et al. (eds) Domain Decomposition Methods in Sciences and Engineering. Proc. 8th Int. Conf. on Domain Decomposition Methods. J. Wiley
3. Blaheta, R. (1988): A multilevel method with overcorrection by aggregation for solving discrete elliptic problems. J. Comp. Appl. Math., **24**, 227–239

4. Blaheta, R.(2002): GPCG - generalized preconditioned CG method and its use with non-linear and non-symmetric displacement decomposition preconditioners. Numer. Linear Algebra Appl., **9**, 527-550
5. Blaheta, R.: Two-level Schwarz preconditioners with interfaces in the coarse grid space. In preparation
6. Blaheta, R., Jakl, O., Krečmer, K., Starý, J. (2002): Large-scale modelling in geomechanics with parallel computing on clusters of PC's. In: Mestat, P. (ed) NUMGE'02 5th European Conference Numerical Methods in Geotechnical Engineering. Presses de l'Ecole Nationale des Ponts et Chausses, Paris, pp.315-320
7. Bramble, J.H., Pasciak, J.E., Wang, J., Xu, J. (1991): Convergence estimates for product iterative methods with applications to domain decomposition. Math. Comp., **57**, 1–21
8. Bramble, J.H., Xu, J. (1991): Some estimates for a weighted L_2 projection, Math. Comput., **56**, 463-476
9. Brenner, S.C. (2000): Lower bounds for two-level additive Schwarz preconditioners with small overlap, SIAM J. Sci. Comput., **21**, 1657-1669
10. Brenner, S.C., Scott, L.R.(2002): The Mathematical Theory of Finite Element Methods, Springer-Verlag, New York Second edition
11. Brezina, M. (1997): Robust iterative methods on unstructured meshes, PhD. thesis, University of Colorado at Denver
12. Cai, X.C. (1995): The use of pointwise interpolation in domain decomposition methods with non-nested meshes. SIAM J. Sci. Comput., **16**, 250–256
13. Chan, T.F., Mathew, T.P. (1994): Domain Decomposition Algorithms. Acta Numerica, 61-143
14. Dryja, M., Widlund, O.B. (1994): Domain decomposition algorithms with small overlap, SIAM J. Sci. Comp., **15**, 604-620
15. Jenkins, E.W., Kelley, C.T., Miller C.T., Kees, C.E. (2001): An aggregation-based domain decomposition preconditioner for groundwater flow. SIAM J. Sci. Comput., **23**, 430–441
16. Lions, P.L. (1988): On the Schwarz alternating methods I. In: Chan, T.F., Glowinski, R., Periaux, J., Widlund, O.B. (eds): First International Symposium on Domain Decomposition Methods for PDE, SIAM, Philadelphia, 1–42
17. Mandel, J. (1994): Hybrid domain decomposition with unstructured subdomains. In: 6th Int. Conf. on Domain Decomposition, AMS Contemporary Mathematics vol. 157, 103-112
18. Saad, Y. (2003): Iterative methods for sparse linear systems. SIAM, Philadelphia
19. Schwarz, H.A. (1870): Gesammelte Mathematische Abhandlungen, volume 2, pages 133-143. Springer, Berlin 1890. First published in 1870
20. Smith, B.F., Bjørstad, P.E., Gropp, W.D. (1996): Domain Decomposition Parallel Multilevel Methods for Elliptic Partial Differential Equations. Cambridge University Press, Cambridge
21. Wolmuth, B. I. (2001): Discretization Methods and Iterative Solvers Based on Domain Decomposition, LNCSE 17, Springer Berlin

Boundary Conditions for Hyperbolic Equations or Systems

Thierry Gallouët

LATP, CMI, 39 rue Joliot-Curie, 13453 Marseille cedex 13 *gallouet@cmi.univ-mrs.fr*

Summary. Different types of boundary conditions in industrial numerical simulators involving the discretization of hyperbolic systems are presented. For some of them, one may determine the problem to which the limit of approximate solutions (as the discretization parameters tend to 0) is the unique solution. In turn, this convergence result may suggest other ways to take into account the boundary conditions.

1 Introduction

In the industrial context, efficient numerical simulators are often developped after a long "trial and error" procedure. The efficiency of the simulators may be evaluated, for instance, by the fact that the solution satisfies some natural constraints and that it is in agreement with experimental data. In some cases, estimates on the approximate solutions allow to obtain the convergence of some sequences of approximate solutions as the discretization size tends to 0. However, it is not easy to give the answer to the following question: "What problem has a unique solution which is the limit of the approximate solutions?".

This paper will focus on the problem of boundary conditions needed in the discretization of nonlinear hyperbolic equations or systems of equations; this problem is not yet clearly understood in many cases. Two different cases will be presented: a two phase flow in a pipeline and a two phase flow in a porous medium.

2 A two phase flow in a pipeline

2.1 Description of the system

A "simple" model for a two phase flow in a pipeline (see [8], for instance) leads to a 3×3 system of conservations laws. The unknown w is a function from $(0,1) \times \mathbf{R}_+$ in $\mathbf{R}^3$, solution of the following system:

$$w_t + (F(w))_x = 0, \ x \in (0,1), \ t \in \mathbf{R}_+, \tag{1}$$

where $(\cdot)_t$ and $(\cdot)_x$ denote the derivatives with respect to t and x variables. The first two equations of (1) give the mass conservation of the 2 phases (gas and liquid) and the third one is the momentum equation for the mixture. The expression of the given function $F: \mathbf{R}^3 \to \mathbf{R}^3$ is quite complicated. It takes into account thermodynamical laws and a hydrodynamical law. System (1) is hyperbolic: for any $w \in \mathbf{R}^3$, the Jacobian matrix $DF(w)$ is diagonalizable in $\mathbf{R}$. The three eigenvalues can be ordered: $\lambda_1(w) < \lambda_2(w) < \lambda_3(w)$. In real situations, the first eigenvalue, $\lambda_1(w)$ is negative and the third, $\lambda_3(w)$, is positive (they correspond to some "pressure waves" which are related to a "sound velocity"). The second eigenvalue, $\lambda_2(w)$, corresponds to some mean velocity between the two phases and can change sign. One can also note that the field related to this second eigenvalue is quite complicated because it is not, in general, a genuinely nonlinear field or a linearly degenerate field. In petroleum engineering, the wave associated to this second eigenvalue is a "void fraction wave"; engineers require a good representation of this wave in the numerical simulations.

Remark 1. In real situations, the function F in System (1) also depends on x, in order to take into account, for instance, the variation in the slope of the pipeline. Moreover, some source terms have to be added to the system, in order to take into account, for instance, some friction terms.

In order to complete System (1), an initial condition is prescribed:

$$w(x,0) = w_0(x), \; x \in (0,1), \tag{2}$$

and it is also necessary to give some boundary conditions. This appears to be not so easy. Indeed, classically, a general principle is that the number of boundary conditions needs to be equal to the number of positive eigenvalues of the Jacobian matrix at $x = 0$ and to the number of negative eigenvalues of the Jacobian matrix at $x = 1$ (and these boundary conditions have to satisfy some compatibility conditions). However, this principle is not so easy to understand when an eigenvalue changes sign during the simulation (or in the case of a null eigenvalue). A very interesting case is the so called "severe slugging" case in a pipeline. For this case, there are always two positive eigenvalues at $x = 0$ and two natural boundary conditions are prescribed at $x = 0$, namely the fluxes of gas and liquid; these boundary conditions can be taken constant in time. At $x = 1$, there is one natural boundary condition, namely the pressure (which is the same for the two phases, in this model), to be prescribed. It can also be constant in time. The true physical solution, which is measured by experiments (and the aim is to modelize these experiments), is periodical in time and it appears that, at $x = 1$, the first eigenvalue is always positive and the third one is always negative but the second eigenvalue changes sign during the simulation. In the sequel, one presents different ways to take into account the boundary conditions and one gives a convergence result in a simplified case.

2.2 Discretization of the problem

In order to discretize Problem (1), (2) and some boundary conditions, which will be introduced later, let $h = \frac{1}{N}$ (with $N \in \mathbf{N}^\star$) be the mesh size and $k > 0$ be the time step (assumed to be constant, for the sake of simplicity). The discrete unknown are the values $w_i^n \in \mathbf{R}^3$ for $i \in \{1, \ldots, N\}$ and $n \in \mathbf{N}$. The discretization of the initial condition leads to

$$w_i^0 = \frac{1}{h} \int_{(i-1)h}^{ih} w_0(x)dx, \ i \in \{1, \ldots, N\}. \tag{3}$$

For the computation of w_i^n for $n > 0$, one uses an explicit, 3-points scheme:

$$\frac{h}{k}(w_i^{n+1} - w_i^n) + F_{i+\frac{1}{2}}^n - F_{i-\frac{1}{2}}^n = 0, \ i \in \{1, \ldots, N\}, \ n \in \mathbf{N}. \tag{4}$$

For $i \in 1, \ldots, N-1$, one takes $F_{i+\frac{1}{2}}^n = g(w_i^n, w_{i+1}^n)$, where g is the numerical flux. It has to satisfy, in particular, the classical consistency condition, namely $g(a,a) = F(a)$, and needs to be chosen in order to obtain some stability properties for the numerical scheme under a so called CFL condition on the time step (see Sect. 3 for the study of a scalar model). In the case of two phase flow in a pipeline, the classical numerical fluxes such as the Godunov flux (see [9]) or the Roe flux (see [11]) may not be implemented, because of computational difficulties. A convenient choice is obtained with a simplified Roe flux, namely $g(a,b) = \frac{g(a)+g(b)}{2} + \frac{1}{2}|A(a,b)|(a-b)$, where $A(a,b)$ is some appoximation of the Jacobian matrix, depending on a and b, but not satisfying the so called Roe condition, see [8].

Remark 2. In fact, for the simulation of a two phase flow in a pipeline, the magnitude of the so-called fast eigenvalues, λ_1 and λ_3, is much greater than that of λ_2; the choice in [8] is to use an implicit scheme with respect to the fast eigenvalues, whereas the eingevalue λ_2, which corresponds to the void fraction wave, is handled with an explicit second order discretization, since the void fraction wave needs to be simulated precisely (see [8] for details).

Let us now define the fluxes $F_{\frac{1}{2}}^n$ and $F_{N+\frac{1}{2}}^n$ at the boundary.

2.3 Boundary conditions for the discretized problem

In order to compute $F_{\frac{1}{2}}^n$ (and similarily $F_{N+\frac{1}{2}}^n$) a good way is to know, or to determine, some artificial value $w_0^n \in \mathbf{R}^3$ (and $w_{N+1}^n \in \mathbf{R}^3$) and to take $F_{\frac{1}{2}}^n = g_0(w_0^n, w_1^n)$ (and $F_{N+\frac{1}{2}}^n = g_1(w_N^n, w_{N+1}^n)$). The numerical fluxes g_0 and g_1 can be chosen equal to g, but this is not at all necessary (see the convergence result of Sect. 3); in fact, there are numerous situations where one should take g_0 and g_1 different from g. Indeed, the scheme is often very sensitive to the computation of the boundary fluxes and it is often worthwhile to use a more precise, but also more expensive numerical flux (such as the Godunov flux, for instance) for the computation of the boundary fluxes than for the computation of the interior fluxes. The difficulty is now to determine these artificial values, w_0^n and w_{N+1}^n.

Remark 3. In some cases, the choice of w_0^n and w_{N+1}^n is quite easy. A well known example is given by the wall-boundary condition for the Euler equations (with a perfect gas state law or a more general state law). For the sake of simplicity, let us mention the one-dimensional case; the generalization to a multi-dimensional case is quite easy. The Euler equations may be written the form (1), corresponding to conservation of mass, momentum and energy, with $w = (\rho, \rho u, E)^t$, where ρ is the density of the fluid, u its velocity, and E its energy. The wall-boundary condition at $x = 0$ is $u = 0$, and the only component to compute for the boundary condition is the second component of $F_{\frac{1}{2}}^n$ which is equal here to the pressure at $x = 0$ (since $u = 0$ at the wall), say $p_{\frac{1}{2}}^n$. The value w_1^n may be computed from the values ρ_1^n, u_1^n and p_1^n. A natural choice for w_0^n is to take $\rho_0^n = \rho_1^n$, $u_0^n = -u_1^n$ and $p_0^n = p_1^n$. The flux $F_{\frac{1}{2}}^n$ (that is the value $p_{\frac{1}{2}}^n$) is then obtained with $F_{\frac{1}{2}}^n = g_0(w_0^n, w_1^n)$ and a convenient choice of the numerical flux g_0. We suggest to choose g_0 as the Godunov flux (or as a linearized Godunov flux, see [3] for instance). Numerical tests which were performed in [3] show that this choice is very satisfactory, even in the difficult case of a strong depressurization at the boundary. These tests also show that the pressure obtained with the Roe flux is not so satisfactory and neither is the choice $p_{\frac{1}{2}}^n = p_1^n$ which may seem natural (in particular, in 2D simulations, using a dual mesh obtained with a finite element primal mesh).

In most cases, however, the choice of w_0^n and w_{N+1}^n is not so easy. A possible method, which is described in [4], is now layed out, for a fixed n and g_0 given:

1. Compute $DF(w_1^n)$, its eigenvalues $\{\lambda_1, \lambda_2, \lambda_3\}$ and a basis of $\mathbf{R}^3$, $\{\varphi_1, \varphi_2, \varphi_3\}$, such that $DF(w_1^n)\varphi_i = \lambda_i \varphi_i$, $i = 1, 2, 3$.
2. Write w_1^n on the basis $\{\varphi_1, \varphi_2, \varphi_3\}$, namely $w_1^n = \alpha_1\varphi_1 + \alpha_2\varphi_2 + \alpha_3\varphi_3$,
3. Let p be the number of positive eigenvalues, compute $w_0^n = \beta_1\varphi_1 + \beta_2\varphi_2 + \beta_3\varphi_3$ and $F_{\frac{1}{2}}^n = g_0(w_0^n, w_1^n)$, where the three unknowns β_1, β_2, and β_3 are determined by the p equations stating the boundary conditions (note that these equations involve the components of $F_{\frac{1}{2}}^n$) and by the $3 - p$ equalities $\beta_i = \alpha_i$ for $\lambda_i < 0$.

This method leads, at each time step, to a nonlinear system of 3 equations with 3 unknowns (except if $\lambda_i = 0$ for some i), namely β_1, β_2 and β_3; note that some compatibility conditions are needed in order that this nonlinear system has a solution. Several variants of this method are possible. For instance, a boundary condition may be imposed on w_0^n rather than $F_{\frac{1}{2}}^n$. A similar method is, of course, possible at point $x = 1$ (changing the role of positive and negative eigenvalues).

This method is not always satisfactory. In the case of severe slugging for the simulation of two phase flow in a pipeline, the method seems to perform well at $x = 0$, where the eigenvalues λ_1 and λ_2 are always positive and the two boundary conditions (gas and liquid fluxes) are convenient. However, at $x = 1$, the second eigenvalue sometimes becomes negative and one needs a second boundary condition (the first one is a condition on the pressure). A natural condition seems to be $Q_l = 0$, where Q_l is the second component of the flux F, that is the liquid flux, but this condition does not lead to good results. Other possible choices of this additional boundary condition at $x = 1$ were tested and did not give good results. A possible interpretation of this problem is the fact that the sign of λ_2 is computed with w_N^n. Roughly speaking, it is "too late" when $\lambda_2(w_N^n)$ becomes negative (see Sect. 3 for the study of a simple scalar case). Indeed, good results (in agreement with experiments) are obtained with the unilateral condition $Q_l \geq 0$ (whatever the sign of $\lambda_2(w_N^n)$). It consists in using the preceeding method (for the boundary condition at $x = 1$) and in replacing, in the numerical scheme (4), the second component of $F^n_{N+\frac{1}{2}}$ by its positive part. Then, if $\lambda_2(w_N^n) < 0$, two boundary conditions are given at $x = 1$ (pressure and $Q_l = 0$) and if $\lambda_2(w_N^n) \geq 0$, one boundary condition is given at $x = 1$ (pressure) but, in (4), the second component of $F^n_{N+\frac{1}{2}}$ is replaced by its positive part.

In the following section, we will try to understand the sense of this boundary condition in a simplified scalar case.

3 The scalar case

A general convergence result is presented here in the case of a scalar equation. Then this result will be applied to understand the sense of the boundary condition, described at $x = 1$ in the previous section, in a simplified scalar case.

3.1 A general convergence result

The unknown is now a function $u : (0,1) \times \mathbf{R}+ \to \mathbf{R}$. The flux is a function $f \in C^1(\mathbf{R},\mathbf{R})$ (or $f : \mathbf{R} \to \mathbf{R}$ Lipschitz continuous) and the initial datum is $u_0 \in L^\infty((0,1))$. Let $A, B \in \mathbf{R}$ be such that $A \leq u_0 \leq B$ a.e.. The problem to solve is:

$$u_t + (f(u))_x = 0,\ x \in (0,1),\ t \in \mathbf{R}_+, \tag{5}$$

with the initial condition:

$$u(x,0) = u_0(x),\ x \in (0,1), \tag{6}$$

and some boundary conditions which will be prescribed later.

As in the previous section, let $h = \frac{1}{N}$ (with $N \in \mathbf{N}^\star$) be the mesh size and $k > 0$ be the time step (assumed to be constant, for the sake of simplicity). The discrete unknowns are now the values $u_i^n \in \mathbf{R}$ for $i \in \{1, \ldots, N\}$ and $n \in \mathbf{N}$. In order to define the approximate solution a.e. in $(0,1) \times \mathbf{R}$, one sets $u_{h,k}(x,t) = u_i^n$ for $x \in ((i-1)h, ih)$, $t \in (nk, (n+1)k)$, $i \in \{1, \ldots, N\}$, $n \in \mathbf{N}$.

The discretization of the initial condition leads to

$$u_i^0 = \frac{1}{h} \int_{(i-1)h}^{ih} u_0(x)dx, \; i \in \{1, \ldots, N\}. \tag{7}$$

For the computation of u_i^n for $n > 0$, one uses, as before, an explicit, 3-points scheme:

$$\frac{h}{k}(u_i^{n+1} - u_i^n) + f_{i+\frac{1}{2}}^n - f_{i-\frac{1}{2}}^n = 0, \; i \in \{1, \ldots, N\}, \; n \in \mathbf{N}. \tag{8}$$

For $i \in 1, \ldots, N-1$, one takes

$$f_{i+\frac{1}{2}}^n = g(u_i^n, u_{i+1}^n), \tag{9}$$

where g is the numerical flux. Sufficient conditions on $g: \; [A,B]^2 \to \mathbf{R}$, in order to have a convergent scheme if $x \in \mathbf{R}$ instead of $(0,1)$, are:

C1: g is non decreasing with respect to its first argument and nonincreasing with respect to its second argument,
C2: $g(s,s) = f(s)$, for all $s \in [A,B]$,
C3: g is Lipschitz continuous.

Let L be a Lipschitz constant for g (on $[A,B]^2$) and $\zeta > 0$. If $(0,1)$ is replaced by $\mathbf{R}$, it is well known (see e.g. [4]) that, if $k \le (1-\zeta)\frac{h}{L}$, the approximate solution $u_{h,k}$, that is the solution defined by (7)-(9) (with $i \in \mathbf{Z}$), takes its values in $[A,B]$ and converges towards the unique entropy weak solution of (5)-(6) in $L^p_{loc}(\mathbf{R} \times \mathbf{R}_+)$ as $h \to 0$.

In the case $x \in (0,1)$ instead of $x \in \mathbf{R}$, one assumes the same conditions on g, namely (C1)-(C3). In order to complete the scheme, one has to define $f_{\frac{1}{2}}^n$ and $f_{N+\frac{1}{2}}^n$.

Let $\overline{u}, \overline{\overline{u}} \in L^\infty(\mathbf{R}_+)$ be such that $A \le \overline{u}, \overline{\overline{u}} \le B$, a.e. on $\mathbf{R}_+$, let $g_0, g_1: \; [A,B]^2 \to \mathbf{R}$, satisfying (C1)-(C3), and define:

$$\begin{aligned} f_{\frac{1}{2}}^n &= g_0(\overline{u}^n, u_1^n); \;\; \overline{u}^n = \tfrac{1}{k} \textstyle\int_{nk}^{(n+1)k} \overline{u}(t)dt \\ f_{N+\frac{1}{2}}^n &= g_1(u_N^n, \overline{\overline{u}}^n,); \;\; \overline{\overline{u}} = \tfrac{1}{k} \textstyle\int_{nk}^{(n+1)k} \overline{\overline{u}}(t)dt, \end{aligned} \tag{10}$$

Then a convergence theorem can be proven as in the case $x \in \mathbf{R}$, see [13]:

Theorem 1. *Let $f \in C^1(\mathbf{R}, \mathbf{R})$ (or $f: \; \mathbf{R} \to \mathbf{R}$ Lipschitz continuous). Let $u_0 \in L^\infty((0,1))$, $\overline{u}, \overline{\overline{u}} \in L^\infty(\mathbf{R}_+)$ and $A, B \in \mathbf{R}$ be such that $A \le u_0 \le$*

B a.e. on $(0,1)$, $A \le \overline{u}, \overline{\overline{u}} \le B$ a.e. on $\mathbf{R}_+$. Let $g_0, g_1 : [A,B]^2 \to \mathbf{R}$, satisfying (C1)-(C3). Let L be a common Lipschitz constant for g, g_0 and g_1 (on $[A,B]^2$) and let $\zeta > 0$. Then, if $k \le (1-\zeta)\frac{h}{L}$, the equations (7)-(10) define an approximate solution $u_{h,k}$ which takes its values in $[A,B]$ and converges towards the unique solution of (11) in $L^p_{loc}([0,1] \times \mathbf{R}_+)$ for any $1 \le p < \infty$, as $h \to 0$:

$$
\begin{aligned}
&u \in L^\infty((0,1) \times (0,\infty)),\\
&\int_0^\infty \int_0^1 [(u-\kappa)^\pm \varphi_t + sign_\pm(u-\kappa)(f(u)-f(\kappa))\varphi_x]dxdt\\
&+M\int_0^\infty (\overline{u}(t)-\kappa)^\pm \varphi(0,t)dt + M\int_0^\infty (\overline{\overline{u}}(t)-\kappa)^\pm \varphi(1,t)dt\\
&\qquad\qquad + \int_0^1 (u_0-\kappa)^\pm \varphi(x,0)dx \ge 0,\\
&\qquad\qquad \forall \kappa \in [A,B],\ \forall \varphi \in C^1_c([0,1] \times [0,\infty), \mathbf{R}_+).
\end{aligned}
\tag{11}
$$

In (11), M is any bound for $|f'|$ on $[A,B]$ (and the solution of (11) does not depends on the choice of M). The definition of $sign_\pm$ is: $sign_+(s) = 1$ if $s > 0$, $sign_+(s) = 0$ if $s < 0$, $sign_-(s) = 0$ if $s > 0$, $sign_-(s) = -1$ if $s < 0$.

Remark 4.

1. It is interesting to remark that this convergence result is also true if the function g depends on i and n, provided that L is a common Lipschitz constant for all these functions.
2. The definition (11) of solution of (5)-(6) with the "weak" boundary conditions $\overline{u}$ and $\overline{\overline{u}}$ at $x = 0$ and $x = 1$ is essentially due to F. Otto, see [10].
3. It is interesting also to remark that if one replaces, in (11), the two entropies $(u-\kappa)^\pm$ by the sole entropy $|u-\kappa|$, one has an existence result (since $|u-\kappa| = (u-\kappa)^+ + (u-\kappa)^-$) but no uniqueness result, see [13] for a counter-example to uniqueness.
4. This convergence result can be generalized to the multidimensional case, see Sect. 5 and [13].

If u, solution of (11), is regular enough (say $u \in C^1([0,1] \times \mathbf{R}_+)$, for instance), u satisfies $u(0,t) = \overline{u}(t)$ and $u(1,t) = \overline{\overline{u}}(t)$ in the weak sense given in [1]. This condition is very simple if f is monotone:
If $f' > 0$, then $u(0,\cdot) = \overline{u}$ and u does not depend on $\overline{\overline{u}}$.
If $f' < 0$, then $u(1,\cdot) = \overline{\overline{u}}$ and u does not depend on $\overline{u}$.

3.2 A very simple example

One considers here Equation (5), with initial condition (6) and weak boundary condition $\overline{u}$ and $\overline{\overline{u}}$ at $x = 0$ and $x = 1$, that is in the sense of (11), in the particular case $f' > 0$. In this case, the main example of numerical flux is

$g = g_0 = g_1$, $g(a,b) = f(a)$, which leads to the well known upstream scheme. With this choice of g_0 and g_1, using the notations of Sect. 3.1, the boundary conditions are taken into account in the form:

$$f^n_{\frac{1}{2}} = f(\overline{u}^n),\ f^n_{N+\frac{1}{2}} = f(u^n_N), \tag{12}$$

with $\overline{u}^n = \frac{1}{k}\int_{nk}^{(n+1)k} \overline{u}(t)dt$. One may apply the general convergence theorem. The approximate solutions converge (as $h \to 0$) towards the solution of (11). In this case, the approximate solutions, as well as the solution of (11), do not depends on $\overline{\overline{u}}$.

In the case $f' < 0$ the main example is $g = g_0 = g_1$, $g(a,b) = f(b)$, which also leads to the upstream scheme. The boundary conditions are taken into account in the following way:

$$f^n_{\frac{1}{2}} = f(u^n_1),\ f^n_{N+\frac{1}{2}} = f(\overline{\overline{u}}^n), \tag{13}$$

with $\overline{\overline{u}}^n = \frac{1}{k}\int_{nk}^{(n+1)k} \overline{\overline{u}}(t)dt$.

These simple cases suggest the following scheme for any f, which is the scalar version of the scheme described in Sect. 2.2 (note that $f'(u)$ is the Jacobian matrix at point $u \in \mathbf{R}$):

– Boundary condition at $x = 0$:

$$\begin{cases} f^n_{\frac{1}{2}} = f(\overline{u}^n), & \text{if } f'(u^n_1) > 0, \\ f^n_{\frac{1}{2}} = f(u^n_1), & \text{if } f'(u^n_1) < 0. \end{cases} \tag{14}$$

– Boundary condition at $x = 1$:

$$\begin{cases} f^n_{N+\frac{1}{2}} = f(\overline{\overline{u}}^n), & \text{if } f'(u^n_N) < 0, \\ f^n_{N+\frac{1}{2}} = f(u^n_N), & \text{if } f'(u^n_N) > 0. \end{cases} \tag{15}$$

This solution is not always satisfactory as can be shown on the following simple example with the Burgers equation:

Let $f(s) = s^2$, $u_0 = 1$ a.e. on $(0,1)$, $\overline{u} = 1$ a.e. on $\mathbf{R}_+$ and $\overline{\overline{u}} = -2$ a.e. on $\mathbf{R}_+$. The exact solution which has to be approached by the numerical scheme is the unique solution of (11) with these values of f, u_0, $\overline{u}$ and $\overline{\overline{u}}$. Computing the approximate solution with (7)-(9), the function g satisfying $(C2)$, and with (14)-(15), leads to an approximate solution which is constant and equal to 1 for any h and k. Then it does not converge (as h and k go to 0) towards the exact solution which is not constant and equal to 1 since, for the exact solution, a shock wave with a negative speed starts from the point $x = 1$ at time $t = 0$. Indeed, one can also remark that this approximate solution is the exact solution of (11) with the same values of f, u_0, $\overline{u}$ and with any $\overline{\overline{u}}$ satisfying $\overline{\overline{u}} \geq -1$ a.e. on $\mathbf{R}_+$. In order to obtain a convergent approximation of the exact solution corresponding to $\overline{\overline{u}} = -2$, a good choice is, instead of (15), $f^n_{N+\frac{1}{2}} = g_1(u^n_N, -2)$ with g_1 satisfying (C1)-(C3).

3.3 A simplifed model for two phase flows in pipelines

It is now possible to understand the treatment of the boundary described in Sect. 2 on a simplified model. This simplifed model for two phase flows in pipelines is given in [12]. For this model, the densities are constant so that there are no longer pressure waves but only the void fraction wave, corresponding to the second eigenvalue of the original system (1). It is also easy to see that for this model, the total flux (that is the sum of the fluxes of the two phases) is constant in space. One also assumes that this total flux is constant in time (and positive). System (1) is then reduced to a scalar equation, Equation (5), where the unknown, $u : (0,1) \times \mathbf{R} \to \mathbf{R}$, is the gas fraction which takes its values between 0 and 1.

The function f can be taken as $f(s) = as - bs^2$, where $a, b \in \mathbf{R}$ are given and such that $0 < b < a < 2b$. In (5), the quantity $f(u)$ is the flux of gas. Then $f(1) - f(u)$ is the flux of liquid. The function f is increasing between 0 and $u_M = a/(2b)$ and decreasing between u_M and 1. An important value is $u_m \in [0, u_M]$ such that $f(u_m) = f(1)$.

One takes $u_0 = 0$ a.e. on $[0,1]$ as an initial condition. At $x = 0$, the gas flux is given (as in the complete model, see Sect. 2), one takes $f(u(0,\cdot)) = \overline{f}$ with $\overline{f}(t) = c$ for $t \le T$ and $\overline{f}(t) = 0$ for $t \ge T$, where c and T are given with $c > f(1)$ and T large enough so that f' changes sign at $x = 1$ during the simulation. Indeed, in this simplified model, it is also necessary to take T not too large in order to avoid a problem at $x = 0$ (for T too large, f' will also changes sign at $x = 0$). The boundary condition at $x = 1$ will be described on the discrete problem below.

The discretization of the problem is performed as before with (7)-(9), with g satisfying (C1)-(C3).

For the discretization of the boundary condition at $x = 0$, the method described in Sect. 2 leads here to

$$f^n_{\frac{1}{2}} = \overline{f}(nk), \tag{16}$$

which is indeed in accordance with the fact that $f'(u^n_1) > 0$ for all n, at least if T is not too large.

For the discretization of the boundary condition at $x = 1$, the first method described in Sect. 2 and given in (15), using the sign of $f'(u^n_N)$ leads to

$$\begin{cases} f^n_{N+\frac{1}{2}} = f(u^n_N), & \text{if } u^n_N < u_M, \\ f^n_{N+\frac{1}{2}} = f(1), & \text{if } u^n_N > u_M, \end{cases} \tag{17}$$

and does not lead to the desired results. Note also that $f^n_{N+\frac{1}{2}}$, given by (17), is a discontinuous function of u^n_N.

The second method, described in Sect. 2, uses the fact that the liquid flux cannot be negative at $x = 1$. Since the liquid flux at $x = 1$ is $f(1) - f_{N+\frac{1}{2}}$ and since $f(u_m) = f(1)$, this method leads to

$$\begin{cases} f^n_{N+\frac{1}{2}} = f(u^n_N), & \text{if } u^n_N \le u_m, \\ f^n_{N+\frac{1}{2}} = f(u_m), & \text{if } u^n_N > u_m, \end{cases} \tag{18}$$

Note that $f^n_{N+\frac{1}{2}}$, given by (18), is a continuous function of u^n_N. We shall apply the convergence theorem, Theorem 1 given in Sect. 3.1, for the boundary conditions (16) and (18), and understand the boundary conditions satisfied by the limit of the approximate solutions. In order to do so, we need to find g_0 and g_1, satisfying (C1)-(C3), and $\overline{u}, \overline{\overline{u}} \in L^\infty(\mathbf{R}_+)$ such that $f^n_{\frac{1}{2}}$ and $f^n_{N+\frac{1}{2}}$, respectively defined by (16) and (18), satisfy (10). Indeed, it is shown in [7] that both boundary fluxes $f^n_{\frac{1}{2}}$ and $f^n_{N+\frac{1}{2}}$ may be expressed with the Godunov flux in the following way:

- Boundary flux at $x = 1$. One takes $\overline{\overline{u}} = 1$ a.e. on $\mathbf{R}_+$ and g_0 equal to the Godunov flux, that is $g_0 = g_G$ with

$$g_G(\alpha, \beta) = \begin{cases} \min\{f(s),\ s \in [\alpha, \beta]\} & \text{if } \alpha \le \beta, \\ \max\{f(s),\ s \in [\beta, \alpha]\} & \text{if } \alpha > \beta. \end{cases}$$

  The formula (18) reads

$$f^n_{N+\frac{1}{2}} = g_G(u^n_N, 1) = \begin{cases} f(u^n_N) & \text{if } u^n_N \le u_m, \\ f(1) & \text{if } u^n_N > u_m. \end{cases} \tag{19}$$

- Boundary flux at $x = 0$. One assumes (for simplicity) that $\frac{T}{k} \in \mathbf{N}$. Let $\alpha, \beta \in (0,1)$ such that $\alpha < \beta$ and $f(\alpha) = f(\beta) = c$. One takes

$$\overline{u}(t) = \begin{cases} \alpha & \text{if } t < T, \\ 0 & \text{if } t > T, \end{cases} \tag{20}$$

  so that, recalling that $\overline{u}^n = \frac{1}{k}\int_{nk}^{(n+1)k} \overline{u}(t)dt$,

$$f(\overline{u}^n) = \begin{cases} c & \text{if } nk < T, \\ 0 & \text{if } nk \ge T, \end{cases}$$

  Then, if $u^n_1 \le \beta$, the formula (16) reads

$$f^n_{\frac{1}{2}} = g_G(\overline{u}^n, u^n_1), \tag{21}$$

  since, in this case, $g_G(\overline{u}^n, u^n_1) = f(\overline{u}^n)$. The fact that $u^n_1 \le \beta$ is true for all n if T is not too large. If T is too large, the convergence result can be applied with (21) instead of (16).

It is now possible to apply Theorem 1. Let L be a common Lipschitz constant for g and g_G (on $[0,1]^2$) and let $\zeta > 0$. If $k \le (1-\zeta)\frac{h}{L}$, the approximate solution $u_{h,k}$, that is the solution defined by (7)-(9), with the boundary fluxes

(19)-(21) (and $u_0 = 0$, $\overline{\overline{u}} = 1$ and $\overline{u}$ given by (20)), takes its values in $[0,1]$ and converges towards the unique solution of (22) in $L^p_{loc}([0,1] \times \mathbf{R}_+)$ for any $1 \leq p < \infty$, as $h \to 0$:

$$\begin{aligned}
&u \in L^\infty((0,1) \times (0,\infty)),\\
&\int_0^\infty \int_0^1 [(u-\kappa)^\pm \varphi_t + \text{sign}_\pm(u-\kappa)(f(u) - f(\kappa))\varphi_x] dx dt\\
&\quad + M \int_0^\infty (\overline{u}(t) - \kappa)^\pm \varphi(0,t) dt + M \int_0^\infty (1-\kappa)^\pm \varphi(1,t) dt \qquad (22)\\
&\qquad + \int_0^1 (0-\kappa)^\pm \varphi(x,0) dx \geq 0,\\
&\qquad \forall \kappa \in [0,1],\ \forall \varphi \in C^1_c([0,1] \times [0,\infty), \mathbf{R}_+).
\end{aligned}$$

If u, solution of (22), is regular enough on $[0,1] \times (0,T)$, then, it is possible to prove that u satisfies the boundary conditions, for $0 < t < T$, in the following sense (see [13] and [7]):

- Boundary condition at $x = 0$ (recall that $\overline{u}$ is given by (20)): $u(0,t) = \alpha$ or $u(0,t) \geq \beta$. In fact, if T is not too large, one has $u(0,t) = \alpha$.
- Boundary condition at $x = 1$: $u(1,t) \leq u_m$ or $u(1,t) = 1$.

Thanks to Theorem 1, it is possible to give other choices for $f^n_{N+\frac{1}{2}}$ for which the approximate solutions obtained with this new choice of $f^n_{N+\frac{1}{2}}$ converge towards the same function u, which is the unique solution of (22). Indeed, let $h: [0,1] \to \mathbf{R}$ be a nondecreasing function such that $h \leq f$ and $h(1) = f(1)$ and take:

$$f^n_{N+\frac{1}{2}} = h(u^n_N). \qquad (23)$$

One may construct a function g_1 satisfying (C1)-(C3) such that $h(s) = g_1(s,1)$, for all $s \in [0,1]$, and then use Theorem 1. Let L be a common Lipschitz constant for g and g_G and g_1 (on $[0,1]^2$) and let $\zeta > 0$. If $k \leq (1-\zeta)\frac{h}{L}$, the approximate solution $u_{h,k}$, that is the solution defined by (7)-(9), with the boundary fluxes (23) and (21) (and $u_0 = 0$, $\overline{\overline{u}} = 1$ and $\overline{u}$ given by (20)), takes its values in $[0,1]$ and converges towards the unique solution of (22) in $L^p_{loc}([0,1] \times \mathbf{R}_+)$ for any $1 \leq p < \infty$, as $h \to 0$.

Turning back to the complete system described in Sect. 2, the analysis of this simplified model for two phase flows in pipelines may also suggest another way to take into account the boundary condition at $x = 1$ (with a given numerical flux g_1):

1. Compute $DF(w^n_N)$, its eigenvalues $\{\lambda_1, \lambda_2, \lambda_3\}$ and a basis of $\mathbf{R}^3$, $\{\varphi_1, \varphi_2, \varphi_3\}$, such that $DF(w^n_N)\varphi_i = \lambda_i \varphi_i$, $i = 1,2,3$.
2. Write w^n_N on the basis $\{\varphi_1, \varphi_2, \varphi_3\}$, namely $w^n_N = \alpha_1\varphi_1 + \alpha_2\varphi_2 + \alpha_3\varphi_3$.

3. Since $\lambda_3 < 0$ and since one wants $Q_l \geq 0$, compute $w^n_{N+1} = \beta_1\varphi_1 + \beta_2\varphi_2 + \beta_3\varphi_3$ and $F^n_{N+\frac{1}{2}} = g_1(w^n_N, w^n_{N+1})$ with the following 3 conditions on the components of w^n_{N+1}: usual condition on the pressure, $\beta_3 = \alpha_3$ and $R^n_{N+1} = 1$ where R^n_{N+1} is the gas fraction computed with w^n_{N+1}.

4 Two phase flow in a porous medium

A second example is given by the modelization of a two phase flow, oil and water (for instance), in a porous medium. Phases are immiscible. Compressibility and capillarity effects are neglected. The model is obtained using the conservation of mass for each phase and Darcy's law. This study is limited to the one dimensional case. In this case the pressure can be eliminated and the problem is reduced to a single equation, namely (5) with:

$$f(u) = \frac{f_1(u)(\alpha + \beta f_2(u))}{f_1(u) + f_2(u)}. \tag{24}$$

The unknown is the saturation of one phase, say water, and is denoted by u. The quantity α is the total flux, which is constant in space, thanks to the incompressibility of the phases. One assumes also that it is constant in time and positive. The quantity β is the difference between the densities of the phases. The functions f_1 and f_2 are the mobilities of the phases. The function f_1 is nondecreasing, regular and satisfies $f_1(0) = 0$. The function f_2 is nonincreasing, regular and satisfies $f_2(1) = 0$. The function $f_1 + f_2$ is bounded from below by a positive number.

Remark 5. For the equivalent two or three dimensional model, the pressure cannot be eliminated and the resulting model is a coupled system of two partial differential equations and two unknowns (pressure and saturation). The problem to which the limit of the approximate solutions is solution is then much more complicated to determine. See [6] for a partial study of this question.

Here again, an initial condition is prescribed, namely (6), with $u_0 \in L^\infty((0,1))$, $0 \leq u_0 \leq 1$ a.e.. The boundary condition will be given later.

The numerical scheme is as in Sect. 3.1; it is given by (7) and (8) with (9). The choice of the numerical flux, g, satisfying (C1)-(C3), is usually given, for this model, using an "upwinding phase by phase", that is (see [2], for instance):

$$\begin{aligned} g(a,b) &= \frac{f_1(a)(\alpha + \beta f_2(a))}{f_1(a) + f_2(a)} \text{ if } -\alpha + \beta f_1(a) \leq 0 \\ g(a,b) &= \frac{f_1(a)(\alpha + \beta f_2(b))}{f_1(a) + f_2(b)} \text{ if } -\alpha + \beta f_1(a) > 0. \end{aligned} \tag{25}$$

Let us then define $f^n_{\frac{1}{2}}$ and $f^n_{N+\frac{1}{2}}$. On considers here the case of an injection of pure water at $x = 0$. Then:

$$f^n_{\frac{1}{2}} = \alpha,\ n \geq 0. \tag{26}$$

At $x = 1$, The boundary condition is quite complicated. A simple example is (see [7] for a more complete study):

$$f^n_{N+\frac{1}{2}} = \frac{f_1(u^n_N)\alpha}{f_1(u^n_N) + f_2(u^n_N)}. \tag{27}$$

Then the approximate solution is given with (7)-(9), g given by (25), and (26)-(27).

In order to prove that the approximate solutions converge, as h and k go to zero, and to determine the problem for which the limit of the approximate solutions is its unique solution, one proceeds as in Sect. 3.3. One has to find g_0 and g_1 satisfying (C1)-(C3) and $\overline{u}, \overline{\overline{u}} \in L^\infty(\mathbf{R}_+)$ such that $f^n_{\frac{1}{2}}$ and $f^n_{N+\frac{1}{2}}$, respectively defined by (26) and (27), satisfy (10). This is again performed in [7]. The most interesting case is obtained for $\beta f_1(1) > \alpha$ and when the function f is increasing on $(0, u_M)$ and decreasing on $(u_M, 1)$, as in Sect. 3.3. In fact, the main point is the existence of a unique $u_m \in (0,1)$such that $f(u_m) = f(1) = \alpha$ and that f is increasing on $[0, u_m]$ and greater or equal to α on $[u_m, 1]$. Then it is quite easy to prove that (26) yields

$$f^n_{\frac{1}{2}} = \alpha = g_G(u_m, u^n_1),$$

where g_G is the Godunov flux given in Sect. 3.3.

For the boundary condition at $x = 1$, it is possible to construct (see [7]) a function $g_1 :\ [0,1]^2 \to \mathbf{R}$ satisfying (C1)-(C3) such that (27) gives:

$$f^n_{N+\frac{1}{2}} = g_1(u^n_N, 1).$$

It is now possible to use Theorem 1.

Let L be a common Lipschitz constant for g (given by (25)), g_G and g_1 (on $[0,1]^2$) and let $\zeta > 0$. If $k \leq (1-\zeta)\frac{h}{L}$, the approximate solution $u_{h,k}$, that is the solution defined by (7)-(9) (with g given by 25), and by the boundary fluxes (26)-(27), takes its values in $[0,1]$ and converges towards the unique solution of (28) in $L^p_{loc}([0,1] \times \mathbf{R}_+)$ for any $1 \leq p < \infty$, as $h \to 0$:

$$\begin{aligned}
&u \in L^\infty((0,1) \times (0,\infty)),\\
&\int_0^\infty \int_0^1 [(u-\kappa)^\pm \varphi_t + \mathrm{sign}_\pm(u-\kappa)(f(u)-f(\kappa))\varphi_x]dxdt\\
&\quad +M\int_0^\infty (u_m-\kappa)^\pm \varphi(0,t)dt + M\int_0^\infty (1-\kappa)^\pm \varphi(1,t)dt\\
&\qquad\qquad + \int_0^1 (u_0-\kappa)^\pm \varphi(x,0)dx \geq 0,\\
&\qquad\qquad \forall \kappa \in [0,1],\ \forall \varphi \in C^1_c([0,1] \times [0,\infty), \mathbf{R}_+),
\end{aligned} \tag{28}$$

where M is a bound for $|f'|$ on $[0,1]$ (f is given by (24). As in Sect. 3.3 it is possible to give the sense of the boundary condition if u is regular enough. Indeed, let u be a regular solution of (28). Then u satisfies the boundary conditions in the sense given by [1], that is:

$$\mathrm{sign}(u(0,t)-u_m)(f(u(0,t))-f(\kappa))\leq 0,\ \forall\kappa\in[u_m,u(0,t)],\ \text{for a.e. } t\in\mathbf{R}_+,$$

$$\mathrm{sign}(u(1,t)-1)(f(u(1,t))-f(\kappa))\geq 0,\ \forall\kappa\in[1,u(1,t)],\ \text{for a.e. } t\in\mathbf{R}_+,$$

with $[a,b]=\{ta+(1-t)b,\ t\in[0,1]\}$ and $\mathrm{sign}(s)=1$ for $s>0$, $\mathrm{sign}(s)=-1$ for $s<0$, $\mathrm{sign}(0)=0$.

This gives $u(0,t)=u_m$ or $u(0,t)=1$ and $u(1,t)\leq u_m$ or $u(1,t)=1$. In particular, at $x=0$, one has $f(u(0,t))=\alpha$ (only water is injected) and, at $x=1$, $f(u(1,t))<\alpha$ if $u(1,t)<u_m$ (which states that there is some oil production).

5 The multidimensional scalar case

In this section, a generalization of Theorem 1 is presented for the multidimensional scalar case together with a rough sketch of proof. For the sake of simplicity, one considers $d=2$ (the extension to $d=3$ is straightforward) and a flux function under the form $v(x,t)f(u)$, with $\mathrm{div}(v(\cdot,t))=0$ (see [13] for the general case of a flux function $f(x,t,u)$). This leads to the following equation:

$$u_t+\mathrm{div}(vf(u))=0,\ \text{in } \Omega\times(0,T), \tag{29}$$

where Ω is a bounded polygonal open set of $\mathbf{R}^2$, $T>0$, $f\in C^1(\mathbf{R},\mathbf{R})$ (or $f:\ \mathbf{R}\to\mathbf{R}$ Lipschitz continuous) and $v\in C^1(\mathbf{R}^2\times[0,T])\to\mathbf{R}^2$ with $\mathrm{div}(v(\cdot,t))=0$ in $\mathbf{R}^2$ for all $t\in[0,T]$. The unknown is $u:\ \Omega\times(0,T)\to\mathbf{R}$.

Let $u_0\in L^\infty(\Omega)$ and $\overline{u}\in L^\infty(\partial\Omega\times(0,T))$. Let $A,B\in\mathbf{R}$ be such that $A\leq u_0\leq B$ a.e. on Ω and $A\leq\overline{u}\leq B$ a.e. on $\partial\Omega\times(0,T)$. Following the work of [10], an entropy weak solution of (29) with the initial condition u_0 and the (weak) boundary condition $\overline{u}$ is a solution of (30):

$$\begin{aligned}
&u\in L^\infty(\Omega\times(0,T)),\\
&\int_0^T\int_\Omega[(u-\kappa)^\pm\varphi_t+\mathrm{sign}_\pm(u-\kappa)(f(u)-f(\kappa))v\cdot\mathrm{grad}\varphi]dxdt\\
&\qquad\qquad+M\int_0^T\int_{\partial\Omega}(\overline{u}(t)-\kappa)^\pm\varphi(x,t)d\gamma(x)dt\\
&\qquad\qquad\qquad+\int_\Omega(u_0-\kappa)^\pm\varphi(x,0)dx\geq 0,\\
&\qquad\forall\kappa\in[A,B],\ \forall\varphi\in C_c^1(\overline{\Omega}\times[0,T),\mathbf{R}_+),
\end{aligned} \tag{30}$$

where $d\gamma(x)$ stands for the integration with respect to the one dimensional Lebesgue measure on the boundary of Ω and M is such that

$$\|v\|_\infty |f(s_1) - f(s_2)| \leq M|s_1 - s_2| \text{ for all } s_1, s_2 \in [A, B],$$

where $\|v\|_\infty = \sup_{(x,t)\in\Omega\times[0,T]} |v(x,t)|$ (and $|\cdot|$ denotes here the Euclidean norm in $\mathbf{R}^2$).

Remark 6.

1. If u satisfies the family of inequalities (30), it is possible to prove that u is a solution of (29) (on a weak form), u satisfies some entropy inequalities in $\Omega\times(0,T)$, namely $|u-\kappa|_t+\mathrm{div}(v(f(\max(u,\kappa))-f(\min(u,\kappa)))) \leq 0$ for all $\kappa \in \mathbf{R}$, but also on the boundary of $\partial\Omega$ and on $t = 0$. u satisfies the initial condition ($u(\cdot, 0) = u_0$) and u satisfies partially the boundary condition. For instance, if $f' > 0$ and u is regular enough, then $u(x,t) = \overline{u}(x,t)$ if $x \in \partial\Omega$, $t \in (0,T)$ and $v(x,t)\cdot n(x,t) < 0$, where n is the outward normal vector to $\partial\Omega$.
2. Let $\overline{M} \geq 1$. It is interesting to remark that u is solution of (30) if and only if u is solution of (30) where the term $\int_\Omega (u_0 - \kappa)^\pm \varphi(x,0)dx$ is replaced by $\overline{M} \int_\Omega (u_0 - \kappa)^\pm \varphi(x,0)dx$.

A sketch of proof of existence and uniqueness of the solution of (30) together with the convergence of numerical approximations is now given, following [13].

STEP 1: APPROXIMATE SOLUTION. With a quite general mesh of Ω (with triangles, for instance), denoted by $\mathcal{T}$, and a time step k, it is possible to define an approximate solution, denoted by $u_{\mathcal{T},k}$, using some numerical fluxes (on the edges of the mesh) satisfying conditions similar to (C1)-(C3) in Sect. 3.1. Under a so called CFL condition (like $k \leq (1-\zeta)\frac{h}{L}$ in Sect. 3.1), it is easy to prove that $A \leq u_{\mathcal{T},k} \leq B$ a.e. on $\Omega \times (0,T)$. Unfortunately, it does not seem easy to obtain directly a strong compactness result on the familly of approximate solutions (alhough this strong compactness result is true, as we shall see below).

STEP 2: WEAK COMPACTNESS. Using only this L^∞ bound on $u_{\mathcal{T},k}$, one can assume (for convenient subsequences of sequences of approximate solutions) that $u_{\mathcal{T},k} \to u$, as the mesh size goes to zero (with the CFL condition), in a "nonlinear weak-$\star$ sense" (similar to the convergence towards young measures, see [4] for instance), that is $u \in L^\infty(\Omega \times (0,T) \times (0,1))$ and

$$\int_0^T \int_\Omega g(u_{\mathcal{T},k}(x,t))\varphi(x,t)dxdt \to \int_0^1 \int_0^T \int_\Omega g(u(x,t,\alpha))\varphi(x,t)dxdtd\alpha,$$
$$\text{for all } \varphi \in L^1(\Omega \times (0,T)).$$

STEP 3: PASSING TO THE LIMIT. Using the monotonicity of the numerical fluxes, the approximate solutions satisfy some discrete entropy inequalities. Passing to the limit in these inequalities gives that u (defined in Step 2) is solution of some inequalities which are very similar to (30), namely:

$$
\begin{aligned}
&u \in L^\infty(\Omega \times (0,T) \times (0,1)),\\
&\int_0^1 \int_0^T \int_\Omega [(u-\kappa)^\pm \varphi_t + \text{sign}_\pm(u-\kappa)(f(u)-f(\kappa))v \cdot \text{grad}\varphi] dxdtd\alpha\\
&\qquad\qquad + M \int_0^T \int_{\partial\Omega} (\overline{u}(t)-\kappa)^\pm \varphi(x,t) d\gamma(x) dt \\
&\qquad\qquad\qquad + \int_\Omega (u_0-\kappa)^\pm \varphi(x,0)dx \geq 0,\\
&\qquad\qquad \forall \kappa \in [A,B],\ \forall \varphi \in C_c^1(\overline{\Omega} \times [0,T), \mathbf{R}_+).
\end{aligned} \tag{31}
$$

For this step, one chooses M not only greater than the Lipschitz constant of $\|v\|_\infty f$ on $[A,B]$, but also greater than the Lipschitz constant (on $[A,B]^2$) of the numerical fluxes associated to the edges of the meshes (the equivalent of L in Theorem 1). This choice of M is possible since the unique solution of (30) does not depend on M provided that M is greater than the Lipschitz constant of $\|v\|_\infty f$ on $[A,B]$ and since it is possible to choose numerical fluxes (namely, Godunov flux, for instance) such as the Lipschitz constant of these numerical fluxes is bounded by the Lipschitz constant of $\|v\|_\infty f$ (then, the present method leads to an existence result with M only greater than the Lipschitz constant of $\|v\|_\infty f$ on $s \in [A,B]$, passing to the limit on approximate solutions given with these numerical fluxes).

STEP 4: UNIQUENESS OF THE SOLUTION OF (31). In this step, the "doubling variables" method of Krushkov is used to prove the uniqueness of the solution of (31). Indeed, if u and w are two solutions of (31), the doubling variables method leads to:

$$
\begin{aligned}
&\int_0^1 \int_0^1 \int_0^T \int_\Omega |u(x,t,\alpha) - w(x,t,\beta)| \varphi_t \, dxdtd\alpha d\beta\\
&+ \int_0^1 \int_0^1 \int_0^T \int_\Omega (f(\max(u,w)) - f(\min(u,w)))v \cdot \text{grad}\varphi \, dxdtd\alpha d\beta \geq 0\\
&\qquad\qquad \forall \varphi \in C_c^1(\overline{\Omega} \times [0,T), \mathbf{R}_+),
\end{aligned} \tag{32}
$$

Taking $\varphi(x,t) = (T-t)^+$ in (32) (which is, indeed, possible) gives that u does not depend on α, v does not depend on β and $u = v$ a.e. on $\Omega \times (0,T)$. As a result, u is also the unique solution of (30).

STEP 5: CONCLUSION. Step 4 gives, in particular, the uniqueness of the solution of (30). It also implies that the nonlinear weak-$\star$ limit of sequences of approximate solutions is a solution of (30) and, therefore, it guarantees the existence of the solution of (30). Furthermore, since the nonlinear weak-$\star$ limit of sequences of approximate solutions does not depend on α, it is quite easy to deduce that this limit is "strong" in $L^p(\Omega \times (0,T))$ for any $p \in [1,\infty)$ (see [4], for instance) and, thanks to the uniqueness of the limit, the convergence holds without extraction of subsequences.

References

1. Bardos, C., LeRoux, A.Y., Nédélec, J.C. (1979): First order quasilinear equations with boundary conditions. Comm. Partial Differential Equations, **9**, 1017–1034
2. Brenier, Y., Jaffré, J. (1991): Upstream differencing for multiphase flow in reservoir simulation. SIAM J. Num. Ana. **28**, 685–696
3. Buffard, T., Gallouët, T., Hérard, J.M. (2000): A sequel to a rough Godunov scheme: Application to real gases. Computers and Fluids, **29**, 813–847
4. Eymard, R., Gallouët, T., Herbin, R. (2000): Finite volume methods. Handbook of numerical analysis, Vol. VII, 713–1020. North-Holland, Amsterdam
5. Kagan, A.M., Linnik, Y.V., Rao, C.R. (1973): Characterization Problems in Mathematical Statistics. Wiley, New York
6. Eymard, R., Gallouët, T. (2003): H-convergence and numerical schemes for elliptic equations. SIAM Journal on Numerical Analysis, **41**, Number 2, 539–562
7. Eymard, R., Gallouët, T., Vovelle, J.: Boundary conditions in the numerical approximation of some physical problems via finite volume schemes. Accepted for publication in "Journal of CAM"
8. Faille, I., Heintzé, E. (1999): A rough finite volume scheme for modeling two phase flow in a pipeline. Computers and Fluids, **28**, 213–241
9. Godunov, S. (1976): Résolution numérique des problèmes multidimensionnels de la dynamique des gaz. Editions de Moscou
10. Otto, F. (1996): Initial-boundary value problem for a scalar conservation law. C. R. Acad. Sci. Paris Sér. I Math. **8**, 729–734
11. Roe, P.L. (1981): Approximate Riemann solvers, parameter vectors, and difference schemes. J. Comp. Phys., **43**, 357–372.
12. Patault, S., Q.-H. Tran, Q.H. (1996): Modèle et schéma numérique du code TACITE-NPW, tech. report, rapport IFP 42415
13. Vovelle, J. (2002): Convergence of finite volume monotone schemes for scalar conservation laws on bounded domains. Num. Math., **3**, 563–596

Fictitious Domain Methods in Shape Optimization with Applications in Free-Boundary Problems

Jaroslav Haslinger[1], Tomáš Kozubek[2], Karl Kunisch[3] and Günter Peichl[3]

[1] Charles University, Prague, Czech Republic, *haslin@apollo.karlov.mff.cuni.cz*
[2] Technical University of Ostrava, Czech Republic, *tomas.kozubek@vsb.cz*
[3] University of Graz, Austria, *karl.kunisch@uni-graz.at, gunther.peichl@uni-graz.at*

Summary. This paper deals with a class of 2D shape optimization problems with a 'flux' cost functional and a fictitious domain formulation of state constraints. These constraints are given by nonhomogeneous Dirichlet boundary problems in bounded, doubly connected domains. This approach is used for the numerical realization of free-boundary problems of Bernoulli type.

1 Introduction

This paper deals with a particular shape optimization problem with a fictitious domain (FD) formulation of the state equation. Solvers which are based on FD formulations represent nowadays one of efficient tools for solving large scale algebraic systems arising from discretizations of state problems. The fact that the new FD problem is solved in a domain Ω with a simple shape (a box, e.g.) enables us to construct uniform partitions of Ω and consequently to use fast solvers and special preconditioning techniques. FD solvers have additional advantages when used in shape optimization. To see that let us recall the standard approach in shape optimization which is based on boundary variations of admissible domains. Let us suppose that a linear state problem is solved by a standard finite element method and a gradient type method is used for the minimization of the cost functional. Then the following steps have to be performed after every change of the shape: (i) *remeshing the new configuration*; (ii) *assembling the new stiffness matrix and the right-hand side of the linear algebraic system*; (iii) *solving this new system.* As a result the computational process is not efficient. As we shall see, FD solvers utilizing nonfitted meshes completely avoid step (i) and partially avoid step (ii) since the stiffness matrix remains the same for every admissible domain. The FD formulation that we use in this paper is based on the dualization of Dirichlet conditions by boundary Lagrange multipliers ([6], [8]). It turns out that this variant is appropriate in shape optimization problems for the following reasons: the Lagrange multiplier being part of the solution is equal to the conormal derivative on the searched boundary of the solution to the original state problem. The conormal derivative of the state appears in expressions for the shape derivative of cost functionals

([18]). In addition, the computed Lagrange multiplier is used when evaluating our particular cost functional.

The paper is organized as follows. In Section 2 the optimal shape design problem with the "flux" cost functional is defined and analyzed. State problems are given by Dirichlet boundary value problems in doubly connected, bounded domains. To avoid difficulties with shape dependent function spaces and for numerical purposes we use the above mentioned FD formulation of the state problems. We introduce appropriate assumptions on the family of admissible domains which guarantee the existence of optimal shapes. Continuity of solutions to the FD formulation with respect to domain variations which plays a key role in the existence analysis is the main result of this section. To avoid the evaluation of the dual trace norm which defines the cost functional we give an alternative setting of the problem in which the standard $H_0^1(\Omega)$-norm is used for expressing the cost functional. Section 3 deals with computational aspects of this approach. We present a finite element discretization of the FD formulation. Then we shortly describe the modified controlled random search (MCRS) method, i.e. the gradient free global minimization method of the stochastic type which will be used for minimizing the cost functional. For more details on computational aspects we refer to [11] and [9]. Finally, in Section 4 several Bernoulli free-boundary problems will be solved using our approach. Both the gradient and MCRS methods will be used for the numerical minimization of the cost functional.

2 Setting of the problem, existence analysis

In what follows we shall consider the following optimal shape design problem:

$$\min_{\omega\in\mathcal{O}} \frac{1}{2}\left\|\frac{\partial u(\omega)}{\partial\nu_A} - Q\right\|^2_{-1/2,\Gamma_f(\omega)}, \; Q\in\mathbb{R}^1, \qquad (\tilde{\mathbb{P}})$$

where:

- $\mathcal{O}$ is a family of admissible domains which consists of doubly connected domains in $\mathbb{R}^2$ contained in a box Ω. The components of the boundary $\partial\omega$ are denoted by Γ_0 and $\Gamma_f(\omega)$. We shall suppose that Γ_0 is fixed and the same for all $\omega\in\mathcal{O}$ while $\Gamma_f(\omega)$ is variable and defines the shape of ω (in our presentation $\Gamma_f(\omega)$ is exterior to Γ_0 but one can consider also the opposite situation);
- $\dfrac{\partial u(\omega)}{\partial\nu_A}$ denotes the conormal derivative of u on $\Gamma_f(\omega)$, where u solves the Dirichlet state problem in ω:

$$\begin{cases} -\operatorname{div}(A\nabla u(\omega)) = f & \text{in } \omega,\\ u(\omega) = g & \text{on } \Gamma_0,\\ u(\omega) = 0 & \text{on } \Gamma_f(\omega), \end{cases} \qquad (\mathcal{P}(\omega))$$

with $f \in L^2(\Omega)$, $g \in H^{1/2}(\Gamma_0)$, $A \in L^\infty(\Omega, \mathbb{R}^2 \times \mathbb{R}^2)$ satisfying:

$$\exists \alpha > 0: \; A(x)\xi \cdot \xi \geq \alpha \|\xi\|^2 \quad \forall \xi \in \mathbb{R}^2, \quad \text{a.e. in } \Omega.$$

The symbol $\| \; \|_{-1/2,\Gamma_f(\omega)}$ stands for the dual norm in $H^{-1/2}(\Gamma_f(\omega))$.

Remark 1. Problem $(\tilde{\mathbb{P}})$ is closely related to Bernoulli free-boundary problems. Indeed, let there exist $\omega^* \in \mathcal{O}$ such that

$$J(\omega^*) := \frac{1}{2} \| \frac{\partial u(\omega^*)}{\partial \nu_A} - Q \|^2_{-1/2,\Gamma_f(\omega^*)} = 0.$$

Then simultaneously $u(\omega^*) = 0$ and $\dfrac{\partial u(\omega^*)}{\partial \nu_A} = Q$ on $\Gamma_f(\omega^*)$ which is a typical system of boundary conditions satisfied on free boundaries. A shape optimization approach can be used for the numerical realization of this type of problems. To illustrate how we proceed let us consider the exterior Bernoulli free-boundary problem (see [3]): for $Q < 0$ given, find $\omega^* \in \mathcal{O}$ and $u^* : \; \omega^* \to \mathbb{R}^1$ such that

$$\begin{cases} \Delta u^* = 0 & \text{in } \omega^*, \\ u^* = 1 & \text{on } \Gamma_0, \\ u^* = 0, \; \dfrac{\partial u^*}{\partial \nu} = Q & \text{on } \Gamma_f(\omega^*). \end{cases} \tag{1}$$

For $\omega \in \mathcal{O}$ given à-priori, problem (1) is ill-posed due to the conditions on $\Gamma_f(\omega)$. A possible way how to solve (1) is to skip one of the conditions on $\Gamma_f(\omega)$ (say $\frac{\partial u^*}{\partial \nu} = Q$). The rest of the problem is now well posed and defines the state problem $(\mathcal{P}(\omega))$. The remaining Neumann condition on $\Gamma_f(\omega)$ will be satisfied by minimizing the cost functional J. If $\omega_{opt} \in \mathcal{O}$ is a solution to $(\tilde{\mathbb{P}})$ such that $J(\omega_{opt}) = 0$ then $\omega^* = \omega_{opt}$ solves (1). Let us mention however that the assumptions on $\mathcal{O}$ which are specified below *do not* automatically ensure $J(\omega_{opt}) = 0$.

Next we shall closely follow [10] where detailed proofs of all results can be found.

2.1 Parametrization of shapes

We shall suppose that the outer components $\Gamma_f(\omega)$ of the boundary $\partial\omega$, $\omega \in \mathcal{O}$ are parametrized by 2π-periodic functions $\gamma : \; [0, 2\pi] \to \mathbb{R}^2$. We shall use the following notations: $\Gamma_f(\omega) := \Gamma_\gamma$ meaning that $\Gamma_f(\omega)$ is the range of γ. Further ω_γ denotes the doubly connected domain between Γ_0 and Γ_γ. The family of admissible domains can be specified by a class $\mathcal{S}$ to which γ belongs.

Definition 1. *A function $\gamma : \; [0, 2\pi] \to \mathbb{R}^2$ belongs to $\mathcal{S}$ if:*

(S_1) $\gamma \in C^2_{2\pi}$, *i.e. γ is 2π-periodic, twice continuously differentiable on $[0, 2\pi]$;*

(S_2) $\exists \alpha, \gamma_1, \gamma_2 > 0\, \forall t \in [0, 2\pi]:\ |\gamma'(t)| \geq \alpha,\ \ \|\gamma'\|_\infty \leq \gamma_1,\ \ \|\gamma''\|_\infty \leq \gamma_2$;

(S_3) γ *is positively oriented;*

(S_4) $\overline{\omega}_\gamma \subset \Omega = (0,1)^2$;

(S_5) $\exists d > 0:\ \operatorname{dist}(\Gamma_0, \Gamma_\gamma) \geq d,\ \operatorname{dist}(\Gamma_\gamma, \partial\Omega) \geq d$;

(S_6) $\exists h > 0 \quad \forall \gamma \in \mathcal{S} \quad \forall t \in [0, 2\pi]$: $\exists B_i,\ B_o$ *open discs of radius* h *such that* $B_i \subset \omega_\gamma,\ B_o \subset \Omega \setminus \overline{\omega}_\gamma,\ \gamma(t) \in \overline{B}_i \cap \overline{B}_o$.

Remark 2. Assumption (S_6) means that there exists a tubular neighborhood of Γ_γ of uniform thickness for all $\gamma \in \mathcal{S}$ (see Fig. 1).

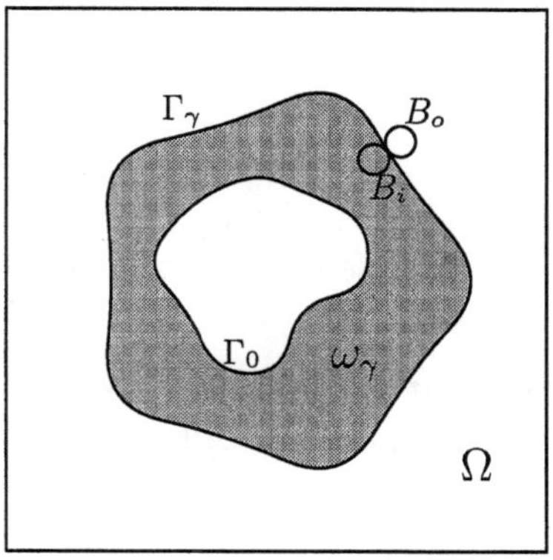

Fig. 1. Geometry of the Bernoulli problem

2.2 The fictitious domain formulation of $(\mathcal{P}(\omega_\gamma))$

The unit square Ω which contains $\overline{\omega}_\gamma$ for all $\gamma \in \mathcal{S}$ will serve as the fictitious domain in the FD formulation. Let us introduce the following spaces:

$$\hat{V}_g = \{\hat{v} \in H_0^1(\Omega) \mid \quad \hat{v} = g \text{ on } \Gamma_0,\ \hat{v} = 0 \text{ on } \Gamma_\gamma\},$$
$$\hat{V}_0 := \hat{V}_g \text{ with } \quad g = 0.$$

Instead of $(\mathcal{P}(\omega_\gamma))$, $\gamma \in \mathcal{S}$ fixed, we formulate the following problem on Ω:

$$\textit{Find } \hat{u} \in \hat{V}_g :\ a(\hat{u}, \hat{v}) = (\tilde{f}, \hat{v})_{0,\Omega} \quad \forall \hat{v} \in \hat{V}_0, \tag{2}$$

where

$$a(\hat{u}, \hat{v}) := \int_\Omega A \nabla \hat{u} \cdot \nabla \hat{v}\, dx$$

and $\tilde{f}$ is the zero extension of f from ω_γ into $\Omega \setminus \overline{\omega}_\gamma$.

Remark 3. It would be possible to use any continuous extension of f outside of ω_γ but the zero extension is of a special significance.

It is easy to see that (2) has a unique solution $\hat{u}$. In addition, $\hat{u}_{|\omega_\gamma}$ solves $(\mathcal{P}(\omega_\gamma))$.

Now we may look at the conditions on Γ_0 and Γ_γ as constraints which will be released by means of the Lagrange multipliers belonging to $H^{-1/2}(\Gamma_0)$, $H^{-1/2}(\Gamma_\gamma)$, respectively. The FD formulation of $(\mathcal{P}(\omega_\gamma))$ reads as follows:

$$(\hat{\mathcal{P}}(\omega_\gamma))\quad \begin{cases} \textit{Find } (\hat{u}, \lambda_0, \lambda_\gamma) \in H_0^1(\Omega) \times H^{-1/2}(\Gamma_0) \times H^{-1/2}(\Gamma_\gamma) \textit{ s.t.} \\ a(\hat{u}, \hat{v}) - \langle \lambda_0, \tau_0 \hat{v} \rangle_{\Gamma_0} - \langle \lambda_\gamma, \tau_\gamma \hat{v} \rangle_{\Gamma_\gamma} = (\tilde{f}, \hat{v})_{0,\Omega} \quad \forall \hat{v} \in H_0^1(\Omega), \\ \langle \mu_0, \tau_0 \hat{u} \rangle_{\Gamma_0} + \langle \mu_\gamma, \tau_\gamma \hat{u} \rangle_{\Gamma_\gamma} = \langle \mu_0, g \rangle_{\Gamma_0} \\ \qquad\qquad \forall (\mu_0, \mu_\gamma) \in H^{-1/2}(\Gamma_0) \times H^{-1/2}(\Gamma_\gamma). \end{cases}$$

Here $\tau_0 : H_0^1(\Omega) \to H^{1/2}(\Gamma_0)$, $\tau_\gamma : H_0^1(\Omega) \to H^{1/2}(\Gamma_\gamma)$ stand for the trace mappings and $\langle\, ,\, \rangle_{\Gamma_0}$, $\langle\, ,\, \rangle_{\Gamma_\gamma}$ denote the respective duality pairings. Using results of [2] one can easily prove

Theorem 1. *There exists a unique solution $(\hat{u}, \lambda_0, \lambda_\gamma)$ to $(\hat{\mathcal{P}}(\omega_\gamma))$. In addition, $u := \hat{u}_{|\omega_\gamma}$ solves $(\mathcal{P}(\omega_\gamma))$ and $\lambda_\gamma = \dfrac{\partial u}{\partial \nu_A}$ on Γ_γ.*

Remark 4. The equality $\lambda_\gamma = \dfrac{\partial u}{\partial \nu_A}$ on Γ_γ is due to the zero extension of f outside of ω_γ. For any other extension one only knows that $\lambda_\gamma = \left[\dfrac{\partial \hat{u}}{\partial \nu_A}\right]_{\Gamma_\gamma}$, where $[\ ]_{\Gamma_\gamma}$ denotes the jump of the corresponding quantity across Γ_γ.

The previous result motivates us to consider $(\tilde{\mathbb{P}})$ in the following form:

$$\min_{\gamma \in \mathcal{S}_L} \frac{1}{2} \|\lambda_\gamma - Q\|^2_{-1/2,\Gamma_\gamma}, \qquad (\mathbb{P})$$

where $\lambda_\gamma \in H^{-1/2}(\Gamma_\gamma)$ is the last component of the solution to $(\hat{\mathcal{P}}(\omega_\gamma))$ and $\mathcal{S}_L \subseteq \mathcal{S}$ is a compact subset of $\mathcal{S}$.

2.3 Periodic function spaces on $[0, 2\pi]$

To prove the existence of optimal shapes in $(\mathbb{P})$ one has to show that the solutions to $(\hat{\mathcal{P}}(\omega_\gamma))$ depend continuously on variations of $\gamma \in \mathcal{S}$. One of the difficulties that we face in any shape optimization problem is the fact that the functions have their own, variable domain of definition. To handle this difficulty we pass to reference, shape independent function spaces. This will be done for the trace space $H^{1/2}(\Gamma_\gamma)$ and its dual $H^{-1/2}(\Gamma_\gamma)$. Using the parametrization of Γ_γ we shall replace $H^{1/2}(\Gamma_\gamma)$ by the reference space $H^{1/2}_{2\pi}$ and $H^{-1/2}(\Gamma_\gamma)$ by $H^{-1/2}_{2\pi}$.

The space of 2π-periodic, square integrable functions will be denoted by $L^2_{2\pi}$. The reference trace space $H^{1/2}_{2\pi}$ is defined as follows (see [13]):

$$H^{1/2}_{2\pi} = \{\varphi \in L^2_{2\pi} \mid \quad \|\varphi\|_{1/2,2\pi} < \infty\},$$

where

$$\|\varphi\|^2_{1/2,2\pi} := \|\varphi\|^2_{L^2_{2\pi}} + \int_0^{2\pi}\int_0^{2\pi} \frac{|\varphi(t)-\varphi(s)|^2}{|\sin((t-s)/2)|^2}\,dt\,ds.$$

Next we shall define the trace mapping $\mathcal{T}_\gamma$ from $H^1_0(\Omega)$ into the reference space $H^{1/2}_{2\pi}$. To this end we introduce the following spaces on Γ_γ, $\gamma \in \mathcal{S}$:

$$H^{1/2}_p(\Gamma_\gamma) = \{\varphi \in L^2(\Gamma_\gamma) \mid \quad \varphi\circ\gamma \in H^{1/2}_{2\pi}\}$$

with the norm

$$\|\varphi\|_{1/2,p} := \|\varphi\circ\gamma\|_{1/2,2\pi}$$

and the standard Sobolev space $H^{1/2}(\Gamma_\gamma)$ endowed with the norm ([15]):

$$\|\varphi\|^2_{1/2,\gamma} := \|\varphi\|^2_{L^2(\Gamma_\gamma)} + \int_{\Gamma_\gamma}\int_{\Gamma_\gamma} \frac{|\varphi(x)-\varphi(y)|^2}{|x-y|^2}\,ds_x\,ds_y = \int_0^{2\pi} |\varphi\circ\gamma|^2|\gamma'|\,dt$$
$$+ \int_0^{2\pi}\int_0^{2\pi} \frac{|\varphi\circ\gamma(t)-\varphi\circ\gamma(s)|^2}{|\gamma(t)-\gamma(s)|^2}|\gamma'(t)||\gamma'(s)|\,dt\,ds$$

making use of the parametrization of Γ_γ. The relation between $H^{1/2}_p(\Gamma_\gamma)$ and $H^{1/2}(\Gamma_\gamma)$ follows from the next lemma.

Lemma 1. *The spaces $H^{1/2}_p(\Gamma_\gamma)$ and $H^{1/2}(\Gamma_\gamma)$ coincide as sets and are topologically equivalent, uniformly with respect to $\gamma \in \mathcal{S}$, i.e. there exist positive constants c_1, c_2 such that*

$$c_1\|\varphi\|_{1/2,\gamma} \le \|\varphi\circ\gamma\|_{1/2,2\pi} \le c_2\|\varphi\|_{1/2,\gamma}$$

holds for every $\varphi \in H^{1/2}(\Gamma_\gamma)$ and every $\gamma \in \mathcal{S}$.

Let $i_\gamma : H^{1/2}(\Gamma_\gamma) \to H^{1/2}_p(\Gamma_\gamma)$ be the identity mapping and $I_\gamma : H^{1/2}_p(\Gamma_\gamma) \to H^{1/2}_{2\pi}$ be an isometry defined by $I_\gamma(\varphi) := \varphi\circ\gamma$. Then the trace mapping $\mathcal{T}_\gamma : H^1_0(\Omega) \to H^{1/2}_{2\pi}$ is given by

$$\mathcal{T}_\gamma := I_\gamma \circ i_\gamma \circ \tau_\gamma.$$

The trace mapping $\mathcal{T}_\gamma$ enjoys properties which are useful in the existence analysis for $(\mathbb{P})$.

Lemma 2. 1) *There exists a constant $c > 0$ such that*

$$\|\mathcal{T}_\gamma\|_{\mathcal{L}(H_0^1(\Omega), H_{2\pi}^{1/2})} \leq c \quad \textit{for all } \gamma \in \mathcal{S}. \tag{3}$$

2) *If $\gamma_n \to \gamma$ in $C_{2\pi}^1$, $\gamma_n, \gamma \in \mathcal{S}$, then*

$$\mathcal{T}_{\gamma_n} \hat{v} \to \mathcal{T}_\gamma \hat{v} \quad \textit{in } H_{2\pi}^{1/2} \tag{4}$$

holds for every $\hat{v} \in H_0^1(\Omega)$.

The mapping $\mathcal{T}_\gamma$ maps $H_0^1(\Omega)$ *onto* $H_{2\pi}^{1/2}$. This follows from

Lemma 3. (*Inverse trace property*) *There exists a continuous extension mapping $\mathcal{E}_\gamma : H_{2\pi}^{1/2} \to H_0^1(\Omega)$ such that*

$$\mathcal{T}_\gamma \mathcal{E}_\gamma \varphi = \varphi \quad \forall \varphi \in H_{2\pi}^{1/2}$$

and

$$\|\mathcal{E}_\gamma\|_{\mathcal{L}(H_{2\pi}^{1/2}, H_0^1(\Omega))} \leq c, \tag{5}$$

where $c > 0$ does not depend on $\gamma \in \mathcal{S}$. In addition, $\operatorname{supp} \mathcal{E}_\gamma \varphi$ *is contained in the h-neighborhood of Γ_γ.*

For the proof of the previous lemmas we refer to [10].

In a similar way one can associate with any $\mu_\gamma \in H^{-1/2}(\Gamma_\gamma)$ a unique element $\tilde{\mu}_\gamma \in H_{2\pi}^{-1/2}$:

$$\langle \mu_\gamma, \varphi \rangle_{\Gamma_\gamma} = \langle \tilde{\mu}_\gamma, I_\gamma i_\gamma \varphi \rangle_{2\pi} = \langle \tilde{\mu}_\gamma, \varphi \circ \gamma \rangle_{2\pi} \quad \forall \varphi \in H^{1/2}(\Gamma_\gamma),$$

where $\tilde{\mu}_\gamma := I_\gamma^{-*} i_\gamma^{-*} \mu_\gamma$ and $\langle \, , \, \rangle_{2\pi}$ denotes the duality pairing between $H_{2\pi}^{-1/2}$ and $H_{2\pi}^{1/2}$.

This enables us to rewrite problem $(\hat{\mathcal{P}}(\omega_\gamma))$ in the following equivalent form in which only shape independent function spaces appear:

$$(\hat{\mathcal{P}}(\omega_\gamma))_{ref} \quad \left\{ \begin{array}{l} \textit{Find } (\hat{u}, \lambda_0, \tilde{\lambda}_\gamma) \in H_0^1(\Omega) \times H^{-1/2}(\Gamma_0) \times H_{2\pi}^{-1/2} \textit{ s.t.} \\ a(\hat{u}, \hat{v}) - \langle \lambda_0, \tau_0 \hat{v} \rangle_{\Gamma_0} - \langle \tilde{\lambda}_\gamma, \mathcal{T}_\gamma \hat{v} \rangle_{2\pi} = (\tilde{f}, \hat{v})_{0,\Omega} \quad \forall \hat{v} \in H_0^1(\Omega), \\ \langle \mu_0, \tau_0 \hat{u} \rangle_{\Gamma_0} + \langle \tilde{\mu}, \mathcal{T}_\gamma \hat{u} \rangle_{2\pi} = \langle \mu_0, g \rangle_{\Gamma_0} \\ \hfill \forall (\mu_0, \tilde{\mu}) \in H^{-1/2}(\Gamma_0) \times H_{2\pi}^{-1/2}. \end{array} \right.$$

Since

$$\langle Q, \varphi \rangle_{\Gamma_\gamma} = Q \int_0^{2\pi} \varphi \circ \gamma \, |\gamma'| \, dt = \langle Q|\gamma'|, \varphi \circ \gamma \rangle_{2\pi},$$

we obtain the following transformation of $(\mathbb{P})$:

$$\min_{\gamma \in \mathcal{S}_L} \frac{1}{2} \|\tilde{\lambda}_\gamma - Q|\gamma'|\,\|^2_{H^{-1/2}_{2\pi}}, \qquad (\mathbb{P})_{ref}$$

where $\tilde{\lambda}_\gamma \in H^{-1/2}_{2\pi}$ is the last component of the solution to $(\hat{\mathcal{P}}(\omega_\gamma))_{ref}$.

2.4 Existence analysis for $(\mathbb{P})_{ref}$

As we have already mentioned, the continuous dependance of solutions to $(\hat{\mathcal{P}}(\omega_\gamma))_{ref}$ with respect to $\gamma \in \mathcal{S}$ plays a central role in the existence analysis. To prove this property we shall need the following auxiliary results.

Lemma 4. *The set of solutions* $\{(\hat{u}_\gamma, \lambda_{0\gamma}, \tilde{\lambda}_\gamma),\ \gamma \in \mathcal{S}\}$ *to* $(\hat{\mathcal{P}}(\omega_\gamma))_{ref}$ *is bounded in* $H^1_0(\Omega) \times H^{-1/2}(\Gamma_0) \times H^{-1/2}_{2\pi}$.

Lemma 5. *(Stability of the Dirichlet boundary condition) Let* $\gamma_n \to \gamma$ *in* $C_{2\pi}$, $\gamma_n, \gamma \in \mathcal{S}$ *and let* $\{\hat{v}_n\}$ *be any sequence in* $H^1_0(\Omega)$ *such that* $\tau_0 \hat{v}_n = g$, $\tau_{\gamma_n} \hat{v}_n = 0$. *If* $\hat{v}_n \rightharpoonup \hat{v}$ *in* $H^1_0(\Omega)$, *then* $\tau_0 \hat{v} = g$ *and* $\tau_\gamma \hat{v} = 0$.

Continuity of solutions to $(\hat{\mathcal{P}}(\omega_\gamma))_{ref}$ with respect to $\gamma \in \mathcal{S}$ now follows from the next theorem.

Theorem 2. *Let* $\gamma_n \to \gamma$ *in* $C^1_{2\pi}$, $\gamma_n, \gamma \in \mathcal{S}$ *and* $(\hat{u}_n, \lambda_{0n}, \tilde{\lambda}_n) \in H^1_0(\Omega) \times H^{-1/2}(\Gamma_0) \times H^{-1/2}_{2\pi}$ *be the solution to* $(\hat{\mathcal{P}}(\omega_{\gamma_n}))_{ref}$. *Then*

$$\begin{cases} \hat{u}_n \to \hat{u} \ \text{in } H^1_0(\Omega), \\ \lambda_{0n} \to \lambda_0 \ \text{in } H^{-1/2}(\Gamma_0), \\ \tilde{\lambda}_n \rightharpoonup \tilde{\lambda}_\gamma \ \text{in } H^{-1/2}_{2\pi} \end{cases} \qquad (6)$$

and $(\hat{u}, \lambda_0, \tilde{\lambda}_\gamma)$ *is a solution to* $(\hat{\mathcal{P}}(\omega_\gamma))_{ref}$. *In addition, if* $\gamma_n \to \gamma$ *in* $C^2_{2\pi}$ *then*

$$\tilde{\lambda}_n \to \tilde{\lambda}_\gamma \ \text{in } H^{-1/2}_{2\pi}. \qquad (7)$$

Proof. By Lemma 4 we may assume that there exists a subsequence of $\{(\hat{u}_n, \lambda_{0n}, \tilde{\lambda}_n)\}$ (denoted by the same symbol) such that

$$(\hat{u}_n, \lambda_{0n}, \tilde{\lambda}_n) \rightharpoonup (\hat{u}, \lambda_0, \tilde{\lambda}_\gamma) \text{ in } H^1_0(\Omega) \times H^{-1/2}(\Gamma_0) \times H^{-1/2}_{2\pi}. \qquad (8)$$

From Lemma 5 it follows that $\tau_0 \hat{u} = g$, $\tau_\gamma \hat{u} = 0$ so that the second equation in $(\hat{\mathcal{P}}(\omega_\gamma))_{ref}$ is satisfied. Let $\hat{v} \in H^1_0(\Omega)$ be fixed. Then

$$\begin{aligned} \lim_{n\to\infty} \langle \tilde{\lambda}_n, \mathcal{T}_{\gamma_n} \hat{v} \rangle_{2\pi} &= \lim_{n\to\infty} \{a(\hat{u}_n, \hat{v}) - \langle \lambda_{0n}, \tau_0 \hat{v} \rangle_{\Gamma_0} - (\tilde{f}, \hat{v})_{0,\Omega}\} \\ &= a(\hat{u}, \hat{v}) - \langle \lambda_0, \tau_0 \hat{v} \rangle_{\Gamma_0} - (\tilde{f}, \hat{v})_{0,\Omega} \end{aligned} \qquad (9)$$

as follows from (8). On the other hand

$$\lim_{n\to\infty}\langle\tilde{\lambda}_n, \mathcal{T}_{\gamma_n}\hat{v}\rangle_{2\pi} = \lim_{n\to\infty}\{\langle\tilde{\lambda}_n, \mathcal{T}_{\gamma_n}\hat{v} - \mathcal{T}_\gamma\hat{v}\rangle_{2\pi} + \langle\tilde{\lambda}_n, \mathcal{T}_\gamma\hat{v}\rangle_{2\pi}\} = \langle\tilde{\lambda}_\gamma, \mathcal{T}_\gamma\hat{v}\rangle_{2\pi},$$

making use of (4) and (8). From this and (9) we see that $(\hat{u}, \lambda_0, \tilde{\lambda}_\gamma)$ solves $\left(\hat{\mathcal{P}}(\omega_\gamma)\right)_{ref}$. In addition, the whole sequence in (8) tends to $(\hat{u}, \lambda_0, \tilde{\lambda}_\gamma)$. Strong convergence of $\{\hat{u}_n\}$ follows from

$$a(\hat{u}_n, \hat{u}_n) = \langle\lambda_{0n}, g\rangle_{\Gamma_0} + (\tilde{f}, \hat{u}_n)_{0,\Omega} \quad\longrightarrow\quad \langle\lambda_0, g\rangle_{\Gamma_0} + (\tilde{f}, \hat{u})_{0,\Omega} = a(\hat{u}, \hat{u})$$

using that $\tau_0\hat{u}_n = g$ and $\mathcal{T}_{\gamma_n}\hat{u}_n = 0$.

Let $\varphi \in H^{1/2}(\Gamma_0)$ be given and $\hat{v} \in H_0^1(\Omega)$ be its continuous extension: $\tau_0\hat{v} = \varphi$, $\|\hat{v}\|_{1,\Omega} \le c\|\varphi\|_{1/2,\Gamma_0}$ with $\operatorname{supp}\hat{v}$ being in a δ-neighborhood of Γ_0, $\delta \le d$. Then

$$|\langle\lambda_{0n} - \lambda_0, \varphi\rangle_{\Gamma_0}| \le \|a\|\,\|\hat{u}_n - \hat{u}\|_{1,\Omega}\|\hat{v}\|_{1,\Omega} \le c\|\hat{u}_n - \hat{u}\|_{1,\Omega}\|\varphi\|_{1/2,\Gamma_0}$$

yielding strong convergence of $\{\lambda_{0n}\}$ to λ_0. For the proof of (7) which is more technical we refer to [10]. □

It remains to specify a compact subset $\mathcal{S}_L$ of $\mathcal{S}$. Let $L > 0$ be given and define

$$\mathcal{S}_L = \{\gamma \in \mathcal{S} \mid \quad |\gamma''(t) - \gamma''(s)| \le L|t - s| \quad \forall t, s \in [0, 2\pi]\}. \tag{10}$$

Theorem 3. *Let $\mathcal{S}_L$ be defined by* (10). *Then* $(\mathbb{P})_{ref}$ *has a solution.*

The proof follows from Theorem 2 and continuity of the cost functional.

The dual norm defining the cost functional is hard to evaluate. For this reason we use another functional which also measures the distance between $\frac{\partial u}{\partial \nu_A}$ and the target Q and which is easy to compute. To this end we introduce the new norm in $H^{1/2}(\Gamma_\gamma)$:

$$[\![\varphi]\!]_{1/2,\Gamma_\gamma} := \inf_{\substack{\hat{v}\in H_0^1(\Omega)\\ \tau_\gamma\hat{v}=\varphi}} |\hat{v}|_{1,\Omega}. \tag{11}$$

It is easy to show that the respective dual norm is given by

$$\|\mu\|_{-1/2,\Gamma_\gamma} = |\hat{z}|_{1,\Omega}, \tag{12}$$

where $\hat{z}$ is the solution of the transmission problem:

$$\begin{cases} \textit{Find } \hat{z} := \hat{z}(\mu) \in H_0^1(\Omega) \textit{ such that} \\ (\nabla\hat{z}, \nabla\hat{v})_{0,\Omega} = \langle\mu, \hat{v}\rangle_{\Gamma_\gamma} \quad \forall\hat{v} \in H_0^1(\Omega). \end{cases} \tag{$\mathcal{A}(\mu)$}$$

This leads to a new definition of the shape optimization problem $(\tilde{\mathbb{P}})$:

$$\min_{\mathcal{O}} \frac{1}{2} |\hat{z}(u(\omega))|^2_{1,\Omega}, \qquad (\overline{\mathbb{P}})$$

where $\hat{z}(u(\omega))$ solves $(\mathcal{A}(\mu))$ with $\mu = \dfrac{\partial u}{\partial \nu_A} - Q$. Observe that $\hat{z}(u(\omega)) = 0$ if and only if $\dfrac{\partial u}{\partial \nu_A} = Q$, i.e. the zero value (if any) of the cost functionals in $(\tilde{\mathbb{P}})$ and $(\overline{\mathbb{P}})$ is realized by the same shapes.

The mathematical analysis for $(\overline{\mathbb{P}})$ is done in [9]. It is worth noticing that the existence of optimal shapes in $(\overline{\mathbb{P}})$ is ensured under weaker assumptions on $\mathcal{O}$. Instead of $C^{2,2}$ in $(\tilde{\mathbb{P}})$ one needs $C^{1,1}$ boundaries in $(\overline{\mathbb{P}})$, only. This is due to the fact that the cost functional in $(\overline{\mathbb{P}})$ is expressed by a volume integral.

2.5 Computational aspects

In this section we briefly describe the discretization of $(\overline{\mathbb{P}})$. For more details we refer to [9]. Instead of the family of admissible domains $\mathcal{O}$ we shall introduce a new family $\mathcal{O}_\kappa$ which contains domains with boundaries defined by a finite number $d := d(\kappa)$ of parameters, e.g. splines. For $\omega_\kappa \in \mathcal{O}_\kappa$ given we define a discretization of $(\hat{\mathcal{P}}(\omega_\gamma))$ as follows:

$$\left(\hat{\mathcal{P}}(\omega_\kappa)\right)_h^H \quad \begin{cases} \textit{Find } (\hat{u}_h, \lambda_{H_0}, \lambda_{H_\gamma}) \in V_h \times \Lambda_{H_0} \times \Lambda_{H_\gamma} \ s.t. \\ a(\hat{u}_h, \hat{v}_h) - [\lambda_{H_0}, \hat{v}_h]_0 - [\lambda_{H_\gamma}, \hat{v}_h]_\gamma = (\tilde{f}, \hat{v}_h)_{0,\Omega} \quad \forall \hat{v}_h \in V_h, \\ [\mu_{H_0}, \hat{u}_h]_0 + [\mu_{H_\gamma}, \hat{u}_h]_\gamma = [\mu_{H_0}, \overline{g}]_0 \\ \qquad \qquad \forall (\mu_{H_0}, \mu_{H_\gamma}) \in \Lambda_{H_0} \times \Lambda_{H_\gamma}. \end{cases}$$

Here V_h is a discretization of $H_0^1(\Omega)$, Λ_{H_0}, Λ_{H_γ} are appropriate discretizations of $H^{-1/2}(\Gamma_0)$, $H^{-1/2}(\Gamma_\gamma)$, respectively, $\overline{g}$ is an approximation of g and $[\,,\,]_0$, $[\,,\,]_\gamma$ are approximations of the duality pairings $\langle\,,\,\rangle_{\Gamma_0}$, $\langle\,,\,\rangle_{\Gamma_\gamma}$, respectively. The symbols h, H_0 and H_γ denote the norms of the respective partitions used for the construction of V_h, Λ_{H_0}, Λ_{H_γ}, respectively and $H := (H_0, H_\gamma)$. In what follows we shall suppose that the partition $\hat{\mathcal{T}}_h$ of $\overline{\Omega}$ characterizing V_h *does not depend* on the geometry of ω_κ.

To ensure the existence and uniqueness of solutions to $\left(\hat{\mathcal{P}}(\omega_\kappa)\right)_h^H$ we shall need the following stability condition:

$$[\mu_{H_0}, \hat{v}_h]_0 + [\mu_{H_\gamma}, \hat{v}_h]_\gamma = 0 \quad \forall \hat{v}_h \in V_h \implies \mu_{H_0} = \mu_{H_\gamma} = 0. \qquad (S)$$

For the detailed mathematical analysis of $\left(\hat{\mathcal{P}}(\omega_\kappa)\right)_h^H$, in particular for the verification of the LBB-condition we refer to [6]. Having λ_{H_γ} at our disposal we define the discretization of $\left(\mathcal{A}(\omega_k)\right)$:

$$\begin{cases} \textit{Find } \hat{z}_h := \hat{z}_h(\lambda_{H_\gamma}) \in V_h \textit{ such that} \\ (\nabla \hat{z}_h, \nabla \hat{v}_h)_{0,\Omega} = [\lambda_{H_\gamma} - Q, \hat{v}_h]_\gamma \quad \forall \hat{v}_h \in V_h. \end{cases} \qquad \left(\mathcal{A}(\omega_\kappa)\right)_h^H$$

Finally, the discretization of $(\overline{\mathbb{P}})$ reads as follows:

$$\min_{\mathcal{O}_\kappa} \frac{1}{2} |\hat{z}_h(\lambda_{H_\gamma})|^2_{1,\Omega}, \qquad (\overline{\mathbb{P}})_{\kappa h}^H$$

where $\hat{z}_h(\lambda_{H_\gamma})$ solves $\left(\mathcal{A}(\omega_\kappa)\right)_h^H$.

To see better the structure of the discrete problem we now present its algebraic form. Any $\omega_\kappa \in \mathcal{O}_\kappa$ is characterized by a vector $\boldsymbol{\kappa} \in \mathbb{R}^d$ of discrete design variables. The family $\mathcal{O}_\kappa$ can be identified with a compact subset $\mathcal{U} \subseteq \mathbb{R}^d$. For a given $\boldsymbol{\kappa} \in \mathcal{U}$ we first solve the saddle-point problem:

$$\begin{cases} \textit{Find } (\mathbf{u}, \boldsymbol{\lambda}_0, \boldsymbol{\lambda}_\gamma) \in \mathbb{R}^n \times \mathbb{R}^{m_1} \times \mathbb{R}^{m_2} \textit{ s.t.} \\ \mathbb{A}\,\mathbf{u} - \mathbb{B}_0^T\,\boldsymbol{\lambda}_0 - \mathbb{B}_\gamma^T(\boldsymbol{\kappa})\boldsymbol{\lambda}_\gamma = \mathbf{f}(\boldsymbol{\kappa}), \\ \mathbb{B}_0\,\mathbf{u} = \mathbf{g}, \quad \mathbb{B}_\gamma(\boldsymbol{\kappa})\mathbf{u} = \mathbf{0}, \end{cases} \qquad (\mathcal{P}(\boldsymbol{\kappa}))$$

then the linear system:

$$\mathbb{A}\,\mathbf{z}(\boldsymbol{\kappa}) = \mathbb{B}_\gamma^T(\boldsymbol{\kappa})(\boldsymbol{\lambda}_\gamma - \mathbf{Q}), \qquad (\mathcal{A}(\boldsymbol{\kappa}))$$

where $\mathbb{A}$ is the stiffness matrix and $\mathbb{B}_0$, $\mathbb{B}_\gamma(\boldsymbol{\kappa})$ are matrices coupling the primal variable $\mathbf{u}$ with the Lagrange multipliers $\boldsymbol{\lambda}_0$, $\boldsymbol{\lambda}_\gamma$, respectively. Notice that only the matrix $\mathbb{B}_\gamma$ and the right hand side $\mathbf{f}$ depend on $\boldsymbol{\kappa}$ but *not* $\mathbb{A}$! Therefore, $\mathbb{A}$ needs to be assembled only once. We finally arrive at the following non-linear mathematical programming problem:

$$\min_{\mathcal{U}} \frac{1}{2} (\mathbb{A}\,\mathbf{z}(\boldsymbol{\kappa}), \mathbf{z}(\boldsymbol{\kappa}))_{\mathbb{R}^n}, \qquad (\mathbb{P})_{\boldsymbol{\kappa}}$$

where $\mathbf{z}(\boldsymbol{\kappa})$ solves $\left(\mathcal{A}(\boldsymbol{\kappa})\right)$.

A traditional way of solving $(\mathbb{P})_{\boldsymbol{\kappa}}$ is based on derivative information. It is known from the theory that fictitious domain methods which use non-fitted meshes may reduce the smoothness of the control-to-state mapping (see [11], [12]). In addition, the classical gradient minimization techniques are local methods. To obtain a global minimizer one should use global minimization methods which do not need any gradient information. For our class of problems we use a stochastic type method, namely the modified controlled random search (MCRS) algorithm [14].

This algorithm uses ideas of the simplex method [16] and the CRS (Controlled Random Search) algorithm [17]. It starts with a population P of N points ($N \gg n$) which are chosen at random in the search space $\mathcal{X}$ ($n := \dim(\mathcal{X})$). A new trial point $\mathbf{x}$ is generated from a simplex S (a set of $n+1$ linearly independent points of a population P in $\mathcal{X}$) by the following operation called reflection:

$$\mathbf{x} = \mathbf{g} - Y(\mathbf{z} - \mathbf{g}),$$

where $\mathbf{z}$ is one (randomly chosen) vertex of the simplex S, $\mathbf{g}$ is the center of gravity of the remaining n vertices of the simplex and Y is a randomized multiplicative factor. Thus the new point $\mathbf{x}$ is obtained from the reflection of the point $\mathbf{z}$ with respect to $\mathbf{g}$. Let $\mathbf{x}_{max}$ be the point with the largest objective function value among the N points currently stored in P. If $f(\mathbf{x}) < f(\mathbf{x}_{max})$ then $\mathbf{x}_{max} := \mathbf{x}$, i.e. the worst point in the population is replaced by the new trial point. The process continues until a stopping criterion is fulfilled.

The main modification of the original CRS algorithm consists in randomizing the multiplicative factor Y. The best results in most tested examples were obtained with Y distributed uniformly in $[0, \alpha[$ with α ranging from 4 to 8 and $N = \max(5n, n^2)$. For more details on this algorithm we refer to [14].

3 Applications in free-boundary problems

In this section we show that the shape optimization problem $(\tilde{\mathbb{P}})$ with the FD solver of the state problem represents an efficient computational tool for solving a large class of free-boundary problems. Since free-boundary problems are very well investigated from the theoretical point of view they may serve as benchmarks for testing the reliability of the method. In what follows we use our approach for solving exterior and interior Bernoulli free-boundary (BFB) problems and a dam problem.

3.1 Bernoulli free boundary problems

We shall be concerned with the exterior as well as interior BFB problem. Unlike to the exterior BFB problem defined by (1), the unknown component $\Gamma_f(\omega^*)$ of the boundary is interior to Γ_0 in the interior BFB problem and the respective boundary conditions on Γ_0 and $\Gamma_f(\omega^*)$ read as follows (see [3]):

$$\begin{aligned} u^* &= 0 \quad \text{on } \Gamma_0, \\ u^* = 1, \ \frac{\partial u^*}{\partial \nu} &= Q \quad \text{on } \Gamma_f(\omega^*), \end{aligned} \tag{13}$$

where Q is a positive constant. Since the Dirichlet condition on $\Gamma_f(\omega^*)$ is nonhomogeneous, one has to introduce the new variable $\tilde{u} := u - 1$ in order to ensure that $\lambda_\gamma = \frac{\partial \tilde{u}}{\partial \nu}$ on $\Gamma_f(\omega^*)$.

The discrete family $\mathcal{O}_\kappa$ consists of all doubly connected domains ω_κ whose variable component of the boundary is realized by piecewise second order Bezier curves. The discretization parameter κ is related to the number of Bezier segments. The space $V_h \subset H_0^1(\Omega)$ is realized by *continuous, piecewise linear* functions over a uniform partition $\hat{\mathcal{T}}_h$ of Ω. Further, $\Lambda_{H_0}, \Lambda_{H_\gamma}$ are spaces of *piecewise constant* functions over partitions $\hat{\mathcal{T}}_{H_0}, \hat{\mathcal{T}}_{H_\gamma}$ of polygonal approximations $\tilde{\Gamma}_0, \tilde{\Gamma}_\gamma$ of Γ_0, Γ_γ, respectively (see Fig. 2). Finally, the duality pairings $[\,,\,]_0$, $[\,,\,]_\gamma$ are realized by the $L^2(\tilde{\Gamma}_0)$, $L^2(\tilde{\Gamma}_\gamma)$-scalar products, respectively. In order to satisfy the LBB-condition, the partitions $\hat{\mathcal{T}}_{H_0}, \hat{\mathcal{T}}_{H_\gamma}$ are constructed in such a way that $H_0 \doteq 3h$, $H_\gamma \doteq 3h$. For more details on the practical realiza-

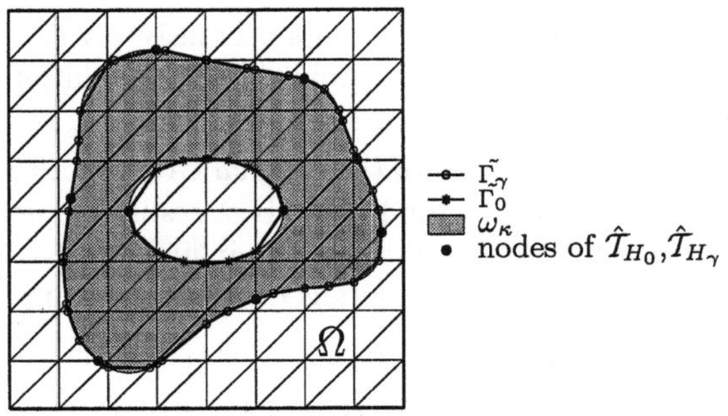

Fig. 2. FE partitions

tion we refer to [9], [11]. Both external and internal problems were computed with the following data: $\Omega = (0, 10)^2$, $h = 10/64$, the number of Bezier segments $\kappa = 10$ and Γ_0 is L-shaped. The examples are computed for different values of the target Q by using two cost functionals, namely:

(a) $J_1(\omega)$ as in the definition of $(\overline{\mathbb{P}})$;

(b) $J_2(\omega) = \frac{1}{2}\|\lambda_{H_\gamma} - Q\|^2_{-1/2,\tilde{\Gamma}_\gamma}$, where

$$\|v\|_{-1/2,\tilde{\Gamma}_\gamma} = \|v \circ \gamma\|_{-1/2,[0,1]} \quad \text{and}$$

$$\|v\|^2_{-1/2,[0,1]} = \sum_{n=-\infty}^{\infty} \frac{|c_n|^2}{1+|n|}, \quad c_n = \int_0^1 v(t)\, e^{-2\pi i n t}\, dt.$$

Here γ denotes a piecewise linear parametrization representing $\tilde{\Gamma}_\gamma$.

The results obtained by the MCRS method after 2000 function evaluations are shown in Figs. 3-4.

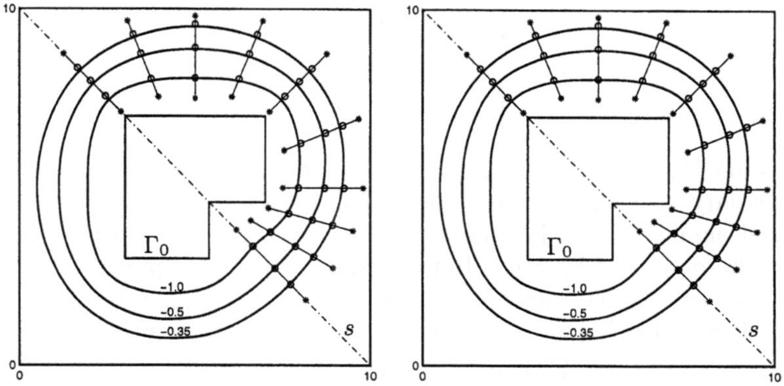

Variant (*a*) Variant (*b*)

Fig. 3. Exterior BFB

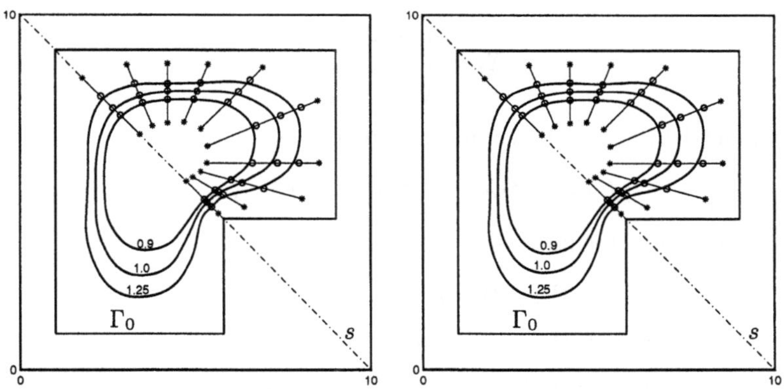

Variant (*a*) Variant (*b*)

Fig. 4. Interior BFB

3.2 A gradient approach

In this subsection we outline a gradient based approach to the optimal shape design problem $(\tilde{\mathbb{P}})$ with the $H^{-1/2}(\Gamma_f)$-norm replaced by the $L^2(\Gamma_f)$-norm. A rigorous foundation of the following arguments as well as a detailed numerical analysis will be carried out in a future paper. For simplicity we consider the following variant

$$\min_{\omega \in \mathcal{O}_\Pi} \frac{1}{2} \int_{\Gamma_f} \left| \frac{\partial u(\omega)}{\partial \nu} - Q \right|^2 d\sigma, \tag{14}$$

where $\dfrac{\partial u(\omega)}{\partial \nu}$ denotes the outer normal derivative of the solution to

$$\begin{cases} -\Delta u = 0 & \text{on } \omega, \\ u = g & \text{on } \Gamma_0, \\ u = 0 & \text{on } \Gamma_f \end{cases} \tag{15}$$

with Γ_0 and Γ_f as in Section 2. We assume that the searched component of the boundary $\Gamma_p := \Gamma_f$ of every admissible domain ω_p, $p \in \Pi$, is the union of a fixed number k of adjoining arcs $\Gamma_{p,i}$, $\Gamma_p = \cup_{i=1}^k \Gamma_{p,i}$, such that each arc $\Gamma_{p,i}$ can be described by a quadratic Bezier curve. More precisely, given an ordered set of distinct control nodes $x_1, \ldots, x_k$, $x_i = (\xi_i, \eta_i)$, $i = 1, \ldots, k$, the i-th Bezier arc $\Gamma_{p,i}$ is determined by the triple (m_{i-1}, x_i, m_i) with $m_i = \frac{1}{2}(x_i + x_{i+1})$, $i = 1, \ldots, k$. Here we set $x_{k+1} = x_0$ and $m_0 = m_k$. Thus $\Gamma_{p,i}$ can be parametrized by

$$\gamma_i(t) = m_{i-1}(1-t)^2 + 2x_i t(1-t) + m_i t^2, \quad t \in [0,1].$$

Defining $\mathcal{B}\colon [0,1] \to \mathbb{R}^{1\times 3}$ (a (1x3) matrix), $t \mapsto \frac{1}{2}[(1-t)^2, 1 + 2t(1-t), t^2]$ and

$$X_i = \begin{pmatrix} \xi_{i-1} & \eta_{i-1} \\ \xi_i & \eta_i \\ \xi_{i+1} & \eta_{i+1} \end{pmatrix}$$

this parametrization can be compactly written as

$$\gamma_i(t) = \mathcal{B}(t) X_i, \quad t \in [0,1], \quad i = 1, \ldots, k.$$

Shifting the parameter interval

$$\begin{aligned} &\mathcal{B}_i(t) = \mathcal{B}(t - (i-1)), \\ &\gamma_i(t) = \mathcal{B}_i(t) X_i, \quad t \in [i-1, i], \quad i = 1, \ldots, k, \end{aligned}$$

one obtains a parametrization $\gamma\colon [0,k] \to \mathbb{R}^2$ of Γ_p defined by $\gamma|_{[i-1,i]} = \gamma_i$. Thus each index $p \in \mathbb{R}^{k\times 2}$ corresponds to an ordered list of coordinates of control nodes and $\Pi \subset \mathbb{R}^{k\times 2}$ describes some a priori assumptions on the location of the control nodes.

Let $J(p)$ denote the value of the cost functional defined in (14) computed by solving (15) on ω_p corresponding to the configuration p of the control nodes. The Gateaux derivative of J at p in the direction δp is defined by

$$J'(p)\delta p = \lim_{s \to 0+} \frac{1}{s}(J(p + s\delta p) - J(p)).$$

Proceeding formally one can derive the following representation

$$J'(p)\delta p = -\int_{\Gamma_p} \frac{\partial u}{\partial \nu}\frac{\partial \mu}{\partial \nu} \alpha \, d\sigma - \frac{1}{2}\int_{\Gamma_p} \kappa((\frac{\partial u}{\partial \nu})^2 - Q^2)\alpha \, d\sigma. \tag{16}$$

Above α describes the normal component of the displacement of Γ_p caused by the perturbation δp of the configuration p and κ stands for the curvature of Γ_p. The auxiliary variable μ is the solution of the adjoint equation

$$\begin{cases} -\Delta\mu = 0 \text{ on } \omega_p, \\ \mu = 0 \text{ on } \Gamma_0, \\ \mu = \dfrac{\partial u}{\partial \nu} - Q \text{ on } \Gamma_p \end{cases} \tag{17}$$

and u is the solution of (15) on ω_p. Let us introduce the shorthand notation

$$J'(p)\delta p = \int_{\Gamma_p} \mathcal{K}\alpha \, d\sigma \tag{18}$$

and let $\delta X_i \in \mathbb{R}^{3\times 2}$ hold the perturbation of the coordinates of the nodes x_{i-1}, x_i, x_{i+1} specified by δp. If we insert

$$\alpha(t) = \mathcal{B}_i(t)\delta X_i \nu(t), \quad t \in [i-1, i], i = 1, \ldots, k,$$

into (18) we obtain

$$J'(p)\delta p = \sum_{i=1}^{k} \int_{i-1}^{i} \mathcal{K}(t)\mathcal{B}_i(t)\delta X_i \nu(t)|\gamma'(t)|\, dt.$$

If we just perturb the j-th coordinate of the l-th control node by $\delta\tau$, $j = 1, 2$, this will affect only the Bezier arcs $\Gamma_{p,l-1}$, $\Gamma_{p,l}$ and $\Gamma_{p,l+1}$ determined by $(m_{l-2}, x_{l-1}, m_{l-1})$, (m_{l-1}, x_l, m_l) and (m_l, x_{l+1}, m_{l+1}), respectively. This results in

$$\begin{aligned} J'(p)\delta p = \tfrac{1}{2}\big[\textstyle\int_{l-2}^{l-1} \mathcal{K}(t)t^2\nu_j(t)|\gamma'(t)|\, dt + \int_{l-1}^{l} \mathcal{K}(t)(1+2t(1-t))\nu_j(t)|\gamma'(t)|\, dt \\ + \textstyle\int_{l}^{l+1} \mathcal{K}(t)(1-t)^2\nu_j(t)|\gamma'(t)|\, dt\big]\, \delta\tau := g_{lj}\delta\tau \end{aligned} \tag{19}$$

with obvious modifications for $l = 1$ and $l = k$. Hence the gradient $J'(p)$ is represented by the matrix $G \in \mathbb{R}^{k\times 2}$ with entries $G_{lj} = g_{lj}$ determined by (19).

We demonstrate the feasibility of this approach by applying this concept to the Bernoulli problem described in Section 3. Figure 5.1 shows the computed free boundaries (solid line) after 13, 12 and 19 optimization steps based on the gradient information supplied by (19) for $Q = -1.0$, $Q = -0.5$ and $Q = -0.35$, respectively. The dotted line indicates the initial configuration, stars the corresponding initial and circles the final positions of the control nodes. For comparison purposes we also show the free boundaries (dashed line) obtained in Section 3 by a global method. Figure 5.2 illustrates the decrease of J on a logarithmic scale.

3.3 A dam problem

We conclude this paper by solving a dam problem [4], [5]. The classical boundary variation approach has been used in [1] and [7]. The vertical wall Ω made

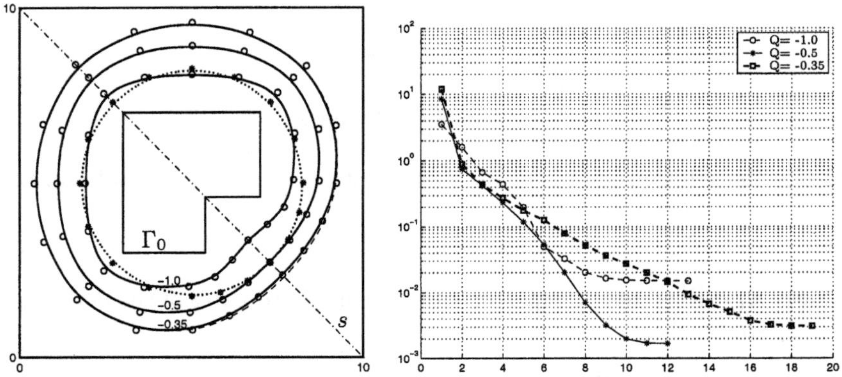

1. Found free boundaries **2.** Minimization history

Fig. 5. Exterior BFB (gradient approach)

of a non-homogeneous material separates two water levels of height y_1, y_2. One wants to find a curve separating the wet and dry part of Ω. The mathematical model leads to the following free-boundary problem:

$$\left\{\begin{array}{rcl} \multicolumn{3}{l}{\textit{Find } \Omega_\varphi \subseteq \Omega \textit{ and } u:\ \Omega_\varphi \to \mathbb{R}^1 \textit{ such that}} \\ -\operatorname{div}(k\nabla u) &=& 0 \ \text{ in } \ \Omega_\varphi, \\ u &=& y_i \ \text{ on } \ \Gamma_i, \quad i=1,2, \\ k\dfrac{\partial u}{\partial \nu} &=& 0 \ \text{ on } \ \Gamma_0 \cup \Gamma_\varphi, \\ u &=& y \ \text{ on } \ \Gamma_\varphi \cup \Gamma_\sigma, \end{array}\right. \tag{20}$$

where $k \in L^\infty(\Omega)$, $k \geq k_0 > 0$ and the partition of $\partial\Omega_\varphi$ into Γ_0, Γ_1, Γ_2, Γ_φ and Γ_σ is shown in Fig. 6. The Neumann condition on Γ_φ will be satisfied by

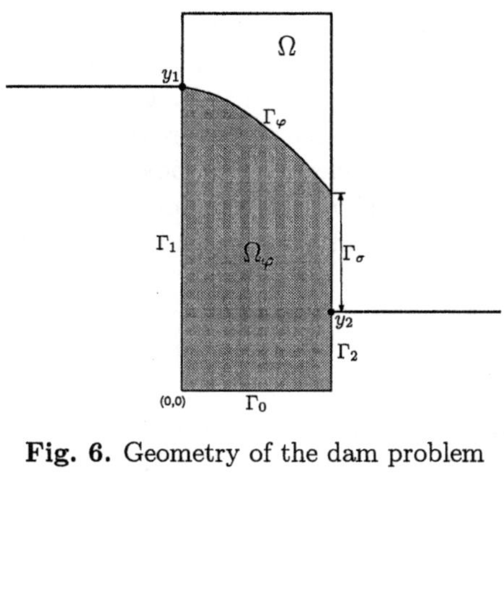

Fig. 6. Geometry of the dam problem

minimizing the cost functional $J(\varphi) = \frac{1}{2}\|k\frac{\partial u}{\partial \nu}\|^2_{-1/2,\Gamma_\varphi}$, whereas the rest of system (20) defines a well-posed state problem $(\mathcal{P}(\varphi))$ for any $\varphi \in \mathcal{U}^{ad}$. If Ω is made of a *homogeneous* material, corresponding to a constant $k > 0$, $\mathcal{U}^{ad}$ consists of smooth, concave and decreasing functions defined over Γ_0. After substituting $\tilde{u} := u - y$ for given $\varphi \in \mathcal{U}^{ad}$ we obtain the new state problem with homogeneous Dirichlet data on $\Gamma_\sigma \cup \Gamma_\varphi$:

$$\begin{cases} -\operatorname{div}(k\nabla\tilde{u}) = \dfrac{\partial k}{\partial y} \text{ in } \Omega_\varphi, \\ \tilde{u} = y_i - y \text{ on } \Gamma_i, \quad i = 1,2, \\ \tilde{u} = 0 \text{ on } \Gamma_\varphi \cup \Gamma_\sigma, \\ \dfrac{\partial \tilde{u}}{\partial \nu} = 1 \text{ on } \Gamma_0. \end{cases} \qquad (\tilde{\mathcal{P}}(\varphi))$$

The FD formulation of $(\tilde{\mathcal{P}}(\varphi))$ is now straightforward:

$$\begin{cases} \textit{Find } (\hat{u}, \lambda_\varphi) \in V_{\hat{g}} \times H^{-1/2}(\Gamma_\varphi) \textit{ such that} \\ \displaystyle\int_\Omega k\nabla\hat{u}\cdot\nabla\hat{v}\,dx\,dy = -\int_\Omega \tilde{k}\frac{\partial \hat{v}}{\partial y}\,dx\,dy + \langle\lambda_\varphi, \hat{v}\rangle_{\Gamma_\varphi} \quad \forall \hat{v} \in V_0, \\ \langle \mu, \hat{u}\rangle_{\Gamma_\varphi} = 0 \quad \forall \mu \in H^{-1/2}(\Gamma_\varphi), \end{cases} \qquad (\hat{\mathcal{P}}(\varphi))$$

where

$$V_{\hat{g}} = \{\hat{v} \in H^1(\Omega) |\ \hat{v} = \hat{g} \text{ on } \partial\Omega \setminus \Gamma_0\},$$

$$V_0 := V_{\hat{g}=0}, \quad \hat{g} = \begin{cases} y_i - y \text{ on } \Gamma_i, \quad i = 1,2, \\ 0 \text{ on } \partial\Omega \setminus (\Gamma_1 \cup \Gamma_2 \cup \Gamma_0) \end{cases}$$

and $\tilde{k}$ is the zero extension of k from Ω_φ to Ω. The cost functional to be minimized now takes the form

$$J(\varphi) = \frac{1}{2}\|\lambda_\varphi\|^2_{-1/2,\Gamma_\varphi}, \tag{21}$$

where ν_y is the y-component of ν. In computations the cost functional (21) is replaced by the $L^2(\Gamma_\varphi)$-norm. As before the spaces $V_{\hat{g}}$ and $H^{-1/2}(\Gamma_\varphi)$ are approximated by continuous, piecewise linear and piecewise constant functions, respectively and the unknown component Γ_φ by piecewise, second order Bezier curves. The example was solved with the following data: $\Omega = (0, 1.62) \times (0, 4)$, $h = 1/32$, $y_1 = 3.22$, $y_2 = 0.84$, $k = 1$ and the number of Bezier segments $\kappa = 6$. The free boundary found by the MCRS method after 2000 function evaluations is shown in Fig. 7.

Conclusions

The variant of the FD method presented in this paper provides an efficient computational tool for the numerical realization of shape optimization problems. Its main features are the following:

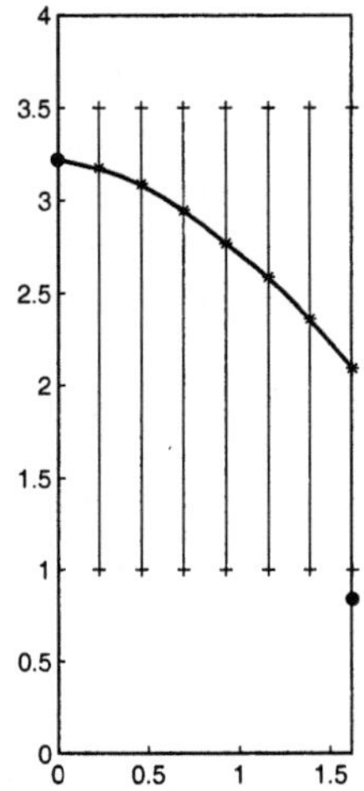

Fig. 7. Found free boundary

- it enables us to solve efficiently state problems;
- the use of non-fitted meshes in discretized problems avoids remeshing of the fictitious domain after any change of the shape of an optimized structure. Consequently, the whole process is more "user friendly" compared with the standard boundary variation technique;
- the (co)-normal derivative of the state which appears in the shape derivative of cost functionals is a part of the solution to the FD-formulation;
- it can be utilized with minor changes for the numerical realization of a large class of free-boundary problems.

Acknowledgement. Research was supported in part by the AKTION Czech Republic-Austria under 36p6, by the grant IAA1075005 (J. Haslinger and T. Kozubek), 201/02/D102 (T. Kozubek) and MSM 113200007.

References

1. D. Begis and R. Glowinski (1975): Application de la méthode des éléments finis à l'approximation d'un problème de domaine optimale. Méthodes de résolution des problèmes approchés, Applied Mathematics & Optimization, **2**, 130–169.
2. F. Brezzi and M. Fortin (1991): Mixed and hybrid finite element methods. Springer-Verlag, New York.
3. M. Flucher and M. Rumpf (1997): Bernoulli's free-boundary problem, qualitative theory and numerical approximation, J. Reine Angew. Math., **486**, 165–204.
4. C. Baiocchi and A. Friedman (1977): A filtration problem in a porous medium with variable permeability, Ann. Mat. Pura Appl., **114**, 377–393.

5. M. Chipot (1984): Variational inequalities and flow in porous media, Springer-Verlag, New York.
6. V. Girault and R. Glowinski (1995): Error analysis of a fictitious domain method applied to a Dirichlet problem, Japan J. Indust. Appl. Math., **12**, 487–514.
7. J. Haslinger, K.H. Hoffmann and R. Mäkinen (1993): Optimal control/dual approach for the numerical solution of a dam problem, Advances in Math. Sciences and Appl., **2**, 189–213.
8. J. Haslinger and A. Klarbring (1995): Fictitious domain / mixed finite element approach for a class of optimal shape design problems. RAIRO M^2AN, **29**, 435–450.
9. J. Haslinger, T. Kozubek, K. Kunisch and G. Peichl (2003): Shape Optimization and Fictitious Domain Approach for Solving Free Boundary Problems of Bernoulli Type, COAP, **26**, 231–251.
10. J. Haslinger, T. Kozubek, K. Kunisch and G. Peichl (2003): An embedding domain approach for a class of 2-D shape optimization problems: mathematical analysis, (to appear in Journal of Mathematical Analysis and Applications)
11. J. Haslinger and R.A.E. Mäkinen (2003): Introduction to Shape Optimization, Theory, Approximation, and Computation. SIAM, Philadelphia.
12. J. Haslinger and P. Neittaanmäki (1996): Finite Element Approximation for Optimal Shape, Material and Topology Design. Second Edition, J. Wiley & Sons, Chichester.
13. R. Kress (1989): Linear Integral Equations, Springer-Verlag.
14. I. Křivý and J. Tvrdík (1995): The controlled random search algorithm in optimizing regression models, Comput. Statist. and Data Anal., **20**, pp. 229–234.
15. J. Nečas (1967): Les Méthodes Directes en Théorie des Equations Elliptiques, Masson, Paris.
16. J.A. Nelder and R. Mead (1964): A simplex method for function minimization, Computer J., **7**, 308–313.
17. W.L. Price (1976): A controlled random search procedure for global optimisation, Computer J., **20**, 367–370.
18. J. Sokolowski and J.P. Zolesio (1992): Introduction to Shape Optimization, Shape Sensitivity Analysis, Springer Series in Computational Mathematics, **16**, Springer-Verlag, Berlin.

Part II

Contributed Papers

Domain Decomposition Method for a Class of Non-Linear Elliptic Equation with Arbitrary Growth Nonlinearity and Data Measure

Nour Eddine Alaa[1] and Jean Rodolphe Roche[2]

[1] Département de Mathématiques et Informatique, Université des Sciences et Techniques Cadi Ayyad, B.P. 618, Guéliz,Marrakech,Maroc, *alaa@fstg-marrakech.ac.ma*

[2] I.E.C.N., Université Henri Poincaré, B.P. 239, 54506 Vandoeuvre lès Nancy, France, *roche@iecn.u-nancy.fr*

Summary. In this paper we show the existence of weak solutions for a nonlinear elliptic equations with arbitrary growth of the non linearity and data measure. A numerical algorithm to compute a numerical approximation of the weak solution is discribed and analysed. In a first step a super-solution is computed using a domain decomposition method. A numerical example is presented and commented.

1 Introduction

This work deals with weak solutions of the following quasi-linear elliptic problem with Dirichlet boundary conditions:

$$\begin{cases} -u''(t) + G(t, u'(t)) = F(t, u(t)) + f(t) \;\; in \;\; (0,1) \\ u(0) = u(1) = 0 \end{cases} \tag{1}$$

where $G, F : [0,1] \times \mathbb{R} \to [0, +\infty[$ are measurable and continuous with respect to u' and u, f is a given finite non negative measure on $(0,1)$.

The main goal is to present a numerical analysis of this weak solutions and to study their existence and uniqueness. Such problems arise from biological, chemical and physical systems and various methods have been proposed to study existence, uniqueness, qualitative properties and numerical simulation of such solutions (see [8]). When f is regular, it is proved in [9] that if (1) has a nonnegative super-solution in $W_0^{1,\infty}$ then (1) has a solution in $W_0^{1,\infty} \bigcap W^{2,p}$. Many authors dealt with this problem when f is irregular and G is sub-quadratic with respect to u' namely:

$$|G(t,r)| \leq c(g(t) + |r|^2), \quad g(t) \in L^1(0,1), c > 0 \tag{2}$$

They showed that, if G satisfies (2), (1) has a solution $u \in H_0^1(0,1)$ provided that (1) has a super-solution in $W^{1,\infty}(0,1)$ see [6], [5]. The case where the super-solution itself is irregular has been treated in [2], if the super-solution belongs to $H_0^1(0,1)$ then (1) has a solution in $H_0^1(0,1)$ provided that G satisfies (2).

In this work we are particularly interested in situations where f is irregular and where the growths of G with respect to u' and F with respect to u are arbitrary. Let us make some precisions about the model problem:

$$\begin{cases} -u''(t) + |u'(t)|^q = |u(t)|^p + f(t) \text{ in } (0,1) \\ u(0) = u(1) = 0 \end{cases} \tag{3}$$

where $p, q \geq 1$ and $f \in M_B^+(0,1)$, the set of nonnegative finite measure on $(0,1)$. We show here that if the semi-linear problem:

$$\begin{cases} -w''(t) = |w(t)|^p + f(t) \text{ in } (0,1) \\ w(0) = w(1) = 0 \end{cases} \tag{4}$$

has a solution then (3) has a solution. Remark here that no restrictions for p and q are imposed. For an elegant study of (4), see the work of Pierre and Baras [4]. If $w'(0) = +\infty$ or $w'(1) = -\infty$ then $w \notin W_0^{1,\infty}$ and obviously the classical approach fails to provide existence of solutions of (3) and new techniques have to be used.

Another approach studied here is the numerical approximation of the solution of problem (1). In this approach the most important difficulties are to determine the uniqueness and the blowup of the solution.

The general algorithm for the numerical approximation of this equation is the application of the Newton method to the discrete version of problem (1):

$$\text{Find } U \in \mathbb{R}^m \text{ such that } AU = H(U) \tag{5}$$

where A is a sparse matrix and $H : \mathbb{R}^m \to \mathbb{R}^m$ is a nonlinear operator.

The Newton algorithm is given by:

$$\begin{cases} \text{chose } U^0 \text{ in a neighbourhood of the solution} \\ \text{and solve until convergence} \\ (A - H'(U^k)\,Id)\,(U^{k+1} - U^k) = -A\,U^k + H(U^k) \end{cases} \tag{6}$$

where $H'(U^k)$ is the Jacobian matrix of the operator H computed in U^k and Id is a identity matrix in $\mathbb{R}^m$. This method converges quadratically when it converges. Convergence depends in particular on the choice of U^0 and on the existence and uniqueness of solutions of the linear system (6). In the case of problem (1) the matrix $A - H'(U^k)Id$ is often singular.

To overcome this difficulty we introduce a domain decomposition to compute an approximation of $\delta u^k = u^{k+1} - u^k$ by the resolution of a sequence of problems of type (1) in subset Ω_i of $(0,1)$, such that $\Omega = \bigcup_{i=1,K} \Omega_i$. The idea of the method comes from the following remark [11]:

Lemma 1. *Let $0 \leq a < b \leq 1$, $a_i \in L^\infty(0,1)$, for $i = 1,2$. If $|b-a|$ is small enough then the operator $-\frac{d^2}{dt^2} - a_1(t)\frac{d}{dt} - a_2(t)Id$ has an inverse in (a,b).*

We have organized this paper in the following manner. In section 2 we give the exact setting of the problem, we present an approximate equation for (1) and we prove that the existence of weak super-solutions implies the existence of weak solutions, without any restriction of the growth of G with respect to u', this result being a generalization of the classical result of [9], [6] and [2].

In section 3 we present an approximation scheme for problem (1) based on the Schwarz overlapping domain decomposition method, combined with finite element method.

This work was supported by the French Grant "Action Intégrée MA/02/33".

2 Mathematical analysis of the problem

Throughout this paper we suppose that f is a nonnegative finite measure on $(0,1)$ and $G, F : [0,1] \times \mathbb{R} \to [0,+\infty)$ are measurable.

$$\text{The functions } r \to G(t,r), F(t,r) \text{ are continuous a. e. } t \tag{7}$$

$$F(t,.) \text{ is nondecreasing and } G(t,.) \text{ is convex,} \tag{8}$$

$$\forall r \in \mathbb{R},\ G(.,r),\ F(.,r) \text{ are integrable on } (0,1) \tag{9}$$

$$G(t,0) = \min\{G(t,r), r \in \mathbb{R}\} = 0 \text{ and } F(t,0) = 0. \tag{10}$$

We introduce the notion of weak solution, super-solution and sub-solution used here.

Definition 1. *A function u is said to be a weak solution of (1) if*

$$\begin{cases} u \in W^{1,\infty}_{loc}(0,1) \bigcap C_0[0,1] \\ -u''(t) + G(t,u'(t)) = F(t,u(t)) + f \ \ in \ \mathcal{D}'(0,1) \end{cases} \tag{11}$$

(replace in (11) = by $\geq$ for a weak super-solution and by $\leq$ for a weak sub-solution)

Remark 1. In (11) $u \in W^{1,\infty}_{loc}(0,1)$, using (9) we have $G(t,u'(t))$ and $F(t,u(t)) \in L^1_{loc}(0,1)$. Hence every term in (11) makes sense.

This enables us to state the main result of this paper.

Theorem 1. *Assume that (7)–(10) and $f \in M^+_B(0,1)$ hold. Assume that there exists a weak solution $\overline{w}$ to the problem,*

$$\begin{cases} \overline{w} \in W^{1,\infty}_{loc}(0,1) \cap C_0[0,1] \\ -\overline{w}'' = F(.,\overline{w}) + f \quad in\ \mathcal{D}'(0,1) \end{cases} \tag{12}$$

Then $\overline{w}$ is a super-solution of (1) and there exists a weak solution u of (1) such that $u \leq \overline{w}$.

Remark 2. 1) It should be noted that there are no growth restrictions on the lower order nonlinearity of F and G w.r.t. u and u' respectively. Hence the present theorem extends some results proposed in [2], [6].

2) For any finite nonnegative measure f, the problem:

$$\begin{cases} \underline{w} \in W_0^{1,\infty}(0,1) \\ -\underline{w}'' + G(t, \underline{w}') = f \quad \text{in } \mathcal{D}'(0,1) \end{cases} \tag{13}$$

has a unique solution $\underline{w}$, see [1], and remark here that $\underline{w}$ is a sub-solution of the problem (11).

In order to prove theorem 1, we consider $G_n(t,.)$, for $n \geq 0$, the Yoshida approximation of $G(t,.)$ which increases a.e. to $G(t,.)$ as n tends to infinity. Note that $G_n(t,.)$ satisfies (7) -(10) and

$$G_n \leq G \ , \ G_n \leq G_{n+1} \tag{14}$$

According to the result given in [1], [6] there exists a sequence (u_n) of solution to the problem:

$$\begin{cases} u_{n+1} \in W_0^{1,\infty}(0,1) \\ -u''_{n+1} + G_{n+1}(t, u'_{n+1}) = F(u_n) + f \quad \text{in } \mathcal{D}'(0,1) \end{cases} \tag{15}$$

where $u_0 = \overline{w}$. To have estimates when passing to the limit, we give the following three lemmas, see [1].

Lemma 2. *Let* $a(t) \in L^1_{loc}(0,1)$, $v \in W^{1,1}_{loc}(0,1) \cap C_0[0,1]$ *such that*

$$\begin{cases} a(t)\, v'(t) \in L^1_{loc}(0,1) \\ -v'' - a\, v' \geq 0 \ \textit{in} \ \mathcal{D}'(0,1) \end{cases} \tag{16}$$

Then $v \geq 0$ *in* $[0,1]$.

Lemma 3. *Let* $u \in W^{1,1}_{loc}(0,1)$, $\underline{v}, \overline{v} \in L^\infty(0,1)$ *and* $\mu \in M_B^+(0,1)$ *such that:*

$$\begin{cases} \underline{v} \leq u \leq \overline{v} \ \textit{in} \ ]0,1[\\ -u'' \leq \mu \ \textit{in} \ \mathcal{D}'(0,1) \\ -\overline{v}'' \geq \mu \ \textit{in} \ \mathcal{D}'(0,1) \end{cases} \tag{17}$$

Then $u \in W^{1,\infty}_{loc}(0,1)$, *and*

$$|u'(t)| \leq \frac{1}{d(t;a,b)}\left(c(a,b) + ||\underline{v}||_{L^\infty} + ||\overline{v}||_{L^\infty} + ||\mu||_{M_B}\right) \tag{18}$$

for all $0 < a < b < 1$, *where* $d(t;a,b) = min(b-t, t-a)$ *and* $c(a,b)$ *is a constant depending on* a *and* b.

Lemma (3), will provide $W^{1,\infty}_{loc}(0,1)$ estimates for the approximate solution u_n. But this estimate does not allow us to pass to the limit in the nonlinear terms. We need the strong convergence of u_n in $W^{1,\infty}_{loc}(0,1)$. We obtain this result from the following Lemma.

Lemma 4. *Let* $(u_n)_n \subset W_0^{1,\infty}(0,1)$ *such that,*

$$u_n \to u \quad \text{strongly in} \quad L^\infty(0,1) \tag{19}$$

$$\begin{cases} \underline{w} \le u \le u_n, \\ -u'' \le \mu \quad \text{in} \quad \mathcal{D}'(0,1) \\ \overline{w}'' \le \mu \quad \text{in} \quad \mathcal{D}'(0,1) \end{cases} \tag{20}$$

Then $u_n' \to u'$ *strongly in* $L^\infty_{loc}(0,1)$

Proof of the theorem (1). In a first step we have by direct application of the maximum principle(see [2]) that

$$\underline{w} \le u_{n+1} \quad \text{for all} \quad n \ge 0 \tag{21}$$

In a second step, since F is monotonous w.r.t. r we can prove by induction that

$$u_{n+1} \le u_n \le \overline{w} \quad \text{in} \quad [0,1] \quad \text{for all} \quad n \ge 0 \tag{22}$$

By lemma (3) u_n is bounded in $W^{1,\infty}_{loc}(0,1) \cap C_0[0,1]$ independently of n. Therefore, there exists a subsequence, still denoted by (u_n) for simplicity, u_n such that converges to u strongly in $L^\infty(0,1)$ if $n \to \infty$. Also u'_{n+1} converges to u' strongly in $L^1_{loc}(0,1)$ and a.e. in $(0,1)$. Then from lemma (3) we have u'_{n+1} converging to u' strongly in $L^\infty_{loc}(0,1)$, and

$$||u_n'||_{L^\infty(a,b)} \le K(a,b)\,(\,c(a,b) + ||\overline{w}||_{L^\infty(0,1)} + ||f||_{M_B} + ||\underline{w}||_{L^\infty(0,1)}) \tag{23}$$

where $K(a,b) = 1/\eta$ and $0 < \eta < a < \eta + b < 1$.

Since $G(t,.)$ and $F(t,.)$ are continuous with respect to the two last arguments, we have for all $0 < a < b < 1$

$$G(t,u'_{n+1})\,,\, F(t,u_n) \text{ converges to } G(t,u')\,,\, F(t,u) \text{ a.e. } t \in (0,1). \tag{24}$$

On the other hand, for a.e $t \in (a,b)$

$$|G(t,u'_{n+1}(t)| \le \max_{|r| \le C'(a,b)} |G(t,r)| = \theta(t) \tag{25}$$

and

$$|F(t,u_n(t))| \le \max_{|s| \le \max(\,||\overline{w}||_{L^\infty(0,1)},\,||\underline{w}||_{L^\infty(0,1)})} |F(t,s)\,| = \widehat{\theta}(t) \tag{26}$$

and $\theta, \widehat{\theta} \in L^1_{loc}\,(0,1)$ from (9). Using Lebesgue's dominate convergence Theorem (see [7]), we also have;

$$G(t,u'_{n+1}),\; F(t,u_n) \text{ converges to } G(t,u'),\, F(t,u) \text{ in } L^1(a,b) \text{ respectively} \tag{27}$$

Now, we can pass to the limit in (15), and if $\varphi \in \mathcal{D}\,(0,1)$ with sup $\varphi \subset [a,b]$ then

$$0 = \lim_{n\to\infty} \langle\, -u''_{n+1} + G(u'_{n+1}) - F(u_n)\,,\,\varphi\rangle = \langle\, -u'' + G(u') - F(u)\,,\varphi\,\rangle \tag{28}$$

where $\langle .,. \rangle$ denotes the duality pairing between $\mathcal{D}'(0,1)$ and $\mathcal{D}(0,1)$. The theorem follows.

3 Numerical method

In this section we present the numerical method to solve equation (1). Formally the iterative method brings out a sequence of numerical solutions of (15) in $H_0^1(0,1)$ with a first guess which is a super-solution of (1), in our case, a solution of problem (12).

Then the algorithm can be formulated in the following way:

1) Find $\overline{w} \in H_0^1(0,1)$ such that:

$$(P_1) \qquad -\overline{w}'' \geq F(.,\overline{w}) + f \tag{29}$$

2)Given $u_0 = \overline{w}$ we compute a sequence, solution in $H_0^1(0,1)$ of the non linear equation:

$$(P_2) \qquad -u''_{n+1} + G_{n+1}(t, u'_{n+1}) = F(u_n) + f \tag{30}$$

Both problems (P_1) and (P_2) are nonlinear, and if (P_1) has a solution, in theorem (1) we prove that the solution of (P_2) is also a solution of the equation (1). Problem (P_2) has a unique solution and the numerical calculation is straightforward by Newton method. To solve the nonlinear equation (P_1), which presents some interesting difficulties, we construct a sequence w^k such that w^k is a solution of a linear problem and w^k converges to $\overline{w}$.

Let $\overline{w}^0 = 0$, we define $\overline{w}^{k+1} = \overline{w}^k + \delta$ where δ is the solution of the following linear problem:

$$(P_3) \qquad \begin{cases} -\delta'' - \frac{\partial F(\overline{w}^k)}{\partial r}\delta = (\overline{w}^k)'' + F(\overline{w}^k) + f \quad \text{in} \quad (0,1) \\ \delta(0) = \delta(1) = 0 \end{cases} \tag{31}$$

Then at each iteration we have to solve the linear problem (P_3). To achieve this, we consider a weak formulation of the problem and use the finite element method.

To simplify the text we reformulate (P_3) in the following way: find $v \in H_0^1(a,b)$ such that:

$$(P_4) \begin{cases} -v(t)'' + c(t)\,v(t) = h(t) \text{ in } (a,b) \\ v(a) = v(b) = 0 \end{cases} \tag{32}$$

where $h \in M_B(a,b)$, the set of finite measure in (a,b), and $c(t) \in L^2(a,b)$, without restriction on its sign. We assume $c_\infty = ||c||_{L^\infty(a,b)}$ is bounded.

According to Lemma 1, problem (P_4) has a solution in a domain (a,b) small enough.

If $V = H_0^1(0,1)$ then the weak formulation of (P_4) is:

$$\text{find } v \in V : a(v,w) \int_a^b v'w' + c(x)\,v\,w\,dx = \int_a^b h\,w\,dx = (h,w) \;\; \forall\, w \,\text{in}\, V \tag{33}$$

Thanks to Poincaré inequality we have:

$$a(w,w) = (w',w') + (c(t)w,w) \geq (\frac{c_0}{|b-a|} - c_\infty)(w,w) \tag{34}$$

Thus the bilinear form $a(w,v)$ is coercive if $|b-a| < \frac{c_0}{c_\infty}$.

This remark is of great interest, because it can be exploited to obtain a numerical solution of (P_4) using a domain decomposition technique. In other words, this means that the domain partition should be determined by the behavior of $\left\| \frac{\partial F(\overline{w}^k)}{\partial r} \right\|_\infty$.

The aim of this paragraph is to introduce the Schwarz overlapping domain decomposition method [10] applied to problem (P_4). To simplify, without lost of generality, we assume that we can consider a two domains decomposition $(a,b) = (a,\beta) \bigcup (\alpha,b)$ such that:

$$\text{with } \alpha < \beta \text{ and } (\beta - a), (b - \alpha) \; < \min(\frac{c_0}{c_\infty}, \frac{\pi}{2\sqrt{c_\infty}}) \tag{35}$$

Then, if v^0 is an initialization function, defined on (a,b) and vanishing at a and b, we define for $k \geq 0$, two sequences v_i^k, $i = 1,2$ solving the following problems:

$$\begin{cases} -(v_1^{k+1})''(t) + c(t)\, v_1^{k+1}(t) = h(t) \text{ in } (a,\beta) \\ v_1^{k+1}(a) = 0; \; v_1^{k+1}(\beta) = v_2^k(\beta) \end{cases} \tag{36}$$

and

$$\begin{cases} -(v_2^{k+1})''(t) + c(t)\, v_2^{k+1}(t) = h(t) \text{ in} (\alpha,b) \\ v_2^{k+1}(\alpha) = v_1^k(\alpha); \; v_2^{k+1}(b) = 0 \end{cases} \tag{37}$$

Now to prove the convergence of the Schwarz overlapping domain decomposition algorithm applied to problem (P_4) we consider two problems:

$$(P_{4.1}) \begin{cases} -v_1(t)'' + c(t)\, v_1(t) = h(t) \in (a,\beta) \\ v_1(a) = 0 \; ; \; v_1(\beta) = v_2(\beta) \end{cases} \tag{38}$$

and:

$$(P_{4.2}) \begin{cases} -v_2(t)'' + c(t)\, v_2(t) = h(t) \text{ in } (\alpha,1) \\ v_2(\alpha) = v_1(\alpha),\, v_2(b) = 0 \end{cases} \tag{39}$$

Let v be $v = v_1$ in (a,β), $v = v_2$ in (α,b), then $v_1 = v_2$ in (α,β). We suppose the existence of a solution of $P_{4.1}$ in $C(a,\beta)$ and a solution of $P_{4.2}$ in $C(\alpha,b)$.

Theorem 2. *Assume a,b,α and β are such that $\alpha < \beta$, $(\beta-a),(b-\alpha) < \min(\frac{c_0}{c_\infty}, \frac{\pi}{2\sqrt{c_\infty}})$. Then the sequence v^k converges to v in $C(a,\beta)$ and $C(\alpha,b)$.*

Proof:

Let $d^k = v_1^k - v$ in (a,β) and $e^k = v_2^k - v$ in (α,b). We prove the following inequality:

$$||d^{k+2}||_\infty \leq \gamma\, ||d^k||_\infty \text{ and } ||e^{k+2}||_\infty \leq \gamma\, ||e^k||_\infty \tag{40}$$

where $\gamma < 1$.

The difference d^k satisfies the following equation.

$$(P_{d.1}) \left\{ \begin{array}{l} -\,d^{k+1}(t)'' + c(t)\, d^{k+1}(t) = 0 \text{ in } (a,\beta) \\ d^{k+1}(a) = 0 \;;\; d^{k+1}(\beta) = v_2^k(\beta) - v(\beta) \end{array} \right. \tag{41}$$

and e^k satisfies a similar equation in (α, b) with boundary conditions
$e^{k+1}(\alpha) = d^k(\alpha)$;
$e^{k+1}(b) = 0$.

If we consider the following equation:

$$(P_\varphi) \left\{ \begin{array}{l} -\,\varphi(t)'' - c_\infty\, \varphi(t) = 0 \text{ in } (a,\beta) \\ \varphi(a) = 0 \;;\; \varphi(\beta) = |e^{k+1}(\beta)| \end{array} \right. \tag{42}$$

then $\varphi(t) = |e^{k+1}(\beta)| \dfrac{\sin(\sqrt{c_\infty}\,(t-a))}{\sin(\sqrt{c_\infty}\,(\beta - a))}$. This solution is unique and positive if $(\beta - a) < \frac{\pi}{2\sqrt{c_\infty}}$. If we consider the difference $z = \varphi - d^{k+2}$ it is easy to prove that $z \geq 0$ and $d^{k+2} \leq \varphi \leq |e^{k+1}(\beta)|$. If now $z = \varphi + d^{k+2}$ we have also $z \geq 0$ and $-|e^{k+1}(\beta)| \leq \varphi \leq d^{k+2}(t)$, $\forall t$ in (a,β). Then the inequality $||d^{k+2}||_\infty \leq |e^{k+1}(\beta)| \leq ||e^{k+1}||_\infty$ holds.

To prove that $|e^{k+1}(\beta)| \leq \gamma\, ||d^k||_\infty$ with $\gamma < 1$, we consider the equation:

$$(P_\phi) \left\{ \begin{array}{l} -\,\phi(t)'' - c_\infty\, \phi(t) = 0 \quad \text{in } (\alpha, b) \\ \phi(\alpha) = |d^k(\alpha)| \textit{ and } \phi(b) = 0 \end{array} \right. \tag{43}$$

The solution of this equation is given by: $\phi(t) = |d^k(\alpha)| \dfrac{\sin(\sqrt{c_\infty}(b-t))}{\sin(\sqrt{c_\infty}\,(b-\alpha))}$. This solution is positive if $(b - \alpha) < \frac{\pi}{2\sqrt{c_\infty}}$ and we have $\phi(t) \geq |e^{k+1}(t)|$ $\forall t \in (\alpha, b)$.

Then $|e^{k+1}(\beta)| \leq \phi(\beta) \leq \gamma\, |d^k(\alpha)|$ with $\gamma = \dfrac{\sin(\sqrt{c_\infty}\,(b-\beta))}{\sin(\sqrt{c_\infty}\,(b-\alpha)}$. The coefficient γ is smaller than 1 only if $\alpha < \beta$.

In conclusion we have $||d^{k+2}||_\infty < ||d^k||_\infty$ if $\alpha < \beta$ and $(\beta - a), (b - \alpha) < \min(\frac{c_0}{c_\infty}, \frac{\pi}{2\sqrt{c_\infty}})$. In the same way we prove that $||e^{k+2}||_\infty < ||e^k||_\infty$ if $\alpha < \beta$ and $(\beta - a), (b - \alpha) < \min(\frac{c_0}{c_\infty}, \frac{\pi}{2\sqrt{c_\infty}})$.

We conclude that the Schwarz overlapping domain decomposition method applied to problem (P_4) converges.

4 Numerical Results

The algorithm introduced in the previous section has been implemented numerically for the model problem (3) with $p = q = 3$ and $f(t)$ a Dirac in $\dfrac{1}{3}$.

To study the convergence history of the numerical simulation plotted in figure 1 we consider two steps. In the first step, where we compute a super-solution, we observe the evolution of the number of sub-domains: it goes from $m = 2$ sub-domains to $m = 10$ sub-domains in five iterations according to criterion (35). Simulation stops after 17 iterations when the residual is of the order 10^{-11}.

In the second step, starting with the super-solution computed in the previous step we perform nine iterations of the Yoshida approximation described in section 2 and the simulation stops when the correction computed is in uniform norm of the order 10^{-11}.

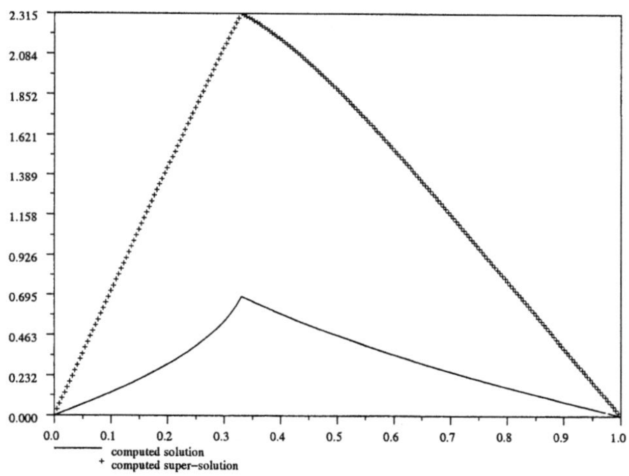

Fig. 1. Example $f(t) = 8. * \delta_t$, $t = \frac{1}{3}$, m=10

References

1. Alaa, N. (1989): Etude d'équations elliptiques non-linéaires à dépendance convexe en le gradient et à données measures, Thèse de Doctorat, Université de Nancy I
2. Alaa, N., Pierre, M. (1993): Weak solution of some quasi-linear elliptic equations with data measures, SIAM J. Math. Anal., **24**, 23–35
3. Alaa, N., Iguernane, M. (2002): Weak periodic solutions of some quasi-linear parabolic equations with data measure, J. Inequal. Pure Appl. Math. **3**
4. Baras, P., Pierre, M. (1984): Critères d'existence de solutions positives pour des équations semi-linéaires, Annales Fourier Grenoble, **24**, 1985–2006
5. Bensoussan, A., Boccardo, L., Murat, F. (1988): On a nonlinear partial differential equation having natural growth terms and unbounded solution, Ann. Inst. Henri Poincaré, **5**, 347–364
6. Boccardo, L., Murat, F., Puel, J.P. (1989): Existence results for some quasi-linear parabolic equations, Nonlinear Analysis Theory Method and Applications, **13**, 373–392
7. Brezis, H. (1983): Analyse fonctionnelle théorie et applications, Masson

8. Levin, S.A., Hallam, Th. G., Gross, L.J. (1989): Applied Mathematical Ecology, Biomathematics **18**, Springer Verlag
9. Lions, P. L.(1980): Résolution de problèmes elliptiques quasilinéaires, Arch. Rational Mech. Anl., **74**, 335-353
10. Quarteroni, A.,Valli, A.(1999): Domain decomposition Methods for Partial Differential Equations, Oxford Science Publications
11. Witomski, P. (1983): Sur la résolution numérique de quelques problèmes nonlinéaires, Thèse d'Etat, Université Scientifique et médicale de Grenoble

Variants of Relaxation Schemes and the Lattice Boltzmann Model Relaxation Systems

Mapundi Kondwani Banda[1]

Darmstadt University of Technology, Schlossgartenstr. 7, D-64289 Darmstadt
banda@mathematik.tu-darmstadt.de

Summary. In the low Mach number limit of the Lattice Boltzmann type models one obtains the incompressible Navier-Stokes equation. This is achieved by asymptotic analysis. Moreover in the course of this analysis, the Lattice Boltzmann Model reduces to a relaxation system which can be discretized using relaxation schemes. We present two variants of the relaxation schemes characterized by local approximation of characteristic speeds and a multidimensional flux approximation. These are applied to relaxation systems. Their performance will be discussed with reference to test cases of isothermal incompressible flow.

1 Introduction

Many kinetic equations or discrete velocity models of kinetic equations yield in the limit for small Knudsen and Mach numbers an approximation of the incompressible Navier Stokes (INS) equations. A classical example is given by the discrete velocity models used for Lattice-Boltzmann methods, see [1]. These discrete velocity models can be viewed as relaxation systems for the INS equations.

Relaxation type schemes have been used successfully to discretize such relaxation systems. In particular, a large number of numerical methods for kinetic equations with stiff relaxation terms have been considered in fluid dynamic or diffusive limits. For these relaxation methods and asymptotic preserving methods, we refer to [2, 3] and for more general applications of relaxation schemes to the recent review paper [4]. Such a multiscale based approach provides an alternative to understanding the numerical transition from kinetic models to the continuum models. This can be used as a platform for developing alternative numerical schemes for INS. In the context of hyperbolic conservation laws relaxation schemes provide efficient high resolution and Riemann solver free numerical methods. Here we introduce an HLL type of relaxation scheme and apply it in the context of INS. Further by choosing a different method for computing cell averages a scheme that can be considered multidimensional is realized. This paper follows closely the work presented in [6].

2 Lattice-Boltzmann type Discrete Velocity Models and Simplified Relaxation Systems

2.1 The Lattice-Boltzmann moment system

Consider kinetic equations with a *diffusive scaling* with a small parameter, ϵ, together with a rescaling of velocity. This scaling describes the small Knudsen and low Mach number limit of kinetic equations, see [7] for details. Under these transformations, we obtain

$$\frac{\partial f}{\partial t} + \frac{1}{\epsilon^{\alpha}} \mathbf{v} \cdot \nabla f = -\frac{1}{\epsilon^{1+\alpha_T}} (f - f^{eq}[\rho, \epsilon^{\alpha} \mathbf{u}]) =: J(f). \quad (1)$$

which describes the evolution of a particle density $f(\mathbf{x}, \mathbf{v}, t)$ with $\mathbf{x} = (x, y) \in \mathbb{R}^2$ and $\mathbf{v} = (v_1, v_2) \in \mathbb{R}^2$. For discrete velocity models in two-dimensions (2-D) consider a model with nine velocities ($N = 9$). In the discrete case, the $\mathbf{v}$-dependence of the particle distribution $f(\mathbf{x}, \mathbf{v}, t)$ is uniquely determined through 9 functions $f_i(\mathbf{x}, t) = f(\mathbf{x}, \mathbf{c}_i, t), \quad i = 0, \ldots, 8.$ where $\mathbf{c}_i$ are discrete velocities.

A discrete moment M of order $m \in \mathbb{N}$ of f is defined by $M(t, \mathbf{x}) = \langle f(\mathbf{x}, \mathbf{v}, t), P(\mathbf{v})\rangle_{\mathbf{v}}$, where P is a $\mathbf{v}$-polynomial of degree $m \in \mathbb{N}$. In the following we denote the components of the velocity by $\mathbf{u} = (u, v)$. The scalar product is defined as $\langle g_1, g_2\rangle_{\mathbf{v}} = \sum_{\mathbf{v}} g_1(\mathbf{v}) g_2(\mathbf{v})$. Equation (1) is transformed into an equivalent set of moment equations (see also [8, 9] for a similar approach) using moments based on $\mathbf{v}$-polynomials [10]. The mass and momentum density are given by the zeroth and first order moments of f, $\langle f, 1\rangle_{\mathbf{v}} = \rho$ and $\langle f, v_1\rangle_{\mathbf{v}} = \rho u$, $\langle f, v_2\rangle_{\mathbf{v}} = \rho v$. The second order moments form a symmetric tensor, $\boldsymbol{\Theta} = (\boldsymbol{\Theta}^x \quad \boldsymbol{\Theta}^y)$, and the remaining moments are set to $\mathbf{q}$ and s. The equations of mass and momentum conservation are

$$\partial_t \rho + \operatorname{div} \rho \mathbf{u} = 0, \qquad \partial_t \rho \mathbf{u} + \operatorname{div} \boldsymbol{\Theta} + \frac{1}{3\epsilon^{2\alpha}} \nabla \rho = 0. \quad (2)$$

Here, the divergence is applied to the rows of $\boldsymbol{\Theta}$. The equation for $\boldsymbol{\Theta}$ is

$$\partial_t \boldsymbol{\Theta} + \frac{2}{3\epsilon^{2\alpha}} \mathbf{S}[\rho \mathbf{u}] + \frac{1}{3} \mathbf{Q}[\mathbf{q}] = -\frac{1}{\epsilon^{1+\alpha_T}} (\boldsymbol{\Theta} - \rho \mathbf{u} \otimes \mathbf{u}), \quad (3)$$

where

$$\mathbf{S}[\mathbf{u}] = \frac{1}{2} \begin{pmatrix} 2\partial_x u & \partial_y u + \partial_x v \\ \partial_y u + \partial_x v & 2\partial_y v \end{pmatrix}.$$

Since $\mathbf{Q}[\mathbf{q}]$, $\mathbf{q}$ and s are not needed to derive INS, they will be ignored.

From the momentum equation in (2) one can deduce that as $\epsilon \to 0$, ρ approaches a constant $\bar{\rho}$ and can be written as $\rho = \bar{\rho}(1 + 3\epsilon^{2\alpha} p)$ to obtain $\mathcal{O}(1)$ terms. Hence

$$\partial_t p + \frac{1}{3\epsilon^{2\alpha}} \operatorname{div} \mathbf{u} = -\operatorname{div} p\mathbf{u}, \quad \partial_t \mathbf{u} + \operatorname{div} \frac{1}{\bar{\rho}} \boldsymbol{\Theta} + \nabla p = -3\epsilon^{2\alpha} \partial_t p \mathbf{u}. \quad (4)$$

For $\epsilon \to 0$, equation (3) yields at the lowest order

$$\frac{1}{\bar{\rho}}\boldsymbol{\Theta} = \mathbf{u} \otimes \mathbf{u} - \frac{2\epsilon^{1-\alpha}\tau}{3}\mathbf{S}[\mathbf{u}]. \tag{5}$$

Since $2\,\mathrm{div}\,\mathbf{S}[\mathbf{u}] = (\Delta + \nabla\,\mathrm{div}\,)\mathbf{u}$, we obtain from (4) and (5) the INS equations as limiting system for $\alpha = 1$

$$\mathrm{div}\,\mathbf{u} = 0, \qquad \partial_t \mathbf{u} + \mathrm{div}\,\mathbf{u} \otimes \mathbf{u} + \nabla p = \frac{\tau}{3}\Delta \mathbf{u}, \tag{6}$$

where the Reynolds number is related to the relaxation time by $Re = 3/\tau$.

For $0 < \alpha < 1$, we obtain the incompressible Euler equations

$$\mathrm{div}\,\mathbf{u} = 0, \qquad \partial_t \mathbf{u} + \mathrm{div}\,(\mathbf{u} \otimes \mathbf{u}) + \nabla p = 0.$$

2.2 Simplified Relaxation Systems

By neglecting the lower order terms in the above equations (3) and (4) and setting $\overline{\rho} \equiv 1$, we can introduce a simplified relaxation system:

$$\begin{aligned} \partial_t p + \frac{1}{\epsilon^{2\alpha}}\,\mathrm{div}\,\mathbf{u} = 0, \quad & \partial_t \mathbf{u} + \mathrm{div}\,\boldsymbol{\Theta} + \nabla p = 0, \\ \partial_t \boldsymbol{\Theta} + \nabla^{\mathbf{a}}[\mathbf{u}] + \frac{2}{\epsilon^{2\alpha}}\mathbf{S}^{\epsilon}[\mathbf{u}] &= -\frac{1}{\epsilon^{1+\alpha}\tau}(\boldsymbol{\Theta} - \mathbf{u} \otimes \mathbf{u}), \end{aligned} \tag{7}$$

where

$$\mathbf{S}^{\epsilon}[\mathbf{u}] = \mathbf{S}[\mathbf{u}] - \frac{\epsilon^{2\alpha}}{2}\nabla^{\mathbf{a}}[\mathbf{u}].$$

We have added and subtracted the term

$$\nabla^{\mathbf{a}}[\mathbf{u}] = (a^2 \partial_x \mathbf{u}, b^2 \partial_y \mathbf{u}) = \begin{pmatrix} a^2\partial_x u & b^2 \partial_y u \\ a^2 \partial_x v & b^2 \partial_y v \end{pmatrix},$$

with $\mathbf{a} = \begin{pmatrix} a \\ b \end{pmatrix} \in \mathbb{R}^2_+$. Obviously the limit equations for this system are again the INS equations with Reynolds number $Re = 1/\tau$.

Considering the nonstiff advection parts in (7) separately for $\mathbf{u}$ and $\boldsymbol{\Theta}$, we obtain a hyperbolic system with characteristic speeds $\pm a$ and $\pm b$ in x and y directions:

$$\partial_t \mathbf{u} + \mathrm{div}\,\boldsymbol{\Theta} = 0, \qquad \partial_t \boldsymbol{\Theta} + \nabla^{\mathbf{a}}[\mathbf{u}] = 0. \tag{8}$$

As we will see in the last section, $\mathbf{a}$ is chosen depending on the local speed. In the x-direction (8) can be written as:

$$\begin{pmatrix} \mathbf{u} \\ \boldsymbol{\Theta}^x \end{pmatrix}_t + \begin{pmatrix} \mathbf{0} & \mathbf{I} \\ a^2\mathbf{I} & \mathbf{0} \end{pmatrix} \begin{pmatrix} \mathbf{u} \\ \boldsymbol{\Theta}^x \end{pmatrix}_x = 0. \tag{9}$$

A variant of this scheme incorporates a simple linear HLL-type Riemann solver. To incorporate the HLL type formulation, equation (9) is written as:

$$\begin{pmatrix} \mathbf{u} \\ \mathbf{\Theta}^x \end{pmatrix}_t + \begin{pmatrix} \mathbf{0} & \mathbf{I} \\ -a^- a^+ \mathbf{I} & (a^- + a^+)\mathbf{I} \end{pmatrix} \begin{pmatrix} \mathbf{u} \\ \mathbf{\Theta}^x \end{pmatrix}_x = 0 \tag{10}$$

where a^- and a^+ are determined by some algorithm [17]. The submatrices $\mathbf{0}$ and $\mathbf{I}$ are 2×2 zero and identity matrices. Observe that if $a^- = -a^+$ the formulation (9) is obtained.

3 Numerical Schemes

3.1 Space Discretizations

To discretize the equations in space a uniform grid in x and y with grid points (x_i, y_j) with spacing h is used. Consider the linear system (8). For the x-direction $\mathbf{\Theta}^x \pm a\mathbf{u}$ are the characteristic variables associated with the characteristic speeds $\pm a$ while for the y-direction the characteristic variables associated with the characteristic speeds $\pm b$ are $\mathbf{\Theta}^y \pm b\mathbf{u}$. Similarly the characteristic variables associated with the x-direction for the HLL type scheme in equation (10) are

$$W^+ = \frac{2a^+}{(a^+ - a^-)}(\mathbf{\Theta}^x - a^-\mathbf{u}) \text{ and } W^- = \frac{-2a^-}{(a^+ - a^-)}(\mathbf{\Theta}^x - a^+\mathbf{u}). \tag{11}$$

The values of the characteristic variables will be determined at cell edges as in [2]. Hence the numerical fluxes at cell boundaries are:

$$\begin{aligned}
\mathbf{\Theta}^x_{i+1/2j} &= \frac{1}{2}\Big(\mathbf{\Theta}^x_{ij} + \mathbf{\Theta}^x_{i+1j}\Big) - \frac{a}{2}\Big(\mathbf{u}_{i+1j} - \mathbf{u}_{ij}\Big) \\
&\quad + \frac{1}{2}\Big(\sigma^1_{ij}(\mathbf{\Theta}^x + a\mathbf{u}) - \sigma^1_{i+1j}(\mathbf{\Theta}^x - a\mathbf{u})\Big), \\
\mathbf{u}_{i+1/2j} &= \frac{1}{2}\Big(\mathbf{u}_{ij} + \mathbf{u}_{i+1j}\Big) - \frac{1}{2a}\Big(\mathbf{\Theta}^x_{i+1j} - \mathbf{\Theta}^x_{ij}\Big) \\
&\quad + \frac{1}{2a}\Big(\sigma^1_{ij}(\mathbf{\Theta}^x + a\mathbf{u}) + \sigma^1_{i+1j}(\mathbf{\Theta}^x - a\mathbf{u})\Big)
\end{aligned}$$

in the case of second order method. If minmod slope limiting is applied then σ^1_{ij} are given by $\sigma^1_{ij}(\mathbf{z}) = \frac{1}{2}\text{minmod}\Big(\mathbf{z}_{ij} - \mathbf{z}_{i-1j}, \mathbf{z}_{i+1j} - \mathbf{z}_{ij}\Big)$.

For the HLL type scheme the numerical fluxes at cell edges are

$$\Theta^x_{i+1/2j} = \frac{1}{(a^+ - a^-)}\Big(a^+\Theta^x_{ij} - a^-\Theta^x_{i+1j}\Big) + \frac{a^+a^-}{a^+ - a^-}\Big(\mathbf{u}_{i+1j} - \mathbf{u}_{ij}\Big)$$
$$+\frac{1}{2}\Big(\sigma^1_{ij}(W^+) - \sigma^1_{i+1j}(W^-)\Big),$$
$$\mathbf{u}_{i+1/2j} = \frac{2}{(a^+ - a^-)}\Big\{\frac{1}{(a^+ - a^-)}\Big(a^+\Theta^x_{ij} + a^-\Theta^x_{i+1j}\Big)$$
$$-\frac{a^+a^-}{a^+ - a^-}\Big(\mathbf{u}_{i+1j} + \mathbf{u}_{ij}\Big) + \frac{1}{2}\Big(\sigma^1_{ij}(W^+) + \sigma^1_{i+1j}(W^-)\Big)\Big\}.$$

We denote by $\mathbf{F}^1_h$, $\mathbf{F}^2_h$ the discretization of the convective parts $\operatorname{div}\Theta$ and $\nabla^{\mathbf{a}}[\mathbf{u}]$ in equation (7), respectively. Using the numerical fluxes above one obtains

$$\mathbf{F}^1_h(\Theta, \mathbf{u}) = \frac{1}{h}\Big(\Theta^x_{i+1/2j} - \Theta^x_{i-1/2j}\Big) + \frac{1}{h}\Big(\Theta^y_{ij+1/2} - \Theta^y_{ij-1/2}\Big),$$
$$\mathbf{F}^2_h(\Theta, \mathbf{u}) = \Big(\frac{1}{h}\Big(a^2\mathbf{u}_{i+1/2,j} - a^2\mathbf{u}_{i-1/2j}\Big), \frac{1}{h}\Big(b^2\mathbf{u}_{ij+1/2} - b^2\mathbf{u}_{ij-1/2}\Big)\Big).$$

And for the HLL type scheme we get the following convective term:

$$\mathbf{F}^2_h(\Theta, \mathbf{u}) = \Big(\frac{1}{h}\Big(-a^-a^+\mathbf{u}_{i+1/2,j} + a^-a^+\mathbf{u}_{i-1/2j}\Big)$$
$$+\frac{1}{h}\Big((a^- + a^+)\Theta^x_{i+1/2,j} - (a^- + a^+)\Theta^x_{i-1/2j}\Big),$$
$$\frac{1}{h}\Big(-b^-b^+\mathbf{u}_{ij+1/2} + b^-b^+\mathbf{u}_{ij-1/2}\Big)$$
$$+\frac{1}{h}\Big((b^- + b^+)\Theta^y_{ij+1/2} - (b^- + b^+)\Theta^y_{ij-1/2}\Big)\Big).$$

To obtain a multidimensional scheme the computation of the cell averages is modified. To take advantages of diagonal points of the cell the trapezoidal approximation is used. The cell averages are thus defined as:

$$\Theta_{ij} = \frac{1}{4}\Big(f(\mathbf{u}^{NE}_{ij}) + f(\mathbf{u}^{NW}_{ij}) + f(\mathbf{u}^{SE}_{ij}) + f(\mathbf{u}^{SW}_{ij})\Big) \tag{12}$$

where $\mathbf{u}^{NE(NW)}_{ij} = p_{ij}(x_{i\pm\frac{1}{2}}, y_{j+\frac{1}{2}})$, $\mathbf{u}^{SE(SW)}_{ij} = p_{ij}(x_{i\pm\frac{1}{2}}, y_{j-\frac{1}{2}})$ and $p_{ij}(x, y) = \mathbf{z}_{ij} + (\mathbf{z}_x)_{ij}(x - x_i) + (\mathbf{z}_y)_{ij}(y - y_j)$. The slopes $(\mathbf{z}_x)_{ij}$ and $(\mathbf{z}_y)_{ij}$ are (at least first order) approximations to derivatives $\mathbf{z}_x$ and $\mathbf{z}_y$, respectively. The flux is the equilibrium flux (5), $f(\mathbf{u}) = \mathbf{u}\otimes\mathbf{u} - (2\epsilon^{1-\alpha}\tau)/3\mathbf{S}[\mathbf{u}]$ at $\epsilon \to 0$.

Denote the discrete gradient by $\mathbf{G}_h$ and the discrete divergence by $\mathbf{D}_h$ which are given by second order centered differences. Second order centered difference approximations of $\mathbf{S}^\epsilon$ and $\mathbf{S}$ are denoted by $\mathbf{S}^\epsilon_h$ and $\mathbf{S}_h$. Hence we obtain the spatial discretization:

$$\dot{p} + \frac{1}{\epsilon^{2\alpha}}\mathbf{D}_h \cdot \mathbf{u} = 0, \qquad \dot{\mathbf{u}} + \mathbf{F}^1_h(\Theta, \mathbf{u}) + \mathbf{G}_h p = 0, \tag{13}$$
$$\dot{\Theta} + \mathbf{F}^2_h(\Theta, \mathbf{u}) + \frac{2}{\epsilon^{2\alpha}}\mathbf{S}^\epsilon_h(\mathbf{u}) = -\frac{1}{\epsilon^{1+\alpha}\tau}\Big(\Theta - \mathbf{u}\otimes\mathbf{u}\Big)$$

or equivalently

$$\mathbf{D}_h \cdot \mathbf{G}_h p - 2\epsilon^{2\alpha}\ddot{p} = -\mathbf{D}_h \cdot \mathbf{F}_h^1(\boldsymbol{\Theta}, \mathbf{u}), \quad \dot{\mathbf{u}} + \mathbf{F}_h^1(\boldsymbol{\Theta}, \mathbf{u}) + \mathbf{G}_h p = 0,$$
$$\dot{\boldsymbol{\Theta}} + \mathbf{F}_h^2(\boldsymbol{\Theta}, \mathbf{u}) = -\frac{1}{\epsilon^{1+\alpha}\tau}\Big(\boldsymbol{\Theta} - \mathbf{u} \otimes \mathbf{u} + 2\epsilon^{1-\alpha}\tau \mathbf{S}_h^{\epsilon}(\mathbf{u})\Big).$$

A corresponding high order upwind based space discretization for the INS equations is obtained considering the limit of the above discretization as $\epsilon \to 0$

$$\mathbf{D}_h \cdot \mathbf{G}_h p = -\mathbf{D}_h \cdot \mathbf{F}_h(\mathbf{u}), \quad \dot{\mathbf{u}} = -\mathbf{F}_h^1\Big(\mathbf{u} \otimes \mathbf{u} - 2\epsilon^{1-\alpha}\tau \mathbf{S}_h(\mathbf{u}), \mathbf{u}\Big) - \mathbf{G}_h p.$$

3.2 Time Discretizations

To treat only the limit equations ($\epsilon = 0$) any explicit high order Runge-Kutta method [11] can be used in combination with a Poisson solver and the limiting (relaxed) spatial discretization.

Further one would want to discretize the relaxation system (13) for all ranges of the parameter ϵ. This allows the study of the numerical passage from the Boltzmann to the INS regime. An implicit-explicit (IMEX) Runge-Kutta method of the type recently developed in [12] is used as suggested in [6]. We denote the time step by k and use superscript n to denote the time iterations. For the second order time discretization a two stage IMEX Runge Kutta method [12] which guarantees second order accuracy in the stiff limit is chosen. For $\epsilon \to 0$ this scheme suggests a formulation for a second order time discretization of INS equations i.e. the projection is taken at every step:
Step 1:

$$\begin{aligned}
\Delta p^{n+1/2} &= \frac{1}{k\gamma}\,\mathrm{div}\left(\mathbf{u}^n - k\gamma \mathrm{div}\boldsymbol{\Theta}^n\right),\\
\mathbf{u}^{n+1/2} &= \mathbf{u}^n - k\gamma\Big(\mathrm{div}\,\boldsymbol{\Theta}^n + \nabla p^{n+1/2}\Big),\\
\boldsymbol{\Theta}^{n+1/2} &= \mathbf{u}^{n+1/2} \otimes \mathbf{u}^{n+1/2} - 2\epsilon^{1-\alpha}\tau \mathbf{S}[\mathbf{u}^{n+1/2}].
\end{aligned}$$

Step 2:

$$\begin{aligned}
\Delta p^{n+1} &= \frac{1}{k\gamma}\,\mathrm{div}\left(\mathbf{u}^n - k\Big(\delta\,\mathrm{div}\,\boldsymbol{\Theta}^n + (1-\delta)\,\mathrm{div}\,\boldsymbol{\Theta}^{n+1/2} + (1-\gamma)\nabla p^{n+1/2}\Big)\right),\\
\mathbf{u}^{n+1} &= \mathbf{u}^n - k\Big(\delta\,\mathrm{div}\,\boldsymbol{\Theta}^n + (1-\delta)\boldsymbol{\Theta}^{n+1/2} + (1-\gamma)\nabla p^{n+1/2} + \gamma\nabla p^{n+1}\Big),\\
\boldsymbol{\Theta}^{n+1} &= \mathbf{u}^{n+1} \otimes \mathbf{u}^{n+1} - 2\epsilon^{1-\alpha}\tau \mathbf{S}[\mathbf{u}^{n+1}].
\end{aligned}$$

with $\gamma = 1 - \sqrt{2}/2$ and $\delta = 1 - 1/2\gamma$, which gives a scheme for the incompressible Euler equations ($0 < \alpha < 1$) and INS equations for $\alpha = 1$.

Guided by the above formulation we obtain the second-order modified Runge-Kutta scheme for INS:
Step 1:

$$\Delta p^{n+1/2} = \frac{1}{k}\operatorname{div}\left(\mathbf{u}^n - k\operatorname{div}\boldsymbol{\Theta}^n\right), \quad \mathbf{u}^{n+1/2} = \mathbf{u}^n - k\Big(\operatorname{div}\boldsymbol{\Theta}^n + \nabla p^{n+1/2}\Big).$$

Step 2:

$$\begin{aligned}\Delta p^{n+1} &= \frac{2}{k}\operatorname{div}\left(\mathbf{u}^n - \frac{k}{2}\Big(\operatorname{div}\boldsymbol{\Theta}^n + \operatorname{div}\boldsymbol{\Theta}^{n+1/2}\Big)\right),\\ \mathbf{u}^{n+1} &= \mathbf{u}^n - \frac{k}{2}\Big(\operatorname{div}\boldsymbol{\Theta}^n + \operatorname{div}\boldsymbol{\Theta}^{n+1/2}\Big) - \frac{k}{2}\nabla p^{n+1}.\end{aligned}$$

In addition, we do not have to compute $\boldsymbol{\Theta}$ at every time step unless it is needed. We rather compute $\operatorname{div}\boldsymbol{\Theta}$ instead. This reduces the number of variables. The usual hyperbolic and parabolic CFL conditions have to be fulfilled to guarantee stability.

4 Numerical Examples and Results

The cases for $\epsilon = 0$ will be tested in this section. The space discretization in equation (13) will be applied. For slope limiting the van Leer limiter is used.

Example 1: Taylor-Vortex Flow - Accuracy test

Relaxation schemes will be first tested on incompressible Euler equations, i.e. $\nu = \tau = 0.0$ augmented with smooth periodic initial data. The test admits the following exact solution [13]:

$$\begin{aligned}u(x,y,t) &= -\cos(2\pi x)\sin(2\pi y)\exp(-2\nu t);\\ v(x,y,t) &= \sin(2\pi x)\cos(2\pi y)\exp(-2\nu t); \quad x,y \in [0,1].\end{aligned} \tag{14}$$

Taking $h = 1/N$ with $N = 32, 64, 128$, and 256, the solution is computed up to $t = 2.0$ and the L^1 and L^2 norms of the errors and convergence rates of u (the velocity component in the x-direction) are listed in Table 1 below.

The following schemes have been tested: $\mathcal{R}^{\epsilon=0,2}_{\text{TVD}}$ (Second-order relaxed scheme with TVD time integration); $\mathcal{R}^{\epsilon=0,2}_{\text{DIRK}}$ (Second-order relaxed scheme with DIRK time integration).

All examples use the CFL number of 0.475 based on the local flow velocity.

$$a_{i+\frac{1}{2},j} := 2\max\{|u_{i,j}|, |u_{i+1,j}|\}; \quad b_{i,j+\frac{1}{2}} := 2\max\{|v_{i,j}|, |v_{i,j+1}|\}.$$

Example 2: Travelling wave - accuracy test

This numerical test was used by Minion and Brown in [14]. The computational domain of unit length is doubly-periodic. The exact solution of the Navier-Stokes equations for this problem is:

Table 1. Error-norms for the Incompressible Euler Problem with the initial condition in 14 at $t = 2.0$ using relaxation schemes, $\nu = 0.0$.

N	Scheme	L^1-error	Rate	L^2-error	Rate
64	$\mathcal{R}^{\epsilon=0,2}_{\text{TVD}}$	$2.99588 \cdot 10^{-3}$	2.5683	$4.10701 \cdot 10^{-3}$	2.4753
	$\mathcal{R}^{\epsilon=0,2}_{\text{DIRK}}$	$2.80480 \cdot 10^{-3}$	2.5351	$3.91299 \cdot 10^{-3}$	2.4382
128	$\mathcal{R}^{\epsilon=0,2}_{\text{TVD}}$	$6.16432 \cdot 10^{-4}$	2.281	$9.55745 \cdot 10^{-3}$	2.1034
	$\mathcal{R}^{\epsilon=0,2}_{\text{DIRK}}$	$5.97678 \cdot 10^{-4}$	2.2305	$9.28992 \cdot 10^{-4}$	2.0745
256	$\mathcal{R}^{\epsilon=0,2}_{\text{TVD}}$	$1.13089 \cdot 10^{-4}$	2.4465	$1.8214 \cdot 10^{-4}$	2.3916
	$\mathcal{R}^{\epsilon=0,2}_{\text{DIRK}}$	$1.05389 \cdot 10^{-4}$	2.5036	$1.64960 \cdot 10^{-4}$	2.4935

$$\begin{aligned} u(x,y,t) &= 1 + 2\cos(2\pi(x-t))\sin(2\pi(y-t))\exp(-8\pi^2\nu t); \\ v(x,y,t) &= 1 - 2\sin(2\pi(x-t))\cos(2\pi(y-t))\exp(-8\pi^2\nu t); \\ p(x,y,t) &= -(\cos(4\pi(x-t)) + \cos(4\pi(y-t)))\exp(-16\pi^2\nu t). \end{aligned} \tag{15}$$

Numerical results are presented in Table 2.

Table 2. Error-norms for the Incompressible Euler Problem with the initial condition in (15) at $t = 0.7$ using relaxation schemes, $\nu = 0.0$.

N	Scheme	L^1-error	Rate	L^2-error	Rate
64	$\mathcal{R}^{\epsilon=0,2}_{\text{TVD}}$	$1.3222 \cdot 10^{-2}$	1.4691	$1.60313 \cdot 10^{-2}$	1.5942
	$\mathcal{R}^{\epsilon=0,2}_{\text{DIRK}}$	$1.29641 \cdot 10^{-2}$	1.5573	$1.57197 \cdot 10^{-2}$	1.5656
128	$\mathcal{R}^{\epsilon=0,2}_{\text{TVD}}$	$2.63745 \cdot 10^{-3}$	2.3257	$3.26585 \cdot 10^{-3}$	2.2954
	$\mathcal{R}^{\epsilon=0,2}_{\text{DIRK}}$	$2.59158 \cdot 10^{-3}$	2.3226	$3.20928 \cdot 10^{-3}$	2.2923
256	$\mathcal{R}^{\epsilon=0,2}_{\text{TVD}}$	$3.18838 \cdot 10^{-4}$	3.0482	$4.2203 \cdot 10^{-4}$	2.952
	$\mathcal{R}^{\epsilon=0,2}_{\text{DIRK}}$	$3.1598 \cdot 10^{-4}$	3.0359	$4.18859 \cdot 10^{-4}$	2.9377

Example 3: Doubly Periodic Shear Layer

This test was introduced by Bell, Colella and Glaz in [15]. In the periodic, two-dimensional computational domain of size $[1 \times 1]$, the following velocity fields are generated as initial conditions:

$$u^{\varrho}(x,y,0) = \begin{cases} \tanh(\varrho(y-1/4)),\ y \le 1/2; \\ \tanh(\varrho(3/4-y)),\ y > 1/2; \end{cases} \quad v^{\delta}(x,y,0) = \delta\sin(2\pi x);$$

where ϱ is the shear layer width parameter and δ is the strength of the initial perturbation. The strength coefficient $\delta = 0.05$ is kept unchanged. The results of vorticity profiles are displayed using 20 equidistant contours.

In figure 1 vorticity profiles of solutions obtained by applying different schemes without recourse to slope limiters are shown.

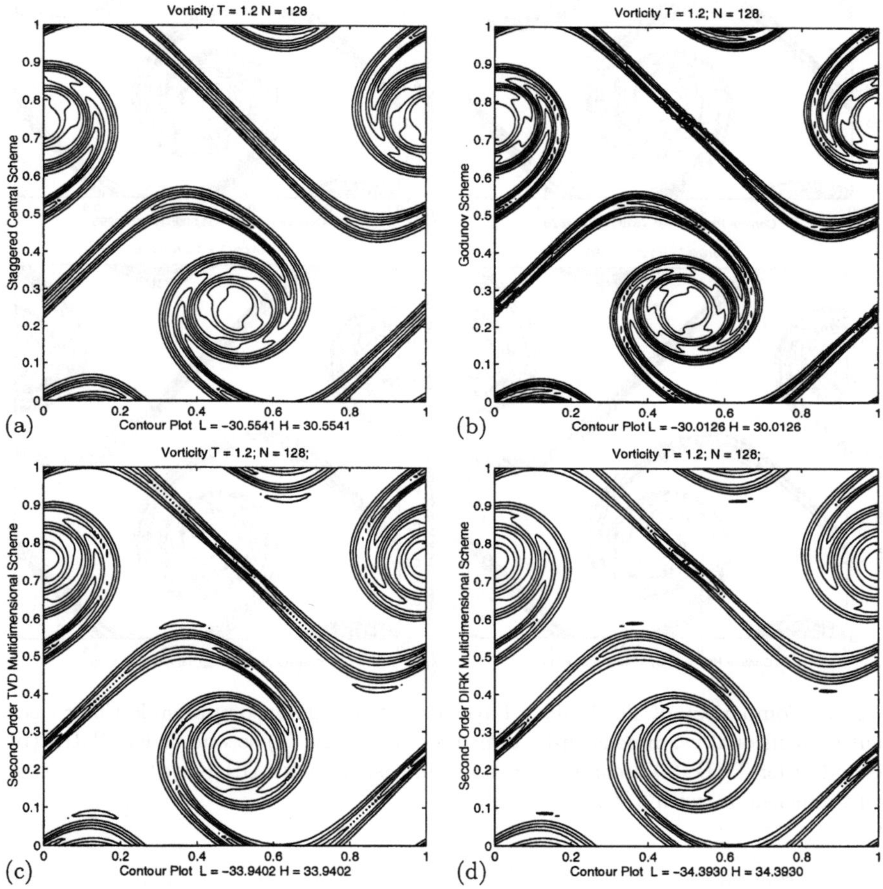

Fig. 1. Thick Shear ($\varrho = 30$) Layer Results for Euler case $\epsilon = 0$: Staggered Central Scheme [16] (a), Godunov Based Scheme [15](b); Relaxation Scheme with TVD time integration (c); and Relaxation Scheme with relaxed DIRK time integration (d).

Further a refinement of the grid to $N = 256$ was made. The same computation was repeated with the van Leer limiter applied to the relaxation-based schemes. The results are shown in figure 2. In figure 3 a v velocity cut at $x = 0.5$ is shown.

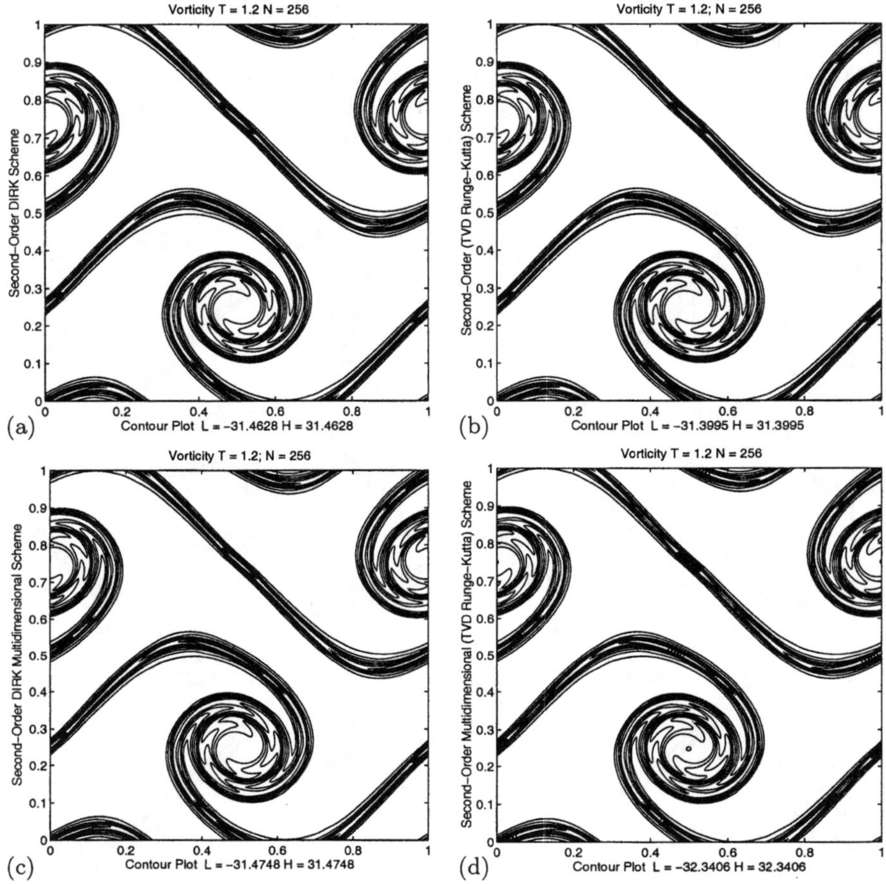

Fig. 2. Comparison of Thick Shear Layer Results for the incompressible Euler case, multidimensional vs. dimension-by-dimension approach: (a) second-order DIRK, (b) second-order TVD, (c) second-order multidimensional with DIRK, (d) second-order multidimensional with TVD.

The relaxation system was also tested on the thin shear layer, $\rho = 80$, problem. The Navier-Stokes equation was considered for $t = 1.0$ using a grid of $N = 256$. A comparison of the time evolution of total kinetic energy and enstrophy in the incompressible Euler equation up to time $t = 2.0$ was made. Figure 4 presents the evolution of the decay of the total kinetic energy of the flow and a history of the mean enstrophy.

In all the three examples we observe that the relaxed schemes ($\epsilon = 0$) perform reasonably well inspite of their simplicity. Much as the DIRK formulation suggests the projection structure, the direct TVD formulation has better qualitative results. The DIRK formulation tends to be more dissipative. Nevertheless they both tend to achieve their expected convergence rate. Further we

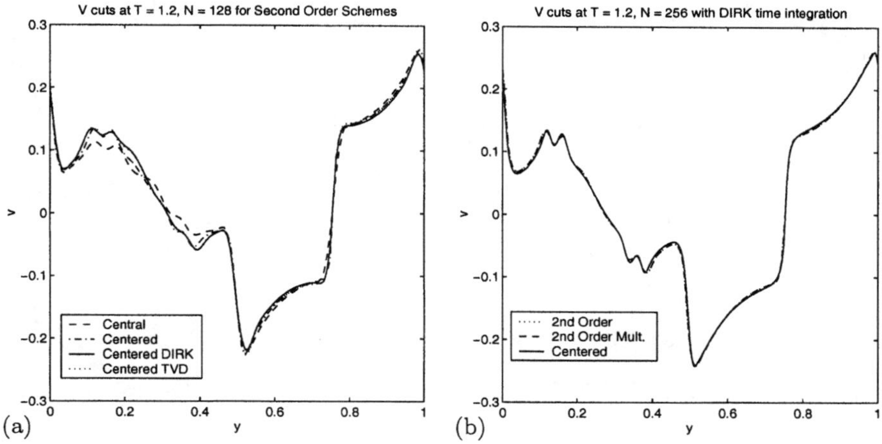

Fig. 3. The cut at $x = 0.5$ of v at $t = 1.2$ for the Thick Shear Layer Problem computed with the staggered central scheme ('Central'), the Godunov projection method ('Centered'), the relaxation-based scheme with DIRK time integration ('Centered DIRK') and the relaxation-based scheme with TVD time integration ('Centered TVD') (Left). The cut at $x = 0.5$ of v at $t = 1.2$ for the Thick Shear Layer Problem computed with the Godunov projection method ('Centered'), the relaxation-based scheme ('2nd Order') and the multidimensional relaxation-based scheme ('2nd Order Mult.') (Right).

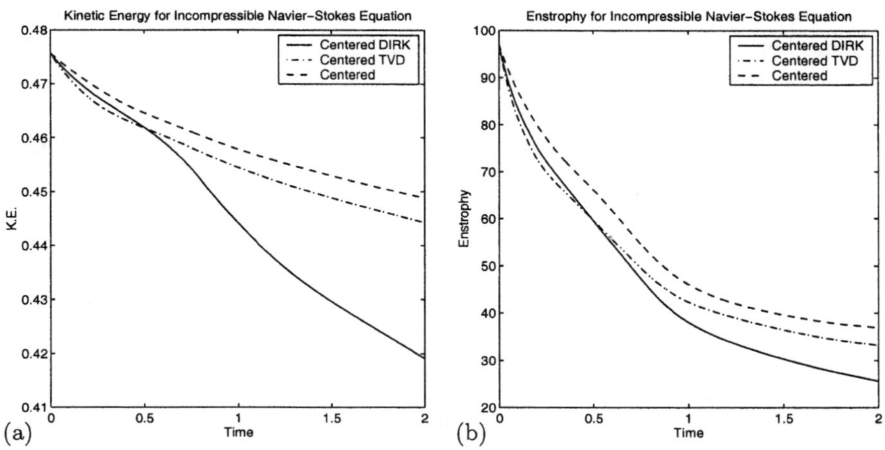

Fig. 4. Comparison of total kinetic energy and enstrophy at $t = 2.0$, $Re = 10000$ for the thin shear layer problem computed with the Godunov projection method ('Centered'), the multidimensional relaxation-based scheme with TVD time integration ('Centered TVD') and the multidimensional relaxation-based scheme with DIRK time integration ('Centered DIRK').

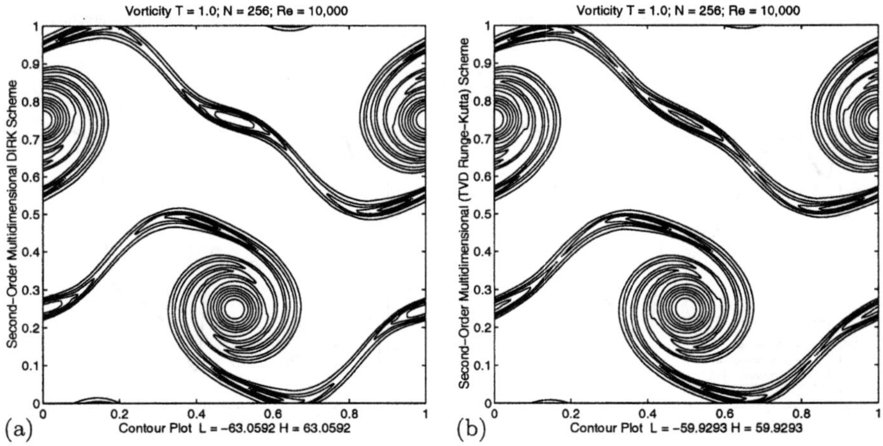

Fig. 5. Comparison of Thin Shear Layer Results for Navier-Stokes case: (a) second-order multidimensional DIRK scheme, (b) second-order multidimensional TVD scheme.

observed that for the DIRK scheme there is no significant difference between the multidimensional and dimension-by-dimension scheme. A comparison with other schemes shows that relaxations schemes are less dissipative than central schemes while on some occasion they are very close to the Godunov scheme. The resolution of the solution for thin shear layer problems shows that the scheme has a lot of potential for improvement. The implementation of the HLL formulation and relaxing schemes ($\epsilon \neq 0$) is underway to investigate how this can be achieved.

Acknowledgement: This work was supported by a DFG Grant KL 1105/9.

References

1. Chen, S., Doolen, G.D. (1998): Lattice Boltzmann Method for fluid Flows. Ann. Rev. Fluid Mech., **30**, 329–364
2. Jin, S., Xin, Z. (1995): The Relaxation Schemes for Systems of Conservation Laws in Arbitrary Space Dimensions. Comm. Pure Appl. Math., **48**, 235–277
3. Klar, A. (1998): An Asymptotic-Induced Scheme for Nonstationary Transport Equations in the Diffusive Limit. SIAM J. Num. Anal., **35**, 1073–1094
4. Naldi, G., Pareschi, L., Toscani, G. (to appear): Relaxation schemes for PDEs and applications to second and fourth order degenerate diffusion problems. Surveys in Mathematics for Industry.
5. LeVeque, R.J., Pelanti, M. (2001): A Class of Approximate Riemann Solvers and their Relation to Relaxation Schemes. J. Comp. Phys., **172**, 572 – 591
6. Banda, M.K., Klar, A., Pareschi, L., Seaïd, M. (2001): Lattice-Boltzmann type relaxation systems and high order relaxation schemes for the incompressible Navier-Stokes equation. preprint.

7. De Masi, A., Esposito, R., Lebowitz, J.L. (1989): Incompressible Navier Stokes and Euler Limits of the Boltzmann Equation. Comm. Pure Appl. Math., **42**, 1189
8. Klar, A. (1999): Relaxation Schemes for a Lattice Boltzmann type discrete velocity model and numerical Navier Stokes limit. J. Comp. Phys., **148**, 1–17
9. d'Humières, D. (1992): Generalized Lattice-Boltzmann Equations. In: AIAA Rarefied Gas Dynamics: Theory and Applications, Progress in Astronautics and Aeoronautics, **159**, 450–458
10. Giraud, L., d'Humieres, D., Lallemand, P. (1998): A lattice Boltzmann model for Jeffreys viscoelastic fluid. Europhys. Lett., **42**, 625–630
11. Shu, C.W., Osher, S. (1988): Efficient implementation of essentially non-oscillatory shock-capturing schemes. J. Comp. Phys., **77**, 439–471
12. Ascher, U., Ruuth, S., Spiteri, R. (1997): Implicit-explicit Runge-Kutta methods for time-dependent partial differential equations. Appl. Numer. Math., **25**, 151–167
13. Chorin, A.J. (1968): Numerical Solution of the Navier-Stokes equations. Math. Comp., **22**, 745–762
14. Brown, D., Minion, M.(1997): Performance of under-resolved two-dimensional incompressible flow simulations, II. J. Comp. Phys., **138**, 734–765
15. Bell, J.B., Colella, P., Glaz, H.M.(1989): A second order projection method for incompressible Navier-Stokes equations. J. Comp. Phys., **85(2)**, 257–283
16. Kupferman, R., Tadmor, E. (1997): A fast high-resolution second-order central scheme for incompressible flows. Proc. Nat. Acad. Sci., **94**, 4848
17. Toro, E. F. (1997): Riemann Solvers and numerical methods for fluid dynamics. Springer, Berlin

A Time Semi-Implicit Relaxation Scheme for Two-Phase Flows in Pipelines

Michaël Baudin[1], Frédéric Coquel[2] and Quang-Huy Tran[1]

[1] Institut Français du Pétrole, Rueil-Malmaison, France, *baudinm@yahoo.com*
[2] Laboratoire Jacques-Louis Lions, Université Pierre et Marie Curie, Paris, France

Summary. The aim of this paper is to present a relaxation scheme designed to approximate the solutions of the system of conservation laws which arises in the modeling of two-phase flows in oil and gas pipelines. The main idea is to relax only the highly non-linear closure laws, thanks to a Lagrangian change of coordinates. By construction, all the fields of the nonlinear hyperbolic relaxation system are linearly degenerated. In addition to the simplicity of the evaluation of the flux function, the relaxation coefficients can be easily devised so as to guaranty the positivity of the mass fractions. We propose to use an integration in time which is explicit with respect to the small eigenvalues and linearly implicit with respect to the large eigenvalues. In the first part, we construct a second-order explicit relaxation scheme based on a Godunov method. In the second part, we present the semi-implicit relaxation scheme. A "stiff" numerical simulation of an industrial case is shown.

Introduction

The aim of this paper is to present a relaxation scheme designed to approximate the solutions of the system of conservation laws which arises in the modeling of two-phase flows in oil and gas pipelines[10]. The system, made up of three conservation laws, is a drift-flux type model and is closed by two thermodynamic and hydrodynamic models. The complexity of these models make classical numerical schemes such as Godunov or Roe schemes very difficult to use. Our feeling is that only a "rough" scheme would be able to successfully meet the challenge of nonlinearities.

An essential property of our system is that it possesses fast characteristic speeds (acoustic waves) and slow ones (corresponding to the mass transport). From the engineer's standpoint, the crucial aspect is the petroleum transport and not the acoustic. That is why we will never use an explicit scheme, the time step of which is limited by the CFL stability condition: the time required for the simulation would be prohibitive due to the large characteristic speeds.

Inspired by [7], I. Faille and E. Heintzé have proposed in [6] a VFRoe-type scheme which is sufficiently rough (diffusive) to handle very stiff industrial cases. The scheme shown in [6] is linearly implicit with respect to the large eigenvalues and explicit with respect to the small eigenvalues. Therefore this semi-implicit scheme will combine accuracy on the waves propagating at the small velocities and reduced CPU-time because the CFL time step limitation only applies to small eigenvalues.

But the VFRoe scheme also suffers from a few drawbacks. The first one is that the scheme is still CPU-time consuming because it needs to compute numerically the eigenvalues of the Jacobian matrix of the system. The second one is that theses eigenvalues are not necessarily real, that is, the system is not always hyperbolic. It is observed in [4] that the system is hyperbolic only if the slip between the phases is not too large.

This is why the authors of [2] designed an explicit relaxation scheme. The approach [2] is close in spirit to Jin & Xin's [11] but differs in the fact that only the genuine non-linearities are relaxed, namely the pressure and the hydrodynamic laws. In a first step, the genuine nonlinearities are shown by means of a Lagrangian change of coordinates. Then, these are relaxed. Finally, the equations are brought back to the Eulerian frame. The resulting relaxation system is automatically hyperbolic and has all its fields linearly degenerated. This scheme is less CPU-time consuming than the VFRoe-type one because the most complex (algorithmically speaking) step is the computation of only two relaxation coefficients. The relaxation coefficients can be easily devised so as to guaranty two properties : first, the stability of the first order asymptotic system computed thanks to the Chapman-Enskog expansion and, second, the positivity of the mass fractions. As a first attempt to design a fast and rough numerical scheme for two-phase flows, these results were encouraging.

The goal of this paper is to work out a semi-implicit version of the explicit scheme [2]. In essence, the extension is possible because of the linear degeneracy property of the relaxation system.

The paper is organized as follows. In section 1, we introduce the two-phase flow model together with the boundary conditions and the characteristics of this model. In section 2, we develop the second order explicit relaxation scheme with the computation of new relaxation coefficients and boundary conditions. In section 3, we present the semi-implicit scheme and section 4 is devoted to the numerical results.

Most of the results presented here are extracted from [1], [2] and [3].

1 Two-phase flow model

1.1 Equations

In the flow, the mixture is characterized by its density ρ, its velocity v and its gas (resp. liquid) mass fraction Y (resp. $X = 1 - Y$). The model is governed by the following system of conservation laws:

$$\begin{cases} \partial_t(\rho) + \partial_x(\rho v) = 0, \\ \partial_t(\rho v) + \partial_x(\rho v^2 + P(\mathbf{u})) = S(\mathbf{u}), \\ \partial_t(\rho Y) + \partial_x(\rho Y v - \sigma(\mathbf{u})) = 0, \end{cases} \tag{1}$$

for all $x \in \mathbb{R}$ and $t \geq 0$ where the unknown is $\mathbf{u} := (\rho,\ \rho v,\ \rho Y)$. Here the source term S includes gravity and wall friction terms which are given functions of the unknown. The two following functions of the unknown $\mathbf{u}$

$$\sigma(\mathbf{u}) = \rho Y(1-Y)\Phi(\mathbf{u}), \quad P(\mathbf{u}) = p(\mathbf{u}) + \rho Y(1-Y)\Phi(\mathbf{u})^2, \tag{2}$$

are highly non-linear closure laws. The natural phase space associated with such variables then reads:

$$\Omega = \left\{\mathbf{u} = (\rho,\ \rho v,\ \rho Y) \in \mathbb{R}^3;\ \rho > 0,\ v \in \mathbb{R},\ Y \in [0,1]\right\}.$$

Considering the pressure p, we will consider a perfect gas and a compressible liquid. The pressure law can be put in the general form $p = p(\rho,\ \rho Y)$ and is a smooth function. We consider a general algebraic hydrodynamic law of the type

$$v_{\mathrm{L}} - v_{\mathrm{G}} = \Phi(\mathbf{u}), \quad \forall\ \mathbf{u} \in \Omega, \tag{3}$$

in order to close (1). In (3), the mapping Φ is assumed to be smooth enough. In practical situations, Φ turns out to be nonlinear in the unknown $\mathbf{u}$ (see [4], [12] for instance).

1.2 Boundary and initial conditions

At the inlet of the pipe ($x = 0$), the mass flowrates are given as functions of time, i.e.

$$(\rho v)_\alpha(0,\ t) = q_\alpha^0(t),\ t \geq 0,\ \alpha = \mathrm{L}, \mathrm{G},$$

where and $(\rho v)_{\mathrm{L}}$ (resp. $(\rho v)_{\mathrm{G}}$) denotes the mass flux of the liquid (resp. the gas). At the outlet of the pipe ($x = L$, the length of the pipeline), the pressure is a given function of time, i.e.

$$p(L,\ t) = p^L(t),\ t \geq 0.$$

We will treat cases in which the flow is induced only by variations of the boundary conditions: in such experiments, the initial condition is the steady state, computed by the values of the boundary conditions at time $t = 0$.

1.3 Characteristics of the model

There is no analytical expression for the physical flux of the considered system, except for very simple hydrodynamic laws. Therefore the eigenvalues of the system (1) are not known in full generality. However, in most common situations, i.e., for usual values of $\mathbf{u}$, the system is hyperbolic and has three real eigenvalues $\lambda_1 < \lambda_2 < \lambda_3$ with $\lambda_1 < 0$ and $\lambda_3 > 0$.

The interesting property of the system is that the eigenvalues $\lambda_{1,3}$ correspond to the acoustic waves (or "pressure waves") and propagate at fast speeds thanks to the compressibility of the fluid. The eigenvalue λ_2, which has a variable sign, corresponds to the kinematic waves and propagates at slow speeds with the fluid. These properties imply that the large eigenvalues are 10-100 times bigger than the small eigenvalue, i.e. $|\lambda_{1,3}| \gg |\lambda_2|$.

2 Explicit relaxation scheme

2.1 The relaxation model

We relax all the *genuine* non-linearities of the equilibrium system (1) which appear by means of a Lagrangian change of coordinates. Then, we introduce two new state variables Σ and Π which are intended to coincide respectively with $\sigma(\mathbf{u})$ and $P(\mathbf{u})$ in the limit of the relaxation parameter λ. Finally, back in Eulerian coordinates, we propose as a relaxation model the following system:

$$\begin{cases} \partial_t \rho & + \ \partial_x (\rho v) & = 0, \\ \partial_t (\rho v) & + \ \partial_x (\rho v^2 + \Pi) & = S(\mathbf{u}), \\ \partial_t (\rho \Pi) & + \ \partial_x (\rho \Pi v + a^2 v) & = \lambda \rho (P(\mathbf{u}) - \Pi), \\ \partial_t (\rho Y) & + \ \partial_x (\rho Y v - \Sigma) & = 0, \\ \partial_t (\rho \Sigma) & + \ \partial_x (\rho \Sigma v - b^2 Y) & = \lambda \rho (\sigma(\mathbf{u}) - \Sigma), \end{cases} \tag{4}$$

where a and b are two real positive parameters that we call "relaxation coefficients". The above relaxation system will be given hereafter the convenient abstract form:

$$\partial_t \mathbf{v} + \partial_x \mathcal{G}(\mathbf{v}) = \lambda \mathcal{R}(\mathbf{v}) + \mathcal{S}(\mathbf{v}), \quad t > 0, \ x \in \mathbb{R}; \tag{5}$$

where the functions $\mathcal{G}$, $\mathcal{R}$ and $\mathcal{S}$ receive clear definitions.

The first order system with no source term extracted from (4) admits five real eigenvalues: v, $v \pm a\tau$, $v \pm b\tau$, $(\tau = 1/\rho)$ and five linearly independent corresponding eigenvectors. Consequently, the first order system with no source term extracted from (4) is hyperbolic. Moreover, each eigenvalue is associated with a linearly degenerated field.

Because of the specific values of the functions P and σ, we always have $a > b$. This is in harmony with the fact that a is associated with pressure waves and b with kinematic waves.

2.2 First order explicit scheme

The pipeline is made of I cells, denoted $(M_i)_{i=1,I}$. Let x_i be the center of the cell and Δx its length. We also denote $x_{i+1/2} = (x_i + x_{i+1})/2$ the interface between two cells, $x_{1/2} = 0$ the inlet boundary interface and $x_{I+1/2} = L$ the outlet boundary interface. Let $\Delta t^n = t^{n+1} - t^n$ be the time step. Let $\mathbf{u}_i^n \approx \mathbf{u}(x_i, t^n)$ be the discrete unknown. The numerical scheme is based on the following splitting method.

1. Relaxation. We take $\lambda = \infty$ and solve the ODE system $\partial_t \mathbf{v} = \lambda \mathcal{R}(\mathbf{v})$ by projecting the variables on the equilibrium variety i.e. we set $\Pi_i^n := P(\mathbf{u}_i^n)$ and $\Sigma_i^n := \sigma(\mathbf{u}_i^n)$.
2. Evolution. We take $\lambda = 0$ and solve the system $\partial_t \mathbf{v} + \partial_x \mathcal{G}(\mathbf{v}) = \mathcal{S}(\mathbf{v})$ on one iteration in order to go to the time $t = t^{n+1}$.

In the evolution step, we consider the system $\partial_t \mathbf{v} + \partial_x \mathcal{G}(\mathbf{v}) = \mathcal{S}(\mathbf{v})$ that we approximate by a classical Godunov finite volume scheme which is based on the update $\mathbf{v}_i^{n+1} = \mathbf{v}_i^n - \frac{\Delta t}{\Delta x}\left(\mathcal{G}(\mathbf{v}_{i+1/2}^\star) - \mathcal{G}(\mathbf{v}_{i-1/2}^\star)\right)$. The Riemann solution $\mathbf{v}_{i+1/2}^\star$ on the interface $i+1/2$ is made of six constant states separated by five contact discontinuities (see [2] for details). It is easily computed and leads to a low cost numerical scheme.

The time step of the relaxation explicit scheme is limited by a CFL condition.

2.3 Relaxation coefficients

We present in this section the computation of the relaxation coefficients a and b. We consider one local couple $(a_{i+1/2}, b_{i+1/2})$ on each interface $x_{i+1/2}$ in order to minimize the numerical dissipation. The relaxation coefficients are designed in order to ensure:

1. the stability of the first order asymptotic equilibrium system thanks to the Chapman-Enskog expansion
$$\Pi = P + \lambda^{-1}\Pi_1 + \mathcal{O}(\lambda^{-2}), \qquad \Sigma = \sigma + \lambda^{-1}\Sigma_1 + \mathcal{O}(\lambda^{-2}). \tag{6}$$
The Chapman-Enskog analysis justifies the following choice:
$$\begin{aligned} \overline{a}_{i+1/2} &= \sqrt{\max(A(\mathbf{u_L}), A(\mathbf{u_R}))}, \qquad & A(\mathbf{u}) &= -P_\tau(\mathbf{u}) + P_v^2(\mathbf{u}), \\ \overline{b}_{i+1/2} &= \sqrt{\max(B(\mathbf{u_L}), B(\mathbf{u_R}))}, \qquad & B(\mathbf{u}) &= \sigma_Y^2(\mathbf{u}). \end{aligned} \tag{7}$$
2. physical properties of the approximate solution.
It is possible to compute the relaxation coefficients a and b in order to satisfy, at the discrete level, the following two basic physical properties: the positivity of the density and, above all, the maximum principle on the gas mass fraction, i.e., $Y_i^n \in [0,1]$. This is done by taking $\tilde{a}_{i+1/2}$ and $\tilde{b}_{i+1/2}$ large enough, which results in a increased, but well-adjusted, amount of numerical dissipation.

Finally, on one interface $x_{i+1/2}$, the relaxation coefficients are $a_{i+1/2} = \max(\overline{a}_{i+1/2}, \tilde{a}_{i+1/2})$ and $b_{i+1/2} = \max(\overline{b}_{i+1/2}, \tilde{b}_{i+1/2})$.

2.4 Second order in space and time

The scheme is extended to second-order accuracy in space by using the classical MUSCL (Monotonic Upstream Scheme for Conservation Laws) technique ([8], [9]). Instead of taking a constant approximate solution on each cell, we construct a linear approximation. The limited slopes are those of the "physical variable" $(p,\ Y,\ v)$ rather than those of the conservative variable $\mathbf{u}$ [6]. We choose the classical minmod slope limiter.

The scheme is extended to second-order accuracy in time thanks to a Runge-Kutta second-order procedure.

3 Semi-implicit relaxation scheme

3.1 First order linearly implicit scheme

The linearly implicit scheme is classically splitted into three steps.

1. In the physical step, one computes a "predictor" using the explicit scheme
$$\mathbf{v}_i^\sharp = \mathbf{v}_i^n - \frac{\Delta t^n}{\Delta x}\left(\mathbf{H}(\mathbf{v}_i^n, \mathbf{v}_{i+1}^n) - \mathbf{H}(\mathbf{v}_{i-1}^n, \mathbf{v}_i^n)\right) + \Delta t^n\ \mathcal{S}(\mathbf{v}_i^n). \tag{8}$$
Here $\mathbf{H}$ is the numerical flux, based on a Godunov method.
2. In the mathematical step, one solves the linear equation in $\delta\mathbf{v}$
$$-\alpha_{i-1}\delta\mathbf{v}_{i-1} + \beta_i\delta\mathbf{v}_i + \gamma_{i+1}\delta\mathbf{v}_{i+1} = \mathbf{v}_i^\sharp - \mathbf{v}_i^n \tag{9}$$
where α_{i-1}, β_i and γ_{i+1} are 5×5 matrices involving partial derivatives of the numerical flux.
This consists, at each time step, in solving a linear system $\mathbb{A}\ \delta\mathbf{v} = b$ thanks to, for example, a Gauss method. The matrix $\mathbb{A}$ is a tridiagonal by block matrix with each block of size 5×5: the resulting matrix is a band matrix with 9 extra-diagonal terms.
3. Finally, we can update the conservative variable of the cell i by $\mathbf{v}_i^{n+1} = \mathbf{v}_i^n + \delta\mathbf{v}_i$.

Now, the remaining question is: how to construct the matrices α, β and γ? It is easy (see [6]) to compute these matrices when the explicit scheme is a Roe-type scheme. The main objective is therefore to put the relaxation explicit scheme under Roe's form. Such a rewriting is based on a shock curve decomposition and is possible when the solution of the Riemann problem is made of shocks and contact discontinuities (which is the case of the relaxation system). This computation is detailed in [1, 3].

3.2 First order semi-implicit relaxation scheme

The relaxation scheme is constructed as to be:

- linearly implicit on the fast waves of speed $v \pm a\tau$ ("pressure waves"),
- explicit on the slow waves of speed v ("kinematic waves") and the associated relaxation waves of speed $v \pm b\tau$.

Accordingly, when one computes the partial derivatives of the numerical flux, only the terms associated with the largest eigenvalues are kept : we nullify the entries of the diagonal matrix implied in the computation of Roe's matrix. The eigenvalues of the Jacobian matrices are modified in the same way. The matrices α, β and γ are then computed with these modified matrices.

Time step of the semi-implicit relaxation scheme At each time step, one computes two auxiliary time steps. The explicit time step is computed on the base of the small eigenvalues of the relaxation system with a CFL number of 0.5. The linearly implicit time step is computed on the base of the large eigenvalues of the relaxation system with a CFL number of 20. The final time step is the minimum of the two last time steps. The CFL number 20, experimentally determined, turns out to be a good compromise between a small CPU-time and an acceptable amount of smearing out for the numerical solution.

Implicit projection We follow the ideas of Chalons [5]. The use of a linearly implicit scheme implies that the projection step is giving steady states that are not accurate. The solution is to link the evolutions of the relaxed variables associated with an implicit field to the evolutions of the equilibrium variables. This modifies the linear system to be solved in the mathematical step. It enables us to reduce the size of each block from 5×5 to 4×4 and therefore reduce the size of the global linear system. Numerical experiments shows that not only the steady solutions are more accurate (as already shown in [5]), but even transient solutions are more accurate.

3.3 Second order semi-implicit scheme

The slope limiters used to build the second order explicit scheme leads to a non differentiable expression. That is why we choose a simplified version in which we do not differentiate the nonlinear operator involved in the second order correction. In the physical step, we first compute the second order correction of the states $\mathbf{u}_{\mathbf{L},\mathbf{R}}$ and evaluate the numerical flux. In the mathematical step, we solve the first order linear system but the derivatives are computed with the corrected states $\mathbf{u}_{\mathbf{L},\mathbf{R}}$.

In order to have a second order accuracy in time, we use again the Runge-Kutta 2 procedure. Since the scheme is only semi-implicit, the order 2 in time is only achieved on the explicit waves and the linearly implicit waves are solved with order 1 in time.

4 Numerical results

In this Section, we show the numerical results of the semi-implicit relaxation scheme: we consider a real-life problem in which the solution is driven by the changes of the boundary conditions.

The details of this test-case are shown in figure (1) and the results are given in figures (2).

In this experiment[6], the inlet gas flow rate is decreased from 0.114 to 0 kg/s. As the mass flowrates are small, the decrease in the inlet gas mass flow rate gives rise to negative oil velocities in the upper part of the riser. Therefore,

Extracted from : [6]
Geometry of the pipe and discretization :
- Length : 80 m, **vertical**
- Diameter : 0.146 m
- Cell size : 1.6 m

Closure laws :
- Thermodynamic : compressible liquid, perfect gas
- Hydrodynamic : like Zuber-Findlay

Source terms : Friction+ Gravity.
Boundary and initial conditions :
- Stabilization period : 10 s
- Transient period : 100 s
- Inlet condition on gas flowrate : 0.114 to 0. kg/s
- Inlet condition on liquid flowrate : 1.628 kg/s
- Outlet condition on pressure : 10^6 Pa

Fig. 1. Detail of the experiment, Zuber-Findlay-like law with boundary conditions

a void fraction between a "single-phase gas" state and a "two-phase" state propagates from the outlet down the pipe. Simultaneously, the change in the inlet gas mass flowrates induces another void fraction front which propagates from the inlet up the pipeline. These two fronts meet around $x = 47$ m at time $t = 260$ s to form a unique discontinuity wave which propagates toward the outlet of the pipe. At time $t = 200$ s, this discontinuity wave reaches the outlet and the pipe turns to a single-phase liquid steady state.

This test-case is quite stiff. During the simulation, the scheme must handle two-phase states, liquid state and gas state. Moreover, the discontinuity propagating at the end on the simulation is between two one-phase states. The classical VFRoe scheme is not enough rough in order to handle this case and that is why the authors of [6] introduced more numerical diffusion in their VFRoe-TACITE scheme. The numerical results show a good agreement between the two schemes.

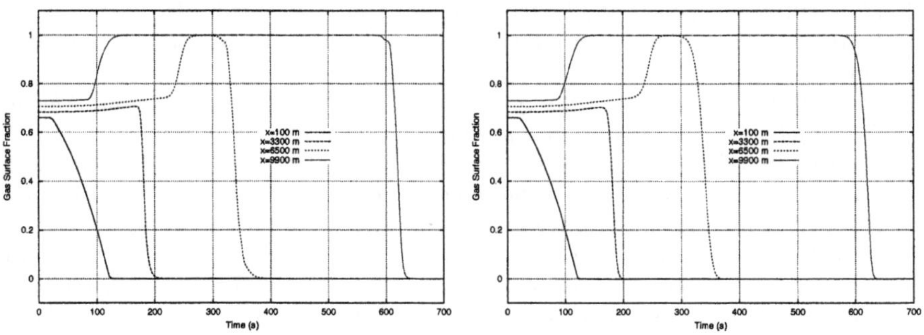

Fig. 2. Experiment 4, gas surface fraction. Left : TACITE results, Right : relaxation.

The difference between the two schemes is mainly the CPU cost of the two simulations. The TACITE scheme requires 22 minutes and the relaxation scheme requires 7 minutes. See [1] for a discussion on this point.

Conclusion

We have presented a second order semi-implicit relaxation scheme. The main difficulty was to express the relaxation Godunov scheme in Roe's form, that is to say, to compute Roe's matrix. The semi-implicit relaxation scheme is then explicit for the slow waves and linearly implicit for the fast waves and enable us to reduce the CPU time as well as to be more accurate on slow waves. Numerical experiments show a good agreement with a VFRoe-type scheme on realistic problems.

The main open issue is the extension of this scheme to the compositional flows which have a very important role in petroleum industry. The energy equation should also be investigated in order to take care of thermal effects. These systems are explored in [1]. Another effort could be made in developing rough boundary conditions: the explicit relaxation only ensures the physical properties inside the domain and not at the boundaries.

Acknowledgments

The authors wish to thank C. Berthon, I. Faille and F. Willien for the numerous helpful discussions we had.

References

1. M. Baudin, *Méthodes de relaxation pour la simulation des écoulements diphasiques dans les conduites petrolières*, PhD thesis, Université Pierre et Marie Curie, 2003.
2. M. Baudin, C. Berthon, F. Coquel, R. Masson, and Q.-H. Tran, *A relaxation method for two-phase flow models with hydrodynamic closure law*, submitted, (2002).
3. M. Baudin, F. Coquel, and Q.-H. Tran, *A semi-implicit relaxation scheme for modeling two-phase flow in a pipeline*, in preparation, (2002).
4. S. Benzoni-Gavage, *Analyse Numérique des Modèles Hydrodynamiques d'Écoulements Diphasiques Instationnaires dans les Réseaux de Production Pérolière*, PhD thesis, Ecole Normale Supérieure de Lyon, 1991.
5. C. Chalons, *Bilans d'entropie discrets dans l'approximation numérique des chocs non classiques. Application aux équations de Navier-Stockes multi-pression 2D et à quelques systèmes visco-capillaires*, PhD thesis, Université Pierre et Marie Curie, Paris VI, novembre 2002.
6. I. Faille and E. Heintzé, *A rough finite volume scheme for modeling two phase flow in a pipeline*, Computers and Fluids, 28 (1999), pp. 213–241.

7. T. GALLOUËT AND J.-M. MASELLA, *A rough godunov scheme*, in Compte Rendus à l'Académie des Sciences, Paris, 1996, p. 77.
8. E. GODLEWSKI AND P.-A. RAVIART, *Hyperbolic systems of conservation laws*, Mathématiques et Applications, SMAI, Ellipses, 1991.
9. R. J. LEVEQUE, *Numerical methods for conservation laws*, Lectures in Mathematics, ETH Zürich, Birkhäuser Verlag, Berlin, (1992).
10. C. PAUCHON, H. DHULÉSIA, G. BINH-CIRLOT, AND J. FABRE, TACITE*: a transient tool for multiphase pipieline and well simulation*, in SPE annual Technical Conference, New Orleans, 1994.
11. Z. XIN AND S. JIN, *The relaxation schemes for systems of conservation laws in arbitrary space dimensions*, Comm. Pure Appl. Math., 48 (1995), pp. 235–276.
12. N. ZUBER AND J. FINDLAY, *Average volumetric concentration in two-phase flow systems*, J. Heat Transfer, C87 (1965), pp. 453–458.

Computational Study of Field Scale BTEX Transport and Biodegradation in the Subsurface

Markus Bause

Institut für Angewandte Mathematik, Universität Erlangen-Nürnberg, Martensstr. 3, 91058 Erlangen, Germany (*bause@am.uni-erlangen.de*)

Summary. In this work we simulate numerically the transport and biodegradation of an organic contaminant (BTEX) in the subsurface. A "real world" contaminated site is considered and realistic laboratory-derived or field-measured input parameters are used. The biodegradation of the dissolved contaminant plume is modelled by Monod type kinetics. For the computations we use an approximation scheme which is based on a higher order finite element method and the two step backward differentiation formula and was proposed and carefully analyzed by the author in a recent work [3]. It was successfully applied to test (benchmark) problems reported in the literature (cf. [10, 13]) as well as complex scenarios with an additional numerical computation of the flow field by solving the parabolic-elliptic degenerate Richards equation; cf. [3]. The higher order approximation scheme has shown to reduce significantly the amount of inherent numerical diffusion compared to lower order ones. Thereby an artificial transverse mixing of the species leading to a strong overestimation of the biodegradation process and wrong prediction is avoided.

1 Introduction

Groundwater contamination by biodegradable organic compounds has become a serious and widespread environmental problem in industrialized countries. Major organic contaminants include petroleum fuels (gasoline, diesel), petroleum byproducts (coal tar, coal-tar creosote), and chlorinated solvents. In many cases, groundwater contains a mixture of organic contaminants, either due to the complex mixture in many non-aqueous phase liquids (NAPLs; e.g., gasoline) or due to co-disposal/co-spillage (e.g., landfill leachates). The degradation of these contaminants is controlled to a large extent by the biological and geochemical conditions in the groundwater. Fortunately, biodegradation tends to attenuate at least some organics during groundwater transport.

The question of whether active remediation is required, or whether natural processes of attenuation (passive remediation) will be sufficient is a critical issue in "real world" situations. Passive or intrinsic remediation is generally preferred, if feasible, due to the potential to, firstly, eliminate permanently contaminants through biogeochemical transformation or mineralization and, secondly, avoid expensive biological, chemical and physical treatments. However, the possible attenuation of organic compounds and the impact of that contamination on a groundwater resource is difficult to predict since field sampling limitations make it difficult to develop an accurate mass balance. Numerical models can be used to help answer these questions, predict the long-term evaluation of the contaminant plume and evaluate factors limiting biodegradation. However, a predictive capability for decision-making can only be found in advanced contaminant transport models which include the full range of the controlling processes and efficient, accurate and reliable numerical methods for solving these equations. Although the understanding of conservative transport and the effect of medium heterogeneity on transport are now well-advanced, methods for modelling bioreactive processes, in particular at the field scale, are less well understood; cf. [6]. Also, the accuracy of numerical techniques in the context of bioreactive transport has been little explored so far.

In a recent work [3], the author proposed a higher order approximation scheme (cf. Sec. 3), based on finite element methods and backward differentiation formulae, for biochemically reacting contaminant transport in the subsurface with Monod type kinetics (cf. Sec. 2). The numerical scheme was carefully compared to a recently published adaptive finite volume approach (cf. [10, 13]) by recomputing some computational experiments of biodegradation processes presented in [10, 13]. The higher order finite element techniques seemingly provided more accurate results than the finite volume methods; cf. [3]. Further, the simulations presented in [3] have clearly borne out that in order to ensure the reliability of the numerical discretization of the bioreactive transport model, it is of importance to use higher order approximation schemes, in particular, for the spatial discretization. Using lower order methods may lead to an overrepresentation of the transverse mixing of the substances and, thereby, to a significant overprediction of the biodegradation process; cf. [6, 13]. Completely wrong solutions are obtained, even if the spatial grid is locally refined and adapted to the solution. Higher order methods help to overcome these difficulties due to their less inherent numerical diffusion.

In this work we use the approximation scheme suggested by the author in [3] to study the long-term evaluation and biodegradation of a "real world" field scale **B**enzene **T**oluene **E**thylbenzene **X**ylene plume in the subsurface. The contaminated site is located in the north part of the city Geretsried in Germany close to Munich and was recently analyzed within the interdisciplinary network project "Sustainable Remediation involving Natural Attenuation" that was supported by the Bavarian State Ministry for Regional Development and Environmental Affairs; cf. Sec. 4 or contact [7] for further information. At this site, large quantities of mineral oil were infiltrated into the soil between 1948 and 1989 by a chemical laundry. Despite the cleaning up performed in 2001, a significant concentration of BTEX is still being measured there. The plan for the paper is now as follows. In Sec. 2 we introduce the mathematical model describing the transport and Monod type biodegradation of organic contaminants in the subsurface. In Sec. 3 the numerical discretization techniques are briefly introduced. The simulated results for the expansion and movement of the contaminant plume are presented in Sec. 4.

2 Governing equations

Microbial activity in the subsurface is dependent on the bioavailability of all substrates utilized by the microorganisms. The main substrates are the electron donor, the electron acceptor, and the primary carbon source. In the standard case of metabolic aerobic degradation, oxygen is the electron acceptor, and the contaminant to be degraded acts both as the electron donor and the primary carbon source; cf. [5]. Here, for the sake of simplicity, only the basic process of aerobic degradation of a single substrate (xylene) is considered. The principles hold equally for multiple electron donors or acceptors.

For this "simple" scenario, biomass growth is assumed to follow the double Monod kinetics, also referred to as the double Michaelis–Menten law; cf. [5, 6, 15]. Then the governing equations for the electron donor c_D $[\mathrm{M\,L^{-3}}]$, electron acceptor c_A $[\mathrm{M\,L^{-3}}]$ and immobile biomass c_X $[\mathrm{M\,L^{-3}}]$ are respectively given by

$$\begin{aligned} \partial_t(\Theta c_i) - \nabla \cdot (D_i \nabla c_i - q c_i) &= -\alpha_i \mu \,, \\ \partial_t c_X + k_d c_X &= \frac{Y}{\Theta}\left(1 - \frac{c_X}{c_{X_{\max}}}\right)\mu \,, \end{aligned} \tag{1}$$

for $i = D, A$, where the Monod term μ $[\mathrm{M\,L^{-3}\,T^{-1}}]$ is defined by

$$\mu = \Theta \mu_{\max}\, c_X \, \frac{c_D}{K_D + c_D}\, \frac{K_{ID}}{K_{ID} + c_D}\, \frac{c_A}{K_A + c_A}\, \frac{K_{IA}}{K_{IA} + c_A} \,. \tag{2}$$

Thus, we have to consider a coupled system of partial and ordinary differential equations. In (1), Θ [–] denotes the volumetric water content, q $[\mathrm{L\,T^{-1}}]$ the Darcy velocity vector (volumetric flux) and D_i, $i = D, A$, $[\mathrm{L^2\,T^{-1}}]$ with

$$D_i = (\Theta d_i + \beta_t |q|) I + (\beta_l - \beta_t)(q \otimes q)/|q| \tag{3}$$

the dispersion tensor following the Scheidegger parametrization (cf. [14]) in which d_i, $i = D, A$, $[\mathrm{L^2\,T^{-1}}]$ is the molecular diffusion, I [–] the identity matrix and β_l [L] and β_t [L] are the longitudinal and transverse dispersivities, respectively. We have $\alpha_D = 1$ [–]. The constant α_A [–] denotes the electron acceptor to donor mass ratio, k_d $[\mathrm{T^{-1}}]$ is the first order decay rate for the biomass, Y [–] is the microbial yield coefficient per unit electron donor consumed (mg biomass per mg electron donor) and $c_{X_{\max}}$ $[\mathrm{M\,L^{-3}}]$ is the maximum biomass concentration. In (2), $\mu_{\max}$ $[\mathrm{T^{-1}}]$ denotes the maximum growth rate, K_i, $i = D, A$, $[\mathrm{M\,L^{-3}}]$ is the half-utilization constant of the electron donor and acceptor, respectively, and K_{Ii}, $i = D, A$, $[\mathrm{M\,L^{-3}}]$ is the Haldane inhibition concentration of the electron donor and acceptor, respectively. The inhibition term $K_{Ii}/(K_{Ii}+c_i)$, $i = D, A$, proposed by Haldane [8] and Andrews [1], yields a slower microbial growth and, therefore, a slower effective electron donor utilization rate at higher concentrations; cf. [15]. In the numerical model, unrealistically high biomass concentrations are avoided by introducing the term $1 - c_X/c_{X_{\max}}$ in the second of the equations (1). If microbial growth is not restricted in the model, simulated microbial concentrations may become very large, especially in source areas with continuous electron donor and acceptor supply. In real aquifers, the size of the biomass is limited, for example, due to a lack of available pore space, production of inhibitory metabolites, lack of nutrients and viral attack. The constant $c_{X_{\max}}$ represents the maximum microbial concentration at which the biomass reaches a quasi-steady state.

We consider solving (1), (2) over $(0,T) \times \Omega$ where $\Omega \subset \mathbb{R}^d$, $d = 2, 3$, is a two- or three-dimensional bounded domain and the system (1)–(2) is supplied with initial conditions

$$c_D(0,\cdot) = c_{D,0}\,,\ c_A(0,\cdot) = c_{A,0}\,,\ c_X(0,\cdot) = c_{X,0} \quad \text{in } \Omega \text{ at } t = 0\,, \tag{4}$$

and nonhomogeneous Dirichlet and Robin boundary conditions for c_i, $i = D, A$,

$$c_i = g_i \quad \text{on } (0,T) \times \Gamma_D\,, \qquad (qc_i - D_i \nabla c_i) \cdot \nu = h_i \quad \text{on } (0,T) \times \Gamma_R\,. \tag{5}$$

Here, ν denotes the outer unit normal to the boundary $\partial\Omega = \Gamma_D \cup \Gamma_R$ of Ω. The existence of a global unique *non-negative* solution $c_D, c_A \in W_p^{1,2}((0,T) \times \Omega)$, with $p > 2$, and $c_X \in C^1([0,T]; C(\overline{\Omega}))$ to (1)–(5) for any given $T \in (0,\infty)$ was recently proved; cf. [12]. In particular, the non-negativeness of c_D, c_A, c_X can be ensured. The proof can be carried over to $d = 3$. For the definition of $W_p^{l/2,l}((0,T) \times \Omega)$ we refer to [11].

In our computational experiment, the velocity vector q $[\mathrm{L\,T}^{-1}]$ in (1), (3) and (5) is prescribed analytically. This is done due to a lack of information and measurements of the flow field for the considered site; cf. Sec. 4. For numerical simulations of contaminant transport and biodegradation scenarios where the flow field is additionally computed numerically by solving the parabolic-elliptic degenerate Richards equation we refer to [3].

The established regularity of solutions to problem (1)–(5) is, by far, too weak to justify the use of higher order approximation schemes. However, one may expect that a higher regularity of the solution still holds in some sense locally. This might be sufficient to get a significant advantage of higher order approximation schemes over lower order ones. Such superiority of the higher order methods was recently confirmed by numerical computations; cf. [3]. Further, higher order regularity results for solutions to equations (1)–(5) are established in a forthcoming paper; cf. [4].

3 Discretization and solution techniques

We shall now briefly describe our numerical methods and solution techniques for solving the equations (1)–(5). For the spatial discretization we use conforming finite element methods. Here, for simplicity, we assume that $\Omega \subset \mathbb{R}^2$ is a polygonal bounded domain. If either the whole boundary of Ω or at least some part of it is curved, we adapt the mesh to the boundary by using the isoparametric counterparts of the finite elements introduced below; cf. [3]. In our computations we will consider non-vanishing Dirichlet boundary values g_i, $i = D, A$, in (5). However, for simplicity, the variational formulation of (1)–(5) is given for homogeneous Dirichlet boundary conditions only. Nonhomogeneous boundary values are incorporated by standard techniques.

Let $\mathcal{T}_h = \{K\}$ be a finite decomposition of mesh size h of $\overline{\Omega}$ into triangles. The decompositions are assumed to be regular, i.e., "face to face". We use standard conforming P_2 elements for the Monod model (1)–(5). The approximation spaces V_h for the electron donor and acceptor C_i, $i = D, A$, and X_h for the biomass C_X are thus defined as $V_h = \{C_i \in C(\overline{\Omega}) \mid C_{i|K} \in P_2(K) \text{ for } K \in \mathcal{T}_h\} \cap W^{1,2}_{0,\Gamma_D}$ and $X_h = \{C_X \in C(\overline{\Omega}) \mid C_{X|K} \in P_2(K) \text{ for } K \in \mathcal{T}_h\}$. By $P_j(K)$, $j \in \mathbb{N}$, we denote the space of all continuous polynomials of maximum degree j. Further, $W^{1,2}_{0,\Gamma_D} = \{c \in W^{1,2}(\Omega) \mid c = 0 \text{ on } \Gamma_D\}$. Hence, the spatial discretization of c_D, c_A and c_X converges formally of third order with respect to the norm in $L^2(\Omega)$. Advection-dominated transport of the mobile species (electron donor and acceptor) introducing local numerical instabilities in the solution can efficiently be captured by the streamline upwind Petrov–Galerkin method (SUPG); cf. [3, 9].

For the temporal discretization of problems (1)–(5) we consider a mesh $\{t_n\}$, $n = 0, \ldots, \mathcal{N}$ with $t_0 = 0$ and $t_{\mathcal{N}} = T$, for the time variable t and define $\tau_n = t_{n+1} - t_n$. Due to the generally high stiffness of semidiscretizations to flow and transport problems, implicit schemes should be preferred in the choice of time-stepping methods for solving these problems. The backward Euler method is robust and has excellent stability properties, but it is inaccurate due to its first convergence order only and also strongly damping. So, it should only be used for nonstationary calculations which aim to iterate towards the steady limit. A scheme having similar stability properties as the backward Euler method but being of second order accuracy is the two step backward differentiation formula BDF_2 which we use in our computations. Further, to increase the efficiency of the calculations, an adaptive time stepping procedure was developed and tested in [3] for the proposed discretization of the transport and biodegradation model (1)–(5).

Now, we suppose that sequences $\{q(t_n), \Theta(t_n)\} \in W^{1,\infty}(\Omega) \times L^\infty(\Omega)$ are explicitly prescribed. Let P_{Z_h} denote the L^2-projection onto the finite element space Z_h. The discretization of the Monod model (1)–(5) by the Galerkin method and the two step backward differentiation formula BDF_2 then reads as follows:

Set $C_i^0 = P_{V_h} c_{i,0}$ and $C_X^0 = P_{X_h} c_{X,0}$. For all time steps $n = 0, \ldots, \mathcal{N} - 2$ compute approximations $C_i^{n+2} \in V_h$, $i = D, A$, and $C_X^{n+2} \in X_h$ by solving the equations

$$
\begin{aligned}
&\gamma_{n+2}\left\langle \Theta(t_{n+2}) C_i^{n+2}, V_i\right\rangle - \gamma_{n+1}\left\langle \Theta(t_{n+1}) C_i^{n+1}, V_i\right\rangle + \gamma_n \left\langle \Theta(t_n) C_i^{n}, V_i\right\rangle \\
&+\tau_{n+1}\langle q(t_{n+2})\cdot \nabla C_i^{n+2}, V_i\rangle + \tau_{n+1}\langle D_i(t_{n+2})\nabla C_i^{n+2}, \nabla V_i\rangle \\
&-\tau_{n+1}\langle (q\cdot\nu) C_i^{n+2}, V_i\rangle_{\Gamma_R} + \tau_{n+1}\langle \nabla\cdot q(t_{n+1}) C_i^{n+2}, V_i\rangle \\
&= -\tau_{n+1}\langle \alpha_i\, \mathcal{U}^{n+2}, V_i\rangle - \tau_{n+1}\langle h_i, V_i\rangle_{\Gamma_R}\,,
\end{aligned} \tag{6}
$$

for all $V_i \in V_h$, $i = D, A$, *and*

$$
\begin{aligned}
&\gamma_{n+2}\, C_X^{n+2} - \gamma_{n+1}\, C_X^{n+1} + \gamma_n\, C_X^{n} + \tau_{n+1} k_d\, C_X^{n+2} = \\
&\tau_{n+1}\frac{Y}{\Theta(t_{n+2})}\left(1 - \frac{C_X^{n+2}}{c_{X_{\max}}}\right)\mathcal{U}^{n+2}
\end{aligned} \tag{7}
$$

for all nodes $(x_j)_{j=1,\dots,M}$ *associated with degrees of freedom of* C_X^{n+2}, *where*

$$
\begin{aligned}
&\mathcal{U}^{n+2} = \\
&\Theta(t_{n+2})\,\mu_{\max}\, C_X^{n+2}\,\frac{C_D^{n+2}}{K_D + C_D^{n+2}}\,\frac{K_{ID}}{K_{ID}+C_D^{n+2}}\,\frac{C_A^{n+2}}{K_A+C_A^{n+2}}\,\frac{K_{IA}}{K_{IA}+C_A^{n+2}}\,.
\end{aligned} \tag{8}
$$

By $\langle\cdot,\cdot\rangle$ and $\langle\cdot,\cdot\rangle_{\Gamma_R}$ we denote the standard L^2 inner product in $L^2(\Omega)$ and $L^2(\Gamma_R)$, respectively. Further, in (6) and (7) we use the notation $\gamma_{n+2} = 1 + \tau_{n+1}/(\tau_{n+1}+\tau_n)$, $\gamma_{n+1} = 1 + \tau_{n+1}/\tau_n$ and $\gamma_n = \tau_{n+1}^2/((\tau_{n+1}+\tau_n)\tau_n)$. The time step sizes τ_{n+1} can be chosen adaptively; cf. [3]. Clearly, identity (7) formulates a pointwise condition for all nodes associated with degrees of freedom of C_X^{n+2}. Using instead of (7) a variational equation, analogously to (6), leads to stability problems and severe oscillations.

As usual, for the test functions V_i we choose the basis functions of V_h. Further, let $C_X^i \in X_h$, $i = n, n+1, n+2$, be represented in terms of the finite element basis functions $\{W_j\}_{j=1}^M$ of X_h, i.e., $X_h = \operatorname{span}\{W_j \mid 1 \le j \le M\}$ and $C_X^i = \sum_{j=1}^M \xi_j^i W_j$, for $i = n, n+1, n+2$, where the vector $\xi^i = (\xi_1^i, \dots, \xi_M^i)$ denotes the degrees of freedom of C_X^i. Then, (7) amounts to solving the system of equations in the unknown vector ξ^{n+2},

$$
\gamma_{n+2}\xi^{n+2} - \gamma_{n+1}\,\xi^{n+1} + \gamma_n\,\xi^n + \tau_{n+1}k_d\,\xi^{n+2} = \tau_{n+1}\frac{Y}{\Theta(t_{n+2})}\left(1 - \frac{\xi^{n+2}}{c_{X_{\max}}}\right)U^{n+2},
$$

where $U^{n+2} = (\mathcal{U}^{n+2}(x_1),\dots,\mathcal{U}^{n+2}(x_M))$.

Since the BDF_2 is a two step method, we need a starting procedure to compute appropriate approximations C_D^1, C_A^1 and C_X^1 of $c_D(t_1,\cdot)$, $c_A(t_1,\cdot)$ and $c_X(t_1,\cdot)$, respectively. Here, the first time step is done by performing M substeps of the backward Euler method with step size τ_0/M. In our computations we use $M = 4$. To solve the resulting nonlinear systems of equations, a damped version of Newton's method is applied. The linear problems of the Newton iteration are solved by standard Krylov space methods like GMRES, for instance, with SSOR preconditioning. For the future we plan to use multigrid methods for the linear solver which is motivated by our former experiences with computing variably saturated subsurface flow; cf. [2]. In the simple case of a single

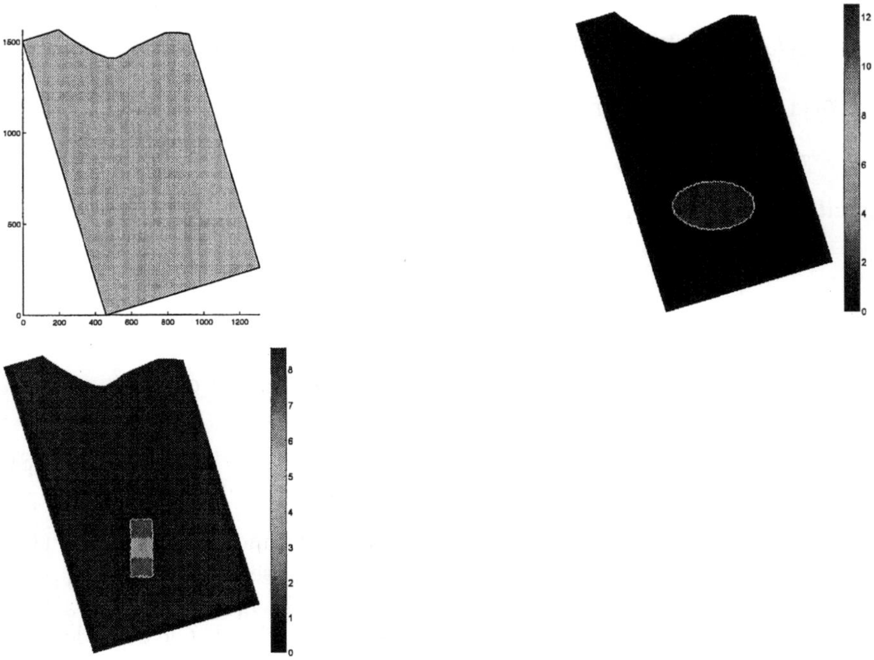

Fig. 1. Computational domain (left) for contaminated site Geretsried and initial concentration of electron donor xylene (middle) and electron acceptor oxygen (right)

electron donor and acceptor and a single biomass an alternative treatment of the transport and biodegradation problem (1)–(5) seems to be possible. After a temporal discretization of (1), one may resolve the time-discrete version of the second of the equations (1) for the biomass concentration c_X and substitute the resulting identity into the time-discrete version of the first of the equations (1). Thus, the biomass concentration is eliminated from the nonlinear system of equations and can be computed in a postprocessing procedure which leads to smaller systems of linear equations to be solved. However, a generalization of such approach to the case of multiple microbial populations and, in particular, its implementation seems to be more complex than an explicit treatment of the ordinary differential equations for the biomass. Therefore, the approach is not considered here.

4 Computational results

We shall now present our computational results obtained for a "real world" residual waste and thereby provide valuable insights into the complex interactions of biological, chemical and physical processes that are involved in natural attenuation phenomena. The contaminated site is located in the city Geretsried in Bavaria (Germany) and was recently analyzed within a network project; contact [7] for details. Large quantities of mineral oil were infiltrated into the soil between 1948 and 1989 by a chemical laundry. Despite some active remediation, a significant concentration of BTEX is still being measured there.

The computational domain Ω in the subsurface is visualized in Fig. 1 with length given in meter. For simplicity and due to a lack of information, a constant groundwater flow field parallel to the left and right boundary of Ω is supposed. The flow direction is from the lower to the upper boundary adjacent to the river Isar. The measured concentration profiles at the current state, assumed to be the initial state of the simulation, are shown in Fig. 1. The shapes within the profiles were slightly idealized. We restrict ourselves to the electron donor xylene which here is the main component of BTEX with 70–90%. The xylene concentration inside the ellipse is 12.5 [mg/l] and 0 [mg/l] elsewhere. Biodegradation of xylene with either oxygen or nitrate as electron acceptor is observed. Here, we consider the first case. We have a small rectangular domain overlapping with the xylene ellipse where the oxygen concentration is 2 [mg/l] and 4.0 [mg/l] in its center. The ambient oxygen concentration is 8.6 [mg/l]. The measured higher oxygen concentration inside the contaminant source puzzles us and has not been completely understood yet. One reason for that might be a lower permeability inside the contaminant source due to a lack of available pore space. Further investigations are necessary. Therefore and due to the lack of information in particular about the spatial variation of the flow field, our simulations have to be considered rather as a qualitative analysis of the biodegradation process than as a quantitative prediction of the xylene degradation. Nevertheless they contribute to a better understanding of natural attenuation phenomena for this site.

In our calculations we used reliable field-measured and laboratory-derived input parameters that were given in [15]. They proved to describe adequately field scale degradation provided that all controlling factors are incorporated in the field scale model. The flow field $q = (0.045, 0.15)^T$, with time given in days, and the diffusion-dispersion parameters were obtained by measurements. Precisely, we put Θ: 0.3, d_D, d_A: 8.64e−5, β_t: 2.0, β_l: 10.2, α_A: 3.16, k_d: 0.001, Y: 0.52, $c_{X_{\max}}$: 1.0, $\mu_{\max}$: 4.13, K_D: 0.79, K_A: 0.1, K_{ID}: 91.7, K_{IA}: ∞. The initial concentration of the biomass was $c_X(0,\cdot) = 0.003$ in $\overline{\Omega}$. We chose homogeneous Neumann boundary conditions at the left, right and upper boundary and a Dirichlet condition at the lower one. The computations were done on an almost uniform grid with 12322 elements. The time step sizes were chosen adaptively; cf. [3]. The calculated concentration profiles of the electron donor (contaminant) xylene, acceptor oxygen and biomass are visualized in Fig. 2 to 4. For comparison, the problem was recomputed on a very fine mesh with 197152 elements. This was done on a Linux cluster with 16 processors. No significant changes in the computed profiles were observed. It shows that the proposed numerical scheme reliably predicts the degradation rates even on

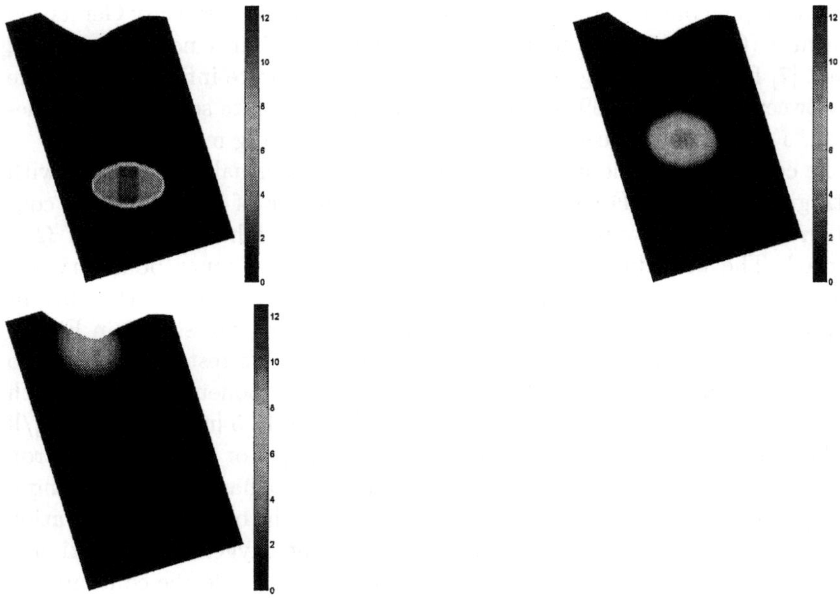

Fig. 2. Concentration of electron donor xylene at T = 20 days, 1.5 years and 4.5 years

relatively coarse meshes which is in agreement with the results presented in [3].

Fig. 2 to 4 show that the contaminant is transported by the flow field to the upper boundary of Ω adjacent to the river. Simultaneously, it is degraded by a reaction between electron donor, acceptor and biomass. The initial oxygen concentration inside the contaminant plume becomes depleted within a few days which is not consistent with the measured profile and might result from an insufficient description of the flow and permeability conditions inside the contaminant source. The reaction between the species is restricted to those regions where their concentrations are sufficiently large. Basically, it is the interface between the electron donor and the surrounding region where still enough acceptor is available. If a numerical method with much artificial diffusion is used, this interface between the species smears out and the reaction takes place in the larger region. Then, the contaminant is degraded too fast; cf. [3, 6, 10, 13]. In particular, this happens if lower order methods are applied on not highly refined meshes; cf. [3, 13].

References

1. J.F. Andrews. A mathematical model for the continuous culture of microorganisms utilizing inhibitory substrates, Biotechnol. Bioeng., **10**:707–723, 1968.

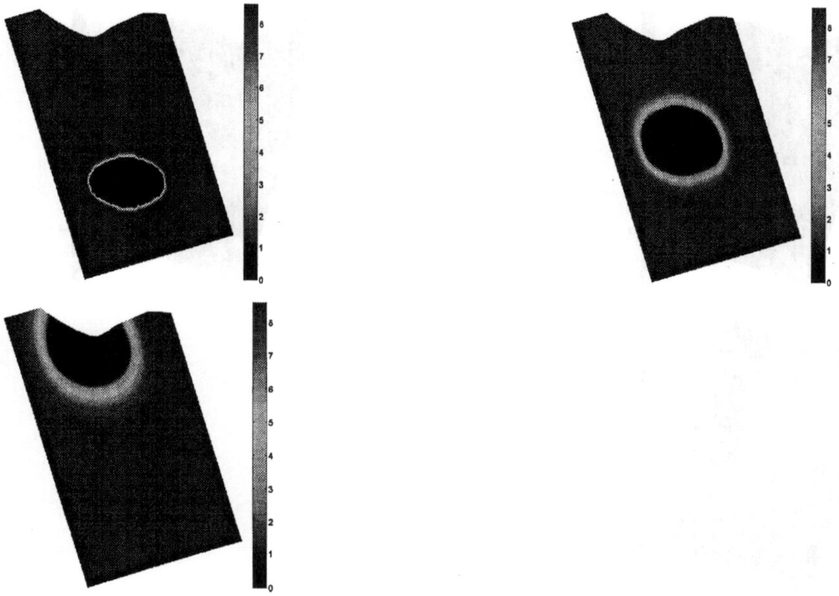

Fig. 3. Concentration of electron acceptor oxygen at T = 20 days, 1.5 years and 4.5 years

2. M. Bause, P. Knabner. Computation of variably saturated subsurface flow by adaptive mixed hybrid finite element methods, Adv. Water Resour., submitted, 2002.
3. M. Bause, P. Knabner. Numerical simulation of contaminant biodegradation by higher order methods and adaptive time stepping, Comp. Vis. Sci., accepted, 2004.
4. M. Bause, W. Merz. Higher order regularity and approximation of solutions to the Monod biodegradation model, Appl. Numer. Math., submitted, 2004.
5. R.C. Borden, P.B. Bedient. Transport of dissolved hydrocarbons influenced by oxygen limited biodegradation: 1. Theoretical development, Water Resour. Res., **22**:1973–1982, 1986.
6. O.A. Cirpka, E.O. Frind, R. Helmig. Numerical simulation of biodegradation controlled by transverse mixing, J. Contam. Hydrol., **40**:159–182, 1999.
7. Gesellschaft zur Altlastensanierung in Bayern mbH (GAB mbH), Innere Wiener Straße 11a/I, 81667 München, Germany (http://www.altlasten-bayern.de).
8. J.B.S. Haldane. Enzymes, M.I.T. Press, Cambridge, MA, 1965.
9. T.J.R. Hughes, A.N. Brooks. A multidimensional upwind scheme with no crosswind diffusion. Finite Element Methods for Convection Dominated Flows, T.J.R. Hughes (ed.), 19–35, ASME, New York, 1979.
10. R. Klöfkorn, D. Kröner, M. Ohlberger. Local adaptive methods for convection dominated problems, Internat. J. Numer. Methods Fluids, **40**:79-91, 2002.
11. O.A. Ladyženskaya, V.A. Solonnikov, N.N. Ural'ceva. Linear and Quasilinear Equations of Parabolic Type, Am. Math. Soc., Providence, RI, 1968.

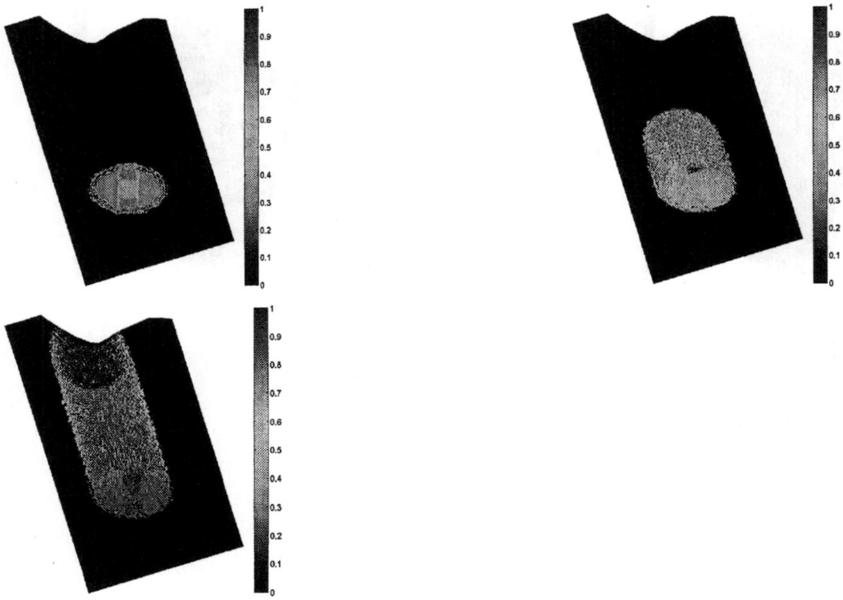

Fig. 4. Concentration of biomass at T = 20 days, 1.5 years and 4.5 years

12. W. Merz. Global existence result of the monod model, Adv. Math. Sci., accepted, 2003.
13. M. Ohlberger, C. Rohde. Adaptive finite volume approximations of weakly coupled convection dominated parabolic systems, IMA J. Numer. Anal., **22**:253–280, 2002.
14. A.E. Scheidegger. General theory of dispersion in porous media, J. Geophys. Res., **66**(10):3273–3278, 1961.
15. M. Schirmer, J.W. Molson, E.O. Frind, J.F. Barker. Biodegradation modelling of a dissolved gasoline plume applying independent laboratory and field parameters, J. Contam. Hydrol., **46**:339–374, 2000.

A Two-Level Stabilization Scheme for the Navier-Stokes Equations

Roland Becker[1] and Malte Braack[2]

[1] Laboratoire de Mathématiques Appliquées
Université de Pau et des Pays de l'Adour
BP 1155, 64013 PAU Cedex, France
roland.becker@univ-pau.fr

[2] Institut für Angewandte Mathematik
Universität Heidelberg
Im Neuenheimer Feld 294, 69120 Heidelberg, Germany
malte.braack@iwr.uni-heidelberg.de

Summary. As an alternative to classical stabilization schemes as, for instance, Galerkin-Least-Squares or streamline diffusion techniques, a stable equal-order finite element scheme for the Navier-Stokes equation is proposed. The approach is based on filtering small-scale fluctuations of pressure and velocities by local projections. For the Stokes system, we prove stability and analyze the arising system matrix. Furthermore, the transport equation is analyzed with respect to stability and an a-priori estimate is given.

1 Introduction

In this note we present a discretization of the stationary Navier-Stokes equations based on equal-order finite elements. We combine the pressure stabilization for the Stokes equations developed in [2] with a similar technique for the nonlinear convection term. The entire approach is based on the use of two discrete spaces W_{2h}, V_h. We use finite elements on quadrilateral meshes. The discrete space V_h corresponds to bilinear finite element functions and W_{2h} to piecewise constant elements on a globally coarser mesh.

Although stabilization by weighted least-squares terms, as for instance Galerkin-Least-Squares (GLS) or streamline-upwind Petrov-Galerkin (SUPG), see [10, 9, 6, 14], is now classical and provides a rather general framework, there is a certain need for different stabilization techniques, see the more recent approaches [4, 8, 2, 5]. One common feature of the new approaches is that they have better local conservation properties then the classical ones. Further, the difficulties of SUPG for higher order polynomials might be overcome. The choice of the stabilization parameter does no longer depend on the constants of an inverse estimate, see [2].

Our motivation for development of new stabilized schemes comes from two fields of application: a) reacting flows with complex chemistry and b) optimal control of incompressible flows. In the first case, SUPG-like stabilization leads to enormous coupling terms which are rather difficult and time-consuming to compute, see [3]. Further, the design of robust Newton-type methods in this context is not evident since one has to decide whether or not to take into account the derivative of each stabilization term in the approximate Jacobian.

Related to the just mentioned problem of computing the derivatives of the stabilization terms is the discretization of optimal control problems. Here, the critical terms directly influence the quality of the computed gradients of the cost functional, see [1] for a discussion.

In Section 2, we describe the two-level discretization of the stationary Navier-Stokes equations. In the following sections, we present some aspects of the analysis: Section 3 deals with the arising algebraic system for the Stokes equations and Section 4 with stabilization of the convective terms.

2 Two-level scheme for the Navier-Stokes equations

Let $\Omega \subset \mathbb{R}^2$ be a polygonal domain. We want to solve the Navier-Stokes equations for an incompressible fluid with fluid velocity v and pressure p,

$$(v\cdot\nabla)v - \nu\Delta v + \nabla p = f \quad \text{in} \quad \Omega\,, \tag{1}$$

$$\operatorname{div} v = 0 \quad \text{in} \quad \Omega, \tag{2}$$

supplied with homogeneous Dirichlet boundary conditions:

$$v = 0 \quad \text{on} \quad \partial\Omega, \tag{3}$$

and the normalization of the pressure

$$\int_\Omega p\,dx = 0\,. \tag{4}$$

In (1), $f \in L^2(\Omega)^2$ represents given data. For the weak formulation of (1)-(4) we introduce the following notations:

$$u := (p,v) \in X := L^2(\Omega)/\mathbb{R} \times H_0^1(\Omega)^2\,,$$

$$a(u)(\phi) := ((v\cdot\nabla)v,\psi) + \nu(\nabla v \nabla\psi) - (p, \operatorname{div}\psi) + (\operatorname{div} v, \xi)\,, \tag{5}$$

for test functions $\phi = (\psi,\xi) \in X$. Now the weak formulation of (1)-(4) reads in compact notation:

$$a(u)(\phi) = (f,\psi) \qquad \forall\phi \in X. \tag{6}$$

We denote by $\mathcal{T}_h$ a shape regular partition of the domain into quadrilaterals. Hanging nodes are allowed with moderation for ease of local mesh refinement. We consider two finite element spaces W_{2h}, V_h, which are constructed in the following way.

V_h consists of bilinear finite elements on $\mathcal{T}_h$. On hanging nodes, the finite element functions are interpolated by the neighbor nodes so that no degrees of freedom are present on those irregular nodes. The space W_{2h} consists of constants on each cell of a coarser mesh $\mathcal{T}_{2h}$ obtained by one global coarsening of $\mathcal{T}_h$: Each quadrilateral $K \in \mathcal{T}_{2h}$ is cut into four new quadrilaterals (dividing all lengths of edges of K by 2) in order to obtain the fine partition $\mathcal{T}_h$. Note, that the functions in W_{2h} are discontinuous across edges of elements of $\mathcal{T}_{2h}$. We indicate the subspaces of discrete functions respecting the Dirichlet condition by an additional subscript $V_{h,0} \subset V_h$. The restriction to a mean value of zero, cf. (4), is denoted by V_h^0. The discrete solution $u_h = (v_h, p_h)$ is searched in the discrete space $X_h := V_h^0 \times V_{h,0}^2$.

Furthermore, we use the L^2-projections on the piecewise constants $P_W : L^2(\Omega) \to W_{2h}$ and the *fluctuation operator* $\pi_h : V_h \to V_h$:

$$\pi_h := I - P_W, \tag{7}$$

where I denotes the identity mapping.

We use the following stabilization terms, defined on $X_h \times X_h$:

$$s(u)(\phi) := (\pi_h[(v\cdot\nabla)v],\ \delta\pi_h[(v\cdot\nabla)\psi]) + (\pi_h\nabla p,\ \alpha\pi_h\nabla\xi). \tag{8}$$

Here, α and δ denote piecewise constant functions which depend among other things on the local cell size h_K. The precise definition is:

$$\alpha|_K := \frac{h_K^2}{\nu}, \qquad \delta|_K := \min\left(\frac{h_K^2}{\nu}, \frac{h_K}{\|v\|_{\infty,K}}\right). \tag{9}$$

The discrete problem reads: Find $u_h \in X_h$ such that

$$a(u_h)(\phi_h) + s(u_h)(\phi_h) = (f, \phi_h) \qquad \forall \phi_h \in X_h. \tag{10}$$

One remarkable feature of (10) is, that the stabilization terms only act on the diagonal of the coupled system. The structure of the stabilization is unchanged, if additional lower order terms are added to the equations.

Our numerical experience shows that the resulting scheme has very similar properties to SUPG concerning stability and accuracy. In the following we present some aspects of the analysis of (10).

3 Structure of the system matrix for the Stokes equations

As the first step, we consider the proposed stabilization in the case of the Stokes equations. It can be easily seen that the stabilization term $(\pi_h\nabla p_h, \alpha\pi_h\nabla\xi)$ in (8) leads to a larger stencil for the pressure then the original one coming from the Galerkin part. The discrete problem for the Stokes equations reads: Find $u_h \in X_h$ such that

$$a(u_h, \phi_h) + s_h u_h, \phi_h) = (f, \phi_h) \quad \forall \phi_h \in X_h \tag{11}$$

holds, where now

$$a(u,\phi) := \nu(\nabla v, \nabla \psi) - (p, \operatorname{div} \psi) + (\operatorname{div} v, \xi), \tag{12}$$

$$s(u,\phi) := (\pi_h \nabla p, \alpha \pi_h \nabla \xi). \tag{13}$$

The analysis presented in [2] gives us the following error estimate in terms of the L^2–norm $\|\cdot\|$ and Sobolev norms $\|\cdot\|_{H^r(\Omega)}$.

Theorem 1. *Let $p \in H^2(\Omega)$ and $v \in H^3(\Omega)^2$. Then, if V_h consists of isoparametric biquadratic functions, the solution (p_h, v_h) of* (10) *for the bilinear forms* (12), (13) *allows for the following error estimate:*

$$\|p - p_h\| + \|\nabla(v - v_h)\| \le Ch^2(\|p\|_{H^2(\Omega)} + \|v\|_{H^3(\Omega)}), \tag{14}$$

where h denotes the maximal mesh size.

A similar error estimate holds true for the classical Taylor-Hood element with biquadratic velocities and bilinear pressure (which does not require any stabilization), see [7]. One might therefore wonder what the additional pressure degrees of freedom in our scheme produce. The answer to this question is provided in the following.

We denote by $\widehat{v}_h \in I\!R^{2n}$ the vector of coefficients of the function v_h with respect to the canonical finite element nodal basis $\{\psi_h^i\}$, i.e., $v_h = \sum_i (\widehat{v}_h)_i \psi_h^i$. Now, we split the pressure p_h into a coarse grid part $\overline{p}_h$ and small-scale fluctuations p_h':

$$p_h = \overline{p}_h + p_h', \qquad \overline{p}_h \in V_{2h}, \qquad p_h' \in V_h' := V_h \setminus V_{2h}. \tag{15}$$

The coefficient vectors $\widehat{\overline{p}}_h$ and $\widehat{p'}_h$ are defined analogously by the nodal basis. Then, the matrix representation of the linear system (11) reads

$$\begin{bmatrix} A & -(\overline{B})^* & -(B')^* \\ \overline{B} & S_1 & S_2^* \\ B' & S_2 & S_3 \end{bmatrix} \begin{bmatrix} \widehat{v}_h \\ \widehat{\overline{p}}_h \\ \widehat{p_h'} \end{bmatrix} = \begin{bmatrix} \widehat{f}_h \\ 0 \\ 0, \end{bmatrix}$$

where $\widehat{f}_h$ has the obvious meaning. The matrix A stands for the Laplacian, $\overline{B}$ and B' for the divergence, and S for the stabilization.

Block elimination of the pressure component $\widehat{p_h'}$ leads to:

$$\begin{bmatrix} \widetilde{A} & -(\widetilde{B})^* \\ \widetilde{B} & \widetilde{S} \end{bmatrix} \begin{bmatrix} \widehat{v}_h \\ \widehat{\overline{p}}_h \end{bmatrix} = \begin{bmatrix} \widehat{f}_h \\ 0 \end{bmatrix},$$

$$\text{where} \quad \widetilde{A} = A + D, \qquad D = -(B')^* S_3^{-1} B',$$
$$\widetilde{S} = S_1 - S_2^* S_3^{-1} S_2,$$
$$\widetilde{B} = \overline{B} - S_2^* S_3^{-1} B'.$$

We provide the following result as an interpretation of our findings.

Remark 1. The additional diagonal block D in the system matrix is suspected to act as an additional stabilization term controlling the discrete divergence.

We denote the l_2 scalar product by $\langle \cdot, \cdot \rangle$. On quasi-regular meshes where all cells are parallelograms, we have a mesh-size independent constant $c > 0$ so that

$$\frac{1}{c} \sum_{K \in \mathcal{T}_h} \|\operatorname{div} v_h\|_K^2 \leq \langle D\widehat{v}_h, \widehat{v}_h \rangle \leq c \sum_{K \in \mathcal{T}_h} \|\operatorname{div} v_h\|_K^2 . \tag{16}$$

Such a term is of common use in stabilized schemes, see [13].

Proof. By definition of D it holds:

$$\langle D\widehat{v}_h, \widehat{v}_h \rangle = \langle S_3^{-1} B'\widehat{v}_h, B'\widehat{v}_h \rangle .$$

We denote by I_f the index set of fine grid nodes $\mathcal{N}_i$, and by $\xi_i \in V_h$ the standard nodal hat functions of node $\mathcal{N}_i$ with support P_i. It holds

$$\langle B'\widehat{v}_h, B'\widehat{v}_h \rangle = \|B'\widehat{v}_h\|^2 = \sum_{i \in I_f} (\operatorname{div} v_h, \xi_i)^2$$
$$\leq \sum_{i \in I_f} (\|\operatorname{div} v_h\|_{P_i} \|\xi_i\|)^2 \leq c \sum_{i \in I_f} h_{P_i}^2 \|\operatorname{div} v_h\|_{P_i}^2 \leq c \sum_{K \in \mathcal{T}_h} h_K^2 \|\operatorname{div} v_h\|_K^2 .$$

Since S_3 scales like h^2, we get $\langle D\widehat{v}_h, \widehat{v}_h \rangle \leq c \sum_{K \in \mathcal{T}_h} \|\operatorname{div} v_h\|_K^2$. The proof for the opposite direction

$$\frac{1}{c} \sum_{K \in \mathcal{T}_h} h_K^2 \|\operatorname{div} v_h\|_K^2 \leq \langle B'\widehat{v}_h, B'\widehat{v}_h \rangle$$

will be given in [12].

4 Analysis for a transport equation

We consider the following transport equation with a given constant transport vector $\beta \in \mathbb{R}^2$ and given continuous data f and g:

$$u + (\beta \cdot \nabla) u = f \quad \text{in} \quad \Omega, \qquad u = g \quad \text{on} \quad \Gamma_-, \tag{17}$$

where Γ_- is the inflow part of the boundary:

$$\Gamma_- := \{x \in \partial\Omega : n(x) \cdot \beta < 0\} .$$

Denoting the L^2-scalar product on the boundary $\partial\Omega$ by $\langle \cdot, \cdot \rangle$, we define the following bilinear form and linear functional:

$$a(u, \phi) := (u, \phi) + ((\beta \cdot \nabla) u, \phi) - \langle (\beta \cdot n)_- u, \phi \rangle$$
$$l(\phi) := (f, \phi) - \langle (\beta \cdot n)_- \tilde{g}, \phi \rangle.$$

Here we have used the notation $x_- := \min(x, 0)$ and $\tilde{g}$ is a prolongation of g to the whole boundary. Later on, we will use the notation $x_+ := \max(x, 0)$. Then, the continuous solution u of equation (17) satisfies the variational equation:

$$a(u, \phi) = l(\phi) \qquad \forall \phi \in V\,, \tag{18}$$

where, for instance, $V = H^1(\Omega)$. The discretization of (17) to be considered here is based on the stabilized version of (18) using the space of continuous piecewise bilinear finite elements V_h as before:

$$a(u_h, \phi) + s(u_h, \phi) = l(\phi) \qquad \forall \phi \in V_h. \tag{19}$$

The stabilization term is defined similar to one part used for the Navier-Stokes equations (8):

$$s(u_h, \phi) := (\pi_h[(\beta\cdot\nabla)u_h],\ \delta\pi_h[(\beta\cdot\nabla)\phi]).$$

From the following stability result we obtain existence and uniqueness of the discrete solution:

Lemma 1. *We have*

$$a(u_h, u_h) + s(u_h, u_h) = |||u_h|||^2, \tag{20}$$

with

$$|||u_h||| := \left(\|u_h\|^2 + \frac{1}{2}\|\sqrt{|\beta\cdot n|}\,u_h\|^2_{\partial\Omega} + \|\sqrt{\delta}\,\pi_h[(\beta\cdot\nabla)u_h]\|^2 \right)^{1/2}. \tag{21}$$

Proof. Follows from integration by parts.

In the following we give an error estimate. The proof is very similar to the classical one in [11] for the same equation supplied with SUPG stabilization. However, notice that in contrast to the proof for SUPG, we only have control over the streamline derivative of the *fluctuations* $\pi_h[(\beta\cdot\nabla)u_h]$ in (21).

Theorem 2. *Let u be the continuous solution to* (17) *satisfying $u \in H^2(\Omega)$ and u_h the discrete solution of* (19). *Then we have the following estimate:*

$$\|u - u_h\| \le Ch^{3/2}\|u\|_{H^2(\Omega)}. \tag{22}$$

The estimate is similar to the standard estimate for SUPG or the discontinuous Galerkin method, [11]. With respect to the interpolation error we loose a power of 1/2.

Proof. By $j_h : V \to V_h$ we denote the modified Scott-Zhang interpolation operator introduced in [2] which has the following orthogonality property:

$$(u - j_h u, \phi) = 0, \qquad \forall \phi \in W_{2h}\,, \tag{23}$$

and allows for optimal interpolation in $L^2(\Omega)$ and $H^1(\Omega)$. That is, there exist a constant C such that

$$\|u - j_h u\| \le Ch^2\|u\|_{H^2(\Omega)},$$
$$\|\nabla(u - j_h u)\| \le Ch\|u\|_{H^2(\Omega)}.$$

We split the error as $u - u_h = \eta - \xi$, $\xi := u_h - j_h u$, $\eta := u - j_h u$. Due to this interpolation result, it is sufficient to show

$$\|\xi\| \leq C h^{3/2} \|u\|_{H^2(\Omega)} \tag{24}$$

with (another) constant C for proving the assertion. We have:

$$\begin{aligned} |||\xi|||^2 &= a(u_h, \xi) + s(u_h, \xi) - a(j_h u, \xi) - s(j_h u, \xi) \\ &= l(\xi) - a(j_h u, \xi) - s(j_h u, \xi) \\ &= a(u, \xi) - a(j_h u, \xi) - s(j_h u, \xi) \\ &= a(\eta, \xi) - s(j_h u, \xi) \\ &= (\eta, \xi) + ((\beta \cdot \nabla)\eta, \xi) - \langle (\beta \cdot n)_- \eta, \xi \rangle - s(j_h u, \xi)\,. \end{aligned}$$

The only critical terms are $((\beta \cdot \nabla)\eta, \xi)$ and $s(j_h u, \xi)$. We use partial integration to obtain

$$\begin{aligned} ((\beta \cdot \nabla)\eta, \xi) &= -(\eta, (\beta \cdot \nabla)\xi) + \langle (\beta \cdot n)\eta, \xi \rangle \\ &= -(\eta, \pi_h(\beta \cdot \nabla)\xi) + \langle (\beta \cdot n)\eta, \xi \rangle\,. \end{aligned}$$

In the last line, we have used the orthogonality property (23) of the interpolation operator j_h. Furthermore, the stabilization term can be bounded as follows:

$$\begin{aligned} |s(j_h u, \xi)| &\leq \sum_{K \in \mathcal{T}_{2h}} h_K^{1/2} \|\pi_h(\beta \cdot \nabla) j_h u\|_K \sqrt{\delta_P} \|\pi_h(\beta \cdot \nabla)\xi\|_K \\ &\leq h^{1/2} \left(\sum_{K \in \mathcal{T}_{2h}} \|\pi_h(\beta \cdot \nabla) j_h u\|_K \right) |||\xi||| \\ &\leq h^{3/2} \|u\|_{H^2(\Omega)} |||\xi||| \end{aligned}$$

Here, the last line is obtained by stability of the L^2 projection P_W and the interpolation property of j_h:

$$\begin{aligned} &\sum_{K \in \mathcal{T}_{2h}} \|\pi_h(\beta \cdot \nabla) j_h u\|_K \leq \sum_{K \in \mathcal{T}_{2h}} \left(\|\pi_h(\beta \cdot \nabla) u\|_K + \|\pi_h(\beta \cdot \nabla)(j_h u - u)\|_K \right) \\ &\leq h\|u\|_{H^2(\Omega)} + \sum_{K \in \mathcal{T}_{2h}} \left(\|(\beta \cdot \nabla)(j_h u - u)\|_K + \|P_W(\beta \cdot \nabla)(j_h u - u)\|_K \right) \\ &\leq Ch\|u\|_{H^2(\Omega)}\,. \end{aligned}$$

Now we get:

$$\begin{aligned} |||\xi|||^2 &= (\eta, \xi) - (\eta, \pi_h[(\beta \cdot \nabla)\xi]) + \langle (\beta \cdot n)_+ \eta, \xi \rangle - s(j_h u, \xi) \\ &\leq \|\eta\|\,\|\xi\| + \|\sqrt{\delta^{-1}}\,\eta\|\,\|\sqrt{\delta}\pi_h[(\beta \cdot \nabla)\xi]\| \\ &\quad + \|\sqrt{(\beta \cdot n)_+}\,\eta\|_{\partial\Omega} \|\sqrt{(\beta \cdot n)_+}\,\xi\|_{\partial\Omega} + Ch^{3/2}\|u\|_{H^2(\Omega)} |||\xi||| \\ &\leq \left(\|\eta\| + \|\sqrt{\delta^{-1}}\,\eta\| + \|\sqrt{(\beta \cdot n)_+}\,\eta\|_{\partial\Omega} + Ch^{3/2}\|u\|_{H^2(\Omega)} \right) |||\xi|||\,, \end{aligned}$$

and hence by the trace theorem:

$$
\begin{aligned}
|||\xi||| &\le \|\eta\| + \|\sqrt{\delta^{-1}}\,\eta\| + \|\sqrt{(\beta\cdot n)_+}\,\eta\|_{\partial\Omega} + Ch^{3/2}\|u\|_{H^2(\Omega)} \\
&\le Ch^{-1/2}\|\eta\| + Ch^{3/2}\|u\|_{H^2(\Omega)} \\
&\le Ch^{3/2}\|u\|_{H^2(\Omega)} ,
\end{aligned}
$$

which shows (24) since $\|\xi\| \le |||\xi|||$.

References

1. R. Becker. Adaptive Finite Elements for optimal control problems. Habilitationsschrift, Institut für Angewandte Mathematik, Universität Heidelberg, 2001.
2. R. Becker and M. Braack. A finite element pressure gradient stabilization for the Stokes equations based on local projections. *Calcolo*, 38(4):173–199, 2001.
3. M. Braack. An Adaptive Finite Element Method for Reactive Flow Problems. PhD Dissertation, Universität Heidelberg, 1998.
4. R. Codina and J. Blasco. A finite element formulation for the Stokes problem allowing equal velocity-pressure interpolation. *Comput. Methods Appl. Mech. Engrg.*, 143:373–391, 1997.
5. P. Hansbo E. Burman. Edge stabilization for galerkin approximations of the generalized stokes' problem. submitted to M2AN, 2004.
6. L.P. Franca and S.L. Frey. Stabilized finite element methods: II. The incompressible Navier-Stokes equations. *Comput. Methods Appl. Mech. Engrg.*, 99:209–233, 1992.
7. V. Girault and P.-A. Raviart. *Finite Elements for the Navier Stokes Equations.* Springer, Berlin, 1986.
8. J.-L. Guermond. Stabilization of Galerkin approximations of transport equations by subgrid modeling. *Modél. Math. Anal. Numér.*, 33(6):1293–1316, 1999.
9. P. Hansbo and A. Szepessy. A velocity-pressure streamline diffusion finite element method for the incompressible Navier-Stokes equations. *Comput. Methods Appl. Mech. Engrg.*, 84:175–192, 1990.
10. T.J.R. Hughes, L.P. Franca, and M. Balestra. A new finite element formulation for computational fluid dynamics: V. circumvent the Babuska-Brezzi condition: A stable Petrov-Galerkin formulation for the Stokes problem accommodating equal order interpolation. *Comput. Methods Appl. Mech. Engrg.*, 59:89–99, 1986.
11. C. Johnson. *Numerical Solution of Partial Differential Equations by the Finite Element Method.* Cambridge University Press, Cambridge, UK, 1987.
12. M. Braack R. Becker. A two-level finite element stabilization for Navier-Stokes. in preparation, 2004.
13. M.A. Olshanskii T. Gelhard, G. Lube. Stabilized finite element schemes with LBB-stable elements for incompressible flows. submitted, 2003.
14. L. Tobiska and R. Verfürth. Analysis of a streamline diffusion finite element method for the Stokes and Navier-Stokes equations. *SIAM J. Numer. Anal.*, 33(1):107–127, 1996.

A Posteriori Error Estimates for Parameter Identification

Roland Becker[1] and Boris Vexler[2]

[1] Laboratoire de Mathematiques Appliquees, Université de Pau et des Pays de l'Adour,
BP 1155, 64013 Pau Cedex, France
roland.becker@univ-pau.fr

[2] Institut fü Angewandte Mathematatik, Universität Heidelberg, Im Neuenheimer Feld 294,
69120 Heidelberg, Germany
boris.vexler@iwr.uni-heidelberg.de

Summary. In this paper we present an a posteriori error estimator for parameter identification problems governed by partial differential equations. This estimator aims to control the error in parameters due to the discretization by finite elements. It is used in an adaptive mesh refinement algorithm generating a sequence of locally refined meshes for efficient computation of the parameters. Comparison with some heuristic mesh refinement algorithms is done for a simple example inverse problem.

1 Introduction

We consider parameter identification problems involving a finite number of unknown parameters in the following form: The *state variable* u in an appropriate Hilbert space V is determined by a partial differential equation (*state equation*) in weak form:

$$a(u,q)(\phi) = f(\phi) \quad \forall \phi \in V, \tag{1}$$

where $q \in Q = \mathbb{R}^{n_p}$ denotes the unknown parameters. The form a is defined on the Hilbert space $V \times Q \times V$ and the linear functional $f \in V'$ represents the right hand side of the state equation, where V' denotes the dual space of V. Further, we are given an observation operator $C : V \to Z$, which maps the state variable u to the space of measurements $Z = \mathbb{R}^{n_m}$, where we assume $n_m \geq n_p$. We denote by $\langle \cdot, \cdot \rangle_Z$ the scalar product of Z and by $\| \cdot \|_Z$ the corresponding norm. Similar notations are used for the scalar product and norm in the space Q.

The values of the parameters are estimated from a given set of measurements $\bar{C} \in Z$ using a least squares approach such that we obtain a constrained optimization problem with the cost functional $J : V \to \mathbb{R}$:

$$\text{Minimize} \quad J(u) := \frac{1}{2}\|C(u) - \bar{C}\|_Z^2 \tag{2}$$

under the constraint (1). Here, the cost functional is the squared norm of the residual R^{LS} defined by

$$R^{LS}(u) := \bar{C} - C(u). \tag{3}$$

The state equation is discretized by conforming finite elements on a regular mesh $\mathcal{T}_h$, resulting in a finite element space $V_h \subset V$, see e.g. Ciarlet [8] for the standard construction. In order to ease mesh refinement, the cells are allowed to have nodes, which lie on midpoints of faces of neighboring cells. But at most one such *hanging node* is permitted for each face, see Carey & Oden [7] for implementation details.

The discrete state $u_h \in V_h$ and parameter $q_h \in Q$ are determined by:

$$\text{Minimize} \quad J(u_h) \tag{4}$$

under the constraint

$$a(u_h, q_h)(\phi_h) = f(\phi_h) \quad \forall \phi_h \in V_h. \tag{5}$$

Due to the finite dimension of Q, we suppose the parameter q_h in (4) to be sought in the space Q.

The paper is organized as follows: In the next section we describe an optimization algorithm for solving the problem (4, 5) on a fixed mesh $\mathcal{T}_h$. Section 3 is devoted to a posteriori error estimation. Here, we present an a posteriori error estimator for the error in parameter $E(q) - E(q_h)$ for a given error functional $E : Q \to \mathbb{R}$. This error estimator is developed in [5]. It is based on the optimal control approach to a posteriori error estimation from Becker & Rannacher [4]. However, a direct application of the techniques described in [4] leads to an estimator which controls the error in the cost functional J (2). In general, such an estimator does not provide useful error bounds for the parameters, in contrast to the approach presented here. In Section 4 we discuss a numerical example illustrating the usage of the error estimator. The presented approach is compared with some heuristic methods with respect to the quality of generated meshes. Conclusions are given in the last section.

2 Optimization algorithm

In this section we discuss an optimization algorithm for solving the problem (4, 5) on a fixed mesh $\mathcal{T}_h$.

Under the assumption of regularity of the partial derivative a'_u, the implicit function theorem in Banach spaces implies the existence of an open set $Q_0 \subset Q$, containing the optimal parameter q, and a continuously differentiable solution operator $S : Q_0 \to V$, $q \to S(q)$, so that (1) is fulfilled for $u = S(q)$. This allows us to reformulate the problem (1, 2) as an unconstrained optimization problem:

$$\text{Minimize} \quad j(q) := \frac{1}{2}\|c(q) - \overline{C}\|_Z^2, \quad q \in Q, \tag{6}$$

where the reduced observation operator c is given by $c(q) = C(S(q))$. Denoting by $G = c'(q)$ the Jacobian matrix of the reduced observation operator c, the first-order necessary condition $j'(q) = 0$ for (6) reads:

$$G^*(c(q) - \overline{C}) = 0\,, \tag{7}$$

where G^* denotes the transpose of G. The Jacobian matrix G can be obtained using tangent solutions $w_j \in V$ determined by:

$$a'_u(u,q)(w_j,\phi) = -a'_{q_j}(u,q)(1,\phi) \quad \forall \phi \in V, \quad j = 1 \ldots n_p. \tag{8}$$

Then, one simply proves that the entries of the matrix G are given by $G_{ij} = C'_i(u)(w_j)$.

Similarly to the continuous case, we introduce a discrete solution operator $S_h : Q \to V_h$ for the discretized state equation (5) and obtain an unconstrained formulation of the discretized problem (4, 5) by:

$$\text{Minimize} \quad j_h(q_h) := \frac{1}{2}\|c_h(q_h) - \overline{C}\|_Z^2, \quad q_h \in Q, \tag{9}$$

where c_h is the discrete reduced observation operator defined by $c_h(q_h) = C(S_h(q_h))$. The corresponding Jacobian matrix G_h can be computed similarly to the continuous case using the discrete tangent solution $w_{j,h} \in V_h$ determined by the discrete version of (8).

The problem (9) is solved iteratively starting with an initial guess q_h^0 and using the recursive setting $q_h^{k+1} = q_h^k + \delta q_h$. The update δq_h is obtained using a symmetric approximation H_k of the hessian $\nabla^2 j_h(q_h^k)$ as the solution of the system of linear equations:

$$H_k\,\delta q_h = G_h^*(\bar{C} - c_h(q_h^k)), \tag{10}$$

where $G_h = c'_h(q_h^k)$. The most widely used choice of the matrix $H_k = G_h^*\,G_h$ leads to the Gauss-Newton algorithm, see e.g. Nocedal & Wright [10].

For one step of the Gauss-Newton algorithm the state equation and n_p tangent problems (8) have to be solved which originate from the same linear operator but with different right-hand sides. Due to the small dimension n_p of the parameter space Q the solution of (10) is uncritical.

3 A posteriori error estimation

In this section we present an error estimator for the error with respect to a given error functional $E : Q \to \mathbb{R}$. The precise error representation is given in the following theorem. Here, we use an interpolation operator $i_h : V \to V_h$, see e.g. Clément [9].

Theorem 1. *Let (u, q) be a solution of the parameter identification problem (1, 2) and (u_h, q_h) the corresponding discrete solution of the problem (4, 5). Then, for a given error functional E, we have that:*

$$E(q) - E(q_h) = \frac{1}{2}\rho(u_h)(y - i_h y) + \frac{1}{2}\rho^*(u_h, y_h)(u - i_h u) + P + R, \tag{11}$$

where $y \in V$ is the solution of the adjoint problem:

$$a'_u(u, q)(\phi, y) = -\langle G(G^*G)^{-1}\nabla E(q), C'(u)(\phi)\rangle \quad \forall \phi \in V \tag{12}$$

and $\rho(\cdot)(\cdot)$ and $\rho^*(\cdot)(\cdot)$ are the residuals of the state and adjoint equation defined by:

$$\begin{aligned} \rho(u_h)(\phi) &:= f(\phi) - a(u_h, q_h)(\phi) \\ \rho^*(u_h, y_h)(\phi) &:= -\langle G_h(G_h^*G_h)^{-1}\nabla E(q_h), C'(u_h)(\phi)\rangle - a'_u(u_h, q_h)(\phi, y_h). \end{aligned} \tag{13}$$

The remainder term R (due to linearization) is quadratic in the error and the additional remainder term P admits the estimate:

$$|P| \le \widetilde{C}\left(\|e_u\|_V + \|e_q\|_Q + \|\delta_h v\|_V + \|\delta_h \bar{z}\|_V\right)\|R^{LS}(u)\|_Z, \tag{14}$$

where $e_u := u - u_h$, $e_q := q - q_h$ and $\delta_h \phi := \phi - i_h \phi$ is an interpolation error operator. The mean tangent solution $v \in V$ is given by

$$v = -\sum_{j=1}^{n_p}\left((G^*G)^{-1}\nabla E(q)\right)_j w_j \tag{15}$$

and the normalized adjoint solution $\bar{z} \in V$ is determined by:

$$a'_u(u, q)(\phi, \bar{z}) = \langle -\frac{R^{LS}(u)}{\|R^{LS}(u)\|}, C'(u)(\phi)\rangle_Z \quad \forall \phi \in V, \tag{16}$$

if the least squares residual $R^{LS}(u)$ does not vanish; otherwise we set $\bar{z} = 0$. The constant $\widetilde{C}$ does not depend on the mesh parameter h nor on the measurements $\bar{C}$.

Proof. For proof we refer to [5].

For evaluation of the error estimator, denoted by η_h, the local interpolation errors $y - i_h y$ and $u - i_h u$ have to be approximated. In our numerical examples, we use interpolation of the computed bilinear finite element solutions y_h and u_h on the space of biquadratic finite elements on patches of cells. The main computational cost for the a posteriori error estimator described above is the solution of one auxiliary equation (12). This is cheap, even in comparison with only one Gauss-Newton step, which includes solution of the state (nonlinear) and of the several (linear) tangent equations.

In order to illustrate the typical use of the error estimator η_h, we sketch a generic adaptive mesh refinement algorithm. Such an algorithm generates a sequence of locally refined meshes and corresponding finite element spaces until the estimated error with respect to E is below a given tolerance TOL.

Adaptive Mesh Refinement Algorithm

1. Choose an initial mesh $\mathcal{T}_{h_0}$ and set $k = 0$
2. Construct the finite element space V_{h_k}
3. Compute $u_{h_k} \in V_{h_k}, q_{h_k} \in Q$ solving (4,5)
4. Evaluate the a posteriori error estimator η_{h_k}
5. If $\eta_{h_k} \leq TOL$ quit
6. Refine $\mathcal{T}_{h_k} \to \mathcal{T}_{h_{k+1}}$ using information from η_{h_k}
7. Increment k and go to 2.

Remark 1. In step 3, the parameter identification problem is solved on a fixed mesh. As initial data, we use the values from the computation on the previous mesh. This allows us to avoid unnecessary iterations of the optimization loop on fine meshes.

4 Numerical result

In this section we compare our general approach to mesh refinement for parameter identification problems with some heuristic methods. We consider three types of heuristic approaches for mesh refinement: a strategy based only on the information obtained from the computed state variable, a strategy based only on the a priori knowledge of the structure of the observation operator and a strategy, which combines both types of information.

We consider the following diffusion-reaction equation with unknown coefficient q in the unit square $\Omega = (0,1)^2$:

$$\begin{aligned} -q\Delta u + su &= 2 \qquad \text{in } \Omega, \\ u &= 0 \qquad \text{on } \partial\Omega, \end{aligned} \tag{17}$$

where s is chosen as $s = 200$. The parameter q is estimated using measurements given by the values of the state variable at nine different points ξ_i, see Figure 1. The exact value of the parameter is $q = 1$.

The components of the corresponding observation operator C have the following form:

$$C_i(v) = v(\xi_i), \tag{18}$$

and the parameter identification problem is formulated as follows: For $(u,q) \in V \times Q$ with $V = H_0^1(\Omega)$ and $Q = \mathbb{R}$

$$\text{Minimize } \tfrac{1}{2}\sum_{i=1}^{9}(u(\xi_i) - \bar{C}_i)^2 \tag{19}$$

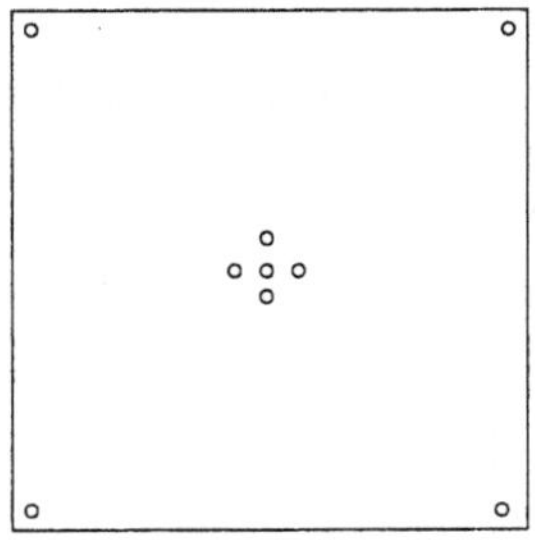

Fig. 1. The computational domain with measurement points marked by circles

under the constraint (17), where $\bar{C}_i$ denote the components of the measurement vector $\bar{C} \in Z = \mathbb{R}^9$ and are given by the values of the state variable u for the exact parameter q, i.e. $\bar{C}_i = u(\xi_i)$.

First, we compare the quality of meshes generated by our a posteriori error estimator for this problem with a typical strategy based on a posteriori information obtained by the state variable, i.e with the mesh refinement guided by one of the well-known "energy" type error estimators for uncontrolled equation, see e.g. Bank & Weiser [2] and Babuška & Miller [1]. This estimator aims to control the error $u - u_h$ in H^1-norm, but they do not take care of the structure of the parameter identification problem. As seen from Figure 2, adaptive refinement based on the "energy" estimator leads to a similar reduction of the error as global refinement. However, the strategy based on our error estimator leads to an obvious saving in the number of unknowns necessary to achieve a prescribed accuracy level.

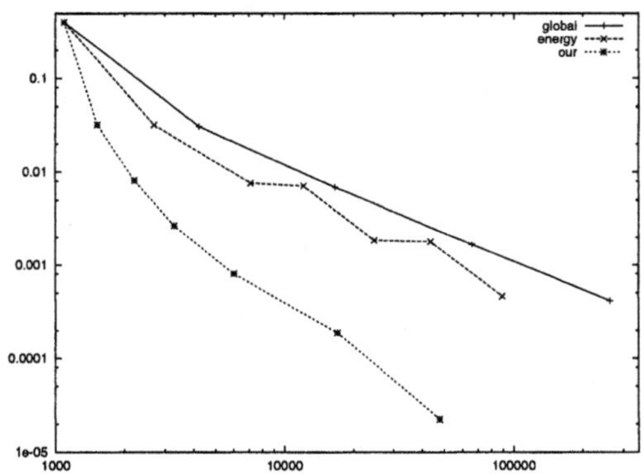

Fig. 2. Errors in q for different refinement strategies vs. number of nodes (global refinement, "energy"-based refinement and refinement resulting from our a posteriori error estimator)

Next, we compare our strategy for mesh refinement with the following heuristic approach: In each iteration of the mesh refinement we refine the cells, which lie close to one of the measurement points, i.e. in the set

$$\bigcup_i \{x \mid \|x - \xi_i\| \leq r\}, \tag{20}$$

where $r \in \mathbb{R}_+$ is a given number. In contrast to our approach, this strategy is unable to weight the relative importance of the measurement points. The corresponding comparison is done in Figure 3 for two choices of r ($r = 0.04$ and $r = 0.1$).

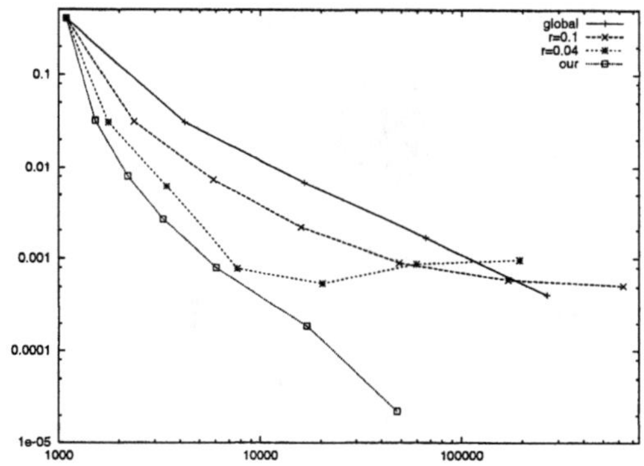

Fig. 3. Errors in q for different refinement strategies vs. number of nodes (global refinement, refinement across measurement points and refinement resulting from our a posteriori error estimator)

After several steps one does not observe any error reduction despite increasing the number of nodes for both choices of r in the described strategy, as could be expected. Typical meshes resulting from application of our a posteriori error estimator, "energy" estimator and the last strategy are shown in Figure 4.

We also compare our mesh refinement procedure with a combination of the last heuristic methods. By this strategy both the cells marked by the "energy" estimator and the cells across the measurement points (20) are refined. The corresponding comparison with our mesh refinement procedure is made in Figure 5 for two choices of r ($r = 0.04$ and $r = 0.1$).

The typical meshes resulting from application of this strategy for two choices of r ($r = 0.04$ and $r = 0.1$) are shown in Figure 6.

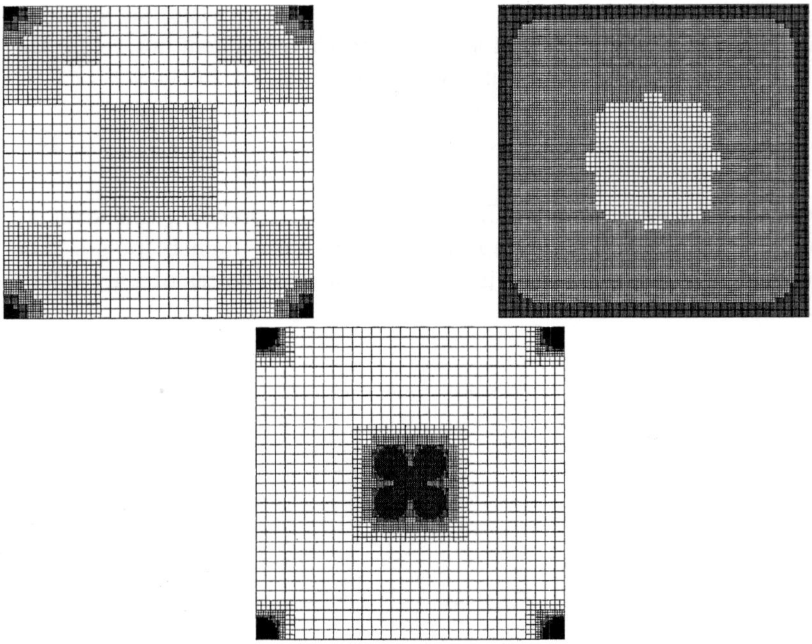

Fig. 4. Typical meshes produced by our a posteriori error estimator (left), energy error estimator (right) and the refinement across the measurement points

5 Conclusions

We presented an a posteriori error estimator for finite element discretization of parameter identification problems. This error estimator is cheap to evaluate

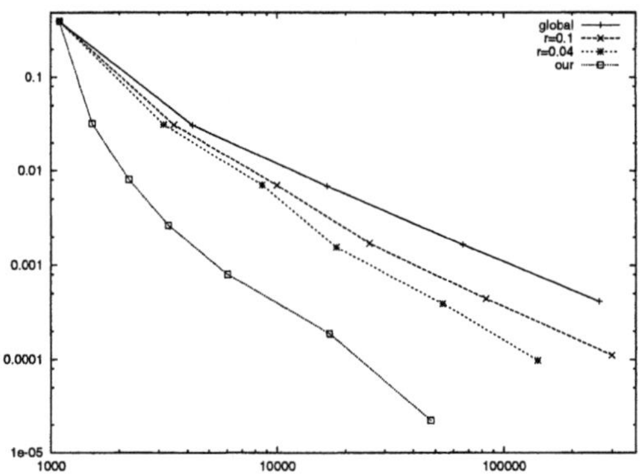

Fig. 5. Errors in q for different refinement strategies vs. number of nodes (global refinement, refinement produced by combing the refinement across measurement points and "energy"-based refinement, and refinement resulting from our a posteriori error estimator)

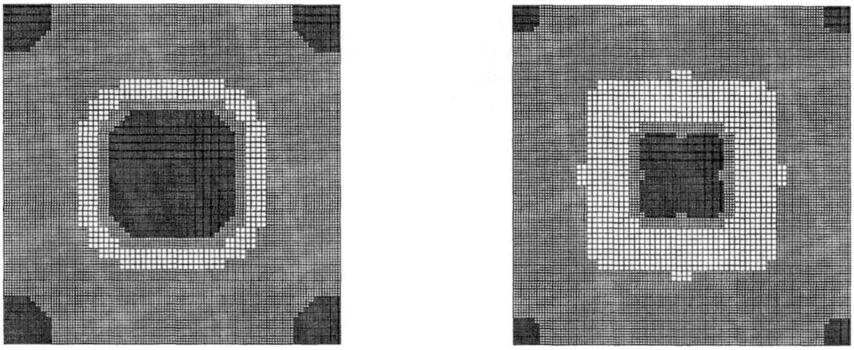

Fig. 6. Typical meshes produced by combining the refinement according to "energy" error estimator and the refinement across the measurement points for $r = 0.1$ (left) and $r = 0.04$ (right)

and assess the error we are interested in, i.e. the error in parameters. We compared our approach with some heuristic methods with respect to the quality of the generated meshes. The presented error estimator is successfully applied to parameter identification in CFD problems, see Becker & Vexler [6] and to estimation of chemical models in multidimensional reactive flows, see Becker, Vexler & Braack [3].

References

1. Babuška, I., Miller, A.D. (1987): A feedback finite element method with a posteriori error estimation. Comput. Methods Appl. Mech. Engrg, 61:1-40
2. Bank, R.E., Weiser, A. (1985): Some a posteriori error estimators for elliptic partial differential equations. Math. Comp., 44:283-301
3. Becker, R., Braack, M., Vexler, B. (2003): Numerical parameter estimation for chemical models in multidimensional reactive flows. Combustion Theory and Modelling, submitted
4. Becker, R., Rannacher, R. (2001): An optimal control approach to a posteriori error estimation in finite element methods. In Acta Numerica 2001 (A. Iserles, ed.), Cambridge University Press, Cambridge
5. Becker, R., Vexler, B. (2003): A posteriori error estimation for finite element discretization of parameter identification problems. Numerische Mathematik, published online
6. Becker, R., Vexler, B. (2003): Calibration of PDE Models and Sensitivity Analysis with Adaptive Finite Elements: Application to CFD Problems. Journal of Computational Physics, submitted
7. Carey, C.F., Oden J.T. (1984): Finite Elements, Computational Aspects. Vol. III. Prentice-Hall
8. Ciarlet, P.G. (1978): The Finite Element Method for Elliptic Problems. North-Holland Publishing Company, Amsterdam

9. Clément, Ph. (1975): Approximation by finite element functions using local regularization. [J] Revue Franc. Automat. Inform. Rech. Operat. 9(R-2), 77–84
10. Nocedal, J., Wright, S.J. (1999): Numerical Optimization. Springer Series in Operations Research, Springer New York

On a Phase-Field Model with Advection

Michal Beneš

Department of Mathematics, Faculty of Nuclear Sciences and Physical Engineering, Czech Technical University in Prague, Trojanova 13, 120 00 Prague 2, Czech Republic *Michal.Benes@fjfi.cvut.cz*

Summary. In this contribution, we present a phase-field model of advected mean-curvature flow and of advected pattern formation in solidification. The model is based on the approach presented in [3], where an extensive literature list on methods treating mean-curvature problems can be found. The model represents a step towards simulation of solidification processes where the melt motion is important. We give a basic mathematical information concerning the weak solution of the model equations, introduce a numerical scheme based on the Finite-Difference Method, and show several numerical studies demostrating basic qualitative effects of advection in the given context.

Mean curvature flow with advection. The problem of mean curvature flow of hypersurfaces (see [7, 6, 5]) is usually set as follows:

$$\textit{normal velocity} = -\textit{mean curvature} + \textit{forcing},$$

in the normal direction to a closed hypersurface Γ. In this work, we consider the law modified by an imposed velocity field $\mathbf{V}$, which means that the hypersurface is advected by the vector field, as well. Using the notations $\mathbf{n}_\Gamma$ for the Euclidean normal vector to Γ, v_Γ for the normal velocity, κ_Γ for the mean curvature, and F for the forcing term, we formulate the motion law for Γ as follows

$$v_\Gamma - \mathbf{V}\cdot\mathbf{n}_\Gamma = -\kappa_\Gamma + F. \tag{1}$$

The equation (1) has origin in the modified Stefan problem with advection describing the solidification of crystalline materials where the bulk mixture of liquid and solid is carried by an imposed vector field $\mathbf{V}$:

$$\begin{aligned} \frac{\mathrm{d}_m u}{\mathrm{d}t} &= \Delta u \text{ in } \Omega_s \text{ and } \Omega_l, \qquad (2)\\ \left.\frac{\partial u}{\partial n_\Gamma}\right|_s - \left.\frac{\partial u}{\partial n_\Gamma}\right|_l &= L(v_\Gamma - \mathbf{V}\cdot\mathbf{n}_\Gamma) \text{ on } \Gamma(t),\\ F(u) &= \kappa_\Gamma + \alpha(v_\Gamma - \mathbf{V}\cdot\mathbf{n}_\Gamma) \text{ on } \Gamma(t), \end{aligned}$$

where u denotes the temperature field, u^* the temperature of melting point, Ω_s, Ω_l the solid and liquid subdomains of Ω, L the latent heat per unit volume, α a material parameter, $F(u)$ a coupling term ($\cong u - u^*$), and $\frac{\mathrm{d}_m}{\mathrm{d}t} = \frac{\partial}{\partial t} + \mathbf{V}\cdot\nabla$ the material derivative.

For details of physical context, we refer the reader to [8, 10]. Obviously, the above given physical problem simplifies the problem of density driven flow around the solidifying structure in a real material. In this general case, the law of momentum conservation would be needed, and the boundary conditions should also correspond to the conservation of quantities in question. Our purpose is to study the problem (2) by a diffuse-interface approach which allows us to design a suitable numerical algorithm and to perform qualitative numerical studies.

Allen-Cahn equation with advection. The law (1) can be treated by a particular method of levelset type (see [3]), which relies on solution properties of a reaction-diffusion equation of Allen-Cahn type (see [9]). We refer the reader to [3], a sample of literature resources on this topic. Introducing the thickness $\xi > 0$ of the diffuse layer surrounding Γ and the polynomial $f_0(s) = ap(1-p)(p-\frac{1}{2})$ as the minus derivative of a double-well function w_0, we present **the phase-field approximation of the advected mean curvature flow** as follows

$$\alpha\xi^2 \frac{\mathrm{d}_m p}{\mathrm{d}t} = \xi^2 \Delta p + f_0(p) + F\xi^2|\nabla p|. \tag{3}$$

In Figure 1, we illustrate the expected fact, that the $\frac{1}{2}$-levelset of a solution to (3) converges to the set $\Gamma(t)$ evolved by (1) - see [2] for the no-advection case.

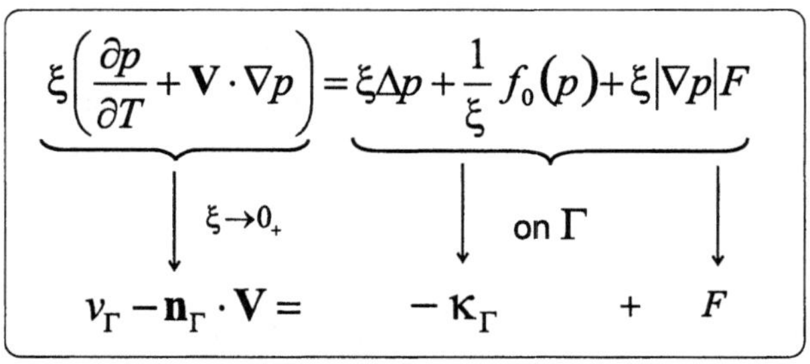

Fig. 1. Schematic relationship to the original motion law

Phase-field equations with advection. The above indicated approach can be used to approximate (or regularize) the physical problem (2). In this case, we obtain a complete system of phase-field equations with advection reading as

$$\begin{aligned} \frac{\mathrm{d}_m u}{\mathrm{d}t} &= \Delta u + L\chi'(p)\frac{\mathrm{d}_m p}{\mathrm{d}t}, \\ \alpha\xi \frac{\mathrm{d}_m p}{\mathrm{d}t} &= \xi \Delta p + \frac{1}{\xi} f_0(p) + F(u)\xi|\nabla p|, \end{aligned} \tag{4}$$

with the initial conditions $u\ |_{t=0} = u_{ini}$, $p\ |_{t=0} = p_{ini}$, and with the homogeneous Dirichlet boundary conditions (set for the sake of simplicity). Additionally, we assume that $F(u)$ is a bounded Lipschitz-continuous function.

The enthalpy of the system $\mathcal{H}(u) = u - L\chi(p)$ is expressed by means of a focusing function χ, which is monotone with bounded, Lipschitz-continuous derivative:

$$\chi(0) = 0,\ \chi(0.5) = 0.5,\ \chi(1) = 1,\ supp(\chi') \subset \langle 0, 1\rangle.$$

In the following theorem, we set $\chi(p) = p$ for simplicity, although the computations are performed for a nontrivial χ. The general case is investigated in [2]. Considering a bounded domain $\Omega \subset \mathbb{R}^2$ with $\mathcal{C}^2$ boundary, we can state the following basic property of the system (4):

Theorem 1. *If $u_{ini}, p_{ini} \in \mathrm{H}_0^1(\Omega)$, $\mathbf{V} \in \mathrm{L}_\infty(\Omega; \mathbb{R}^2)$, and ξ remains fixed, then there is a unique solution $u, p \in \mathrm{L}_2(0, T; \mathrm{H}_0^1(\Omega))$ of the weak problem*

$$\begin{aligned}
&\forall v, q \in \mathcal{D}(\Omega),\ a.e.\ in\ (0, T):\\
&\frac{\mathrm{d}}{\mathrm{d}t}(u - Lp, v) + (\mathbf{V} \cdot \nabla(u - Lp), v) + (\nabla u, \nabla v) = 0, \qquad (5)\\
&\alpha\xi^2 \left(\frac{\mathrm{d}}{\mathrm{d}t}(p, q) + (\mathbf{V} \cdot \nabla p, v)\right) + \xi^2(\nabla p, \nabla q) = (f_0(p), q) + \xi^2(F(u)|\nabla p|, q),\\
&\quad u(0) = u_0,\ p(0) = p_0,
\end{aligned}$$

for which

$$p \in \mathrm{L}_2(0, T; \mathrm{H}^2(\Omega) \cap \mathrm{H}_0^1(\Omega)),\ \frac{\partial u}{\partial t}, \frac{\partial p}{\partial t} \in \mathrm{L}_2(0, T; \mathrm{L}_2(\Omega)).$$

Proof is an extension of the result stated in [2]. We concentrate ourselves on issues closely related to the advection terms in both equations. By means of a total set in $\mathrm{L}_2(\Omega)$ (e.g., consisting of eigenvectors of $-\Delta$) denoted as $\{v_i\}_{i\in\mathbb{N}}$, we define a finite-dimensional subspace

$$V_m = span\{v_i\}_{i\in\mathbb{N}_m} \text{ where } \mathbb{N}_m = [1, \dots, m],$$

and consider the projector $\mathcal{P}_m : \mathrm{L}_2(\Omega) \to V_m$. By means of the Faedo-Galerkin method, we derive a semi-discrete scheme $\forall v, q \in V_m$

$$\begin{aligned}
&(\partial_t(u^m - Lp^m), v) + (\mathbf{V} \cdot \nabla(u^m - Lp^m), v) + (\nabla u^m, \nabla v) = 0 \text{ a.e. in } (0, T),\\
&\alpha\xi^2 ((\partial_t p^m, q) + (\mathbf{V} \cdot \nabla p^m, v)) + \xi^2(\nabla p^m, \nabla q) \qquad (6)\\
&\qquad\qquad = (f_0(p^m), q) + \xi^2(F(u^m)|\nabla p^m|, q),\\
&\quad u^m(0) = \mathcal{P}_m u_{ini}, \quad p^m(0) = \mathcal{P}_m p_{ini}.
\end{aligned}$$

The approximate solution is given by basis functions of V_m

$$u^m(t) = \sum_{i\in\mathbb{N}_m} \gamma_i^m(t) v_i,\ p^m(t) = \sum_{i\in\mathbb{N}_m} \beta_i^m(t) v_i.$$

In fact, the semi-discrete scheme is a system of ODEs for unknown functions of time: γ_i^m, β_i^m . Next step is derivation of energy estimates. We multiply the equations by $\dot{\gamma}_i^m(t)$, $\dot{\beta}_i^m(t)$, respectively, and sum over $i \in \mathbb{N}_m$.

$$\|\partial_t u^m\|^2 + \frac{1}{2}\frac{\mathrm{d}}{\mathrm{d}t}\|\nabla u^m\|^2 + (\mathbf{V}\cdot\nabla(u^m - Lp^m), \partial_t u^m) = L(\partial_t p^m, \partial_t u^m),$$

$$\alpha\xi^2\|\partial_t p^m\|^2 + \frac{\xi^2}{2}\frac{\mathrm{d}}{\mathrm{d}t}\|\nabla p^m\|^2 + \alpha\xi^2(\mathbf{V}\cdot\nabla p^m, \partial_t p^m)$$
$$= (f_0(p^m), \partial_t p^m) + \xi^2(F(u^m)|\nabla p^m|, \partial_t p^m).$$

Consequently, we obtain the inequalities

$$\|\partial_t u^m\|^2 + \frac{\mathrm{d}}{\mathrm{d}t}\|\nabla u^m\|^2 \leq 3L^2C_\chi^2\|\partial_t p^m\|^2 + 3V^2\|\nabla u^m\|^2 + 3V^2L^2C_\chi^2\|\nabla p^m\|^2,$$

$$\frac{1}{2}\alpha\xi^2\|\partial_t p^m\|^2 + \frac{\xi^2}{2}\frac{\mathrm{d}}{\mathrm{d}t}\|\nabla p^m\|^2 + \frac{\mathrm{d}}{\mathrm{d}t}(w_0(p^m), 1) \leq \left(\frac{C_F^2}{\alpha}\xi^2 + \alpha\xi^2V^2\right)\|\nabla p^m\|^2,$$

where $w_0' = -f_0$, $|F(u)| \leq C_F$, $V = \|\mathbf{V}\|_\infty$. They allow us to use the compactness method in the same manner as it is presented in [2]. Namely, the theorem assumptions together with the above estimates processed by the Gronwall lemma give that, independently of m, ∇u^m, ∇p^m are bounded in $\mathrm{L}_\infty(0,T;\mathrm{L}_2(\Omega))$, and p^m are bounded in $\mathrm{L}_\infty(0,T;\mathrm{L}_s(\Omega))$ for each finite time $T > 0$, and for any $s \geq 1$. Repeated integration says that $\frac{\partial u^m}{\partial t}$, $\frac{\partial p^m}{\partial t}$ are bounded in $\mathrm{L}_2(0,T;\mathrm{L}_2(\Omega))$ for each finite time $T > 0$, independently of m.

Therefore, we are able to pass to a weak limit $u^{m'} \rightharpoonup u$ in $\mathrm{L}_2(0,T;\mathrm{H}_0^1(\Omega))$ $p^{m'} \rightharpoonup p$ in $\mathrm{L}_2(0,T;\mathrm{H}_0^1(\Omega)\cap\mathrm{L}_4(\Omega))$ via a subsequence m', and additionally, thanks to the compact-imbedding theorem with the assumptions

$$\{u^m\}_{m=1}^\infty \text{ bounded in } \mathrm{L}_2(0,T;\mathrm{H}_0^1(\Omega)), \{\frac{\partial p^m}{\partial t}\}_{m=1}^\infty \text{ bounded in } \mathrm{L}_2(0,T;\mathrm{L}_2(\Omega)),$$

$$\{p^m\}_{m=1}^\infty \text{ bounded in } \mathrm{L}_4(0,T;\mathrm{H}_0^1(\Omega)\cap\mathrm{L}_4(\Omega)),$$
$$\{\frac{\partial p^m}{\partial t}\}_{m=1}^\infty \text{ bounded in } \mathrm{L}_2(0,T;\mathrm{L}_2(\Omega)),$$

to the strong limits u in $\mathrm{L}_2(0,T;\mathrm{L}_2(\Omega))$, p in $\mathrm{L}_4(0,T;\mathrm{L}_4(\Omega))$.

The passage to the limit in the semi-discrete scheme (6) can be accomplished due to the following facts:

1. $\nabla p^{m'}$ converges strongly in $\mathrm{L}_2(0,T;\mathrm{L}_2(\Omega;\mathbb{R}^2))$ to ∇p (Lemma 3.4 of [2]), $\nabla u^{m'}$ converges weakly in $\mathrm{L}_2(0,T;\mathrm{L}_2(\Omega;\mathbb{R}^2))$ to ∇u,
2. $\mathbf{V}\cdot\nabla(u^{m'} - Lp^{m'})$ converges weakly in $\mathrm{L}_2(0,T;\mathrm{L}_2(\Omega))$ to $\mathbf{V}\cdot\nabla(u - Lp)$, $\mathbf{V}\cdot\nabla p^{m'}$ converges weakly in $\mathrm{L}_2(0,T;\mathrm{L}_2(\Omega))$ to $\mathbf{V}\cdot\nabla p$,
3. $f_0(p^{m'})$ converges weakly in $\mathrm{L}_{\frac{4}{3}}(0,T;\mathrm{L}_{\frac{4}{3}}(\Omega))$ (polynomial nonlinearity and Aubin lemma),
4. $F(u^m)|\nabla p^m|$ converges weakly to $F(u)|\nabla p|$ in $\mathrm{L}_2(0,T;\mathrm{L}_2(\Omega))$ (Lemma 3.5 of [2]),

5. $\mathcal{P}_{m'} p_{ini}$, $\mathcal{P}_{m'} u_{ini}$ converge strongly to p_{ini}, u_{ini} in $\mathrm{L}_2(\Omega)$,
6. $p^{m'}(0) = \mathcal{P}_{m'} p_{ini}$, $u^{m'}(0) = \mathcal{P}_{m'} u_{ini}$.

Additionally, we observe that the function p belongs to $\mathrm{L}_2(0,T;\mathrm{H}^1_0(\Omega)\cap\mathrm{H}^2(\Omega))$ (Lemma 3.3 of [2]). The weak solution satisfies the initial condition (again see [2] for details).

Due to Lipschitz-continuity of F with the Lipschitz constant denoted by L_F, we prove **uniqueness** of the solution of (5). We consider two solutions of the problem (5), denoted by $[u_1, p_1]$ and $[u_2, p_2]$. Subtracting corresponding systems of equations and denoting $[u_{12}, p_{12}] = [u_1 - u_2, p_1 - p_2]$, multiplying the first equation by $u_{12} - Lp_{12}$ and the second equation by p_{12}, we have

$$\begin{aligned}
&\frac{1}{2}\frac{\mathrm{d}}{\mathrm{d}t}\|u_{12} - Lp_{12}\|^2 + \|\nabla(u_{12} - Lp_{12})\|^2 + L(\nabla p_{12}, \nabla(u_{12} - Lp_{12})) \\
&\quad +(\mathbf{V}\cdot\nabla(u_{12} - Lp_{12})), u_{12} - Lp_{12}) = 0 \text{ in } (0,T), \\
&\frac{1}{2}\alpha\xi^2\frac{\mathrm{d}}{\mathrm{d}t}\|p_{12}\|^2 + \xi^2\|\nabla p_{12}\|^2 + \alpha\xi^2(\mathbf{V}\cdot\nabla p_{12}), p_{12}) = (f_0(p_1) - f_0(p_2), p_{12}) \\
&\quad +\xi^2(F(u_1)|\nabla p_1| - F(u_2)|\nabla p_2|, p_{12}) \text{ in } (0,T), \\
&(u_{12} - Lp_{12})(0) = 0, \quad p_{12}(0) = 0.
\end{aligned}$$

Due to the form of f_0, we have that

$$(f_0(p_1) - f_0(p_2), p_{12}) \le \frac{7a}{4}\|p_{12}\|^2,$$

Using the Schwarz inequality, we get

$$\begin{aligned}
&\frac{\mathrm{d}}{\mathrm{d}t}\|u_{12} - Lp_{12}\|^2 \le V^2\|u_{12} - Lp_{12}\|^2 + L^2\|\nabla p_{12}\|^2, \\
&\frac{1}{2}\alpha\xi^2\frac{\mathrm{d}}{\mathrm{d}t}\|p_{12}\|^2 + \xi^2\|\nabla p_{12}\|^2 \le \\
&\qquad \frac{7a}{4}\|p_{12}\|^2 + \xi^2 L_F\|u_{12}\|\|\nabla p_1\|_{\mathrm{L}_4(\Omega)}\|p_{12}\|_{\mathrm{L}_4(\Omega)} \\
&\qquad +(\xi^2 C_F + \alpha\xi^2 V)\|\nabla p_{12}\|\|p_{12}\|,
\end{aligned} \tag{7}$$

in $(0,T)$. Due to the Young inequality, we obtain

$$\begin{aligned}
&\frac{\mathrm{d}}{\mathrm{d}t}\|u_{12} - Lp_{12}\|^2 \le L^2\|\nabla p_{12}\|^2 + V^2\|u_{12} - Lp_{12}\|^2 \text{ in } (0,T), \\
&\frac{1}{2}\alpha\xi^2\frac{\mathrm{d}}{\mathrm{d}t}\|p_{12}\|^2 + \frac{\xi^2}{2}\|\nabla p_{12}\|^2 \le \\
&\qquad \frac{3}{2}(C_F^2\xi^2 + \frac{7a}{6} + \alpha^2V^2\xi^2 + 2L^2\xi^2L_F^2C_4^2\|\nabla p_1\|^2_{\mathrm{L}_4(\Omega)})\|p_{12}\|^2 \\
&\qquad +3\xi^2L_F^2C_4^2\|\nabla p_1\|^2_{\mathrm{L}_4(\Omega)}\|u_{12} - Lp_{12}\|^2.
\end{aligned} \tag{8}$$

Combining these inequalities, we have in $(0,T)$:

$$\frac{\mathrm{d}}{\mathrm{d}t}\left(\frac{1}{2}\alpha\xi^2\|p_{12}\|^2 + \frac{\xi^2}{2L^2}\|u_{12} - Lp_{12}\|^2\right) \leq \qquad (9)$$

$$M(t)\left(\frac{1}{2}\alpha\xi^2\|p_{12}\|^2 + \frac{\xi^2}{2L^2}\|u_{12} - Lp_{12}\|^2\right)$$

with

$$M(t) = V^2 + 6L^2L_F^2C_4^2\|\nabla p_1\|^2_{\mathrm{L}_4(\Omega)} + \frac{3C_F^2}{\alpha} + \frac{7a}{2\alpha\xi^2} + 3\alpha V^2 + \frac{6L^2}{\alpha}L_F^2C_4^2\|\nabla p_1\|^2_{\mathrm{L}_4(\Omega)}.$$

Considering the fact that there is a constant C_p for which

$$\int_0^T \|\nabla p_1\|^2_{\mathrm{L}_4(\Omega)} dt \leq C_4^2 \int_0^T \|p_1\|^2_{\mathrm{H}^2(\Omega)} dt \leq C_p^2,$$

where C_4 is the norm of the imbedding $\mathrm{H}_0^1(\Omega)$ into $\mathrm{L}_4(\Omega)$, and $p_1 \in \mathrm{L}_2(0,T;\mathrm{H}^2(\Omega))$, we have that

$$p_{12}(t) = u_{12}(t) = 0 \text{ in } \mathrm{L}_2(\Omega),\ \forall t \in (0,T).$$

as follows from the Gronwall lemma. □

Asymptotic behaviour. A priori estimates in the above given proof imply that

$$E_\xi[p](t) \leq E_\xi[p](0)\exp\{\frac{C_F^2}{2\alpha}t\} \quad t \in (0,T),$$

where we denoted

$$E_\xi[p_\xi](t) = \int_\Omega [\xi\frac{1}{2}|\nabla p_\xi|^2 + \frac{1}{\xi}w_0(p_\xi)]\mathrm{d}x,$$

and additionally, there is an estimate for the time derivative given by

$$\frac{1}{2}\alpha\xi\int_0^T \|\partial_t p\|^2 dt + E_\xi[p](T) \leq C_T E_\xi[p](0).$$

We therefore can state that the function $p = p(t,x;\xi)$ tends to a stepwise constant function as in the Theorem 2.2 of [4]. As in [2], the matching procedure recovers the Gibbs-Thompson law $\alpha(v_\Gamma - \mathbf{V}\cdot\mathbf{n}_\Gamma) = -\kappa_\Gamma + F(u^\xi) + \mathcal{O}(\xi^2)$, as well as the Stefan condition at Γ.
Remark. As indicated in [1], the model (4) can be modified by incorporating anisotropy into coefficients of (1). The above given analysis is applicable again, namely for weak anisotropies.

Method of lines in 2D. In the computations, we take $\Omega = (0, L_1) \times (0, L_2)$ and use a uniform rectangular grid. The following notations are introduced:

$$\mathbf{h} = (h_1, h_2), \; h_1 = \frac{L_1}{N_1}, \; h_2 = \frac{L_2}{N_2}, \; u_{ij} = u(ih_1, jh_2),$$

$$\omega_{\mathbf{h}} = \{[ih_1, jh_2] \mid i = 1, \ldots, N_1 - 1; \; j = 1, \ldots, N_2 - 1\},$$

$$\bar{\omega}_{\mathbf{h}} = \{[ih_1, jh_2] \mid i = 0, \ldots, N_1; \; j = 0, \ldots, N_2\}, \quad \gamma_{\mathbf{h}} = \bar{\omega}_{\mathbf{h}} - \omega_{\mathbf{h}},$$

$$u_{\bar{x}_1,ij} = \frac{u_{ij} - u_{i-1,j}}{h_1}, \quad u_{x_1,ij} = \frac{u_{i+1,j} - u_{ij}}{h_1},$$

$$u_{\bar{x}_2,ij} = \frac{u_{ij} - u_{i,j-1}}{h_2}, \quad u_{x_2,ij} = \frac{u_{i,j+1} - u_{ij}}{h_2},$$

$$u_{\bar{x}_1 x_1,ij} = \frac{1}{h_1^2}\left(u_{i+1,j} - 2u_{ij} + u_{i-1,j}\right),$$

and

$$\bar{\nabla}_h u = [u_{\bar{x}_1}, u_{\bar{x}_2}], \quad \nabla_h u = [u_{x_1}, u_{x_2}], \quad \Delta_h u = u_{\bar{x}_1 x_1} + u_{\bar{x}_2 x_2}.$$

The set of grid functions is denoted by $\mathcal{H}_{\mathbf{h}}$. The semi-discrete scheme has the following form

$$\begin{aligned}
&(\frac{\mathrm{d}}{\mathrm{d}t} + \mathbf{V} \cdot \bar{\nabla}_h)(u^h - L\chi(p^h)) = \Delta_h u^h, \\
&\alpha\xi^2(\frac{\mathrm{d}}{\mathrm{d}t} + \mathbf{V} \cdot \bar{\nabla}_h)p^h = \xi^2 \Delta_h p^h + f_0(p^h) + \xi^2 |\nabla_h p^h| F(u^h) \text{ on } \omega_{\mathbf{h}}, \\
&u^h \mid_{\gamma_{\mathbf{h}}} = 0, \quad p^h \mid_{\gamma_{\mathbf{h}}} = 0, \\
&u^h(0) = \mathcal{P}_h u_{ini}, \quad p^h(0) = \mathcal{P}_h p_{ini},
\end{aligned}$$

where its solution is a map $u^h, p^h :< 0, T > \longrightarrow \mathcal{H}_{\mathbf{h}}$ and $\mathcal{P}_h : \mathcal{C}(\bar{\Omega}) \to \mathcal{H}_{\mathbf{h}}$ is a restriction operator. The stability and convergence of the scheme can be investigated in a way very similar to the proof of the Theorem 1. The scheme is designed to meet real conditions, where the diffusion and growth process dominates the advection.

Computational results. We present several results of advected curve dynamics and advected pattern formation in solidification. Figure 2 shows how the advection field influences particular situations of curve dynamics. In Figure 2a, the circle of critical radius is carried down, and when it interacts with the domain boundary, it is brought to shrinking. Obviously, such circle remains unchanged, when no advection is imposed. In Figure 2b, the initial circle is converted to an anisotropic pattern due to the anisotropy incorporated into the model. When it expands, it interacts with the boundary. In Figure 2c, a circle is shrinking when being carried around by advection. In Figure 2d, a circle at critical radius is carried around by advection. As there is no interaction with boundary, it remains unchanged.

In Figure 3, we observe a single-dendrite growth. The pattern falls downwards. In Figure 4, three nucleation sites were imposed. The dendrites growth, interact and fall downwards, where they touch the domain boundary.

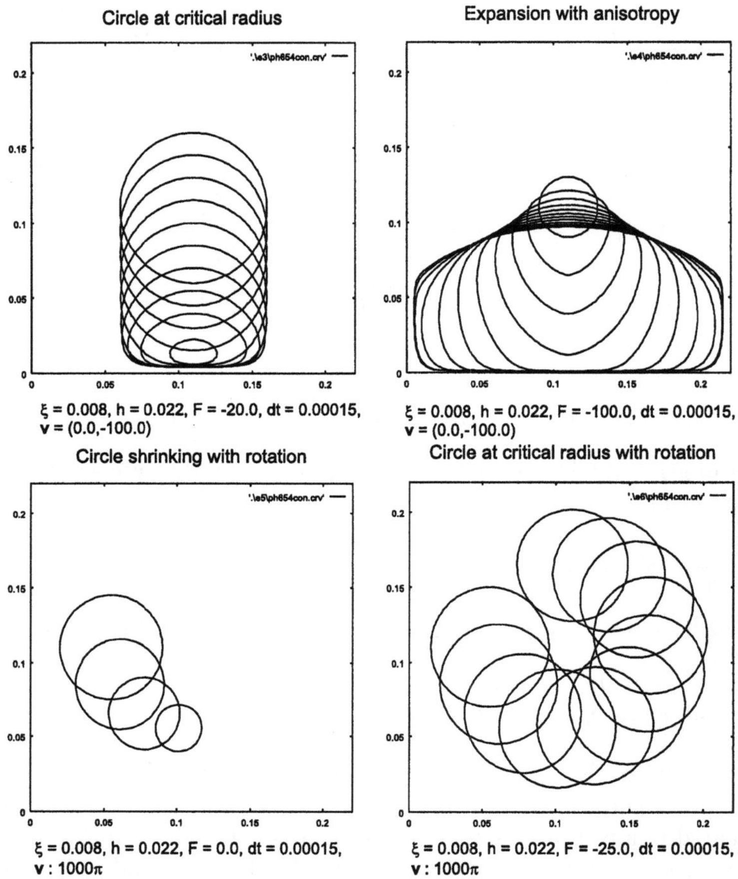

Fig. 2. Various situations of the advected mean curvature flow

Acknowledgement. The author was partly supported by the project of the Czech Grant Agency No. 201/01/0676. Some computations were performed under the support of the Institute of Computer Science, Czech Academy of Sciences within the project No. A1030103 of the Grant Agency of the Czech Academy of Sciences.

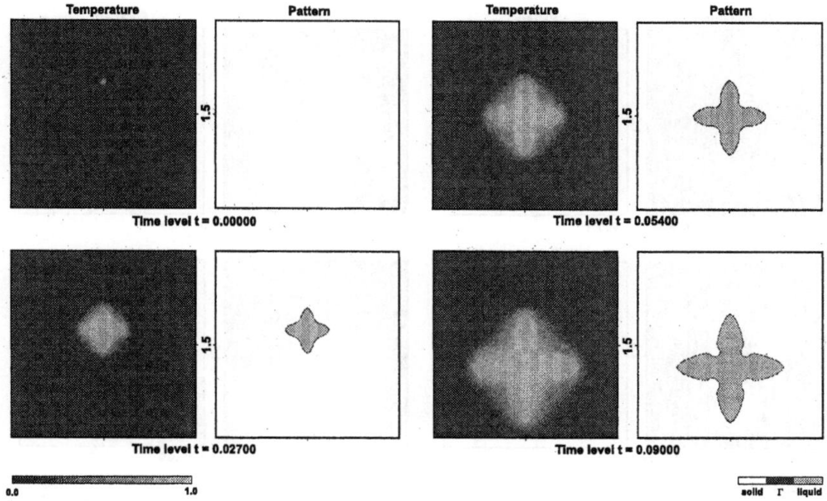

Fig. 3. Solidification of a single falling dendrite - parameters are $u^* = 1.0$, $u_{ini} = 0.0$, $L = 2.0$, $\beta = 300$, $a = 4.0$, $\alpha = 3$, $L_1 = L_2 = 3.0$, initial radius $= 0.025$, $N_1 = N_2 = 300$, $\xi = 0.015$

References

1. M. Beneš, *Anisotropic phase-field model with focused latent-heat release*, FREE BOUNDARY PROBLEMS: Theory and Applications II (Chiba, Japan), 2000, GAKUTO International Series Mathematical Sciences and Applications, Vol.14, pp. 18–30.
2. ———, *Mathematical analysis of phase-field equations with numerically efficient coupling terms*, Interfaces and Free Boundaries **3** (2001), 201–221.
3. ———, *Mathematical and computational aspects of solidification of pure substances*, Acta Mathematica Universitatis Comenianae **70, No. 1** (2001), 123–152.
4. M. Beneš and K. Mikula, *Simulation of anisotropic motion by mean curvature–comparison of phase-field and sharp-interface approaches*, Acta Math. Univ. Comenianae **67, No. 1** (1998), 17–42.
5. Y.-G. Chen, Y. Giga, and S. Goto, *Uniqueness and existence of viscosity solutions of generalized mean curvature flow equations*, J. Diff. Geom. **33** (1991), 749–786.
6. L.C. Evans, H.M. Soner, and P.E. Souganidis, *Phase transitions and generalized motion by mean curvature*, Comm. Pure Appl. Math. **45** (1992), 1097–1123.
7. L.C. Evans and J. Spruck, *Motion of level sets by mean curvature I*, J. Diff. Geom. **33** (1991), 635–681.
8. M. Gurtin, *On the two-phase Stefan problem with interfacial energy and entropy*, Arch. Rational Mech. Anal. **96** (1986), 200–240.
9. T. Ohta, M. Mimura, and R. Kobayashi, *Higher-dimensional localized patterns in excitable media*, Physica D **34** (1989), 115–144.
10. A. Visintin, *Models of phase transitions*, Birkhäuser, Boston, 1996.

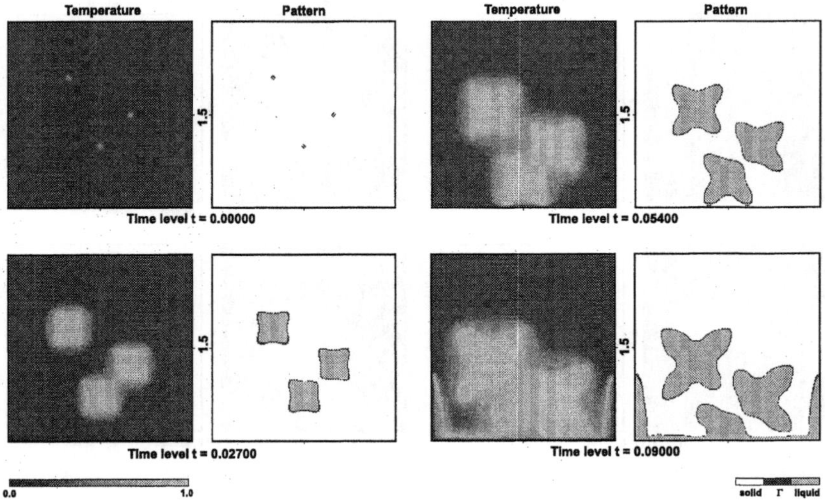

Fig. 4. Solidification of three falling dendrites - parameters are $u^* = 1.0$, $u_{ini} = 0.0$, $L = 2.0$, $\beta = 300$, $a = 4.0$, $\alpha = 3$, $L_1 = L_2 = 3.0$, initial radius $= 0.025$, $N_1 = N_2 = 300$, $\xi = 0.015$

Fast Evaluation of Eddy Current Integral Operators

Steffen Börm

Max-Planck-Institute für Mathematik in den Naturwissenschaften
Inselstraße 22–26, 04103 Leipzig, Germany
sbo@mis.mpg.de

Summary. Boundary element formulations for eddy current problems are based on non-local operators. Discretizing these operators by standard Galerkin techniques leads to large dense matrices. In order to treat the discretized system efficiently, we cannot store these dense matrices directly, but use data-sparse approximations.

We present an approach based on piecewise polynomial interpolation of the underlying kernel functions. The resulting $\mathcal{H}^2$*-matrix approximation* can be stored using only $\mathcal{O}(nm^3)$ units of storage, where n is the number of degrees of freedom and m is the order of the interpolation. Construction and evaluation of the approximated matrix requires only $\mathcal{O}(nm^3)$ operations.

This paper presents joint work with Jörg Ostrowski.

1 Introduction

1.1 Problem

The eddy-current model introduced in [6] in combination with impedance boundary conditions [9] leads to the variational equation

$$\begin{aligned} a(\mathbf{v},\mathbf{u}) \quad -b(\mathbf{v},\phi) \quad &= f(\mathbf{v}) \quad \text{for all } \mathbf{v}\in\mathcal{V} \\ -b(\mathbf{u},\psi) \quad -q(\psi,\phi) \quad &= \zeta(\psi) \quad \text{for all } \psi\in\mathcal{W} \end{aligned} \tag{1}$$

for the unknown vector field $\mathbf{u}\in\mathcal{V}$ and the unknown potential $\phi\in\mathcal{W}$, where the bilinear forms have the form

$$\begin{aligned} a(\mathbf{v},\mathbf{u}) &= \int_\Gamma\int_\Gamma \langle \operatorname{curl}_\Gamma \mathbf{v}(\mathbf{x}), \operatorname{curl}_\Gamma \mathbf{u}(\mathbf{y})\rangle \Phi(\mathbf{x},\mathbf{y})\,d\mathbf{y}\,d\mathbf{x} + \text{sparse}, \\ q(\psi,\phi) &= \int_\Gamma\int_\Gamma \langle \mathbf{curl}_\Gamma \psi(\mathbf{x}), \mathbf{curl}_\Gamma \phi(\mathbf{y})\rangle \Phi(\mathbf{x},\mathbf{y})\,d\mathbf{y}\,d\mathbf{x}, \\ b(\mathbf{v},\phi) &= \int_\Gamma\int_\Gamma \langle \mathbf{curl}_\Gamma \phi(\mathbf{y}), \mathbf{v}(\mathbf{x})\rangle \langle \mathbf{grad}_x \Phi(\mathbf{x},\mathbf{y}), \mathbf{n}(\mathbf{x})\rangle\,d\mathbf{y}\,d\mathbf{x} \\ &\quad - \int_\Gamma\int_\Gamma \langle \mathbf{curl}_\Gamma \phi(\mathbf{y}), \mathbf{n}(\mathbf{x})\rangle \langle \mathbf{grad}_x \Phi(\mathbf{x},\mathbf{y}), \mathbf{v}(\mathbf{x})\rangle\,d\mathbf{y}\,d\mathbf{x} \\ &\quad + \text{sparse}. \end{aligned}$$

Here, Φ denotes the singularity function for the Laplace equation, i.e.,

$$\Phi(\mathbf{x}, \mathbf{y}) = \frac{1}{4\pi \|\mathbf{x} - \mathbf{y}\|}, \tag{2}$$

and the surface differential operators are defined by

$$\mathbf{curl}_\Gamma\, \mathbf{u} := \langle \mathbf{n}, \mathbf{curl}\, \mathbf{u} \rangle, \qquad \mathrm{curl}_\Gamma\, \psi := (\mathbf{grad}\, \psi) \times \mathbf{n}.$$

The sparse parts of the bilinear forms can be treated by standard techniques, so we consider only the non-local components of the system.

1.2 Compression

Discretizing the non-local operators leads to large densely populated matrices that cannot be handled efficiently by standard techniques, therefore we use the $\mathcal{H}^2$-matrix approximation technique [2] to reduce the storage requirements and the complexity of the discretization and the matrix-vector multiplication. The application of $\mathcal{H}^2$-matrices to the problem of the eddy current model was investigated in [7] and [3].

We remark that there is a close relationship of $\mathcal{H}^2$-matrices to the panel clustering technique [5] and the fast multi-pole method for integral operators [8, 4]. While the multi-pole technique applies an expansion specially designed for the kernel function under investigation in order to reach the, in some sense optimal, complexity of $\mathcal{O}(n \log^2 n)$, the $\mathcal{H}^2$-matrix technique has a complexity of $\mathcal{O}(n \log^3 n)$ but can be applied to any asymptotically smooth (cf. (4)) kernel function.

2 Approximation

2.1 Discretization

We approximate the surface Γ by a polyhedron Γ_h described by a conforming triangulation. Its triangles, edges and nodes are denoted by $\mathcal{T}$, $\mathcal{E}$ and $\mathcal{N}$.

For each edge $e \in \mathcal{E}$, we denote the surface edge element basis function (cf. [3]) by $\mathbf{b}_e$ and define the finite-dimensional subspace $\mathcal{V}_h := \mathrm{span}\{\mathbf{b}_e \ : \ e \in \mathcal{E}\}$ of $\mathcal{V}$. For each node $v \in \mathcal{N}$, the surface nodal basis function is denoted by ψ_v, and $\mathcal{W}_h := \mathrm{span}\{\psi_v \ : \ v \in \mathcal{N}\}$ is a finite-dimensional subspace of $\mathcal{W}$.

The standard Galerkin approach with these basis functions leads to

$$\begin{pmatrix} \mathsf{A} & -\mathsf{B} \\ -\mathsf{B}^\top & -\mathsf{Q} \end{pmatrix} = \mathrm{rhs}$$

defined by

$$\mathsf{A}_{ef} = a(\mathbf{b}_e, \mathbf{b}_f), \quad \mathsf{Q}_{vw} = q(\psi_v, \psi_w) \quad \text{and} \quad \mathsf{B}_{ew} = b(\mathbf{b}_e, \psi_w)$$

for $e, f \in \mathcal{E}$ and $v, w \in \mathcal{N}$, where all matrices are densely populated.

2.2 Simplification

For $\mathbf{b} \in \mathcal{V}_h$ and $\psi \in \mathcal{W}_h$, the functions $\operatorname{curl}_\Gamma \mathbf{b}$ and $\mathbf{curl}_\Gamma \psi$ are piecewise constant. Therefore we introduce the auxiliary space $\mathcal{X}_h$ spanned by piecewise constant basis functions χ_t for $t \in \mathcal{T}$ and the auxiliary matrix G defined by

$$\mathsf{G}_{ts} := \int_\Gamma \int_\Gamma \chi_t(\mathbf{x}) \Phi(\mathbf{x}, \mathbf{y}) \chi_s(\mathbf{y}) d\mathbf{y} d\mathbf{x}$$

and observe that

$$\mathsf{A} = \mathsf{L}_1^\top \mathsf{G} \mathsf{L}_1 + \text{sparse} \qquad \text{and} \qquad \mathsf{Q} = \mathsf{L}_2^\top \begin{pmatrix} \mathsf{G} & & \\ & \mathsf{G} & \\ & & \mathsf{G} \end{pmatrix} \mathsf{L}_2 \tag{3}$$

hold for sparse matrices L_1 and L_2. This representation is more efficient than the original one, but it is still based on a densely populated matrix.

3 Approximation

3.1 Interpolation

The idea is to replace the kernel function Φ on a sub-domain $\tau \times \sigma \subseteq \Gamma \times \Gamma$ by

$$\tilde{\Phi}^{\tau,\sigma}(\mathbf{x}, \mathbf{y}) := \sum_{\nu=1}^{k} \sum_{\mu=1}^{k} \Phi(\mathbf{x}_\nu^\tau, \mathbf{x}_\mu^\sigma) \mathcal{L}_\nu^\tau(\mathbf{x}) \mathcal{L}_\mu^\sigma(\mathbf{y}),$$

where $(\mathbf{x}_\nu^\tau)_{\nu=1}^k$ and $(\mathbf{x}_\mu^\sigma)_{\mu=1}^k$ are interpolation points and $(\mathcal{L}_\nu^\tau)_{\nu=1}^k$ and $(\mathcal{L}_\mu^\sigma)_{\mu=1}^k$ are the corresponding Lagrange polynomials.

This results in an approximation of the local matrix

$$\begin{aligned}
\mathsf{G}_{ts}^{\tau,\sigma} &:= \int_\tau \int_\sigma \chi_t(\mathbf{x}) \Phi(\mathbf{x}, \mathbf{y}) \chi_s(\mathbf{y})\, d\mathbf{y}\, d\mathbf{x} \approx \int_\tau \int_\sigma \chi_t(\mathbf{x}) \tilde{\Phi}^{\tau,\sigma}(\mathbf{x}, \mathbf{y}) \chi_s(\mathbf{y})\, d\mathbf{y}\, d\mathbf{x} \\
&= \sum_{\nu=1}^{k} \sum_{\mu=1}^{k} \underbrace{\Phi(\mathbf{x}_\nu^\tau, \mathbf{x}_\mu^\sigma)}_{=:\mathsf{S}_{\nu\mu}^{\tau,\sigma}} \underbrace{\int_\tau \chi_t(\mathbf{x}) \mathcal{L}_\nu^\tau(\mathbf{x})\, d\mathbf{x}}_{=:\mathsf{V}_{t\nu}^\tau} \underbrace{\int_\sigma \chi_s(\mathbf{y}) \mathcal{L}_\mu^\sigma(\mathbf{y})\, d\mathbf{y}}_{=:\mathsf{V}_{s\mu}^\sigma} \\
&= (\mathsf{V}^\tau \mathsf{S}^{\tau,\sigma} (\mathsf{V}^\sigma)^\top)_{ts}.
\end{aligned}$$

The approximation requires only $2nk + k^2$ units of storage. For typical interpolation schemes, k will be much smaller than n, so the factorized representation will by much more efficient (cf. Figure 1).

Typical interpolation schemes, e.g., tensor-product Chebyshev interpolation, work only for smooth functions. Since the function Φ is not globally

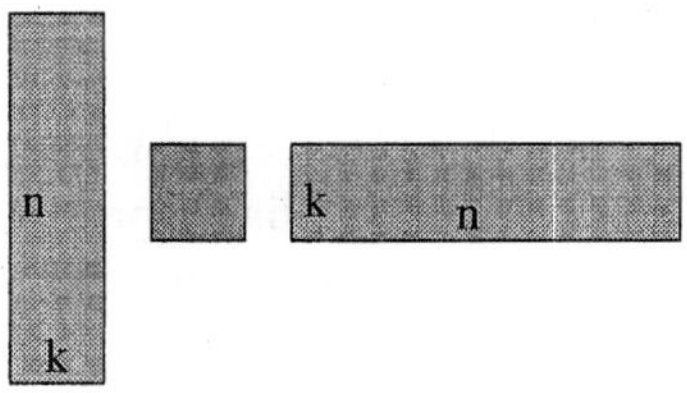

Fig. 1. Compressed representation of local matrices

smooth, we cannot hope to find a factorized representation for the entire matrix G. Instead, we make use of the fact that Φ is *asymptotically smooth*, i.e., that there are constants $C_{\mathrm{asymp}}, c_0, d \in \mathbb{R}_{>0}$ such that

$$|\partial_x^\alpha \partial_y^\beta \Phi(\mathbf{x}, \mathbf{y})| \leq C_{\mathrm{asymp}} c_0^{|\alpha|+|\beta|} (\alpha+\beta)! \|\mathbf{x}-\mathbf{y}\|^{-d-|\alpha|-|\beta|} \tag{4}$$

holds for $\mathbf{x}, \mathbf{y} \in \mathbb{R}^3$ with $\mathbf{x} \neq \mathbf{y}$.

Combining this inequality with standard estimates for the interpolation error, we find that there is a polynomial C_{apx} such that

$$\|\tilde{\Phi}^{\tau,\sigma} - \Phi\|_{L^\infty(B^\tau \times B^\sigma)} \leq \frac{C_{\mathrm{apx}}(m)}{\operatorname{dist}(B^\tau, B^\sigma)^d} \left(1 + \frac{2\operatorname{dist}(B^\tau, B^\sigma)}{\operatorname{diam}(B^\tau \times B^\sigma)}\right)^{-(m+1)}$$

holds for tensor-product Chebyshev interpolation of order $m \in \mathbb{N}$ (corresponding to a rank of $k = (m+1)^d$) on axis-parallel boxes B^τ and B^σ satisfying $\tau \subseteq B^\tau$ and $\sigma \subseteq B^\sigma$.

To ensure a uniform rate of convergence, the *admissibility condition*

$$\operatorname{diam}(B^\tau \times B^\sigma) \leq 2\eta \operatorname{dist}(B^\tau, B^\sigma). \tag{5}$$

has to hold for a constant $\eta \in \mathbb{R}_{>0}$ (cf. Figure 2).

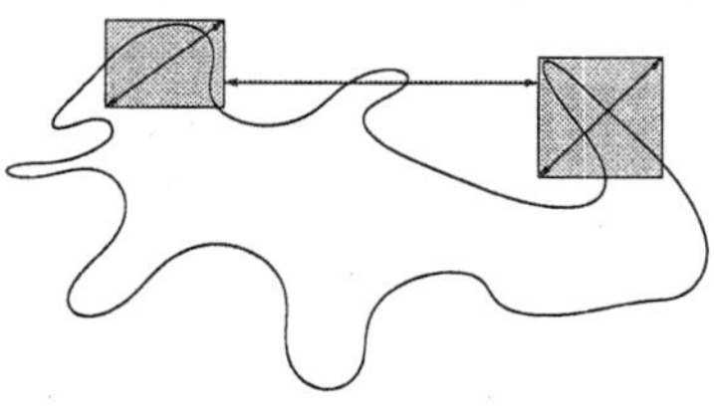

Fig. 2. Admissibility condition

3.2 Cluster tree and block partition

Since we can apply our interpolation scheme only to sub-domains $\tau \times \sigma$ satisfying the admissibility condition (5), we have to split the entire domain $\Gamma_h \times \Gamma_h$

into sub-domains that either fulfil this condition or are so small that we can store the corresponding local matrix block $\mathsf{G}^{\tau,\sigma}$ in the standard format.

We construct the splitting of $\Gamma_h \times \Gamma_h$ using a hierarchical splitting of the domain Γ_h that is called a *cluster tree*:

Definition 1. *A tree $\mathcal{C}$ is called a* cluster tree *for a set Γ_h if*

- *the set Γ_h is the root of $\mathcal{C}$, i.e.,* $\mathrm{root}(\mathcal{C}) = \Gamma_h$, *and*
- *if a node $\tau \in \mathcal{C}$ is not a leaf, then it is the (up to sets of measure zero) disjoint union of its sons, i.e.,*

$$\tau = \dot{\bigcup}\{\tau' \ : \ \tau' \in \mathrm{sons}(\tau)\}.$$

Each node $\tau \in \mathcal{C}$ is called a cluster.

A cluster tree can be constructed from an arbitrary set of triangles by *binary space partitioning*: We start with the root cluster containing all the triangles, corresponding to the entire domain, split it into two son clusters and repeat the procedure recursively until the clusters contain less than k triangles.

Using the cluster tree, we can construct a partition

$$P \subseteq \{\tau \times \sigma \ : \ \tau, \sigma \in \mathcal{C}\}$$

of $\Gamma_h \times \Gamma_h$ containing only admissible and small blocks (cf. [1]).

3.3 Construction of matrices

The matrix V^τ for each cluster τ is given by

$$\mathsf{V}^\tau_{t\nu} := \int_\tau \chi_t(\mathbf{x})\mathcal{L}^\tau_\nu(\mathbf{x})\, d\mathbf{x}.$$

Since χ_t and $\mathcal{L}^\tau_\nu$ are polynomials, the entries of the matrix can be computed by standard quadrature.

Since the coefficient matrix $\mathsf{S}^{\tau,\sigma}$ is defined by

$$\mathsf{S}^{\tau,\sigma}_{\nu\mu} := \Phi(x^\tau_\nu, x^\sigma_\mu)$$

for each admissible sub-domain $\tau \times \sigma$, it can be constructed by simply evaluating the expression (2).

Setting

$$P_{\mathrm{far}} := \{\tau \times \sigma \in P \ : \ \tau \times \sigma \text{ is admissible}\}, \qquad P_{\mathrm{near}} := P \setminus P_{\mathrm{far}},$$

the approximation of the matrix G is defined by

$$\widetilde{\mathsf{G}} := \sum_{\tau\times\sigma\in P_{\mathrm{far}}} \mathsf{V}^\tau \mathsf{S}^{\tau,\sigma} \mathsf{V}^{\sigma\top} + \sum_{\tau\times\sigma\in P_{\mathrm{near}}} \mathsf{G}^{\tau,\sigma}. \tag{6}$$

The approximation (6) requires $\mathcal{O}(nk \log n) = \mathcal{O}(nm^3 \log n)$ units of storage, and the approximation error decreases exponentially in m.

4 Nested cluster basis

Now, we aim to use the special structure of the matrices V^τ, the *cluster bases*, in order to reduce the storage complexity for the matrix $\widetilde{\mathsf{G}}$.

4.1 Nested approximation spaces

Let τ be a cluster and τ' one of its sons. Since we use the same order of interpolation for all clusters, we have

$$\mathcal{L}_\nu^\tau = \sum_{\nu'=1}^{k} \underbrace{\mathcal{L}_\nu^\tau(x_{\nu'}^{\tau'})}_{=:\mathsf{T}_{\nu'\nu}^{\tau',\tau}} \mathcal{L}_{\nu'}^{\tau'}, \quad \text{which implies} \quad \mathsf{V}^\tau = \sum_{\tau' \in \mathrm{sons}(\tau)} \mathsf{V}^{\tau'} \mathsf{T}^{\tau',\tau}, \tag{7}$$

so we need to store the $n \times k$-matrix V^τ only for clusters without sons, since we can reconstruct it for all other clusters by using the $k \times k$-matrices $\mathsf{T}^{\tau',\tau}$ (cf. Figure 3). This reduces the storage complexity to $\mathcal{O}(nk) = \mathcal{O}(nm^3)$.

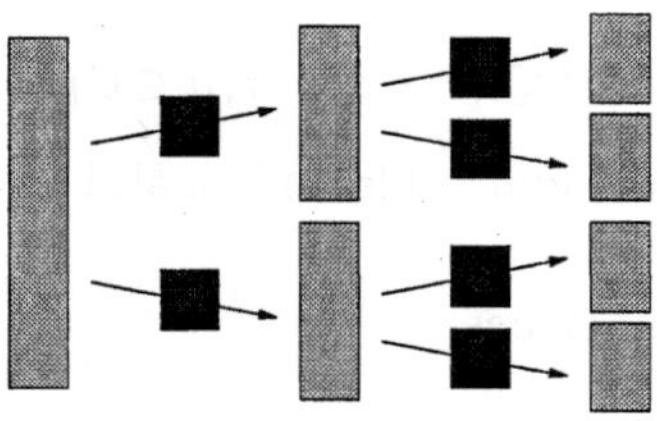

Fig. 3. Representation of cluster basis matrices by transfer matrices

4.2 Fast matrix-vector multiplication

The equation (7) can also be used to perform the matrix-vector multiplication efficiently: We introduce the auxiliary variables

$$\mathbf{u}^\sigma := \mathsf{V}^{\sigma\,\top}\mathbf{u}, \qquad\qquad \mathbf{v}^\tau := \sum_{\sigma, \tau\times\sigma \in P_{\mathrm{far}}} \mathsf{S}^{\tau,\sigma}\mathbf{u}^\sigma$$

for clusters $\tau, \sigma \in \mathcal{C}$ and find

$$\begin{aligned}
\tilde{\mathsf{G}}\mathbf{u} &= \sum_{(\tau,\sigma)\in P_{\mathrm{near}}} \mathsf{S}^{\tau,\sigma}\mathbf{u} + \sum_{(\tau,\sigma)\in P_{\mathrm{far}}} \mathsf{V}^\tau \mathsf{S}^{\tau,\sigma}\mathsf{V}^{\sigma\,\top}\mathbf{u} \\
&= \sum_{(\tau,\sigma)\in P_{\mathrm{near}}} \mathsf{G}^{\tau,\sigma}\mathbf{u} + \sum_{(\tau,\sigma)\in P_{\mathrm{far}}} \mathsf{V}^\tau \mathsf{S}^{\tau,\sigma}\mathbf{u}^\sigma \\
&= \sum_{(\tau,\sigma)\in P_{\mathrm{near}}} \mathsf{G}^{\tau,\sigma}\mathbf{u} + \sum_{\tau\in\mathcal{C}} \mathsf{V}^\tau \mathbf{v}^\tau = \mathbf{v}.
\end{aligned}$$

For typical domain splittings, the set $\{\sigma : \tau \times \sigma \in P_{\text{far}}\}$ contains only a small number of elements, so the computation of $\mathbf{v}^\tau$ can be done in complexity $\mathcal{O}(nk)$ for all $\tau \in \mathcal{C}$. This leaves us with the task of evaluating $\mathbf{u}^\sigma = \mathsf{V}^{\sigma\top}\mathbf{u}$ and $\mathsf{V}^\tau\mathbf{v}^\tau$ efficiently. Due to (7), we have

$$\mathbf{u}^\sigma = \mathsf{V}^{\sigma\top}\mathbf{u} = \sum_{\sigma' \in \text{sons}(\sigma)} \mathsf{T}^{\sigma',\sigma\top}\mathsf{V}^{\sigma'\top}\mathbf{u} = \sum_{\sigma' \in \text{sons}(\sigma)} \mathsf{T}^{\sigma',\sigma\top}\mathbf{u}^{\sigma'},$$

so we can use a recursive procedure to compute the vectors $\mathbf{u}^\sigma$ for all $\sigma \in \mathcal{C}$ in $\mathcal{O}(nk)$ operations. A similar recursion can be used to construct $\sum_{\tau\in\mathcal{C}} \mathsf{V}^\tau\mathbf{v}^\tau$ in $\mathcal{O}(nk)$ operations.

5 Approximation of the double layer potential

Now that we know how to store and multiply by the matrices A and Q in complexity $\mathcal{O}(nk)$, we only have to handle the matrix B efficiently.

We replace Φ by the local approximations $\tilde{\Phi}^{\tau,\sigma}$ and $\mathbf{grad}_x\,\Phi$ by $\mathbf{grad}_x\,\tilde{\Phi}^{\tau,\sigma}$ and recall the sparse lifting matrix L_2 from (3) in order to find that

$$\widetilde{\mathsf{B}} = \sum_{\tau\times\sigma\in P_{\text{far}}} \mathsf{L}_2^\top \begin{pmatrix} \mathsf{V}^{\tau\top}\mathsf{S}^{\tau,\sigma}\mathsf{W}^{\sigma,1} \\ \mathsf{V}^{\tau\top}\mathsf{S}^{\tau,\sigma}\mathsf{W}^{\sigma,2} \\ \mathsf{V}^{\tau\top}\mathsf{S}^{\tau,\sigma}\mathsf{W}^{\sigma,3} \end{pmatrix} + \sum_{\tau\times\sigma\in P_{\text{near}}} \mathsf{B}^{\tau,\sigma}$$

is a good approximation of B, where

$$\mathsf{W}^{\sigma,\ell}_{j\mu} = \int_\sigma (\mathbf{b}_j)_\ell(\mathbf{x})\langle \mathbf{grad}\,\mathcal{L}^\sigma_\mu(\mathbf{x}), \mathbf{n}(\mathbf{x})\rangle - \mathbf{n}_\ell(\mathbf{x})\langle \mathbf{grad}\,\mathcal{L}^\sigma_\mu(\mathbf{x}), \mathbf{b}_j(\mathbf{x})\rangle\, dx$$

for $j \in \mathcal{E}$, $\mu \in \{1,\ldots,k\}$ and $\ell \in \{1,\ldots,3\}$.

For clusters with sons, the matrices $\mathsf{W}^{\sigma,\ell}$ can again be expressed by the transfer matrices $\mathsf{T}^{\sigma',\sigma}$, so we can re-use most of the matrices involved in the approximation of G and require additional storage only for clusters without sons and for the near-field.

6 Numerical experiment

We approximate the auxiliary matrix G on the unit sphere using a local interpolation operator of $m = 2$. We use the admissibility condition (5) for $\eta = 2$ and find the results given in Table 1. We can see that the approximation error is stable, while the memory and time requirements grow linearly.

Table 1. Approximation of the auxiliary matrix G

n	Mem[MB]	Mem[KB]/n	Build[s]	MVM[ms]	Error
512	3.9	7.68	1	4	2.3_{-4}
2048	21.6	10.54	5	46	4.2_{-4}
8192	113.9	13.90	18	269	4.3_{-4}
32768	461.4	14.08	79	1138	4.2_{-4}
131072	2024.4	15.45	305	4990	–
524288	7976.1	15.21	1224	19772	–
2097152	32023.8	15.27	4974	83181	–

References

1. S. Börm, L. Grasedyck, and W. Hackbusch, *Introduction to hierarchical matrices with applications*, Engineering Analysis with Boundary Elements, 27 (2003), pp. 405–422.
2. S. Börm and W. Hackbusch, *$\mathcal{H}^2$-matrix approximation of integral operators by interpolation*, Applied Numerical Mathematics, 43 (2002), pp. 129–143.
3. S. Börm and J. Ostrowski, *Fast evaluation of boundary integral operators arising from an eddy current problem*, Tech. Rep. 33, Max Planck Institute for Mathematics in the Sciences, 2003. To appear in Journal of Computational Physics.
4. L. Greengard and V. Rokhlin, *A fast algorithm for particle simulations*, Journal of Computational Physics, 73 (1987), pp. 325–348.
5. W. Hackbusch and Z. P. Nowak, *On the fast matrix multiplication in the boundary element method by panel clustering*, Numerische Mathematik, 54 (1989), pp. 463–491.
6. R. Hiptmair, *Symmetric coupling for eddy current problems*, SIAM J. Numer. Anal., 40 (2002), pp. 41–65.
7. J. Ostrowski, *Boundary Element Methods for Inductive Hardening*, PhD thesis, University of Tübingen, Germany, 2003. available under `http://w210.ub.uni-tuebingen.de/dbt/volltexte/2003/672`.
8. V. Rokhlin, *Rapid solution of integral equations of classical potential theory*, Journal of Computational Physics, 60 (1985), pp. 187–207.
9. C. Schwab and O. Sterz, *A scalar BEM for time harmonic eddy current problems with impedance boundary conditions*, in Scientific Computing in Electrical Engineering, M. Günther, D. Hecht, and U. van Rienen, eds., vol. 18 of Lecture Notes in Computational Science and Engineering, Springer, Berlin, 2001, pp. 129–136.

Adaptive Computation of Reactive Flows with Local Mesh Refinement and Model Adaptation

Malte Braack[1] and Alexandre Ern[2]

[1] Institut für Angewandte Mathematik, Universität Heidelberg
malte.braack@iwr.uni-heidelberg.de

[2] Cermics, Ecole nationale des ponts et chaussées, Marne la Vallée, France
ern@cermics.enpc.fr

Summary. An adaptive method for reactive flows involving locally refined meshes and different types of diffusion models is proposed. Starting with a less exact diffusion model, the model is changed locally throughout the computational domain to a more accurate and much more expensive model. An a posteriori error estimator provides reliable information on where to refine the mesh and where to adapt the model. Discretization and modeling errors are equilibrated.

1 Introduction

The underlying equations describing reactive flows are well-known but may involve models of different complexity, scales and accuracy. In various cases, the most accurate and validated model cannot be chosen in numerical simulations because of the large amount of computational costs. Simpler models usually need less computing time and involve less couplings between variables. For instance, the choice of diffusion models in gas mixtures is not straightforward. Although multicomponent diffusion models are accepted to be accurate [8], simpler and less accurate models, e.g. Fick's law, are widely used in practice for two- and three-dimensional simulations. While simple diffusion models, as for instance Fick's law, may involve only diagonal diffusion, multicomponent diffusion models leads to couplings between all chemical variables. For implicit solvers, these couplings lead to a huge fill-in in the (sparse) Jacobians. Due to the resulting high numerical cost of complex models, it is desirable to apply the complex diffusion model only in those regions of the computational domain where necessary; for instance in the flame front where a complex balance of reaction, convection and diffusion phenomena takes place. However, it is not a priori known, where an accurate diffusion model is necessary. In this work, we present an adaptive method which automatically detects the regions where an accurate diffusion model is important.

In the previous work [6], the mathematical background for a posteriori control of modeling errors and discretization errors is given. Other work addressing the estimation of modeling error includes [9, 13, 14]. For measuring the modeling error the variational formulation of the partial differential equation together with a duality argument is used. The a posteriori error estimator for the discretization error is obtained by using Galerkin orthogonality of a finite element discretization [3]. These two aspects allow to measure the overall error in terms of user-defined output functionals of the numerical solution. Both types of adaptivity (mesh-size and diffusion model) are merged in a such a way that both sources of errors are equilibrated and subsequently reduced. This strategy avoids the situation that the accuracy of the solution is affected by a poor diffusion model although very fine meshes are used, and vice-versa.

As an extension of the theoretical fundamentals in [6], here we present the application of mesh size adaptation and diffusion model adaptation and show numerical results for a combustion problem with two types of diffusion models.

We use the following notations: The usual L^2–scalar-product in a subdomain $\omega \subset \Omega$ will be denoted by $(\cdot,\cdot)_\omega$. By $(u,v) = (u,v)_\Omega$ we denote the integration over the entire computational domain $\Omega \subset I\!R^d, d = 2,3$. We denote the velocity by v, pressure by p, temperature by T, the n_s species mass fractions by y_k, $k = 1,\dots,n_s$, and density by ρ.

2 Variational formulation of the underlying equations

We start from the basic equations in variational formulation for steady-state reactive viscous flow describing the conservation of mass, momentum, energy and species mass fractions. For this we assemble all variables in the vector $u = (p, v, T, y_1, \dots, y_s)$ which is an element of a functional Hilbert space V. For test functions $\phi = (\xi, \psi, \sigma, \tau_0, \dots \tau_s) \in V$, $s := n_s - 1$, the following nonlinear forms are used:

$$
\begin{aligned}
a_1(u)(\phi) &:= (\operatorname{div}(\rho v), \xi)\,,\\
a_2(u)(\phi) &:= (\rho(v\cdot\nabla)v, \psi) + (\pi, \nabla\psi) - (p, \operatorname{div}\psi)\,,\\
a_3(u)(\phi) &:= (\rho c_p v\cdot\nabla T, \sigma) + (\lambda\nabla T, \nabla\sigma) + \sum_{k=1}^{n_s}(h_k m_k \dot{\omega}_k, \sigma)\,,\\
a_4(u)(\phi) &:= \sum_{k=1}^{s}\{(\rho v\cdot\nabla y_k, \tau_k) - (m_k\dot{\omega}_k, \tau_k)\}\,,\\
a(u)(\phi) &:= \sum_{i=1}^{4} a_i(u)(\phi)\,.
\end{aligned}
$$

The mass fraction y_{n_s} of the last species is set to $y_{n_s} = 1 - \sum_{i=1}^{s} y_i$, in order to ensure that the sum over all y_k is equal to 1. The density is considered as a coefficient determined by the gas law

$$
\rho = \frac{p\overline{m}}{RT}\,, \quad \overline{m} = \left(\sum_{k=1}^{n_s}\frac{y_k}{m_k}\right)^{-1}\,, \tag{1}
$$

with the universal gas constant R and the mean molar weight $\overline{m}$. Coefficients are the viscosity μ, the heat capacity c_p at constant pressure, and the head conductivity λ. For each species k, we have the molecular weight m_k, the specific enthalpy h_k, and the molecular production rate $\dot{\omega}_k$. The viscous tensor π is given as usual for compressible Newtonian fluids. For ease of presentation, in the form $a_3(u)(\phi)$ the effect of temperature flux due to diffusion fluxes of species with different specific heat capacities is neglected.

The form $a_3(u)(\phi)$ does not yet include species diffusion. We consider the following two models for the mass diffusion fluxes $\mathcal{F}_k$:

Fick's law: diagonal diffusion driven by the gradient of mole fractions $x_k = y_k\overline{m}/m_k$,

$$\mathcal{F}_k^F = -\rho D_k^* \frac{y_k}{x_k} \nabla x_k \,. \tag{2}$$

The diffusion coefficients $D_k^* = D_k^*(y)$ are given by an empirical law which is about 10% accurate, see [12].

Multicomponent diffusion: a full diffusion matrix driven by the gradients of species mole fractions:

$$\mathcal{F}_k^M = -\rho y_k \left\{ \sum_{l=1}^{n_s} D_{kl} \nabla x_l + \theta_k \nabla (\log T) \right\} . \tag{3}$$

The diffusion coefficients D_{kl} are given only implicit as solutions of linear systems. Therefore, the computational costs are higher as for the previous diffusion model. However, this form of diffusion flux can be derived by the theory of gases, see [8, 10].

Recent DNS computations [7] investigate the differences of these two models and show a legibly impact of model (ii) especially for lean and rich hydrogen flames.

Partial integration of the multicomponent model (3) leads to the semi-linear form

$$d(u)(\phi) := \sum_{k=1}^{s} (-\mathcal{F}_k^M, \nabla \tau_k) \,.$$

The residual of the problem will be denoted by

$$\varrho(u)(\phi) := (f, \phi) - a(u)(\phi) - d(u)(\phi) \,.$$

Now, the solution u fulfills the equation

$$u \in \widehat{u} + V : \quad \varrho(u)(\phi) = 0 \quad \forall \phi \in V \,,$$

where non-homogeneous Dirichlet conditions for u are described by $\widehat{u}$. Homogeneous Dirichlet conditions are already included in the choice of the space V.

3 Discretization

The discrete counterpart of the equations is the basis for the a posteriori error estimator we need later for the adaptive procedure.

In order to switch locally between the diffusion models (2) and (3), we introduce a (symbolic) parameter m and a subdomain Ω_m of Ω. In Ω_m, we use multicomponent diffusion, and in $\Omega_f := \Omega \setminus \Omega_m$ the simpler Fick's law. We define the diffusion part

$$d_m(u)(\phi) := \sum_{k=0}^{s} -\{(\mathcal{F}_k^M, \nabla \tau_k)_{\Omega_m} + (\mathcal{F}_k^F, \nabla \tau_k)_{\Omega_f}\},$$

and formulate a perturbed solution $u_m \in V$ solving

$$a(u_m)(\phi) + d_m(u_m)(\phi) = (f, \phi) \quad \forall \phi \in V.$$

The way to determine the subdomain Ω_m will be explained later on.

The discretization is done by conforming finite elements on a triangulation $\mathcal{T}_h$ of Ω. We denote the corresponding space by $V_h \subset V$. In order to handle the convective terms and the stiff pressure-velocity coupling one should add stabilization terms $s_h(\cdot,\cdot)$ to the discrete systems. Such stabilization can be done in different ways. We use the stabilization concept for the v,p-coupling proposed in [1] and for the convective terms in [11] and [2]. To this purpose, we need certain restrictions on the meshes used. We assume that the triangulation $\mathcal{T}_h$ is organized patch-wise: $\mathcal{T}_h$ results from a global refinement of a mesh $\mathcal{T}_{2h}$. Note that $\mathcal{T}_h$ contains in two dimensions twice as much hanging nodes as $\mathcal{T}_{2h}$. The same construction is possible in three dimensions.

By $i_{2h}^h : V_h \to V_{2h}$ we denote the nodal interpolation to the coarse grid, and by $\pi_h := i - i_{2h}^h : V_h \to V_h$ the projection operator which filters the small-scale fluctuations The stabilization form reads now:

$$\begin{aligned} s(u)(\phi) = {} & (\nabla \pi_h p, \delta_p \nabla \pi_h \xi) + ((\beta_v \cdot \nabla)\pi_h v, \delta_v (\beta_v \cdot \nabla)\pi_h \psi) \\ & + (\beta_T \cdot \nabla \pi_h T, \delta_T \beta_T \cdot \nabla \pi_h \sigma) + \sum_{k=1}^{s} (\beta_v \cdot \nabla \pi_h y_k, \delta_k \beta_v \cdot \nabla \pi_h \tau), \end{aligned}$$

with $\beta_v := \rho v$, $\beta_T := (\rho c_p v + \alpha)$ and piece-wise constant functions $\delta_p, \delta_v, \delta_T, \delta_1, \ldots, \delta_s$, depending on the mesh-size h. For further details, we refer to [2].

This leads us to the definition of the discrete residual

$$\varrho_{hm}(u)(\phi) := (f, \phi) - a(u)(\phi) - d_m(u)(\phi) - s_h(u)(\phi).$$

The reduced discrete system to be solved reads

$$u_{hm} \in V_h : \quad \varrho_{hm}(u_{hm})(\phi) = 0 \quad \forall \phi \in V_h.$$

The difference between the continuous and the reduced discrete residual is

$$\begin{aligned} g_{hm}(u)(\phi) &:= \varrho(u)(\phi) - \varrho_{hm}(u)(\phi) \\ &= -d(u)(\phi) + d_m(u)(\phi) + s_h(u)(\phi) \\ &= \sum_{k=0}^{s} (\mathcal{F}_k^M - \mathcal{F}_k^F, \nabla \tau_k)_{\Omega_f} + s_h(u)(\phi). \end{aligned}$$

This is the difference of the two diffusion models in the domain Ω_f, where Fick's law is used, and the additional stabilization terms, which are usually small. Since, in this work we do not focus on adaptively chosen stabilization terms, we neglect the contribution $s_h(u)(\phi)$ in the expression $g_{hm}(u)(\phi)$.

4 A posteriori control

For the local refinement of the mesh and the adaptivity respect to the diffusion model, we need an a-posteriori error estimator which gives us information of the two error contributions. We aim to measure the error respect to an arbitrary output functional $j : V \to I\!R$. We formulate the following dual residuals

$$\varrho^*(u)(z,\phi) := j(\phi) - a'(u)(\phi,z) - d'(u)(\phi,z)\,,$$
$$\varrho^*_{hm}(u)(z,\phi) := j(\phi) - a'(u)(\phi,z) - d'_m(u)(\phi,z) - s'_h(u)(\phi,z)\,.$$

We use the dual solution $z \in V$ to capture the influence of the error to the functional:

$$z \in V: \quad \varrho^*(u)(z,\phi) := 0 \quad \forall \phi \in V\,.$$

The corresponding reduced discrete version reads:

$$z_{hm} \in V_h: \quad \varrho^*_{hm}(u_{hm})(z_{hm},\phi) := 0 \quad \forall \phi \in V_h$$

The errors will be denoted by $e_u = u - u_{hm}$ and $e_z = z - z_{hm}$. We recall the error representation in [6] wherein a proof is also given.

Theorem 1. *If the semi-linear forms $a(u)(\cdot)$, $d(u)(\cdot)$, $s_h(u)(\phi)$ and the functional $j(u)$ are sufficiently differentiable with respect to u, then it holds*

$$j(u) - j(u_{hm}) = g_{hm}(u_{hm})(z_{hm}) + \frac{1}{2}\{g_{hm}(u_{hm})(e_z) + g'_{hm}(u_{hm})(e_u, z_{hm})\}$$
$$+ \frac{1}{2}\{\varrho_{hm}(u_{hm})(z - i_h z) + \varrho^*_{hm}(u_{hm}, z_{hm})(u - i_h u)\} + R\,,$$

where $i_h : V \to V_h$ is an arbitrary interpolation operator and a remainder R which is cubic in the error $e = \{e_u, e_z\}$.

For the specific form of the remainder R and a proof of this Theorem, we refer to [6].

The error representation of the Theorem stated above cannot be directly used numerically, because it involves the unknown primal and dual solutions u and z, respectively. However, the first term $g_{hm}(u_{hm})(z_{hm})$ can be easily computed, because it depends only on the reduced discrete solutions u_{hm} and z_{hm}. Furthermore, the terms $g_{hm}(u_{hm})(e_z)$ and $g'_{hm}(u_{hm})(e_u, z_{hm})$ are quadratic in the modeling error since they involve both, e as an argument and the difference of the two diffusion models in the expression g_{hm}. We neglect these contributions in the estimator. However, if more accuracy of the estimator is required, these terms can also be approximated. Evaluation of the remainder R is not worthwhile, because it is cubic in the error.

The terms $\varrho_{hm}(u_{hm})(z - i_h z)$ and $\varrho^*_{hm}(u_{hm}, z_{hm})(u - i_h u)$ describe the discretization error and have to be approximated by a numerical evaluation of the interpolation errors $z - i_h z$ and $u - i_h u$. An efficient possibility to do this, is the recovery process of the computed quantities by higher-order polynomials, see [4]. For instance, in the case of quadrilaterals and piecewise bilinear elements (so called Q_1 elements), the interpolation can be done on biquadratic elements. Let $i_{2h}^{(2)} : V_h \to V_{2h}^{(2)}$ be the quadratic interpolation of piecewise bilinears on $\mathcal{T}_h$ onto biquadratic finite elements on $\mathcal{T}_{2h}$.

The interpolation errors of z, for instance, will be numerically approximated by

$$z - i_h z \approx i_{2h}^{(2)} z_{hm} - z_{hm} .$$

Taking into account that the residuals $\varrho_{hm}(u_{hm})(\phi)$ and $\varrho^*_{hm}(u_{hm}, z_{hm})(\phi)$ with respect to a discrete test function $\phi \in V_h$ vanish, leads to the following estimator η consisting of two parts

$$j(u) - j(u_{hm}) \approx \eta := \eta_h + \eta_m , \tag{4}$$

$$\eta_h := \frac{1}{2}\{\varrho_{hm}(u_{hm})(i_{2h}^{(2)} z_{hm}) + \varrho^*_{hm}(u_{hm}, z_{hm})(i_{2h}^{(2)} u_{hm})\} , \tag{5}$$

$$\eta_m := g_{hm}(u_{hm})(z_{hm}) . \tag{6}$$

The part η_h of the estimator can be considered as contributions of the discretization, and the part η_m measures the influence of the model. For multicomponent diffusion, the evaluation of η_m is expensive. However, the gain of an adaptive algorithm with local model modification becomes substantial, since we do not need to include the global detailed model neither in each residual evaluation nor in the Jacobian for solving the equations.

On the basis of the estimators η_h and η_m, local error indicators can be derived. A standard method is partial integration of the diffusive parts and the application of Cauchy-Schwarz on each element. We proceed in a different way by filtering coarse grid contributions. We obtain nodal quantities $\eta_{h,i}$ and $\eta_{m,i}$ for each node $\mathcal{N}_i$ of the mesh:

$$|\eta| \le \sum_{i=1}^{n} (\eta_{h,i} + \eta_{m,i}) .$$

A proof is given in [6]. Further details, especially concerning the adaptation strategy for equilibrating both error contributions are given in [5].

5 Computation of an ozone flame

We investigate the methodology presented above for an ozone decomposition flame in a two-dimensional geometry with a moderate impact of the diffusion model. Moreover, the complexity of the problem size is moderate enough to obtain a reference solution on a very fine mesh. This reference solution is used for validation of the error estimators. A more involved example can be found in [5], where an hydrogen flame is investigated.

The flame under consideration is modeled with three chemical species, namely O_3, O_2, and O-atoms. The reaction mechanism consists of six reactions, see [3]. At the inflow ($x = 0$), we prescribe 20% mass fraction for ozone and 80% mass fraction for oxygen. Furthermore, the inflow velocity profile is parabolic with maximum velocity $v = 35\,\text{cm/s}$. The computational domain corresponds to a Cartesian tube, $\Omega = [0, 2\,\text{cm}] \times [0, 5\,\text{mm}]$. The temperature is fixed at the inflow, bottom, and top boundary according to

$$T = T_{min} + \Delta T \cdot \exp\{-\sigma(x - x_0)^2\},$$

with $x_0 = 5\,\text{mm}$, $\sigma = 10^5$, $T_{\text{min}} = 298\,K$, and $\Delta T = 502\,K$. A homogeneous Neumann condition is imposed at the outflow boundary.

Figure 1 shows the computed profile of the mass fractions of O-atoms when the multicomponent diffusion model is used over the whole computational domain. For this type of flame, Fick's law is a good approximation of the multicomponent diffusion model. Therefore, the difference between both diffusion models is small, and the computation with Fick's law yields very similar pictures. However, the difference in the model is in the same range as the discretization error on the coarsest mesh.

Now, we report on the adaptive algorithm described before. The initial mesh is an equidistant mesh with 585 nodes. We start with Fick's law (2) for the species mass diffusion fluxes over the whole computational domain. After computing the stationary solution u_h on this mesh and with the crude diffusion model, we compute the associate dual solution z_h for the functional $j(u) = c\int_\Omega y_O dx$, with a constant scaling factor c. This functional gives the mean value of mass fractions of O-atoms in the computational domain Ω. The error indicators η_h and η_m are obtained from (5) and (6). On the basis of η_h we change the mesh-size locally by bisection of element edges. According to η_m, we change the diffusion model locally from Fick's diffusion to multicomponent diffusion.

This procedure is iterated various times. Results are listed in Table 1. The first column contains the number of nodes of the mesh, the second column contains the relative amount of cells (in percent) where the multicomponent diffusion model is used. The two parts of the estimator η_h and η_m are listed in columns 3 and 4, respectively. The sum of these two terms yields the error estimator η (column 5). The effectivity index $I_{\text{eff}} = \eta / j(u - u_{hm})$ is listed in the last column and is obtained by using the reference solution computed on a very fine mesh and with the accurate diffusion model. For an exact error estimator, the efficiency index would be equal to 1. Our values are in the range of $1.5 - 2.6$, which means that the estimator η slightly overestimates the error.

Fig. 1. O-atom mass fraction for the ozone flame. The maximum value is $1.24 \cdot 10^{-3}$.

Table 1. Results for the ozone flame: number of nodes (#nodes); fraction of cells flagged for multicomponent diffusion (% of multi.); estimator of the discretization error η_h; estimator of the model error η_m; their sum η; the true error $j(u - u_{hm})$; the effectivity index I_{eff}

#nodes	% of multi.	η_h	η_m	η	$j(u - u_{hm})$	I_{eff}
585	0	2.168	1.043	3.210	2.031	1.58
1 047	21.1	1.250	9.953e-2	1.350	5.385e-1	2.51
2 085	37.4	1.584e-1	7.729e-2	2.356e-1	1.378e-1	1.71
4 871	48.9	7.488e-2	3.830e-2	1.132e-1	5.351e-2	2.12
12 421	52.4	5.605e-2	1.602e-2	7.206e-2	3.186e-2	2.26
30 013	66.3	5.029e-2	9.372e-3	5.966e-2	2.757e-2	2.16
81 021	79.4	2.160e-2	6.017e-3	2.761e-2	1.065e-2	2.59

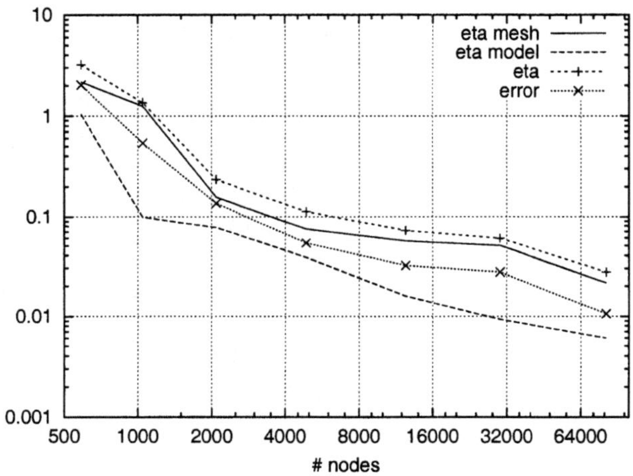

Fig. 2. Ozone flame: estimator η_m for the modeling error; estimator η_h for the discretization error; their sum η; true error $j(u - u_h)$ as a function of the number of nodes (mesh points)

However, the error estimator is reliable since it provides an upper bound for the actual error.

The adaptive algorithm balances both types of errors by adapting the mesh-size and the model. Figure 2 illustrates how the two sources of error are equilibrated. We plot the estimators η_h and η_m, their sum η, and the true error $j(u - u_{hm})$ as a function of the number of mesh nodes. The estimator and the true error clearly show the same asymptotic behavior.

The sequence of locally refined meshes with 1 047, 2 085, and 4 871 nodes is shown in Figure 3. The darker areas indicate the part of the computational domain where the multicomponent diffusion model is used. In the remaining (light) part, the simple Fick law is used. We observe that the estimator de-

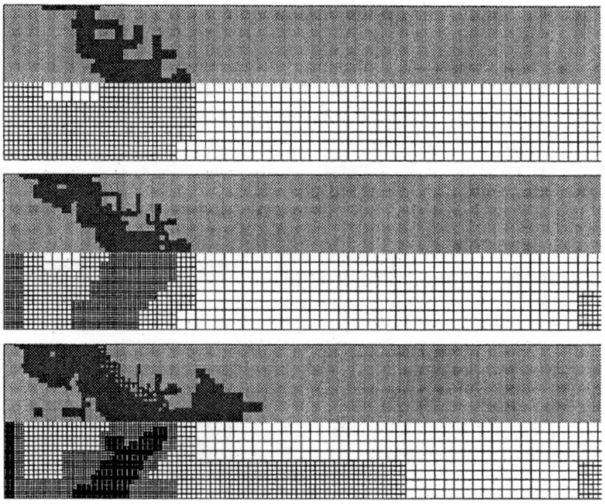

Fig. 3. On the upper half of each picture, the areas where multicomponent diffusion is used (dark/red areas) and where Fick's law is used (light areas) are indicated; the lower half shows the corresponding locally refined mesh

tects quite well the reaction area where a difference in both diffusion models influences the accuracy of the output functional $j(u)$.

References

1. R. Becker and M. Braack. A finite element pressure gradient stabilization for the Stokes equations based on local projections. *Calcolo*, 38(4):173–199, 2001.
2. R. Becker and M. Braack. A two-level stabilization scheme for the Navier-Stokes equations. In Feistauer, editor, *Enumath Proceedings*, Berlin, submitted, 2003. Springer.
3. R. Becker, M. Braack, and R. Rannacher. Numerical simulation of laminar flames at low Mach number with adaptive finite elements. *Combust. Theory Modelling*, 3:503–534, 1999.
4. R. Becker and R. Rannacher. An optimal control approach to a posteriori error estimation in finite element methods. In A. Iserles, editor, *Acta Numerica 2001*. Cambridge University Press, 2001.
5. M. Braack and A. Ern. Coupling multimodeling with local mesh refinement for the numerical solution of laminar flames. in preparation, 2003.
6. M. Braack and A. Ern. A posteriori control of modeling errors and discretization errors. *Multiscale Model. Simul.*, 1(2):221–238, 2003.
7. J. de Charentenay and A. Ern. Multicomponent transport impact on turbulent premixed H_2/O_2 flames. *Combust. Theory Modelling*, 6:439–462, 2002.
8. A. Ern and V. Giovangigli. *Multicomponent Transport Algorithms.* Lecture Notes in Physics, m24, Springer, 1994.
9. L. Fatone, P. Gervasio, and A. Quarteroni. Multimodels for incompressible flows: iterative solutions for the navier-stokes/oseen coupling. *M2AN, Math. Model. Numer. Anal.*, 35(3):549–574, 2001.

10. V. Giovangigli. *Multicomponent Flow Modeling.* Birkhäuser, Boston, 1999.
11. J.-L. Guermond. Stabilization of Galerkin approximations of transport equations by subgrid modeling. *Modél. Math. Anal. Numér.*, 33(6):1293–1316, 1999.
12. J. O. Hirschfelder and C. F. Curtiss. *Flame and Explosion Phenomena.* Williams and Wilkins Cp., Baltimore, 1949.
13. J. T. Oden and S. Prudhomme. Estimation of modeling error in computational mechanics. *J. Comput. Phys.*, 182:496–515, 2002.
14. E. Stein and S. Ohnimus. Anisotropic discretization- and model-error estimation in. *Comput. Methods Appl. Mech. Engrg.*, 176:363–385, 1999.

An Alternative to the Least-Squares Mixed Finite Element Method for Elliptic Problems

Jan Brandts[1] and Yanping Chen[2]

[1] Korteweg-de Vries Institute for Mathematics, Faculty of Science, University of Amsterdam, Plantage Muidergracht 24, 1018 TV Amsterdam, Netherlands. *brandts@science.uva.nl*

[2] Department of Mathematics, Xiangtan University, Xiangtan 411105, China. *ypchen@xtu.edu.cn*

Summary. In this paper we derive a strengthened Cauchy-Schwarz inequality that enables us to formulate a short and transparant proof of the coercivity of a Least Squares Mixed Finite Element bilinear form. Also, it shows that the coupling between $H_0^1(\Omega)$ and $H(\text{div}; \Omega)$ is weak enough to be neglected. This results in an alternative way to compute approximations of both the scalar variable and its gradient for second order elliptic problems.

1 Least squares mixed finite elements

We consider the application of the least-squares mixed finite element method to the following second order elliptic problem. Let Ω be a bounded domain of arbitrary dimension with Lipschitz continuous boundary. Given $f \in H^{-1}(\Omega)$, find $u \in H_0^1(\Omega)$ such that

$$-\text{div}\,(A\nabla u) = f \text{ in } \Omega, \quad u = 0 \text{ on } \partial\Omega, \tag{1}$$

where A is uniformly symmetric positive definite with Lipschitz continuous coefficients and all eigenvalues in the interval $[\beta^2, \beta^{-2}]$ for some $\beta \in (0,1]$. The least-squares mixed finite element method first writes the second order differential equation as a system of two first order equations,

$$\mathbf{p} = -A\nabla u \text{ in } \Omega, \quad \text{div}\,\mathbf{p} = f \text{ in } \Omega. \tag{2}$$

Denoting the L^2 inner product and norm by $(\cdot,\cdot)_0$ and $|\cdot|_0$, the following quadratic functional $J : H_0^1(\Omega) \times \mathbf{H}(\text{div}\,; \Omega) \to \mathbb{R}$

$$J(v, \mathbf{q}) = |f - \text{div}\,\mathbf{q}|_0^2 + (\mathbf{q} + A\nabla v, A^{-1}(\mathbf{q} + A\nabla v))_0, \tag{3}$$

is minimized over suitable subspaces $V_h \times \mathbf{\Gamma}_h \subset H_0^1(\Omega) \times \mathbf{H}(\text{div}\,; \Omega)$. Setting the first variation in (3) to zero leads to the following discrete problem to solve,

$$\forall (v_h, \mathbf{q}_h) \in V_h \times \mathbf{\Gamma}_h, \quad B(u_h, \mathbf{p}_h | v_h, \mathbf{q}_h) = (f, \text{div}\,\mathbf{q}_h)_0, \tag{4}$$

where the bilinear form $B : [H_0^1(\Omega) \times \mathbf{H}(\mathrm{div}\,;\Omega)] \times [H_0^1(\Omega) \times \mathbf{H}(\mathrm{div}\,;\Omega)] \to \mathbb{R}$ is defined by

$$B(w, \mathbf{r}; v, \mathbf{q}) = (\mathrm{div}\,\mathbf{r}, \mathrm{div}\,\mathbf{q})_0 + (\mathbf{r} + A\nabla w, A^{-1}(\mathbf{q} + A\nabla v))_0. \tag{5}$$

It was proved in [4] that B is continuous and coercive. The proof for coercivity was in our opinion somewhat tedious and may also be derived as a simple corollary of the following lemma. This lemma, which may be called a strengthened Cauchy-Schwarz inequality, may also serve useful in a different context.

Firstly, define norms on $\mathbf{H}(\mathrm{div}\,;\Omega)$ and $H_0^1(\Omega)$ and on their product space by

$$\|\mathbf{q}\|^2_{\mathrm{div},A} = (A^{-1}\mathbf{q}, \mathbf{q})_0 + |\mathrm{div}\,\mathbf{q}|_0^2, \quad |v|^2_{1,A} = (A\nabla v, \nabla v)_0, \tag{6}$$

$$\|(v, \mathbf{q})\|^2_{1\times\mathrm{div},A} = |v|^2_{1,A} + \|\mathbf{q}\|^2_{\mathrm{div},A}. \tag{7}$$

If $A = I$, they reduce to the usual norms on those spaces. They are equivalent to them in case $A \neq I$. In particular, we will explicitly use that for all $v \in H_0^1(\Omega)$,

$$\beta^2 |v|_1^2 \le |v|^2_{1,A} \le \beta^{-2}|v|_1^2. \tag{8}$$

Lemma 1. *For all* $\mathbf{q} \in \mathbf{H}(\mathrm{div}\,;\Omega)$ *and* $v \in H_0^1(\Omega)$ *we have*

$$|(\mathbf{q}, \nabla v)_0| \le \gamma |v|_{1,A} \|\mathbf{q}\|_{\mathrm{div},A} \tag{9}$$

where

$$0 < \gamma = \frac{d}{\sqrt{d^2 + \beta^2}} < 1 \quad \textit{with} \quad 0 < d = \sup_{0 \neq v \in H_0^1(\Omega)} \frac{|v|_0}{|v|_1}. \tag{10}$$

Proof. Let $\mathbf{q} \in \mathbf{H}(\mathrm{div}\,;\Omega)$ and $v \in H_0^1(\Omega)$ be given. Then

$$|(\mathbf{q}, \nabla v)_0| = |(A^{-1/2}\mathbf{q}, A^{1/2}\nabla v)_0| \le |v|_{1,A}\sqrt{(A^{-1}\mathbf{q}, \mathbf{q})_0}. \tag{11}$$

By definition of d and in view of (8), we obtain by using Green's formula that

$$|(\mathbf{q}, \nabla v)_0| = |(\mathrm{div}\,\mathbf{q}, v)_0| \le |v|_0 |\mathrm{div}\,\mathbf{q}|_0 \le \frac{d}{\beta}|v|_{1,A}|\mathrm{div}\,\mathbf{q}|_0. \tag{12}$$

Multiply (11) by d/β and square the result, and add it to the square of (12). This gives

$$\left(1 + \frac{d^2}{\beta^2}\right) |(\mathbf{q}, \nabla v)_0|^2 \le \frac{d^2}{\beta^2}|v|^2_{1,A}\left((A^{-1}\mathbf{q}, \mathbf{q})_0 + |\mathrm{div}\,\mathbf{q}|_0^2\right) = \frac{d^2}{\beta^2}|v|^2_{1,A}\|\mathbf{q}\|^2_{\mathrm{div},A}. \tag{13}$$

This proves the statement. ◊

Remark 1. Instead of using $|v|_0 \le d|v|_1 \le d/\beta |v|_{1,A}$ as in (12), we might also have chosen to define just one single constant η giving rise to a possibly smaller constant γ in (9) as follows,

$$0 < \gamma = \frac{\eta}{\sqrt{\eta^2 + 1}} \quad \text{with} \quad 0 < \eta = \sup_{0 \neq v \in H_0^1(\Omega)} \frac{|v|_0}{|v|_{1,A}}. \tag{14}$$

Remark 2. If we define $\|\cdot\|_{1,A}^2 = |\cdot|_0^2 + |\cdot|_{1,A}^2$, then following the same lines as in the proof of Lemma 1 we get,

$$|(\mathbf{q}, \nabla v)_0| \leq \frac{1}{2}\|v\|_{1,A}\|\mathbf{q}\|_{\mathrm{div},A}. \tag{15}$$

This seems a more symmetric and in a sense even stronger result. However, the norm $\|\cdot\|_{1,A}$ is too strong for the applications to be discussed below.

Corollary 1. *With γ as in* (10)*, for all non-zero $(v, \mathbf{q}) \in H_0^1(\Omega) \times \mathbf{H}(\mathrm{div}\,;\Omega)$ we have*

$$B(v, \mathbf{q}|v, \mathbf{q}) \geq (1-\gamma)\|(v, \mathbf{q})\|_{1\times\mathrm{div},A}^2 > 0. \tag{16}$$

Proof. By definition of B we find

$$B(v, \mathbf{q}|v, \mathbf{q}) = \|(v, \mathbf{q})\|_{1\times\mathrm{div},A}^2 + 2(\mathbf{q}, \nabla v)_0 \geq \|(v, \mathbf{q})\|_{1\times\mathrm{div},A}^2 - 2|(\mathbf{q}, \nabla v)_0|. \tag{17}$$

Lemma 1 and the inequality $2|ab| \leq a^2 + b^2$ give that

$$2|(\mathbf{q}, \nabla v)_0| \leq 2\gamma|v|_{1,A}\|\mathbf{q}\|_{\mathrm{div},A} \leq \gamma\|(v, \mathbf{q})\|_{1\times\mathrm{div},A}^2. \tag{18}$$

Since $\gamma < 1$, the statement follows. ◊

Remark 3. Using Lemma 1, it can also easily be shown that B is continuous in the sense that

$$|B(w, \mathbf{r}; v, \mathbf{q})| \leq (1+\gamma)\|(w, \mathbf{r})\|_{1\times\mathrm{div},A}\|(v, \mathbf{q})\|_{1\times\mathrm{div},A},$$

which, in view of the constant for the coercivity in Corollary 1, is an appealing result.

The continuity and coercivity of B give that the Lax-Milgram lemma assures the existence of a unique pair $(u_h, \mathbf{p}_h) \in V_h \times \mathbf{\Gamma}_h$ of approximations of $(u, \mathbf{p})$. Céa's lemma now implies quasi-optimality in the sense that for all $(v_h, \mathbf{q}_h) \in H_0^1(\Omega) \times \mathbf{H}(\mathrm{div}\,;\Omega)$,

$$\|(u - u_h, \mathbf{p} - \mathbf{p}_h)\|_{1\times\mathrm{div},A} \leq \frac{1+\gamma}{1-\gamma}\|(u - v_h, \mathbf{p} - \mathbf{q}_h)\|_{1\times\mathrm{div},A}. \tag{19}$$

From this perspective, it may seem as if it is not wise to employ spaces V_h and $\mathbf{\Gamma}_h$ which have different approximation qualities, because the approximation quality of the product space is not better than that of the worst of the two. However, it can be shown that under rather weak conditions, there is still a certain amount of independence present between the two approximations. In terms of the linear algebraic problem that arises from (4) this means that the two diagonal blocks of the system matrix, which correspond to the interactions of V_h with itself and of $\mathbf{\Gamma}_h$ with itself, are only weakly coupled by the two off-diagonal blocks, which represent the interactions between V_h and $\mathbf{\Gamma}_h$. In fact, this coupling is so weak that solving just two well-chosen linear systems with the diagonal blocks only, leads to approximations u_h^0 and $\mathbf{p}_h^0$ for which

$$\|(u_h - u_h^0, \mathbf{p}_h - \mathbf{p}_h^0)\|_{1\times\mathrm{div},A} \leq Ch\|(u - u_h, \mathbf{p} - \mathbf{p}_h)\|_{1\times\mathrm{div},A}. \quad (20)$$

This defines an approximation method for $(u, \mathbf{p})$ in $V_h \times \mathbf{\Gamma}_h$ that gives approximations that are *superclose* to the least-squares mixed approximations $(u_h, \mathbf{p}_h)$, but which is computationally much more efficient.

2 Solving the system of linear equations

Let $(v_j)_{j=1}^M$ be a basis for V_h and $(\mathbf{q}_j)_{j=1}^N$ a basis for $\mathbf{\Gamma}_h$. Define matrices $S = (s_{ij}), C = (c_{ij}), D = (d_{ij})$ and $G = (g_{ij})$ by means of their entries

$$s_{ij} = (A\nabla_i, \nabla v_j),\ c_{ij} = (v_i, \mathrm{div}\,\mathbf{q}_j),\ d_{ij} = (A^{-1}\mathbf{q}_i, \mathbf{q}_j),\ g_{ij} = (\mathrm{div}\,\mathbf{q}_i, \mathrm{div}\,\mathbf{q}_j).$$

Let $f_N \in \mathbb{R}^N$ have entries $(f, \mathrm{div}\,\mathbf{q}_j)$, and $f_M \in \mathbb{R}^M$ entries (f, v_j). Define $\mathbf{p}_N \in \mathbb{R}^N$ and $u_M, u_M^s \in \mathbb{R}^M$ as the solutions of the systems

$$\begin{pmatrix} D+G & C \\ C^* & S \end{pmatrix} \begin{pmatrix} \mathbf{p}_N \\ u_M \end{pmatrix} = \begin{pmatrix} f_N \\ 0 \end{pmatrix} \quad \text{and} \quad Su_M^s = f_M. \quad (21)$$

Let e_k^j be the jth row of the $k \times k$ identity matrix, then

$$\mathbf{p}_h = \sum_{j=1}^N (e_N^j \mathbf{p}_N)\mathbf{q}_j \in \mathbf{\Gamma}_h \quad \text{and} \quad u_h = \sum_{j=1}^M (e_M^j u_M) v_j \in V_h \quad (22)$$

are the solution of (4), whereas

$$u_h^s = \sum_{j=1}^M (e_M^j u_M^s) v_j \in V_h$$

yields the standard finite element approximation of u in V_h.

Now, given an approximation u_M^0 of u_M, we may substitute it in the first (block)-equation of (21) to obtain an approximation $\mathbf{p}_N^0$ of $\mathbf{p}_N$ from

$$(D+G)\mathbf{p}_N^0 = f_N - Cu_M^0. \quad (23)$$

Write u_h^0 and $\mathbf{p}_h^0$ for the finite element functions corresponding to the vectors u_M^0 and $\mathbf{p}_N^0$, respectively. Then the counterpart of (23) in terms of the finite element spaces is to write out (4) and substitute u_h^0 for u_h, resulting in

$$(A^{-1}\mathbf{p}_h^0, \mathbf{q}_h) + (\mathrm{div}\,\mathbf{p}_h^0, \mathrm{div}\,\mathbf{q}_h) = (f, \mathrm{div}\,\mathbf{q}_h) + (u_h^0, \mathrm{div}\,\mathbf{q}_h). \quad (24)$$

Notice that

$$(A^{-1}\mathbf{q}_h, \mathbf{q}_h) + (\operatorname{div}\mathbf{q}_h, \operatorname{div}\mathbf{q}_h) = \|\mathbf{q}_h\|^2_{\operatorname{div},A}, \tag{25}$$

hence $D+G$ is invertible and the approximation $\mathbf{p}_h^0$ well-defined. It can be used to compute a hopefully better approximation u_M^1 of u_M by substituting $\mathbf{p}_N^0$ into the second (block)-equation in (21), and to solve $Su_M^1 = -C^*p_N^0$. If we continue like this, we get the Block Gauss-Seidel iteration for the linear system (21) as follows,

$$\text{given } u_M^0, \text{ iterate} \quad \begin{cases} (D+G)\mathbf{p}_N^j = f_N - Cu_M^j, \\ Su_M^{j+1} = -C^*\mathbf{p}_N^j. \end{cases} \tag{26}$$

If we write u_h^j and $\mathbf{p}_h^j$ for the finite element functions that correspond to the vectors u_M^j and $\mathbf{p}_N^j$, respectively, this iteration reads as

$$\text{given } u_h^0, \text{ iterate} \begin{cases} (A^{-1}\mathbf{p}_h^j, \mathbf{q}_h) + (\operatorname{div}\mathbf{p}_h^j, \operatorname{div}\mathbf{q}_h) = (f, \operatorname{div}\mathbf{q}_h) + (u_h^j, \operatorname{div}\mathbf{q}_h), \\ (A\nabla u_h^{j+1}, \nabla v_h) = -(v_h, \operatorname{div}\mathbf{p}_h^j). \end{cases} \tag{27}$$

We will now study its convergence.

Corollary 2. *With γ as in* (10), *the iterates $(u_h^j, \mathbf{p}_h^j)$ defined in* (27) *satisfy*

$$\gamma^{-1}\|u_h - u_h^{j+1}|_{1,A} \le \|\mathbf{p}_h - \mathbf{p}_h^j\|_{\operatorname{div},A} \le \gamma|u_h - u_h^j|_{1,A}. \tag{28}$$

Proof. Subtracting (27) from the least-squares mixed discrete equations (4) and testing with $\mathbf{q}_h = \mathbf{p}_h - \mathbf{p}_h^j$ and with $v_h = u_h - u_h^{j+1}$ results in

$$\|\mathbf{p}_h - \mathbf{p}_h^j\|^2_{\operatorname{div},A} = (u_h - u_h^j, \operatorname{div}(\mathbf{p}_h - \mathbf{p}_h^j)) \tag{29}$$

and

$$|u_h - u_h^{j+1}|^2_{1,A} = -(u_h - u_h^{j+1}, \operatorname{div}(\mathbf{p}_h - \mathbf{p}_h^j)). \tag{30}$$

Applying Green's formula and Lemma 1 gives the statement. ◇

This proves convergence of the Block Gauss-Seidel iteration (21) with a convergence factor that is independent of the dimension of the subspace. In each step of the Block Gauss-Seidel iteration, a system with S and one with $D+G$ should be solved. Even though for many finite element spaces this can be done in optimal computational complexity with MultiGrid solvers [1], there is no need to do so more than once if the right start vector is chosen. Indeed, choosing $u_h^0 := u_h^s$, the standard finite element approximation of the problem defined via the linear system $Su_M^s = f_M$, and computing $\mathbf{p}_h^0$ is already sufficient in most situations. This procedure costs only one solve with S and one with $D+G$.

Theorem 1. *Let $u_h^0 = u_h^s$ be the standard finite element approximation of* (1) *resulting from solving the linear system $Su_M^s = f_M$. Solve $\mathbf{p}_h^0$ from $(D + G)\mathbf{p}_h^0 = f_N - Cu_M^s$. Then, under the assumtion that V_h and Γ_h satisfy*

– $(\mathbf{\Gamma}_h, \operatorname{div} \mathbf{\Gamma}_h)$ *is a Babuska Brezzi stable mixed finite element pair,*
– *For all* $f \in L^2(\Omega)$, *the solution* u *of (1) satisfies* $\|u\|_2 \leq C|f|_0$,
– $\forall v \in H_0^1(\Omega) \cap H^2(\Omega),\ \exists v_h \in V_h,\ |v - v_h|_1 \leq Ch|v|_2$,
– $\forall \mathbf{q} \in [H^1(\Omega)]^2,\ \exists \mathbf{q}_h \in \mathbf{\Gamma}_h,\ |\mathbf{q} - \mathbf{q}_h|_0 \leq Ch|\mathbf{q}|_1$,
– $\forall \mathbf{q} \in [H^2(\Omega)]^2,\ \exists \mathbf{q}_h \in \mathbf{\Gamma}_h,\ |\operatorname{div} \mathbf{q} - \operatorname{div} \mathbf{q}_h|_0 \leq Ch|\mathbf{q}|_2$,

we have that

$$\|(u_h - u_h^0, \mathbf{p}_h - \mathbf{p}_h^0)\|_{1\times\mathrm{div},A} \leq Ch\|(u - u_h, \mathbf{p} - \mathbf{p}_h)\|_{1\times\mathrm{div},A}. \tag{31}$$

Proof. The assumptions above imply that $|u_h - u_h^s|_1 \leq Ch\|(u - u_h, \mathbf{p} - \mathbf{p}_h)\|_{1\times\mathrm{div}}$, which is a supercloseness result proved in [2]. By equivalence of the norms defined in terms of A with their counterparts defined by taking $A = I$, we may switch to A norms. Corollary 2 then shows that $\|\mathbf{p}_h - \mathbf{p}_h^0\|_{\mathrm{div},A}$ shares the same upper bound. Hence, the statement follows. ◊

3 Conclusions

The above shows that under rather weak conditions, $u_h^s, \mathbf{p}_h^0$ are higher order perturbations of the least-squares mixed finite element solutions $u_h, \mathbf{p}_h$. Since the computation of $u_h^s, \mathbf{p}_h^0$ requires only one solve with S and one with $D + G$, they are much cheaper to compute than $u_h, \mathbf{p}_h$, whereas they cannot be distinguished from one another. Hence, it does not make sense to apply the least-squares mixed method under these conditions.

In fact, instead of putting energy into solving for $\mathbf{p}_h^0$, it is also possible[1] to construct yet another approximation for $\mathbf{p}_h$ in $\mathbf{\Gamma}_h$ by means of a simple local post-processing u_h^s. It is however not clear when or if such a postprocessed approximation will also be a higher order perturbation of $\mathbf{p}_h$, as $\mathbf{p}_h^0$ is.

Acknowledgments

J.H. Brandts was supported by a Research Fellowship of the Royal Netherlands Academy of Arts and Sciences. Y.P. Chen was supported by National Science Foundation of China, the Foundation of China State Education Commission and the Special Funds for Major State Basic Research Projects. Both authors gratefully acknowledge this support, and thank the referee for valuable comments on an earlier version of this paper.

References

1. Arnold, D.N., Falk, R.S, Winther, R. (1997). Preconditioning in $\mathbf{H}(\mathrm{div}\,;\Omega)$ and applications. Math. Comp., **66**, 957–984.

[1] We thank Prof. R. Rannacher for pointing this out.

2. Brandts, J.H., Chen, Y., Yang, J. (2003). Analysis of least-squares mixed finite elements in terms of standard and mixed elements. UvA Numerica Preprint **08**, University of Amsterdam, Netherlands. Submitted.
3. Pehlivanov, A.I., Carey, G.F., Lazarov, R.D. (1994). Least-squares Mixed Finite Elements for Second Order Elliptic Problems. SIAM J. Numer. Anal., **31**, 1368–1377.
4. Pehlivanov, A.I., Carey, G.F., Vassilevski, P.S. (1996): Least-squares mixed finite element methods for non-selfadjoint elliptic problems. I. Error estimates. Numer. Math., **72**, 501–522.

Limit Analysis Method in Electrostatics

Igor A. Brigadnov

North-Western State Technical University, Millionnaya 5, St. Petersburg, 191186, Russia *brigadnov@nwpi.ru*

Summary. The limit analysis problem (LAP) for estimation of electric durability for a dielectric in a powerful electric field is examined. The appropriate dual problem is formulated. After the standard piecewise linear continuous finite-element approximation the dual LAP is transformed into the problem of mathematical programming with linear equality constraints. This finite dimension problem is effectively solved by the standard method of gradient projection.

1 Introduction

Investigation of the electrostatic boundary-value problems (BVPs) for dielectrics in a powerful electric field is of particular interest in both theory and practice. The current research is motivated by significance and practical interests in Electrical Engineering and Microelectronics.

The electric state of a medium in a given domain $\Omega \subset \mathbb{R}^3$ is characterized by the bulk and surface density of charges and by vectors of electric field intensity $\mathbf{E} = \{E_i\} \in \mathbb{R}^3$, electric induction $\mathbf{D} = \{D_i\} \in \mathbb{R}^3$ and electric current density $\mathbf{J} = \{J_i\} \in \mathbb{R}^3$. Vector $\mathbf{D}$ is introduced by the relation $\mathbf{D} = \varepsilon_0\mathbf{E} + \mathbf{P}$, where $\varepsilon_0 \approx 8.85{\cdot}10^{-12}$ is the electric permittivity of a vacuum and $\mathbf{P} \in \mathbb{R}^3$ is the vector of polarization density [11, 14, 16]. For the electric field intensity the electrostatic potential u is introduced such that $\mathbf{E}(\mathbf{x}) = -\nabla u(\mathbf{x})$ for almost every $\mathbf{x} \in \Omega$, where $\nabla = \partial/\partial\mathbf{x}$ is the differential vector-operator.

In weak electric fields the conductivity current in a dielectrical medium is practically absent and the simplest linear constitutive relation $\mathbf{E} \mapsto \mathbf{D}$ is used [11, 14, 16]. As a result, for the solution of the appropriate *linear* BVPs various effective analytical and numerical methods have been worked out, for example, in [15].

The classic method of estimation of puncture conditions is based on the *point criteria.* Namely, it is assumed that the electric puncture sets in when $\max\{|\mathbf{E}(\mathbf{x})| : \mathbf{x} \in \Omega\} \geq E_0$, where $\mathbf{E}$ is the solution of the linear electrostatic BVP and $E_0 > 0$ is the critical value, which is measured in physical experiments on thin plates in a homogeneous electric field [15, 16]. Unfortunately, for dielectrics with a complex shape in nonhomogeneous electric fields this method introduces a large error.

In powerful electric fields the essentially *nonlinear* phenomena of polarization saturation ($|\mathbf{P}| \leq P_* < +\infty$) and powerful growth of the electric current

$(\partial|\mathbf{J}|/\partial|\mathbf{E}| \gg \varepsilon_0)$ must be taken into account [9, 11]. As a result, the complimentary physical parameter of the dielectrical medium $\lambda > 0$ always exists such that $|\mathbf{D} - \mathbf{J}| \leq \lambda < \infty$.

Within the framework of the variational method, the existence of the limit electrostatic load (such external charges with no solution of the electrostatic BVP) was pointed out in [6, 7]. From the physical point of view this effect is treated as a loss of electrostatic balance, i.e. as the beginning of the electric puncture of dielectric. For calculation of the limit electrostatic load the original variational *limit analysis problem* (LAP) was formulated. From the mathematical point of view this problem needs a relaxation because its solution belongs to the space $BV(\Omega)$ of scalar functions with bounded variations having the generalized gradient as the bounded Radon's measure [1, 13, 17].

Unfortunately no clear physical interpretation can be provided for the fully relaxed LAP. Therefore, the original partial relaxation of LAP was proposed in [6, 7]. This relaxation is based on the special discontinuous finite-element approximation (FEA), which was proposed earlier by the author for LAP in non-linear elasticity [4, 5]. But after relaxation the appropriate finite dimensional problem becomes ill-conditioned and thus needs special preconditioned numerical methods as, for example, presented by the author in [3].

In this paper the *dual* LAP in electrostatics is formulated. It has a clear physical interpretation and after the standard piecewise linear continuous FEA is transformed into a problem of mathematical programming with linear equality constraints. This finite dimensional problem is effectively solved by the standard method of gradient projection, which is easily adapted for parallel computations.

The numerical results show that the proposed limit analysis method has a qualitative advantage over the classic technique of estimation of puncture conditions. This method can be used in Electrical Engineering and Microelectronics.

2 LAP in Electrostatics

Let a dielectrical medium occupy a domain $\Omega \subset \mathbb{R}^3$. The polarization and ionization properties of the dielectrical medium are described by two constitutive relations $\mathbf{D} = \hat{\mathbf{D}}(\mathbf{x}, \mathbf{E}) : \Omega \times \mathbb{R}^3 \to \mathbb{R}^3$ and $\mathbf{J} = \hat{\mathbf{J}}(\mathbf{x}, \mathbf{E}) : \Omega \times \mathbb{R}^3 \to \mathbb{R}^3$, as shown, for example in [11, 14, 16]. In practice the relations $D_i = \varepsilon_{ij}(\mathbf{x}, \mathbf{E})E_j$ and $J_i = \sigma_{ij}(\mathbf{x}, \mathbf{E})E_j$ are used, where $\{\varepsilon_{ij}\}$ and $\{\sigma_{ij}\}$ are the symmetric second-order tensors of dielectric permittivity and conductivity, respectively. Here and in what follows over repeated indices the summation rule applies. For an isotropic medium $\varepsilon_{ij} = \varepsilon\delta_{ij}$ and $\sigma_{ij} = \sigma\delta_{ij}$, where $\varepsilon = \varepsilon(\mathbf{x}, |\mathbf{E}|)$ and $\sigma = \sigma(\mathbf{x}, |\mathbf{E}|)$ are scalar functions and δ_{ij} is the Kronecker symbol. For a homogeneous medium $\{\varepsilon_{ij}\}, \{\sigma_{ij}\} = const(\mathbf{x})$.

Let the following quasi-static electric influences act on the dielectric: a bulk charge with density ϱ in Ω, a surface charge with density g on a portion Γ^2

of the boundary, and a portion Γ^1 of the boundary is grounded, i.e. $u \equiv 0$ on Γ^1. Here $\Gamma^1 \cup \Gamma^2 = \partial\Omega$, $\Gamma^1 \cap \Gamma^2 = \emptyset$ and $|\Gamma^1| > 0$. Point charges are absent.

In electrostatics it is assumed that an external source of the electric field compensates the work of the electric current in the dielectric. In accordance with the classical Thomson principle the true electrostatic potential is the solution of the following variational problem:

$$u_* = \arg\inf \{ \Pi(u) - A(u) : \ u \in V \}, \tag{1}$$

$$\Pi(u) = \int_\Omega \Phi(\mathbf{x}, \nabla u(\mathbf{x}))\, d\Omega, \qquad A(u) = \int_\Omega \varrho u \, d\Omega + \int_{\Gamma^2} g u \, d\gamma,$$

where $V = \{u : \overline{\Omega} \to \mathbb{R};\ u(\mathbf{x}) = 0,\ \mathbf{x} \in \Gamma^1\}$ is the set of admissible electrostatic potentials, Φ is the specific and $\Pi(u)$ is the full potential energy of the electric field, $A(u)$ is the work of an external source on a transference of charges from infinity to Ω.

In compliance with the Thompson and Joule–Lenz laws the function $\Phi(\mathbf{x}, \mathbf{E})$ is calculated as

$$\Phi(\mathbf{x}, \mathbf{E}) = E_i \int_0^1 \left[\hat{D}_i(\mathbf{x}, p\mathbf{E}) - \hat{J}_i(\mathbf{x}, p\mathbf{E})\right] dp.$$

In Fig. 1 experimental constitutive relations $|\mathbf{E}| \mapsto |\mathbf{D}|$ (line 1) and $|\mathbf{E}| \mapsto |\mathbf{J}|$ (line 2) for real isotropic dielectrical media are presented [9, 11, 16]. The appropriate function of *effective induction* is shown by the line 3. It is easily seen that for every dielectrical medium the scalar $\lambda > 0$ always exists such that for every $\mathbf{E} \in \mathbb{R}^3$ and almost every $\mathbf{x} \in \Omega$ the following estimation is true:

$$\Phi(\mathbf{x}, \mathbf{E}) \leq \lambda(\mathbf{x})|\mathbf{E}|, \tag{2}$$

where

$$\lambda(\mathbf{x}) = \max\left\{ \left|\hat{\mathbf{D}}(\mathbf{x}, \mathbf{E}) - \hat{\mathbf{J}}(\mathbf{x}, \mathbf{E})\right| : \ \mathbf{E} \in \mathbb{R}^3 \right\}.$$

From the physical point of view λ is *the electric saturation.* In what follows, we consider a homogeneous dielectrical medium for which $\lambda = const(\mathbf{x})$.

From the estimation (2) it follows [12, 17] that the set of admissible electrostatic potentials is defined as the following subspace

$$V = \{u \in W^{1,1}(\Omega) : \ u(\mathbf{x}) = 0,\ \mathbf{x} \in \Gamma^1\}. \tag{3}$$

From the mathematical point of view the variational problem (1) can have no solution because the functional $\Pi(u) - A(u)$ can be unbounded from below on the set V. In particular, after the point $E_0 = |\mathbf{E}^*|$ (see Fig. 1) the full potential energy of the electric field $\Pi(u)$ has growth in $\|u\|_{1,1}$ *less than linear.* But the work of the electric field on the external charges $A(u)$ always has *linear* growth in $\|u\|_{1,1}$. As a result, for an admissible minimizing sequence $u_m \in V$

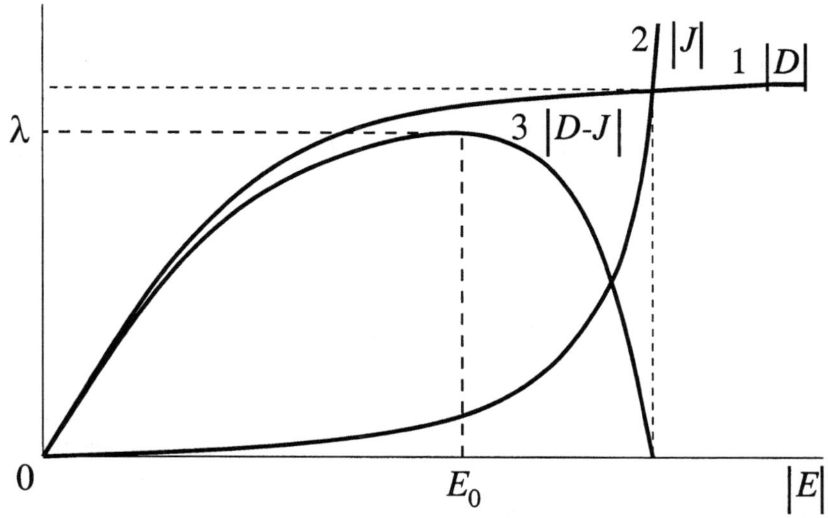

Fig. 1. Experimental (lines 1,2) and effective (line 3) constitutive relations

with $\|u_m\|_{1,1} \to \infty$ we have $\Pi(u) - A(u) \to -\infty$ as $m \to \infty$ (details see in [7]), i.e. the electrostatic variational problem (1) is not well-posed. Namely, the limit electrostatic load exists, i.e. external charges (ϱ, g) with no solution of the problem (1). From the physical point of view this effect can be treated as the beginning of the electric puncture of dielectric because it corresponds to a loss of the electrostatic balance between an external source of charges and the dielectrical medium. Here we have the full analogy with some problems of global stability and fracture in Mechanics of Solids [2, 4, 5, 17].

For a definition of the limit electrostatic load we introduce the set of admissible external charges for which the minimized functional in problem (1) is bounded from below on V and, therefore, a solution of this problem exists:

$$B = \{ \, (\varrho, g) \in L^\infty(\Omega) \times L^\infty(\Gamma^2) : \ \inf \, (\, \Pi(u) - A(u) : u \in V \,) > -\infty \, \} \, .$$

The set is non-empty because for small external charges the problem (1) is transformed into the classic variational problem of linear electrostatics, which always has a solution [9].

For arbitrary external charges $(\varrho_0, g_0) \in B$ we examine the sequence of charges, which are proportional to the real parameter $t \geq 0$.

Definition 1. *The number $t_* \geq 0$ is defined as the limit parameter of electrostatic loading and $(t_*\varrho_0, t_*g_0)$ is the limit electrostatic load, if $(t\varrho_0, tg_0) \in B$ for $0 \leq t \leq t_*$ and $(t\varrho_0, tg_0) \notin B$ for $t > t_*$.*

As a result, the analysis of the electrostatic balance in a dielectric comes to the investigation of the set of positive parameters t for which the one-parametric functional

$$I_t(u) = \Pi(u) - t\,A(u)$$

is bounded from below on the set of admissible electrostatic potentials V from (3). The following basic result has been proven recently by the author in [6, 7].

Theorem 1. *The finite limit parameter of electrostatic loading exists. It is calculated as the solution of the following limit analysis problem:*

$$t_* = \lambda \inf \left\{ \int_\Omega |\nabla u(\mathbf{x})|\, d\Omega : \; u \in V,\; A(u) = 1 \right\}, \tag{4}$$

where λ *is the electric saturation from* (2).

From the Definition 1 it follows that for $t_* < 1$ the electrostatic variational problem (1) has no solution. This phenomenon corresponds to the beginning of the electric puncture of the dielectrical medium. Therefore, the limit analysis problem (4) is the main problem for the estimation of puncture conditions for dielectrics of complex shape in powerful nonhomogeneous electric fields that closes one of the modern fundamental problems [11, 14, 16].

It was pointed out in [6, 7] that the solution of LAP (4) belongs to the space of scalar functions with bounded variations $BV(\Omega) \supset W^{1,1}(\Omega)$. This space consists of functions $u \in L^1(\Omega)$ having the bounded total variation

$$\int_\Omega |\partial u| = \sup \left\{ \int_\Omega (\nabla \cdot \mathbf{D})\, u\, d\Omega : \; \mathbf{D} \in C_0^1(\Omega, R^3),\; |\mathbf{D}| \le 1 \right\},$$

where ∂u is the vector Radon's measure denoting the gradient of the function u in the sense of distribution theory [1, 13, 17]. In this case $\int_\Omega |\partial u| = \int_\Omega |\nabla u(x)|\, d\Omega$ for every $u \in W^{1,1}(\Omega)$. Here and in what follows the point between vectors denotes the scalar product of these vectors in $\mathbb{R}^3$.

The Banach space $BV(\Omega)$ is weak$_*$ sequentially compact with the norm $\|u\|_{BV} = \|u\|_1 + \int_\Omega |\partial u|$, therefore, the mathematically correct and fully relaxed LAP has the following form:

$$t_* = \lambda \inf \left\{ \int_\Omega |\partial u| : \; u \in BV(\Omega),\; u_{|_{\Gamma 1}} = 0,\; A(u) = 1 \right\}.$$

Unfortunately, at present this problem has no a clear physical interpretation.

Within the framework of the classic approach the original partial relaxation of LAP can be used [6, 7]. This relaxation is based on the special discontinuous FEA [4, 5]. But after relaxation the appropriate finite dimensional problem becomes ill-conditioned and thus needs special preconditioned numerical methods as, for example, presented by the author in [3].

3 Dual LAP

We construct here the dual LAP for a homogeneous dielectric using methods from duality theory [10]. Thus, we introduce the set of admissible fields of the effective induction compensating for the electric field of external charges in the weak sense [12],

$$G = \left\{ \mathbf{D} \in L^\infty(\Omega, \mathbb{R}^3) : \int_\Omega (\nabla \cdot \mathbf{D} + \varrho) u \, d\Omega = 0, \right.$$

$$\left. \int_{\Gamma^2} (\mathbf{n} \cdot \mathbf{D} - g) u \, d\gamma = 0, \ \forall u \in V \right\}, \tag{5}$$

where $\mathbf{n}$ is the unit normal vector on the boundary Γ^2 and $\nabla \cdot \mathbf{D}$ is the distribution.

The dual LAP is formulated as the problem of finding an admissible effective induction of minimal intensiveness

$$\tau_* = \inf \{ \|\mathbf{D}\|_\infty : \mathbf{D} \in G \}, \tag{6}$$

where $\|\mathbf{D}\|_\infty = \inf\{ \sup(|\mathbf{D}(\mathbf{x})| : \mathbf{x} \in \Omega \setminus \omega) : |\omega| = 0 \}$.

Theorem 2. *For solutions of the problems* (4) *and* (6) *the relation* $t_* \tau_* = \lambda$ *is true.*

Proof. It is easily verified that dual LAP (6) is equivalent to the problem

$$\tau_*^{-1} = \sup \{ \nu > 0 : \mathbf{D} \in G, \ \|\nu \mathbf{D}\|_\infty \leq 1 \}. \tag{7}$$

For every value $\nu > 0$ and field $\mathbf{D} \in G$ the following equality is true:

$$\nu = \nu + \nu \inf \left\{ \int_{\Gamma^2} (\mathbf{n} \cdot \mathbf{D} - g) u \, d\gamma - \int_\Omega (\nabla \cdot \mathbf{D} + \varrho) u \, d\Omega : \ u \in V \right\}.$$

After integration by parts and taking into account the boundary condition on Γ^1, we find

$$\nu = \inf \left\{ \int_\Omega \nu \mathbf{D} \cdot \nabla u \, d\Omega : \ u \in V, \ A(u) = 1 \right\}. \tag{8}$$

We introduce the bilinear functional $L(\mathbf{D}, u) = \int_\Omega \mathbf{D} \cdot \nabla u \, d\Omega$ on $L^\infty(\Omega, \mathbb{R}^3) \times W^{1,1}(\Omega)$ [12], then from (8) it follows that the problem (7) has the form

$$\tau_*^{-1} = \sup \{ \inf(L(\mathbf{D}, u) : \ u \in V, \ A(u) = 1) : \ \mathbf{D} \in G, \ \|\mathbf{D}\|_\infty \leq 1 \}.$$

For the bilinear functional $L(\mathbf{D}, u)$ the classic equality

$$\sup_{\mathbf{D}} \inf_{u} L(\mathbf{D}, u) = \inf_{u} \sup_{\mathbf{D}} L(\mathbf{D}, u)$$

is fulfilled [10], therefore,

$$\tau_*^{-1} = \inf \{ \sup (L(\mathbf{D}, u) : \ \mathbf{D} \in G, \ \|\mathbf{D}\|_\infty \leq 1) : \ u \in V, \ A(u) = 1 \}.$$

For every vectors $\mathbf{E}, \mathbf{D} \in \mathbb{R}^3$ the relation $|\mathbf{E}| = \sup\{\mathbf{E} \cdot \mathbf{D} : \ |\mathbf{D}| \leq 1\}$ is true. As a result, for every $u \in V$ we have the equality completing the proof:

$$\sup\{L(\mathbf{D}, u) : \ \mathbf{D} \in G, \ \|\mathbf{D}\|_\infty \leq 1\} = \int_\Omega |\nabla u(\mathbf{x})| \, d\Omega \,.$$

We showed that the estimation of the electric durability of a homogeneous dielectric is equivalent to finding an effective induction of minimal intensiveness, which compensates the electric field of external charges. As a consequence of Theorem 2, if $\tau_* > \lambda$ then the electrostatic variational problem (1) has no solution. The problem (6) is fully correct because the admissible effective induction has *three* independent components satisfying only *one* differential equation, i.e. a minimization in *two* independent components is possible.

4 Finite-element approximation of dual LAP

By the standard FEA for the domain $\Omega \subset \mathbb{R}^n$ $(n = 1, 2, 3)$ the sets $\Omega_h = \cup T_h$ and $\Gamma_h = \partial\Omega_h$ are constructed such that $|\Omega \backslash \Omega_h| \to 0$ and $|\Gamma \backslash \Gamma_h| \to 0$ for $h \to +0$, where h is the characteristic step of approximation and T_h is a simplex [8]. Every FEA is described by the set of nodes $\{\mathbf{x}^k\}_{k=1}^m$.

For the admissible effective induction the piecewise linear continuous approximation is used [8]:

$$\mathbf{D}_h(\mathbf{x}) = \mathbf{D}^k \Psi_k(\mathbf{x}) \qquad (k = 1, 2, \ldots, m),$$

where $\mathbf{D}^k = \{D_i^k\} \in \mathbb{R}^3$ is the admissible effective induction in the node $\mathbf{x}^k$, $\Psi_k : \Omega_h \to \mathbb{R}$ is continuous and linear on every simplex scalar function such that $\Psi_k(\mathbf{x}^r) = \delta_{kr}$ $(k, r = 1, 2, \ldots, m)$. The *supp* (Ψ_k) consists of simplices having $\mathbf{x}^k$ as a common node.

After the standard piecewise linear continuous FEA of external charges (ϱ_h, g_h) and normal vector $\mathbf{n}_h$ on the boundary Γ^2, the set of admissible effective inductions (5) is approximated by the set

$$G_h = \{\mathbf{D}^k \in \mathbb{R}^3 : \ \mathbf{D}^k \cdot \nabla\Psi_k(\mathbf{x}) + \varrho_h(\mathbf{x}) = 0, \ \mathbf{x} \in \Omega_h; \\ \mathbf{n}_h(\mathbf{x}^r) \cdot \mathbf{D}^k \Psi_k(\mathbf{x}^r) = g_h(\mathbf{x}^r), \ \mathbf{x}^r \in \Gamma_h^2\},$$

which is the *convex set with piecewise linear boundaries*, i.e. it is a simplex in the space of global variables $\mathbb{R}^{3m}$.

As a result, the dual LAP (6) is approximated by the problem of mathematical programming with linear equality constraints

$$\tau_h = \min \left\{ \max \left(|\mathbf{D}^k| : \ k = 1, 2, \ldots, m \right) : \ \mathbf{D}^k \in G_h \right\}. \tag{9}$$

If the number of finite elements equals m_1 and the number of nodes on Γ_h^2 equals m_2 then the number of free variables in the problem (9) equals $3m - (m_1 + m_2)$. It is easily verified that the minimum number of variables equals $2n + 1$ $(n = 1, 2, 3)$, which is reached for the domain coinciding with the simplest n-dimension simplex because in this case $m = n + 1$, $m_1 = 1$, $m_2 = n + 1$.

The objective functional in the problem (9) is a combination of convex hypercones in the space $\mathbb{R}^{3m}$. Therefore, due to the linearity of the constraints in the set G_h this finite dimensional problem is effectively solved by the standard *method of gradient projection*, which is easily adapted for parallel computations.

5 Numerical results

In the numerical experiment the following electrostatic BVP was considered: an isotropic and homogeneous dielectric has the form of a finite cylindrical rod with the radius of section a and length $2l$. The small round blocks of dielectric are covered by conductors having charges $\pm Q$.

In view of the axial symmetry, the initial LAP (4) has the following form:

$$t_* = \lambda \inf\{I(u) : \ u \in V, \ u(r, 1) = 1\},$$

$$I(u) = 2 \int_0^1 \int_0^1 \left[\eta^2 \left(\frac{\partial u}{\partial r} \right)^2 + \left(\frac{\partial u}{\partial z} \right)^2 \right]^{1/2} r dr dz,$$

$$V = \left\{ u \in W^{1,1}((0,1)\times(0,1)) : u(r,0) = 0, \ \frac{\partial u}{\partial z}(r,0) = 0, \ \frac{\partial u}{\partial r}(0,z) = 0 \right\},$$

where λ is the electric saturation from (2) (see Fig. 1), $\eta = l/a$ is the geometric parameter and $r \in [0, 1]$, $\varphi \in [0, 2\pi)$, $z \in [0, 1]$ are the reduced cylindrical coordinates [4, 6]. Here $\Omega = (0,1)\times(0,1)$ and $\Gamma^2 = \{r \in [0,1], z = 1\}$. It is easily verified that puncturing charge $Q_* = \pi a l t_*$.

From Theorem 2 we have $t_* = \lambda / \tau_*$, where the parameter τ_* is the solution of the appropriate dual LAP (6) on the following set of admissible effective inductions:

$$G = \left\{ \mathbf{D} \in L^\infty(\Omega, \mathbb{R}^3) : \ \nabla \cdot \mathbf{D} = 0 \ \text{ in } \ \Omega, \ D_z = 1 \ \text{ on } \ \Gamma^2, \ D_\varphi = 0 \ \text{ in } \ \overline{\Omega} \right\}.$$

In the computer experiment a uniform $N \times N$ triangulation of the domain Ω was used. As a result, the problem of mathematical programming (9) was solved for $3(N+1)^2$ variables satisfying $2N^2+N+1$ linear equality constraints. The experimental results for $\lambda = 1$ are shown in Table 1. It is easily seen that $\tau_h \searrow \tau_* = 1$ as $h \to +0$ that fully coincides with results presented in [6, 7].

Table 1. Numerical results.

N	5	10	20	40	80
τ_h	2.17	1.83	1.36	1.06	1.01

The above analytical and numerical results are new. They are of practical interest, but more theoretical and experimental research is desirable. For example, the presented limit analysis method can be used for the design of electric isolators.

References

1. Ambrosio, L., Fusco, N., Pallera, D. (2000): Functions of Bounded Variations and Free Discontinuity Problems. Oxford Uni. press, New York
2. Brigadnov, I.A. (1996): Existence theorems for boundary value problems of hyperelasticity. Sbornik: Mathematics, **187**, 1–14
3. Brigadnov, I.A. (1996): Numerical methods in non-linear elasticity. In: Desideri, J.-A., Le Tallec, P., Onate, E., Periaux, J., Stein, E. (eds.) Numerical methods in engineering. Wiley, Chichester, 158–163
4. Brigadnov, I.A. (1998): Discontinuous solutions and their finite element approximation in non-linear elasticity. In: Van Keer, R., Verhegghe, B., Hogge, M., Noldus, E. (eds.) Advanced computational methods in engineering. Shaker Publishing B.V., Maastricht, 141–148
5. Brigadnov, I.A. (1999): The limited static load in finite elasticity. In: Dorfmann, Al, Muhr, A. (eds.) Constitutive models for rubber. A.A.Balkema, Rotterdam, 37–43
6. Brigadnov, I.A. (2001): Numerical analysis of dielectrics in powerful electrical fields. Computer Assisted Mech. Eng. Sci., **8**, 227–234
7. Brigadnov, I.A. (2002): Variational-difference method for estimation of electrical durability of dielectrics. Mathematical Modeling. **14**(4), 57–66 (in Russian)
8. Ciarlet, Ph.G. (1980): The Finite Element Method for Elliptic Problems, North-Holland Publ. Co., Amsterdam
9. Duvaut, G., Lions, J.-L. (1972): Les Inéquations en Mécanique et en Physique. Dunod, Paris
10. Ekeland, I., Temam, R. (1976): Convex Analysis and Variational Problems. North-Holland Publ. Co., Amsterdam.
11. Eringen, A.C., Maugin, G. (1989): Electrodynamics of Continua, Vol.I and II. Springer, New York.
12. Fučik, S., Kufner, A. (1980): Nonlinear Differential Equations, Elsevier Sci. Publ. Co., Amsterdam–Oxford–New York

13. Giusti, E. (1984): Minimal Surfaces and Functions of Bounded Variations. Birkhäuser, Boston
14. Landau, L.D., Lifshitz, E.M. (1957): Electrodynamics of Continuum Media. Gostekhizdat, Moscow (in Russian)
15. Mirolyubov, N.M., Kostenko, M.V., Levinstein, M.L., Tikhodeev, N.N. (1963): Solving Methods for Electrostatic Fields. Vischaya Shkola, Moscow (in Russian)
16. Tamm, I.E. (1989): Bases of the Electricity Theory. Nauka, Moscow (in Russian)
17. Temam, R. (1983): Problèmes Mathématiques en Plasticité. Gauthier-Villars, Paris

Finite Element Mesh Adjusted to Singularities Applied to Axisymmetric and Plane Flow

Pavel Burda[1], Jaroslav Novotný[2], Bedřich Sousedík[3] and Jakub Šístek[1]

[1] Department of Mathematics, Faculty of Mechanical Engineering, Czech Univ. Technology, Karlovo nám. 13, CZ-121 35 Praha 2, Czech Republic
burda@fsik.cvut.cz, *sistek@seznam.cz*

[2] Institute of Thermomechanics AS CR, Dolejškova 5, CZ-18200 Prague 8, Czech Republic *novotny@bivoj.it.cas.cz*

[3] Department of Mathematics, Faculty of Civil Engineering, Czech University of Technology, Thákurova 7, CZ-166 29 Praha 6, Czech Republic
sousedik@seznam.cz

Summary. We consider the Navier-Stokes equations for incompressible flow in axisymmetric tubes with abrupt changes of diameter. This paper is based on results for the asymptotics of the solution in the vicinity of nonconvex internal angles where the velocities possess an expansion $u(\rho, \vartheta) = \rho^{\gamma}\varphi(\vartheta) + \ldots$, where ρ, ϑ are local spherical coordinates. Problems with corners, edges, etc. are often successfully solved by the finite element method using an adaptive strategy usually based on a posteriori error estimates. In our paper we suggest an alternative approach for the mesh refinement near the corners, which makes use of the above expansion. It gives very precise results in a cheap way. We give numerical results and show the pros and cons of this approach.

1 Introduction

We consider the Navier-Stokes equations for incompressible fluid flow in axisymmetric tubes with abrupt changes of diameter. At present, problems with corners, edges, etc. are often successfully solved by the finite element method using an adaptive strategy based on a posteriori error estimates. Concerning linear elliptic equations let us mention the work of I. Babuška and W. C. Rheinboldt [2], concerning the Stokes problem the paper of M. Ainsworth and J.T. Oden [1]. Other references can be found in [5]. In [5] we derived an a posteriori error estimate for the Stokes problem in a 2D polygonal domain, and in [6] also for 3D case.

In this paper we apply an alternative approach for the mesh refinement near the corners, which makes use of the knowledge of the local behaviour of the solution near the corners.

For the stationary Navier-Stokes equations we proved in [4] that for nonconvex internal angles the velocities near the corners possess an expansion $u(\rho, \vartheta) = \rho^{\gamma}\varphi(\vartheta) + \ldots$ (smoother terms), where ρ, ϑ are local spherical coordinates. E.g. for the angle $\alpha = \frac{3}{2}\pi$ we have $\gamma = 0.5444837$. It is well-known that using the standard finite element method on triangles with polynomials of degree $p = 1, 2, 3$ we have the a priori error estimate

$$\|(u - u_h)\|_{H^1(\Omega)} \leq C\ h^{\gamma-\epsilon}\|(u)\|_{H^{\gamma+1-\epsilon}(\Omega)},$$

which cannot be improved by increasing the degree of the approximating polynomials.

In this paper the local behaviour of the solution near the singular point is used to design a priori, a mesh which is adjusted to the shape of the solution.

The first part of the paper is devoted to the behaviour of the singularity near the corner. The second part deals with the impact of the singularity on the refinement of the mesh. We show an example of the mesh with quadratic polynomials for velocity. Then we use this mesh for the numerical solution of flow in the channel with corners.

2 Steady Navier-Stokes Equations Near the Corner

The asymptotic behaviour of *plane flow* with corner singularities has been studied e.g. by Kondratiev [11], Ladevéze, Peyret [12].

In this section we deal with *pipe flow* (axially symmetric). To study the asymptotic behaviour of the solution of the Navier-Stokes equations for an incompressible fluid, we utilize the stream function - vorticity formulation, which in cylindrical geometry reads

$$\frac{\partial \omega}{\partial t} + u_1\frac{\partial \omega}{\partial z} + u_2\frac{\partial \omega}{\partial r} + u_2\frac{\omega}{r} = \nu\Big(\frac{\partial^2 \omega}{\partial z^2} + \frac{\partial^2 \omega}{\partial r^2} + \frac{1}{r}\frac{\partial \omega}{\partial r} - \frac{\omega}{r^2}\Big), \tag{1}$$

$$-r\omega = \frac{\partial^2 \psi}{\partial z^2} + \frac{\partial^2 \psi}{\partial r^2} - \frac{1}{r}\frac{\partial \psi}{\partial r}, \tag{2}$$

$$u_1 = \frac{1}{r}\frac{\partial \psi}{\partial r}, \qquad u_2 = -\frac{1}{r}\frac{\partial \psi}{\partial z}, \tag{3}$$

where r, z are cylindrical coordinates, $u_1 = V_z$, $u_2 = V_r$ are velocity components in the z, r directions, respectively, ω is the vorticity, ψ is the stream function, and ν is the viscosity. We assume that all derivatives exist here at least in the generalized sense.

In [4] we studied the *stationary Stokes flow.* There, substituting ω, u_1, u_2 from (2) – (3) into (1) we got

$$\begin{aligned} 0 \;=\; \nu\, \{&\frac{1}{r}(\frac{\partial^4 \psi}{\partial z^4} + 2\frac{\partial^4 \psi}{\partial z^2 \partial r^2} + \frac{\partial^4 \psi}{\partial r^4}) - \\ -\frac{1}{r^2}(\frac{\partial^3 \psi}{\partial z^3} + \frac{\partial^3 \psi}{\partial z^2 \partial r} &+ \frac{\partial^3 \psi}{\partial z \partial r^2} + \frac{\partial^3 \psi}{\partial r^3}) + \frac{3}{r^3}\frac{\partial^2 \psi}{\partial z^2} - \frac{3}{r^4}\frac{\partial \psi}{\partial z}\}. \end{aligned} \tag{4}$$

We are interested in the asymptotic behaviour of the solution near the corners. One example of our solution domain is shown in Figure 1, where the corners are the points P, Q. The lower edge of the domain coincides with the axis of symmetry.

Fig. 1. The solution domain Ω

Investigating the equation (4) and using the technique of Kondratiev [11], we showed in [4] that near the corners such as P or Q, the solution ψ possesses the expansion

$$\psi(x,y) = \sum_j \sum_{s=0}^{p_j-1} a_{js}\rho^{-i\lambda_j} \ln^s \rho \cdot \psi_{sj}(\vartheta) + w(x,y), \tag{5}$$

where ρ, ϑ are polar coordinates, w is smooth, and λ_j are the poles of multiplicity p_j of the corresponding resolvent $R(\lambda)$. More specifically, for the internal angle $\alpha = \frac{3}{2}\pi$, we proved that the leading term of the expansion for the velocity components is as follows

$$u_l(\rho,\vartheta) = \rho^{0.54448374}\varphi_l(\vartheta) + \ldots, \quad l = 1,2, \tag{6}$$

Similar results have been proved for the Navier-Stokes equations.

3 Finite Element Solution to Steady Navier-Stokes Equations

We are concerned with the finite element solution of the stationary Navier-Stokes equations. We would like to use the information about the asymptotics of the flow near the singular point, in order to suggest adequate local mesh refinement.

In [4] we showed that the behaviour near the singular point, of the axisymmetric flow and of the plane flow, are the same. So in what follows, for simplicity we deal with plane flow, in the domain Ω which has the same shape as in Figure 1. We investigate the Navier-Stokes equations in primitive variables $\mathbf{u}$ (velocity vector) and p (pressure)

$$(\nabla \mathbf{u}) \cdot \mathbf{u} - \nu \Delta \mathbf{u} + \nabla p = 0 \quad \text{in } \Omega. \tag{7}$$

For the finite element approximation we take Ω to be a polygon in $\boldsymbol{R}^2$, for simplicity. Let $\{\mathcal{T}_h\}_{h\to 0}$ be a regular (cf. [10]) family of triangulations of Ω.

Let P^m be the set of all polynomials of degrees $m > 0$. Let X^h, M^h be the finite element spaces of Hood-Taylor elements (cf. e.g. F. Brezzi, M. Fortin [3]), i.e.

$$X^h = \{\mathbf{u} \in H_0^1(\Omega)^2 \cap C(\Omega), \mathbf{u}/_K \in P^2(K)^2, K \in \mathcal{T}_h\}, \tag{8}$$
$$M^h = \{p \in L_0^2(\Omega), p/_K \in P^1(K), K \in \mathcal{T}_h\},$$

cf. [7]. Velocity values are given in corner nodes and midside nodes of the quadrilateral or the triangle, and pressure values only in corner nodes, in order to satisfy Babuška-Brezzi condition [3]. Then the velocity components and the pressure are approximated as continuous functions of spatial variables.

4 Refinement of FEM Mesh Adjusted to Singularity

Near the corner P where the angle is $\frac{3}{2}\pi$, velocities have the leading term in the expansion as given in (6). There ρ is the distance from the corner, ϑ the angle. Note that $\frac{\partial u_i(\rho,\vartheta)}{\partial \rho} \rightarrow \infty$ for $\rho \rightarrow 0$.

Now, in this section we assume the Stokes flow, for simplicity. A priori estimate of the finite element error is (cf. [10], [8])

$$\|\nabla(\mathbf{u}-\mathbf{u_h})\|_0 + \|p-p_h\|_0 \leq$$
$$\leq C\Big[\Big(\sum_T h_T^{2k} \mid \mathbf{u} \mid^2_{H^{k+1}(T)}\Big)^{1/2} + \Big(\sum_T h_T^{2k} \mid p \mid^2_{H^k(T)}\Big)^{1/2}\Big], \tag{9}$$

where $k = 2$. Taking into account the expansion (6), and following Johnson's idea [9], we derived in [4] the estimate

$$\mid \mathbf{u} \mid^2_{H^{k+1}(T)} \approx C \int_{r_T-h_T}^{r_T} \rho^{2(\gamma-k-1)}\ \rho\, d\rho \approx C\ r_T^{2(\gamma-k)}, \tag{10}$$

where h_T is the diameter of the triangle T of a triangulation $\mathcal{T}_h$, and r_T is the distance of the element T from the corner. So, in order to get an error estimate of the order $O(h^k)$, we should guarantee

$$h_T^{2k}\ r_T^{2(\gamma-k)} \approx h^{2k}. \tag{11}$$

This lead us in [4] to an algorithm for generating the mesh near the corner:

Algorithm. *Let r_1 be the distance of the large element from the corner. For given auxiliary stepsize h we compute recursively:*

for $i = 1, 2, \ldots, N$:

$$h_i = h \cdot (r_i)^{1-\frac{\gamma}{k}},$$
$$r_{i+1} = r_i - h_i$$

5 Model Problem

Consider two-dimensional flow of a viscous, incompressible fluid described by the Navier-Stokes equations in a domain with corner singularity, cf. Fig. 2.

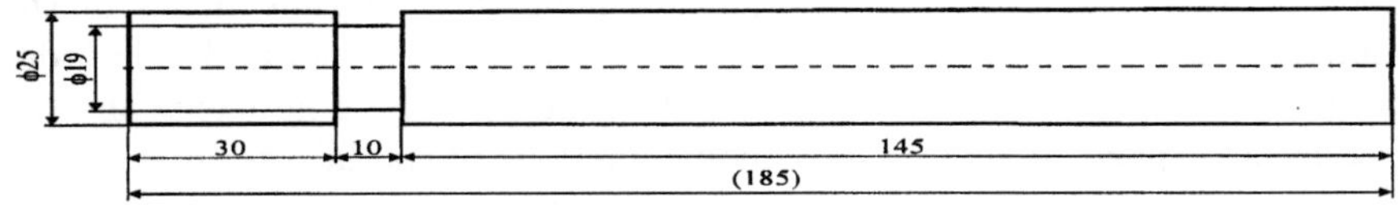

Fig. 2. Geometry of the channel

Due to symmetry, we solve the problem only on the upper half of the channel, cf. Figs 1, 3. On the inflow we consider a parabolic velocity profile, at the outflow 'do nothing' boundary condition. On the upper wall no-slip condition is imposed and on the lower wall a condition of symmetry is assumed (i.e. only the $y-$ component of velocity equals zero). We consider the following parameters: $\nu = 0.0001$ m^2/s, u_{in} (max.) $= 1$ m/s.

The algorithm for mesh refinement described in the previous section is applied to the corner where the channel or tube suddenly decreases in diameter (forward step, corresponding to point A in Fig. 1).
We start with $r_1 = 0,25mm$, $h = 0,1732\,mm$, $k = 2$, $\gamma = 0,5444837$. This corresponds to the contribution of cca 3% of individual elements to the global error. This way we get ten diameters of elements, cf. Tab. 1.

Tab. 1. Resulting refinement

i	$r_i(mm)$	$h_i(mm)$
1	0.25000	0.06316
2	0.18685	0.05110
3	0.13575	0.04050
4	0.09526	0.03129
5	0.06396	0.02342
6	0.04054	0.01681
7	0.02374	0.01138
8	0.01235	0.07077
9	0.00528	0.00381
10	0.00147	0.00147

6 Design of the mesh detail near the corner

Using the parameters from Table 1, J. Šístek [13] suggested three variants of mesh refinement near the corner (Figures 3-5). Mesh No. 1 was a classical mesh used before. He suggested two other variants in order to fit better to the algorithm of mesh refinement, especially to its polar coordinate nature (Figures 4-5).

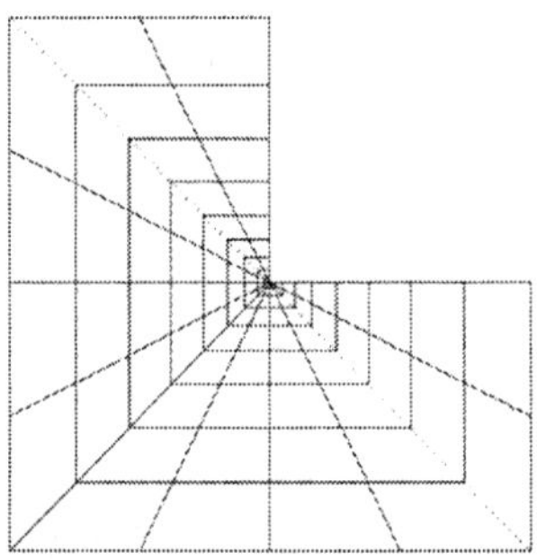

Fig. 3. Mesh No. 1

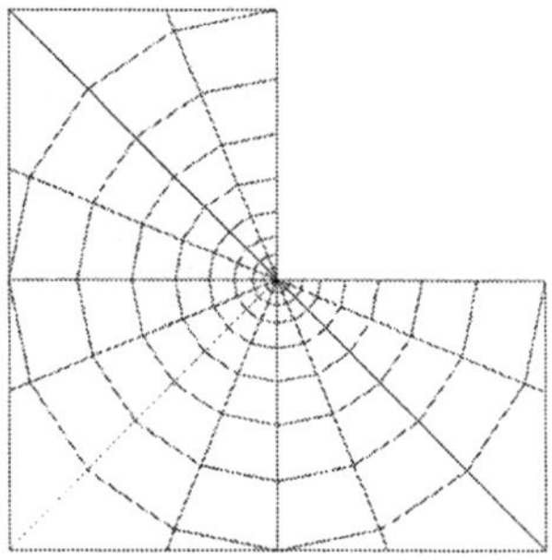

Fig. 4. Mesh No. 2

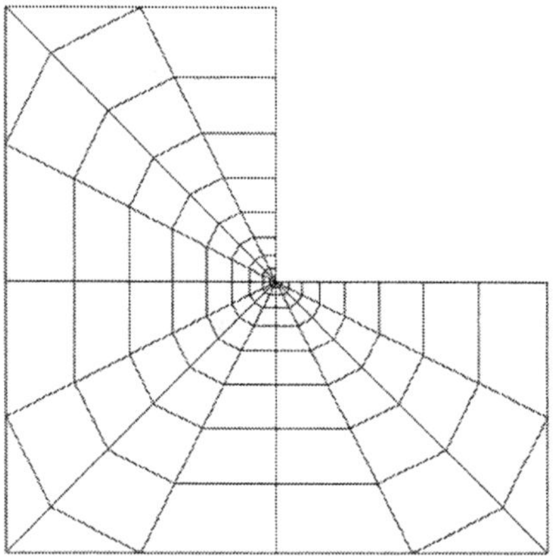

Fig. 5. Mesh No. 3

In Figs. 6-7 we present the whole computational mesh and its detail, using Mesh 3.

Fig. 6. The whole computational mesh

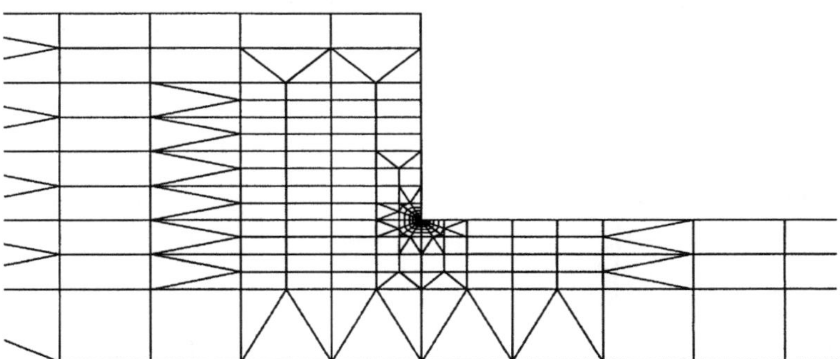

Fig. 7. The mesh - detail

7 Evaluation of the Approximation Error

To evaluate the error we use an a posteriori error estimate derived for the Stokes problem e.g. in [6]. Here $\{\mathbf{u}, p\} = (u, v, p)$ is the vector of the exact solution, $\{\mathbf{u_h}, p_h\} = (\overline{u}, \overline{v}, \overline{p})$ is the approximate solution computed by the

FEM. The error $\mathbf{e} = (e_u, e_v, e_p) = (u - \overline{u}, v - \overline{v}, p - \overline{p})$. The a posteriori estimate is:

$$\mathcal{U}^2(u - \overline{u}, v - \overline{v}, p - \overline{p}, \Omega) \leq \mathcal{E}^2(\overline{u}, \overline{v}, \overline{p}, \Omega), \tag{12}$$

where

$$\mathcal{U}^2(u - \overline{u}, v - \overline{v}, p - \overline{p}, \Omega) = \sum_l \|(e_u, e_v)\|^2_{1,\Omega_l} + \|e_p\|^2_{0,\Omega_l},$$

$$\mathcal{E}^2(\overline{u}, \overline{v}, \overline{p}, \Omega) = C \sum_l \left(h_l^2 \int_{\Omega_l} (r_1^2(\overline{u}, \overline{v}, \overline{p}) + r_2^2(\overline{u}, \overline{v}, \overline{p}))\, d\Omega + \int_{\Omega_l} r_3^2(\overline{u}, \overline{v}, \overline{p}) d\Omega \right).$$

where r_1, r_2 a r_3 denote the residuals with respect to the 1-st and the 2-nd N-S equation and the continuity equation (cf. [7], concerning choice of the constant C). To evaluate the error on elements we use the modified absolute error, defined as

$$\mathcal{A}_m^2(\overline{u}, \overline{v}, \overline{p}, \Omega_l) = \frac{|\Omega| \mathcal{E}^2(\overline{u}, \overline{v}, \overline{p}, \Omega_l)}{|\overline{\Omega}_l| \mathcal{U}^2(\overline{u}, \overline{v}, \overline{p}, \Omega)}, \tag{13}$$

where $|\Omega|$ is the area of the whole domain and $|\overline{\Omega}_l|$ is the mean area of elements obtained as $|\overline{\Omega}_l| = \frac{|\Omega|}{n}$. Here n denotes the number of all elements in the domain.

8 Numerical results

On Figures 8 - 11 we present the graphical output of entities that chartacterize the flow in the channel. On Figs. 8, 9 we observe how strong the singularity is both for the velocity and for the pressure (note that here the flow is from the right to the left, to have better view). Figure 8 shows that the location of the peak of the singularity of velocity is outside the patch where the refinement was done. One can see that, again it is the behaviour of the pressure that is decisive, cf. Fig. 9.

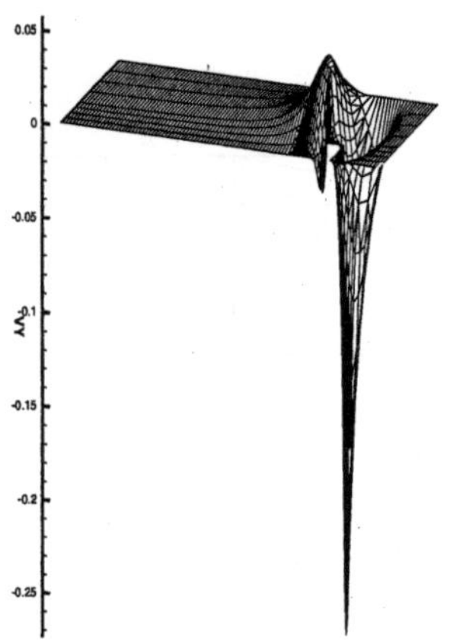

Fig. 8. Velocity component v (Mesh 3)

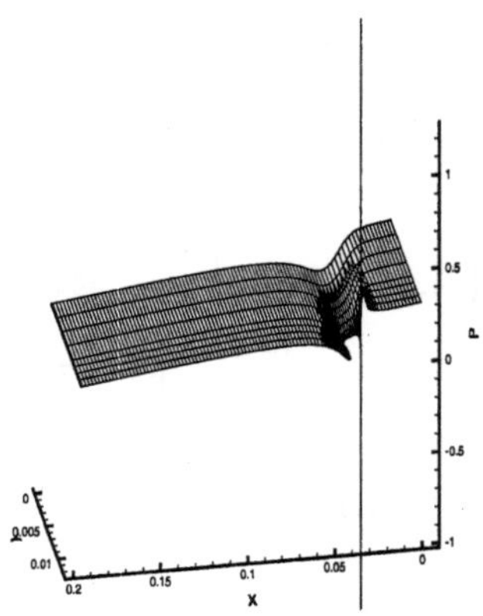

Fig. 9. Pressure p (Mesh 3)

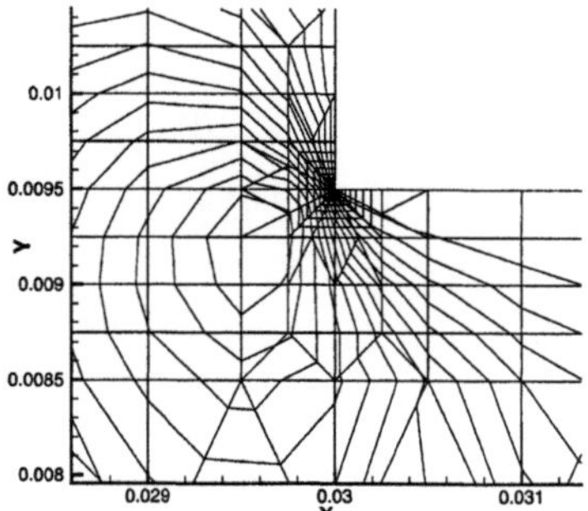

Fig. 10. Isolines of v (Mesh 1)

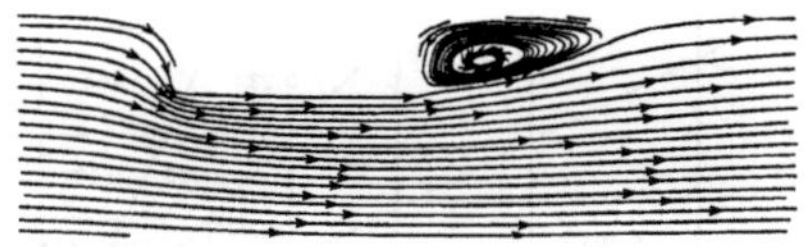

Fig. 11. Streamlines near the forward and backward step.

On Figures 12-13 we present errors on elements near the corner (forward step) for Mesh 1 and Mesh 2.

9 Conclusions

Pros:

- distribution of the error on elements is quite uniform (esp. for Mesh 2);
- strength of singularity (both for the velocity and the pressure) is well captured;
- the algorithm of adjusted mesh refinement has been confirmed;
- the efficiency of the algorithm: to achieve the desired precision one needs to carry out only one computation (compared with an adaptive approach, the same precision would require approximately 10 successive refinements).

Cons:

- suitable only for sigularities due to "geometry";
- adaptive approach is much more robust.

Nevertheless, efficient refinement of the mesh near the corners still remains a challenge.

References

1. Ainsworth, M., Oden, J., T. (1997): A posteriori error estimators for the Stokes and Oseen problems. SIAM J. Numer. Anal., **34**, 228 - 245
2. Babuška, I., Rheinboldt, W., C. (1978): A posteriori error estimates for the finite element method. Internat. J. Numer. Meth. Engrg., **12**, 1597 - 1615
3. Brezzi, F., Fortin, M. (1991): Mixed and Hybrid Finite Element Methods. Springer, Berlin
4. Burda, P. (1998): On the F.E.M. for the Navier-Stokes equations in domains with corner singularities. In: Křížek, M., et al. (eds) Finite Element Methods, Supeconvergence, Post-Processing and A Posteriori Estimates, Marcel Dekker, New York, 41-52

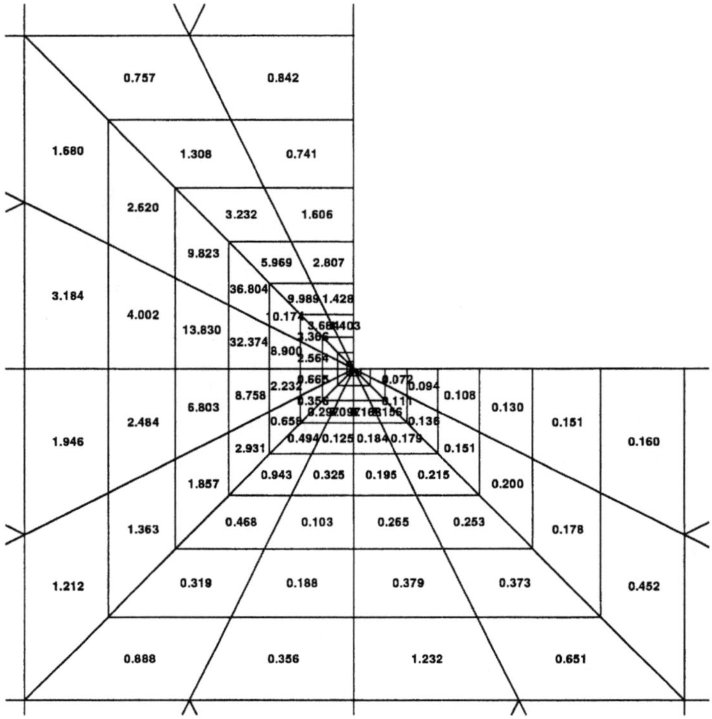

Fig. 12. Errors on elements (Mesh 1)

5. Burda, P. (2000): An a posteriori error estimate for the Stokes problem in a polygonal domain using Hood-Taylor elements. In: Neittaanmäki, et al. (eds) ENUMATH 99, Proc. Conf. Numerical Mathematics and Advanced Applications. World Scientific, Singapore, 448 - 455
6. Burda, P. (2001): A posteriori error estimates for the Stokes flow in 2D and 3D domains, In: Neittaanmäki, P., Křížek, M. (eds) Finite Element Methods, 3D. (GAKUTO Internat. Series, Math. Sci. and Appl., vol. 15) Gakkotosho, Tokyo, pp. 34-44
7. Burda, P., Novotný, J., Sousedík, B. (2003): Adaptive mesh refinement based on a posteriori error estimates for the Stokes flow in a 2D Problem, in: F. Brezzi et al. (ed), ENUMATH 2001, Proc. Conf. Numerical Mathematics and Advanced Applications, Springer Verlag, pp. 681 - 690
8. Douglas, J., Jr., Wang, J. (1989): An absolutely stabilized Finite Element Method for the Stokes problem, Math. Comp., **52**, 495-508
9. Johnson, C. (1994): Numerical Solution of Partial Differential Equations by the Finite Element Method, Cambridge University Press
10. Girault, V., Raviart, P. G. (1986): Finite Element Method for Navier- Stokes Equations. Springer, Berlin
11. Kondratiev, V. A. (1967): Asimptotika reshenija uravnienija Nav'je-Stoksa v okrestnosti uglovoj točki granicy, Prikl. Mat. i Mech., **1**, 119-123

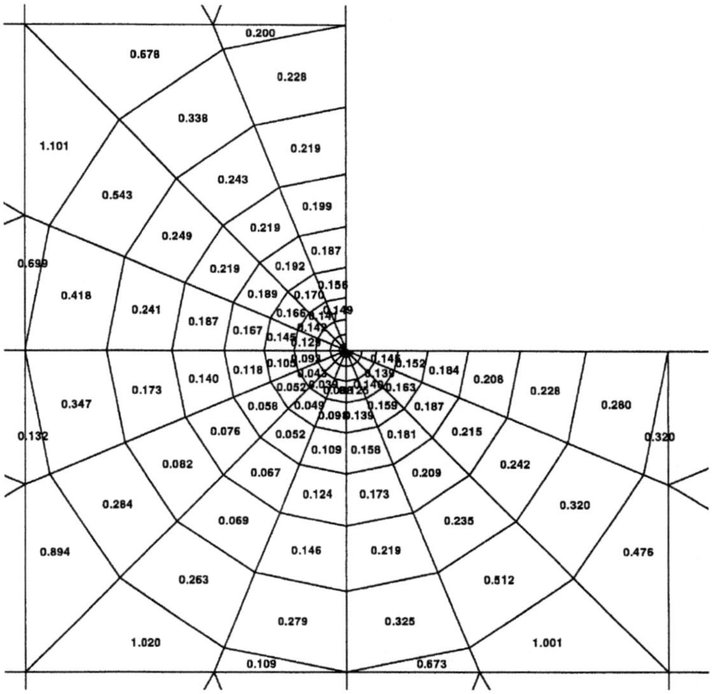

Fig. 13. Errors on elements (Mesh 2)

12. Ladevéze J., Peyret, R. (1974): Calcul numérique d'une solution avec singularit des équations de Navier-Stokes: écoulement dans un canal avec variation brusque de section, Journal de Mécanique, 13, 367-396
13. Šístek, J. (2003): Solution of the fluid flow in a tube with singularity using suitable mesh refinement (in Czech), Student Conference STČ, ČVUT Praha

Acknowledgements

This research has been supported partly by the GACR Grant No. GA 101/02/0391, and partly by the State Research Project No. J04/98/210000010.

The Edge Stabilization Method for Finite Elements in CFD

Erik Burman[1] and Peter Hansbo[2]

[1] Department of Mathematics, Ecole Polytechnique Federale de Lausanne, Switzerland, *erik.burman@epfl.ch*

[2] Department of Applied Mechanics, Chalmers University of Technology, SE–41296 Göteborg, Sweden, *hansbo@am.chalmers.se*

Summary. We give a brief overview of our recent work on the edge stabilization method for flow problems. The application examples are convection–diffusion, with small diffusion parameter, and a generalized Stokes model.

1 Introduction

In order for the standard Galerkin finite element method to be stable for problems in CFD, some care must be taken. For convection-dominated flow problems, stabilization must be introduced, while for mixed methods the approximations of velocities and pressure must either be carefully balanced or, again, stabilized. Examples of stabilization methods are the SUPG/SD-method [8], the discontinuous Galerkin method [9], the residual free bubbles [2], subviscosity models [7], and pressure projection methods for the Stokes problem [5]. The relation between the different approaches is also well understood in most cases. However for complex flow problems, most of these methods have drawbacks. The SUPG stabilization becomes non-symmetric and the formulation does not permit lumped mass; the residual free bubbles and discontinuous Galerkin method add additional degrees of freedom; the projection methods introduce the need of hierarchical meshes for the projection or the sub-viscosity model. In this paper we give a brief review of our recent work, [3, 4], on an alternative method originally proposed by Douglas and Dupont [6]. This method stabilizes convection-diffusion-reaction problems, as well as equal order interpolation methods for the generalized Stokes problem, by adding a least-squares term based on the jump in the gradient of the discrete solution over element boundaries. With this simple concept we obtain stability for convection-reaction-diffusion problems also in the vanishing viscosity limit as well as for the generalized Stokes problem with equal order interpolation.

The advantage of this method, in comparison with the others mentioned, is that no additional degrees of freedom are added, no hierarchical meshes are needed, the formulation remains symmetric, and the mass can be lumped for efficient time marching and treatment of stiff source terms. The drawback is an increased number of non-zero elements in the stiffness matrix due to the fact that the gradient jump term couple neighboring elements. The implementation also requires an element neighbor data structure that is not necessarily available in standard finite element codes.

2 Model problems

As a first model problem, we consider, in $\Omega \subset \mathbb{R}^d$, $d = 2, 3$, the problem of solving

$$\sigma u + \boldsymbol{\beta} \cdot \nabla u - \nabla \cdot (\varepsilon \nabla u) = f \quad \text{in } \Omega \tag{1}$$

with, for simplicity, $u = 0$ on $\partial\Omega$. Here, f is a given source term, $\boldsymbol{\beta}$ is a given smooth velocity field, satisfying $\nabla \cdot \boldsymbol{\beta} = 0$, and σ and ε are bounded positive functions.

The weak form of this problem is to find $u \in H_0^1(\Omega)$ such that

$$A(u, v) = (f, v) \quad \forall v \in H_0^1(\Omega), \tag{2}$$

where

$$A(u, v) := \int_\Omega (\sigma\, u\, v + \varepsilon \nabla u \cdot \nabla v + \boldsymbol{\beta} \cdot \nabla u\, v)\, \mathrm{d}x \quad \text{and} \quad (f, v) := \int_\Omega f\, v\, \mathrm{d}x.$$

The second model problem that we consider is a generalized Stokes problem, given by the partial differential equation

$$\begin{aligned} \sigma \boldsymbol{u} - \nu \Delta \boldsymbol{u} + \nabla p &= \boldsymbol{f} \quad \text{in } \Omega, \\ \nabla \cdot \boldsymbol{u} &= g \quad \text{in } \Omega, \\ \boldsymbol{u} \cdot \boldsymbol{n} &= 0 \quad \text{on } \partial\Omega \\ \nu \boldsymbol{u} \cdot \boldsymbol{t} &= 0 \quad \text{on } \partial\Omega. \end{aligned} \tag{3}$$

where Ω is bounded polygonal domain in $\mathbb{R}^d$ with boundary $\partial\Omega$, $d = 2, 3$ and σ and ν are two positive parameters, that are not allowed to vanish simultaneously, and $\boldsymbol{f}$ is a force term. By $\boldsymbol{n}$ we denote the outward pointing normal to Ω and $\boldsymbol{t}$ is a vector orthogonal to $\boldsymbol{n}$. This problem can be written in weak form as follows: Find

$$\boldsymbol{u} \in V = \{\boldsymbol{v} \in [H^1(\Omega)]^d : \; \boldsymbol{v}|_{\partial\Omega} = 0\}$$

when $\nu > 0$, alternatively

$$\boldsymbol{u} \in V = \{\boldsymbol{v} \in [L^2(\Omega)]^d, \nabla \cdot \boldsymbol{v} \in L^2(\Omega) : \; \boldsymbol{v} \cdot \boldsymbol{n}|_{\partial\Omega} = 0\}$$

for $\nu = 0$, and $p \in Q = L_2(\Omega)/\mathbb{R}$ when $\nu > 0$, alternatively $p \in H^1(\Omega)$ for $\nu = 0$, such that

$$a(\boldsymbol{u}, \boldsymbol{v}) + b(p, \boldsymbol{v}) - b(q, \boldsymbol{u}) = L(\boldsymbol{v}, q), \quad \forall (\boldsymbol{v}, q) \in V \times Q, \tag{4}$$

where

$$a(\boldsymbol{u},\boldsymbol{v}) = \int_\Omega \sum_{i=1}^{d} \sigma u_i v_i + \nu \nabla u_i \cdot \nabla v_i \, \mathrm{d}x, \quad b(p,\boldsymbol{v}) = -\int_\Omega p \, \nabla \cdot \boldsymbol{v} \, \mathrm{d}x,$$

and

$$L(\boldsymbol{v},q) = \int_\Omega \boldsymbol{f} \cdot \boldsymbol{v} \, \mathrm{d}x - \int_\Omega gq \, \mathrm{d}x.$$

In the following, we shall denote the L_2-scalar product by $(\cdot,\cdot)$ and the corresponding norm by $\|\cdot\|$.

3 The finite element methods

For our first model problem, the finite element method consists of seeking piecewise polynomial approximation $\boldsymbol{u}^h \in V^h \subset H_0^1(\Omega)$, and for our second model problem $\boldsymbol{u}_h$ and p_h, where $\boldsymbol{u}_h \in W^h \subset V$ and $p_h \in Q^h \subset Q$, with V^h, W^h, and Q^h built from continuous functions.

Consider a partitioning of Ω into a conforming triangulation T_h of affine simplicies K. We shall be concerned with the approximation

$$V^h = \{v \in H_0^1 \cap C^0(\Omega) : \; v|_K \in P^1(K) \quad \forall K \in T_h\}, \quad W^h = [V^h]^d,$$

and a continuous pressure space,

$$Q^h = \{q \in Q \cap C^0(\Omega) : \; q|_K \in P^1(K) \quad \forall K \in T_h\}.$$

It is well known that Galerkin methods, based on V^h, for convection dominated problems produce severe oscillations in the presence of rapidly varying features of the solution, and that the combination $W^h \times Q^h$ is unstable for the generalized Stokes problem. The stabilization method is similar in both cases

Our finite element methods are defined as follows.

Model problem I: find $u_h \in V^h$ such that

$$A(u_h,v) + J(u_h,v) = (f,v) \quad \forall v \in V^h, \tag{5}$$

where

$$\begin{aligned} J(u,v) &= \sum_K \frac{1}{2} \int_{\partial K} \gamma h_{\partial K}^2 [\nabla u] \cdot [\nabla v] \, \mathrm{d}s \\ &= \sum_K \frac{1}{2} \int_{\partial K} \gamma h_{\partial K}^2 [\boldsymbol{n}_K \cdot \nabla u] \, [\boldsymbol{n}_K \cdot \nabla v] \, \mathrm{d}s. \end{aligned} \tag{6}$$

Here, $h_{\partial K}$ is the size of ∂K, $[q]$ denotes the jump of q across ∂K for $\partial K \cap \partial\Omega = \emptyset$, $[q] = 0$ on $\partial K \cap \partial \Omega \neq \emptyset$, $\boldsymbol{n}_K$ is the outward pointing unit normal to K, and γ is a constant. We also introduce the local mesh size h_K as the largest of the $h_{\partial K}$ associated with element K, and we will assume that $h_K / h_{\partial K} < C$ where C is a fixed constant.

Model problem II: Find $(\boldsymbol{u}^h, p^h) \in W^h \times Q^h$ such that

$$
\begin{aligned}
a(\boldsymbol{u}_h, \boldsymbol{v}) + b(p_h, \boldsymbol{v}) + \tilde{\jmath}(\boldsymbol{u}_h, \boldsymbol{v}) &= L(\boldsymbol{v}, 0) \quad \text{in } \Omega, \\
b(q, \boldsymbol{u}) - \jmath(p_h, q) &= L(0, q) \quad \text{in } \Omega,
\end{aligned}
\tag{7}
$$

for all $(\boldsymbol{v}, q) \in W^h \times Q^h$, where

$$
\jmath(p, q) := \sum_K \frac{1}{2} \int_{\partial K} \gamma h_K^{s+1} \left[\boldsymbol{n}_K \cdot \nabla p_h\right] \left[\boldsymbol{n}_K \cdot \nabla q\right] \mathrm{d}s \tag{8}
$$

and

$$
\tilde{\jmath}(\boldsymbol{u}, \boldsymbol{v}) := \sum_K \frac{1}{2} \int_{\partial K} \gamma h_K^{s+1} \left[\nabla \cdot \boldsymbol{u}\right] \left[\nabla \cdot \boldsymbol{v}\right] \mathrm{d}s, \tag{9}
$$

For these methods we have, under additional regularity conditions, consistency in that the exact solution fulfills the discrete equations (because the jumps in the continuous variables vanish). What we also need to show is stability. The stability estimates obtained using edge stabilization is less immediate than that obtained in the case of streamline diffusion or discontinuous Galerkin. For our first model problem, it is well known (cf. [8]) that the crucial term to control is $\|h_K^{1/2} \boldsymbol{\beta} \cdot \nabla u_h\|$. For a Galerkin method, no control whatsoever of this term exists. In a discontinuous Galerkin method one exploits the fact that $h_K \boldsymbol{\beta} \cdot \nabla u_h$ is in the finite element test space and hence can be chosen as test function, while in a streamline diffusion method it is simply included in the test space. In the case of edge stabilization we do not have $h_K \boldsymbol{\beta} \cdot \nabla u_h$ in the finite element space. However, something which is close is there (e.g., its interpolant), and the point is that the difference is controlled by the edge stabilization. Thus, thanks to the term $J(u_h, v)$, we get the necessary control of $\|h_K^{1/2} \boldsymbol{\beta} \cdot \nabla u_h\|$.

For the second model problem the jump terms similarly give us control of $\|h_K^{1/2} \nabla p_h\|$ necessary in the Stokes case and also $\|h_K^{1/2} \nabla \cdot \boldsymbol{u}_h\|$ needed in the Darcy case. We then need to separate the Darcy case, where we use $s = 1$, and the Stokes problem, where we use $s = 2$. This is similar to streamline diffusion stabilization in the Navier-Stokes case, where the streamline diffusion stabilization parameter must change from $O(h_K^2)$ to $O(h_K)$ when going from viscosity-dominated to convection-dominated situations.

The details are given in [3, 4]; here we just give an indication of where the stability comes from, exemplified using model problem I. More precisely, we shall sow that there exists some $\zeta \geq \zeta_0 > 0$ such that

$$
\|h_K^{1/2} (\pi_h \boldsymbol{\beta} \cdot \nabla u_h - \boldsymbol{\beta} \cdot \nabla u_{)}\|^2 \leq \zeta J(u_h, u_h)
$$

for $\boldsymbol{\beta}$ constant. Extensions to $\boldsymbol{\beta} \in W^1_\infty$ are straighforward. To this end, let $\mathcal{N}_i$ be the set of all triangles K^i containing node i and assume that the cardinality of $\mathcal{N}_i$ is bounded uniformly in i. Let $\mathcal{F}_K$ be the set of all test functions φ_i such that $K \subset \operatorname{supp}\varphi_i$ and $\Omega_i = \bigcup_{\mathcal{N}_i} K^i$. We will consider a function $w \in P_0(K)$, and its representation in the finite element basis $\tilde{w}$ defined by

$$\tilde{w}|_K = w|_K \sum_{i \in \mathcal{F}_K} \varphi_i. \tag{10}$$

It follows that $\tilde{w} = w$ everywhere except on elements adjacent to Dirichlet boundaries where the boundary nodes are not included in the finite element space. We have:

Lemma 1. *Suppose that K is an element with at least one node on a Dirichlet boundary then*

$$\|w\|_K^2 = \frac{d+1}{n_i}(w, \tilde{w}), \tag{11}$$

where n_i denotes the number of interior nodes of the element.

Proof. The proof is immediate noting that

$$(w, \tilde{w}) = |w_K|^2 \int_K \sum_{i \in \mathcal{F}_K} \varphi_i \mathrm{d}x = \frac{n_i}{d+1} |w_K|^2 m(K).$$

We will now proceed to prove that

$$\|h^{s/2}(\tilde{w} - \pi_h w)\|^2 \leq C \tilde{J}_s(w, w) = C \sum_K \int_{\partial K} h^{s+1} [w]^2 \mathrm{d}s.$$

The operator $\pi_h : P_0(K) \to V^h$, which denotes the lowest order Clément operator is constructed as follows.

$$\pi_h w = \sum_i w_i \varphi_i, \quad \text{where} \quad w_i = \frac{1}{m(\Omega_i)} \sum_{\mathcal{N}_i} w|_{K^i} m(K^i). \tag{12}$$

In the following we will also write $w|_{K^i} - w|_K = \sum_{K_i}^{K} [w]$, with $[w]$ denoting the jump across element boundaries and the sum is taken over the shortest "path" from element K^i to element K.

It is now straightforward to show that the projection error is controlled by the operator $\tilde{J}_s(w, w)$

$$\begin{aligned}
\|h_K^{s/2}(\pi_h w - \tilde{w})\|^2 &= \sum_K \int_K h_K^s \Big(\sum_{i \in \mathcal{F}_K} \big(\frac{1}{m(\Omega_i)} \sum_{K^i \in \mathcal{N}_i} w|_{K^i}\, m(K^i) \big) \varphi_i - \tilde{w} \Big)^2 \mathrm{d}x \\
&= \sum_K \int_K h_K^s \Big(\sum_{i \in \mathcal{F}_K} \frac{1}{m(\Omega_i)} \sum_{K^i \in \mathcal{N}_i} (w|_{K^i} - w|_K) m(K^i) \varphi_i \Big)^2 \mathrm{d}x \\
&\leq C \sum_K \int_K h_K^s \sum_{i \in \mathcal{F}_K} \frac{1}{m(\Omega_i)^2} \sum_{K^i \in \mathcal{N}_i} \Big\{ \sum_{K_i}^{K} [w] \Big\}^2 m(K^i)^2 \mathrm{d}x \\
&\leq C \sum_K \int_{\partial K} h_K^{s+1} [w]^2 \mathrm{d}s \leq C \tilde{J}_s(w, w),
\end{aligned}$$

where we used the upper bound on the number of triangles neighboring to a node and a scaling argument. We have proved the following:

Lemma 2. *If w is some piecewise constant function, $\tilde{w}$ is defined by* (10) *and π_h is the Clément interpolant on V^h, then the edge stabilization term satisfies*

$$\|h_K^{s/2}(\pi_h w - \tilde{w})\|^2 \leq \gamma \tilde{J}_s(w,w) \tag{13}$$

for some $\gamma \geq \gamma_0 > 0$ independent of h_K.

Note that by the construction of $\tilde{w}$ we get less stabilization in elements adjacent to Dirichlet boundaries than in the interior of the domain, hence we expect to get poorer stabilizing properties close to sharp out flow layers (when diffusion is present), something which is confirmed by the numerical experiments.

4 Numerical examples

4.1 Convection–diffusion

We consider the domain $(0,1)\times(0,1)$, with $\sigma = 0$, $\varepsilon = 10^{-6}$, and $\boldsymbol{\beta} = (1-y, x)$. The boundary condition on the inflow boundary is $u = 0$ at $y = 0$ and $u = 0$ at $x = 0$ except for $0.3 < y < 0.5$, where $u = 1$. We compare the the numerical solutions using the present method and the streamline diffusion method in Figure 1. The results are of comparable quality.

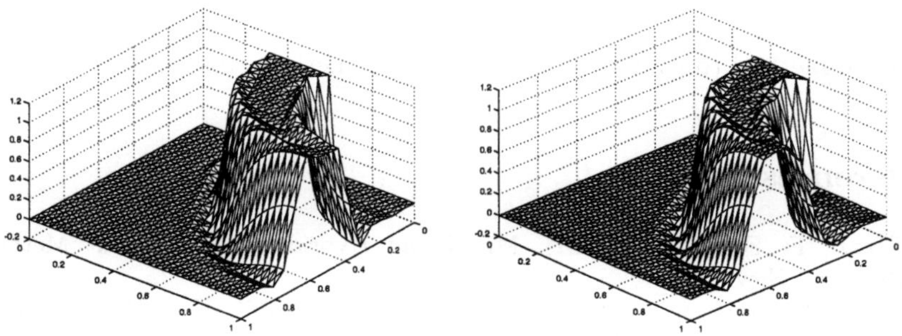

Fig. 1. Streamline diffusion (left) and gradient jump stabilization (right) for convection–diffusion

4.2 Stokes problem

To illustrate the absence of boundary layers in the pressure using the gradient jump method, we consider the Poiseulle flow problem on $(0,4) \times (0,1)$ on an unstructured mesh. The inflow boundary velocity is $\boldsymbol{u} = (y * (y-1), 0)$, we have set $\sigma = 0$ and $\nu = 1$. At $x = 4$, we apply a natural boundary condition. We solve this problem using P1P1-approximations with the streamline diffusion method and the gradient jump method on the mesh shown in Figure 2. Also shown is the computed velocity field (indistinguishable for the different methods). Finally, in Figure 3, we show the pressure isolines for the streamline

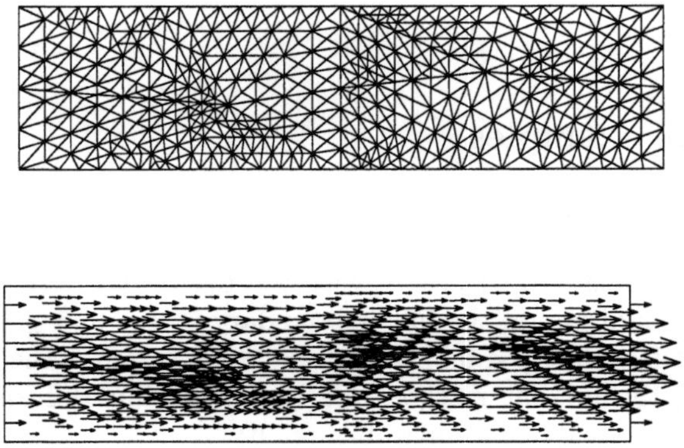

Fig. 2. Mesh and computed velocity field

diffusion method (top) and the gradient jump method (bottom).

References

1. Brezzi, F, Fortin, M. (1991): Mixed and Hybrid Finite Element Methods. Springer, New York
2. Brezzi, F., Hughes, T.J.R., Marini, L.D., Russo, A., Süli, E.A. (1999): A priori error analysis of residual-free bubbles for advection-diffusion problems. SIAM J. Numer. Anal., **36**, 1933–1948
3. Burman, E., Hansbo, P. (2002): Edge stabilization for Galerkin approximations of convection-diffusion problems. Chalmers Finite Element Center Preprint 2002-17
4. Burman, E., Hansbo, P. (2003): Edge stabilization for the generalized Stokes problem: a continuous interior penalty method. Chalmers Finite Element Center Preprint 2003-16
5. Codina, R., Blasco, J. (2000): Analysis of a pressure-stabilized finite element approximation of the stationary Navier-Stokes equations. Numer. Math., **87**, 59–81

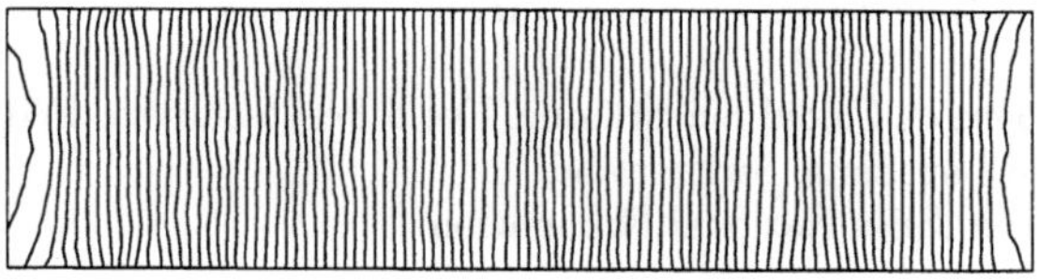

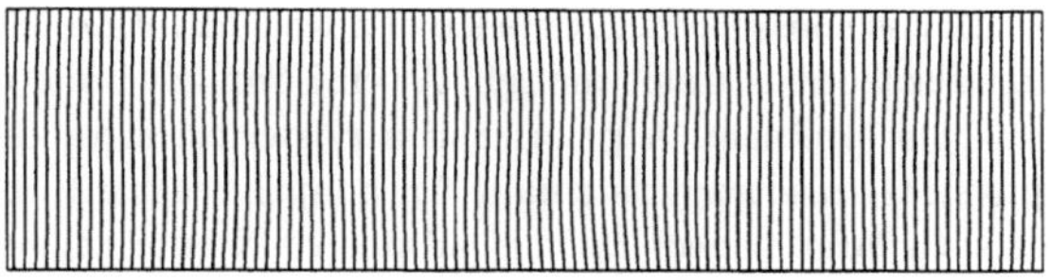

Fig. 3. Pressure isolines for streamline diffusion (top) and the gradient jump method (bottom)

6. Douglas, J., Dupont, T. (1976): Interior penalty procedures for elliptic and parabolic Galerkin methods, in: R. Glowinski, R., Lions, J. L. (eds) Computing Methods in Applied Sciences. Springer Berlin
7. Guermond, J.L. (1999): Stabilization of Galerkin approximations of transport equations by subgrid modeling. M2AN Math. Model. Numer. Anal., **33**, 1293–1316
8. Johnson, C., Nävert, U., Pitkäranta, J. (1984): Finite element methods for linear hyperbolic equations. Comput. Methods Appl. Mech. Engrg., **45**, 285–312
9. Johnson, C., Pitkäranta, J. (1986): An analysis of the discontinuous Galerkin method for a scalar hyperbolic equation, Math. Comp., **46**, 427–444.

Analysis and Computation of Dendritic Growth in Binary Alloys Using a Phase-Field Model

Eric Burman[1,2], Marco Picasso[1] and Jacques Rappaz[1]

[1] Institut d'Analyse et Calcul Scientifique, Ecole Polytechnique Fédérale de Lausanne, 1015 Lausanne, Switzerland
[2] supported by the Swiss National Foundation

Summary. A solutal, anisotropic phase-field model for dendritic growth of an isothermal binary alloy is considered. Existence of a weak solution is established provided the physical anisotropy is small enough. A semi-implicit finite element method is proposed to solve the problem. A priori and a posteriori error estimates are derived when the physical anisotropy is small. An adaptive algorithm which aims at producing meshes with large aspect ratio is proposed. Numerical results show that accurate solutions can be obtained, even when the physical anisotropy is large.

1 Introduction

A phase-field model for the dendritic growth in a domain Ω of an isothermal binary alloy is considered. The mathematical model contains the physical descriptions of [22, 20] and consists in a parabolic system of nonlinear equations set in Ω. The unknowns are the phase-field ϕ, which is an order parameter taking the value 1 in the solid phase and 0 in the liquid phase (see Fig. 1) and the concentration c of the binary alloy. In the phase-field approach, ϕ is regularized, both ϕ and c vary rapidly but smoothly across the thin solid-liquid interface. A typical profile of ϕ and c across the horizontal abscissa x_1 is presented in Fig. 2.

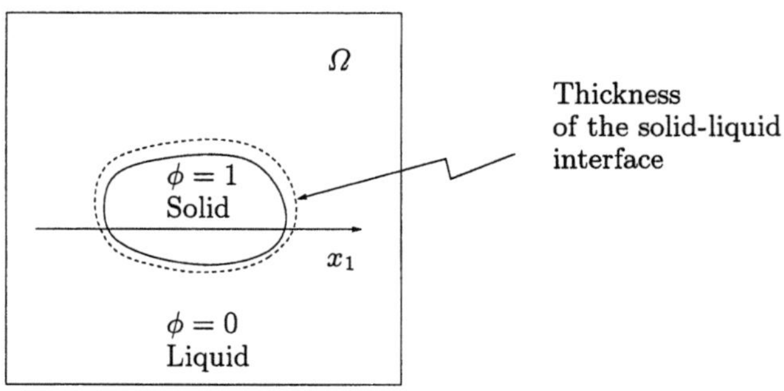

Fig. 1. The phase-field ϕ

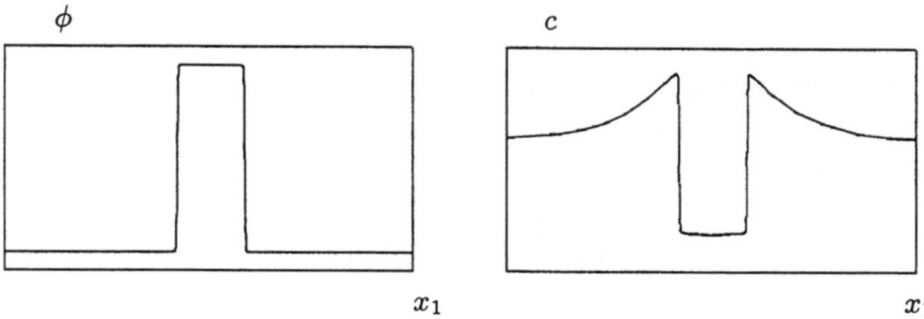

Fig. 2. Typical profiles of the phase field (left) and concentration field (right). The phase-field has values zero or one, except in the phase change region. The concentration field changes rapidly across the phase change region, but may also vary outside the phase change region.

A general mathematical formulation of this solidification problem is the following. Given a bounded domain Ω of $\mathbb{R}^2$ with boundary $\partial\Omega$ and outer unit normal $\mathbf{n}$, two functions $c_0, \phi_0 : \Omega \to \mathbb{R}$ and a time interval $(0, T)$ we consider the problem of finding $\phi, c : \Omega \times (0, T) \to \mathbb{R}$ such that

$$\alpha\frac{\partial\phi}{\partial t} - \operatorname{div}\Big(A(\nabla\phi)\nabla\phi\Big) - S(c,\phi) = 0 \qquad \text{in } \Omega \times (0,T), \tag{1}$$

$$\frac{\partial c}{\partial t} - \operatorname{div}\Big(D_1(\phi)\nabla c + D_2(c,\phi)\nabla\phi\Big) = 0 \qquad \text{in } \Omega \times (0,T), \tag{2}$$

$$A(\nabla\phi)\nabla\phi \cdot \mathbf{n} = 0 \qquad \text{on } \partial\Omega \times (0,T), \tag{3}$$

$$\Big(D_1(\phi)\nabla c + D_2(c,\phi)\nabla\phi\Big) \cdot \mathbf{n} = 0 \qquad \text{on } \partial\Omega \times (0,T), \tag{4}$$

$$\phi(0) = \phi_0, \quad c(0) = c_0 \qquad \text{in } \Omega. \tag{5}$$

Here $\alpha \in \mathbb{R}$ is a positive parameter and $A(\cdot)$ is the matrix defined for $\xi \in \mathbb{R}^2 \setminus \{0\}$ by

$$A(\xi) = \begin{pmatrix} a^2(\theta(\xi)) & -a(\theta(\xi))a'(\theta(\xi)) \\ a(\theta(\xi))a'(\theta(\xi)) & a^2(\theta(\xi)) \end{pmatrix}, \tag{6}$$

where $\theta(\xi)$ denotes the angle between the vector ξ and a preferential direction, a is the real-valued function defined by

$$a(\theta) = 1 + \bar{a}\cos(\kappa\theta), \tag{7}$$

with $\bar{a} \geq 0$ (the anisotropy parameter) and κ a positive integer corresponding to the number of dendritic branching directions. Without loss of generality, the horizontal abscissa x_1 can be choosed as preferential direction and we have

$$\cos\theta(\xi) = \frac{\xi \cdot e_1}{\|\xi\|},$$

where e_1 is the unit vector in the horizontal direction. From the physical point of view, (1) is nothing but a generalization of a mean curvature - or Allen-Cahn

- equation, see for instance [9, 21] for a general presentation. This equation is obtained by taking the derivative of a free energy functional accounting for phase transformation, double well barriers and an interfacial anisotropic energy contribution

$$E(\nabla\phi) = \frac{1}{2}\int_{\Omega} a^2(\theta(\nabla\phi))|\nabla\phi|^2 dx.$$

It can be noticed that the derivative DE of E with respect to $\nabla\phi$ is given by

$$< DE(\nabla\phi), \nabla\psi >= \int_{\Omega} A(\nabla\phi)\nabla\phi \cdot \nabla\psi dx,$$

which is the weak form corresponding to the second term of (1). Due to the double well barriers, the function $S(\cdot,\cdot)$ in (1) contains small parameters that force the phase-field ϕ to take values 0 or 1, except in a small region, see again Fig. 1. Finally, equation (2) corresponds to solute conservation. We refer to [22, 20] for details about the physical derivation of the model.

2 Existence

Mathematical and numerical studies corresponding to (1)-(5) have already been presented in the isotropic case, that is when $\bar{a} = 0$ thus $A(\cdot) = I$. More precisely, existence has been proved in [17] using a Faedo-Galerkin method, a finite element procedure together with a priori error estimates have been proposed in [13] and adaptive finite elements have been presented in [14]. Existence of a weak solution in the anisotropic case $\bar{a} \neq 0$ has been proved in [8]. More precisely, if $\phi_0, c_0 \in L^2(\Omega)$, if S, D_1 and D_2 are continuous, bounded Lipschitz functions satisfying, for all ϕ in $\mathbb{R}$:

$$0 < D_s \leq D_1(\phi) \leq D_l, \tag{8}$$

and if $\bar{a} \leq \frac{1}{\kappa^2-1}$, then problem (1)-(5) has a weak solution

$$\phi, c \in L^2(0,T;H^1(\Omega)) \cap H^1(0,T;(H^1(\Omega))')$$

such that

$$\alpha\left\langle \frac{\partial\phi}{\partial t}, v\right\rangle + \int_{\Omega} A(\nabla\phi)\nabla\phi\cdot\nabla v - \int_{\Omega} S(c,\phi)v = 0,$$
$$\left\langle \frac{\partial c}{\partial t}, w\right\rangle + \int_{\Omega} D_1(\phi)\nabla c\cdot\nabla w + \int_{\Omega} D_2(c,\phi)\nabla\phi\cdot\nabla w = 0,$$

for all $v, w \in H^1(\Omega)$, a.e. $0 < t \leq T$. Hereabove, $< \cdot,\cdot >$ denotes the duality pairing between $H^1(\Omega)$ and its dual. Moreover, if the functions S and D_2 satisfy, for all ϕ, c in $\mathbb{R}$

$$S(c,0) = S(c,1) = 0,$$
$$D_2(0,\phi) = D_2(1,\phi) = 0,$$

they can be extended by zero outside the interval $(0,1)$, and a maximum principle holds for ϕ and c, that is, if $0 \le \phi_0, c_0 \le 1$, then $0 \le \phi(t), c(t) \le 1$, for all t.

In order to prove the existence result, the implicit Euler scheme is considered. Given an integer N, we set $\tau = T/N$ the time step, $t^n = n\tau$, $n = 0, ..., N$, $\phi^0 = \phi_0$ and $c^0 = c_0$. For $n = 1, 2, ..., N$, ϕ^{n-1} and c^{n-1} being known, let ϕ^n and c^n be approximations of $\phi(t^n)$ and $c(t^n)$ such that

$$\alpha\frac{\phi^n - \phi^{n-1}}{\tau} - \operatorname{div}\Big(A(\nabla\phi^n)\nabla\phi^n\Big) - S(c^{n-1},\phi^n) = 0 \qquad \text{in } \Omega, \tag{9}$$

$$\frac{c^n - c^{n-1}}{\tau} - \operatorname{div}\Big(D_1(\phi^n)\nabla c^n + D_2(c^{n-1},\phi^n)\nabla\phi^n\Big) = 0 \qquad \text{in } \Omega. \tag{10}$$

Existence of ϕ^n is obtained rewriting the first equation as a minimization problem. The functional to be minimized is defined by the potential

$$I(\phi) = \int_\Omega \left(\frac{\alpha}{2}\frac{(\phi - \phi^{n-1})^2}{\tau} + \frac{1}{2}a^2(\theta(\nabla\phi))|\nabla\phi|^2 + f(c^{n-1},\phi)\right),$$

where the function f is such that $\frac{\partial f}{\partial\phi}(c,\phi) = -S(c,\phi)$. The existence of a unique minimizer is a consequence of the fact that the Ginzburg-Landau potential $E(\cdot)$ defined by

$$E(\xi) = \frac{1}{2}\int_\Omega a^2(\theta(\xi))|\xi|^2 dx$$

is strictly convex $\forall\xi \in L^2(\Omega)^2$, whenever $\bar{a} < \frac{1}{\kappa^2 - 1}$.

3 Finite elements and error estimates

3.1 A priori error estimates

For any $0 < h < 1$, let $\mathcal{T}_h$ be a conforming triangulation of $\overline{\Omega}$ into triangles K with diameter h_K less than h. Let V_h be the usual finite element space of continuous, piecewise linear functions on the triangles of $\mathcal{T}_h$. The finite element scheme corresponding to (9) (10) is considered. For each $n = 1..., N$, we are first looking for ϕ_h^n in V_h such that

$$\int_\Omega \alpha\frac{\phi_h^n - \phi_h^{n-1}}{\tau}v_h + \int_\Omega A(\nabla\phi_h^n)\nabla\phi_h^n \cdot \nabla v_h - \int_\Omega S(c_h^{n-1},\phi_h^n)v_h = 0, \tag{11}$$

for all $v_h \in V_h$ and then for c_h^n in V_h such that

$$\int_\Omega \frac{c_h^n - c_h^{n-1}}{\tau}w_h + \int_\Omega D_1(\phi_h^n)\nabla c_h^n \cdot \nabla w_h + \int_\Omega D_2(c_h^{n-1},\phi_h^n)\nabla\phi_h^n \cdot \nabla w_h = 0, \tag{12}$$

for all $w_h \in V_h$.

A priori error estimates in the $L^2(0,T;H^1(\Omega))$ norm have been proved in [5] in the case when $\bar{a} < \frac{1}{\kappa^2-1}$. The convergence proof strongly relies on the strong monotonicity on the operator A. More precisely, the following result is used to prove convergence.

Lemma 1. *Let $A(\cdot)$ be the operator defined by* (6) *and let the convexity condition $\bar{a} < \frac{1}{\kappa^2-1}$ hold. Then there exists μ (depending on $\bar{a}$) such that, for all ϕ, $\psi \in H^1(\Omega)$, we have*

$$\mu \|\nabla(\phi - \psi)\|^2_{L^2(\Omega)} \leq \Big(A(\nabla\phi)\nabla\phi - A(\nabla\psi)\nabla\psi, \nabla(\phi - \psi)\Big). \quad (13)$$

Then, assuming that the solution (ϕ, c) is smooth the following result is proved in [5].

Theorem 1. *Let ϕ, c be the weak solution of* (1)-(5)*, let ϕ_h^n, c_h^n be defined by* (11)-(12)*. Assume that*

$$(\phi, c) \in \Big(L^2(0,T;H^3(\Omega)) \cap L^\infty(0,T;H^2(\Omega) \cap W^{1,\infty}(\Omega)) \cap H^1(0,T;H^1(\Omega))\Big)^2.$$

Then, there is a constant C such that, for all $h, \tau > 0$, we have

$$\sum_{n=1}^{N} \tau \|\nabla(\phi_h^n - \phi(t^n))\|^2_{L^2(\Omega)} + \sum_{n=1}^{N} \tau \|\nabla(c_h^n - c(t^n))\|^2_{L^2(\Omega)} \leq C\left(\tau^2 + h^2 + \frac{h^4}{\tau}\right). \quad (14)$$

3.2 A posteriori error estimates for meshes with high aspect ratio

As in [15, 16], our goal is to perform adaptive finite elements with high aspect ratio for solving numerically (1)-(5) in the general case when $\bar{a} \neq 0$. The general framework of [10, 11] is considered.

For any triangle K of the mesh, let $T_K : \hat{K} \to K$ be the affine transformation which maps the reference triangle $\hat{K}$ onto K. Let M_K be the Jacobian of T_K that is

$$\mathbf{x} = T_K(\hat{\mathbf{x}}) = M_K \hat{\mathbf{x}} + \mathbf{t}_K.$$

Since M_K is invertible, it admits a singular value decomposition $M_K = R_K^T \Lambda_K P_K$, where R_K and P_K are orthogonal and where Λ_K is diagonal with positive entries. In the following we set

$$\Lambda_K = \begin{pmatrix} \lambda_{1,K} & 0 \\ 0 & \lambda_{2,K} \end{pmatrix} \qquad \text{and} \qquad R_K = \begin{pmatrix} \mathbf{r}_{1,K}^T \\ \mathbf{r}_{2,K}^T \end{pmatrix},$$

with the choice $\lambda_{1,K} \geq \lambda_{2,K}$. A simple example of such a transformation is $x_1 = H\hat{x}_1$, $x_2 = h\hat{x}_2$, with $H \geq h$, thus

$$M_K = \begin{pmatrix} H & 0 \\ 0 & h \end{pmatrix} \quad \lambda_{1,K} = H, \quad \lambda_{2,K} = h, \quad \mathbf{r}_{1,K} = \begin{pmatrix} 1 \\ 0 \end{pmatrix}, \quad \mathbf{r}_{2,K} = \begin{pmatrix} 0 \\ 1 \end{pmatrix},$$

see Fig. 3.

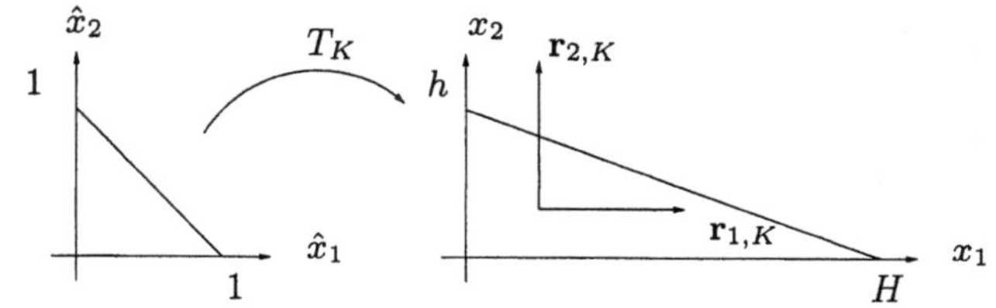

Fig. 3. A simple example of transformation from element $\hat{K}$ to K

In the framework of meshes with high aspect ratio, the classical minimum angle condition must be avoided. However, it is required that, for each vertex, the number of neighbouring vertices is bounded above, uniformly with respect to the mesh size h. Also, for any patch Δ_K (the set of triangles having a vertex common with K), the diameter of the corresponding reference patch $\Delta_{\hat{K}}$, that is $\Delta_{\hat{K}} = T_K^{-1}(\Delta_K)$, must be uniformly bounded independently of h. This latter hypothesis excludes some distorted patches, see Fig. 4. Let $I_h : H^1(\Omega) \to V_h$ be a Clément or Scott-Zhang like interpolation operator. We now recall some interpolation results due to [10, 11].

Lemma 2. *There is a constant $\hat{C}$ depending only on the reference element $\hat{K}$ such that, for all $v \in H^1(\Omega)$, for all $K \in \mathcal{T}_h$, we have*

$$\|v - R_h v\|^2_{L^2(K)} + \lambda_{2,K} \|v - R_h v\|^2_{L^2(\partial K)} \leq \hat{C} \omega_K^2(v).$$

Here we have set

$$\omega_K^2(v) = \lambda_{1,K}^2 \left(\mathbf{r}_{1,K}^T G_K(v) \mathbf{r}_{1,K} \right) + \lambda_{2,K}^2 \left(\mathbf{r}_{2,K}^T G_K(v) \mathbf{r}_{2,K} \right),$$

and where $G_K(v)$ denotes the 2×2 matrix defined by

$$G_K(v) = \begin{pmatrix} \displaystyle\int_{\Delta K} \left(\frac{\partial v}{\partial x_1} \right)^2 dx & \displaystyle\int_{\Delta K} \frac{\partial v}{\partial x_1} \frac{\partial v}{\partial x_2} dx \\ \displaystyle\int_{\Delta K} \frac{\partial v}{\partial x_1} \frac{\partial v}{\partial x_2} dx & \displaystyle\int_{\Delta K} \left(\frac{\partial v}{\partial x_2} \right)^2 dx \end{pmatrix}.$$

For $0 \leq t \leq T$, we denote $\phi_h(t)$ and $c_h(t) \in V_h$ the semi-discrete finite element approximation corresponding to (11) and (12), respectively. In order to prove a posteriori error estimates we need to assume that the error in the $L^2(0,T; L^2(\Omega))$ norm converges faster than the error in the $L^2(0,T; H^1(\Omega))$

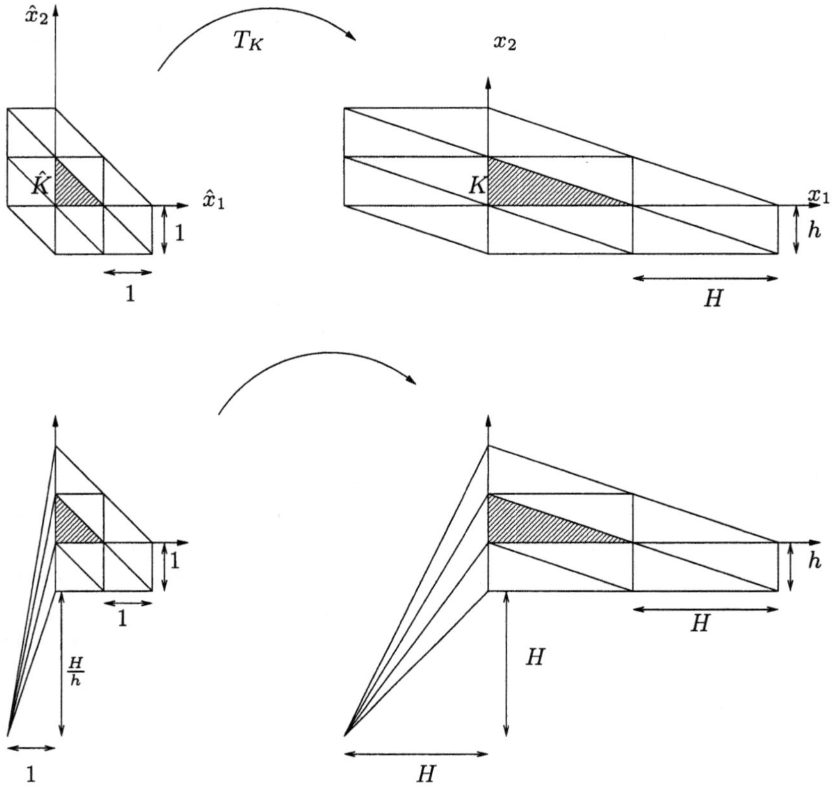

Fig. 4. Example of an acceptable patch (top): the size of the reference patch $\Delta_{\hat{K}}$ does not depend on the aspect ratio H/h. Example of a nonacceptable patch (bottom): the size of the reference patch $\Delta_{\hat{K}}$ now depends on the aspect ratio H/h.

norm, that is, there are two constant $C > 0$ and $s \in]0,1]$ such that, for all mesh $\mathcal{T}_h$ we have

$$\int_0^T \left(\|\phi - \phi_h\|^2_{L^2(\Omega)} + \|c - c_h\|^2_{L^2(\Omega)} \right)$$
$$\leq C \Big(\max_{K \in \mathcal{T}_h} \lambda_{2,K} \Big)^{2s} \int_0^T \left(\|\nabla(\phi - \phi_h)\|^2_{L^2(\Omega)} + \|\nabla(c - c_h)\|^2_{L^2(\Omega)} \right). \quad (15)$$

Let us comment this assumption in the frame of meshes with small aspect ratio, that is to say when $\lambda_{1,K}$ and $\lambda_{2,K}$ are of order h, for all $K \in \mathcal{T}_h$. According to the results of section 3.1, the error in the $L^2(0,T;H^1(\Omega))$ norm is shown to be $O(h)$ whenever the time step is small. On the other hand, $O(h^2)$ convergence in the $L^\infty(0,T;L^2(\Omega))$ norm is proved for in [13], but only in the case when $\bar{a} = 0$. Therefore, we expect the following assumption

$$\int_0^T \left(\|\phi - \phi_h\|^2_{L^2(\Omega)} + \|c - c_h\|^2_{L^2(\Omega)} \right) \leq Ch^2 \int_0^T \left(\|\nabla(\phi - \phi_h)\|^2_{L^2(\Omega)} + \|\nabla(c - c_h)\|^2_{L^2(\Omega)} \right)$$

to hold for meshes with small aspect ratio, provided optimal convergence results hold in both $L^2(0,T;H^1(\Omega))$ and $L^2(0,T;L^2(\Omega))$ norms. Note that this assumption has already been used in [14] to obtain a posteriori error estimates in the case when $\bar{a} = 0$. When considering meshes with high aspect ratio, h in the above estimate should be replaced by $\max_{K\in\mathcal{T}_h} \lambda_{2,K}$, which yields (15). This assumption is checked numerically in section 4.

For each interior edge of $\mathcal{T}_h$, let us choose an arbitrary normal direction $\mathbf{n}$, let $[\xi]$ denote the jump of ξ across the edge. For each edge of $\mathcal{T}_h$ lying on the boundary $\partial\Omega$, we set $[\xi]$ to twice the inner side value of ξ. The following result is proved in [6].

Theorem 2. *Let ϕ, c be the weak solution of* (1)-(5), *let ϕ_h, c_h be the semi-discrete approximation corresponding to* (11) (12). *Assume that $\phi, c \in L^\infty(0,T;H^2(\Omega))$ and that* (15) *holds. Then, there is a constant C depending only on the interpolation constants of Lemma 2 such that, for all mesh $\mathcal{T}_h$ such that $\max_{K\in\mathcal{T}_h} \lambda_{2,K}$ is sufficiently small, we have*

$$\begin{aligned}
&\frac{\alpha}{2}\|(\phi - \phi_h)(T)\|^2_{L^2(\Omega)} + \frac{\mu}{2}\int_0^T \|\nabla(\phi - \phi_h)\|^2_{L^2(\Omega)} \\
&+ \frac{\mu D_s}{8M_2^2}\|(c - c_h)(T)\|^2_{L^2(\Omega)} + \frac{\mu D_s^2}{8M_2^2}\int_0^T \|\nabla(c - c_h)\|^2_{L^2(\Omega)} \\
&\leq \frac{\alpha}{2}\|(\phi - \phi_h)(0)\|^2_{L^2(\Omega)} + \frac{\mu D_s}{8M_2^2}\|(c - c_h)(0)\|^2_{L^2(\Omega)} \\
&+ C\int_0^T \sum_{K\in\mathcal{T}_h} \left(\left\| \alpha\frac{\partial \phi_h}{\partial t} - div\,(A(\nabla\phi_h)\nabla\phi_h) - S(c_h, \phi_h) \right\|_{L^2(K)} \right. \\
&\qquad \left. + \frac{1}{2\lambda_{2,K}^{1/2}} \|[A(\nabla\phi_h)\nabla\phi_h \cdot \mathbf{n}]\|_{L^2(\partial K)} \right) \times \omega_K(\phi - \phi_h) \\
&+ C\frac{\mu D_s^2}{4M_2^2}\int_0^T \sum_{K\in\mathcal{T}_h} \left(\left\| \frac{\partial c_h}{\partial t} - div\left(D_1(\phi_h)\nabla c_h + D_2(c_h, \phi_h)\nabla\phi_h \right) \right\|_{L^2(K)} \right. \\
&\qquad \left. + \frac{1}{2\lambda_{2,K}^{1/2}} \left\| \left[D_1(\phi_h)\frac{\partial c_h}{\partial n} + D_2(c_h, \phi_h)\frac{\partial \phi_h}{\partial n} \right] \right\|_{L^2(\partial K)} \right) \times \omega_K(c - c_h). \qquad (16)
\end{aligned}$$

Here μ is the constant of Lemma 1, D_s is the constant of (8) *and $M_2 = \|D_2\|_{L^\infty(\mathbb{R}^2)}$. Moreover, $\omega_K(\cdot)$ is defined as in Lemma 2.*

Estimate (16) is not a usual a posteriori error estimate since ϕ and c are still involved in the right hand side. We then proceed as in [15, 16] and introduce an estimator based on superconvergent recovery, namely a Zienkiewicz-Zhu (Z-Z) like estimator [23, 3, 24]. More precisely, we consider the simplest Z-Z error estimator as defined in [19, 1]. The Z-Z error estimator corresponding to $\nabla(c - c_h)$ is defined by the difference between ∇c_h and an approximate L^2 projection of ∇c_h onto V_h^2, namely :

$$\eta^{ZZ}(c_h) = \begin{pmatrix} \eta_1^{ZZ}(c_h) \\ \eta_2^{ZZ}(c_h) \end{pmatrix} = \begin{pmatrix} (I - \Pi_h)\left(\dfrac{\partial c_h}{\partial x_1}\right) \\ (I - \Pi_h)\left(\dfrac{\partial c_h}{\partial x_2}\right) \end{pmatrix}. \tag{17}$$

Here $\Pi_h : g \in L^2(\Omega) \to (\Pi_h g) \in V_h$ is defined by

$$\int_\Omega r_h((\Pi_h g) v_h) = \int_\Omega g v_h \qquad \forall v_h \in V_h,$$

where r_h denotes the usual Lagrange interpolant. In other words, from constant values of ∇c_h on triangles, we build values at vertices P using the formula

$$\begin{pmatrix} \Pi_h\left(\dfrac{\partial c_h}{\partial x_1}\right)(P) \\ \Pi_h\left(\dfrac{\partial c_h}{\partial x_2}\right)(P) \end{pmatrix} = \frac{1}{\sum\limits_{\substack{K\in\mathcal{T}_h \\ P\in K}} |K|} \begin{pmatrix} \sum\limits_{\substack{K\in\mathcal{T}_h \\ P\in K}} |K| \left(\dfrac{\partial c_h}{\partial x_1}\right)_{|K} \\ \sum\limits_{\substack{K\in\mathcal{T}_h \\ P\in K}} |K| \left(\dfrac{\partial c_h}{\partial x_2}\right)_{|K} \end{pmatrix}.$$

From [2, 19] we know that for a certain class of meshes (namely parallel meshes) and for smooth solutions, Z-Z like error estimators are asymptotically exact (i.e. the Z-Z error estimator converges to the true error when h goes to zero). Our error indicator corresponding to $c - c_h$ is then obtained by replacing the matrix $G_K(c - c_h)$ present in the definition of $\omega_K(c - c_h)$ in (16) by the matrix $\tilde{G}_K(c_h)$ defined by

$$\tilde{G}_K(c_h) = \begin{pmatrix} \displaystyle\int_K (\eta_1^{ZZ}(c_h))^2 dx & \displaystyle\int_K \eta_1^{ZZ}(c_h)\eta_2^{ZZ}(c_h) dx \\ \displaystyle\int_K \eta_1^{ZZ}(c_h)\eta_2^{ZZ}(c_h) dx & \displaystyle\int_K (\eta_2^{ZZ}(c_h))^2 dx \end{pmatrix}. \tag{18}$$

Finally, our simplified error indicator corresponding to the concentration error $c - c_h$ is defined on each triangle K by

$$\int_0^T \frac{1}{2\lambda_{2,K}^{1/2}} \left\| \left[\frac{\partial c_h}{\partial n} \right] \right\|_{L^2(\partial K)} \times \left(\lambda_{1,K}^2 \left(\mathbf{r}_{1,K}^T \tilde{G}_K(c_h) \mathbf{r}_{1,K} \right) + \lambda_{2,K}^2 \left(\mathbf{r}_{2,K}^T \tilde{G}_K(c_h) \mathbf{r}_{2,K} \right) \right)^{1/2}. \tag{19}$$

4 Numerical study of the effectivity index for small time steps

In this section, the quality of our error indicator (19) is investigated numerically. For details we refer to [6]. Let us consider ϕ_h^n and $c_h^n \in V_h$ the solutions of (11) and (12), respectively. In practice, ϕ_h^n is obtained by performing only one Newton iteration at each time step. Proceeding as in [16] we introduce $c_{h\tau}$ the continuous, piecewise linear approximation in time defined by

$$c_{h\tau}(x,t) = \frac{t-t^{n-1}}{\tau}c_h^n(x) + \frac{t^n-t}{\tau}c_h^{n-1}(x) \qquad t^{n-1} \le t \le t^n, \quad x \in \Omega. \quad (20)$$

and the simplified error indicator for each time interval $[t^{n-1}, t^n]$ and triangle K by

$$\left(\eta_{n,K}(c_{h\tau})\right)^2 = \int_{t^{n-1}}^{t^n} \frac{1}{2\lambda_{2,K}^{1/2}} \left\| \left[\frac{\partial c_{h\tau}}{\partial n}\right] \right\|_{L^2(\partial K)}$$
$$\times \left(\lambda_{1,K}^2\left(\mathbf{r}_{1,K}^T \tilde{G}_K(c_{h\tau})\mathbf{r}_{1,K}\right) + \lambda_{2,K}^2\left(\mathbf{r}_{2,K}^T \tilde{G}_K(c_{h\tau})\mathbf{r}_{2,K}\right)\right)^{1/2}, \quad (21)$$

where $\tilde{G}_K(c_{h\tau})$ is defined as in (18).

We consider the model of [20], the notations being those of [12]. The source term S in (1) is defined by

$$S(c,\phi) = -\frac{1}{\lambda^2}\phi(1-\phi)(1-2\phi) + \frac{5m_l}{\lambda\Gamma}\phi^2(1-\phi)^2\left(\frac{c}{1-\phi+k\phi} - c_l\right) \qquad \text{if } 0 \le \phi \le 1,$$

whereas $S(c,\phi) = 0$ if $\phi < 0$ or $\phi > 1$. Here λ is the thickness of the solid-liquid interface, m_l is the liquid slope in the phase diagram, Γ is the Gibbs-Thomson isotropic coefficient, c_l is the liquid concentration in the phase diagram, and k is the phase diagram partition coefficient (thus $c_s = kc_l$, where c_s is the solid concentration in the phase diagram). The coefficient α in (1) equals $\frac{1}{\mu_k \Gamma}$, where μ_k is the interface kinetic coefficient. The first term in the definition of $S(\cdot,\cdot)$ is nothing but the derivative of a double well potential that forces ϕ to values zero or one, except in the phase change region. An asymptotic expansion of the phase-field equation at first order in λ links the normal velocity of the solid-liquid interface, some anisotropic measure of the interface curvature, and the concentration field. The function D_1 in (2) is given by

$$D_1(\phi) = D_s + \frac{1-\phi}{1-\phi+k\phi}(D_l - D_s) \qquad \text{if } 0 \le \phi \le 1,$$

whereas $D_1(\phi) = D_l$ if $\phi < 0$ and $D_1(\phi) = D_s$ if $1 < \phi$. Here D_s and D_l are the solid and liquid diffusion coefficients. Finally, the function D_2 in (2) is given by

$$D_2(c,\phi) = D_1(\phi)\frac{(1-k)c(1-c)}{1-\phi+k\phi} \qquad \text{if } 0 \le c \le 1,$$

whereas $D_2(c,\phi) = 0$ if $c < 0$ or $c > 1$. All the physical parameters are given below in the international MKSA unit system.

Our first goal is to validate numerically assumption (15) in the context of meshes with high aspect ratio. For this purpose, we set the computational domain to $\Omega = [-0.0002, 0.0002]^2$, we add source terms in (1) (2) so that ϕ and c are given by

$$\phi(x_1, x_2, t) = c(x_1, x_2, t) = \frac{1 - \tanh\left(\dfrac{x_1 - vt}{\delta}\right)}{2},$$

where $v = 2\ 10^{-4}$ and $\delta = 10^{-5}$. Dirichlet boundary conditions are prescribed on the vertical sides of Ω, homogeneous Neumann boundary conditions on the horizontal sides. The physical parameters involved in the definition of S, D_1 and D_2 are given in Tab. 1, the time step is $\tau = 5\ 10^{-5}$ and is small enough in order to overkill the error due to time discretization. Meshes with high aspect ratio are used to validate assumption (15). In Tab. 2, errors in the $L^2(0,T;L^2(\Omega))$ and $L^2(0,T;H^1(\Omega))$ norms (resp. e_{L^2} and e_{H^1}) are reported when using distorded meshes ($h1 - h2$ denotes the mesh size in horizontal and vertical directions). Also, the effectivity indices (the ratio between the error estimator and the true error) ei^{ZZ} and ei^{A} corresponding to the Zienkiewicz-Zhu error estimator (17) and our simplified error indicator (21) are shown. Clearly, when the mesh is refined in the horizontal direction, assumption (15) holds with $s = 1$ since the $L^2(0,T;L^2(\Omega))$ error converges at order two and the $L^2(0,T;H^1(\Omega))$ error at rate one. However, when the mesh is refined in the wrong (vertical) direction, then the error does not decrease and (15) does not hold.

Table 1. Test case with exact solution : parameters used for the computations.

λ	m_l	Γ	c_s	c_l	D_s	D_l	μ_k
10^{-5}	-260	0.1	0.015	0.0238	$5\ 10^{-10}$	$5\ 10^{-9}$	0.0015

5 An adaptive algorithm generating meshes with high aspect ratio

We now present the adaptive algorithm of [6]. Given a time step τ, the goal is to build triangulations $\mathcal{T}_h^n$, $n = 1, ..., N$ having high aspect ratio such that the relative estimated error in the $L^2(0,T;H^1(\Omega))$ norm is close to a preset tolerance TOL. Our adaptive algorithm aims at building triangulations $\mathcal{T}_h^n$, $n = 1, ..., N$ such that

Table 2. Various convergence results for the travelling wave solution.

$h1 - h2$	e_{L^2}	e_{H^1}	ei^{ZZ}	ei^A
$0.000005 - 0.0001$	$1.1\ 10^{-6}$	0.29	1.01	1.85
$0.0000025 - 0.00005$	$3.2\ 10^{-7}$	0.13	1.01	1.74
$0.00000125 - 0.000025$	$8.8\ 10^{-8}$	0.066	1.01	1.74

Anisotropic meshes refined in both horizontal and vertical directions

$h1 - h2$	e_{L^2}	e_{H^1}	ei^{ZZ}	ei^A
$0.000005 - 0.0001$	$1.1\ 10^{-6}$	0.29	1.01	1.85
$0.0000025 - 0.0001$	$4.4\ 10^{-7}$	0.14	1.00	1.79
$0.00000125 - 0.0001$	$1.2\ 10^{-8}$	0.061	1.00	1.79

Anisotropic meshes refined in horizontal direction only

$h1 - h2$	e_{L^2}	e_{H^1}	ei^{ZZ}	ei^A
$0.00005 - 0.00001$	$4.0\ 10^{-5}$	4.3	0.85	1.67
$0.00005 - 0.000005$	$3.2\ 10^{-5}$	5.3	0.86	1.87
$0.00005 - 0.0000025$	$4.2\ 10^{-5}$	8.48	0.86	1.54

Anisotropic meshes refined in vertical direction only

$$0.75\ TOL \leq \frac{\sum_{n=1}^{N} \sum_{K \in \mathcal{T}_h} \left(\eta_{n,K}(c_{h\tau})\right)^2}{\int_0^T \int_\Omega |\nabla c_{h\tau}|^2} \leq 1.25\ TOL, \tag{22}$$

where $\eta_{n,K}(c_{h\tau})$ defined by (21). A sufficient condition to satisfy (22) is to build, for each $n = 1, ..., N$, an anisotropic triangulation $\mathcal{T}_h^n$ such that

$$\frac{0.75^2 TOL^2}{NV_h^n} \int_{t^{n-1}}^{t^n} \int_\Omega |\nabla c_{h\tau}|^2 \leq \left(\eta_{n,K}(c_{h\tau})\right)^2 \leq \frac{1.25^2 TOL^2}{NV_h^n} \int_{t^{n-1}}^{t^n} \int_\Omega |\nabla c_{h\tau}|^2$$

for all triangle $K \in \mathcal{T}_h^n$, where NV_h^n is the number of vertices of the mesh $\mathcal{T}_h^n$. We then proceed as in [15, 16] to build such an anisotropic mesh, using the BL2D mesh generator [4].

5.1 Computations with small anisotropy

We now consider the following physical situation. At initial time, the computational domain is liquid, with homogeneous concentration 0.02. Then, a circular solid seed of diameter 2.5 10^{-6} and concentration 0.015 is placed at the center of Ω. The physical parameters are now given in Tab. 3 and are taken from [12], table 1, column B, except c_s and c_l.

Table 3. Parameters used for the computations.

λ	m_l	Γ	c_s	c_l	D_s	D_l	μ_k
0.5 10^{-6}	−260	5 10^{-7}	0.015	0.0238	5 10^{-10}	5 10^{-9}	0.0015

We first present computations in the case when the anisotropy parameter $\bar{a}$ is small. We set the number of dendrite arms $\kappa = 4$ and choose $\bar{a} = 0.04$ so that $\bar{a} < \frac{1}{\kappa^2-1} \simeq 0.0667$. The time step is $\tau = 5\ 10^{-4}$ and the final time is $T = 1$, making the total number of time steps 2000.

In Fig. 5, the adapted meshes, concentration and phase fields corresponding to an adaptive computation with tolerance $TOL = 0.0625$ (6.25% estimated relative error) are reported. The concentration c and phase ϕ appear to be smooth, but exhibit strong gradients across the solid to liquid transition zone, therefore the mesh is strongly refined in the neighbourhood of the solid-liquid interface. Zooms of the results are shown in Fig. 6. The adaptive algorithm generates 400 meshes from initial to final time. The computation takes about 4 hours on a Pentium III 1.2 Ghz PC, with a required memory of less than 300 Mb. The maximum aspect ratio of the generated meshes is approximately 30, without any a priori upper bound imposed by the adaptive method.

5.2 Computations with large anisotropy

We now choose the anisotropy parameter $\bar{a} > \frac{1}{\kappa^2-1} \simeq 0.0667$, namely $\bar{a} = 0.1$. In this case there are no known existence results for the system in $L^2(0,T;H^1(\Omega))$.

In Fig. 7, the concentration fields corresponding to an adaptive computation with tolerance $TOL = 0.0625$ are reported. A zoom of the results at final time, Fig. 8, shows that the gradient is discontinuous in some regions. This phenomenon is explained in details in [6].

6 Conclusions and perspectives

A phase-field model corresponding to the isothermal, dendritic growth of a binary alloy is considered. Existence is proved when the physical anisotropy is small. A finite element method is proposed and a priori error estimates are obtained, again when the physical anisotropy is small. A posteriori error estimates are derived for meshes with large aspect ratio and a numerical study of

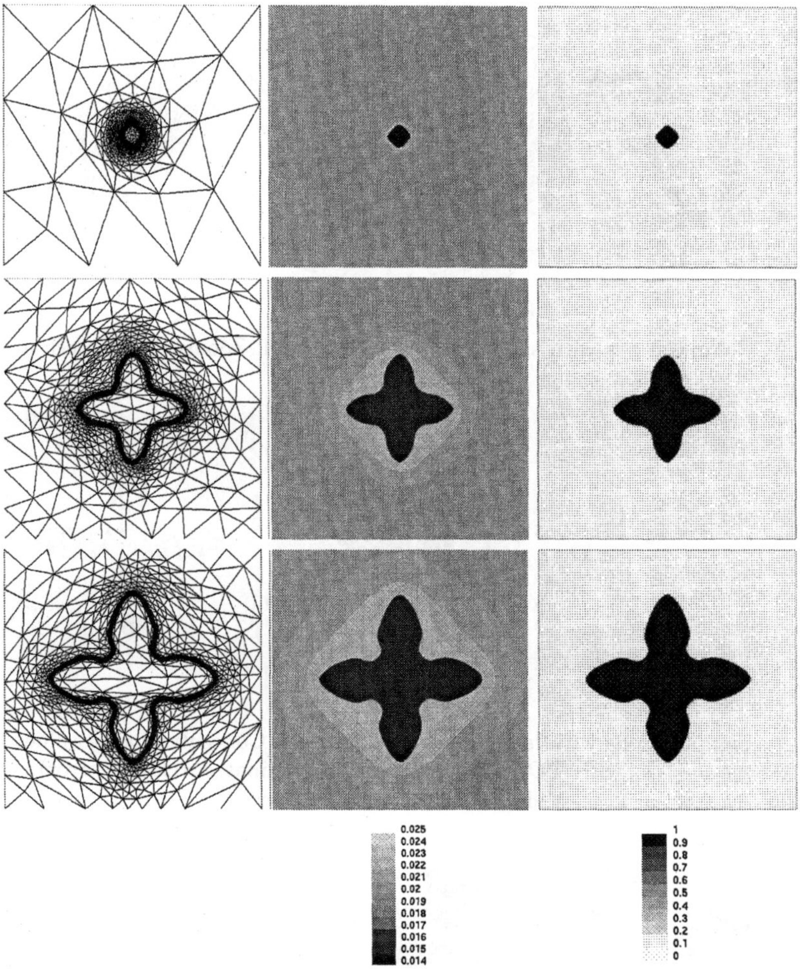

Fig. 5. Computations with small anisotropy, $\bar{a} = 0.04$. Adapted meshes (left column), concentration isovalues (middle column) and phase isovalues (right column), from $t = 0$ to $t = 1$ s, with $TOL = 0.0625$ (6.25% estimated relative error). Row 1: $t = 0.05$ s, 6874 vertices. Row 2: $t = 0.5$ s, 17170 vertices. Row 3: $t = 1$ s, 24441 vertices.

the effectivity index is proposed. Finally, an adaptive algorithm that generates successive meshes with high aspect ratio is presented.

Numerical results corresponding to dendritic growth are then presented. When the physical anisotropy is small, the numerical solution is smooth whereas, when the physical anisotropy exceeds the predicted value, then ir-

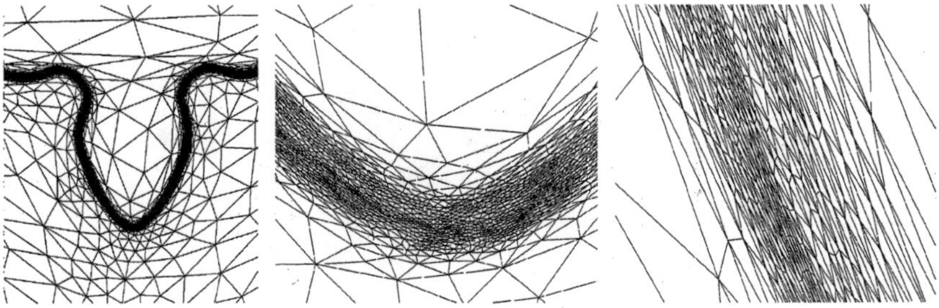

Fig. 6. Computations with small anisotropy, $\bar{a} = 0.04$. Zooms of the mesh at final time.

Fig. 7. Computations with large anisotropy, $\bar{a} = 0.1$. Concentration isovalues from $t = 0$ to $t = 1$ s, with $TOL = 0.0625$ (6.25% estimated relative error). Left : $t = 0.05$ s, 5138 vertices. Middle : $t = 0.5$ s, 20580 vertices. Right : $t = 1$ s, 29971 vertices.

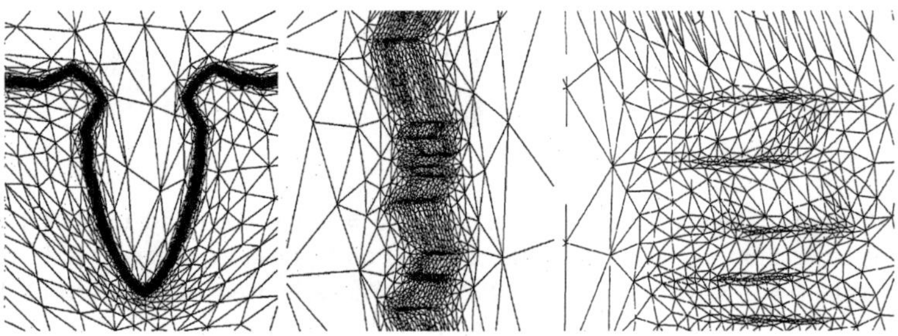

Fig. 8. Computations with large anisotropy, $\bar{a} = 0.1$. Zooms of the mesh at final time.

regular dendritic shapes are obtained. We are looking forward to obtaining similar theoretical results for the multiphase-field model described in [18, 7].

References

1. M. Ainsworth and J. T. Oden. A posteriori error estimation in finite element analysis. *Comput. Methods Appl. Mech. Engrg.*, 142(1-2):1–88, 1997.
2. M. Ainsworth and J.T. Oden. A unified approach to a posteriori error estimation using finite element residual methods. *Numer. Math.*, 65:23–50, 1993.
3. M. Ainsworth, J. Z. Zhu, A. W. Craig, and O. C. Zienkiewicz. Analysis of the Zienkiewicz-Zhu a posteriori error estimator in the finite element method. *Internat. J. Numer. Methods Engrg.*, 28(9):2161–2174, 1989.
4. H. Borouchaki and P. Laug. The bl2d mesh generator: Beginner's guide, user's and programmer's manual. Technical Report RT-0194, Institut National de Recherche en Informatique et Automatique (INRIA), Rocquencourt, 78153 Le Chesnay, France, 1996.
5. E. Burman, D. Kessler, and J. Rappaz. Convergence of the finite element method for an anisotropic phase-field model. Technical report, Départernent de Mathématiques, Ecole Polytechnique Fédérale de Lausanne, 1015 Lausanne, Switzerland, 2002.
6. E. Burman and M. Picasso. Anisotropic, adaptive finite elements for the computation of a solutal dendrite. *Interfaces Free Bound.*, 5(2):103–127, 2003.
7. E. Burman, M. Picasso, and A. Jacot. Adaptive finite elements with high aspect ratio for the computation of coalescence using a phase-field model. *J. Comput. Phys.*, accepted, 2003.
8. E. Burman and J. Rappaz. Existence of solutions to an anisotropic phase-field model. *Math. Methods Appl. Sci.*, 26(13):1137–1160, 2003.
9. C. M. Elliott. Approximation of curvature dependent interface motion. In *The state of the art in numerical analysis (York, 1996)*, pages 407–440. Oxford Univ. Press, New York, 1997.
10. L. Formaggia and S. Perotto. New anisotropic a priori error estimates. *Numer. Math.*, 89:641–667, 2001.
11. L. Formaggia and S. Perotto. Anisotropic error estimates for elliptic problems. *Numer. Math.*, 94(1):67–92, 2003.
12. A. Jacot and M. Rappaz. A pseudo front tracking technique for the modelling of solidification microstrucures in multicomponent alloys. *Acta Materialia*, 50:1909–1926, 2002.
13. D. Kessler and J.-F. Scheid. A priori error estimates for a phase-field model for the solidification process of a binary alloy. *IMA J. Numer. Anal.*, 22:281–305, 2002.
14. O. Kruger, M. Picasso, and J.-F. Scheid. A posteriori error estimates and adaptive finite elements for a nonlinear parabolic problem arising from solidification. *Comput. Methods Appl. Mech. Engrg.*, 192:535–558, 2001.
15. M. Picasso. Numerical study of the effectivity index for an anisotropic error indicator based on zienkiewicz-zhu error estimator. *Comm. Numer. Methods Engnrg.*, 19:13–23, 2002.
16. M. Picasso. An anisotropic error indicator based on zienkiewicz-zhu error estimator : application to elliptic and parabolic problems. *SIAM J. Sci. Comp.*, 24:1328–1355, 2003.
17. J. Rappaz and J.-F. Scheid. Existence of solutions to a phase-field model for the isothermal solidification process of a binary alloy. *Math. Meth. Appl. Sci.*, 23:491–513, 2000.

18. M. Rappaz, A. Jacot, and W.J. Boettinger. Last stage solidification of alloys : a theoretical study of dendrite arm and grain coalescence. *Met. Trans. A*, 34:467–479, 2003.
19. R. Rodríguez. Some remarks on Zienkiewicz-Zhu estimator. *Numer. Methods Partial Differential Equations*, 10(5):625–635, 1994.
20. J. Tiaden, B. Nestler, H. J. Diepers, and I. Steinbach. The multiphase-field model with an integrated concept for modelling solute diffusion. *Physica D: Nonlinear Phenomena*, 115(1-2):73–86, 1998.
21. A. Visintin. *Models of phase transitions.* Birkhäuser Boston Inc., Boston, MA, 1996.
22. J. A. Warren and W. J. Boettinger. Prediction of dendritic growth and microsegregation patterns in a binary alloy using the phase-field model. *Acta metall. mater.*, 43(2):689–703, 1995.
23. O. Zienkiewicz and J. Zhu. A simple error estimator and adaptive procedure for practical engineering analysis. *Internat. J. Numer. Methods Engrg.*, 24(2):337–357, 1987.
24. O. C. Zienkiewicz and J. Z. Zhu. The superconvergent patch recovery and a posteriori error estimates. I. The recovery technique. *Internat. J. Numer. Methods Engrg.*, 33(7):1331–1364, 1992.

Discontinuous Galerkin Methods for Timoshenko Beams

Fatila Celiker[1*], Bernardo Cockburn[1], Sukru Güzey[2], Ramdev Kanapady[3], Sew-Chew Soon[2], Henrik K. Stolarski[2], and Kummar Tamma[3]

1. School of Mathematics, University of Minnesota, 206 Church St SE Minneapolis, MN 55455
2. Department of Civil Engineering, University of Minnesota, 500 Pillsbury Drive SE Minneapolis, MN 55455
3. Department of Mechanical Engineering, University of Minnesota, 111 Church Street S.E. Minneapolis, MN 55455
**Corresponding Author: e-mail: celiker@math.umn.edu*

Summary. We devise a family of discontinuous Galerkin methods for the Timoshenko beam problem. Sufficient conditions for the existence and uniqueness of the approximation are given. The method allows arbitrary meshes and arbitrary polynomial degrees within the mesh, and hence is suitable for *hp* adaptivity. Numerical results showing optimal and exponential convergence are provided. These features of the method render it appealing for other problems in structure mechanics such as, plates, shells etc.

1 Introduction

In this paper, we introduce and numerically study discontinuous Galerkin (DG) for Timoshenko beams. The Timoshenko beam model, see [1] and [2], can be written as

$$\frac{dw}{dx} = \theta(x) - \frac{T(x)}{(GA)(x)}, \quad \frac{d\theta}{dx} = \frac{M(x)}{(EI)(x)}, \quad \frac{dM}{dx} = T(x), \quad \frac{dT}{dx} = q(x), \quad (1)$$

where $x \in \Omega = (0, L)$. Here, the unknowns are the transverse displacement w, the rotation of the transverse cross-section of the beam θ, the bending moment M, and the shear force T. The material and geometrical properties of the beam are characterized by the shear modulus G, the cross-section area A, the Young modulus E, and the moment of inertia I. The transverse load, q, is part of data of the problem. To complete the model and ensure the existence and uniqueness of its solution, we must impose suitable boundary conditions; we take, for example,

$$w(0) = w_0 \qquad M(0) = M_0 \qquad w(L) = w_L \qquad M(L) = M_L. \qquad (2)$$

Our long-term goal is to investigate the possible advantages of DG methods in computational structural mechanics. In this paper, we begin our efforts by studying how to properly devise DG methods for the Timoshenko beam. In a forthcoming paper, we give a complete error analysis of these methods and show that they can easily overcome the so-called shear locking.

In [4] Arnold analyzed the continuous version of the Galerkin method. He proved error estimates which degenerate as the aspect ration of the beam tends to zero, and hence the method suffers from shear locking. In the same paper he proved that locking is overcome if we use the so-called reduced integration technique. These findings are verified by numerical experiments. For the relationship between mixed finite element methods and reduced integration techniques we refer to [5]. In [6], Li analyzed the p and hp versions of the same method and proved error estimates independent of the aspect ratio of the beam. This is consistent with the well known fact that locking can be overcome by increasing the polynomial degree of the approximations. Our preliminary results indicate that the DG methods overcome locking even if we approximate all the unknowns with piecewise constants and do not use reduced integration. This and other features render DG appealing for other problems in structure mechanics, such as plates, shells and elasticity. Of course, there are additinal issues involved in these problems; Reissner-Mindlin plates have boundary layers, elasticity problems have volumetric locking and shells exhibit membrane locking. For a recent DG method for the Reissner-Mindlin plate see [7]. Arnold and Falk provided a theoretical analysis of the boundary layers in [8], and a uniformly accurate finite element method in [9]. For a locking-free finite element method for shells see [10].

The paper is organized as follows. In section 2, we introduce the weak formulation we are going to use to define the DG methods. Then, in section 3, we introduce their general form. In section 4, we introduce what we call a discrete energy identity which we use, in section 5, to establish conditions that ensure existence and uniqueness of the approximate solution. In section 6, we present numerical results showing that the method can achieve optimal h-convergence as well as exponential convergence. We end in section 7 with some concluding remarks.

2 The weak formulation for the continuous case

To display the weak formulation we use to define the DG methods, we need to introduce some notations. Let $\mathcal{T} = \{I_j = (x_{j-1}, x_j), j = 1, \ldots, N\}$ be a triangulation of the computational domain $\Omega = (0, L)$; we assume that the nodes x_i are such that $0 = x_0 < x_1 < \cdots < x_{N-1} < x_N = L$. Then, we write

$$(\varphi, \psi) := \sum_{j=1}^{N} \int_{I_j} \varphi\psi \quad \text{and} \quad \langle R, [\![\varphi n]\!]\rangle := \sum_{j=0}^{N} R(x_j)\, [\![\varphi n]\!](x_j).$$

Here, R is a function defined on the set of nodes $\mathscr{E}_h := \{x_0, x_1, \ldots, x_N\}$. The *jump* of the function φ, $[\![\varphi n]\!]$, is defined as follows. If the node e is in

$\mathscr{E}_h^o := \{x_1, x_2, \ldots, x_{N-1}\}$, then we take $[\![\varphi n]\!](e) = \varphi(e^+)n_e^+ + \varphi(e^-)n_e^-$, where $\varphi(e^\pm) := \lim_{\epsilon\downarrow 0} \varphi(e - \epsilon\, n_e^\pm)$ and $n_e^\pm = \mp 1$. For the boundary nodes, we take $[\![\varphi n]\!](0) = -\varphi(0^+)$, $[\![\varphi n]\!](L) = \varphi(L^-)$. These jumps are well defined for φ in $H^1(\Omega_h)$, where $\Omega_h = \cup_{j=1,\ldots,N} I_j$.

It is now easy to see that if we assume that $(T, M, \theta, w) \in [H^1(\Omega)]^4$, we have

$$-(w, \frac{d}{dx}v^1) + \langle w, [\![v^1 n]\!]\rangle = (\theta, v^1) - (\frac{1}{GA}T, v^1), \tag{3a}$$

$$-(\theta, \frac{d}{dx}v^2) + \langle \theta, [\![v^2 n]\!]\rangle = (\frac{1}{EI}M, v^2), \tag{3b}$$

$$-(M, \frac{d}{dx}v^3) + \langle M, [\![v^3 n]\!]\rangle = (T, v^3), \tag{3c}$$

$$-(T, \frac{d}{dx}v^4) + \langle T, [\![v^4 n]\!]\rangle = (q, v^4). \tag{3d}$$

for all $v^1, v^2, v^3, v^4 \in H^1(\Omega_h)$. This is the weak formulation we use to define the DG methods.

3 The DG Methods

The approximate solution $(T_h, M_h, \theta_h, w_h)$ given by the DG method will be sought in the finite dimensional space $V_h^{k_1} \times V_h^{k_2} \times V_h^{k_3} \times V_h^{k_4}$; here,

$$V_h^k := \{v : \Omega_h \mapsto \mathbb{R} : \quad v|_{I_j} \in P^k(I_j), j = 1, \ldots, N\},$$

where $P^k(K)$ is the set of all polynomials on K of degree not exceeding k. The approximate solution is determined by requiring that

$$-(w_h, \frac{d}{dx}v^1) + \langle \widehat{w}_h, [\![v^1 n]\!]\rangle = (\theta_h, v^1) - (\frac{1}{GA}T_h, v^1), \tag{4a}$$

$$-(\theta_h, \frac{d}{dx}v^2) + \langle \widehat{\theta}_h, [\![v^2 n]\!]\rangle = (\frac{1}{EI}M_h, v^2), \tag{4b}$$

$$-(M_h, \frac{d}{dx}v^3) + \langle \widehat{M}_h, [\![v^3 n]\!]\rangle = (T_h, v^3), \tag{4c}$$

$$-(T_h, \frac{d}{dx}v^4) + \langle \widehat{T}_h, [\![v^4 n]\!]\rangle = (q, v^4), \tag{4d}$$

hold for all $v^i \in V_h^{k_i}$ for $i = 1, 2, 3, 4$. To complete the definition of the method, we have to define the numerical traces $(\widehat{T}_h, \widehat{M}_h, \widehat{\theta}_h, \widehat{w}_h)$ at the nodes. It is through them that the interaction between the degrees of freedom of different intervals is introduced and the boundary conditions are actually imposed. Moreover, their choice is crucial as it affects both the stability and the accuracy of the method; see [3] for a detailed discussion of this issue for some other problems.

Extending to our framework what have been already done for fluid flow problems, we assume that the form of these traces is as follows. For an interior node e $\in \mathscr{E}_h^o$, we take

$$\widehat{w}_h = \{\!\{ w_h \}\!\} + C_{11}[\![w_h n]\!] + C_{12}[\![\theta_h n]\!] + C_{13}[\![M_h n]\!] + C_{14}[\![T_h n]\!], \tag{5a}$$

$$\widehat{\theta}_h = \{\!\{ \theta_h \}\!\} + C_{21}[\![w_h n]\!] + C_{22}[\![\theta_h n]\!] + C_{23}[\![M_h n]\!] + C_{24}[\![T_h n]\!], \tag{5b}$$

$$\widehat{M}_h = \{\!\{ M_h \}\!\} + C_{31}[\![w_h n]\!] + C_{32}[\![\theta_h n]\!] + C_{33}[\![M_h n]\!] + C_{34}[\![T_h n]\!], \tag{5c}$$

$$\widehat{T}_h = \{\!\{ T_h \}\!\} + C_{41}[\![w_h n]\!] + C_{42}[\![\theta_h n]\!] + C_{43}[\![M_h n]\!] + C_{44}[\![T_h n]\!], \tag{5d}$$

where $\{\!\{\varphi\}\!\}(e) = \frac{1}{2}(\varphi(e^+) + \varphi(e^-))$. At $x = 0$, we take

$$\widehat{w}_h(0) = w_0, \tag{6a}$$

$$\widehat{\theta}_h(0) = \theta_h(0^+) + C_{21}(0)(w_0 - w_h(0^+)) + C_{23}(0)(M_0 - M_h(0^+)), \tag{6b}$$

$$\widehat{M}_h(0) = M_0, \tag{6c}$$

$$\widehat{T}_h(0) = T_h(0^+) + C_{41}(0)(w_0 - w_h(0^+)) + C_{43}(0)(M_0 - M_h(0^+)). \tag{6d}$$

and at $x = L$,

$$\widehat{w}_h(L) = w_L, \tag{7a}$$

$$\widehat{\theta}_h(L) = \theta_h(L^-) + C_{21}(L)(w_h(L^-) - w_L) + C_{23}(L)(M_h(L^-) - M_L), \tag{7b}$$

$$\widehat{M}_h(L) = M_L, \tag{7c}$$

$$\widehat{T}_h(L) = T_h(L^-) + C_{41}(L)(w_h(L^-) - w_L) + C_{43}(L)(M_h(L^-) - M_L). \tag{7d}$$

Note how the boundary conditions are incorporated into the DG method through the definition of the numerical traces at the border. Note also that the parameters C_{ij} defining the numerical traces can have different values at different nodes. In the next section, we investigate the role of these parameters. In particular, we show that out of these sixteen parameters, six can be expressed in terms of the remaining ten and that only four of them have an impact on the "energy" of the discretization.

4 The Discrete Energy Identity

To see this, we consider a classical energy argument. It is not difficult to see that if we take $v^1 = T$, $v^2 = -M$, $v^3 = -\theta$, and $v^4 = w$ in the equations (3), and add them, we obtain the energy identity

$$(\frac{1}{EI}M, M) + (\frac{1}{GA}T, T) = (q, w) + bc(T, M, \theta, w),$$

where

$$bc(T, M, \theta, w) = M_L\,\theta(L^-) - M_0\,\theta(0^+) - w_L\,T(L^-) + w_0\,T(0^+).$$

Since this identity captures an essential feature of the problem under consideration, we would like to obtain a similar energy identity for the DG method. Such an identity is obtained in the following result.

Proposition 1 (Discrete Energy Identity). *Assume that* $(T_h, M_h, \theta_h, w_h)$ *is a solution of the DG method given by* (4), *with numerical traces given by* (5), (6), *and* (7). *Moreover, assume that for* all *nodes* $e \in \mathscr{E}_h$ *we have*

$$C_{21} = C_{43},\ -C_{22} = C_{33},\ C_{24} = C_{13},\ C_{31} = C_{42},\ C_{34} = C_{12},\ -C_{11} = C_{44}. \tag{8}$$

Then, we have

$$(\frac{1}{EI}M_h, M_h) + (\frac{1}{GA}T_h, T_h) + \Theta_{jumps} = (q, w_h) + bc(T_h, M_h, \theta_h, w_h) + \Theta_{bc}, \tag{9}$$

Here, setting $C_{14} = C_{32} = 0$ *at the boundary nodes, we have*

$$\begin{aligned}\Theta_{jumps} &= \sum_{e\in\mathscr{E}_h} \left(C_{14}[\![T_h n]\!]^2 - C_{23}[\![M_h n]\!]^2 - C_{32}[\![\theta_h n]\!]^2 + C_{41}[\![w_h n]\!]^2\right)(e),\\ \Theta_{bc} &= w_0[C_{41}w(0^+) - C_{21}M(0^+)] + M_0[C_{43}w(0^+) - C_{23}M(0^+)]\\ &\quad + w_L[C_{41}w(L^-) - C_{21}M(L^-)] + M_L[C_{43}w(L^-) - C_{23}M(L^-)].\end{aligned}$$

Proof. The proof of the above result follows by mimicking what was done for the continuous case, that is, by taking $v^1 = T_h$, $v^2 = -M_h$, $v^3 = -\theta_h$, and $v^4 = w_h$ in the definition of the DG method (4), adding the resulting equations, and carrying out some simple algebraic manipulations. □

It is now clear that if we take

$$C_{14},\ -C_{23},\ -C_{32},\ C_{41} \geq 0, \tag{10}$$

then each of the terms of Θ_{jumps} can be considered to be an *energy* associated with the discontinuous nature of the discretization. Thus, the above condition ensures that the appearance of the jumps in the DG approximation is accompanied by an increase of the total energy of the system. Since this can also be thought of as being a stabilizing effect, the above parameters are called the *stabilization* parameters. None of the remaining parameters appear in the expression for the energy of the approximation, as we can see in the above result.

5 Existence and uniqueness of the DG approximation

Our main theoretical result is the following.

Theorem 1 (Existence and uniqueness of the DG approximation). *Consider the DG method defined by the weak formulation* (4) *and the numerical traces* (5), (6) *and* (7). *Assume that the constants C_{ij} satisfy* (8) *and* (10). *Then the method has a unique solution in the following cases:*

Case 1: *$C_{41} > 0$ on $\mathscr{E}_h$, $-C_{32} > 0$ on $\mathscr{E}_h^o$, $k_2 \geq k_3 - 1$, and $k_1 \geq k_4 - 1$.*
Case 2: *$C_{ij} = 0$ on $\mathscr{E}_h^o$, except $C_{11} = C_{22} = -C_{33} = -C_{44} = 1/2$, $C_{41}(L) > 0$, $k_2 \geq k_3$ and $k_1 \geq k_4$.*
Case 3: *$k_2 \geq k_3 + 1$ and $k_1 \geq k_4 + 1$.*

From the first case, we see that the stabilization parameters associated to θ_h and w_h, namely, C_{32} and C_{41}, respectively, have a stronger influence on the existence and uniqueness of the solution of the method than the ones associated to T_h and M_h, namely, C_{14} and C_{23}, respectively.

From the second and third cases, we see that when the stabilization effect of a jump is turned off (by setting the corresponding stabilization parameter equal to zero), the existence and uniqueness of the approximate solution can still be guaranteed by a suitable definition of the other parameters and/or by modifying the polynomial degree of the approximate solutions. Roughly speaking, the more stabilization parameters are equal to zero, the more the spaces for θ_h and w_h have to be in relation to the spaces of M_h and T_h, respectively.

Proof. Due to the linearity of the problem, it is enough to show that the only solution to (4) with $q = 0$, $w_0 = w_L = M_0 = M_L = 0$ is $w_h = \theta_h = M_h = T_h = 0$. In this case, (9) takes the form

$$(\frac{1}{EI} M_h, M_h) + (\frac{1}{GA} T_h, T_h) + \Theta_{jumps} = 0. \tag{11}$$

By the assumptions on the parameters C_{ij}, this implies $T_h = 0$, $M_h = 0$. On the other hand, taking $v^1 = 1$ in (4a), we get that $(\theta_h, 1) = 0$.

Case 1. In the first case, from the discrete energy identity (11) we see that $[\![\theta_h\, n]\!] = 0$ on $\mathscr{E}_h^o$, and $[\![w_h\, n]\!] = 0$ on $\mathscr{E}_h$. As a consequence, if $k_3 = 0$, θ_h is a constant and since $(\theta_h, 1) = 0$, it is equal to zero. If $k_3 > 0$, a similar conclusion can be reached as follows: We have, by (5b), that $\widehat{\theta}_h = \theta_h$ and, by (4b), that $(\frac{d}{dx}\theta_h, v^2) = 0$ for all $v^2 \in V_h^{k_2}$. Since $\theta_h \in V_h^{k_3}$ and $k_2 \geq k_3 - 1$, this implies that θ_h is a constant and since $(\theta_h, 1) = 0$, that it is equal to zero. A similar argument shows that w_h is also zero if $k_1 \geq k_4 - 1$.

Case 2. In the second case, by the definition of the numerical trace of θ_h, we have that $\widehat{\theta}_h(0) = \theta_h(0^+)$ and that $\widehat{\theta}_h(e) = \theta_h(e^-)$ for all the remaining nodes e. Taking v^2 with support (x_0, x_1) we get, by equation (4b), that $-\int_{x_0}^{x_1} \theta_h \frac{d}{dx} v^2 + (\theta_h\, v^2)(x_1^-) - (\theta_h\, v^2)(x_0^+) = \int_{x_0}^{x_1} v^2 \frac{d}{dx}\theta_h = 0$. Since $\theta_h \in V_h^{k_3}$ and $k_2 \geq k_3$ this implies that $\theta_h = 0$ on that interval. Now, if we take v^2 with support on (x_1, x_2), equation (4b) becomes $\int_{x_1}^{x_2} \theta_h \frac{d}{dx} v^2 = \theta_h(x_2^-)\, v^2(x_2^-)$. Taking $v^2 = 1$, we get that $\theta_h(x_2^-) = 0$. Also, since $k_2 \geq k_3$, we can take

$v^2 = \theta_h$ and obtain that $\theta_h(x_1^+) = 0$. Hence, $\int_{x_0}^{x_1} v^2 \frac{d}{dx}\theta_h = 0$, and so $\theta_h = 0$ in (x_1, x_2). By repeating this argument, we obtain that $\theta_h = 0$. Similarly, it is easy to show that $w_h = 0$ outside the last interval (x_{N-1}, x_N). By the equation (4a), we have $\int_{x_{N-1}}^{x_N} v^1 \frac{d}{dx} w_h = 0$, and hence w_h is a constant on the last interval since $w_h \in V_h^{k_4}$ and $k_1 \geq k_4$. Finally, since $C_{41}(L) > 0$, by the discrete energy identity (11), we have that $[\![w_h\, n]\!](L) = 0$ and so $w_h(L^-) = 0$. This implies that $w_h = 0$.

Case 3. Finally, consider the third case. Since $k_2 \geq 1$, taking $v^2 = x/L$ in (4b), we get that $\widehat{\theta}_h(L) = (\theta_h, 1) = 0$. Then, taking $v^2 = 1$ on (x_j, x_{j+1}) and $v^2 = 0$ for the rest of the domain, the equation (4b) yields $\widehat{\theta}_h(x_j) = \widehat{\theta}_h(x_{j+1})$. This implies that $\widehat{\theta}_h = 0$ on all the nodes. By (4b), this implies that $(\theta_h, \frac{d}{dx}v^2) = 0$ for all $v^2 \in V_h^{k_2}$, and since $\theta_h \in V_h^{k_3}$ and $k_2 \geq k_3 + 1$, this implies that $\theta_h = 0$. A similar argument shows that $w_h = 0$ if $k_1 \geq k_4 + 1$.

This completes the proof. □

6 Numerical Results

In this section, we display preliminary numerical experiments showing that DG methods for the Timoshenko beam can be constructed so as to achieve optimal rates of convergence as well as exponential convergence. We consider two test problems. In both problems, same-degree polynomial approximations are used for all the unknowns, that is, we take $k_1 = k_2 = k_3 = k_4 = k$.

Test problem 1. We solve (1)-(2) in $\Omega = (0, 1)$ with the boundary conditions $w_0 = w_L = 0$ and $M_0 = M_L = 0$. We take $q(x) = \sin(\pi x)$, for all $x \in \Omega$. The transverse cross-section area of the beam at the point x, $A(x)$, is assumed to be rectangular with width and height $a(x)$ and $b(x)$; the moment of inertia is given by the formula $I(x) = a(x)b^3(x)/12$. We take $a(x) = b(x) = 0.01$ The Young modulus is $E = 10^7$ whereas the shear modulus is $G = (0.35)E$.

We pick the numerical traces by setting $C_{ij}(x) = 0$ for $i \neq j$ except $-C_{23}(0) = C_{41}(L) = 1/h$. And $C_{11}(x) = C_{22}(x) = -C_{33}(x) = -C_{44}(x) = 0.5$. The existence and uniqueness of the approximate solution is guaranteed by the second case of Theorem 1.

Test problem 2. The purpose of this test problem is to show that the method can easily handle other boundary conditions and discontinuities in the material properties and the load. Thus, we take the boundary conditions $w_0 = 0$, $\theta_0 = 0$, $M_L = 0$ and $T_L = 0$. We also take $q(x) = \sin(\pi x)$ if $0 \leq x \leq 1/2$ and $q(x) = -16(x - 3/4)^2 + 1$ if $1/2 < x \leq 1$. $a(x) = b(x) = 0.02$ if $0 \leq x \leq 1/2$ and 0.05 if $1/2 < x \leq 1$.

We use the following numerical traces. To capture the boundary conditions, we take $\widehat{w}_h(0) = 0, \widehat{\theta}_h(0) = 0, \widehat{M}_h(L) = 0$, and $\widehat{T}_h(L) = 0$. To define the remaining numerical traces, we simply take $C_{ij} = 0$ except for $C_{11} = C_{22} = -C_{33} = -C_{44} = 1/2$. The existence and uniqueness of the approximate solution is *not* guaranteed by Theorem 1, but can be easily proven.

in Fig. 1, we see that in both test problems, exponential convergence is achieved. In Figs. 2, and 3, the solid line represents the exact solution and "+" represents the numerical trace of the approximate solution at the nodes of the mesh. We display the results obtained on a uniform mesh of 10 elements. In the Tables 1 and 2, we display convergence orders up to polynomial degree $k = 5$. Mesh number i means a uniform mesh of 2^i elements. We can see that optimal orders of convergence are achieved for all the unknowns even in the case of piecewise constant approximation.

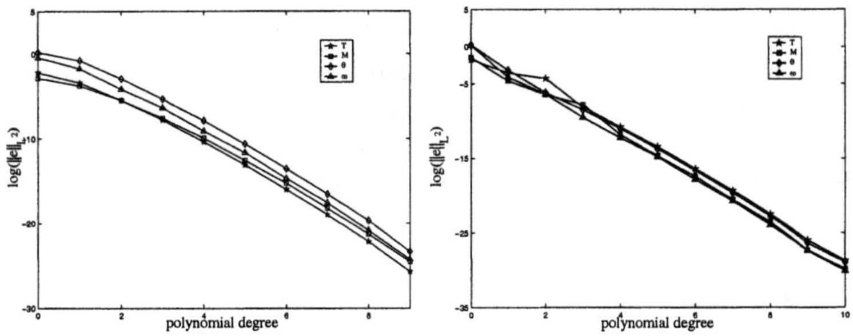

Fig. 1. Exponential convergence: Test problem 1 (left) and 2 (right)

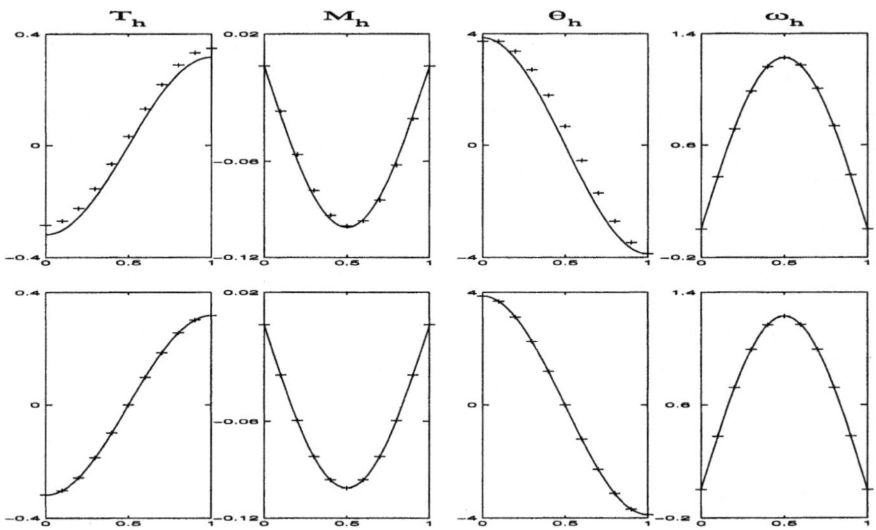

Fig. 2. Piecewise constant (top) and linear (bottom) approximations for test problem 1

Table 1. Convergence rates for same-degree approximation for test problem 1.

degree k	mesh number	$\|\|e_T\|\|_{L^2(\Omega)}$ error	order	$\|\|e_M\|\|_{L^2(\Omega)}$ error	order	$\|\|e_\theta\|\|_{L^2(\Omega)}$ error	order	$\|\|e_w\|\|_{L^2(\Omega)}$ error	order
0	4	1.6e-02	0.99	9.3e-03	0.98	1.6e-01	1.00	9.5e-02	0.99
	5	8.0e-03	1.00	4.7e-03	1.00	7.9e-02	1.00	4.8e-02	1.00
	6	4.0e-03	1.00	2.3e-03	1.00	3.9e-02	1.00	2.4e-02	1.00
1	4	5.3e-04	2.00	2.5e-04	2.70	6.4e-03	1.99	2.0e-03	2.01
	5	1.3e-04	2.00	4.6e-05	2.47	1.6e-03	1.99	5.1e-04	2.00
	6	3.3e-05	2.00	1.1e-05	2.11	4.0e-04	2.00	1.3e-04	2.00
2	4	8.3e-06	3.00	2.7e-06	3.24	1.0e-04	2.99	3.2e-05	3.00
	5	1.0e-06	3.00	3.3e-07	3.04	1.3e-05	3.00	4.0e-06	3.00
	6	1.3e-07	3.00	4.1e-08	3.00	1.6e-06	3.00	5.0e-07	3.00
3	4	1.0e-07	4.00	4.9e-08	4.92	1.2e-06	3.97	3.9e-07	4.00
	5	6.3e-09	4.00	2.2e-09	4.49	7.6e-08	3.99	2.4e-08	4.00
	6	3.9e-10	4.00	1.3e-10	4.11	4.8e-09	4.00	1.5e-09	4.00
4	4	9.8e-10	5.00	3.2e-10	5.31	1.2e-08	5.00	3.8e-09	5.00
	5	3.1e-11	5.00	9.7e-12	5.04	3.7e-10	5.00	1.2e-10	5.00
	6	9.5e-13	5.00	3.0e-13	5.00	1.2e-11	5.00	3.7e-12	5.00
5	3	5.1e-10	5.99	4.9e-10	6.83	6.1e-09	5.93	2.0e-09	6.01
	4	7.9e-12	6.00	3.9e-12	6.99	9.6e-11	5.98	3.1e-11	6.00
	5	1.2e-13	5.99	4.3e-14	6.50	1.5e-12	6.00	4.8e-13	6.00

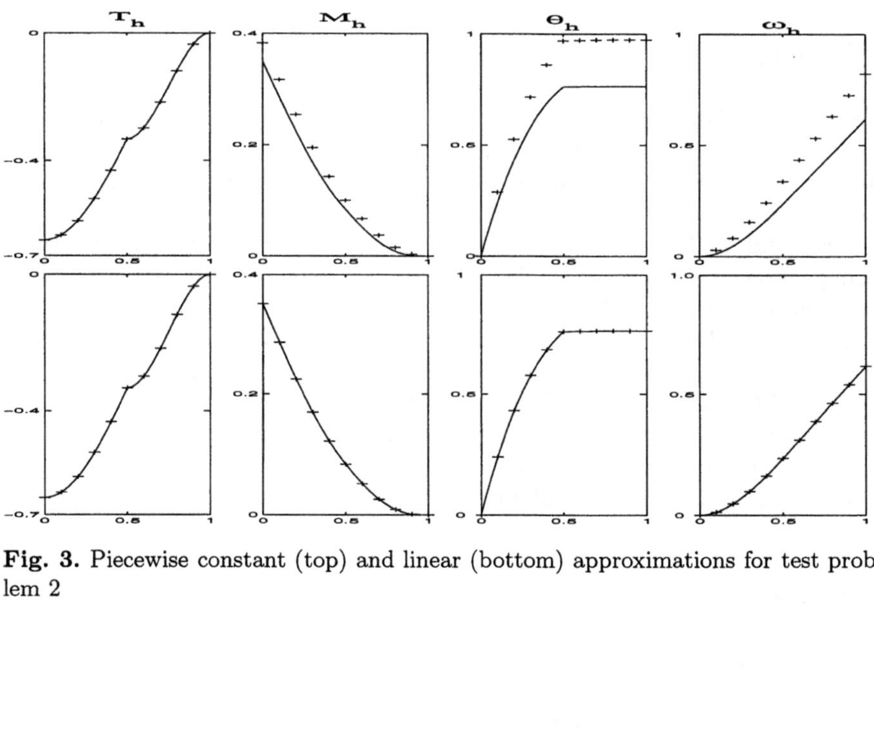

Fig. 3. Piecewise constant (top) and linear (bottom) approximations for test problem 2

Table 2. Convergence rates for same-degree approximation for test problem 2.

degree k	mesh number	$\|\|e_T\|\|_{L^2(\Omega)}$ error	order	$\|\|e_M\|\|_{L^2(\Omega)}$ error	order	$\|\|e_\theta\|\|_{L^2(\Omega)}$ error	order	$\|\|e_w\|\|_{L^2(\Omega)}$ error	order
0	2	9.7e-2	0.89	1.1e-1	1.04	5.7e-1	1.14	4.7e-1	1.32
	3	5.1e-2	0.94	5.6e-2	1.04	2.7e-1	1.09	2.1e-1	1.19
	4	2.6e-2	0.98	2.7e-2	1.03	1.3e-1	1.05	9.6e-2	1.10
1	2	1.3e-2	1.25	2.7e-3	1.94	1.1e-2	1.99	4.2e-3	2.04
	3	3.4e-3	1.89	6.7e-4	1.99	2.8e-3	2.00	1.1e-3	1.98
	4	8.6e-4	1.98	1.7e-4	1.99	6.9e-4	2.00	2.7e-4	1.98
2	2	1.8e-3	3.00	2.6e-4	2.61	2.8e-4	2.92	2.1e-4	2.95
	3	2.2e-4	3.00	3.4e-5	2.93	3.5e-5	2.97	2.7e-5	2.98
	4	2.8e-5	3.00	4.3e-6	2.99	4.5e-6	2.99	3.4e-6	2.99
3	2	1.8e-5	4.01	2.7e-5	3.99	1.4e-5	3.96	4.5e-6	4.06
	3	1.1e-6	4.01	1.7e-6	4.00	8.6e-7	3.99	2.8e-7	4.02
	4	7.1e-8	4.00	1.1e-7	4.00	5.4e-8	4.00	1.7e-8	4.01
4	2	6.9e-7	4.94	2.3e-7	5.05	5.3e-7	4.96	1.7e-7	4.89
	3	2.2e-8	4.98	7.1e-9	5.02	1.7e-8	4.99	5.3e-9	4.97
	4	6.9e-10	4.99	2.2e-10	5.01	5.3e-10	5.00	1.7e-10	4.99
5	2	2.3e-8	6.01	7.1e-9	5.91	1.7e-8	5.97	5.7e-9	6.04
	3	3.6e-10	6.00	1.1e-10	5.98	2.7e-10	5.99	8.9e-11	6.01
	4	5.6e-12	6.00	1.8e-12	5.99	4.3e-12	6.00	1.4e-12	6.01

7 Conclusion

We devised a family of DG methods for the Timoshenko beam problem and provided sufficient conditions for the existence and uniqueness of its solution. We then displayed numerical results showing that the method converges to the exact solution with optimal order even in the case of discontinuities of the material and the load were presented. We also provided numerical results indicating exponential convergence. In a forthcoming paper, we provide a complete error analysis of the methods and show that they are free from shear locking.

Acknowledgement. The authors are pleased to acknowledge support of this research by the Army High Performance Computing Research Center(AHPCRC) under the auspices of the Department of Army, Army Research Laboratory(ARL) under cooperative agreement DAAD19-01-2-0014. The content of which does not necessarily reflect the position or the policy of the government, and no official endorsement should be inferred.

References

1. S.P.Timoshenko (1921) *On the correction for shear of the differential equation for transverse vibrations of prismatic bars,*Philosophical Magazine,**41**, 744-746.
2. S.P.Timoshenko (1922) *On the transverse vibrations of bars of uniform cross section,* Philosophical Magazine,**43**,125-131.
3. B. Cockburn(2003) *Discontinuous Galerkin Methods,* ZAMM Z. Angew. Math. Mech.,**83**,731–754.
4. D.N.Arnold(1981) *Discretization by Finite Elements of a model Parameter Dependent Problem* , Numer.Math.37, 405-421, (1981).

5. Malkus, D.S., Hughes, T.J.R (1978) *Mixed finite element methods-reduced integration and selective integration techniques: a unification of concepts.* Comput. Methods Appl. Mech. Engrg. **15**, 63-81
6. Likang Li(1990) *Discretization of the Timoshenko Beam Problem by the p and the $h-p$ Versions of the Finite Element Method,* Numer.Math.57, 413-420.
7. D. N. Arnold, F.Brezzi and D. Marini *A family of discontinuous Galerkin finite elements for the Reissner-Mindlin plate*, submitted to Journal of Scientific Computing.
8. D.N.Arnold, R.Falk (1990) *The Boundary Layer for the Reissner-Mindlin Plate Model,* SIAM J. Math. Anal., Vol 21, No.2, pp. 281-312.
9. D.N.Arnold, R.Falk (1989) *A Uniformly Accurate Finite Element Method for the Reissner-Mindlin Plate,* SIAM J. Numer. Anal., Vol 26, No.6, pp. 1276-1290.
10. D.N.Arnold, F.Brezzi (1997) *Locking-free Finite Element Methods for Shells* Math.Comp., Vol.66, Number 217, pp. 1-14.

Numerical Algorithms for Solving Elliptic–Parabolic Problems

Raimondas Čiegis

Vilnius Gediminas Technical University, Saulėtekio Str. 11, LT-2040 Vilnius, Lithuania rc@fm.vtu.lt

Summary. This paper deals with numerical algorithms for solving elliptic–parabolic problems. An example of such problem is given by the Richards equation for modeling the saturated-unsaturated water flow in porous media. We consider a linear model problem and investigate the convergence of two finite–volume schemes. The first one uses the implicit approximation in the whole domain, and the second scheme is constructed using the splitting method. Results of numerical experiments are also given.

1 Introduction

In $(x,t) \in \Omega_l \times (0,T]$ we solve a problem, describing the two–phase flow in a porous layered media (for a more detailed description we refer to the book of Helmig [Hel97]):

$$\begin{cases} \varepsilon_l \rho_l \dfrac{\partial s_l}{\partial t} = \nabla \cdot \big(\lambda_l(s_l) K_l \nabla (p_l(s_l) - x_1)\big), \\ p_l(x,t) = q_0, \quad (x,t) \in \partial\Omega_D \times [0,T], \\ \nabla(p_l - x_1) = 0, \quad (x,t) \in \partial\Omega \backslash \partial\Omega_D \times [0,T], \\ s_l(x,0) = s_{init}(x), \end{cases}$$

where s_l is the water saturation in the l-th layer, p_l is the pressure, $\lambda_l(s_l)K_l$, ε_l, ρ_l are the permeability, porosity of the porous media, and density of the fluid, respectively.

The equation becomes elliptic in the region of saturation, where $p_l > 0$ and $s_l = s_l^s$:

$$\nabla \cdot (\lambda_l(s_l^s) K_l \nabla (p_l - x_1)) = 0,$$

here the pressure p_l is the primary unknown, and s_l^s denotes the water content of a water–saturated medium.

The actual determination of the discrete solution may require large computational resources due to the strong nonlinearity of the problem, the dominance of the convective process, discontinuity of the porous medium properties. The formulated problem looks like a system of parabolic partial differential equations, but its type can become either nonlinear hyperbolic or degenerate parabolic, depending on the influence of capillary pressure (see, e.g. Helmig [Hel97], Eymard *et al.* [EGH00]).

An additional difficulty arises due to degeneracy of the parabolic problem in the unsaturated region to the elliptic problem in the saturation region of the porous medium. Fully implicit discrete finite volume and finite difference schemes are usually used for the approximation of the flow equations in the whole region (see the papers of Alt and Luckhaus [ALu83], Chen and Ewing [ChE97], Eymard *et al.* [EGH00], Jäger and Kačur [JaK91]).

It is well–known that most effective numerical algorithms for solving multi–dimensional parabolic problems are based on splitting methods. New time splitting schemes for the time integration of the problems describing the two–phase flow in the porous medium are proposed by Čiegis *et al.* [CPZ00]. The parallel version of this algorithm is considered in [CCZ99].

In Section 2 we formulate the model linear 3D elliptic–parabolic initial–boundary value problem and define its approximation in space by the finite–volume method. In Section 3 we formulate and investigate the fully implicit scheme for integration in time. The additive integration scheme, which is based on the Douglas algorithm, is presented and investigated in Section 4. Finally, the results of numerical experiments are presented in Section 5.

2 Problem Formulation

We consider a linear elliptic–parabolic initial–boundary value problem, which is defined in the region $Q = [0,1]^3 \times [0,T]$:

$$\begin{cases} \dfrac{\partial u}{\partial t} = \sum\limits_{j=1}^{3} \dfrac{\partial}{\partial x_j}\Big(k_j \dfrac{\partial u}{\partial x_j}\Big) - q(t)u + f(\mathbf{X},t), \quad (\mathbf{X},t) \in Q_{par}, \\ -\sum\limits_{j=1}^{3} \dfrac{\partial}{\partial x_j}\Big(k_j \dfrac{\partial u}{\partial x_j}\Big) + q(t)u = f(x,t), \quad (\mathbf{X},t) \in Q_{ell}, \\ u(\mathbf{X},t) = \mu_0, \quad (\mathbf{X},t) \in \partial\Omega \times [0,T], \\ u(\mathbf{X},0) = u_0(\mathbf{X}), \quad \mathbf{X} \in [0,1]^3. \end{cases} \tag{1}$$

We assume that:

$$Q = Q_{par} \bigcup Q_{ell}, \quad Q_{ell} = [a,b]^3 \times [0,T].$$

Let $Q_{h\tau} = Q_h \times Q_\tau$ be the discrete uniform mesh:

$$Q_h = \{\, X_{ijk} = (x_{1i}, x_{2j}, x_{3k}) : \; x_{lm} = mh, \;\; 0 \le m \le M \,\},$$
$$Q_\tau = \{\, t^n : \; t^n = n\tau, \;\; 0 \le n \le N \,\}.$$

We use the notation $U^n = U(X_{ijk}, t^n)$. Using the finite-difference method we approximate a part of the differential operator (1) by the following discrete operators

$$\tilde{A}_j U^n = \left(k_j U^n_{x_j}\right)_{\bar{x}_j} - q_j(t^n)U^n, \quad A_j U^n = \tilde{A}_j U^n + f_j(X, t^n), \quad j = 1, 2, 3,$$

here the discrete difference operators are defined by

$$U^n_{x_j} = \frac{U^n(x_j + h) - U^n}{h}, \quad U^n_{\bar{x}_j} = \frac{U^n - U^n(x_j - h)}{h}.$$

The selection of a time–stepping procedure is a non trivial task, since the stability and robustness of the algorithm on the one hand must be balances with the computational efficiency on the other hand. In the following section we investigate two different numerical integration schemes.

3 Fully Implicit Difference Scheme

We approximate the differential problem by the following modification of the backward Euler scheme:

$$c(X)\,\frac{U^{n+1} - U^n}{\tau} = \sum_{j=1}^{3} A_j U^{n+1}, \tag{2}$$

$$c(X) = \begin{cases} 1 & \text{if } X \in Q_{par}, \\ 0 & \text{if } X \in Q_{ell}. \end{cases}$$

At each time level t^n we get a system of linear equations:

$$\Big(\frac{c}{\tau}\mathbf{I} - \sum_{j=1}^{3} \tilde{A}_j\Big)U^{n+1} = \frac{c}{\tau}U^n + \sum_{j=1}^{3} f_j(X, t^{n+1}),$$

or

$$\mathbf{A}U^{n+1} = F^{n+1}. \tag{3}$$

We note that even if the problem is parabolic in the whole region of its definition (i.e., the flow in the porous media is unsaturated) the $3D$ parabolic problem is approximated by the backward Euler scheme.

The system of linear equations (3) is solved by some iterative method, e.g. the *Conjugate Gradient* method. In the case of the nonlinear two–phase flow problem the matrix of the obtained system is non–symmetric, thus some special methods such as $GMRES$ should be used to solve a system of linearized equations.

The convergence rate of iterative methods depends essentially on the stiffness of the matrix $\mathbf{A}$ and on the distribution of its eigenvalues. It is easy to prove that the following spectral estimates are valid:

$$\left(\frac{meas(Q_{par})}{\tau}+\lambda_{A,\min}\right)\mathbf{I}\le\mathbf{A}\le\left(\frac{meas(Q_{par})}{\tau}+\lambda_{A,\max}\right)\mathbf{I},$$

here $\lambda_{A,\min}, \lambda_{A,\max}$ are the smallest and largest eigenvalues of the matrix $\mathbf{A}$, respectively, and $meas(Q)$ denotes the volume of Q.

Now we can investigate the stiffness number of the matrix $\mathbf{A}$:

$$\kappa(\mathbf{A})=\frac{meas(Q_{par})/\tau+\lambda_{A,\max}}{meas(Q_{par})/\tau+\lambda_{A,\min}}.$$

Taking into account that $\lambda_{A,\min}=O(1)$ and $\lambda_{A,\max}=O(h^{-2})$, and assuming that $meas(Q_{par})>0$, we obtain the following asymptotic estimates:

$$\kappa(\mathbf{A})=\begin{cases}\dfrac{c}{meas(Q_{par})\,h^{3/2}} & \text{if } \tau=O(\sqrt{h}),\\ \dfrac{c}{meas(Q_{par})\,h} & \text{if } \tau=O(h),\\ \dfrac{c}{meas(Q_{par})} & \text{if } \tau=O(h^2).\end{cases}\tag{4}$$

Thus for sufficiently large time steps τ the number of iterations of the CG method will be approximately the same as for solving the pure elliptic problem.

4 Finite Difference Scheme 2

We approximate the differential problem by the following modification of the *stability–correction* scheme

$$\begin{cases}\dfrac{U_s^{n+1/3}-U_s^n}{\tau}=A_1U_s^{n+1/3}+A_2U_s^n+A_3U_s^n,\\ \dfrac{U_s^{n+2/3}-U_s^{n+1/3}}{\tau}=A_2U_s^{n+2/3}-A_2U_s^n,\\ \dfrac{U_s^{n+1}-U_s^{n+2/3}}{\tau}=A_3U_s^{n+1}-A_3U_s^n.\end{cases}\tag{5}$$

Here s denotes the *elliptic* iteration number, and the initial condition at time level t^n is recalculated after each iteration as

$$U_{s+1}^n=\begin{cases}U^n & \text{if } x\in\Omega_{h,par},\\ U_s^{n+1} & \text{if } x\in\Omega_{h,ell},\end{cases}$$

$$U_0^n=U^n,\quad \text{for } (x,t^n)\in Q_{h\tau}.$$

The proposed algorithm coincides with the Douglas splitting method if the problem (1) is parabolic in the whole region of definition. The stability analysis of the Douglas method is done by Hundsdorfer [Hun96]. It is interesting to note that the Douglas splitting method is unconditionally stable if all operators A_j have negative real parts of all eigenvalues and only one operator can have complex valued eigenvalues (see, also [CPZ00]).

Stability analysis of SCS

In this section we will investigate the stability of the proposed iterative algorithm (5). Let $\lambda_1, \lambda_2, \lambda_3$ be eigenvalues of the discrete operators A_1, A_2, A_3, respectively:

$$\lambda_j < 0, \quad j = 1, 2, 3.$$

Let us denote the error of the iterative solution by

$$Z_s^{n+1} = U_s^{n+1} - U^{n+1},$$

$$Z_s^n = \begin{cases} 0, & \text{if } x \in \Omega_{h,par}, \\ U_s^n - U^{n+1}, & \text{if } x \in \Omega_{h,ell}. \end{cases}$$

Theorem 1. *The stability–correction scheme is unconditionally stable for the three dimensional linear elliptic–parabolic problem and the following convergence estimate is valid:*

$$\|Z_s^{n+1}\| \le q \; \|Z_{s-1}^{n+1}\|, \quad q < 1 \, .$$

Proof. Let us consider the Fourier series:

$$Z_s^{n+1} = \sum_{i,j,m} q_{i,j,m}^{n+1} \, c_{i,j,m} \, X_{i,j,m} \, ,$$

where $q_{i,j,m}$ are the *stability* functions, or the growing factors

$$q_{i,j,m} = \frac{1 + \tau^2(\lambda_1\lambda_2 + \lambda_1\lambda_3 + \lambda_2\lambda_3) - \tau^3\lambda_1\lambda_2\lambda_3}{(1-\tau\lambda_1)(1-\tau\lambda_2)(1-\tau\lambda_3)} \, .$$

Thus for $\lambda_j < 0$ the inequalities

$$|q_{i,j,m}| \le q_1 < 1$$

hold unconditionally. Recall, that we have the following initial condition:

$$Z_s^n = \begin{cases} 0, & \text{if } x \in \Omega_{h,par}, \\ Z_{s-1}^{n+1}, & \text{if } x \in \Omega_{h,ell}. \end{cases}$$

Hence $\|Z_s^n\|$ can be bounded as follows

$$\|Z_s^n\| \le q_2 \, \|Z_{s-1}^{n+1}\|, \quad q_2 \le 1 \, ,$$

which implies that

$$\|Z_s^{n+1}\| \le q_1 \, \|Z_s^n\| \le q_1 q_2 \, \|Z_{s-1}^{n+1}\| \, .$$

This completes the proof.

5 Numerical Experiments

In this section we present some results of numerical simulations for problem (1) with the following discrete operators

$$A_j U^n = \left(U^n_{x_j}\right)_{\bar{x}_j} + f_j(x, t^n), \quad j = 1, 2, 3\,.$$

Two cases of elliptic regions are investigated:

$$Q^1_{ell} = [0.4, 0.7] \times [0.4, 0.7] \times [0.4, 0.7]\,,$$

$$Q^2_{ell} = [0.4, 0.5] \times [0.4, 0.5] \times [0.4, 0.5]\,.$$

The function $f(x,t)$ is defined using an exact solution of the differential problem

$$u(X, t) = \exp(t) \sin(\pi x_i) \sin(\pi x_j) \sin(\pi x_k)\,.$$

A stopping condition of the iterative algorithm is defined as:

$$\|\sum_{j=1}^{3} A_j U^{n+1}_s\|_{\Omega_{h,ell}} \leq \varepsilon\,.$$

In Table 1 we present the averaged numbers of iterations at one time level, obtained by the CG method applied for the realization of the fully implicit finite–difference scheme (2) with $T = 0.5$.

Table 1. The averaged numbers of iterations for the CG method

τ	$N = 20$		$N = 40$		$N = 80$	
	Q^1_{ell}	Q^2_{ell}	Q^1_{ell}	Q^2_{ell}	Q^1_{ell}	Q^2_{ell}
0.1	21.8	7.8	48.8	20.8	102.8	53.8
0.05	19.8	7.0	44.1	18.9	96.1	49.9
0.025	17.6	6.15	38.3	16.9	83.5	43.7
0.0125	14.95	5.02	32.32	14.65	68.45	37.0

In Table 2 we present the averaged numbers of iterations at one time level, obtained by the splitting scheme (5). Here τ_0 denotes the value of parameter τ used in the elliptic region Q^1_{ell}.

6 Conclusions

In this paper we have discussed two numerical algorithms for solving a three–dimensional elliptic–parabolic problem. The main difference between these

Table 2. The averaged number of iterations for one time step

τ	$N=20$	τ_0	$N=40$	τ_0
0.1	15.8	0.005	26.2	0.002
0.05	14.35	0.006	26.5	0.002
0.025	12.85	0.007	24.3	0.0025

methods is that in the first one the differential problem is treated as an elliptic problem in the whole region of the definition and it is integrated in time by the backward Euler scheme, whereas the second method treats the problem as parabolic and the integration is done by the splitting–type method. The advantage of the latter approach is that the linear algebra algorithm is reduced to simple one–dimensional subproblems. The advantage of the first method is that the fully implicit approximation leads to a very robust algorithm.

References

[ALu83] Alt, H.W., Luckhaus, S.: Quasilinear elliptic-parabolic differential equations. Math. Z., **183**(4), 311–341 (1983)

[ChE97] Chen, Z., Ewing, R.: Fully discrete finite element analysis of multiphase flow in groundwater hydrology. SIAM J. Numer. Anal., **34**, 2228–2253 (1997)

[CPZ00] Čiegis, R., Papastavrou, A., Zemitis, A.: Additive splitting methods for elliptic-parabolic problems. Annalli del Universiteta di Ferrara, Sez. VII, Sc. Mat. Vol. XLVI, 291–306 (2000)

[CCZ99] Čiegis, Raim., Čiegis, Rem., Zemitis, A.: Parallel numerical methods for the elliptic–parabolic problem, *Progress in Industrial Mathematics at ECMI98*, B.G.Teubner, Stuttgart, Leipzig, 206–213 (1999)

[EGH00] Eymard, R., Gallouet, T., Gutnic, M., Herbin, R., Hilhorst, D.: Numerical approximation of an elliptic–parabolic equation arising in environment. Comput. Visual Sci., **3**, 33–38 (2000)

[Hel97] Helmig, R.: Multiphase Flow and Transport Processes in the Subserface. *Modelling of Hydrosystems.* Springer–Verlag (1997)

[Hun96] Hundsdorfer, W.: A note on stability of the Douglas splitting method, CWI Report NM-R9606, Amsterdam, 1996.

[JaK91] Jäger, W., Kačur, J.: Solution of porous medium systems by linear approximation scheme, Numer. Math., **60**, pp.407–427 (1991)

Stochastic Relaxation of Variational Integrals with Non-attainable Infima

Dennis D. Cox[1], Petr Klouček[2], Daniel R. Reynolds[3] and Pavel Šolín[2]

[1] Department of Statistics, Rice University, 6100 Main Street, Houston, TX 77005, USA *dcox@rice.edu*
[2] Department of Computational and Applied Mathematics, Rice University, 6100 Main Street, Houston, TX 77005, USA *kloucek@rice.edu*
[3] Center for Applied Scientific Computing, Lawrence Livermore National Laboratory, P.O. Box 808, L-551, Livermore, CA 94551, USA *reynoldd@llnl.gov*

Summary. We provide an example of a stochastic approach to relaxation of the variational integrals with non-attainable infima in one dimension. We provide an approximation for the coefficients of the Laplace transformation of the Probability Density Function. This approaximation yields the relaxing microstructures.

1 Variational Formulation of Non-attainable Differential Inclusions

We have reported in [6] application of the *Subgrid Projection Method* to the problem of finding an approximation to solutions of non-attainable differential inclusions. In this contribution, we describe how this approach leads to a stochastic variational formulation of this problem. Consequently, we can approach solutions to such problems by stochastic gradient flows. We recall that we consider the following

Problem 1.1 *Let $f \in W^{1,\infty}(0,1)$, $|f'(x)| < 1$ for a. a. $x \in [0,1]$. Find a function $u \in W^{1,\infty}(0,1)$ such that*

$$\begin{aligned} &u'(x) \in \{\pm 1\}, &&\text{for almost all } x \in (0,1), \quad \text{and}\\ &u(x) = f(x), &&\text{for all } x \in (0,1). \end{aligned}$$

□

This problem cannot be solved in $W^{1,1}(0,1)$, much less in $W^{1,\infty}(\Omega)$, but if we relax any of the two contradictory requirements in (1.1) a bit, the set of the solutions is enormous. In fact it is dense in the sense of Baire category, [8]. Namely, for any $\epsilon > 0$ there exists $u_\epsilon \in W^{1,\infty}(0,1)$ such that

$$\begin{aligned} &u'_\epsilon \in \{\pm 1\}, \qquad \text{a. e. in } (0,1), \quad \text{and}\\ &\|u_\epsilon - f\|_{L^\infty(0,1)} < \epsilon. \end{aligned} \tag{1}$$

Moreover, for any continuous function $h = h(x)$ such that $h(u'_\epsilon) \rightharpoonup g$, weakly in $L^1(0,1)$, as $\epsilon \to 0_+$, we have

$$g(x) = \int_{-\infty}^{+\infty} h(y) d\mu_{x,u'_\epsilon}(y), \tag{2}$$

where $\mu_{x,u'_\epsilon} = \lambda(x)\delta_{-1} + (1-\lambda(x))\delta_{+1}$; $\delta_{\pm 1}$ denotes the Dirac measures on $\mathbb{R}$ giving unit mass to the points ± 1; and $\lambda(x) = \frac{1}{2}(1 - f'(x))$ a. e. in $[0,1]$. A functional with non-attainable infimum compatible with Problem 1.1 is the popular potential

$$I(u) \stackrel{\text{def}}{=} \int_0^1 \frac{1}{4}\left| |u'(x)|^2 - 1 \right|^2 + \frac{1}{2}|u(x) - f(x)|^2 \, dx. \tag{3}$$

Of course, the choice of this particular form is somewhat arbitrary. It is obvious that

$$\inf_{u_h \in V_h} I(u_h) > 0 \tag{4}$$

for any conforming approximation V_h of $W^{1,4}(\Omega)$. Hence it follows from (1) that the set of local minimizers of (4) is large. In particular it is shown in [4] that if $E_h = \inf_{V_h} I(u_h)$ then there exists a family K_h consisting of $(1/h)^{1/h}$ local minimizers of the same discretized problem such that

$$\begin{aligned} &I(v_h) \leq (1 + 24\sqrt{h})E_h, \qquad \text{for any } v_h \in K_h, \\ &\sup_{t\in[0,1]} I(tv_h^1 + (1-t)v_h^2) > \tfrac{13}{h}E_h, \quad v_h^1, v_h^2 \in K_h. \end{aligned} \tag{5}$$

Moreover, we have shown in [7] that if we consider a problem similar to Problem 1.1, corresponding to tetragonal-to-cubic phase transformations, and if we apply any Descent Algorithm with a pseudo-gradient as descent navigation then such an algorithm converges strongly regardless of the initial guess (that is a weakly differentiable function) even if the infimum cannot be attained! The situation is even more complicated by the fact that this class of problems suffers from the so-called Lavrentiev phenomenon which makes the minimizers dependent on the choice of the functional space, [3]. In summary: classical minimization algorithms applied to (3)-like potentials with non-attainable infima fail. There exists a huge number of local minimizers on the discrete level, computationally the minimizers will depend on the polynomial approximation, the large energy barriers makes the solution dependent on the initial guess. We conclude that, in the light of these difficulties, it is hopeless to expect a reasonable outcome by including into the variational formulation just the information about the averaged states and the crystallographic constraints for the derivatives. We have shown in [5] and [7] that the variational formulation should have a form

$$\int W_{macro} + W_{micro} + W_{stochastic}. \tag{6}$$

Our conclusion is motivated by the observation that we can achieve very good numerical results by building minimizing sequences which become asymptotically (weak) white noise in their derivatives, c.f. [5], [7]. Note that such sequences cannot be periodic! Hence, we conclude that the appropriate microscale dynamical system for approaching the solutions of (1.1) ought to be given by a Langevin dynamics, i.e. by a stochastic gradient flow. Assuming that the Helmholtz free energy has a form (6), then we obtain the Langevin system as an Euler-Lagrange equation corresponding to this functional. We present such an approach in this paper. We refer to [5], [6] and [7] for two possible constructions of $W_{stochastic}$.

2 Two Microscale Finite Dimensional Langevin Models

We consider two models: one with a fixed potential and one with a time varying potential. Our objective is to find appropriate local minima of the potential

$$I(u) = W_{macro}(u) + \lambda W_{micro}(u), \tag{7}$$

where

$$W_{macro}(u) = \int_0^1 |f(x) - u(x)|^2 \, dx \tag{8}$$

$$W_{micro}(u) = \int_0^1 \left|u'(x)^2 - 1\right|^2 \, dx. \tag{9}$$

As $\lambda \to \infty$, the minimizer of I will converge to the solution of the constrained minimization problem

$$\text{minimize } W_{macro}(u) \qquad\qquad \text{subject to } |u'| = 1.$$

We will generally consider piecewise affine approximate solutions to this minimization problem. Thus

$$u'(x) = \sum_{i=1}^{n} v_i \phi_{A_i}(x), \tag{10}$$

where $v = (v_1, \ldots, v_n)$, $\phi_A(x) = \begin{cases} 1 & \text{if } x \in A, \\ 0 & \text{if } x \notin A, \end{cases}$ is the characteristic function of A and $A_i = [(i-1)/n, i/n)$. For a given vector v of values of u' on the intervals A_i, we will recover u via the formula

$$u(x) = \int_0^x \sum_{i=1}^{n} v_i \phi_{A_i}(y) \, dy + \int_0^1 \left[f(y) - \sum_{i=1}^{n} v_i \phi_{A_i}(y) \right] dy. \tag{11}$$

With this convention, we may write $u = Hv$ where H is the linear operator implicit in (11) and think of I as a function of v rather than u, say

$$\tilde{I}(v) = I(Hu).$$

2.1 Time Independent Energy Density

Our first approach is to consider a Stochastic Differential Equation (SDE) for v of the Langevin form

$$\begin{aligned} d\boldsymbol{v}(t) &= -\nabla \tilde{I}(\boldsymbol{v})\,dt + d\eta(t), \\ \boldsymbol{v}(0) &= 0, \end{aligned} \tag{12}$$

where η is an n-dimensional Wiener process satisfying

$$E[\eta(t)] = 0, \quad E[\eta(t) \otimes \eta(s)] = A \min\{s, t\}.$$

The matrix A is a given $n \times n$ positive definite matrix, usually taken as an identity matrix. Note that $d\eta(t)$ is a white noise process. In other words, we assume that the energy corresponding to Problem 1.1 has the form (6), and (12) is the underlying stochastic gradient flow enforcing the competition in the weak topology to provide the bridge between the atomic nano-scale and the specimens microscale ($\sim \mu m$).

The choice of the initial condition is somewhat arbitrary. We will have more to say on this point in our second approach. This SDE describes a particle moving under a force (defined by the gradient of the potential) subject to "thermal" agitation (defined by the white noise term). The resulting stochastic process is a diffusion.

For large values of λ, the behavior of the diffusion process defined by (12) is wander for a short period of time until it "falls" into one of the "wells" corresponding to the local minima of W_{micro}. Our Monte Carlo experiments indicate that there is a slight preference for the particle to be initially attracted to the deeper wells (which correspond to smaller local minima of W because the corresponding u is a better approximation to f in that it makes the W_{macro} term smaller), but this preference is very weak. For this reason, a second approach was considered, which we believe better reflects the physical process of the material phase changes we are modeling.

2.2 Time Dependent Energy Density

Consider a time dependent potential

$$I(u, t) = [1 - \alpha(t)] W_{macro}(u) + \alpha(t) \lambda W_{micro}(u). \tag{13}$$

The SDE remains of the same Langevin form. Namely,

$$d\boldsymbol{v}(t) = -\nabla \tilde{I}(\boldsymbol{v}, t)\,dt + d\eta(t). \tag{14}$$

Here, the gradient is taken only with respect to the first variable v. The strategy for specifying α is to set

$$\alpha(t) = 0, \quad 0 \le t \le t_1, \tag{15}$$

where t_1 is chosen large enough that $\boldsymbol{v}(t)$ has converged to a steady state solution of the Langevin equation similar to (13) but containing only W_{macro}. Then, $\alpha(t)$ is increased linearly to its maximum value of 1

$$\alpha(t) = \max\{a(t - t_1), \alpha_{max}\}, \quad (16)$$

until the particle drops into one of the wells of W_{micro}.

The actual implementation requires selection of a time interval $\triangle t$ and then we approximate the SDE with the Stochastic Difference Equation (S$\varDelta$E):

$$\boldsymbol{v}((k+1)\triangle t) - \boldsymbol{v}(k \triangle t) = -\nabla \tilde{W}(\boldsymbol{v}(k \triangle t), k \triangle t) \triangle t + Z_{k+1} \triangle t^{1/2}, \quad (17)$$

where $Z_1, Z_2, \ldots$ are independent and identically distributed Gaussian random vectors having mean 0 and covariance A. For given W_{macro} and W_{micro} the simulation requires the following inputs:

1. λ which controls the weight given to the W_{micro} term in the potential.
2. A which is the covariance matrix for the noise; we have taken this to be a multiple of the identity $A = \sigma^2 I$, so only specification of σ^2 is necessary.
3. t_1 which is the "burn in time" so that the process V is in a steady state determined by W_{macro}. This requires some experimentation depending on f.
4. a which controls how quickly the potential transforms from W_{macro} to λW_{micro}. This was controlled by choosing the maximum number of time steps and choosing a so that α_{max} was achieved at the maximum number of time steps.
5. α_{max}, the maximum change in α.
6. $\triangle t$, the time step.

The actual numerical values we used are given in Table 1.

3 Meso-scale Fokker-Planck Equation

The Probability Density Function (PDF) $g : \mathbb{R}^+ \times \mathbb{R}^{\dim \boldsymbol{v}} \mapsto \mathbb{R}$ can be used to obtain any statistical information contained in the microscale Langevin systems for $\boldsymbol{v}$ at a fixed time point t. Namely,

$$E[h(t,\boldsymbol{v})] = \int h(t,\boldsymbol{v}) g(t,\boldsymbol{v})\, d\boldsymbol{v}, \qquad \text{for any } h \in C^0\left(\mathbb{R}^+, \mathbb{R}^1\right). \quad (18)$$

The PDF g is obtained by solving the meso-scale deterministic Fokker-Planck equation, [10]. The Fokker-Planck equation for the Langevin system in Section 2.1 has the form

$$\frac{\partial g(t,\boldsymbol{v})}{\partial t} = -\mathrm{div}_{\boldsymbol{v}} \left[DW_{density}(\boldsymbol{v}) g(t,\boldsymbol{v}) \right] + \frac{\sigma^2}{2} \Delta_{\boldsymbol{v}} g(t,\boldsymbol{v}), \quad \boldsymbol{v} \in \mathbb{R}^1,\, t > 0, \quad (19)$$

where σ^2 is the white noise standard deviation. Hence, for our specific form of the energy density given by (7), the Fokker-Planck Equation (19) becomes

$$\frac{\partial g(t,\boldsymbol{v})}{\partial t} = \left(\int_0^1 \tilde{f}(x)\boldsymbol{\beta}(x)\,\mathrm{d}x\right)\cdot\nabla g(t,\boldsymbol{v}) - \underbrace{\left(\int_0^1 \boldsymbol{\beta}(x)\otimes\boldsymbol{\beta}(x)\,\mathrm{d}x\right)}_{\{B_{ij}\}_{i,j=1}^n}\boldsymbol{v}\cdot\nabla g(t,\boldsymbol{v})$$
$$+\frac{\lambda}{n}\sum_{i=1}^n v_i(v_i^2-1)\frac{\partial g(t,\boldsymbol{v})}{\partial v_i} - \left(2\lambda+\sum_{i=1}^n B_{ii}\right)g(t,\boldsymbol{v}) + \frac{\sigma^2}{2}\triangle g(t,\boldsymbol{v}), \tag{20}$$

where

$$\tilde{f}(x) = f(x) - \int_0^1 f(y)\,\mathrm{d}y,$$
$$B_{ij} = \int_0^1 \beta_i(x)\beta_j(x)\,\mathrm{d}x,$$
$$\beta_i(x) = \begin{cases} -\frac{n-i+1/2}{n^2}, & x\in(0,(i-1)/n),\\ x-(i-1)/n-\frac{n-i+1/2}{n^2}, & x\in[(i-1)/n,i/n),\\ 1/n-\frac{n-i+1/2}{n^2}=\frac{i-1/2}{n^2}, & x\in(i/n,1). \end{cases}$$

Since the microscale process given by the Langevin equation is a diffusion, we expect, in the sense of the convergence in measure,

$$\lim_{\sigma^2\to0_+}\lim_{\lambda\to+\infty}\lim_{t\to+\infty} g(t,\mathbf{v}) = \sum_{i=1}^{2^n} q_i\delta_{\mathbf{v}_i^*}(\mathbf{v}) := g(\mathbf{v}), \qquad \sum_{i=1}^{2^n} q_i = 1, \quad q_i\ge 0, \tag{21}$$

where $\{\mathbf{v}_i^*\}_{i=1}^{2^n}$ represent all 2^n states with $\mathbf{v}_{ij}^*\in\{\pm1\}$. Note that

1. $t\to+\infty$ corresponds to finding the equilibrium distribution of the states, which is given by the Gibbs distribution [9],
2. $\lambda\to+\infty$ represents imposing the ±1 constraint,
3. $\sigma^2\to0_+$ corresponds to cooling,
4. $n\to\infty$ gives the continuum case.

The Laplace transform of $g(\boldsymbol{v})$ has the form

$$(Lg)(\boldsymbol{y}) = \sum_{j=1}^{2^n} q_j e^{\boldsymbol{y}^T\boldsymbol{v}_j^*}, \tag{22}$$

where $\boldsymbol{v}_j^*$ denotes one of the 2^n possible distributions of ± 1, having the probability q_j. Our goal now is to compute the coefficients q_j in (22). In principle, we have two options:

1. Take the Laplace transform of the Fokker-Planck equation (20) with the aim to obtain a dense linear $2^n \times 2^n$-system for the unknown coordinates q_j. The system can be obtained by inserting 2^n various vectors $\boldsymbol{y}$ into (20) together with the representation (22). One may then use a reduction technique, [2], [1], to obtain a sparse system and solve it.
2. Enumerate all the 2^n states $\boldsymbol{v}_j^*$, determine their probabilities q_j, and pick the state with highest probability as the most likely one to which the system will relax.

Here, we show only the second approach which is based on the following conjecture.

Conjecture 3.1

$$
\begin{aligned}
q_i &= \prod_{j=1}^{n} q^{\frac{1}{2}(1+\mathrm{v}_{ij}^*)}(1-q)^{1-\frac{1}{2}(1+\mathrm{v}_{ij}^*)}, \qquad \textit{where,} \\
q &= \frac{1}{2}\left(1 - \frac{f(j/n) - f((j-1)/n)}{1/n}\right).
\end{aligned}
\tag{23}
$$

This conjecture has the important application that it can be used to compute directly the volume fraction. We believe that it is probably not exact, but in fact provides an excellent approximation. It is in part based on our previous work [5] where we found that the computed approximate solutions of (1.1) had a white noise property. Since

$$
f'(x) = \int_R y \, d\mu_x(y), \tag{24}
$$

where $\mu_x = a(x)\delta_{-1} + (1 - a(x))\delta_{+1}$, $a(x) = \frac{1}{2}(1 - f'(x))$. Then if we let

$$
a_j \stackrel{\text{def}}{=} \sum_{\mathrm{v}_{ij}^* = -1} q_i, \tag{25}
$$

and if a_n were defined by $a_n \in C^0(0,1)$ such that $a_n(x_j) = a_j$, then

$$
a_n \to a, \qquad \text{as } n \to \infty \quad \text{in } C^0(0,1). \tag{26}
$$

4 Monte Carlo Simulations Based on the Langevin Equation

We use the model (17) and we apply Monte Carlo simulations to obtain sequences which come close to approximating Problem 1.1. The target function for this simulation is $f(x) = x(1-x)$ with $\int_0^1 f(x)\,dx = 1/16$. The volume fraction is computed by averaging over the replications. Namely,

$$a_j \stackrel{\text{def}}{=} \frac{1}{2}\left(\frac{1}{m}\sum_{i=1}^{m} v_{ij} + 1\right). \tag{27}$$

Here, m is the number of replications and v_i is the vector of the derivatives for the i–th replication. The index j refers to the interval A_j, i.e., $[(j-1)/n, j/n]$, where n is the number of intervals in $(0,1)$. The result shown in this section is based on twenty independent simulations for the mesh with $h = 1/200$. To deduce the macroscopic shape, we average over all twenty replications, c.f. Figure 1.

Table 1. Data for the model (17) used in this section

Variable	Value
spatial resolution	$\frac{1}{200}$
number of independent replications	20
max no. of steps per replication	10^7
no. of steps for burn in	10^5
time increment per step	0.05
standard deviation for white noise	0.02
maximum change in alpha	0.5

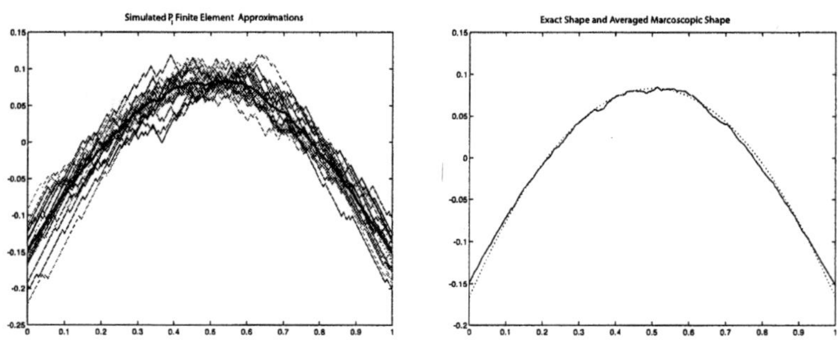

Fig. 1. Macroscopic approximations. The left picture shows all twenty replications (each replication has different color) and the averaged state (thick smoother line). The right picture shows the difference between the target function $f = x(1-x)$ and its computed shape. The spatial resolution is $h = 1/200$.

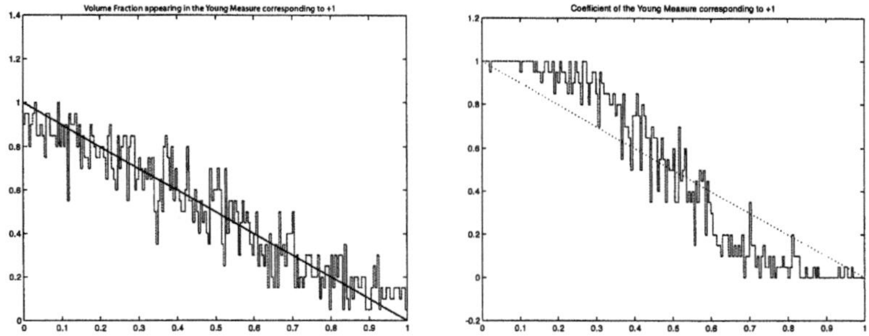

Fig. 2. Volume fractions. The volume fraction on the left is computed with $\sigma = 0.02$. The volume fraction on the right is computed with the data given by Table 1 but with much smaller deviation. Namely, with $\sigma = 0.005$. This shows that particles need a higher values of the deviation to discover states with lower energy.

5 Simulations Based on the Analysis of the Fokker-Planck Equation

We use in this section the formulae and convergence results to investigate the approximation properties based on the formulae (23) and (25). We chose the spatial resolution to be $h = 1/16$. With 16 elements on the $(0, 1)$ segment we have $2^{16} = 65536$ coefficients in the formula (23) and each state vector has 16 components. We evaluate the values of q_i and we select the state for which q_i is the biggest. In other words we select the state with the highest probability to exhibit the microscopic structure of the solution. The volume fraction is computed using the formula (25). It seems that the discrete solution corresponding to the maximum probability state minimizes the L^2–distance to the target function. We chose $f(x) = \frac{1}{13}\sin(13x)$ for the calculations in this section. The results are plotted in Figure 3.

6 Acknowledgment

Dennis Cox was supported in part by the grant NSF DMS-0204723. Petr Klouček was supported in part by the grant NSF DMS-0107539, by the grant from TRW Foundation and by the grant NASA SECTP. The work of Daniel Reynolds was performed in part under the auspices of the U.S. Department of

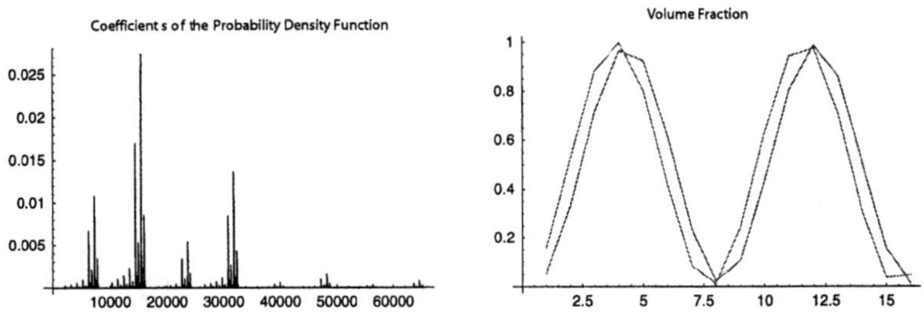

Fig. 3. The picture on the left shows the distributions of the coefficients of the Probability Density Function given by (23). It is clear that the more the target function deviates from a constant the more isolated is the maximum among the rest of the coordinates. The right picture shows the reconstruction of the volume fraction using the formula (25). The darker line corresponds to the computed volume fraction, the lighter line correspond to the P_1-projection of the exact function representing the volume fraction.

Energy by the University of California, Lawrence Livermore National Laboratory under Contract No. W–7405–Eng–48. Pavel Šolín was supported in part by the Grant Agency of the Czech Republic under Grant No. GP102/01/D114. This material is based upon work supported in part by Contract No. 74837-001-0349 from the Regents of University of California (Los Alamos National Laboratory) to William Marsh Rice University.

References

1. A.C. Antoulas and D.C. Sorensen. Approximation of large-scale dynamical systems: An overview. *Special Issue in Numerical Analysis and System Theory, Edited by S.L. Campbell, International J. of Applied Mathematics and Computational Science*, 11:1093–1121, 2001.
2. A.C. Antoulas and D.C. Sorensen. The sylvester equation and approximate balanced reduction. *Special Issue in Numerical Analysis and System Theory; Edited by V. Blondel, D. Hinrichsen, J. Rosenthal, and P.M. van Dooren, Linear Algebra and It's Applications*, 2002. to appear.
3. J. M. Ball. Singularities and computation of minimizers for variational problems. *Oxford FoCM, Lecture Notes*, 1999.
4. C. Carstensen. Numerical analysis of non-convex minimization problems alllowing microstructures. *Zeitschrift für Angewandte mathematik und Mechanik*, 76(S2):437–438, 1996.
5. D. Cox, P. Klouček, and D. R. Reynolds. The non-local relaxation of nonattainable differential inclusions using a subgrid projection method: One dimensional theory and computations. *Technical Report 10, École Polytechnique Fédérale de Lausanne, July 2001.* to appear in: SIAM J. Sci. Comp., (2003).

6. D. Cox, P. Klouček, and D. R. Reynolds. A subgrid projection mathod for nonattainable differential inclusions. pages 575–584, 2001. in proceedings: ENUMATH 2001, Numerical Mathematics and Advanced Applications, F. Brezzi, A. Buffa, S. Corsaro, A. Murli (Eds).
7. D. Cox, P. Klouček, and D. R. Reynolds. On the asymptotically stochastic computational modeling of microstructures. *Future Generation Computer Systems*, 1080:1–16, 2003.
8. B. Dacorogna and P. Marcellini. *Implicit Partial Differential Equations*. Birkhäuser, 2000.
9. R. Jordan, D. Kinderlehrer, and F. Otto, The variational formulation of the Fokker-Planck equation, *SIAM J. Math. Anal.*, 29:1-17, 1998.
10. H. Risken. *The Fokker-Planck equation: methods of solution and applications*. Springer–Verlag, Berlin-New York, 1984. Springer series in synergetics: 18.

A Pressure-Weighted Upwind Scheme in Unstructured Finite-Element Grids

Masoud Darbandi[1], Kiumars Mazaheri-Body[2] and Shidvash Vakilipour[3]

[1] Sharif University of Technology, Tehran, P.O. Box 11365-8639, Iran
darbandi@sharif.edu

[2] Tarbiat-Modares University, Tehran, P.O. Box 14115-111, Iran
kmb@modares.ac.ir

[3] Sharif University of Technology, Tehran, Iran *vakilipour@mehr.sharif.edu*

Summary. Today the finite element method is known as a powerful tool capable of solving complex flow in complex geometries. Additionally, the unstructured grid topology is a complementary tool which effectively increases computational efficiencies. On the other hand, the finite element volume methods incorporate the advantages of conserving the conservative quantities within elements. However, the accurate conservation statements need utilizing suitable approximation at cell faces. In convection dominated flows, upwind-based schemes are strongly utilized. However, these schemes do not suffice to incorporate the details of pressure field in the approximation. Therefore, the pressure-weighted upwind scheme is a better choice for a flow field with high pressure gradients. In this work, a pressure-weighted upwind scheme is suitably extended for solving incompressible flow on unstructured grids. Subsequently, a remedy is given for the problem associated with using equal-order pressure and velocity interpolations. Eventually, the extended formulations are validated against suitable benchmark problems involving small and large scale recirculation zones. Comparing with the benchmark solutions, the current results are excellent.

1 Introduction

The rapid progress in unstructured grid generation techniques has encouraged the CFD code developers to extend their formulations in order to solve the fluid flow and heat transfer problems on unstructured grid distributions [1]. In fact, computational methods always require to improve their accuracies in solving more complex realistic configurations, of course, at lower computational cost. The limit in computer memory storage enforces the computational methods to use limited number of nodes. This limit can in turn degrade the achieved accuracy. The unstructured grid can be used as a powerful tool to compensate the degraded accuracy by suitable grid clustering within the zones with high flow field gradients.

Basically, there are two potential numerical instabilities associated with Galerkin-based methods [2]. The first instability is spurious oscillation in the flow field due to the presence of advection terms. The second instability is due to using inappropriate combinations of velocity and pressure interpolations. Generally speaking, the standard Galerkin-based formulations need to be modified when convection is the dominant physics. The Petrov-Galerkin [3], Taylor-Galerkin [4], Galerkin/Least-squares [5] ... are a number of effective remedies to resolve the problem.

Although the advantage of using unstructured grid in finite element volume methods is many [6, 7], the instability issues still persist. In fact, the key point is in the correct approximation of the cell face velocities which is still challenging. Normally, the upwind-biased interpolations do not take into account the weight of pressure-field in flow acceleration/deceleration. Therefore, the pressure-weighted upwinding scheme is used as an important alternative [6]. Unfortunately, this scheme cannot be taken as a solution to the consequences of using equal-order interpolations [8]. In this work, a physical-based pressure-weighted upwind scheme is introduced and suitably extended for using in appropriate combinations of velocity and pressure interpolations. Meantime, a robust prescription for accommodating equal-order interpolations is given.

2 Governing Equations

In the present study, we are concerned with the two-dimensional incompressible steady flow. The governing equations consist of the conservation statements for mass and momentums. The non-dimensional vector form of the governing equations is given by

$$\nabla^* \cdot \mathbf{V}^* = 0 \tag{1}$$

$$Re\ [\nabla^* \cdot (\mathbf{V}^*\mathbf{V}^*) + \nabla^* p^*] = \nabla^{*2}\mathbf{V}^* \tag{2}$$

where the lengths x & y, velocity V, and pressure p variables are nondimensionalized with respect to a characteristic length L_∞ (e.g., $x^*{=}x/L_\infty$, $y^*{=}y/L_\infty$), a reference velocity V_∞ (e.g., $\mathbf{V}^*{=}\mathbf{V}/V_\infty$), and a reference density ρ_∞ (e.g., $p^*{=}p/(\rho_\infty V_\infty^2)$).

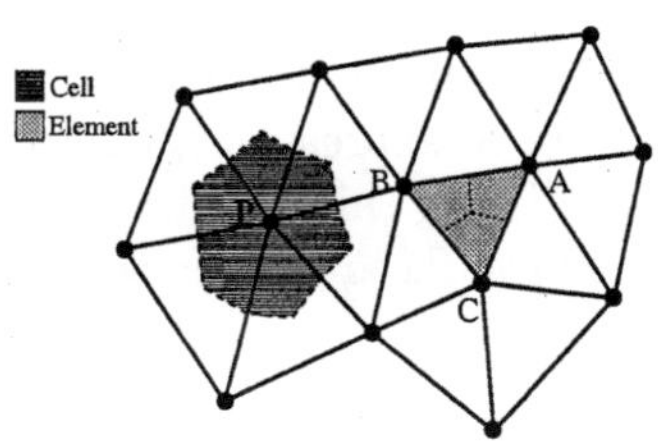

Fig. 1. A part of unstructured grid representing the elements and a constructed cell

The solution domain is broken into a huge number of triangular elements which are distributed in an unstructured form. The elements fully cover the solution domain with no overlapping. Figure 1 shows a small part of the solution domain. Nodes are located at the triangle vertices. They are the locations of the unknown variables. There are three main neighbors around each triangle, see element ABC in Fig. 1. There are no limits in the number of elements intersect at a node. For example, the shaded area in Fig. 1 shows six triangles which encompass node P. Therefore, to utilize the benefits of cell-centered schemes, each element is divided into three quadrilaterals by the help of its three medians. The medians are demonstrated by dashlines in Fig. 1. The cells are then constructed from the proper assemblage of these sub-quadrilaterals. As is seen, irrespective of the shape and distribution of the elements, each node is surrounded by a number of sub-quadrilaterals. The proper assemblage of neighboring sub-quadrilaterals around any non-boundary node creates a polygon cell. In case of unstructured grid, it is possible to have a hybrid mesh composed of polygons having different number of sides. It is because the number of elements which visit an specific node is not fixed.

3 Computational Modelling

To utilize the advantages of finite element volume methods, the governing equations are initially integrated over the shaded area or the cell shown in Fig. 1. The employment of Gauss divergence theorem to the dimensional form of the governing equations leads to

$$\int_A \mathbf{V} \cdot \mathbf{dA} = 0 \tag{3}$$

$$\int_A u(\rho\mathbf{V}) \cdot \mathbf{dA} = -\int_A p\, dA_x + \int_A (\mu\nabla u) \cdot \mathbf{dA} \tag{4}$$

$$\int_A v(\rho\mathbf{V}) \cdot \mathbf{dA} = -\int_A p\, dA_y + \int_A (\mu\nabla v) \cdot \mathbf{dA} \tag{5}$$

where $\mathbf{V} = u\mathbf{i} + v\mathbf{j}$, p, ρ, and μ represent velocity, pressure, density, and the molecular viscosity, respectively. The above integrals are evaluated over the surface which encloses each cell. The cell area is indicated by A. The above equations are suitably discretized using finite difference scheme and finite element interpolations. In the above expressions, $\mathbf{dA}{=}dA_x\mathbf{i} - dA_y\mathbf{j}{=}\Delta y\mathbf{i} - \Delta x\mathbf{j}$ is calculated on each cell face. Using this definition, the above integrals can be evaluated by summation over the faces that enclose the cell, i.e.,

$$\sum_{i=1}^{ns}[\rho(u\, dA_x + v\, dA_y)]_i = 0 \tag{6}$$

$$\sum_{i=1}^{ns}[\rho\bar{u}\ (u\, dA_x + v\, dA_y)]_i = -\sum_{i=1}^{ns}(p\, dA_x)_i + \sum_{i=1}^{ns}\left[\mu\left(\frac{\partial u}{\partial x}\, dA_x + \frac{\partial u}{\partial y}\, dA_y\right)\right]_i \tag{7}$$

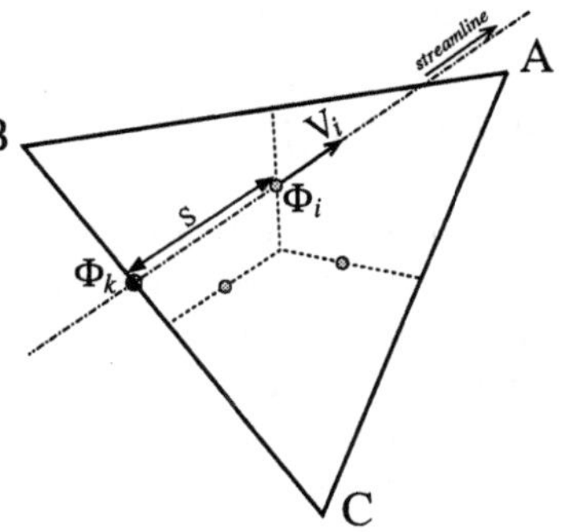

Fig. 2. The velocity upwinding strategy within an element

$$\sum_{i=1}^{ns}[\rho\bar{v}\ (u\ dA_x + v\ dA_y)]_i = -\sum_{i=1}^{ns}(p\ dA_y)_i + \sum_{i=1}^{ns}\left[\mu\left(\frac{\partial v}{\partial x}\ dA_x + \frac{\partial v}{\partial y}\ dA_y\right)\right]_i \tag{8}$$

where i counts the number of cell faces from 1 to ns. The number of cell faces around node P in Fig. 1 is 12. The bar over $\bar{u}$ and $\bar{v}$ indicates that the variables are approximated from the known magnitudes of the preceding iteration. These estimations are necessary in order to linearize the nonlinear convection terms. The rest of procedure is to relate the cell face magnitudes (identified by lower case letters such as u, v, and p variables) directly to the nodal magnitudes (identified by upper case letters such as U, V, and P variables) where the unknown variables are located. A simple idea for treating the right-hand-side terms is the use of finite element shape functions $N_{j=1\dots3}$, i.e.,

$$p_i = \sum_{j=1}^{3} N_{ij} P_j \tag{9}$$

$$\left.\frac{\partial\phi}{\partial z}\right|_i = \sum_{j=1}^{3}\frac{\partial N_{ij}}{\partial z}\Phi_j \tag{10}$$

where p_i identifies the magnitude of p at the mid-point of ith cell face. The j notation counts the node numbers of an element where the ith cell face is located inside it. Additionally, the variable z represents either x or y coordinates and ϕ represents either u or v velocity components. In the above expressions, lower and upper case letters represent cell face and nodal magnitudes, respectively.

The above treatments end the pressure and diffusion term calculations at cell face i. However, more sophisticated expressions are required to treat the convection terms. In fact, the treatment should not disregard the convection-diffusion physics and concept. To mimic the correct physics of the convection, the convection term in the left-hand-side is upwinded. Considering the ith cell face in Fig. 2, one inclusive suggestion is given by

$$\phi_i = \phi_k + \left(\frac{\partial \phi}{\partial s}\right)_i \Delta S_{ki} \tag{11}$$

which has been written in the streamwise direction at mid-point of the ith cell face. The length ΔS is a geometry sensitive parameter shown in Fig. 2 as s. Then, we need to determine the gradient of ϕ along the streamline. We try to approximate this gradient using the original governing PDE's. In another words, one meaningful approximation can be obtained by writing the revised momentum equations in the streamwise direction, i.e.,

$$\rho \bar{\forall} \frac{\partial \phi}{\partial s} = \nabla \cdot \mu \nabla \phi + S_\phi \tag{12}$$

where $\forall = \sqrt{\bar{u}^2 + \bar{v}^2}$ is the total velocity at the cell mid-point and the source term S_ϕ represents either $\partial p / \partial x$ in treating x-momentum or $\partial p / \partial y$ in treating y-momentum. The substitution of Eq.(12) in Eq.(11) results in

$$\phi_i = \phi_k + \left[\frac{1}{\rho \bar{\forall}} \left(\nabla \cdot \mu \nabla \phi + S_\phi\right)\right]_i \Delta S_{ki} \tag{13}$$

As is observed, the influence of pressure has been considered in calculating the correction part of Eq.(11) now. Using the finite-element context, this statement can be revised to

$$\phi_i = \sum_{j=1}^{3} N_{kj} \Phi_j + \frac{1}{\rho \bar{\forall}_i} \left(\frac{\sum_{j=1}^{3} N_{ij} \Phi_j - \phi_i}{L_i^2} - \sum_{j=1}^{3} \frac{\partial N_{ij}}{\partial z} P_j \right) \Delta S_{ki} \tag{14}$$

where L_i is an appropriate diffusion length scale [6]. This length can be estimated in an specified triangle by discretizing the diffusion terms using central differencing.

Equation (14) shows that ϕ_i appears in both sides of equation. As is seen, considering a lagged role for $\bar{\phi}$ in the diffusion term results in a passive role of diffusion term in the formulations. To switch it to an active role, it is not lagged and the impact of this term is taken to the left-hand-side of Eq.(14). A suitable rearrangement of the new equation in terms of our major dependent variables, i.e., Φ_j and P_j, yields

$$\phi_i = \sum_{j=1}^{3} \alpha_{ij} \Phi_j + \sum_{j=1}^{3} \beta_{ij} P_j + \gamma_i \tag{15}$$

where α, β, and γ represent matrix, matrix, and vector coefficients, respectively. The above statement indicates that ϕ ($\equiv u, v$) at cell face can be approximated by the proper assemblage of Φ and P influences. In fact, this approximation can be regarded as a pressure-weighted upwind scheme.

As is known, one major disaster with the continuity equation is the lack of having any explicit pressure term despite representing the pressure field. The past investigation has shown that the ignorance of this important point can result in non-physical wavy solution [9]. The most important reason for this non-physical wavy solution is the employment of equal order pressure and velocity interpolations in Eqs. (6)-(8). The use of unequal interpolations is known as a general remedy to suppress the non-physical solution. However, the current innovate idea suggests the use of more sophisticated interpolations such as Eq.(15) which enforces the direct role of pressure field in the continuity equation. Therefore, using Eq.(15) in the continuity equation may eliminate the need for unequal order interpolations. In another words, Eq.(3)should no longer permit the occurrence of a non-physical solution in the domain. Although this strategy theoretically seems to work well, some deficits has been practically encountered. For example, defining $m = \rho(u\ dA_x + v\ dA_y)$, Eqs.(6)-(8) for Euler flow can be re-written as

$$\sum_{i=1}^{ns} m_i = 0 \tag{16}$$

$$\sum_{i=1}^{ns} m_i u_i = -\sum_{i=1}^{ns} (p\ dA_x)_i \tag{17}$$

$$\sum_{i=1}^{ns} m_i v_i = -\sum_{i=1}^{ns} (p\ dA_y)_i \tag{18}$$

In a one-dimensional context, Reference [9] shows that the above equations can still result in undesirable wavy solution under special circumstances. However, the wavy solution can be eliminated if Eq.(14) is suitably modified. Therefore, instead of using unequal order interpolation functions, we introduce and utilize new ϕ's at cell faces. Considering the generic form of upwinding given by Eq.(11), the new velocity gradients are given by

$$\rho\bar{\forall}\frac{\partial\phi}{\partial s} = \nabla\cdot\mu\nabla\phi + S_\phi - \phi(\frac{\partial u}{\partial x} + \frac{\partial v}{\partial y}) \tag{19}$$

where the new additional gradient terms in the parentheses provide new statements for the velocity components. Considering the new definitions, Eq.(13) is revised to

$$\tilde{\phi}_i = \phi_k + \left\{\frac{1}{\rho\bar{\forall}}\left[\nabla\cdot\mu\nabla\phi + S_\phi - \phi\left(\frac{\partial u}{\partial x} + \frac{\partial v}{\partial y}\right)\right]\right\}_i \Delta S_{ki} \tag{20}$$

The new additional gradient terms are approximated using the approach employed in Eq.(10). The rest of equation is treated similar to Eq.(14). The tilde over $\tilde{\phi}$ indicates that these velocity statements are used in the continuity equation.

4 Results and Discussion

The derived formulations are tested using standard squared cavity [10] and triangular cavity [11] benchmark problems which involve many complex recirculation zones. The first problem is the flow in a squared cavity driven by its upper lid. The problem is tested in Re=3200. Figure 3(left) demonstrates a typical non-uniform unstructured grid distribution in the cavity. The grid generator has been developed by the current authors. It is capable of generating different types of unstructured grid topology. As is observed, the grid has been properly refined in regions with high flow field gradients. Figure 3(right) depicts the streamlines in the cavity. The recirculation zones in bottom corners and top-left corner resemble the complexity of this flow field. They have been detected successfully and accurately.

To investigate the advantages of using the extended formulations on an unstructured grid, the cavity is tested on both uniform and non-uniform unstructured grids. The grid resolution is 71×71 for uniform grid. The total number of nodes in non-uniform grid is 5508 which is close to that of uniform grid, i.e., 5684. The number of cell for them is 10734 and 11086, respectively.

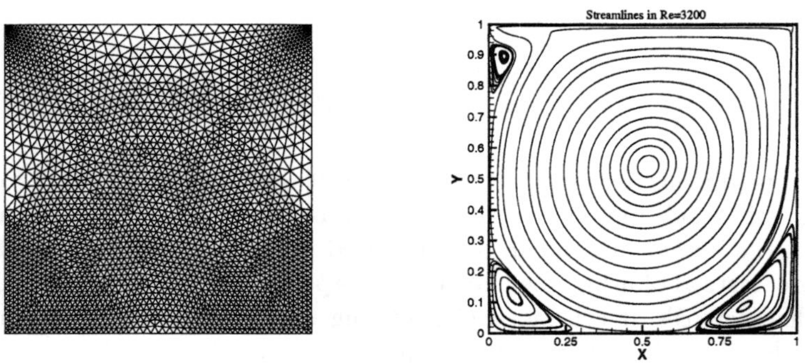

Fig. 3. A typical unstructured grid and the obtained streamlines in the cavity

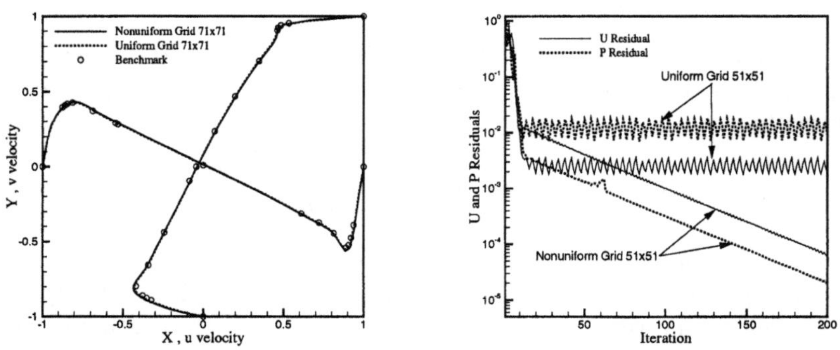

Fig. 4. The centerline velocities in the cavity and the convergence histories

Figure 4(left) demonstrates the centerline u and v velocities for both uniform and non-uniform grid types. The current results are compared with each other and those of benchmark [10]. The figure shows that the results of non-uniform grid distribution are similar to that of uniform grid. Additionally, they are in good agreement with the benchmark solution. Figure 4(right) shows the residual histories for both U velocity component and P fields. The V velocity component history is very similar to that of U. The histories are presented for both uniform and non-uniform grids. Despite performing equal accuracies in Fig. 4(left), the performances are different. As is seen, the convergence instabilities are dominant in uniform grid. However, the non-uniform unstructured grid resembles a smooth and stable residual reduction.

The second test problem is the steady recirculating viscous flow in an equilateral triangular cavity. A primary eddy and several secondary eddies at different regions indicate the complexity of the flow field. Figure 5 shows the geometry of the cavity with two typical unstructured grid distributions. The top horizontal wall is sliding with a constant velocity. The non-uniform topol-

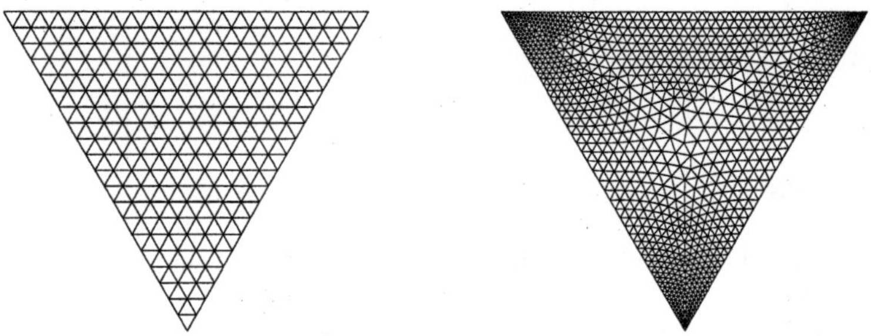

Fig. 5. Using uniform and non-uniform unstructured grid topologies in triangular cavity

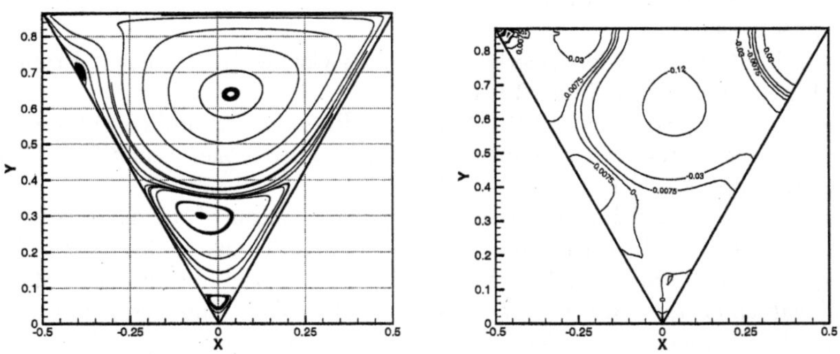

Fig. 6. The streamlines and iso-bars in triangular cavity

ogy suitably clusters the grid around the corners. Of course, the refinement helps to achieve more accurate results for an insufficient number of mesh nodes.

Figure 6 shows the streamline patterns (left) and isobar lines using non-uniform grid distribution. The Reynolds number is 500. The total number of nodes is 1135. As is observed, there are one primary eddy on top and two secondary eddies under it. The eddies shrink in lower levels. Additionally, there is one irregular recirculation on the left edge which has been successfully detected. Unfortunately, there is no quantitative report of velocity magnitudes for this benchmark case. However, a qualitative comparison has been done with other references such as Refs. [11, 12]. Table 1 shows that the x and y magnitudes of the eddy center locations are in good agreement with those obtained from the benchmark solutions.

Table 1. A comparison on the (x, y) locations of vorticities in triangular cavity

	top vortex	mid vortex	bottom vortex	edge vortex
Current results	0.038, 0.639	-0.047, 0.298	0.000, 0.064	-0.389, 0.706
Guermond [12]	0.042, 0.632	-0.049, 0.294	0.000, 0.062	-0.395, 0.701

References

1. Mavriplis, D.J. (1997): Unstructured Grid Techniques, Annu. Rev. Fluid Mech., **29**, 473–514.
2. Tezduyar, T.E. (1992): Finite Element Computation of Unsteady Incompressible Flows Involving Moving Boundaries and Interfaces and Iterative Solution Strategies, AGARD-R-787, France, Chap. 3.
3. Hughes, T.J.R., Franca, L.P., and Balestra, M. (1986): A New Finite Element Formulation for Computational Fluid Dynamics: V, Comp. Methods Appl. Mech. Eng., **59** 85–99.
4. Donea, J. (1984): A Taylor-Galerkin Method for Convective Transport Problems, Int. J. Numer. Methods Eng., **20**, 101–119.
5. Hansbo, P., and Szepessy (1990): A Velocity-Pressure Streamline Diffusion Finite Element Method for the Incompressible Navier-Stokes Equations, Comp. Methods Appl. Mech. Eng., **84**, pp.175–192.
6. Darbandi, M., and Schneider, G.E. (1999): Application of an All-Speed Flow Algorithm to Heat Transfer Problems, Numer. Heat Trans. A, **35**, pp.695–715.
7. Baliga, B.R., and Patankar, S.V. (1987): Elliptic Systems: Finite-Element Method II, *Handbook of Numerical Heat Transfer*, Edited by W.J. Minkowycz, E.M. Sparrow, G.E. Schneider, R.H. Pletcher, John Wiley, New York, pp.421–461.
8. Prakash, C. (1986): An Improved Control Volume Finite-Element Method for Heat and Mass Transfer, and for Fluid Flow Using Equal-Order Velocity-Pressure Interpolation, Numer. Heat Trans., **9**, 253–276.

9. Darbandi, M., Schneider, G.E., and Bostandoost, S.M. (2003): Improvement of Velocity Role in Coupling of Mass and Momentum Equations, AIAA Paper 2003-0856, 41st AIAA Aero. Sciences Meeting, Reno, NV, Jan. 6-9.
10. Ghia, U., Ghia, K.N., and Shin, C.T. (1982): High-Re Solutions for Incompressible Flow Using the Navier-Stokes Equations and a Multigrid Method, J. Comp. Physics, **48**, 387–411.
11. Ribbens, C.J., Watson, L.T., and Wang, C.Y., (1994): Steady Viscous Flow in a Triangular Cavity, J. Comp. Physics, **112**, 173–181.
12. Guermond, J.L., and Quartapelle, L. (1997): Calculation of Incompressible Viscous Flows by an Unconditionally Stable Projection FEM, J. Comp. Phys., **132** 12–33.

Discontinuous Galerkin Finite Element Method for the Numerical Solution of Viscous Compressible Flows

Vít Dolejší

Charles University Prague, Faculty of Mathematics and Physics, Sokolovská 83, 186 75 Prague, Czech Republic *dolejsi@karlin.mff.cuni.cz*

Summary. We deal with the numerical solution of the compressible Navier-Stokes equations with the aid of the discontinuous Galerkin finite element (DG FE) approach with the nonsymmetric interior penalty terms. The linearization of diffusive terms and the treatment of the boundary conditions are discussed. Several numerical examples demonstrating the efficiency of the numerical method are presented.

1 Introduction

We deal with the numerical solution of the compressible Navier-Stokes equations with the aid of the discontinuous Galerkin finite element method (DGFEM). This method has become quite popular and it is discussed in a number of papers. For a review of DG methods, see [4] or [5]. Let us mention the papers [1] and [2] dealing with the numerical simulation of compressible flows, where the mixed formulation is applied to the treatment of the viscous terms.

We develop the so-called DGFEM with nonsymmetric interior penalty terms. This method was applied to the solution of a scalar nonlinear convection-diffusion equation in [8] where a complete numerical analysis is presented. The extension of DGFEM to the system of the Navier-Stokes equations is straightforward (see preliminary results in [6]) but some suitable linearization of diffusive terms has to be performed. That is the subject of this paper.

In Section 2 the continuous problem describing compressible flow is formulated. DGFE discretization is introduced in Section 3, where also the linearization of diffusive terms is discussed. Several numerical examples demonstrating the efficiency of the method are presented in Section 4.

2 Continuous problem

Let $\Omega \subset \mathbb{R}^2$ be a bounded plain domain and $T > 0$. We set $Q_T = \Omega \times (0, T)$ and by $\partial\Omega$ we denote the boundary of Ω which consists of several disjoint parts. We distinguish inlet Γ_I, outlet Γ_O and impermeable walls Γ_W on $\partial\Omega$. We want to find a vector-valued function $\boldsymbol{w} : Q_T \rightarrow \mathbb{R}^4$, such that

$$\frac{\partial \boldsymbol{w}}{\partial t} + \sum_{s=1}^{2} \frac{\partial \boldsymbol{f}_s(\boldsymbol{w})}{\partial x_s} = \sum_{s=1}^{2} \frac{\partial \boldsymbol{R}_s(\boldsymbol{w}, \nabla \boldsymbol{w})}{\partial x_s} \quad \text{in } Q_T, \tag{1}$$

where

$$\begin{aligned}
\boldsymbol{w} &= \boldsymbol{w}(x,t), \ x \in \Omega, \ t \in (0,T), \\
\boldsymbol{w} &= (\rho, \rho v_1, \dots, \rho v_N, E)^{\mathrm{T}} \in I\!R^4, \\
\boldsymbol{f}_i(\boldsymbol{w}) &= \left(\rho v_i, \rho v_1 v_i + \delta_{1i}\, p, \rho v_2 v_i + \delta_{2i}\, p, (E+p) v_i\right)^{\mathrm{T}}, \ i = 1,2 \\
\boldsymbol{R}_i(\boldsymbol{w}, \nabla \boldsymbol{w}) &= \left(0, \tau_{i1}, \tau_{i2}, \tau_{i1}\, v_1 + \tau_{i2}\, v_2 + \frac{\gamma}{Re\, Pr} \partial\theta / \partial x_i\right)^{\mathrm{T}}, \ i = 1,2 \\
\tau_{ij} &= \frac{1}{Re}\left(\frac{\partial v_i}{\partial x_j} + \frac{\partial v_j}{\partial x_i} - \frac{2}{3}\operatorname{div}\boldsymbol{v}\delta_{ij}\right), \ i,j = 1,2.
\end{aligned} \tag{2}$$

To system (1) we add the thermodynamical relations

$$p = (\gamma - 1)(E - \rho |\boldsymbol{v}|^2/2), \quad \theta = \left(\frac{E}{\rho} - \frac{1}{2}|\boldsymbol{v}|^2\right). \tag{3}$$

We use the following notation: $\boldsymbol{v} = (v_1, v_2)^{\mathrm{T}}$ – velocity vector, ρ – density, p – pressure, θ – temperature, E – total energy, γ – Poisson adiabatic constant, Re – Reynolds number, Pr – Prandtl number.

System (1) is equipped with the initial condition

$$\boldsymbol{w}(x,0) = \boldsymbol{w}^0(x), \quad x \in \Omega, \tag{4}$$

and the following set of boundary conditions on appropriate parts of the boundary:

$$\begin{aligned}
&\text{a)}\ \rho|_{\Gamma_I \times (0,T)} = \rho_D, \quad \text{b)}\ \boldsymbol{v}|_{\Gamma_I \times (0,T)} = \boldsymbol{v}_D = (v_{D1}, v_{D2})^{\mathrm{T}}, \\
&\text{c)}\ \sum_{j=1}^{2}\left(\sum_{i=1}^{2} \tau_{ij} n_i\right) v_j + \frac{\gamma}{Re\, Pr}\frac{\partial \theta}{\partial n} = 0 \quad \text{on } \Gamma_I \times (0,T);
\end{aligned} \tag{5}$$

$$\text{a)}\ \boldsymbol{v}|_{\Gamma_W \times (0,T)} = 0, \quad \text{b)}\ \frac{\partial \theta}{\partial n}|_{\Gamma_W \times (0,T)} = 0; \tag{6}$$

$$\text{a)} \sum_{i=1}^{2} \tau_{ij} n_i = 0, \quad j = 1, \dots, 2, \quad \text{b)}\ \frac{\partial \theta}{\partial n} = 0 \quad \text{on } \Gamma_O \times (0,T); \tag{7}$$

The problem to solve the compressible Navier–Stokes equations, equipped with the above initial and boundary conditions will be denoted by (CFP) (compressible flow problem).

3 DGFE discretization

3.1 Triangulation

By Ω_h we denote a polygonal approximation of the domain Ω. Let $\mathcal{T}_h$ $(h > 0)$ denote a standard triangulation of the closure $\overline{\Omega}_h$ of the domain Ω_h into a finite number of closed triangles.

We set $h_K = \mathrm{diam}(K)$, $h = \max_{K \in \mathcal{T}_h} h_K$. All elements of $\mathcal{T}_h$ will be numbered so that $\mathcal{T}_h = \{K_i\}_{i \in I}$, where $I \subset Z^+ = \{0, 1, 2, \ldots\}$ is a suitable index set. If two elements K_i, $K_j \in \mathcal{T}_h$ have a common edge, we call them *neighbours* and put $\Gamma_{ij} = \partial K_i \cap \partial K_j$. For $i \in I$ we set $s(i) = \{j \in I; K_j$ is a neighbour of $K_i\}$. The boundary $\partial \Omega_h$ is formed by a finite number of edges of elements K_i adjacent to $\partial \Omega_h$. We denote all these boundary faces by S_j, where $j \in I_b \subset Z^- = \{-1, -2, \ldots\}$ and set $\gamma(i) = \{j \in I_b; S_j$ is an edge of $K_i\}, \Gamma_{ij} = S_j$ for $K_i \in \mathcal{T}_h$ such that $S_j \subset \partial K_i$, $j \in I_b$. For K_i not containing any boundary edge S_j we set $\gamma(i) = \emptyset$. Obviously, $s(i) \cap \gamma(i) = \emptyset$ for all $i \in I$. Now, if we write $S(i) = s(i) \cup \gamma(i)$, we have

$$\partial K_i = \bigcup_{j \in S(i)} \Gamma_{ij}, \qquad \partial K_i \cap \partial \Omega_h = \bigcup_{j \in \gamma(i)} \Gamma_{ij}. \tag{8}$$

Moreover, for $i \in I$, by $\gamma_D(i)$ we denote the subset of $\gamma(i)$ formed by such indexes j that the faces Γ_{ij} approximate the parts of $\partial \Omega$, where the Dirichlet boundary condition is prescribed at least for one component of $\boldsymbol{w}$. Then, for the Navier-Stokes equations, with respect of (5)–(7) we have

$$\bigcup_{i \in I} \bigcup_{j \in \gamma_D(i)} \Gamma_{ij} = \Gamma_I \cup \Gamma_W. \tag{9}$$

Moreover, we set

$$\gamma_N(i) = \gamma(i) \setminus \gamma_D(i), \tag{10}$$

where a Neumann type of boundary condition is prescribed for all components of $\boldsymbol{w}$.

Furthermore, we use the following notation: $\boldsymbol{n}_{ij} = ((n_{ij})_1, (n_{ij})_2) =$ unit outer normal to ∂K_i on the edge Γ_{ij} ($\boldsymbol{n}_{ij}$ is a constant vector on Γ_{ij}) and $|\Gamma_{ij}| =$ length of the edge Γ_{ij}. Over the triangulation $\mathcal{T}_h$ we define the *broken Sobolev space*

$$H^k(\Omega, \mathcal{T}_h) = \{v; v|_K \in H^k(K)\ \forall\, K \in \mathcal{T}_h\}. \tag{11}$$

For $v \in H^1(\Omega, \mathcal{T}_h)$ we set

$$\langle v \rangle_{\Gamma_{ij}} = \frac{1}{2}\left(v\big|_{\Gamma_{ij}} + v\big|_{\Gamma_{ji}}\right) \quad \text{and} \quad [v]_{\Gamma_{ij}} = v\big|_{\Gamma_{ij}} - v\big|_{\Gamma_{ji}}, \tag{12}$$

denoting the *average* and *jump of the traces* of v on $\Gamma_{ij} = \Gamma_{ji}$, respectively. Obviously, $\langle v \rangle_{\Gamma_{ij}} = \langle v \rangle_{\Gamma_{ji}}$, $[v]_{\Gamma_{ij}} = -[v]_{\Gamma_{ji}}$ and $[v]_{\Gamma_{ij}} \boldsymbol{n}_{ij} = [v]_{\Gamma_{ji}} \boldsymbol{n}_{ji}$.

3.2 Approximate solution

We derive the discretization of problems (CFP) with the aid of the DGFEM. The approximate solution $\boldsymbol{w}_h$ as well as test functions $\boldsymbol{\varphi}_h$ are elements of the finite dimensional space of vector-valued functions

$$\boldsymbol{S}_h = [S_h]^4, \tag{13}$$

where

$$S_h = S^{p,-1}(\Omega, \mathcal{T}_h) = \{v; v|_K \in P^p(K)\ \forall\, K \in \mathcal{T}_h\}, \tag{14}$$

$p \in Z^+$ and $P^p(K)$ denotes the space of all polynomials on K of degree $\leq p$.

Assuming that $\boldsymbol{w}$ is a classical sufficiently regular solution of problem (CFP) and $\boldsymbol{\varphi} \in [H^2(\Omega, \mathcal{T}_h)]^4$, we multiply equation (1) by $\boldsymbol{\varphi}$, integrate over $K_i \in \mathcal{T}_h$, apply Green's theorem, sum over all $K_i \in \mathcal{T}_h$ and with the aid of (8) we arrive at the identity

$$\begin{aligned}
&\int_\Omega \frac{\partial \boldsymbol{w}}{\partial t} \cdot \boldsymbol{\varphi}\, \mathrm{d}x + \sum_{i\in I} \sum_{j\in S(i)} \int_{\Gamma_{ij}} \sum_{s=1}^{2} \boldsymbol{f}_s(\boldsymbol{w})\, (n_{ij})_s \cdot \boldsymbol{\varphi}|_{\Gamma_{ij}} \mathrm{d}S \\
&- \sum_{i\in I} \int_{K_i} \sum_{s=1}^{2} \boldsymbol{f}_s(\boldsymbol{w}) \cdot \frac{\partial \boldsymbol{\varphi}}{\partial x_s}\, \mathrm{d}x + \sum_{i\in I} \int_{K_i} \sum_{s=1}^{2} \boldsymbol{R}_s(\boldsymbol{w}, \nabla \boldsymbol{w}) \cdot \frac{\partial \boldsymbol{\varphi}}{\partial x_s}\, \mathrm{d}x \\
&- \sum_{i\in I} \sum_{\substack{j\in s(i)\\ j<i}} \int_{\Gamma_{ij}} \sum_{s=1}^{2} \langle \boldsymbol{R}_s(\boldsymbol{w}, \nabla \boldsymbol{w}) \rangle\, (n_{ij})_s \cdot [\boldsymbol{\varphi}]\, \mathrm{d}S \\
&- \sum_{i\in I} \sum_{j\in\gamma(i)} \int_{\Gamma_{ij}} \sum_{s=1}^{2} \boldsymbol{R}_s(\boldsymbol{w}, \nabla \boldsymbol{w})\, (n_{ij})_s \cdot \boldsymbol{\varphi}\, \mathrm{d}S = 0.
\end{aligned} \tag{15}$$

In order to obtain a stable numerical method we add to (15) some stabilization terms which vanish for a smooth solution $\boldsymbol{w}$. In order to define a well-posed scheme we have to linearize the viscous terms $\boldsymbol{R}_s(\boldsymbol{w}, \nabla \boldsymbol{w})$. From (2) we obtain

$$\boldsymbol{R}_1(\boldsymbol{w}, \nabla \boldsymbol{w}) \tag{16}$$
$$= \begin{pmatrix}
0 \\
\frac{2}{3} \frac{1}{Re\, w_1} \left[2 \left(\frac{\partial w_2}{\partial x_1} - \frac{w_2}{w_1} \frac{\partial w_1}{\partial x_1} \right) - \left(\frac{\partial w_3}{\partial x_2} - \frac{w_3}{w_1} \frac{\partial w_1}{\partial x_2} \right) \right] \\
\frac{1}{Re\, w_1} \left[\left(\frac{\partial w_3}{\partial x_1} - \frac{w_3}{w_1} \frac{\partial w_1}{\partial x_1} \right) + \left(\frac{\partial w_2}{\partial x_2} - \frac{w_2}{w_1} \frac{\partial w_1}{\partial x_2} \right) \right] \\
\frac{w_2}{w_1} \boldsymbol{R}_1^{(2)} + \frac{w_3}{w_1} \boldsymbol{R}_1^{(3)} + \frac{\gamma}{Re\, Pr\, w_1} \left[\frac{\partial w_4}{\partial x_1} - \frac{w_4}{w_1} \frac{\partial w_1}{\partial x_1} - \frac{1}{w_1} \left(w_2 \frac{\partial w_2}{\partial x_1} + w_3 \frac{\partial w_3}{\partial x_1} \right) \right. \\
\qquad \left. + \frac{1}{w_1^2} (w_2^2 + w_3^2) \frac{\partial w_1}{\partial x_1} \right]
\end{pmatrix},$$

$$
\boldsymbol{R}_2(\boldsymbol{w}, \nabla \boldsymbol{w}) = \begin{pmatrix} 0 \\ \frac{1}{Re\, w_1}\left[\left(\frac{\partial w_3}{\partial x_1} - \frac{w_3}{w_1}\frac{\partial w_1}{\partial x_1}\right) + \left(\frac{\partial w_2}{\partial x_2} - \frac{w_2}{w_1}\frac{\partial w_1}{\partial x_2}\right)\right] \\ \frac{2}{3}\frac{1}{Re\, w_1}\left[2\left(\frac{\partial w_3}{\partial x_2} - \frac{w_3}{w_1}\frac{\partial w_1}{\partial x_2}\right) - \left(\frac{\partial w_2}{\partial x_1} - \frac{w_2}{w_1}\frac{\partial w_1}{\partial x_1}\right)\right] \\ \frac{w_2}{w_1}\boldsymbol{R}_2^{(2)} + \frac{w_3}{w_1}\boldsymbol{R}_2^{(3)} + \frac{\gamma}{Re\, Pr\, w_1}\left[\frac{\partial w_4}{\partial x_1} - \frac{w_4}{w_1}\frac{\partial w_1}{\partial x_2} - \frac{1}{w_1}\left(w_2\frac{\partial w_2}{\partial x_2} + w_3\frac{\partial w_3}{\partial x_2}\right) \right. \\ \left. \qquad + \frac{1}{w_1^2}\left(w_2^2 + w_3^2\right)\frac{\partial w_1}{\partial x_2}\right] \end{pmatrix},
$$

where $\boldsymbol{R}_s^{(r)} = \boldsymbol{R}_s^{(r)}(\boldsymbol{w}, \nabla \boldsymbol{w})$ denotes the r-th component of $\boldsymbol{R}_s$ $(s = 1, 2,\ r = 2, 3)$.

Now for $\boldsymbol{w} = (w_1, \ldots, w_4)^{\mathrm{T}}$ and $\boldsymbol{\varphi} = (\varphi_1, \ldots, \varphi_4)^{\mathrm{T}}$ we define the vector-valued functions

$$
\boldsymbol{D}_1(\boldsymbol{w}, \nabla \boldsymbol{w}, \boldsymbol{\varphi}, \nabla \boldsymbol{\varphi}) = \begin{pmatrix} 0 \\ \frac{2}{3}\frac{1}{Re\, w_1}\left[2\left(\frac{\partial \varphi_2}{\partial x_1} - \frac{\varphi_2}{w_1}\frac{\partial w_1}{\partial x_1}\right) - \left(\frac{\partial \varphi_3}{\partial x_2} - \frac{\varphi_3}{w_1}\frac{\partial w_1}{\partial x_2}\right)\right] \\ \frac{1}{Re\, w_1}\left[\left(\frac{\partial \varphi_3}{\partial x_1} - \frac{\varphi_3}{w_1}\frac{\partial w_1}{\partial x_1}\right) + \left(\frac{\partial \varphi_2}{\partial x_2} - \frac{\varphi_2}{w_1}\frac{\partial w_1}{\partial x_2}\right)\right] \\ \frac{w_2}{w_1}\boldsymbol{D}_1^{(2)} + \frac{w_3}{w_1}\boldsymbol{D}_1^{(3)} + \frac{\gamma}{Re\, Pr\, w_1}\left[\frac{\partial \varphi_4}{\partial x_1} - \frac{\varphi_4}{w_1}\frac{\partial w_1}{\partial x_1} - \frac{1}{w_1}\left(w_2\frac{\partial \varphi_2}{\partial x_1} + w_3\frac{\partial \varphi_3}{\partial x_1}\right) \right. \\ \left. \qquad + \frac{1}{w_1^2}\left(w_2\varphi_2 + w_3\varphi_3\right)\frac{\partial w_1}{\partial x_1}\right] \end{pmatrix}, \tag{17}
$$

$$
\boldsymbol{D}_2(\boldsymbol{w}, \nabla \boldsymbol{w}, \boldsymbol{\varphi}, \nabla \boldsymbol{\varphi}) = \begin{pmatrix} 0 \\ \frac{1}{Re\, w_1}\left[\left(\frac{\partial \varphi_3}{\partial x_1} - \frac{\varphi_3}{w_1}\frac{\partial w_1}{\partial x_1}\right) + \left(\frac{\partial \varphi_2}{\partial x_2} - \frac{\varphi_2}{w_1}\frac{\partial w_1}{\partial x_2}\right)\right] \\ \frac{2}{3}\frac{1}{Re\, w_1}\left[2\left(\frac{\partial \varphi_3}{\partial x_2} - \frac{\varphi_2}{w_1}\frac{\partial w_1}{\partial x_2}\right) - \left(\frac{\partial \varphi_2}{\partial x_1} - \frac{\varphi_2}{w_1}\frac{\partial w_1}{\partial x_1}\right)\right] \\ \frac{w_2}{w_1}\boldsymbol{D}_2^{(2)} + \frac{w_3}{w_1}\boldsymbol{D}_2^{(3)} + \frac{\gamma}{Re\, Pr\, w_1}\left[\frac{\partial \varphi_4}{\partial x_2} - \frac{\varphi_4}{w_1}\frac{\partial w_1}{\partial x_2} - \frac{1}{w_1}\left(w_2\frac{\partial \varphi_2}{\partial x_2} + w_2\frac{\partial \varphi_3}{\partial x_2}\right) \right. \\ \left. \qquad + \frac{1}{w_1^2}\left(w_2 + \varphi_2 + w_3\varphi_2\right)\frac{\partial w_1}{\partial x_2}\right] \end{pmatrix},
$$

where $\boldsymbol{D}_s^{(r)}$ denotes the r-th component of $\boldsymbol{D}_s$ $(s = 1, 2,\ r = 2, 3)$. Obviously, $\boldsymbol{D}_1$ and $\boldsymbol{D}_2$ are linear with respect to $\boldsymbol{\varphi}$ and $\nabla\boldsymbol{\varphi}$ and

$$
\boldsymbol{D}_s(\boldsymbol{w}, \nabla \boldsymbol{w}, \boldsymbol{w}, \nabla \boldsymbol{w}) = \boldsymbol{R}_s(\boldsymbol{w}, \nabla \boldsymbol{w}), \quad s = 1, 2. \tag{18}
$$

The definition of $\boldsymbol{D}_s$, $s = 1, 2$ can be given in other forms. We only require that they are linear with respect $\boldsymbol{\varphi}_h$ and satisfy (18). The natural way, how to perform the linearization of the diffusion terms, follows from (16) where the space derivatives of $\boldsymbol{w}$ are simply replaced by the derivatives of $\boldsymbol{\varphi}$. This linearization gives also the dependence on the gradient of the first component of $\boldsymbol{\varphi}$. However, numerical experiments carried out with the aid of this linearization do not yield satisfactory results. Therefore we use (17) which gives terms $\boldsymbol{D}_s$, $s = 1, 2$ independent of $\nabla\varphi_1$. For more detail see [7].

We add the following terms to the left-hand side of (15):

$$\sum_{i\in I}\sum_{\substack{j\in s(i)\\ j<i}}\int_{\Gamma_{ij}}\sum_{s=1}^{2}\langle \boldsymbol{D}_s(\boldsymbol{w},\nabla\boldsymbol{w},\boldsymbol{\varphi},\nabla\boldsymbol{\varphi})\rangle\,(n_{ij})_s\cdot[\boldsymbol{w}]\,\mathrm{d}S \tag{19}$$

$$+\sum_{i\in I}\sum_{j\in\gamma(i)}\int_{\Gamma_{ij}}\sum_{s=1}^{2}\boldsymbol{D}_s(\boldsymbol{w},\nabla\boldsymbol{w},\boldsymbol{\varphi},\nabla\boldsymbol{\varphi})\,(n_{ij})_s\cdot\boldsymbol{w}\,\mathrm{d}S.$$

In the second term we use the zero natural Neumann boundary conditions (7), a)–b) and the Dirichlet conditions are taken into account with the aid of additional terms on the right-hand side of (15).

Moreover, to the left-hand side of (15) we add the vanishing *interior penalty* terms

$$\sum_{i\in I}\sum_{\substack{j\in s(i)\\ j<i}}\int_{\Gamma_{ij}}\sigma[\boldsymbol{w}]\cdot[\boldsymbol{\varphi}]\,\mathrm{d}S \tag{20}$$

with $\sigma|_{\Gamma_{ij}} = (Re\,|\Gamma_{ij}|)^{-1}$ and boundary penalty terms balanced by additional right-hand side terms containing the Dirichlet boundary data.

We arrive at the definition of the following form

$$\begin{aligned}
\boldsymbol{A}_h(\boldsymbol{w},\boldsymbol{\varphi}) = &\sum_{i\in I}\int_{K_i}\sum_{s=1}^{2}\boldsymbol{R}_s(\boldsymbol{w},\nabla\boldsymbol{w})\cdot\frac{\partial\boldsymbol{\varphi}}{\partial x_s}\,\mathrm{d}x\\
&-\sum_{i\in I}\sum_{\substack{j\in s(i)\\ j<i}}\int_{\Gamma_{ij}}\sum_{s=1}^{2}\langle\boldsymbol{R}_s(\boldsymbol{w},\nabla\boldsymbol{w})\rangle\,(n_{ij})_s\cdot[\boldsymbol{\varphi}]\,\mathrm{d}S\\
&+\sum_{i\in I}\sum_{\substack{j\in s(i)\\ j<i}}\int_{\Gamma_{ij}}\sum_{s=1}^{2}\langle\boldsymbol{D}_s(\boldsymbol{w},\nabla\boldsymbol{w},\boldsymbol{\varphi},\nabla\boldsymbol{\varphi})\rangle\,(n_{ij})_s\cdot[\boldsymbol{w}]\,\mathrm{d}S\\
&-\sum_{i\in I}\sum_{j\in\gamma_D(i)}\int_{\Gamma_{ij}}\sum_{s=1}^{2}\boldsymbol{R}_s(\boldsymbol{w},\nabla\boldsymbol{w})\,(n_{ij})_s\cdot\boldsymbol{\varphi}\,\mathrm{d}S\\
&+\sum_{i\in I}\sum_{j\in\gamma_D(i)}\int_{\Gamma_{ij}}\sum_{s=1}^{2}\boldsymbol{D}_s(\boldsymbol{w},\nabla\boldsymbol{w},\boldsymbol{\varphi},\nabla\boldsymbol{\varphi})\,(n_{ij})_s\,(\boldsymbol{w}-\boldsymbol{w}_B)\,\mathrm{d}S\\
&+\sum_{i\in I}\sum_{\substack{j\in s(i)\\ j<i}}\int_{\Gamma_{ij}}\sigma[\boldsymbol{w}]\cdot[\boldsymbol{\varphi}]\mathrm{d}S+\sum_{i\in I}\sum_{j\in\gamma_D(i)}\int_{\Gamma_{ij}}\sigma\,(\boldsymbol{w}-\boldsymbol{w}_B)\cdot\boldsymbol{\varphi}\,\mathrm{d}S.
\end{aligned}$$

The boundary state $\boldsymbol{w}_B$ will be defined later. The convective terms are represented by the form

$$\boldsymbol{B}_h(\boldsymbol{w}_h, \boldsymbol{\varphi}_h) = -\sum_{i\in I} \int_{K_i} \sum_{s=1}^{N} \boldsymbol{f}_s(\boldsymbol{w}_h) \cdot \frac{\partial \boldsymbol{\varphi}_h}{\partial x_s} \, \mathrm{d}x \tag{21}$$

$$+\sum_{i\in I} \sum_{j\in S(i)} \int_{\Gamma_{ij}} \boldsymbol{H}(\boldsymbol{w}_h|_{\Gamma_{ij}}, \boldsymbol{w}_h|_{\Gamma_{ji}}, \boldsymbol{n}_{ij}) \cdot \boldsymbol{\varphi}_h \mathrm{d}S, \quad \boldsymbol{w}_h, \boldsymbol{\varphi}_h \in H^1(\Omega, \mathcal{T}_h)^4,$$

where $\boldsymbol{H}$ is a suitable numerical flux commonly used in the finite volume method. We use the numerical flux based on the direct solution of the local Riemann problem, see [10]. If $\Gamma_{ij} \subset \partial\Omega_h$, then there is no neighbour K_j of K_i adjacent to Γ_{ij} and the values of $\boldsymbol{w}_h|_{\Gamma_{ij}}$ must be determined on the basis of "inviscid" boundary conditions, see [9].

The boundary state $\boldsymbol{w}_B = (w_{B1}, \ldots, w_{B4})^{\mathrm{T}}$ is determined in the following way: We set

$$w_{Br}|_{\Gamma_{ij}} = w_r^*|_{\Gamma_{ij}}, \tag{22}$$

if the r-th component w_r of $\boldsymbol{w}$ is prescribed on Γ_{ij}. Here w_r^* is the r-th component of $\boldsymbol{w}^* : Q_T \to I\!R^4$, which is a function satisfying the boundary conditions (5) – (7). Otherwise, we set

$$w_{Br}|_{\Gamma_{ij}} = w_r|_{\Gamma_{ij}}, \tag{23}$$

which means that we use the "extrapolation" of w_r onto Γ_{ij} from $K_i \in \mathcal{T}_h$. In particular, we have

$$\boldsymbol{w}_B = (\rho_{ij}, 0, 0, \rho_{ij}\theta_{ij}) \quad \text{on } \Gamma_W, \tag{24}$$

$$\boldsymbol{w}_B = \left(\rho_D, \rho_D v_{D1}, \rho_D v_{D2}, \rho_{ij}\theta_{ij} + \frac{1}{2}\rho_D|\boldsymbol{v}_D|^2\right) \quad \text{on } \Gamma_I,$$

where ρ_D and $\boldsymbol{v}_D = (v_{D1}, v_{D2})$ are the given density and velocity from the boundary conditions (5) – (7) and ρ_{ij}, θ_{ij} are the values of the density and absolute temperature extrapolated from K_i onto Γ_{ij}.

Now the *discrete DGFE Navier-Stokes problems* read:

Definition 1. *An approximate DGFE solution of the compressible Navier–Stokes problem (CFP) is defined as a vector-valued function $\boldsymbol{w}_h$ such that*

$$a)\ \boldsymbol{w}_h \in C^1([0,T]; \boldsymbol{S}_h) \tag{25}$$

$$b)\ \frac{\mathrm{d}}{\mathrm{d}t}(\boldsymbol{w}_h(t), \boldsymbol{\varphi}_h) + \boldsymbol{B}_h(\boldsymbol{w}_h(t), \boldsymbol{\varphi}_h) + \boldsymbol{A}_h(\boldsymbol{w}_h(t), \boldsymbol{\varphi}_h) = 0 \qquad \forall \boldsymbol{\varphi}_h \in \boldsymbol{S}_h,\ t \in (0,T),$$

$$c)\ \boldsymbol{w}_h(0) = \boldsymbol{w}_h^0,$$

where $\boldsymbol{w}_h^0$ is an $\boldsymbol{S}_h$-approximation of $\boldsymbol{w}^0$.

The problem (25) exhibits a system of ordinary differential equations which can be solved with the aid of a suitable ODE solver.

4 Numerical examples

We implemented the numerical scheme (25), a)–c) with the aid of a piecewise linear approximation on regular triangular grids. The system of ODE was solved by the explicit Euler method.

Now we consider four cases of viscous flow around the profile NACA0012 with the following data, see [3]:

case	M_{in}	α	Re	case	M_{in}	α	Re
C1	0.80	10°	500	C3	0.85	0°	500
C2	2.00	10°	106	C4	0.85	0°	2000

where M_{in} is the far field Mach number, α the angle of attack and Re the Reynolds number. We compare our results with the numerical results presented in [3], where ten methods were applied. The following table contains our computed lift c_L and drag c_R coefficients in comparison with [3] ($\#\mathcal{T}_h$ denotes the number of elements of the mesh $\mathcal{T}_h$)

	computed values			reference values from [3]	
case	$\#\mathcal{T}_h$	c_L	c_D	c_L (range); mean value	c_D (range); mean value
C1	4563	0.4985	0.1938	(0.4199 – 0.5170); 0.4526	(0.1597 – 0.2868); 0.2559
C2	5640	0.3969	0.4172	(0.3063 – 0.4059); 0.3443	(0.4120 – 0.4910); 0.4660
C3	4946	0.0003	0.2304	(0.0000 – 0.0007); 0.0001	(0.1790 – 0.2420); 0.2192
C4	4946	0.0001	0.1179	(0.0000 – 0.0002); 0.0001	(0.1012 – 0.1360); 0.1171

The computed values of drag and lift correspond to the reference values from [3]. Figures 1 shows the employed triangulation and the computed isolines of the Mach number for the case C2.

References

1. Bassi, F., Rebay, S.(1997): A high-order accurate discontinuous finite element method for the numerical solution of the compressible Navier–Stokes equations. J. Comput. Phys, **131**, 267–279
2. Bassi, F., Rebay, S.(2000): A high order discontinuous Galerkin method for compressible turbulent flow. In Cockburn, B., Karniadakis, G. E., Shu, C.-W. (eds) Discontinuous Galerkin Method: Theory, Computations and Applications, Lecture Notes in Computational Science and Engineering 11, pages 113–123. Springer-Verlag
3. Bristeau, M. O., Glowinski, R., Periaux, J., Viviand, H. eds (1987): *Numerical Simulation of Compressible Navier-Stokes Flows*, volume 18 of *Notes on Numerical Fluid Mechanics.* Vieweg, Braunschwig
4. Cockburn B. (1999): Discontinuous Galerkin methods for convection dominated problems. In Barth, T.J., Deconinck, H., (eds), High–Order Methods for Computational Physics, Lecture Notes in Computational Science and Engineering 9, pages 69–224. Springer, Berlin

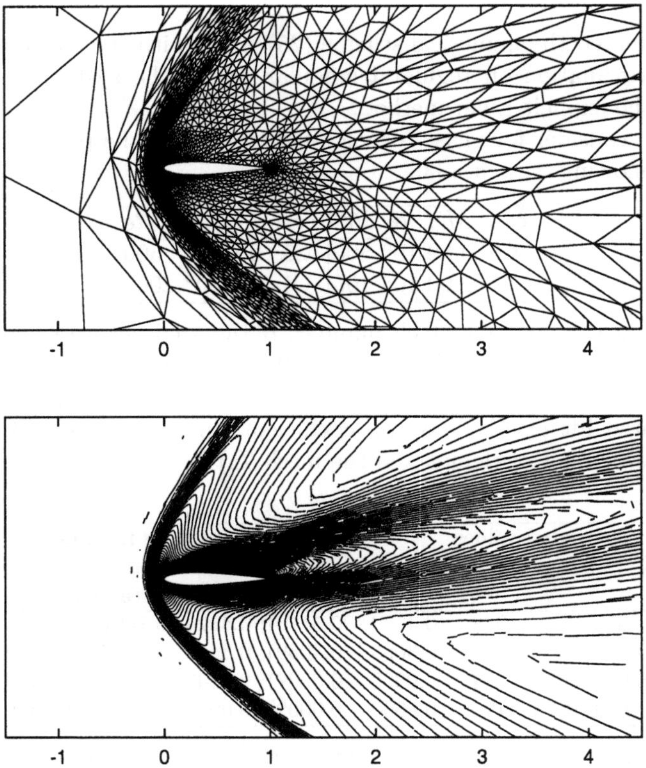

Fig. 1. Viscous flow along NACA 0012, case C2, triangulation (top), isolines of the Mach number (bottom)

5. Cockburn, B., Karniadakis, G. E., Shu, C.-W. eds (2000) Discontinuous Galerkin methods. Lecture Notes in Computational Science and Engineering 11., Springer, Berlin
6. Dolejší, V. (2002): A higher order scheme based on the finite volume approach. In R. Herbin and D. Kröner, (eds), Finite Volumes for Complex Applications III (Problems and Perspectives), pages 333–340. Hermes
7. Dolejší, V.: On the discontinuous Galerkin method for the numerical solution of the Navier–Stokes equations. Int. J. Numer. Methods Fluids, (submitted)
8. Dolejší, V., Feistauer, M., Sobotíková, V.: A discontinuous Galerkin method for nonlinear convection–diffusion problems. Comput. Methods Appl. Mech. Eng. (submitted)
9. Feistauer, M., Felcman, J., Straškraba, I. (2003): Mathematical and Computational Methods for Compressible Flow. Oxford University Press, Oxford
10. Toro, E.F. (1997): Riemann Solvers and Numerical Methods for Fluid Dynamics. Springer-Verlag

A Finite Volume Scheme on General Meshes for the Steady Navier-Stokes Equations in Two Space Dimensions

Robert Eymard[1] and Raphaĺe Herbin[2]

[1] Université de Marne-la-Vallée, Paris, France *eymard@math.univ-mlv.fr*
[2] Université de Provence, Marseille, France *herbin@cmi.univ-mrs.fr*

Summary. We introduce a new finite volume scheme for the discretization of the incompressible Navier-Stokes equations on general meshes, for which we prove convergence without any condition on the regularity of the solution. Numerical results are presented.

1 The incompressible Navier-Stokes equations

Numerical schemes for the Navier-Stokes equations (1) have been extensively studied: see [7, 11, 12, 13, 8, 6, 15] and references therein. An advantage of the finite volume schemes is that the unknowns are approximated by piecewise constant functions: this makes it easy to take into account additional nonlinear phenomena or the coupling with algebraic or differential equations, for instance in the case of reactive flows. In [11] is presented the classical finite volume scheme on rectangular meshes, which is the basis of many industrial applications. A convergence proof of the so-called MAC scheme is given in [10] in the case of a uniform rectangular grid. However, the use of rectangular grids limits the type of domain which can be gridded, and more recently, finite volume schemes for the Navier-Stokes equations on triangular grids have been presented: see for example [9] where the vorticity formulation is used, and [2] where primal variables are used with a Chorin type projection method (but no proof of convergence is known). Here, we propose a new method using the primitive variables and enforcing the divergence condition directly, using quite general meshes such as mixed rectangular-triangular or Voronoï meshes, and for which we are able to prove convergence under general conditions (in particular, no regularity of the exact solution is required). An error estimate in the case of the linear Stokes equations was presented in [4].

We seek an approximation of $u = (u^{(1)}, u^{(2)})^t \in H_0^1(\Omega) \times H_0^1(\Omega)$ and $p \in L^2(\Omega)$, weak solution to the incompressible generalized Navier-Stokes equations:

$$\begin{aligned} &\eta u^{(i)} - \nu \Delta u^{(i)} + \partial_i p + u^{(1)} \partial_1 u^{(i)} + u^{(2)} \partial_2 u^{(i)} = f^{(i)} \text{ in } \Omega, \quad \text{for } i = 1, 2, \\ &\partial_1 u^{(1)} + \partial_2 u^{(2)} = 0 \text{ in } \Omega. \end{aligned} \tag{1}$$

where $\eta \geq 0$, $u^{(1)}$ and $u^{(2)}$ are the two components of the velocity, p denotes the pressure, ν the viscosity of the fluid, under the following assumptions:

$$\Omega \text{ is a polygonal open bounded connected subset of } \mathbb{R}^2, \tag{2}$$

$$\nu \in (0, +\infty), \ \eta \in [0, +\infty), \tag{3}$$

$$f^{(i)} \in L^2(\Omega), \quad \text{for } i = 1, 2. \tag{4}$$

The terms $\eta u^{(i)}$ appear when considering an implicit time discretization of the unsteady Stokes or Navier-Stokes equations (with η as the inverse of the time step, $\eta = 0$ yields in the steady-state.

We prescribe for both problems a homogeneous Dirichlet boundary condition on the velocity $(u^{(1)}, u^{(2)})$. Let us denote by $x = (x^{(1)}, x^{(2)})$ any point of Ω and by dx the 2-dimensional Lebesgue measure $dx = dx^{(1)}dx^{(2)}$.

Definition 1 (Weak solution). *Under hypotheses* (2)-(4), $u = (u^{(1)}, u^{(2)})^t$ *is called a weak solution of* (1) *if and only if*

$$\begin{cases} u = (u^{(1)}, u^{(2)})^t \in E(\Omega), \\ \eta \displaystyle\sum_{i=1,2} \int_\Omega u^{(i)}(x) v^{(i)}(x) dx + \nu \sum_{i=1,2} \int_\Omega \nabla u^{(i)} \cdot \nabla v^{(i)}(x) dx + b(u, u, v) = \\ \displaystyle\sum_{i=1,2} \int_\Omega f^{(i)}(x) v^{(i)}(x) dx, \quad \forall v = (v^{(1)}, v^{(2)})^t \in E(\Omega), \end{cases} \tag{5}$$

where the trilinear form b is defined for all $u, v, w \in (H_0^1(\Omega))^2$ by

$$b(u, v, w) = \sum_{k=1,2} \sum_{i=1,2} \int_\Omega u^{(i)}(x) \partial_i v^{(k)}(x) w^{(k)}(x) dx, \tag{6}$$

which classically satisfies, for all $u \in E(\Omega)$,

$$b(u, v, w) = \sum_{k=1,2} \sum_{i=1,2} \int_\Omega \partial_i (u^{(i)} v^{(k)})(x) w^{(k)}(x) dx.$$

2 The finite volume scheme

Definition 2. [Admissible discretization] *Let Ω be an open bounded polygonal subset of $\mathbb{R}^2$, and $\partial\Omega = \overline{\Omega} \setminus \Omega$ its boundary. An admissible finite volume discretization of Ω, denoted by $\mathcal{D}$, is given by $\mathcal{D} = (\mathcal{M}, \mathcal{E}, \mathcal{P}, \mathcal{V})$, where:*

- *$\mathcal{M}$ is a finite family of non empty open polygonal convex disjoint subsets of Ω (the "control volumes") such that $\overline{\Omega} = \cup_{K \in \mathcal{M}} \overline{K}$. For any $K \in \mathcal{M}$, let $\partial K = \overline{K} \setminus K$ be the boundary of K and $\mathrm{m}(K) > 0$ denote the area of K.*

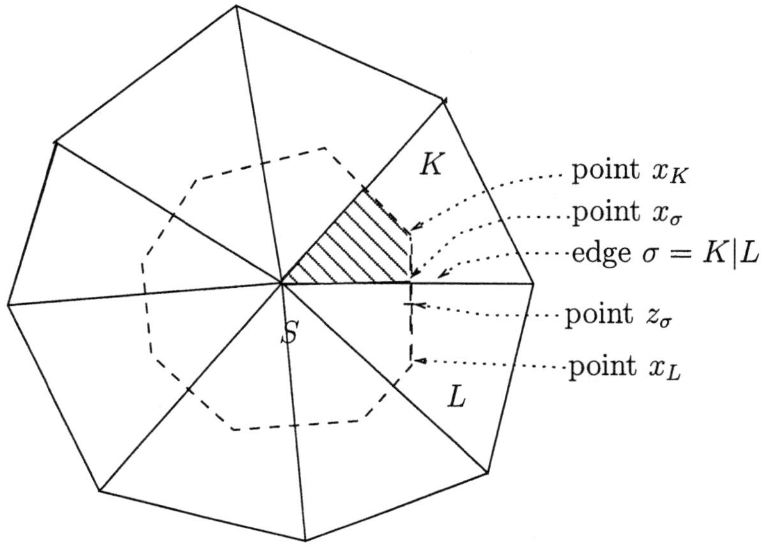

Fig. 1. Example of an admissible triangular discretization

- $\mathcal{E}$ *is a finite family of disjoint subsets of* $\overline{\Omega}$ *(the "edges" of the mesh), such that, for all* $\sigma \in \mathcal{E}$, *there exists a hyperplane* E *of* $\mathbb{R}^2$ *and* $K \in \mathcal{M}$ *with* $\overline{\sigma} = \partial K \cap E$ *and* σ *is a non empty open subset of* E. *We then denote by* $m_\sigma > 0$ *the 1-dimensional measure of* σ. *We assume that, for all* $K \in \mathcal{M}$, *there exists a subset* $\mathcal{E}_K$ *of* $\mathcal{E}$ *such that* $\partial K = \cup_{\sigma \in \mathcal{E}_K} \overline{\sigma}$. *It then results from the previous hypotheses that, for all* $\sigma \in \mathcal{E}$, *either* $\sigma \subset \partial\Omega$ *or there exists* $(K, L) \in \mathcal{M}^2$ *with* $K \neq L$ *such that* $\overline{K} \cap \overline{L} = \overline{\sigma}$; *we denote in the latter case* $\sigma = K|L$.
- $\mathcal{P}$ *is a family of points of* Ω *indexed by* $\mathcal{M}$, *denoted by* $\mathcal{P} = (x_K)_{K \in \mathcal{M}}$. *The coordinates of* x_K *are denoted by* $x_K^{(i)}$, $i = 1, 2$. *The family* $\mathcal{P}$ *is such that, for all* $K \in \mathcal{M}$, $x_K \in K$. *Furthermore, for all* $\sigma \in \mathcal{E}$ *such that there exists* $(K, L) \in \mathcal{M}^2$ *with* $\sigma = K|L$, *it is assumed that the straight line* (x_K, x_L) *going through* x_K *and* x_L *is orthogonal to* $K|L$. *For all* $K \in \mathcal{M}$ *and all* $\sigma \in \mathcal{E}_K$, *let* z_σ *be the orthogonal projection of* x_K *on* σ. *We suppose that* $z_\sigma \in \sigma$.
- $\mathcal{V}$ *is a finite family of non empty open polygonal disjoint subsets of* Ω *(constituting the "dual mesh" of* $\mathcal{M}$*), which are centered around the vertices* $(x_s)_{s=1,N_V}$ *in the following way (*N_V *is the number of vertices):*
  for $1 \leq s \leq N_V$, *let* $\mathcal{M}_s \subset \mathcal{M}$ *be the set of control volumes to which* x_s *is a vertex. For* $K \in \mathcal{M}_s$, *denote by* $\sigma_{K,s,1}$ *and* $\sigma_{K,s,2} \in \mathcal{E}_K$ *the two edges of* K *with vertex* x_s. *Define* K_s *as the convex hull of the four points*

$$\left(x_s, x_K, z_{\sigma_{K,s,1}}, z_{\sigma_{K,s,2}}\right).$$

The dual cell around x_s, denoted by S, is then defined as (also see Figure 1):

$$\overline{S} = \cup_{K \in \mathcal{M}_s} K_s.$$

Since there is a one-to-one mapping between the set $\{1, \ldots, N_V\} \subset \mathbb{N}$ and the set $\mathcal{V}$, we shall replace all subscripts s by S when dealing with the dual mesh. Let $\mathcal{V}_K$ denote the set of vertices of a given control volume K. Note that:

$$\overline{K} = \cup_{x_s \in \mathcal{V}_K} K_s, \quad \textit{and } K_s = \overline{K \cap S}.$$

The size of the discretization is defined by:

$$\text{size}(\mathcal{D}) = \sup\{\text{diam}(K), K \in \mathcal{M}\}.$$

The regularity of the mesh is defined by

$$\text{angle}(\mathcal{D}) = \inf\left\{|\widehat{z_\sigma x_K x_S}|,\ |\widehat{z_\sigma x_S x_K}|, K \in \mathcal{M},\ S \in \mathcal{V}_K,\ \sigma \in \mathcal{E}_K \cap \mathcal{E}_S\right\}, \quad (7)$$

where $|\widehat{xyz}|$ designates the absolute value of the measure of the angle $\widehat{xyz}$ (note that $\widehat{z_\sigma x_K x_S} = \frac{\pi}{2} - \widehat{z_\sigma x_S x_K}$.).

For all $K \in \mathcal{M}$ and $\sigma \in \mathcal{E}_K$, we denote by $\mathbf{n}_{K,\sigma}$ the unit vector normal to σ outward to K. We denote by $d_{K,\sigma}$ the Euclidean distance between x_K and σ. We then define

$$\tau_{K,\sigma} = \frac{m_\sigma}{d_{K,\sigma}}.$$

The set of interior (resp. boundary) edges is denoted by $\mathcal{E}_{\text{int}}$ (resp. $\mathcal{E}_{\text{ext}}$), that is $\mathcal{E}_{\text{int}} = \{\sigma \in \mathcal{E};\ \sigma \not\subset \partial\Omega\}$ (resp. $\mathcal{E}_{\text{ext}} = \{\sigma \in \mathcal{E};\ \sigma \subset \partial\Omega\}$). For any $\sigma \in \mathcal{E}_{\text{int}}, \sigma = K|L$ (resp. $\mathcal{E}_{\text{ext}},\ \sigma \in \mathcal{E}_K$), let x_σ be the center point of the line segment $[x_K x_L]$ (resp. $[x_K z_\sigma]$), and $x_\sigma^{(1)}$ and $x_\sigma^{(2)}$ its coordinates.

For all $K \in \mathcal{M}$ and all $S \in \mathcal{V}_K$, let σ_1 and $\sigma_2 \in \mathcal{E}_K \cap \mathcal{E}_S$ numbered such that $(x_{\sigma_1}^{(2)} - x_S^{(2)})(x_{\sigma_2}^{(1)} - x_S^{(1)}) - (x_{\sigma_2}^{(2)} - x_S^{(2)})(x_{\sigma_1}^{(1)} - x_S^{(1)}) > 0$.

Let $A_{K,S}^{(1)} = x_{\sigma_1}^{(2)} - x_{\sigma_2}^{(2)}$ and $A_{K,S}^{(2)} = x_{\sigma_2}^{(1)} - x_{\sigma_1}^{(1)}$.

Definition 3. *Let Ω be an open bounded polygonal subset of $\mathbb{R}^N$, with $N \in \mathbb{N}_*$. Let $\mathcal{D} = (\mathcal{M}, \mathcal{E}, \mathcal{P}, \mathcal{V})$ be an admissible finite volume discretization of Ω in the sense of Definition 2. We denote by $H_\mathcal{D}(\Omega) \subset L^2(\Omega)$ the space of functions which are piecewise constant on each control volume $K \in \mathcal{M}$. For all $w \in H_\mathcal{D}(\Omega)$ and for all $K \in \mathcal{M}$, we denote by w_K the constant value of w in K and we define $(w_\sigma)_{\sigma \in \mathcal{E}}$ by:*

$$w_\sigma = 0,\ \forall \sigma \in \mathcal{E}_{\text{ext}} \quad (8)$$

and

$$\tau_{K,\sigma}(w_\sigma - w_K) + \tau_{L,\sigma}(w_\sigma - w_L) = 0,\ \forall \sigma \in \mathcal{E}_{\text{int}}, \sigma = K|L. \quad (9)$$

Let $L_\mathcal{D}(\Omega)$ be the space of functions which are piecewise constant on the domains S, for all $S \in \mathcal{V}$. Let $\mathrm{div}_\mathcal{D} : (H_\mathcal{D}(\Omega))^2 \to L_\mathcal{D}(\Omega)$ be defined by:

$$\mathrm{div}_\mathcal{D}(u)(x) = \frac{1}{\mathrm{m}(S)} \sum_{K \in \mathcal{M}_S} \sum_{i=1,2} A_{K,S}^{(i)} u_K^{(i)}, \quad \textit{for a.e. } x \in S, \forall S \in \mathcal{V}.$$

We then set $E_\mathcal{D}(\Omega) = \{u \in (H_\mathcal{D}(\Omega))^2, \mathrm{div}_\mathcal{D}(u) = 0\}$. For $(v,w) \in (H_\mathcal{D}(\Omega))^2$, we denote by

$$[v,w]_\mathcal{D} = \sum_{K \in \mathcal{M}} \sum_{\sigma \in \mathcal{E}_K} \tau_{K,\sigma}(v_\sigma - v_K)(w_\sigma - w_K), \tag{10}$$

Remark that thanks to (9), *one has:*

$$[v,w]_\mathcal{D} = \sum_{\sigma \in \mathcal{E}_{\mathrm{int}}, \sigma = K|L} \tau_\sigma (v_K - v_L)(w_K - w_L) + \sum_{\sigma \in \mathcal{E}_{\mathrm{ext}}} \tau_\sigma v_{K_\sigma} w_{K_\sigma},$$

where K_σ denotes the control volume to which σ is an edge. We define a norm in $H_\mathcal{D}(\Omega)$ (thanks to the discrete Poincaré inequality (11) *given below) by*

$$|w|_\mathcal{D} = ([w,w]_\mathcal{D})^{1/2}.$$

Similarly, for $u = (u^{(1)}, u^{(2)})^t \in (H_\mathcal{D}(\Omega))^2$, $v = (v^{(1)}, v^{(2)})^t \in (H_\mathcal{D}(\Omega))^2$ and $w = (w^{(1)}, w^{(2)})^t \in (H_\mathcal{D}(\Omega))^2$, we define:

$$|u|_\mathcal{D} = \left(\sum_{i=1,2} [u^{(i)}, u^{(i)}]_\mathcal{D} \right)^{1/2}.$$

and

$$[v,w]_\mathcal{D} = \sum_{i=1,2} [v^{(i)}, w^{(i)}]_\mathcal{D}.$$

The discrete Poincaré inequality (see [3]) writes:

$$\|w\|_{L^2(\Omega)} \le \mathrm{diam}(\Omega)|w|_\mathcal{D}, \ \forall w \in H_\mathcal{D}(\Omega). \tag{11}$$

We only present here a centered finite volume scheme, and refer to [5] for the upstream version. Under hypotheses (2)-(4), let $\mathcal{D}$ be an admissible discretization of Ω. Let $\lambda \in (0, +\infty)$. The finite volume scheme for the approximation of the solution (1) writes: find u such that

$$\begin{aligned} & u \in E_\mathcal{D}(\Omega), \\ & \eta \int_\Omega u(x) \cdot v(x) dx + \nu [u,v]_\mathcal{D} + b_\mathcal{D}(u,u,v) = \int_\Omega f(x) \cdot v(x) dx, \ \forall v \in E_\mathcal{D}(\Omega), \end{aligned} \tag{12}$$

where, for u, v and $w \in H_\mathcal{D}(\Omega)$,

$$\begin{aligned}
b_{\mathcal{D}}(u,v,w) &= \sum_{K\in\mathcal{M}} \sum_{k=1,2} w_K^{(k)} \sum_{S\in\mathcal{V}_K} v_S^{(k)} \sum_{i=1,2} A_{K,S}^{(i)}\ u_K^{(i)} \\
v_S^{(k)} &= \frac{1}{\mathrm{m}(S)} \sum_{K\in\mathcal{M}_S} \mathrm{m}(K\cap S)\, v_K^{(k)},\ \forall S\in\mathcal{V},\ k=1,2.
\end{aligned} \tag{13}$$

The trilinear form $b_{\mathcal{D}}(u,v,w)$ satisfies some continuity properties in $(H_{\mathcal{D}}(\Omega))^3$ (see [5] for the proof).

Lemma 1. [Continuity of the trilinear form in discrete H^1 space] *Under Hypothesis (2), let $\mathcal{D}$ be an admissible discretization in the sense of Definition* 2, *let $\alpha > 0$ be such that* angle$(\mathcal{D}) \geq \alpha$, *let $H_{\mathcal{D}}(\Omega)$ be the space of piecewise constant functions defined in 3 and let $b_{\mathcal{D}}$ be the trilinear form defined by* (13).

Then there exists $C_1 > 0$, only depending on α, such that:

$$|b_{\mathcal{D}}(u,v,w)| \leq C_1\, |u|_{\mathcal{D}} |v|_{\mathcal{D}} |w|_{\mathcal{D}}. \tag{14}$$

As in the case of the linear problem (see [4]), we use the following penalized approximation of (12):

$$\begin{aligned}
&(u,p)\in (H_{\mathcal{D}}(\Omega))^2 \times L_{\mathcal{D}}(\Omega),\\
&\nu\left([u,v]_{\mathcal{D}}\right) - \int_\Omega p(x)\mathrm{div}_{\mathcal{D}}(v)(x)dx + b_{\mathcal{D}}(u,u,v) = \int_\Omega f(x)\cdot v(x)dx,\\
&\qquad\qquad\qquad\qquad\qquad\qquad\qquad\qquad \forall v\in (H_{\mathcal{D}}(\Omega))^2,\\
&\mathrm{div}_{\mathcal{D}}(u) = -\lambda\ \mathrm{size}(\mathcal{D})\ p,
\end{aligned} \tag{15}$$

3 Convergence of the scheme

The following proposition gives a sufficient condition for the existence and uniqueness of a solution to the scheme (with or without penalization), under the classical assumption that the data are small, or the viscosity is large enough (see [14] Theorem 1.3 page 167 for the continuous case). Note that in the continuous case, the "small data" assumption is only required to prove uniqueness, not existence. Here, however, this assumption is also required for the existence of a discrete solution. Moreover, uniqueness is only proven for "small enough" solutions.

Proposition 1 [Existence and uniqueness of small discrete solutions in the small data case, with or without a penalization] *Under hypotheses* (2)-(4), *let $\mathcal{D}$ be an admissible discretization of Ω in the sense of Definition 2 and let $\alpha > 0$ with* angle$(\mathcal{D}) \geq \alpha$. *Let C_1 be the real value which only depends on α, given by (14) of Lemma* 1. *Assume that the condition*

$$\frac{1}{\nu^2}\left(\sum_{i=1,2} \|f^{(i)}\|_{L^2(\Omega)}\right) < C_2 := \frac{1}{4\mathrm{diam}(\Omega)C_1} \tag{16}$$

is fulfilled. Then there exists one and only one function $u \in (H_{\mathcal{D}}(\Omega))^2$ such that

$$|u|_{\mathcal{D}} \le C_3 := \frac{1}{2C_1}\left[\nu - \left(\nu^2 - 4\left(\sum_{i=1,2}\|f^{(i)}\|_{L^2(\Omega)}\right)\operatorname{diam}(\Omega)C_1\right)^{1/2}\right], \tag{17}$$

and u is solution to (12) *and* (13) *(no penalization), or u is such that there exists a function p with (u,p) solution to* (15) *and* (13) *for a given $\lambda \in (0,+\infty)$. Furthermore, in the latter case, the following inequality holds:*

$$\lambda \operatorname{size}(\mathcal{D})\ \|p\|^2_{L^2(\Omega)} \le \operatorname{diam}(\Omega)\left(\sum_{i=1,2}\|f^{(i)}\|_{L^2(\Omega)}\right) + C_1 C_3{}^2. \tag{18}$$

and the function p is unique too.

Proposition 2 [Convergence of the centered penalized scheme in the nonlinear case] *Under Hypotheses* (2)-(4)*, let $\alpha > 0$ be given and let $C_2 > 0$ be given by Proposition 1. We assume that the property (16) holds. Let $\lambda \in (0,+\infty)$ be given and let $(\mathcal{D}^{(n)})_{n\in\mathbb{N}}$ be a sequence of admissible discretization of Ω in the sense of Definition 2, such that $\lim_{n\to\infty} \operatorname{size}(\mathcal{D}^{(n)}) = 0$ and $\operatorname{angle}(\mathcal{D}^{(n)}) \ge \alpha$, for all $n \in \mathbb{N}$. Let $(u^{(n)}, p^{(n)}) \in (H_{\mathcal{D}^{(n)}}(\Omega))^2 \times L_{\mathcal{D}^{(n)}}(\Omega)$ be a solution to* (15), (13), (17)*. Then there exists a subsequence of the sequence $(u^{(n)})_{n\in\mathbb{N}}$ which converges in $L^2(\Omega)^2$ to u, weak solution of the Navier-Stokes problem in the sense of* (5)*. If C_2 is taken small enough, the uniqueness property of the solution entails the convergence of the whole sequence.*

4 Numerical results

Experiments with an analytical solution were performed. For the centered scheme, the results indicate a rate of convergence of h^2 for the velocities, and better than $h^{0.5}$ for the pressures in the case of unstructured triangular meshes. In the case of rectangular meshes or structured triangular meshes, we obtain an order h^2 for the velocities and better than h for the pressures. For the upstream weighting scheme on structured meshes, we obtain an order $h^{0.8}$ for the velocities.

Some experiments were also carried out for the classical example of the lid driven cavity, using triangular meshes. We refer to [5] for these, and shall only give here some results on the backward facing step, for a Reynolds number equal to 800. This is a well documented case in the literature (see e.g. [1], and allows to test the performance of methods with respect to the precision on the zones of recirculating flow. The geometrical data of the backward step is taken from [1]. We computed the streamlines using a reconstruction of a discrete potential Φ_σ, located at the edges $\sigma \in \mathcal{E}$ of the mesh (see [5]). We present in Figure 2 the streamlines in three different cases: starting form the top, the first figure is obtained with the centered scheme, using a 25200 rectangular grid blocks mesh, the second one with the centered scheme using a 2800 rectangular grid blocks mesh, the third one with the upstream scheme using a 2800 rectangular grid blocks mesh, and the two last ones with respectively the centered and the upstream scheme for 847 cells. It is clear from these figures that the centered scheme is, as one could expect, more precise, but that it becomes unstable for coarser meshes. In fact, for a mesh of 700 cells, the Newton iterations do not converge, even when using an under-relaxation procedure.

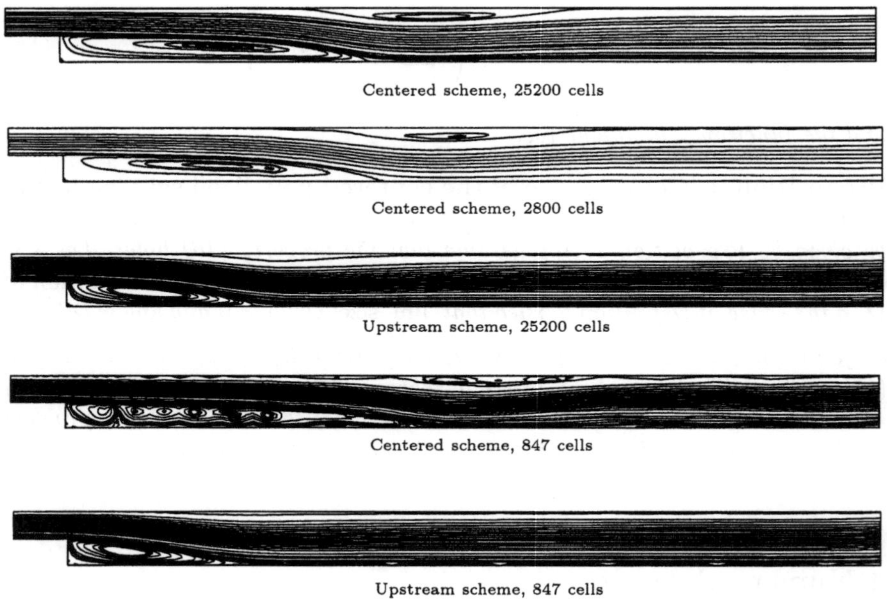

Fig. 2. Streamlines for the backward step

The numerical solution obtained with the centered scheme, using a 25200 rectangular grid blocks mesh seems to be precise enough (comparing the separation and reattachment lengths with those of the literature, see [5]) to be used as a reference solution for experiments carried out on coarser meshes. This allows to compute a rate of convergence of h^2.

We conclude from these numerical tests that the upstream scheme is too diffusive and cannot be used for accurate results, although it has the advantage of remaining stable even on coarse meshes. The centered scheme yields accurate results for a reasonable number of Newton iterations (typically between 5 and 15).

Future developments will concentrate on the extension to three-dimensional meshes and to the time-dependent case.

References

1. B.F. Armaly, F. Durst, J.C.F. Pereira and B. Schönung, "Experimental and Theoretical investigation of backward-facing step flow" J. Fluid Mech. (1983) vol. 127, pp 473–496.
2. S. Boivin, F. Cayré, J.M. Hérard, A finite volume method to solve the Navier-Stokes equations for incompressible flows on unstructured meshes, Int. J. Therm. Sci., 38, 806-825, 2000.
3. R. Eymard, T. Gallouët and R. Herbin, Finite Volume Methods, Handbook of Numerical Analysis, Vol. VII, pp. 713-1020. Edited by P.G. Ciarlet and J.L. Lions (North Holland).
4. R. Eymard and R. Herbin, A cell-centered finite volume scheme on general meshes for the Stokes equations in two dimensions, 125–128, t.337, 2, 2003. CRAS, Mathématiques.
5. R. Eymard and R. Herbin, A finite volume scheme on general meshes for the steady Navier-Stokes problem in two space dimensions, LAPT Report n0 , submitted
6. J.H. Ferziger, M. Peric, Computational Methods for Fluid Dynamics. Springer, Berlin, 1996.
7. V. Girault, P.-A. Raviart, Finite element methods for the Navier-Stokes equations: Theory and algorithms, Springer, Berlin, 1986.
8. M.D. Gunzburger, Finite element methods for viscous incompressible flows, A guide to thoery, practice, and algorithms, Computer Science qnd Scientific Computing, Academic Press 1989.
9. M.D. Gunzburger and R.A Nicolaïdes Incompressible computational fluid dynamics, Cambridge University Press, 1993.
10. R.A Nicolaïdes and X. Wu, Analysis and convergence of the MAC scheme II, Navier-Stokes equations, Math. Comp. 65 (1996), 29-44.
11. S.V. Patankar, (1980), Numerical Heat Transfer and Fluid Flow, Series in Computational Methods in Mechanics and Thermal Sciences, Minkowycz and Sparrow Eds. (Mc Graw Hill).
12. R. Peyret and T. Taylor, Computational methods for for fluid flow, Springer, New-York, 1893.
13. O. Pironneau, Finite element methods for fluids, John Wiley and sons, 1989.
14. R. Temam, Navier-Stokes Equations, Studies in mathematics and its applications, J.L. Lions, G. Papanicolaou, R.T. Rockafellar Editors, North-Holland, 1977.
15. P. Wesseling, Principles of Computational Fluid Dynamics, Springer, Berlin, 2001.

Existence and Uniqueness of a Weak Solution to a Stratigraphic Model

Robert Eymard[1], Thierry Gallouët[2], Véronique Gervais[3] and Roland Masson[3]

[1] Dépt de Mathématiques, Université de Marne-La-Vallée, Marne-La-Vallée, France; *eymard@math.univ-mlv.fr*
[2] LATP, Université de Provence, Marseille, France; *Thierry.Gallouet@cmi.univ-mrs.fr*
[3] Institut Français du Pétrole, Rueil-Malmaison, France; *Veronique.Gervais@ifp.fr, Roland.Masson@ifp.fr*

Summary. In this paper, we study a multi-lithology diffusion model used to simulate the evolution through time of a sedimentary basin composed of several lithologies such as sand or shale. It is a simplified model for which the surficial flux in lithology i is taken proportional to the slope and to a lithology fraction c_i^s in lithology i at the top of the basin with a unitary diffusion coefficient. Thus, the sediment thickness variable satisfies a linear parabolic problem and decouples from the other unknowns. The remaining equations couple, for each lithology, a first order linear equation for the surface concentration c_i^s with a linear advection equation for the basin concentration, for which c_i^s appears as an input boundary condition at the top of the basin in case of sedimentation. The existence and uniqueness of a weak solution in L^∞ is proved for this problem.

1 Introduction

Thanks to recent progress in geosciences, the process of sedimentary basin infill is generally well understood today, and is often considered as the response to the interaction between three main processes : the available space created in the basin by sea level variations, tectonic, compaction ...; the sediment supply (boundary fluxes, sediment production); and the transport of the sediments at the surface of the basin. This interaction is essentially treated in a qualitative manner using field data, such as seismic or well data, but these informations can be difficult and expensive to get. Thus, a 4D numerical model of the basin appears as a powerful tool to solve the problem, and stratigraphic models are developed to answer the need for quantifying the sedimentary basin infill.

A stratigraphic model describes the evolution through time of sedimentary basins in terms of geometry and rock properties. These models are suited for large scales in time and space (greater that 10 km and 10.000 yr), and thus average several geological processes such as transport processes (river transport, creep, slumps,...). Descriptions of such models are given in [4], [6], [7] and [8].

We consider here the stratigraphic model detailed in [2], in which sediments are modeled as a mixture of several lithologies characterized by their grain size population (sand or shale for example). The surficial flux of lithology i, $i = 1, \ldots, L$, is taken as in [7] proportional to the slope of the topography h and to a lithology fraction c_i^s defined at the surface of the basin. In this paper, the diffusion coefficients are taken equal to one, leading to a simplified model in the sense that the sediment thickness variable h decouples from the other unknowns, the L concentrations u_i in lithology i inside the basin and the L surface concentrations c_i^s in lithology i at the top of the basin, and satisfies a linear parabolic problem. The remaining equations, accounting for the mass conservation of the lithologies, couple for all $i = 1, \ldots, L$ a first order linear equation for the surface concentration variable c_i^s and a linear advection equation for the basin concentration variable u_i, for which c_i^s appears as an input boundary condition at the top of the basin. A weak formulation has been introduced for this coupled problem (see Definition 1).

The aim of this paper is then to study the problem satisfied by the concentration variables, and more especially to state the existence and uniqueness of a weak solution in L^∞. This result is given below in Theorem 1. The proof of existence has already been achieved in [1] and will be briefly recalled in section 3. It derives from the convergence of an implicit finite volume scheme. The uniqueness will be obtained using the linearity of the coupled problem in the concentration variables, the existence of a weak solution to the adjoint system and two integration by part formulae for the non smooth solutions of the direct and adjoint problems. Then, the paper outlines as follows : the mathematical model and its weak formulation are described in section 2, and the proof of existence and uniqueness of a weak solution is achieved in section 3.

2 Mathematical Model

We consider in this paper the model defined by Eymard et al. in [2] in a simplified case for which the diffusion coefficients of the lithologies are taken equal to one. Furthermore, the sea level variations and the ground distortions are not taken into account.

Let us denote by $h(x,t)$ the sediment thickness variable, function of time $t > 0$ and of $x \in \Omega \subset \mathbb{R}^d$, $d = 1$ or 2, the horizontal extension of the basin. The sediments are modeled as a mixture of L immiscible lithologies, such as sand or shale, characterized by their grain size population, and considered as incompressible materials of constant grain density and null porosity. Inside the basin, the mixture is described by its composition given by the L concentrations $c_i(x,z,t) \geq 0$ in lithology i defined on the domain $\mathcal{B} = \{(x,z,t) \,|\, x \in \Omega, t > 0, z < h(x,t)\}$, and satisfying $\sum_{i=1}^{L} c_i = 1$. The sediments transported by the surficial fluxes, i.e. deposited at the surface in case of sedimentation and passing through it in case of erosion, are characterized by their concentrations $c_i^s(x,t) \geq 0$, defined on $\Omega \times \mathbb{R}_+^*$, and also satisfying $\sum_{i=1}^L c_i^s = 1$.

Since the sediment fluxes are non zero only at the surface of the basin, no change of the sediment composition occurs inside the basin : $\partial_t c_i = 0$ on $\mathcal{B}$. The evolution of c_i is then only governed by the boundary condition at the top of the basin stating that $c_i|_{z=h} = c_i^s$ in case of sedimentation ($\partial_t h > 0$). An initial condition to the basin concentrations is also prescribed : $c_i|_{t=0} = c_i^0$ on $\{(x,z) \,|\, x \in \Omega,\ z < h^0(x)\}$.

Let us now consider these equations in the new coordinate system $(x, \xi, t) = (x', h(x',t') - z, t')$ in which the vertical position of a point is measured downward from the top of the basin, and let us define $u_i(x,\xi,t) = c_i(x, h(x,t) - \xi, t)$ on $\Omega \times \mathbb{R}_+^* \times \mathbb{R}_+^*$, $u_i^0(x,\xi) = c_i^0(x, h^0(x) - \xi, t)$ on $\Omega \times \mathbb{R}_+^*$. Then, we get the new problem :

$$\begin{cases} \partial_t u_i + \partial_t h\, \partial_\xi u_i = 0 & \text{on } \Omega \times \mathbb{R}_+^* \times \mathbb{R}_+^*, \\ u_i|_{\xi=0} = c_i^s & \text{on } \mathcal{D}^+ = \{(x,t) \in \Omega \times \mathbb{R}_+^* \,|\, \partial_t h(x,t) > 0\}, \\ u_i|_{t=0} = u_i^0 & \text{on } \Omega \times \mathbb{R}_+^*. \end{cases} \tag{1}$$

The surficial transport process is the multi-lithology diffusive model introduced in [7] for which the flux of lithology i is taken proportional to the slope and to the surface concentration : $\varphi_i = -\, c_i^s\, k_i\, \nabla h$. The coefficient $k_i > 0$ is the diffusion coefficient of the lithology i, chosen equal to one in this paper for all $i = 1, \ldots, L$. Therefore, the sediment thickness variable decouples from the other unknowns and satisfies a linear parabolic equation as we shall see later. Then, the model accounts for the conservation of the fraction $\mathcal{M}_i(x,t) = \int_0^{h(x,t)} u_i(x,\xi,t)\, d\xi$ in lithology i, stating that

$$\begin{cases} u_i|_{\xi=0}\, \partial_t h + \operatorname{div}(-c_i^s\, \nabla h) = 0 & \text{on } \Omega \times \mathbb{R}_+^*, \\ \sum_{i=1}^L c_i^s = 1 & \text{on } \Omega \times \mathbb{R}_+^*. \end{cases} \tag{2}$$

In this equation, $u_i|_{\xi=0}\, \partial_t h$ is formally equal to $\partial_t \mathcal{M}_i$ thanks to (1). A Neumann boundary condition is imposed to h on $\partial\Omega \times \mathbb{R}_+^*$: $\nabla h \cdot \mathbf{n}|_{\partial\Omega \times \mathbb{R}_+^*} = g$, with $\mathbf{n}$ the unit normal vector to $\partial\Omega$ outward to Ω, as well as the initial condition $h|_{t=0} = h^0$ on Ω. Finally, Dirichlet input boundary conditions are prescribed to the surface concentrations : $c_i^s|_{\Sigma^+} = \tilde{c}_i$ on $\Sigma^+ = \{(x,t) \in \partial\Omega \times \mathbb{R}_+^* \,|\, g(x,t) > 0\}$.

Summing the first equation of (2) for all $i = 1, \ldots, L$, it appears that, for this simplified model, the sediment thickness variable decouples from the other unknowns and satisfies the linear parabolic equation

$$\begin{cases} \partial_t h - \Delta h = 0 & \text{on } \Omega \times \mathbb{R}_+^*, \\ \nabla h \cdot \mathbf{n}|_{\partial\Omega \times \mathbb{R}_+^*} = g & \text{on } \partial\Omega \times \mathbb{R}_+^*, \\ h|_{t=0} = h^0 & \text{on } \Omega. \end{cases} \tag{3}$$

The solution of this problem is then used in the remaining equations (4) accounting for the mass conservation of the lithologies. They couple for each lithology a first order linear equation for the surface concentration c_i^s with a linear advection equation for the concentration u_i, for which c_i^s appears as an input boundary condition at the top of the basin in case of sedimentation :

$$\begin{cases} u_i|_{\xi=0}\,\partial_t h + \operatorname{div}(-c_i^s \nabla h) = 0 & \text{on } \Omega \times \mathbb{R}_+^*, \\ c_i^s|_{\Sigma^+} = \tilde{c}_i & \text{on } \Sigma^+ = \{(x,t) \in \partial\Omega \times \mathbb{R}_+^* \,|\, g(x,t) > 0\}, \\ \partial_t u_i + \partial_t h\,\partial_\xi u_i = 0 & \text{on } \Omega \times \mathbb{R}_+^* \times \mathbb{R}_+^*, \\ u_i|_{\xi=0} = c_i^s & \text{on } \mathcal{D}^+ = \{(x,t) \in \Omega \times \mathbb{R}_+^* \,|\, \partial_t h(x,t) > 0\}, \\ u_i|_{t=0} = u_i^0 & \text{on } \Omega \times \mathbb{R}_+^*. \end{cases} \tag{4}$$

In the sequel, the following hypothesis are made on the data :

Hypothesis 1
(i) Ω is an open bounded subset of $\mathbb{R}^d$, of class $\mathcal{C}^\infty$,
(ii) $h^0 \in \mathcal{C}^2(\bar{\Omega})$, $g \in \mathcal{C}^1(\partial\Omega \times \mathbb{R}_+) \cap L^2(\partial\Omega \times \mathbb{R}_+)$, and g, h^0 are chosen so that the unique solution h of (3) is in $\mathcal{C}^2(\bar{\Omega} \times [0,T])$ for all $T > 0$,
(iii) $\tilde{c}_i \in L^\infty(\Sigma^+)$ with $\tilde{c}_i \geq 0$ for $i = 1, \ldots, L$, and $\sum_{i=1}^L \tilde{c}_i = 1$,
(iv) $u_i^0 \in L^\infty(\Omega \times \mathbb{R}_+^*)$, $u_i^0 \geq 0$ for $i = 1, \ldots, L$, and $\sum_{i=1}^L u_i^0 = 1$,
(v) For all $T > 0$, the boundaries $\partial\Sigma_T^+$ and $\partial\Sigma_T^-$ of the sets $\Sigma_T^+ = \{(x,t) \in \partial\Omega \times (0,T) \,|\, g(x,t) > 0\}$ and $\Sigma_T^- = \{(x,t) \in \partial\Omega \times (0,T) \,|\, g(x,t) < 0\}$ are the union of a finite number of $\mathcal{C}^1$ manifolds of dimension at most $d-1$,
(vi) For all $T > 0$, the boundaries $\partial\mathcal{D}_T^+$ and $\partial\mathcal{D}_T^-$ of the sets $\mathcal{D}_T^+ = \{(x,t) \in \Omega \times (0,T) \,|\, \partial_t h(x,t) > 0\}$, and $\mathcal{D}_T^- = \{(x,t) \in \Omega \times (0,T) \,|\, \partial_t h(x,t) < 0\}$ are the union of a finite number of $\mathcal{C}^1$ manifolds of dimension at most d.

We shall also denote by $\mathcal{L}$ the operator $\mathcal{L} = \partial_t + \partial_t h\,\partial_\xi$ and by $\mathcal{C}_c^\infty(\mathbb{R}^n)$ the space of real valued functions $\{\varphi \in \mathcal{C}^\infty(\mathbb{R}^n) \,|\, \mathrm{Supp}(\varphi) \text{ bounded in } \mathbb{R}^n\}$.

To cope with the difficulty to define the trace of the basin concentration at the top of the basin, we are looking for weak solutions defined as follows :

Definition 1. *Let us assume that Hypothesis 1 holds, and let h denote the solution of (3). Then $(u_i, c_i^s) \in L^\infty(\Omega \times \mathbb{R}_+^* \times \mathbb{R}_+^*) \times L^\infty(\Omega \times \mathbb{R}_+^*)$ is said to be a weak solution of problem (4) if it satisfies :*
(i) for all $\varphi \in \{\phi \in \mathcal{C}_c^\infty(\mathbb{R}^{d+2}) \,|\, \phi(.,0,.) = 0 \text{ on } \Omega \times \mathbb{R}_+^* \setminus \mathcal{D}^+\}$,

$$\int_\Omega \int_{\mathbb{R}_+} \int_{\mathbb{R}_+} \left[\partial_t \varphi(x,\xi,t) + \partial_t h(x,t)\,\partial_\xi \varphi(x,\xi,t)\right] u_i(x,\xi,t)\,dt\,d\xi\,dx$$
$$+ \int_\Omega \int_{\mathbb{R}_+} u_i^0(x,\xi)\varphi(x,\xi,0)\,d\xi\,dx + \int_\Omega \int_{\mathbb{R}_+} \partial_t h(x,t)\,c_i^s(x,t)\varphi(x,0,t)\,dt\,dx = 0, \tag{5}$$

(ii) for all $\psi \in \{\phi \in \mathcal{C}_c^\infty(\mathbb{R}^{d+2}) \,|\, \phi(.,0,.) = 0 \text{ on } \partial\Omega \times \mathbb{R}_+^* \setminus \Sigma^+\}$,

$$
\begin{aligned}
-\int_{\Omega}\int_{\mathbb{R}_+}\int_{\mathbb{R}_+} \left[\partial_t\psi(x,\xi,t) + \partial_t h(x,t)\,\partial_\xi\psi(x,\xi,t)\right] u_i(x,\xi,t)\,dt\,d\xi\,dx \\
-\int_{\Omega}\int_{\mathbb{R}_+} u_i^0(x,\xi)\psi(x,\xi,0)\,d\xi\,dx + \int_{\mathbb{R}_+}\Bigg(\int_{\Omega} c_i^s(x,t)\,\nabla h(x,t)\cdot\nabla\psi(x,0,t)\,dx \\
-\int_{\partial\Omega} \tilde{c}_i(x,t)\,g(x,t)\,\psi(x,0,t)d\gamma(x)\Bigg)dt = 0.
\end{aligned}
\tag{6}
$$

Then, the main result of this paper is the following Theorem. Its proof will be developed in the next section.

Theorem 1. *Assuming that Hypothesis 1 holds, there exists a weak solution* $(u_i, c_i^s) \in L^\infty(\Omega \times \mathbb{R}_+^* \times \mathbb{R}_+^*) \times L^\infty(\Omega \times \mathbb{R}_+^*)$ *for all* $i \in \{1,\dots,L\}$ *to problem* (4) *in the sense of Definition 1, and* u_i *is unique.*

3 Existence and Uniqueness of a Weak Solution

The aim of this section is to prove Theorem 1. The proofs of Lemmae 1 and 2 used in the sequel are technical and will be detailed in a forthcoming paper.

The existence of a weak solution $(u_i, c_i^s) \in L^\infty(\Omega\times\mathbb{R}_+^*\times\mathbb{R}_+^*)\times L^\infty(\Omega\times\mathbb{R}_+^*)$ to (4) in the sense of Definition 1 is obtained by convergence of an implicit finite volume scheme for the model, and has already been proved in [1]. Let us just recall the main stages of this proof.

In the sequel, we shall consider admissible finite volume meshes defined as follows :

Definition 2. *Let Ω be a bounded domain of $\mathbb{R}^d$, $d = 1$ or 2. An admissible finite volume mesh of Ω for the discretization of problem (3)–(4) is given by a family of "control volumes", denoted by $\mathcal{K}$, which are open disjoint subsets of Ω, and a family of points of Ω, denoted by $\mathcal{P}$, satisfying the following properties :*

1. *The closure of the union of all the control volumes of $\mathcal{K}$ is $\bar{\Omega}$.*
2. *For any κ, $\kappa' \in \mathcal{K}$ with $\kappa \neq \kappa'$, either the $(d-1)$-dimensional measure of $\bar{\kappa}\cap\bar{\kappa}'$, denoted by $m(\bar{\kappa}\cap\bar{\kappa}')$, is null, or it is strictly positive and $\bar{\kappa}\cap\bar{\kappa}'$ is included in an hyperplane of $\mathbb{R}^d$. In the following, we will denote by Σ_{int} the family of subsets σ of Ω contained in hyperplanes of $\mathbb{R}^d$ with strictly positive measures, and such that there exist $\kappa, \kappa' \in \mathcal{K}$ with $m(\bar{\kappa}\cap\bar{\kappa}') > 0$ and $\bar{\sigma} = \bar{\kappa}\cap\bar{\kappa}'$. We shall also denote by $\kappa|\kappa' \in \Sigma_{int}$ the edge between the cells κ and κ'.*
3. *The family $\mathcal{P} = (x_\kappa)_{\kappa\in\mathcal{K}}$ is such that $x_\kappa \in \bar{\kappa}$ (for any $\kappa \in \mathcal{K}$), and, if $\sigma = \kappa|\kappa'$, it is assumed that $x_\kappa \neq x_{\kappa'}$ and that the straight line going through x_κ and $x_{\kappa'}$ is orthogonal to the edge σ. We shall denote by $d(\kappa,\kappa')$ the distance between the points x_κ and $x_{\kappa'}$.*
4. *For any $\kappa \in \mathcal{K}$, there exists a subset Σ_κ of Σ_{int} such that $\partial\kappa \setminus \partial\Omega =$*

$\bar{\kappa} \setminus (\kappa \cup \partial\Omega) = \cup_{\sigma \in \Sigma_\kappa} \bar{\sigma}$.
We shall denote by $(\mathcal{K}, \Sigma_{int}, \mathcal{P})$ *this mesh, and by* $\delta\mathcal{K} = \sup\{diam(\kappa), \kappa \in \mathcal{K}\}$ *its size.*

Let $(\mathcal{K}, \Sigma_{int}, \mathcal{P})$ be an admissible mesh of Ω. The time discretization is denoted by t^n, $n \in \mathbb{N}$, such that $t^0 = 0$ and $\Delta t^{n+1} = t^{n+1} - t^n > 0$. The superscript n, $n \in \mathbb{N}$, will be used to denote that the unknowns are considered at time t^n. For each control volume $\kappa \in \mathcal{K}$ and each time t^n, $n \geq 0$, h^n_κ shall denote the approximation of the sediment thickness on κ at time t^n, $c^n_{i,\kappa}(z)$, $z \in (-\infty, h^n_\kappa)$, the approximation of the basin concentration c_i in lithology i in the column $\{(x,z) \,|\, x \in \kappa, z < h(x,t^n)\}$, and $c^{s,n+1}_{i,\kappa}$ the approximation of the surface concentration in lithology i at time t^{n+1} on κ. Then, (3)–(4) is discretized by a fully implicit time integration and a finite volume method with cell centered variables. For the computation of the fluxes at the edges of the control volumes, the discretization uses an upstream weighted evaluation of the surface concentrations. The approximate concentration $c^{n+1}_{i,\kappa}$ is the solution at time t^{n+1} of the conservation equation

$$\begin{cases} \partial_t c_{i,\kappa}(z,t) = 0, \quad \text{for all } t^n < t < t^{n+1}, z < h_\kappa(t) = h^n_\kappa + (t - t^n)\frac{h^{n+1}_\kappa - h^n_\kappa}{\Delta t^{n+1}}, \\ c_{i,\kappa}(h_\kappa(t), t) = c^{s,n+1}_{i,\kappa} \quad \text{if } h^{n+1}_\kappa > h^n_\kappa, \\ c_{i,\kappa}(z, t^n) = c^n_{i,\kappa}(z) \quad \text{for all } z < h^n_\kappa, \end{cases}$$

and $u^n_{i,\kappa}(\xi) = c^n_{i,\kappa}(h^n_\kappa - \xi)$ for all $\xi > 0$. One can refer to [2] or [1] for the complete numerical scheme.

Then, the approximate sediment thickness $(h^n_\kappa)_{\kappa \in \mathcal{K}}$ satisfies an implicit finite volume numerical scheme for the parabolic problem (3) for which existence, uniqueness and error estimates have already been proved in [3]. Concerning the concentration variables, we show the existence of solutions to the discrete problem bounded in the interval $[0,1]$, which are unique except for the surface concentration variables $c^{s,n+1}_{i,\kappa}$. For any admissible mesh $(\mathcal{K}, \Sigma_{int}, \mathcal{P})$ of Ω, any time step $\Delta t > 0$, and $i = 1, \ldots, L$, let us define the piecewise constant functions $h_{\mathcal{K},\Delta t}$, $c^s_{i,\mathcal{K},\Delta t}$ on $\Omega \times \mathbb{R}^*_+$, and $u_{i,\mathcal{K},\Delta t}$ on $\Omega \times \mathbb{R}^*_+ \times \mathbb{R}^*_+$ by

$$h_{\mathcal{K},\Delta t}(x,t) = h^{n+1}_\kappa, \quad u_{i,\mathcal{K},\Delta t}(x,\xi,t) = u^{n+1}_{i,\kappa}(\xi), \quad c^s_{i,\mathcal{K},\Delta t}(x,t) = c^{s,n+1}_{i,\kappa} \tag{7}$$

for all $x \in \kappa$, $\kappa \in \mathcal{K}$, $t \in (t^n, t^{n+1}]$, $n \geq 0$, $\xi \in \mathbb{R}^*_+$, where h^n_κ, $c^{s,n+1}_{i,\kappa}$ and $u^n_{i,\kappa}$ are any given solution of the discrete problem with $c^{s,n+1}_{i,\kappa}$ bounded in the interval $[0,1]$. For all $m \in \mathbb{N}$, let $(\mathcal{K}_m, \Sigma^m_{int}, \mathcal{P}_m)$ be an admissible mesh of Ω and $\Delta t_m > 0$, and let us assume that $\Delta t_m \to 0$, $\delta\mathcal{K}_m/\sqrt{\Delta t_m} \to 0$ as $m \to \infty$ and that there exists $\alpha > 0$ such that, for all $m \in \mathbb{N}$, $\max_{\substack{\sigma \in \Sigma^m_{int} \\ \sigma = \kappa|\kappa'}} \frac{\delta\mathcal{K}_m}{d(\kappa,\kappa')} \leq \alpha$. Let $h_{\mathcal{K}_m,\Delta t_m}$, $u_{i,\mathcal{K}_m,\Delta t_m}$, $c^s_{i,\mathcal{K}_m,\Delta t_m}$ defined by (7) with $\mathcal{K} = \mathcal{K}_m$ and $\Delta t = \Delta t_m$. Then the stability of the discrete concentrations gives the convergence, up to a subsequence, of $(u_{i,\mathcal{K}_m,\Delta t_m}, c^s_{i,\mathcal{K}_m,\Delta t_m})_{m \in \mathbb{N}}$ in L^∞ for the weak-$\star$ topology as $m \to \infty$. To prove that the limit (u_i, c^s_i) is a weak solution of (4), we finally use an interpolation in time of the approximate basin concentration $u_{i,\mathcal{K}_m,\Delta t_m}$ which converges towards u_i in L^∞, and a weak-BV estimate for the flux terms, which is an adaptation to the coupling of a parabolic and an hyperbolic equation of the result proved in [3] for the coupling of an elliptic and an hyperbolic equation in the case of a two-phase Darcy flow. The detailed proof is given in [1].

Let us now show the uniqueness of u_i. It is achieved using mainly the linearity of (4) in the concentration variables and the adjoint problem on $\Omega \times (0,T)$, $T > 0$.

For any given surface concentration $c_i^s \in L^\infty(\Omega \times \mathbb{R}_+^*)$, we have first studied the weak formulation (5) of the linear advection equation $\mathcal{L}u_i = 0$ with input boundary condition c_i^s on $\mathcal{D}^+$ and initial condition u_i^0. Using the characteristic solution of this problem (see [5]), we have proved the following Lemma, in which an integration by part formula for the solutions of the advection equation and its adjoint problem is stated :

Lemma 1. *Hypothesis* 1 *is assumed to hold. Then, for any time* $T > 0$, *any functions* $f \in L^\infty(\Omega \times \mathbb{R}_+^* \times (0,T))$, $l^s \in L^\infty(\mathcal{D}_T^+)$, *and* $v^0 \in L^\infty(\Omega \times \mathbb{R}_+^*)$, *the equation*

$$\begin{cases} \mathcal{L}v = f \quad on\ \Omega \times \mathbb{R}_+^* \times (0,T), \\ v|_{\xi=0} = l^s \quad on\ \mathcal{D}_T^+, \\ v|_{t=0} = v^0 \quad on\ \Omega \times \mathbb{R}_+^*, \end{cases} \tag{8}$$

has a unique weak solution in $L^\infty(\Omega \times \mathbb{R}_+^* \times (0,T))$ *in the sense that for all* $\varphi \in \{\phi \in C_c^\infty(\mathbb{R}^{d+2}) \,|\, \phi(.,0,.) = 0 \text{ on } \Omega \times (0,T) \setminus \mathcal{D}_T^+ \text{ and } \phi(.,.,T) = 0 \text{ on } \Omega \times \mathbb{R}_+^*\}$, *one has*

$$\int_\Omega \int_{\mathbb{R}_+} \int_0^T \Big((\mathcal{L}\varphi)(x,\xi,t)\, v(x,\xi,t) + f(x,\xi,t)\, \varphi(x,\xi,t) \Big)\, dt\, d\xi\, dx$$
$$+ \int_\Omega \int_{\mathbb{R}_+} v^0(x,\xi)\, \varphi(x,\xi,0)\, d\xi\, dx + \int_\Omega \int_0^T \partial_t h(x,t)\, l^s(x,t)\, \varphi(x,0,t)\, dt\, dx = 0. \tag{9}$$

The weak solution v *of* (8) *has a trace on* $t = T$ *in* $L^\infty(\Omega \times \mathbb{R}_+^*)$, *and the function* $v\, \partial_t h$ *has a trace on* $\xi = 0$ *in* $L^\infty(\Omega \times (0,T))$, *such that for any* $\varphi \in C_c^\infty(\mathbb{R}^{d+2})$ *one has*

$$\int_\Omega \int_{\mathbb{R}_+} \int_0^T \Big((\mathcal{L}\varphi)\, v + f\, \varphi \Big)(x,\xi,t)\, dt\, d\xi\, dx + \int_\Omega \int_{\mathbb{R}_+} \Big(v(x,\xi,0)\, \varphi(x,\xi,0)$$
$$- v(x,\xi,T)\, \varphi(x,\xi,T) \Big)\, d\xi\, dx + \int_\Omega \int_0^T \partial_t h(x,t)\, v(x,0,t)\, \varphi(x,0,t)\, dt\, dx = 0. \tag{10}$$

Let $T > 0$ *and* w *be the weak solution in* $L^\infty(\Omega \times \mathbb{R}_+^* \times (0,T))$ *of the adjoint equation*

$$\begin{cases} -\mathcal{L}w = r & on\ \Omega \times \mathbb{R}_+^* \times (0,T), \\ w|_{\xi=0} = q^s & on\ \mathcal{D}_T^-, \\ w|_{t=T} = w^T & on\ \Omega \times \mathbb{R}_+^*, \end{cases} \tag{11}$$

defined in a similar way as above with $r \in L^\infty(\Omega \times \mathbb{R}_+^* \times (0,T))$ *a compactly supported function on* $\bar{\Omega} \times \mathbb{R}_+ \times [0,T]$, $w^T \in L^\infty(\Omega \times \mathbb{R}_+^*)$ *a compactly supported function on* $\bar{\Omega} \times \mathbb{R}_+$, *and* $q^s \in L^\infty(\Omega \times (0,T))$. *Then, one has*

$$\int_\Omega \int_{\mathbb{R}_+} \int_0^T \Big(v\,(\mathcal{L}w) + (\mathcal{L}v)\,w \Big)(x,\xi,t)\,dt\,d\xi\,dx - \int_\Omega \int_{\mathbb{R}_+} \Big(v(x,\xi,T)\,w(x,\xi,T) \\ -v(x,\xi,0)\,w(x,\xi,0) \Big)\,d\xi\,dx + \int_\Omega \int_0^T \partial_t h(x,t)\,v(x,0,t)\,w(x,0,t)\,dt\,dx = 0. \tag{12}$$

Let us denote by (v_i, d_i^s) the difference between any two weak solutions of (4). From the linearity of the set of equations (4) in the concentration variables, the functions (v_i, d_i^s) satisfy the weak formulation (5)–(6) with homogeneous boundary and initial conditions.

Let $T > 0$. From Lemma 1, the function $v_i\,\partial_t h$ has a trace at $\xi = 0$ in $L^\infty(\Omega \times (0,T))$ denoted by $v_i|_{\xi=0}\,\partial_t h$. Then, from the integration by part formula (10) of Lemma 1 and the weak formulation (6), it results that for all $\varphi \in \{\phi \in C_c^\infty(\mathbb{R}^{d+1}) \,|\, \phi(x,t) = 0 \text{ on } \partial\Omega \times (0,T) \setminus \Sigma_T^+, \text{ and } \phi(x,T) = 0 \text{ on } \Omega\}$, one has

$$\int_\Omega \int_0^T \Big(v_i(x,0,t)\,\partial_t h(x,t)\,\varphi(x,t) + d_i^s(x,t)\,\nabla h(x,t) \cdot \nabla\varphi(x,t) \Big)\,dt\,dx = 0. \tag{13}$$

We easily deduce, using (13) and $\partial_t h - \Delta h = 0$, that

$$\operatorname{div}(-d_i^s\,\nabla h) = -v_i|_{\xi=0}\,\partial_t h \in L^\infty(\Omega \times (0,T)), \tag{14}$$

$$\nabla h \cdot \nabla d_i^s = (v_i|_{\xi=0} - d_i^s)\,\partial_t h \in L^\infty(\Omega \times (0,T)). \tag{15}$$

Let us now consider the adjoint system

$$\begin{cases} -w_i|_{\xi=0}\,\partial_t h + \operatorname{div}(q_i^s \nabla h) = 0 & \text{on } \Omega \times (0,T), \\ q_i^s|_{\Sigma_T^-} = 0 & \text{on } \Sigma_T^-, \\ -\mathcal{L}w_i = v_i & \text{on } \Omega \times \mathbb{R}_+^* \times (0,T), \\ w_i|_{\xi=0} = q_i^s & \text{on } \mathcal{D}_T^-, \\ w_i|_{t=T} = v_i|_{t=T} & \text{on } \Omega \times \mathbb{R}_+^*. \end{cases} \tag{16}$$

The direct and adjoint problems are very close, apart from the non vanishing right hand side $v_i \in L^\infty(\Omega \times \mathbb{R}_+^* \times \mathbb{R}_+^*)$ in the advection equation of (16). Then, the existence of a weak solution $(w_i, d_i^s) \in L^\infty(\Omega \times \mathbb{R}_+^* \times \mathbb{R}_+^*) \times L^\infty(\Omega \times \mathbb{R}_+^*)$ to the adjoint problem, defined similarly as in Definition 1, can be obtained in a very close way as in [1] by the convergence of a finite volume numerical scheme, adapted to the non vanishing right hand side in L^∞ in the advection equation.

Considering such a weak solution, the following equation is derived as above

$$\operatorname{div}(q_i^s \nabla h) = w_i|_{\xi=0}\, \partial_t h \in L^\infty(\Omega \times (0,T)). \tag{17}$$

From (15) and (17), the function $\operatorname{div}(q_i^s\, d_i^s\, \nabla h)$ is defined in $L^\infty(\Omega \times (0,T))$ and hence the vector field $q_i^s d_i^s \nabla h$ has a normal trace in $L^\infty(0,T; H^{-\frac{1}{2}}(\partial\Omega))$. As formally d_i^s vanishes on Σ_T^+, q_i^s vanishes on Σ_T^-, and the normal trace g of ∇h vanishes on $\partial\Omega \times (0,T) \setminus (\Sigma_T^+ \cup \Sigma_T^-)$, the normal trace of $q_i^s d_i^s \nabla h$ vanishes on the boundary $\partial\Omega \times (0,T)$. We can prove this result stated by the following Lemma :

Lemma 2. *Hypothesis* 1 *is assumed to hold. Then, for any* $T > 0$, *any weak solutions* (w_i, q_i^s) *of the adjoint problem* (16), *and* (v_i, d_i^s) *of problem* (4) *with homogeneous boundary and initial conditions, one has*

$$\int_\Omega \int_0^T \operatorname{div}(q_i^s\, d_i^s\, \nabla h)\, dt\, dx = 0.$$

According to the definition of the characteristic solution of (11) (see [5]) and since the velocity $\partial_t h$ is uniformly bounded on $\bar{\Omega} \times [0,T]$ for any time $T > 0$, the function v_i (resp. its trace $v_i|_{t=T}$) is compactly supported in $\bar{\Omega} \times \mathbb{R}_+ \times [0,T]$ (resp. in $\bar{\Omega} \times \mathbb{R}_+$). Applying the integration by part formula (12) of Lemma 1 to $v = v_i$ and $w = w_i$, we get that for any time $T > 0$

$$\begin{aligned} \int_\Omega \int_{\mathbb{R}_+} \int_0^T |v_i|^2(x,\xi,t)\, dt\, d\xi\, dx + \int_\Omega \int_{\mathbb{R}_+} |v_i|^2(x,\xi,T)\, d\xi\, dx \\ = \int_\Omega \int_0^T \partial_t h(x,t)\, v_i(x,0,t)\, w_i(x,0,t)\, dt\, dx. \end{aligned} \tag{18}$$

From Lemma 2 and the integration over $\Omega \times (0,T)$ of (17) multiplied by d_i^s, we obtain

$$\int_\Omega \int_0^T \Big[d_i^s(x,t)\, w_i(x,0,t)\, \partial_t h(x,t) + q_i^s(x,t)\, \nabla d_i^s(x,t) \cdot \nabla h(x,t) \Big]\, dt\, dx = 0. \tag{19}$$

Also, multiplying (15) by q_i^s and integrating over $\Omega \times (0,T)$, we get

$$\begin{aligned} \textstyle\int_\Omega \int_0^T \Big[\big(v_i(x,0,t) - d_i^s(x,t)\big)\, q_i^s(x,t)\, \partial_t h(x,t) - \\ - q_i^s(x,t) \nabla d_i^s(x,t) \cdot \nabla h(x,t) \Big]\, dt\, dx = 0. \end{aligned} \tag{20}$$

Summing (19) and (20) and taking into account the boundary conditions $w_i|_{\xi=0} = q_i^s$ on $\mathcal{D}_T^-$, $v_i|_{\xi=0} = d_i^s$ on $\mathcal{D}_T^+$, and that $\partial_t h = 0$ on $\Omega \times (0,T) \setminus (\mathcal{D}_T^+ \cup \mathcal{D}_T^-)$, we obtain

$$\begin{aligned} \int_\Omega \int_0^T \Big(w_i(x,0,t)\, d_i^s(x,t) + v_i(x,0,t)\, q_i^s(x,t) - d_i^s(x,t)\, q_i^s(x,t) \Big)\, \partial_t h(x,t) dt\, dx \\ = \int_\Omega \int_0^T v_i(x,0,t)\, w_i(x,0,t)\, \partial_t h(x,t) dt\, dx = 0. \end{aligned} \tag{21}$$

Equation (21) together with (18) conclude the proof of Theorem 1.

This theorem, together with the convergence result on the solutions of the implicit finite volume numerical scheme seen previoulsy, also give the convergence of the full sequence of approximate solutions $(u_{i,\mathcal{K}_m,\Delta t_m})_{m\in\mathbb{N}}$ towards the weak solution u_i of (4).

The proof will be detailed in a forthcoming paper, and particularly Lemmae 1, 2, and the existence of a weak solution to the adjoint problem.

References

[1] Eymard, R., Gallouët, T., Gervais, V., Masson, R.(submitted 2003) : Convergence of a numerical scheme for stratigraphic modeling. SIAM J. Num. Anal.

[2] Eymard, R., Gallouët, T., Granjeon, D., Masson, R., Tran, Q.(2003) : Multi-lithology stratigraphic model under maximum erosion rate constraint. Int. J. of Num. Methods in Eng., (to appear)

[3] Eymard, R., Gallouët, T., Herbin, R.(2000) : The Finite Volume Method. In : Ciarlet P. and Lions J.(eds.) Handbook of Numerical Analysis, **7**. Elsevier

[4] Flemings, P.B., Jordan, T.E.(1989) : A Synthetic Stratigraphic Model of Foreland Basin Development. J. of Geophysical Research, Vol **94**, B4, 3851–3866

[5] Godlewski, E., Raviart, P.(1996) : Numerical Approximation of Hyperbolic Systems of Conservation Laws. Springer

[6] Granjeon, D.,(1997) : Modélisation stratigraphique déterministe; conception et applications d'un modèle diffusif 3D multilithologique. Ph. D. dissertation, Gosciences Rennes, Rennes, France

[7] Rivenaes, J.C.(1992) : Application of a dual lithology, depth-dependent diffusion equation in stratigraphic simulation. Basin Research, **4**, 133–146

[8] Tucker, G.E., Slingerland, R.L.(1994) : Erosional dynamics, flexural isostasy, and long-lived escarpments: A numerical modeling study. J. of Geophysical Research, Vol **99**, B6, 12,229–12,243

Combined Nonconforming/Mixed-hybrid Finite Element–Finite Volume Scheme for Degenerate Parabolic Problems

Robert Eymard[1], Danielle Hilhorst[2] and Martin Vohralík[2,3]

[1] Département de Mathématiques, Université de Marne-la-Vallée, 5, boulevard Descartes Champs-sur-Marne, 77454 Marne-la-Vallée, France *eymard@math.univ-mlv.fr*

[2] Laboratoire de Mathématiques, Analyse Numérique et EDP, Université de Paris-Sud et CNRS, Bât. 425, 91405 Orsay, France *danielle.hilhorst@math.u-psud.fr*

[3] Department of Mathematics, Faculty of Nuclear Sciences and Physical Engineering, Czech Technical University in Prague, Trojanova 13, 120 00 Prague 2, Czech Republic *vohralik@km1.fjfi.cvut.cz*

Summary. We propose and analyze an efficient numerical scheme for nonlinear degenerate parabolic convection–reaction–diffusion equations. We discretize the diffusion term, which generally involves a full matrix diffusion tensor, by means of piecewise linear nonconforming (Crouzeix–Raviart) finite elements over a triangulation of the space domain, or using the stiffness matrix of the hybridization of the lowest order Raviart–Thomas mixed finite element method. The other terms are discretized by means of a finite volume scheme on a dual mesh, where the dual volumes are constructed around the sides of the original triangulation. Checking the local Péclet number, we set up the exact necessary amount of upstream weighting to avoid spurious oscillations in the velocity dominated case. Under the regularity condition for the triangulation, using a priori estimates and Kolmogorov's relative compactness theorem, the convergence of the scheme is proved.

1 Introduction

The contaminant transport equation writes in the form

$$\frac{\partial \beta(c)}{\partial t} - \nabla \cdot (\mathbf{D} \nabla c) + \nabla \cdot (c\mathbf{v}) + F(c) = q\,, \tag{1}$$

where c is the unknown concentration of the contaminant, the function $\beta(\cdot)$ represents time evolution and equilibrium adsorption reaction, $\mathbf{v}$ is the velocity field, $\mathbf{D}$ is the diffusion–dispersion tensor, the function $F(\cdot)$ represents the changes due to chemical reactions, and q stands for the sources. The main features of equation (1) are its degeneracy since β' may be unbounded, the possible dominance of the convection term, and the presence of a heterogeneous and anisotropic diffusion–dispersion tensor.

The convergence of a finite volume scheme for the equation (1) with $\mathbf{D} = Id$ and $F = 0$ has been shown in [7]. Finite volumes with upstream weighting techniques are unconditionally stable; however, there are geometrical restrictions on the mesh for the discretization of the diffusion term and there is no general prescription how to discretize full tensors. The finite element method for degenerate parabolic problems has been studied e.g. in [2]. One can discretize full tensors and there are no restrictions on the mesh. However, spurious oscillations may appear in the velocity dominated case or in the presence of a reaction term. Hence a quite intuitive idea is to combine finite volume and finite element methods, trying to use the "best of both worlds". In [1], the authors introduce a combined scheme for a convection–diffusion equation with a nonlinear convection term in two space dimensions. In the presented paper, we prove the convergence of this scheme for the equation (1) in two or three space dimensions. We extend the techniques used in [7] for a scheme with negative transmissibilities, general meshes satisfying only the regularity assumption, and cases when the discrete maximum principle is not satisfied.

2 The degenerate parabolic problem

We consider the equation (1) in a polygonal domain $\Omega \subset \mathbb{R}^d$, $d = 2, 3$ and on a time interval $(0,T)$, $0 < T < \infty$. We set $Q_T = \Omega \times (0,T)$. We impose the initial condition by

$$c(\mathbf{x}, 0) = c_0(\mathbf{x}) \qquad \mathbf{x} \in \Omega\,, \tag{2}$$

and a homogeneous Dirichlet boundary condition by

$$c(\mathbf{x}, t) = 0 \qquad \mathbf{x} \in \partial\Omega\,, t \in (0,T)\,. \tag{3}$$

We make the following assumption on the data:

Assumption (A)

(A1) $\beta \in C(\mathbb{R})$, $\beta(0) = 0$ *is a strictly increasing function such that*

$$|\beta(a) - \beta(b)| \geq c_\beta |a - b| \qquad \forall a, b \in \mathbb{R}\,, c_\beta > 0\,,$$

or

(A2) (A1) is satisfied and there in addition exists $P \in \mathbb{R}, P > 0$ *such that* $|\beta(x)| \leq C_\beta$ *in* $[-P, P]$, $C_\beta > 0$ *and Lipschitz continuous with a constant* L_β *on* $(-\infty, P]$ *and* $[P, +\infty)$;

(A3) $\mathbf{D}_{ij} \in L^\infty(Q_T)$, $|\mathbf{D}_{ij}| \leq \dfrac{C_\mathbf{D}}{d}$ *a.e. in* Q_T, $1 \leq i, j \leq d$, $C_\mathbf{D} > 0$, $\mathbf{D}$ *is a symmetric and uniformly positive definite tensor for almost all* $t \in (0,T)$ *with a constant* $c_\mathbf{D} > 0$,

$$\mathbf{D}(\mathbf{x}, t)\boldsymbol{\eta} \cdot \boldsymbol{\eta} \geq c_\mathbf{D} \boldsymbol{\eta} \cdot \boldsymbol{\eta} \qquad \forall \boldsymbol{\eta} \in \mathbb{R}^d\,, \textit{for a.e. } (\mathbf{x}, t) \in Q_T\,;$$

(A4) $\mathbf{v} \in \mathbf{L}^2(0,T;\mathbf{H}(div,\Omega)) \cap \mathbf{L}^\infty(Q_T)$ *satisfies* $\nabla \cdot \mathbf{v} = q_S \geq 0$ *a.e. in* Q_T, $|\mathbf{v} \cdot \mathbf{n}| \leq C_\mathbf{v}$, $C_\mathbf{v} > 0$ *a.e. on* $\ell \times (0,T)$ *for each hyperplane* $\ell \subset \Omega$ *with normal vector* $\mathbf{n}$*;*

(A5) $F(0) = 0$, *F is a non decreasing, Lipschitz continuous function with a constant* L_F

or

(A6) $F(0) = 0$, *F is a Lipschitz continuous function with a constant* L_F *and* $xF(x) \geq 0$ *for* $x < 0$ *and* $x > M$, $M > 0$*;*

(A7) $q \in L^2(Q_T)$, *where* $q = q_S c_S$ *with* $c_S \in L^\infty(Q_T)$, $0 \leq c_S \leq M$ *a.e. in* Q_T*;*

(A8) $c_0 \in L^\infty(\Omega)$, $0 \leq c_0 \leq M$ *a.e. in* Ω.

We now give the definition of a weak solution of the problem (1) – (3).

Definition 1. (Weak solution) *We say that a function c is a weak solution of* (1)–(3), *if* $c \in L^2(0,T;H_0^1(\Omega))$, $\beta(c) \in L^\infty(0,T;L^2(\Omega))$, *and*

$$-\int_0^T \int_\Omega \beta(c)\varphi_t \,\mathrm{d}\mathbf{x}\,\mathrm{d}t - \int_\Omega \beta(c_0)\varphi(\cdot,0)\,\mathrm{d}\mathbf{x} + \int_0^T \int_\Omega \mathbf{D}\nabla c \cdot \nabla\varphi \,\mathrm{d}\mathbf{x}\,\mathrm{d}t -$$
$$-\int_0^T \int_\Omega c\mathbf{v} \cdot \nabla\varphi \,\mathrm{d}\mathbf{x}\,\mathrm{d}t + \int_0^T \int_\Omega F(c)\varphi \,\mathrm{d}\mathbf{x}\,\mathrm{d}t = \int_0^T \int_\Omega q\varphi \,\mathrm{d}\mathbf{x}\,\mathrm{d}t \qquad (4)$$

for all $\varphi \in L^2(0,T;H_0^1(\Omega))$ *with* $\varphi_t \in L^\infty(Q_T)$, $\varphi(\cdot,T) = 0$.

3 The combined finite element–finite volume scheme

We suppose a family of triangulations $\{\mathcal{T}_h\}_h$ of the domain Ω, where each $\mathcal{T}_h$ consists of closed simplices (triangles in the case $d = 2$, tetrahedrons when $d = 3$) such that $\overline{\Omega} = \bigcup_{K\in\mathcal{T}_h} K$. We define $h \equiv \max_{K\in\mathcal{T}_h} \mathrm{diam}(K)$ and suppose that $\{\mathcal{T}_h\}_h$ is regular:

Assumption (B)

(B1) There exists a positive constant $C_\mathcal{T}$ *such that*

$$\max_{K\in\mathcal{T}_h} \frac{\mathrm{diam}(K)}{\rho_K} \leq C_\mathcal{T} \qquad \forall h > 0\,,$$

where ρ_K *is the diameter of the largest ball inscribed in the simplex* K.

We also use a dual partition $\mathcal{D}_h$ of Ω, such that $\overline{\Omega} = \bigcup_{D\in\mathcal{D}_h} D$. The dual volume D associated to the side σ_D is constructed by connecting the barycentres of every $K \in \mathcal{T}_h$ that contains σ_D through the vertices of σ_D. For σ_D from the boundary, the contour is completed by σ_D itself, see Fig. 1. We denote by Q_D the barycentre of σ_D, by $\mathcal{D}_h^{int}$ the set of all interior and by $\mathcal{D}_h^{ext}$

the set of all boundary dual volumes, and by $\mathcal{N}(D)$ the set of all adjacent volumes to the volume D. For $E \in \mathcal{N}(D)$, we finally set $\sigma_{D,E} = \partial D \cap \partial E$.

We suppose the partition of the time interval $(0,T)$ such that $0 = t_0 < \ldots < t_n < \ldots < t_N = T$ and define $\Delta t_n \equiv t_n - t_{n-1}$, $\Delta t \equiv \max_{1\leq n\leq N} \Delta t_n$. When Assumption *(A5)* is satisfied, we do not impose any restriction on Δt. When only *(A6)* holds, we suppose:

Assumption (C)

(C1) The maximum time step condition $\Delta t < \dfrac{c_\beta}{L_F}$ *is satisfied.*

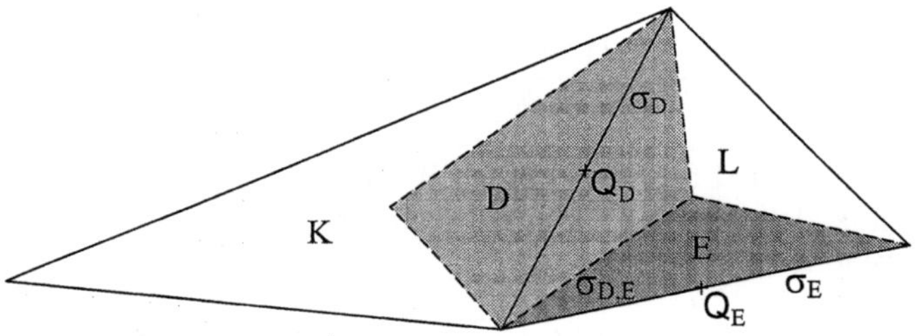

Fig. 1. Triangles $K, L \in \mathcal{T}_h$ and dual volumes $D, E \in \mathcal{D}_h$ associated with edges σ_D, σ_E

We define the following finite-dimensional spaces:

$$X_h \equiv \{\varphi_h \in L^2(\Omega);\ \varphi_h|_K \text{ is linear } \forall K \in \mathcal{T}_h,$$
$$\varphi_h \text{ is continuous at } Q_D, D \in \mathcal{D}_h^{int}\},$$
$$X_h^0 \equiv \{\varphi_h \in X_h;\ \varphi_h(Q_D) = 0 \quad \forall D \in \mathcal{D}_h^{ext}\}.$$

The basis of X_h is spanned by the shape functions φ_D, $D \in \mathcal{D}_h$, such that $\varphi_D(Q_E) = \delta_{DE}$, $E \in \mathcal{D}_h$, δ being the Kronecker delta. We equip X_h^0 with the norm

$$\|c_h\|_{X_h}^2 \equiv \sum_{K\in\mathcal{T}_h} \int_K |\nabla c_h|^2 \,\mathrm{d}\mathbf{x}. \tag{5}$$

Definition 2. (Combined scheme) *The fully implicit combined nonconforming/mixed-hybrid FE–FV scheme reads: find the values* c_D^n, $n \in \{0,1,\ldots,N\}$, $D \in \mathcal{D}_h$, *such that*

$$c_D^0 = \frac{1}{|D|}\int_D c_0(\mathbf{x})\,\mathrm{d}\mathbf{x} \qquad D \in \mathcal{D}_h^{int}, \tag{6}$$

$$c_D^n = 0 \qquad D \in \mathcal{D}_h^{ext},\ n \in \{0,1,\ldots,N\}, \tag{7}$$

$$\frac{\beta(c_D^n) - \beta(c_D^{n-1})}{\Delta t_n}\,|D| - \sum_{E\in\mathcal{N}(D)} \mathbb{D}_{D,E}^n (c_E^n - c_D^n) + \sum_{E\in\mathcal{N}(D)} \mathbf{v}_{D,E}^n\,\overline{c_{D,E}^n} +$$
$$+F(c_D^n)\,|D| = q_D^n\,|D| \qquad D \in \mathcal{D}_h^{int},\ n \in \{1,2,\ldots,N\}. \tag{8}$$

In (6)–(8), $\mathbf{v}^n_{D,E} = \frac{1}{\Delta t_n}\int_{t_{n-1}}^{t_n}\int_{\sigma_{D,E}} \mathbf{v}(\mathbf{x},t)\cdot \mathbf{n}_{D,E}\,\mathrm{d}\gamma(\mathbf{x})\,\mathrm{d}t$ with $\mathbf{n}_{D,E}$ the unit normal vector of the side $\sigma_{D,E}$, outward to D, and

$$q^n_D = \frac{1}{\Delta t_n |D|}\int_{t_{n-1}}^{t_n}\int_D q(\mathbf{x},t)\,\mathrm{d}\mathbf{x}\,\mathrm{d}t.$$

Finally,

$$\begin{array}{lll} \text{if} & \mathbf{v}^n_{D,E} \geq 0 & \overline{c^n_{D,E}} \equiv c^n_D + \alpha^n_{D,E}(c^n_E - c^n_D) \\ \text{if} & \mathbf{v}^n_{D,E} < 0 & \overline{c^n_{D,E}} \equiv c^n_E + \alpha^n_{D,E}(c^n_D - c^n_E) \end{array}. \tag{9}$$

Here, $\alpha^n_{D,E}$ is the coefficient of the amount of upstream weighting, defined by

$$\alpha^n_{D,E} \equiv \frac{\max\left\{\min\left\{\mathbb{D}^n_{D,E}, \frac{1}{2}|\mathbf{v}^n_{D,E}|\right\}, 0\right\}}{|\mathbf{v}^n_{D,E}|}\quad , \quad \mathbf{v}^n_{D,E} \neq 0\,. \tag{10}$$

Remark 1. (Numerical flux) We can easily see that $0 \leq \alpha^n_{D,E} \leq 1/2$, i.e. the numerical flux defined by (9) ranges from the centered scheme to the full upstream weighting.

Diffusion matrix from the nonconforming method We set

$$\mathbb{D}^n_{D,E} \equiv -\sum_{K\in\mathcal{T}_h} (\mathbf{D}^n \nabla\varphi_E, \nabla\varphi_D)_{0,K} \qquad D,E\in\mathcal{D}_h\,,\ n\in\{1,2,\dots,N\}\,,$$

where

$$\mathbf{D}^n(\mathbf{x}) \equiv \frac{1}{\Delta t_n}\int_{t_{n-1}}^{t_n} \mathbf{D}(\mathbf{x},t)\,\mathrm{d}t \qquad n\in\{1,2,\dots,N\}\,,\ \mathbf{x}\in\Omega\,. \tag{11}$$

Diffusion matrix from the mixed-hybrid method Let us consider the problems

$$\begin{aligned} -\nabla\cdot(\mathbf{D}^n\nabla p) &= g \qquad \text{in } \Omega\,,\\ p &= 0 \qquad \text{in } \partial\Omega\,, \end{aligned}$$

at each discrete time t_n, with $g\in L^2(\Omega)$. Then using the hybridization of the lowest order Raviart–Thomas mixed finite element method, one ends up with a linear system $\mathbb{M}^n\Lambda = G$ for the Lagrange multipliers Λ located in barycentres of sides, see [4, Section V.1.2]. Using the analytic form of $\mathbb{M}^n$, we define

$$\mathbb{D}^n_{D,E} \equiv \mathbb{M}^n_{D,E} = -\sum_{K\in\mathcal{T}_h} (\widetilde{\mathbf{D}}^n \nabla\varphi_E, \nabla\varphi_D)_{0,K} \qquad D,E\in\mathcal{D}_h\,,\ n\in\{1,2,\dots,N\}$$

where

$$\widetilde{\mathbf{D}}^n(\mathbf{x}) = \left(\frac{1}{|K|}\int_K (\mathbf{D}^n)^{-1}\,\mathrm{d}\mathbf{x}\right)^{-1} \qquad \mathbf{x}\in K\,,\ K\in\mathcal{T}_h\,,\ n\in\{1,2,\dots,N\}\,. \tag{12}$$

In the sequel, we shall consider apart the following special case, satisfied e.g. when $\mathbf{D} = Id$ and when there is a maximal angle condition:

Assumption (D)

(D1) All non-diagonal terms of the diffusion matrix are nonnegative, i.e.

$$\mathbb{D}^n_{D,E} \geq 0 \qquad \forall D, E \in \mathcal{D}_h^{int}, \, D \neq E \quad \forall n \in \{1, 2, \ldots, N\}.$$

4 Existence, uniqueness, and discrete properties

Lemma 1. (Conservativity of the scheme) *The scheme* (6)–(8) *is conservative with respect to the dual mesh.*

One can easily verify that $\mathbb{D}^n_{D,D} = -\sum_{E \in \mathcal{N}(D)} \mathbb{D}^n_{D,E}$ for all $D \in \mathcal{D}_h$ and $n \in \{1, 2, \ldots, N\}$. Adding the finite volume discretization of the other terms, the assertion follows.

Lemma 2. (Coercivity of the bilinear diffusion form) *We have*

$$-\sum_{D \in \mathcal{D}_h} c_D \sum_{E \in \mathcal{D}_h} \mathbb{D}^n_{D,E} c_E \geq c_{\mathbf{D}} \|c_h\|^2_{X_h} \quad \forall c_h = \sum_{D \in \mathcal{D}_h} c_D \varphi_D \in X_h, \, \forall n \in \{1, 2, \ldots, N\}.$$

The assertion follows immediately from Assumption *(A3)* and the subsequent uniform positive definiteness of the diffusion tensors (11) or (12).

Lemma 3. (Estimate on the convection term) *We have*

$$\sum_{D \in \mathcal{D}_h^{int}} c_D \sum_{E \in \mathcal{N}(D)} \mathbf{v}^n_{D,E} \, \overline{c_{D,E}} \geq 0 \quad \forall c_h = \sum_{D \in \mathcal{D}_h} c_D \varphi_D \in X_h^0, \, \forall n \in \{1, 2, \ldots, N\}.$$

The proof is similar to that in [6] for pure finite volume schemes.

Lemma 4. (A priori estimate for an extended scheme) *Let $u \in [0, 1]$. We define an extended scheme by*

$$c_D^0 = \frac{1}{|D|} \int_D c_0(\mathbf{x}) \, \mathrm{d}\mathbf{x} \qquad D \in \mathcal{D}_h^{int}, \tag{13}$$

$$c_D^n = 0 \qquad D \in \mathcal{D}_h^{ext}, \, n \in \{0, 1, \ldots, N\}, \tag{14}$$

$$u \, \frac{\beta(c_D^n) - \beta(c_D^{n-1})}{\Delta t_n} \, |D| - \sum_{E \in \mathcal{D}_h^{int}} \mathbb{D}^n_{D,E} \, c_E^n + u \sum_{E \in \mathcal{N}(D)} \mathbf{v}^n_{D,E} \, \overline{c^n_{D,E}} +$$
$$+ u \, F(c_D^n) \, |D| = u \, q_D^n \, |D| \qquad D \in \mathcal{D}_h^{int}, \, n \in \{1, 2, \ldots, N\}. \tag{15}$$

Then $\sum_{D \in \mathcal{D}_h} (c_D^n)^2 |D| < C_{\mathrm{es}}$ *for all* $n \in \{1, 2, \ldots, N\}$ *with* $C_{\mathrm{es}} > 0$.

PROOF:

We multiply (15) by $\Delta t_n c_D^n$ and sum over $D \in \mathcal{D}_h^{int}$ and $n \le k$ to have

$$u \sum_{n=1}^{k} \sum_{D \in \mathcal{D}_h^{int}} [\beta(c_D^n) - \beta(c_D^{n-1})] c_D^n |D| + c_{\mathrm{D}} \sum_{n=1}^{k} \Delta t_n \|c_h^n\|_{X_h}^2 \le \tag{16}$$

$$\le u \sum_{n=1}^{k} \Delta t_n \sum_{D \in \mathcal{D}_h^{int}} c_D^n q_D^n \, |D| + u L_F M^2 \sum_{n=1}^{k} \sum_{D \in \mathcal{D}_h^{int}} \Delta t_n |D| ,$$

considering $u \ge 0$, Lemmas 2 and 3, the fact that for $c_D^n < 0$ or $c_D^n > M$, $F(c_D^n) c_D^n \ge 0$ follows from Assumption *(A5)* or *(A6)*, and that when $0 \le c_D^n \le M$, $-F(c_D^n) c_D^n \le |F(c_D^n)||c_D^n| \le L_F M^2$. Let us now introduce a function B, $B(s) \equiv \beta(s)s - \int_0^s \beta(\tau)\,\mathrm{d}\tau$, $s \in \mathbb{R}$. One then can derive

$$B(c_D^n) - B(c_D^{n-1}) = [\beta(c_D^n) - \beta(c_D^{n-1})] c_D^n - \int_{c_D^{n-1}}^{c_D^n} [\beta(\tau) - \beta(c_D^{n-1})] \,\mathrm{d}\tau \,. \tag{17}$$

Using that β is non decreasing, one can easily show that $\int_{c_D^{n-1}}^{c_D^n} [\beta(\tau) - \beta(c_D^{n-1})] \,\mathrm{d}\tau \ge 0$. In view of this and (17), one has

$$\sum_{n=1}^{k} \sum_{D \in \mathcal{D}_h^{int}} [B(c_D^n) - B(c_D^{n-1})] |D| \le \sum_{n=1}^{k} \sum_{D \in \mathcal{D}_h^{int}} [\beta(c_D^n) - \beta(c_D^{n-1})] c_D^n |D| ,$$

which yields

$$\sum_{D \in \mathcal{D}_h^{int}} B(c_D^k)|D| - \sum_{D \in \mathcal{D}_h^{int}} B(c_D^0)|D| \le \sum_{n=1}^{k} \sum_{D \in \mathcal{D}_h^{int}} [\beta(c_D^n) - \beta(c_D^{n-1})] c_D^n |D| .$$

Using the growth condition on β from Assumption *(A1)*, one can derive $B(s) \ge \frac{c_\beta}{2} s^2$ for all $s \in \mathbb{R}$. Thus, using in addition Assumption *(A8)*,

$$\frac{c_\beta}{2} \sum_{D \in \mathcal{D}_h^{int}} (c_D^k)^2 |D| - M\beta(M)|\Omega| \le \sum_{n=1}^{k} \sum_{D \in \mathcal{D}_h^{int}} [\beta(c_D^n) - \beta(c_D^{n-1})] c_D^n \, |D| .$$

Using the Cauchy-Schwarz inequality, extending the summation over all $n \in \{1, 2, \dots, N\}$ and $D \in \mathcal{D}_h$ in the first right term of (16), and using the Young inequality, we have

$$\sum_{n=1}^{k} \Delta t_n \sum_{D \in \mathcal{D}_h^{int}} c_D^n q_D^n \, |D| \le \Big(\sum_{n=1}^{N} \Delta t_n \sum_{D \in \mathcal{D}_h} (c_D^n)^2 |D| \Big)^{\frac{1}{2}} \|q\|_{0,Q_T} \le$$

$$\le \frac{\varepsilon}{2} \sum_{n=1}^{N} \Delta t_n \sum_{D \in \mathcal{D}_h} (c_D^n)^2 |D| + \frac{1}{2\varepsilon} \|q\|_{0,Q_T}^2 \,.$$

Substituting all the above estimates into (16), we obtain

$$u\frac{c_\beta}{2}\max_{n\in\{1,2,\dots,N\}}\sum_{D\in\mathcal{D}_h}(c_D^n)^2|D|+c_{\mathbf{D}}\sum_{n=1}^{k}\triangle t_n\|c_h^n\|_{X_h}^2\le uM\beta(M)|\Omega|+ \qquad (18)$$

$$+u\frac{\varepsilon}{2}T\max_{n\in\{1,2,\dots,N\}}\sum_{D\in\mathcal{D}_h}(c_D^n)^2|D|+u\frac{1}{2\varepsilon}\|q\|_{0,Q_T}^2+uL_FM^2T|\Omega|\,,$$

considering also (14) and the fact that k was arbitrarily chosen. We now choose $\varepsilon=\frac{c_\beta}{2T}$. When $u\neq 0$, this already leads the assertion of the lemma. When $u=0$, it follows from (18) that $c_D^n=0$ for all $D\in\mathcal{D}_h$ and all $n\in\{1,2,\dots,N\}$, since $\|\cdot\|_{X_h}$ is a norm on X_h^0. Thus the assertion of the lemma is trivially satisfied in this case. □

Theorem 1. (Existence of the solution to the discrete problem) *The problem* (6)–(8) *has at least one solution.*

The proof makes use of an induction argument. At each time level, Lemma 4 is employed. Consequently, on can use the (Brouwer) topological degree argument (see [5]).

Theorem 2. (Uniqueness of the solution to the discrete problem) *The solution to the problem* (6)–(8) *is unique.*

The assertion follows from Assumption *(A1)* and *(A5)* or *(A6)*.

Theorem 3. (Discrete maximum principle) *Under Assumption (D1), the solution of the problem* (6)–(8) *satisfies, for all* $D\in\mathcal{D}_h$ *and* $n\in\{1,2,\dots,N\}$,

$$0\le c_D^n\le M\,. \qquad (19)$$

One sets a transmissibility $\mathbb{E}_{D,E}^n\equiv\mathbb{D}_{D,E}^n-|\mathbf{v}_{D,E}^n|\alpha_{D,E}^n$, $E\in\mathcal{N}(D)$. In view of Assumption *(D1)* and (10), one has $\mathbb{E}_{D,E}^n\ge 0$ for all $D\in\mathcal{D}_h^{int}$, $E\in\mathcal{N}(D)$, and hence one can prove the assertion as in [6].

5 A priori estimates

Theorem 4. (A priori estimates) *The solution of the scheme (6)–(8) satisfies*

$$c_\beta\max_{n\in\{1,2,\dots,N\}}\sum_{D\in\mathcal{D}_h}(c_D^n)^2|D|\le C_{\mathrm{ae}}\,, \qquad (20)$$

$$\max_{n\in\{1,2,\dots,N\}}\sum_{D\in\mathcal{D}_h}[\beta(c_D^n)]^2|D|\le C_{\mathrm{ae}}\,, \qquad (21)$$

$$c_{\mathbf{D}}\sum_{n=1}^{N}\triangle t_n\|c_h^n\|_{X_h}^2\le C_{\mathrm{ae}} \qquad (22)$$

with $c_h^n = \sum_{D \in \mathcal{D}_h} c_D^n \varphi_D$ *and* C_{ae} *a constant independent of* h *and* Δt.

The a priori estimates follow from (18) with $\varepsilon = \dfrac{c_\beta}{2T}$ and $u = 1$ and using Assumption *(A1)* or Assumption *(A2)*.

Definition 3. (Approximate solution) *As the approximate solution of* (1)–(3) *by means of the combined nonconforming/mixed-hybrid FE–FV scheme, we understand:*

(*i*) *a function* $c_{h,\Delta t}$ *given by* $c_h^n \in X_h^0$, $c_h^n = \sum_{D \in \mathcal{D}_h} c_D^n \varphi_D$ *with* c_D^n, $D \in \mathcal{D}_h$, $n \in \{0, 1, \ldots, N\}$, *solutions to* (6–8), *such that* $c_{h,\Delta t}$ *is piecewise constant in time;*

(*ii*) *a function* $\tilde{c}_{h,\Delta t}$ *given by the values* c_D^n, $D \in \mathcal{D}_h$, $n \in \{0, 1, \ldots, N\}$, *solutions to* (6–8), *and piecewise constant on the dual volumes* $D \in \mathcal{D}_h$ *and in time.*

The function $c_{h,\Delta t}$ is piecewise linear on $\mathcal{T}_h$ and continuous at the barycentres of the interior sides, whereas $\tilde{c}_{h,\Delta t}$ is piecewise constant on $\mathcal{D}_h$. From the a priori estimate (22), it follows immediately that

$$\|c_{h,\Delta t} - \tilde{c}_{h,\Delta t}\|_{0,Q_T} \longrightarrow 0 \quad as \quad h \to 0\,. \tag{23}$$

Hence, to show the convergence, we can work with $\tilde{c}_{h,\Delta t}$ as in finite volume methods.

Lemma 5. (Time translate estimate) *There exists a constant* $C_{\mathrm{tt}} > 0$, *such that*

$$\int_0^{T-\tau} \int_\Omega \Big(\tilde{c}_{h,\Delta t}(\mathbf{x}, t+\tau) - \tilde{c}_{h,\Delta t}(\mathbf{x}, t)\Big)^2 \,\mathrm{d}\mathbf{x}\,\mathrm{d}t \le C_{\mathrm{tt}}(\tau + \Delta t) \qquad \forall \tau \in (0,T)\,.$$

The proof is an adaptation of the technique used in [8] for degenerate parabolic equations discretized by a finite volume scheme. It uses the equation (8) and the priori estimate (22).

Lemma 6. (Space translate estimate) *Let us define* $\tilde{c}_{h,\Delta t}$ *by zero outside of* Ω. *Then there exists a constant* $C_{\mathrm{st}} > 0$, *such that*

$$\int_0^{T} \int_\Omega \Big(\tilde{c}_{h,\Delta t}(\mathbf{x}+\boldsymbol{\xi}, t) - \tilde{c}_{h,\Delta t}(\mathbf{x}, t)\Big)^2 \,\mathrm{d}\mathbf{x}\,\mathrm{d}t \le C_{\mathrm{st}}|\boldsymbol{\xi}|(|\boldsymbol{\xi}| + h) \qquad \forall \boldsymbol{\xi} \in \mathbb{R}^d\,.$$

The proof is again an adaptation of a technique used to investigate finite volume schemes.

6 Convergence

Theorem 5. (Strong convergence in $L^2(Q_T)$) *There exist subsequences of $\tilde{c}_{h,\triangle t}$ and $c_{h,\triangle t}$ which converge strongly in $L^2(Q_T)$ to some function $u \in L^2(0,T;H_0^1(\Omega))$.*

PROOF:

From Lemmas 5 and 6 and (20), the sequence $\tilde{c}_{h,\triangle t}$ verifies the assumptions of Kolmogorov's theorem [3, Theorem IV.25], and thus $\tilde{c}_{h,\triangle t}$ converges strongly in $L^2(Q_T)$ to some function $u \in L^2(Q_T)$. Moreover, due to Lemma 6, [6, Theorem 3.10] gives that this $u \in L^2(0,T;H_0^1(\Omega))$. Finally, considering (23), $c_{h,\triangle t}$ converges to the same u. □

Theorem 6. (Convergence to a weak solution) *There exist subsequences of $\tilde{c}_{h,\triangle t}$ and $c_{h,\triangle t}$ which converge strongly in $L^2(Q_T)$ to a weak solution given by (4). If the weak solution is unique, then the whole sequences $\tilde{c}_{h,\triangle t}$, $c_{h,\triangle t}$ converge to the weak solution.*

The strong convergence of Theorem 5 permits to pass to the limit in the nonlinear terms. For slightly more restrictive assumptions than *(A)*, the uniqueness follows from [9].

References

1. Angot, A., Dolejší, V., Feistauer, M., Felcman, J. (1998): Analysis of a combined barycentric finite volume–nonconforming finite element method for nonlinear convection–diffusion problems. Appl. Math. Praha, **43**, 263–310
2. Barrett, J. W., Knabner, P. (1997): Finite element approximation of transport of reactive solutes in porous media. Part II: Error estimates for the equilibrium adsorption processes. SIAM J. Numer. Anal., **34**, 455–479
3. Brezis, H. (1983): Analyse fonctionnelle, théorie et applications. Masson, Paris
4. Brezzi, F., Fortin, M. (1991): Mixed and Hybrid Finite Element Methods. Springer-Verlag, New York
5. Deimling, K. (1985): Nonlinear Functional Analysis. Springer-Verlag, Berlin-Heidelberg
6. Eymard, R., Gallouët, T., Herbin, R. (2000): The finite volume method. In: Ciarlet P.G., Lions J.L. (eds) Handbook of numerical analysis, vol. 7, 715–1022, Elsevier Science Publishers B.V. (North-Holland), Amsterdam
7. Eymard, R., Gallouët, T., Herbin, R., Michel, M. (2002): Convergence of a finite volume scheme for nonlinear degenerate parabolic equations. Numer. Math., **92**, 41–82
8. Eymard, R., Gallouët, T., Hilhorst, D., Naït Slimane, Y. (1998): Finite volumes and nonlinear diffusion equations. Modél. Math. Anal. Numér., **32** , 747–761
9. Knabner, P., Otto, F. (2000): Solute transport in porous media with equilibrium and nonequilibrium multiple-site adsorption: uniqueness of weak solutions. Nonlinear Anal., **42** , 381–403

Discrete Maximum Principle for Galerkin Finite Element Solutions to Parabolic Problems on Rectangular Meshes

István Faragó[1], Róbert Horváth[2] and Sergey Korotov[3]

[1] Department of Applied Analysis, Eötvös Loránd University, H-1518, Budapest, Pf. 120, Hungary, *faragois@cs.elte.hu*, supported by the National Scientific Research Found (OTKA) N. T043765

[2] Institute of Mathematics and Statistics, University of West-Hungary, Erzsébet u. 9, H-9400, Sopron, Hungary, *rhorvath@ktk.nyme.hu*, supported by the National Scientific Research Found (OTKA) N. T043765

[3] Department of Mathematical Information Technology, University of Jyväskylä, P.O. Box 35, FIN–40014 Jyväskylä, Finland, *korotov@mit.jyu.fi*, supported by the Agora Center under Grant InBCT of TEKES.

Summary. One of the most important problems in numerical simulation is the preservation of qualitative properties of solutions of mathematical models. For problems of parabolic type, one of such properties is the maximum principle. In [5], Fujii analyzed the discrete analogue of the (continuous) maximum principle for the linear parabolic problems, and derived sufficient conditions guaranteeing its validity for the Galerkin finite element approximations built on simplicial meshes. In our paper, we present the sufficient conditions for the validity of the discrete maximum principle for the case of bilinear finite element space approximations on rectangular meshes.

1 Introduction

Consider a two-dimensional linear parabolic problem in the classical setting: Find a function $u \in C^{1,2}((0,T) \times \Omega) \cap C([0,T) \times \bar{\Omega})$ such that

$$\frac{\partial u}{\partial t} - \alpha \sum_{k=1}^{2} \frac{\partial^2 u}{\partial x_k^2} = f \quad \text{in} \quad (0,T) \times \Omega, \tag{1}$$

$$u = g \quad \text{on} \quad [0,T) \times \partial\Omega, \quad \text{and} \quad u|_{t=0} = u_0 \quad \text{in} \quad \Omega, \tag{2}$$

where Ω is a polygonal domain in $\mathbb{R}^2$ with a boundary $\partial\Omega$, $T > 0$ and α is a positive constant. In order to guarantee the existence and uniqueness of the classical solution $u = u(t, x_1, x_2) = u(t, x)$, we assume that the functions $u_0 : \Omega \to \mathbb{R}$, $f : (0,T) \times \Omega \to \mathbb{R}$ and $g : [0,T) \times \partial\Omega \to \mathbb{R}$ are sufficiently smooth.

The problem (1)–(2) serves as the mathematical model of various physical, chemical or even ecological phenomena. It is well-known that the estimation

$$\min\{0; \min_{\Gamma_t} u\} + t\min\{0; \min_{Q_t} f\} \le u(t,x) \le \max\{0; \max_{\Gamma_t} u\} + t\max\{0; \max_{Q_t} f\} \qquad (3)$$

is valid for the solution u (see [7, Theorem 2.1] and [9, p. 79]). Here Q_t with $t \in [0,T]$ stands for the cylinder $(0,t) \times \Omega$ and Γ_t is the union of its lateral surface and its bottom. Formula (3) is called the *continuous maximum principle.*

To solve problem (1)–(2) numerically, we use certain discretizations, both in spatial and in time coordinates. It is obvious that the validity of the discrete analogue of the maximum principle, the so-called *discrete maximum principle* (DMP), is a natural requirement for having an adequate numerical solution.

The topic of validity of various discrete maximum principles arose already 30 years ago. Thus, in the works [3] and [4], DMP was formulated and proved for the finite difference and finite element approximations, respectively, for the second order linear elliptic equations. In particular, in [4], DMP was proved in 2D case for the continuous piecewise linear finite element approximations under the following geometrical conditions: the angles of triangles in the used triangular meshes are not greater than $\pi/2$ (*nonobtuse type condition*), or less than $\pi/2$ (*acute type condition*). More results on DMP for the elliptic problems can be found in [6] and [8].

In [5], the validity of the DMP for the linear parabolic problem is analyzed: the finite element discretization was performed with linear elements on triangular (simplicial) meshes and the so-called θ-method was applied to the time discretization. The discrete analogue of (3) and sufficient conditions guaranteeing its validity were obtained, where one of such conditions was the acuteness of the triangulations used.

To the authors' knowledge, there is no similar result on the validity of DMP for parabolic problems solved with the help of bilinear finite elements in space. As far as the validity of DMP on rectangular meshes is concerned, we mention the only work [2] in this respect, where the authors considered the simplest elliptic problem and showed that the corresponding DMP may not hold if the rectangular elements are chosen arbitrarily. They also derived sufficient conditions for the validity of DMP.

In our paper, we give sufficient conditions for the validity of the discrete maximum principle for the Galerkin finite element solutions based on the bilinear elements on rectangular meshes for parabolic problems. Similarly to the case of triangular meshes, we obtain certain geometrical condition on the shape of elements. Namely, we introduce the notion of *non-narrow rectangular element*, which represents an analogue of nonobtuse triangular element for the case of triangular meshes.

2 Discretization

2.1 Galerkin Finite Element Discretization with θ-method

Let Ω be a rectangular domain covered by the rectangular mesh $\mathcal{R}_h$, where h stands for the discretization parameter. Let $P_1, ..., P_N$ denote the interior nodes, and $P_{N+1}, \dots, P_{\bar{N}}$ the boundary ones in $\mathcal{R}_h$. We also define $N_\partial := \bar{N} - N$.

Let $\phi_1, \dots, \phi_{\bar{N}}$ be basis functions defined as follows: each ϕ_i is required to be continuous piecewise bilinear such that $\phi_i(P_j) = \delta_{ij}$, $i, j = 1, \dots, \bar{N}$, where δ_{ij} is the Kronecker's symbol. It is obvious that these basis functions have the properties

$$(a) \;\; \phi_i \geq 0, \; i = 1, ..., \bar{N}, \qquad (b) \;\; \sum_{i=1}^{\bar{N}} \phi_i \equiv 1 \;\; \text{in } \overline{\Omega}. \tag{4}$$

We denote the space of all possible linear combinations of the basis functions by V^h, and define its subspace $V_0^h = \{v \in V^h \mid v|_{\partial\Omega} = 0\}$. Based on the usual weak formulation of the original problem, the semidiscrete form for (1)–(2) reads: Find a function $u_h = u_h(t, x)$ such that

$$u_h(0, x) = u_0^h(x), \quad x \in \Omega, \tag{5}$$

$$u_h(t, x) - g_h(t, x) \in V_0^h, \quad t \in (0, T), \tag{6}$$

and

$$\int_\Omega \frac{\partial u_h}{\partial t} v_h \, dx + B(u_h, v_h) = \int_\Omega f v_h \, dx, \quad \forall v_h \in V_0^h, \quad t \in (0, T), \tag{7}$$

where $B(u_h, v_h) = \alpha \int_\Omega \text{grad } u_h \cdot \text{grad } v_h \, dx$. In the above, $u_0^h(x)$ and $g_h(t, x)$ (for any fixed t) are linear interpolants in V^h, i.e.,

$$u_0^h(x) = \sum_{i=1}^{\bar{N}} u_0(P_i)\phi_i(x) \quad \text{and} \quad g_h(t, x) = \sum_{i=1}^{N_\partial} g(t, P_{N+i})\phi_{N+i}(x).$$

We notice that from the consistency of the initial and the boundary conditions ($g(0, s) = u_0(s)$, $s \in \partial\Omega$), we observe that $g(0, P_{N+i}) = u_0(P_{N+i})$, $i = 1, \dots, N_\partial$.

We search for the semidiscrete solution in the form

$$u_h(t, x) = \sum_{i=1}^{N} u_i^h(t)\phi_i(x) + g_h(t, x),$$

and notice that it is sufficient that u_h satisfies (7) for $v_h = \phi_i$, $i = 1, \dots, N$, only.

Introducing the denotation

$$\mathbf{v}^h(t) = [u_1^h(t), \ldots, u_N^h(t), g(t, P_{N+1}), \ldots, g(t, P_{N+N_\partial})]^T,$$

we arrive at the Cauchy problem for the system of ordinary differential equations

$$\mathbf{M}\frac{\mathrm{d}\mathbf{v}^h}{\mathrm{d}t} + \mathbf{K}\mathbf{v}^h = \mathbf{f}, \qquad (8)$$
$$\mathbf{v}^h(0) = [u_0(P_1), \ldots, u_0(P_N), g(0, P_{N+1}), \ldots, g(0, P_{N+N_\partial})]^T$$

for the solution of the semidiscrete problem, where

$$\mathbf{M} = [M_{ij}]_{N\times\bar{N}}, \qquad \mathbf{K} = [K_{ij}]_{N\times\bar{N}}, \qquad \mathbf{f} = [f_i]_{N\times 1}, \qquad (9)$$

$$M_{ij} = \int_\Omega \phi_j\phi_i \, dx, \qquad K_{ij} = B(\phi_j, \phi_i), \qquad f_i = \int_\Omega f\phi_i \, dx. \qquad (10)$$

The above defined matrices $\mathbf{M}$ and $\mathbf{K}$ are called *mass* and *stiffness matrices*, respectively.

In order to get a fully discrete numerical scheme, we choose a time-step Δt and denote the approximation to $\mathbf{v}^h(n\Delta t)$ and $\mathbf{f}(n\Delta t)$ by $\mathbf{v}^n$ and $\mathbf{f}^n$, $n = 0, 1, ..., n_T$ ($n_T\Delta t = T$), respectively. To discretize (9), we apply the θ-method ($\theta \in [0, 1]$ is a given parameter) and obtain the system of linear algebraic equations

$$(\mathbf{M}+\theta\Delta t\mathbf{K})\mathbf{v}^{n+1} = (\mathbf{M}-(1-\theta)\Delta t\mathbf{K})\mathbf{v}^n+\Delta t\, \mathbf{f}^{(n,\theta)}, \quad n = 0, 1, \ldots, n_T-1, \quad (11)$$

where $\mathbf{v}^0 = \mathbf{v}^h(0)$ and $\mathbf{f}^{(n,\theta)} = \theta\mathbf{f}^{n+1} + (1-\theta)\mathbf{f}^n$.

Further, let the matrices $\mathbf{M} + \theta\Delta t\mathbf{K}$ and $\mathbf{M} - (1-\theta)\Delta t\mathbf{K}$ be denoted by $\mathbf{A}$ and $\mathbf{B}$ respectively. In what follows, we shall use the partitions

$$\mathbf{A} = [\mathbf{A}_0|\mathbf{A}_\partial], \quad \mathbf{B} = [\mathbf{B}_0|\mathbf{B}_\partial], \quad \mathbf{v}^n = \begin{bmatrix}\mathbf{u}^n\\ \mathbf{g}^n\end{bmatrix}, \qquad (12)$$

where $\mathbf{A}_0$ and $\mathbf{B}_0$ are square matrices from $\mathbb{R}^{N\times N}$; $\mathbf{A}_\partial, \mathbf{B}_\partial \in \mathbb{R}^{N\times N_\partial}$, $\mathbf{u}^n = [u_1^n, ..., u_N^n]^T \in \mathbb{R}^N$, and $\mathbf{g}^n = [g_1^n, ..., g_{N_\partial}^n]^T \in \mathbb{R}^{N_\partial}$. (Similar partition is used for the matrices $\mathbf{M}$ and $\mathbf{K}$.) Then, the iteration (11) can be also written as

$$[\mathbf{A}_0|\mathbf{A}_\partial]\begin{bmatrix}\mathbf{u}^{n+1}\\ \mathbf{g}^{n+1}\end{bmatrix} = [\mathbf{B}_0|\mathbf{B}_\partial]\begin{bmatrix}\mathbf{u}^n\\ \mathbf{g}^n\end{bmatrix} + \Delta t\, \mathbf{f}^{(n,\theta)}. \qquad (13)$$

The iteration is well-defined, because $\mathbf{A}_0$ is a Gram-matrix, thus it is invertible.

2.2 Entries of the mass and stiffness matrices

To calculate the entries of the mass matrix $\mathbf{M}$, we first calculate the integral of the product $\phi_j\phi_i$ only on a single rectangle R (denoted by $M_{ij}|_R$). Obviously, it is sufficient to do that on the reference rectangle defined by the vertices $P_1 = (0,0)$, $P_2 = (a,0)$, $P_3 = (a,b)$ and $P_4 = (0,b)$. The four basis functions corresponding to the vertices of the reference rectangle are:

$$\phi_1(x_1,x_2) = \tfrac{1}{ab}(x_1 - a)(x_2 - b), \qquad \phi_2(x_1,x_2) = -\tfrac{1}{ab}x_1(x_2 - b),$$

$$\phi_3(x_1,x_2) = \tfrac{1}{ab}x_1x_2, \qquad \phi_4(x_1,x_2) = -\tfrac{1}{ab}(x_1 - a)x_2.$$

A simple calculation leads to $(M_{ij}|_R = M_{ji}|_R)$

$$M_{11}|_R = M_{22}|_R = M_{33}|_R = M_{44}|_R = \frac{ab}{9},$$
$$M_{12}|_R = M_{14}|_R = M_{23}|_R = M_{34}|_R = \frac{ab}{18},$$
$$M_{13}|_R = M_{24}|_R = \frac{ab}{36},$$

where ab denotes the area of the rectangle R.

The elements of the stiffness matrix (integrating only on R) can be obtained similarly. Thus,

$$K_{11}|_R = \frac{\alpha(a^2 + b^2)}{3ab}, \quad K_{12}|_R = -\frac{\alpha(2b^2 - a^2)}{6ab}, \tag{14}$$

$$K_{13}|_R = -\frac{\alpha(a^2 + b^2)}{6ab}, \quad K_{14}|_R = -\frac{\alpha(2a^2 - b^2)}{6ab},$$

and any other value is equal to one of the above four numbers.

2.3 Non-narrow rectangular meshes

In paper [5], a geometrical condition, the acuteness of the triangular meshes, guaranteed the nonpositivity of the off-diagonal entries of **K**. The situation is similar for the case of the bilinear elements. The nonpositivity of the off-diagonal entries of **K** is fulfilled for so-called non-narrow rectangular meshes.

Let $\mathcal{R}_h$ be a rectangular mesh and let us introduce the notation

$$\mu = \max_{R \in \mathcal{R}_h} \frac{\max^2\{a_R, b_R\} - 2\min^2\{a_R, b_R\}}{a_R b_R},$$

where a_R and b_R denote the length of the edges of the rectangle R. A rectangular mesh $\mathcal{R}_h$ is called *non-narrow* if $\mu \leq 0$. It is called *strictly non-narrow* if $\mu < 0$. Hence, the non-narrowness of a mesh means that the longest edge of each rectangle is not greater than $\sqrt{2}$ times the shortest one. The non-narrowness of the mesh will imply the nonpositivity of the off-diagonal elements of the stiffness matrix (see [1], page 254).

3 The Discrete Maximum Principle

Let us define the values

$$g^n_{\min} = \min\{0, g^n_1, \ldots, g^n_{N_\partial}\}, \quad g^n_{\max} = \max\{0, g^n_1, \ldots, g^n_{N_\partial}\},$$

$$v^n_{\min} = \min\{0, g^n_{\min}, u^n_1, \ldots, u^n_N\}, \quad v^n_{\max} = \max\{0, g^n_{\max}, u^n_1, \ldots, u^n_N\},$$

for $n = 0, ..., n_T$, and

$$f^{(n,n+1)}_{\min} = \min\{0, \min_{x\in\Omega, \tau\in(n\Delta t,(n+1)\Delta t)} f(\tau, x)\},$$

$$f^{(n,n+1)}_{\max} = \max\{0, \max_{x\in\Omega, \tau\in(n\Delta t,(n+1)\Delta t)} f(\tau, x)\},$$

for $n = 0, ..., n_T - 1$.

The discrete analogue, the so-called discrete maximum principle (DMP), for the continuous maximum principle (3) can be written in the form (cf. [5, p. 100])

$$\min\{0, g^{n+1}_{\min}, v^n_{\min}\} + \Delta t f^{(n,n+1)}_{\min} \le u^{n+1}_i \le \max\{0, g^{n+1}_{\max}, v^n_{\max}\} + \Delta t f^{(n,n+1)}_{\max}, \tag{15}$$

$i = 1, \ldots, N$; $n = 0, ..., n_T - 1$.

Let us introduce the denotations

$$\mathbf{e} = [1, \ldots, 1]^T \in \mathbb{R}^{\bar N}, \ \mathbf{e}_0 = [1, \ldots, 1]^T \in \mathbb{R}^N, \ \mathbf{e}_\partial = [1, \ldots, 1]^T \in \mathbb{R}^{N_\partial},$$

$$\mathbf{f}^{(n,n+1)}_{\max} = f^{(n,n+1)}_{\max}\mathbf{e} \in \mathbb{R}^{\bar N}, \ \mathbf{f}^{(n,n+1)}_0 = f^{(n,n+1)}_{\max}\mathbf{e}_0 \in \mathbb{R}^N,$$
$$\mathbf{f}^{(n,n+1)}_\partial = f^{(n,n+1)}_{\max}\mathbf{e}_\partial \in \mathbb{R}^{N_\partial},$$

$$\mathbf{v}^n_{\max} = v^n_{\max}\mathbf{e} \in \mathbb{R}^{\bar N}, \qquad \mathbf{v}^n_0 = v^n_{\max}\mathbf{e}_0 \in \mathbb{R}^N, \qquad \mathbf{v}^n_\partial = v^n_{\max}\mathbf{e}_\partial \in \mathbb{R}^{N_\partial}.$$

For simplicity, we denote zero matrices and zero vectors by the symbol $\mathbf{0}$, whose size is always chosen according to the context. The ordering relation is meant elementwise.

Before proving the sufficient condition of the DMP, in the next auxiliary lemmas, some important properties of the matrices $\mathbf{M}, \mathbf{K}$, $\mathbf{A}$ and $\mathbf{B}$ are summarized.

Lemma 1. *Let the rectangular mesh $\mathcal{R}_h$ for Ω be of non-narrow type ($\mu \le 0$). Then $K_{ij} \le 0$ ($i \ne j$, $i = 1, \ldots, N$, $j = 1, \ldots, \bar N$).*

PROOF. We denote $\mathrm{supp}\phi_i \cap \mathrm{supp}\phi_j$ by S_{ij} and calculate K_{ij} ($i \ne j$):

$$K_{ij} = \alpha \int_\Omega \mathrm{grad}\phi_j \cdot \mathrm{grad}\phi_i \, dx = \alpha \sum_{R\subseteq S_{ij}} \int_R \mathrm{grad}\phi_j \cdot \mathrm{grad}\phi_i \, dx = \sum_{R\subseteq S_{ij}} K_{ij}|_R.$$

Because $K_{ij}|_R$ ($i \ne j$) is nonpositive for any non-narrow rectangle, we observe that the off-diagonal entries of the stiffness matrix are nonpositive. ■

Lemma 2. *Let the rectangular mesh $\mathcal{R}_h$ for Ω be of non-narrow type ($\mu \leq 0$) and condition*

$$B_{ii} = M_{ii} - (1-\theta)\Delta t K_{ii} \geq 0, \quad i = 1, \ldots, N, \tag{16}$$

be satisfied. Then $\mathbf{B} \geq 0$.

PROOF. The matrix $\mathbf{M}$ is nonnegative, because the basis functions are nonnegative. Moreover, the previous lemma guarantees the nonpositivity of K_{ij} $(i \neq j)$ that implies that the off-diagonal entries of $\mathbf{B}$ are nonnegative. The nonnegativity of the diagonal entries of $\mathbf{B}$ follows from the condition (16). ■

Lemma 3. *The relations*

$$(P1)\ \ \mathbf{Ke} = \mathbf{0}, \qquad (P2)\ \ \mathbf{f}^{(n,\theta)} \leq \mathbf{Af}_{\max}^{(n,n+1)}$$

are valid.

PROOF. (P1) For the i-th coordinate of the vector $\mathbf{Ke}$, we have

$$(\mathbf{Ke})_i = \sum_{j=1}^{\bar{N}} B(\phi_j, \phi_i) = B\left(\sum_{j=1}^{\bar{N}} \phi_j, \phi_i\right) = B(1, \phi_i) =$$

$$= \alpha \int_\Omega \operatorname{grad} 1 \cdot \operatorname{grad} \phi_i \, dx = 0,$$

which proves the statement.

(P2) For the i-th element of $\mathbf{f}^{(n,\theta)}$, we observe that

$$(\mathbf{f}^{(n,\theta)})_i = \int_\Omega ((1-\theta) f(n\Delta t, x) + \theta f((n+1)\Delta t, x)) \phi_i(x) \, dx \leq$$

$$\leq \int_\Omega f_{\max}^{(n,n+1)} \phi_i(x) \, dx = f_{\max}^{(n,n+1)} \int_\Omega \left(\sum_{j=1}^{\bar{N}} \phi_j(x)\right) \phi_i(x) \, dx =$$

$$= f_{\max}^{(n,n+1)} \sum_{j=1}^{\bar{N}} M_{ij} = \left(\mathbf{Mf}_{\max}^{(n,n+1)}\right)_i = \left((\mathbf{M} + \theta \Delta t \mathbf{K}) \mathbf{f}_{\max}^{(n,n+1)}\right)_i = \left(\mathbf{Af}_{\max}^{(n,n+1)}\right)_i.$$

In the above, we used the facts that the basis functions are nonnegative, their sum equals to the constant one function, and property (P1). ■

Lemma 4. *If*

$$A_{ij} = M_{ij} + \theta \Delta t K_{ij} \leq 0, \quad i \neq j, \ i = 1, \ldots, N, \ j = 1, \ldots, \bar{N}, \tag{17}$$

then $\mathbf{A}_0^{-1} \geq 0$. *Furthermore, the relations*

$$-\mathbf{A}_0^{-1} \mathbf{A}_\partial \geq \mathbf{0}, \quad -\mathbf{A}_0^{-1} \mathbf{A}_\partial \, \mathbf{e}_\partial \leq \mathbf{e}_0 \tag{18}$$

are valid.

PROOF. Combining the positive definite property of $\mathbf{A}_0$ with the assumption of the lemma, we obtain that $\mathbf{A}_0$ is a non-singular M-matrix. This yields that the inverse of $\mathbf{A}_0$ is nonnegative. Because of relation (17), the matrix $\mathbf{A}_\partial$ is nonpositive, which implies $-\mathbf{A}_0^{-1}\mathbf{A}_\partial \geq \mathbf{0}$. The matrix $\mathbf{M}$ is nonnegative, because the basis functions are nonnegative. Thus,

$$\mathbf{0} \leq \mathbf{Me} = (\mathbf{M} + \theta \Delta t \mathbf{K})\mathbf{e} = \mathbf{Ae} = \mathbf{A}_0\mathbf{e}_0 + \mathbf{A}_\partial \mathbf{e}_\partial,$$

and the last statement of the lemma can be obtained by multiplying both sides by $\mathbf{A}_0^{-1}$. ∎

Now, we prove the main result of our paper, which presents a sufficient condition for the validity of DMP (cf. Theorem 1 in [5]).

Theorem 1. *Let the rectangular mesh $\mathcal{R}_h$ of Ω be of non-narrow type and let the time increment Δt satisfy* (16) *and* (17). *Then the discrete maximum principle* (*relation* (15)) *is valid.*

PROOF. Using (13), property ($P2$), the relation $\mathbf{Bv}^n_{\max} = \mathbf{Av}^n_{\max}$ (it follows from ($P1$)) and Lemma 2, we have

$$\mathbf{Av}^{n+1} = \mathbf{Bv}^n + \Delta t\, \mathbf{f}^{(n,\theta)} \leq \mathbf{Bv}^n_{\max} + \Delta t \mathbf{A} \mathbf{f}^{(n,n+1)}_{\max} = \mathbf{Av}^n_{\max} + \Delta t \mathbf{A}\mathbf{f}^{(n,n+1)}_{\max}. \tag{19}$$

From (19), using the partition (12), multiplying both sides by $\mathbf{A}_0^{-1}$ ($\geq \mathbf{0}$, see Lemma 4), and regrouping the inequality, we get

$$\mathbf{u}^{n+1} - \mathbf{v}^n_0 - \Delta t \mathbf{f}^{(n,n+1)}_0 \leq -\mathbf{A}_0^{-1}\mathbf{A}_\partial(\mathbf{g}^{n+1} - \mathbf{v}^n_\partial - \Delta t \mathbf{f}^{(n,n+1)}_\partial).$$

Obviously,

$$e(\mathbf{g}^{n+1} - \mathbf{v}^n_\partial - \Delta t \mathbf{f}^{(n,n+1)}_\partial)_i = g_i^{n+1} - v^n_{\max} - \Delta t f^{(n,n+1)}_{\max} \leq$$

$$\leq \max\{0, \max_j \{g_j^{n+1} - v^n_{\max}\}\}. \tag{20}$$

Therefore, using (19)–(20) and Lemma 4, we get

$$\mathbf{u}^{n+1} - \mathbf{v}^n_0 - \Delta t \mathbf{f}^{(n,n+1)}_0 \leq \max\{0, \max_j\{g_j^{n+1} - v^n_{\max}\}\}\mathbf{e}_0. \tag{21}$$

Writing (21) for the i-th component, and expressing u_i^{n+1}, we obtain the right-hand side inequality in (15). The left-hand side inequality in (15) can be proved in a similar manner. ∎

The previous theorem does not say anything about the choice of the rectangulation and the choice of the parameters θ and Δt in order to guarantee the DMP. The validity of the DMP can be checked only after the direct calculation of the elements of the matrices $\mathbf{A}$ and $\mathbf{B}$ by testing the two inequalities in the above theorem. The next theorem can guarantee the DMP *a priori*.

Theorem 2. *The finite element solution of* (1)–(2)*, using bilinear basis functions on a strictly non-narrow rectangular mesh $\mathcal{R}_h$ of a rectangular domain Ω, satisfies the discrete maximum principle* (15) *if the conditions*

$$\Delta t \geq \frac{\lambda^2}{3\,\theta|\mu|\alpha}, \tag{22}$$

and

$$\Delta t \leq \frac{\gamma^2}{3\,(1-\theta)\alpha}, \tag{23}$$

are fulfilled, where

$$\lambda = \max_{R\in\mathcal{R}_h}\{\sqrt{ab}\}, \quad \gamma = \min_{R\in\mathcal{R}_h}\left\{\frac{a_R b_R}{\sqrt{a_R^2+b_R^2}}\right\}.$$

Proof. It is easy to show that under conditions (22)–(23), the sufficient conditions of the DMP are satisfied in Theorem 1. The two inequalities in the theorem can be proven using condition (23) and (22), respectively. ■

4 Final comments

In this paper *a priori* sufficient conditions for the validity of the discrete maximum principle have been given for the Galerkin finite element methods based on bilinear finite elements in space. We close the paper with some remarks regarding our results.

- As it usually happens in the qualitative analysis of finite element approximations, there are both, upper and lower bounds for the time-step, which means that Δt cannot be chosen neither too small nor too large.
- A square mesh with the mesh-size h is, obviously, strictly non-narrow. On such meshes, the sufficient conditions for the DMP are

$$\Delta t \geq \frac{h^2}{3\theta\alpha} \tag{24}$$

and

$$\Delta t \leq \frac{h^2}{6(1-\theta)\alpha}. \tag{25}$$

This shows that the time step can be chosen only for the values $\theta \geq 2/3$, i.e., the Crank-Nicolson scheme is not included.
- The results of Theorem 1 are similar to Theorem 1 in [5]. The only difference is the application of the condition of the strict non-narrowness instead of the acute type condition.

References

1. O. Axelsson, V.A. Barker, *Finite Element Solution of Boundary Value Problems*, Theory and Computation. Academic Press, Inc. 1984.
2. I. Christie, C. Hall, *The Maximum Principle for Bilinear Elements*, Internat. J. Numer. Methods Engrg. 20 (1984), pp. 549–553.
3. P. G. Ciarlet, *Discrete Maximum Principle for Finite Difference Operators*, Aequationes Math. 4 (1970), pp. 338–352.
4. P. G. Ciarlet, P. A. Raviart, *Maximum Principle and Uniform Convergence for the Finite Element Method*, Comput. Methods Appl. Mech. Engrg. 2 (1973), pp. 17–31.
5. H. Fujii, *Some Remarks on Finite Element Analysis of Time-Dependent Field Problems*, Theory and Practice in Finite Element Structural Analysis, Univ. Tokyo Press, Tokyo (1973), pp. 91–106.
6. S. Korotov, M. Křížek, P. Neittaanmäki, *Weakened Acute Type Condition for Tetrahedral Triangulations and the Discrete Maximum Principle*, Math. Comp. 70, (2001), pp. 107-119.
7. O. A. Ladyzenskaja, V. A. Solonnikov, N. N. Ural'ceva, *Linear and Quasilinear Equations of Parabolic Type*, Translations of Mathematical Monographs, Vol. 23, American Mathematical Society, Providence, R.I., 1968.
8. V. Ruas Santos, *On the Strong Maximum Principle for Some Piecewise Linear Finite Element Approximate Problems of Non-Positive Type*, J. Fac. Sci. Univ. Tokyo Sect. IA Math. 29 (1982), pp. 473–491.
9. J. Smoller, *Shock Waves and Reaction-Diffusion Equations*, Springer Verlag, 1981.

Cubature-Differences Method for Singular Integro-differential Equations

Alexander I. Fedotov

Chebotarev Institute of Mathematics & Mechanics, Kazan, Russia fedotov@mi.ru

Summary. In the papers [1] - [4] the quadrature-differences methods for the various classes of the 1-dimensional periodic singular integro-differential equations with Hilbert kernels were justified. The convergence of the methods was proved and error estimates were obtained. Here we propose and justify the cubature-differences method for 2-dimensional [1] linear periodic singular integro-differential equations. Such equations appear in the theory of elastity (see [5]) and in some problems of diffraction of electromagnetic waves (see e.g. [6]) The convergence of the method is proved and error estimate is obtained.

1 Statement of the problem

Let's define the sets $\mathbf{N} = \mathrm{N}^2$, $\mathbf{Z} = \mathrm{Z}^2$, $\mathbf{R} = \mathrm{R}^2$, $\boldsymbol{\Delta} = [-\pi;\pi]^2$. For the elements of this sets (2-components vectors) beside the usual operations we'll define the following operations

$$\mathbf{l}\cdot\mathbf{k} = l_1k_1 + l_2k_2,\ \mathbf{l}*\mathbf{k} = (l_1k_1, l_2k_2),\ \mathbf{l}^2 = l_1^2 + l_2^2,\ \mid\mathbf{l}\mid = l_1 + l_2,\ [\mathbf{l}] = l_1l_2,$$

and the partial order

$$\mathbf{l} < \mathbf{k} \equiv (l_1 < k_1)\&(l_2 < k_2), \quad \mathbf{l} = (l_1, l_2), \quad \mathbf{k} = (k_1, k_2).$$

For the fixed $s \in \mathrm{R}$ let's denote by H^s the Sobolev space of 2-dimensional 2π-periodic complex-valued functions with the norm

$$\|u\|_s = \|u\|_{H^s} = (\sum_{\mathbf{k}\in\mathbf{Z}} (1+\mathbf{k}^2)^s \mid \widehat{u}(\mathbf{k}) \mid^2)^{1/2},$$

where

$$\widehat{u}(\mathbf{k}) = (2\pi)^{-2}\int_{\boldsymbol{\Delta}} u(\boldsymbol{\tau})\bar{e}_{\mathbf{k}}(\boldsymbol{\tau})d\boldsymbol{\tau}$$

are the Fourier coefficients of the function $u(\boldsymbol{\tau})$ to the system of trigonometric monomials

$$e_{\mathbf{k}}(\boldsymbol{\tau}) = \exp(i\mathbf{k}\cdot\boldsymbol{\tau}), \qquad \mathbf{k}\in\mathbf{Z}, \boldsymbol{\tau}\in\boldsymbol{\Delta}.$$

[1] 2-dimensional case is considered only in sake of simplicity. All results could be easily generalised to the case of m ($m \geq 3$) dimensions.

For the following we'll asume that $s > 1$ providing (see e.g. [8]) the embedding of H^s into the space of continuous functions.

Consider the linear singular integro-differential equation

$$ABu + Tu = f, \tag{1}$$

where A is a 2-dimensional singular integral operator

$$Au \equiv a_{00}(\mathbf{t})u(\mathbf{t}) + a_{01}(\mathbf{t})(J_{01}u)(\mathbf{t}) + a_{10}(\mathbf{t})(J_{10}u)(\mathbf{t}) + a_{11}(\mathbf{t})(J_{11}u)(\mathbf{t}),$$

$$A : H^s \to H^s,$$

with the singular integrals

$$(J_{01}u)(\mathbf{t}) = (2\pi)^{-1} \int_{-\pi}^{\pi} u(t_1, \tau_2) \cot \frac{\tau_2 - t_2}{2} d\tau_2,$$

$$(J_{10}u)(\mathbf{t}) = (2\pi)^{-1} \int_{-\pi}^{\pi} u(\tau_1, t_2) \cot \frac{\tau_1 - t_1}{2} d\tau_1,$$

$$(J_{11}u)(\mathbf{t}) = (2\pi)^{-2} \int_{-\pi}^{\pi} \int_{-\pi}^{\pi} u(\tau_1, \tau_2) \cot \frac{\tau_1 - t_1}{2} \cot \frac{\tau_2 - t_2}{2} d\tau_2 d\tau_1$$

which are to be interpreted as the Cauchy-Lebesgue principal value, B is an elliptic differential operator

$$Bu \equiv (Bu)(\mathbf{t}) = \sum_{|\boldsymbol{\alpha}|=|\boldsymbol{\beta}|=m} b_{\boldsymbol{\alpha\beta}}(\mathbf{t})(D^{\boldsymbol{\alpha}+\boldsymbol{\beta}}u)(\mathbf{t}), \ B : H^{s+2m} \to H^s, \ m \in \mathrm{N},$$

with derivatives

$$D^{\boldsymbol{\alpha}}u = \frac{\partial^{|\boldsymbol{\alpha}|}u}{\partial {t_1}^{\alpha_1} \partial {t_2}^{\alpha_2}} \qquad \text{of order } \boldsymbol{\alpha} = (\alpha_1, \alpha_2) \in \mathbf{N},$$

and $T : H^{s+2m} \to H^s$ is known linear operator. The coefficients $a_{kl}(\mathbf{t}),\ k, l = 0, 1$, $b_{\boldsymbol{\alpha\beta}}(\mathbf{t})$, $| \boldsymbol{\alpha} |=| \boldsymbol{\beta} |= m$, and the right-hand side $f(\mathbf{t})$ of equation (1) are assumed to belong to C^∞.

2 Calculation scheme

Let's fix $\mathbf{n} = (n_1, n_2) \in \mathbf{N}$, denote by

$$\mathbf{I_n} = I_{n_1} \times I_{n_2},\ I_{n_j} = \{k_j \mid k_j \in Z, | k_j | \le n_j\},\ j = 1, 2,$$

the index set and difine the grid

$$\boldsymbol{\Delta}_{\mathbf{n}} = \{\mathbf{t_k} = (t_{k_1}, t_{k_2}) \mid \mathbf{k} \in \mathbf{I_n}, t_{k_j} = k_j h_j, h_j = 2\pi/(2n_j + 1), j = 1, 2\}.$$

on $\boldsymbol{\Delta}$. The approximate solution of equation (1) we'll seek as a periodic grid function (vector of values) $u_{\mathbf{n}} = u_{\mathbf{n}}(\mathbf{t})$ defined on $\boldsymbol{\Delta}_{\mathbf{n}}$.

The differential operators $D^{\boldsymbol{\alpha}+\boldsymbol{\beta}}$ of equation (1) we'll approximate by the operators

$$D_{\mathbf{n}}^{\boldsymbol{\alpha}+\boldsymbol{\beta}} u_{\mathbf{n}} = \frac{1}{2}(\partial^{\boldsymbol{\alpha}}\bar{\partial}^{\boldsymbol{\beta}} + \bar{\partial}^{\boldsymbol{\alpha}}\partial^{\boldsymbol{\beta}})u_{\mathbf{n}}, \tag{2}$$

where

$$\partial^{\boldsymbol{\alpha}} u_{\mathbf{n}} = \partial_1^{\alpha_1}\partial_2^{\alpha_2} u_{\mathbf{n}}, \qquad \bar{\partial}^{\boldsymbol{\alpha}} u_{\mathbf{n}} = \bar{\partial}_1^{\alpha_1}\bar{\partial}_2^{\alpha_2} u_{\mathbf{n}},$$

$$\partial_j u_{\mathbf{n}} = h_j^{-1}(u_{\mathbf{n}}(\mathbf{t} + h_j\boldsymbol{\delta}_j) - u_{\mathbf{n}}(\mathbf{t})), \quad \bar{\partial}_j u_{\mathbf{n}} = h_j^{-1}(u_{\mathbf{n}}(\mathbf{t}) - u_{\mathbf{n}}(\mathbf{t} - h_j\boldsymbol{\delta}_j)),$$

$$\boldsymbol{\delta}_j = (\delta_{j1}, \delta_{j2}), \ j = 1, 2,$$

and δ_{jk} is Kronecker symbol.

Singular integrals are to be approximated by cubatures and quadratures. To do this we'll integrate interpolative Lagrange polynomials

$$(P_{\mathbf{n}} u_{\mathbf{n}})(\boldsymbol{\tau}) = \sum_{\mathbf{k}\in\mathbf{I_n}} u_{\mathbf{n}}(\mathbf{t_k})\xi_{\mathbf{n}}(\boldsymbol{\tau}, \mathbf{t_k}),$$

$$\xi_{\mathbf{n}}(\boldsymbol{\tau}, \mathbf{t_k}) = \prod_{j=1,2} \frac{\sin((2n_j+1)(\tau_j - t_{k_j})/2)}{(2n_j+1)\sin((\tau_j - t_{k_j})/2)},$$

$$\boldsymbol{\tau} = (\tau_1, \tau_2) \in \boldsymbol{\Delta}, \ \mathbf{t_k} = (t_{k_1}, t_{k_2}) \in \boldsymbol{\Delta}_{\mathbf{n}}.$$

Then the integrals will take the form

$$(J_{01} P_{\mathbf{n}} u_{\mathbf{n}})(\mathbf{t_k}) = (2n_2+1)^{-1} \sum_{l_2\in I_{n_2}} \gamma_{k_2-l_2}^{(n_2)} u_{\mathbf{n}}(t_{k_1}, t_{l_2}), \tag{3}$$

$$(J_{10} P_{\mathbf{n}} u_{\mathbf{n}})(\mathbf{t_k}) = (2n_1+1)^{-1} \sum_{l_1\in I_{n_1}} \gamma_{k_1-l_1}^{(n_1)} u_{\mathbf{n}}(t_{l_1}, t_{k_2}),$$

$$(J_{11} P_{\mathbf{n}} u_{\mathbf{n}})(\mathbf{t_k}) = [2\mathbf{n}+\mathbf{1}]^{-1} \sum_{\mathbf{l}\in\mathbf{I_n}} \gamma_{k_1-l_1}^{(n_1)}\gamma_{k_2-l_2}^{(n_2)} u_{\mathbf{n}}(\mathbf{t_l}), \quad \mathbf{t_k} \in \boldsymbol{\Delta}_{\mathbf{n}}, \quad \mathbf{1} = (1,1),$$

and the coefficients $\gamma_r^{(q)}$ are

$$\gamma_r^{(q)} = \{\tan\frac{r\pi}{2(2q+1)}, \ r - \text{even}, \ -\cot\frac{r\pi}{2(2q+1)}, \ r - \text{odd}\}.$$

The operator T we'll approximate by any covergent operator $T_{\mathbf{n}}$.

Substituting numerical differential formulas (2), cubature and quadrature sums (3), values of the coefficients $a_{kl}(\mathbf{t})$, $k, l = 0, 1$, $b_{\boldsymbol{\alpha\beta}}(\mathbf{t})$, $|\boldsymbol{\alpha}| = |\boldsymbol{\beta}| = m$, of the operator $(T_{\mathbf{n}} u_{\mathbf{n}})(\mathbf{t})$ and right-hand side $f(\mathbf{t})$ at the nodes of the grid $\boldsymbol{\Delta}_{\mathbf{n}}$ in equation (1) we'll obtain a system of linear algebraic equations

$$a_{00}(\mathbf{t_k}) \sum_{|\boldsymbol{\alpha}|=|\boldsymbol{\beta}|=m} b_{\boldsymbol{\alpha\beta}}(\mathbf{t_k})(D_{\mathbf{n}}^{\boldsymbol{\alpha}+\boldsymbol{\beta}} u_{\mathbf{n}})(\mathbf{t_k}) + \tag{4}$$

$$+\frac{a_{01}(\mathbf{t_k})}{2n_2+1}\sum_{l_2\in I_{n_2}}\gamma_{k_2-l_2}^{(n_2)}\sum_{|\boldsymbol{\alpha}|=|\boldsymbol{\beta}|=m}b_{\boldsymbol{\alpha\beta}}(t_{k_1},t_{l_2})(D_{\mathbf{n}}^{\boldsymbol{\alpha}+\boldsymbol{\beta}}u_{\mathbf{n}})(t_{k_1},t_{l_2})+$$

$$+\frac{a_{10}(\mathbf{t_k})}{2n_1+1}\sum_{l_1\in I_{n_1}}\gamma_{k_1-l_1}^{(n_1)}\sum_{|\boldsymbol{\alpha}|=|\boldsymbol{\beta}|=m}b_{\boldsymbol{\alpha\beta}}(t_{l_1},t_{k_2})(D_{\mathbf{n}}^{\boldsymbol{\alpha}+\boldsymbol{\beta}}u_{\mathbf{n}})(t_{l_1},t_{k_2})+$$

$$+\frac{a_{11}(\mathbf{t_k})}{[2\mathbf{n}+1]}\sum_{\mathbf{l}\in\mathbf{I_n}}\gamma_{k_1-l_1}^{(n_1)}\gamma_{k_2-l_2}^{(n_2)}\sum_{|\boldsymbol{\alpha}|=|\boldsymbol{\beta}|=m}b_{\boldsymbol{\alpha\beta}}(t_{l_1},t_{l_2})(D_{\mathbf{n}}^{\boldsymbol{\alpha}+\boldsymbol{\beta}}u_{\mathbf{n}})(t_{l_1},t_{l_2})+$$

$$+(T_{\mathbf{n}}u_{\mathbf{n}})(\mathbf{t_k})=f(\mathbf{t_k}),\qquad \mathbf{t_k}\in\boldsymbol{\Delta}_{\mathbf{n}},$$

of the cubature-differences method.

3 Preliminaries

Let's denote by $H_{\mathbf{n}}^s$ the set of grid functions (vectors of values) on $\boldsymbol{\Delta}_{\mathbf{n}}$ with the norm

$$\|u_{\mathbf{n}}\|_{s,\mathbf{n}}=\|u_{\mathbf{n}}\|_{H_{\mathbf{n}}^s}=\Big(\sum_{\mathbf{k}\in\mathbf{I_n}}(1+\mathbf{k}^2)^s\mid\widehat{u}_{\mathbf{n}}(\mathbf{k})^{(\mathbf{n})}\mid^2\Big)^{1/2},$$

where

$$\widehat{u}_{\mathbf{n}}(\mathbf{k})^{(\mathbf{n})}=[2\mathbf{n}+\mathbf{1}]^{-1}\sum_{\mathbf{l}\in\mathbf{I_n}}u_{\mathbf{n}}(\mathbf{t_l})\bar{e}_{\mathbf{k}}(\mathbf{t_l}),\qquad \mathbf{k}\in\mathbf{I_n},$$

are Fourier-Lagrange coefficients of the function $u_{\mathbf{n}}(\mathbf{t})$ belonging to the grid $\boldsymbol{\Delta}_{\mathbf{n}}$.

The sets H^s and $H_{\mathbf{n}}^s$ will be mapped onto each other by the operators

$$p_{\mathbf{n}}u=(u(\mathbf{t_k}))_{\mathbf{k}\in\mathbf{I_n}},\ p_{\mathbf{n}}:H^s\to H_{\mathbf{n}}^s,$$

$$(P_{\mathbf{n}}u_{\mathbf{n}})(\boldsymbol{\tau})=\sum_{\mathbf{k}\in\mathbf{I_n}}u_{\mathbf{n}}(\mathbf{t_k})\xi_{\mathbf{n}}(\boldsymbol{\tau},\mathbf{t_k}),\quad P_{\mathbf{n}}:H_{\mathbf{n}}^s\to H^s,$$

by $E_{\mathbf{n}}(u)_s$ we denotethe best approximation of the function $u\in H^s$ by the trigonometrical polynomials of order not higher than $\mathbf{n}$.

Lemma 1. *For any* $u\in H^s$, $s\in\mathrm{R}$, $s>1$ *and* $\mathbf{n}\in\mathbf{N}$ *the following estimations are valid*

$$\|P_{\mathbf{n}}\|_{H_{\mathbf{n}}^s\to H^s}=1,\qquad \|p_{\mathbf{n}}\|_{H^s\to H_{\mathbf{n}}^s}\le 2M(\mathbf{n},s)\sqrt{\zeta(2s-1)},$$

$$\|P_{\mathbf{n}}p_{\mathbf{n}}u-u\|_s\le(1+2M(\mathbf{n},s)\sqrt{\zeta(2s-1)})E_{\mathbf{n}}(u)_s,$$

where $M(\mathbf{n},s)=\left(\frac{\sqrt{n_1^2+n_2^2}}{\min(n_1,n_2)}\right)^s$, $\mathbf{n}\in\mathbf{N}$, *and* $\zeta(t)$ *is Riemann's* ζ*-function.*

To prove the convergence of the method we need the function $M(\mathbf{n}, s)$ to be bounded. Let's for some $c, s \in R$ define the set

$$\mathbf{N}(c, s) = \{\mathbf{n} \mid \mathbf{n} \in \mathbf{N},\ M(\mathbf{n}, s) \le c\}.$$

Obviously, $\mathbf{N}(c, s) = \emptyset$ for $c < 2^{s/2}$ and $\mathbf{N}(c, s) = \{\mathbf{n} \mid \mathbf{n} = (j, j),\ j \in N\}$ for $c = 2^{s/2}$. For the following we'll mean that all indices $\mathbf{n},\ \mathbf{n}_0,\ \mathbf{n}_1$ mentioned below belong to $\mathbf{N}(c, s)$ for some $c \ge 2^{s/2}$.

Lemma 2. *For any $s \le p$, $u \in H^p$*

$$E_{\mathbf{n}}(u)_s \le (1 + \mathbf{n}^2)^{(s-p)/2} E_{\mathbf{n}}(u)_p.$$

4 Justification

Theorem. *Let for some $c, s \in R$, $s > 1$, $c \ge 2^{s/2}$ equation (1) and calculation scheme (2)–(4) of the method satisfy the following conditions:*

1) *for any $\mathbf{n}$ the operator A maps the set of all trigonometric polynomials of order not higher than $\mathbf{n}$ to itself,*

2) *B is an elliptic operator,*

3) *the operator $T : H^{s+2m} \to H^{s+\varepsilon}$ is bounded for some $\varepsilon \in R$, $\varepsilon > 0$,*

4) *the sequence of the the operators $T_{\mathbf{n}}$ approximates operator T with respect to $p_{\mathbf{n}}$, i.e. for any function $u \in H^s$:*

$$\|T_{\mathbf{n}} p_{\mathbf{n}} u - p_{\mathbf{n}} T u\|_{s,\mathbf{n}} = \eta_{\mathbf{n}} \|u\|_{s+2m} \quad \text{with} \quad \eta_{\mathbf{n}} \to 0 \text{ for } \mathbf{n} \to \infty,$$

5) *equation (1) has a unique solution $u^* \in H^{s+2m}$ for any right-hand side $f \in H^s$.*

Then for all $\mathbf{n}$, beginning from some $\mathbf{n}_0$, the system of equations (4) is uniquely solvable and approximate solutions $u^*_{\mathbf{n}}$ converge to exact solution u^* of equation (1)

$$\|u^*_{\mathbf{n}} - p_{\mathbf{n}} u^*\|_{s+2m,\mathbf{n}} \to 0,\ \mathbf{n} \to \infty.$$

If, in addition, $u^* \in H^{s+2m+2}$, then the error estimate

$$\|u^*_{\mathbf{n}} - p_{\mathbf{n}} u^*\|_{s+2m,\mathbf{n}} \le C(\mathbf{h}^2 + \eta_{\mathbf{n}}), \quad \mathbf{h} = (h_1, h_2),\ h_j = 2\pi/(2n_j + 1),\ j = 1, 2,$$

is valid.

Proof. Let's take an arbitrary constant $r \in R$ which is not an eigenvalue of problem $Bu + ru = 0$, $u \in H^{s+2m}$ and substitute into equation (1)

$$v = Bu + ru, \qquad v \in H^s. \tag{5}$$

The existence of such a constant follows from the properties of the spectrum of elliptic operators (see e.g. [7]). Then

$$u = Gv,\ Bu = v - rGv, \tag{6}$$

where G is the inverse to $Bu + ru$ and equation (1) will take the form

$$Kv \equiv Av - rAGv + TGv = f, \qquad K : H^s \to H^s, \tag{7}$$

being still equivalent to the original one. The equivalence hear means, that solvability of one of the equations yields solvability of the another, and their solutions are related by the relationships (5), (6). Now let's rewrite the system of equations (4) as an operator equation

$$A_{\mathbf{n}}B_{\mathbf{n}}u_{\mathbf{n}} + T_{\mathbf{n}}u_{\mathbf{n}} = f_{\mathbf{n}}, \tag{8}$$

$$A_{\mathbf{n}} = p_{\mathbf{n}}AP_{\mathbf{n}}, \ f_{\mathbf{n}} = p_{\mathbf{n}}f,$$

$$(B_{\mathbf{n}}u_{\mathbf{n}})(\mathbf{t_k}) = \sum_{|\boldsymbol{\alpha}|=|\boldsymbol{\beta}|=m} b_{\boldsymbol{\alpha\beta}}(\mathbf{t_k})(D_{\mathbf{n}}^{\boldsymbol{\alpha}+\boldsymbol{\beta}}u_{\mathbf{n}})(\mathbf{t_k}), \qquad \mathbf{t_k} \in \boldsymbol{\Delta}_{\mathbf{n}},$$

and make the substitution

$$v_{\mathbf{n}} = B_{\mathbf{n}}u_{\mathbf{n}} + ru_{\mathbf{n}}, \qquad v_{\mathbf{n}} \in H_{\mathbf{n}}^s. \tag{9}$$

As it is shown in [10] equation (9) is uniquely solvable for all $\mathbf{n}$, beginning from some $\mathbf{n_1}$, and for $v_{\mathbf{n}} = p_{\mathbf{n}}v$ solutions $u_{\mathbf{n}} = G_{\mathbf{n}}v_{\mathbf{n}} = G_{\mathbf{n}}p_{\mathbf{n}}v$ converge to the solution $u = Gv$ of equation (5). Here $G_{\mathbf{n}}$ is inverse to operator $B_{\mathbf{n}}u_{\mathbf{n}} + ru_{\mathbf{n}}$ and

$$u_{\mathbf{n}} = G_{\mathbf{n}}v_{\mathbf{n}}, \qquad B_{\mathbf{n}}u_{\mathbf{n}} = v_{\mathbf{n}} - rG_{\mathbf{n}}v_{\mathbf{n}}. \tag{10}$$

By substitution (9) we'll get equation

$$K_{\mathbf{n}}v_{\mathbf{n}} \equiv A_{\mathbf{n}}v_{\mathbf{n}} - rA_{\mathbf{n}}G_{\mathbf{n}}v_{\mathbf{n}} + T_{\mathbf{n}}G_{\mathbf{n}}v_{\mathbf{n}} = f_{\mathbf{n}}, \qquad K_{\mathbf{n}} : H_{\mathbf{n}}^s \to H_{\mathbf{n}}^s, \tag{11}$$

which is equivalent to equation (8). As before, the equivalence here means, that solvability of one of equations yields solvability of the another and their solutions are related by the relationships (9), (10).

The invertibility of the operators $K_{\mathbf{n}} : H_{\mathbf{n}}^s \to H_{\mathbf{n}}^s$ we'll prove following [9]. To do this we have to establish the following:

a) $\|P_{\mathbf{n}}f_{\mathbf{n}} - f\|_s \to 0$ for $\mathbf{n} \to \infty$;

b) the sequence of operators $(K_{\mathbf{n}})$ approximates the operator K compactly;

c) K is invertible.

The validity of a) follows immediately from Lemma 1 [2]

$$\|P_{\mathbf{n}}f_{\mathbf{n}} - f\|_s = \|P_{\mathbf{n}}p_{\mathbf{n}}f - f\|_s \le CE_{\mathbf{n}}(f)_s.$$

To check b) we have to show first that the sequence $(K_{\mathbf{n}})$ approximates the operator K with respect to $P_{\mathbf{n}}$, and then that for any bounded sequence $(v_{\mathbf{n}})$, $v_{\mathbf{n}} \in H_{\mathbf{n}}^s$ the sequence $(P_{\mathbf{n}}K_{\mathbf{n}}v_{\mathbf{n}} - KP_{\mathbf{n}}v_{\mathbf{n}})$ is compact in H^s.

For arbitrary $v_{\mathbf{n}} \in H_{\mathbf{n}}^s$ we'll write

$$\|P_{\mathbf{n}}K_{\mathbf{n}}v_{\mathbf{n}} - KP_{\mathbf{n}}v_{\mathbf{n}}\|_s \le \|P_{\mathbf{n}}A_{\mathbf{n}}v_{\mathbf{n}} - AP_{\mathbf{n}}v_{\mathbf{n}}\|_s + \tag{12}$$

[2] Here and further C denotes generic real positive constants, independent from $\mathbf{n}$.

$$+ \mid r \mid \|P_{\mathbf{n}}A_{\mathbf{n}}G_{\mathbf{n}}v_{\mathbf{n}} - AGP_{\mathbf{n}}v_{\mathbf{n}}\|_s + \|P_{\mathbf{n}}T_{\mathbf{n}}G_{\mathbf{n}}v_{\mathbf{n}} - TGP_{\mathbf{n}}v_{\mathbf{n}}\|_s$$

and estimate each summand of the right-hand side. From the definition of the operator $A_{\mathbf{n}}$ and condition 1) of Theorem it follows that the first summand is equal to zero. For the second summand, using once more the definition of the operator $A_{\mathbf{n}}$, condition 1) of Theorem and the boundness of the operators A and $P_{\mathbf{n}}$, we'll have

$$\mid r \mid \|P_{\mathbf{n}}A_{\mathbf{n}}G_{\mathbf{n}}v_{\mathbf{n}} - AGP_{\mathbf{n}}v_{\mathbf{n}}\|_s \leq C\|P_{\mathbf{n}}p_{\mathbf{n}}AP_{\mathbf{n}}G_{\mathbf{n}}v_{\mathbf{n}} - AGP_{\mathbf{n}}v_{\mathbf{n}}\|_s \leq$$

$$\leq C(\|G_{\mathbf{n}}v_{\mathbf{n}} - p_{\mathbf{n}}GP_{\mathbf{n}}v_{\mathbf{n}}\|_{s+2m,\mathbf{n}} + E_{\mathbf{n}}(GP_{\mathbf{n}}v_{\mathbf{n}})_{s+2m}).$$

For the third summand, using Lemma 1 and the boundness of the operators $T_{\mathbf{n}}$, we'll obtain

$$\|P_{\mathbf{n}}T_{\mathbf{n}}G_{\mathbf{n}}v_{\mathbf{n}} - TGP_{\mathbf{n}}v_{\mathbf{n}}\|_s \leq C(\|G_{\mathbf{n}}v_{\mathbf{n}} - p_{\mathbf{n}}GP_{\mathbf{n}}v_{\mathbf{n}}\|_{s+2m,\mathbf{n}} +$$

$$+\|T_{\mathbf{n}}p_{\mathbf{n}}GP_{\mathbf{n}}v_{\mathbf{n}} - p_{\mathbf{n}}TGP_{\mathbf{n}}v_{\mathbf{n}}\|_{s,\mathbf{n}} + E_{\mathbf{n}}(TGP_{\mathbf{n}}v_{\mathbf{n}})_s).$$

Finally, estimation (12) will take the form

$$\|P_{\mathbf{n}}K_{\mathbf{n}}v_{\mathbf{n}} - KP_{\mathbf{n}}v_{\mathbf{n}}\|_s \leq C(\|G_{\mathbf{n}}v_{\mathbf{n}} - p_{\mathbf{n}}GP_{\mathbf{n}}v_{\mathbf{n}}\|_{s+2m,\mathbf{n}} +$$

$$+\|T_{\mathbf{n}}p_{\mathbf{n}}GP_{\mathbf{n}}v_{\mathbf{n}} - p_{\mathbf{n}}TGP_{\mathbf{n}}v_{\mathbf{n}}\|_{s,\mathbf{n}} + E_{\mathbf{n}}(GP_{\mathbf{n}}v_{\mathbf{n}})_{s+2m} + E_{\mathbf{n}}(TGP_{\mathbf{n}}v_{\mathbf{n}})_s),$$

which, taking into account condition 4) of Theorem, convergence of operators $(G_{\mathbf{n}})$ and convergence to zero of the best approximations of functions $GP_{\mathbf{n}}v_{\mathbf{n}}$ and $TGP_{\mathbf{n}}v_{\mathbf{n}}$, means that

$$\|P_{\mathbf{n}}K_{\mathbf{n}}v_{\mathbf{n}} - KP_{\mathbf{n}}v_{\mathbf{n}}\|_s \to 0,$$

and thus the approximation of operator K by sequence of operators $(K_{\mathbf{n}})$ with respect to $P_{\mathbf{n}}$.

Let's assume now, that sequence $(v_{\mathbf{n}})$, $v_{\mathbf{n}} \in H^s_{\mathbf{n}}$ is bounded $\|v_{\mathbf{n}}\|_{s,\mathbf{n}} \leq 1$, and prove that sequence $(P_{\mathbf{n}}K_{\mathbf{n}}v_{\mathbf{n}} - KP_{\mathbf{n}}v_{\mathbf{n}})$ is compact in H^s. We'll write

$$P_{\mathbf{n}}K_{\mathbf{n}}v_{\mathbf{n}} - KP_{\mathbf{n}}v_{\mathbf{n}} = rAGP_{\mathbf{n}}v_{\mathbf{n}} - TGP_{\mathbf{n}}v_{\mathbf{n}} - rAP_{\mathbf{n}}G_{\mathbf{n}}v_{\mathbf{n}} + P_{\mathbf{n}}T_{\mathbf{n}}G_{\mathbf{n}}v_{\mathbf{n}},$$

and prove compactness of each summand of the right-hand side. Operators $G : H^s \to H^{s+2m}$, $T : H^{s+2m} \to H^{s+\varepsilon}$, $A : H^{s+2m} \to H^{s+2m}$ are bounded, so sequences $(rAGP_{\mathbf{n}}v_{\mathbf{n}})$ and $(TGP_{\mathbf{n}}v_{\mathbf{n}})$ are bounded in $H^{s+\gamma}, \gamma = \min(2m, \varepsilon)$ and thus compact in H^s. Operators $G_{\mathbf{n}} : H^s_{\mathbf{n}} \to H^{s+2m}_{\mathbf{n}}$ and $T_{\mathbf{n}}G_{\mathbf{n}} : H^s_{\mathbf{n}} \to H^{s+\varepsilon}_{\mathbf{n}}$ are also bounded so polynomials $P_{\mathbf{n}}G_{\mathbf{n}}v_{\mathbf{n}}$ and $P_{\mathbf{n}}T_{\mathbf{n}}G_{\mathbf{n}}v_{\mathbf{n}}$ are bounded in $H^{s+\gamma}$ and thus, due to Riesz theorem, sequences $(rAP_{\mathbf{n}}G_{\mathbf{n}}v_{\mathbf{n}})$ and $(P_{\mathbf{n}}T_{\mathbf{n}}G_{\mathbf{n}}v_{\mathbf{n}})$ are also compact in H^s, which gives the compactness of sequence $(P_{\mathbf{n}}K_{\mathbf{n}}v_{\mathbf{n}} - KP_{\mathbf{n}}v_{\mathbf{n}})$.

Validity of c) follows from condition 5) of Theorem and equivalence of equations (1) and (7).

Therefore, according to Theorem 6.1 [9], for all $\mathbf{n}$, beginning from some $\mathbf{n}_0$, $\mathbf{n}_0 \geq \mathbf{n}_1$, equations (11), (8), and thus system of equations (4) are uniquely solvable and approximate solutions $(u^*_{\mathbf{n}})$ of system of equations (4) converge to the exact solution u^* of equation (1) with a rate

$$\|u^*_{\mathbf{n}} - p_{\mathbf{n}}u^*\|_{s+2m,\mathbf{n}} \leq C\|p_{\mathbf{n}}(ABu^* + Tu^*) - (A_{\mathbf{n}}B_{\mathbf{n}}p_{\mathbf{n}}u^* + T_{\mathbf{n}}p_{\mathbf{n}}u^*)\|_{s,\mathbf{n}} \leq$$

$$\leq C(E_{\mathbf{n}}(Bu^*)_s + \|p_{\mathbf{n}}Bu^* - B_{\mathbf{n}}p_{\mathbf{n}}u^*\|_{s,\mathbf{n}} + \|p_{\mathbf{n}}Tu^* - T_{\mathbf{n}}p_{\mathbf{n}}u^*\|_{s,\mathbf{n}}).$$

If, moreover, $u^* \in H^{s+2m+2}$, then $Bu^* \in H^{s+2}$ and as it is shown in [9],

$$\|p_{\mathbf{n}}Bu^* - B_{\mathbf{n}}p_{\mathbf{n}}u^*\|_{s,\mathbf{n}} \leq C\mathbf{h}^2.$$

On the other hand, according to Lemma 2, and using the inequality $(1 + \mathbf{n}^2)^{-q} \leq C(\mathbf{h}^2)^q$, $q \in R$, $q > 0$, we'll have

$$E_{\mathbf{n}}(Bu^*)_s \leq (1 + \mathbf{n}^2)^{-1}E_{\mathbf{n}}(Bu^*)_{s+2} \leq C(\mathbf{h}^2),$$

which, together with condition 4) of the Theorem gives desired estimation

$$\|u^*_{\mathbf{n}} - p_{\mathbf{n}}u^*\|_{s+2m,\mathbf{n}} \leq C(\mathbf{h}^2 + \eta_{\mathbf{n}}).$$

Hence, the Theorem is proved.

References

1. Fedotov, A. I., On convergence of quadrature-differences method for linear singular integrodifferential equations, Zh. Vychisl. Mat. i. Mat. Fiz. 29 (1989), N9, 1301-1307 (in Russian).
2. Fedotov, A. I., On convergence of quadrature-differences method for one class of singular integro-differential equations, Izv. Vyssh. Uchebn. Zaved. Mat. (1989), N8, 64-68 (in Russian).
3. Fedotov, A. I., On convergence of quadrature-differences method for linear singular integrodifferential equations with discontinuous coefficients, Zh. Vychisl. Mat. i. Mat. Fiz. 31 (1991), N2, 261-271 (in Russian).
4. Fedotov, A. I., On convergence of quadrature-differences method for nonlinear singular integrodifferential equations, Zh. Vychisl. Mat. i. Mat. Fiz. 31 (1991), N5, 781-787 (in Russian).
5. Parton, V. Z., Perlin P. I., Mathematical methods of the theory of elastity. Vol 1., "MIR", Moscow 1984.
6. Kas'anov V. I., Suchov R. N. Galerkin's methods for a solution of a problem of diffraction of electromagnetic waves on a dielectric wedge, Progress in Electromagnetic Research Symposium, July 5-14, 2000, Cambridge, MA, USA.
7. John, F., Partial differential equations, 4th ed., Springer-Verlag, New York 1981.
8. Taylor, M. E., Pseudodifferential operators, Princeton University Press, Princeton 1981.
9. Vainikko, G. M., The compact approximation of the operators and the approximate solution of the equations, Tartu University Press, Tartu 1970 (in Russian).
10. Vainikko, G. M., Tamme, E. E., Convergence of the differences method in a problem of periodic solutions of elliptical type equations, coefficients, Zh. Vychisl. Mat. i. Mat. Fiz. 16 (1976), N3, 261-271 (in Russian).

Nonconforming Discretization Techniques for Overlapping Domain Decompositions

Bernd Flemisch, Michael Mair, and Barbara Wohlmuth

Inst. for Appl. Analysis and Num. Simulation, Pfaffenwaldring 57, 70569 Stuttgart, Germany, *{flemisch,mair,wohlmuth}@ians.uni-stuttgart.de*. Supported in part by DFG, SFB 404, C12.

Summary. For the numerical solution of coupled problems on two nested domains, two meshes are used which are completely independent to each other. Especially in the case of a moving subdomain, this leads to a great flexibility for employing different meshsizes, discretizations or model equations on the two domains. We present a general setting for these problems in terms of saddle point formulations, and investigate one- and bi-directionally coupled applications.

1 Introduction

We consider coupled problems on two nested domains, the global domain Ω and the subdomain ω, see Figure 1. In order to approximate the involved solution components on Ω and ω, two meshes are used which are completely independent to each other. We like to be able to deal with different meshsizes,

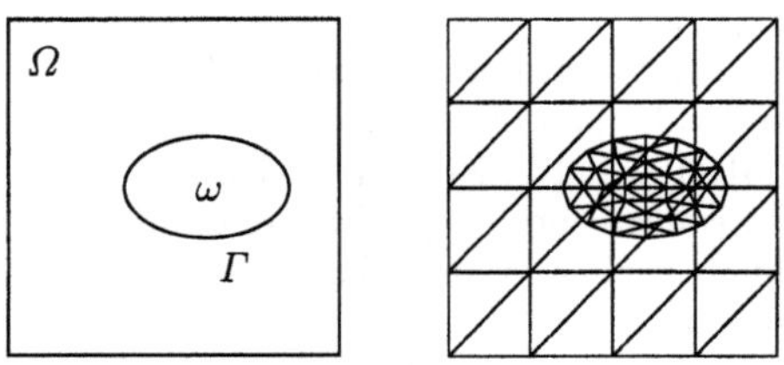

Fig. 1. Two nested domains (left), independent grids (right)

discretizations and model equations on the two domains. Our approach is useful especially for a moving subdomain, i.e., when ω changes its position inside the global domain. In this case, no remeshing will be necessary and only the matrices responsible for the coupling have to be reassembled. In Section 2, we start with the general variational setting in terms of a saddle point formulation. A one-directionally coupled model problem is investigated in Section 3. In Section 4, we consider bi-directionally coupled formulations on the examples of a linear elasticity problem and an eddy current simulation.

2 Variational Setting

Generalized saddle point problems. Our goal is to find a solution $u = (u_\Omega, u_\omega)$ consisting of two components defined on the global domain Ω and on the subdomain ω, respectively. We denote by V_Ω and V_ω the appropriate weak function spaces for the solution components as well as for the test functions. Without taking into account any coupling between the two solution components, the involved differential operators are in general described by continuous bilinear forms $a_\Omega(\cdot,\cdot)$ acting on $V_\Omega \times V_\Omega$, and $a_\omega(\cdot,\cdot)$ acting on $V_\omega \times V_\omega$. Indicating by V the product space $V_\Omega \times V_\omega$, a composed bilinear form $a(\cdot,\cdot) : V \times V \to \mathbb{R}$ is obtained by

$$a(w,v) = a_\Omega(w_\Omega, v_\Omega) + a_\omega(w_\omega, v_\omega), \quad w, v \in V.$$

The coupling between the two solution components is realized via the Lagrange multiplier space M in terms of two continuous bilinear forms $b_1(\cdot,\cdot)$ and $b_2(\cdot,\cdot)$ acting on $V \times M$. For the applications in Section 3 and Subsection 4.2, M is the dual of the trace space $H^{1/2}(\Gamma)$, i.e., $M = H^{-1/2}(\Gamma)$, whereas in Subsection 4.1 $M = H^{-1/2}(\Gamma)^2$, with $\Gamma = \partial\omega$ indicating the subdomain boundary.

Solving additionally for the Lagrange multiplier $p \in M$, the following generalized saddle point problem is derived: find $(u,p) \in V \times M$ such that

$$\begin{aligned} a(u,v) + b_1(v,p) &= \langle f, v\rangle_{V'\times V}, \quad && v \in V, \\ b_2(u,q) &= \langle g, q\rangle_{M'\times M}, \quad && q \in M, \end{aligned} \tag{1}$$

where $\langle\cdot,\cdot\rangle_{V'\times V}$ and $\langle\cdot,\cdot\rangle_{M'\times M}$ denote the usual duality pairings. We point out that for $b_1(\cdot,\cdot) = b_2(\cdot,\cdot)$, problem (1) has the usual symmetric structure, which is encountered for example in the framework of mixed [1] and mortar [2] finite element methods. Moreover, if $b_1(\cdot,q)$ acts only either on V_Ω or V_ω, one-directionally coupled problems are derived.

The bilinear forms $b_i(\cdot,\cdot)$ define coupling operators $B_i : V \to M'$ and $B_i^{\mathrm{T}} : M \to V'$ by $\langle B_i v, q\rangle_{M'\times M} = \langle v, B_i^{\mathrm{T}} q\rangle_{V\times V'} = b_i(v,q)$ for $v \in V$ and $q \in M$. The validation of the following coercivity- and inf-sup-conditions guarantees the unique solvability of problem (1) in $V \times M/KerB_1^{\mathrm{T}}$, [5]:

$$\exists \alpha_0 > 0 : \quad \sup_{v_0 \in \operatorname{Ker} B_1} \frac{a(w_0, v_0)}{\|w_0\|_V \, \|v_0\|_V} \geq \alpha_0, \quad w_0 \in \operatorname{Ker} B_2, \tag{2}$$

$$\sup_{w_0 \in \operatorname{Ker} B_2} \frac{a(w_0, v_0)}{\|w_0\|_V \, \|v_0\|_V} \geq \alpha_0, \quad v_0 \in \operatorname{Ker} B_1. \tag{3}$$

$$\exists k_0 > 0 : \quad \inf_{q \in M} \sup_{v \in V} \frac{b_i(v,q)}{\|v\|_V \|q\|_{M/\operatorname{Ker} B_i^{\mathrm{T}}}} \geq k_0, \quad i = 1,2. \tag{4}$$

We note that the above conditions can be more relaxed [3, 11].

Discretization. We use two different shape regular quasi-uniform triangulations $\mathcal{T}_H$ on Ω and $\mathcal{T}_h$ on ω, as illustrated in Figure 1, with H and h indicating the corresponding maximum element diameter. The function spaces V_Ω, V_ω, and M are replaced by discrete approximations $V_H \subset V_\Omega$, $V_h \subset V_\omega$, and $M_h \subset M$, respectively. We denote an element (v_H, v_h) of the product space $V_h^H = V_H \times V_h$ by v_h^H. It may become necessary to involve approximate bilinear forms $a_h(\cdot,\cdot)$, $b_{1,h}(\cdot,\cdot)$, and $b_{2,h}(\cdot,\cdot)$. The discrete saddle point formulation reads: find $(u_h^H, p_h) \in V_h^H \times M_h$ such that

$$\begin{aligned} a_h(u_h^H, v) + b_{1,h}(v, p_h) &= \langle f, v\rangle_{V'\times V}, \quad && v \in V_h^H, \\ b_{2,h}(u_h^H, q) &= \langle g, q\rangle_{M'\times M}, \quad && q \in M_h, \end{aligned} \tag{5}$$

If the conditions corresponding to (2)–(4) hold for $a_h(\cdot,\cdot)$, $b_{1,h}(\cdot,\cdot)$, and $b_{2,h}(\cdot,\cdot)$ with constants independent of the meshsizes H and h, it is possible to derive optimal a priori estimates. This is a consequence of the next lemma, which follows from [3, Thm. 2.2].

Lemma 1. *Under conditions* (2)–(4)*, the following estimate holds with a constant C depending on α_0, k_0 and the continuity constants of the involved bilinear forms:*

$$\|u - u_h^H\|_V + \|p - p_h\|_M \le C \inf_{v\in V_h^H} \|u - v\|_V + C \inf_{q\in M_h} \|p - q\|_M$$

$$+C \sup_{v\in V_h^H} \frac{(a - a_h)(u, v)}{\|v\|_V} + C \sup_{v\in V_h^H} \frac{(b_1 - b_{1,h})(v, p)}{\|v\|_V} + C \sup_{q\in M_h} \frac{(b_2 - b_{2,h})(u, q)}{\|q\|_M}.$$

The most delicate step for the quality of the discretization and the computational complexity is the information transfer between the two grids via the discrete Lagrange multiplier space M_h. In all our considered applications, we essentially couple between the global grid on Ω and the subdomain boundary Γ. On Γ, dual Lagrange multipliers [13] are used to approximate M, which have optimal stability and approximation properties. Moreover, they have local support and satisfy a biorthogonality relation with the basis functions of the trace space $V_{h|\Gamma}$. Therefore, the implementation of the corresponding operators $B_{1,h}$ and $B_{2,h}$ can be performed with low computational costs.

3 A one-directionally coupled model problem

We apply the framework presented in the last section to a one-directionally coupled model problem. We present a uniqueness proof and an a priori error estimate, which we confirm by a numerical example.

Continuous formulation. Consider the problem

$$-\Delta u_\Omega = f_\Omega \text{ in } \Omega, \quad u_{\Omega|\partial\Omega} = 0, \tag{6}$$

with its associated bilinear form $a_\Omega(w_\Omega, v_\Omega) := \int_\Omega \nabla w_\Omega \nabla v_\Omega$, for $w_\Omega, v_\Omega \in H_0^1(\Omega)$. We want to solve an additional problem for the subdomain ω, namely,

$$-\Delta u_\omega = f_\omega \text{ in } \omega, \quad u_{\omega|\Gamma} = u_{\Omega|\Gamma}, \tag{7}$$

with $f_\omega = f_{\Omega|\omega}$. Using Green's formula, we obtain

$$a_\omega(u_\omega, v_\omega) + \langle v_\omega, \frac{\partial u_\omega}{\partial n} \rangle_{M' \times M} = (f_\omega, v_\omega)_\omega, \quad v_\omega \in H^1(\omega),$$

with the obvious meanings for $a_\omega(\cdot,\cdot)$ and $(\cdot,\cdot)_\omega$, and where $M = H^{-1/2}(\Gamma)$. Introducing the Lagrange multiplier $p = \frac{\partial u_\omega}{\partial n}$, we find that $b_1(v,q) = \langle v_\omega, q\rangle_{M'\times M}$ for $v = (v_\Omega, v_\omega) \in V = H_0^1(\Omega) \times H^1(\omega)$, $q \in M$. We realize the continuity requirement along the boundary Γ in (7) by the bilinear form $b_2(v,q) = \langle v_\Omega - v_\omega, q\rangle_{M'\times M}$, $v \in V$, $q \in M$ and obtain the saddle point formulation of (6), (7) given in (1) with $g = 0$.

It is obvious that problem (1) has a unique solution, since the problem on the global domain Ω is not influenced by the problem on the subdomain ω, and its solution u_Ω yields the boundary data for a well posed subdomain problem. Nevertheless, we provide a complete proof within the saddle point setting.

Theorem 1. *With the above definitions, problem* (1) *is uniquely solvable.*

Proof. We first show the unique solvability of problem (1) by validating the conditions (2)–(4). Our main tool is the harmonic extension operator $\mathcal{H} : M' \to V_\omega$, defined by

$$a_\omega(\mathcal{H}w, v_\omega) = 0, \quad v_\omega \in V_\omega^0 = H_0^1(\omega), \quad (\mathcal{H}w)_{|\Gamma} = w. \tag{8}$$

We observe that the trace of V_ω onto Γ is the space M, and that $b_1((0,v),q) = b_2((0,-v),q)$. Taking $v = (0, \pm\mathcal{H}w)$, $w \in M'$, condition (4) is a consequence of the definition of the $H^{-1/2}$-norm and of the fact that $\|\mathcal{H}w\|_{1,\omega} \le C\|w\|_{M'}$.

Let us focus on condition (2). The kernels of the coupling operators are $\operatorname{Ker} B_1 = V_\Omega \times V_\omega^0$, and $\operatorname{Ker} B_2 = \{v \in V : \operatorname{tr} v_\Omega = v_{\omega|\Gamma}\}$, where $\operatorname{tr} : H^1(\Omega) \to H^{1/2}(\Gamma)$ denotes the trace operator. We uniquely decompose $v_\omega \in V_\omega$ into $v_B + v_I$ such that $v_B = \mathcal{H}(v_{\omega|\Gamma})$ and $v_I \in V_\omega^0$. For an arbitrary $w_0 = (w_\Omega, w_B + w_I) \in \operatorname{Ker} B_2$, we consider $v_0 = (w_\Omega, w_I) \in \operatorname{Ker} B_1$. By using the properties of the harmonic extension, we get

$$\|w_\Omega\|_{1,\Omega}^2 \ge c\left(\textstyle\int_\Gamma \operatorname{tr} w_\Omega\right)^2 + c|w_\Omega|_{1,\Omega}^2 \ge c\left(\textstyle\int_\Gamma w_B\right)^2 + c|w_B|_{1,\omega}^2 \ge c\|w_B\|_{1,\omega}^2. \tag{9}$$

Condition (2) follows from (9):

$$a(w_0, v_0) = a_\Omega(w_\Omega, w_\Omega) + a_\omega(w_I, w_I) \ge c\|w_\Omega\|_{1,\Omega}^2 + c\|w_I\|_{1,\omega}^2 \ge c\|w_0\|_1\|v_0\|_1 \tag{10}$$

The proof of condition (3) is similar. For an arbitrary $v_0 = (v_\Omega, v_I) \in \operatorname{Ker} B_1$, we set $w_0 = (v_\Omega, \mathcal{H}(\operatorname{tr} v_\Omega) + v_I) \in \operatorname{Ker} B_2$, and obtain (10).

Discretization. We use standard conforming finite elements of order r and s on $\mathcal{T}_H$ and $\mathcal{T}_h$, respectively. The associated discrete spaces with no boundary conditions are denoted by $S_H^r(\Omega)$ and $S_h^s(\omega)$, and we set $S_{0,H}^r(\Omega) = S_H^r(\Omega) \cap H_0^1(\Omega)$ and $S_{0,h}^s(\omega) = S_h^s(\omega) \cap H_0^1(\omega)$ to be the spaces taking into account homogeneous Dirichlet conditions on $\partial\Omega$ and Γ, respectively. For the discrete Lagrange multiplier space M_h, we propose the use of dual basis functions [13] adapted to the order s of elements from the trace space of $S_h^s(\omega)$, which is indicated by $W_h^s(\Gamma)$. Setting $V_h^H = S_{0,H}^r(\Omega) \times S_h^s(\omega)$, the discrete saddle point problem (5) is obtained. The unique solvability of problem (5) can be shown by replacing the harmonic extension $\mathcal{H}$ and the trace operator tr in the proof of Theorem 1 by discrete operators $\mathcal{H}_h : W_h^s(\Gamma) \to S_h^s(\omega)$ and $\mathrm{tr}_h : S_{0,H}^r(\Omega) \to W_h^s(\Gamma)$, respectively. In order to obtain the estimate (9), these operators have to satisfy certain stability and extension properties with respect to the $H^{1/2}$-norm. The discrete harmonic extension $\mathcal{H}_h$ is naturally obtained by taking $S_{0,h}^s(\omega)$ as a test function space in (8), and the operator tr_h is given by the mortar projection associated with the discrete Lagrange multiplier space M_h, in particular, for this choice, we find $\int_\Gamma w_\Omega = \int_\Gamma \mathrm{tr}_h w_\Omega$.

We intend to use a smaller meshsize $h < H$ or a higher order $s > r$ on the subdomain, and, therefore, expect a better solution u_h compared to $u_{H|\omega}$. Thus, the finite element solution u_{FE} is defined by

$$u_{\mathrm{FE}} := \begin{cases} u_H \text{ in } \omega^c = \Omega \setminus \overline{\omega}, \\ u_h \text{ in } \omega. \end{cases}$$

Lemma 1 only provides a global estimate, which is not sufficient here, since we like to disregard the approximate solution component u_H on the subdomain ω. The necessary tools for a more local analysis can be found in [12], resulting in the following estimate which is proved in [7].

Theorem 2. *Let $B \supset \omega^c$ such that $d = \mathrm{dist}\,(\partial B \setminus \partial\Omega, \partial\omega^c \setminus \partial\Omega) > 0$. Then for H small enough and u regular enough, there exists a constant C depending on d such that*

$$\|u - u_H\|_{1,\omega^c} + \|u - u_h\|_{1,\omega} \leq Ch^s |u|_{s+1,\omega} + CH^r |u|_{r+1,B} + CH^{r+1} |u|_{r+1,\Omega}. \tag{11}$$

We note that the last term in (11) is the fundamental difference of our approach to the estimates obtained by standard adaptive finite element methods. It is due to the fact that in our one-directionally coupled approach no pollution effect is taken into account.

Numerical test. Consider the model problem (6) on $\Omega := (0,1)^2$ with source term f derived from the exact solution $u(x,y) := \exp(-100((x-0.6)^2/a^2 + (y-0.5)^2/b^2))$. An elliptic patch with radii 0.25 and 0.15 is placed in the domain Ω with its center at $(0.6, 0.5)$, as illustrated in Figure 1. Since the solution goes to zero with an exponential decay, we may have a coarser triangulation far enough away from $(0.6, 0.5)$. Therefore, we choose an initial triangulation with $h/H = 1/4$. We use $P1$ elements on $\mathcal{T}_H$, whereas on $\mathcal{T}_h$, we consider two different cases and use $P1$ elements for one test, and $P2$ elements for another test. Figure 2 shows the decay of the errors $e_H = u - u_H$ and $e_{\mathrm{FE}} = u - u_{\mathrm{FE}}$ in the H^1-norm under uniform refinement. The errors e_H and e_{FE} both satisfy the

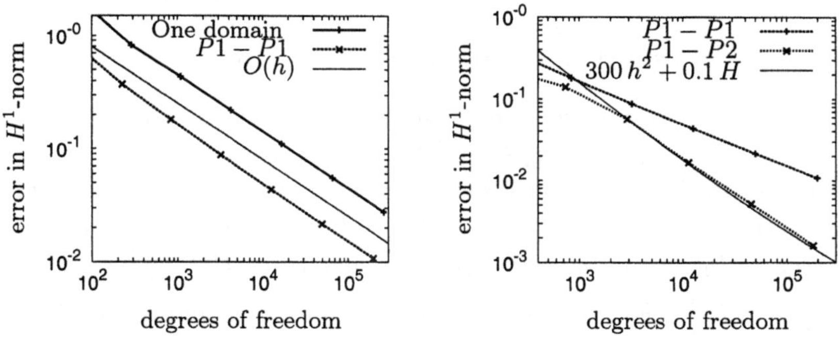

Fig. 2. Error decay in the H^1-norm of $P1$-$P1$ and $P1$-$P2$ coupling

a priori estimates. Choosing the same number of unknowns for the standard and the overlapping method, the solution obtained by the $P1$-$P1$ coupling is significantly better than the solution obtained by the standard method. For the $P2$-$P1$ coupling, the error decay is almost optimal with respect to the piecewise quadratic finite elements used on $\mathcal{T}_h$. In agreement with (11), the error behaves like $c_1 h^2 + c_2 H$, and, numerically, $c_2 \ll c_1$.

4 Bi-directionally coupled problems

We present two applications which result in bi-directionally coupled problems. The first one illustrates a complementary coupling procedure, the second one considers an eddy current problem.

4.1 Natural boundary conditions at the hole ω

We want to solve a boundary value problem on the domain $\Omega \setminus \overline{\omega} =: \omega^c$ with natural boundary conditions on the hole boundary Γ. The solution on the global domain Ω yields the solution on the domain with hole ω^c. This problem is analyzed for the linear elasticity setting in [9]. In addition, we show an application for rotating "holes", see Figure 4.

Saddle point formulation for the linear elasticity problem. We consider the problem: find $u_c \in \left(H^1(\omega^c)\right)^2$ such that

$$
\begin{aligned}
-\mathsf{Div}\,\sigma(u_c) &= f_c && \text{in } \omega^c, \\
\sigma(u_c)\, n_c &= t && \text{on } \Gamma,
\end{aligned}
$$

with $u_c = 0$ on Γ_0 and $\sigma(u_c)\, n_c = g$ on Γ_1 where $\Gamma_0 \subset \partial\Omega$ has a positive measure and $\partial\Omega = \Gamma_0 \cup \Gamma_1$, $\Gamma_0 \cap \Gamma_1 = \emptyset$, with body forces $f_c \in \left(L^2(\omega^c)\right)^2$ and surface tractions $g \in \left(L^2(\Gamma_1)\right)^2$, $t \in \left(L^2(\Gamma)\right)^2$. An equivalent formulation is given by

$$\begin{aligned} -\mathsf{Div}\,\sigma(u_\Omega) &= f_\Omega && \text{in } \Omega, \\ -\mathsf{Div}\,\sigma(u_\omega) &= f_\omega && \text{in } \omega, \\ u_\Omega &= u_\omega && \text{on } \Gamma, \\ \left[\sigma(u_\Omega)\,n_c\right] &= -\sigma(u_\omega)\,n_c + t && \text{on } \Gamma, \end{aligned}$$

with $u_\Omega = 0$ on Γ_0 and $\sigma(u_\Omega)\,n_c = g$ on Γ_1, and where $[w_\Omega] := w_{\Omega|_{\omega^c}} - w_{\Omega|_\omega}$, $f_\omega = f_{\Omega|_\omega}$ and $f_\Omega \in \left(L^2(\Omega)\right)^2$ is an extension of f_c to Ω. It can be written in its weak formulation as saddle point problem with the bilinear forms $a_\Omega(u_\Omega, v_\Omega) = \int_\Omega \sigma(u_\Omega) : \epsilon(v_\Omega)$, $a_\omega(u_\omega, v_\omega) = \int_\omega \sigma(u_\omega) : \epsilon(v_\omega)$, $b_1(v,q) = \langle v_\omega + v_\Omega, q\rangle_{M\times M'}$, $b_2(v,q) = \langle v_\omega - v_\Omega, q\rangle_{M\times M'}$, and $V = \left(H^1(\Omega)\right)^2 \times \left(H^1(\omega)\right)^2$, $M = \left(H^{-\frac{1}{2}}(\Gamma)\right)^2$. For the discretization, we use linear and quadratic finite elements on quasi-uniform and shape regular triangulations. The conditions (2) – (4) hold for h/H small enough in the discrete setting as well as in the continuous setting, yielding unique solvability. For details, we refer to [9].

The realization of this approach allows for an easy shift of the hole without having to remesh and can be used in shape optimization algorithms to determine an optimal hole position. Note that the quantity we pass back from the hole to the background is the jump in the fluxes, i.e., in general the solution u_Ω is only $H^{\frac{3}{2}-\varepsilon}$-regular, $\varepsilon > 0$. Another application of this complementary coupling technique are time dependent problems where the hole is an object moving through the domain Ω emanating some flux into ω^c.

Numerical examples

Beam with one hole. We consider the problem domain $\omega^c = (-5,5) \times (0,1) \setminus \{(x,y) \in \mathbf{R}^2 \mid \|(x,y) - (-1,0.5)\| \le 0.3\}$ with $u_c = 0$ for $x = \pm 5$, $\sigma(u_c)\,n_c = (0,-1)$ for $y = 1$, $\sigma(u_c)\,n_c = 0$ elsewhere and $f = 0$. We use Young's modulus $E = 200$ and Poisson ratio $\nu = 0.3$. The stress σ_{xx} is monitored as a graphical quantity, see Figures 3. The iterative solver is based on a block Gauß-Seidel method for the symmetric positive definite system arising from static condensation of the Lagrange multiplier. The convergence rates are level independent, see [9].

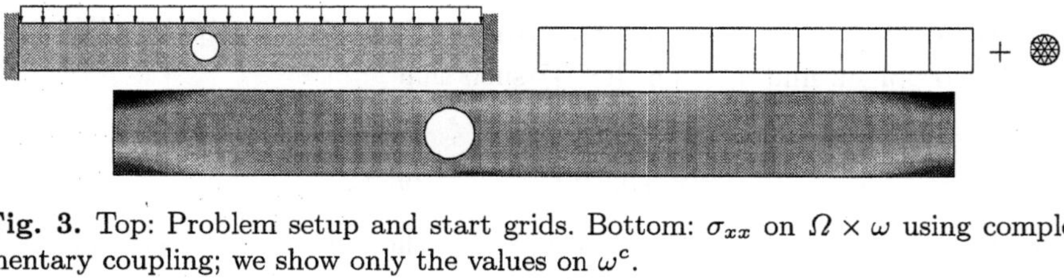

Fig. 3. Top: Problem setup and start grids. Bottom: σ_{xx} on $\Omega \times \omega$ using complementary coupling; we show only the values on ω^c.

Rotating hole. The hole domain is a smooth star with five points described by $(x,y) = (m_x, m_y) + (r_i + \Delta r)\cos(10\pi\lambda)\,(\cos(\alpha(\lambda,t)), \sin(\alpha(\lambda,t)))$ with $\lambda, t \in [0,1]$ and $\alpha(\lambda,t) = 2\pi(\lambda - t - cos(10\pi(\lambda + \frac{1}{8})/51))$, and a center $(m_x, m_y) = (1.1, 0.65))$, the medium radius $r_i = 0.2$ and the radius change amplitude $\Delta r = 0.05$. We solve the Poisson problem. The boundary segment described by $\lambda \in (0, 0.1)$ carries non-zero natural data. The background region is $\Omega = (0,3) \times (0,1)$. The situation and the solution at $t = 0, 0.15, 0.3, 0.45, 0.65, 0.85$ can be found in Figure 4. For each new position of the star, no remeshing procedure has to be carried out, simply the existent grid has to be rotated. Moreover, only the coupling matrix has to be reassembled, all other involved matrices stay the same throughout the whole computation.

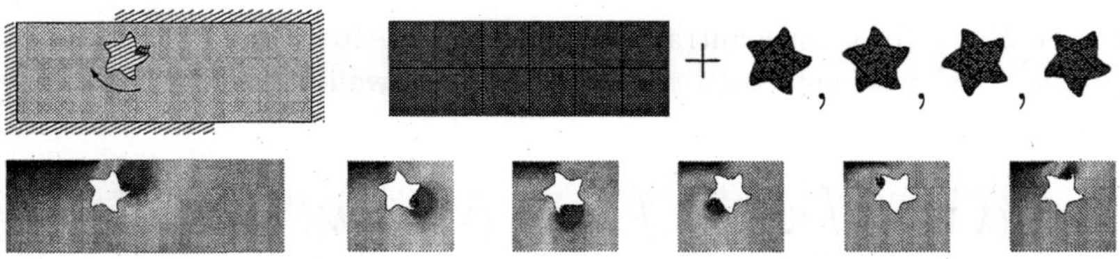

Fig. 4. Problem setup (top left): Zero natural b.c. in hatched areas, 0 and 1 Dirichlet b.c. at the top left and bottom right side, respectively, influx of 5 (natural b.c.) at the back side of the first wing. Initial background grid plus rotated star grid at $t = 0, 0.15, 0.3, 0.45$ (top right). Bottom: Solutions at different times t: $t = 0$ (complete), $t = 0.15, 0.3, 0.45, 0.65, 0.85$ (partly displayed).

4.2 Eddy current simulation

We want to approximate the eddy currents inside a conductor ω which is exposed to a time dependent electromagnetic field acting in the global domain Ω. A detailed problem description and analysis concerning the statically condensated elliptic system can be found in [8], numerical results are available in [6]. Here, we present an alternative approach which fits into the saddle point framework presented in Section 2.

Saddle point formulation. Elimination of the other involved field quantities from the quasistationary Maxwell equations yields for the magnetic field H

$$\operatorname{div} \mu H = 0 \quad \text{in } \Omega, \tag{12}$$

$$\partial_t H + \frac{1}{\sigma\mu} \operatorname{curl} \operatorname{curl} H = 0 \quad \text{in } \omega, \tag{13}$$

with positive material parameters μ and σ, and μ constant in ω. We assume knowing a source vector potential T_s such that $\operatorname{curl} T_s = J_s$ in ω^c, with

J_s a given source current density. The magnetic field H is decomposed into $T - \operatorname{grad}\phi$ on ω and $T_s - \operatorname{grad}\phi$ on ω^c, where $T \in H_0^{\text{curl}}(\omega)$ is a vector valued potential defined on the conductor ω, and $\phi \in H_0^1(\Omega)$ is a scalar valued potential defined on the global domain Ω. Moreover, we use the Coulomb gauge, i.e., T is chosen to be solenoidal. From (12), we obtain

$$a_\Omega(\phi, v) - \int_\Gamma (Tn)v = \int_{\omega^c} \beta T_s \operatorname{grad} v, \quad v \in V_\Omega = H_0^1(\Omega), \tag{14}$$

with $a_\Omega(w,v) = \int_\Omega \beta \operatorname{grad} w \operatorname{grad} v$ and β depending on μ. Taking $v \in H_0^1(\omega)$, (14) implies that ϕ is harmonic on ω, thus, there exists $\gamma = \phi_{|\Gamma} \in H^{1/2}(\Gamma)$ such that $\phi_{|\omega} = \mathcal{H}\gamma$ with the harmonic extension operator $\mathcal{H} : H^{1/2}(\Gamma) \to H^1(\omega)$. Furthermore, due to the solenoidality of T, it holds that $\int_\Gamma (Tn)\lambda - \int_\omega T \operatorname{grad} \mathcal{H}\lambda = 0$ for an arbitrary $\lambda \in H^{1/2}(\Gamma)$. After time discretization by an implicit Euler scheme with time step size Δt, we obtain from (13) at each time step:

$$a_\omega((\gamma, T), (\lambda, W)) + \int_\Gamma (Tn)\lambda = \int_\omega f_\omega W, \quad (\lambda, W) \in V_\omega = H^{1/2}(\Gamma) \times H_0^{\text{curl}}(\omega), \tag{15}$$

with

$$a_\omega((\gamma, T), (\lambda, W)) = \int_\omega \alpha \operatorname{curl} T \operatorname{curl} W + TW - W \operatorname{grad} \mathcal{H}\gamma - T \operatorname{grad} \mathcal{H}\lambda, \tag{16}$$

where $\alpha = \Delta t/(\mu\sigma)$, and f_ω contains the information from the preceding time step.

This suggests the introduction of the Lagrange multiplier $p = Tn \in M = H^{-1/2}(\Gamma)$, and of the coupling bilinear form $b((v, \lambda, W), q) = \langle \lambda - v, q \rangle_{M' \times M}$ for $(v, \lambda, W) \in V = V_\Omega \times V_\omega$ and $q \in M$. Setting $g = 0$ and $b_1(\cdot,\cdot) = b_2(\cdot,\cdot) = b(\cdot,\cdot)$, problem (1) is obtained. By choosing $(\lambda, W) = (0, \operatorname{grad} v)$, $v \in H_0^1(\omega)$, in (15), it is easy to see that the solenoidality of T is guaranteed provided that f_ω is divergence free. The unique solvability of the statically condensated formulation of problem (1) is proved in [8].

Discretization. For the approximation of ϕ, piecewise linear finite elements are used on $\mathcal{T}_H$. Concerning the vector potential T, we employ curl-conforming edge elements [10] on $\mathcal{T}_h$, which are ideally suited for the approximation of $H_0^{\text{curl}}(\omega)$, whereas for γ, we use piecewise linear finite elements on Γ. As before, we approximate the Lagrange multiplier space M by dual basis functions. In contrast to the preceding applications, the bilinear form $a(\cdot,\cdot)$ cannot be implemented directly, and we need an approximation $a_h(\cdot,\cdot)$. Therefore, the harmonic extension operator $\mathcal{H}$ in (16) is replaced by its discrete analogue $\mathcal{H}_h$ corresponding to piecewise linear finite elements on $\mathcal{T}_h$. The gradient operator can be easily realized by the node-to-edge incidence matrix G which acts on the degrees of freedom associated with the linear elements on $\mathcal{T}_h$, and gives the gradient as a linear combination of the basis functions for the edge element space [4]. An optimal a priori estimate based on Lemma 1 for the finite element solution (ϕ_H, T_h) of the statically condensated form of (5) is obtained in [8], provided that the ratio h/H is small enough.

References

1. Babuska, I. (1973): The finite element method with Lagrangian multipliers. Numer. Math., **20**, 179–192.
2. Ben Belgacem, F. (1999): The mortar finite element method with Lagrange multipliers. Numer. Math., **84**, 173–197.
3. Bernardi, C., Canuto, C., Maday, Y. (1988): Generalized inf-sup conditions for Chebyshev spectral approximation of the Stokes problem. SIAM J. Numer. Anal., **25**, 1237–1271.
4. Bossavit, A. (1998): Computational electromagnetism: variational formulations, complementarity, edge elements. Academic Press, New York.
5. Brezzi, F., Fortin, M. (1991): Mixed and hybrid finite element methods. Springer, New York.
6. Flemisch, B., Maday, Y., Rapetti, F., Wohlmuth, B.I. (2003): Coupling scalar and vector potentials on nonmatching grids for eddy currents in a moving conductor. To appear in J. Comput. Appl. Math.
7. Flemisch, B., Wohlmuth, B.I. (2003): A domain decomposition method on nested domains and nonmatching grids. To appear in Numer. Methods Partial Differ. Equations.
8. Maday, Y., Rapetti, F., Wohlmuth, B.I. (2003): Mortar element coupling between global scalar and local vector potentials to solve eddy current problems. In: Brezzi, F. et al (eds), Numerical Mathematics and Advanced Applications, Proceedings of ENUMATH 2001, Springer, Berlin, 847–865.
9. Mair, M., Wohlmuth, B.I. (2003): A domain decomposition method for domains with holes using a complementary decomposition. Report SFB 404 2003/38.
10. Nédélec, J.-C. (1980): Mixed finite elements in $\mathbb{R}^3$. Numer. Math., **35**, 315–341.
11. Nicolaides, R.A. (1982): Existence, uniqueness and approximation for generalized saddle point problems. SIAM J. Numer. Anal., **19**, 349–357.
12. Wahlbin, L.B. (1991): Local behavior in finite element methods. In: Ciarlet, P.G., Lions, J.L. (eds.), Handbook of Numerical Analysis, Vol. II, Elsevier Science Publishers B.V., 1991.
13. Wohlmuth, B.I. (2000): A mortar finite element method using dual spaces for the Lagrange multiplier. SIAM J. Numer. Anal., **38**, 989–1012.

On the Use of Implicit Updates in Minimum Curvature Multi-step Quasi-Newton Methods

John A. Ford[1] and Issam A. Moghrabi[2]

[1] Department of Computer Science, University of Essex, Wivenhoe Park, Colchester, Essex, United Kingdom *fordj@essex.ac.uk*
[2] Beirut Arab University, Beirut, Lebanon *i_moghrabi@yahoo.com*

Summary. Multi-step quasi-Newton methods for optimization employ, at each iteration, an interpolating polynomial in the variable space to construct a multi-step version of the well-known Secant Equation (the relation which constrains the updating of the Hessian approximation). There is some freedom in the choice of the interpolating polynomial and this freedom is exploited, in the case of two-step methods, by the so-called "Minimum Curvature" algorithms, which produce the 'smoothest' interpolation, in the sense of obtaining the polynomial with the smallest possible second derivative (measured in some suitable norm). Typically, these norms are defined by a positive-definite matrix and, in this paper, we will consider and compare the use of different matrices in defining the norm. In particular, we will describe the construction of implicit methods, in which, as we will demonstrate, there is no requirement to compute the matrix defining the norm explicitly.

1 Introduction

We consider two-step quasi-Newton methods for the unconstrained optimization problem

$$\min f(\underline{x}) \quad (\underline{x} \in R^n).$$

If we denote the gradient and Hessian of f by $\underline{g}$ and G, respectively, then such methods are organised in a manner that closely resembles the structure of standard quasi-Newton methods, except that the approximation B_{i+1} to the Hessian $G(\underline{x}_{i+1})$ is required to satisfy a condition of the form (where γ_i is a scalar determined by the precise variant of the method under consideration)

$$B_{i+1}(\underline{s}_i - \gamma_i \underline{s}_{i-1}) = (\underline{y}_i - \gamma_i \underline{y}_{i-1}), \text{ or} \tag{1}$$

$$B_{i+1}\underline{r}_i = \underline{w}_i, \text{ say,} \tag{2}$$

in place of the usual condition (known as the Secant Equation)

$$B_{i+1}\underline{s}_i = \underline{y}_i. \tag{3}$$

(In equations (1) and (3), the step-vectors $\underline{s}_j$ and $\underline{y}_j$ are defined by

$$\underline{s}_j \stackrel{\text{def}}{=} \underline{x}_{j+1} - \underline{x}_j\ , \tag{4}$$

$$\underline{y}_j \stackrel{\text{def}}{=} \underline{g}(\underline{x}_{j+1}) - \underline{g}(\underline{x}_j) = \underline{g}_{j+1} - \underline{g}_j\ , \text{ say,} \tag{5}$$

where $\{\underline{x}_j\}$ are the successive iterates produced by the method.) A matrix satisfying (1)/(2) can be constructed by appropriately modifying, for example, the $BFGS$ update formula, as follows:

$$B_{i+1} = B_i - \frac{B_i\underline{r}_i\underline{r}_i^T B_i}{\underline{r}_i^T B_i\underline{r}_i} + \frac{\underline{w}_i\underline{w}_i^T}{\underline{w}_i^T\underline{r}_i} \tag{6}$$

$$\stackrel{\text{def}}{=} BFGS(B_i, \underline{r}_i, \underline{w}_i), \text{ say.} \tag{7}$$

The derivation of the condition (1) is described by Ford and Moghrabi [5, 4]. In short, quadratic curves $\underline{x}(\tau)$ and $\hat{\underline{g}}(\tau)$ in R^n (where $\tau \in R$) are constructed which interpolate respectively, for the same set of values of τ, the three most recent iterates $\underline{x}_{i-1}$, $\underline{x}_i$ and $\underline{x}_{i+1}$, and the three associated gradient evaluations (assumed to be known). The derivatives of these two curves (at $\tau = \tau_2$, where τ_2 is the value of τ corresponding to $\underline{x}_{i+1}$ and $\underline{g}(\underline{x}_{i+1})$ on the respective curves) are then substituted into the relation

$$G(\underline{x}_{i+1})\underline{x}'(\tau_2) = \underline{g}'(\underline{x}(\tau_2)) \;, \tag{8}$$

derived by applying the Chain Rule to the function $\underline{g}(\underline{x}(\tau))$. (In (8), primes denote differentiation with respect to τ.) Of course,

$$\underline{w}_i \stackrel{\text{def}}{=} \hat{\underline{g}}'(\tau_2) \tag{9}$$

will, in general, only be an *approximation* to the vector $\underline{g}'(\underline{x}(\tau_2))$ required in (8), whereas

$$\underline{r}_i \stackrel{\text{def}}{=} \underline{x}'(\tau_2) \tag{10}$$

may be computed exactly. Nevertheless, on making these substitutions into (8) and removing a common scaling factor, a relation of the form (1) for $B_{i+1} \approx G(\underline{x}_{i+1})$ is obtained.

The remainder of this paper is organised as follows: in Section 2, we review the "minimum curvature" approach to determining a suitable set of parameters $\{\tau_j\}_{j=0}^2$, while Section 3 describes the concept of implicit updates. Section 4 then develops the use of implicit updates within "minimum curvature" methods. Finally, we present the results of numerical experiments in Section 5, and Section 6 draws conclusions on the basis of the results.

2 Minimum curvature methods

If the values of τ corresponding to $\underline{x}_{i-1}/\underline{g}(\underline{x}_{i-1})$ and $\underline{x}_i/\underline{g}(\underline{x}_i)$ are denoted by τ_0 and τ_1 respectively, then Ford and Moghrabi [6] observed that, without loss of generality, the values $\tau_0 = 0$ and $\tau_2 = 1$ could be specified, leaving the remaining value τ_1 to be chosen according to suitable criteria. Defining the related quantity

$$\delta = (\tau_2 - \tau_1)/(\tau_1 - \tau_0) = (1 - \tau_1)/\tau_1 \Rightarrow \tau_1 = (1 + \delta)^{-1}, \tag{11}$$

they then discussed choosing the parameter δ to minimize the norm $\| \underline{x}''(\tau) \|_M$, where

$$\| \underline{z} \|_M \stackrel{\text{def}}{=} \{\underline{z}^T M \underline{z}\}^{1/2} \tag{12}$$

and where M is a given symmetric positive-definite matrix, thus producing an interpolating curve $\underline{x}(\tau)$ which is the "smoothest" in the sense of the norm $\| . \|_M$. It was shown in [6] that fulfilling this criterion leads to the requirement of solving the cubic polynomial

$$\psi(\delta) \equiv \delta^3 - \mu\delta^2 + \mu\delta - \sigma \tag{13}$$

at each iteration, where

$$\sigma_j \stackrel{\text{def}}{=} \| \underline{s}_j \|_M^2; \quad \mu_j \stackrel{\text{def}}{=} \underline{s}_j^T M \underline{s}_{j-1}; \tag{14}$$

$$\sigma \stackrel{\text{def}}{=} \sigma_i / \sigma_{i-1}; \quad \mu \stackrel{\text{def}}{=} \mu_i / \sigma_{i-1}. \tag{15}$$

In [6], the properties of the polynomial ψ were analyzed and it was shown that its zeroes may be determined efficiently. In addition, the circumstances under which ψ has three real zeroes (rather than only one) were identified and it was shown (in that case) which zero should be selected to yield the lowest curvature. It was observed that, in general, this approach was capable (depending on the relative dispositions of the three iterates $\underline{x}_{i-1}$, $\underline{x}_i$ and $\underline{x}_{i+1}$) of producing all three essentially different orderings of the iterates on the interpolating curve.

In their original paper [6] on these "minimum curvature" methods, the authors reported the results of numerical experiments conducted with the version of the algorithm (called **A**) obtained by making the straightforward choice $M = I$, and showed that this yielded a substantial improvement in performance, when compared with the standard *BFGS* method. In a subsequent paper [7], they investigated the performance of the algorithm, called **B**, arising from choosing $M = B_i$ (this choice being motivated by the previous success of other multi-step methods employing the same matrix), and showed that a further improvement in performance was thereby obtained. In this paper, we will pursue this avenue of investigation further by considering related choices for the matrix M. In particular, we will consider the use of implicit updates - that is, updated forms of the matrix B_i which are not calculated explicitly.

3 Implicit updates

Since the choice $M = B_i$ produces an algorithm with good numerical performance, a natural question to consider is whether related matrices might yield further gains. An obvious line of enquiry to pursue in answering this question is the use of updated versions of B_i, where the update employs data from the most recent iteration(s). Because this updated matrix (call it $\widehat{B}_i$ for the present) will be used to compute $\{\tau_j\}_{j=0}^2$ and hence γ_i, $\underline{r}_i$ and $\underline{w}_i$, it cannot be the matrix B_{i+1} which will be produced via equation (6). Thus, it appears that use of such a matrix would necessitate carrying out a *second* update (requiring $\mathcal{O}(n^2)$ operations) at each iteration. However, we observe that explicit knowledge of the updated matrix is not our real goal - rather, we only need this matrix to enable us to calculate σ_i, σ_{i-1} and μ_i and hence (by solving the cubic polynomial) δ, γ_i, $\underline{r}_i$ and $\underline{w}_i$. Therefore, if it can be shown that the expressions required in equations (14) and (15) may be computed at low cost without explicit calculation of $\widehat{B}_i$, then we will have gained our objective of using an updated matrix, while avoiding most of the computational expense. Because explicit computation of $\widehat{B}_i$ is avoided, methods which use this technique were termed *implicit methods* in [3].

4 Implicit updates in minimum curvature methods

For the purposes of simplicity and of easing the computational requirements, we will only consider here *single-step* implicit updates of B_i (although use of two-step updates is an obvious further line of research). We therefore propose the following matrices for use in the norm $\| \cdot \|_M$, denoting the algorithms thus defined by **C** and **D**:

$$\mathbf{C}\text{: } \widetilde{B}_{i-1} \stackrel{\text{def}}{=} BFGS(B_i, \underline{s}_{i-1}, \underline{y}_{i-1}) \tag{16}$$

$$\mathbf{D}\text{: } \widetilde{B}_{i+1} \stackrel{\text{def}}{=} BFGS(B_i, \underline{s}_i, \underline{y}_i). \tag{17}$$

In order to avoid the explicit computation of these updated matrices (for the reasons explained above), it is necessary to show (in the context of "minimum curvature" methods) how the quantities σ_i, σ_{i-1} and μ_i can be calculated without explicit knowledge of the matrix. We consider this issue for each update, in turn.

4.1 Update C

Since the matrix $\widetilde{B}_{i-1}$ is constructed by means of a standard single-step BFGS update, it follows immediately from (16) that

$$\widetilde{B}_{i-1}\underline{s}_{i-1} = \underline{y}_{i-1}. \tag{18}$$

On the other hand, we can use equations (6) and (7) to show that

$$\begin{aligned}\widetilde{B}_{i-1}\underline{s}_i &= \left\{ B_i - \frac{B_i \underline{s}_{i-1}\underline{s}_{i-1}^T B_i}{\underline{s}_{i-1}^T B_i \underline{s}_{i-1}} + \frac{\underline{y}_{i-1}\underline{y}_{i-1}^T}{\underline{y}_{i-1}^T \underline{s}_{i-1}} \right\} \underline{s}_i \\ &= -t_i \underline{g}_i + t_i \left\{ \frac{\underline{s}_{i-1}^T \underline{g}_i}{\underline{s}_{i-1}^T B_i \underline{s}_{i-1}} \right\} (B_i \underline{s}_{i-1}) + \left\{ \frac{\underline{y}_{i-1}^T \underline{s}_i}{\underline{y}_{i-1}^T \underline{s}_{i-1}} \right\} \underline{y}_{i-1}. \end{aligned} \tag{19}$$

(In deriving (19), we have assumed that $\underline{x}_{i+1}$ is obtained by some form of search along the 'quasi-Newton' direction

$$\underline{p}_i = -B_i^{-1}\underline{g}_i, \tag{20}$$

which implies that

$$B_i\underline{s}_i = -t_i\underline{g}_i, \tag{21}$$

for some positive scalar t_i.) Thus we are able to derive the following expressions for the quantities σ_i, σ_{i-1} and μ_i:

$$\sigma_{i-1} = \underline{s}_{i-1}^T\underline{y}_{i-1}; \tag{22}$$

$$\sigma_i = -t_i\underline{s}_i^T\underline{g}_i - t_i^2\left\{\frac{(\underline{s}_{i-1}^T\underline{g}_i)^2}{\underline{s}_{i-1}^T B_i\underline{s}_{i-1}}\right\} + \left\{\frac{(\underline{y}_{i-1}^T\underline{s}_i)^2}{\underline{y}_{i-1}^T\underline{s}_{i-1}}\right\}; \tag{23}$$

$$\mu_i = \underline{s}_i^T\underline{y}_{i-1}. \tag{24}$$

4.2 Update D

In a similar manner, we can derive the following expressions for σ_i, σ_{i-1} and μ_i in the case when the implicit update (17) is applied:

$$\sigma_{i-1} = \underline{s}_{i-1}^T B_i\underline{s}_{i-1} + t_i\left\{\frac{(\underline{s}_{i-1}^T\underline{g}_i)^2}{\underline{s}_i^T\underline{g}_i}\right\} + \left\{\frac{(\underline{s}_{i-1}^T\underline{y}_i)^2}{\underline{s}_i^T\underline{y}_i}\right\}; \tag{25}$$

$$\sigma_i = \underline{s}_i^T\underline{y}_i; \tag{26}$$

$$\mu_i = \underline{s}_{i-1}^T\underline{y}_i. \tag{27}$$

It is evident, from equations (23) and (25), that there remains one obstacle to be overcome, in each case, in achieving our goal of avoiding terms requiring $\mathcal{O}(n^2)$ computation in the calculation of $\sigma = \sigma_i/\sigma_{i-1}$. That difficulty resides in the computation of the product $\underline{s}_{i-1}^T B_i\underline{s}_{i-1}$. This problem has been tackled in two ways - by alternation of updates and by use of a recurrence.

4.3 Alternation

Alternation [8] involves the repeated application of a *cycle* consisting (in its basic form) of two iterations, the first of which is a standard single-step BFGS iteration and the second of which is the required two-step method. The consequence of this arrangement is that each *two-step* iteration can be implemented in the knowledge that the relation

$$B_i\underline{s}_{i-1} = \underline{y}_{i-1} \tag{28}$$

holds (because of the preceding *single-step* iteration). This implies that

$$\underline{s}_{i-1}^T B_i\underline{s}_{i-1} = \underline{s}_{i-1}^T\underline{y}_{i-1}, \tag{29}$$

which means that $\underline{s}_{i-1}^T B_i\underline{s}_{i-1}$ can be computed in $\mathcal{O}(n)$ operations.

4.4 Recurrence

In [9], the authors showed how the following recurrence for the efficient calculation of the quantity $\lambda_i \stackrel{\text{def}}{=} \underline{s}_{i-1}^T B_i \underline{s}_{i-1}$ could be derived:

$$\pi_{i-1} = t_{i-1}\underline{s}_{i-1}^T\underline{g}_{i-1} + \gamma_{i-1}^2\lambda_{i-1} - 2t_{i-1}\underline{r}_{i-1}^T\underline{g}_{i-1} ;$$

$$\lambda_i = -t_{i-1}\underline{s}_{i-1}^T\underline{g}_{i-1} + \frac{(\underline{s}_{i-1}^T\underline{w}_{i-1})^2}{\underline{r}_{i-1}^T\underline{w}_{i-1}} - \frac{t_{i-1}^2(\underline{r}_{i-1}^T\underline{g}_{i-1})^2}{\pi_{i-1}} , \tag{30}$$

where

$$\underline{r}_{i-1} = \underline{s}_{i-1} - \gamma_{i-1}\underline{s}_{i-2} \text{ , and} \tag{31}$$

$$\underline{w}_{i-1} = \underline{y}_{i-1} - \gamma_{i-1}\underline{y}_{i-2}. \tag{32}$$

Again, only $\mathcal{O}(n)$ operations are required in order to obtain $\lambda_i = \underline{s}_{i-1}^T B_i \underline{s}_{i-1}$.

4.5 New methods

In fact, the algorithm denoted by **B** was developed *before* the derivation of the recurrence (30), so we have a total of five new algorithms to compare with the existing methods *BFGS*, **A** (using $M = I$) and **B** (using $M = B_i$ and alternation). For consistency of notation, we will now rename **B** as $\mathbf{B}_{alt}$, since it uses alternation. The five new algorithms are therefore

1. $\mathbf{B}_{recur}$ (using $M = B_i$ and the recurrence (30))
2. $\mathbf{C}_{alt}$ (using $M = \widetilde{B}_{i-1}$ and alternation)
3. $\mathbf{C}_{recur}$ (using $M = \widetilde{B}_{i-1}$ and the recurrence (30))
4. $\mathbf{D}_{alt}$ (using $M = \widetilde{B}_{i+1}$ and alternation)
5. $\mathbf{D}_{recur}$ (using $M = \widetilde{B}_{i+1}$ and the recurrence (30)).

5 Numerical experiments

The algorithms $\mathbf{C}_{alt}$, $\mathbf{C}_{recur}$, $\mathbf{D}_{alt}$ and $\mathbf{D}_{recur}$ derived from the new implicit updates were compared with each other, in our first set of experiments. All the multi-step algorithms tested in these and the following experiments employed the BFGS formula to update the inverse Hessian approximations $H_i = B_i^{-1}$, but with the usual vectors $\underline{s}_i$ and $\underline{y}_i$ replaced by $\underline{r}_i$ and $\underline{w}_i$:

$$H_{i+1} = H_i + \left(1 + \frac{\underline{w}_i^T H_i \underline{w}_i}{\underline{r}_i^T \underline{w}_i}\right)\frac{\underline{r}_i \underline{r}_i^T}{\underline{r}_i^T \underline{w}_i} - \left(\frac{H_i \underline{w}_i \underline{r}_i^T + \underline{r}_i \underline{w}_i^T H_i}{\underline{r}_i^T \underline{w}_i}\right). \tag{33}$$

The line-search employed by all the algorithms was an implementation of safeguarded cubic interpolation and was required to produce a point $\underline{x}_{i+1}$ satisfying the following standard stability conditions (see Fletcher [2], for example):

$$f(\underline{x}_{i+1}) \le f(\underline{x}_i) + 10^{-4}\underline{s}_i^T \underline{g}(\underline{x}_i) \; ; \tag{34}$$

$$\underline{s}_i^T \underline{g}(\underline{x}_{i+1}) \ge 0.9\underline{s}_i^T \underline{g}(\underline{x}_i) \; . \tag{35}$$

The suite of test functions used in the experiments is as described in Ford and Moghrabi [4], with a small number of modifications to starting-points and convergence criteria. The suite comprises 107 functions, in all, and includes many test functions which are well-documented in the literature (for example, Moré , Garbow and Hillstrom [11], Conn et al. [1] and Toint [12]), such as the Penalty I and II functions, the Chained Wood function, the Discrete Boundary-Value function, the Engvall function, the Discrete Integral Equation function and the Gragg-Levy function. The dimensions of these test problems ranged from 2 to 80. For each function, four starting-points were used, giving a total of 428 test problems. For convenience, the functions were notionally classified into those of 'low' ($2 \le n \le 15$), 'medium' ($16 \le n \le 45$) and 'high' ($46 \le n \le 80$) dimension. In total, there were 29 functions in the 'low' set, 43 in the 'medium' set and 35 in the 'high' set, giving 116, 172 and 140 test problems in the respective sets.

Table 1. Comparison of four new minimum curvature methods

Problem set	Low	Medium	High	Combined
$\mathbf{C}_{alt}$	20739 (15148)	32634 (28769)	25085 (23239)	78458 (67156)
Scores	**45**	**84**	**71**	**200**
$\mathbf{C}_{recur}$	21820 (15659)	35036 (30606)	24561 (22926)	81417 (69191)
Scores	**38**	**73**	**64**	**175**
$\mathbf{D}_{alt}$	20898 (15560)	37327 (33663)	29485 (27933)	87710 (77156)
Scores	**31**	**17**	**8**	**56**
$\mathbf{D}_{recur}$	21172 (15299)	36247 (32237)	27571 (26001)	84990 (73537)
Scores	**31**	**10**	**6**	**47**

Results from these first experiments are summarized in Table 1. Each of the tables which we present is divided into five columns, three of which correspond to the subsets of functions referred to above, while the last refers to the complete set. The main entry for a method in each column gives the total number of function / gradient evaluations required by that method to solve all the problems in the specified set, followed by the total number of iterations (in brackets). A 'best performance' for each problem was decided on the basis of the lowest number of evaluations, with ties resolved by the number of iterations. The row labelled 'Scores' shows the number of best performances by each method for the relevant set.

It is evident, from Table 1, that methods based on the implicit update **D** are not competitive with methods based on **C**. (Comparison with Table 2 below shows that they are not even competitive, on the problems with highest

Table 2. Comparison of new and old minimum curvature methods

Problem set	**Low**	**Medium**	**High**	**Combined**
BFGS	*21269 (16073)*	*42202 (38478)*	*33549 (32192)*	*97020 (86743)*
Ratios	*100.0% (100.0%)*	*100.0% (100.0%)*	*100.0% (100.0%)*	*100.0% (100.0%)*
A	21053 (15300)	32769 (28898)‡	25721 (23908)	79543 (68106)‡
Ratios	99.0% (95.2%)	77.6% (75.1%)‡	76.7% (74.3%)	82.0% (78.5%)‡
Scores	**30**	**17**	**12**	**59**
$\mathbf{B}_{alt}$	21201 (15557)	34848 (30625)	25083 (23306)	81132 (69488)
Ratios	99.7% (96.8%)	82.6% (79.6%)	74.8% (72.4%)	83.6% (80.1%)
Scores	**31**	**55**	**53**	**139**
$\mathbf{B}_{recur}$	22015 (15869)	36646 (31840)	24853 (23152)	83514 (70861)
Ratios	103.5% (98.7%)	86.8% (82.7%)	74.1% (71.9%)	86.1% (81.7%)
Scores	**18**	**21**	**16**	**55**
$\mathbf{C}_{alt}$	20739 (15148)	32634 (28769)	25085 (23239)	78458 (67156)
Ratios	97.5% (94.2%)	77.3% (74.8%)	74.8% (72.2%)	80.9% (77.4%)
Scores	**35**	**77**	**53**	**165**
$\mathbf{C}_{recur}$	21820 (15659)	35036 (30606)	24561 (22926)	81417 (69191)
Ratios	102.6% (97.4%)	83.0% (79.5%)	73.2% (71.2%)	83.9% (79.8%)
Scores	**29**	**56**	**48**	**133**

dimension, with the original "minimum curvature" method **A**.) Although this result may be somewhat surprising (it might have been expected, instead, that an update employing the 'latest' data would be more successful still than the method using $M = B_i$, let alone the method **A**), we point out that it is consistent with the results obtained by Ford and Tharmlikit [10] when using a similar implicit update. On the basis of the results reported in Table 1, the methods $\mathbf{D}_{alt}$ and $\mathbf{D}_{recur}$ will not be considered further here.

A second set of experiments (using the same test functions) was then conducted, in order to compare the more successful new methods $\mathbf{C}_{alt}$ and $\mathbf{C}_{recur}$ with the existing "minimum curvature" methods **A** and $\mathbf{B}_{alt}$, and the new version $\mathbf{B}_{recur}$ of **B**. These results are reported in Table 2. For comparison, we have also included the results returned by the standard single-step *BFGS* method. In this Table, the entries in each 'Ratios' row give the proportions of evaluations and (in brackets) iterations for that method, expressed as percentages of the corresponding figures for the *BFGS* method. Finally, scores (indicating 'best performances') are recorded again for this set of experiments, but we point out that the results returned by the *BFGS* method were not included in assessing these best performances, since our primary purpose is to compare the "minimum curvature" methods. (The notation ‡ placed against two of the results for the method **A** indicate that there was one failure for this method [for a test problem in the **Medium** category], where it was unable to converge to an acceptable minimum within the permitted limit of 5000 iterations. The evaluations and iterations incurred on this failure have *not* been included in the relevant totals.)

6 Summary and conclusions

It has been shown how the technique of "implicit updates" may be embedded within the "minimum curvature" approach to determining smooth interpolatory polynomials for use in two-step quasi-Newton methods and, further, how this embedding can be carried out in a computationally inexpensive manner (without, for example, requiring any additional $\mathcal{O}(n^2)$ quasi-Newton updates). Numerical experiments have demonstrated that methods based on the implicit update $\mathbf{D}$ are not competitive with other "minimum curvature" methods. Further experiments have shown that some "minimum curvature" methods are capable of out-performing the standard $BFGS$ method on higher-dimension problems by as much as 27 - 29% in terms of both function / gradient evaluations and iterations. On the basis of total evaluations, total iteration and scores, the most successful method is the alternating implicit method $\mathbf{C}_{alt}$ introduced in this paper. Its nearest competitors are $\mathbf{B}_{alt}$ and $\mathbf{C}_{recur}$. We also note that methods based on the recurrence (30) tend to perform a little less effectively than the corresponding methods employing alternation. The alternating methods are, in addition, a little cheaper to operate, because they do not need to compute the recurrence and because they only need to solve the "minimum curvature" sub-problem once in every second iteration, instead of on each iteration.

References

1. Conn, A.R., Gould, N.I.M., Toint, Ph.L. (1986): Testing a class of methods for solving minimization problems with simple bounds on the variables. Research Report CS–86–45, University of Waterloo
2. Fletcher, R. (1987): Practical Methods of Optimization (2nd ed.). Wiley, New York
3. Ford, J.A. (2001): Implicit updates in multistep quasi-Newton methods. Comput. Math. Appl., **42**, 1083–1091
4. Ford, J.A., Moghrabi, I.A. (1993): Alternative parameter choices for multi-step quasi-Newton methods. Optimization Methods and Software, **2**, 357–370
5. Ford, J.A., Moghrabi, I.A. (1994): Multi-step quasi-Newton methods for optimization. J. Comput. Appl. Math., **50**, 305–323
6. Ford, J.A., Moghrabi, I.A. (1996): Minimum curvature multistep quasi-Newton methods. Comput. Math. Appl., **31**, 179–186
7. Ford, J.A., Moghrabi, I.A. (1997): Further development of minimum curvature multi-step quasi-Newton methods. In: Bainov, D. (ed) Proceedings of the Seventh International Colloquium on Differential Equations. VSP Utrecht Tokyo
8. Ford, J.A., Moghrabi, I.A. (1997): Alternating multi-step quasi-Newton methods for unconstrained optimization. J. Comput. Appl. Math., **82**, 105–116.

9. Ford, J.A., Moghrabi, I.A. (2002): On the Use of Alternation and Recurrences in Two-Step Quasi-Newton Methods. Submitted to Comput. Math. Appl.
10. Ford, J.A., Tharmlikit, S. (2003): New implicit methods in multi-step quasi-Newton methods for unconstrained optimisation. J. Comput. Appl. Math., **152**, 133–146
11. Moré, J.J., Garbow, B.S., Hillstrom, K.E. (1981): Testing unconstrained optimization software. ACM Trans. Math. Software, **7**, 17–41
12. Toint, Ph.L. (1987): On large scale nonlinear least squares calculations. SIAM J. Sci. Stat. Comput., **8**, 416–435

A Boundary Movement Identification Method for a Parabolic Partial Differential Equation

Tom P. Fredman[1]

Heat Engineering Laboratory, Åbo Akademi University, Biskopsgatan 8, FIN-20500 Åbo, Finland *tfredman@abo.fi*

Summary. We study boundary movement identification for a parabolic partial differential equation describing a dynamic diffusion process, on basis of internally recorded data. Formulated as a sideways diffusion equation, the problem is treated by a spatial continuation technique to extend the solution to a known boundary condition at the desired boundary position. Recording the positions traversed in the continuation for each time instant yields the boundary position trajectory and hence the solution of the identification problem. As the problem is ill-posed, a hyperbolic approximation approach is used to regularize the computation and recast the equations into a form amenable to analysis.

1 Introduction

A common feature of *inverse* problems is their objective of determining the "cause from the effects". A consequence of this is the ill-posed mathematical nature [1] of such formulations. This means that they do *not satisfy* Hadamard's definition of well-posedness: (i) For all admissible data, a solution exists. (ii) For all admissible data, the solution is unique. (iii) The solution depends continuously on the data. Practical consequences of ill-posedness are strong error growth in the computation and the fact that approximation errors as well as measurement noise cause blow up of results. Hence, special regularization techniques are necessary for stabilizing the computations.

Here, estimation of boundary position from real process data is considered based on a regularized, slowly divergent space marching [6] method. Possible approaches for this have been adopted for the sideways heat equation [2], including sequential and entire time-domain computation by output-norm minimization as well as direct methods. An entire time-domain direct computation method will be demonstrated by application to simulated input data contaminated by random noise. If necessary, nonlinearities through variation (temperature, time or geometry) of material properties and mixed (Robin type) boundary conditions are readily included. The integrated regularization and straightforward space marching algorithm makes the method especially suited to industrial applications. To our knowledge, the proposed approach has not previously been used to tackle dynamical boundary identification problems, although the main component in the calculation - the sideways diffusion equation - has been extensively investigated.

The sideways diffusion equation is a useful formulation in boundary estimation when the boundary shape follows a certain solution value or conforms to a specified flux. Then, an iterative scheme can be devised for identifying the boundary. Solution of the sideways diffusion equation is feasible through discretization of the time variable, using a stabilizing approximation, and solution of the resulting system of ordinary differential equations [8, 10]. Possible stabilizing approximations are central and forward differences [9, 10], Fourier transform, wavelets or mollification filtering [11]. Our approach follows that of Weber [22], who approximated the sideways heat equation by a well-posed hyperbolic partial differential equation. In fact, such a model of heat conduction (frequently termed *the telegraph equation*) had been proposed by Morse and Feshbach [16], who pointed out the nonphysical nature of instantaneous heat transfer in space permitted by the conventional parabolic heat conduction model.

2 Hyperbolic regularization

Consider the hyperbolic system in the static and bounded domain

$$\frac{\partial^2 u(x,t)}{\partial x^2} = \gamma^2 \frac{\partial^2 u(x,t)}{\partial t^2} + \frac{\partial u(x,t)}{\partial t} , \quad (x,t) \in]0,S[\times]0,t_f[. \tag{1}$$

$$u(0,t) = u_m(t) , \quad \left(\frac{\partial u(x,t)}{\partial x}\right)_{x=0} = q_m(t) \tag{2}$$

$$u(x,0) = u_0(x) , \quad u(x,t_f) = u_f(x) , \tag{3}$$

Here, $\gamma > 0$ is a small regularization parameter and u_f an arbitrary function. For consistency, it is presumed that $u_m(0) = u_0(0)$ and $u_m(t_f) = u_f(0)$. The characteristics of (1) are straight lines with slopes $\pm\gamma$ and since the domain of dependence for the solution at any point is bounded by the characteristics through that point, influence of the endpoint condition $u(x,t_f) = u_f(x)$ becomes negligible as γ becomes small. Since (1, 2, 3) is well-posed and, as will be seen, a regularization of the conventional sideways diffusion problem, it can be used as an approximation for analysis of the conventional problem. There is a well-established theory for this type of partial differential equations, ensuring existence, uniqueness and stability of the solution for realistic measurement data $u_m(t)$, $q_m(t)$. In the case of a moving boundary, $(x,t) \in]0,s(t)[\times]0,t_f[$, to be determined from the measurements, *a priori* conditions on the solution together with the hyperbolic regularization ensure a meaningful solution.

Conceptually (1, 2, 3) is an admissible regularization of the conventional sideways diffusion problem, where it is desired to obtain the solution and flux at $x = S$, if it is required that $\gamma \downarrow 0$ as the measurement error diminishes to zero. Assuming the solution and flux at the origin in (2) can be approximated by solution of a direct parabolic diffusion equation in $x \in]-\infty, 0[$, the solution at $x = S$ takes a particularly simple form. Then, the solution trajectory $f(t)$ at the position S can be solved from the Volterra integral equation

$$(Kf)(t) = u_m(t) \ , \quad 0 \leq t < \infty \ , \tag{4}$$

where the integral operator is defined by

$$(Kf)(t) = \frac{S}{2\sqrt{\pi}} \int_0^t \frac{f(\tau)}{(t-\tau)^{3/2}} \exp\left[\frac{-S^2}{4(t-\tau)}\right] d\tau. \tag{5}$$

The integral equation (4) is inherently ill-posed although exact knowledge of $u_m(t)$ on any finite time horizon uniquely determines the sought temperature trajectory $f(t)$ on the same time interval. The following Lemma was proved by Carasso [5], using auxiliary results in [13].

Lemma 1 *Let $t_f > 0$. Then, $\forall\, p$, $1 \leq p \leq \infty$, K is a compact, linear operator in $L^p\,[0, t_f]$. $Kf = 0$ implies $f = 0$. Thus, K^{-1} exists and is unbounded.*

In the literature, various additional regularization methods are reported in conjunction with iterative solution of the sideways diffusion equation. Filtering has been done, either directly in the frequency domain [18], through higher-order finite differencing [20] or mollification [14, 15, 2]. Wavelets and spline approximation have also been applied in some cases to ensure well-behaved computation [3, 11, 2]. For our purposes, regularization beyond hyperbolic approximation is unnecessary and would only require further tradeoffs in solution accuracy.

3 Boundary identification

3.1 Existence, Uniqueness and Stability

If the space and time variables of (1, 2, 3) are interchanged, a conventional hyperbolic initial-boundary value problem is obtained, to which a solution $u(x,t)$, $0 < x \leq S$ can be obtained using Riemann functions [21, 19, 22]. From such a solution, existence, uniqueness and continuity with respect to initial data, here the measurements $u_m(t)$ and $q_m(t)$ in (2), can be verified. Furthermore, as the solution and distance are connected through the flux, stability of the boundary identification problem follows whenever a nonzero flux and parameter γ may be presumed. More interesting is the error estimate of how well the $\gamma > 0$ approximates the case $\gamma = 0$. Such an estimate was obtained by Eldén [7], indicating the same log-convex behavior as for stability of the sideways heat conduction equation [9, 11].

A different approach to investigating existence, uniqueness and stability of the boundary identification is to convert the problem to a coefficient estimation problem by introducing the spatial variable $y(t) = x/s(t)$. For $\gamma = 0$ the objective is then to find the solution $(v(y,t), 1/s^2(t))$ to

$$\frac{1}{s^2(t)}\frac{\partial^2 v(y,t)}{\partial y^2} = \frac{\partial v(y,t)}{\partial t}\ , \quad (y,t) \in]0,1[\times]0,t_f[. \tag{6}$$

$$v(0,t) = u_m(t)\ , \quad \left(\frac{\partial v(y,t)}{\partial y}\right)_{y=0} = q_m(t)\ , \tag{7}$$

$$v(y,0) = v_0(y)\ , \quad v(y,t_f) = v_f(y(t_f))\ . \tag{8}$$

With certain restrictions on the input data $u_m(t), q_m(t)$, this problem can be solved "in a well-posed manner", quoting [4], where the problem was treated in a modern way applied to a multidimensional geometry. For the one-dimensional situation in (6, 7, 8) the input data must be differentiable and at least one of the measured functions must be monotone [12]. For our purposes this formulation is too restrictive, as in particular $u_m(t)$ may be an oscillating function and all inputs may contain noise. The coefficient estimation setting is useful, however, in demonstrating the requirements on the data to achieve a well-posed computation and in illustrating the degree to which our problem can be considered ill-posed.

An alternative strategy for stabilization at $\gamma = 0$ with respect to input noise can be formulated in a case where $s(t) \ll S$ and an *a priori* bound $u(S,t) \le M$ as in (9) exists with $u(s(t),t) \ll u(S,t)$. When the solution and flux at the origin in (2) can be approximated by solution of a direct parabolic diffusion equation in $x \in]-\infty, 0[$, a bound on the measurement error amounting to ϵ results in a log-convex stability bound proportional to $M^{x/S}\epsilon^{1-x/S}$ for the solution [9, 11]. For $\gamma > 0$, a similar slightly less tight bound has been established [7].

Discretization of the time variable in (1, 2, 3) and solution of the resulting system of ordinary differential equations in the space variable has a stabilizing effect as such, since this prevents blow-up of high-frequency components in the solution [9]. Discretization represents the unbounded time-differentiation operator on the left-hand side of (1) by a bounded matrix and although of high condition number it limits magnification of measurement noise. The recommended step size for time discretization is $t_f(\log(M/\epsilon))^{-2}/2$ [9, 10].

3.2 Preliminaries

Consider now the situation depicted in the (x,t)-plane in Fig. 1 for (1) having a moving boundary $s(t) > 0$ and a specified solution function on this boundary, i.e., $u(s(t),t) = u_s(t)$. It is presumed that this function is such that there exists an intermediate value solution $s(t)$, i.e.,

$$u_m(t) \ \le\ u_s(t) \ \le\ u(S,t) \ \forall \ 0 \le t \le t_f . \tag{9}$$

Furthermore, for solution uniqueness it is presumed that the desired boundary trajectory is found when reaching the correct boundary solution function the first time according to the procedure:

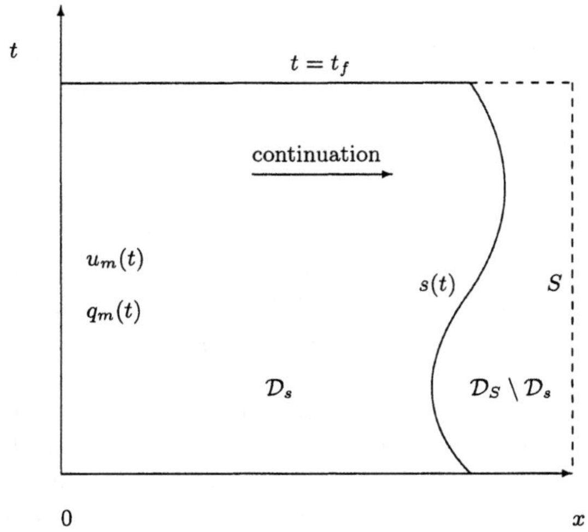

Fig. 1. The relation between domains for the classical sideways diffusion problem and the boundary identification problem for a finite time horizon

- As continuation proceeds and the solution boundary condition is reached, the solution in $\mathcal{D}_S \setminus \mathcal{D}_s$ is put equal to $u_s(t)$.
- As continuation proceeds and the solution boundary condition is reached, the flux $\frac{\partial u(x,t)}{\partial x}$ in $\mathcal{D}_S \setminus \mathcal{D}_s$ is put equal to zero.

Keeping in mind the regularized formulation of the problem as a damped wave equation (1), it is clear that the speed of boundary movement $s'(t)$ must not exceed the wave velocity $1/\gamma$ in absolute value. The rather restrictive assumption (9) together with the above procedure ensures (by a simple intermediate value argument) that the condition $u(s(t), t) = u_s(t)$ can be satisfied for almost all times yielding a unique $s(t)$. Hence, conditions (i) and (ii) of the introduction can be considered fulfilled through formulation, noting that (9) is realistic for the class of industrial boundary identification problems we have in mind for application of the method proposed in this paper. Remains condition (iii), which is tackled as follows: accept possible invalidity of (iii) (for $\gamma = 0$) and devise a method diverging slowly enough to amplify moderate measurement errors in $u_m(t)$, $q_m(t)$ to within bounds of resolution. A Lax-Richtmyer analysis is given below for theoretical discussion of this issue.

4 Computational method

4.1 Cauchy Problem

Transform (1, 2, 3) to an abstract Cauchy-problem applying *the method of lines* by using subscript x for spatial differentiation

$$\begin{bmatrix} u \\ u_x \end{bmatrix}_x = \begin{bmatrix} 0 & I \\ \frac{\partial}{\partial t} + \gamma^2 \frac{\partial^2}{\partial t^2} & 0 \end{bmatrix} \begin{bmatrix} u \\ u_x \end{bmatrix}. \tag{10}$$

$$\begin{bmatrix} u \\ u_x \end{bmatrix}(0) = \begin{bmatrix} u_m(t) \\ q_m(t) \end{bmatrix}. \tag{11}$$

Here, the initial and endpoint conditions (3) will be incorporated into the differential operator in the matrix of (10).

4.2 Solution Procedure

Adopting a difference approximation to the differential operator

$$\left(\frac{\partial}{\partial t} + \gamma^2 \frac{\partial^2}{\partial t^2}\right) u(x,t) \approx \Delta_t u(x,t), \tag{12}$$

the sampled measurement data

$$\begin{aligned} U(0) &= [u_m(0)\ u_m(\Delta t)\ \ldots\ \ u_m(N\Delta t)]^T \\ U_x(0) &= [q_m(0)\ q_m(\Delta t)\ \ldots\ \ q_m(N\Delta t)]^T \end{aligned} \tag{13}$$

can be used for integration of (10) to obtain the discrete solution profile

$$U(x) = [u(x,0)\ u(x,\Delta t)\ \ldots\ \ u(x,N\Delta t)]^T\ ,\quad 0 \leq x \leq s(t), \tag{14}$$

Direct numerical integration yields solution trajectories (functions of time in vector form) at positions $0 < x \leq s(t)$. When an element of the solution vector reaches the boundary condition $u(s(t),t) = u_s(t)$, this element is excluded and the time interval split into two subintervals for which the computation is repeated. To obtain the boundary position at the element, record the integration distance. Depending on the shape of the initial-value vector (13), a small number of such interval divisions are necessary to satisfy the boundary condition for most of the nodes. For each subinterval, starting point and endpoint conditions are not known (except for the first interval starting point, specified by the initial condition at $t = 0$). Therefore, it is reasonable to assume that the solution extends smoothly across these points and to introduce a corresponding linear extrapolation [10] into the difference scheme (12).

4.3 Lax-Richtmyer Analysis

Applying the Lax-Richtmyer theory [17] the discretized solution operator of (10), viewed as the mapping from the noisy sensor data to the unknown boundary, can be investigated [6]. The system (10), discretized through (12) can hence be considered as a space marching scheme

$$\begin{bmatrix} u \\ u_x \end{bmatrix}(x+\Delta x) = \left(\Delta x \begin{bmatrix} 0 & I \\ \Delta_t & 0 \end{bmatrix} + I\right) \begin{bmatrix} u \\ u_x \end{bmatrix}(x) := C(\Delta x, \Delta t) \begin{bmatrix} u \\ u_x \end{bmatrix}(x). \tag{15}$$

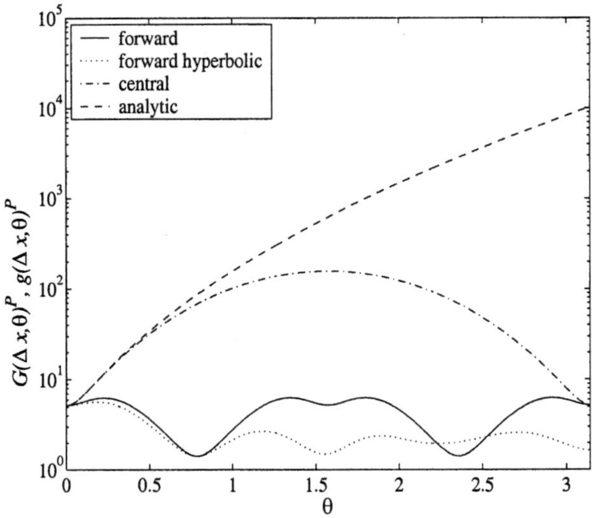

Fig. 2. Amplification factors for the analytical case, two-point central (S3 in [6]) and four-point forward difference approximations ((17) for $\gamma = 0$) and the forward difference hyperbolic regularization in (17)

Introducing the Fourier image $G(\Delta x, \theta)$, in terms of the normalized frequency θ, of $C(\Delta x, \Delta t)$, one has the relation

$$\|C^P\| = \max_{0 \leq \theta \leq \pi} |G(\theta)^P|_{l^2}, \tag{16}$$

where $\|\cdot\|$ denotes the L^2 norm in the time variable, $|\cdot|_{l^2}$ the Euclidean norm and P is the maximum number of spatial steps required to reach the desired boundary. The matrix $G(\theta)$ is 2×2 and, e.g., a four-step forward difference approximation to $\frac{\partial}{\partial t}$ combined with a central difference approximation to $\frac{\partial^2}{\partial t^2}$ in (12) yields

$$G(\theta) = \begin{bmatrix} 1 & \Delta x \\ \frac{\Delta x}{4\Delta t}\left(e^{4i\theta} - 1\right) + \gamma^2 \frac{\Delta x}{(\Delta t)^2}\left(e^{i\theta} - 2 + e^{-i\theta}\right) & 0 \end{bmatrix}. \tag{17}$$

For comparison, the corresponding analytical solution operator for one space step Δx is a convolution kernel with the Fourier image $e^{\Delta x \sqrt{i\theta/\Delta t}}$. The magnitude of this is $g(\Delta x, \theta) = e^{\Delta x \sqrt{|\theta|/2\Delta t}}$ to which $|G(\Delta x, \theta)|_{l^2}$ is a discrete approximation. Thus, marching the input data P spatial steps to reach the desired boundary, amplifies the noise by a factor $g(\Delta x, \theta)^P$. Alternative discretizations (12) are treated extensively by Carasso [6], including the cases with $\gamma = 0$ and central difference as well as the forward difference approximation of $\frac{\partial}{\partial t}$ in (17). For the sake of illustration, an example computation with exaggerated parameter values was performed, results of which are depicted in Fig. 2 together with the analytic amplification factor $g(\Delta x, \theta)^P$ for the parameter values $\Delta t = 0.5$, $\Delta x = 5 \cdot 10^{-4}$, $P = 10^4$ and $\gamma = 0.5$. From Fig. 2 it appears that the hyperbolic regularization in (17) performs well with respect to error growth.

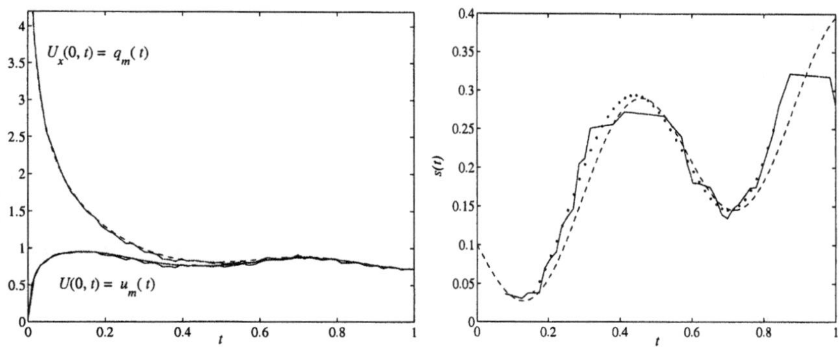

Fig. 3. (Left) Diffusion profile and flux measured at the origin (solid) together with their theoretical trajectories (dashed), for increasing and oscillating thickness trajectory. (Right) Corresponding identified (solid) and theoretical (dashed) thickness trajectories

5 Example simulation

A simulation was carried out in order to test the ability of the outlined computation method to identify different types of smooth thickness variation. To generate corresponding simulated measurements at $x = 0$, the direct parabolic problem ((1) with $\gamma = 0$) was solved with the theoretical trajectories for $s(t)$ and $u_s(t)$ as boundary conditions. After superposition with a noise (normally distributed, relative amplitude 0.1 %) signal the full curves $u_m(t)$ and $q_m(t)$, depicted in Fig. 3, were obtained. Subsequently, the system (10, 13) was solved with the "measurement signals" as initial conditions using a simple explicit Euler-method *for space marching* with $\Delta x = 10^{-3}$, $\Delta t = 7.9 \cdot 10^{-3}$ and $\gamma = 0.01$. Hence, the applicable Courant-Friedrichs-Lewy condition $\gamma \Delta x / \Delta t \leq 1$ is fulfilled. Identification of the boundary trajectory from the direct integration as described above, yielded the results on the right hand of Fig. 3.

The results confirm the feasibility of the used method for situations with noisy measurement data. As seen in the figures, the solutions are stable and consistent with the theoretical (dashed) curves. It should be noted that there is a natural time lag involved before the variation induced by boundary movement reaches the sensor location at $x = 0$. In the direct integration method, the entire time domain is treated simultaneously without account of this causality aspect. Causality of the problem would require measurement data for the interval $(1, 1.1]$ for the solution of the final time interval $(0.9, 1.0]$ to be entirely trustworthy, despite the extrapolation made for the initial and final solution nodes (cf. the discussion in Eldén [8]). Thus, if heads and tails of the function trajectories are disregarded the results are encouraging.

6 Conclusions

A simple one-dimensional, dynamic model for boundary identification was formulated on basis of direct integration of the sideways diffusion equation. Simulated, noisy input signals were used to illustrate stability against measurement errors and model response to different types of boundary variation. The results indicate model feasibility by both of these criteria.

References

1. Alifanov, O. M. (1994): Inverse Heat Transfer Problems. International Series in Heat and Mass Transfer, Springer-Verlag, Berlin
2. Berntsson, F. (2001): Numerical methods for solving a non-characteristic Cauchy problem for a parabolic equation. Technical Report LiTH-MAT-R-2001-17, Department of Mathematics, Linköping University, Linköping, Sweden
3. Berntsson, F., Eldén, L., Loyd, D., Garcia-Padrón, R. (1997): A comparison of three numerical methods for an inverse heat conduction problem and an industrial application. In: Proc. Tenth Int. Conf. for Numerical Methdods in Thermal Problems, volume X, Institute for Numerical Methods in Engineering, University of Wales Swansea, Pineridge Press, Swansea, UK
4. Cannon, J. R., Rundell, W. (1991): Recovering a time dependent coefficient in a parabolic differential equation. Journal of Mathematical Analysis and Applications, 572–582
5. Carasso, A. (1982): Determining surface temperatures from interior observations. SIAM J. Appl. Math., **42**, 558–574
6. Carasso, A. S. (1992): Space marching difference schemes in the nonlinear inverse heat conduction problem. Inverse Problems, **8**, 25–43
7. Eldén, L. (1988): Hyperbolic approximations for a Cauchy problem for the heat equation. Inverse Problems, **4**, 59–70
8. Eldén, L. (1995): Numerical solution of the sideways heat equation. In: Engl, H. W., Rundell, W. (eds.) Proc. GAMM-SIAM Symp., Gesellschaft für Angewandte Mathematik und Mechanik, GAMM-SIAM, Regensburg, Germany
9. Eldén, L. (1995): Numerical solution of the sideways heat equation by difference approximation in time. Inverse Problems, **11**, 913–923
10. Eldén, L. (1997): Solving an inverse heat conduction problem by a "method of lines". Transactions of the ASME, **119**, 406–412

11. Eldén, L., Berntsson, F., Regińska, T. (2000): Wavelet and Fourier methods for solving the sideways heat equation. SIAM J. Sci. Comput., **21**, 2187–2205
12. Jones, Jr, B. F. (1963): Various methods for finding unknown coefficients in parabolic differential equations. Communications on Pure and Applied Mathematics, **XVI**, 33–44
13. Kato, T. (1984): Perturbation Theory for Linear Operators, 2nd edition. A Series of Comprehensive Studies in Mathematics, Springer-Verlag, Berlin
14. Mejía, C. E., Murio, D. A. (1993): Mollified hyperbolic method for coefficient identification problems. Computers Math. Aplic., **26**, 1–12
15. Mejía, C. E., Murio, D. A. (1996): Numerical solution of generalized IHCP by discrete mollification. Computers Math. Aplic., **32**, 33–50
16. Morse, P. M., Feshbach, H. (1953): Methods of Theoretical Physics, volume I of *International Series in Pure and Applied Physics*. McGraw-Hill Book Company, Inc., New York
17. Richtmyer, R. D., Morton, K. W. (1967): Difference Methods for Initial-Value Problems, 2nd edition. Tracts in Pure and Applied Mathematics, Wiley Interscience, New York
18. Seidman, T. I., Eldén, L. (1990): An 'optimal filtering' method for the sideways heat equation. Inverse Problems, **6**, 681–696
19. Sobolev, S. L. (1989): Partial Differential Equations of Mathematical Physics. Dover Publications Inc., New York
20. Taler, J., Duda, P. (2001): Solution of non-linear inverse heat conduction problems using the method of lines. Heat and Mass Transfer, **37**, 147–155
21. Tikhonov, A. N., Samarskii, A. A. (1990): Equations of Mathematical Physics. Dover Publications Inc., New York
22. Weber, C. F. (1981): Analysis and solution of the ill-posed inverse heat conduction problem. Int. J. Heat Mass Transfer, **24**, 1783–1792

On Computational Properties of a Posteriori Error Estimates Based upon the Method of Duality Error Majorants

Maxim Frolov[1a], Pekka Neittaanmäki[1b] and Sergey Repin[2]

[1] Department of Mathematical Information Technology, University of Jyväskylä, P.O. Box 35, FIN–40014, Finland *frolov@it.jyu.fi*[a], *pn@mit.jyu.fi*[b]
[2] V.A. Steklov Institute of Mathematics in St.-Petersburg, 191011, Fontanka 27, St.-Petersburg, Russia *repin@pdmi.ras.ru*

Summary. In the present paper, we analyze computational properties of the functional type a posteriori error estimates that have been derived for elliptic type boundary-value problems by duality theory in calculus of variations. We are concerned with the ability of this type of a posteriori estimates to provide accurate upper bounds of global errors and properly indicate the distribution of local ones. These questions were analyzed on a series of boundary-value problems for linear elliptic operators of 2nd and 4th order. The theoretical results are confirmed by numerical tests in which the duality error majorant for the classical diffusion problem is compared with the standard error indicator used in the MATLAB PDE Toolbox. Numerical tests performed show that the meshes generated on the basis of the majorant are very close to those that would be computed if on each step of the mesh refinement process we knew the exact error distribution. At the same time, meshes generated by the MATLAB code may considerably differ from them.

1 Introduction

For several decades the attention of a number of authors has been focused on questions of reliability and efficiency of calculations in computational engineering. These questions are closely related to the progress in the theory of a posteriori error control. On the one hand, it is necessary to have guaranteed upper bounds on the errors computed in a suitable norm. On the other hand, it is very desirable to also have qualitative indication of their local behavior. These efforts are aimed to decrease computational costs while ensuring accurate and reliable modelling of physical phenomena.

Nowadays, in the framework of Finite Element Methods several approaches to error control are used widely. The first of these was formulated at the end of 70th in the works of Babuška and Rheinboldt (see [2, 3]). Further investigations of this subject were pursued by a number of authors and the amount of the corresponding literature is very large. The most complete description of the methods and the associated literature are given, for instance, in [1, 4, 11].

The main idea of the method which we investigate in this paper was introduced in [8, 9]. An important property of this method (which makes it different from other approaches) is that the Duality Error Majorants (DEM) allow us to estimate the accuracy of any conforming approximate solution independently of the type of approximation (i.e., majorants are also suitable for methods other than FEM). The main advantage of this technique is calculation to *guaranteed* upper bounds of the energy norm of the error. In principle, bounds can be computed as accurately as required (subject to the implied computational effort). This fact has been mathematically justified in [6] and [10]. As has been shown in [10], the method also provides local indication of the error. Both results were confirmed by a considerable amount of numerical testing.

In view of the above-mentioned properties, it is natural to expect that a combination of the DEM error estimator with standard packages will lead to highly effective numerical procedures. This expectation is confirmed by the tests performed. In this paper, we use the MATLAB PDE Toolbox for mesh generation and adaptive refinements. Besides, we compare the majorant with the standard error indicator of the toolbox.

In the very last example, we also present the results of numerical experiments for the biharmonic problem.

2 Duality Error Majorant for a 2nd Order Model Problem

In this section, we take the classical diffusion problem with a Dirichlet type boundary condition as a basis of our investigations (we state it in the variational form and call it the *primal* problem).
Problem $\mathcal{P}$. Find $u \in V$ such that

$$J(u) = \inf_{v \in V} J(v)\ , \quad J(v) := \int\limits_{\Omega} \left(\frac{1}{2} A \nabla v \cdot \nabla v - f v \right) dx\ ,$$

where $V := \{v \in H^1(\Omega) \mid v = w + u_0\ , \quad w \in V_0\}$ and $V_0 := \overset{0}{H}{}^1(\Omega)$.
It is assumed that Ω is a bounded connected domain in $\mathbb{R}^n$ with a Lipschitz continuous boundary $\partial\Omega$, $f \in \mathbb{L}^2(\Omega)$, $A \in \mathbb{M}_s^{n \times n}$, and there exist positive constants α_1, α_2 such that

$$\alpha_1 |\mathbf{q}|^2 \le A\mathbf{q} \cdot \mathbf{q} \le \alpha_2 |\mathbf{q}|^2 \quad \forall \mathbf{q} \in \mathbb{R}^n\ , \quad |\mathbf{q}|^2 := \mathbf{q} \cdot \mathbf{q}\ .$$

From the theory of the Calculus of Variations, it is well-known that Problem $\mathcal{P}$ has the *dual* counterpart Problem $\mathcal{P}^*$ (see, e.g., [5]).
Problem $\mathcal{P}^*$. Find $\mathbf{p}^* \in Q^*_f$ such that

$$I^*(\mathbf{p}^*) = \sup_{\mathbf{q}^* \in Q^*_f} I^*(\mathbf{q}^*)\ , \quad I^*(\mathbf{q}^*) := \int\limits_{\Omega} \left(\nabla u_0 \cdot \mathbf{q}^* - \frac{1}{2} A^{-1} \mathbf{q}^* \cdot \mathbf{q}^* - f u_0 \right) dx\ ,$$

where $Q_f^* := \left\{ \mathbf{q}^* \in \mathbb{L}^2(\Omega, \mathbb{R}^n) \mid \int\limits_\Omega \mathbf{q}^* \cdot \nabla w dx = \int\limits_\Omega f w dx \quad \forall w \in V_0 \right\}$.
Solutions of Problem $\mathcal{P}$ and Problem $\mathcal{P}^*$ satisfy the relations

$$J(u) = \inf_{v \in V} J(v) = \sup_{\mathbf{q}^* \in Q_f^*} I^*(\mathbf{q}^*) = I^*(\mathbf{p}^*) \,, \tag{1}$$

$$\mathbf{p}^* = A \nabla u \,, \quad -\nabla \cdot \mathbf{p}^* = f \quad \text{a.e. in } \Omega \,. \tag{2}$$

For any conforming approximate solution v, using the relation (1), we immediately arrive at the a posteriori error estimate for the energy norm of the error $e = v - u$

$$|\!|\!| e |\!|\!|^2 := \int\limits_\Omega A \nabla e \cdot \nabla e dx \leq \int\limits_\Omega (A \nabla v - \mathbf{q}^*) \cdot (\nabla v - A^{-1} \mathbf{q}^*) dx \,,$$

where $\mathbf{q}^*$ is any element of Q_f^*.
This estimate is mainly of a theoretical value. In many cases, it is practically hard to construct conforming approximations that belong to Q_f^*. But, as was originally shown in [8, 9] (see, also, [10]), this difficulty can be overcome by extending the set of admissible functions for the dual variable. Hence we arrive at the estimate:

$$|\!|\!| e |\!|\!|^2 \leq M(v, \beta, \mathbf{y}^*) := M_D(v, \beta, \mathbf{y}^*) + M_R(\beta, \mathbf{y}^*) \,, \tag{3}$$

$$M_D(v, \beta, \mathbf{y}^*) := (1 + \beta) \int\limits_\Omega (A \nabla v - \mathbf{y}^*) \cdot (\nabla v - A^{-1} \mathbf{y}^*) dx \,,$$

$$M_R(\beta, \mathbf{y}^*) := (1 + 1/\beta) \, \mathbb{C}^2_{\nabla \Omega} / \alpha_1 \int\limits_\Omega (\nabla \cdot \mathbf{y}^* + f)^2 dx \,,$$

where $\mathbf{y}^*$ is an element of $Q^*_\nabla := \{ \mathbf{q}^* \in \mathbb{L}^2(\Omega, \mathbb{R}^n) \mid \nabla \cdot \mathbf{q}^* \in \mathbb{L}^2(\Omega) \}$, β is a positive number, and $\mathbb{C}_{\nabla \Omega}$ is a constant in the Poincaré–Friedrichs inequality.

The functional M is called *the Duality Error Majorant*. It is defined on the pair of free variables $(\beta, \mathbf{y}^*)$. For any v, *any* values of β and $\mathbf{y}^*$ from $\mathbb{R}_+ \times Q^*_\nabla$ provide a *guaranteed* upper bound on the error. However, we should minimize $M(v, \beta, \mathbf{y}^*)$ with respect to β and $\mathbf{y}^*$ to compute a sharp estimate.

Let us consider a sequence $\{Q^*_{\nabla k}\}_{k=1}^{+\infty}$ of finite dimensional subspaces of Q^*_∇ that possess the limit density property, i.e., for any $\delta > 0$ and any $\mathbf{q}^* \in Q^*_\nabla$ there exists a positive integer k_δ such that

$$\inf_{\mathbf{y}^* \in Q^*_{\nabla k}} \| \mathbf{q}^* - \mathbf{y}^* \|_{Q^*_\nabla} \leq \delta \quad \forall k \geq k_\delta \,, \quad \| \mathbf{q}^* \|^2_{Q^*_\nabla} := \| \mathbf{q}^* \|^2 + \| \nabla \cdot \mathbf{q}^* \|^2$$

(we denote by $\| \cdot \|$ two different $\mathbb{L}^2$-norms).

Let us introduce the functional

$$M_k(v) := M(v, \beta_k, \mathbf{y}_k^*) = \inf_{\beta>0, \mathbf{y}^* \in Q^*_{\nabla k}} M(v, \beta, \mathbf{y}^*) ,$$

whose value can be computed by solving an auxiliary finite-dimensional problem on the subspace $Q^*_{\nabla k}$.

Theorem 1. *If the sequence* $\{Q^*_{\nabla k}\}_{k=1}^{+\infty}$ *possess the limit density property in* Q^*_∇, *then*

$$M_k(v) \to \| e \|^2 \qquad as\ k \to \infty .$$

The proof of this theorem is based on the relations (2) and the above-mentioned property.
It is also important to emphasize the following result concerning the convergence of the optimal sequence $\{\mathbf{y}_k^*\}_{k=1}^{+\infty}$ to $\mathbf{p}^*$ in Q^*_∇ (see [6]).

Theorem 2. *If the sequence of pairs* $(\beta_k, \mathbf{y}_k^*)$ *that minimizes* $M(v, \beta, \mathbf{y}^*)$ *on* $\mathbb{R}_+ \times Q^*_{\nabla k}$ *is such that* $\beta_k \to 0$, *then*

$$\mathbf{y}_k^* \to \mathbf{p}^* \ \ in\ Q^*_\nabla , \quad M_D(v, \beta_k, \mathbf{y}_k^*) \to \| e \|^2 ,$$

and

$$M_R(\beta_k, \mathbf{y}_k^*) \to 0 \qquad as\ k \to \infty .$$

The theorems state properties of the duality error majorant as a global estimator.

Let us denote by $\varepsilon(x)$ and $\mu_k(x)$ the *integrands* of the error and majorant $M(v, \beta_k, \mathbf{y}_k^*)$, respectively:

$$\varepsilon(x) := A\nabla e(x) \cdot \nabla e(x) ,$$

$$\mu_k(x) := (1+\beta_k)(A\nabla v(x) - \mathbf{y}_k^*(x)) \cdot (\nabla v(x) - A^{-1}\mathbf{y}_k^*(x)) + \\ +(1+1/\beta_k)\, \mathrm{C}^2_{\nabla\Omega}/\alpha_1 \, (\nabla \cdot \mathbf{y}_k^*(x) + f(x))^2 .$$

For any positive σ, we define the set

$$\Omega_\sigma := \{x \in \Omega : |\mu_k(x) - \varepsilon(x)| \geq \sigma\} .$$

Theorem 3. *Under the assumptions of Theorem 2*

$$\mathrm{meas}\,(\Omega_\sigma) \to 0 \ \ for\ any\ given\ \sigma > 0 \qquad as\ k \to \infty .$$

The proof of Theorem 3 can be found in [10].

Theorem 3 states that the Duality Error Majorant also provides effective indication of the distribution of local errors, and, therefore, this proposition has great importance for any mesh refinement processes.

3 Duality Error Majorant for a 4th Order Model Problem

In this section, we consider the *primal* problem related to the biharmonic operator and the corresponding *dual* problem.
Problem $\mathcal{P}$. Find $u \in W_0$ such that

$$J(u) = \inf_{v \in W_0} J(v) \, , \quad J(v) := \int_\Omega \left(\frac{1}{2} B \nabla\nabla v : \nabla\nabla v - fv \right) dx \, ,$$

where $W_0 := \overset{0}{H}{}^2(\Omega)$. As in Section 2, we assume that $f \in \mathbb{L}^2(\Omega)$, the tensor $B = \{b_{ijsl}\}$ possess the symmetry property $b_{(ij)(sl)} = b_{(sl)(ij)}$ for $i, j, s, l = 1, ..., n$, and there exist positive constants α_1, α_2 such that

$$\alpha_1 |\varkappa|^2 \leq B\varkappa : \varkappa \leq \alpha_2 |\varkappa|^2 \quad \forall \varkappa \in \mathbb{M}_s^{n\times n} \, , \quad |\varkappa|^2 := \varkappa : \varkappa \, . \tag{4}$$

Problem $\mathcal{P}^*$. Find $\mathsf{m}^* \in N_f^*$ such that

$$I^*(\mathsf{m}^*) = \sup_{\mathsf{n}^* \in N_f^*} I^*(\mathsf{n}^*) \, , \quad I^*(\mathsf{n}^*) := \int_\Omega \left(-\frac{1}{2} B^{-1} \mathsf{n}^* : \mathsf{n}^* \right) dx \, ,$$

where $N_f^* := \left\{ \mathsf{n}^* \in \mathbb{L}^2(\Omega, \mathbb{M}_s^{n\times n}) \mid \int_\Omega \mathsf{n}^* : \nabla\nabla w dx = \int_\Omega f w dx \, , \quad \forall w \in W_0 \right\}$.
The relationship between Problem $\mathcal{P}$ and Problem $\mathcal{P}^*$ here is similar to that presented by (1) and (2):

$$J(u) = \inf_{v \in W_0} J(v) = \sup_{\mathsf{n}^* \in N_f^*} I^*(\mathsf{n}^*) = I^*(\mathsf{m}^*) \, ,$$

$$\mathsf{m}^* = B\nabla\nabla u \, , \quad \nabla\cdot\nabla\cdot\mathsf{m}^* = f \quad \text{a.e. in } \Omega \, . \tag{5}$$

In this case, the respective majorant was derived in [7] and has the form

$$|\!|\!| e |\!|\!|^2 := \int_\Omega B\nabla\nabla e : \nabla\nabla e dx \leq M(v, \beta, \varkappa^*) := M_D(v, \beta, \varkappa^*) + M_R(\beta, \varkappa^*) \, , \tag{6}$$

$$M_D(v, \beta, \varkappa^*) := (1+\beta) \int_\Omega (B\nabla\nabla v - \varkappa^*) : (\nabla\nabla v - B^{-1}\varkappa^*) dx \, ,$$

$$M_R(\beta, \varkappa^*) := (1 + 1/\beta) \, \mathbb{C}^2_{\nabla\nabla\Omega} / \alpha_1 \int_\Omega (\nabla\cdot\nabla\cdot\varkappa^* - f)^2 dx \, ,$$

where $\varkappa^* \in N^*_{\nabla\nabla} := \{\mathsf{n}^* \in \mathbb{L}^2(\Omega, \mathbb{M}_s^{n\times n}) \mid \nabla\cdot\nabla\cdot\mathsf{n}^* \in \mathbb{L}^2(\Omega)\}$, $\beta \in \mathbb{R}_+$, $\mathbb{C}_{\nabla\nabla\Omega}$ is a constant in the inequality

$$\| w \| \leq \mathbb{C}_{\mathbb{W}\Omega} \, \| \nabla\nabla w \| \qquad \forall w \in W_0 \;,$$

and α_1 is as in (4).

In general terms, the techniques used for deriving estimates (3) and (6) are rather close. As in Sect. 2, we consider a sequence $\{N^*_{\mathbb{W}k}\}_{k=1}^{+\infty}$ of finite dimensional subspaces of $N^*_{\mathbb{W}}$ and introduce the corresponding functionals

$$M_k(v) := \inf_{\beta>0, \varkappa^* \in N^*_{\mathbb{W}k}} M(v, \beta, \varkappa^*) \;.$$

Let us formulate also analogues of Theorems 1–3.

Theorem 4. *If* $\{N^*_{\mathbb{W}k}\}_{k=1}^{+\infty}$ *possess the limit density property in* $N^*_{\mathbb{W}}$*, then*

$$M_k(v) \to \| e \|^2 \qquad \text{as } k \to \infty \;.$$

Proof. By the limit density property, for the exact solution m^* of the dual problem and any given $\delta > 0$ we can find k_δ such that, for $k \geq k_\delta$, there exists an element $\mathsf{m}^*_k \in N^*_{\mathbb{W}k}$ satisfying the inequality

$$\| \mathsf{e}^*_k \|_{N^*_{\mathbb{W}}} \leq \delta \;, \quad \| \mathsf{e}^*_k \|^2_{N^*_{\mathbb{W}}} := \| \mathsf{e}^*_k \|^2 + \| \nabla\cdot\nabla\cdot \mathsf{e}^*_k \|^2 \;,$$

where $\mathsf{e}^*_k := \mathsf{m}^*_k - \mathsf{m}^*$. Let $k \geq k_\delta$; then

$$M_k(v) = \inf_{\beta>0, \varkappa^* \in N^*_{\mathbb{W}k}} M(v, \beta, \varkappa^*) \leq M(v, \delta, \mathsf{m}^*_k) = M_D(v, \delta, \mathsf{m}^*_k) + M_R(\delta, \mathsf{m}^*_k) \;.$$

Given the relations (5), we consider parts of the majorant $M(v, \delta, \mathsf{m}^*_k)$:

$$\int_\Omega (B\nabla\nabla v - \mathsf{m}^*_k) : (\nabla\nabla v - B^{-1}\mathsf{m}^*_k) dx =$$

$$= \int_\Omega (B\nabla\nabla e - \mathsf{e}^*_k) : (\nabla\nabla e - B^{-1}\mathsf{e}^*_k) dx =$$

$$= \| e \|^2 - 2\int_\Omega \nabla\nabla e : \mathsf{e}^*_k dx + \int_\Omega B^{-1}\mathsf{e}^*_k : \mathsf{e}^*_k dx \leq \| e \|^2 + 2\delta \, \| e \| / \sqrt{\alpha_1} + \delta^2/\alpha_1 \;;$$

$$\int_\Omega (\nabla\cdot\nabla\cdot \mathsf{m}^*_k - f)^2 dx = \int_\Omega (\nabla\cdot\nabla\cdot \mathsf{e}^*_k)^2 dx \leq \delta^2 \;.$$

Taking into account the multipliers $(1+\delta)$ and $(1+1/\delta)\,\mathbb{C}^2_{\mathbb{W}\Omega}/\alpha_1$, we arrive at the estimate $\| e \|^2 \leq M_k(v) \leq \| e \|^2 + \delta\, C$. Therefore,

$$M_k(v) \to \| e \|^2 \qquad \text{as } k \to \infty \;.$$

□

Theorem 5. *If the sequence of pairs* $(\beta_k, \varkappa_k^*)$ *that minimizes* $M(v, \beta, \varkappa^*)$ *on* $\mathbb{R}_+ \times N^*_{\nabla k}$ *is such that* $\beta_k \to 0$, *then*

$$\varkappa_k^* \to \mathsf{m}^* \ \ in\ N^*_{\nabla}\ , \quad M_D(v, \beta_k, \varkappa_k^*) \to \| e \|^2\ , \quad M_R(\beta_k, \varkappa_k^*) \to 0\ ,$$

and

$$\operatorname{meas}(\Omega_\sigma) \to 0 \ \ for\ any\ given\ \sigma > 0 \quad as\ k \to \infty\ .$$

4 Numerical Results

In this section, we justify the method of Duality Error Majorants as an efficient tool for numerical simulations. During the last few years we have performed various types of tests (see [6, 10] and other papers cited therein). In these tests, it was observed that the method is accurate and robust. The next step in our analysis, which is considered in the present paper, is to investigate the behavior of the DEM in the process of adaptive mesh refinement. Certainly, in a short note it is impossible to describe all of the tests that we have performed. Therefore, we select only one interesting and representative example.

The main purpose of our investigations is to compare the efficiency of different error indicators in the process of mesh adaptation. It is important to emphasize that all refinements are based on the same principle (the marking strategy is quite standard: we flag elements if the corresponding local error is greater than half of the maximum local error). Therefore, any difference in the final results is due only to the differences in the efficiencies of the approaches.

We have used three error indicators. The first indicator is computed by comparing an approximate solution v with the exact solution and provides an objective judgement of the quality of mesh adaptations. The value of this reference error indicator on an element T is denoted by η_T^{E}:

$$\left(\eta_T^{\mathrm{E}}\right)^2 := \int_T A(\nabla u - \nabla v) \cdot (\nabla u - \nabla v)dx\ .$$

Two further indicators to be compared with η_T^{E} are defined as follows: a local indicator based on the Duality Error Majorant that arises from (3):

$$\left(\eta_T^{\mathrm{D}}\right)^2 := (1+\beta) \int_T (A\nabla v - \mathbf{y}_k^*) \cdot (\nabla v - A^{-1}\mathbf{y}_k^*)dx\ + \\ + (1 + 1/\beta)\, \mathbb{C}^2_{\nabla\Omega}/\alpha_1 \int_T (\nabla \cdot \mathbf{y}_k^* + f)^2 dx\ ,$$

and the standard local indicator of the MATLAB PDE Toolbox:

$$\eta_T^{\mathrm{S}} := C_1 \, \| h_T \, f \, \|_T + C_2 \left\{ \frac{1}{2} \sum_{E \in \partial T \setminus \partial \Omega} h_E^2 [n_E \cdot (A \nabla u_h)]^2 \right\}^{1/2} .$$

In general, the numerical tests performed can be classified into two groups. For those in the first group, it was observed that the standard error indicator provides adaptively refined meshes of suitable quality. For this group, the DEM is also preferable but its advantage is not very considerable. However, for the second group of examples, the distinction is rather more obvious. Below, we present such a case.

Example. Let us consider the classical problem

$$\begin{cases} -\nabla \cdot (A \nabla u) = f & \text{in } \Omega \,, \\ u = 0 & \text{on } \partial\Omega \,, \end{cases}$$

where the values of A and f are given in Table 1 (the subdomains of Ω are depicted in Fig. 1). From Fig. 2, we conclude that the local errors computed by the DEM reproduce the actual distribution on the initial mesh with high accuracy. We also observe a disadvantage of the standard approach – overes-

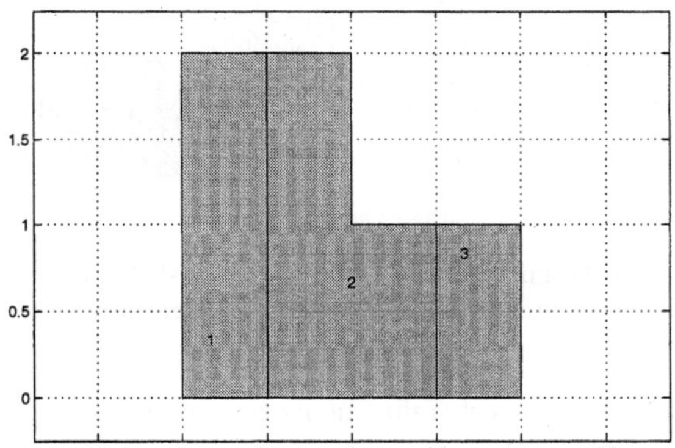

Fig. 1. Domain Ω

Table 1. The matrix A and in the right-hand side f

Subdomain	1	2	3
A	$\begin{bmatrix} 10 & 0 \\ 0 & 1 \end{bmatrix}$	$\begin{bmatrix} 1 & 0 \\ 0 & 1 \end{bmatrix}$	$\begin{bmatrix} 10 & 0 \\ 0 & 1 \end{bmatrix}$
f	1	0	1

timation of local errors in subdomains, where $f \neq 0$. From this point of view, it is natural to predict that the DEM will lead to a more effective adaptation than the one obtained by the indicator η_T^{S}.

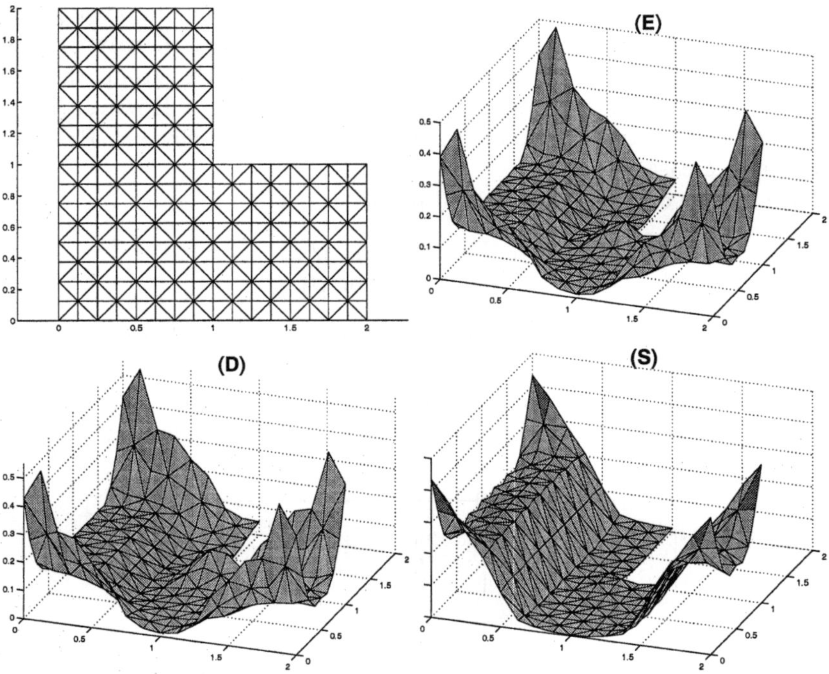

Fig. 2. Indication on the initial mesh: **(E)** exact error distribution; **(D)** error indication by the DEM; **(S)** error indication by the standard indicator

The results of several mesh refinements are collected in Table 2; we denoted by N the corresponding number of degrees of freedom for the meshes generated. The quantity % presents percentage of the relative error for the approximate solutions. Let us compare similar steps of adaptation computed by the reference technique (the 1st block of columns) and DEM (the 2nd block). Final meshes obtained on the basis of the indicators are also depicted on Fig. 3. For every step, the DEM gives accuracy of approximations, which is very close to the optimal. It is worth outlining, that the effectivity index

$$I_{eff} := \sqrt{M} / \, |\!|\!| e |\!|\!|$$

of the computed upper bounds is very close to 1 in each step of the mesh refinement. Eventually, the mesh (D15) almost coincides with the optimal mesh (E13) (see Fig. 3). At the same time, the standard approach leads to essentially worse results (see the 3rd block of columns). The difference is clearly observed for meshes (D16) and (S11). The same accuracy of computed solutions is pro-

Table 2. Mesh adaptation performed by the considered indicators

(E) Iter.	N	%	(D) Iter.	N	%	I_{eff}	(S) Iter.	N	%
0	225	19.02	0	225	19.02	1.09	0	225	19.02
5	862	6.99	7	876	7.12	1.07	4	1022	7.79
7	1422	5.42	9	1428	5.50	1.07	5	1929	5.70
11	3268	3.45	13	3376	3.50	1.06	7	3989	4.01
(13)	5552	2.70	(15)	5698	2.74	1.06	9	7417	3.01
			(16)	7142	2.38	1.06	(11)	13560	2.37

vided by the DEM-based technology of adaptation with approximately half the number of degrees of freedom.

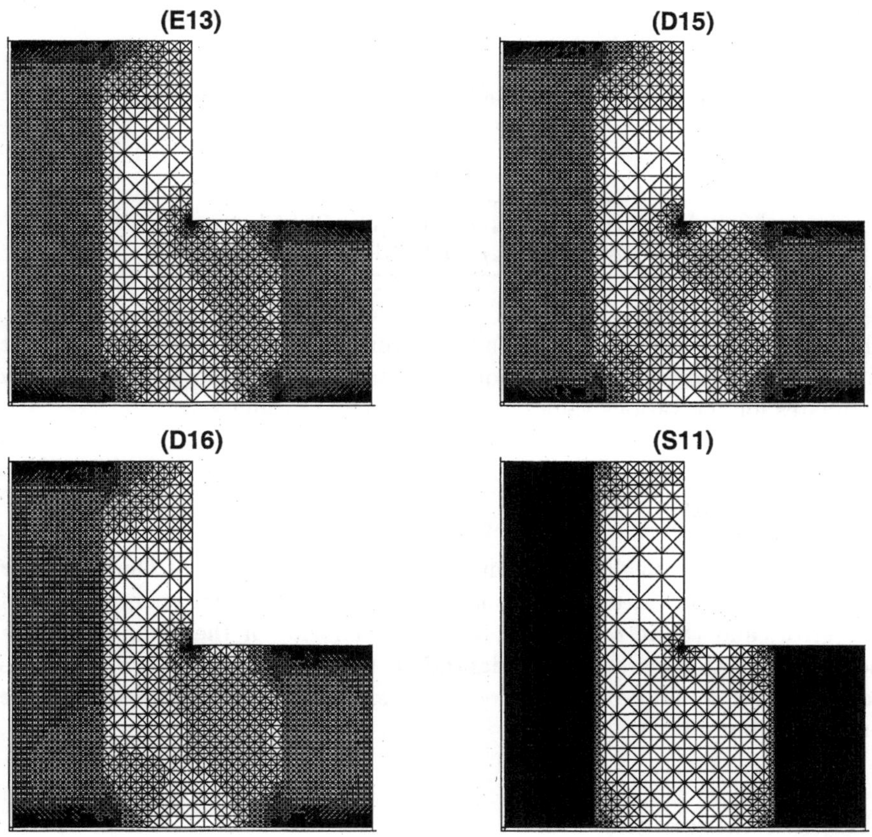

Fig. 3. Final meshes computed by different indicators

In the very last example, we show that the DEM that stems from (6) also leads to error estimates of high quality.
Example. Let us consider the biharmonic problem with homogeneous boundary conditions:

$$\begin{cases} \Delta^2 u = f & \text{in } \Omega\,, \\ u = \partial_n u = 0 & \text{on } \partial\Omega\,, \end{cases}$$

where Ω is the unit square. In this example, we aim to show that the DEM provides effective error control not only in the framework of Finite Element Methods but more widely. For this purpose, we choose the exact solution of the following form: $u = \phi(x_1)\,\phi(x_2)$, where $\phi(x) = (1-x)^2\,x^2$. An approximate solution is taken in the form $v = u + cw$, where $w = \psi(x_1, k_1)\,\psi(x_2, k_2)$ and $\psi(x, k) = e^{10x^k}\,\phi(x)$. We select such a value of the constant c that the accuracy of v is about of 5%. An approximation of the dual variable $\varkappa^*$ is also constructed as a combination of global basis functions (the total number of which is denoted by N_b). The effectivity of the corresponding error bounds for

Table 3. Efficiency of the DEM for the biharmonic problem

k_1		0.1	0.2	0.5	0.5
k_2		0.1	0.3	0.5	1.0
$N_b = 144$	I_{eff}	1.42	1.49	1.52	1.53
$N_b = 196$	I_{eff}	1.11	1.13	1.13	1.14

the various cases is presented in Table 3. From these results, we conclude that a combination of 196 basis functions is quite enough to provide high-quality estimates for the various values of the parameters k_1 and k_2.

5 Conclusions

We justified theoretically and numerically that the method of Duality Error Majorants (a) provides sharp upper bounds on the global energy norm of the error, and (b) it reproduces the local behavior of the error with high accuracy. For this reason, mesh adaptations based on the DEM are very close to those that would be obtained on the basis of the exact knowledge on the error distribution.

Acknowledgement

The research was partially supported by grant N201728 of the Academy of Sciences of Finland.

References

1. Ainsworth, M., Oden, J.T. (2000): A posteriori error estimation in finite element analysis. Wiley, New York
2. Babuška, I., Rheinboldt, W.C. (1978): A-posteriori error estimates for the finite element method. Int. J. Numer. Methods Eng., **12**, 1597–1615
3. Babuška, I., Rheinboldt, W.C. (1978): Error estimates for adaptive finite element computations. SIAM J. Numer. Anal., **15**, no. 4, 736–754
4. Babuška, I., Strouboulis, T. (2001): The finite element method and its reliability. Numerical Mathematics and Scientific Computation, The Clarendon Press, Oxford University Press, New York
5. Ekeland, I., Temam, R. (1976): Convex analysis and variational problems. Studies in Mathematics and its Applications, Vol. 1., North-Holland Publishing Co., Amsterdam-Oxford, American Elsevier Publishing Co., Inc., New York
6. Frolov, M., Neittaanmäki, P., Repin, S. (2003): On the reliability, effectivity and robustness of a posteriori error estimation methods. In: Kuznetsov, Y., Neittaanmäki, P., Pironneau, O. (eds.) Numerical Methods for Scientific Computing. Variational problems and applications. CIMNE Barcelona (in press)
7. Neittaanmäki, P., Repin, S.I. (2001): A posteriori error estimates for boundary-value problems related to the biharmonic operator. East-West J. Numer. Math., **9**, no. 2, 157–178.
8. Repin, S.I. (1997): A posteriori error estimation for nonlinear variational problems by duality theory. Zap. Nauchn. Semin. POMI, **243**, 201–214 ((2000): english translation in J. Math. Sci., New York, **99**, no. 1, 927–935)
9. Repin, S.I. (2000): A posteriori error estimation for variational problems with uniformly convex functionals. Math. Comp., **69**, no. 230, 481–500
10. Repin, S.I, Sauter, S., Smolianski, A. (2003): A posteriori error estimation for the Dirichlet problem with account of the error in the approximation of boundary conditions. Computing, **70**, 205–233
11. Verfürth, R. (1996): A review of a posteriori error estimation and adaptive mesh-refinement techniques. Wiley-Teubner Series Advances in Numerical Mathematics

Efficient Algorithm for Local-Bound-Preserving Remapping in ALE Methods

Rao Garimella[1], Milan Kuchařík[2] and Mikhail Shashkov[3]

[1] Los Alamos National Laboratory, T-7, MS-284, Los Alamos, NM 87545, USA *rao@lanl.gov*
[2] Czech Technical University in Prague, Břehová 7, 115 19 Prague 1, Czech Republic *kucharik@karkulka.fjfi.cvut.cz*
[3] Los Alamos National Laboratory, T-7, MS-284, Los Alamos, NM 87545, USA *shashkov@lanl.gov*

Summary. The remapping algorithm is an essential part of the ALE (Arbitrary Lagrangian-Eulerian) method. In this talk we present such an algorithm based on linear function reconstruction, approximate integration and mass redistribution.

1 Introduction

Conservative remapping is an essential part of the ALE (Arbitrary Lagrangian-Eulerian) method for fluid dynamics computations. This method tries to use advantages of both the Lagrangian and Eulerian approaches.

At first, several time steps of the pure Lagrangian computation are used. As the grid moves together with the fluid, it may happen that the grid becomes distorted or tangled due to shear flow. Now comes the Eulerian part of the algorithm. We prepare a new rezoned grid and recompute (remap) the quantities from the distorted grid to the rezoned one.

We have several conditions this remapping step must satisfy. It must be efficient to be usable in real computations. Total sum of the conservative quantities must be preserved – the algorithm must be conservative. We do not want to create new local extrema, we want it to be local-bound preserving. It must be stable and applicable to general unstructured meshes in 2D and 3D. In this article we introduce a 3D algorithm, which satisfies these conditions. A similar procedure in 2D is described in [1].

2 Algorithm Description

2.1 Problem statement

Suppose, we have two grids: Lagrangian $C = \{c\}$ and rezoned $\tilde{C} = \{\tilde{c}\}$. The grids have the same topology. The rezoned grid is created from the original one just by small movement of the grid nodes. There exists some underlying function $g(\mathbf{r})$, $\mathbf{r} = (x, y, z)$ in the Lagrangian cells (for example $g = \rho$, $g = \rho u$, $g = \rho v$, $g = \rho w$, $g = \rho\left(\varepsilon + |\mathbf{U}|^2/2\right)$, where ρ is mass density, $\mathbf{U} = (u, v, w)$ is a vector of velocities and ε is the internal energy). We do not know the function itself, we know just the mean values in the grid cells and their masses and volumes

$$\bar{g}_c = \frac{\int_c g(\mathbf{r})\, dV}{V(c)} = \frac{m(c)}{V(c)}, \quad m(c) \equiv \int_c g(\mathbf{r})\, dV, \quad V(c) \equiv \int_c 1\, dV\,. \tag{1}$$

Total mass (momentum, energy) in the computational domain Ω can be computed as

$$M \equiv \int_\Omega g(\mathbf{r})\, dV = \sum_{\forall c} \int_c g(\mathbf{r})\, dV = \sum_{\forall c} m(c)\,. \tag{2}$$

We want to compute new masses $m^*(\tilde{c})$ and corresponding mean values in the rezoned cells

$$\bar{\bar{g}}^*_{\tilde{c}} = \frac{m^*(\tilde{c})}{V(\tilde{c})} \tag{3}$$

and we want them to be as close to the exact values as possible ($m^*(\tilde{c}) \approx m(\tilde{c}) = \int_{\tilde{c}} g(\mathbf{r})\, dV$). We also want not to create new local extrema

$$g_c^{\max} \geq \bar{\bar{g}}^*_{\tilde{c}} \geq g_c^{min}\,, \quad g_c^{\max} = \max_{c_n \in \mathcal{C}(c)} g_{c_n}\,, \quad g_c^{\min} = \min_{c_n \in \mathcal{C}(c)} g_{c_n}\,, \tag{4}$$

where $\mathcal{C}(c) \subset C$ is neighborhood of cell c, and to be conservative (total mass must be the same)

$$\sum_{\forall c} m^*(\tilde{c}) = M\,.$$

If the underlying function is a linear function, we want our method to be exact

$$m^*(\tilde{c}) = m(\tilde{c}) = \int_{\tilde{c}} g(\mathbf{r})\, dV \quad \text{for} \quad g(\mathbf{r}) = a + b\,x + c\,y + d\,z\,.$$

2.2 Remapping Algorithm

We design our algorithm in three stages. In the first stage, we make a piecewise linear reconstruction of the underlying function on the original mesh. This can be done using different methods, with or without limiters. In the second stage, we integrate this reconstructed function to obtain means on the new grid. The most natural approach would be exact integration, but it needs computation of the intersections of the Lagrangian grid with the rezoned one. This intersection is very time consuming in 2D and almost unfeasible in 3D, so we use numerical quadrature - swept integration. It does not require finding these intersections so it is much faster. The problem is that it is an approximate method and it may happen that the local extrema are violated, so we need also the third stage - repair - which ensures us this local-bound preservation.

2.3 Stage 1 – Piecewise Linear Reconstruction

We want to reconstruct the underlying function in the form

$$\rho_c(\mathbf{r}) = \rho_c(x,y,z) = \bar{\rho}_c + S_c^x\,(x - x_c) + S_c^y\,(y - y_c) + S_c^z\,(z - z_c)\,, \tag{5}$$

where

$$x_c = \frac{\int\limits_c x\,dV}{V(c)}\,, \quad y_c = \frac{\int\limits_c y\,dV}{V(c)}\,, \quad z_c = \frac{\int\limits_c z\,dV}{V(c)} \tag{6}$$

are coordinates of the cell center and $V(c)$ is the volume of the cell defined in (1).

For computation of slopes we use the limited form

$$S_c^x = \Phi_c\,S_c^{x\ \mathrm{unlim}}\,,\ S_c^y = \Phi_c\,S_c^{y\ \mathrm{unlim}}\,,\ S_c^z = \Phi_c\,S_c^{z\ \mathrm{unlim}}\,, \tag{7}$$

where $S_c^{\{x,y,z\}\ \mathrm{unlim}}$ are unlimited slopes and Φ_c is Barth-Jasperson limiter, which must be computed firstly.

Unlimited Slopes In 1D we can use just the central difference as the unlimited slope. To compute unlimited slopes in 2D we construct a contour surrounding the cell and use Green's Theorem. In 3D this would require computing intersections of this neighborhood with the original grid, which would be too slow. So we must use another method.

Let's construct the functional

$$F(S_c^x, S_c^y, S_c^z) = \sum_{c_n \in C(c)} \left(\bar{\rho}_{c_n} - \frac{\int\limits_{c_n} \rho_c(x,y,z)\,dx\,dy\,dz}{V(c_n)} \right)^2 \tag{8}$$

for each cell, which measures the sum of differences between the mean values in the neighboring cells and average values of the reconstructed function from the original cell in the same neighboring cell. We want to minimize this functional, so we want the reconstructed function to be as close to the mean values in the neighboring cells as possible.

We easily compute derivative of this functional with respect to all three variables and let them be equal to zero. This gives us a linear system

$$\frac{\partial F(S_c^x, S_c^y, S_c^z)}{\partial S_c^{\{x,y,z\}}} = 0\,, \tag{9}$$

which can be easily solved and gives us our unlimited slopes $S_c^{\{x,y,z\}\ \mathrm{unlim}}$.

Limited Slopes For computation of the slopes $S_c^{\{x,y,z\}}$ we use the Barth-Jasperson limiter at each cell vertex n and than the minimum of them as a cell limiter

$$\Phi_n = \begin{cases} \min\left(1, \frac{\rho_n^{\max} - \bar{\rho}_c}{\rho_n^{\text{unlim}} - \bar{\rho}_c}\right) & \text{for } \rho_n^{\text{unlim}} - \bar{\rho}_c > 0 \\ \min\left(1, \frac{\rho_n^{\min} - \bar{\rho}_c}{\rho_n^{\text{unlim}} - \bar{\rho}_c}\right) & \text{for } \rho_n^{\text{unlim}} - \bar{\rho}_c < 0 \\ 1 & \text{for } \rho_n^{\text{unlim}} - \bar{\rho}_c = 0\,, \end{cases} \qquad \Phi_c = \min_{n \in \mathcal{N}(c)} \Phi_n\,, \tag{10}$$

which ensures us preservation of local extrema and also preservation of a linear function. Here ρ_n^{unlim} is the value of the reconstructed function (using the unlimited slopes) in the node n. It is described in details in [2].

Integration over an Arbitrary Polyhedron The only part in the functional, we don't know, is the integral

$$\int_{c_n} \rho_c(x, y, z)\, dx\, dy\, dz\,. \tag{11}$$

We also need to compute the integrals in the definition of cell centers x_c, y_c, z_c (6) and cell volumes V_{c_n} (1). So we need a method for integration of the linear function over an arbitrary polyhedron. We note, that the boundary of the polyhedron is uniquely defined , we know just the vertices of each face. If the face vertices do not lie in one plane, the face is curved and the boundary is not uniquely defined.

We demonstrate our integration procedure for the example of the cell volume, the integration of an arbitrary linear function is similar. The cell volume can be written in the form

$$V(c) = \int_c \mathbf{1}\, dV = \frac{1}{3} \int_c \operatorname{div}(x, y, z)\, dV \tag{12}$$

and using the Divergence Theorem we can rewrite it as an integral over the boundary ∂c

$$V(c) = \frac{1}{3} \int_{\partial c} (x, y, z)^{\mathsf{T}} \cdot \mathbf{S}\, dA\,. \tag{13}$$

Here the superscript ^{T} means the transposition of a vector and $\mathbf{S}$ is the vector normal to the boundary. The boundary integral can be split into the sum over all faces Π of the face integrals

$$V(c) = \frac{1}{3} \sum_{\Pi \in \partial V} \int_{\Pi} (x, y, z)^{\mathsf{T}} \cdot \mathbf{S}\, dA \tag{14}$$

Now just by averaging the coordinates of vertices of each face we compute its center, connect it with all face vertices and split these face integrals to the integrals over such defined triangles Δ. On each this triangle the face normal $\mathbf{S}$ is constant so it can go in front of the integral

$$V(c) = \frac{1}{3} \sum_{\Pi \in \partial V} \sum_{\Delta \in \Pi} \left(S^x \int_{\Delta} x \, dA + S^y \int_{\Delta} y \, dA + S^z \int_{\Delta} z \, dA \right) \tag{15}$$

Now we project all triangles to the coordinate planes. For each triangle we select the coordinate plane in which the triangle has the biggest area. This ensures us that we do not get into trouble due to numerical problems. Using Green's Theorem we reduce these integrals over triangles to 1D edge integrals, which can be computed directly from vertex coordinates. This algorithm gives us a method for computing the integral of the arbitrary linear function over an arbitrary polyhedron. More details can be seen in [3].

2.4 Stage 2 – Swept Integration

Swept region quadrature concept has been explained in detail in [1].

The swept region is created by the movement of the face from the original grid to the new position. It is bordered by the old face, the new face, and by not necessarily flat quadrilaterals connecting each edge from the original face to the edge of the new face. We can compute the volume and mass of a such region - we talk about swept volume and swept mass. We use these terms in their signed sense. Suppose we have a cell on the original mesh and we move just one face as illustrated on the Fig. 1. In this case, the right face moves

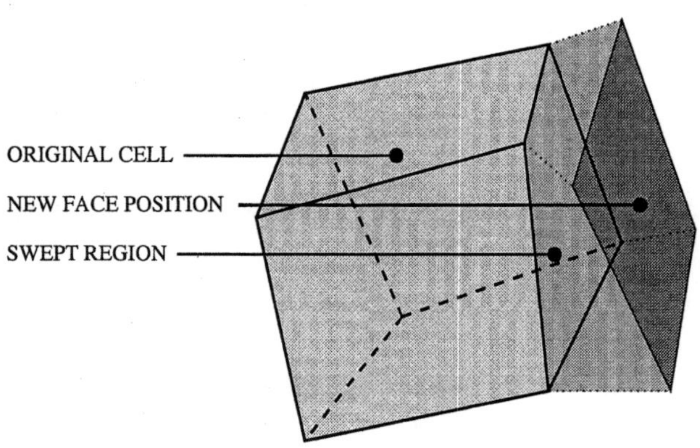

Fig. 1. One swept region

outward from the original cell and the middle part is the swept region. In fact, all faces can move in different ways and swept regions can be tangled. If most of the swept region goes outward from the original cell, the swept volume and swept mass are positive, otherwise they are negative. The mass of the swept region is computed by integration of the reconstructed function over the cell, in which the most swept region lies. The new cell mass can be composed from the mass of the original cell and the masses of all swept regions

$$m^*(\tilde{c}) = m(c) + \sum_{f \in \mathcal{F}(c)} \delta m_f \,. \tag{16}$$

Here f means a swept region from the set $\mathcal{F}(c)$ of all swept regions of the cell c. The new mean value can be than computed as

$$\bar{\bar{\rho}}_{\tilde{c}} = \frac{m^*(\tilde{c})}{V(\tilde{c})} \tag{17}$$

and as noticed before, it can violate the local bounds due to the approximation of the integration. So the third stage is necessary to enforce local-bound preservation.

2.5 Stage 3 – Repair

The repair stage works as the conservative redistribution of a conserved quantity. It corrects the overshoots back to their local bounds. At first, we must compute these local extrema. For each cell c we define a bound-determining neighborhood $\mathcal{C}(c)$, which is a piece of the original grid fully covering the new cell. Usually we use the original cell plus its nearest neighbors. We compute the local extrema in this neighborhood

$$\rho_c^{\min} = \min_{c_n \in \mathcal{C}(c)} \bar{\rho}_{c_n} \,, \quad \rho_c^{\max} = \max_{c_n \in \mathcal{C}(c)} \bar{\rho}_{c_n} \,. \tag{18}$$

We show the repair for the example of violation of the lower bound

$$\bar{\bar{\rho}}_{\tilde{c}} < \rho_c^{\min} \,, \tag{19}$$

upper bound is done similarly. At first we compute mass, which is needed in the cell to bring the mean value back to the local minimum

$$\delta m_{\tilde{c}}^{\text{needed}} = (\rho_c^{\min} - \bar{\bar{\rho}}_{\tilde{c}})\, V(\tilde{c}) \,. \tag{20}$$

We want our algorithm to be conservative, so we do not just add this mass to the wrong cell, but we look for available mass in the bound-determining neighborhood. For each neighboring cell we compute the mass

$$\delta m_{\tilde{c}_n}^{\text{avail}} = \max\Big((\bar{\bar{\rho}}_{\tilde{c}_n} - \rho_{c_n}^{\min})\, V(\tilde{c}), 0 \Big) \tag{21}$$

which can safely be taken from the cell without violating the local bound also. The total available mass in the neighborhood is

$$\delta m_{\mathcal{C}(c)}^{\text{avail}} = \sum_{c_n \in \mathcal{C}(c)} \delta m_{\tilde{c}_n}^{\text{avail}} . \tag{22}$$

If the available mass is too small ($\delta m^{\text{avail}}_{C(c)} < \delta m^{\text{needed}}_{\tilde{c}}$), we extend the stencil and look for the available mass in a larger area. If there is enough mass available, we perform the repair. We bring the wrong value back to the local minimum

$$m'(\tilde{c}) = \rho_c^{\min} V(\tilde{c}) \tag{23}$$

and we take the mass from the neighborhood proportionally to the mass available

$$m'(\tilde{c}_n) = m(\tilde{c}_n) - \frac{\delta m^{\text{avail}}_{\tilde{c}_n}}{\delta m^{\text{avail}}_{C(\tilde{c})}} \, \delta m^{\text{needed}}_{\tilde{c}} \, . \tag{24}$$

In [1] we proved that this algorithm succeeds in a finite number of steps and the repair stage corrects all local-bound violations.

3 Numerical Tests

3.1 Orthogonal Uniform Grid

In the first example the underlying function is equal to zero everywhere, only in a spherical region around the center of the computation domain $\langle 0,1\rangle^3$ it is equal to 1

$$g(x,y,z) = \begin{cases} 1 & \text{for } r \le 0.25 \\ 0 & \text{else} \end{cases} , \qquad r = \sqrt{\left(x-\frac{1}{2}\right)^2 + \left(y-\frac{1}{2}\right)^2 + \left(z-\frac{1}{2}\right)^2} \, . \tag{25}$$

We define the uniform orthogonal grid in the computational domain, the initial

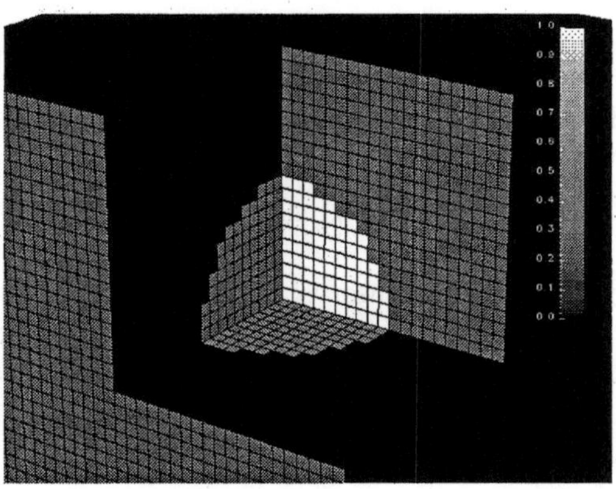

Fig. 2. Initial spherical function on 40^3 orthogonal uniform grid

function is shown on the Fig. 2. We move the grid as the tensor product movement

$$x_n^{\text{new}} = (1-\alpha)\, x_n + \alpha\, x_n^{1.5}\,,\, y_n^{\text{new}} = (1-\alpha)\, y_n + \alpha\, y_n^{2.0}\,,\, z_n^{\text{new}} = (1-\alpha)\, z_n + \alpha\, z_n^{2.5}\,, \tag{26}$$

where

$$\alpha = 0.5\,\sin(4\,\pi\, t)\,, \qquad t = N/N_{\max}\,, \quad t \in \langle 0,1\rangle \text{ – time of } N\text{th timestep.} \tag{27}$$

We make $N_{\max} = 200$ remappings to obtain accumulated errors and to have the problematic regions visible. On the Fig. 3 we can see this spherical function

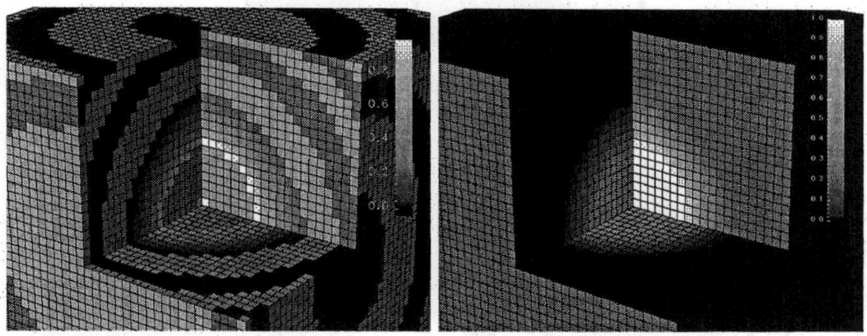

a) b)

Fig. 3. 200times remapped spherical function using only unlimited reconstruction a) without repair b) with repair

remapped using only unlimited slopes. This causes more errors, so the effect of the repair stage is more obvious. In the a) part of the figure we can see the function without the repair stage. The light gray cells show areas where the extrema are violated. In the b) part we see the same remapping with repair, no values violate the bounds.

3.2 Tetrahedral Grid with Random Movement

The second numerical example shows the same cubical computational domain with tetrahedral mesh inside. It includes about 9000 tetrahedrons. We use the same spherical function as before, we can see it on the Fig. 4. Now, we shake the grid randomly 10 times and remap between these grids. In the last time step we remap back to the original grid. On the Fig. 5 we can see the situation with the usage of the Barth-Jasperson limiter with and without the repair stage. Again, we can see several white cells in the a) part, where the bounds are violated. In the b) part, the repair stage corrects everything and no problem with bound preservation is observed.

Fig. 4. Initial spherical function on about 9000 tetrahedrons

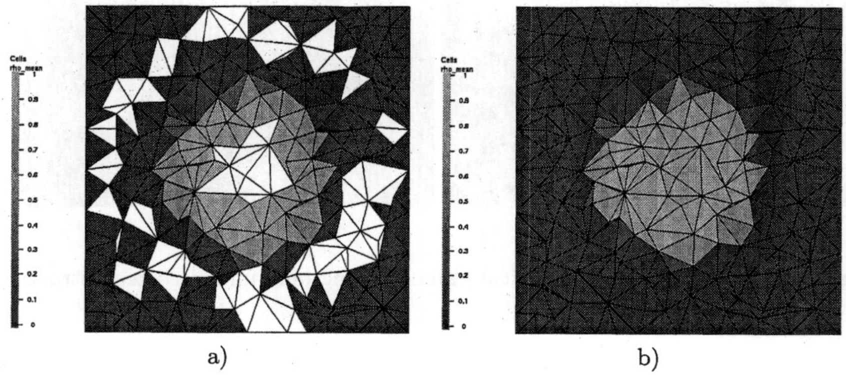

Fig. 5. 10times remapped spherical function using Barth-Jasperson limiter a) without repair b) with repair

4 Conclusion

In this article we constructed an efficient algorithm for function remapping between two similar grids. It is face-based and usable in 3D unlike the most natural exact integration algorithm, which is not feasible in 3D. The algorithm is conservative (total mass remains constant), local-bound preserving (does not create new extrema), stable and linearity preserving. We presented several numerical examples to show, that we can use it for different types of grids and grid movements.

5 Acknowledgments

This work was performed under the auspices of the US Department of Energy at Los Alamos National Laboratory, under contract W-7405-ENG-36. The authors acknowledge the partial support of the DOE/ASCR Program in the

Applied Mathematical Sciences and the Laboratory Directed Research and Development program (LDRD). R. Garimella and M. Shashkov acknowledge the partial support of DOE's Accelerated Strategic Computing Initiative (ASCI). M. Kuchařík also acknowledges the partial support of the Czech Technical University grant CTU0310614. The authors thank L. Margolin, B. Wendroff, B. Swartz, R. Liska, M. Berndt and V. Dyadechko for fruitful discussions and constructive comments.

References

1. M. Kuchařík, M. Shashkov, and B. Wendroff. An efficient linearity-and-bound-preserving remapping method. *Journal of Computational Physics*, 188(2):462–471, 2003.
2. T. J. Barth. Numerical methods for gasdynamic systems on unstructured meshes. In C. Rohde D. Kroner, M. Ohlberger, editor, *An introduction to Recent Developments in Theory and Numerics for Conservation Laws, Proceedings of the International School on Theory and Numerics for Conservation Laws*, Berlin, 1997. Lecture Notes in Computational Science and Engineering, Springer.
3. B. Mirtich. Fast and accurate computation of polyhedral mass properties. *Journal of Graphics Tools*, 1(2):31–50, 1996.

Mimetic Finite Difference Methods for Diffusion Equations on Unstructured Triangular Grid

Victor Ganzha[1], Richard Liska[2], Mikhail Shashkov[3] and Christoph Zenger[1]

[1] Department of Informatics, Technical University of Munich, Boltzmannstr. 3, D-85748 Garching, Germany *ganzha@in.tum.de, zenger@in.tum.de*

[2] Faculty of Nuclear Sciences and Physical Engineering, Czech Technical University in Prague, Břehová 7, 115 19 Prague 1, Czech Republic *liska@siduri.fjfi.cvut.cz*

[3] Group T-7, Los Alamos National Laboratory, Los Alamos, NM 87544, USA *shashkov@lanl.gov*

Summary. A finite difference algorithm for solution of stationary diffusion equation on unstructured triangular grid has been developed earlier by a support operator method. The support operator method first constructs a discrete divergence operator from the divergence theorem and then constructs a discrete gradient operator as the adjoint operator of the divergence. The adjointness of the operators is based on the continuum Gauss theorem which remains valid also for discrete operators. Here we extend the method to general Robin boundary conditions, generalize it to time dependent heat equation and perform the analysis of space discretization. One parameter family of discrete vector inner products, which produce exact gradients for linear functions, is designed. Our method works very well for discontinuous diffusion coefficient and very rough or very distorted grids which appear quite often e.g. in Lagrangian simulations.

1 Introduction

The mimetic finite difference methods preserve fundamental properties of the original continuum differential operators and allow the discrete approximations of partial differential equations (PDEs) to mimic critical properties including conservation laws and symmetries in the solution of the underlying physical problem [1]. The discrete analogs of differential operators satisfy the identities and theorems of vector and tensor calculus [2] and provide new reliable algorithms for a wide class of PDEs. In [3] the mimetic method for parabolic diffusion equation has been developed on 2D quadrilateral logically rectangular grids and in [4] the method has been developed for stationary diffusion equation on unstructured triangular grid. In this paper we apply our ideas to the construction of mimetic methods for the solution of parabolic diffusion problems in strongly heterogeneous materials on unstructured triangular computational grids in 2D, capable to treat arbitrary computational region.

1.1 Continuous problem

We consider heat equation with general Robin boundary conditions

$$u_t - \operatorname{div} K\, \mathbf{grad}\, u = f \text{ on } \Omega \qquad (1)$$
$$\alpha u + \beta(K\, \mathbf{grad}\, u, \mathbf{n}) = \psi \text{ on } \partial\Omega$$

on an arbitrary 2D region Ω with boundary $\partial\Omega$ with possible discontinuous diffusion coefficient K and unknown function u. The goal of the paper is to develop a numerical method for this problem on given triangular grid. The method should work well also on bad quality meshes typically appearing in Lagrangian hydrodynamics computations, where we need to treat the parabolic part of the model as heat conductivity of fluid. The spatial dicretization will be the same as in [4]. For derivation of discretization we will use the first order coordinate invariant operators div and **grad**, so we first transform the heat equation (1) into the first order system $u_t + \operatorname{div} \mathbf{w} = f,\ \ \mathbf{w} = -K\, \mathbf{grad}\, u$ where $\mathbf{w}$ is the heat flux.

For further analysis we introduce the generalized gradient operator

$$\mathbf{G}u = -K\, \mathbf{grad}\, u \qquad (2)$$

and extended divergence operator

$$\mathbf{D}\,\mathbf{w} = \begin{cases} \operatorname{div} \mathbf{w} & \text{on } \Omega \\ -(\mathbf{w}, \mathbf{n}) & \text{on } \partial\Omega \end{cases} \qquad (3)$$

and look at some integral properties of these operators.

First we note that the divergence Green formula

$$\int_\Omega \operatorname{div} \mathbf{w}\, d\,\Omega - \oint_{\partial\Omega} (\mathbf{w}, \mathbf{n})\, d\, S = 0 \qquad (4)$$

can be written as $(\mathbf{D}\ \mathbf{w}, 1)_H = 0$, where the inner product of scalar functions $(.,.)_H$ on the space H of sufficiently smooth scalar functions on Ω is defined by $(u, v)_H = \int_\Omega u\ v\ d\ \Omega + \oint_{\partial\Omega} u\ v\ d\ S$. On the other hand the Gauss theorem

$$\int_\Omega u \operatorname{div} \mathbf{w}\ d\ \Omega - \oint_{\partial\Omega} u(\mathbf{w}, \mathbf{n})\ d\ S + \int_\Omega (\mathbf{w}, K^{-1} K \mathbf{grad}\, u) d\ \Omega = 0 \qquad (5)$$

can be written as

$$(\mathbf{D}\mathbf{w}, u)_H = (\mathbf{w}, \mathbf{G}u)_{\mathbf{H}}, \qquad (6)$$

where the inner product of vector functions $(.,.)_{\mathbf{H}}$ on the space of vector functions $\mathbf{H}$ is defined by $(\mathbf{A}, \mathbf{B})_{\mathbf{H}} = \int_\Omega (K^{-1}\mathbf{A}, \mathbf{B}) d\ \Omega$.

Our operators $\mathbf{G}, \mathbf{D}$ are acting between spaces H and $\mathbf{H}$ as $\mathbf{G} : H \to \mathbf{H}, \mathbf{D} : \mathbf{H} \to H$. The Gauss theorem (6) implies that the operator $\mathbf{G}$ is the adjoint operator of the operator $\mathbf{D}$ in the sense of defined inner products $\mathbf{G} = \mathbf{D}^*$.

1.2 Semidiscrete problem

The heat equation problem (1) is discretized in time by fully implicit scheme

$$\frac{u^{n+1}-u^n}{\Delta t} - \operatorname{div} K\,\mathbf{grad}\,u^{n+1} = f \text{ on } \Omega, \qquad (7)$$
$$\alpha u^{n+1} + \beta(K\,\mathbf{grad}\,u^{n+1}, \mathbf{n}) = \psi \text{ on } \partial\Omega,$$

which can be written in operator form as $\mathbf{A}u^{n+1} = \mathbf{F}u^n$ where the operators $\mathbf{A}$ and $\mathbf{F}$ are given by

$$\mathbf{A}u = \begin{cases} u/\Delta t - \operatorname{div} K\,\mathbf{grad}\,u \text{ on } \Omega \\ \alpha u + \beta(K\,\mathbf{grad}\,u, \mathbf{n}) \text{ on } \partial\Omega \end{cases}, \qquad \mathbf{F}u = \begin{cases} u/\Delta t + f \text{ on } \Omega \\ \psi \qquad \text{on } \partial\Omega \end{cases}$$

For simplicity we assume Neumann boundary conditions in the semidiscrete problem (7), so $\alpha = 0, \beta = 1$. In this case the global operator A is given by

$$\mathbf{A} = (\mathbf{I}/\Delta t - \mathbf{DG}) \qquad (8)$$

where the operator $\mathbf{I}$ is the identity inside the region Ω and zero on the boundary $\partial\Omega$, $\mathbf{G}$ is the generalized gradient operator (2) and $\mathbf{D}$ is the extended divergence (3) operator. One can quite easily [4, 3] show ($\mathbf{G}$ is the adjoint operator of $\mathbf{D}$, $\mathbf{G} = \mathbf{D}^*$), that the global operator $\mathbf{A}$ is self adjoint and positive definite $\mathbf{A} = \mathbf{A}^* > 0$. The same conclusion can be reached also for the case of Dirichlet and Robin boundary conditions, [5].

2 Spatial discretization

We first describe the approximation of scalar and vector functions on the given unstructured triangulation of the region Ω. Triangles of the grid are numbered by index i, vertexes by index j and edges by index k with boundary edges ordered first. The scalar function u is approximated by the piecewise constant discrete function with constant values U_i inside each triangle i and with constant values U_k on each boundary edge k. The vector heat flux function $\mathbf{w}$ is discretized by point values at the center of each edge by the projection W_k of $\mathbf{w}$ to the normal to the edge as shown in Fig. 1 The normal flux is continuous across the edges. We define space HC of discrete scalar functions (piecewise constant functions inside each triangle and on each boundary edge) with natural inner product

$$(U,V)_{HC} = \sum_{i=1}^{N_t} U_i\, V_i\, VC_i + \sum_{k=1}^{N_{eb}} U_k\, V_k\, S_k,$$

where N_t is the number of triangles, N_{eb} is the number of boundary edges, VC_i is the area of triangle i and S_k is the length of the boundary edge k. The space $\mathbf{HL}$ of discrete vector functions has the natural inner product

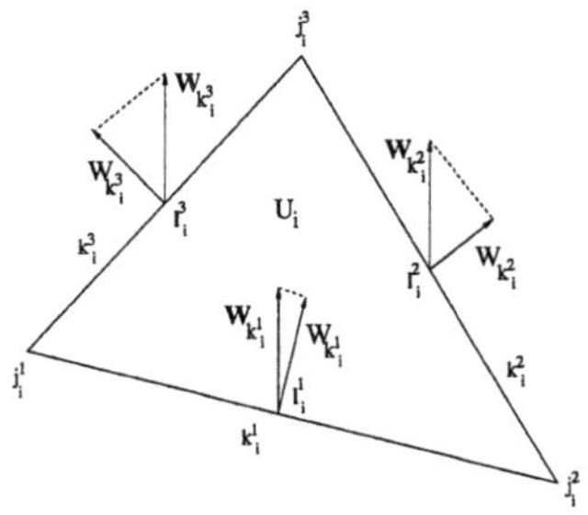

Fig. 1. Projection of vector function $\mathbf{w}$ to the edges normals at centers of the edges

$$(\mathbf{A}, \mathbf{B})_{\mathbf{HL}} = \sum_{i=1}^{N_t} (\mathbf{A}, \mathbf{B})_i \frac{VC_i}{\mathbf{K}_i}.$$

Both natural inner products of discrete functions are approximations of inner products of continuum functions defined above in section 1.1.

The traditional definition of the discrete inner product $(\mathbf{A}, \mathbf{B})_i$ of vector functions at triangle i is

$$(\mathbf{A}, \mathbf{B})_i = \frac{1}{3} \sum_{J=1}^{3} (\mathbf{A}, \mathbf{B})_{j_i^J}, \tag{9}$$

part of which at the vertex j_i^J of the triangle i is

$$(\mathbf{A}, \mathbf{B})_{j_i^J} = \frac{A_{k_i^J} B_{k_i^J} + A_{k_i^{J-1}} B_{k_i^{J-1}} + (A_{k_i^J} B_{k_i^{J-1}} + A_{k_i^{J-1}} B_{k_i^J}) \cos \phi_{k_i^J}^{k_i^{J-1}}}{\sin^2 \phi_{k_i^J}^{k_i^{J-1}}}. \tag{10}$$

Such inner product gives the exact gradient of linear functions.

2.1 General vector local inner product

The general discrete inner product of vector functions at a triangle i can be defined by a symmetric positive definite matrix M

$$(\mathbf{A}, \mathbf{B})_i^M = (M \cdot \mathbf{A}) \cdot \mathbf{B}, \quad M = \begin{pmatrix} m_{11} & m_{12} & m_{13} \\ m_{12} & m_{22} & m_{23} \\ m_{13} & m_{23} & m_{33} \end{pmatrix}.$$

Any triangle can be transformed into triangle with vertexes $(0,0), (1,0)$, (x, y) by moving, rotation and scaling. We continue the explanation here on this triangle. The Gauss theorem (5) is applied to our triangle with linear scalar function u and arbitrary vector function $\mathbf{w}$. We require the inner product to define the exact gradient which results in a system of 6 linear equations for 6 variables, elements m_{kl} of the matrix M. Only 5 equations from the system are independent and their solution is

$$m_{11} = \frac{((x-2)x+y^2+1)(x+1)s_2 + 3m_{12}s_1y^2}{3s_2y^2}$$

$$m_{13} = \frac{-(x^2+y^2-1)s_1s_2 + 3m_{12}y^2}{3s_2y^2}$$

$$m_{22} = \frac{-(x^2+y^2)(x-2)s_1 + 3m_{12}s_2y^2}{3s_1y^2}$$

$$m_{23} = \frac{-(x^2-2x+y^2)s_1s_2 + 3m_{12}y^2}{3s_1y^2}$$

$$m_{33} = \frac{((x-1)x+y^2+1)s_1s_2 + 3m_{12}y^2}{3s_1s_2y^2}$$

where s_1, s_2 are lengths of the triangle edges $s_1 = \sqrt{(x-1)^2+y^2}, s_2 = \sqrt{x^2+y^2}$ and m_{12} remains as a free parameter.

The Sylvester criterion for positive definiteness of matrix M results in a set of 3 inequalities which reduces to the constraint on $m_{12} > m_{12}^{\min} = s_1s_2(1 + (x+1)(x-2)/y^2)/9$ after simplification by using quantifier elimination. We have found a family of inner products depending on the parameter m_{12} which produce exact gradient for linear functions. The traditional scalar product defined by (9)-(10) belongs to this family. The free parameter m_{12} can be used to improve some properties of our numerical algorithm as its accuracy and condition number of the local matrix M [6].

2.2 Divergence and gradient discretization

Before proceeding to divergence and gradient discretizations we need to define formal inner products. The formal inner product for scalar discrete functions is given by

$$[U,V]_{HC} = \sum_{i=1}^{N_t} U_iV_i + \sum_{k=1}^{N_{eb}} U_kV_k, \qquad (U,V)_{HC} = [MU,V]_{HC}$$

and for vector discrete functions by

$$[\mathbf{A},\mathbf{B}]_{\mathbf{HL}} = \sum_{k=1}^{N_e} A_kB_k, \qquad (\mathbf{A},\mathbf{B})_{\mathbf{HL}} = [L\mathbf{A},\mathbf{B}]_{\mathbf{HL}}$$

where we have introduced the operators M, L which connect the natural and formal inner products. Note that the formal inner products are plain sums of discrete values products while the natural inner products approximate inner products on the spaces of continuum functions.

The discrete operators divergence D and gradient G act between discrete functions spaces as $D : \mathbf{HL} \to HC, G : HC \to \mathbf{HL}$. The divergence Green formula (4) can be written in the discrete case as $(DW, 1)_{HC} = 0$ and when applied to one triangle i it gives us the discretization of the divergence

$$(DW)_i = \frac{1}{VC_i} \sum_{J=1}^{3} W_{k_i^J} S_{k_i^J} \mathrm{sign}(j_i^{J+1} - j_i^J)$$

inside the triangle i ($\mathrm{sign}(j_i^{J+1} - j_i^J)$ distinguishes the unique direction on the edge k_i^J connecting vertexes j_i^{J+1} and j_i^J [4]) and $(DW)_k = -W_k n_k$ on the boundary edge k.
The Gauss theorem (6) can be rewritten in the discrete case as $(DW, U)_{HC} = (W, GU)_{\mathbf{HL}}$ so that the discrete gradient is the adjoint of discrete divergence $G = D^*$. When we transform this inner products equality into formal inner products using the operators L and M we get

$$[\mathbf{W}, D^{\otimes} MU]_{\mathbf{HL}} = [\mathbf{W}, LD^* U]_{\mathbf{HL}} \Rightarrow LD^* = D^{\otimes} M.$$

Now the formal adjoint $D^{\otimes}$ can be constructed [4] and to get the gradient $\mathbf{W} = GU$ (which is the natural adjoint D^*U) of the scalar grid function U the system

$$L\mathbf{W} = D^{\otimes} MU \tag{11}$$

has to be solved. The gradient constructed by this way as adjoint to divergence has global stencil.

The discrete approximation of the global operator (8) is symmetric and positive definite and for its inversion, which is needed in each time step of the implicit method, we employ the conjugate gradient method. The numerical gradient evaluated on every iteration of the conjugate gradient method by solving (11) for $\mathbf{E}$ is computed by the standard Gauss-Seidel method. Our method is exact on piecewise constant or piecewise linear solutions, otherwise it is second order accurate.

2.3 Boundary conditions

The shortly outlined mimetic discretization incorporates a different kind of boundary conditions, namely Dirichlet, Neumann and Robin ones. Each type of boundary conditions is treated differently. For Dirichlet boundary conditions the value of the scalar function u on the boundary is known, the gradient on the boundary is computed and boundary conditions are fulfilled exactly. For Neumann boundary conditions the gradient on the boundary is known, we do not solve for it, and the boundary flux (gradient) is moved to the right hand side of the global system, which is the discretization of the semidiscrete system (7). The value of the scalar function u on the Neumann boundary is not needed and again boundary conditions are fulfilled exactly.

The situation is different for the Robin boundary conditions when both value of the scalar function u and the boundary flux (gradient of u) are unknown on the boundary. The discrete form of the boundary conditions in (7) is included in the global discrete system and the boundaries with Robin conditions are included in scalar products used for computing conjugate gradient coefficients and residuals. More details on numerical treatment of boundary conditions by mimetic methods can be found in [5].

3 Numerical tests

In this section we provide several tests of the developed mimetic method. For all tests we use initial conditions $u = 0$ everywhere and compute till time sufficiently large for the solutions to reach the steady state. Exact discrete solutions used in error evaluation are given by the point values of exact continuous solution at the median (average of its vertexes) of each triangle.

3.1 Piecewise linear and quadratic tests

Both piecewise linear and quadratic tests are solved on the region $(x, y) \in (0,1) \times (0,1)$ with a discontinuous piecewise constant diffusion coefficient

$$k = \begin{cases} k_1, \, 0 < x < 0.5, \\ k_2, \, 0.5 < x < 1. \end{cases}$$

We solve these problems for particular values of diffusion coefficient $k_1 = 1$, $k_2 = 2$ till time $t = 10$ when the solution converges to the stationary one. Of course the triangulation is done in such a way that the whole discontinuity line $x = 0.5$ is covered by the edges so that inside each triangle the diffusion coefficient is constant.

The stationary exact solution of the piecewise linear test, coming e.g. from [7, 3], is a piecewise linear function

$$u = \begin{cases} \frac{k_2 x + 2k_1 k_2}{0.5(k_1+k_2)+4k_1k_2} & 0 < x < 0.5 \\ \frac{k_1 x + 2k_1k_2 + 0.5(k_2-k_1)}{0.5(k_1+k_2)+4k_1k_2} & 0.5 < x < 1, \end{cases} .$$

The maximal numerical errors for this test are shown in Table 1(a) and are close to machine precision, showing that our method is exact for piecewise linear solutions.

The stationary exact solution of the piecewise quadratic test, coming e.g. from [8, 3], is a piecewise quadratic function

$$u = \begin{cases} a_1 \frac{x^2}{2} + b_1 x & 0 \le x \le 0.5 \\ a_2 \frac{x^2}{2} + b_2 x + c_2 & 0.5 \le x \le 1 \end{cases}$$

where $a_i = -1/k_i$, $b_1 = (3a_2 + a_1)k_2/(4(k_1 + k_2))$, $b_2 = k_2 b_1/k_1$, $c_2 = -b_2 - a_2/2$. The convergence analysis for this test is presented by maximum errors and numerical order of convergence in Table 1(a) and confirms that our method is second order for non-linear solutions, even in the case of discontinuous diffusion coefficients.

3.2 Anisotropic triangulation

One of our aims was to develop a method working well also for bad quality, rather distorted triangular grid which appears quite often in Lagrangian meshes moving with the fluid. To show how our mimetic method works on bad quality grids including triangles with big angles we choose the initial grid on the region $(x, y) \in (-1, 1) \times (0, 1)$ as shown in Fig. 2 (a) and stretch this grid by a parameter a producing bad quality grids on the region $(x, y) \in (-a, a) \times (0, 1)$. The initial grid stretched by parameter $a = 5$ is shown in Fig. 2 (b). We solve one problem on series of grids obtained by stretching by increasing parameter a.

(a)

(b)

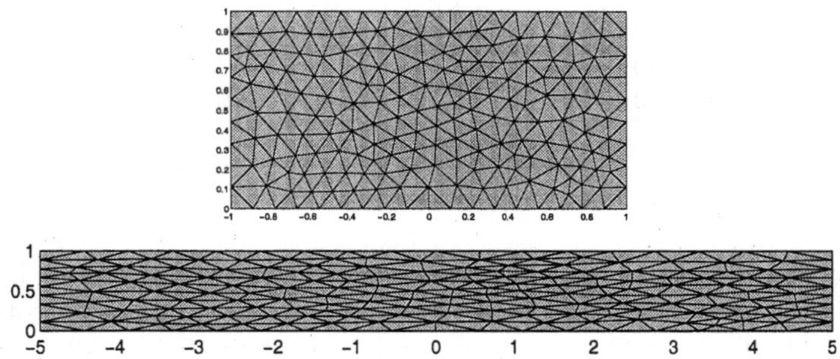

Fig. 2. Grid used for stretching the triangulation; (a) for parameter $a = 1$, i.e. $x \in (-1, 1)$, (b) for parameter $a = 5$, i.e. $x \in (-5, 5)$

The problem with anisotropic triangulation is heat equation (1) with diffusion coefficient $K = 1$, right hand side $f = -2/a^2$ and zero Dirichlet boundary conditions on left and right and zero Neumann boundary condition on top and down. The stationary solution of this problem is $u = x^2/a^2 - 1$. This problem is solved till time $t = 10a^2$ when the solution reaches the stationary state. We compare the results of our mimetic support operator method with the results of standard linear finite element method for which we use its implementation in Partial Differential Equation Toolbox in Matlab. This comparison is presented by maximal errors in Table 1(b) for stretching parameter a growing from 1 to 10 000. Our mimetic method keeps the accuracy well for all values of a while finite element method is loosing accuracy already for $a = 100$. The minimal numerical value of numerical solution, which should be minus one, is also presented in Table 1 (b). One can note that the finite element method is not resolving this minimum for big a, its solution remains flat and very close to initial condition $u = 0$. The reason for this behavior of the finite element method is that for big a linear interpolation introduces zig-zagging in y direction (there are no edges close to be parallel to y axis) for solution with curvature in x direction. This zig-zagging is eating too much of the overall energy and the parabola in the x direction is not resolved well.

Table 1. (a) Convergence tables of maximum errors $E_{\max}$ for piecewise linear and piecewise quadratic tests, for quadratic test also numerical order of convergence q is shown; (b) Maximum errors and minima of numerical solution of problem with stationary solution $u = x^2/a^2 - 1$ on grids stretched by parameter a by mimetic support operator method (MSOM) and standard linear finite element method (FEM).

Nr. of triangles	piecewise linear test $E_{\max}$	piecewise quadratic test $E_{\max}$	q
126	$1.4 \cdot 10^{-10}$	0.0024	1.96
504	$2.5 \cdot 10^{-11}$	0.00063	1.98
2016	$8.9 \cdot 10^{-12}$	0.00016	1.99
8064	$9.4 \cdot 10^{-12}$	0.000040	

(a)

	MSOM		FEM	
a	E_{max}	$min(u)$	E_{max}	$min(u)$
1	0.011	-1.0087	0.0039	-1.0028
10	0.0055	-1.0032	0.072	-0.95
100	0.0055	-1.0031	0.88	-0.12
1000	0.0055	-1.0032	1.0	-0.0013
10000	0.0074	-1.0034	1.0	-0.00001

(b)

Acknowledgment: This research has been partially supported by International Bureau of the BMBF and Czech Ministry of Education in the framework of the German-Czech join research project nr. CZE-00-010 and by the project nr. ME 436/2001 from the program Kontakt of the Czech Ministry of Education. The work of M. Shashkov was performed under auspices of the US Department of Energy under contract W-7405-ENG-36.

References

1. M. Shashkov. *Conservative Finite-Difference Methods on General Grids.* CRC Press, Boca Raton, Florida, 1996.
2. L. Margolin, M. Shashkov, and P. Smolarkiewicz. A discrete operator calculus for finite difference approximations. *Comput. Methods Appl. Mech. Engrg.*, 187:365–383, 2000.
3. M. Shashkov and S. Steinberg. Solving diffusion equation with rough coefficients in rough grids. *J. Comp. Phys.*, 129:383–405, 1996.
4. Victor Ganzha, Richard Liska, Mikhail Shashkov, and Christoph Zenger. Support operator method for laplace equation on unstructured triangular grid. *Selcuk Journal of Applied Mathematics*, 3(1):21–48, 2002.

5. J.M. Hyman and M. Shashkov. Approximation of boundary conditions for mimetic finite-difference methods. *Computers Math. Applic.*, 36:79–99, 1998.
6. R. Liska, M. Shashkov, and V. Ganzha. Analysis and optimization of inner products for mimetic finite difference methods on triangular grid. *Mathematics and Computers in Simulation*, Accepted, 2004.
7. J.M. Morel, J.E. Dendy Jr, M.L. Hall, and S.W. White. A cell-centered lagrangian-mesh diffusion differencing scheme. *J. Comp. Phys.*, 103, 1992.
8. R. J. MacKinnon and G. F. Carey. Analysis of material interface discontinuities and superconvergent fluxes in finite difference theory. *J. Comp. Phys.*, 75, 1988.

On Computational Glaciology: FE-Simulation of Ice Sheet Dynamics

Günter Gödert[1] and Franz-Theo Suttmeier[2]

[1] *Guenter.Goedert@zf.com*
[2] University Dortmund *info@fem2m.de*

Summary. The main focus of this paper is on stable FE-discretisations for treating systems of partial differential equations arising in glaciology. The systems are coupled ones, consisting of a flow problem determining stress, pressure and velocity and evolution problems for temperature and mean orientation densities, describing anisotropic material behaviour. The proposed strategies are applied to a standard model for describing ice sheet dynamics and an enhanced one, taking into account the developement of certain fabrics in the structure of the ice.

1 Introduction

Climate, climate history and climate forecast have become more and more important. If one wants to conceive changes of climate in the past and to make precdictions for the future, the climate of an ice-age has to be studied.

In the context of climate simulations the flow of polar ice masses represents an essential part. In so far there emerges a need for appropriate climate boundary conditions, Greve e.a. [11], Huybrechts [13], Fabre e.a. [3]. Furthermore there is a need of an appropriate description of thermo-mechanical material behaviour. For instance simulations considering future climate need reliable constitutive relations to generate reliable predictions about the global hydrobalance. In addition problems emerged from cold region structural engineering, e.g. Calov e.a. [1], are often become more reliable through such flow simulations. On the other hand, because it may record the past history of ice and climatic changes and because it is sensible to ice sheets deformation history, the microstructure of polar ice is worth studying.

The growth and retreat of inland ice masses is governed by the snowfall onto the surface, the melting and calving of the ice close to and at the outer ice boundaries. Owing to its own weight, the ice deforms with velocities of typically 100 meter per year causing a transport of ice towards the ice sheet boundaries where the ice melts and calves. This process, in turn, is influenced by the temperature distribution within the ice, implying a delicate balance between the thermal and mechanical regimes that are established by the climate input and the geothermal conditions of the substrate. The thermomechanically coupled ice dynamics together with the mass flux due to snowfall and mass loss in the vicinity of ice boundaries determine the thickness distribution of a particular ice sheet.

The deformation of an ice sheet and the variation of its temperature distribution depends to a large extent on its thermomechanical constitutive modelling. Here, we treat ice as a rheologically nonlinear, thermally coupled, viscous fluid, i.e., we asume its fluidity (inverse viscosity) to be temperature-dependent, the latter according to a power law with exponent 3, the former essentially following an Arrhenius-type relationship.

Ice sheets are *cold*, i.e. below their melting point, except at parts of their base, where the temperature may reach the melting point. For instance, this can be concluded from radio-echo sounding of sub-Antarctic lakes (c.f. Oswald and Robin [16]). The usual fluid no-slip boundary conditions only apply when basal ice is cold, but at the melting temperature, ice can slide over its base (see Paterson [17]). In Fowler [4] it is mentioned, that several models are based on the assumption of non-zero sliding velocity, and in some cases this is even required in order to obtain a solution (see Morland and Johnson [15] and Hutter e.a. [12]). On the other hand, basal topographie in Antarctica is so rough, that for the sliding law we should expect $\mathbf{v} \approx 0$ (c.f. Paterson [17], Richardson [18]).

The underlying mathematical model is a coupled system of partial differential equations for describing the distributions of velocities, temperature and evolution of the geometry of the ice sheets. This system is solved numerically by employing the Finite Element (FE) method. The choice of an appropriate discretisation is involved to some extent:

1) assuming the temperature to be known, we have to choose stable elements for the saddle-point problem determining stress, pressure and velocity simultaneously.

2) assuming the velocity field to be known, we have to choose stable elements for the convection-diffusion problem determining the temperature.

The standard fluid model mentioned above is necessarily isotrop and thus cannot describe stress-induced anisotropies evident in specimens from boreholes. The extreme anisotropy of the ice single crystal leads to heterogeneous intra-granular deformation modes within the polycrystal and hence to the development of a certain fabric. The expectation is, that the climate becomes to some extent reconstructable from analyzing ice-core textures, e.g., Thorsteinsson [19], in combination with the numerical solution of the ice-sheet flow problem. Therefore considerations for an enhanced model are based on two geometric scales.

2 Mathematical model

The ice in large ice masses is generally polytherm, i.e., the ice mass consists of disjoint regions in which the ice is either cold (i.e., its temperature is below the melting point) or temperate (i.e., it is at the pressure melting point), but except for a few recent cases theoretical formulations are restricted to cold ice. For such a case the continuum mechanical postulate *ice is a slow, gravity-driven, incompressible, heat-conducting, nonlinear viscous fluid* yields the following balance laws of mass, momentum and temperature as well as constitutive relations:

$$\begin{aligned} C\sigma - \varepsilon(v) &= 0 && \text{Material} \\ -\operatorname{div} v &= 0 && \text{Incompressibility} \\ -\operatorname{div}\sigma + \nabla p - f &= 0 && \text{Equilibrium} \\ \dot{T} - \Delta T - \sigma\varepsilon(v) &= 0 && \text{Temperature} \end{aligned}$$

which hold on a bounded domain Ω in $\mathbb{R}^2$. The right-hand-side $f = (0, -g)^T$ is defined by the gravity force g. C describes the constitutive relation between the deviatoric stress σ and the strain rate $\varepsilon(v) = (\nabla \mathbf{v} + (\nabla \mathbf{v})^T)/2$ and may be identified with the 4-th order fluidity tensor. The velocity field is denoted by $v = (v_1, v_2)^T$, the related pressure by p and the temperate by T.

Usually the relation $C\sigma = \varepsilon(v)$ is given by

$$\varepsilon(v) = A(T)G(\sigma)\sigma\,,$$

where $A(T)$ denotes the Arrhenius law and $G(\sigma) = |\sigma|^2$ according to Glenn.

The basis for applying the Finite Element (FE) method to the classical system is the formulation in the variational setting

$$\begin{array}{lrl} (C\sigma, \tau) & -\ (\varepsilon(v), \tau) = & 0 \\ & (\operatorname{div} v, q) = & 0 \\ (\sigma, \varepsilon(\varphi)) - (p, \operatorname{div}\varphi) & = & (f, \varphi) \end{array}$$

and

$$(\partial_t T, w) + ((v \cdot \nabla)T, w) + (\nabla T, \nabla w) = (\sigma\varepsilon(v), w)\,,$$

for arbitrary τ, q, φ, w choosen from suitable function spaces described below. In order to introduce a more compact setting, we define

$$U = \begin{pmatrix} \sigma \\ p \\ v \\ T \end{pmatrix}, \Phi = \begin{pmatrix} \tau \\ q \\ \varphi \\ w \end{pmatrix}. \tag{1}$$

With these definitions, we consider the problem of finding $U \in \mathbf{V} := \Sigma \times Q \times V \times L$ with
$\mathbf{V} \subset (L_2)^3 \times L_2 \times (H^1)^2 \times H^1$, fulfilling

$$\mathcal{B}(U; \Phi) = 0 \qquad \forall\, \Phi \in \mathbf{V}\,, \tag{2}$$

with

$$\begin{aligned}\mathcal{B}(U;\Phi) := (C\sigma,\tau) - (\varepsilon(v),\tau) - (\operatorname{div} v, q) \\ + (\sigma,\varepsilon(\varphi)) - (p, \operatorname{div}\varphi) - (f,\varphi) \\ + (\partial_t T, w) + ((v\cdot\nabla)T, w) + (\nabla T, \nabla w) - (\sigma\varepsilon(v), w)\,.\end{aligned}$$

Here and in what follows, $(.,.)$ represents the L^2 inner product of a bounded domain Ω in $\mathbb{R}^2$ and $\|.\|$ the corresponding norm. Furthermore $H^m = H^m(\Omega)$ denotes the standard Sobolev space of L^2-functions with derivatives in $L^2(\Omega)$ up to the order m, and $H_0^1 \subset H^1$ is the subspace of H^1-functions vanishing on $\Gamma := \partial\Omega$.

2.1 Boundary conditions

Sliding (basal slip), as opposed to fabric or temperature-enhanced basal shear, certainly occurs on ice sheets, particularly at the margins. We consider it unlikely to occur where basal ice is cold, and where shear stresses are close to zero. In addition large scale basal roughness may mean that any sliding which does occur inland will be very small. Following these remarks we consider in test configurations sketched in Figure 1 no-slip boundary conditions, i.e. $v_1 = v_2 = 0$ on Γ_x, and, exploiting symmetry, $v_1 = 0$ on Γ_y, for the components of the velocity field $\mathbf{v} = (v_1, v_2)^T$.

Furthermore, we have free boundary conditions for the temperatute T on Γ_x and Γ_y (due to symmetry) and for $\mathbf{v}$ on $\Gamma_S := \partial\Omega \setminus (\Gamma_x \cup \Gamma_y)$. Moreover T is prescribed on Γ_S by data from climate input.

Now, we calculate the deformation under the assumption of isotropic material behaviour.A plot of the velocity field is depicted in Figure 1. We observe that Γ_S is devided into two parts, the inflow and outflow boundary, defined by

$$\Gamma_- = \{x \in \Gamma_S \mid v\cdot n < 0\}\,, \tag{3}$$

$$\Gamma_+ = \{x \in \Gamma_S \mid v\cdot n \geq 0\}\,, \tag{4}$$

where n denotes the outward unit normal of Γ_S. This notation will be used at the end of Section 4.

3 Discretisation

The full discretisation for the system (2) is derived in two steps. First, we perform a discretisation with respect to the time variable, yielding a sequence of problems continuous with respect to the space variable. In the second step, these problems are approximated by the finite element method.

3.1 Semi-discretisation

For discretization, the time interval $[0, t_M]$ is decomposed like $0 = t_0 < t_1 < \ldots < t_M$ into subintervals $I_m := (t_{m-1}, t_m]$ of length $k_m := t_m - t_{m-1}$.

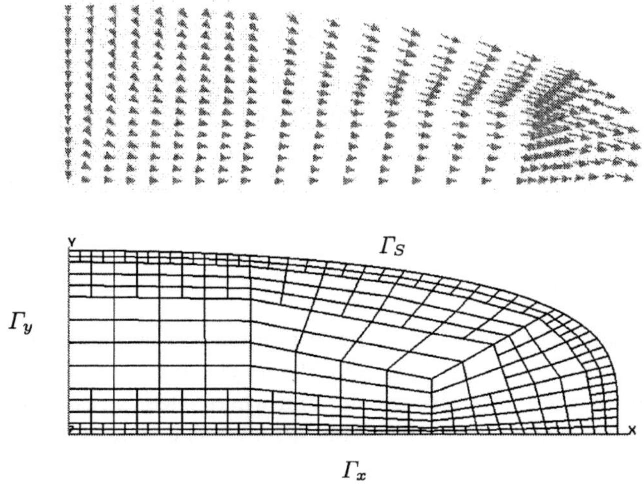

Fig. 1. Sketch of computational domains for the test examples including structure of the FE-meshes for the benchmark problem

Integrating in (2) over I_m and approximating the integrals by quadrature formulas of the type

$$\int_{I_m} w(t)\,dt = k_m\{\alpha w^m + (1-\alpha)w^{m-1}\}\,, \tag{5}$$

with some $\alpha \in (0,1]$, yields the time dicrete schemes. The choice of $\alpha = 1$ corresponds to the backward Euler scheme, while for $\alpha = \frac{1}{2}$, we obtain the Crank-Nicolson scheme. Here, we only consider the simple Euler scheme which reads

$$\begin{aligned} 0 = B(U^m;\Phi) &:= (T^m - T^{m-1}, w) \\ &+ (C\sigma^m, \tau) - (\varepsilon(v^m), \tau) - (\operatorname{div} v^m, q) \\ &+ (\sigma^m, \varepsilon(\varphi)) - (p^m, \operatorname{div}\varphi) - (f^m, \varphi) \\ &+ k_m\Big(((v^m\cdot\nabla)T^m, w) + (\nabla T^m, \nabla w) - (\sigma^m\varepsilon(v^m), w)\Big)\,. \end{aligned} \tag{6}$$

In what follows the superscript m is omitted.

3.2 Nonlinear solution process

In this section, we describe the algorithm (see e.g. Geiger and Kanzow [7]) we employ to solve the problems arising in (6) having the general structure

$$B(U;\Phi) = 0\,. \tag{7}$$

1. Calculate correction D^j by a linear problem

$$L_B(D^j, \Phi) = B(U^j; \Phi) . \tag{8}$$

2. Perform a damped update

$$U^{j+1} = U^j - \alpha^j D^j , \tag{9}$$

where α^j is choosen s.t.

$$B(U^{j+1}; \Phi) \leq B(U^j; \Phi) + \alpha^j \delta D^j B(U^j; \Phi) \tag{10}$$

with a constant $\delta \geq 0$.
3. Set $j = j + 1$ and goto 1.

Remark: The Newton scheme is defined by the choice

$$L_B(D^j, \Phi) = B'(U^j; D^j, \Phi) , \tag{11}$$

where $B'(U^j; \cdot, \cdot)$ is the derivative of $B(\cdot; \cdot)$ in U^j.

The problem $B(U, \Phi) = 0$ has in cases under consideration the special form

$$B(U)U - F(U) = 0 ,$$

omitting the argument Φ to simplify notation. A full Newton-scheme is determined by

$$\begin{aligned} L_B(D) &= \partial_U \left(B(U)U - F(U)\right) D \\ &= B(U)D + B'(U)DU - F'(U)D . \end{aligned}$$

Neglecting the last two terms, our nonlinear iteration is given by

$$\begin{aligned} U^{j+1} &= U^j - \alpha^j B^{-1}(U^j) \left(B(U^j)U^j - F(U^j)\right) \\ &= (1 - \alpha^j)U^j + \alpha^j B^{-1}(U^j) F(U^j) , \end{aligned}$$

which allows for the interpretation as a damped variant of Kachanov's method (*freezing of coefficients*).

3.3 Spatial discretisation

In order to obtain approximate solutions of the time discrete problems (6) we will apply the finite element method on decompositions $\mathbb{T}_h = \{T_i \mid 1 \leq i \leq N_h\}$ of Ω consisting of quadrilateral elements T, satisfying the usual condition of shape regularity. The width of the mesh $\mathbb{T}_h$ is characterised in terms of a piecewise constant mesh size function $h = h(x) > 0$, where $h_T := h_{|T} = \mathrm{diam}(T)$. With these notations the discrete solutions U_h of (6) are defined by

$$B(U_h; \Phi) = 0 \qquad \forall \Phi \in \mathbf{V}_h , \tag{12}$$

where stable discretisations are established by a cell-wise constant pressure approximation and discontinuous and continuous bilinear functions ($Q1$) for stress and velocity respectively. Temperature is approximated by continuous Q_1-elements as well, where the corresponding transport dominated equation requires further stabilisation, which is provided by the streamline diffusion method. For a detailed discussion of this method we refer to the textbook written by C. Johnson [14].

3.4 Linear solution process

1) Assuming the temperature to be known, in each nonlinear iteration step we have to solve a linear system of the form

$$\begin{pmatrix} A & 0 & -B^T \\ 0 & 0 & C^T \\ -B & C & 0 \end{pmatrix} \begin{pmatrix} \sigma \\ p \\ v \end{pmatrix} = \begin{pmatrix} 0 \\ 0 \\ r \end{pmatrix}$$

with the relations

$$\begin{aligned} A &\approx (C(a)\sigma, \tau)\,, \quad & B &\approx (\sigma, \varepsilon(\varphi))\,, \\ C &\approx (p, \operatorname{div}\varphi)\,, \quad & r &\approx -(f, \varphi)\,. \end{aligned}$$

Identifying $\mathcal{U} = (\sigma, p)^T$, $\mathcal{P} = v$, $b = 0$, $c = r$ and

$$\mathcal{A} = \begin{pmatrix} A & 0 \\ 0 & 0 \end{pmatrix}, \quad \mathcal{B} = (-B, C)\,,$$

we have to treat a general discrete saddle point problem of the form

$$\begin{pmatrix} \mathcal{A} & \mathcal{B}^T \\ \mathcal{B} & 0 \end{pmatrix} \begin{pmatrix} \mathcal{U} \\ \mathcal{P} \end{pmatrix} = \begin{pmatrix} b \\ c \end{pmatrix}$$

which can be done by employing augmented Lagrange algorithms (see e.g. Glowinski [8]).

2) assuming the velocity field to be known, in each nonlinear iteration step we have to solve an unsymmetric linear system determining the temperature. This can be done by using a standard `bicgstab`-algorithm.

4 Enhanced Model

An important by-product of the simulation of the dynamics of large ice masses is the determination of the age of the ice as a function of its position where it is located today. Hereby, the age of an ice particle is defined as the time that elapsed since it fell as a snow flake on the free surface. The problem of the ice age - depth correlation has indeed become a central question in the reconstruction of the past climate from many ice cores in Greenland and Antarctica. The thermomechanically coupled ice-sheet models yield this information on the ice motion through spatial and temporal integration, however, the resulting motion depends on the underlying constitutive behaviour usually assumed to obey an isotropic, non-linearly viscous flow law, as described above.

More precisely, such laws should be based upon the microscopic structure of the material, which is for ice, as for metals, a polycrystalline aggregate consisting of hexagonal single crystal grains. The ice in glaciers is not isotropic; at closer examination it is seen to be built by a large number of differently oriented, strongly anisotropic ice crystals. As structural elements one may distinguish two building stones defining the crystal: first the basal plane along which the crystals may relatively easily slide, and second, the unit normal vector perpendicular to it, defining the so-called c-axis.

The anisotropy of ice crystallites is given by these axis, the fine-scale is given by the crystallites orientation defined on the sphere S^2, whereas the large-scale is to be identified with the space of daily experience say $\mathbb{R}^d$. Following Gödert [10] the fine-scale-structure is actually considered via the second-order structure tensor $\mathbf{A}$ denoting mean orientation densities and yielding an anisotropic material behaviour.

Therefore our classical system is enhanced by

$$\dot{\mathbf{A}} = \mathcal{E}[\mathbf{v}, \mathbf{A}] \tag{13}$$

From now on, C depends on the orientation

$$\mathbf{A} = (a_{11}, a_{22}, a_{12})^T$$

of the crystallites the ice consists of. The corresponding evolution of $\mathbf{A}$ is determined by $\mathcal{E}[\mathbf{v}, \mathbf{A}]$. Following Gödert [10], $\mathcal{E}$ is given by

$$\mathcal{E}[\mathbf{v}, \mathbf{A}] = ((\alpha_1 - 1)I_{\text{dev}} - 2\alpha_1 \mathcal{P}_{\mathbf{a}})\varepsilon(\mathbf{v}) \\ + \lambda_0(Id - (d+1)\mathbf{A}) + \mathbf{WA} - \mathbf{AW}\,,$$

where $\mathbf{W} = (\nabla \mathbf{v} - (\nabla \mathbf{v})^T)/2$. The value α_1 defined by

$$\alpha_1 := \alpha \frac{d}{d-1}(I_{\mathbf{A}^2} - \frac{1}{d}) \tag{14}$$

denotes a measure of alignment, $\alpha_1 \to 0$ for randomly distributed c-axes orientations, whereas $\alpha_1 \to \alpha$ if all c-axes are parallel. Furthermore $\alpha = 1.2$ is determined via field data (c.f. Gödert [9]) and $I_{\mathbf{A}^2}$ denotes the inner product, $I_{\mathbf{A}^2} := (Id, \mathbf{A}^2)$. The macro-space dimension is given by d, λ_0 controls the diffusion due to recrystallisation, e.g. Gödert [10].

The operator $((\alpha_1 - 1)I_{\text{dev}} - 2\alpha_1 \mathcal{P}_{\mathbf{a}})$ is realised as a 3×3-matrix, applied to the vector

$$\varepsilon(\mathbf{v}) = (\partial_x v_1, \partial_y v_2, (\partial_y v_1 + \partial_x v_2)/\sqrt{2})^T \,.$$

$\mathcal{P}_{\mathbf{a}}$ is symmetric and in this note given by

$$\begin{pmatrix} a_{11}a_{22} & -a_{11}a_{22} & c \\ -a_{11}a_{22} & a_{11}a_{22} & -c \\ c & -c & 1/2 - 2a_{12} \end{pmatrix},$$

with $c = a_{12}(2a_{22} - 1)$. The material behaviour is characterised by

$$C = \mu \left[\frac{\beta}{2} I_{\text{dev}} + (1 - \beta)\mathcal{P}_{\mathbf{a}} \right] \tag{15}$$

where in accordance to Gagliardini and Meyssonnier [5] $\beta = 0.25$ can be interpreted as the ratio of the prismatic and the basal fluidity. Furthermore μ is given by

$$\mu = \mu_0 \frac{5}{3\beta + 2} \,,$$

with a constant $\mu_0 > 0$. In test calculations below we simply choose $\mu_0 = 1$. For the term $\mathbf{WA} - \mathbf{AW}$, one obtains

$$\mathbf{WA} - \mathbf{AW} = \frac{(\partial_y v_1 - \partial_x v_2)}{2} \left(2a_{12} \;\; -2a_{12} \;\; (a_{22} - a_{11})\right)^T \,.$$

Remark: The evolution for $\mathbf{A}$ has a hyperbolic character. Consequently, we can prescribe inflow boundary data for $\mathbf{A}$ on Γ_-. On Γ_+ $\mathbf{A}$ is left free. The discretisation for $\mathbf{A}$ is performed using the same strategies as described for the numerical treatment of the temperature T.

5 Numerical results

The numerical results presented throughout this work are obtained by FE-implementations based on the DEAL-library [2].

5.1 Benchmark problem

Here, we consider our model on a bounded domain Ω described by

$$1 - \tfrac{2}{3}\varepsilon x^2 - y^4 = 0 \qquad \varepsilon = 1.83 \cdot 10^{-2} \,.$$

The structure of the FE-meshes is shown in Figure 1. Especially, we focus on the evaluation of the velocity field along axis parallel to the $y-$axis.

In order to check our discretisation, we compare the computed solution of v_1 along the vertical line $x = 1$ for the isotropic case to Vialov's profile (c.f. Vialov [20]), an analytical description for such a situation. The result is depicted in Figure 2, demonstrating good agreement between computed and the exact solution, denoted by u and $f(x)$ respectively.

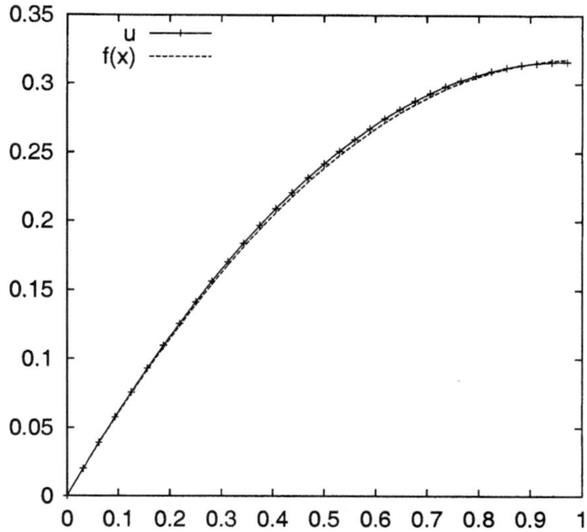

Fig. 2. Evaluation of u along vertical line

5.2 Enhanced model

The computations are performed using about 23365 degrees of freedom. The time step is choosen adaptively via

$$k_m = 0.9k_{m-1}\sqrt{\delta/\|U_1 - U_2\|}\,.$$

with $\delta = 0.001$. Here U_1 and U_2 denote the solution at time step $m-1$ obtained by employing Euler and Crank-Nicolson scheme respectively. The typical developement of the local step size is depicted in Figure 3 showing k_m to increase in the stationary limit. The considerations are restricted to isothermal flow.

Now, we compare the horizontal velocity along the vertical line $x = 1$ for the isotropic ($\beta = 1$) and the enhanced model ($\beta = 0.25$). In Figure 4 it is shown, that the enhanced model becomes faster, after the stationary limes was reached. This is in agreement with results found in Gagliardini and Meyssonnier [6].

Eventually, we investigate the influence of λ_0, which controls the effect of recrystallisation, on the orientation of the c-axis described by the parameter α_1 in (14). The results are shown in Figure 5. As predicted from theory the maximum value of α_1 is controlled by λ_0, and we observe $\alpha_1 \to \alpha$ as λ_0 tends to zero.

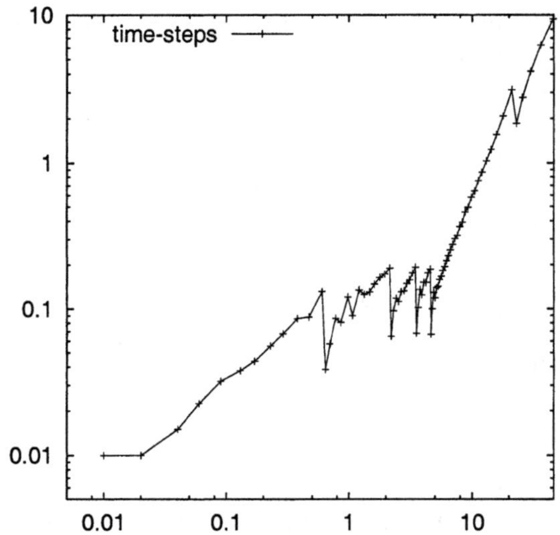

Fig. 3. Typical developement of the local step size during the simulation, showing k_m to increase in the stationary limit

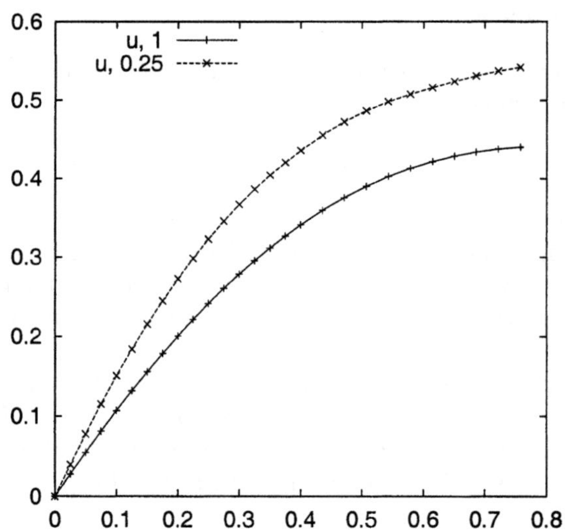

Fig. 4. Evaluation of u along vertical line, demonstrating the enhanced model ($\beta = 0.25$) to become faster compared to the isotropic model ($\beta = 1$)

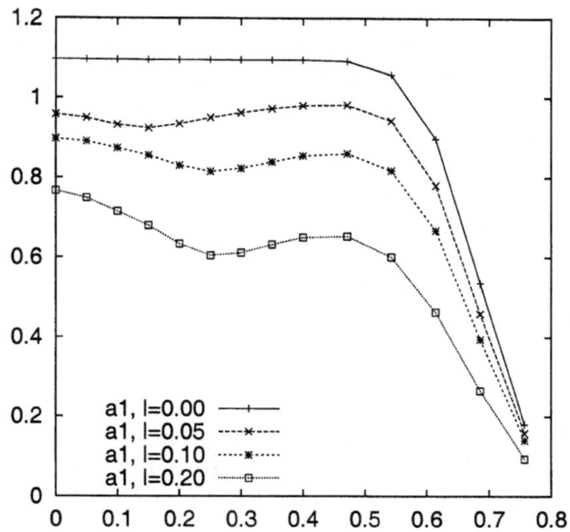

Fig. 5. Evaluation of α_1 along vertical line for different values of the parameter λ_0

6 Conclusion

We have presented a stable FE-discretisation for treating systems of partial differential equations arising in glaciology. The system is a coupled one, consisting of a flow problem determining stress, pressure and velocity and evolution problems for temperature and mean orientation densities, describing anisotropic material behaviour. The time discretisation is done by the stable backward Euler scheme. Stable discretisations with respect to space are established by a cell-wise constant pressure approximation and discontinuous and continuous Q_1-elements for stress and velocity respectively. Temperature and orientation densities are approximated by continuous bilinear functions as well, where the corresponding transport dominated equations require further stabilisation, which is provided by the streamline diffusion method. The proposed strategies are applied to a standard model for describing ice sheet dynamics and an enhanced one, taking into account the developement of certain fabrics in the structure of the ice.

References

1. R. Calov, A.A. Savvin, R. Greve, I. Hansen, and K. Hutter. Simulation of the antarctic ice sheet with a three-dimensional polythermal ice sheet model. *Annals of Glaciology*, 27:201–206, 1998.
2. DEAL. differential equations analysis library. *available via http://www-lsx.mathematik.uni-dortmund.de/user/lsx/suttmeier/deal.html*, 1995.

3. A. Fabre, A. Letrguilly, and C. Ritz. Sensitivity of a greenland ice sheet model to ice flow and ablation parameters: Consequences on the evolution through the last climatic cycle. *Climate Dynamics*, 13:11–24, 1997.
4. A.C. Fowler. Modelling ice sheet dynamics. *Geophys. Astrophys. Fluid Dynamics*, 63:29–65, 1992.
5. O. Gagliardini and J. Meyssonnier. Plane flow of an ice sheet exhibiting strain-induced anisotropy. In Y. Wang K. Hutter and H. Beer, editors, *Advances in Cold-Region Thermal Engineering and Sciences*, pages 171–182. Springer, 1999.
6. O. Gagliardini and J. Meyssonnier. About the condition to apply on the lateral boundary of a model for local flow of anisotropic ice. *to appear*, 2001.
7. C. Geiger and C. Kanzow. *Numerische Verfahren zur Lösung unrestringierter Optimierungsaufgaben.* Springer, 1999.
8. R. Glowinski. *Numerical methods for nonlinear variational problems.* Springer Series in Comp. Physics. Springer, 1983.
9. G. Gödert. Meso-macro model for the description of induced anisotropy of natural ice, including grain interaction. In K. Hutter, Y. Wang, and H. Beer, editors, *Advances in Cold-Region Thermal Engineering and Sciences*, pages 183–196, 1999.
10. G. Gödert. The use of structure tensors to model the evolution of textural anisotropy of polar ice. *Ann. Glaciol.*, 2002. submitted.
11. R. Greve, M. Weis, and K. Hutter. Palaeoclimatic and present conditions of the greenland ice sheet in the vicinity of summit: An approach by large-scale modelling. *Paleoclimates*, 2:133–161, 1998.
12. K. Hutter, S. Yakowitz, and F. Szidarovsky. A numerical study of plane ice sheet flow. *J. Glaciol.*, 32:139–160, 1986.
13. P. Huybrechts. The present evolution of the greenland ice sheet: an assessment by modelling. *Global Planet. Change*, 9:39–51, 1995.
14. C. Johnson. *Numerical solution of partial differential equations by the finite element method.* Studentlitteratur, 1987.
15. L.W. Morland and I.R. Johnson. Steady motion of ice sheets. *J. Glaciol.*, 28:229–246, 1980.
16. G.K.A. Oswald and G.de Q. Robin. Lakes beneath the antartic ice sheet. *Nature*, 245:251–254, 1973.
17. W.S.B. Paterson. *The physics of glaciers.* Pergamon, Oxford, 1981.
18. S. Richardson. On the no-slip boundary condition. *J. Fluid Mech.*, 59:707–719, 1973.
19. T. Thorsteinsson. Textures and fabrics in the grip ice core, in relation to climate history and ice deformation. Technical Report 206, Reports on Polar Research, 1996. ISSN 0176-5027.
20. S.S. Vialov. Regularities of glacial ice shields movements and the the theory of plastic viscous flow. *Physics of the movements of ice, IAHS*, 47:266–275, 1958.

Nonreflecting Boundary Conditions for Multiple Domain Wave Scattering in Unbounded Media

Marcus J. Grote[1], Christoph Kirsch[1] and Patrick Meury[2]

[1] Department of Mathematics, University of Basel, Rheinsprung 21, CH-4051 Basel, Switzerland.
[2] Seminar for Applied Mathematics, ETH Zurich, Switzerland

Summary. A nonreflecting boundary condition is presented, which generalizes the well-known Dirichlet-to-Neumann (DtN) approach for time-harmonic scattering in unbounded domains to multiple scattering problems. Because this boundary condition allows each scatterer to be enclosed by a separate artificial boundary, the size of the computational domain, and hence the computational cost, are greatly reduced.

1 Introduction

For the numerical solution of wave scattering problems in unbounded media, a well-known approach is to enclose all obstacles, inhomogeneities and nonlinearities with an artificial boundary B. A boundary condition is then imposed on B, which leads to a numerically solvable boundary-value problem in a finite computational domain Ω. The boundary condition should be chosen such that the solution of the problem in Ω coincides with the restriction to Ω of the solution in the original unbounded region. Otherwise spurious reflections will appear at B, which will travel back into the interior computational region and spoil the numerical solution throughout Ω.

Dirichlet-to-Neumann (DtN) maps yield exact nonreflecting boundary conditions and thus avoid spurious reflections from B. They are explicitly known for various equations or geometries [1, 2, 3, 4, 5]. Once combined with a finite difference or finite element discretization inside Ω, they lead to a highly accurate and efficient numerical scheme.

Here we extend the DtN approach to multiple scattering problems, where every scatterer is enclosed by a separate artificial boundary B_j. See [6] for an introduction to multiple scattering. Hence Ω consists of multiple disjoint components, Ω_j. We derive an exact DtN boundary condition on B, the disjoint union of all B_j, by combining multiple contributions from purely outgoing wave fields. We present theoretical results that show existence and uniqueness of the solution to the (artificial) boundary value problem, as well as numerical results that demonstrate the accuracy and efficiency of our method.

2 Acoustic waves on two scatterers

We consider two disjoint scatterers in unbounded two-dimensional space. The scatterers may contain obstacles, inhomogeneities and nonlinearity. Let Γ denote the boundary of all obstacles and Ω_∞ the free space outside Γ. The scattered field $u = u(r, \theta)$ is a solution of the exterior boundary value problem problem

$$\Delta u + k^2 u = f \quad \text{in } \Omega_\infty \subset \mathbb{R}^2, \tag{1}$$

$$u = g \quad \text{on } \Gamma, \tag{2}$$

$$\lim_{r \to \infty} \sqrt{r} \left(\frac{\partial}{\partial r} - ik \right) u = 0. \tag{3}$$

The wave number k and the source term f can vary in space, while f may be nonlinear. We have chosen the Dirichlet-type condition (2) for simplicity. The Sommerfeld radiation condition (3) ensures that the scattered field corresponds to a purely outgoing wave at infinity.

We assume that both scatterers have compact support and that they are well separated. In this case, they can be surrounded by two non-overlapping disks centered at c_1, c_2 with radii R_1, R_2, respectively, such that in the unbounded domain D outside these disks, the scattered field satisfies the homogeneous Helmholtz equation with a constant wave number $k > 0$, together with the Sommerfeld radiation condition:

$$\Delta u + k^2 u = 0 \quad \text{in } D, \quad k > 0 \text{ constant} \tag{4}$$

$$\lim_{r \to \infty} \sqrt{r} \left(\frac{\partial}{\partial r} - ik \right) u = 0 \tag{5}$$

Let $\Omega = \Omega_\infty \setminus D$ denote the finite domain inside the disks. Ω consists of two disjoint components, Ω_1 and Ω_2. Ω is bounded by $\partial\Omega = \Gamma \cup B$, where $\Gamma = \Gamma_1 \cup \Gamma_2$ and $B = \partial D$ consists of two circles B_1 and B_2. To solve the scattering problem (1)–(3) in the finite domain Ω, a boundary condition is needed at the artificial boundary B, which ensures that the solution in Ω, with that boundary condition imposed on B, coincides with the restriction to Ω of the solution in the original unbounded region Ω_∞.

2.1 Derivation of the DtN map

Let D_1 denote the unbounded domain outside B_1 and D_2 the unbounded domain outside B_2. We split the scattered field u in the unbounded exterior region D into two purely outgoing wave fields u_1, u_2 which solve the following problems:

$$\Delta u_1 + k^2 u_1 = 0 \quad \text{in } D_1, \tag{6}$$

$$\lim_{r\to\infty} \sqrt{r}\left(\frac{\partial}{\partial r} - ik\right) u_1 = 0, \tag{7}$$

$$\Delta u_2 + k^2 u_2 = 0 \quad \text{in } D_2, \tag{8}$$

$$\lim_{r\to\infty} \sqrt{r}\left(\frac{\partial}{\partial r} - ik\right) u_2 = 0. \tag{9}$$

Either wave field is influenced by a single scatterer and therefore completely oblivious to the other. The following proposition shows that solutions to (4)–(9) are uniquely determined by the values of u on B, u_1 on B_1, and u_2 on B_2, respectively.

Proposition 1. *Let $K \subset \mathbb{R}^2$ be a compact set with smooth boundary. Then the exterior Dirichlet problem*

$$\Delta u + k^2 u = 0 \quad \text{in } \mathbb{R}^2 \setminus K, \quad k > 0 \text{ constant} \tag{10}$$

$$u = 0 \quad \text{on } \partial K \tag{11}$$

$$\lim_{r\to\infty} \sqrt{r}\left(\frac{\partial}{\partial r} - ik\right) u = 0 \tag{12}$$

has only the trivial solution.

Proof. Without loss of generality, we can assume $0 \in K$. Because K is compact, there exists an $R_0 > 0$, such that every circle B_R centered at 0 with radius R satisfies $B_R \subset \mathbb{R}^2 \setminus K$, for all $R \geq R_0$. Let u now be a solution to (10)–(12). A direct computation shows

$$\int_{B_R} \left|\sqrt{r}\left(\frac{\partial}{\partial r} - ik\right) u\right|^2 ds = R \int_{B_R} \left|\frac{\partial u}{\partial r}\right|^2 + k^2 |u|^2 \, ds - ikR \int_{B_R} \left(u\frac{\partial \overline{u}}{\partial r} - \frac{\partial u}{\partial r}\overline{u}\right) ds. \tag{13}$$

From (12) we observe that the left side of (13) tends to zero as $R \to \infty$. Next, we use Green's formula and (10) to conclude that

$$\int_{B_R} \left(u\frac{\partial \overline{u}}{\partial r} - \frac{\partial u}{\partial r}\overline{u}\right) ds = \int_{\partial K} \left(u\frac{\partial \overline{u}}{\partial n} - \frac{\partial u}{\partial n}\overline{u}\right) ds. \tag{14}$$

Here $\partial/\partial n$ is the derivative in the direction of the normal vector on ∂K pointing away from K. The right hand side of (14) vanishes because of (11). Hence (13) implies that

$$\lim_{R\to\infty} \int_{B_R} |u|^2 \, ds = 0. \tag{15}$$

By Rellich's theorem (see for example [7], Lemma 2.11), we conclude

$$u \equiv 0 \quad \text{in } \mathbb{R}^2 \setminus K, \tag{16}$$

which completes the proof of Proposition 1. □

The solutions u_1 and u_2 to (6)–(9) can be explicitly written, in local polar coordinates (r_1, θ_1), (r_2, θ_2), as a Fourier series

$$u_j(r_j, \theta_j) = \frac{1}{\pi} {\sum_{n=0}^{\infty}}' \frac{H_n^{(1)}(kr_j)}{H_n^{(1)}(kR_j)} \int_0^{2\pi} u_j(R_j, \theta') \cos n(\theta_j - \theta')\, d\theta', \quad r_j \geq R_j, \tag{17}$$

for $j = 1, 2$. The prime after the sum indicates that the term for $n = 0$ is multiplied by $1/2$, while $H_n^{(1)}$ denotes the n-th order Hankel function of the first kind. Now let $u \in C^2(\Omega_\infty)$ be a given function which satisfies (4), (5). Let u_1 and u_2 be the solutions to (6), (7) and (8), (9), respectively, together with the following matching condition on B:

$$u_1 + u_2 = u \quad \text{on } B = B_1 \cup B_2. \tag{18}$$

Both u and $u_1 + u_2$ satisfy (4), (5) in $D = D_1 \cap D_2$. Since u and $u_1 + u_2$ coincide on B, they coincide everywhere in the exterior region D. We summarize this result in the following proposition.

Proposition 2. *Let $u \in C^2(\Omega_\infty)$ be a function which satisfies (4), (5). Then*

$$u \equiv u_1 + u_2, \quad \text{in } B \cup D, \tag{19}$$

where u_1 and u_2 are solutions to the problems (6), (7) and (8), (9), respectively, together with the matching condition (18). The decomposition of u into the two purely outgoing wave fields u_1 and u_2 is unique.

Proof. Uniqueness: The uniqueness of the decomposition follows from Proposition 1 and from the linear independence of Hankel and Bessel functions – see [8] for details.
Existence: We define the functions

$$v_1 := u|_{B_1} \in C^0(B_1), \quad v_2 := u|_{B_2} \in C^0(B_2) \tag{20}$$

Then we introduce the propagation operators

$$P_1 : C^0(B_1) \to C^0(B_2), \quad u_1|_{B_1} \mapsto u_1|_{B_2}, \tag{21}$$

$$P_2 : C^0(B_2) \to C^0(B_1), \quad u_2|_{B_2} \mapsto u_2|_{B_1}. \tag{22}$$

Explicit formulas for P_j, $j = 1, 2$, in local polar coordinates are given by (17), with some coordinate transformations between (r_1, θ_1) and (r_2, θ_2). The matching condition (18) can be written as an operator equation

$$\begin{pmatrix} u_1 \\ u_2 \end{pmatrix} + K\left[\begin{pmatrix} u_1 \\ u_2 \end{pmatrix}\right] = \begin{pmatrix} v_1 \\ v_2 \end{pmatrix} \tag{23}$$

on the Banach space $X = C^0(B_1) \times C^0(B_2)$, where the operator $K : X \to X$ is given by

$$K\left[\begin{pmatrix} w_1 \\ w_2 \end{pmatrix}\right] = \begin{pmatrix} P_2[w_2] \\ P_1[w_1] \end{pmatrix}, \quad \begin{pmatrix} w_1 \\ w_2 \end{pmatrix} \in X. \tag{24}$$

Note that the operator equation (23) with vanishing right-hand side admits only the trivial solution, by the uniqueness proof above. Therefore, if K is a compact operator, existence and uniqueness of a solution to (23) for an arbitrary right hand side follows from Fredholm's alternative. When the solution to (23) is found, it can be extended by (17) to a solution u_1 of (6), (7) and a solution u_2 of (8), (9), respectively. In their common domain $D = D_1 \cap D_2$, the superposition $u_1 + u_2$ then satisfies (4), (5), and (18) on B. By Proposition 1, $u \equiv u_1 + u_2$ in D.

It remains to show that the operator K is compact. The propagation operator P_1 is defined by

$$P_1[u](\theta_2) = \frac{1}{\pi} {\sum_{n=0}^{\infty}}' \frac{H_n^{(1)}(kr_1)}{H_n^{(1)}(kR_1)} \int_0^{2\pi} u(\theta') \cos n(\theta_1 - \theta')\, d\theta'. \tag{25}$$

In (25), $r_1 = r_1(\theta_2)$ and $\theta_1 = \theta_1(\theta_2)$ denote the polar coordinates of the points on B_2 relative to the center of B_1. The truncated version of P_1, denoted by P_1^N, $N \in \mathbb{N}$, is defined as in (25), with the infinite sum truncated at N.

Lemma 1. *The propagation operators have the following properties:*

1. *$P_1^N : C^0(B_1) \to C^0(B_2)$ is a bounded linear operator of finite rank*
2. *$P_1 : C^0(B_1) \to C^0(B_2)$ is a bounded linear operator*
3. *$\|P_1^N - P_1\| \to 0$, $N \to \infty$*

From Lemma 1, compactness of P_1 follows (see for example [9], Corollary II.3.3).

Proof. (Lemma 1)

1. This property is obvious from the definition of P_1^N.

2. The linearity of P_1 is obvious from the definition. We shall now show the boundedness of P_1.
   Because the scatterers are well separated, there exists an $r_{\min} > R_1$, such that $r_1(\theta_2) \geq r_{\min}$ for all $\theta_2 \in [0, 2\pi]$. Because the function $x|H_n^{(1)}(x)|^2$ is monotone decreasing for $x > 0$ and $n > 1/2$ [10], we therefore have

   $$kr_1|H_n^{(1)}(kr_1)|^2 \leq kr_{\min}|H_n^{(1)}(kr_{\min})|^2 \leq kR_1|H_n^{(1)}(kR_1)|^2, \quad n \geq 1, \tag{26}$$

   from which we derive

   $$\left|\frac{H_n^{(1)}(kr_1)}{H_n^{(1)}(kR_1)}\right| \leq \left|\frac{H_n^{(1)}(kr_{\min})}{H_n^{(1)}(kR_1)}\right| =: \alpha_n, \quad n \geq 1. \tag{27}$$

   We will now show the convergence of the series $\sum_{n=1}^{\infty} \alpha_n$. From the asymptotic behavior of the modulus of the Hankel functions for large orders [11],

$$|H_n^{(1)}(z)|^2 \sim \frac{1}{2\pi n}\left(\left(\frac{ez}{2n}\right)^{2n} + 4\left(\frac{ez}{2n}\right)^{-2n}\right), \quad n \to \infty,\ z \in \mathbb{R}, \tag{28}$$

we derive the following asymptotic behavior for the ratio $|\alpha_{n+1}/\alpha_n|$:

$$\left|\frac{\alpha_{n+1}}{\alpha_n}\right| \sim \frac{R_1}{r_{\min}}\left(\frac{\left(\gamma_{n+1}^{4(n+1)}+4\right)\left(\Gamma_n^{4n}+4\right)}{\left(\Gamma_{n+1}^{4(n+1)}+4\right)\left(\gamma_n^{4n}+4\right)}\right)^{1/2}, \quad n \to \infty, \tag{29}$$

with

$$\gamma_n := \frac{ekr_{\min}}{2n}, \quad \Gamma_n := \frac{ekR}{2n}. \tag{30}$$

Because $\gamma_n \to 0$, $n \to \infty$ and $\Gamma_n \to 0$, $n \to \infty$, we find that the expression on the right hand side of (29) converges, and therefore

$$\limsup_{n\to\infty}\left|\frac{\alpha_{n+1}}{\alpha_n}\right| = \lim_{n\to\infty}\left|\frac{\alpha_{n+1}}{\alpha_n}\right| = \frac{R_1}{r_{\min}} < 1. \tag{31}$$

The ratio test now ensures convergence of the series $\sum_{n=1}^{\infty}\alpha_n$. We define

$$\alpha_0 := \max_{0\le\theta_2\le 2\pi}\left|\frac{H_0^{(1)}(kr_1)}{H_0^{(1)}(kR_1)}\right| < \infty \tag{32}$$

and estimate

$$\|P_1 u\|_\infty = \max_{0\le\theta_2\le 2\pi}\left|\frac{1}{\pi}{\sum_{n=0}^{\infty}}' \frac{H_n^{(1)}(kr_1)}{H_n^{(1)}(kR_1)}\int_0^{2\pi} u(\theta')\cos n(\theta_1-\theta')\,d\theta'\right| \tag{33}$$

$$\le \max_{0\le\theta_2\le 2\pi}\left(2{\sum_{n=0}^{\infty}}'\left|\frac{H_n^{(1)}(kr_1)}{H_n^{(1)}(kR_1)}\right|\right)\|u\|_\infty \tag{34}$$

$$\le 2\|u\|_\infty \sum_{n=0}^{\infty}\alpha_n. \tag{35}$$

We conclude $\|P_1\| \le 2\sum_{n=0}^{\infty}\alpha_n < \infty$.

3. We use the definitions of P_1 and P_1^N to estimate

$$\|P_1^N - P_1\| \le 2\sum_{n=N+1}^{\infty}\alpha_n. \tag{36}$$

The right hand side tends to zero as $N \to \infty$. We conclude $\|P_1^N - P_1\| \to 0$, $N \to \infty$.

This completes the proof of Lemma 1, which implies the compactness of the operator P_1. Compactness of P_2 is shown similarly and compactness of K then follows. This completes the proof of Proposition 2. □

As a consequence, for any given function $u \in C^2(\Omega_\infty)$ satisfying (4), (5), we can determine an explicit relation between the values of $u|_B$ on the artificial boundary (Dirichlet data) and the values of the normal derivative $(\partial_n u)|_B$ (Neumann data). This DtN map for u is given by

$$\partial_n u = M[u_1] + T[u_2] \quad \text{on } B_1, \tag{37}$$
$$\partial_n u = M[u_2] + T[u_1] \quad \text{on } B_2, \tag{38}$$
$$u_1 + P[u_2] = u \quad \text{on } B_1, \tag{39}$$
$$P[u_1] + u_2 = u \quad \text{on } B_2. \tag{40}$$

The operators M, T and P are defined by

$$M[u_j](\theta_j) := \frac{1}{\pi} {\sum_{n=0}^{\infty}}' \frac{kH_n^{(1)'}(kR_j)}{H_n^{(1)}(kR_j)} \int_0^{2\pi} u_j(R_j, \theta') \cos n(\theta_j - \theta')\, d\theta', \tag{41}$$

$j = 1, 2$ (standard single scatterer DtN operator),

$$T[u_1](\theta_2) := \frac{\partial u_1}{\partial r_2}(R_2, \theta_2), \quad T[u_2](\theta_1) := \frac{\partial u_2}{\partial r_1}(R_1, \theta_1) \quad \text{(transfer operator)} \tag{42}$$

$$P[u_1](\theta_2) := u_1(R_2, \theta_2), \quad P[u_2](\theta_1) := u_2(R_1, \theta_1) \quad \text{(propagation operator)} \tag{43}$$

The expressions on the right hand sides of (37), (38) and on the left hand sides of (39), (40) are evaluated explicitly by using (41)–(43) and the exact Fourier representation (17), which involves some straightforward but technical coordinate transformations.

The matching condition (39), (40) cannot be inverted explicitly, and thereby u_1 and u_2 eliminated in the DtN map (37)–(40). Instead, one also needs to compute the values of u_1 on B_1 and u_2 on B_2.

With the DtN condition (37)–(40), we are now able to solve a boundary value problem in the finite domain Ω:

$$\Delta u + k^2 u = f \quad \text{in } \Omega, \tag{44}$$
$$u = g \quad \text{on } \Gamma, \tag{45}$$
$$\partial_n u = M[u_1] + T[u_2] \quad \text{on } B_1, \tag{46}$$
$$\partial_n u = M[u_2] + T[u_1] \quad \text{on } B_2, \tag{47}$$
$$u_1 + P[u_2] = u \quad \text{on } B_1, \tag{48}$$
$$P[u_1] + u_2 = u \quad \text{on } B_2. \tag{49}$$

In [8] we prove the following theorem, which ensures existence and uniqueness of a solution to the DtN problem (44)–(49).

Theorem 1. *Assume that the free space problem* (1)–(3) *has a unique classical solution* $u \in C^2(\Omega_\infty)$ *which satisfies* (4), (5). *Then the double scattering boundary value problem* (44)–(49) *has a unique solution in* Ω, *which coincides with the restriction of* u *to* Ω.

2.2 Combination with a numerical scheme

The boundary value problem (44)–(49) can be discretized by any numerical scheme suited for the solution of elliptic boundary value problems, for example by a finite difference or finite element scheme. Equations (46), (47) is a Robin-type boundary condition, combined with (48), (49). The discretization of (44)–(49) will lead to a large system of linear equations for the unknowns, which are the values of the solution on the grid points, for example. The matching condition (48), (49) requires the storage of additional values, namely the values of the purely outgoing wave fields u_j on the boundary components B_j, $j = 1, 2$. These auxiliary values are useful during post-processing for the evaluation of u outside the computational domain and in the far-field [8]. Details on the finite difference and finite element implementation of the multi-DtN condition (46)–(49) can be found in [8].

3 Numerical example

We consider scattering of an incident plane wave impinging on three sound-soft obstacles with kidney-shaped boundaries. The wave number is $k = 8\pi$. For the numerical solution with a second-order finite difference scheme we generalize the DtN condition presented above, from two to three scatterers. This generalization is straightforward and explicitly described in [8]. The contour lines of the real part of the total field are shown in Fig. 1. In [8] we compared our multi-DtN condition with the standard DtN condition applied to one single very large domain and showed that the two solutions coincide.

4 Conclusion

We have generalized the Dirichlet-to-Neumann boundary conditions for single scattering to multiple scattering problems. To do so we have used an expansion of the scattered field into multiple purely outgoing wave fields. We have shown that the multi-DtN condition is exact, which implies that no spurious reflections appear at the artificial boundary. When used in a numerical scheme, the multi-DtN condition allows for much smaller computational domains than the single-DtN condition, especially when the scatterers are far away from each other. Moreover, the amount of work does not increase with increasing distance between the obstacles. We have illustrated the use of the multi-DtN condition via a numerical example.

References

1. Keller, J.B., Givoli, D. (1989): Exact non-reflecting boundary conditions. J. Comp. Phys. **82**, 172–192

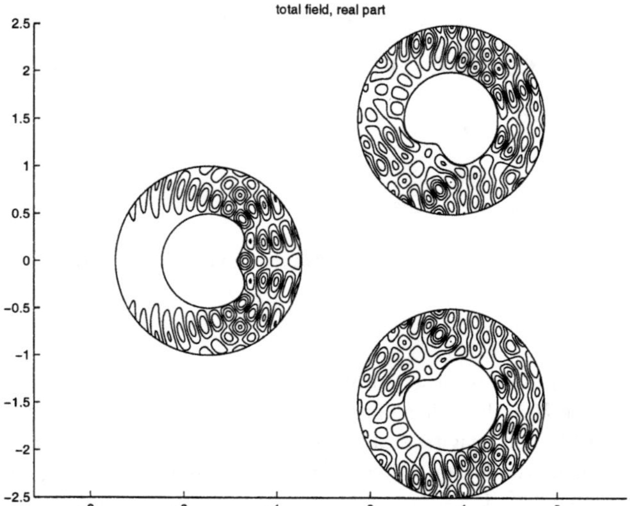

Fig. 1. Contour lines of the real part of the total field for plane wave scattering on three kidney-shaped obstacles with sound-soft boundaries. The plane wave is incident from the right and the wave number is $k = 8\pi$. The multi-DtN condition is used at the artificial boundary components, combined with a second-order finite difference scheme in the interior.

2. Givoli, D. (1992): Numerical Methods for Problems in Infinite Domains. Elsevier
3. Grote, M.J., Keller, J.B. (1995): On Nonreflecting Boundary Conditions. J. Comp. Phys. **122**, 231–243
4. Givoli, D. (1999): Recent Advances in the DtN FE Method. Archives of Comput. Meth. Engin. **6**, 71–116
5. Gächter, G.K., Grote, M.J. (2003): Dirichlet-to-Neumann Map for Three-Dimensional Elastic Waves. Wave Motion **37**, 293–311
6. Martin, P.A. (1995): Multiple Scattering: an Invitation. In: Cohen, G. (ed) Proc. Third Int. Conf. on Math. and Numerical Aspects of Wave Propagation, SIAM Philadelphia, 3–16
7. Colton, D., Kress, R. (1992): Inverse Acoustic and Electromagnetic Scattering Theory. Springer, Berlin, Heidelberg
8. Grote, M.J., Kirsch, C.: Dirichlet-to-Neumann Boundary Conditions for Multiple Scattering Problems, J. Comp. Phys., submitted
9. Werner, D. (1995): Funktionalanalysis. Springer
10. Ryzhik, I.M., Gradshteyn, I.S. (1994): Table of Integrals, Series, and Products. Academic Press
11. Abramowitz, M., Stegun, I.A. (1970): Handbook of mathematical functions. Dover Publications

On the Choice of the Regularization Parameter in the Case of the Approximately Given Noise Level of Data*

Uno Hämarik and Toomas Raus

University of Tartu, Institute of Applied Mathematics, ESTONIA
uno.hamarik@ut.ee, toomas.raus@mtk.ut.ee

Summary. We consider ill-posed problems $Au = f$ with operator $A \in \mathcal{L}(H,H)$, $A = A^* \geq 0$, where H is the Hilbert space and range $\mathcal{R}(A)$ is non-closed. Regularized solutions u_r are obtained by a general regularization scheme, including the Lavrentiev method, iteration methods and others. We assume that instead of $f \in \mathcal{R}(A)$ noisy data $\tilde{f}$ are available with the approximately given noise level δ: it holds $\|\tilde{f} - f\|/\delta \leq$ const for $\delta \to 0$. We propose a new a-posteriori rule for the choice of the regularization parameter $r = r(\delta)$ guaranteeing $u_{r(\delta)} \to u_*$ for $\delta \to 0$, where u_* is solution of problem $Au = f$. The error estimates are given.

1 Introduction

We consider an operator equation

$$Au = f\,, \quad f \in R(A)\,, \tag{1}$$

where $A \in L(H,H)$, $A = A^* \geq 0$ is the linear continuous self-adjoint and non-negative operator; u and f are elements of the real Hilbert space H. We do not suppose that the range $\mathcal{R}(A)$ is closed and so in general our problem is ill-posed. The kernel $N(A)$ may be non-trivial. As usual in the treatment of ill-posed problems, we suppose that instead of the exact right-hand side f we have only an approximation $\tilde{f} \in H$.

The approximate solution u_r of the ill-posed problem $Au = f$ is found by some regularization method and depends on the regularization parameter r. If the noise level δ with $\|f - \tilde{f}\| \leq \delta$ is known, then the proper parameter choice $r = r(\delta)$ guarantees $u_{r(\delta)} \to u_*$ for $\delta \to 0$, where u_* is solution of $Au = f$, the nearest to the initial approximation u_0 (see Section 2; often $u_0 = 0$).

In this paper we are interested in the case of approximately given noise level δ: it is unknown, holds the inequality $\|\tilde{f} - f\| \leq \delta$ or not. Instead of this inequality we assume that $\|\tilde{f} - f\|/\delta \leq$ const for $\delta \to 0$ and we give a new rule for the parameter choice $r = r(\delta)$ guaranteeing $u_{r(\delta)} \to u_*$ for $\delta \to 0$. The error estimates are presented as well.

* This work was supported by the Estonian Science Foundation, Grant No. 5785.

2 Regularization methods

We consider the regularization methods in the general form (see [1, 2])

$$u_r = (I - Ag_r(A))u_0 + g_r(A)\tilde{f}\,, \tag{2}$$

where u_r is the approximate solution, u_0 – initial approximation, r – regularization parameter, I – the identity operator and the function $g_r(\lambda)$ satisfies the conditions (3) and (4):

$$\sup_{0\le\lambda\le a} |g_r(\lambda)| \le \gamma r\,, \quad r \ge 0\,, \tag{3}$$

$$\sup_{0\le\lambda\le a} \lambda^p |1 - \lambda g_r(\lambda)| \le \gamma_p r^{-p}\,, \quad r \ge 0\,, 0 \le p \le p_0\,. \tag{4}$$

Here p_0, γ and γ_p are positive constants, $a \ge \|A\|$, $\gamma_0 \le 1$ and the greatest value of p_0, for which the inequality (4) holds is called the qualification of method.

The following regularization methods are special cases of the general method (2).

M1 The Lavrentiev method $u_\alpha = (\alpha I + A)^{-1}\tilde{f}$. Here $u_0 = 0$, $r = \alpha^{-1}$, $g_r(\lambda) = (\lambda + r^{-1})^{-1}$, $p_0 = 1$.

M2 The iterative variant of the Lavrentiev method. Let $m \in \mathbf{N}$, $m \ge 1$, $u_0 = u_{0,\alpha} \in H$ – initial approximation and $u_{m,\alpha} = (\alpha I + A)^{-1}(\alpha u_{m-1,\alpha} + \tilde{f})$. Here $r = \alpha^{-1}$, $g_r(\lambda) = \frac{1}{\lambda}\left(1 - \left(\frac{1}{1+r\lambda}\right)^m\right)$, $p_0 = m$.

M3 Explicit iteration scheme (the Landweber's method). Let $0 < \mu < 1/\|A\|$ be a constant and

$$u_n = u_{n-1} - \mu(Au_{n-1} - \tilde{f})\,, \quad n = 1, 2, \ldots\,.$$

Here $r = n$, $g_r(\lambda) = \frac{1}{\lambda}(1 - (1 - \mu\lambda)^r)$, $p_0 = \infty$.

M4 Implicit iteration scheme. Let $\alpha > 0$ be a constant and

$$\alpha u_n + Au_n = \alpha u_{n-1} + \tilde{f}\,, \quad n = 1, 2, \ldots\,.$$

Here $r = n$, $g_r(\lambda) = \frac{1}{\lambda}\left(1 - \left(\frac{\alpha}{\alpha+\lambda}\right)^r\right)$, $p_0 = \infty$.

M5 The method of the Cauchy problem. We take the solution of the Cauchy problem

$$u'(r) + Au(r) = \tilde{f}\,, \quad u(0) = u_0$$

for the approximation u_r to the solution of problem (1). Here $g_r(\lambda) = \frac{1}{\lambda}(1 - e^{-r\lambda})$, $p_0 = \infty$.

3 Parameter choice in case of the known noise level of data

In regularization methods (2) the important problem is the choice of a proper regularization parameter r. If r is too big, the numerical implementation will be unstable and u_r will be useless; if r is too small, the approximation u_r is dominated by the initial guess u_0. The rules of the parameter choice can be devided into two groups, where the noise level is used in the rules of the first group and not used in the rules of the second group.

If the noise level δ with $\|\tilde{f} - f\| \leq \delta$ is known the most prominent rule for methods M2–M5 is the Morozov's discrepancy principle.

The Morozov's discrepancy principle [3]. In this rule the regularization parameter $r = r_D$ is chosen as the solution of the equation

$$\left\|Au_r - \tilde{f}\right\| \approx b\delta \quad \text{with} \quad b = \text{const} > 1\,.$$

The second rule of the first group is the **modification of the discrepancy principle (MD rule)** [4]. In this rule the regularization parameter $r = r_{MD}$ is chosen as the solution of the equation

$$\left\|B_r(Au_r - \tilde{f})\right\| \approx b\delta \quad \text{with} \quad b = \text{const} > 1\,,$$

where the operator

$$B_r = \begin{cases} I & \text{for } p_0 = \infty\,, \\ (I - Ag_r(A))^{1/p_0} & \text{for } p_0 = \infty \end{cases}$$

depends on the qualification p_0 of the method.

The discrepancy principle and its modification coincide for regularization methods M3–M5 where $p_0 = \infty$, but these rules differ for the Lavrentiev method and its iterative variant, where $\|B_r(Au_{\alpha,m} - \tilde{f})\| = \|Au_{\alpha,m+1} - \tilde{f}\|$.

Some useful properties of the MD rule which are important for our further study are:

1) convergence: $\|u_{r_{MD}} - u_*\| \to 0$ for $\delta \to 0$; here u_* is nearest to u_0 solution of problem $Au = f$;

2) order-optimality: if $u_0 - u_* = A^p v$, $v \in H$, $\|v\| \leq \varrho$, $p > 0$, then $\|u_{r_{MD}} - u_*\| \leq C_p \varrho^{\frac{1}{\varrho+1}} \delta^{\frac{p}{p+1}}$, $0 < p \leq p_0$;

3) quasioptimality: there exists a constant c such that

$$\|u_{r_{MD}} - u_*\| \leq c \inf_{r \geq 0} \{\|(I - Ag_r(A))(u_0 - u_*)\| + \gamma r\delta\}\,. \tag{5}$$

The MD rule has some advantages over the discrepancy principle since

(i) the MD rule can also be used for the Lavrentiev method;

(ii) the method M2 with $r = r_D$ is order-optimal only for the range $p \in (0, p_0 - 1]$, but not order optimal for $p > p_0 - 1$;

(iii) it can be shown that the method M2 with $r = r_D$ is not quasi-optimal.

However, the discrepancy principle and its modification have an essential disadvantage. Namely, these choices are unstable in this sense that if the actual error of the right-hand side is larger than $b\delta$, then the error of the approximate solution may be arbitrarily large, independently of the value of the ratio of the actual and supposed noise level. For example, if $b = 2$ and the actual noise level is three times larger than the noise level δ, which we use in the rule, then the error of the approximate solution may be arbitrarily large.

There are also parameter choice rules, which do not use the noise level delta. Sometimes these choices are called as heuristic or delta-free choices. The first such parameter choice rule was the quasioptimality criterion [5, 6]. According to this rule the parameter was chosen for which the function $k(r) = r\|B_r(Au_r - \tilde{f})\|$ has the minimum. Other popular delta-free rules are the Wahba's cross-validation rule [7, 8] and the Hansen's L-curve criterion [9, 10]. Some heuristic rules are also proposed in [11].

Although these rules often work well (or even better than the discrepancy principle and the MD rule), it was shown by Bakushinskii [12], that one cannot prove the convergence of the approximate solution for heuristic parameter choice rules.

4 Parameter choice rule in the case of the approximately given noise level of data

In applied ill-posed problems the exact noise level is often unknown. Therefore in the following we assume that the actual noise level is unknown and only some guesses about this level can be made. It means that the supposed error level $\delta > 0$ is given, but we do not know exactly, if $\|\tilde{f} - f\| \le \delta$ holds or not. Our aim is to present a rule for the stable parameter choice which guarantees the convergence of the approximate solution to the exact solution if only the ratio $\|\tilde{f} - f\|/\delta$ is bounded in the process $\delta \to 0$, and to give some error estimates of the approximate solution.

In the following the function

$$\varphi(r) = \sqrt{r}\|A^{1/2}B_r^{3/2}(Au_r - \tilde{f})\| = \sqrt{r}\langle B_r(Au_r - \tilde{f}), AB_r^2(Au_r - \tilde{f})\rangle^{1/2}$$

plays an important role.

Note that for the Lavrentiev method and its iterative variant $B_r = (I + rA)^{-1}$ and $\varphi(r) = \varphi(\alpha^{-1}) = \frac{1}{\sqrt{\alpha}}\langle Au_{m+1,\alpha} - \tilde{f}, A(Au_{m+2,\alpha} - \tilde{f})\rangle^{1/2}$; for iterative methods $\varphi(r) = \varphi(n) = \sqrt{n}\langle Au_n - \tilde{f}, Au_n - \tilde{f})\rangle^{1/2}$.

Rule P. Let $0 \le s \le 1$ and b_1, b_2 be the constants such that $b_2 \ge b_1 > C_m$, where $C_m = 1/2$, $C_m = 1/\sqrt{2m+3}$, $C_m = 1/\sqrt{2\mu e}$, $C_m = \sqrt{\alpha/2}$ and $C_m =$

$1/\sqrt{2e}$ for methods M1–M5 respectively. If $\varphi(1) \le b_2\delta$ then choose $r(\delta) = 1$. In the contrary case we find at first $r_2(\delta) > 1$ such that

$$\varphi(r_2(\delta)) \le b_2\delta\,, \tag{6}$$

$$\varphi(r) \ge b_1\delta \qquad \forall r \in [1, r_2(\delta)]\,. \tag{7}$$

For the regularization parameter $r(\delta)$ we choose the parameter r, for which the function $t(r) = r^s\|B_r(Au_r - \tilde{f})\|$ has the global minimum on the interval $[1, r_2(\delta)]$.

In iterative methods the following rule P' can be used for the choice of the stopping index $n(\delta)$ as the parameter r. For the rule P' the analogous results hold as for the rule P.

Rule P'. Let $0 \le s \le 1$ and b be the constant such that $b > C_m$. Find $n_2(\delta)$ as the first $n = 1, 2, \ldots$, for which $\varphi(n) \le b\delta$. For the regularization parameter $n(\delta)$ we choose $n \in \mathbf{N}$, for which the function $t(n) = n^s\|Au_n - \tilde{f}\|$ has the global minimum on the interval $[1, n_2(\delta)]$.

In [13] the rule for the choice of the parameter was considered, in which for the regularization parameter the parameter $r_2(\delta)$ was taken. We can consider the rule P as the generalization of this rule, since in case $s = 0$ these rules coincide due to the fact that the function $\|B_r(Au_r - \tilde{f})\|$ is monotonically decreasing with respect to r. On the other hand, in case $s = 1$ the rule P is similar to the non-selfadjoint analogue of the parameter choice rule by the quasioptimality criterion, since we choose the minimum point of the function $r\|B_r(Au_r - \tilde{f})\|$ for the regularization parameter. The only difference between the rule P and the analogue of the quasioptimality criterion is the interval, on which the function $r\|B_r(Au_r - \tilde{f}\|$ is minimized: the intervals are $[1, r_2(\delta)]$ and $[1, \infty)$ respectively.

In [13] the following results are proven for methods M1–M5:

(i) for each $\tilde{f} \in H$ we have $\lim_{r\to\infty} \varphi(r) = 0$;

(ii) if $\|\tilde{f} - f\| \le \delta$, $\|u_0 - u_*\| \le M$, $b \ge C_m$, then for each r, $r \ge R_{M,\delta} = \overline{C}_m M/(b - C_m)\delta$ we have $\varphi(r) \le b\delta$; here $\overline{C}_m = 12\sqrt{15}/125$, $\overline{C}_m = (3/2)^{(3/2)} m^m/(m+3/2)^{m+3/2}$, $\overline{C}_m = (3/(2\mu e))^{3/2}$, $\overline{C}_m = (3\alpha/2)^{(3/2)}$, $\overline{C}_m = (3/(2e))^{3/2}$ for methods M1–M5 respectively;

(iii) if $\frac{\|\tilde{f}-f\|}{\delta} \le$ const for $\delta \to 0$ then $\|u_{r_2(\delta)} - u_*\| \to 0$ for $\delta \to 0$.

Due to the continuity of the function $\varphi(r)$ from the property (i) follows that the choice of finite parameters $r_2(\delta)$ and $r(\delta) \le r_2(\delta)$ according to Rule P is possible. The property (ii) says that if we know a constant $M > 0$ such that $\|u_0 - u_*\| \le M$, then it is sufficient to search the parameter $r_2(\delta)$ in the finite interval $[1, R_{M,\delta}]$. Note that the function $\varphi(r)$ is non-monotone and therefore in Rule P we must use the conditions (6)–(7) instead of inequalities $b_1\delta \le \varphi(r) \le b_2\delta$.

Note that the analogues of the results of the paper [13] for non-selfadjoint problems are presented in [14].

Theorem 1. *Let $A \in L(H,H)$, $A = A^* \geq 0$, $f \in R(A)$. Let the parameter $r(\delta)$ be chosen according to Rule P. If $\frac{\|\tilde{f}-f\|}{\delta} \leq \mathrm{const}$ in the process $\delta \to 0$, then in methods M1–M5*

$$\left\|u_{r(\delta)} - u_*\right\| \to 0 \quad \textit{for} \quad \delta \to 0 .$$

Proof. Denote $G_r := I - Ag_r(A)$. Then we have

$$u_r - u_* = G_r(u_0 - u_*) + g_r(A)(\tilde{f} - f) ,$$

from which with (3) follows that

$$\left\|u_{r(\delta)} - u_*\right\| \leq \left\|G_{r(\delta)}(u_0 - u_*) + \gamma C r(\delta)\delta . \tag{8}$$

To prove the theorem, it suffices to show the convergence of the right-hand side of (8). In [13] it is proven that

$$r_2(\delta)\delta \to 0 \quad \text{for} \quad \delta \to 0 . \tag{9}$$

From inequality $r(\delta) \leq r_2(\delta)$ and from (9) follows the convergence of the second term of (8). To show the convergence of the first term of (8), we prove at first that

$$r_2^s(\delta)\left\|B_{r_2(\delta)}(Au_{r_2(\delta)} - \tilde{f})\right\| \to 0 \quad \text{for} \quad \delta \to 0 . \tag{10}$$

We have

$$B_r(Au_r - \tilde{f}) = AB_rG_r(u_0 - u_*) - B_rG_r(\tilde{f} - f) , \tag{11}$$

from which with regard to the inequality $\|B_rG_r(\tilde{f} - f)\| \leq \|\tilde{f} - f\| \leq C\delta$ follows that

$$r_2^s(\delta)\left\|B_{r_2(\delta)}(Au_{r_2(\delta)} - \tilde{f})\right\| \leq r_2^s(\delta)\left\|AB_{r_2(\delta)}G_{r_2(\delta)}(u_0 - u_*)\right\| + r_2^s(\delta)C\delta . \tag{12}$$

To show the convergence

$$r_2^s(\delta)\left\|AB_{r_2(\delta)}G_{r_2(\delta)}(u_0 - u_*)\right\| \to 0 \quad \text{for} \quad \delta \to 0 , \tag{13}$$

we consider the cases a) $r_2(\delta) \to \infty$, b) $r_2(\delta) \leq \overline{r} = \mathrm{const}$ separately. If $r_2(\delta) \to \infty$ in the process $\delta \to 0$ then using the Banach-Steinhaus theorem we can prove similarly as in [2] (p. 43) that $r^p\|AB_rG_r(u_0 - u_*)\| \to 0$ if $r \to \infty$ $(0 \leq p \leq 1)$. Now we consider the case $r_2(\delta) \leq \overline{r} = \mathrm{const}$. Using (11), (4) we get

$$\begin{aligned} r_2(\delta)^{1/2}\left\|A^{3/2}B^{3/2}_{r_2(\delta)}G_{r_2(\delta)}(u_0 - u_*)\right\| &\leq r_2(\delta)^{1/2}\left\|A^{1/2}B^{3/2}_{r_2(\delta)}(Au_{r_2(\delta)} - \tilde{f})\right\| \\ + r_2(\delta)^{1/2}\left\|A^{1/2}B^{3/2}_{r_2(\delta)}G_{r_2(\delta)}(\tilde{f} - f)\right\| &\leq b_2\delta + \gamma_{1/2}\left\|\tilde{f} - f\right\| \leq (b_2 + C\gamma_{1/2})\delta , \end{aligned}$$

from which follows that

$$\left\|A^{3/2}B^{3/2}_{r_2(\delta)}G_{r_2(\delta)}(u_0-u_*)\right\| \to 0 \quad \text{for} \quad \delta \to 0\,.$$

In [2] (p. 66) the implication

$$AG_{r_n}(u_0-u_*) \to 0 \quad (n\to\infty) \quad \Longrightarrow \quad G_{r_n}(u_0-u_*) \to 0 \quad (n\to\infty) \quad (14)$$

is proven. Similarly we can show that if $A^{3/2}B^{3/2}_{r_n}G_{r_n}(u_0-u_*) \to 0$ $(n\to\infty)$, then $AB_{r_n}G_{r_n}(u_0-u_*) \to 0$ $(n\to\infty)$ which proves the convergence (13) in this case. Now the convergence (10) follows from (12),(13) and (9).

Let us remind that the parameter $r(\delta)$ is the global minimum point of the function $t(r)=r^s\|B_r(Au_r-\tilde{f})\|$ in $[1,r_2(\delta)]$. Therefore from (10) follows the convergence

$$r^s(\delta)\left\|B_{r(\delta)}(Au_{r(\delta)}-\tilde{f})\right\| \to 0 \quad \text{for} \quad \delta\to 0\,.$$

Using (11) we get

$$r^s(\delta)\left\|AB_{r(\delta)}G_{r(\delta)}(u_0-u_*)\right\| \le r^s(\delta)\left\|B_{r(\delta)}(Au_{r(\delta)}-\tilde{f})\right\| + r^s(\delta)C\delta \to 0 \quad \text{if} \quad \delta\to 0\,.$$

From this relation with the implication of type (14) we get the convergence $\|G_{r(\delta)}(u_0-u_*)\| \to 0$ for $\delta\to 0$, which with (8) proves the theorem.

In the next two theorems we give estimates for the error of the approximate solution in cases $\|\tilde{f}-f\|\le\delta$ and $\|\tilde{f}-f\|\ge\delta$ respectively. The proofs of these theorems will be presented in another forthcoming paper.

Theorem 2. *Let $A\in L(H,H)$, $A=A^*\ge 0$, $f\in R(A)$, $\|\tilde{f}-f\|\le\delta$. Let the parameter $r(\delta)$ be chosen according to Rule P with $s\in(0,1)$ and let the function $t(r)=r^s\|B_r(Au_r-\tilde{f})\|$ be monotonically increasing on the interval $[r(\delta),r_2(\delta)]$. Then for methods M1–M5 the error estimation*

$$\left\|u_{r(\delta)}-u_*\right\| \le c(b_1,b_*)\frac{1}{1-s}\inf_{r\ge 0}\left\{\left\|(I-Ag_r(A))(u_0-u_*)\right\| + \gamma r\delta\right\} \quad (15)$$

holds, where $b_*=\max\limits_{r(\delta)\le r\le R(\delta)}\varphi(r)/\delta \ge b_2$ and $R(\delta)$ is the greatest parameter for which $\varphi(r)=b_2\delta$.

Theorem 3. *Let $A\in L(H,H)$, $A=A^*\ge 0$, $f\in R(A)$. Let the parameter $r(\delta)$ be chosen according to Rule P with $s\in(0,1)$. Then in case $\|\tilde{f}-f\|>\delta$ for methods M1–M5 the following error estimations hold:*

a) if $\delta\le\|\tilde{f}-f\|\le\delta_0$, where $\delta_0:=\|B_{r(\delta)}(Au_{r(\delta)}-\tilde{f})\|$, then

$$\left\|u_{r(\delta)}-u_*\right\| \le c_1(b_1,b_*)\frac{1}{1-s}\inf_{r\ge 0}\left\{\left\|(I-Ag_r(A))(u_0-u_*)\right\| + \gamma r\left\|\tilde{f}-f\right\|\right\}; \quad (16)$$

b) if $\|\tilde{f}-f\|>\delta_0$, then

$$\left\|u_{r(\delta)}-u_*\right\| \le \quad (17)$$

$$\le c_2(b_1,b_2)\left(\frac{\|\tilde{f}-f\|}{\delta_0}\right)^{1/s}\inf_{r\ge 0}\left\{\left\|(I-Ag_r(A))(u_0-u_*)\right\| + \gamma r\left\|\tilde{f}-f\right\|\right\}.$$

5 Numerical experiments

Discretization of ill-posed problems often leads to the linear systems of algebraic equations with large condition numbers. In our numerical experiments we solved many linear systems of equations $Au = f$, where A was the 100×100 diagonal matrix with the diagonal elements $\lambda_1, \dots, \lambda_{100}$. For generating eigenvalues λ_k of A, solution $u = (u_1, \dots, u_{100})^T$, noise in $\tilde{f}$ and supposed noise level δ different schemes were used and the concrete scheme was chosen randomly by computer. Schemes for eigenvalues λ_k of A and for solution u were

$$\lambda_k = \mathrm{rand}\,(0,1)a^{k-1}\,, \quad a = 1.3; 2; 5; 10\,;$$

$$\lambda_k = \mathrm{rand}\,(0,1)k^{-a}\,, \quad a = 1; 2; 3; 4; 10\,;$$

$$\lambda_k = P_k(\mathrm{rand}\,(0,1))\lambda_{k-1}\,, \quad \lambda_1 = 1; \quad P_k(x) = x; x^2; e^{-x}; x/\sqrt{k}\,; \qquad k = 2, 3, \dots, 100\,;$$

$$u_k = \mathrm{rand}\,(0,1)k^{-a}\,, \quad a = -1; -0.5; 0; 0.5; 1; 2; 4\,;$$

$$u_k = R_k(\mathrm{rand}\,(0,1))u_{k-1}\,, \quad u_1 = 1; \quad R_k(x) = x; x^2; e^{-x}, x^3\,; \qquad k = 2, 3, \dots, 100\,;$$

where rand $(0,1)$ is a random number in interval $(0,1)$. The exact right hand side f was found by formula $f = Au$ and perturbations satisfied the relation $\|\tilde{f} - f\| = 10^{-j/2}\|f\|$, $j = 1, 2, \dots, 13$ (Euclidean norms). Concrete random perturbations were distributed uniformly by the formula

$$\tilde{f}_k = f_k + 2(\mathrm{rand}\,(0,1) - 0.5)\|\tilde{f} - f\| / \left[\frac{1}{n}\sum_{i=1}^{n}(2(\mathrm{rand}\,(0,1) - 0.5))^2\right]^{1/2}, \qquad k = 1, 2, \dots, n$$

or all the noise was concentrated on one eigenvalue, chosen randomly:

$$\tilde{f}_{k_0} = f_{k_0} + \sqrt{n}(\mathrm{rand}\,(0,1) - 0.5)\|\tilde{f} - f\| / |\mathrm{rand}\,(0,1) - 0.5|\,,$$

$$\tilde{f}_k = f_k \quad \text{for} \quad k \neq k_0\,.$$

For the supposable noise level $\delta = \|\tilde{f} - f\|/d$ was taken, where values of d were 1, 3, 5, 10, 20, 50, 100. By the Lavrentiev method 6000 different variants of the problem $Au = f$ were solved, parameter r was chosen by the rule P. The ratios

$$V = \frac{\|u_{r(\delta)} - u_*\|}{\inf\limits_{r \geq 0}\{\|(I - Ag_r(A))(u_0 - u_*)\|^2 + (\gamma r \max(\|\tilde{f} - f\|, \delta))^2\}^{1/2}}$$

were computed, characterizing the coefficients of the quasioptimality (compare with formula (5)). The ratio V also shows how well the rule P works in comparison with the MD rule, which uses the actual noise level. Namely, we solved the same problems also by MD rule using the actual noise level and then the ratio V was almost 1 (average of V was 0.95, maximum of V was 1.24).

The results of numerical experiments are given in Table 1. The results show that in the case of the exactly estimated noise level ($d \equiv \|\tilde{f} - f\|/\delta = 1$) rule P works nearly as well as the MD rule (averages of ratio V were 0.991 and 0.95 respectively). As expected, in the case of the underestimated noise level ($d > 1$) the error of the approximate solution is larger than for the MD rule which uses the exact noise level. But the ratio of errors of these approximate solutions is relatively small in comparison with the error, made by estimating the noise level. For example, if the noise level was $d = 100$ times smaller than the real value, the error of the approximate solution for rule P was only 7% larger than for the MD rule which uses the exact noise level (the corresponding averages of V were 1.022 and 0.95), and for 96% of problems the ratio V for rule P was smaller than 1.5. Numerical experiments also showed that if the actual noise level is larger than the noise level used in the MD rule, then this rule is not good. For example, if the actual noise level was 3 times larger than the noise level used in the MD rule, the ratio V was in most cases larger than 10.

Table 1. The Lavrentiev method, Rule P, $s = 0.75$, $b_1 = b_2 = 1.5C_m$

$d = \|\tilde{f} - f\|/\delta$	Number of problems	Average V	Maximal V	% problems for which $V < 1.5$	% problems for which $V < 3$
1	851	0.991	3.95	97.70	99.19
3	901	0.982	4.74	97.44	99.36
5	826	0.984	7.13	96.51	98.88
10	873	0.956	5.99	97.89	99.60
20	861	0.953	6.83	97.32	99.46
50	874	0.974	11.30	97.10	99.08
100	813	1.022	18.50	96.03	98.58
All cases	6000	0.980	18.50	97.15	99.17

References

1. Vainikko G. (1982): The discrepancy principle for a class of regularization methods. U.S.S.R. Comput. Math. Math. Phys., **22**, 3, 1–19.
2. Vainikko G., Veretennikov A. (1986): Iteration Procedures in Ill-Posed Problems. Nauka, Moscow, (in Russian).
3. Morozov V. (1966): On the solution of functional equations by the method of regularization. Soviet Math. Dokl., **7**, 414–417.
4. Raus T. (1984): Residue principle for ill-posed problems. Acta et Comment. Univ. Tartuensis, **672**, 16–26 (in Russian).
5. Tikhonov A. N., Glasko V. B. (1965): Use of the regularization method in non-linear problems. USSR Comput. Math. Math. Phys., **5**, 3, 93–107.

6. Tikhonov A., Arsenin V. (1977): Solution of Ill-Posed Problems. Wiley, New York.
7. Wahba G. (1977): Practical approximate solutions to linear operator equations when the data are noisy. SIAM J. Numer.Anal. **14**, 651–657.
8. Golub G. H., von Matt Urs. (1997): Generalized cross-validation for large-scale problems. J. Comput. Graph. Statist., **6**, 1, 1–34.
9. Hansen P.C. (1992): Analysis of discrete ill-posed problems by means of the L-curve. SIAM Rev., **34**, 561–580.
10. Calvetti D., Morigi S., Reichel L., Sgallari F. (2000): Tikhonov regularization and the L-curve for large discrete ill-posed problems. Numerical analysis 2000, Vol. III. Linear algebra. J. Comput. Appl. Math., **123**, 1–2, 423–446.
11. Hanke M., Raus T. (1996): A general heuristic for choosing the regularization parameter in ill-posed problems. SIAM J. Sci. Comput., **17**, 954–972.
12. Bakushinskii A. (1984): Remarks on choosing a regularization parameter using the quasi-optimality and ratio criterion. U.S.S.R. Comput. Math. Math. Physics., **24**, 181–182.
13. Raus T. (1990): An a posteriori choice of the regularizartion parameter in case of approximately given error bound of data. Acta et Comment. Univ. Tartuensis, **913**, 73–87.
14. Raus T. (1992): About regularization parameter choice in case of approximately given error bounds of data. Acta Comment. Univ. Tartuensis, **937**, 77–89.

Adaptive Discontinuous Galerkin Finite Element Methods with Interior Penalty for the Compressible Navier–Stokes Equations

Ralf Hartmann[1] and Paul Houston[2]

[1] Institute of Aerodynamics and Flow Technology, German Aerospace Center, Lilienthalplatz 7, 38108 Braunschweig, Germany. Email: *Ralf.Hartmann@dlr.de*

[2] Department of Mathematics, University of Leicester, Leicester LE1 7RH, UK. Email: *Paul.Houston@mcs.le.ac.uk*. Supported by the EPSRC (Grant GR/R76615)

Summary. In this article we consider goal–oriented *a posteriori* error estimation for the symmetric interior penalty discontinuous Galerkin finite element discretization of the compressible Navier–Stokes equations. Numerical experiments demonstrating the accuracy of the error estimation and the performance of the adaptive mesh refinement strategy will be presented.

1 Introduction

In recent years there has been tremendous interest in the design of discontinuous Galerkin finite element methods (DGFEM) for the discretization of compressible fluid flow problems; see, for example, [2, 3, 6, 7] and the references cited therein. The key advantages of these schemes are that DGFEM provide robust and high-order accurate approximations, particularly in transport-dominated regimes, and that they are considerably flexible in the choice of mesh design. Indeed, DGFEM can easily handle non-matching grids and non-uniform, even anisotropic, polynomial approximation degrees.

In this paper we introduce the symmetric version of the interior penalty DGFEM for the numerical approximation of the compressible Navier–Stokes equations. We then consider the a posteriori error analysis and adaptive mesh design for the underlying discretization method. In particular, here we focus on so–called 'goal–oriented' a posteriori error estimation which bounds the error measured in terms of certain target functionals of real or physical interest. Typical examples that we shall consider here include the drag and lift coefficients of a body immersed in a viscous fluid. By employing a duality argument we derive a weighted (Type I) a posteriori error bound which reflects the error creation and error propagation mechanisms inherent in viscous compressible fluid flows. On the basis of this a posteriori estimate, we design and implement the corresponding adaptive algorithm to ensure both the reliable and efficient control of the error in the prescribed target functional. The superiority of the proposed approach over standard mesh refinement algorithms which employ empirical error indicators will be demonstrated. This paper represents a continuation of our previous work presented in the articles [10, 11].

2 The compressible Navier-Stokes equations

Writing ρ, $\mathbf{v} = (v_1, v_2)^T$, p, E and T to denote the density, velocity vector, pressure, specific total energy and temperature, respectively, the equations are given by

$$\nabla \cdot (\mathcal{F}^c(\mathbf{u}) - \mathcal{F}^v(\mathbf{u}, \nabla\mathbf{u})) \equiv \sum_{i=1}^{2} \frac{\partial}{\partial x_i} \mathbf{f}_i^c(\mathbf{u}) - \sum_{i=1}^{2} \frac{\partial}{\partial x_i} \mathbf{f}_i^v(\mathbf{u}, \nabla\mathbf{u}) = \mathbf{0} \quad \text{in } \Omega, \quad (1)$$

where Ω is an open bounded domain in $\mathbb{R}^2$. Here, the vector of conservative variables $\mathbf{u}$, the convective fluxes $\mathbf{f}_i^c$, $i = 1, 2$, and the viscous fluxes $\mathbf{f}_i^v$, $i = 1, 2$, are defined by $\mathbf{u} = [\rho, \rho v_1, \rho v_2, \rho E]^T$, $\mathbf{f}_i^c(\mathbf{u}) = [\rho v_i, \rho v_1 v_i + \delta_{1i} p, \rho v_2 v_i + \delta_{2i} p, \rho H v_i]^T$ and $\mathbf{f}_i^v = [0, \tau_{1i}, \tau_{2i}, \tau_{i1} v_1 + \tau_{i2} v_2 + \mathcal{K} T_{x_i}]^T$, respectively. Here, $\mathcal{K}$ is the thermal conductivity coefficient and H is the total enthalpy defined by $H = E + p/\rho$. The pressure is determined by the equation of state of an ideal gas, i.e., $p = (\gamma - 1)\rho(E - \frac{1}{2}\mathbf{v}^2)$, where $\gamma = c_p/c_v$ is the ratio of specific heat capacities; for dry air, $\gamma = 1.4$. For a Newtonian fluid, the viscous stress tensor is given by $\tau = \mu\left(\nabla\mathbf{v} + (\nabla\mathbf{v})^T - \frac{2}{3}(\nabla\cdot\mathbf{v})I\right)$, where μ is the dynamic viscosity coefficient; the temperature T is given by $\mathcal{K}T = \frac{\mu\gamma}{Pr}\left(E - \frac{1}{2}\mathbf{v}^2\right)$, where $\mathrm{Pr} = 0.72$ is the Prandtl number.

The non-dimensionalized form of the Navier–Stokes equations (1) are given by

$$\nabla \cdot \mathcal{F}^c(\mathbf{u}) - \nabla \cdot (G_{1j}\partial\mathbf{u}/\partial x_j, G_{2j}\partial\mathbf{u}/\partial x_j) = \mathbf{0} \quad \text{in } \Omega, \quad (2)$$

where repeated indices are summed through their range. Here, the matrices $G_{ij} = \partial\mathbf{f}_i^v(\mathbf{u}, \nabla\mathbf{u})/\partial u_{x_j}$, for $i, j = 1, 2$, i.e., $\mathbf{f}_i^v(\mathbf{u}, \nabla\mathbf{u}) = \sum_{j=1}^{2} G_{ij}\partial\mathbf{u}/\partial x_j$, $i = 1, 2$, where

$$G_{11} = \frac{1}{\rho\,\mathrm{Re}} \begin{pmatrix} 0 & 0 & 0 & 0 \\ -\frac{4}{3}v_1 & \frac{4}{3} & 0 & 0 \\ -v_2 & 0 & 1 & 0 \\ -\left(\frac{4}{3}v_1^2 + v_2^2 + \frac{\gamma}{\mathrm{Pr}}\left(E - \mathbf{v}^2\right)\right) & \left(\frac{4}{3} - \frac{\gamma}{\mathrm{Pr}}\right)v_1 & \left(1 - \frac{\gamma}{\mathrm{Pr}}\right)v_2 & \frac{\gamma}{\mathrm{Pr}} \end{pmatrix},$$

$$G_{12} = \frac{1}{\rho\,\mathrm{Re}} \begin{pmatrix} 0 & 0 & 0 & 0 \\ \frac{2}{3}v_2 & 0 & -\frac{2}{3} & 0 \\ -v_1 & 1 & 0 & 0 \\ -\frac{1}{3}v_1v_2 & v_2 & -\frac{2}{3}v_1 & 0 \end{pmatrix}, G_{21} = \frac{1}{\rho\,\mathrm{Re}} \begin{pmatrix} 0 & 0 & 0 & 0 \\ -v_2 & 0 & 1 & 0 \\ \frac{2}{3}v_1 & -\frac{2}{3} & 0 & 0 \\ -\frac{1}{3}v_1v_2 & -\frac{2}{3}v_2 & v_1 & 0 \end{pmatrix},$$

$$G_{22} = \frac{1}{\rho\,\mathrm{Re}} \begin{pmatrix} 0 & 0 & 0 & 0 \\ -v_1 & 1 & 0 & 0 \\ -\frac{4}{3}v_2 & 0 & \frac{4}{3} & 0 \\ -\left(v_1^2 + \frac{4}{3}v_2^2 + \frac{\gamma}{\mathrm{Pr}}\left(E - \mathbf{v}^2\right)\right) & \left(1 - \frac{\gamma}{\mathrm{Pr}}\right)v_1 & \left(\frac{4}{3} - \frac{\gamma}{\mathrm{Pr}}\right)v_2 & \frac{\gamma}{\mathrm{Pr}} \end{pmatrix},$$

and Re denotes the Reynolds number.

Given that $\Omega \subset \mathbb{R}^2$ is a bounded region, with boundary Γ, the system of conservation laws (2) must be supplemented by appropriate boundary conditions. For simplicity of presentation, we assume that Γ may be decomposed as follows $\Gamma = \Gamma_{\mathrm{D}} \cup \Gamma_{\mathrm{N,sup}} \cup \Gamma_{\mathrm{N,sub}} \cup \Gamma_{\mathrm{W}}$, where Γ_{D}, $\Gamma_{\mathrm{N,sup}}$, $\Gamma_{\mathrm{N,sub}}$ and Γ_{W} are distinct subsets of Γ representing the Dirichlet (inflow), Neumann (supersonic-outflow), Neumann (subsonic-outflow) and solid wall boundaries, respectively. Thereby, we may specify the following boundary conditions: $\mathbf{u} = \mathbf{u}_D$ on Γ_{D}, $\mathcal{F}^v(\mathbf{u}, \nabla\mathbf{u}) \cdot \mathbf{n} = \mathbf{0}$ on $\Gamma_{\mathrm{N,sup}} \cup \Gamma_{\mathrm{N,sub}}$; note that on $\Gamma_{\mathrm{N,sub}}$ an additional condition which imposes a given pressure p_{out} is enforced. For solid wall boundaries, we consider the distinction between *isothermal* and *adiabatic* conditions. To this end, decomposing $\Gamma_W = \Gamma_{\mathrm{W,iso}} \cup \Gamma_{\mathrm{W,adia}}$, we set $\mathbf{v} = \mathbf{0}$ on Γ_{W}, $T = T_{\mathrm{wall}}$ on $\Gamma_{\mathrm{W,iso}}$, $\mathbf{n} \cdot \nabla T = 0$ on $\Gamma_{\mathrm{W,adia}}$, where T_{wall} is a given wall temperature; we refer to [2, 3, 5, 6] and the references cited therein for further details concerning the imposition of suitable boundary conditions.

3 Discontinuous Galerkin Discretization

In this section we introduce the discontinuous Galerkin method with interior penalty for the discretization of the compressible Navier–Stokes equations (2).

We assume that Ω can be subdivided into shape-regular meshes $\mathcal{T}_h = \{\kappa\}$ consisting of quadrilateral elements κ. For each $\kappa \in \mathcal{T}_h$, we denote by $\mathbf{n}_\kappa$ the unit outward normal vector to the boundary $\partial\kappa$, and by h_κ the elemental diameter. An interior edge of $\mathcal{T}_h$ is the (non-empty) one-dimensional interior of $\partial\kappa^+ \cap \partial\kappa^-$, where κ^+ and κ^- are two adjacent elements of $\mathcal{T}_h$. Similarly, a boundary edge of $\mathcal{T}_h$ is the (non-empty) one-dimensional interior of $\partial\kappa \cap \Gamma$ which consists of entire edges of $\partial\kappa$. We denote by $\mathcal{E}_\mathcal{I}$ the union of all interior edges of $\mathcal{T}_h$, by $\mathcal{E}_\Gamma$ the union of all boundary edges, and set $\mathcal{E} = \mathcal{E}_\mathcal{I} \cup \mathcal{E}_\Gamma$.

Next, we define average and jump operators. To this end, let κ^+ and κ^- be two adjacent elements of $\mathcal{T}_h$ and $\mathbf{x}$ be an arbitrary point on the interior edge $e = \partial\kappa^+ \cap \partial\kappa^- \subset \mathcal{E}_\mathcal{I}$. Moreover, let $\mathbf{v}$ and $\underline{\tau}$ be vector- and matrix-valued functions, respectively, that are smooth inside each element $\kappa^\pm$. By $(\mathbf{v}^\pm, \underline{\tau}^\pm)$ we denote the traces of $(\mathbf{v}, \underline{\tau})$ on e taken from within the interior of $\kappa^\pm$, respectively. Then, we define the averages at $\mathbf{x} \in e$ by $\{\!\{\mathbf{v}\}\!\} = (\mathbf{v}^+ + \mathbf{v}^-)/2$ and $\{\!\{\underline{\tau}\}\!\} = (\underline{\tau}^+ + \underline{\tau}^-)/2$. Similarly, the jumps at $\mathbf{x} \in e$ are given by $[\![\mathbf{v}]\!] = \mathbf{v}^+ \otimes \mathbf{n}_{\kappa^+} + \mathbf{v}^- \otimes \mathbf{n}_{\kappa^-}$ and $[\![\underline{\tau}]\!] = \underline{\tau}^+ \cdot \mathbf{n}_{\kappa^+} + \underline{\tau}^- \cdot \mathbf{n}_{\kappa^-}$. For matrices $\underline{\sigma}, \underline{\tau} \in \mathbb{R}^{m\times n}$, $m, n \geq 1$, we use the standard notation $\underline{\sigma} : \underline{\tau} = \sum_{k=1}^m \sum_{l=1}^n \sigma_{kl}\tau_{kl}$; additionally, for vectors $\mathbf{v} \in \mathbb{R}^m$, $\mathbf{w} \in \mathbb{R}^n$, the matrix $\mathbf{v} \otimes \mathbf{w} \in \mathbb{R}^{m\times n}$ is defined by $(\mathbf{v} \otimes \mathbf{w})_{kl} = v_k\, w_l$.

Given a polynomial degree $p \geq 1$, we define the finite element space $\mathbf{V}_h = \{\mathbf{v} \in [L_2(\Omega)]^4 : \mathbf{v}|_\kappa \in [\mathcal{Q}_p(\kappa)]^4,\ \kappa \in \mathcal{T}_h\}$, where $\mathcal{Q}_p(\kappa)$ denotes the space of tensor product polynomials on κ of degree p in each coordinate direction. We consider the following interior penalty discontinuous Galerkin discretization of equations (2): find $\mathbf{u}_h \in \mathbf{V}_h$ such that

$$\begin{aligned}
\mathcal{N}(\mathbf{u}_h, \mathbf{v}_h) \equiv & -\int_\Omega \mathcal{F}^c(\mathbf{u}_h) : \nabla_h \mathbf{v}_h \,\mathrm{d}\mathbf{x} + \sum_{\kappa\in\mathcal{T}_h} \int_{\partial\kappa\setminus\Gamma} \mathcal{H}(\mathbf{u}_h^+, \mathbf{u}_h^-, \mathbf{n}_\kappa)\cdot \mathbf{v}_h^+ \,\mathrm{d}s \\
& + \int_\Omega \mathcal{F}^v(\mathbf{u}_h, \nabla_h \mathbf{u}_h) : \nabla_h \mathbf{v}_h \,\mathrm{d}\mathbf{x} - \int_{\mathcal{E}_\mathcal{I}} \{\!\{\mathcal{F}^v(\mathbf{u}_h, \nabla_h \mathbf{u}_h)\}\!\} : \underline{[\![\mathbf{v}_h]\!]} \,\mathrm{d}s \\
& - \int_{\mathcal{E}_\mathcal{I}} \{\!\{(G_{i1}^T \partial_h \mathbf{v}_h/\partial x_i, G_{i2}^T \partial_h \mathbf{v}_h/\partial x_i)\}\!\} : \underline{[\![\mathbf{u}_h]\!]})\,\mathrm{d}s \\
& + \int_{\mathcal{E}_\mathcal{I}} \Sigma \underline{[\![\mathbf{u}_h]\!]} : \underline{[\![\mathbf{v}_h]\!]} \,\mathrm{d}s + \mathcal{N}_\Gamma(\mathbf{u}_h, \mathbf{v}_h) = 0
\end{aligned} \tag{3}$$

for all $\mathbf{v}_h$ in $\mathbf{V}_h$. Here, the subscript h on the operators ∇_h and $\partial_h/\partial x_i$, $i = 1, 2$, is used to denote the discrete counterparts of ∇ and $\partial/\partial x_i$, $i = 1, 2$, respectively, taken elementwise. Furthermore, $\mathcal{H}(\cdot,\cdot,\cdot)$ denotes a *numerical* (*convective*) *flux* function, assumed to be Lipschitz continuous, consistent and conservative. The function $\Sigma \in L^\infty(\mathcal{E})$ denotes the so–called discontinuity penalization function; defining $\mathbf{h} \in L^\infty(\mathcal{E})$ by $\mathbf{h}(\mathbf{x}) = \min\{h_{\kappa^+}, h_{\kappa^-}\}$ when $\mathbf{x} \in e = \partial\kappa^+ \cap \partial\kappa^- \subset \mathcal{E}_\mathcal{I}$, and $\mathbf{h}(\mathbf{x}) = h_\kappa$ when $\mathbf{x} \in e = \partial\kappa \cap \Gamma \subset \mathcal{E}_\Gamma$, we set $\Sigma = C_\Sigma/(\mathbf{h}\,\mathrm{Re})$, with a parameter $C_\Sigma > 0$ that is independent of $\mathbf{h}$. Finally, we write

$$\begin{aligned}
\mathcal{N}_\Gamma(\mathbf{u}_h, \mathbf{v}_h) = & \int_\Gamma \mathcal{H}(\mathbf{u}_h^+, \mathbf{u}_\Gamma(\mathbf{u}_h^+), \mathbf{n})\cdot \mathbf{v}_h^+ \,\mathrm{d}s + \int_{\Gamma_\mathrm{D}\cup\Gamma_\mathrm{W}} \Sigma\,(\mathbf{u}_h^+ - \mathbf{u}_\Gamma(\mathbf{u}_h^+))\cdot \mathbf{v}_h^+ \,\mathrm{d}s, \\
& - \int_{\Gamma_\mathrm{D}\cup\Gamma_{\mathrm{W,iso}}} \mathcal{F}^v(\mathbf{u}_h, \nabla_h \mathbf{u}_h) : \mathbf{v}_h^+ \otimes \mathbf{n}\,\mathrm{d}s - \int_{\Gamma_{\mathrm{W,adia}}} \hat{\mathcal{F}}^v(\mathbf{u}_h, \nabla_h \mathbf{u}_h) : \mathbf{v}_h^+ \otimes \mathbf{n}\,\mathrm{d}s \\
& - \int_{\Gamma\setminus\Gamma_{\mathrm{N,sup}}} (G_{i1}^T(\mathbf{u}_h^+)\partial_h \mathbf{v}_h^+/\partial x_i, G_{i2}^T(\mathbf{u}_h^+)\partial_h \mathbf{v}_h^+/\partial x_i) : (\mathbf{u}_h^+ - \mathbf{u}_\Gamma(\mathbf{u}_h^+)) \otimes \mathbf{n}\,\mathrm{d}s,
\end{aligned}$$

where the boundary function $\mathbf{u}_\Gamma(\mathbf{u})$ is given according to the type of boundary condition imposed. We set $\mathbf{u}_\Gamma(\mathbf{u}) = \mathbf{u}_D$ on Γ_D, $\mathbf{u}_\Gamma(\mathbf{u}) = \mathbf{u}$ on $\Gamma_{\mathrm{N,sup}}$, and $\mathbf{u}_\Gamma(\mathbf{u}) = (\rho, \rho v_1, \rho v_2, \frac{p_{\mathrm{out}}}{\gamma-1} + \frac{1}{2}\rho\mathbf{v}^2)^T$ on $\Gamma_{\mathrm{N,sub}}$. Furthermore, we set $\mathbf{u}_\Gamma(\mathbf{u}) = (\rho, 0, 0, \rho c_v T_{\mathrm{wall}})^T$ on $\Gamma_{\mathrm{W,iso}}$, and $u_\Gamma(\mathbf{u}) = (\rho, 0, 0, \frac{p}{\gamma-1})^T = (\rho, 0, 0, \rho E - \frac{1}{2}\rho\mathbf{v}^2)^T$ on $\Gamma_{\mathrm{W,adia}}$. Finally, we note that on the adiabatic boundary $\Gamma_{\mathrm{W,adia}}$, we define $\hat{\mathcal{F}}^v(\mathbf{u}, \nabla\mathbf{u})$ such that

$$\hat{\mathcal{F}}^v(\mathbf{u}, \nabla\mathbf{u})\cdot\mathbf{n} = (0, \tau_{11}n_1 + \tau_{12}n_2, \tau_{21}n_1 + \tau_{22}n_2, 0)^T.$$

Remark 1. We remark that the discretization of the viscous terms has been done by employing the symmetric version of the interior penalty method, cf. [1], and the references cited therein. In particular, we note that this scheme is derived by first re-writing (2) as a system of first–order partial differential equations through the introduction of appropriate auxiliary variables. By defining suitable numerical flux functions, these additional variables are subsequently eliminated; see [1, 8] for details. We note that within this process the transpose of the matrices G_{ij}, $i,j=1,2$, naturally arise in the definition of the DGFEM. Moreover, this is necessary to ensure the adjoint consistency of the resulting method which is essential for the approximation of functionals of the solution, cf. [1, 8]. As a final remark, we note that the discontinuity penalization parameter Σ must be chosen sufficiently large in order to guarantee the stability of the underlying method.

4 Goal–oriented a posteriori error estimation

In this section, we shall be concerned with controlling the error in the numerical solution measured in terms of a given target functional $J(\cdot)$; for a detailed discussion, we refer to the review articles [4, 13]. Assuming that $J(\cdot)$ is differentiable, we write

$$\bar{J}(\mathbf{u},\mathbf{u}_h;\mathbf{u}-\mathbf{u}_h) = J(\mathbf{u}) - J(\mathbf{u}_h) = \int_0^1 J'[\theta\mathbf{u}+(1-\theta)\mathbf{u}_h](\mathbf{u}-\mathbf{u}_h)\,\mathrm{d}\theta, \quad (4)$$

where $J'[\mathbf{w}](\cdot)$ denotes the Fréchet derivative of $J(\cdot)$ evaluated at some $\mathbf{w}$ in $\mathbf{V}$. Here, $\mathbf{V}$ is some suitably chosen function space such that $\mathbf{V}_h \subset \mathbf{V}$. Analogously, we write

$$\begin{aligned}\mathcal{M}(\mathbf{u},\mathbf{u}_h;\mathbf{u}-\mathbf{u}_h,\mathbf{v}) &= \mathcal{N}(\mathbf{u},\mathbf{v}) - \mathcal{N}(\mathbf{u}_h,\mathbf{v}) \\ &= \int_0^1 \mathcal{N}'_{\mathbf{u}}[\theta\mathbf{u}+(1-\theta)\mathbf{u}_h](\mathbf{u}-\mathbf{u}_h,\mathbf{v})\,\mathrm{d}\theta \qquad (5)\end{aligned}$$

for all $\mathbf{v}$ in $\mathbf{V}$. Here, $\mathcal{N}'_{\mathbf{u}}[\mathbf{w}](\cdot,\mathbf{v})$ denotes the Fréchet derivative of $\mathbf{u}\mapsto\mathcal{N}(\mathbf{u},\mathbf{v})$, for $\mathbf{v}\in\mathbf{V}$ fixed, at some $\mathbf{w}$ in $\mathbf{V}$. We remark that the linearization defined in (5) is only a *formal* calculation, in the sense that $\mathcal{N}'_{\mathbf{u}}[\mathbf{w}](\cdot,\cdot)$ may not in general exist. Instead, a suitable approximation to $\mathcal{N}'_{\mathbf{u}}[\mathbf{w}](\cdot,\cdot)$ must be determined, for example, by computing appropriate finite difference quotients of $\mathcal{N}(\cdot,\cdot)$, cf. [9, 10]. Given a suitable linearization, we introduce the following *dual* problem: find $\mathbf{z}\in\mathbf{V}$ such that

$$\mathcal{M}(\mathbf{u},\mathbf{u}_h;\mathbf{w},\mathbf{z}) = \bar{J}(\mathbf{u},\mathbf{u}_h;\mathbf{w}) \quad \forall\mathbf{w}\in\mathbf{V}. \quad (6)$$

We assume that (6) possesses a unique solution. Clearly, the validity of this assumption depends on both the definition of $\mathcal{M}(\mathbf{u},\mathbf{u}_h;\cdot,\cdot)$ and the choice of the target functional under consideration, cf. [10]. For the proceeding error analysis, we must therefore *assume* that the dual problem (6) is well–posed.

Proposition 1. *Let* $\mathbf{u}$ *and* $\mathbf{u}_h$ *denote the solutions of* (2) *and* (3), *respectively, and suppose that the dual problem* (6) *is well–posed. Then,*

$$J(\mathbf{u}) - J(\mathbf{u}_h) \le \sum_{\kappa\in\mathcal{T}_h} |\eta_\kappa|, \quad (7)$$

where

$$
\begin{aligned}
\eta_\kappa = &\int_\kappa \mathbf{R}_h \cdot \boldsymbol{\omega}_h \, d\mathbf{x} + \int_{\partial\kappa\setminus\Gamma} (\mathcal{F}^c(\mathbf{u}_h)\cdot \mathbf{n}_\kappa - \mathcal{H}(\mathbf{u}_h^+, \mathbf{u}_h^-, \mathbf{n}_\kappa)) \cdot \boldsymbol{\omega}_h^+ \, ds \\
&+ \int_{\partial\kappa\cap\Gamma} (\mathcal{F}^c(\mathbf{u}_h)\cdot \mathbf{n}_\kappa - \mathcal{H}(\mathbf{u}_h^+, \mathbf{u}_\Gamma(\mathbf{u}_h^+), \mathbf{n}_\kappa)) \cdot \boldsymbol{\omega}_h^+ \, ds \\
&- \int_{\partial\kappa\setminus\Gamma} \Sigma \underline{[\![\mathbf{u}_h]\!]} : \boldsymbol{\omega}_h^+ \otimes \mathbf{n}_\kappa \, ds \\
&+ \frac{1}{2}\int_{\partial\kappa\setminus\Gamma} ((G_{i1}^T \partial_h \boldsymbol{\omega}_h/\partial x_i, G_{i2}^T \partial_h \boldsymbol{\omega}_h/\partial x_i) : \underline{[\![\mathbf{u}_h]\!]} - [\mathcal{F}^v(\mathbf{u}_h, \nabla_h \mathbf{u}_h)] \cdot \boldsymbol{\omega}_h^+) \, ds \\
&- \int_{\partial\kappa\cap(\Gamma_D\cup\Gamma_W)} \Sigma \left(\mathbf{u}_h^+ - \mathbf{u}_\Gamma(\mathbf{u}_h^+)\right) \cdot \boldsymbol{\omega}_h^+ \, ds \\
&- \int_{\partial\kappa\cap(\Gamma_{N,sub}\cup\Gamma_{N,sup})} (\mathcal{F}^v(\mathbf{u}_h^+, \nabla_h \mathbf{u}_h^+)\cdot \mathbf{n}_\kappa) \cdot \boldsymbol{\omega}_h^+ \, ds \\
&- \int_{\partial\kappa\cap\Gamma_{W,adia}} \left(\mathcal{F}^v(\mathbf{u}_h^+, \nabla_h \mathbf{u}_h^+) - \hat{\mathcal{F}}^v(\mathbf{u}_h^+, \nabla_h \mathbf{u}_h^+)\right) : \boldsymbol{\omega}_h^+ \otimes \mathbf{n} \, ds \\
&+ \int_{\partial\kappa\cap(\Gamma\setminus\Gamma_{N,sup})} (G_{i1}^T(\mathbf{u}_h^+)\partial_h \boldsymbol{\omega}_h^+/\partial x_i, G_{i2}^T(\mathbf{u}_h^+)\partial_h \boldsymbol{\omega}_h^+/\partial x_i) : \\
&\qquad\qquad : (\mathbf{u}_h^+ - \mathbf{u}_\Gamma(\mathbf{u}_h^+)) \otimes \mathbf{n} \, ds,
\end{aligned}
$$

and $\boldsymbol{\omega}_h = \mathbf{z} - \mathbf{z}_h$ *for all* $\mathbf{z}_h$ *in* $\mathbf{V}_h$*. Here,* $\mathbf{R}_h|_\kappa = -\nabla_h \cdot \mathcal{F}^c(\mathbf{u}_h) + \nabla_h \cdot \mathcal{F}^v(\mathbf{u}_h, \nabla_h \mathbf{u}_h)$*,* $\kappa \in \mathcal{T}_h$*, denotes the elementwise residual.*

Proof. Choosing $\mathbf{w} = \mathbf{u} - \mathbf{u}_h$ in (6), recalling the linearization performed in (4), and exploiting the Galerkin orthogonality property of the DGFEM, we get

$$
\begin{aligned}
J(\mathbf{u}) - J(\mathbf{u}_h) &= \bar{J}(\mathbf{u}, \mathbf{u}_h; \mathbf{u} - \mathbf{u}_h) = \mathcal{M}(\mathbf{u}, \mathbf{u}_h; \mathbf{u} - \mathbf{u}_h, \mathbf{z}) \\
&= \mathcal{M}(\mathbf{u}, \mathbf{u}_h; \mathbf{u} - \mathbf{u}_h, \mathbf{z} - \mathbf{z}_h) = -\mathcal{N}(\mathbf{u}_h, \mathbf{z} - \mathbf{z}_h) \quad \forall \mathbf{z}_h \in \mathbf{V}_h.
\end{aligned}
$$

Equation (7) now follows by the divergence theorem together with the triangle inequality.

We end this section by noting that the Type I a posteriori error bound (7) depends on the unknown analytical solution to the primal and dual problems. Thus, in order to render these quantities computable, both $\mathbf{u}$ and $\mathbf{z}$ must be replaced by suitable approximations. Here, the linearizations leading to $\mathcal{M}(\mathbf{u}, \mathbf{u}_h; \cdot, \cdot)$ and $\bar{J}(\mathbf{u}, \mathbf{u}_h; \cdot)$ are performed about $\mathbf{u}_h$ and the dual solution $\mathbf{z}$ is replaced by a DGFEM approximation $\hat{\mathbf{z}}$ computed on the same mesh $\mathcal{T}_h$ used for $\mathbf{u}_h$, but with a higher degree polynomial.

5 Numerical example

In this section we present a numerical example to highlight the advantages of designing an adaptive finite element algorithm based on the weighted er-

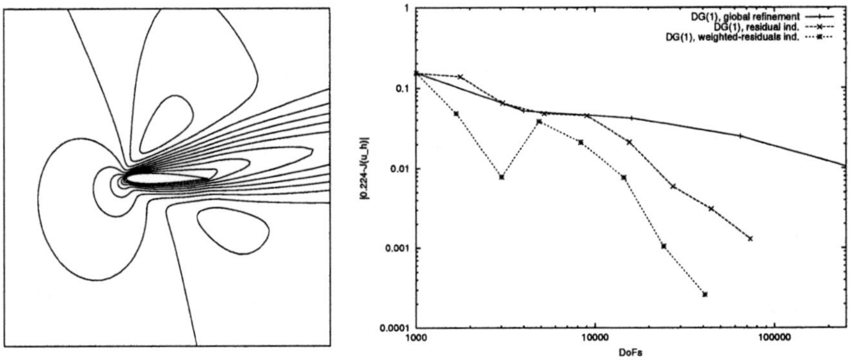

(a) (b)

Fig. 1. (a) Mach isolines, $\mathrm{Ma} = \frac{i}{10}, i = 1, \dots, 10$, of the flow around the NACA0012 airfoil; (b) Convergence of the error $|J(\mathbf{u}) - J(\mathbf{u}_h)|$ using each mesh refinement strategy

Table 1. Adaptive algorithm based on the weighted error indicator $|\eta_\kappa|$.

Elements	DOF	$J(\mathbf{u}) - J(\mathbf{u}_h)$	$\sum_\kappa \eta_\kappa$	θ_1	$\sum_\kappa \|\eta_\kappa\|$	θ_2
63	1008	-1.534e-01	-1.299e-01	0.85	9.2e-01	6.02
105	1680	-4.838e-02	-1.114e-02	0.23	7.4e-01	15.35
189	3024	7.814e-03	5.574e-03	0.71	7.9e-01	100.60
306	4896	-3.819e-02	-1.296e-02	0.34	8.1e-01	21.11
522	8352	-2.107e-02	-9.249e-03	0.44	8.5e-01	40.25
909	14544	-7.686e-03	-3.746e-03	0.49	7.9e-01	102.9
1512	24192	-1.049e-03	-9.130e-04	0.87	1.0e+00	962
2559	40944	-2.628e-04	-2.395e-04	0.91	6.2e-01	2374

ror indicator $|\eta_\kappa|$ in comparison with both uniform mesh refinement, as well as an adaptive algorithm based on an empirical refinement indicator which does not require the solution of an auxiliary (dual) problem; for simplicity, we employ a Type II residual indicator of the form derived in [10]. Throughout this section, we employ the Vijayasundaram flux for the discretization of the convective terms and set $p = 1$ (bilinear elements). Finally, for both adaptive refinement strategies, we use the fixed fraction refinement algorithm with refinement and derefinement fractions set to 20% and 10%, respectively; we also note that for the computation of $|\eta_\kappa|$, the dual solution is approximated using piecewise biquadratic polynomials.

We consider a Mach 0.8 flow at an angle of attack $\alpha = 10°$ with Reynolds number $\mathrm{Re} = 73$ and constant temperature on the profile, cf. [3, 12]. The solution to this problem consists of a flow that is mainly subsonic with a small supersonic region above the airfoil, see Fig. 1(a). Here, we consider the evalu-

Table 2. Nonlinear Newton residuals (Res.) and convergence rates on a sequence of uniformly refined meshes. *: On the coarsest mesh only the last 4 out of 6 iteration steps are displayed.

Mesh 1		Mesh 2		Mesh 3		Mesh 4		Mesh 5	
Res.	Rate	Res.	Rate	Res.	Rate	Res.	Rate	Res.	Rate
6.5-02*	-	8.1-01	-	4.4-01	-	2.4-01	-	1.1-01	
2.5-02	3	4.2-02	19	3.3-02	13	1.6-02	16	5.9-03	19
6.7-04	37	1.2-03	36	1.6-03	20	2.6-04	62	1.4-04	43
6.6-07	1021	1.4-06	808	1.8-05	92	2.6-07	976	1.7-04	804
3.7-10	1755	3.1-10	4538	2.9-09	6199	5.5-11	4750	1.7-10	1021

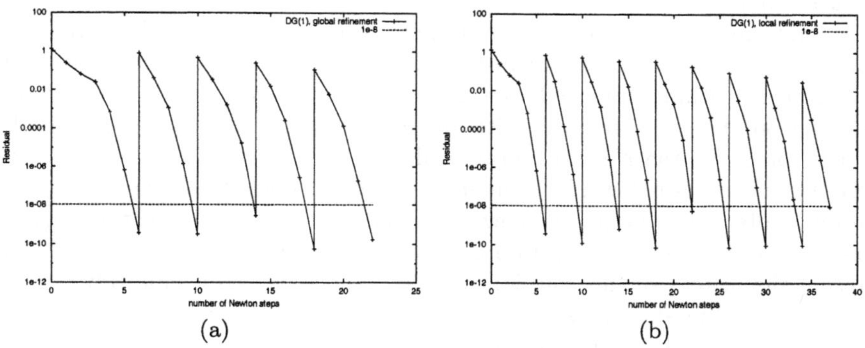

Fig. 2. Convergence of the nonlinear residual with the number of Newton steps employed: (a) Uniform mesh refinement; (b) Adaptive mesh refinement using the empirical error indicator

ation of the inviscid drag coefficient (c_{dp}) on the surface of the airfoil. On the basis of a fine grid computation, the reference value of the functional is given by $J(\mathbf{u}) \approx 0.224$.

In Fig. 1(b) we compare the true error in the computed target functional $J(\cdot)$ using all three mesh refinement strategies. Here, we clearly observe the superiority of the weighted a posteriori error indicator; at all refinement steps the error in the computed functional is less than the corresponding quantity when either uniform refinement or adaptive refinement based on an empirical residual indicator is employed. Indeed, on the final mesh the true error in $J(\cdot)$ is over an order of magnitude smaller than $|J(\mathbf{u}) - J(\mathbf{u}_h)|$ computed on the sequence of meshes generated by the empirical indicator.

In Table 1 we collect the data of the adaptive algorithm when employing the weighted indicators. Here, we show the number of elements and degrees of freedom (DOF) in $\mathbf{V}_h$, the true error in the functional $J(\mathbf{u}) - J(\mathbf{u}_h)$, the computed error representation formula, the approximate a posteriori error bound and their respective effectivity indices $\theta_1 = \sum_\kappa \eta_\kappa / (J(\mathbf{u}) - J(\mathbf{u}_h))$ and

$\theta_2 = \sum_\kappa |\eta_\kappa|/|J(\mathbf{u}) - J(\mathbf{u}_h)|$. First we note that on all refinement steps the correct sign of the error is predicted by the computed error representation formula. Furthermore, whereas on very coarse meshes the quality of $\sum_\kappa \eta_\kappa$ is rather poor, in the sense that θ_1 is noticeable smaller than one, we see that the effectivity indices θ_1 slowly tend towards unity as the mesh is refined. On the other hand, even just the application of the triangle inequality leads to significant over–estimation of the true error in the computed functional; indeed, here we see that θ_2 slowly increases.

As a final remark, we note that in all of our computations, we employed a (damped) Newton iteration to solve the system of nonlinear equations arising from the discontinuous Galerkin discretization of the compressible Navier-Stokes equations. Within each Newton step, GMRES with Block-Gauss-Seidel preconditioning is exploited to solve the resulting linearized problem involving the Jacobian matrix, cf. [9], for further details. For each of the three mesh refinement algorithms employed above, the nonlinear solver proceeds as follows: starting on the coarsest mesh with free-flow conditions the nonlinear problem is solved by the Newton iteration described above. When the nonlinear residual converges below 10^{-8}, the mesh is refined once; then the discrete solution is interpolated onto the new mesh and is thereby taken as the starting solution for the Newton iteration on the newly refined mesh. In Table 2 we present the history of this solution process on a sequence of uniformly refined computational meshes; these results are also summarized in Fig. 2(a). On the coarsest mesh, with free-flow values, the Newton iteration requires a few steps until the iterative solution reaches the range of quadratic convergence. Indeed, after only 6 Newton steps the nonlinear residual is below the given tolerance of 10^{-8} and the mesh is refined; on subsequent meshes, the Newton iteration requires only 4 steps to reduce the residual below the given tolerance.

The convergence behaviour of the Newton iteration is analogous even when locally refined meshes are employed. Indeed, in Fig. 2(b) we plot the nonlinear residuals against the number of Newton steps employed for each of the meshes generated using the empirical error indicator. Here, we again see that on the first mesh 6 Newton steps are required to satisfy the convergence criterion, while only 4 are necessary on subsequent meshes. Analogous behaviour is also observed on the adaptive meshes generated using the weighted error indicator $|\eta_\kappa|$; for brevity, these results have been omitted.

References

1. Arnold, D, Brezzi, F., Cockburn, B., Marini, D. (2002): Unified analysis of discontinuous Galerkin methods for elliptic problems. SIAM J. Numer. Anal., **39(5)**, 1749–1779
2. Baumann, C.E., Oden, J.T. (1999): A discontinuous *hp* finite element method for the solution of the Euler and Navier–Stokes equations. Int. J. Numer. Meth. Fluids, **31**, 79–95

3. Bassi, F., Rebay, S. (1997): A high–order accurate discontinuous finite element method for the numerical solution of the compressible Navier–Stokes equations. J. Comput. Phys., **131**, 267–279
4. Becker, R., Rannacher R. (2001): An optimal control approach to a-posteriori error estimation in finite element methods. In: Iserles, A. (ed), Acta Numerica, pp. 1–102, CUP
5. Capon, P.J. (1995): Adaptive Stable Finite Element Methods for the Compressible Navier–Stokes Equations. Ph.D. Thesis, University of Leeds
6. Dolejsi, V. (2002): On the discontinuous Galerkin method for the numerical solution of the Euler and the Navier–Stokes equations. Int. J. Numer. Meth. Fluids, (submitted)
7. Dolejsi, V., Feistauer, M. Schwab, Ch. (2002): On discontinuous Galerkin methods for nonlinear convection–diffusion problems and compressible flow. Math. Bohem., **127(2)**, 163–179
8. Harriman, K., Houston, P., Senior, B., Süli, E. (2003): *hp*–Version discontinuous Galerkin methods with interior penalty for partial differential equations with nonnegative characteristic form. In: Shu, C.-W., Tang, T., Cheng, S.-Y. (eds), Recent Advances in Scientific Computing and Partial Differential Equations. Contemporary Mathematics, **330**, 89–119
9. Hartmann, R. (2004): The role of the Jacobian in the adaptive discontinuous Galerkin method for the compressible Euler equations. In: Warnecke, G. (ed), Analysis and Numerical Methods for Conservation Laws, Springer–Verlag (submitted)
10. Hartmann, R., Houston, P. (2002): Adaptive discontinuous Galerkin finite element methods for nonlinear hyperbolic conservation laws. SIAM J. Sci. Comp., **24(3)**, 979–1004
11. Hartmann, R., Houston, P. (2002): Adaptive discontinuous Galerkin finite element methods for the compressible Euler equations. J. Comp. Phys., **183**, 508–532
12. Numerical Simulation of Compressible Navier-Stokes Equations-External 2D Flows Around a NACA0012 Airfoil. GAMM Workshop, Dec. 4-6, 1985, Nice, France (Edt. INRIA, Centre de Rocquefort, de Rennes et de Sophia-Antipolis, 1986)
13. Süli, E., Houston, P. (2002): Adaptive finite element approximation of hyperbolic problems. In: Barth, T., Deconinck, H.(eds) Error Estimation and Adaptive Discretization Methods in Computational Fluid Dynamics. Lecture Notes in Computational Science and Engineering, Volume 25, pp. 269-344, Springer-Verlag

On a Novel Technique for Parallel Unstructured Mesh Generation in 3D

Jan Haškovec[1] and Pavel Šolín[2]

[1] Faculty of Mathematics and Physics, Charles University, Sokolovská 83, Prague, Czech Republic *jan.haskovec@st.mff.cuni.cz*

[2] Department of Computational and Applied Mathematics, Rice University, 6100 Main Street, Houston, TX 77251-1892 *solin@rice.edu*

Summary. We present a novel approach to unstructured tetrahedral mesh generation in 3D with minimum geometrical input. The boundary of the domain is identified adaptively with adjustable precision. The grid points of the mesh are distributed by means of an algorithm based on the analogy with a system of electrically charged particles. A-priori local refinements of the mesh are possible. Surface and volume meshes are built using advancing-front-type algorithms. All steps of the algorithm are suitable for efficient parallelisation. An application of the presented approach is shown.

1 Motivation

Nowadays, computational mathematics becomes part of new challenging interdisciplinary computer-assisted technologies in medicine, natural sciences, industry and elsewhere[1]. Specifically for mesh generation this means that simplification and automation of geometrical inputs becomes a crucial issue. Traditional algorithms, among which the most popular ones are based on Solid Modeling Techniques (SMT) and CAD modellers, have not been designed to process machine-generated information such as outputs of MRI scans, image recognition devices, physical measurements, geographic databases etc. It is our aim to design a fully automatic parallel generator of unstructured tetrahedral meshes with this capability.

The example geometry in Fig. 1 is given by the formula

$$\Omega = \left\{ (x,y,z); (x - R(\phi)cos(\phi))^2 + (y - R(\phi)sin(\phi))^2 + (z - \tfrac{\phi}{2\pi})^2 \le r^2(\phi)\,, \quad \phi \in (0, 4\pi) \right\}$$

with

[1] The authors acknowledge the financial support of the Grant Agency of the Czech Republic under Grant No. GP102/01/D114.

Fig. 1. Example of a 3D geometry whose optimal computer representation is a non-trivial issue

$$R(\phi) = 1 - \frac{\phi}{8\pi}, \; r(\phi) = 0.3\left(1 - \frac{\phi}{8\pi}\right).$$

The minimum information necessary for the definition of the geometry of a 3D domain Ω is its characteristic function χ,

$$\chi_\Omega = \begin{cases} 1, & x \in \Omega \\ 0, & x \notin \Omega. \end{cases} \quad (1)$$

In this paper we introduce a first working version of a C++ mesh generator XGEN3D that has been designed to process this form of geometrical input.

The algorithm consists of four steps,

step 1: adaptive identification of the boundary $\partial\Omega$,
step 2: iterative distribution of grid points,
step 3: generation of surface mesh,
step 4: generation of volume mesh,

that will be discussed in the following sections.

Mesh generation is a scientific discipline with a very long tradition, and due to its variety and diversity it is difficult to select a few references on which our work is exactly based. One of the best places where mesh-generation-related information of virtually any kind can be found, that we used intensively, is Robert Schneiders' web page [3].

2 Adaptive identification of the boundary $\partial\Omega$

In order to keep the computer implementation at a reasonable level of complexity, the user is requested to provide the following parameters in addition to the characteristic function χ_Ω: co-ordinates of a (reasonably small) cube $C \subset \mathbb{R}^3$ (*inital cube*) such that $\Omega \subset C$ and tolerances TOL_Ω, $TOL_{\partial\Omega}$, $TOL_{\partial\Omega} \leq TOL_\Omega$, with which the geometry of Ω will be approximated. The meaning of these parametes will be explained later. Moreover the user defines a mesh density function $h(x) : C \to \mathbb{R}$ that at every point $x \in \Omega$ indicates the desired mean edge length.

The algorithm produces a continuous, piecewise triangular approximation of the boundary $\partial\Omega$. It proceeds recursively, starting by dividing the initial cube into eight smaller sub-cubes (hence *oct-tree*), and using the function χ_Ω to determine whether a sub-cube intersects with the boundary $\partial\Omega$ or not. The adaptive splitting process is continued until the size of the sub-cubes reaches the parameter TOL_Ω. The resulting sub-cubes are called *basic cubes*. Splitting of the basic cubes further continues until $TOL_{\partial\Omega}$ is reached. The resulting sub-cubes are called *boundary cubes*.

At this point the complete (discrete) information about the boundary $\partial\Omega$ is contained in the boundary cubes. A simple algorithm allows for accurate location of points (*boundary points*) where the edges of the boundary cubes intersect with the boundary $\partial\Omega$. For $TOL_{\partial\Omega}$ sufficiently small, the boundary $\partial\Omega$ can be locally approximated by a plane. In other words, for each boundary cube the corresponding boundary points form a planar polygon with 3 to 7 vertices $P_1, P_2, \ldots, P_k$.

If $k = 3$, we already have a portion of the desired triangular approximation of $\partial\Omega$. If $k \geq 4$, we need to construct a triangulation of the polygon. This is most easily achieved by calculating its center of gravity,

$$P_c = \frac{1}{k}(P_1 + P_2 + \ldots + P_k),$$

and defining k triangles $(P_1, P_2, P_c), (P_2, P_3, P_c), \ldots, (P_{k-1}, P_k, P_c)$.

Summarized, the recursive algorithm of approximating the boundary $\partial\Omega$ can be written as follows:

Boundary identification algorithm:

1. Begin with the initial cube C.
2. Split the actual cube into 8 sub-cubes.
3. With each sub-cube:
   - 3.1. Does it intersect with the boundary?
     - NO: Is the size of its edge smaller than TOL_Ω?
       - YES: STOP.
       - NO: Continue with step 2.
     - YES: Is the size of its edge smaller than $TOL_{\partial\Omega}$?
       - YES: Build an approximation to $\partial\Omega$ inside of the cube and STOP.

NO: Continue with step 2.

4. STOP.

The output of this algorithm is a piecewise triangular approximation to the boundary $\partial\Omega$. The algorithm can analyse as complicated geometry as desired and approximate it as accurately as desired. The only limiting factor is the amount of memory and CPU-time consumed by the software. Let us point out that the resulting set of triangles cannot be used as a surface mesh yet – the size and shape of the triangles varies irregularly and their diameter has nothing in common with the user-specified mesh density function $h(x)$.

Fig. 2, 3 and 4 show the approximation of a part of the domain Ω from Fig. 1 with three different values of $TOL_{\partial\Omega}$.

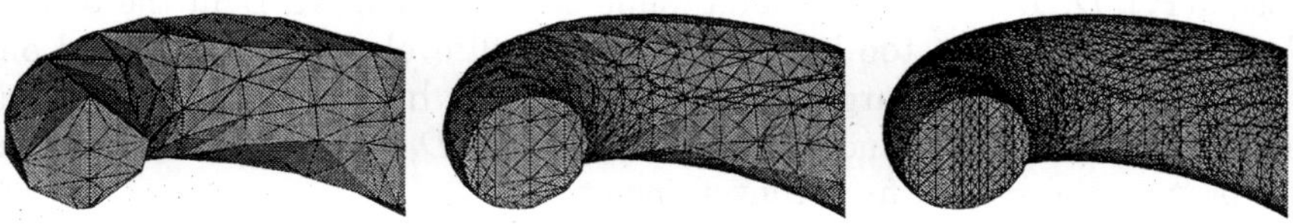

low accuracy optimal accuracy high accuracy

Fig. 2. Approximation of $\partial\Omega$ with various levels of accuracy: $TOL_{\partial\Omega} = 0.1$, 0.03 and 0.015, respectively

3 Distribution of grid points

First we need to estimate the total number N of grid points. With the shape of Ω known from the previous step, we numerically integrate the mesh density function $h(x)$ over the domain. An average edge length is defined as

$$h_a = \frac{\int_\Omega h(x)\,\mathrm{d}x}{|\Omega|}.$$

Finally we use the information about the volume of an equilateral tetrahedron with the edge length h_a to compute the approximate number of tetrahedra in the mesh and the optimal number of grid points N.

Next we place N geometrical points with pseudo-random positions into the domain Ω. By pseudo-random we mean that the positions are generated randomly, but we do not allow any two points to lie too close to each other. When all of the N points are generated, we start optimizing their positions. For this we adopt the approach [4] that minimizes a suitable global potential Φ defined on the set of all grid points. One way to choose Φ is to look for analogy with a system of electrically charged particles. From a potential one can derive forces that act on the particles, and one can apply a time stepping procedure in order to let the system converge to a state with minimum energy.

With a suitable choice of the repulsing force F_{ij} of particles P_i, P_j, steady state of the system is usually achieved after a few iterations. An example of the repulsing force F that respects the variable mesh density h is

$$F_{ij} = F(P_i, P_j) = h\left(\frac{P_i + P_j}{2}\right)\frac{P_i - P_j}{|P_i - P_i|^3}.$$

The total force working on each particle P_i is computed from contributions of particles that lie in a limited distance from P_i. In this way we use the oct-tree structure available from Step 1 to avoid quadratic complexity of the algorithm. One further uses the magnitude of forces working at each particle together with the mesh density function in order to calculate a suitable time step. The time step is defined in such a way that any particle P_i displaces at most $h(P_i)/D$, $D > 0$. The default value of the parameter D in the code is $D = 5$. If D is chosen too large, the system will evolve very slowly and one will have to perform a large number of iterations before the steady state is reached. On the other hand, a too small value of D would cause instabilities and prevent the system from converging at all.

In addition to that one has to withdraw the kinetic part of the total energy from the system after each time step by resetting the velocities of all particles to zero. This action is analogous to cooling of a physical system that is necessary in order to reach the bottom of the potential well.

When moving a particle, we have to check whether it crosses the boundary $\partial\Omega$ on its way. If so, we compute the intersection of the boundary with its trajectory and position the particle to the point of intersection. In this moment the particle becomes a *boundary particle* and it is not allowed to leave the boundary anymore. From now on the particle is being moved along the boundary - we compute the force acting on it but only consider the projection onto the plane tangential to $\partial\Omega$. The outline of the algorithm is as follows:

Algorithm for the distribution of grid points:

1. Compute the total number of grid points N.
2. Place the points into Ω in a pseudo-random way.
3. Until steady state is reached, repeat:
   3.1. For each particle P_i do:
      3.1.1. Calculate the force F_i acting on P_i.
      3.1.2. Does P_i lie on the boundary $\partial\Omega$?
         YES: Compute the projection of F_i to the boundary.
            Move the particle according to the projection.
         NO: Does the last trajectory of P_i intersect with $\partial\Omega$?
            YES: Place P_i to the point of intersection.
            NO: Move the point P_i according to F_i.
4. STOP.

Let us remark that it is desired that the mesh density function $h(x)$ does not vary too fast, otherwise tetrahedra with obtuse angles would arise in the correspondent areas of the domain, which would lead to low quality of the resulting mesh. It is difficult to state any quantitative conditions, but we can roughly say that the function $h(x)$ should be Lipschitz-continous with the Lipschitz constant not greater than $1/3$. However, this value depends on the domain geometry. Generally speaking, the function $h(x)$ should be chosen as smooth as possible.

4 Surface meshing

The set of boundary grid points is used to generate a surface mesh – this is a standard task that virtually all mesh generators have in common. We apply an Advancing Front (AF) algorithm [4] for this purpose. See, e.g., [2] for a more general description of AF algorithms.

Because of some temporary technical difficulties that we hope to overcome soon we request the user to partition the surface of the domain into M subregions $\partial\Omega_1, \partial\Omega_2, \ldots, \partial\Omega_M$ of simpler shapes. Function `Numbering(x,y,z)` then returns the index k of a subregion $\partial\Omega_k$ the point $[x, y, z]$ belongs to. Ordered chains of abscissas that separate a subregion $\partial\Omega_k$ from the rest of $\partial\Omega$ are called *frontiers*.

We generate the surface triangulation for each subregion $\partial\Omega_k$ separately using an AF technique. The k-th frontier is the starting point. We copy all its abscissas to the list of abscissas A. Then, for the first abscissa (P, Q) in A, we search a boundary grid point $Z = [x, y, z]$ such that `Numbering(x,y,z)`$=k$ and the angle (P, Z, Q) of abscissas (P, Z) and (Q, Z) is maximal. In other words, we look for the maximizer Z of the angle (P, Z, Q) in the set of all boundary grid points such that `Numbering`$=n$. The fact that the angle (P, Z, Q) is maximal ensures that there is no boundary grid point lying inside the triangle (P, Q, Z). If the mesh density function $h(x)$ is Lipschitz-continous with reasonably small Lipschitz constant and we have reached an equilibrium steady state in the process of iterative grid points distribution, we can always find a maximizer Z such that the triangle (P, Q, Z) obeys the minimal-angle-rule and this results in a high quality mesh.

When the point Z is found, we add the newly created triangle into the list of surface triangles τ_s and remove the abscissa (P, Q) from the list A. Then we check if the abscissas (P, Z) and (Q, Z) are already contained in A; if so, we remove them from the list. In the opposite case we add them to A. We repeat the search of a maximizer for the next abscissa in the list, until the list becomes empty – this means that the subregion $\partial\Omega_k$ is covered with a surface triangulation. We can sketch the algorithm as follows:

Surface triangulation algorithm:
1. For each subregion $\partial\Omega_k \subset \partial\Omega$:

1.1. Copy its frontier abscissas into a list A.
1.2. Repeat until A is empty:
1.2.1. Find a boundary grid point $Z \subset \partial\Omega_k$ that maximizes the angle (P, Z, Q) for the first abscissa $(P, Q) \in A$.
1.2.2. Add the triangle (P, Q, Z) to the surface mesh τ_s.
1.2.3. Remove the abscissa (P, Q) from A
1.2.4. Is abscissa (P, Z) already contained in A?
YES: Remove it from A.
NO: Add it to A.
1.2.5. Is abscissa (Q, Z) already contained in A?
YES: Remove it from A.
NO: Add it into A.
2. STOP

The result of this algorithm is the surface triangulation list τ_s. Examples of two different surface meshes on the domain Ω from Fig. 1 are shown in Fig. 3.

Obviously the quality of the mesh strongly depends on the distribution of grid points. In general, it is extremely difficult to give a proof of stability of the meshing algorithm or statements about the quality of the mesh. According to our experience, if the disribution of grid points is reasonable, the meshing algorithm is stable and produces meshes of very good quality, containing prevalently almost-equilateral triangles.

5 Volume meshing

We use a generalization of a two-dimensional algorithm [4] for the construction of the volume mesh τ_v. The input parameters are the list τ_s of surface triangles and the set of inner grid points. The idea of the volume meshing algorithm is similar as in the case of the surface mesh, again based on the AF technique.

We start with a first triangle (P, Q, R) in the list τ_s. For this triangle we find a grid point Z such that the sum of the angles $(P, Z, Q) + (Q, Z, R) + (R, Z, P)$ is maximal. According to our experience, this criterion turned out to be optimal for the meshing algorithm. Next we add the newly created tetrahedron (P, Q, R, Z) into the list τ_v of tetrahedra and delete the triangle (P, Q, R) from the list τ_s. We check if the triangles (P, Q, Z), (P, R, Z) and (Q, R, Z) are contained in τ_s. If so, we remove them from the list. In the opposite case we add them to the list. Then we proceed to the next triangle in τ_s and so on, until τ_s becomes empty. The algorithm is written as follows:

Volume meshing algorithm:
1. Repeat until τ_s is empty:
1.1. Find a maximizer Z for the first triangle $(P, Q, R) \in \tau_s$.
1.2. Add the tetrahedron (P, Q, R, Z) into the list τ_v.
1.3. Remove the triangle (P, Q, R) from τ_s.

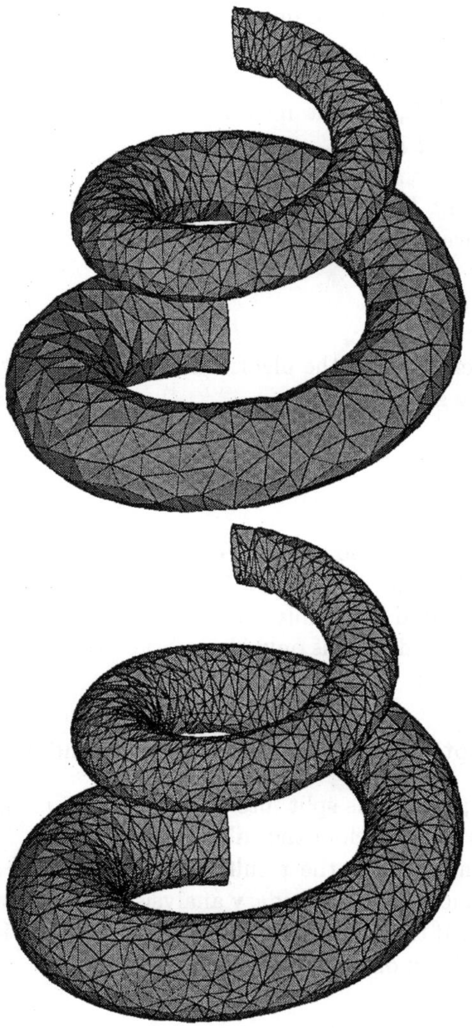

Fig. 3. Meshes on the example geometry from Fig. 1. The first one is gradually refined towards the thinner end of the spiral and has 1460 surface elements. The other consists of 1872 uniformly-sized elements. In both cases geometry analysis parameters $TOL_{\Omega} = 0.15$ and $TOL_{\partial\Omega} = 0.03$ were used.

1.4. Is triangle (P, Q, Z) contained in τ_s?
YES: Remove it from τ_s.
NO: Add it to τ_s.
1.4. Is triangle (P, R, Z) contained in τ_s?
YES: Remove it from τ_s.
NO: Add it to τ_s.
1.4. Is triangle (Q, R, Z) contained in τ_s?
YES: Remove it from τ_s.
NO: Add it to τ_s.
2. STOP

According to our experience, the algorithm is stable and it produces tetrahedral meshes of very good quality.

6 Parallelisation

All algorithms utilized for the adaptive oct-tree analysis of the boundary $\partial\Omega$, distribution of grid points and generation of the surface and volume meshes are extremely well suited for running in parallel. Since we are in the middle of parallelization of the code, let us only give a brief description of the techniques we use.

6.1 Parallel adaptive identification of the boundary $\partial\Omega$

The *initial cube* C, $\Omega \subset C$, is split into eight subcubes in the initial step. The algorithm is then run on each of the subcubes recursively, fully independently of the other instances. Since the resulting subdomains of Ω are disjoint, ne can run eight instances of the boundary analysis algorithm without any risk of potential collisions. If there are more than eight CPUs, obviously it is possible to start the parallel run on further levels of recursion.

6.2 Parallel distribution of grid points

When distributing the grid points, the most time-consuming part of the algorithm is the computation of the total force acting on the particles. With N_{CPU} processors at our disposal, one divides Ω into N_{CPU} subdomains. No domain decomposition method is needed since splitting the *initial cube* C does the job. An independent instance of the algorithm is run for each of the resulting subdomains.

6.3 Parallel generation of surface and volume meshes

The surface of the domain is again split into subregions of simpler shapes in the same spirit as in Section 4. The construction of the surface triangulation in every subregion is a task independent of events in the other subregions. For the sake of technical simplicity, in our conception the number of CPUs is limited by the number of physical areas. This simplification will be eliminated later.

A similar idea as above is applied to the volume meshing – the initial cube is split and independent instances of the volume meshing algorithm are run on the subcubes. After the algorithm finishes independent runs in the subcubes, one final sweep is done to generate tetrahedra that lie across the boundaries of the subcubes. Since these are not many, a single process is sufficient to accomplish this task.

References

1. Cheung, Y.K., Lo, S.H., Leung, A.Y.T. (1996): Finite Element Implementation, Blackwell Science, Oxford
2. George, P.L. (1991): Automatic Mesh Generation: Application to Finite Element Methods, Wiley, New York
3. Schneiders, R.: `http://www-users.informatik.rwth-aachen.de/~roberts/meshgeneration.html`
4. Šolín, P., (2000): On a Mesh Generation Technique Based on a Special Smoothing Procedure for Uniform Inner Point Distribution. Acta Technica ASCR, **45**, 397–417.

Adaptive Finite Element Methods for Turbulent Flow

Johan Hoffman[1] and Claes Johnson[2]

[1] Courant Institute of Mathematical Sciences, New York University, 251 Mercer Street, New York, NY 10012-1185, USA, *hoffman@cims.nyu.edu*
[2] Department of Computational Mathematics, Chalmers University, SE 412 96, Göteborg, Sweden, *claes@math.chalmers.se*

Summary. We present recent results using adaptive finite element methods, based on a posteriori error estimates, to compute various output functionals for incompressible flow problems in 3d, for both laminar and turbulent flows. The a posteriori error estimates are based on the solution of an associated dual problem with data connected to the output functional we want to compute.

1 Introduction

We present recent results from [15, 14], extending earlier results in [17, 13], where we use adaptive finite element methods, based on *a posteriori error estimates*, to compute various output functionals in incompressible flow problems in 3d, for both laminar and turbulent flows. The a posteriori error estimates are based on the solution of an associated linearized dual problem that contains information about error propagation in space-time.

The idea of using duality arguments in a posteriori error estimation goes back to Babuška and Miller [2] in the context of postprocessing 'quantities of physial interest' in elliptic model problems. A framework for more general situations has since then been systematically developed by in particular Eriksson & Johnson and Becker & Rannacher, with coworkers, see e.g. [6, 4, 19, 20]. Applications to incompressible flow have been increasingly advanced with computation of functionals such as the drag coefficient for 2d stationary benchmark problems in [3, 9], and drag and lift coefficients and pressure differences for 3d stationary benchmark problems in [15]. In [17] time dependent problems in 3d are considered, and the extension to *Large Eddy Simulation LES* of turbulent flow is investigated in [13]. In [14] a temporal mean of the drag coefficient of a surface mounted cube in a turbulent channel flow is computed using a LES.

If we use a subgrid model in a LES, the subgrid modeling error is included in the a posteriori error estimates, which opens the possibility of comparing the error using different subgrid models. Altogether, the a posteriori error estimates open the possibility of adaptively choosing both an optimal mesh and an optimal subgrid model. This approach to a posteriori error estimation with respect to the averaged solution, using duality teqniques, in terms of a modeling error and a discretization error was developed for convection-diffusion-reaction equations in [10, 18, 16, 11, 12]. Related approaches with a posteriori error estimates in terms of a modeling and a discretization contribution to the total error have been suggested. For example, more recently in [5] similar ideas are presented with applications to 2d convection-diffusion-reaction problems.

Due to the local nature of turbulence in many applications, in particular for the problem in [14], we stress the possibilities of adaptive mesh refinement for such problems. In [14] we are able to locally resolve scales of motion corresponding to a Reynolds number of about 1000, and in theory we would be able to locally resolve scales corresponding to a Reynolds number of about 10^5, using an ordinary PC or laptop computer.

2 Turbulent flow and LES

The incompressible Navier-Stokes equations for a spatial domain $\Omega \subset \mathbb{R}^3$ take the form:

$$\dot{u} + (u \cdot \nabla)u - \nu \Delta u + \nabla p = f, \quad \nabla \cdot u = 0, \quad \text{in } \Omega \times I, \tag{1}$$

where $u(x,t) = (u_i(x,t))$ is the *velocity* vector and $p(x,t)$ the *pressure* of the fluid at (x,t), f is a given driving force, ν is the *kinematic viscosity*, and $I = (0,T)$ is a time interval. We assume that (1) is normalized so that the reference velocity and typical length scale are both equal to one. The Reynolds number Re is then equal to ν^{-1}.

For low Re we may have time independent solutions that satisfies the *stationary Navier-Stokes equations*, where we simply drop the time derivative in (1). For higher Re we have time dependent solutions, and for sufficiently high Re we get turbulent solutions.

In a turbulent flow we are typically not able to resolve all scales of motion computationally. We may instead aim at computing a *running average* u^h of u on a scale h, defined by

$$u^h(x,t) = \frac{1}{h^3} \int_{Q_h} u(x+y,t)\, dy, \tag{2}$$

where $h = h(x,t)$ is a parameter related to the local resolution of the problem and $Q_h = \{y \in \mathbb{R}^3 : |y_i| \leq h/2\}$. In the LES literature it is common to define the averaging operator through convolution by a certain filter function, and there is a multitude of filter functions being used. Though we only consider the case of the filter corresponding to (2) in this paper, the teqniques for a posteriori error estimation are general and apply to other filters, possibly with modifications for commutation errors associated with such filters.

By an extension of (u,p,f) to $\mathbb{R}^3$ by reflection for all $x \notin \overline{\Omega}$, the averaging operator (2) commutes with space and time differentiation. If we take the running average of the equations (1), corresponding to a LES, we obtain the following equations for u^h:

$$\dot{u}^h + (u^h \cdot \nabla)u^h - \nu \Delta u^h + \nabla p^h + \nabla \cdot \tau^h(u) = f^h, \quad \nabla \cdot u^h = 0, \quad \text{in } \Omega \times I, \tag{3}$$

where $\tau_{ij}^h(u) = (u_i u_j)^h - u_i^h u_j^h$ is the *Reynolds stress tensor*. The closure problem of LES is how to model $\tau^h(u)$ in terms of u^h in a subgrid model $\hat{\tau}^h(u^h)$. In this paper we focus on the computation of chosen output functionals for the problem (3) using adaptive finite element methods, and we refer to [8, 24] and the references therin for work on subgrid modeling for LES, and we refer to [14] for details on the mathematical formulation of the problems (1) and (3).

3 Adaptive finite element methods

An adaptive algorithm includes feed-back from computation to achieve the computational goal with minimal computational cost. In an adaptive finite element method this feed-back from computation relies on a posteriori error estimates.

In [15, 14] we compute approximations $\hat{g}(U,P)$ of functionals $g(u,p)$, where (U,P) is a numerical approximation of (u,p), and we prove a posteriori error estimates of the form

$$|g(u,p) - \hat{g}(U,P)| \leq \sum_{K \in \mathcal{T}_k} \mathcal{E}_K^k, \tag{4}$$

where

$$\mathcal{E}_K^k = \sum_i \left| \int_K R_i^k \cdot \omega_i^k \right| \tag{5}$$

is an *error indicator* for element K in the mesh $\mathcal{T}_k$, with R_i^k residuals, and ω_i^k dual weights from the solution of an associated linearized dual problem, at iteration k. An adaptive algorithm for computing approximations $\hat{g}(U,P)$, to a tolerance TOL, then takes the form:

Algorithm 1 (Adaptive mesh refinement) *Start at $k = 0$, then do*

(1) *compute approximation to the primal problem on $\mathcal{T}_k$*
(2) *compute approximation to the dual problem on $\mathcal{T}_k$*
(3) *if $\sum_{K \in \mathcal{T}_k} \mathcal{E}_K^k < TOL$ then STOP, since $|g(u,p) - \hat{g}(U,P)| \leq TOL$, else*
(4) *refine a fixed fraction of the elements in $\mathcal{T}_k$ with largest $\mathcal{E}_K^k \rightarrow \mathcal{T}_{k+1}$*
(5) *set $k = k + 1$, then goto (1)*

4 Numerical examples

We now present two different applications of Algorithm 1 to incompressible flow in 3d; first a stationary flow from [15], and then a turbulent flow from [14]. For details of the computations and the a posteriori error estimates we refer to [15, 14].

4.1 Stationary benchmark problems in 3d

In [25], computational results for a collection of benchmark problems for laminar flow around a cylinder in 2d and 3d are presented, with contributions from 17 research groups. We consider the case of 3d stationary flow around a cylinder with square cross-section $D \times D$, with $D = 0.1$, centered at $(0.5, 0.2, 0.205)$ aligned in the x_3-direction, in a channel of dimensions $2.5 \times H \times H$, with $H = 0.41$. We have no slip boundary conditions on the cylinder and the channel walls. At the outflow boundary we use a transparant outflow condition, see [23], and the inflow condition is given by $u(0, x_2, x_3) = (16U_m x_2(H - x_2)x_3(H - x_3)/H^4, 0, 0)$. The kinematic viscosity is $\nu = 10^{-3}$ and $U_m = 0.45$, which gives a Reynolds number $Re = \bar{U}D/\nu = 20$, with $\bar{U} = 4U(0, H/2, H/2)/9$.

We consider the computation of the drag coefficient, and the computation of a pressure difference upstream and downstream of the cylinder, using a cG(1) method (continuous piecewise linear trial and test functions) on tetrahedral meshes for both the primal and the dual problems.

To evaluate the performance of the duality based error indicator (5) as a refinement criterion in Algorithm 1, we compare with a commonly used alternative error indicator

$$\mathcal{E}_K^k = \sum_i \|R_i^k\|_K, \tag{6}$$

where $\|\cdot\|_K$ is a norm on the element K, based only on the size of the residuals, coupling to energy estimates (see e.g. [1]).

Computation of the drag coefficient The computational goal is to approximate the drag coefficient c_D, defined in [25] by

$$c_D \equiv \frac{2F_D(u,p)}{\bar{U}^2 DH}, \tag{7}$$

where $F_D(u, p)$ is the drag force on the cylinder. Based the results in [25] we choose $c_D = 7.6$ as our reference value.

In Figure 1 we compare the convergence rates of the two error indicators (5) and (6) with respect to the reference value $c_D = 7.6$. It is obvious that the refinement criterion (5), based on both the residual and the solution to the dual problem, does a better job than the refinement criterion (6), solely based

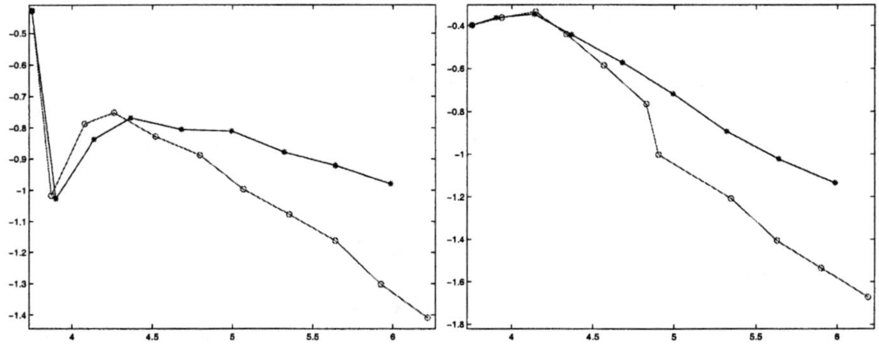

Fig. 1. Convergence rates for the computation of the drag coefficient c_D (left), and the pressure difference Δp (right), for duality based refinement ('o') and residual based refinement ('*'), as a log-log plot of number of unknowns versus relative errors.

on the residuals without any information from the dual problem relating the residual to the error in the drag coefficient c_D.

We then evaluate the a posteriori error estimates as a stopping criterion for the adaptive algorithm by introducing the notion of an effectivity index I_{eff}, defined by

$$I_{eff} = estimated\ error/true\ error, \tag{8}$$

and in Table 1 we present I_{eff} as a function of the number of unknowns. The a posteriori error estimates in this case are quite sharp. After a few initial refinements the error estimates is off by less than a factor 2, and may thus be useful as a stopping criterion.

Table 1. Effectivity indices $I_{eff} = estimated\ error/true\ error$ for computing the drag coefficient c_D (left), and the pressure difference Δp (right), as functions the number of unknowns.

# dof	$c_D : I_{eff}$
5.656	3.36
7.456	13.54
11.996	4.32
18.336	2.53
33.120	2.26
62.252	1.41
116.616	1.27
225.588	0.92
436.444	0.76
844.956	0.66

# dof	$\Delta p : I_{eff}$
5.656	0.85
8.620	0.99
14.044	0.62
21.636	0.71
36.872	0.88
67.412	0.92
80.392	1.11
222.756	1.14
426.612	1.20
797.940	1.08

Computation of a pressure difference Next we consider the problem of computing the pressure difference in two points upstream and downstream of the cylinder respectively, defined by $\Delta p = p(x^d) - p(x^u)$, with $x^u = (0.45, 0.20, 0.205)$ and $x^d = (0.55, 0.20, 0.205)$. Based on the results in [25] we use $\Delta p = 0.176$ as a reference value.

In Figure 1 we compare the error indicators (5) and (6), and we find that the duality based approach again is the better. In Table 1 we present effectivity indices for the a posteriori error estimates, which we find to be quite sharp with I_{eff} close to unity.

4.2 Turbulent flow around a surface mounted cube

We now use Algorithm 1 to compute the temporal mean of the drag coefficient $\bar{c}_D$ over a time interval $I = [T_0, T]$, with $T_0 = 10$ and $T = 20$, defined by

$$\bar{c}_D = \frac{1}{|T - T_0|} \int_{T_0}^{T} c_D(t)\, dt, \tag{9}$$

where $c_D(t)$ is the drag coefficient at time t for a surface mounted cube in a turbulent channel flow, using a cG(1)cG(1) method (continuous piecewise linears in space-time) for both the primal and the dual problem, on tetrahedral meshes $\mathcal{T}_k$ that we choose to be constant in time for each iteration k. In the definition of the LES for the adaptive step k we let $h = h(x)$ be defined to be the piecewise constant function that equals the diameters of the finite elements in the computational mesh $\mathcal{T}_k$.

In our computational model we use the Navier-Stokes equations to model the incompressible fluid around a cubic body of dimension $H \times H \times H$ that sits on the floor of a rectangular channel of length $15H$, height $2H$, and width $7H$, centered at $(3.5H, 0.5H, 3.5H)$. At the inlet we use a velocity profile interpolated from experiments, we use no slip boundary conditions on the body and the vertical boundaries, slip boundary conditions on the lateral boundaries, and a transparent outflow boundary condition. The viscosity ν is chosen to give a Reynolds number $Re = U_b H/\nu = 40.000$, where we have used $U_b = 1.0$.

We use no subgrid model in the computations, but we use the following scale similarity subgrid model

$$\tau_{ij}^h(u) \approx C_L \tau_{ij}^{2h}(u^h), \tag{10}$$

from [22] to estimate the *modeling residual*, see [14], measuring the small scale influence on the resolved scales.

In Figure 2 we plot the mean drag coefficient as a function of number of degrees of freedom. We find that even though we do not reach full convergence using the avaliable number of degrees of freedom, the value for the mean drag coefficient seems to asymptotically approach a value between 1.45-1.5. We know of no experimental reference values of $\bar{c}_D$, but in [21] $\bar{c}_D$ is approximated

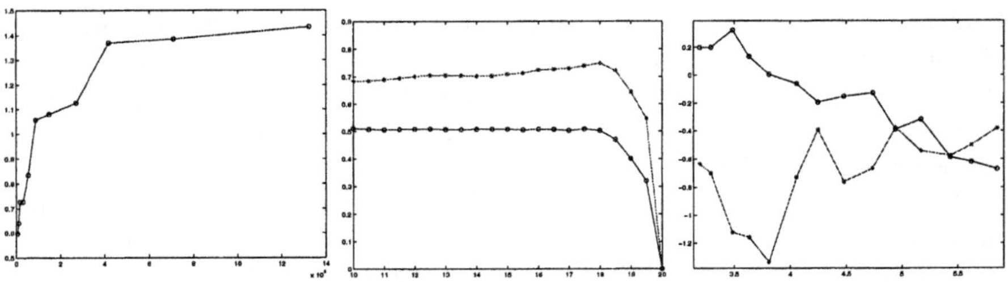

Fig. 2. Mean drag coefficient $\bar{c}_D$ over the time interval [10, 20] as a function of the number of degrees of freedom (left), discretization error e_D ('o') and modeling error e_M ('*') after 13 adaptive mesh refinements as functions of the length of the time interval $[T_0, T]$, with T fix and T_0 varying, assuming $U_h(T_0) = u^h(T_0)$ (middle), and a posteriori error estimates of the discretization error e_D ('o') and the modeling error e_M ('*') for the time interval [10, 20], as functions of the number of degrees of freedom in a log_{10}-log_{10} plot (right).

computationally. The computational setup is similar to the one in [14] except the numerical method, the length of the time interval, and that we in [14] use a channel of length $15H$, compared to a channel of length $10H$ in [21]. Using different meshes and subgrid models, approximations of $\bar{c}_D$ in the interval [1.14, 1.24] are presented in [21].

The diameter of the smallest element in the mesh $\mathcal{T}_{14}$ is about 10^{-3} (with $H = 0.1$), which corresponds to a local Reynolds number $Re_{loc} \approx (2H/h)^{4/3} \approx 1200$ (with channel height $2H$), using standard Kolmogorov arguments of turbulent flow [7], or $Re_{loc} \approx h^{-1} = 1000$, assuming the numerical viscosity of the cG(1)cG(1) method is acting as a term $h(\nabla U_h, \nabla U_h)$. That is, we are locally able to resolve scales corresponding to a Reynolds number of about 1000, even though it would be impossible globally to a similar computational cost. Since turbulence often is a local phenomena, adaptive methods are ideal for computation of turbulence. In theory, if we refine the same elements in each step of the algorithm we would get a finest $h \approx H \times (1/2)^{13} \approx 10^{-5}$, corresponding to $Re_{loc} \approx 10^5$. *That is, we would be able to locally resolve flows corresponding to a Reynolds number of* 10^5 *in a Direct Numerical Simulation using an ordinary PC or laptop computer.*

After 13 adaptive mesh refinements we plot the a posteriori error estimates of the discretization and the modeling errors in Figure 2 as functions of the length of the time interval $[T_0, T]$ (T fix, T_0 varying), where we have assumed that the initial solution is exact for each T_0, so that $U_h(T_0) = u^h(T_0)$. We find that the error at first increases with the length of the time interval, but when the interval exceeds a certain length the error does not increase significally beyond a certain level, and thus the computational cost of computing $\bar{c}_D$ is relatively constant for time intervals longer than a certain length.

In Figure 2 we also plot the discretization and the modeling errors for $\bar{c}_D$ over the time interval [10, 20] as functions of the number of degrees of freedom, where we note an expected decrease in the estimates of the discretization error as we refine the mesh. We also note that the estimates of the modeling error on the other hand increases. This might at first seem alarming, but is in fact to be expected since in this case we have used the simple model (10) to estimate the Reynolds stresses in the modeling residual. Even though the true Reynolds stresses are smaller for a finer resolution h of the problem, the model (10) will in fact first increase as we resolve more scales of motion since it is solely based on the resolved velocity fluctuations on the scale $2h$. This is of course a problem, and in a continuation of this study we seek sharper estimates of the Reynolds stresses based on scale extrapolation.

Remark 1. The use of a stabilized Galerkin finite element method in the computations may be viewed as a type of subgrid model in itself, since we then in fact solve a modified set of equations using a standard Galerkin method. We will further investigate this relation between numerical stabilization and subgrid modeling in a continuation of this work. In this paper we only consider the stabilization to be part of the numerical method and not an explicit subgrid model.

5 Summary

In this paper we have presented results from [15, 14], extending earlier results in [17, 13], where we use adaptive finite element methods based on a posteriori error estimates to compute approximations of output functionals in incompressible fluids, for both laminar and turbulent flow. The a posteriori error estimates are based on the solution of an associated linearized dual problem, and are used as error indicators for the adaptive mesh refinement algorithm.

In the problem of computing the mean drag coefficient in a turbulent channel flow, we emphasize the local nature of turbulence that makes adaptive methods ideal for efficient and accurate computations. Due to the computational goal of approximating the mean drag coefficient we refine the mesh according to the corresponding a posteriori error estimate, resolving scales of motion corresponding to local Reynolds numbers of about 1000, and in theory we would be able to resolve local scales of motion corresponding to local Reynolds numbers of the order 10^5 to a similar computational cost.

In continuations of this study we will address methods for sharp estimation of the modeling residual based on scale extrapolation, as well as adaptive strategies to combine numerical stabilization with subgrid modeling for LES.

Acknowledgments

The first author would like to aknowledge the support by DOE grant DE-FG02-88ER25053.

References

1. M. Ainsworth and J. T. Oden, *A posteriori error estimation in finite element analysis*, Computat. Meth. Appl. Mech. Eng., 142 (1997), pp. 1–88.
2. I. Babuška and A. D. Miller, *The post-processing approach in the finite element method, iii: A posteriori error estimation and adaptive mesh selection*, Int. J. Numer. Meth. Eng., 20 (1984), pp. 2311–2324.
3. R. Becker and R. Rannacher, *A feed-back approach to error control in adaptive finite element methods: Basic analysis and examples*, East-West J. Numer. Math., 4 (1996), pp. 237–264.
4. ——, *A posteriori error estimation in finite element methods*, Acta Numerica, 10 (2001), pp. 1–103.
5. M. Braack and A. Ern, *A posteriori control of modeling errors and discretization errors*, SIAM J. Multiscale Modeling and Simulation, (2003).
6. K. Eriksson, D. Estep, P. Hansbo, and C. Johnson, *Introduction to adaptive method for differential equations*, Acta Numerica, 4 (1995), pp. 105–158.
7. C. Foias, O. Manley, R. Rosa, and R. Temam, *Navier-Stokes Equations and Turbulence*, Cambridge University Press, 2001.
8. T. Gatski, M. Y. Hussaini, and J. Lumley, *Simulation and Modeling of Turbulent Flow*, Oxford University Press, 1996.
9. M. Giles, M. Larson, M. Levenstam, and E. Süli, *Adaptive error control for finite element approximations of the lift and drag coefficients in viscous flow*, Technical Report NA-76/06, Oxford University Computing Laboratory, (1997).
10. J. Hoffman, *On dynamic computational subgrid modeling*, Numerical Analysis Group Preprint, Oxford University, (1999).
11. ——, *Dynamic Subgrid Modeling for Scalar Convection-Diffusion-Reaction Equations with Fractal Coefficients*, Multiscale and Multiresolution Methods: Theory and Application (Ed. T. J. Barth, T. Chan and R. Haimes), Lecture Notes in Computational Science and Engineering, Springer-Verlag Publishing, Heidelberg, 2001.
12. ——, *Dynamic subgrid modeling for time dependent convection-diffusion-reaction equations with fractal solutions*, International Journal for Numerical Methods in Fluids, Vol 40, (2001), pp. 583–592.
13. ——, *Adaptive finite element methods for les: Duality based a posteriori error estimation in various norms and linear functionals*, submitted to SIAM Journal of Scientific Computing, (2002).
14. ——, *Adaptive finite element methods for les: Computation of the drag coefficient in a turbulent flow around a surface mounted cube*, submitted to SIAM Journal of Scientific Computing, (2003).
15. ——, *Computation of functionals in 3d incompressible flow for stationary benchmark problems using adaptive finite element methods*, submitted to Mathematical Models and Methods in Applied Sciences (M3AS), (2003).
16. ——, *Subgrid modeling for convection-diffusion-reaction in 2 space dimensions using a haar multiresolution analysis*, to appear in Mathematical Models and Methods in Applied Sciences, (2003).
17. J. Hoffman and C. Johnson, *Adaptive finite element methods for incompressible fluid flow*, Error Estimation and Solution Adaptive Discretization in Computational Fluid Dynamics (Ed. T. J. Barth and H. Deconinck), Lecture Notes in Computational Science and Engineering, Springer-Verlag Publishing, Heidelberg, 2002.

18. J. Hoffman, C. Johnson, and S. Bertoluzza, *Subgrid modeling for convection-diffusion-reaction in 1 space dimension using a haar multiresolution analysis*, to appear in Computer Methods in Applied Mechanics and Engineering, (2003).
19. C. Johnson and R. Rannacher, *On error control in CFD*, Int. Workshop Numerical Methods for the Navier-Stokes Equations (F.K. Hebeker et.al. eds.), Vol. 47 of *Notes Numer. Fluid Mech.*, Vierweg, Braunschweig, 1994, pp. 133–144.
20. C. Johnson, R. Rannacher, and M. Boman, *Numerics and hydrodynamic stability: Toward error control in cfd*, SIAM J. Numer. Anal., 32 (1995), pp. 1058–1079.
21. S. Krajnović and L. Davidson, *Large-eddy simulation of the flow around a bluff body*, AIAA Journal, 40 (2002), pp. 927–936.
22. S. Liu, C. Meneveau, and J. Katz, *On the properties of similarity subgrid-scale models as deduced from measurements in turbulent jet*, J. Fluid Mech., 275 (1994), pp. 83–119.
23. R. Rannacher, *Finite element methods for the incompressible navier stokes equations*, Preprint Intsitute of Applied Mathematics, Univ. of Heidelberg, (1999).
24. P. Sagaut, *Large Eddy Simulation for Incompressible Flows*, Springer-Verlag, Berlin, Heidelberg, New York, 2001.
25. M. Schäfer and S. Turek, *Benchmark computations of laminar flow around a cylinder*, Flow Simulation with High-Performance Computers II: Notes on Numerical Fluid Mechanics, 52 (1996), pp. 547–566.

Numerical Solution of a Nonlinear Evolution Equation Describing Amorphous Surface Growth of Thin Films

Ronald H.W. Hoppe[1,2] and Eva Nash[3]

[1] Department of Mathematics, University of Houston, Houston TX 772004-3008, U.S.A.
[2] Institute for Mathematics, University of Augsburg, D-86159 Augsburg, Germany
[3] Infineon Technologies AG, D-81541 Munich, Germany

Summary. We consider a nonlinear parabolic partial differential equation that describes the evolution of the surface morphology in the deposition of thin glassy films by molecular beam epitaxy. The dynamics of the growth process exhibits some unexpected initial linear behavior, before the nonlinear dynamics sets in. Therefore, for the numerical solution we suggest a combined spectral element/finite element approach. Results of numerical simulations are given that show a good agreement with experimental measurements.

1 Introduction

We consider the deposition of thin glassy films on the surface of substrates such as silicon by molecular beam epitaxy. Such processes play an important role in materials science with regard to the coating of surfaces in order to obtain specific surface properties (cf., e.g., [11]).
In particular, we assume that the particle beam is impinging perpendicularly to the surface of the substrate (cf. Fig. 1).
Denoting by $\Omega := [0, L]^2 \subset \mathbb{R}^2$ the surface of the substrate, the deposition process can be described by the temporal and spatial distribution of the height profile $u(x,t)$, $x \in \Omega$, $t \geq 0$, as given by

$$u(x,t) \;=\; H(x,t) \;-\; F\,t \;, \tag{1}$$

where $H(x,t)$ is the absolute height and F refers to the deposition rate which is assumed to be constant.
As far as the development of an appropriate mathematical model is concerned, the deposition evolves according to

$$\frac{\partial u}{\partial t}(x,t) \;=\; g(u(x,t)) \quad , \quad x \in \Omega \;,\; t \geq 0 \tag{2}$$

where the right-hand side g describes the surface growth. There have been many attempts to establish appropriate models for the morphology of deposition processes featuring amorphous surface growth (cf., e.g., [1, 4, 13]). Here,

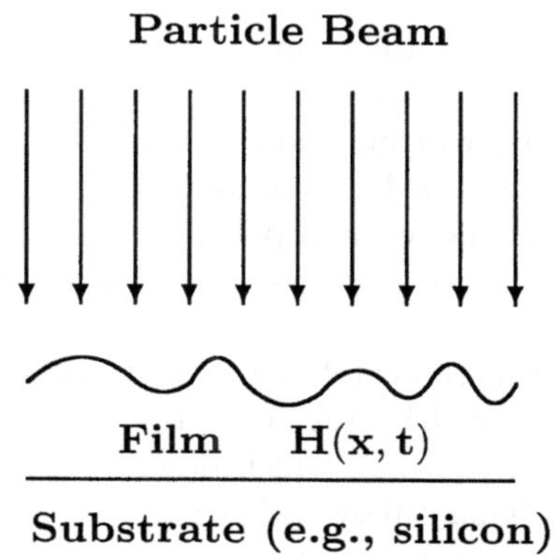

Fig. 1. Schematic representation of the deposition of thin films

following [5, 10] and [11], we take three major growth mechanisms into account:
The first one describes surface growth due to particle interaction

$$g_1(u) \;:=\; -\, S\,(1\,+\,|\nabla u|^{3/2})^{-1}\,\Delta u\;, \tag{3}$$

where S stands for the influence of interatomic and van der Waals forces.
The second one is curvature induced surface relaxation

$$g_2(u) \;:=\; -\,D\,\nabla\,\left[\,(1\,+\,|\nabla u|^2)^{-1/2}\,\nabla\,((1\,+\,|\nabla u|^{3/2})^{-1}\,\Delta u)\,\right]. \tag{4}$$

Here, D refers to the material dependent diffusion coefficient.
Finally, the third mechanism is due to structure coarsening in the sense that particles at locations with high gradients of the height profile move to locations with lower gradients. Here, we follow the model suggested by Moske (see [11])

$$g_3(u) \;:=\; C\,\left[\,(1\,+\,|\nabla u|^2)^{-1/2}\,\nabla\,(1\,+\,|\nabla u|^2)^{-1/2}\,\right], \tag{5}$$

where C denotes the mean surface mobility.
As long as the particle beam impinges perpendicularly onto the surface of the substrate, we have $|\nabla u| \ll 1$. In this case, the functions $g_i = g_i(u), 1 \le i \le 3$, in (3),(4), and (5) simplify to

$$g_1(u) \;:=\; a_1\,\Delta^2 u\;,\; g_2(u)\;:=\;a_2\,\Delta u\;,\; g_3(u)\;:=\;a_3\,\Delta(|\nabla u|^2)\quad,$$

where $a_i < 0\;,\;1 \le i \le 3$. The evolution equation takes the form

$$\frac{\partial u}{\partial t} \;=\; \Delta(a_1 u\;+\;a_2\Delta u\;+\;a_3|\nabla u|^2)\quad\text{in } Q := \Omega \times [0,\infty) \tag{6}$$

with an initial condition $u(x,0) = u_0(x)\;,\; x \in \Omega$, and either periodic boundary conditions or homogeneous Neumann boundary conditions on $\Gamma = \partial\Omega$.

The constants a_1, a_2 and a_3 in the evolution equation are usually determined by parameter identification with respect to experimentally obtained measurements by using Auger spectroscopy and scanning electron microscopy .
However, if the particle beam does not impinge perpendicularly, there are overhangs in the profile and even topological changes due the formation of inclusions. In this case, we must use the original form of the functions $g_i = g_i(u)$, $1 \leq i \leq 3$, as given by (3), (4), (5), and resort to other techniques as, for instance, level set methods (cf., e.g., [8, 9]; see also [3]).
We note that the nonlinear 4th order evolution equation (6) resembles the well-known Cahn-Hilliard equation which describes spinodal decomposition, i.e., phase separation in binary alloys (cf., e.g., [6]).
The paper is organized as follows: In section 2, we will briefly address the dynamics of the growth process which features some unexpected initial linear behavior. This motivates the use of a combined spectral element/finite element approach for the numerical solution of the nonlinear evolution equation (6) that is described in sections 3 and 4. Finally, in section 5 we will give some simulation results in terms of visualizations of the height profile for different film thicknesses.

2 Dynamics of the growth process

The solution of the nonlinear evolution equation exhibits some unexpected initial linear behavior. This can be explained by an appropriate decomposition of the spectrum $\sigma \subset \mathbb{R}$ of the associated linearized operator which is self-adjoint and sectorial. In particular, we specify three constants

$$\gamma^- < 0 < \gamma^+ < \gamma^{++} < 1$$

such that, referring to λ_{max} as the maximum eigenvalue, the spectrum is decomposed into the four parts

$$\begin{array}{rclcrcl} \sigma^{--} & := & (-\infty, \gamma^- \lambda_{max}) & , & \sigma^- & := & (\gamma^- \lambda_{max}, \gamma^+ \lambda_{max}) \quad , \\ \sigma^+ & := & (\gamma^+ \lambda_{max}, \gamma^{++} \lambda_{max}) & , & \sigma^{++} & := & (\gamma^{++} \lambda_{max}, +\infty) \quad . \end{array}$$

We further denote by X^{--} , X^- , X^+ and X^{++} the subspaces spanned by the corresponding eigenfunctions:

$$\begin{array}{rclcrcl} X^{--} & := & \text{span } \{\varphi(\lambda) \mid \lambda \in \sigma^{--}\} & , & X^- & := & \text{span } \{\varphi(\lambda) \mid \lambda \in \sigma^-\} \, , \\ X^+ & := & \text{span } \{\varphi(\lambda) \mid \lambda \in \sigma^+\} & , & X^{++} & := & \text{span } \{\varphi(\lambda) \mid \lambda \in \sigma^{++}\} \, . \end{array}$$

Then, the direct sum of X^+ and X^{++} can be shown to be a dominant subspace which determines the dynamical behavior of solutions to the nonlinear evolution equation in the following sense:

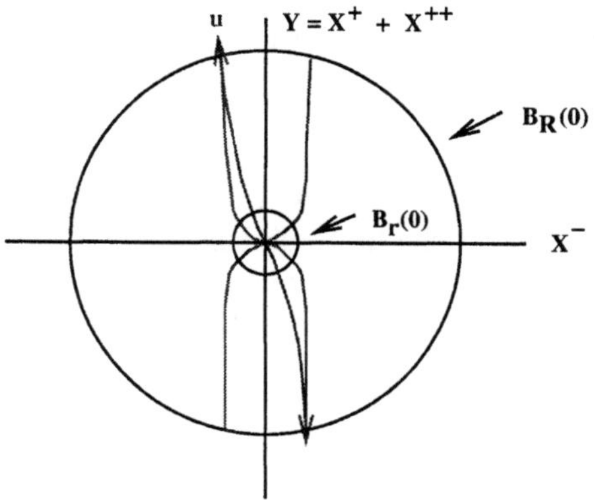

Fig. 2. Illustration of the initial linear behavior of the solution

Theorem 1. *Assume* $u(\cdot,0) \in H^3(\Omega)$ *with*

$$\|u(\cdot,0)\|_{2,\Omega} \ \leq \ r \ := \ C\ L^{3-\alpha}\ ,\ \alpha > 0 \quad ,$$

and set

$$\bar{u}^0 \ := \ L^{-2} \int\limits_{\Omega} u_0(x)\ dx \quad .$$

Then there exists $t^* > 0$ *such that the solution* $u = u(x,t), x \in \Omega, t \in [0,t^*)$, *of* (6) *stays with probability 1 in a vicinity of the dominant subspace*

$$Y \ := \ \bar{u}^0 \ + \ X^+ \oplus X^{++} \tag{7}$$

until at $t = t^*$ *it leaves a ball* $B_R(0)$ *with radius* $R > r$.

Proof. For a proof of this result we refer to [2].

Figure 2 illustrates the initial linear behavior of the solution in case $\bar{u}_0 = 0$. We note that a related result for the Cahn-Hilliard equation has been established in [12].

3 Spectral Galerkin approximation

For the spectral Galerkin approximation we consider the weak formulation of the implicitly in time discretized nonlinear evolution equation which involves the Sobolev space $V := H^2_{per}(\Omega)$ in case of periodic boundary conditions and $V := H^2(\Omega)$, if homogeneous Neumann boundary conditions are imposed. Using the backward Euler scheme and denoting by $u^m \in V$ an approximation of $u(\cdot,t_m)$ at time t_m and by $\tau_m := t_m - t_{m-1}$ the time step from level $m-1$ to level m, the problem is as follows: Find $u^m \in V$ such that for all $\chi \in V$

$$\int_\Omega u^m \chi dx = \int_\Omega u^{m-1} \chi dx + \tau_m \int_\Omega [a_1 u^m + a_2 \Delta u^m + f(u^m)] \Delta \chi dx \ , \qquad (8)$$

where the nonlinearity $f(u^m)$ is given by $f(u^m) := a_3 |\nabla u^m|^2$.
We refer to λ_i , $1 \le i \le N$, as the first N eigenvalues of $-\Delta$ and denote by $V_N :=$ span $\{\varphi_i \mid 1 \le i \le n\}$ the finite dimensional subspace of V spanned by the associated orthonormal eigenfunctions.
The spectral Galerkin approximation is then a linear combination $u_N^m = \sum_{k=1}^{N} u_{N,i}^m \varphi_i \in V_N$ so that (8) with V replaced by V_N gives rise to the nonlinear system

$$g_i(u_N^m) = (1 + a_1 \tau_m \lambda_i - a_2 \tau_m \lambda_i^2) u_{N,i}^m + \qquad (9)$$

$$+ \tau_m \lambda_i \int_\Omega f(\sum_{k=1}^{N} u_{N,k}^m \varphi_k) \varphi_i \, dx - u_{N,i}^{m-1} = 0 \quad , \quad 1 \le i \le N \quad .$$

The solution of that nonlinear system by Newton's method would be quite expensive, since it requires the computation of the Jacobian at each iteration step. It turns out that it is sufficient to use the method of successive iterations which corresponds to the approximation of the original problem by the semi-implicit Euler Scheme: For $\nu \ge 0$ compute $u_N^{m,\nu} \in V_N$ as the solution of

$$\int_\Omega u_N^{m,\nu} \chi_N \, dx = \int_\Omega u_N^{m-1} \chi_N \, dx + \tau_m \int_\Omega [a_1 u_N^{m,\nu} + \qquad (10)$$

$$+ a_2 \Delta u_N^{m,\nu} + f(u_N^{m,\nu-1})] \Delta \chi_N \, dx \quad , \quad \chi_N \in V_N \quad ,$$

where $u_N^{m,-1} := u_N^{m-1}$. In this case, we can explicitly solve for the components of the new iterate

$$u_{N,i}^{m,\nu} = \frac{u_{N,i}^{m-1} - \tau_m \lambda_i f_i^{m,\nu-1}}{1 + a_1 \tau_m \lambda_i - a_2 \tau_m \lambda_i^2} \quad , \quad 1 \le i \le N \quad , \qquad (11)$$

where

$$f_i^{m,\nu-1} := \int_\Omega f(\sum_{k=1}^{N} u_{N,k}^{m,\nu-1} \varphi_k) \, \varphi_i \, dx \quad , \quad 1 \le i \le N \quad .$$

We only have to evaluate the nonlinear terms $f_i^{m,\nu-1}$ which can be efficiently done by the Fast Fourier Transform in case of periodic boundary conditions and by the Fast Cosine Transform for homogeneous Neumann boundary conditions. In both cases, this is done with respect to an equidistant grid consisting of M^2 grid points where M has to be chosen larger than the dimension of the trial space V_N in order to avoid aliasing effects.

The semi-implicit Euler scheme is carried out with an automatic step-size control. The error due to the time discretization is estimated by

$$\gamma \; \|\tilde{u}^m \; - \; u^m\| \; \le \; \|u(t_m) - u^m\| \; \le \; \Gamma \; \|\tilde{u}^m \; - \; u^m\| \quad ,$$

where $\tilde{u}^m \in V$ is the solution of the semi-implicit trapezoidal rule

$$\int\limits_{\Omega} \tilde{u}^m \chi dx = \int\limits_{\Omega} u^{m-1} \chi dx \; + \; \frac{\tau_m}{2} \int\limits_{\Omega} \left[a_1(\tilde{u}^m + u^{m-1}) \; + \right. \tag{12}$$
$$\left. + \; a_2 \Delta(\tilde{u}^m + u^{m-1}) \; + \; (f(u^m) + f(u^{m-1}))) \right] \Delta \chi dx \; , \; \chi \in V \; .$$

If $V = V_N$, the solution of (12) can be easily computed according to

$$(\tilde{u}_N^m)_i = (u_N^{m-1})_i - \frac{\tau_m}{2} \; [a_1 \; \lambda_i \; ((\tilde{u}_N^m)_i + (u_N^{m-1})_i) \; -$$
$$- \; a_2 \; \lambda_i^2 \; ((\tilde{u}_N^m)_i + (u_N^{m-1})_i) - \lambda_i \; (f_i^m + f_i^{m-1})] \; , \; 1 \le i \le N \; .$$

In the practical realization of the step-size control, we additionally take into account the error due to the discretization in space. The error $\|u^m - u_N^m\|_{2,\Omega}$ is caused by the negligence of those eigenmodes $i > N$ that have not been considered in the spectral Galerkin approximation, but are relevant for the exact computation of the nonlinearities. Therefore, for sufficiently large P, we set

$$e_{u_N^m} \; = \; \sum_{N<j<N+P} \frac{\tau_m \; \lambda_j \; f_j^m}{1 \; + \; a_1 \; \tau_m \; \lambda_j \; - \; a_2 \; \tau_m \; \lambda_j^2}$$

Given a tolerance $tol > 0$, we check for convergence:

$$\frac{\|\tilde{u}_N^m \; - \; u_N^m\|_{2,\Omega} \; + \; (\sigma^{-1} \; - \; 1) \; |e_{u_N^m}|}{\|\tilde{u}_N^m\|_{2,\Omega}} \; \le \; tol \quad , \tag{13}$$

where $\sigma < 1$ is an appropriate weighting factor. If (13) is satisfied we proceed with the new time-step

$$\bar{\tau}_m \; := \; \sqrt{\frac{\sigma \; tol \; \|\tilde{u}_N^m\|_{2,\Omega} \; - \; |e_{u_N^m}|}{\|\tilde{u}_N^m \; - \; u_N^m\|_{2,\Omega}}} \quad . \tag{14}$$

Otherwise, we repeat the previous time-step with $\bar{\tau}_m$.

4 The finite element method

The spectral Galerkin method becomes inefficient when the nonlinear dynamics sets in, i.e., when the solution leaves the dominant subspace Y as given by (7). Although a theoretical bound for the exit time t^* is known (cf., e.g., [2]),

this bound is an overestimation and hence not practicable. Therefore, we stop the spectral approach and switch to a finite element method, if the convergence test (13) fails for several consecutive time-steps.
The finite element approximation is based on a reformulation of the 4th order equation (6) as a system of two 2nd order equations

$$\left.\begin{aligned} \partial u/\partial t &= \Delta w \\ w &= a_1 u + a_2 \Delta u + a_3 |\nabla u|^2 \end{aligned}\right\} \quad \text{in } Q := \Omega \times [0,\infty) \qquad (15)$$

We discretize in time by the implicit Euler method and in space by continuous, piecewise linear finite elements with respect to a simplicial triangulation $\mathcal{T}_h$ of Ω. Denoting by $S_1(\Omega;\mathcal{T}_h)$ the associated finite element space, for each time-step we have to solve the nonlinear system of equations:
Find $(u_h^m, w_h^m) \in S_1(\Omega;\mathcal{T}_h) \times S_1(\Omega;\mathcal{T}_h)$ such that

$$\int_\Omega \frac{u_h^m - u_h^{m-1}}{\tau_m} \chi_h \, dx + \int_\Omega \nabla w_h^m \cdot \nabla \chi_h \, dx = 0 \; , \; \chi_h \in S_1(\Omega;\mathcal{T}_h) \; , \qquad (16)$$

$$a_1 \int_\Omega u_h^m \psi_h \, dx - a_2 \int_\Omega \nabla u_h^m \cdot \nabla \psi_h \, dx + a_3 \int_\Omega |\nabla u_h^m|^2 \psi_h \, dx - \qquad (17)$$
$$- \int_\Omega w_h^m \psi_h \, dx = 0 \; , \; \psi_h \in S_1(\Omega;\mathcal{T}_h) \; .$$

Providing a hierarchy $(\mathcal{T}_{h_i})_{i=0}^{\ell}$ of triangulations, we solve the nonlinear system (16), (17) on the finest grid by Newton-Multigrid:
Given $(u_\ell^{m,\nu}, w_\ell^{m,\nu}) \in S_1(\Omega;\mathcal{T}_\ell) \times S_1(\Omega;\mathcal{T}_\ell)$, we compute

$$u_\ell^{m,\nu+1} = u_\ell^{m,\nu} + \delta_{u_\ell}^{m,\nu} \quad , \quad w_\ell^{m,\nu+1} = w_\ell^{m,\nu} + \delta_{w_\ell}^{m,\nu} \quad ,$$

where the Newton increment $(\delta_{u_\ell}^{m,\nu}, \delta_{w_\ell}^{m,\nu})$ is the solution of the linear system

$$\int_\Omega \delta_{u_\ell}^{m,\nu} \chi_\ell dx + \tau_m \int_\Omega \nabla \delta_{w_\ell}^{m,\nu} \cdot \nabla \chi_\ell dx = \int_\Omega u_\ell^{m-1} \chi_\ell dx - \qquad (18)$$
$$- \int_\Omega u_\ell^{m,\nu} \chi_\ell dx - \tau_m \int_\Omega \nabla w_\ell^{m,\nu} \cdot \nabla \chi_\ell dx \quad , \quad \chi_\ell \in S_1(\Omega;\mathcal{T}_{h_\ell} \; ,$$

$$\int_\Omega \delta_{w_\ell}^{m,\nu} \psi_\ell dx - a_1 \int_\Omega \delta_{u_\ell}^{m,\nu} \psi_\ell dx + a_2 \int_\Omega \nabla \delta_{u_\ell}^{m,\nu} \cdot \nabla \psi_\ell dx - \qquad (19)$$
$$- \int_\Omega f'(u_\ell^{m,\nu}) \delta_{u_\ell}^{m,\nu} \psi_\ell dx = - \int_\Omega w_\ell^{m,\nu} \psi_\ell dx + a_1 \int_\Omega u_\ell^{m,\nu} \psi_\ell dx -$$
$$- a_2 \int_\Omega \nabla u_\ell^{m,\nu} \cdot \nabla \psi_\ell dx + \int_\Omega f(u_\ell^{m,\nu}) \psi_\ell dx \quad , \quad \psi_\ell \in S_1(\Omega;\mathcal{T}_{h_\ell} \; .$$

The system (18),(19) is solved by linear multigrid using incomplete LU decomposition both as smoother on all levels $1 \leq i \leq \ell$ as well as an iterative solver for the coarse grid correction equation on level $i = 0$.
For the finite element approach we use a similar step-size control as in case of the spectral Galerkin method, except that we replace the estimation of the error due to the discretization in space by a residual-type a posteriori error estimator.

5 Simulation results

We have used the combined spectral element/finite element approach for the numerical simulation of the deposition of the metallic glassy film $ZrAlCu$ on silicon substrates.
The computational domain Ω has been chosen as a square of length $L = 200nm$ in each direction. Periodic boundary conditions have been imposed and the initial height profile $u_0(x)$, $x \in \Omega$ has been determined randomly with $\bar{u}_0 = 0$.
In the spectral element approach we have used $125, 200$, and 250 modes per dimension, whereas for the computation of the nonlinearities by the Fast Fourier Transform a uniform grid with $M = 400$ grid points in each direction has been employed which is sufficiently large to avoid aliasing effects.
In the finite element method we used a hierarchy $(\mathcal{T}_\langle)_{k=0}^4$ of five simplicial triangulations with $h_0 = 1/25$ and $h_4 = 1/400$.

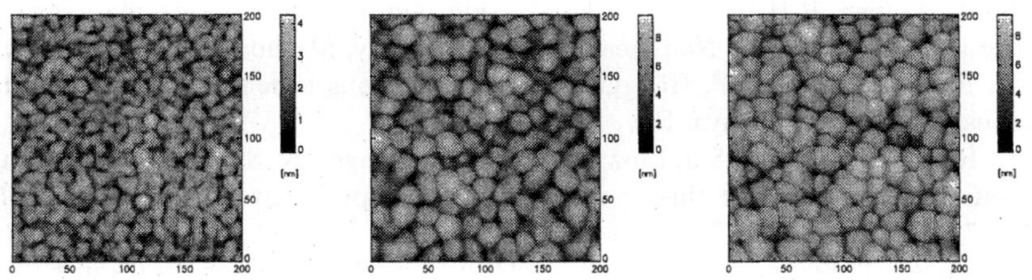

Fig. 3. Computed height profile for different film thicknisses [100 nm (left), 360 nm (middle), and 480 nm (right)]

Figure 3 displays the computed height profiles for different film thicknesses in a grey scale ranging from black (0 nm) to white (4 nm). One clearly observes the effect of structure coarsening: a surface pattern with a mesa-like structure evolves featuring hills with flat plateaus that are separated by narrow deep valleys. We note that already for 125 modes per dimension the computed profiles are both qualitatively and quantitatively in good agreement with experimentally obtained data (cf., e.g., [11]).

Acknowledgement. The work of the authors has been supported by the German National Science Foundation (DFG) within the Collaborative Research Center SFB 438 and the Graduate School GK 283.

References

1. A.L. Barabasi, and H.E. Stanley (1995): Fractal Concepts in SurfaceGrowth. Cambridge Univ. Press Cambridge
2. D. Blömker (2000): Stochastic Partial Differential Equations and Surface Growth. Wissner Augsburg
3. R.H.W. Hoppe, W.G. Litvinov, S.J. Linz (2003): On solutions of certain classes of evolution equations for surface morphologies. Journal of Nonlinear Phenomena in Complex Systems **6**, 582–591
4. M. Kardar, G. Parisi, Y.-C. Zhang (1986): Dynamic scaling of growing interfaces. Phys. Rev. Lett. **56**, 889–892
5. S.J. Linz, M. Raible, and P. Hänggi (2000): Stochastic field equation for amorphous surface growth. Lecture Notes in Physics **557**, 473–483
6. S. Maier-Paape, T. Wanner (2000): Spinodal decomposition for the Cahn-Hilliard equation in higher dimensions: Nonlinear dynamics. Arch. Rat. Mech. Anal. **151**, 187–219
7. E.M. Nash (2001): Finite-Elemente und Spektral-Galerkin Verfahren zur numerischen Lösung der Cahn-Hilliard Gleichung und verwandter nichtlinearer Evolutionsgleichungen. Shaker Aachen
8. Ch.-D. Nguyen (2003): Level set methods for nonlinear deposition equations. Dissertation. Institute for Mathematics, University of Augsburg
9. Ch.-D. Nguyen, R.H.W. Hoppe (2003): Amorphous surface growth via a level set approach. submitted to Nonlinear Analysis: Theory, Methods, and Applications
10. M. Raible, S.J. Linz, P. Hänggi (2000): Amorphous thin film growth: Minimal deposition equation. Phys. Rev. E **62**, 1691–1705
11. M. Raible, S.G. Mayr, S.J. Linz, M. Moske, P. Hänggi, K. Samwer (2000): Amorphous thin film growth: theory compared with experiment. Europhys. Lett. **50**, 61–67
12. E. Sander, T. Wanner (2000): Unexpectedly linear behavior for the Cahn-Hilliard equation. SIAM J. Appl. Math. **60**, 2182–2202
13. D.E. Wolf, J. Villain (1990): Growth with surface diffusion. Europhys. Lett. **13**, 389–394

Constrained Mountain Pass Algorithm for the Numerical Solution of Semilinear Elliptic Problems

Jiří Horák

University of Basel, Department of Mathematics, Rheinsprung 21, 4051 Basel, Switzerland
Current address: University of Cologne, Department of Mathematics, Weyertal 86-90, 50923 Cologne, Germany *jhorak@math.uni-koeln.de*

Summary. A new numerical algorithm for solving semilinear elliptic problems is presented. A variational formulation is used and critical points of a C^1-functional subject to a constraint given by a level set of another C^1-functional (or an intersection of such level sets of finitely many functionals) are sought. First, constrained local minima are looked for, then constrained mountain pass points. The approach is based on the mountain pass theorem in a constrained setting.

Weak solutions of semilinear elliptic partial differential equations can typically be represented as critical points of nonlinear functionals. The easiest approach in constructing critical points is via a minimizing sequence. It leads to the method of the steepest descent which yields *local minimizers.* Another tool used to prove existence of critical points is the mountain pass theorem of Ambrosetti and Rabinowitz [1], repeated in Sec. 1. Choi and McKenna [4] introduced a method based on a constructive form of this theorem – the mountain pass algorithm. It is able to find numerical approximations to critical points of *mountain pass* type (typically, saddle points at which the second derivative of the functional has exactly one negative eigenvalue, i.e., there is just one direction in the function space at which the functional decreases on both sides of the critical point).

The main objective of the work presented in this contribution is to design a numerical method (constrained mountain pass algorithm) that can find numerical approximations of *more complicated* saddle type critical points (typically, with more negative eigenvalues of the second derivative of the functional, i.e., more "directions of decrease"). A similar question was posed in [5]. The "high-linking algorithm" presented by the authors can, however, be only applied to a narrow family of problems. The current work presents a different approach that appears to be more universal.

We illustrate the main idea on a simple example that was used in both [4] and [5]:

$$\begin{aligned} -\Delta u &= u^3 \qquad \text{in } \Omega = (0,1) \times (0,1) \ , \\ u &= 0 \qquad \text{on } \partial\Omega \ . \end{aligned} \tag{1}$$

Weak solutions of this problem correspond to critical points of the functional

$$I(u) = \int_\Omega \left[\frac{1}{2}|\nabla u|^2 - \frac{1}{4}u^4\right] \mathrm{d}x$$

defined on $H_0^1(\Omega)$. It is not difficult to see that $u \equiv 0$ is not only a solution of (1) but also a local minimum of I. Since the steepest descent method would likely yield this trivial solution (or diverge since I is not bounded from below), a different method needs to be used to approximate nontrivial solutions. In [4] a mountain pass solution (Fig. 1(a)) was found numerically using the mountain pass algorithm. Later, the high-linking algorithm of [5] provided a more complicated saddle type solution (Fig. 1(b)).

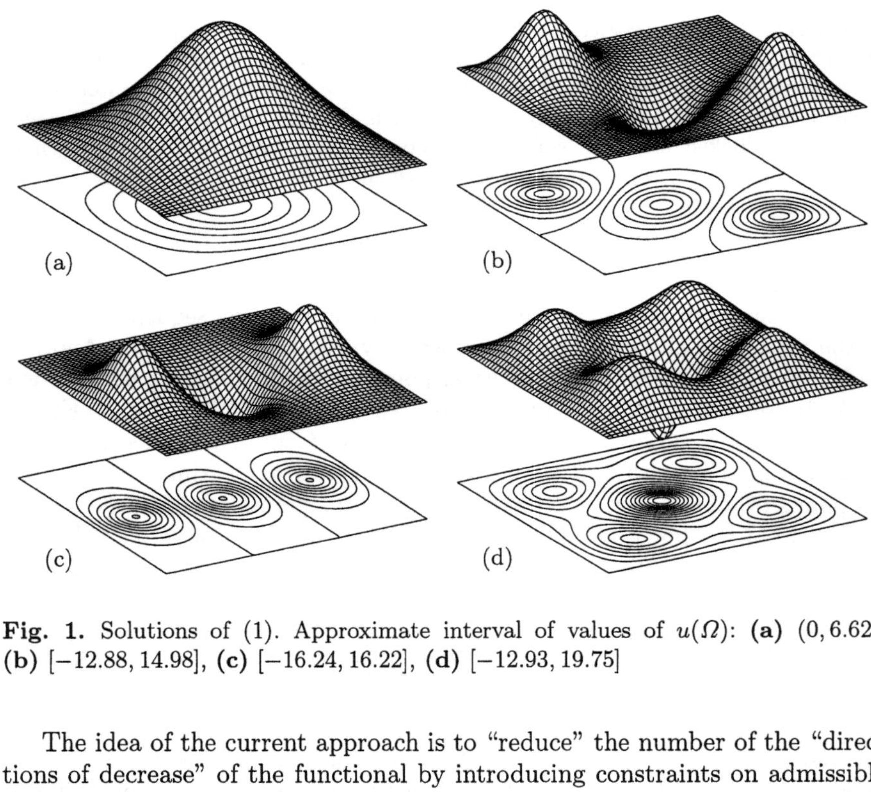

Fig. 1. Solutions of (1). Approximate interval of values of $u(\Omega)$: **(a)** $(0, 6.62]$, **(b)** $[-12.88, 14.98]$, **(c)** $[-16.24, 16.22]$, **(d)** $[-12.93, 19.75]$

The idea of the current approach is to "reduce" the number of the "directions of decrease" of the functional by introducing constraints on admissible functions. Roughly speaking, we could expect to reduce mountain pass points to constrained local minima, or more complicated saddle type points to, for example, constrained mountain pass points.

Define a constraint given by a new functional J. Let $S = \{u \in H_0^1(\Omega) \setminus \{0\} \mid J(u) := \int_\Omega \left[|\nabla u|^2 - u^4\right] \mathrm{d}x = 0\}$. By testing (1) with u we find that all nontrivial weak solution of (1) belong to S. Instead of looking for critical points of I we will look for critical points of I with respect to S, i.e., we will solve

$I'(u) - \lambda J'(u) = 0$, where $\lambda \in \mathbb{R}$ is a Lagrange multiplier. It can be shown [6] that any solution (u, λ), $u \not\equiv 0$ of this equation satisfies $\lambda = 0$, hence u also solves (1). It should be noted that it is not always possible or even desirable to find constraints, such that the Lagrange multipliers vanish. As applications in Sec. 3 show, these multipliers can be a part of the formulation of the problem.

In the constrained setting, the solution in Fig.1(a) can also be found as a local minimum of I with respect to the constraint S by the method of Sec. 2.1. Similarly, the solution of [5] in Fig. 1(b) can also be found as a mountain pass point of I with respect to S by the method of Sec. 2.2. The solutions shown in Figs. 1(c,d) are also constrained mountain pass points found by our method, they, however, were not found by the method of [5]. Hence already this simple example shows advantages of the current approach. But it is the problems to which the high-linking method cannot be applied (e.g., those in Sec. 3) that make the approach unique.

The outline of the rest of the paper: Section 1 presents a summary of known theoretical results – the mountain pass theorem and its constrained version. In Section 2 a description of the constrained steepest descent method (CSDM) and the constrained mountain pass algorithm (CMPA) is given. Finally, Section 3 shows the application of the method to two problems which cannot be handled by the high-linking algorithm of [5]: a second order problem with two constraints, a fourth order problem on an unbounded domain.

1 Theoretical Background

The mountain pass algorithm of [4] is based on the classical mountain pass theorem of [1]. The constrained mountain pass algorithm presented in this contribution is based on the constrained mountain pass theorem. We review both theorems in this section.

Let B be a real Banach space and $I \in C^1(B, \mathbb{R})$ a continuously Fréchet differentiable functional.

Definition 1. *I satisfies the Palais-Smale condition at the level $a \in \mathbb{R}$ if any sequence $\{u_n\} \subset B$ such that $I(u_n) \to a$ and $I'(u_n) \to 0$ possesses a convergent subsequence.*

Theorem 1 (mountain pass). *Let e_1, e_2 be two distinct points in B. Define*

$$c = \inf_{\gamma \in \Gamma} \max_{u \in \gamma([0,1])} I(u) \; , \tag{2}$$

where $\Gamma = \{\gamma \in C([0,1], B) \mid \gamma(0) = e_1, \gamma(1) = e_2\}$.

If $c > \max\{I(e_1), I(e_2)\}$ and I satisfies the Palais-Smale condition at the level c, then c is a critical value of I, i.e., there exists $u \in B$ such that $I'(u) = 0$ and $I(u) = c$.

Let us now introduce k constraints. Assume that $J_i \in C^1(B,\mathbb{R})$, $i \in \{1,\dots,k\}$. We are interested in finding real numbers $\lambda_1,\dots,\lambda_k$ and a function $u \in B$ that satisfy

$$I'(u) - \sum_{i=1}^{k} \lambda_i J_i'(u) = 0 \; . \tag{3}$$

Such a function u is a critical point of I with respect to the set

$$S = \{u \in B \,|\, J_1(u) = 1, \dots, J_k(u) = 1\} \; .$$

Equation (3) is a general formulation of problems that we want to solve numerically in this contribution. We assume further that if $u \in S$, then $J_i'(u) \neq 0$ for all i and $\{J_i'(u)\}_{i=1}^k$ are linearly independent.

The constrained mountain pass theorem [6] is based on the work of Bonnet [2]. In order to use his definition we denote $\|I'|_{S_u}\| = \inf_{\alpha_1,\dots,\alpha_k\in\mathbb{R}} \|I'(u) - \sum_{i=1}^k \alpha_i J_i'(u)\|$ and $tg(u) = \inf_{\alpha_1^2+\dots+\alpha_k^2=1} \|\sum_{i=1}^k \alpha_i J_i'(u)\|$.

Definition 2. *I and $J_1,\dots,J_k$ satisfy the Palais-Smale condition at the level $a \in \mathbb{R}$ if for any sequence $\{u_n\} \subset B$ such that $J_1(u_n) \to 1,\dots,J_k(u_n) \to 1$ and $I(u_n) \to a$ and either $\|I'|_{S_{u_n}}\| \to 0$ or $tg(u_n) \to 0$ there exists a convergent subsequence.*

Theorem 2 (constrained mountain pass). *Let $e_1 \neq e_2$ belong to a path-connected component of S. Define c by (2), where $\Gamma = \{\gamma \in C([0,1],S) \,|\, \gamma(0) = e_1, \gamma(1) = e_2\}$.*

If $c > \max\{I(e_1), I(e_2)\}$ and I and $J_1,\dots,J_k$ satisfy the Palais-Smale condition at the level c, then c is a critical value of I on S, i.e., there exists a solution of equation (3) with $u \in S$ and $I(u) = c$.

2 Description of the Algorithm

2.1 Constrained Steepest Descent Method

Let from now on $B = H$ be a Hilbert space. In order to apply Theorem 2, convenient points $e_1, e_2 \in S$ need to be found. Choosing e_1 and e_2 as local minima of I on S seems reasonable. They can be found using a *constrained steepest descent method.*

The method solves numerically the following initial value problem:

$$\frac{\mathrm{d}}{\mathrm{d}t}\zeta(t) = -P_{\zeta(t)} \nabla I(\zeta(t)) \; , \qquad \zeta(0) = \zeta_0 \in S \; , \tag{4}$$

where $\nabla I(u)$ is defined as the Riesz representation of the Fréchet derivative $I'(u)$ and P_u is the orthogonal projection on the tangent space of S at $u \in S$ and is given by

$$P_u v = v - \sum_{j=1}^{k} \alpha_j \nabla J_j(u) \qquad v \in H \ ,$$

where the coefficients α_j solve the linear algebraic system

$$\sum_{j=1}^{k} \langle \nabla J_i(u), \nabla J_j(u) \rangle \alpha_j = J_i'(u)v \text{ for } i \in \{1, \dots, k\}.$$

Properties of (4) are studied in [6]. If ζ is a solution, then $\zeta(t) \in S$ on its interval of existence and $I(\zeta(t))$ is a nonincreasing function of t.

Equation (4) is discretized in two steps:

Step 1 Choose $\Delta t_n > 0$ small, define $\bar{u}_{n+1} = u_n - \Delta t_n P_{u_n} \nabla I(u_n)$.

Finding the gradients of I and $J_1, \dots, J_k$ at u_n usually means solving a linear partial differential equation by a convenient numerical method (e.g., finite element method).

Step 2 – Projection Since $\bar{u}_{n+1}$ lies in the tangent space of S at u_n but not necessarily on S we need to approximate it by some $u_{n+1} \in S$. This is usually accomplished by scaling (cf. examples of Sec. 3).

The size of Δt_n in Step 1 is chosen to be smaller than a prescribed maximum. After Step 2 we check whether $I(u_{n+1}) < I(u_n)$. If not, we halve the size of Δt_n and repeat both steps. If the value of I cannot be decreased any further, we stop.

2.2 Constrained Mountain Pass Algorithm

Assume we work in a finite dimensional approximating subspace of H (conveniently chosen in 2.1). We take a discretized path in S connecting e_1 and e_2, two local minima obtained by the constrained steepest descent method. We find the maximum of I along the path. The point at which the maximum occurs is moved a small distance in the direction of the projection of the steepest descent of I to the tangent space to S, and then projected back to S. Hence the path has been deformed in S and the maximum of I lowered. The deforming of the path is repeated until the maximum along the path cannot be lowered any more – a critical point is reached.

Path Initialization A path in S connecting e_1 and e_2 is represented by a collection of P points $z_0, \dots, z_P \in S$. There is no general rule for choosing these points but in many cases the following obvious choice suffices:

$$\bar{z}_j = e_1 + \tfrac{j}{P}(e_2 - e_1) \quad j \in \{0, \dots, P\} \ ,$$

$z_j \in S$ is an approximation of $\bar{z}_j$ as in Step 2 of 2.1.

Main Loop First, find the maximum of I on the path, i.e., find j_m with $I(z_{j_m}) \geq I(z_j)\ \forall j$. Use interpolation to improve the maximum by moving z_{j_m} closer to z_{j_m+1} or z_{j_m-1}.

Second, update z_{j_m} by moving it in the direction of the steepest descent of I projected to the tangent space of S at z_{j_m} to decrease the value of $I(z_{j_m})$. In fact, this amounts to the application of the two steps of the constrained steepest descent method 2.1 (with z_{j_m} instead of u_n, u_{n+1} is then the updated z_{j_m}).

Repeat these two steps until one of the following occurs:

1. the value of $I(z_{j_m})$ cannot be decreased any further,
2. in several recent consecutive repetitions of the loop the index j_m has always been the same – infinite loop.

In the first situation $\|P_{z_{j_m}} \nabla I(z_{j_m})\| = \|I'|_{S_{z_{j_m}}}\|$ is too small, i.e., z_{j_m} is an approximation of the desired critical point, the algorithm stops. In the second situation the path needs to be refined.

Refining the Path We prescribe a number of points to be inserted between z_{j_m} and $z_{j_m \pm 1}$ (for example two). At the same time we remove the same number of points from both ends of the path so that the number P of points on the path stays the same.

It is sometimes useful and more efficient to design a Newton scheme for problem (3) and to use the solution obtained by CMPA as an initial guess for this scheme. The approximate constrained mountain pass solution does not need to be very precise. This means we can stop the CMPA early and do not need to refine the path that many times.

3 Examples

3.1 Fučík Spectrum of the Laplacian

$$\begin{aligned} -\Delta u &= \mu u^+ - \nu u^- && \text{in } \Omega \subset \mathbb{R}^2\,, \\ u &= 0 && \text{on } \partial\Omega \end{aligned} \tag{5}$$

with $\int_\Omega u^2 \mathrm{d}x = 1$, where $\Omega = \{(x_1, x_2) \in \mathbb{R}^2 \,|\, x_2 > 0, x_2 < 2 - 4x_1, x_2 < 2 + 4x_1\}$ is an isosceles triangle with base 1 and height 2, $u^\pm = \max\{\pm u, 0\}$. A point $(\mu, \nu) \in \mathbb{R}^2$ is called a Fučík eigenvalue if (5) has a weak solution $u \in H_0^1(\Omega)$. The Fučík spectrum is then the collection of all Fučík eigenvalues. It has been studied for various domains both analytically and numerically in [8]. The numerical investigation was based on a continuation method starting at some known solution. Such a solution can be numerically obtained by CSDM and CMPA.

For $t \in (0, 1)$ define a variational problem with two constraints:

$$I(u) = \int_\Omega |\nabla u|^2 \mathrm{d}x \ , \quad J_1(u) = \int_\Omega (u^+)^2 \mathrm{d}x \ , \quad J_2(u) = \int_\Omega (u^-)^2 \mathrm{d}x \ ,$$
$$S = \{u \in H_0^1 \mid J_1(u) = t, J_2(u) = 1 - t\} \ .$$

Critical points of I on S are solutions of (5) with (μ, ν) obtained as Lagrange multipliers. It is shown in [6] that I and J_1, J_2 satisfy the Palais-Smale condition at any level.

We apply the steepest descent method of Sec. 2.1. The scaling in the projection step is performed in the following way: for $\bar{u} \in H_0^1, \bar{u}^+ \neq 0, \bar{u}^- \neq 0$ find $t_+, t_- > 0$ such that $u = t_+\bar{u}^+ - t_-\bar{u}^- \in S$. Figure 2 shows two local minimizers of I on S for $t = 0.2$.

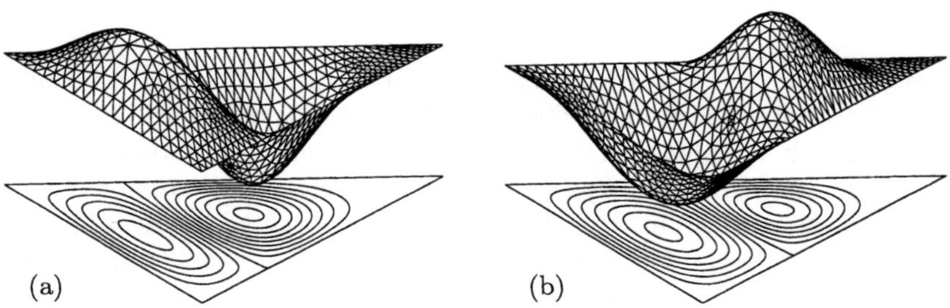

Fig. 2. Solutions of (5) found by CSDM. Approximate values of (μ, ν) and $u(\Omega)$: **(a)** $(52.6, 47.8)$, $[-2.50, 1.66]$, **(b)** $(65.4, 34.3)$, $[-2.19, 1.91]$

These local minimizers are used as endpoints of the path in CMPA. The algorithm converges to the solution in Fig. 3(b). During the run of CMPA the size of $\|I'|_{S_{z_{jm}}}\| = \|P_{z_{jm}} \nabla I(z_{jm})\|$ is checked and if it is small for a number of iterations but later grows again, we may use the point z_{jm} as an initial guess in Newton's method. If this method converges, we obtain a numerical solution that is most likely different from the one to which CMPA eventually converges. The solution in Fig. 3(a) was obtained this way.

A finite element method with piecewise linear functions on a triangular grid with 11,097 nodes and 21,760 triangles was used, the path in CMPA had $P = 50$ points.

3.2 Fourth Order Problem in $\mathbb{R}^2$

$$\Delta^2 u + c^2 \frac{\partial^2 u}{\partial x_1^2} + u + g(u) = 0 \qquad x \in \mathbb{R}^2 \ , \tag{6}$$
$$u(x) \to 0 \text{ as } |x| \to \infty \ ,$$

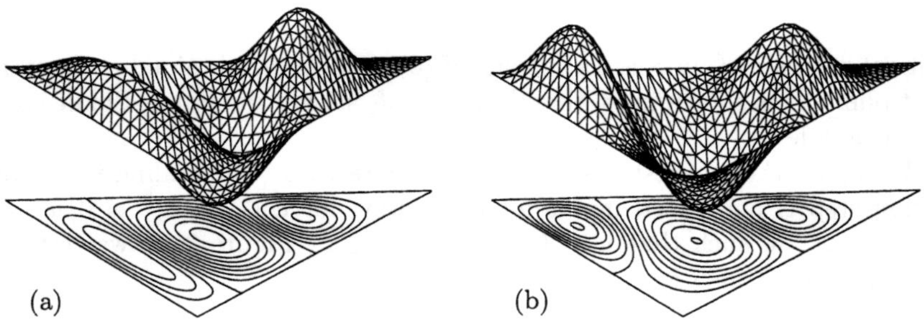

Fig. 3. Solutions of (5) found by CMPA. Approximate values of (μ, ν) and $u(\Omega)$: **(a)** $(103.2, 45.8)$, $[-2.55, 1.90]$, **(b)** $(105.7, 43.4)$, $[-2.41, 1.84]$

where $g(u) = \mathrm{e}^u - 1 - u$, $c^2 \in (0,2)$. It has been studied in [7], its solutions represent traveling waves in the direction of the x_1-axis with speed c in a model of a nonlinearly supported plate. The authors present a variational existence proof based on the mountain pass theorem for a certain class of nonlinearities g. Weak solutions of (6) in $H^2 = W^{2,2}(\mathbb{R}^2)$ are constructed as critical points of a functional. Since the functional does not satisfy the Palais-Smale condition of Theorem 1, additional work has to be done in order to recover some form of compactness. This turns out to be difficult for the above mentioned exponential nonlinearity and hence the proof does not cover this case.

The numerical results of [7] give, however, a strong evidence of existence of mountain pass solutions even with the exponential nonlinearity. Moreover, a comparison is made with a one-dimensional ODE version of problem (6), for which a wide variety of numerical solutions was computed by a shooting method [3]. Hence a question arisen whether there is a way of finding more numerical solutions of the two-dimensional PDE problem than just those in [7] found using the mountain pass algorithm. We will show that the constrained mountain pass algorithm yields additional numerical solutions.

Define functionals $I, J \in C^1(H^2, \mathbb{R})$ by

$$I(u) = \frac{1}{2}\int_{\mathbb{R}^2} [(\Delta u)^2 + u^2]\,\mathrm{d}x + \int_{\mathbb{R}^2} G(u)\mathrm{d}x\ , \qquad G(\xi) = \int_0^\xi g(t)\mathrm{d}t\ ,$$

$$S = \left\{ u \in H^2 \;\middle|\; J(u) := \frac{1}{2}\int_{\mathbb{R}^2} \left(\frac{\partial u}{\partial x_1}\right)^2 \mathrm{d}x = C \right\}\ ,$$

where $C > 0$ is a constant. Critical points of I on S are weak solutions of (6) with c^2 obtained as a Lagrange multiplier. Let the inner product on H^2 be $\langle \phi, \psi \rangle = \int_{\mathbb{R}^2} [(\Delta\phi)(\Delta\psi) + \phi\psi]\mathrm{d}x$. It is shown in [7] that the corresponding norm is equivalent to the standard norm on H^2.

Although the domain is the whole plane $\mathbb{R}^2$, for the numerical purposes we can work on a large enough (but bounded) rectangle because our solutions decay to zero as $|x| \to \infty$. Further, any translation of a solution of (6) is also a solution with the same values of I and J. These translations can be prevented by assuming symmetries of solutions (as in [3, 7]): $u(x_1, x_2) = u(-x_1, x_2)$ and $u(x_1, x_2) = u(x_1, -x_2)$ $\forall (x_1, x_2) \in \mathbb{R}^2$. Hence we can work on a rectangle $[-K_1, 0] \times [-K_2, 0]$. On the boundaries $x_1 = -K_1$ and $x_2 = -K_2$ the conditions $u = 0$ and $\Delta u = 0$ are implemented, on $x_1 = 0$ and $x_2 = 0$ the symmetry conditions.

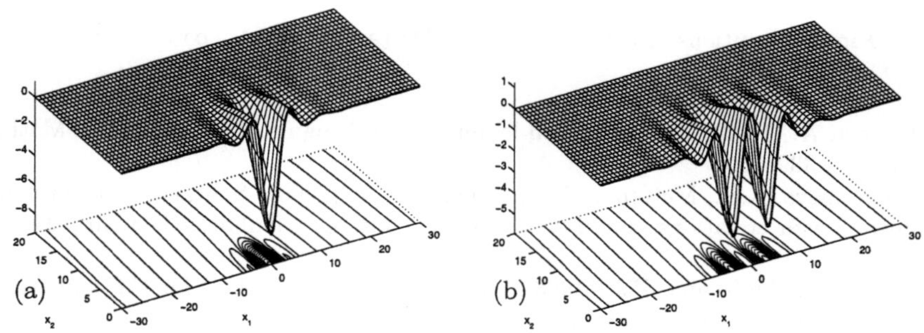

Fig. 4. Solutions of (6) found by CSDM: **(a)** $c \approx 1.247$, **(b)** $c \approx 1.313$

CSDM with $C = 150$ yields numerical solutions shown in Fig. 4 (the profiles of the waves have been highlighted, only a part of the computational domain is shown). These types of solutions were found in [7] by the unconstrained mountain pass algorithm. The projection step (Sec. 2.1) in the numerical methods is again accomplished by scaling: for $\bar{u} \in H^2 \backslash \{0\}$ find $t > 0$ such that $u = t\bar{u} \in S$.

The two numerical local minima can be then used as end points of the path in the constrained mountain pass algorithm. The algorithm converges to the solution shown in Fig. 5(b). As noted in Sec. 3.1 already, by stopping the algorithm early, if $\|I'|_{S_{z_{jm}}}\|$ stays small for a number of iterations, and by applying Newton's method, a new numerical solution may be obtained – here the function in Fig. 5(a).

A finite difference discretization was used with $K_1 = 70$, $K_2 = 50$ and the step size $\Delta x_1 = \Delta x_2 = 0.2$. The number of points on the path in CMPA was $P = 50$.

References

1. Ambrosetti, A., Rabinowitz, P.H. (1973): Dual variational methods in critical point theory and applications. J. Functional Analysis, **14**, 349–381

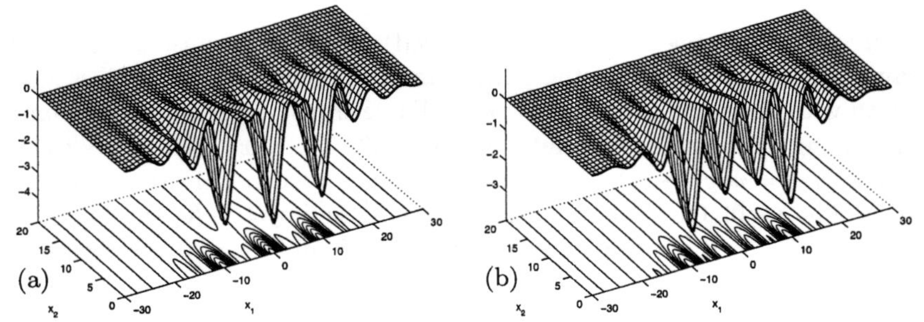

Fig. 5. Solutions of (6) found by CMPA: **(a)** $c \approx 1.384$, **(b)** $c \approx 1.365$

2. Bonnet, A. (1993): A deformation lemma on a C^1 manifold. Manuscripta Math., **81**(3-4), 339–359
3. Champneys, A.R., McKenna, P.J., Zegeling, P.A. (2000): Solitary waves in nonlinear beam equations: stability, fission and fusion. Nonlinear Dynam., **21**(1), 31–53
4. Choi, Y.S., McKenna, P.J. (1993): A mountain pass method for the numerical solution of semilinear elliptic problems. Nonlinear Anal., **20**(4), 417–437
5. Ding, Z., Costa, D., Chen, G. (1999): A high-linking algorithm for sign-changing solutions of semilinear elliptic equations. Nonlinear Anal., **38**(2, Ser. A: Theory Methods), 151–172
6. Horák, J. (2003): Constrained mountain pass algorithm for the numerical solution of semilinear elliptic problems. Preprint 2003-14, University of Basel
7. Horák, J., McKenna, P.J. (2003): Traveling waves in nonlinearly supported beams and plates. In: Nonlinear Equations: Methods, Models and Applications, volume 54 of Progress in Nonlinear Differential Equations and their Applications, pages 197–215. Birkhäuser Verlag, Basel
8. Horák, J., Reichel, W. (2003): Analytical and numerical results for the Fučík spectrum of the Laplacian. To appear in J. of Computational and Applied Mathematics

Optimal Shape Design of Diesel Intake Ports with Evolutionary Algorithm*

András Horváth[1] and Zoltán Horváth[2]

[1] Department of Physics, Széchenyi István University, Győr, Hungary, *horvatha@sze.hu*

[2] Department of Mathematics, Széchenyi István University, Győr, Hungary, *horvathz@sze.hu*

Summary. Intake port shape affects the quality of an engine significantly. In this paper, we present a method for improving an existing geometry with evolutionary algorithm type optimization. Characteristic parameters of different port shapes are calculated with a self-developed CFD program. However only small deformations allowed on original design, significant improvement achieved. Proposed robust parallel evolutionary algorithm seems to be a suitable for other optimization problems on heterogeneous non-reliable cluster of workstations.

1 Introduction

Design parameters (geometrical structure, injection parameters, etc.) of Diesel engines affects the value of it in a very complex way. A large developing effort is taken into optimizing these parameters by engine designers. Modeling a complete engine is a very difficult task both experimentally and numerically, therefore a complex optimization can not be performed for the whole system. Hence the traditional way is to split the engine into several main parts, find some parameters that characterizes the value of each parts and perform optimization processes only on these parts separately. However, after these optimizations we have to put the parts together and check the whole system whether the independently optimized parts can be assembled together or not and check the overall behaviour of system.

One of these main parts is the intake port system. Its geometry highly affects the main parameters (e.g. power, efficiency, pollution emission) of the engine. (See Fig. 1 for illustrations of intake port shape.) The essential reason of this strong dependence is that geometry of intake port biases the amount and initial velocity distribution of drawn in air into the cylinder. The amount of fresh air influes the maximum mass of burnt fuel in one cycle, while the initial velocity distribution affects the air-fuel mixture formation process which has a high impact to the quality of burning.

* This paper was supported under the Hungarian Grant for Scientific Research OTKA T43177.

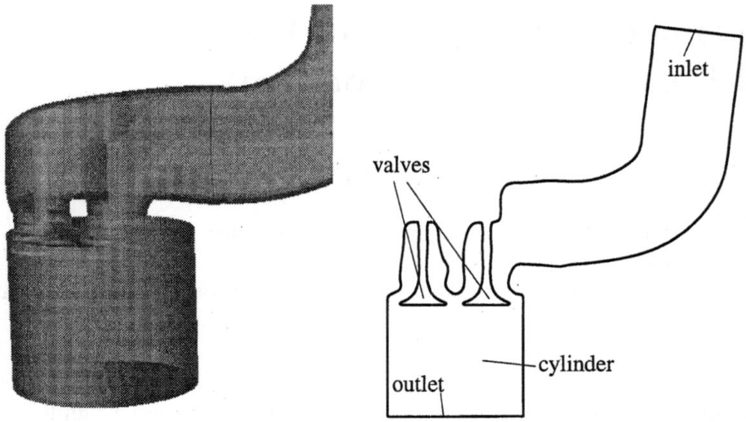

Fig. 1. An opaque view of surface and a cross section of intake port shape

It is obvious that the more air drawn in is better. The dependence on velocity distribution is more complex, but the engineering practice gives us a plausible aspect: the most important property of velocity distribution in the cylinder is the angular momentum per unit mass which characterizes the global rotation. Both two small and too high global rotation are wrong for efficient air-fuel mixture formation. There are commonly accepted optimal rotation ranges for different type of engines coming from experiments.

The traditional way to describe this two important attribute of an intake port is to calculate two non-dimensional parameters: the flux-coefficient and the swirl-coefficient, denoted by C_f and C_s respectively; for a more detailed definition see Section 2.2.

Our task was to improve an existing intake port geometry, namely enhance C_f and keeping C_s near to the original value. Because of avoiding a complete redesign of existing and working engine head structure, only small deformations of the original geometry were allowed. However, this constraint causes that only small improvement is to be expected, even 1% improvement in C_f can be important in engineering point of view.

2 The air flow in the intake port

2.1 CFD calculations

In the core of optimization process we needed a reliable and accurate CFD software which can calculate the two characteristic flow parameters for port shapes that are derived from original shape with small deformations. We used a self-developed software based on a classical FVM method with some improvements for this purpose. A detailed description can be found in [3]. Here we give only a short summary of the properties:

- FVM method with flux vector splitting (Viyajasundaram-type, [7])
- solving compressible Euler or Navier-Stokes equations
- conservative in mass, momenta and energy
- explicit (first order) time steps
- local time-steps for faster calculations
- uses unstructured, conform tetrahedral mesh
- capable to handle small deformations of mesh

A tetrahedral mesh with 147 775 elements was generated with a CAD-software and was used as the discretization of the original shape. (See Fig. 2.)

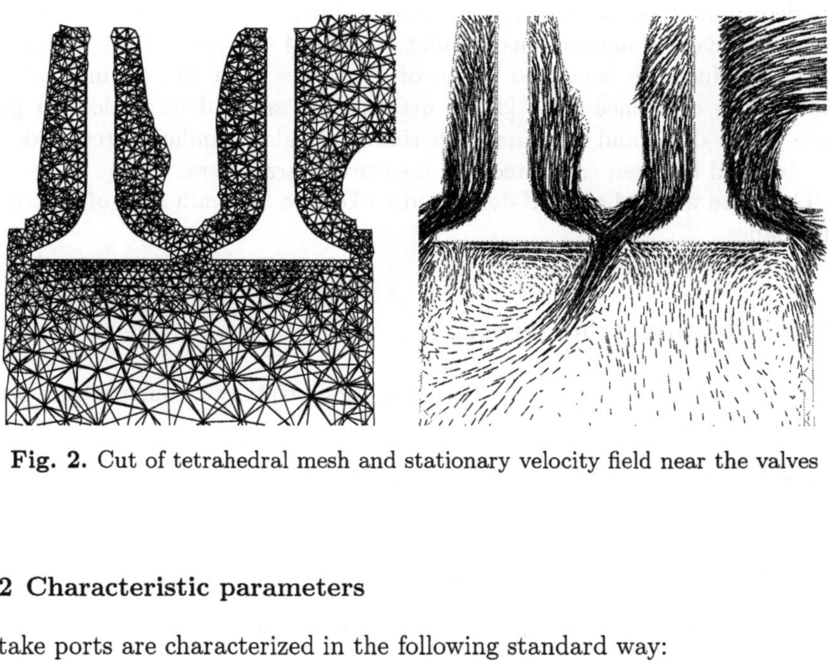

Fig. 2. Cut of tetrahedral mesh and stationary velocity field near the valves

2.2 Characteristic parameters

Intake ports are characterized in the following standard way:

Let the pressure constant at inlet and outlet. (p_{in} and p_{out}) Let us measure the mass flux ($\dot{m}$) and the flux of angular momenta (T) at outlet (in the cylinder) in stationary case.

The characteristic parameters of intake ports are the flux coefficient (C_f) and the swirl coefficient (C_s): (see [4])

$$C_f = \frac{\dot{m}/\rho_0}{A\,v_0}, \qquad C_s = \frac{8T}{\dot{m}B\,v_0} \tag{1}$$

where ρ_0 is the density of air at the inlet, A is the approximate area of smallest intake cross section (two times the valve inner seat area), $v_0 = \sqrt{2(p_{in} - p_{out})/\rho_0}$ is the characteristic velocity based on pressure drop and B is the cylinder bore.

The larger C_f the better, while values of C_s has an "optimum interval" coming from engineering experience. (See the introduction.)

$\dot{m}$ and T can be calculated from numerical model by approximating the following integrals on outlet face (S_{out}):

$$\dot{m} = \int_{S_{out}} \rho \mathbf{v} d\mathbf{S}, \qquad T = \int_{S_{out}} \rho(\mathbf{v} \times (\mathbf{r} - \mathbf{r}_0)) d\mathbf{S}$$

where $\mathbf{r}_0$ is an arbitrary point on symmetr axis of cylinder, ρ and $\mathbf{v}$ are the calculated density and velocity of air.

We found that calculations with this model can reproduce the experimentally determined characteristic parameters within acceptable relative difference even in the case of non-viscous calculations. For 4 different values of p_{out} we could reproduce the measured values of C_f in less than 1%, C_s in less than 10% relative difference. (See [3] for details.) Six small deformation on port shape was realized and measured experimentally also: similar correspondence was observed between computed and measured parameters.

Therefore we used our self-developed CFD code for evaluation of deformed port shapes.

3 The optimization strategy

The main steps of finding a shape optimization strategy in our problem were the followings:

1. Find and parametrize a set of small deformations.
2. Find a suitable object function which measures the quality of a deformed shape.
3. Find and implement method to maximize the object function.

In the following subsections we will go through these steps.

3.1 Parametrization of small deformations

We used local, smooth deformations on the original grid. This way we did not have to generate tetrahedral mesh for each deformed shape. Instead of this very time consuming step we modified only the coordinates of vertices and recalculated the geometrical parameters (volume, face normals, etc.) of tetrahedra.

The overall deformation of the shape was put together from elementary deformations. Each elementary deformation shifted the vertices only inside a specified cylinder with a parallel axis with surface normal. (See Fig. 3.) The shift of vertices is a fourth order polynomial of the distance from the axis of cylinder and the depth relative to deformation center.

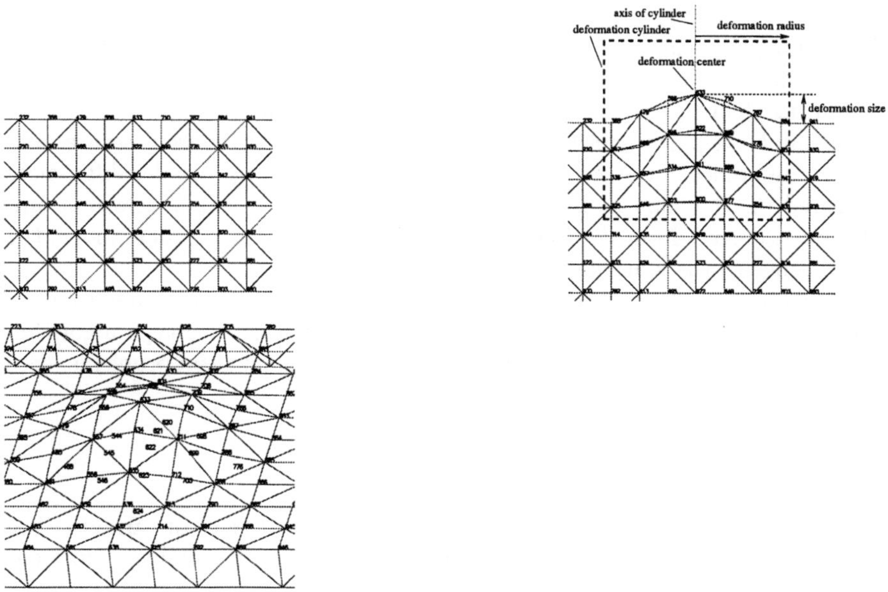

Fig. 3. A smooth local deformation on a simple tetrahedral mesh

This choice fits to the engineering requirements also: we could assure that the deformations keep small and keep some places (e.g. the valves, the top of the cylinder) unchanged.

We used combinations of such elementary deformations. This way parametrization of deformations consists from a list of parameters of elementary deformation, i.e. a list of deformation centers, radii and size. (Height of deformation cylinder is not a free parameter: it does not affect the shape but the tetrahedral mesh. We have to choose its value carefully depending on deformation parameters.)

3.2 The object function

Our goal is to maximize C_f while keeping C_s close to the original value. Therefore we calculated the parameters with no deformations and used them as reference values C_f^{ref} and C_s^{ref}.

Our object function was:

$$Obj = \frac{C_f}{C_f^{ref}} - D\left(1 - \frac{C_s}{C_s^{ref}}\right)^2 \tag{2}$$

Namely we have a one-variable object function with a penalty term depending on change in C_s. We found $D = 1.0$ as adequate in our case. Thus 10% change in C_s matches 1% change in C_f. (Remember that C_s has no strict optimal value rather an optimal interval, that is a few percent change is not significant.) For example: $Obj = 1.02$ means at least 2% improvement in mass flux. (Depending on swirl number change.)

During the CFD calculations we observed significant "noise" in C_f and C_s. For example we ported our C code to two different hardware architectures and found a difference in the order of 10^{-5} in C_f and 10^{-3} in C_s. Similar differences appeared on the same architecture in the case of very small deformations. This noise derives from the rounding errors and the nature of calculations: the convergence to the stationary flow is slow in C_s.

It means that object function has a lot of false local maxima as deformation parameters vary. For this reason we decided to use a genetic type algorithm for optimization. It is a common decision in shape optimization problems. (See e.g. [6])

3.3 Optimization method

Genetic (GA) or evolutionary (EA) type algorithm have a lot of variants. (See [2], [8]) Depending on the problem and the hardware possibilities different versions of GA/EA should be used.

The main peculiarities of our problem and possibilities are the following:

- The workstations we can use have different CPU-speed.
- Calculation time of object function (a complete CFD simulation) depends on geometry and takes 3–4 ours on our fastest workstations.
- We can use approximately 20–30 workstations for a few weeks therefore 2000–4000 object function evaluations is possible.
- There is significant probability of hardware errors during the calculations.

It is obvious that a classical GA/EA with master-worker type parallelization is not suitable in our circumstances. One reason is the significant probability of hardware errors other is the different evaluation time. Both can lead to a significant loss off efficiency due to the synchronization stages.

Furthermore, an island type parallelization with one island per workstation is not a good choice either, because each workstation can calculate only 2–4 full generation with reasonable population size. Since new chromosomes from other islands can be included only at the end of a complete generation cycle, frequency of communication between different subpopulations is too small for efficient parallelization.

A Robust Parallel Evolutionary Algorithm The specialties of our problem and hardware possibilities led us to a special type of parallelization strategy. The main idea is to use a separate population on each workstation with an evolutionary algorithm which tries to send and get messages about evaluated chromosomes after each object function evaluation and if it got a new evaluated chromosome from other machine, incorporates it into the population immediately. The communication is executed in two stages: the EA-process writes the evaluated chromosomes and object function values to the local hard disk and checks it for evaluated chromosomes arrived from other machines. "Good" chromosomes are moved between EA-units by "agent" programs which are running independently from EA-processes.

This way we achieved stability against hardware failures and strong connection between the EA-units with zero synchronization loss. It is clearly seen that if an EA-unit fails, the other can work further, if the connection between units is lost (network error), the units can continue on evaluating object functions and if the workstation executing the agent program fails the agent can be restarted on an arbitrary workstation. Even after a global failure the EA-units could read back the evaluated chromosomes and continued working. However each failure decreases the efficiency, the whole calculation will not stop and after reparation the calculation can continue at full capacity.

Certeanly we need a special EA algorithm to guarantee that the new results coming from other EA-units are used as soon as possible. In a classical GA/EA cycle it is not possible. Therefore we were searching for a more flexible approach and found the concept of "Flexible Evolution Agents" (FEA-s, see [9]).

FEA-s has not a strictly prescribed sequence of different genetic operators but a central "decision engine" decides about which genetic operator (mutation, crossover, etc.) will be executed in the next step. This kind of flexibility is used to get adaptivity property of EA: a learning engine collects statistics about success of operators and decision engine uses this information to choose the operators to be execute next time. (See [9])

We did not implemented the adaptivity of FEA because each EA-unit executes only 100–200 genetic operators in our case and it is too small for a reliable statistics. However the non-deterministic order of genetic operators allowed us to read and use the results of object function evaluation of other units provided by communication agent program.

The skeleton of a robust and parallel EA-unit:

```
1. Population=empty
2. sort the population with niching
3. if ( there are new chromosomes on disk )
      read new values
4. if ( Population.size < Size_min )
      generate and evaluate a random chromosome
      --> step 2
5. if ( Population.size > Size_max )
      truncate Population
6. if ( the best of Population changed )
      try one line search step between
            old and new best value
      --> step 2
7. find a new chromosome by a random elementary step
8. evaluate the new chromosome
```

```
9. add/replace new chromosome to Population
10. --> step 2
```

The possible elementary steps we used:

- mutate a randomly chosen chromosome
- single point crossover between two different chromosomes selected with binary tournament
- examine a random chromosome: if it is closer to a better chromosome by a specified distance then mutate it (constrained mutation)

The set of elementary steps was pieced together using the experience of test calculations on classical test problems. Hill climbing was not used because it would hinder to immediately read and use results of other units. Instead, we implemented step 6 and "constrained mutation" (see above) which are searching near the existing good chromosomes and affect similar property.

The probability of elementary steps is chosen so that in a long calculation the number of them will be equal with the number of such steps in a classical GA-algorithm.

Test results We performed several test calculations with different agent strategies on classical test problems. (Rastrigin and Keane-functions, see [5]) At first we present results with a very simple agent strategy, called "uniform distribution" which means that the agent program collects the best two chromosomes from each unit and sends them to all the other units periodically. We found that the time period of collecting and distributing best results should be more than 5–10 object function evaluation time but should be less than one tenth part of total time of calculations. Within this wide range we found that the convergence of best object function value does not depend on best results redistribution frequency.

On Figure 4 we present the results of 20 variable Keane-function problem with 1, 4 and 16 EA-units with uniform distribution strategy. One can observe decreased efficiency in 16 EA-unit case. Such a decay is a well known property of parallel algorithms.

We tested the robustness of our parallel EA-strategy in test problems. In a test calculation with 16 EA-units we randomly stopped the agent program for significant intervals to simulate network error. It is clearly seen that during the "network errors" increasing best object function values slowed down, but when the communication was restored, the results became rapidly increasing. This shows that with no communication the EA-units was working further producing a high variety of chromosomes and when the communication was repaired, the units could use them to produce good entities. (See Fig. 4 on right.)

To circumvent the efficiency loss mentioned above, we divided EA-units into 5–10 element groups and used uniform distribution within them. An another agent program was applied to realize the communication between groups. This

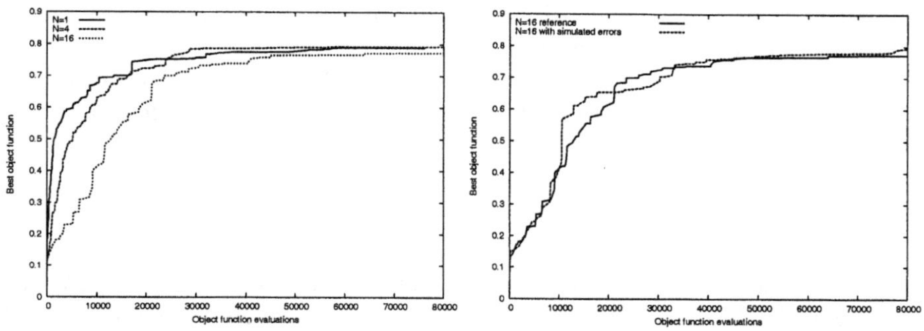

Fig. 4. Test calculations with 20-variable Keane-function.

multilevel strategy has a close relationship with island model; in our case a group of EA-units matches with an island.

In the real CFD shape optimization problem we could use at most 32 workstations. Following the multilevel agent strategy, we divided them into 3 groups (A, B and C) with approximately equal members and implemented a non-symmetric data flow at the top level agent program: the best results of B and C groups was sent to group A periodically. This way group B and C evolved separately from other groups which is a good strategy to maintain diversity, while group A could combine all the best chromosomes.

4 Results of intake port optimization

4.1 About technical aspects

Both CFD and EA code was written in standard C. The agents were Bourne shell scripts. The calculations were performed on Linux workstations at Széchenyi István University. The maximum number of workstations was 32. The CPU-speeds were between 1.5 and 2.4 GHz.

A typical evaluation took 3–4 hours (on 2.4 GHz Pentium4 machines)

There were 6 significant hardware-problems during calculations. (Power-outs, hard disk problems, network problems, etc.) This fact proves that the robustness of EA method had critical importance in our case.

4.2 Preparation

Based on previously mentioned calculations the most sensitive parts of intake system geometry were chosen. (E.g. large pressure gradients on surface indicate large resistance, high velocity values near the boundaries indicates important parts, etc.)

Small deformations of sensitive parts were selected with 5–90 mm radius and (−15)–(+15) mm maximum size.

The goal of optimization process was to find the optimal deformation size values in fixed locations with fixed radius. This way we used floating point values of deformation sizes as genes.

4.3 Calculations

In the first calculation 20 deformation points were used. On each EA-unit Size.min=40, Size.max=120 were set. The optimal value we found in 2000 function evaluation was 1.016. However it is significant, we wanted to get a higher improvement.

The study of optimal chromosomes had important conclusions:

- There were 7 genes where the optimal value was at lower or upper limit of that deformation size (extremal point)
- There were 5 genes where the absolute value of optimal deformation size was less than 1 mm (irrelevant points)

Using this result a new set of deformations with new limits of deformation sizes was chosen and a completely new optimization process was started: The 5 irrelevant points was dropped and 3 new (hopefully not irrelevant) was added. The limits for the 7 extremal points was modified. We present the results of the optimization of these 18 parameters. Figure 5 shows the object function values during the calculations. One can observe very small improvements in best values and decreasing diversity at the end of calculations. These symptoms indicate that there is no reason to continue the calculations further.

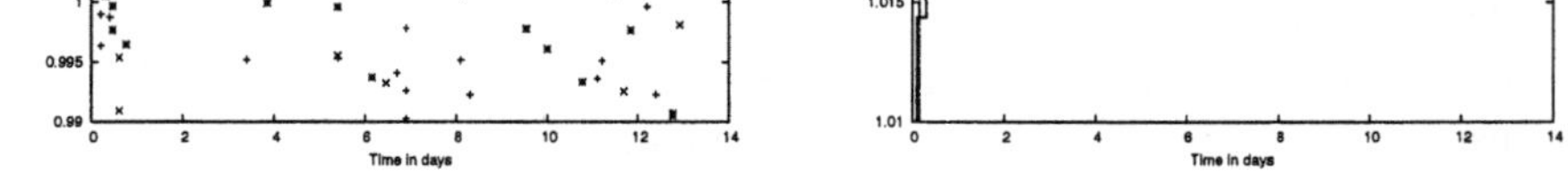

Fig. 5. Object function values during optimization process

The best chromosome we found has object function value 1.024. It means a 2.5% improvement in C_f and 3% decrease of C_s. In practice it may result in e.g. aproximately 2.5% extra power with similar quality of burning.

We present the most important differences between original and optimal shapes on Figure 6. The changes are plausible, but the deformation sizes cannot be figured out by hand.

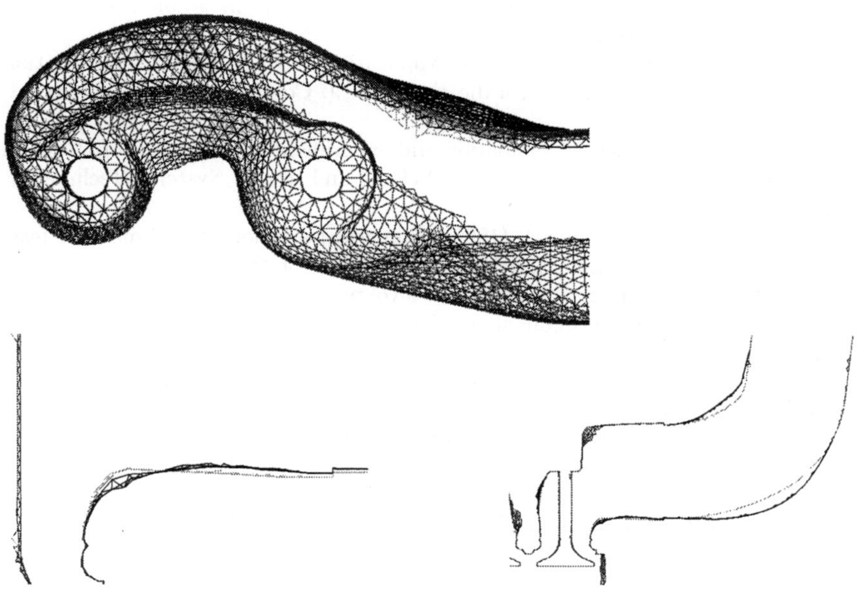

Fig. 6. The resulted "optimal" shape. Black color: modified, optimal shape, light gray: original shape, mid gray: unchanged parts

We examined the flow in optimal shape and found significant differences. For example, the pressure gradient at surface decreased notably.

5 Conclusions

The robust parallel EA-method appeared to be useful in non-reliable hardware circumstances. On a further work we will focus on optimizing EA-units and agent strategies.

With small deformations (less than 12 mm size) a significant improvement was achieved. With a larger set of deformations some further improvement to be expected, but a much higher improvement probably requires major redesign.

We can conclude that our system is suitable for real 3D compressible fluid flow shape optimization tasks.

References

1. Feistauer, M. (1993) Mathematical Methods in Fluid Dynamics. Pitman Monographs and Surveys in Pure and Applied Mathematics, Longman Scientific & Technical, UK.
2. Goldberg, D. (1989): Genetic algorithms in search, optimization and machine learning. Addision-Wesley Publ.
3. Horváth, A. and Horváth, Z. (2003): Application of CFD numerical simulation for intake port shape design of a diesel engine. J. Comput. and App. Mech., **4**, 129–146.
4. Krüs, H. (1993): Numerical simulation and experimental verification of DI diesel intake port designs 4th Int. Conf. on Vehicle and Traffic Systems Technology, Strasbourg.
5. Marco-Blaszka, N., Désidéri, J. (1999): Numerical solution of optimization test-cases by genetic algorithms. INRIA research report 3622.
6. Marco, N. and Lanteri, S. (2000): A two-level parallelization strategy for genetic algorithms applied to optimum shape design. Parallel Computing, **26**, 377–397.
7. Vijayasundaram, G. (1986): Transonic flow simulations using an upstream centered scheme of Godunov in finite elements. J. Comput. Phys. **63**, 416–433.
8. Whitley, D. (2001): An overview of evolutionary algorithms: practical issues and common pitfalls. Information and Software Technology, **43**, 817–831.
9. Winter, G., Galván, B., Alonso, S., González, B. (2002): Evolving form genetic algorithms to flexible evolution agents. In: Late Breaking Papers at the Genetic and Evolutionary Computation Conference (GECCO-2002).

Numerical Simulation of Compressible Fluids with Moving Boundaries: An Effective Method with Applications*

Zoltán Horváth[1] and András Horváth[2]

[1] Department of Mathematics, Széchenyi István University, Győr, Hungary
horvathz@sze.hu

[2] Department of Physics, Széchenyi István University, Győr, Hungary
horvatha@sze.hu

Summary. In this paper we present a numerical algorithm to the solution of the equations of compressible nonviscous fluids on domains with moving (translating) boundaries. Our moving mesh algorithm, defined on special tetrahedral meshes, avoids global interpolation and re-meshing, thus it works quite efficiently on problems with strongly deforming domains. As an illustration we give some computational results when our algorithm is applied to the simulation of gas flow in a high-voltage circuit breaker.

1 The engineering problems and the scope of the paper

The simulation of many industrial processes requires numerical solution of compressible fluids on moving domains. For example we can mention airfoil oscillation, mixture formation in the combustion chamber and the cylinder of a Diesel engine and flow development in a circuit breaker. With several problems the flow domain is strongly compressed and/or stretched in a certain interval of the simulation time. So, to avoid small time step-sizes due to distorted cells, re-meshing of the flow domain and, accordingly, interpolation of the state variables are necessary from time to time, at least in case of explicit methods. However, global re-meshing and the interpolation are time-consuming, moreover, the latter introduces additional numerical and non-conservativity errors. Further, even if the state variables are interpolated in a conservative way the conservative errors of some other important conservative quantities, such as total angular momentum, usually increase, see e.g. [2].

In this paper we would like to present our method that we applied successfully to real-life problems, such as the simulation of the flow in a Diesel engine and in a high-voltage circuit breaker (this can be considered a domain with several pistons and valves). With both problems we have to compute compressible (multicomponent) gas flow in a strongly deforming 3D domain where the

* This paper was supported under the Hungarian Scientific Research Fund OTKA T43177.

deformation is induced by translating boundary parts. In Section 2 we pose our mathematical model, the multicomponent compressible Euler equations on moving domains. Then in Section 3 we introduce a first order finite volume method based on Vijayasundaram's numerical flux function with explicit time stepping. The core of the numerical algorithm is the moving tetrahedral mesh algorithm called snapper, the basic idea of which goes back to the snapper algorithm for hexahedral meshes given in [2]. This results in a method which requires re-meshing and interpolation only in the very close neighbourhood of the moving objects and, moreover, these are done very efficiently.

This numerical method is coded and we show applications to a 3D (academic) test problem and the industrial problem of gas flow in a high-voltage circuit breaker in Section 4. We conclude that comparisons of the computations with the exact solution of the test problem and with actual physical measurements for the latter problem show good agreement.

2 The mathematical model

We have selected the multicomponent Euler equations of gas dynamics on moving domains as our mathematical model of the fluid flow problems described in Section 1. This model is suitable for the modelling of flows where the effect of viscosity is not significant comparing to that of convection.

$$\frac{\partial u}{\partial t} + \operatorname{div} f(u) = 0 \qquad \forall t \in [0, t_{\max}],\ \mathbf{x} \in \Omega(t) \tag{1}$$

$$u(0, \mathbf{x}) = u_0(\mathbf{x}) \qquad \forall \mathbf{x} \in \Omega(0) \tag{2}$$

$$+\ \text{BC (boundary conditions)} \tag{3}$$

$$+\ \text{EOS (equations of states)} \tag{4}$$

where

- $u : \cup_{t\in[0,t_{\max}]}\{t\} \times \Omega(t) \to \mathbb{R}^{K+4}$ is the state variable with $u = (\rho_1, ..., \rho_K, \rho\mathbf{v}^T, e)^T$ where $\rho_k = \rho_k(t, \mathbf{x})$ $(k = 1, ..., K)$, $\rho := \sum \rho_m$, $\mathbf{v} = \mathbf{v}(t, \mathbf{x}) = (v_1, v_2, v_3)^T$, $e = e(t, \mathbf{x})$ are respectively the density of the fluid components, the density of the fluid, the fluid velocity and the total energy density (i.e. the total energy per unit volume of the fluid);
- $f = (f_1, f_2, f_3)^T$ is the flux vector with $f_i(u) = (\rho_1 v_i, \ldots, \rho_K v_i, (\rho v_i \mathbf{v} + p\mathbf{e}_i)^T, v_i(e+p)^T$ $(i = 1, 2, 3)$ where $\mathbf{e}_i$ is the ith coordinate unit vector; $\operatorname{div} f(u) = \sum_{i=1}^{3} \frac{\partial f_i(u)}{\partial x_i}$;
- EOS denotes the set of the equations of states of a non-ideal gas mixture:

$$p = RT \sum_{m=1}^{K} \frac{\rho_m}{W_m}, \qquad e = \sum_{m=1}^{K} \rho_m I_m(T) \tag{5}$$

where $T = T(t, \mathbf{x})$ denotes the temperature of the fluid; $I_m(T)$ is the specific internal energy given a priori by interpolation formulas based on tabulated values, R is the universal gas constant and W_m is the molecule weight of the mth fluid component;

- BC: linearly consistent boundary conditions formulated by the help of ghost cells (for more details see e.g. [3] pp. 457-460, [11] pp. 222-224 with an emphasis on moving boundaries; see also [2]);
- $\Omega(t)$ is the time dependent flow domain defined in the following way: $\Omega(0)$ is a given initial domain, which is deforming according to the mapping $\varphi : [0, t_{\max}] \times \Omega(0) \to \mathbb{R}^3$, i.e. $\Omega(t) := \{\mathbf{x} = \varphi(t, \xi) \in \mathbb{R}^3 \,|\, \xi \in \Omega(0)\}$; we suppose that $\varphi_t := \varphi(t, .)$ is one-to-one for all t; then the velocity of the points of Ω is given by $\kappa(t, \mathbf{x}) := \frac{\partial \varphi}{\partial t}(t, \varphi_t^{-1}(\mathbf{x}))\ \forall t \in [0, t_{\max}]$ and $\forall \mathbf{x} \in \Omega(t)$.

Note that φ is not unique if we prescribe only the deformation of the boundary of Ω, which is the case in the situations we are focussing on in this paper.

The following lemma of the calculus called Reynolds' transport theorem is the basic tool to obtain an integral formulation for the fluid flow on a moving domain.

Lemma 1. *Let V be a moving subdomain of Ω, i.e. $V(0) \subset \Omega(0)$ and $V(t) := \varphi_t(V(0))$. Then for any $\psi : \mathbb{R}^4 \to \mathbb{R}$ differentiable function we have*

$$\int_{V(t)} \frac{\partial \psi}{\partial t}(t, \mathbf{x}) \mathrm{d}\mathbf{x} = \frac{\mathrm{d}}{\mathrm{d}t} \int_{V(t)} \psi(t, \mathbf{x}) \mathrm{d}\mathbf{x} - \int_{\partial V(t)} \psi(t, s) \kappa(t, s) \cdot \mathbf{n}(t, s)\, \mathrm{d}s.$$

Integrating (1) over a moving subdomain V we get, by applying Lemma 1, a weak formulation of the Euler equations on moving domains as

$$\frac{\mathrm{d}}{\mathrm{d}t} \int_{V(t)} u \mathrm{d}\mathbf{x} + \int_{\partial V(t)} f(u) \cdot \mathbf{n} - \kappa \cdot \mathbf{n}\, u\, \mathrm{d}s = 0 \qquad \forall V \subset \Omega \text{ moving subdomain of } \Omega \tag{6}$$

or, integrating in time and using the notation

$$\overline{u} = \overline{u}_V(t) := \frac{1}{|V(t)|} \int_{V(t)} u(t, \mathbf{x})\, \mathrm{d}\mathbf{x},$$

$$|V(t_b)|\overline{u}(t_b) - |V(t_a)|\overline{u}(t_a) + \int_{t_a}^{t_b} \int_{\partial V(t)} f(u) \cdot \mathbf{n} - \kappa \cdot \mathbf{n}\, u\, \mathrm{d}s\, \mathrm{d}t = 0 \qquad \forall [t_a, t_b] \subset [0, t_{\max}],\ \forall V \subset \Omega. \tag{7}$$

3 The numerical methods

For the construction of a numerical method to the problem posed in Section 2 we consider a moving tetrahedral mesh of $\Omega(t)$. This means that we have a face-to-face partition of Ω into tetrahedra T_j, $j = 1, ..., N$ such that each tetrahedron is moving according to φ, the transformation function of Ω. We allow that at certain time-points called snapping (or re-meshing) points of time the structure of the partition and even N can be changed.

Further, we denote by S_{jl} the lth side of T_j ($l = 1, ..., 4$) according to a certain agreement, which does not change between the subsequent snapping points. Then the tetrahedron neighbouring T_j with sharing the face S_{jl} will be denoted by the local indexing T_{jl} as well. (T_{jl} denotes a "ghost" tetrahedron if $S_{jl} \subset \partial\Omega$.) Further, let $\mathbf{n}_{jl}$ denote the outer unit normal vector of T_j corresponding to the face S_{jl}.

For the time discretization we suppose that an adaptively defined subdivision of $[0, t_{\max}]$ as $0 = t^0 < \ldots < t^{n-1} < t^n < \ldots < t^m = t_{\max}$ is given, where each snapping point belongs to $\{t^n\}_{j=0}^m$; further $\tau_n := t^n - t^{n-1}$ is the nth time step-size.

Now we are in a position to introduce our numerical method. Suppose that u is a weak solution of the problem posed in Section 2. Applying (7) with $V = T_j$ on $[t^{n-1}, t^n]$ we obtain the explicit scheme for the $u_j^n \approx \dfrac{1}{|T_j(t_n)|}\displaystyle\int_{T_j(t^n)} u(t^n, \mathbf{x})\mathrm{d}\mathbf{x}$ values

$$|T_j(t^n)|u_j^n - |T_j(t^{n-1})|u_j^{n-1} + \tau_n \sum_l \tilde{\nu}_{jl}\, g(u_j^{n-1}, u_{jl}^{n-1}, \tilde{\mathbf{n}}_{jl}, \tilde{\kappa}_{jl}) = 0, \quad n = 1, 2, ... \tag{8}$$

where g is the numerical flux function on sides and $\tilde{\mathbf{n}}_{jl}$, $\tilde{\kappa}_{jl}$, $\tilde{\nu}_{jl}$ approximate in some sense $\mathbf{n}_{jl}$, κ_{jl} and $|S_{jl}|$, respectively, such that

$$\int_{t^{n-1}}^{t^n} \int_{S_{jl}(t)} (f(u)\cdot\mathbf{n} - \kappa\cdot\mathbf{n}u)\, \mathrm{d}s\, \mathrm{d}t \approx \tau_n \tilde{\nu}_{jl}\, g(u_j^{n-1}, u_{jl}^{n-1}, \tilde{\mathbf{n}}_{jl},\, \tilde{\kappa}_{jl}). \tag{9}$$

The initial values for (8) are defined by $u_j^0 := \dfrac{1}{|T_j(0)|}\displaystyle\int_{T_j(0)} u_0(\mathbf{x})\mathrm{d}\mathbf{x}$. For a time-stepping scheme we have to define the moving mesh algorithm, the numerical flux function g and the geometrical parameters. We devote the following three subsections to the definition of these.

3.1 The moving mesh algorithm: the snapper

Here we give an algorithm for an efficient discretization of the moving domain $\Omega = \Omega(t)$ which is strongly deforming due to some translating parts of the boundary. At first we divide Ω into non-overlapping blocks (moving subdomains) according to the type of the moving/deformation and discretize the blocks separately, taking care of the face-to-face property on the common parts of the boundary of the blocks; the union of the tetrahedra of the blocks finally constitutes the tetrahedral mesh $\{T_j\}$. The type of a block B can be fixed (if $\kappa \equiv \mathbf{0}$), shifting (if $\kappa(t, \mathbf{x}) \equiv \beta(t)\mathbf{b}$) or deforming. We call B of the latter type if B is translation invariant in the sense that there exists a surface $S_0 \subset \mathbb{R}^3$ such that $B_0 := \cup_t B(t)$, the frame of B, equals the volume swept by shifts of S_0, i.e. $B_0 = \cup_{\theta \in [0,1]}(S_0 + \theta \mathbf{b})$, and, moreover, there exists a moving part of the boundary $M \subset \partial B$ such that $\kappa(t, \mathbf{x}) = \beta(t)\mathbf{b}$ whenever $\mathbf{x} \in M(t)$ and $\kappa \cdot \mathbf{n} = 0$ otherwise.

The discretization of the blocks of the first or the second type can be an arbitrary tetrahedral mesh fitting the geometry at time $t = 0$ and this is left unchanged in time (first type) or simply shifted by $\beta(t)\mathbf{b}$ (second type). Let us now assume that B is of deforming type and for the ease of presentation suppose that $S_0 \subset \partial B(0)$ and $M(0) \subset S_0$ (e.g. the block of a valve in its bottom dead center). The task we have to solve is the discretization of B at time points t^n given by the flow calculation (the "hydrocode") such that $B(t^n)$ and its mesh is derived by the mapping φ from the given $B(t^{n-1})$ and its mesh ($n = 1, 2, \ldots$); this step is called mesh modification. But first we need the discretization of the frame $B_0 = B(0)$.

Layered tetrahedral mesh generation The steps of the discretization of B_0 are the following. (For an illustration see Figure 1.)

1. Triangulate $M(0) \subset S_0$ and extend this triangulation to that of S_0.
2. Translate this triangulation in the direction of $1/\ell\mathbf{b}$ where ℓ is a positive integer to obtain a layer of prisms.
3. Divide the prisms into 3 tetrahedra each to get a face-to-face tetrahedral mesh of this layer.
4. Translate the last layer with its meshing with $\theta\mathbf{b}$ while necessary.

Note that Step 3. is not a trivial task, for a solution see [6].

We remark that this algorithm was succesfully applied to the discretization of non-deforming and non-regular blocks either in such a way that first we enframed the block to be meshed into a layered mesh (in these steps we allowed "biased" translations), omitted tetrahedra not intersecting the block and dragged the boundary nodes of the union of tetrahedra to the boundary of the block.

The algorithm of the mesh modification Suppose that we are given the mesh of $B(t^{n-1})$, φ and τ_n. We shall call the layer in the direction of $\mathbf{b}$ from $M(t^{n-1})$ the moving layer.

1. Compute first the new position of the moving part, i.e. $M(t^{n-1} + \tau_n)$.
2. If the height of the moving layer is smaller than half of the original height (i.e. $1/\ell|\mathbf{b}|$) or even worse: $M(t^{n-1} + \tau_n)$ does not belong to the moving layer, reduce τ_n so that in the new position of the moving part the height of the layer is exactly the half of the original height and take $M(t^n) = M(t^{n-1} + \tau_n)$.
3. Inherit the topological structure of the mesh of $B(t^{n-1})$ to the mesh of $B(t^n)$; only update the coordinates of nodes of $M(t^n)$ according to the prescribed

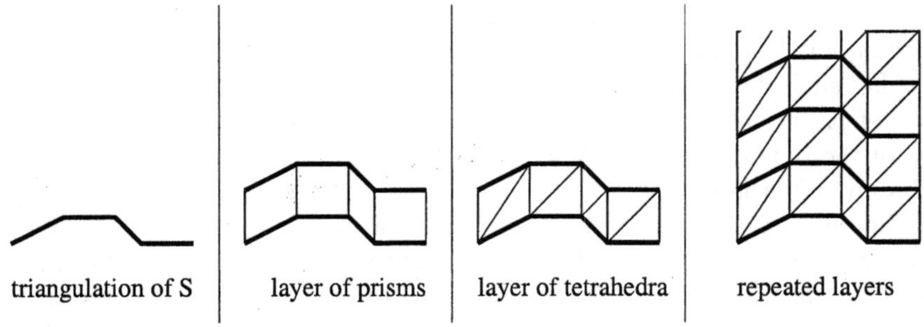

Fig. 1. Layered tetrahedral mesh generation

mapping φ, re-calculate the geometrical data (area of sides, etc.). In this case we assign the node velocities as mean velocities, for example for the node A on the moving part we take $\kappa(t, \mathbf{x}_A) = (\mathbf{x}_A(t^n) - \mathbf{x}_A(t^{n-1})/\tau_n$.

4. If the height of the moving layer is exactly half of the original, we shall call the neighbouring layer (in the direction of $\mathbf{b}$) the new moving layer, and its nodes back in direction $\mathbf{b}$ are snapped to the corresponding nodes on $M(t^{n-1})$. The nodes of $M(t^{n-1})$ are snapped back to the initial position of the former active layer and are deactivated (i.e. signed that these points do not belong from now to the flow domain). Of course, the geometrical data have to be updated and since the tetrahedra in the new moving layer corresponding to the moving part are derived by joining two layers the state variables have to be interpolated (discussed below).

Step 4. in the algorithm above is called the snapping step and the whole algorithm is the snapper (cf. [2]). For an illustration of this algorithm in 2D case see Figure 2. A whole 3D tetrahedral mesh has a too complex structure to illustrate the snapping on it, however the idea can be understood in 2D and on Figure 3 we present the basic blocks of a 3D mesh, before and after a snapping.

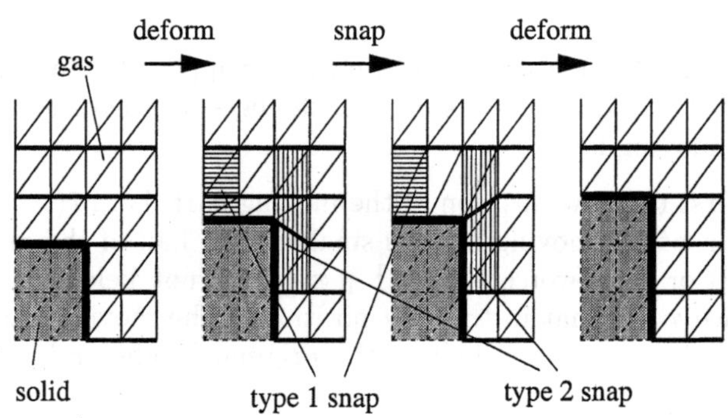

Fig. 2. Sketch for the snapper algorithm

Due to the mesh generation and the mesh modification algorithms given above, tetrahedral meshes before and after snapping have good properties which enables us to do the interpolation of state variables in an effective and conservative way. One can observe that grids before and after snapping can be divided into "snapping blocks" of 2 or 3 prisms such way that these blocks before and after snapping are coincident.

It is obvious that snappings must be handled in a different way if the base triangle of a prism of the moving layer is in the interior of $M(t^{n-1})$ (type 1 snap, 2 prism snapping blocks) or if there are only one or two vertices of the base triangle on $M(t^{n-1})$ (type 2 snap, 3 prism snapping blocks). Type 1 and 2 snapping blocks in 2D are marked in Figure 2. In Figure 3 we present the structure of type 1 and type 2 snapping blocks in 3D. We have to remark that depending on which vertices of base triangle are on $M(t^{n-1})$, there are 6 different cases of type 2 snappings.

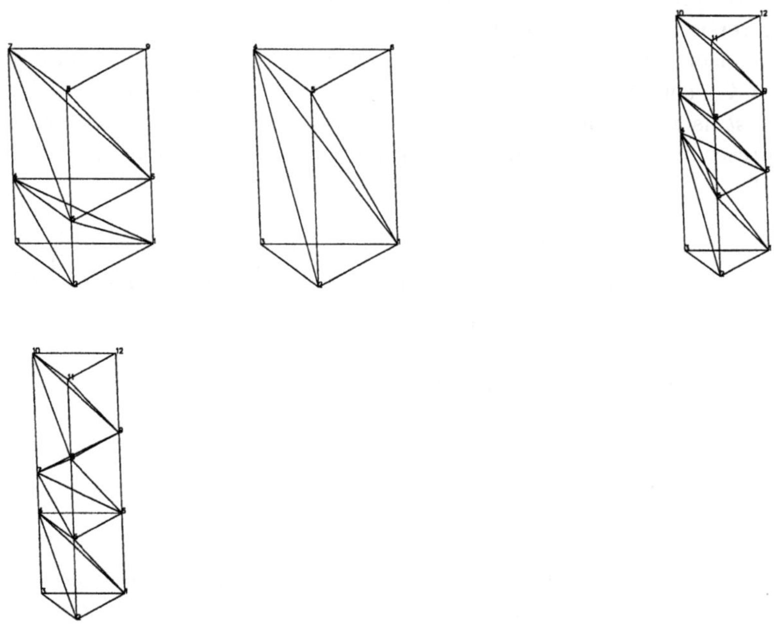

Fig. 3. Corresponding snapping blocks in type 1 (left) and type 2 (right) cases

The interpolation of the state variables can be done locally, inside the snapping blocks. This means interpolation between 3 new and 6 old tetrahedra in type 1, between 9 new and 9 old tetrahedra in type 2 snap. Since our method is of first order and state variables are of density nature assigned to tetrahedra, these interpolation can be implemented in a natural way: state variables after snapping are linear combinations of pre-snapping values and the elements of transformation matrices can be calculated by determinig overlapping volume ratios of tetrahedra. For example in type 1 snap the interpolation matrix from 3 to 6 tetrahedra is the following: (if we use an appropriate order of tetrahedra)

$$\begin{pmatrix} 1/3 & 2/9 & 4/27 & 8/27 & 0 & 0 \\ 0 & 1/9 & 4/27 & 8/27 & 4/9 & 0 \\ 0 & 0 & 1/27 & 2/27 & 2/9 & 2/3 \end{pmatrix}$$

Notice that this snapping algorithm, including the elements of the interpolatory matrices, is independent of the triangulation of the moving surface.

3.2 The geometrical parameters for the scheme

In order to obtain appropriate geometrical parameters, $\tilde{\mathbf{n}}_{jl}$, $\tilde{\kappa}_{jl}$ and $\tilde{\nu}_{jl}$, the discrete version of the geometric conservation law (GCL) gives us a guideline. For the concept and importance of the discrete GCL condition consult e.g. [4].

The GCL condition is derived from the fact that the constant flow $u(t, \mathbf{x}) \equiv u^* = const.$ is a solution of (1) and also its weak form (7), whenever a suitable BC is prescribed. Hence we have for all moving subdomains V and $[t^{n-1}, t^n]$ (c.f. (7))

$$\left(|V(t^n)| - |V(t^{n-1})|\right) u^* + \int_{t^{n-1}}^{t^n} \int_{\partial V(t)} \mathbf{n}\,\mathrm{d}s\,\mathrm{d}t\, f(u^*) - \left(\int_{t^{n-1}}^{t^n} \int_{\partial V(t)} \kappa \cdot \mathbf{n}\,\mathrm{d}s\,\mathrm{d}t \right) u^* = 0$$

i.e., employing the identity $\int_{\partial V(t)} \mathbf{n}\,\mathrm{d}s = \mathbf{0}$,

$$|V(t^n)| - |V(t^{n-1})| - \int_{t^{n-1}}^{t^n} \int_{\partial V(t)} \kappa \cdot \mathbf{n}\,\mathrm{d}s\,\mathrm{d}t = 0. \tag{10}$$

It is a natural requirement that the discretization of the problem should preserve this property called discrete GCL, i.e. the deformation of the mesh alone should not change a constant flow, which means that, for all $u^* = const.$, if $u_j^{n-1} \equiv u^*$ for all j then $u_j^n \equiv u^*$ for all j.

Lemma 2. *If the numerical flux function g is conservative and consistent, i.e. for all $u, v, \mathbf{n}, \kappa$ there holds $g(u, v, \mathbf{n}, \kappa) = g(v, u, -\mathbf{n}, \kappa)$ and $g(u, u, \mathbf{n}, \kappa) = f(u) \cdot n - \kappa \cdot \mathbf{n}\, u$, respectively, then the discrete GCL holds for the method* (8) *whenever*

$$\sum_l \tilde{\nu}_{jl} \tilde{\mathbf{n}}_{jl} = \mathbf{0} \qquad \textit{and} \qquad |T_j^n| - |T_j^{n-1}| = \tau_n \sum_l \tilde{\nu}_{jl} \tilde{\kappa}_{jl} \cdot \tilde{\mathbf{n}}_{jl} \qquad \forall j. \tag{11}$$

Proof. Substituting $u_j^{n-1} = u_j^n = u^*$ into (8) the statement follows from the consistency and conservatitity of the method. □

Lemma 3. *The method* (8) *with a conservative and consistent numerical flux* g *and the "snapper" mesh deformation respects the discrete GCL whenever*

$$\tilde{\mathbf{n}}_{jl} = \mathbf{n}_{jl}(\frac{t^{n-1}+t^n}{2}), \quad \tilde{\nu}_{jl} = |S_{jl}(\frac{t^{n-1}+t^n}{2})|,$$
$$\tilde{\kappa}_{jl} := \frac{1}{3} \sum_{A:\ node\ of\ S_{jl}} \frac{\mathbf{x}_A^n - \mathbf{x}_A^{n-1}}{\tau_n}. \tag{12}$$

Proof. The first relation of (11) follows from the fact that the left hand side of the equation equals the integral of the outer normal vector over the surface of $T_j(t^{n-1}+t^n)/2)$, which is the zero vector. To prove the second relation of (11) it is enough to check by (10) that $\tau_n \sum_l \tilde{\nu}_{jl}\tilde{\kappa}_{jl} \cdot \tilde{\mathbf{n}}_{jl} = \int_{\partial V(t)} \kappa \cdot \mathbf{n} \, ds \, dt$; but this follows now from the actual choice of parameters and the definition of κ. □

In our actual algorithms applied to problems reported in this paper we used (12).

We remark that besides considering the discrete GCL condition is natural to hold true it is proven to guarantee first order accuracy of the scheme (8) provided the method is in addition accurate on fixed meshes, see [4].

3.3 The numerical flux function: Vijayasundaram's function

For the numerical flux function g we employ Vijayasundaram's numerical flux function (see [12] and also [3], [7]), which was proven accurate in our former applied problems as well (see [5]).

Lemma 4. *Let* $u \in \mathbb{R}^{K+4}$, $\mathbf{n} \in \mathbb{R}^3$ *be arbitrary and* f *defined in Section 2. Then we have* $f(u)\,\mathbf{n} = C(u,\mathbf{n})\,u$ *with* $C(u,\mathbf{n}) := \sum_{i=1}^3 f_i'(u)\,n_i$. *Moreover, the eigenvalues of* $C(u,\mathbf{n})$ *are*

$$\lambda_1, \ldots, \lambda_{K+2} = \mathbf{v}\cdot\mathbf{n}, \qquad \lambda_{K+3} = \mathbf{v}\cdot\mathbf{n} + \sqrt{c}, \qquad \lambda_{K+4} = \mathbf{v}\cdot\mathbf{n} - \sqrt{c}$$

where $Y := (\rho_1/\rho, ..., \rho_K/\rho)^T$, $c := Y^T \dfrac{\partial p}{\partial(\rho_1, \ldots, \rho_K)} + (H - \|\mathbf{v}\|^2)\dfrac{\partial p}{\partial e}$, $H := (e+p)/\rho$.

The eigenvectors of $C(u,\mathbf{n})$ *corresponding to* λ_{K+3} *and* λ_{K+4} *are*

right: $\begin{pmatrix} Y \\ \mathbf{v} \pm \sqrt{c}\mathbf{n} \\ H \pm \sqrt{c}\mathbf{v}\cdot\mathbf{n} \end{pmatrix}$

left: $\left(\frac{\partial p}{\partial \rho_1} \mp \sqrt{c}\mathbf{v}\mathbf{n}, \ ..., \ \frac{\partial p}{\partial \rho_K} \mp \sqrt{c}\mathbf{v}\mathbf{n}, \ -\frac{\partial p}{\partial e}\mathbf{v}^T \pm \sqrt{c}\mathbf{n}, \ \frac{\partial p}{\partial e} \right)$.

In the formulas above we have, with the specific heat capacity at constant volume $c_{V,m}(T) = \partial I_m(T)/\partial T$,

$$\frac{\partial p}{\partial e} = \frac{R\rho_m/W_m}{\sum\limits_m \rho_m c_{V,m}(T)}, \qquad \frac{\partial p}{\partial \rho_m} = \frac{RT}{W_m} - p_e\, I_m(T).$$

Proof. The results follow from tedious computations and can be validated by checking the definitions. □

In (9) we have $f(u)\cdot\mathbf{n} - \kappa\cdot\mathbf{n}u = \tilde{C}(u,\mathbf{n},\kappa)u$ with $\tilde{C}(u,\mathbf{n},\kappa) := C(u,\mathbf{n}) - \kappa\cdot\mathbf{n}\,I$ (I is the $(K+4)$-by-$(K+4)$ identity matrix). Then we chose for g the function

$$g(u,v,\mathbf{n},\kappa) = g_{\text{Vijaya}} := \tilde{C}(\frac{u+v}{2},\mathbf{n},\kappa)^+\,u + \tilde{C}(\frac{u+v}{2},\mathbf{n},\kappa)^-\,v. \tag{13}$$

Note that the computation of $\tilde{C}^+$ (and $\tilde{C}^-$) is done by using diagonalization $\tilde{C}^\pm = Q^{-1}\,\tilde{D}^\pm\,Q$ where Q can be computed from the eigenvectors of C and $\tilde{D} = \mathrm{diag}(\lambda_i(C) - \mathbf{n}\cdot\kappa)$.

It is clear that Vijayasundaram's flux function is conservative and consistent, therefore, as a consequence of Lemma 2 the method (8) with the snapper mesh deformation and geometrical parameters (12) respects the discrete GCL condition.

4 Applications

We implemented the numerical algorithm, certainly including the snapper mesh generation, in ANSI C programming language and applied to several problems. Here we display some results of two problems. Based on our experiences we may say that the code performs well for the considered test and real-life problems.

4.1 Testproblem: rectangular block with an oscillating wall

We consider a rectangular block $\Omega(t) = [0,1]\times[0,0.2]\times[0,0.5-0.32\cos(100t)]$, $t\in[0,0.25]$ and a one-component flow with initial data: $\mathbf{v}_0 = \mathbf{0}$, $\rho_0 = 1.3$, $T_0 = 293K$; ideal gas EOS: $\gamma = 1.404$, $p = (\gamma-1)(e - 1/(2\rho)|\rho\mathbf{v}|^2)$ and slip BC. For reference we computed exactly the total energy: $E(t) = \int_{\Omega(t)} e\mathrm{d}\mathbf{x} \cong 9759\left(\frac{0.18}{0.5-0.32\cos(100t)}\right)^{0.404}$ and the components of the total momentum: $P_i(t) := \int_{\Omega(t)} \rho v_i\,\mathrm{d}\mathbf{x}$, which appeared identically 0 for $i = 1,2$ and $P_3 \cong -0.7488\sin 100t$.

We tested our method, which respects the GCL property and the explicit Euler method for time stepping (i.e. all geometrical data are evaluated at the beginning of the time step; we know that this method does not respect GCL). We had a mesh of $6\cdot(20\times5\times20) = 12000$ tetrahedra (at most).

We found numerically that there is a very small error in P_3 and $E(t)$ (with relative error at most 10^{-3}). Moreover, our method produced the fifth of the error of the explicit Euler method in P_2, underlining the importance of the discrete GCL property.

4.2 Application: gas flow in a high voltage circuit breaker

We investigated the gas flow in a high voltage circuit breaker. In these investigations the gas flow was induced by mechanical constraints of a configuration of pistons and valves only, i.e. the current was taken identically zero. The fluid flow domain is slightly not axisymmetric but employing its rotational symmetry of 90 degrees and plane symmetry it is sufficient to compute the flow in its eighth part. Hence we did not assume a priori the usual axisymmetric formulations (c.f. [8]). The gas was a mixture of two components that were originally separated. For an illustration of our results see Figure 4. Our code performed very well: the computed and measured pressure and density values were compared at two control points and there were at most 5% relative errors in the measured and computed quantities. However, at certain small parts of the fluid domain we experienced spurious pressure oscillations but these occured only for a short period during the simulation time. This was somehow expected (see e.g. [1], [9]), and clipping negative values from the energy approximations cured the code. Finally we remark that the flow was proven significantly not axisymmetric (c.f. Figure 4), which justifies our model.

Fig. 4. Graph of $Y_1 := \rho_1/\rho$ at two points of time in two perpendicular plane sections (the plane of the cross section and the direction of movement of parts is marked)

References

1. Abgrall, R. (1996): How to prevent pressure oscillations in multicomponent flow calculations: A quasi conservative approach. J. Comput. Phys., **125**, 150-160.
2. Amsden, A.A., O'Rourke, P.J., Butler, T.D. (1989): KIVA-II: A Computer Program for Chemically Reactive Flows with Sprays. Report LA-11560-MS, Los Alamos National Laboratory.
3. Feistauer, M. (1993): Mathematical Methods in Fluid Dynamics. Longman, London
4. Guillard, H., Farhat, C. (2000): On the significance of the geometric conservation law for computations on moving meshes. Comput. Methods Appl. Engrg., **190**, 1467-1482.
5. Horváth, A., Horváth, Z. (2003): Application of CFD numerical simulation for intake port shape design of a diesel engine. Journal of Computational and Applied Mechanics, **4**, 129–146
6. Horváth, A., Horváth, Z., Křížek, M.: Tetrahedralization of partitions formed by prisms. *(in preparation)*

7. Kröner, D. (1997): Numerical Schemes for Conservation Laws. Wiley and Teubner, Chichester Stuttgart
8. Yan, J.D., Fang, M.T.C. and Hall, W. (1999): The development of PC based CAD tools for auto-expansion circuit-breaker design. IEEE Trans. on Power Delivery, **14**, 176-181.
9. Liu, X.D., Fedkiw, R.P., Osher, S., (1998): A quasi-conservative approach to the multiphase Euler equations without spurious pressure oscillations. CAM Reports, 98-11, UCLA
10. Schenk, K., Bader, G., Berti, G. (1998): Analysis and Approximation of Multicomponent Gas Mixtures. Technical Report, TU-Cottbus
11. Toro, E.F. (1999): Riemann Solvers and Numerical Methods for Fluid Dynamics: a Practical Introduction. Springer, Berlin Heidelberg
12. Vijayasundaram, G. (1986): Transonic flow simulations using an upstream centered scheme of Godunov in finite elements. J. Comput. Phys., **63**, 416–433

Discontinuous Galerkin Methods for the Time-Harmonic Maxwell Equations

Paul Houston[1], Ilaria Perugia[2], Anna Schneebeli[3] and Dominik Schötzau[4]

[1] Department of Mathematics, University of Leicester, Leicester LE1 7RH, United Kingdom, email: *Paul.Houston@mcs.le.ac.uk*. Supported by the EPSRC (Grant GR/R76615).
[2] Dipartimento di Matematica, Università di Pavia, Via Ferrata 1, 27100 Pavia, Italy, email: *perugia@dimat.unipv.it*.
[3] Department of Mathematics, University of Basel, Rheinsprung 21, 4051 Basel, Switzerland, email: *schneeba@math.unibas.ch*. Supported by the SNSF (Project 21-068126.02).
[4] Mathematics Department, University of British Columbia, Vancouver, BC V6T 1Z2, Canada, email: *schoetzau@math.ubc.ca*.

Summary. Interior penalty discontinous Galerkin methods for the time-harmonic Maxwell equations in frequency-domain, together with their stability and convergence properties, are reviewed. A new set of numerical tests carried out on a model problem with a singular analytical solution validates the theoretical error estimates of the presented method for the high-frequency case.

1 Introduction

In this paper, we review recent work on discontinuous Galerkin (DG) methods for the discretization of the time-harmonic Maxwell equations: find the electric field $\mathbf{u}$ such that

$$\nabla \times (\mu^{-1}\nabla \times \mathbf{u}) - \omega^2(\varepsilon - i\omega^{-1}\sigma)\mathbf{u} = \mathbf{j} \qquad \text{in } \Omega, \tag{1}$$

$$\mathbf{n} \times \mathbf{u} = \mathbf{0} \qquad \text{on } \Gamma = \partial\Omega. \tag{2}$$

Here, Ω is a simply-connected Lipschitz polyhedron in $\mathbb{R}^3$ with connected boundary $\Gamma = \partial\Omega$ and outward normal unit vector $\mathbf{n}$. The function $\mathbf{j}$ is a given source term in $L^2(\Omega)$. The temporal frequency is denoted by $\omega > 0$. The real-valued functions μ, ε and σ are the magnetic permeability, the electric permittivity, and the electric conductivity, respectively.

The main motivation for using a DG approach for the numerical approximation of the above problem is that DG methods, being based on discontinuous finite element spaces, can easily handle meshes with hanging nodes and local spaces of different orders. This renders DG methods ideally suited for *hp*-adaptive algorithms. Moreover, the implementation of discontinuous elements can be based on standard shape functions; a convenience that is particularly advantageous for high-order elements and that is not straightforwardly shared by standard edge elements commonly used in computational electromagnetics (see [1, 3, 14] and the references therein for *hp*-adaptive edge element methods). On the other hand, in the *hp*-context, since most of the degrees of freedom are in the interior of the elements, the increase in the total number of degrees of freedom with respect to the corresponding conforming method is not dramatic.

In this paper, we focus on interior penalty DG discretizations for (1)–(2) in the low-frequency and high-frequency cases (the term $\omega^2\varepsilon$ in (1) is neglected in the former case, whereas the term $i\omega\sigma$ is neglected in the latter).

In the low-frequency case, problem (1)–(2) has to be completed by a divergence-free constraint in the subdomain $\Omega_0 \subseteq \Omega$ covered by insulating material where $\sigma = 0$ (additional scalar constraints arise if $\partial\Omega_0$ is not connected, see, e.g., [13]). This results in the following system:

$$\nabla\times(\mu^{-1}\nabla\times\mathbf{u})+i\omega\sigma\,\mathbf{u}=\mathbf{j}\quad\text{in }\Omega,\qquad \nabla\cdot(\varepsilon\mathbf{u})=0\quad\text{in }\Omega_0,\qquad \mathbf{n}\times\mathbf{u}=\mathbf{0}\quad\text{on }\Gamma. \tag{3}$$

The main difficulty here is the incorporation of the divergence constraint in the DG framework. Following [8], we show that this can be achieved using a mixed approach where the constraint is accounted for by a suitable Lagrange multiplier. We present the underlying theoretical properties, as well as the energy norm *a priori* and *a posteriori* estimates that were derived in [8] and [6]. For further numerical tests, we also refer to [9].

In the high-frequency case, the problem that consists in finding the electric field $\mathbf{u}$ such that

$$\nabla\times(\mu^{-1}\nabla\times\mathbf{u})-\omega^2\varepsilon\mathbf{u}=\mathbf{j}\quad\text{in }\Omega,\qquad \mathbf{n}\times\mathbf{u}=\mathbf{0}\quad\text{on }\Gamma. \tag{4}$$

Here, we assume that ω^2 is not an eigenvalue of the underlying Maxwell eigenproblem. While the design of interior penalty DG methods is straightforward, the key difficulty for (4) arises in the numerical analysis of the methods due to the indefiniteness caused by the zero order term. We present the main results of a novel error analysis that was recently developed in [4]. These results show that DG methods for (4) yield optimal rates of convergence in the energy norm and the L^2-norm. We further present a new set of numerical results on a test problem with a singular solution.

To simplify the presentation in this article, we assume that μ and ε are constants. However, the analysis in [8] covers the case of piecewise smooth material coefficients, whereas the theoretical results in [4] hold for smooth coefficients μ and ε only.

2 Discontinuous Galerkin Discretizations

In this section, we introduce interior penalty DG methods for the two model problems in (3) and (4), and review their theoretical properties.

2.1 Meshes, Trace Operators, Finite Element Spaces and DG Norms

We consider shape-regular affine meshes $\mathcal{T}_h$ that partition the domain Ω into tetrahedra $\{K\}$; the parameter h denotes the mesh size of $\mathcal{T}_h$ given by $h = \max_{K\in\mathcal{T}_h} h_K$, where h_K is the diameter of the element $K \in \mathcal{T}_h$. We denote by $\mathcal{F}_h^{\mathcal{I}}$ the set of all interior faces of elements in $\mathcal{T}_h$, by $\mathcal{F}_h^{\mathcal{B}}$ the set of all boundary faces, and set $\mathcal{F}_h := \mathcal{F}_h^{\mathcal{I}} \cup \mathcal{F}_h^{\mathcal{B}}$. We define the *local meshsize* $\mathtt{h}$ on $\mathcal{F}_h$ by setting $\mathtt{h}(\mathbf{x}) := \max\{h_{K^+}, h_{K^-}\}$, if $\mathbf{x}$ is in the interior of $\partial K^+ \cap \partial K^-$, and by $\mathtt{h}(\mathbf{x}) := h_K$ if $\mathbf{x} \in \partial K$ is on the boundary.

For piecewise smooth vector-valued and scalar-valued functions $\mathbf{v}$ and q, respectively, we introduce the following trace operators. On an interior face $f \in \mathcal{F}_h^{\mathcal{I}}$ shared by two neighboring elements K^+ and K^- with unit outward normal vectors $\mathbf{n}^{\pm}$, respectively, denoting by $\mathbf{v}^{\pm}$ and $q^{\pm}$ the traces of $\mathbf{v}$ and q taken from within $K^{\pm}$, respectively, we define the *jumps* and *averages* across f by $[\![\mathbf{v}]\!]_T := \mathbf{n}^+ \times \mathbf{v}^+ + \mathbf{n}^- \times \mathbf{v}^-$, $[\![q]\!]_N = q^+\mathbf{n}^+ + q^-\mathbf{n}^-$, $\{\!\{\mathbf{v}\}\!\} := (\mathbf{v}^+ + \mathbf{v}^-)/2$ and $\{\!\{q\}\!\} := (q^+ + q^-)/2$, respectively. On a boundary face $f \in \mathcal{F}_h^{\mathcal{B}}$, we set $[\![\mathbf{v}]\!]_T := \mathbf{n} \times \mathbf{v}$, $\{\!\{\mathbf{v}\}\!\} := \mathbf{v}$ and $[\![q]\!]_N = q\mathbf{n}$.

For a given partition $\mathcal{T}_h$ of Ω and an approximation order $\ell \geq 1$, we introduce the following discontinuous finite element spaces:

$$\begin{aligned} \mathbf{V}_h &:= \{\mathbf{v} \in L^2(\Omega)^3 : \mathbf{v}|_K \in \mathcal{P}^{\ell}(K)^3 \ \ \forall K \in \mathcal{T}_h\}, \\ Q_h &= \{q \in L^2(\Omega)^3 : q|_K \in \mathcal{P}^{\ell+1}(K) \ \ \forall K \in \mathcal{T}_h\}, \end{aligned} \tag{5}$$

where $\mathcal{P}^k(K)$ denotes the space of polynomials of total degree at most k on K.

Finally, denoting by $\|\cdot\|_{s,D}$, for $s \geq 0$ and D a bounded domain in $\mathbb{R}^2$ or $\mathbb{R}^3$, the standard norm in the Sobolev space $H^s(D)^d$, $d \geq 1$, we define the following DG norms with which we will measure the approximation errors:

$$\begin{aligned} \|\mathbf{v}\|^2_{\mathbf{V}(h)} &= \|\varepsilon^{\frac{1}{2}}\mathbf{v}\|^2_{0,\Omega} + \|\mu^{-\frac{1}{2}}\nabla\times\mathbf{v}\|^2_{0,\Omega} + \|\mu^{-\frac{1}{2}}\mathtt{h}^{-\frac{1}{2}}[\![\mathbf{v}]\!]_T\|^2_{0,\mathcal{F}_h}, \\ \|q\|^2_{Q(h)} &= \|\varepsilon^{\frac{1}{2}}\nabla_h q\|^2_{0,\Omega} + \|\varepsilon^{\frac{1}{2}}\mathtt{h}^{-\frac{1}{2}}[\![q]\!]_N\|^2_{0,\mathcal{F}_h}. \end{aligned}$$

2.2 DG Discretization of the Low-Frequency Problem (Insulating Materials)

We consider the case of insulating materials, i.e., $\Omega_0 = \Omega$ since all the key difficulties in the numerical treatment of (3) are already present in this particular case. The DG method for the discretization of (3) with $\Omega_0 = \Omega$ and divergence-free source term $\mathbf{j}$ is based on the following mixed formulation of the problem:

$$\begin{aligned} \nabla\times(\mu^{-1}\nabla\times\mathbf{u}) - \varepsilon\nabla p &= \mathbf{j} && \text{in } \Omega, \\ \nabla\cdot(\varepsilon\mathbf{u}) &= 0 && \text{in } \Omega, \\ \mathbf{n}\times\mathbf{u} = \mathbf{0},\ p &= 0 && \text{on } \Gamma. \end{aligned} \tag{6}$$

Here, p is the Lagrange multiplier related to the divergence constraint. The standard variational formulation of (6) is well-posed in $H_0(\text{curl};\Omega) \times H_0^1(\Omega)$; see, e.g., [3, 14].

The mixed DG method for (6) then reads as follows: find $(\mathbf{u}_h, p_h)$ in $\mathbf{V}_h \times Q_h$ such that

$$\begin{aligned} a_h(\mathbf{u}_h, \mathbf{v}) + b_h(\mathbf{v}, p_h) &= (\mathbf{j}, \mathbf{v}), \\ b_h(\mathbf{u}_h, q) - c_h(p_h, q) &= 0 \end{aligned} \tag{7}$$

for all $(\mathbf{v}, q) \in \mathbf{V}_h \times Q_h$, where the discrete forms $a_h(\cdot,\cdot)$, $b_h(\cdot,\cdot)$ and $c_h(\cdot,\cdot)$ are defined, respectively, by

$$\begin{aligned} a_h(\mathbf{u},\mathbf{v}) &= (\mu^{-1}\nabla_h \times \mathbf{u}, \nabla_h \times \mathbf{v}) - \int_{\mathcal{F}_h} [\![\mathbf{u}]\!]_T \cdot \{\!\{\mu^{-1}\nabla_h \times \mathbf{v}\}\!\}\, ds \\ &\quad - \int_{\mathcal{F}_h} [\![\mathbf{v}]\!]_T \cdot \{\!\{\mu^{-1}\nabla_h \times \mathbf{u}\}\!\}\, ds + \int_{\mathcal{F}_h} \mathtt{a}\, \mu^{-1} [\![\mathbf{u}]\!]_T \cdot [\![\mathbf{v}]\!]_T\, ds, \\ b_h(\mathbf{v}, p) &= -(\varepsilon\mathbf{v}, \nabla_h p) + \int_{\mathcal{F}_h} \{\!\{\varepsilon\mathbf{v}\}\!\} \cdot [\![p]\!]_N\, ds, \\ c_h(p,q) &= \int_{\mathcal{F}_h} \mathtt{c}\,\varepsilon [\![p]\!]_N \cdot [\![q]\!]_N\, ds. \end{aligned}$$

Here, and in the following, we denote by $(\cdot,\cdot)$ the standard inner product in $L^2(\Omega)^d$, $d \geq 1$, and use ∇_h to denote the elementwise application of the operator ∇. Further, we use the notation $\int_{\mathcal{F}_h} \varphi\, ds := \sum_{f\in\mathcal{F}_h} \int_f \varphi\, ds$. The form $a_h(\cdot,\cdot)$ corresponds to the interior penalty discretization of the curl-curl operator, the form $b_h(\cdot,\cdot)$ discretizes the divergence operator in a DG fashion, and the form $c_h(\cdot,\cdot)$ is the interior penalty form that weakly enforces the continuity of p_h.

The parameters $\mathtt{a}$ and $\mathtt{c}$ in $L^\infty(\mathcal{F}_h)$ are the usual *interior penalty stabilization functions* defined by

$$\mathtt{a} := \alpha\, \mathtt{h}^{-1}, \qquad \mathtt{c} := \gamma\, \mathtt{h}^{-1}, \tag{8}$$

where α and γ are positive parameters independent of the mesh size.

The results contained in the following theorem have been proven and numerically validated in [8].

Theorem 1. *There is a parameter $\alpha_{\min} > 0$ only depending on the shape regularity of the mesh and the polynomial approximation degree ℓ such that, for parameters $\alpha > \alpha_{\min}$ and $\gamma > 0$ in (8), the DG method (7) possesses a unique solution.*

Moreover, assume that the analytical solution $(\mathbf{u}, p)$ of (6) satisfies the smoothness assumptions $\varepsilon\mathbf{u} \in H^s(\Omega)^3$, $\mu^{-1}\nabla\times\mathbf{u} \in H^s(\Omega)^3$ and $p \in H^{s+1}(\Omega)$, for an exponent $s > 1/2$, and let $(\mathbf{u}_h, p_h)$ be the DG approximation on conforming meshes defined by (7), with $\alpha > \alpha_{\min}$ and $\gamma > 0$. Then we have the optimal a priori error bound

$$\|\mathbf{u}-\mathbf{u}_h\|_{\mathbf{V}(h)} + \|p-p_h\|_{Q(h)} \le C\, h^{\min\{s,\ell\}} \left[\|\varepsilon\mathbf{u}\|_{s,\Omega} + \|\mu^{-1}\nabla\times\mathbf{u}\|_{s,\Omega} + \|p\|_{s+1,\Omega}\right],$$

with a constant C independent of the mesh size.

The analysis developed in [8] is valid for *piecewise* smooth coefficients μ and ε and the error estimate in Theorem 1 holds for piecewise smooth solutions. On the other hand, it requires the assumption of conformity of the meshes, although numerical tests in [9] have shown that the method is robust on meshes with hanging nodes as well.

Moreover, the following energy norm *a posteriori* error estimate has been established and tested in [6].

Theorem 2. *We assume that $\nabla\cdot\mathbf{j}=0$ holds, so that $p\equiv 0$. Let $(\mathbf{u}_h,p_h)$ be the DG approximation on conforming meshes defined by (7), with $\alpha>\alpha_{\min}$ and $\gamma>0$. Then there is a constant $C>0$ independent of the mesh size, such that*

$$\|\mathbf{u}-\mathbf{u}_h\|_{\mathbf{V}(h)} + \|p-p_h\|_{Q(h)} \le C\Big(\sum_{K\in\mathcal{T}_h}\eta_K^2\Big)^{1/2},$$

where the elemental error indicator η_K is given by

$$\begin{aligned}\eta_K^2 &= h_K^{2\sigma}\|\mathbf{j}-\nabla\times(\mu^{-1}\nabla\times\mathbf{u}_h)+\varepsilon\nabla p_h\|_{0,K}^2 + h_K^{2\sigma-1}\|\widehat{\tau}_K(\mathbf{u}_h)-\tau_K(\mathbf{u}_h)\|_{0,\partial K}^2\\ &\quad + h_K^{-1}\|\mu^{\frac12}[\![\mathbf{u}_h]\!]_T\|_{0,\partial K}^2 + h_K\|[\![\varepsilon\mathbf{u}_h]\!]_N\|_{0,\partial K\setminus\Gamma}^2 + h_K^2\|\nabla\cdot(\varepsilon\mathbf{u}_h)\|_{0,K}^2\\ &\quad + \|\varepsilon^{\frac12}\nabla p_h\|_{0,K}^2 + h_K^{-1}\|\varepsilon^{\frac12}[\![p_h]\!]_N\|_{0,\partial K}^2,\end{aligned}$$

$\sigma\in(1/2,1]$ is the parameter of the embeddings $H_0(\mathrm{curl};\Omega)\cap H(\mathrm{div};\Omega)\hookrightarrow H^\sigma(\Omega)^3$ and $H(\mathrm{curl};\Omega)\cap H_0(\mathrm{div};\Omega)\hookrightarrow H^\sigma(\Omega)^3$ (see [2]), and $\widehat{\tau}_K(\mathbf{v})$ is the numerical flux *defined by*

$$\widehat{\tau}_K(\mathbf{v}) = \begin{cases}\mathbf{n}_K\times(\{\!\{\mu^{-1}\nabla\times\mathbf{v}\}\!\} - \mu^{-1}\mathtt{a}\,[\![\mathbf{v}]\!]_T) & \text{on } \partial K\setminus\Gamma,\\ \mathbf{n}_K\times(\mu^{-1}\nabla\times\mathbf{v} - \mu^{-1}\mathtt{a}\,(\mathbf{n}_K\times\mathbf{v})) & \text{on } \partial K\cap\Gamma.\end{cases}$$

2.3 DG Discretization of the High-Frequency Problem

For problem (4), the interior penalty DG method is given by: find $\mathbf{u}_h\in\mathbf{V}_h$ such that

$$a_h(\mathbf{u}_h,\mathbf{v}) - \omega^2(\varepsilon\mathbf{u}_h,\mathbf{v}) = (\mathbf{j},\mathbf{v}) \tag{9}$$

for all $\mathbf{v}\in\mathbf{V}_h$, where the discrete form $a_h(\cdot,\cdot)$ is the same as in Section 2.2. The *interior penalty stabilization function* $\mathtt{a}\in L^\infty(\mathcal{F}_h)$ is defined again by (8), with α chosen independently of the mesh size and the frequency.

The following *a priori* error estimates in the energy norm and in the L^2-norm have been proven in [4]. Their proof is based on techniques similar to those of [12] and [11, Section 7.2], for the energy error bound, and of [10, Theorem 3.2], for the L^2-error bound, combined with novel results that allow one to approximate a discontinuous function by a conforming one. This result is instrumental in controlling the non-conformity of the DG method.

Theorem 3. *Assume that the analytical solution* $\mathbf{u}$ *of* (4) *satisfies the regularity assumptions* $\varepsilon\mathbf{u} \in H^s(\Omega)^3$ *and* $\mu^{-1}\nabla \times \mathbf{u} \in H^s(\Omega)^3$, *for* $s > \frac{1}{2}$, *and let* $\mathbf{u}_h$ *be the DG approximation on conforming meshes defined by (9). Then there is a parameter* $\alpha_{\min} > 0$ *only depending on the shape regularity of the mesh and on the polynomial approximation degree* ℓ, *and a mesh size* $h_0 > 0$ *such that, for* $\alpha \geq \alpha_{\min}$ *and* $0 < h \leq h_0$, *we have the optimal a priori error bound*

$$\|\mathbf{u} - \mathbf{u}_h\|_{\mathbf{V}(h)} \leq C\, h^{\min\{s,\ell\}} \Big[\|\varepsilon\mathbf{u}\|_{s,\Omega} + \|\mu^{-1}\nabla \times \mathbf{u}\|_{s,\Omega} \Big],$$

with a constant $C > 0$ *independent of the mesh size.*

Consequently, for $\alpha \geq \alpha_{\min}$, *the DG method* (9) *admits a unique solution* $\mathbf{u}_h \in \mathbf{V}_h$, *provided that* $h \leq h_0$.

Finally, assume that the analytical solution $\mathbf{u}$ *of* (4) *satisfies the additional regularity assumption* $\mathbf{u} \in H^{s+\sigma}(\Omega)^3$ *for* $s > \frac{1}{2}$ *and* $\sigma \in (1/2, 1]$ *the parameter of the embeddings* $H_0(\mathrm{curl};\Omega) \cap H(\mathrm{div};\Omega) \hookrightarrow H^\sigma(\Omega)^3$ *and* $H(\mathrm{curl};\Omega) \cap H_0(\mathrm{div};\Omega) \hookrightarrow H^\sigma(\Omega)^3$ *(see* [2]*). Then, for* $\alpha \geq \alpha_{\min}$, *there is a mesh size* $h_2 > 0$ *such that, for* $0 < h \leq h_2$, *we have*

$$\|\varepsilon^{\frac{1}{2}}(\mathbf{u} - \mathbf{u}_h)\|_{0,\Omega} \leq C\, h^{\min\{s,\ell\}+\sigma} \Big[\|\varepsilon\mathbf{u}\|_{s+\sigma,\Omega} + \|\mu^{-1}\nabla \times \mathbf{u}\|_{s,\Omega} \Big],$$

with a constant $C > 0$ *independent of the mesh size.*

The analysis of [4] is based on duality arguments; thus, the result of Theorem 3 can easily be extended to smooth material coefficients μ and ε. However, the extension to piecewise smooth coefficients requires alternative mathematical tools; this is the subject of ongoing research.

3 Numerical Example

In this section we present a numerical example to highlight the practical performance of the DG method introduced and analyzed in this article for the numerical approximation of the high-frequency indefinite time-harmonic Maxwell equations in (4), considering a model problem with a singular solution. Throughout this section, we take $\mu = \mu_0$ and $\varepsilon = \varepsilon_0$, the permeability and permittivity of the free space, respectively, and select the interior penalty parameter α in (8) as follows: $\alpha = 10\,\ell^2$. As is standard in electromagnetic computations, we scale the electric field by $\mathbf{u} \to \mu_0\mathbf{u}$ and obtain a problem for the scaled field (that we again denote by $\mathbf{u}$) of the form

$$\nabla \times \nabla \times \mathbf{u} - k^2\mathbf{u} = \mathbf{j} \quad \text{in } \Omega, \qquad \mathbf{n} \times \mathbf{u} = \mathbf{0} \quad \text{on } \Gamma, \tag{10}$$

with a rescaled right-hand side (again denoted by $\mathbf{j}$) and the wave number $k = \omega\sqrt{\mu_0\varepsilon_0}$.

For simplicity, we restrict ourselves to the two-dimensional analogue of (10). To this end, we let Ω be the L-shaped domain $(-1,1)^2 \setminus [0,1) \times (-1,0]$ and select $\mathbf{j}$ (and suitable non-homogeneous boundary conditions for $\mathbf{u}$) so that the analytical solution $\mathbf{u}$ to the two-dimensional analogue of (10) is given, in terms of the polar coordinates (r, ϑ), by

$$\mathbf{u}(x,y) = \nabla S(r,\vartheta), \quad \text{where} \quad S(r,\vartheta) = (kr)^{2/3}\sin(2\vartheta/3). \tag{11}$$

Here, the boundary conditions are enforced in the usual DG manner by adding boundary terms in the formulation (9); see [5, 7] for details. The analytical solution given by (11) then contains a singularity at the re-entrant corner located at the origin of Ω; in particular, we note that $\mathbf{u}$ lies in the Sobolev space $H^{2/3-\varepsilon}(\Omega)^2$, $\varepsilon > 0$. This example represents a slight modification of the numerical experiment presented in [4]; cf., also, [1].

We investigate the asymptotic convergence of the DG method on a sequence of successively finer (quasi-uniform) unstructured triangular meshes for $\ell = 1, 2, 3$ as the wave number k increases. To this end, in Tables 1, 2, 3 and 4 we present numerical experiments for $k = 1, 2, 4, 6$, respectively. In each case we show the number of elements in the computational mesh, the corresponding DG-norm of the error $\|\mathbf{u}-\mathbf{u}_h\|_{\mathbf{V}(h)}$ and the numerical rate of convergence r. In view of the scaling we introduced, we have taken $\|\mathbf{u}-\mathbf{u}_h\|_{\mathbf{V}(h)}$ as $\left(\|\mathbf{u}-\mathbf{u}_h\|^2_{0,\Omega}+\|\nabla\times(\mathbf{u}-\mathbf{u}_h)\|^2_{0,\Omega}+\|\mathrm{h}^{-\frac{1}{2}}[\![\mathbf{u}-\mathbf{u}_h]\!]_T\|^2_{0,\mathcal{F}_h}\right)^{\frac{1}{2}}$. We observe that (asymptotically) $\|\mathbf{u}-\mathbf{u}_h\|_{\mathbf{V}(h)}$ converges to zero at the optimal rate $\mathcal{O}(h^{2/3-\varepsilon})$, for each fixed ℓ and each k, as h tends to zero, as predicted by Theorem 3. In particular, we make two key observations: firstly, we note that for a given fixed mesh and fixed polynomial degree, an increase in the wave number k leads to an increase in the DG-norm of the error in the approximation to $\mathbf{u}$ As pointed out in [1], where curl-conforming finite element methods were employed for the numerical approximation of (10), the pre-asymptotic region increases as k increases; this is particularly evident when $k = 6$, cf. Table 4. Secondly, we observe that the DG-norm of the error decreases when either the mesh is refined, or the polynomial degree is increased as we would expect; this is also the case when the DG-norm of the error is compared with the total number of degrees of freedom employed in the underlying finite element space, for each fixed k; for brevity these results have been omitted.

Finally, we end this section by considering the rate of convergence of the error in the approximation to $\mathbf{u}$ measured in the L^2-norm. While for smooth solutions the optimal L^2-order has been confirmed numerically (see [4]), the additional regularity assumptions for the L^2 estimate in Theorem 3 do not hold in the example considered here. Notwithstanding this, in Figure 1 we plot the L^2-norm of the error in the approximation to $\mathbf{u}$, with the square root of the number of degrees of freedom in the finite element space $\mathbf{V}_h$, for $k = 1$ and $k = 6$. We observe that (asymptotically) $\|\mathbf{u}-\mathbf{u}_h\|_{0,\Omega}$ converges to zero at the rate $\mathcal{O}(h^{2/3})$, for each fixed ℓ and k, as in the case of the DG-norm of the error.

Table 1. Convergence of $\|\mathbf{u}-\mathbf{u}_h\|_{\mathbf{V}(h)}$ with $k = 1$.

	$\ell = 1$		$\ell = 2$		$\ell = 3$	
Elements	$\|\mathbf{u}-\mathbf{u}_h\|_{\mathbf{V}(h)}$	r	$\|\mathbf{u}-\mathbf{u}_h\|_{\mathbf{V}(h)}$	r	$\|\mathbf{u}-\mathbf{u}_h\|_{\mathbf{V}(h)}$	r
24	1.525e-1	-	8.881e-2	-	6.078e-2	-
96	8.875e-2	0.78	5.374e-2	0.73	3.744e-2	0.70
384	5.393e-2	0.72	3.331e-2	0.69	2.337e-2	0.68
1536	3.348e-2	0.69	2.085e-2	0.68	1.467e-2	0.67
6144	2.096e-2	0.68	1.310e-2	0.67	9.227e-3	0.67

Table 2. Convergence of $\|\mathbf{u}-\mathbf{u}_h\|_{\mathbf{V}(h)}$ with $k = 2$.

	$\ell = 1$		$\ell = 2$		$\ell = 3$	
Elements	$\|\mathbf{u}-\mathbf{u}_h\|_{\mathbf{V}(h)}$	r	$\|\mathbf{u}-\mathbf{u}_h\|_{\mathbf{V}(h)}$	r	$\|\mathbf{u}-\mathbf{u}_h\|_{\mathbf{V}(h)}$	r
24	2.272e-1	-	1.366e-1	-	9.490e-2	-
96	1.369e-1	0.73	8.428e-2	0.70	5.904e-2	0.69
384	8.464e-2	0.69	5.262e-2	0.68	3.700e-2	0.67
1536	5.290e-2	0.68	3.303e-2	0.67	2.326e-2	0.67
6144	3.322e-2	0.67	2.078e-2	0.67	1.464e-2	0.67

Table 3. Convergence of $\|\mathbf{u}-\mathbf{u}_h\|_{\mathbf{V}(h)}$ with $k = 4$.

	$\ell = 1$		$\ell = 2$		$\ell = 3$	
Elements	$\|\mathbf{u}-\mathbf{u}_h\|_{\mathbf{V}(h)}$	r	$\|\mathbf{u}-\mathbf{u}_h\|_{\mathbf{V}(h)}$	r	$\|\mathbf{u}-\mathbf{u}_h\|_{\mathbf{V}(h)}$	r
24	8.727e-1	-	3.900e-1	-	2.274e-1	-
96	3.637e-1	1.26	1.853e-1	1.07	1.156e-1	0.98
384	1.793e-1	1.02	9.786e-2	0.92	6.462e-2	0.84
1536	9.670e-2	0.89	5.622e-2	0.80	3.845e-2	0.75
6144	5.611e-2	0.79	3.395e-2	0.73	2.363e-2	0.70

4 Conclusions

In this paper, we have reviewed two interior penalty discontinuous Galerkin methods for the numerical approximation of the time-harmonic Maxwell equations in both the low-frequency and high-frequency regimes. The predicted performance of the DG method for the indefinite problem in the high-frequency case has been confirmed in a new set of numerical experiments carried out on a model problem with a singular solution.

Table 4. Convergence of $\|\mathbf{u}-\mathbf{u}_h\|_{\mathbf{V}(h)}$ with $k=6$.

	$\ell=1$		$\ell=2$		$\ell=3$	
Elements	$\|\mathbf{u}-\mathbf{u}_h\|_{\mathbf{V}(h)}$	r	$\|\mathbf{u}-\mathbf{u}_h\|_{\mathbf{V}(h)}$	r	$\|\mathbf{u}-\mathbf{u}_h\|_{\mathbf{V}(h)}$	r
24	2.535	-	8.535e-1	-	6.884e-1	-
96	7.305e-1	1.79	5.972e-1	0.52	2.048e-1	1.75
384	3.230e-1	1.18	1.641e-1	1.86	1.002e-1	1.03
1536	1.879e-1	0.78	8.383e-2	0.97	5.461e-2	0.88
6144	8.275e-2	1.18	4.722e-2	0.83	3.207e-2	0.77

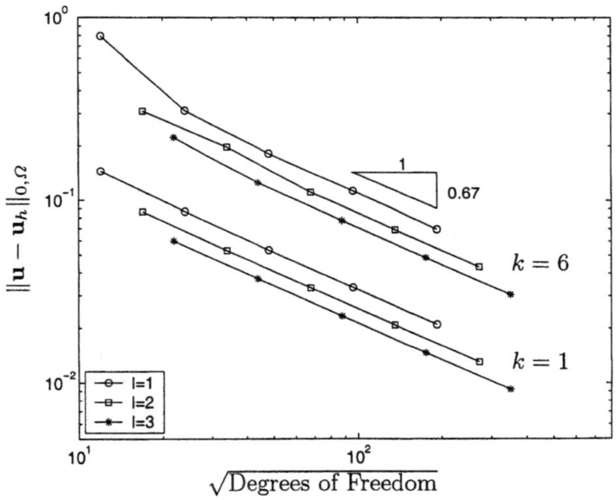

Fig. 1. Convergence of $\|\mathbf{u}-\mathbf{u}_h\|_{0,\Omega}$ for $k=1$ and $k=6$

References

1. M. Ainsworth and J. Coyle. Hierarchic *hp*-edge element families for Maxwell's equations on hybrid quadrilateral/triangular meshes. *Comput. Methods Appl. Mech. Engrg.*, 190:6709–6733, 2001.
2. C. Amrouche, C. Bernardi, M. Dauge, and V. Girault. Vector potentials in three-dimensional non-smooth domains. *Math. Models Appl. Sci.*, 21:823–864, 1998.
3. L. Demkowicz and L. Vardapetyan. Modeling of electromagnetic absorption/scattering problems using *hp*–adaptive finite elements. *Comput. Methods Appl. Mech. Engrg.*, 152:103–124, 1998.
4. P. Houston, I. Perugia, A. Schneebeli, and D. Schötzau. Interior penalty method for the indefinite time-harmonic maxwell equations. Technical Report PIMS-03-15, Pacific Institute for the Mathematical Sciences, Universify of British Columbia, 2003.

5. P. Houston, I. Perugia, and D. Schötzau. Mixed discontinuous Galerkin approximation of the Maxwell operator. Technical Report 02-16, University of Basel, Department of Mathematics, 2002. Accepted for publication in *SIAM J. Numer. Anal.*
6. P. Houston, I. Perugia, and D. Schötzau. Energy norm a posteriori error estimation for mixed discontinuous Galerkin approximations of the Maxwell operator. Technical Report 2003/16, University of Leicester, Department of Mathematics, 2003.
7. P. Houston, I. Perugia, and D. Schötzau. *hp*-DGFEM for Maxwell's equations. In F. Brezzi, A. Buffa, S. Corsaro, and A. Murli, editors, *Numerical Mathematics and Advanced Applications ENUMATH 2001*, pages 785–794. Springer-Verlag, 2003.
8. P. Houston, I. Perugia, and D. Schötzau. Mixed discontinuous Galerkin approximation of the Maxwell operator: Non-stabilized formulation. Technical Report 2003/17, University of Leicester, Department of Mathematics, 2003.
9. P. Houston, I. Perugia, and D. Schötzau. Nonconforming mixed finite element approximations to time-harmonic eddy current problems. Technical Report 2003/15, University of Leicester, Department of Mathematics, 2003.
10. P. Monk. A finite element method for approximating the time-harmonic Maxwell equations. *Numer. Math.*, 63:243–261, 1992.
11. P. Monk. *Finite element methods for Maxwell's equations.* Oxford University Press, New York, 2003.
12. P. Monk. A simple proof of convergence for an edge element discretization of Maxwell's equations. In C. Carstensen, S. Funken, W. Hackbusch, R. Hoppe, and P. Monk, editors, *Computational electromagnetics*, volume 28 of *Lect. Notes Comput. Sci. Engrg.*, pages 127–141. Springer–Verlag, 2003.
13. I. Perugia and D. Schötzau. The *hp*-local discontinuous Galerkin method for low-frequency time-harmonic Maxwell equations. *Math. Comp.*, 72:1179–1214, 2003.
14. L. Vardapetyan and L. Demkowicz. *hp*-adaptive finite elements in electromagnetics. *Comput. Methods Appl. Mech. Engrg.*, 169:331–344, 1999.

MIX
Papier aus verantwortungsvollen Quellen
Paper from responsible sources
FSC® C105338
FSC
www.fsc.org

If you have any concerns about our products,
you can contact us on
ProductSafety@springernature.com

In case Publisher is established outside the EU,
the EU authorized representative is:
Springer Nature Customer Service Center GmbH
Europaplatz 3, 69115 Heidelberg, Germany

Printed by Libri Plureos GmbH
in Hamburg, Germany

Numerical Mathematics and Advanced Applications

M. Feistauer • V. Dolejší • P. Knobloch • K. Najzar
Editors

Numerical Mathematics and Advanced Applications

Proceedings of ENUMATH 2003
the 5th European Conference
on Numerical Mathematics and
Advanced Applications

Prague, August 2003

Springer

Editors

Miloslav Feistauer
Vít Dolejší
Petr Knobloch
Karel Najzar

Charles University Prague
Faculty of Mathematics and Physics
Department of Numerical Mathematics
Sokolovská 83, 186 75 Praha 8

Czech Republic

email:
feist, dolejsi, knobloch, knaj@karlin.mff.cuni.cz

Library of Congress Control Number: 2004107015
Mathematics Subject Classification (2000): 65-XX, 74-XX, 76-XX, 78-XX

ISBN 978-3-642-62288-5 ISBN 978-3-642-18775-9 (eBook)
DOI 10.1007/978-3-642-18775-9

Originally published by Springer-Verlag Berlin Heidelberg New York in 2004
Softcover reprint of the hardcover 1st edition 2004

Cover Design: *design & production GmbH*, Heidelberg
Typesetting: Computer to film by author's data
Printed on acid-free paper 40/3142XT 5 4 3 2 1 0

Preface

These proceedings collect the major part of the lectures given at ENUMATH2003, the European Conference on Numerical Mathematics and Advanced Applications, held in Prague, Czech Republic, from 18 August to 22 August, 2003.

The importance of numerical and computational mathematics and scientific computing is permanently growing. There is an increasing number of different research areas, where numerical simulation is necessary. Let us mention fluid dynamics, continuum mechanics, electromagnetism, phase transition, cosmology, medicine, economics, finance, etc. The success of applications of numerical methods is conditioned by changing its basic instruments and looking for new appropriate techniques adapted to new problems as well as new computer architectures.

The ENUMATH conferences were established in order to provide a forum for discussion of current topics of numerical mathematics. They seek to convene leading experts and young scientists with special emphasis on contributions from Europe. Recent results and new trends are discussed in the analysis of numerical algorithms as well as in their applications to challenging scientific and industrial problems.

The first ENUMATH conference was organized in Paris in 1995, then the series continued by the conferences in Heidelberg 1997, Jyväskylä 1999 and Ischia Porto 2001. It was a great pleasure and honour for the Czech numerical community that it was decided at Ischia Porto to organize the ENUMATH2003 in Prague. It was the first time when this conference crossed the former Iron Courtain and was organized in a postsocialist country.

The ENUMATH2003 was organized by the Faculty of Mathematics and Physics of the Charles University in cooperation with the Department of Mathematics of the Institute of Chemical Technology in Prague. The Charles University, the oldest university in the Middle Europe, was founded in 1348. In the middle ages, mathematics was studied in Prague at the Artistic Faculty, later at the Philosophical Faculty and in the 20th century it belonged to the Faculty of Natural Sciences till 1952, when the Faculty of Mathematics and Physics was founded. As follows from historical sources, already in the 15th century the students of the Charles University had the opportunity to be trained in "Computational Mathematics". Křišťan from Prachatice, who is in the Czech history well–known as a Church reformer (he was a friend of the famous reformer Jan Hus), or a personal doctor of medicine of the Czech and German-Roman King Wenceslas IV, was a professor of astronomy and mathematics at the Charles University. He wrote lecture notes with the title "Algoritmus Prosaicus", where he describes among other approximate methods for the realization of various mathematical operations. The contemporary Czech school of numerical and applied mathematics was born much later in the second part of the 20th century. It is connected mainly with the names of Ivo Babuška, who can be considered as a father of the Czech numerical mathematics, Karel Rektorys, Milan Práger, Emil Vitásek, Miloš Zlámal, Alexander Ženíšek, Ivan Hlaváček, a famous specialist in PDE's Jindřich Nečas (who influenced a number of Czech numerical analysts) and Jan Polášek (who founded the Prague school of CFD).

These proceedings contain a selection of invited plenary lectures, papers presented in minisymposia and works communicated within the sessions. All contributions of these proceedings have been reviewed by members of the Scientific Committee. At this occasion we want to thank the members of the Program Committee (F. Brezzi, M. Feistauer, R. Glowinski, R. Jeltsch, Yu. Kuznetsov, J. Periaux, R. Rannacher) and the members of the Scientific Committee (O. Axelsson, C. Bernardi, C. Canuto, M. Griebel, R. Hoppe, G. Kobelkov, M. Křížek, P. Neittaanmäki, O. Pironneau, A. Quarteroni, C. Schwab, E. Süli, W. Wendland) for their scientific support. We are grateful to the plenary speakers R. Blaheta, A. Bermúdez, T. Gallouët, J. Haslinger, R. Hiptmair, T. Hughes, J. Rappaz, A. Russo, A. Tveito and V. Schulz for coming to Prague and richly contributing to the success of the conference.

We are very much obliged to A. Klíč, the head of the Department of Mathematics of the Institute of Chemical Technology, and our colleagues J. Felcman, J. Segethová and E. Plandorová from the local organizing committee for the cooperation in the organization of the conference. Finally, we are gratefully indebted to O. Ulrych for the TEX editing and the preparation of the camera-ready manuscript of the proceedings. Lastly, we thank all participants for coming and animating the meeting.

We believe that this volume will be an invaluable instrument for obtaining an overview of the latest and newest results and aspects of numerical mathematics and scientific computing and their applications.

M. Feistauer
V. Dolejší
P. Knobloch
K. Najzar
editors

Table of Contents

Mixed *hp*-Discontinuous Galerkin Finite Element Methods for the Stokes Problem in Polygons

Paul Houston[1], Dominik Schötzau[2] and Thomas P. Wihler[3]

[1] Department of Mathematics, University of Leicester
Leicester LE1 7RH, UK, Email: *Paul.Houston@mcs.le.ac.uk*

Supported by the EPSRC (Grant GR/R76615)

[2] Mathematics Department, University of British Columbia
Vancouver BC V6T 1Z2, Canada, Email: *schoetzau@math.ubc.ca*

[3] School of Mathematics, University of Minnesota
Minneapolis MN 55455, USA, Email: *wihler@math.umn.edu*

Supported by the Swiss National Science Foundation under project PBEZ2-102321

Summary. We consider mixed *hp*–discontinuous Galerkin finite element methods (DGFEM) for Stokes flow in general polygons. In particular, we show that, on geometrically refined meshes, the *hp*-DGFEM yields exponential rates of convergence for problems with piecewise analytic input data. Numerical results confirming the exponential convergence rates are presented.

1 Introduction

Over the last few years, several mixed discontinuous Galerkin finite element methods (DGFEM) have been proposed for the discretization of incompressible fluid flow problems; see, e.g., [4, 6, 7, 9, 12, 16] and the references therein. The main motivations that led to these schemes are that mixed DGFEM provide robust and high-order accurate approximations, particularly in transport-dominated regimes, and that they are considerably flexible in the choice of velocity-pressure combinations, without excessive numerical stabilization. For example, no extra stabilization is required to use optimally matched combinations where the approximation degree for the pressure is of one order lower than that of the velocity; this result was first established in [11] in the context of linear elasticity of nearly incompressible materials.

The work in [13] presented a unifying framework for the analysis of mixed *hp*-DGFEM for Stokes flow. For discontinuous $\mathcal{Q}^p - \mathcal{Q}^{p-1}$ elements on hexahedral meshes, the results there ensure (slightly suboptimal) error bounds for the *p*-version of the DGFEM, where convergence is obtained by increasing the polynomial approximation order p on a fixed mesh. However, these bounds result in algebraic rates of convergence and are restricted to piecewise smooth solutions – an assumption that is unrealistic in general polygons, due to the presence of corner singularities.

In this note, we report on the recent results in [14] that extend the approach of [13] to mixed *hp*-DGFEM for Stokes flow in general polygons. In particular, we show that, on geometrically refined meshes, the *hp*-DGFEM yields exponential rates of convergence for problems with piecewise analytic input data; see [18] for similar results in the context of diffusion problems. We further present numerical results that confirm the exponential convergence rates as predicted in [14].

2 The Stokes problem in polygons

In this section, we introduce the Stokes problem and use the results of [10] to describe the regularity of its solution for piecewise analytic input data.

2.1 The Stokes problem

Let $\Omega \subset \mathbb{R}^2$ be a bounded polygonal domain with outward unit normal vector $\mathbf{n}$ on the boundary $\partial\Omega$. Then, for a given forcing term $\mathbf{f} \in H^{-1}(\Omega)^2$ and a Dirichlet datum $\mathbf{g} \in H^{\frac{1}{2}}(\partial\Omega)^2$ satisfying the compatibility condition $\int_{\partial\Omega} \mathbf{g} \cdot \mathbf{n}\, ds = 0$, the Stokes problem is to find a velocity field $\mathbf{u} \in H^1(\Omega)^2$ and a pressure $p \in L^2_0(\Omega) := L^2(\Omega)/\mathbb{R}$ such that

$$\begin{aligned} -\Delta \mathbf{u} + \nabla p &= \mathbf{f} \quad \text{in } \Omega, \\ \nabla \cdot \mathbf{u} &= 0 \quad \text{in } \Omega, \\ \mathbf{u} &= \mathbf{g} \quad \text{on } \partial\Omega. \end{aligned} \tag{1}$$

This system is uniquely solvable; see, e.g., [5, 8] for details.

2.2 Analytic regularity in polygons

In [10] the regularity of the solution $(\mathbf{u}, p)$ to the Stokes equations with piecewise analytic data $\mathbf{f}$ and $\mathbf{g}$ was described in terms of the countably normed Sobolev spaces introduced by Babuška and Guo for closely related diffusion and elasticity problems (see [2, 3, 15], and the references cited therein). To define these spaces, let $\{A_i\}_{i=1}^M$ denote the vertices of the domain Ω. To each vertex A_i we assign a weight $\beta_i \geq 0$ and store these numbers in the M-tuple $\beta = (\beta_1, \ldots, \beta_M)$. We define $\beta \pm j := (\beta_1 \pm j, \ldots, \beta_M \pm j)$ and use the shorthand notation $C_1 > \beta > C_2$ to mean $C_1 > \beta_i > C_2$ for $i = 1, \ldots, M$. Writing $r_i(\mathbf{x}) = \min\{1, |\mathbf{x} - A_i|\}$, we define the weight function $\Phi_\beta(\mathbf{x}) := \prod_{i=1}^M r_i(\mathbf{x})^{\beta_i}$, and introduce the semi-norms

$$|u|^2_{H^{k,l}_\beta(\Omega)} := \sum_{|\alpha| \geq l}^{k} \|\Phi_{\beta+|\alpha|-l} D^\alpha u\|^2_{L^2(\Omega)}, \qquad k \geq l \geq 0.$$

We denote by $H_\beta^{k,l}(\Omega)$ the completion of $C^\infty(\overline{\Omega})$ with respect to the norm

$$\|u\|^2_{H_\beta^{k,l}(\Omega)} := \|u\|^2_{H^{l-1}(\Omega)} + |u|^2_{H_\beta^{k,l}(\Omega)}, \qquad l \geq 1,$$

$$\|u\|^2_{H_\beta^{k,0}(\Omega)} := \sum_{|\alpha|=0}^{k} \|\Phi_{\beta+|\alpha|} D^\alpha u\|^2_{L^2(\Omega)}.$$

Here, we denote by $\|\cdot\|_{L^2(\Omega)}$ the usual L^2-norm. Similarly, $\|\cdot\|_{H^l(\Omega)}$ is the norm on the standard Sobolev space $H^l(\Omega)$.

Definition 1. *For an M-tuple $\beta = (\beta_1, \ldots, \beta_M)$ and $l \geq 0$, the countably normed space $B_\beta^l(\Omega)$ consists of all functions u for which $u \in H_\beta^{k,l}(\Omega)$ for $k \geq l$ and*

$$\|\Phi_{\beta+k-l} D^\alpha u\|_{L^2(\Omega)} \leq C d^{(k-l)} (k-l)!, \qquad |\alpha| = k \geq l,$$

for constants $C > 0$, $d \geq 1$ independent of k. Moreover, for $l \geq 1$, the space $B_\beta^{l-\frac{1}{2}}(\partial\Omega)$ is the space of traces of functions in $B_\beta^l(\Omega)$.

Functions in $B_\beta^l(\Omega)$ (or their traces) are referred to as *piecewise analytic functions*. Indeed, they are analytic in any interior domain $\Omega_{\text{int}} \subset \Omega$ with $\overline{\Omega}_{\text{int}} \subset \overline{\Omega} \setminus \{A_i\}_{i=1}^M$, but develop singularities at the corners $\{A_i\}_{i=1}^M$. Globally, we have $B_\beta^l(\Omega) \subset H^{l-1}(\Omega)$, but $B_\beta^l(\Omega) \not\subset H^l(\Omega)$. The following regularity result was proved in [10].

Theorem 1. *There exists a weight vector $0 \leq \beta_{\min} < 1$ depending on the opening angles of Ω at the vertices $\{A_i\}_{i=1}^M$ such that for weight vectors β with $\beta_{\min} < \beta < 1$ and piecewise analytic data $(\mathbf{f}, \mathbf{g}) \in B_\beta^0(\Omega)^2 \times B_\beta^{\frac{3}{2}}(\partial\Omega)^2$, the solution $(\mathbf{u}, p)$ of the Stokes system (1) satisfies $(\mathbf{u}, p) \in B_\beta^2(\Omega)^2 \times B_\beta^1(\Omega)$.*

In the rest of the paper, we assume that $(\mathbf{f}, \mathbf{g}) \in B_\beta^0(\Omega)^2 \times B_\beta^{\frac{3}{2}}(\partial\Omega)^2$ for a weight vector β with $\beta_{\min} < \beta < 1$, in order to ensure the piecewise analyticity of the solution $(\mathbf{u}, p)$, as stated in Theorem 1.

3 Discontinuous Galerkin methods

In this section, we introduce discontinuous Galerkin methods for the Stokes problem (1) and review their well-posedness, using the recent results in [13].

3.1 Meshes

Throughout, we assume that the domain Ω can be subdivided into shape-regular affine meshes $\mathcal{T}_h = \{K\}$ consisting of parallelograms K. For each $K \in \mathcal{T}_h$, we denote by $\mathbf{n}_K$ the outward unit normal vector to the boundary ∂K, and by h_K the elemental diameter. Further, we assign to each element $K \in \mathcal{T}_h$ an approximation order $k_K \geq 1$. The local quantities h_K and k_K are stored in the vectors $\mathbf{h} = \{h_K\}_{K \in \mathcal{T}_h}$ and $\mathbf{k} = \{k_K\}_{K \in \mathcal{T}_h}$, respectively. An interior edge of $\mathcal{T}_h$ is the (non-empty) one-dimensional interior of $\partial K^+ \cap \partial K^-$, where K^+ and K^- are two adjacent elements of $\mathcal{T}_h$. Similarly, a boundary edge of $\mathcal{T}_h$ is the (non-empty) one-dimensional interior of $\partial K \cap \partial \Omega$ which consists of entire edges of ∂K. We denote by $\mathcal{E}_\mathcal{I}$ the union of all interior edges of $\mathcal{T}_h$, by $\mathcal{E}_\mathcal{D}$ the union of all boundary edges, and set $\mathcal{E} = \mathcal{E}_\mathcal{I} \cup \mathcal{E}_\mathcal{D}$. We allow for meshes with 1-irregular hanging nodes, and further assume that there is a constant $\kappa > 0$ such that $\kappa k_K \leq k_{K'} \leq \kappa^{-1} k_K$, whenever K and K' share a common edge.

3.2 Averages and jumps

Next, we define average and jump operators. To this end, let K^+ and K^- be two adjacent elements of $\mathcal{T}_h$ and $\mathbf{x}$ be an arbitrary point on the interior edge $e = \partial K^+ \cap \partial K^- \subset \mathcal{E}_\mathcal{I}$. Moreover, let q, $\mathbf{v}$, and $\underline{\tau}$ be scalar-, vector-, and matrix-valued functions, respectively, that are smooth inside each element $K^\pm$. By $(q^\pm, \mathbf{v}^\pm, \underline{\tau}^\pm)$ we denote the traces of $(q, \mathbf{v}, \underline{\tau})$ on e taken from within the interior of $K^\pm$, respectively. Then, we define the averages at $\mathbf{x} \in e$ by:

$$\{\!\{q\}\!\} = (q^+ + q^-)/2, \qquad \{\!\{\mathbf{v}\}\!\} = (\mathbf{v}^+ + \mathbf{v}^-)/2, \qquad \{\!\{\underline{\tau}\}\!\} = (\underline{\tau}^+ + \underline{\tau}^-)/2.$$

Similarly, the jumps at $\mathbf{x} \in e$ are given by

$$\begin{aligned} [\![q]\!] &= q^+ \mathbf{n}_{K^+} + q^- \mathbf{n}_{K^-}, & [\![\mathbf{v}]\!] &= \mathbf{v}^+ \cdot \mathbf{n}_{K^+} + \mathbf{v}^- \cdot \mathbf{n}_{K^-}, \\ \underline{[\![\mathbf{v}]\!]} &= \mathbf{v}^+ \otimes \mathbf{n}_{K^+} + \mathbf{v}^- \otimes \mathbf{n}_{K^-}, & [\![\underline{\tau}]\!] &= \underline{\tau}^+ \mathbf{n}_{K^+} + \underline{\tau}^- \mathbf{n}_{K^-}. \end{aligned}$$

On boundary edges $e \subset \mathcal{E}_\mathcal{D}$, we set $\{\!\{q\}\!\} = q$, $\{\!\{\mathbf{v}\}\!\} = \mathbf{v}$, $\{\!\{\underline{\tau}\}\!\} = \underline{\tau}$, as well as $[\![q]\!] = q\mathbf{n}$, $[\![\mathbf{v}]\!] = \mathbf{v} \cdot \mathbf{n}$, $\underline{[\![\mathbf{v}]\!]} = \mathbf{v} \otimes \mathbf{n}$, and $[\![\underline{\tau}]\!] = \underline{\tau}\mathbf{n}$.

3.3 Mixed *hp*-DGFEM

Given a mesh $\mathcal{T}_h$ and a degree vector $\mathbf{k} = \{k_K\}$, $k_K \geq 1$, we wish to approximate the Stokes problem (1) by finite element functions $(\mathbf{u}_h, p_h) \in \mathbf{V}_h \times Q_h$, where

$$\begin{aligned} \mathbf{V}_h &= \{\, \mathbf{v} \in L^2(\Omega)^2 : \mathbf{v}|_K \in \mathcal{Q}^{k_K}(K)^2, \ K \in \mathcal{T}_h \,\}, \\ Q_h &= \{\, q \in L^2_0(\Omega) : q|_K \in \mathcal{Q}^{k_K - 1}(K), \ K \in \mathcal{T}_h \,\}. \end{aligned}$$

Here, $\mathcal{Q}^k(K)$ is the space of polynomials of degree at most k in each variable on K. Thereby, we consider the following mixed method: find $(\mathbf{u}_h, p_h) \in \mathbf{V}_h \times Q_h$ such that

$$\left\{ \begin{aligned} A_h(\mathbf{u}_h, \mathbf{v}) \ + B_h(\mathbf{v}, p_h) &= F_h(\mathbf{v}), \\ -B_h(\mathbf{u}_h, q) \qquad\qquad &= G_h(q) \end{aligned} \right. \tag{2}$$

for all $(\mathbf{v}, q) \in \mathbf{V}_h \times Q_h$. The forms A_h and B_h are discontinuous Galerkin forms that discretize the Laplacian and the incompressibility constraint, respectively, with corresponding right-hand sides F_h and G_h. These forms are given by

$$\begin{aligned} A_h(\mathbf{u},\mathbf{v}) &= \int_\Omega \nabla_h \mathbf{u} : \nabla_h \mathbf{v}\, d\mathbf{x} - \int_\mathcal{E} \left(\{\!\{\nabla_h \mathbf{v}\}\!\} : \underline{[\![\mathbf{u}]\!]} + \{\!\{\nabla_h \mathbf{u}\}\!\} : \underline{[\![\mathbf{v}]\!]}\right) ds \\ &\quad + \int_\mathcal{E} \mathsf{c}\underline{[\![\mathbf{u}]\!]} : \underline{[\![\mathbf{v}]\!]}\, ds, \\ B_h(\mathbf{v},q) &= -\int_\Omega q\, \nabla_h \cdot \mathbf{v}\, d\mathbf{x} + \int_\mathcal{E} \{\!\{q\}\!\} [\![\mathbf{v}]\!]\, ds, \\ F_h(\mathbf{v}) &= \int_\Omega \mathbf{f}\cdot\mathbf{v}\, d\mathbf{x} - \int_{\mathcal{E}_\mathcal{D}} (\mathbf{g}\otimes\mathbf{n}) : \nabla_h \mathbf{v}\, ds + \int_{\mathcal{E}_\mathcal{D}} \mathsf{c}\mathbf{g}\cdot\mathbf{v}\, ds, \\ G_h(q) &= -\int_{\mathcal{E}_\mathcal{D}} q\,\mathbf{g}\cdot\mathbf{n}\, ds. \end{aligned}$$

Here, ∇_h and $\nabla_h\cdot$ denote the discrete gradient and divergence operator, respectively, taken element-wise. The function $\mathsf{c} \in L^\infty(\mathcal{E})$ is the so-called discontinuity stabilization function that is chosen as follows. Define the functions $\mathsf{h} \in L^\infty(\mathcal{E})$ and $\mathsf{k} \in L^\infty(\mathcal{E})$ by

$$\mathsf{h}(\mathbf{x}) := \begin{cases} \min\{h_K, h_{K'}\}, & \mathbf{x} \in e = \partial K \cap \partial K' \subset \mathcal{E}_\mathcal{I}, \\ h_K, & \mathbf{x} \in e = \partial K \cap \partial\Omega \subset \mathcal{E}_\mathcal{D}, \end{cases}$$

$$\mathsf{k}(\mathbf{x}) := \begin{cases} \max\{k_K, k_{K'}\}, & \mathbf{x} \in e = \partial K \cap \partial K' \subset \mathcal{E}_\mathcal{I}, \\ k_K, & \mathbf{x} \in e = \partial K \cap \partial\Omega \subset \mathcal{E}_\mathcal{D}. \end{cases}$$

Then we set

$$\mathsf{c} = \gamma \mathsf{h}^{-1}\mathsf{k}^2,$$

with a parameter $\gamma > 0$ that is independent of h and k.

Remark 1. The form A_h corresponds to the so-called symmetric interior penalty (IP) discretization of the Laplace operator; see [1] and [13] where the presentation and analysis of several different DG methods were unified for diffusion problems and the Stokes system, respectively. All the results presented in this paper hold true verbatim for all the mixed DG methods investigated in [13].

Remark 2. For piecewise analytic data $\mathbf{f}$ and $\mathbf{g}$ as in Theorem 1, the forms F_h and G_h are well-defined.

3.4 Well-posedness

Well-posedness of the discrete system (2) was established in [13]. Indeed, by endowing $\mathbf{V}_h$ with the broken norm

$$\|\mathbf{v}\|_h^2 := \|\nabla_h \mathbf{v}\|_{L^2(\Omega)}^2 + \int_\mathcal{E} \mathsf{h}^{-1}\mathsf{k}^2 |\underline{[\![\mathbf{v}]\!]}|^2\, ds,$$

the forms A_h and B_h are continuous on $\mathbf{V}_h \times \mathbf{V}_h$ and $\mathbf{V}_h \times Q_h$, respectively, with continuity constants $C > 0$ independent of $\mathbf{h}$ and $\mathbf{k}$. Furthermore, there exists a parameter $\gamma_{\min} > 0$ independent of $\mathbf{h}$ and $\mathbf{k}$ such that for any $\gamma \geq \gamma_{\min}$ there exists a coercivity constant $C > 0$ independent of $\mathbf{h}$ and $\mathbf{k}$ with

$$A_h(\mathbf{v}, \mathbf{v}) \geq C\|\mathbf{v}\|_h^2 \qquad \forall \mathbf{v} \in \mathbf{V}_h.$$

Finally, the following discrete inf-sup condition holds true:

$$\inf_{0 \neq q \in Q_h} \sup_{0 \neq \mathbf{v} \in \mathbf{V}_h} \frac{B_h(\mathbf{v}, q)}{\|\mathbf{v}\|_h \|q\|_{L^2(\Omega)}} \geq C|\mathbf{k}|^{-1} > 0,$$

with a constant $C > 0$ that is independent of $\mathbf{h}$ and $\mathbf{k}$. Here, $|\mathbf{k}| := \max_{K \in \mathcal{T}_h}\{k_K\}$.

The above properties of the forms A_h and B_h, combined with the continuity of the forms F_h and G_h, guarantee the well-posedness of the formulation (2) for $\gamma \geq \gamma_{\min}$.

4 Exponential rates of convergence

In this section, we present the main result of [14], namely exponential rates of convergence for mixed *hp*-DGFEM on geometrically refined meshes.

4.1 Geometric meshes

We first define geometric meshes on $\hat{Q} = (0,1)^2$.

Definition 2. *Fix $n \in \mathbb{N}_0$ and $\sigma \in (0,1)$. On $\hat{Q}$, the geometric mesh $\Delta_{n,\sigma}$ with $n+1$ layers and grading factor σ is created recursively as follows: If $n = 0$, $\Delta_{0,\sigma} = \{\hat{Q}\}$. Given $\Delta_{n,\sigma}$ for $n \geq 0$, $\Delta_{n+1,\sigma}$ is generated by subdividing the square K with $0 \in \overline{K}$ into four smaller rectangles by dividing its sides in a $\sigma : (1-\sigma)$ ratio.*

An example of a geometric mesh $\Delta_{n,\sigma}$ on $\hat{Q}$ is shown in Figure 1. We denote the elements in the basic geometric mesh by $\{K_{ij}\}$, as indicated. We say that the elements K_{1j}, K_{2j} and K_{3j} constitute layer j for $j \geq 2$. The element at the origin is denoted as K_{11}.

Definition 3. *A geometric mesh $\mathcal{T}_{n,\sigma}$ on the polygon $\Omega \subset \mathbb{R}^2$ is obtained by mapping the geometric meshes $\Delta_{n,\sigma}$ on $\hat{Q}$ affinely to a vicinity of each convex corner of Ω. At reentrant corners three suitably scaled copies of $\Delta_{n,\sigma}$ are used (as shown in Figure 2). The remainder of Ω is subdivided with a fixed affine and quasi-uniform partition.*

In Figure 2 this local geometric refinement is illustrated. For ease of exposition, we only consider mesh patches that are identically refined with the same parameters σ and n, although different grading factors and numbers of layers may be used for the different corner patches.

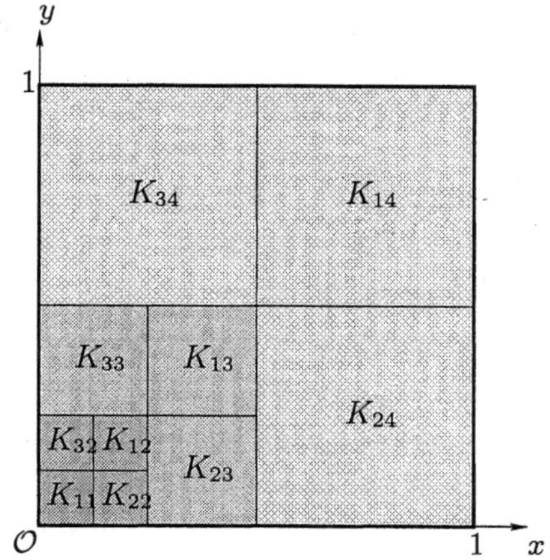

Fig. 1. The basic geometric mesh $\Delta_{n,\sigma}$ with $n = 3$ and $\sigma = 0.5$

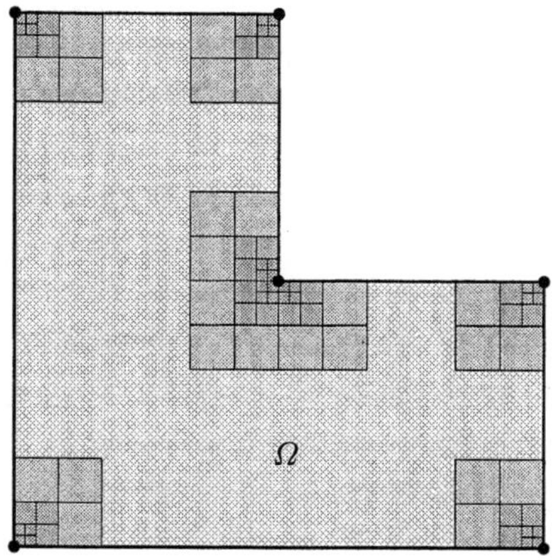

Fig. 2. Local geometric refinement towards the vertices of Ω. In all corners, $n = 3$ and $\sigma = 0.5$

Definition 4. *A polynomial degree distribution* $\mathbf{k}$ *on a geometric mesh* $\mathcal{T}_{n,\sigma}$ *is called linear with slope* $\mu > 0$ *if the elemental polynomial degrees are layerwise constant in the geometric patches and given by* $k_j := \max(2, \lfloor \mu j \rfloor)$ *in layer* j*,* $j = 1, \ldots, n+1$*. In the interior of the domain the elemental polynomial degree is set to* $\max(2, \lfloor \mu(n+1) \rfloor)$.

4.2 Exponential convergence

Our main result is the exponential convergence of the mixed hp-DGFEM for problems with piecewise analytic data. Its detailed proof can be found in [14].

Theorem 2. *Assume that the analytical solution* $(\mathbf{u}, p)$ *of the Stokes problem* (1) *is piecewise analytic as stated in Theorem 1. Further, let* $(\mathbf{u}_h, p_h) \in \mathbf{V}_h \times Q_h$ *denote the DGFEM approximation defined in* (2) *with* $\gamma \geq \gamma_{\min}$ *obtained on geometric meshes* $\mathcal{T}_{n,\sigma}$*. Then there exists a parameter* $\mu_0 = \mu_0(\sigma, \beta) > 0$*, such that for linear degree vectors* $\mathbf{k}$ *with slope* $\mu \geq \mu_0$*, there holds*

$$\|\mathbf{u} - \mathbf{u}_h\|_h + \|p - p_h\|_{L^2(\Omega)} \leq C \exp(-bN^{1/3}),$$

with constants $C, b > 0$ *independent of* $N = \dim(\mathbf{V}_h) \approx \dim(Q_h)$.

Remark 3. If the polynomial degree is chosen to be constant throughout the mesh, i.e., $k_K = k$ for all $K \in \mathcal{T}_h$, exponential convergence is still obtained by choosing k proportional to the number of layers n.

5 Numerical experiment

The goal of this section is to numerically confirm the exponential convergence result stated in Theorem 2. To this end, let Ω be the L-shaped domain shown in Figure 3. As in [17, p. 113], we select the right-hand side $\mathbf{f} = \mathbf{0}$ and take the Dirichlet boundary datum $\mathbf{g}$ in such a way that, in the polar coordinates (r, φ), the exact solution $(\mathbf{u}, p)$ to the Stokes problem (1) is given by

$$\mathbf{u}(r,\varphi) = r^{\lambda} \begin{pmatrix} (1+\lambda)\sin(\varphi)\Psi(\varphi) + \cos(\varphi)\Psi'(\varphi) \\ \sin(\varphi)\Psi'(\varphi) - (1+\lambda)\cos(\varphi)\Psi(\varphi) \end{pmatrix},$$
$$p = -r^{\lambda-1}[(1+\lambda)^2\Psi'(\varphi) + \Psi'''(\varphi)]/(1-\lambda),$$

where

$$\Psi(\varphi) = \sin((1+\lambda)\varphi)\cos(\lambda\omega)/(1+\lambda) - \cos((1+\lambda)\varphi) - \sin((1-\lambda)\varphi)\cos(\lambda\omega)/(1-\lambda) + \cos((1-\lambda)\varphi).$$

Furthermore, $\omega = {}^{3\pi}/_{2}$, and the exponent λ is the smallest positive solution of the equation $\sin(\lambda\omega) + \lambda\sin(\omega) = 0$; thereby, $\lambda \approx 0.54448373678246$.

We emphasize that the solution $(\mathbf{u}, p)$ is in fact the strongest corner singularity of the Stokes operator in the domain Ω; it is piecewise analytic, that is analytic in $\overline{\Omega} \setminus \{\mathcal{O}\}$, but both $\nabla\mathbf{u}$ and p are strongly singular at the origin. Indeed, here $\mathbf{u} \notin H^2(\Omega)^2$ and $p \notin H^1(\Omega)$; thus, this example reflects the typical (singular) behavior that solutions of the Stokes problem exhibit in the vicinity of reentrant corners.

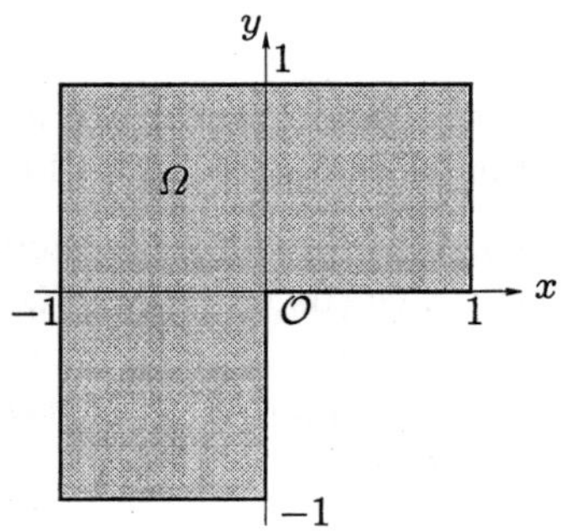

Fig. 3. L-shaped domain Ω

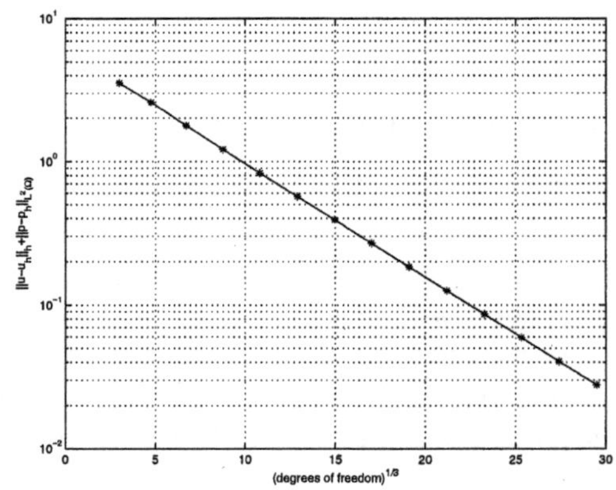

Fig. 4. Performance of the mixed *hp*-DGFEM

Figure 4 shows the performance of the mixed *hp*-DGFEM for the above problem, on meshes that are geometrically refined towards the origin and for polynomial degree distributions that are linearly increasing away from the origin. In our computations we used the grading factor $\sigma = 0.5$ and the linear slope $\mu = 1$. The interior penalty parameter γ was chosen as $\gamma = 10$. The exponential convergence rate, according to Theorem 2, is clearly visible. In addition, we observe that the asymptotic regime is already achieved even with a small number of degrees of freedom.

References

1. D.N. Arnold, F. Brezzi, B. Cockburn, and L.D. Marini. Unified analysis of discontinuous Galerkin methods for elliptic problems. *SIAM J. Numer. Anal.*, 39:1749–1779, 2001.
2. I. Babuška and B.Q. Guo. Regularity of the solution of elliptic problems with piecewise analytic data, I. *SIAM J. Math. Anal.*, 19:172–203, 1988.
3. I. Babuška and B.Q. Guo. Regularity of the solution of elliptic problems with piecewise analytic data, II. *SIAM J. Math. Anal.*, 20:763–781, 1989.
4. G.A. Baker, W.N. Jureidini, and O.A. Karakashian. Piecewise solenoidal vector fields and the Stokes problem. *SIAM J. Numer. Anal.*, 27:1466–1485, 1990.
5. F. Brezzi and M. Fortin. Mixed and hybrid finite element methods. In *Springer Series in Computational Mathematics*, volume 15. Springer–Verlag, New York, 1991.
6. B. Cockburn, G. Kanschat, and D. Schötzau. The local discontinuous Galerkin method for the Oseen equations. *Math. Comp.*, to appear.
7. B. Cockburn, G. Kanschat, D. Schötzau, and C. Schwab. Local discontinuous Galerkin methods for the Stokes system. *SIAM J. Numer. Anal.*, 40:319–343, 2002.
8. V. Girault and P.A. Raviart. *Finite Element Methods for Navier–Stokes Equations.* Springer–Verlag, New York, 1986.
9. V. Girault, B. Rivière, and M.F. Wheeler. A discontinuous Galerkin method with non-overlapping domain decomposition for the Stokes and Navier-Stokes problems. Technical Report 02-08, TICAM, UT Austin, 2002.
10. B.Q. Guo and C. Schwab. Analytic regularity of Stokes flow in polygonal domains. Technical Report 2000-18, SAM, ETH Zürich, 2000.
11. P. Hansbo and M.G. Larson. Discontinuous finite element methods for incompressible and nearly incompressible elasticity by use of Nitsche's method. *Comput. Methods Appl. Mech. Engrg.*, 191:1895–1908, 2002.
12. O.A. Karakashian and W.N. Jureidini. A nonconforming finite element method for the stationary Navier-Stokes equations. *SIAM J. Numer. Anal.*, 35:93–120, 1998.
13. D. Schötzau, C. Schwab, and A. Toselli. Mixed *hp*-DGFEM for incompressible flows. *SIAM J. Numer. Anal.*, 40:2171–2194, 2003.
14. D. Schötzau and T. P. Wihler. Exponential convergence of mixed *hp*-DGFEM for Stokes flow in polygons. *Numer. Math.*, to appear.
15. C. Schwab. *p- and hp-FEM – Theory and Application to Solid and Fluid Mechanics.* Oxford University Press, Oxford, 1998.
16. A. Toselli. *hp*-discontinuous Galerkin approximations for the Stokes problem. *Math. Models Methods Appl. Sci.*, 12:1565–1616, 2002.
17. R. Verfürth. *A Review of a Posteriori Error Estimation and Adaptive Mesh-Refinement Techniques.* B.G. Teubner, Stuttgart, 1996.
18. T.P. Wihler, P. Frauenfelder, and C. Schwab. Exponential convergence of the *hp*-DGFEM for diffusion problems. *Comput. Math. Appl.*, to appear.

A Postprocessing of Hopf Bifurcation Points

Dáša Janovská[1] and Vladimír Janovský[2]

[1] Institute of Chemical Technology, CZ-166 28 Prague 6, Czech Republic, *janovskd@vscht.cz*

[2] Faculty of Mathematics and Physics, Charles University, CZ-186 00 Prague 8, Czech Republic, *janovsky@karlin.mff.cuni.cz*

Summary. The aim is to initialize a branch of periodic orbits emanating from a Hopf bifurcation point. The second-order predictor of the branch is developed. The problem is discussed in the context of the MATLAB toolbox CL_MATCONT.

1 Introduction

We consider dynamical system

$$\dot{u}(t) = F(u(t), \lambda), \tag{1}$$

where $F : \mathbb{R}^n \times \mathbb{R}^1 \to \mathbb{R}^n$ is a sufficiently smooth mapping; $u \in \mathbb{R}^n$ is a vector of state variables and λ is a real parameter. Let (u_H, λ_H) be a *Hopf bifurcation point.* Under generic assumptions, see e.g. [9], Theorem 3.3, there is a branch of periodic orbits (cycles) that emanate from this bifurcation point. One can decide about stability of the cycles provided that the first Lyapunov coefficient l_1 is nonzero.

The packages [4], [10] or [1] can detect a Hopf bifurcation point (u_H, λ_H) and continue the mentioned branch of cycles.

In order to compute T-periodic solutions of (1), one usually rescals time and looks for 1-periodic solutions to

$$-\dot{u}(t) + T\, F(u(t), \lambda) = 0\,, \quad 0 \le t \le 1\,, \tag{2}$$

$$u(0) - u(1) = 0\,. \tag{3}$$

T is the *period* of the actual motion and $\omega \equiv \frac{2\pi}{T}$ is its *frequency.* The quoted packages use *orthogonal collocations*, see [9], to discretize the problem (2) & (3).

The continuation is initialized by a cycle with a small amplitude h, see e.g. [1], function `init_H_LC`:

$$u^0(t) \equiv u_H + h\, \Re\left(\exp\left(2\pi i\, t\right)\, \xi_H\right)\,, \quad 0 \le t \le 1\,, \tag{4}$$

where $\xi_H \in \mathbb{C}^n$ and $\omega_H > 0$ satisfy

$$F_u(u_H, \lambda_H)\xi_H = i\,\omega_H \xi_H\,, \quad \|\xi_H\| = 1\,. \tag{5}$$

It reflects the well known fact that the spectrum of the differential $F_u(u_H, \lambda_H)$ contains a purely imaginary eigenpair. The tuple $\left(u^0(t), T \equiv \frac{2\pi}{\omega_H}, \lambda \equiv \lambda_H\right)$ is considered to be an approximate solution to (2) & (3).

As the predictor for the continuation, it is natural to take

$$v^0(t) \equiv \Re\left(\exp\left(2\pi i\, t\right)\, \xi_H\right)\,, \quad 0 \le t \le 1\,, \tag{6}$$

namely, the differential of function u^0 with respect to h. Following this strategy, differentials of the components T and λ with respect to h are zeroes. Therefore, the predictor for the pathfollowing for the first cycle reads as $\left(v^0(t), 0, 0\right)$.

The aim of this contribution is to suggest a more sophisticated cycle initialization then just (4) and (6). The technique is based on *Lyapunov-Schmidt reduction* at Hopf bifurcation point (u_H, λ_H), see e.g. [11], [5]. We consider the version, which is characterized by making use of *bordering operators*, see [8] and [13].

In Section 2, we review the technique. In Section 3, we give the 2nd-order formula to initiate the branch. Finally, in Section 4, we will report on a numerical experiment.

2 Preliminaries

We review the main points of [8]. Instead of solving (2)&(3) we seek for 2π-periodic solutions of (1); the reasons are just historical. The problem is formulated as functional equation $\Phi\left(u, \lambda, \omega\right) = 0$ on proper function spaces:

$$\Phi\left(u, \lambda, \omega\right) \equiv -\omega \dot{u} + F(u, \lambda)\,, \quad T = \frac{2\pi}{\omega}\,, \tag{7}$$

$$\Phi : U \times \Lambda \times \mathbb{R} \to V\,, \quad U = C^1(S^1, \mathbb{R}^n)\,, \quad V = C^0(S^1, \mathbb{R}^n)\,, \tag{8}$$

where $S^1 \equiv \mathbb{R}/2\pi\mathbb{Z}$. The roots of Φ are, locally and up to a phase shift, one-to-one with the roots of a scalar *bifurcation equation*

$$\phi\left(x, \lambda - \lambda_H, \omega - \omega_H\right) = 0\,; \tag{9}$$

$x \in \mathbb{R}$ is a reduced state variable $u \in U$. The root

$$\left(x = 0, \lambda - \lambda_H = 0, \omega - \omega_H = 0\right)$$

of (9) is related to the root $(u_H, \lambda_H, \omega_H)$ of Φ. The nontrivial solutions $x \neq 0$ of (9) are linked with cycles, the trivial solutions $x = 0$ are linked with steady states.

It is natural to develop the periodic solutions $u = u_H + v \in U$ to Fourier series

$$u = u_H + [v]_0 + \sum_{k=1}^{+\infty}\left([v]_k e^{iks} + [v]_{-k} e^{-iks}\right)\,, \tag{10}$$

where $[v]_k : U \to \mathbb{C}^n$ is the k–th Fourier coefficient of $v \in U$.

The reduction procedure depends on a choice of two vectors $M \in \mathbb{C}^{n\times 1}$ and $L \in \mathbb{C}^{1\times n}$. The requirement is that the bordered matrix

$$\begin{pmatrix} F_u(u_H, \lambda_H) - i\omega_H & M \\ L & 0 \end{pmatrix} \tag{11}$$

is regular. This is generically satisfied.

We will consider two particular variants: The classical Lyapunov-Schmidt reduction, see e.g. [5], which corresponds to the choice

$$M = \xi_H\,, \quad L = \eta_H^*\,, \tag{12}$$

where ξ_H is the unit eigenvector, see (5), and η_H is an adjoint eigenvector,

$$F_u^\top(u_H, \lambda_H)\eta_H = -i\omega_H \eta_H\,, \tag{13}$$

with the scaling $\eta_H^* \xi_H = 1$.

As the second variant, we just set

$$M = \xi_H\,, \quad L = \xi_H^*\,. \tag{14}$$

According to [13], Remark 4, the option (14) leads to an alternative formula for the first Lyapunov index l_1.

Actually, the periodic orbits are linked with solutions $(x, \lambda - \lambda_H, \omega - \omega_H)$ of the factored bifurcation equation

$$x^{-1}\phi\,(x, \lambda - \lambda_H, \omega - \omega_H) = 0\,. \tag{15}$$

We consider the truncated Taylor expansion of (15)

$$\frac{1}{6}\phi_{xxx}\, x^2 + \phi_{x\lambda}\, (\lambda - \lambda_H) + \phi_{x\omega}\, (\omega - \omega_H) + \text{h.o.t.} = 0\,. \tag{16}$$

It consists of low-order terms of the expansion (16), see [5], p.86. The corresponding cycles $u \in U$ are approximated by the accordingly truncated series (10). In (16), it is understood that the differentials of ϕ are evaluated at the origin, i.e. $\phi_{xxx} = \phi_{xxx}\,(0,0,0)$, etc. An algorithm for computing of the chain of relevant differentials is supplied in [8], and in a larger extend in [13].

In next Section 3, we resume the asymptotic formula for a cycle (it is already done in [8]) and provide a formula for differential of the cycle with respect to x. The latter formula is the cycle 'velocity' and serves as predictor step of the cycle continuation.

We rescale time back to period one, see (3).

3 The algorithm

We set $h := x$, where x is the reduced state, see (9).

The solution $(u(t), T, \lambda)$ to (2)&(3) could be represented as

$$u(t) = u_H + 2h\,\Re(\exp(2\pi i t)\,[v_x]_1) + \delta\lambda\,[v_\lambda]_0 + \\ +h^2\left(\frac{1}{2}[v_{xx}]_0 + \Re(\exp(4\pi i t)\,[v_{xx}]_2)\right) + \mathrm{O}(h^3)\,, \tag{17}$$

$$T = \frac{2\pi}{\omega}\,, \quad \omega = \omega_H + \delta\omega + \mathrm{O}(h^4)\,, \tag{18}$$

$$\lambda = \lambda_H + \delta\lambda + \mathrm{O}(h^4) \tag{19}$$

for small parameter h. The increments $\delta\lambda$ and $\delta\omega$ are the solution of a linear system

$$\mathbf{B}\begin{pmatrix}\delta\lambda\\ \delta\omega\end{pmatrix} = -\frac{h^2}{6}\begin{pmatrix}\Re\phi_{xxx}\\ \Im\phi_{xxx}\end{pmatrix}\,, \quad \mathbf{B} \equiv \begin{pmatrix}\Re\phi_{x\lambda}\,, \Re\phi_{x\omega}\\ \Im\phi_{x\lambda}\,, \Im\phi_{x\omega}\end{pmatrix}\,. \tag{20}$$

The objects $[v_x]_1$, $[v_\lambda]_0$, $[v_{xx}]_0$, $[v_{xx}]_2$ are vectors in $\mathbb{C}^n$ and ϕ_{xxx}, $\phi_{x\lambda}$, $\phi_{x\omega}$ are complex constants, see [8], p. 1167. They are computed at a particular Hopf bifurcation point u_H, λ_H, ω_H.

Truncating the higher order terms in (17), (18) and (19) we get the 2nd-oder cycle approximation.

Let us differentiate the cycle $(u(t), T, \lambda)$ with respect to h. The resulting differentials $(\frac{d}{dh}u(t), \frac{d}{dh}\lambda, \frac{d}{dh}T)$ can be considered as cycle velocity. For the components of cycle velocity we obtain the following asymptotic formulae:

$$\frac{d}{dh}u(t) = 2\,\Re(\exp(2\pi i t)\,[v_x]_1) + \delta\lambda\,[v_\lambda]_0 + \\ +2h\left(\frac{1}{2}[v_{xx}]_0 + \Re(\exp(4\pi i t)\,[v_{xx}]_2)\right) + \mathrm{O}(h^2)\,, \tag{21}$$

$$\frac{d}{dh}T = -2\pi\frac{\frac{d}{dh}\omega}{\omega^2}\,, \quad \frac{d}{dh}\omega = \delta\omega + \mathrm{O}(h^3)\,, \tag{22}$$

$$\frac{d}{dh}\lambda = \delta\lambda + \mathrm{O}(h^3)\,, \tag{23}$$

where the increments $\delta\lambda$, $\delta\omega$ satisfy

$$\mathbf{B}\begin{pmatrix}\delta\lambda\\ \delta\omega\end{pmatrix} = -\frac{h}{3}\begin{pmatrix}\Re\phi_{xxx}\\ \Im\phi_{xxx}\end{pmatrix}\,. \tag{24}$$

Truncating the higher order terms in (21), (22) and (23) we obtain the 1st-oder cycle velocity approximation.

Remark 1. Assuming the bordering (12) or (14), it comes out that $[v_x]_1 = \xi_H$. Actually, for a generic bordering, $[v_x]_1$ is a positive multiple of ξ_H.

Truncating in the expansions (17), (18) and (19) the 2nd-order, and in the expansions (21), (22) and (23) the 1st-order terms, we get

$$u^0(t) \equiv u_H + 2h\,\Re(\exp(2\pi i t)\,[v_x]_1)\,, \quad v^0(t) \equiv 2\,\Re(\exp(2\pi i t)\,[v_x]_1)\,. \qquad (25)$$

The claim is that choosing L properly, we get (4) and (6). In fact, due to Remark 1, the choice e.g. $L \equiv 2\xi_H^*$ or $L \equiv 2\eta_H^*$ would do.

4 Numerical tests

We consider

$$\begin{aligned}\dot{x} &= x\mu^+(\frac{\alpha}{\beta + x} - k) - \mu^-\gamma xyz\,,\\ \dot{y} &= y(\epsilon x - \gamma z - \rho)\,,\\ \dot{z} &= z(\sigma\gamma y - \delta)\,,\end{aligned}$$

which depends on ten parameters. This dynamical system models corruption in democratic societies, see [12]. In [6], p. 317, the system was investigated for the parameter setting

$$\alpha \equiv 1.5,\ \beta \equiv 0.5,\ k \equiv 1,\ \rho \equiv 0.1,\ \mu^+ \equiv 1,\ \mu^- \equiv 10,\ \gamma \equiv 1,\ \sigma \equiv 2\,,$$

leaving two parameters ϵ and δ free. There were detected *Generalized Hopf bifurcation point* (GH) as *codim* $= 2$ organizing center and a branch of Hopf bifurcation points containing this GH. We consider the particular Hopf point with ordinate $\epsilon = 0.2$. Coordinates of this point are $u_H = [0.7623; 0.3588; 0.0524]$ and $\delta_H = 0.7176$. On Fig.1, there is a branch of limit cycles emanating from (u_H, δ_H). There is a turning point on the branch (LPC); the detection of bifurcation points was switched off. The zoom of two first cycles is shown on Fig.2. It illustrates the action of initialization procedure `init_H_LC`, see Section 1.

All computations were performed using `CL_MATCONT`, see [1] and [2]. The toolbox works really nicely.

In order to compare the action of `init_H_LC` with the 2nd-oder predictor, we coded the algorithm from Section 3 as an m-file called `init_H_LC_new`. It is supposed to replace `init_H_LC` in the cycle continuation. On Fig.3, there is a sequence of the 2nd-oder cycle approximations due to `init_H_LC_new` for selected h's. The bordering (12) is considered. Fig.4 compares the actions of `init_H_LC` (dotted) with the actions of `init_H_LC_new`. Note that the dotted cycles do *not* advance with parameter δ.

The parameters of both `init_H_LC` and `init_H_LC_new` are the already mentioned h and collocation data setting. The latter will be fixed qua *nstol* $= 10$, *ncol* $= 4$. Let us consider the particular Hopf point (u_H, δ_H).

Using `init_H_LC`, one needs to set $h \leq 0.0003$ to achieve a successful continuation. Choosing h too small, say $h \leq 0.00001$, the continuation is

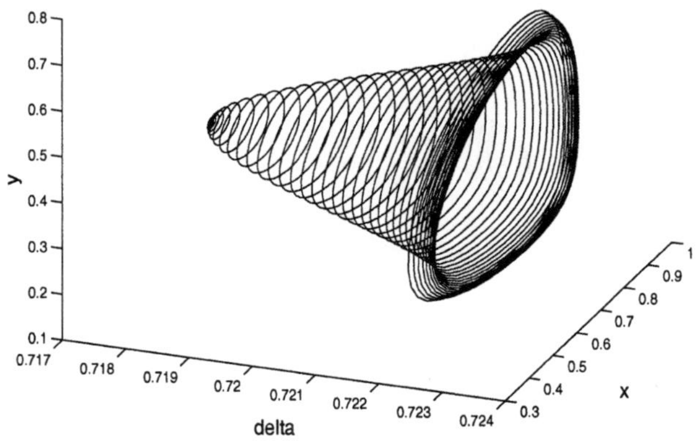

Fig. 1. Continuation of the limit cycle via orthogonal collocations

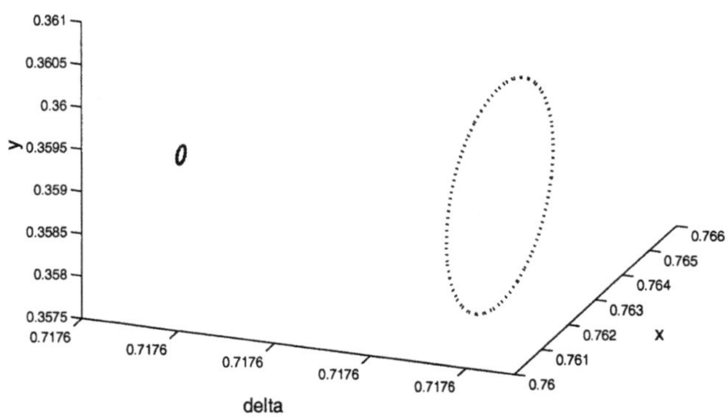

Fig. 2. Zoom - the initial cycle (solid), the next cycle (dotted)

not initialized either due to the round-off errors. On the other hand, applying `init_H_LC_new` with bordering (14), the continuation is successful for $h \leq 0.002$.

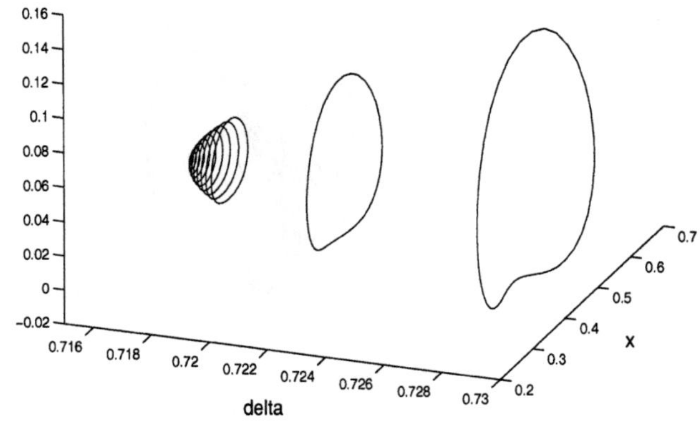

Fig. 3. The 2nd-order approximations of cycles; h=0.01:0.01:0.09, h=0.1:0.1: 0.3

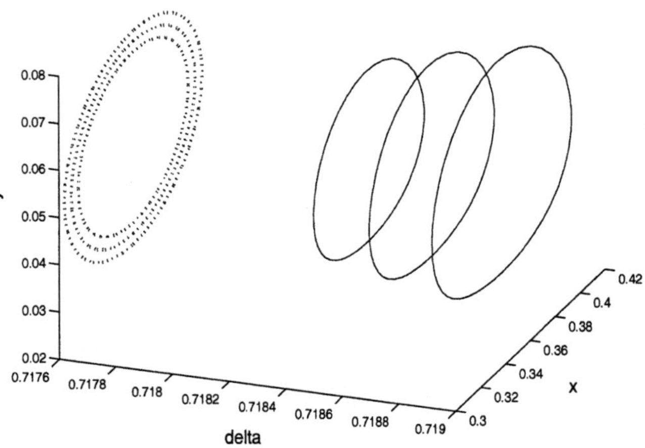

Fig. 4. The 1st-oder (dotted) vs the 2nd-oder (solid) predictors; h=0.08, h=0.09, h=0.1

Conclusion: Using the 2nd-order predictor we can afford to take a larger amplitude h. It means, we can skip over the initial stages of the continuation.

Acknowledgments: The authors acknowledge with pleasure the support of the Grant Agency of the Czech Republic (grant # 201/02/0844, grant # 201/02/0595) and support of the projects of the Czech Ministry of Education: MSM 113200007 and MSM 223400007

References

1. Dhooge, A., Govaerts, W., Kuzetsov, Yu.A., Mestrom, W., Riet, A.M.: CL_MATCONT: A continuation toolbox in Matlab. Gent University, http://allserv.rug.ac.be/ ajdhooge/
2. Dhooge, A., Govaerts, W., Kuzetsov, Yu.A. (2003): MATCONT: A Matlab package for numerical bifurcation analysis of ODEs. ACM Transactions on Mathematical Software, **31**, 141–164
3. Doedel, E.J., Govaerts, W., Kuzetsov, Yu.A.: Computation of periodic solution bifurcations in ODEs using bordered systems. SIAM J. Numer. Anal., to appear
4. Doedel, E.J., Kernévez, J. (1986): AUTO: Software for continuation problems in ordinary differential equations with applications. California Institute of Technology, Applied Mathematics
5. Golubitsky, M., Schaeffer, D.G. (1985): Singularities and Groups in Bifurcation Theory. Vol. I. Springer-Verlag, New York
6. Govaerts, W. (2000): Numerical methods for bifurcation of dynamical equilibria. SIAM, Philadalphia
7. Govaerts, W., Kuzetsov, Yu.A., Sijnave, B. (2000): Numerical methods for the generalized Hopf bifurcation. SIAM J. Numer. Anal., **38**, 329–346
8. Janovský, V., Plecháč, P. (1996): Local numerical analysis of Hopf bifurcation. SIAM J. Numer. Anal., **33**, 1150–1168
9. Kuzetsov, Yu.A. (1995): Elements of Applied Bifurcation Theory. Springer Verlag, New York
10. Kuzetsov, Yu.A., Levitin, V.V. (1977): CONTENT: Integrated environment for analysis of dynamical systems. CWI, Amsterdam, ftp://ftp.cwi.nl/pub/CONTENT
11. Marsden, J.E., McCracken, M. (1976): The Hopf Bifurcation and Its Applications. Springer-Verlag, New York
12. Rinaldi, S., Feichtlinger, G., Wirl, F. (1998): Corruption dynamics in democratic societies. Complexity **3**, 53–64
13. Xu, H., Janovský, V., Werner, B. (1998): Numerical computation of degenerate Hopf points. ZAMM Z. Angew. Mech., **78**, 807–821

Givens' Reduction of Quaternion-Valued Matrices to Upper Hessenberg Form

Drahoslava Janovská[1] and Gerhard Opfer[2]

[1] Institute of Chemical Technology, CZ-166 28 Prague 6, Czech Republic, *janovskd@vscht.cz*
[2] University of Hamburg, Department of Mathematics, Bundesstraße 55 D-20146 Hamburg, Germany, *opfer@math.uni-hamburg.de*

Summary. The number of real operations necessary to reduce a quaternion-valued matrix into a similar upper Hessenberg matrix by making use of quaternion-valued Givens' transformation matrices is relatively large. Two possibilities of how to reduce the columns of a quaternion-valued matrix more effectively are presented.

1 Introduction

In the literature, several applications of quaternions and quaternion-valued matrices can be found. In most of them, the quaternion-valued matrix has to be factored. One conceivable reason is that factorizations yield more time-efficient processing schemes. Another reason is that factorizations can be chosen so as to provide necessary geometric relationships in certain applications.

Let us give an example, which arises in quantum mechanics, see [1], [2]. In particular, solving the Secular Equation, we have to find an eigensystem of a matrix $\mathbf{A}$. If the nonrelativistic or spin-orbitless problem is considered, the matrix $\mathbf{A}$ is real and symmetric. The effect of including the spin-orbit coupling into the model is to replace each scalar matrix element with a 2×2 complex matrix of the form

$$a = \begin{pmatrix} \alpha & \beta \\ -\overline{\beta} & \overline{\alpha} \end{pmatrix}.$$

As a consequence, the matrix $\mathbf{A}$ becomes complex, in general nonhermitean and its size is doubled, i.e. the smallest eigenvalue ($\sim 10^{-9}$) occurs twice. In numerical calculations, computational noise is dramatically increased.

Let $\widetilde{\mathbb{H}}$ be the space of such complex 2×2 matrices:

$$\widetilde{\mathbb{H}} := \{\, \widetilde{\mathbf{h}} = \begin{pmatrix} \alpha & \beta \\ -\overline{\beta} & \overline{\alpha} \end{pmatrix} : \alpha = a_1 + \mathrm{i}a_2,\ \beta = a_3 + \mathrm{i}a_4,\ a_1, a_2, a_3, a_4 \in \mathbb{R}\}$$

and let $\mathbb{H}$ be the space of quaternions:

$$\mathbb{H} := \{h = (a_1, a_2, a_3, a_4) :\ a_1, a_2, a_3, a_4 \in \mathbb{R}\}.$$

Since $\mathbb{H}$ and $\widetilde{\mathbb{H}}$ are isomorphic, see [4], we can replace 2×2 complex matrices with corresponding quaternions and use quaternion arithmetic in calculations. The arithmetic of quaternions is, of course, more complicated, but it allows us to economize the storage and to increase the accuracy of results, see [1].

In this paper, our aim is to develop an algorithm for a reduction of a given quaternion-valued matrix into its upper Hessenberg form. At first, we shortly review the algebra of quaternions, see [6]. We also briefly mention the algorithm of Given's reduction of a quaternion-valued vector $\mathbf{x} \in \mathbb{H}^2$. It can be found in [3], where also all necessary facts are proved in detail. Then, Givens' transformation matrices for a vector $\mathbf{x} \in \mathbb{H}^n$ are defined. We discuss some possibilities of reducing the number of necessary real operations. Finally, the number of real operations for the reduction of quaternion-valued matrices into a similar matrix in upper Hessenberg form will be calculated. Our aim is to show in particular that it is necessary to develop an algorithm, which reduces the large number of operations. A numerical example of Givens' reduction of a quaternion-valued matrix into a similar upper Hessenberg form is presented.

2 Short review of the algebra of quaternions

Let $\mathbb{H} = \mathbb{R}^4$ be equipped with the ordinary vector space structure and with an additional multiplicative operation $\mathbb{H} \times \mathbb{H} \longrightarrow \mathbb{H}$ which most easily can be defined by a multiplication of the four basis elements

$$(1,0,0,0) = \mathbf{1}, \quad (0,1,0,0) = \mathbf{i}, \quad (0,0,1,0) = \mathbf{j}, \quad (0,0,0,1) = \mathbf{k}:$$

$$\mathbf{i}^2 = \mathbf{j}^2 = \mathbf{k}^2 = \mathbf{ijk} = -\mathbf{1}. \tag{1}$$

An element $x = (x_1, x_2, x_3, x_4) \in \mathbb{H}$ has the representation

$$x = x_1\mathbf{1} + x_2\mathbf{i} + x_3\mathbf{j} + x_4\mathbf{k}, \tag{2}$$

where $x_1, x_2, x_3, x_4 \in \mathbb{R}$, $\Re x = x_1$ is the *real part* of x. We will identify the quaternion $x = (x_1, 0, 0, 0)$ with the real number x_1, the quaternion $x = (x_1, x_2, 0, 0)$ will be identified with the complex number $x_1 + \mathrm{i}x_2$. For $x = (x_1, x_2, x_3, x_4) \in \mathbb{H}$, $y = (y_1, y_2, y_3, y_4) \in \mathbb{H}$ it follows from (1) that

$$\begin{aligned} xy &= (x_1y_1 - x_2y_2 - x_3y_3 - x_4y_4)\,\mathbf{1} + (x_1y_2 + x_2y_1 + x_3y_4 - x_4y_3)\,\mathbf{i} \\ &\quad + (x_1y_3 - x_2y_4 + x_3y_1 + x_4y_2)\,\mathbf{j} + (x_1y_4 + x_2y_3 - x_3y_2 + x_4y_1)\,\mathbf{k}. \end{aligned}$$

Obviously, in general, the multiplication is not commutative.

Given x according to (2), the conjugate $\overline{x}$ of x is defined to be

$$\overline{x} = (x_1, -x_2, -x_3, -x_4). \tag{3}$$

Let us note that conjugation obeys the rules

$$\overline{xy} = \overline{y}\,\overline{x}, \qquad \overline{\overline{x}} = x.$$

We define the absolute value of x by

$$|x| = \sqrt{x_1^2 + x_2^2 + x_3^2 + x_4^2}. \tag{4}$$

We will use the following properties of $|x|$:

$$|x|^2 = x\,\overline{x} = \overline{x}\,x, \qquad |xy| = |yx| = |x||y|. \tag{5}$$

Here, we use the convention that a real number x may be identified with the quaternion $(x, 0, 0, 0)$.

The space $\mathbb{H}$ is a normed vector space over $\mathbb{H}$, where the norm is introduced in (4).

Let us remark, that for any $x \in \mathbb{H}\backslash\{0\}$ an inverse quaternion x^{-1} is defined,

$$x^{-1} = \frac{\overline{x}}{|x|^2}. \tag{6}$$

Let $x = (x_1, x_2, x_3, x_4)$, $y = (y_1, y_2, y_3, y_4) \in \mathbb{H}$ be two quaternions. Then

$$\Re(xy) = x_1y_1 - x_2y_2 - x_3y_3 - x_4y_4 = \Re(yx), \tag{7}$$

an equation which will be used later.

Two quaternions x and y are called equivalent, denoted by $x \sim y$, if $y = \alpha^{-1}x\alpha$, for some $\alpha \in \mathbb{H}\backslash\{0\}$. For fixed $x \in \mathbb{H}$ the set

$$[x] = \{y \in \mathbb{H} : \ y = \alpha^{-1}x\alpha \text{ for } \alpha \in \mathbb{H}\backslash\{0\}\} \tag{8}$$

is called equivalence class of x.

Lemma 1. *Two quaternions x and y are equivalent if and only if*

$$\Re x = \Re y \quad \text{and} \quad |x| = |y|. \tag{9}$$

Corollary 1. *Let $x = (x_1, x_2, x_3, x_4)$. Then*

$$y = (x_1, \sqrt{x_2^2 + x_3^2 + x_4^2}, 0, 0) \in [x]$$

is the only complex element in $[x]$ with non negative imaginary part.

Let $\mathbf{x} = (x_1, x_2, \ldots, x_n)^{\mathrm{T}} \in \mathbb{H}^n$ and define the norm $\|\mathbf{x}\|$ of $\mathbf{x}$ by

$$\|\mathbf{x}\| = \sqrt{\sum_{j=1}^{n} |x_j|^2}\,. \tag{10}$$

The space $\mathbb{H}^n$ becomes a normed vector space over $\mathbb{H}$ with the norm defined in (10). For $\mathbf{x} \in \mathbb{H}^n$, we denote by $\mathbf{x}^*$ the transpose of the entrywise conjugate of $\mathbf{x}$.

We define the conjugate transposition of the matrix $\mathbf{B} = (b_{ij}) \in \mathbb{H}^{m\times n}$ as $\mathbf{B}^* = (\overline{b_{ji}}) \in \mathbb{H}^{n\times m}$. The square quaternion valued matrix $\mathbf{B} \in \mathbb{H}^{n\times n}$ is called Hermitean if $\mathbf{B} = \mathbf{B}^*$, and $\mathbf{B}$ is positive definite if it is Hermitean and

$$\mathbf{x}^* \mathbf{B}\, \mathbf{x} > 0 \qquad \forall\, \mathbf{x} \in \mathbb{H}^n \backslash \{\mathbf{0}\}\,.$$

A matrix $\mathbf{B} \in \mathbb{H}^{n\times n}$ is said to be unitary if $\mathbf{B}^*\mathbf{B} = \mathbf{I}$.

Theorem 1. *A matrix $\mathbf{B} \in \mathbb{H}^{n\times n}$ is unitary if and only if $\|\mathbf{B}\mathbf{x}\| = \|\mathbf{x}\|$ for all $\mathbf{x} \in \mathbb{H}^n$.*

Definition 1. Let $\mathbf{B} \in \mathbb{H}^{n\times n}$. If there exist a vector $\mathbf{x} \in \mathbb{H}^n\backslash\{\mathbf{0}\}$ and a quaternion $\lambda \in \mathbb{H}$ such that

$$\mathbf{B}\mathbf{x} = \mathbf{x}\lambda, \tag{11}$$

we call λ an eigenvalue of $\mathbf{B}$ and $\mathbf{x}$ an eigenvector corresponding to λ.

The number of eigenvalues of a quaternion valued matrix $\mathbf{B} \in \mathbb{H}^{n\times n}$ is, in general, not finite. If λ is an eigenvalue, one can easily show that the whole equivalence class $[\lambda]$ consists of eigenvalues. If λ is real, then $[\lambda] \equiv \{\lambda\}$, if λ is not real, then according to Corollary 1 there exists exactly one complex number $\widetilde{\lambda} \in [\lambda]$ with positive imaginary part. It can also be proved that the number of non equivalent eigenvalues is at most n.

Theorem 2. *Let $\mathbf{A} \in \mathbb{H}^{n\times n}$ be Hermitean. Then $\mathbf{A}$ has only real eigenvalues and their number is n.*

Theorem 3. *For any unitary quaternion valued matrix $\mathbf{A}$, the eigenvalues λ satisfy*
$|\lambda| = 1\,.$

3 Givens' transformation of a vector $\mathbf{x} \in \mathbb{H}^2$

Let $\mathbf{x} \in \mathbb{H}^2\backslash\{\mathbf{0}\}$ be given. Let $\mathbf{G}$ be a matrix of the form

$$\mathbf{G} = \begin{pmatrix} \overline{c} & s \\ -\overline{s} & c \end{pmatrix} \in \mathbb{H}^{2\times 2}. \tag{12}$$

Our aim is to find c and s such that

(i) $\mathbf{G}$ is unitary ,
(ii) $\mathbf{G}^*\mathbf{x} = u\mathbf{e}_1$, where $u \in \mathbb{H}\backslash\{0\}$, $\mathbf{e}_1 = (\mathbf{1}, 0)^{\mathrm{T}}$.

The solution is contained in the following theorem.

Theorem 4. *Let $\mathbf{x} = (x_1, x_2)^{\mathrm{T}} \in \mathbb{H}^2\backslash\{\mathbf{0}\}$ be given. Define*

$$\mathbf{G} = \begin{pmatrix} \overline{c} & s \\ -\overline{s} & c \end{pmatrix} \in \mathbb{H}^{2\times 2},$$

$$s = -\sigma \frac{\overline{x_2}}{||\mathbf{x}||}, \quad c = \sigma \frac{\overline{x_1}}{||\mathbf{x}||}, \quad |\sigma| = 1,$$

where σ is arbitrary in case $x_1 = \alpha\ x_2, \alpha \in (\mathbb{R})\backslash\{0\}$. Otherwise we have to choose $\sigma \in \Sigma$, where

$$\Sigma = \left\{ \sigma \in \mathbb{H} : \sigma = \frac{\alpha x_1 + \beta x_2}{|\alpha x_1 + \beta x_2|}, \ \alpha, \beta \in \mathbb{R}, \ |\alpha| + |\beta| > 0 \right\}.$$

Then $\mathbf{G}$ is a unitary matrix,

$$\mathbf{G}^* \mathbf{x} = \mathbf{u} = \sigma ||\mathbf{x}|| (\mathbf{1}, 0)^{\mathrm{T}}$$

and there are no other unitary quaternion-valued matrices satisfying this condition.

Proof. See [3].

Remark 1. The set Σ is not empty, since it contains the subset $\{\pm \mathrm{sgn}\overline{x_1}, \pm \mathrm{sgn}\overline{x_2}\}$. But it is different from the whole unit sphere, since ± 1 do not, in general, belong to Σ. Let us note that s is real for the choice $\sigma = \pm \mathrm{sgn}\overline{x_2}$ and c is real if $\sigma = \pm \mathrm{sgn}\overline{x_1}$.

4 Givens' reduction of a vector $\mathbf{x} \in \mathbb{H}^n$

Let $\mathbf{x} = (x_1, \ldots, x_n)^{\mathrm{T}} \in \mathbb{H}^n \backslash \{\mathbf{0}\}$. For $1 \le i < j \le n$, we define the Givens' rotation matrix $\mathbf{G}_j^i \in \mathbb{H}^{n \times n}$ as follows:

$$\mathbf{G}_j^i = \begin{pmatrix} 1 & & & & & & & & 0 \\ & \ddots & & & & & & & \\ & & 1 & & & & & & \\ & & & \overline{c}_i & \cdots & & s_i & & \\ & & & \vdots & 1 & & \vdots & & \\ & & & \vdots & & \ddots & \vdots & & \\ & & & \vdots & & 1 & \vdots & & \\ & & & -\overline{s}_i & \cdots & & c_i & & \\ & & & & & & & 1 & \\ 0 & & & & & & & & \ddots \; 1 \end{pmatrix} \begin{matrix} \\ \\ \\ \longleftarrow i \\ \\ \\ \\ \longleftarrow j \\ \\ \\ \end{matrix}, \tag{13}$$

$$\qquad\qquad\qquad\qquad\uparrow i \qquad\qquad \uparrow j$$

where

$$s_i = \frac{-\sigma_i \overline{x}_j}{\sqrt{|x_i|^2 + |x_j|^2}}, \qquad c_i = \frac{\sigma_i \overline{x}_i}{\sqrt{|x_i|^2 + |x_j|^2}}, \tag{14}$$

$\sigma_i \in \mathbb{H}$, $|\sigma_i| = 1$, σ_i is arbitrary if $x_i = \alpha x_j$, $\alpha \in \mathbb{R}\backslash\{0\}$. Otherwise, $\sigma_i \in \Sigma_i$,

$$\Sigma_i = \{\sigma \in \mathbb{H} : \sigma = \frac{\alpha\, x_i + \beta\, x_j}{|\alpha\, x_i + \beta\, x_j|}, \ \alpha, \beta \in \mathbb{R}, \ |\alpha| + |\beta| > 0\}. \tag{15}$$

Then $\mathbf{G}_j^i$ is a unitary matrix and

$$(\mathbf{G}_j^i)^*\mathbf{x} = (x_1, \dots, x_{i-1}, u_i, x_{i+1}, \dots, x_{j-1}, 0, x_{j+1}, \dots, x_n)^{\mathrm{T}},$$

$$u_i = \sigma_i \sqrt{|x_i|^2 + |x_j|^2}\,.$$

Let us suppose that our aim is to reduce all components of a given quaternion-valued vector except the first one. We will discuss two possibilities of the ordering of such a reduction.

Let us reduce components in the order from bottom to top. In order to decrease the number of operations, all matrices necessary for the reduction can be multiplied to obtain just one matrix of the reduction.

Theorem 5. *For a given vector* $\mathbf{x} = (x_1, \dots, x_n)^{\mathrm{T}} \in \mathbb{H}^n$, $x_n \neq 0$, *let*

$$\mathbf{G} = \mathbf{G}_n^{n-1}\mathbf{G}_{n-1}^{n-2} \cdots \mathbf{G}_3^2\mathbf{G}_2^1. \tag{16}$$

Then, there exists $\sigma \in \mathbb{H}$, $|\sigma| = 1$ *such that*

$$\mathbf{G}^*\mathbf{x} = (\sigma||\mathbf{x}||, 0, \dots, 0)^{\mathrm{T}} \in \mathbb{H}^n\,.$$

The unitary matrix

$$\mathbf{G}^* = (\mathbf{G}_2^1)^*(\mathbf{G}_3^2)^* \cdots (\mathbf{G}_{n-1}^{n-2})^*(\mathbf{G}_n^{n-1})^*$$

has upper Hessenberg form.

Proof. All matrices $\mathbf{G}_j^i$ in (16) are unitary, so is their product.

Let us set $\varrho_i = \sqrt{\sum_{j=i}^{n} |x_j|^2}$, $i = 1, \dots, n-1$. Since $x_n \neq 0$, also $\varrho_i \neq 0$ for all i.

Let us start from the first reduction. We construct $\mathbf{G}_n^{n-1}$ to reduce the last component of $\mathbf{x} \in \mathbb{H}^n$:

$$(\mathbf{G}_n^{n-1})^* \left(x_1, x_2, \dots, x_{n-2}, x_{n-1}, x_n\right)^{\mathrm{T}} = \left(x_1, x_2, \dots, x_{n-2}, u_{n-1}, 0\right)^{\mathrm{T}},$$

where

$$u_{n-1} = \sigma_{n-1}\sqrt{|x_{n-1}|^2 + |x_n|^2}\,,$$

$$\sigma_{n-1} = \frac{\alpha_{n-1}x_{n-1} + \beta_{n-1}x_n}{|\alpha_{n-1}x_{n-1} + \beta_{n-1}x_n|}, \quad \alpha_{n-1}, \beta_{n-1} \in \mathbb{R},\ |\alpha_{n-1}| + |\beta_{n-1}| > 0.$$

In the next step we continue:

$$(\mathbf{G}_{n-1}^{n-2})^* \left(x_1, x_2, \dots, x_{n-2}, u_{n-1}, 0\right)^{\mathrm{T}} = \left(x_1, x_2, \dots, u_{n-2}, 0, 0\right)^{\mathrm{T}},$$

where

$$u_{n-2} = \sigma_{n-2}\sqrt{|x_{n-2}|^2 + |u_{n-1}|^2} = \sigma_{n-2}\varrho_{n-2}\,,$$

$$\sigma_{n-2} = \frac{\alpha_{n-2}x_{n-2} + \beta_{n-2}u_{n-1}}{|\alpha_{n-2}x_{n-2} + \beta_{n-2}u_{n-1}|} = \frac{\alpha_{n-2}x_{n-2} + \beta_{n-2}\sigma_{n-1}\varrho_{n-1}}{|\alpha_{n-2}x_{n-2} + \beta_{n-2}\sigma_{n-1}\varrho_{n-1}|}$$

$$\alpha_{n-2}, \beta_{n-2} \in \mathbb{R}, \quad |\alpha_{n-2}| + |\beta_{n-2}| > 0.$$

Now, we form the product $(\mathbf{G}_{n-1}^{n-2})^*(\mathbf{G}_n^{n-1})^*$ and continue by reducing u_{n-2}. By induction we obtain the following explicit formulae for the matrix $\mathbf{G}^* = (\widetilde{g_{kj}}) \in \mathbb{H}^{n\times n}$:

$$\begin{aligned}
\widetilde{g}_{1,j} &= \frac{\sigma_1 \overline{x}_j}{||\mathbf{x}||}, \qquad j = 1, \ldots, n;\\
\widetilde{g}_{i+1,i} &= -\frac{\varrho_{i+1}}{\varrho_i}\sigma_{i+1}\overline{\sigma_i}, \quad i = 1, \ldots, n-2;\\
\widetilde{g}_{k,j} &= -\frac{x_{k-1}\overline{\sigma_{k-1}}\sigma_k \overline{x}_j}{\varrho_{k-1}\varrho_k}, \; k = 2, \ldots, n-1, \quad j = k, \ldots, n;\\
\widetilde{g}_{n,n-1} &= -\frac{x_n \overline{\sigma_{n-1}}}{\varrho_{n-1}};\\
\widetilde{g}_{n,n} &= \frac{x_{n-1}\overline{\sigma_{n-1}}}{\varrho_{n-1}};\\
\widetilde{g}_{i,j} &= 0, \qquad i = 3, \ldots, n, \quad j = 1, \ldots, i-2;
\end{aligned}$$

where

$$\sigma_{n-1} = \frac{\alpha_{n-1}\, x_{n-1} + \beta_{n-1}\, x_n}{|\alpha_{n-1}\, x_{n-1} + \beta_{n-1}\, x_n|}, \quad \alpha_{n-1}, \beta_{n-1} \in \mathbb{R}, \; |\alpha_{n-1}| + |\beta_{n-1}| > 0,$$

$$\sigma_i = \frac{\alpha_i\, x_i + \beta_i \sigma_{i+1}\varrho_{i+1}}{|\alpha_i\, x_i + \beta_i \sigma_{i+1}\varrho_{i+1}|}, \; \alpha_i, \beta_i \in \mathbb{R}, \; |\alpha_i|+|\beta_i| > 0 \; \text{ for all } \; i = n-2, \ldots, 1,$$

i.e., the matrix $\mathbf{G}^*$ has upper Hessenberg form and

$$\mathbf{G}^*\mathbf{x} = (\sigma||\mathbf{x}||, 0, \ldots, 0)^{\mathrm{T}} \in \mathbb{H}^n .$$

Remark 2. For a given quaternion-valued vector $\mathbf{x} \in \mathbb{H}^n$ we can construct a Householder transformation matrix $\mathbf{H}$, which also reduces all components of $\mathbf{x}$ except the first one: $\mathbf{Hx} = u\mathbf{e}_1$, $u \in \mathbb{H}$, $\mathbf{e}_1 = (\mathbf{1}, 0, \ldots, 0)^{\mathrm{T}}$, but this matrix $\mathbf{H}$ is full, in general.

For a given $\mathbf{x} = (x_1, \ldots, x_n)^{\mathrm{T}} \in \mathbb{H}^n$, $x_n \neq 0$, let us choose $\sigma_{i-1} = \mathrm{sgn}\overline{x}_i$ at each step of the reduction. Such σ_i always belongs to the set Σ_i, see Remark 1. If we set $\sigma = \sigma_{n-1}$ we obtain

$$\sigma_i = \sigma \quad \text{for} \quad i = 1, \ldots, n-1$$

and $\quad \mathbf{G}^*\mathbf{x} = (\sigma||\mathbf{x}||, 0, \ldots, 0)^{\mathrm{T}} = (\mathrm{sgn}\overline{x_n}\; ||\mathbf{x}||, 0, \ldots, 0)^{\mathrm{T}}, \quad$ where

$$\mathbf{G}^* = \begin{pmatrix} \frac{\sigma\overline{x}_1}{||\mathbf{x}||} & \frac{\sigma\overline{x}_2}{||\mathbf{x}||} & \frac{\sigma\overline{x}_3}{||\mathbf{x}||} & \frac{\sigma\overline{x}_4}{||\mathbf{x}||} & \frac{\sigma\overline{x}_5}{||\mathbf{x}||} & \cdots & \frac{\sigma\overline{x}_n}{||\mathbf{x}||} \\ -\frac{\varrho_2}{\varrho_1} & \frac{x_1\overline{x_2}}{\varrho_1\varrho_2} & \frac{x_1\overline{x_3}}{\varrho_1\varrho_2} & \frac{x_1\overline{x_4}}{\varrho_1\varrho_2} & \frac{x_1\overline{x_5}}{\varrho_1\varrho_2} & \cdots & \frac{x_1\overline{x_n}}{\varrho_1\varrho_2} \\ 0 & -\frac{\varrho_3}{\varrho_2} & \frac{x_2\overline{x_3}}{\varrho_2\varrho_3} & \frac{x_2\overline{x_4}}{\varrho_2\varrho_3} & \frac{x_2\overline{x_5}}{\varrho_2\varrho_3} & \cdots & \frac{x_2\overline{x_n}}{\varrho_2\varrho_3} \\ \vdots & & & & & & \vdots \\ 0 & \cdots & 0 & -\frac{\varrho_{n-1}}{\varrho_{n-2}} & \frac{x_{n-2}\overline{x_{n-1}}}{\varrho_{n-2}\varrho_{n-1}} & & \frac{x_{n-2}\overline{x_n}}{\varrho_{n-2}\varrho_{n-1}} \\ 0 & \cdots & & 0 & -\frac{x_n\overline{\sigma}}{\varrho_{n-1}} & & \frac{x_{n-1}\overline{\sigma}}{\varrho_{n-1}} \end{pmatrix}.$$

The elements below the main diagonal are real.

If we reduce $\mathbf{x}$ in the order from top to bottom, a theorem similar to Theorem 5 can be proved. In particular,

$$\mathbf{G}^*\mathbf{x} = (\mathbf{G}_n^1)^*(\mathbf{G}_{n-1}^1)^* \cdots (\mathbf{G}_3^1)^*(\mathbf{G}_2^1)^*\mathbf{x} = (\sigma||\mathbf{x}||, 0, \dots, 0)^{\mathrm{T}},$$

the resulting $\mathbf{G}^*$ is not an upper Hessenberg matrix, but it has "nearly triangular form":

$$\begin{pmatrix} * & * & * & * & \dots & * \\ * & * & 0 & 0 & & 0 \\ * & * & * & 0 & & 0 \\ \vdots & & & \ddots & \ddots & \vdots \\ \vdots & & & & \ddots & 0 \\ * & * & * & \dots & & * \end{pmatrix};$$

the diagonal elements except the first one are real.

5 A reduction of an arbitrary quaternion-valued matrix into upper Hessenberg form

In order to reduce an arbitrary matrix $\mathbf{Y} = (y_{ij}) \in \mathbb{H}^{n\times n}$ to upper Hessenberg form there are $(n-2)(n-1)/2$ elements to be reduced. They are usually reduced step by step for example in the order

$$y_{31}, y_{41}, \dots, y_{n1}; y_{42}, y_{52}, \dots, y_{n2}; \dots; y_{nn-2}.$$

Corresponding Givens transformations $\mathbf{G}_j^i$, see (13), are then applied to k-th column of $\mathbf{Y}$, $k = 1, \dots, n-2$:

$$\mathbf{x} := (y_{1k}, \dots, y_{ik}, \dots, y_{jk}, \dots, y_{nk})^{\mathrm{T}}, \quad i = k+1; \quad j = k+2, \dots, n.$$

Let $1 \leq i_0 < j_0 \leq n$ be fixed. Let k–th column of the matrix $\mathbf{Y}$ plays the role of $\mathbf{x}$ in (14), (15).

The corresponding Givens' matrix for the reduction of the element $y_{j_0,k}$ has the form

$$\mathbf{G}_{j_0}^{i_0} = (g_{ij}) \in \mathbb{H}^{n\times n}, \quad \begin{pmatrix} g_{i_0 i_0} & g_{i_0 j_0} \\ g_{j_0 i_0} & g_{j_0 j_0} \end{pmatrix} = \begin{pmatrix} \overline{c}_{i_0} & s_{i_0} \\ -\overline{s}_{i_0} & c_{i_0} \end{pmatrix},$$
$$g_{ij} = \delta_{ij} \quad \text{otherwise}.$$

Here (for a properly chosen σ_{i_0}),

$$s_{i_0} = \frac{-\sigma_{i_0}\overline{y}_{j_0 k}}{\sqrt{|y_{i_0 k}|^2 + |y_{j_0 k}|^2}}, \qquad c_{i_0} = \frac{\sigma_{i_0}\overline{y}_{i_0 k}}{\sqrt{|y_{i_0 k}|^2 + |y_{j_0 k}|^2}}.$$

Let us count the number of real operations necessary for the reduction of an arbitrary matrix $\mathbf{Y} \in \mathbb{H}^{n\times n}$ to upper Hessenberg form. Moreover, let our resulting upper Hessenberg matrix be similar to the original one, so that we can use the reduced form to compute the eigenvalues of the matrix $\mathbf{Y}$. Let us remark that 1 addition of two quaternions needs 4 real flops, 1 multiplication of two quaternions needs 28 real flops. For the reduction of one element of $\mathbf{Y}$ we have to perform two matrix multiplications:

$$\begin{array}{rcl} \mathbf{V} & = & (\mathbf{G}_{j_0}^{i_0})^*\mathbf{Y} \end{array} \qquad \begin{array}{l} 4n \text{ multiplications} \\ 2n \text{ additions} \end{array} \quad \text{of quaternions}$$

$$\mathbf{V}\,\mathbf{G}_{j_0}^{i_0} \qquad \begin{array}{l} 4n \text{ multiplications} \\ 2n \text{ additions} \end{array} \quad \text{of quaternions}$$

All together, the reduction of one element of $\mathbf{Y}$ needs 240 n real flops. We can substantially reduce the number of operation (to 144 n real flops) if s or c is real.

To obtain the upper Hessenberg form we have to reduce $(n-2)+(n-3)+\cdots+2+1 = \dfrac{(n-2)(n-1)}{2}$ elements. Let $\mathbf{\Gamma}$ be the product of the corresponding Givens' matrices:

$$\mathbf{\Gamma} = \mathbf{G}_3^1\mathbf{G}_4^1\cdots\mathbf{G}_n^1\mathbf{G}_4^2\mathbf{G}_5^2\cdots\mathbf{G}_n^2\cdots\mathbf{G}_n^{n-2}.$$

We denote the resulting upper Hessenberg matrix by $\mathbf{W}$. After all reductions we obtain

$$\mathbf{W} = \mathbf{\Gamma}^*\mathbf{Y}\mathbf{\Gamma}, \quad \mathbf{W} \sim \mathbf{Y}, \quad \mathbf{W} \text{ is an upper Hessenberg matrix.}$$

Example 1. The elements of the following 5×5 quaternion-valued matrix $\mathbf{Y}$ are chosen randomly (integer elements in $\langle -5, 5\rangle$). Let $\mathbf{Y} =$

$$\left(\begin{array}{rrrrrrrrrrrrrrrrrrrrrrrrr}
y_{11} & y_{12} & y_{13} & y_{14} & y_{15} & y_{21} & y_{22} & y_{23} & y_{24} & y_{25} & y_{31} & y_{32} & y_{33} & y_{34} & y_{35} & y_{41} & y_{42} & y_{43} & y_{44} & y_{45} & y_{51} & y_{52} & y_{53} & y_{54} & y_{55} \\
\hline
5 & 4 & 3 & -3 & -2 & 0 & 0 & 1 & 4 & -5 & 3 & 3 & -4 & -2 & 2 & 4 & 1 & -1 & 3 & 3 & -5 & -2 & 5 & -2 & -5 \\
0 & 2 & 3 & 4 & 0 & 0 & 0 & -4 & -1 & 0 & -3 & -3 & 1 & 0 & -4 & -5 & 4 & -1 & 4 & -1 & -1 & -2 & -2 & -4 & 0 \\
-4 & 5 & 4 & 2 & -4 & -3 & 3 & -4 & 2 & 4 & 2 & -3 & -3 & 3 & 3 & -4 & 0 & -1 & 3 & -1 & 3 & 0 & 3 & -2 & 1 \\
-4 & 3 & -4 & -1 & 0 & 4 & 2 & 1 & -4 & 3 & -3 & 0 & 4 & -2 & 0 & 2 & 0 & 4 & 4 & -3 & -2 & -3 & 4 & 2 & 2
\end{array}\right).$$

Then $\mathbf{W} := \mathbf{\Gamma}^* \mathbf{Y} \mathbf{\Gamma} =$

w_{11}	w_{12}	w_{13}	w_{14}	w_{15}	w_{21}	w_{22}	w_{23}	w_{24}	w_{25}
5.0000	1.3205	−2.9935	1.7707	2.3056	10.0000	0.0833	−0.8560	−3.3842	−0.4310
0	2.7692	1.6199	−5.2047	1.0482	2.0000	3.5534	4.3072	4.6697	1.8606
−4.0000	6.5897	−0.8168	0.7781	3.7071	−6.0000	1.9352	−3.9121	−1.1416	−3.9023
−4.0000	1.5513	5.8488	0.7602	0.8553	4.0000	−0.1335	4.6311	−0.2449	0.6140

w_{31}	w_{32}	w_{33}	w_{34}	w_{35}	w_{41}	w_{42}	w_{43}	w_{44}	w_{45}	w_{51}	w_{52}	w_{53}	w_{54}	w_{55}
0	−4.6098	−4.4234	−0.9866	−1.7903	0	0	−2.4908	−2.4512	1.5053	0	0	0	−2.0766	0.7912
0	−3.7619	4.5482	−2.9750	−2.0253	0	0	−2.2011	2.3583	−2.0050	0	0	0	2.9598	0.6768
0	−3.4546	−2.0860	0.5185	3.4334	0	0	0.7713	3.6406	−2.9308	0	0	0	−4.8980	4.6884
0	−6.2755	1.5631	−0.8724	−2.2988	0	0	6.8124	1.1480	−1.3262	0	0	0	−5.2663	−0.0303

Thus, the structure of $\mathbf{W}$ is

$$\begin{pmatrix} * & * & * & * & * \\ * & * & * & * & * \\ 0 & * & * & * & * \\ 0 & 0 & * & * & * \\ 0 & 0 & 0 & * & * \end{pmatrix},$$

just an upper Hessenberg matrix.

6 Conclusions

We have developed an algorithm for reducing a quaternion-valued matrix into a similar upper Hessenberg matrix by making use of quaternion-valued Givens' transformation matrices. The algorithm was written and tested in MATLAB. It is a generalization of algorithm given in [3]. The number of real operations necessary to perform this reduction is still too large. We have suggested to reduce the number per each column in Section 4.

We are convinced that a good way to reduce the number of real operations is to try to apply Fast Givens' algorithm, which is used in the real case.

A complex version of Fast Givens' algorithm can be found in [5] (written in Chinese). This is the only paper dealing with Fast Givens' transformations applied to complex vectors, which we were able to find.

Acknowledgement

The authors acknowledge with pleasure the support of the Grant Agency of the Czech Republic (grant # 201/02/0844, grant # 201/02/0595) and support of the project MSM 223400007 of the Czech Ministry of Education.

References

1. Dongarra, J.J., Gabriel, J.R., Koelling, D.D., Wilkinson, J. H. (1984): Solving the Secular Equation Including Spin Orbit Coupling for Systems with Inversion and Time Reversal Symmetry. J. Comput. Phys., **54**, 278–288

2. Dongarra, J.J., Gabriel, J.R., Koelling, D.D., Wilkinson, J. H. (1984): The Eigenvalue Problem for Hermitian Matrices with Time Reversal Symmetry. Linear Algebra Appl., **60**, 27–42
3. Janovská, D., Opfer, G., (2003): Givens' transformation applied to quaternion-valued vectors. BIT
4. Van der Waerden, B.L. (1960): Algebra I, 5. Aufl. Springer, Berlin, Göttingen, Heidelberg
5. Xu, L. (1988): Eine schnelle Givens Transformation für komplexe Matrizen (in Chinese). Journal of East China Normal University (Natural Science), **3**, 15–21
6. Zhang, F. (1997): Quaternions and matrices of quaternions. Linear Algebra Appl. **251**, 21-57

Model of Compressible Flow and Transport in a Time-Dependent Domain*

Pavel Jiránek, Jiří Maryška and Jan Šembera

Technical University of Liberec, Hálkova 6, 461 17 Liberec, Czech Republic
jan.sembera@vslib.cz

Summary. In the contribution, an introduction of the model of compressible flow, transport of mass and energy, and production of energy in a time-dependent domain, which is being developed at Technical University of Liberec, is done. The main focus is concentrated on the finite volume model of advection-diffusion mass transport. The explicit upwind scheme is used, therefore the computational time step is restricted by the stability condition and the phenomenon of numerical diffusion may be significant. A scheme for reduction of numerical diffusion is proposed and numerical tests are presented.

1 Introduction

The final aim of our modelling is to predict production of nitrogen oxides from an internal combustion engine. For this purpose, a precise chemical reaction model should be developed. Its main inputs would be flow, temperature, and pressure fields and their development in time. This contribution is devoted to a model of relevant physical processes, which is being developed at Technical University of Liberec and whose outputs would form such inputs to the precise chemical reaction model. It is being built as a model of compressible flow, transport of mass and energy, and production of energy in a time-dependent domain.

A short overview of the model is done in the next section. The rest of the paper a bit more precisely discusses the finite volume mass transport model, a method of reduction of numerical diffusion, and its testing.

2 Overview of the model

The volume and shape of the cylinder of engine changes in time. To simplify the problem, we discretize it in time and in each time step we split the solution into two stages, isochoric and adiabatic one. All modelled processes are computed in the isochoric stage, supposed to take place in a fixed domain:

* This work was supported with the subvention from Ministry of Education of the Czech Republic, project code 242200001.

- production of mass and energy by chemical reactions,
- compressible flow of gas mixture,
- mass and energy transport.

All of them are discretized in time and space and solved by either finite volume or finite element method. The spatial computational mesh, built up of trilateral prismatic elements/volumes in layers, is common to all models (see Fig. 1).

The adiabatic stage models an immediate change of volume. Its key procedure is change of computational mesh. A more detailed description of the setting of the model can be found in [5].

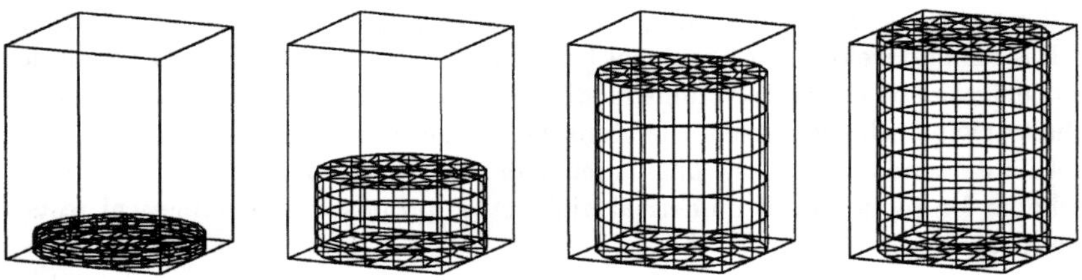

Fig. 1. A schematic example of computational mesh in four time steps

2.1 Model of chemical processes

All chemical processes are described by the set of stechiometric equations where each one can be written as

$$a_1 A_1 + a_2 A_2 + \ldots + a_{n_R} A_{n_R} \rightarrow b_1 B_1 + b_2 B_2 + \ldots + b_{n_P} B_{n_P} + q,$$

where a_i and b_j are stechiometric coefficients of reagents A_i and products B_j, respectively. During this reaction, the heat q is produced. It can be positive or negative depending on the type of such reaction. Local kinetic equations for a computation of mass and energy production can be expressed by linear ordinary differential equations

$$\frac{dc_i}{dt} = -R m_i a_i, \qquad \frac{dT}{dt} = R \frac{q}{C_V},$$

where m_i is the molar mass of reagent A_i and C_V is the heat capacity of the gas mixture. There are an application of the simplifying assumption that the reaction rate R depends only on the mass fraction of a chosen gas component and results of its calibration presented in [4]. One calibration result is illustrated on Fig. 2.

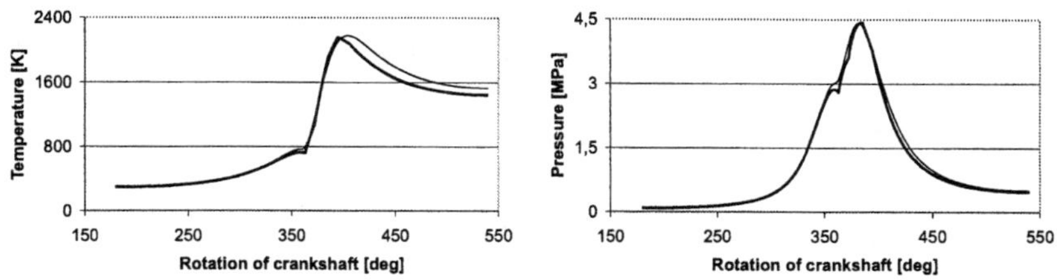

Fig. 2. Results of calibration of simplified reaction model. Bold lines are computational results, thin lines are measured data.

2.2 Fluid flow model

The model of flow of compressible gas mixture is governed by the set of Navier-Stokes equations, the continuity equation, and the state equation of perfect gas:

$$\begin{aligned}
&\tfrac{\partial \mathbf{v}}{\partial t} + (\mathbf{v}\cdot\nabla)\mathbf{v} = \nu\Delta\mathbf{v} + \tilde{\nu}\nabla(\nabla\cdot\mathbf{v}) - \tfrac{1}{\varrho}\nabla p,\\
&\tfrac{\partial \varrho}{\partial t} + \nabla\cdot(\varrho\mathbf{v}) = 0,\\
&p = \hat{R}T\varrho.
\end{aligned}$$

Here, $\mathbf{v}$ is the velocity vector, p pressure of gas, ϱ its density, and ν and $\tilde{\nu}$ are viscosity coefficients, $\hat{R}$ is the molar gas constant, and T is temperature.

Nonlinear system is discretized by mixed hybrid finite element method and linearized. Formulation of the model is set in [5], global behaviour tests were succesfully performed, further testing is being in process.

2.3 Model of mass and energy transport

Let $\Omega \subset \mathrm{R}^3$ represents the domain of the interior of the engine cylinder (since all processes are solved as isochoric processes in the fixed domain, we also omit the time dependence of Ω in our notation). The boundary of Ω is divided into two disjoint parts Γ_{in} and Γ_{ex}. The part Γ_{in} represents the inlet part of the boundary, the remaining part is denoted by Γ_{ex}. The problem is solved in the time interval $(0, \bar{t})$

The mass transport is governed by the set of mass balance equations for each component of a gas mixture [1]:

$$\frac{\partial(\varrho c_i)}{\partial t} + \nabla\cdot(\mathbf{f}c_i + \mathbf{j}_i) + c_i\gamma_- = c_{i,*}\gamma_+ \text{ in } \Omega,\ i = 1,\ldots,N \tag{1}$$

in a given flow field $\mathbf{f}(\mathbf{x}, t)$ for unknown functions $\varrho(\mathbf{x}, t)$, $c_i(\mathbf{x}, t)$ $(i = 1, \ldots, N)$. N is the number of components of gas mixture, γ_+ denotes the density of sources (with defined mass fractions $c_{i,*}$) and γ_- is the (positive) density of sinks. The diffusion flux $\mathbf{j}_i$ is given using the effective diffusivity $\mathcal{D}_i$ by the Fick's-law-like relation $\mathbf{j}_i = -\varrho\mathcal{D}_i\nabla c_i$ slightly corrected as proposed by Sutton and Gnoffo in [6].

The energy transport is derived from the balance of the internal energy, which can be written with respect to the temperature as [1]

$$C_V\left[\frac{\partial(\varrho T)}{\partial t}+\nabla\cdot(\mathbf{f}T)\right]-\nabla\cdot(k\nabla T)+TC_V\gamma_-=T_*C_V\gamma_++C_V(\mathcal{R}^{\mathrm{rev}}+\mathcal{R}^{\mathrm{irr}})\text{ in }\Omega.$$

Here, T_* is the temperature of inflowing fluid. The terms $\mathcal{R}^{\mathrm{rev}}$ and $\mathcal{R}^{\mathrm{irr}}$ express reversible and irreversible rates of conversion between mechanical and internal energy and can be written as

$$\mathcal{R}^{\mathrm{rev}}=-\frac{\hat{R}\varrho T}{C_V}(\nabla\cdot\mathbf{v}),\qquad \mathcal{R}^{\mathrm{irr}}=\frac{\mu}{C_V}\sum_{i,j=1}^{3}\left[\frac{\partial v_j}{\partial x_i}+\frac{\partial v_i}{\partial x_j}-\frac{2}{3}(\nabla\cdot\mathbf{v})\right]\frac{\partial v_j}{\partial x_i},$$

where μ is viscosity coefficient. In these formulas, the ideal behaviour of gas and the Newtonian fluid are supposed.

The boundary conditions are set of Dirichlet type at inlet part of the boundary:

$$\left.\begin{array}{l} c_i(\mathbf{x},t)=c_{i,D}(\mathbf{x},t),\ i=1,\dots,N,\\ \varrho(\mathbf{x},t)=\varrho_D(\mathbf{x},t),\\ T(\mathbf{x},t)=T_D(\mathbf{x},t),\end{array}\right\}\ \mathbf{x}\in\Gamma_{\mathrm{in}},\ t\in(0,\bar{t}) \tag{2}$$

and of Neumann type at the rest part of boundary:

$$\left.\begin{array}{l} \nabla c_i(\mathbf{x},t)\cdot\mathbf{n}(\mathbf{x})=0,\ i=1,\dots,N\\ \nabla T(\mathbf{x},t)\cdot\mathbf{n}(\mathbf{x})=0,\end{array}\right\}\ \mathbf{x}\in\Gamma_{\mathrm{ex}},\ t\in(0,\bar{t}), \tag{3}$$

where $\mathbf{n}(\mathbf{x})$ is the outward normal to Γ_{ex}.

The initial conditions are set as

$$\varrho(\mathbf{x},0)=\varrho_0(\mathbf{x}),\qquad c_i(\mathbf{x},0)=c_{0,i}(\mathbf{x}),i=1,\dots,N,\qquad T(\mathbf{x},0)=T_0(\mathbf{x}). \tag{4}$$

3 Numerical model of mass transport

In this section, we focus on the numerical model of mass transport. The energy transport model is very similar.

Before starting with the numerical scheme, let us make a note about the space decomposition. The presented model use the meshes consisting of trilateral prisms with parallel bases. The base of the cylider is decomposed into the set of triangles and then is the triangulation extruded along the z axis up to the height of the cylinder. It should be noted that the 2-D mesh must satisfy some conditions needed for a consistency of the numerical scheme. These conditions can be found e.g. in [2] as the definition of admissible mesh.

Using the notation from [2], let us introduce the set of control volumes $\mathcal{T}$, the set of their faces $\mathcal{E}$ and the set of points $\mathcal{P}$ such that each point $\mathbf{x}_K\in\mathcal{P}$ can be uniquely assigned to one control volume $K \in \mathcal{T}$. Suppose the mesh $\{\mathcal{T}, \mathcal{E}, \mathcal{P}\}$ to approximate domain Ω and to be admissible. Let us denote $m(K)$ volume of $K \in \mathcal{T}$, $m(\sigma)$ area of $\sigma \in \mathcal{E}$, $\mathcal{N}(K)$ the set of adjacent control volumes of $K \in \mathcal{T}$, $\mathcal{E}_K$ the set of sides common to the volume K. For adjacent control volumes K and L let $K|L \in \mathcal{E}_K$ be their common side. Let $d_{K,\sigma}$ be the distance between the point $\mathbf{x}_K$ and the side σ. Let $d_{K|L}$ denote the distance between $\mathbf{x}_K$ and $\mathbf{x}_L$.

The time discretization is realized by the ascending sequence of time values $(t_n)_{n\in\mathrm{N}_0}$, $t_0 = 0$ with the time step $\Delta t_n = t_{n+1} - t_n$.

Explicit finite volume scheme of the problem (1)–(4) can be written as follows:

$$\begin{aligned} m(K)\frac{\varrho^{n+1}_{i,K} - \varrho^n_{i,K}}{\Delta t_n} + \sum_{\sigma\in\mathcal{E}_K} F^n_{i,K,\sigma} + \sum_{\sigma\in\mathcal{E}_K} f^n_{K,\sigma} c^n_{i,\sigma,+} \\ = m(K)\gamma^n_K c^n_{i,K,+}, \ i = 1,\dots,N, \ K \in \mathcal{T}, \end{aligned} \tag{5}$$

where $\varrho^n_{i,K} = \varrho^n_K c^n_{i,K}$ means partial density of ith specie in the finite volume K. We use such a notation that e.g. ϱ^n_K approximates $\varrho(\mathbf{x}_K, t_n)$. Mass fractions in the advective and source term are expressed following the upwind scheme as

$$c^n_{i,\sigma,+} = \begin{cases} c^n_{i,K} & \text{if } f^n_{K,\sigma} \geq 0, \\ c^n_{i,L} & \text{if } f^n_{K,\sigma} < 0, \ \sigma = K|L \not\subset \partial\Omega, \\ c^n_{i,\sigma,D} & \text{if } f^n_{K,\sigma} < 0, \ \sigma \subset \Gamma_{\text{in}}, \end{cases} \qquad c^n_{i,K,+} = \begin{cases} c^n_{i,K,*} & \text{if } \gamma^n_K \geq 0, \\ c^n_{i,K} & \text{if } \gamma^n_K < 0. \end{cases} \tag{6}$$

The diffusion term is approximated by

$$F_{i,K,\sigma} = \begin{cases} -\frac{m(\sigma)}{d_{K|L}}\varrho^n_\sigma \mathcal{D}^n_{i,\sigma}(c^n_{i,L} - c^n_{i,K}) & \text{for } \sigma = K|L \not\subset \partial\Omega, \\ -\frac{m(\sigma)}{d_{K,\sigma}}\varrho^n_\sigma \mathcal{D}^n_{i,\sigma}(c^n_{i,\sigma,D} - c^n_{i,K}) & \text{for } \sigma \subset \Gamma_{\text{in}}, \\ 0 & \text{for } \sigma \subset \Gamma_{\text{ex}}. \end{cases} \tag{7}$$

Final decomposition into the density and mass fraction is computed as follows:

$$\varrho^{n+1}_K = \sum_{i=1}^N \varrho^{n+1}_{i,K}, \qquad c^{n+1}_{i,K} = \frac{\varrho^{n+1}_{i,K}}{\varrho^{n+1}_K}.$$

The scheme is conservative. The main disadvantage of the proposed model is the strong restriction for the selection of the time step Δt_n. A simple analysis shows that the condition

$$\Delta t_n \leq \frac{\varrho m(K)}{\sum\limits_{\substack{\sigma\in\mathcal{E}_K \\ f^n_{K,\sigma}\geq 0}} f^n_{K,\sigma} + \varrho \sum\limits_{L\in\mathcal{N}(K)} \frac{m(K|L)}{d_{K|L}}\mathcal{D}^n_{i,K|L} + \varrho \sum\limits_{\substack{\sigma\in\mathcal{E}_K \\ \sigma\subset\Gamma_{\text{in}}}} \frac{m(\sigma)}{d_{K,\sigma}}\mathcal{D}^n_{i,\sigma}}$$

for all $K \in \mathcal{T}$ and $i = 1, \dots, N$ assures stability of the scheme in incompressible stationary flow field. It leads to the h^2-stability of the scheme (h has meaning of the largest diameter of finite volumes of the mesh).

One more disadvantage of the scheme arises from using of the upwind approximation of the advective term: numerical diffusion. The analysis of 1-D scheme in [3] shows that the 1-D upwind numerical solution of the advection-diffusion equation with diffusivity $\mathcal{D}$ with equidistant finite volume mesh approximates the exact solution with the diffusivity $\mathcal{D} + \mathcal{D}_{\text{num}}$, where

$$\mathcal{D}_{\text{num}} = \frac{1}{2} v h \left(1 - v \frac{\Delta t}{h}\right). \tag{8}$$

Here v is the velocity and h is the length of finite volumes.

Our approach to the elimination of numerical diffusion is just to subtract the estimated numerical diffusivity from all physical coefficients. The numerical diffusivity is estimated on each inter-volume face of the mesh separately. The substraction cannot be performed arbitrarily but a non-negativity of the resulting coefficients must be achieved, i.e. in the scheme, the coefficient

$$\mathcal{D}_{\text{red}} = \max\{\mathcal{D} - \mathcal{D}_{\text{num}}, 0\} \tag{9}$$

is applied instead of $\mathcal{D}$.

Additionally, the estimate (8) of $\mathcal{D}_{\text{num}}$ should be extended to more-dimensional case. It is made by using the same formula (8) and estimating the 1-D parameters v and h in the following way: On each face $\sigma = K|L$, v is set equal to the magnitude of the dominant velocity in small neighbourhood and h is estimated as the length of the projection of the vector $(\mathbf{x}_L - \mathbf{x}_K)$ to the direction of the dominant velocity.

4 Tests of numerical diffusion

4.1 1-D test

Our 1-D test problem of advection-diffusion transport of coloured fluid is

$$\frac{\partial c}{\partial t}(x,t) + v\frac{\partial c}{\partial x}(x,t) - \mathcal{D}\frac{\partial^2 c}{\partial x^2}(x,t) = 0, (x,t) \in (-\infty, +\infty) \times (0, \bar{t}) \tag{10}$$

$$c(x,0) = M\delta(x - x_0), \tag{11}$$

where $\delta(x)$ denotes the Dirac function. The analytic solution of the problem is

$$c_{\text{an}}(x,t) = \frac{M}{2\sqrt{\pi \mathcal{D} t}} \exp\left[-\frac{(x - x_0 - vt)^2}{4\mathcal{D}t}\right], \tag{12}$$

where M is the initial mass of coloured fluid in the domain.

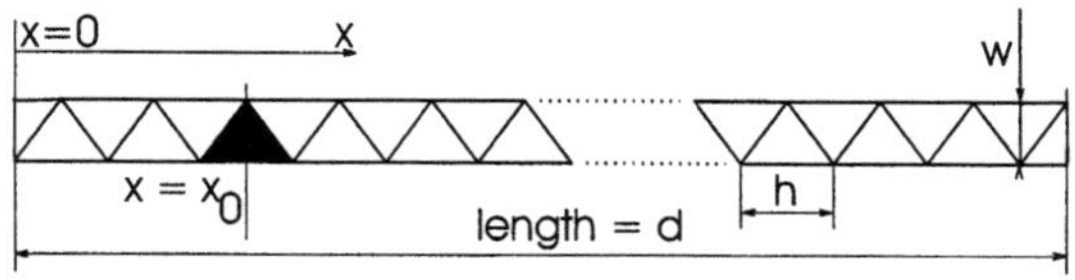

Fig. 3. 1-D test mesh

The solution of the problem (10)–(11) is compared with numerical solution of its 2-D finite-domain approximation:

$$\frac{\partial c}{\partial t}(x,y,t)+\mathbf{v}\nabla c(x,y,t)-\mathcal{D}\Delta c(x,y,t)=0,\,(x,y,t)\in(0,d)\times(0,w)\times(0,\bar{t})\quad(13)$$

$$\begin{array}{l}\frac{\partial c(x,0,t)}{\partial y}=\frac{\partial c(x,w,t)}{\partial x}=0,x\in(0,d),t\in(0,\bar{t})\\ c(0,y,t)=\frac{\partial c(d,y,t)}{\partial x}=0,y\in(0,w),t\in(0,\bar{t})\end{array}\quad c(x,y,0)=\begin{cases}1,\ (x,y)\in K_0\\ 0,\ (x,y)\notin K_0\end{cases},\quad(14)$$

where K_0 is the triangle including point $(x_0, w/2)$. The problems are comparable, if parameter M in (11) has meaning of area of the finite volume K_0 (see Fig. 3).

The tests were performed using the following parameters: $d = 20$, $x_0 = d/4 = 5$, $h = d/N$ (where N is the number of control volumes). The mesh consisted of equilateral triangles, therefore $w = h\cos\frac{\pi}{6}$.

In Tab. 1, there are results of several computations for comparison summarized. The error E is computed as the L_1 norm of the difference of analytical solution (12) and the numerical solution of (13)–(14): $E = \|c_{\mathrm{num}}(x,y,t) - c_{\mathrm{an}}(x,t)\|_{L_1((0,d)\times(0,w))}$, where c_{num} was computed numerically either without [E(without)] or with [E(with)] the proposed reduction of numerical diffusion (9).

4.2 2-D test

The other presented test problem is a 2-D extension of the previous one. It is given as

$$\frac{\partial c}{\partial t}(\mathbf{x},t)+\nabla\cdot(c(\mathbf{x},t)\mathbf{v})-\mathcal{D}\nabla^2 c(\mathbf{x},t)=0,\,(\mathbf{x},t)\in\mathrm{R}^2\times(0,\bar{t})\quad(15)$$

$$c(\mathbf{x},0)=M\delta(|\mathbf{x}-\mathbf{x}_0|)\quad(16)$$

with the analytical solution $c_{\mathrm{an}}(\mathbf{x},t)=\frac{M}{4\pi\mathcal{D}t}\exp\left[-\frac{|\mathbf{x}-\mathbf{x}_0-\mathbf{v}t|^2}{4\mathcal{D}t}\right]$. The solution is compared with numerical solution of the finite-domain approximation of the problem (15), (16) defined as follows: Governing equation (15) is supposed to hold in the domain Ω shown on Fig. 4. The domain is decomposed to the set of 800 equilateral triangles (their side length is 1). The point $\mathbf{x}_0 = (5, 5\sqrt{3}/2)$. The boundary and initial conditions of the finite-domain problem approximation are:

Table 1. Comparison of numerical results of 1-D tests for various input parameters N (number of elements), v (velocity) and $\mathcal{D}$ (diffusivity) performed without and with the proposed reduction of numerical diffusion

N	v	$\mathcal{D}$	E(without)	E(with)	$\frac{E(\text{without})}{E(\text{with})}$
20	1	0.1	$8.49 \cdot 10^{-3}$	$7.61 \cdot 10^{-4}$	11.15
20	1	0.2	$6.54 \cdot 10^{-3}$	$5.86 \cdot 10^{-4}$	11.15
20	1	0.4	$4.40 \cdot 10^{-3}$	$3.58 \cdot 10^{-4}$	12.28
40	1	0.1	$3.21 \cdot 10^{-3}$	$2.06 \cdot 10^{-4}$	15.58
40	1	0.2	$2.23 \cdot 10^{-3}$	$1.33 \cdot 10^{-4}$	16.78
40	1	0.4	$1.35 \cdot 10^{-3}$	$7.24 \cdot 10^{-5}$	18.68
80	1	0.1	$1.11 \cdot 10^{-3}$	$4.83 \cdot 10^{-5}$	22.96
80	1	0.2	$6.97 \cdot 10^{-4}$	$2.35 \cdot 10^{-5}$	29.61
80	5	0.2	$1.73 \cdot 10^{-3}$	$8.98 \cdot 10^{-5}$	19.22
80	5	0.4	$1.26 \cdot 10^{-3}$	$5.76 \cdot 10^{-5}$	21.87
80	10	0.4	$1.73 \cdot 10^{-3}$	$8.98 \cdot 10^{-5}$	19.22

$$\begin{aligned} \nabla c(\mathbf{x},t)\cdot\mathbf{n}(\mathbf{x}) &= 0, \mathbf{x}\in\Gamma_{\text{ex}}, t\in(0,\bar{t}) \\ c(\mathbf{x},t) &= 0, \mathbf{x}\in\Gamma_{\text{in}}, t\in(0,\bar{t}) \end{aligned} \qquad c(\mathbf{x},0)=\begin{cases}1, \mathbf{x}\in K_0\\ 0, \mathbf{x}\notin K_0\end{cases}, \quad (17)$$

where K_0 is the finite volume including point $\mathbf{x}_0$ and the boundary is split into inlet and rest part due to direction of velocity.

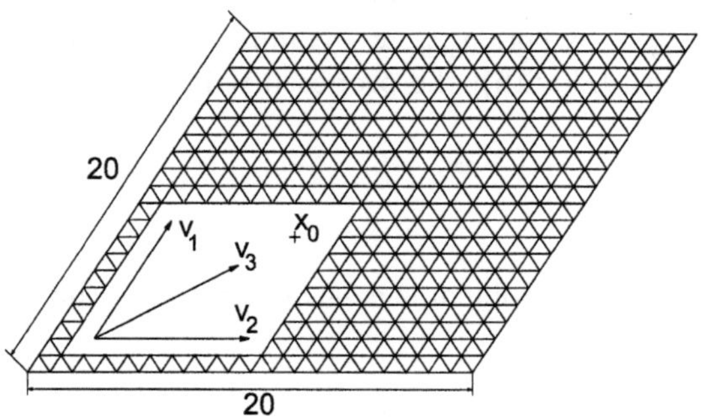

Fig. 4. 2-D test mesh

The solution is computed for three different homogeneous velocity fields differing by direction of the velocity (on Fig. 4, they are denoted as $\mathbf{v}_1$, $\mathbf{v}_2$ and $\mathbf{v}_3$), magnitude of velocity and diffusion coefficients. The error is computed analogically to the 1-D case: $E = \|c_{\text{num}}(\mathbf{x},t) - c_{\text{an}}(\mathbf{x},t)\|_{L_1(\Omega)}$. The comparison results are collected in Tab. 2. The notation is the same as in 1-D case.

From Tab. 2 it can be seen that for advection-dominated transport, our method of reduction of numerical diffusion is not very efficient: In case of the

Table 2. Numerical results of 2-D tests for various magnitudes and directions of velocity **v** and diffusion coefficients $\mathcal{D}$

$\|\mathbf{v}\|$	direction	$\mathcal{D}$	E(without)	E(with)	$\frac{E(\text{without})}{E(\text{with})}$
1	1 or 2	0.2	$5.66 \cdot 10^{-5}$	$7.81 \cdot 10^{-6}$	7.25
1	3	0.2	$7.83 \cdot 10^{-5}$	$1.74 \cdot 10^{-5}$	4.50
1	1 or 2	0.1	$9.21 \cdot 10^{-5}$	$1.59 \cdot 10^{-5}$	5.79
1	3	0.1	$1.26 \cdot 10^{-4}$	$5.10 \cdot 10^{-5}$	2.47
5	1 or 2	0.2	$1.42 \cdot 10^{-4}$	$6.97 \cdot 10^{-5}$	2.04 (34% red.)
5	3	0.2	$2.05 \cdot 10^{-4}$	$1.65 \cdot 10^{-4}$	1.24 (63% red.)

last two tests, the numerical diffusion coefficients respective to 34% or 63% mesh sides were smaller than the physical diffusion coefficient and could be fully substracted, the resting diffusion coefficients were set to 0 due to (9).

5 Conclusions

In the contribution, especially the advection-diffusion model of mass transport of gas mixture was discussed. The stability condition and the simple method for reduction of numerical diffusion were presented. Numerical results show that the proposed method of reduction of numerical diffusion works quite well in 1-D and 2-D cases. Since it is not efficient in advection-dominated transport problems, we are looking for a more sophisticated and successful one. Therefore we plan to concentrate on development of a more acurate numerical scheme for this purpose.

Also other parts of the model are being developed. Especially calibration tests of production of energy are continuing with more natural conditions of ignition and reaction parameters and local tests of the flow model are being performed.

References

1. Bird, R.B., Stewart, W.E., Lightfoot, E.N. (2002): Transport phenomena. Wiley, New York
2. Eymard, R., Gallouët, T., Herbin R. (1994): Finite volume methods. In: Ciarlet P.G., Lions, J.L.(eds) Handbook of Numerical Analysis, Vol. VII, 713–1020. North Holland, Amsterdam
3. Odman, M.T. (1997): A Quantitative Analysis of Numerical Diffusion Introduced by Advection Algorithms in Air Quality Models. Atmospheric Environment, **31**, No. 13, 1933–1940

4. Šembera, J., Maryška, J. (2002): On the Local Model of Energy Production Inside a Combustion Engine. In: Příhoda, J., Kozel, K. (eds) Proceedings of Seminat Topical Problems of Fluid Mechanics 2002, 71–74. Institute of Thermomechanics AS CR, Prague
5. Šembera, J., Maryška, J., Novák, J. (2003): FEM/FVM Modelling of Processes in Combustion Engine. In: Chen, Z., Glowinski, R., Li, K. (eds) Current Trends in Scientific Computing. Contemporary Mathematics, **329**.
6. Sutton, K., Gnoffo, P.A. (1998): Multi-Component Diffusion with Application To Computational Aerothermodynamics. 7th AIAA/ASME Joint Thermophysics and Heat Transfer Conference, Albuquerque, NM. AIAA **98-2575**.

Numerical Study of Convection of Multi-Component Fluid in Porous Medium

Olga Kantur and Vyacheslav Tsybulin

Rostov State University, Department of Mathematics and Mechanics, ul. Zorge 5, Rostov-on-Don, 344090 Russia
kantur_rsu@mail.ru, tsybulin@math.rsu.ru

Summary. Planar problem of convective filtration of incompressible multi-component fluid saturated a porous medium is studied. The combined spectral and finite-difference approach is applied to compute nonstationary regimes and continuous families of steady states with variative spectrum.

1 Introduction

Convective flows in a porous medium is a subject of many works [1]. For the convective filtration of viscous fluid in porous medium (Darcy model) it was observed an appearance of one-parameter family of steady states with the spectrum, which varies along the family [2]. V. Yudovich shown that these families cannot be the orbits of any symmetry group and derived the theory of cosymmetry [3, 4]. Investigation of continuous families of steady states may be performed only experimentally or numerically. Computer modelling of Darcy convection was done by spectral method [5], finite-difference method [6], and combined spectral and finite-difference approach [7, 8]. It was found that keeping a cosymmetry in finite-dimensional approximation of differential equation is extremely important to correct computation of continuous families of steady states [6, 7]. In the present work we give an extension of approach [7] for the convection of multi-component fluid obeying Darcy law.

2 Darcy convection problem

The equations of filtrational convection of multi-component fluid [9] in dimensionless form may be written as

$$\beta_r \theta^r_t = \kappa_r \Delta\theta^r + \lambda_r \psi_x + \beta_r J(\psi, \theta^r) \equiv F^r, \quad r = 1, \dots, S. \tag{1}$$

$$0 = \Delta\psi - \sum_{r=1}^{S} \theta^r_x \equiv G. \tag{2}$$

$$\theta^r = \phi^r, \quad r = 1, \dots, S, \quad \psi = 0 \quad \text{on} \quad \partial\mathcal{D}. \tag{3}$$

Here, x and y are Cartesian coordinates on a plane, t is time, Δ is the Laplacian, $J(\psi,\theta) = \psi_x\theta_y - \psi_y\theta_x$ denotes the Jacobian operator, ψ is the streamfunction, θ^1 is the temperature, $\theta^r \quad (r = 2,\dots,S)$ are concentrations. This problem is characterized by the following parameters: $\kappa_r = \chi_r/\nu$, β_r, $\lambda_r = g\eta_r A_r l^2 K/\nu^2, r = 1,\dots,S$, where λ_1 is Rayleigh number and λ_r – parameters for concentrations ($r = 2,\dots,S$). Here for each r we have expansion coefficient η_r, gradient of distribution for species A_r, diffusivity coefficient χ_r, kinematic coefficient β_r, and g is the acceleration due to gravity, l – height of enclosure, K – permeability, ν – viscosity. We consider the enclosure $\mathcal{D} = [0,a] \times [0,b]$ and ϕ_r are given and do not depend on time.

Cosymmetry for underlying system is given by $(\psi, -\kappa_1\theta_1, \dots -\kappa_S\theta_S)$. Really, multiply (1) by ψ and (2) by $-\kappa_r\theta_r$, sum and integrate over domain $\mathcal{D}$. Then, using integration by parts and Green's formula we derive

$$\sum_{r=1}^{S} \int_{\mathcal{D}} (F^r\psi - G\sum_{r=1}^{S} \kappa_r\theta_r)dxdy = 0. \tag{4}$$

3 Method of solution

Spectral and spectral-difference methods are the powerful tools for solving problems in mathematical physics [10]. We apply here the spectral-finite-difference method in the form derived in [7]. Solution to the problem (1)–(3) is seeking in the form:

$$\{\theta^1, \dots, \theta^S, \psi\} = \sum_{j=1}^{m} \{\theta_j^1(x,t), \dots, \theta_j^S(x,t), \psi_j(x,t)\} \sin\frac{\pi j y}{b}. \tag{5}$$

Substituting (5) into (1)–(3) and performing projections, we obtain [7]:

$$\beta_s\dot{\theta}_j^s = \kappa_s\theta_j^{\prime\prime s} - c_j\kappa_s\theta_j^s + \lambda_s\psi_j' - \beta_r J_j \equiv F_j^s, \quad j = 1,\dots,m, \tag{6}$$

$$0 = \psi_j'' - c_j\psi_j - \theta_j^{\prime s} \equiv G_j, \quad j = 1,\dots,m, \tag{7}$$

$$\theta_j^s(t,0) = \theta_j^s(t,a) = \phi^s, \quad \psi_j(t,0) = \psi_j(t,a) = 0, \quad j = 1,\dots,m. \tag{8}$$

Then, we define the grid $\omega = \{x_k = kh, k = 0,\dots,n, h = a/(n+1)\}$ on the segment $[0,a]$ and introduce the notion: $\theta_{j,k}^r = \theta_j^r(x_k,t)$, $\psi_{j,k} = \psi_j(x_k,t)$, $J_{j,k}^r = J_j^r(x_k,t)$. Using the centered finite-difference operators of second order accuracy on the three-point stencil we obtain a system of ordinary differential equations

$$\begin{aligned} \beta_r\dot{\theta}_{jk} = \kappa_r \frac{\theta_{j,k+1}^r - 2\theta_{j,k}^r + \theta_{j,k-1}^r}{h^2} - c_j\kappa_r\theta_{jk}^r + \\ +\lambda_r \frac{\psi_{j,k+1} - \psi_{j,k-1}}{2h} - \beta_r J_{j,k}^r \equiv \phi_{1jk}^r, \end{aligned} \tag{9}$$

$$0 = \frac{\psi_{j,k+1} - 2\psi_{j,k} + \psi_{j,k-1}}{h^2} - c_j\psi_{jk} + \tag{10}$$
$$+\frac{\theta^r_{j,k+1} - \theta^r_{j,k-1}}{2h} \equiv \phi_{2jk}, \quad r = 1, \ldots, S.$$

Henceforth, the dot is a derivative with respect to t, $c_j = j^2\pi^2/b^2$, and $J^r_{j,k}$ is expressed as

$$J^r_{jk} = \frac{2\pi}{b}\left(\sum_{i=1}^{m-j} \xi^r_{j,i,k} + \sum_{i=1}^{j-1} \zeta^r_{j,i,k}\right),$$

$$\xi^r_{j,i,k} = \frac{2i+j}{2}[d_{s,k}(\theta^r_{i+j}, \psi_i) - d_{s,k}(\theta^r_i, \psi_{i+j})] - \frac{j}{2}[d_{a,k}(\theta^r_{i+j}, \psi_i) + d_{a,k}(\theta^r_i, \psi_{i+j})],$$

$$\zeta^r_{j,i,k} = \frac{j-i}{2}[d_{s,k}(\theta^r_i, \psi_{j-i}) + d_{a,k}(\theta^r_i, \psi_{j-i}) - d_{s,k}(\theta^r_{j-i}, \psi_i) + d_{a,k}(\theta^r_{j-i}, \psi_i)],$$

where operators $d_{a,k}$ and $d_{s,k}$ are derived in [7] using the requirement that discrete version of (4) took place

$$d_{a,k}(\theta^r, \psi) = \frac{\theta^r_{k+1} - \theta^r_{k-1}}{2h}\psi_k - \theta^r_k\frac{\psi_{k+1} - \psi_{k-1}}{2h},$$

$$d_{s,k}(\theta^r, \psi) = \frac{2\theta^r_{k+1}\psi_{k+1} + \psi_k(\theta^r_{k+1} - \theta^r_{k-1}) + \theta^r_k(\psi_{k+1} - \psi_{k-1}) - 2\theta^r_{k-1}\psi_{k-1}}{6h}.$$

Excluding stream function $\psi_{j,k}$ we may rewrite the system (9)–(10) in a vector form

$$\beta_r\frac{d\Theta^r}{dt} = (\kappa_r A + \lambda_r BA^{-1}B)\Theta^r + L(\Theta^r, A^{-1}B\Theta^r), \quad r = 1, \ldots, S, \tag{11}$$

where

$$\Theta^r = (\theta^r_{11}, \ldots, \theta^r_{n1}, \ldots, \theta^r_{1m}, \ldots, \theta^r_{nm}).$$

The matrix A consists of m three-diagonal submatrices, the nonzero entries of the skew-symmetric matrix $B = \{b_{sr}\}^{nm}_{s,r=1}$ are given by $b_{s,s+1} = -b_{s+1,s} = h/2, s = 1, \ldots, nm-1$, and L presents the nonlinear terms in (9).

The system (11) has a trivial solution with zero velocity and linear temperature and concentration profiles. The integration is performed by the fourth order Runge-Kutta method, and the family is calculated by using the algorithm [5, 7]. Starting from the vicinity of unstable zero equilibrium we integrate the system (11) up to a point close to a stable equilibrium on the family. Then of the algorithm to family computation may be formulated as the sequence of the following steps. Correct the point using the modified Newton method. Determine the kernel of the matrix of linearization at given point and predict the next point on the family. Repeat these steps until a closed curve is obtained.

4 Numerical Results

We explored the derived technique to calculate the families consisting of both stable and unstable equilibria. The narrow enclosure ($b/a \geq 1, a = 1$) is considered and the case of $\beta_r = 1$, $\kappa_r = 1$ is analyzed. We found that a scenario of the appearance of a family of the steady states in the case of multi-component fluid with co-directed temperature and concentration gradients is similar to the case of one-component fluid [11]. Firstly a family consisted of stable steady states

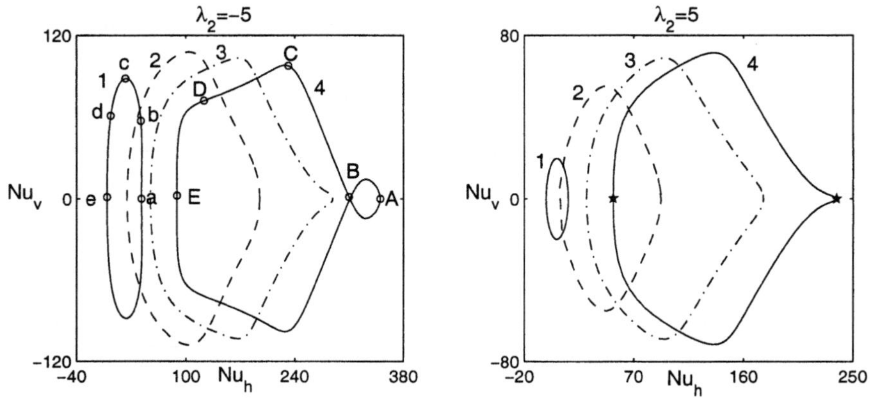

Fig. 1. Evolution of stationary regimes family; $S = 2$, $b = 2$, $a = 1$, left ($\lambda_2 = -5$): $\lambda_1 = 72$ (curve 1), 100 (2), 120 (3), 138.1 (4), right ($\lambda_2 = 5$): $\lambda_1 = 40$ (curve 1), 60 (2), 80 (3), 95.5 (4)

branched off from the state of rest (zero equilibrium) as result of monotonic instability. With increasing the Rayleigh parameter λ_1 the family deforms, and then on it the arches of unstable equilibria occur. In Fig. 1 the families for the case of two-component fluid and the container with $b = 2$ are given as the projections on a plane Nu_h and Nu_v:

$$Nu_h = \int_0^b \theta_x^1(\frac{a}{2}, y, t)\, dy, \quad Nu_v = \int_0^a \theta_y^1(x, 0, t)\, dx.$$

At the critical value λ_u two unstable points appear on the family. For instance, for $\lambda_2 = -5$ we found this instability at $\lambda_u = 138.1$ (circles A and E on Fig. 1 at the left), and for $\lambda_2 = 5$ – respectively at $\lambda_u = 95.5$ (asterisk on Fig. 1 on the right). We present in Figs. 2 and 3 the streamlines, isotherms and isolines for concentration corresponding the letters on Fig. 1.

At large negative gradient of concentration and small diffusive coefficients the state of rest loses stability by oscillatory manner. We have observed here a new scenario of convective regimes development. Convective transitions are

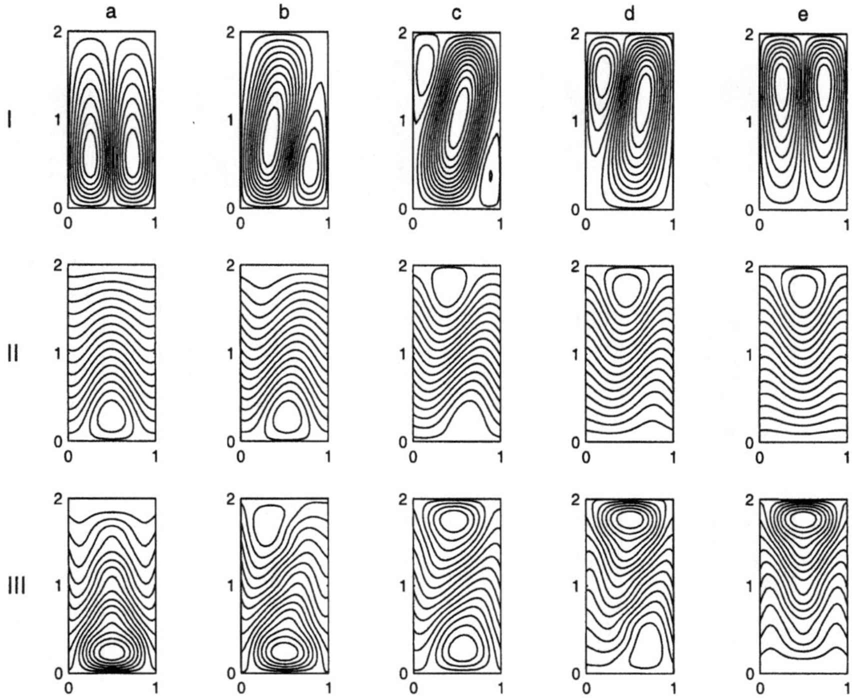

Fig. 2. Stream functions (I), temperature (II), concentration (III) for some regimes, $S = 2$, $b = 2$, $a = 1$, $\lambda_1 = 72$, $\lambda_2 = -5$

organized using nonstationary regime branched off from zero equilibrium (state of rest), and also two families of the stationary solutions were born 'from air'. The given scenario is realized for the cases of two-component and three-component fluids. We draw the development of convective regimes for two-component fluid in Fig. 4; the parameters are the following $\lambda_2 = -10$, $\kappa_2 = 0.3$ and $b = 2$.

The state of rest is stable up to $\lambda_1 \approx 77$ and simultaneously there exist two families of steady states originated via 'out of thin air' bifurcation [12]. Each of these families consists of stable and unstable arches (here we mean transversal stability or instability with respect to a family). State of rest loses its stability via Poincare–Andronov–Hopf bifurcation (oscillatory instability) and the stable limit cycle is formed (curve 3 in Fig. 4). Increasing λ_1 both families become more complicated and at $\lambda_1 \approx 79.4$ collide one with another. Then two new families are appeared: wholly unstable (curve 4) and partially stable (curve 5). Further increasing of λ_1 leads to the reduction of unstable arches and their disappearance. After that, on the interval $80 < \lambda_1 < 156.5$ given family is wholly stable. Limit cycle on small interval of λ_1 undergoes

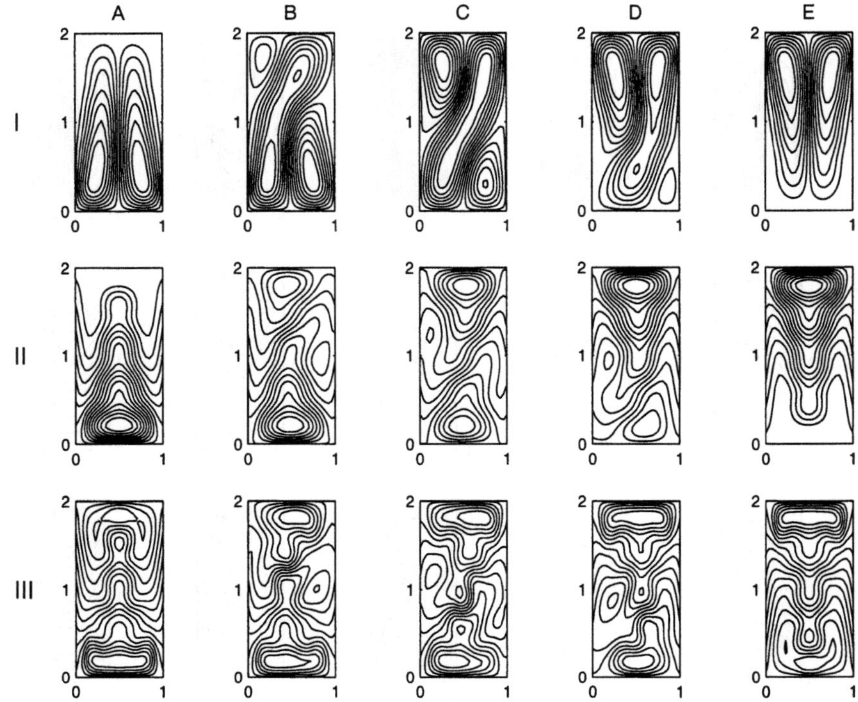

Fig. 3. Stream functions (I), temperature (II), concentration (III) for some regimes, $S = 2$, $b = 2$, $a = 1$, $\lambda_1 = 138.1$, $\lambda_2 = -5$

a sequence of bifurcations leading to the chaotic regime. The last collides with the unstable family (curve 4) and disappears. This family reduces and vanishes at $\lambda_1 = 85$ in the result of collision with the state of rest. It corresponds to the transition of two eigenvalues from left half plane to right one along real axis. Then we observe the growing of main family and an appearance of four arches of unstable equilibria as a result of oscillatory instability at $\lambda_1 = 156.6$.

5 Conclusion

The combined spectral and finite-difference approach gives us an opportunity to compute continuous families steady states in the planar problem of convective filtration of incompressible multi-component fluid saturated a porous medium. We study the scenario of transformation of convective regimes and paricularly the onset of instability on the family of steady states with variative spectrum.

Acknowledgements

This work was partially supported by the Program "Russian Universities - Fundamental Studies" (# UR.04.01.063) and Russian Foundation for Basic Research (# 02-01-00337).

References

1. Nield, D.A., Bejan, A. (1999): Convection in Porous Media. Springer, New York
2. Lyubimov, D. V. (1975): On the convective flows in the porous medium heated from below. J. Appl. Mech. Techn. Phys., **16**, 257–261.
3. Yudovich, V. I. (1991): Cosymmetry, degeneration of solutions of operator equations, and the onset of filtration convection. Mat. Zametki, **49**, 142–148
4. Yudovich, V. I. (1995): Secondary cycle of equilibria in a system with cosymmetry, its creation by bifurcation and impossibility of symmetric treatment of it. Chaos, **5**, 402–441
5. Govorukhin V. N. (1998): Numerical simulation of the loss of stability for secondary steady regimes in the Darcy plane-convection problem. Doklady Akademii Nauk, **363**, 806-808
6. Karasözen B., Tsybulin V.G. (1999): Finite-difference approximation and cosymmetry conservation in filtration convection problem. Physics Letters A, **262**, 321–329
7. Kantur O. Yu., Tsybulin V.G. (2002): Spectral-difference method to the computation of convective motion of the fluid in porous medium and preservation of cosymmetry. J. Comput. Mathematics and Math. Physics, **42**, 878–888
8. Kantur O. Yu., Tsybulin V.G. (2002): Filtration-convection problem: spectral-difference method and preservation of cosymmetry. ICCS, LNCS 2330, Springer-Verlag Berlin Heidelberg, 432–441
9. Yudovich V. I. (2001): Cosymmetry and convection of multi-component fluid in porous medium. Izvestiya Vuzov, SKNTs Vsh., Special Issue, Mat. model., 174–178
10. Canuto C., Hussaini M.Y., Quarteroni A., Zang T.A. (1988): Spectral methods in fluid dynamics. Springer-Verlag
11. Kantur O. Yu., Tsybulin V.G. (2003): Computation of steady states family of filtration convection in tall enclosure. Applied. Mechanics and Technical Physics, **44**, 228–235
12. Kurakin L.G., Yudovich V.I. (2000): Bifurcations accompanying monotonic instability of an aquilibrium of a cosymmetric dynamical system. Chaos, **10**, 311–330

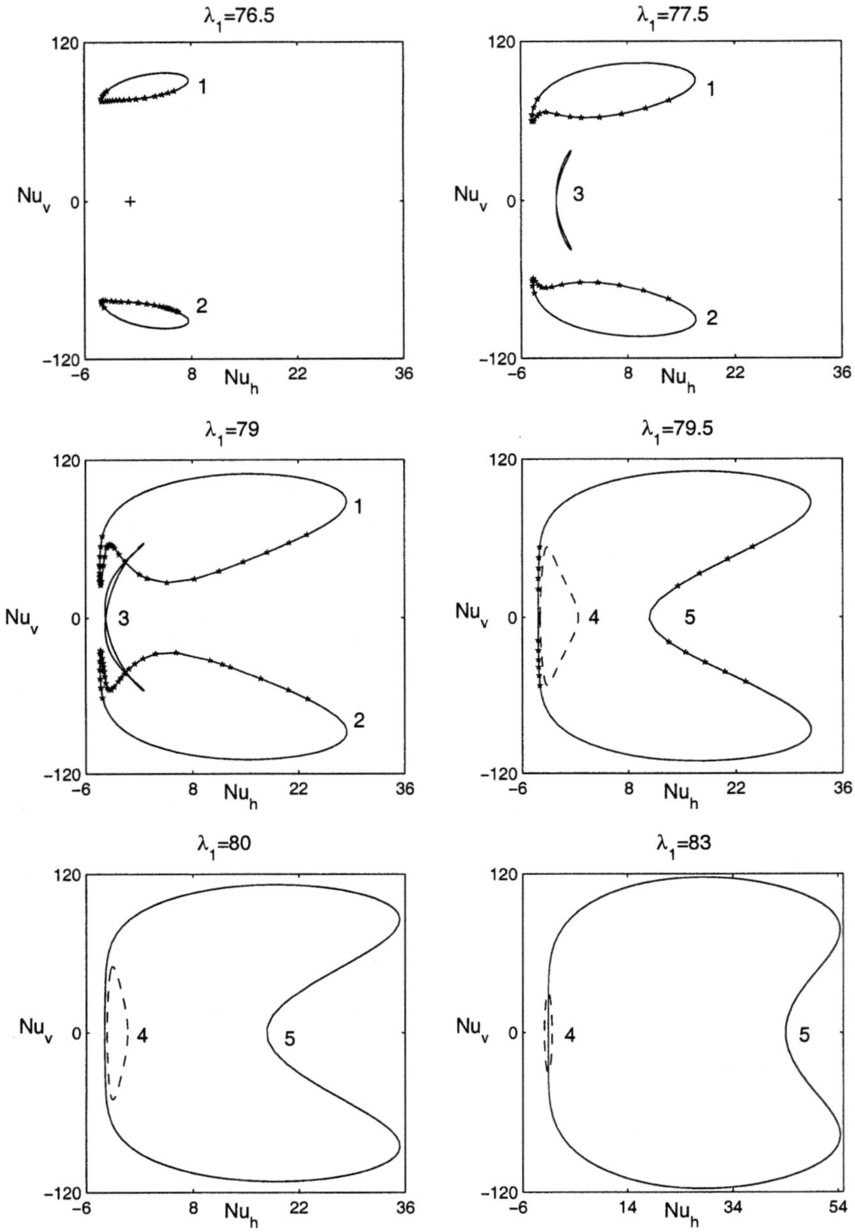

Fig. 4. The convective flows scenario: $\lambda_2 = -10$, $S = 2$, $b = 2$, $\kappa_2 = 0.3$; cross – state of rest, curves 1 and 2 – families with stable and unstable equilibria, curve 3 – limit cycle, dashed line (curve 4) – unstable family, curve 5 – joint family with stable states (solid line) and unstable ones (stars)

Multi-yield Elastoplastic Continuum–Modeling and Computations

Johanna Kienesberger, Jan Valdman

Special Research Program SFB F013
'Numerical and Symbolic Scientific Computing'
Johannes Kepler University Linz
{johanna.kienesberger, jan.valdman}@sfb013.uni-linz.ac.at

Summary. The quasi-static evolution of an elastoplastic body with a multi-surface constitutive law of linear kinematic hardening type allows the modeling of curved stress-strain relations. It generalises classical small-strain elastoplasticity from one to various plastic phases. Firstly, we briefly recall a mathematical model represented by an initial-boundary value problem in the form a variational inequality. Then, the main concern of this paper is focused on an efficient numerical implementation of a one time-step problem. Based on the minimisation problem we describe an iterative non-linear algorithm whose linear subsystems are solved by a geometrical multigrid method. Finally, the numerical computations in 2D and 3D are presented.

1 Introduction

In this paper we consider the quasi-static initial-boundary value problem for small strain elastoplasticity with a multi-surface constitutive law. We treat here a Prandtl-Ishlinskii model of a play type which goes back in the 1D case to PRANDTL [Pra28] and ISHLINSKII [Ish54] and in the multidimensional case to BESSELING [Bes58] and IWAN [Iwa66]. The model extends the classical linear kinematic harding model (single-yield model), that goes back to MELAN [Mel38] and PRAGER [Pra49] in the sense, that is operates with more plastic strains (multi-yield model). Hysteresis properties have been intensively studied by VISINTIN [Vis94] or KREJČÍ [Kre96] amongst others. Our functional formulation of the model and its analysis is based on a direct extention of the work of HAN AND REDDY [HR99] for the linear kinematic hardening model in terms of a time dependent variational inequality. Our numerical approximation for one time-step problem uses the formulation of ALBERTY, CARSTENSEN, AND ZARRABI [ACZ99] extended for a two-yield model, where the solution parameters, i.e., the displacement and two plastic strains, are sought as minimisers of a convex but non-smooth functional. For our approach we regularise this functional, thus standard methods can be applied to the quadratic optimisation problem. The main idea for the algorithm is the use of the Schur-Complement form of the discretised problem in the displacements. The arising linear system is solved by a multi-grid preconditioned conjugate gradient solver.

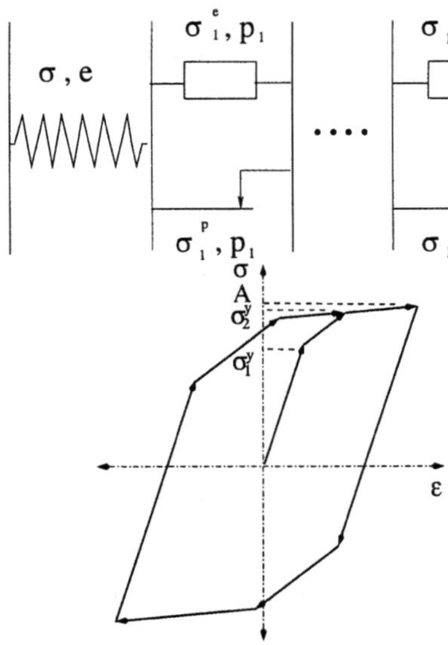

Fig. 1. Prandtl-Ishlinskii model of play type (left) and its $\sigma - \varepsilon$ hysteresis type behaviour for a periodical stress $\sigma(t) = A\sin(t), t \in (0, 2\pi)$ (right)

The paper is organised as follows: In Section 2, the local material model is presented, which is the basis for the boundary value problem in Section 3. The numerical algorithm is designed in Section 4, the numerical experiments are presented in Section 5. Finally, an outlook on the work still to do is given.

2 The Local Material Model

The constitutive law furnishes the relationship between the stress tensor σ and the strain tensor ε. The model discussed here is the Prandtl-Ishlinskii model of play type described by VISINTIN [Vis94] and KREJČÍ [Kre96] among others. It contains finitely many surfaces and its rheological structure and typical hysteresis behaviour are depicted in Figure 1. It is local in the sense that for any given material point x it involves only the time histories $\sigma = \sigma(t)$ and $\varepsilon = \varepsilon(t)$ at that point. It is given by the following system of equations and an evolution variational inequality:

$$\begin{aligned} &\varepsilon = e + p \\ &p = \sum_{r \in I} p_r \\ &\sigma = \sigma_r^b + \sigma_r^p, \quad r \in I \end{aligned} \tag{1}$$

$$\sigma = \mathbb{C}e \tag{2}$$

$$\sigma_r^b = \mathbb{H}_r p_r, \quad r \in I \tag{3}$$

$$\sigma_r^p \in Z, \quad \dot{p}_r : (\tau_r - \sigma_r^p) \leq 0 \quad \text{for all } \tau_r \in Z_r, r \in I, \tag{4}$$

Equation (1) represents the additive decomposition of the strain ε into its elastic part e and its plastic part p as well as of the stress σ into the backstresses σ_r^b and the plastic stress σ_r^p, where $r \in I = \{1, \dots, M\}$. The plastic strain p is additively decomposed to internal plastic strains p_r. The equation (2) denotes a linear elastic law, in the isotropic case one has

$$\mathbb{C}\varepsilon = 2\mu\varepsilon + \lambda(\operatorname{tr}\varepsilon)\mathbb{I}, \tag{5}$$

where the (positive) coefficients μ and λ are called Lamé coefficients. Here $\mathbb{I}$ denotes the second order identity tensor (an identity matrix) and $\operatorname{tr} : \mathbb{R}^{d\times d} \to \mathbb{R}$ defines the trace of a matrix, $\operatorname{tr}\varepsilon := \sum_{j=1}^{d} \varepsilon_{jj}$, for $\varepsilon \in \mathbb{R}^{d\times d}$, where d is the problem dimension. Equation (3) couples the backstresses σ_r^b and the plastic strains p_r through linear mappings with positive definite hardening matrices $\mathbb{H}_r, r \in I$. A typical choice will be $\mathbb{H}_r = h_r\mathbb{I}$, where $h_r > 0, r \in I$ are hardening coefficients. Variational inequality (4) formalises the Prandtl-Reuß normality law, also called the principle of maximal dissipation. The sets $Z_r \subset \mathbb{R}^{d\times d}_{sym}, r \in I$ describe the admissible (plastic) stresses, their boundaries ∂Z_r are called the yield surfaces. We will exclusively use the standard von Mises cylinder with yield stress σ^y

$$Z_r = \{\sigma \in \mathbb{R}^{d\times d}_{sym} : ||\operatorname{dev}\sigma|| \leq \sigma_r^y\}. \tag{6}$$

Here, $||a||^2 = a : a, \quad a : b = \sum_{i,j=1}^{d} a_{ij}b_{ij}$ defines the (Frobenius) norm and the corresponding scalar product, and the deviator of σ is defined as $\operatorname{dev}\sigma := \sigma - \frac{1}{d}(\operatorname{tr}\sigma)\mathbb{I}$. Since this model is described by more (namely M) yield stresses σ_r^y, we classify the model as a multi-yield model or as M-yield model in order to express the number of yield stresses. If $M = 1$ then we speak about a single-yield model, which represents a classical linear kinematic hardening model.

3 The Boundary Value Problem

The elastoplastic continuum is assumed to occupy a bounded domain $\Omega \subset \mathbb{R}^d$, with a Lipschitz boundary $\Gamma = \partial\Omega$. The boundary Γ is split into a Dirichlet boundary Γ_D, a closed subset of Γ with a positive surface measure, and the remaining (relatively open and possibly empty) Neumann part $\Gamma_N := \Gamma \setminus \Gamma_D$. We pose essential and static boundary conditions, namely

$$u = 0 \quad \text{on } \Gamma_D \qquad \text{and} \qquad \sigma \cdot n = g \quad \text{on } \Gamma_N,$$

where g is a given applied surface force and n denotes the outer normal to the boundary Γ_N. Our analysis will be restricted to the study of a boundary value problem defined in these functional spaces:

$$H_D^1(\Omega) = \{v \in H^1(\Omega)^d | v = 0 \text{ on } \Gamma_D\},$$
$$Q = \{q : q \in \operatorname{dev}\mathbb{R}^{d\times d}_{sym}, q_{ij} \in L^2(\Omega)\},$$

where $H^1(\Omega)$ and $L^2(\Omega)$ are the usual Sobolev and Lebesgue spaces. The condition $q \in \operatorname{dev} \mathbb{R}^{d\times d}_{sym}$ in the definition of Q implies that $\operatorname{tr} q = 0$, i.e., q is a trace free matrix. It is shown by BROKATE, CARSTENSEN, VALDMAN [BCV03] that the combination of the system (1)-(4) describing the Prandtl-Ishlinskii model of the play type together with (quasi-static) equilibrium between external (denoted as f) and internal forces, i.e.,

$$\operatorname{div} \sigma(x,t) + f(x,t) = 0, \quad x \in \Omega, t \in (0,T) \tag{7}$$

results in the time-dependent variational inequality for the state variable $w = (u, (p_r)_{r\in I})$:

$$a(w(t), z - \dot{w}(t)) + \psi(z) - \psi(\dot{w}(t)) \geq \langle \ell(t), z - \dot{w}(t) \rangle, \quad \text{for all } z \in \mathcal{H}. \tag{8}$$

w is considered to be an element of the Hilbert space $\mathcal{H} = H^1_D(\Omega) \times \prod_{r\in I} Q$ and to satisfy the zero initial condition $w(0) = 0$. Writing $z = (v, (q_r)_{r\in I})$, a bilinear form $a(\cdot,\cdot)$, a linear functional $\ell(\cdot)$ and a nonlinear functional $\psi(\cdot)$ are defined as:

$$\begin{aligned} a : \mathcal{H} \times \mathcal{H} \to \mathbb{R}, \quad a(w,z) = &\int_\Omega \mathbb{C}(\varepsilon(u) - \sum_{r\in I} p_r) : (\varepsilon(v) - \sum_{r\in I} q_r)\, \mathrm{d}x + \\ &+ \sum_{r\in I} \int_\Omega \mathbb{H} p_r : q_r \, \mathrm{d}x, \end{aligned} \tag{9}$$

$$\ell(t) : \mathcal{H} \to \mathbb{R}, \quad \langle \ell(t), z \rangle = \int_\Omega f(t) \cdot v \, \mathrm{d}x + \int_{\Gamma_N} g(t) \cdot v \, \mathrm{d}S(x), \tag{10}$$

$$\psi : \mathcal{H} \to \mathbb{R}, \quad \psi(z) = \sum_{r\in I} \int_\Omega \sigma^y_r ||q_r|| \, \mathrm{d}x. \tag{11}$$

Thus we can formulate the following formulation of the boundary value problem of quasi-static elastoplasticity.

Problem 1 (BVP of quasi-static multi-surface elastoplasticity). For given $l \in H^1(0,T;\mathcal{H}^*)$ with $\ell(0) = 0$, find $w \in H^1(0,T;\mathcal{H})$ with $w(0) = 0$, such that (8) holds for almost all $t \in (0,T)$.

The unique solvability of Problem 1 under the assumption that the elastic and hardening tensors are symmetric and positive definite bases on the extension on the proof of HAN AND REDDY [HR95, HR99] and can be found in works of VALDMAN [Val02] or BROKATE, CARSTENSEN, VALDMAN [BCV03]:

Theorem 1. *Let $l \in H^1(0,T;\mathcal{H}^*)$ with $\ell(0) = 0$. Then there exists a unique solution $w \in H^1(0,T;\mathcal{H})$ of Problem 1.*

4 Numerical Algorithm

The starting point for the finite element method is the time-discretised form of the variational problem. Problem 1 is solved by an implicit time discretisation, we use the implicit Euler scheme with equidistant time intervals.

It was shown by ALBERTY, CARSTENSEN, AND ZARRABI [ACZ99] that in the single-yield case, i.e., $M = 1$, the time-discretised dual formulation in each time step is equivalent to an optimisation problem depending only on the displacement u and the plastic strain p. This result was obtained by using functional analytic arguments, as the variational inequality is regarded as a sub-differential, for which the dual sub-differential exists and can be reformulated. The resulting objective depends on the chosen hardening law (linear kinematic hardening or isotropic hardening), though the structure remains the same. In KIENESBERGER [Kie03] an algorithm solving the single-yield problems was developed using the results and notation of ALBERTY, CARSTENSEN, AND ZARRABI [ACZ99]. Since the multi-yield hardening model structurally generalises the linear kinematic hardening model, authors managed to extend the original code using templates in C++ effectively in the way, that the multi-yield hardening model becomes a new hardening model. For computational reasons new parameters α_r are introduced, which are internal hardening parameters of the the same dimension as the plastic strains p_r and are defined by $\alpha_r = \mathbb{H}_r p_r$.

The notation is as follows: For given variables with index 0 of an initial time step t^0, the upgrades of the variables at the time step $t^1 = t^0 + \Delta t$ have to be determined. The already time-discretised generalised optimisation problem for the multi-yield case in each time step, subject to the modifications for fitting to the single-yield algorithm, reads as:

$$\begin{aligned} f(u,p_1,\ldots,p_M) := \frac{1}{2}\int_\Omega \mathbb{C}(\varepsilon(u) - \sum_{r\in I} p_r) : (\varepsilon(u) - \sum_{r\in I} p_r)\,\mathrm{d}x \\ +\frac{1}{2}\int_\Omega \sum_{r\in I} |\alpha_r^0|^2\,\mathrm{d}x + \frac{1}{2}\int_\Omega \sum_{r\in I} |p_r - p_r^0|^2\,\mathrm{d}x + \int_\Omega \sum_{r\in I} \alpha_r^0 : (p_r - p_r^0)\,\mathrm{d}x \\ + \int_\Omega \sum_{r\in I} \sigma_r^y |p_r - p_r^0|\,\mathrm{d}x - \int_\Omega f u\,\mathrm{d}x \to \min, \end{aligned} \tag{12}$$

where α^0 is the internal hardening variable from the initial time step.

The basic idea idea for solving the quasi-static problem is using a uniform time discretisation and iterate in each time step until the minimisers, i.e., the displacement u and the plastic strains p_r are determined. Then these values and the separately calculated α_r are used as the reference values with index 0 for the next time step t^2.

The fifth term in (12) contains a norm the sharp bend of which may cause trouble, as the function f is not differentiable. To apply standard methods, the objective is desired to be differentiable and quadratic, thus the function is regularised as follows: The term $|.|$ is regularised by smoothing the norm function, i.e.,

$$|.|_\epsilon := \begin{cases} |.| & \text{if } |.| \geq \epsilon, \\ \frac{1}{2\epsilon}|.|^2 + \frac{\epsilon}{2} & \text{if } |.| < \epsilon. \end{cases} \tag{13}$$

For small ϵ, the quadratised function $f(u, p_1, \ldots, p_M)$ is very similar to the original one, but its properties change enormously. Therefore, it will be referred to by the new symbol $\bar{f}$.

Another simplification is defining the change of p_r by $\tilde{p}_r = p_r - p_r^0$, and using it as an argument of the objective instead of p_r:

The spatial discretisation is carried out by the standard finite element method using linear triangular, resp. tetrahedral finite elements. For reasons of better readability and coherence, the name of the vector denoting the discretised displacement u is again u. The same is valid for $\tilde{p}_r$, p_r^0, furthermore the symmetric matrices are transformed to vectors, e.g. in 2D

$$\begin{pmatrix} \tilde{p}_r^{11} & \tilde{p}_r^{12} \\ \tilde{p}_r^{12} & \tilde{p}_r^{22} \end{pmatrix} \Longrightarrow \begin{pmatrix} \tilde{p}_r^{11} \\ \tilde{p}_r^{22} \\ \tilde{p}_r^{12} \end{pmatrix},$$

such that the objective and other equations can be written in a matrix and vector notation.

For the derivation of the algorithm and numerical experiments we will consider only the two-yield case, i.e., $M = 2$, as it shows the characteristics of the multi-yield problem and can be extended easily. Now, the objective reads as

$$\begin{aligned} \bar{f}(u, \tilde{p}_1, \tilde{p}_2) = \frac{1}{2} \begin{pmatrix} u \\ \tilde{p}_1 \\ \tilde{p}_2 \end{pmatrix}^T \begin{pmatrix} B^T\mathbb{C}B & -B^T\mathbb{C} & -B^T\mathbb{C} \\ -\mathbb{C}B & \mathbb{C} + \mathcal{D}^1 & \mathbb{C} \\ -\mathbb{C}B & \mathbb{C} & \mathbb{C} + \mathcal{D}^2 \end{pmatrix} \begin{pmatrix} u \\ \tilde{p}_1 \\ \tilde{p}_2 \end{pmatrix} \\ + \begin{pmatrix} -f - B^T\mathbb{C}(p_1^0 + p_2^0) \\ \mathbb{C}(p_1^0 + p_2^0) + \mathbb{Q}\alpha_1^0 \\ \mathbb{C}(p_1^0 + p_2^0) + \mathbb{Q}\alpha_2^0 \end{pmatrix}^T \begin{pmatrix} u \\ \tilde{p}_1 \\ \tilde{p}_2 \end{pmatrix} \\ + \frac{1}{2}\mathbb{C}p_1^0 : p_1^0 + \frac{1}{2}\mathbb{C}p_2^0 : p_2^0 + \frac{1}{2}|\alpha_1^0|^2 + \frac{1}{2}|\alpha_2^0|^2 \rightarrow \min, \end{aligned} \tag{14}$$

where Bu denotes the discretised strain $\varepsilon(u)$, and $\mathbb{Q}$ is the result of regarding $\tilde{p}_r$ as vectors, i.e., the matrix norm is defined by $|p| = (p^T\mathbb{Q}p)^{\frac{1}{2}}$.

$\mathcal{D}^1 = \mathbb{Q}(1 + \frac{2\sigma_1^y}{|\tilde{p}_1|_\epsilon})$ is the non-linear iteration matrix of $\bar{f}$ with respect to $\tilde{p}_1$, and analogous for $\mathcal{D}^2$ and $\tilde{p}_2$. These matrices are computed in every iteration step using the current p_r, but apart from that the dependencies on $|\tilde{p}_r|_\epsilon$ will be neglected. This is not an exact method for determining the change of the plastic strain, but its error will be corrected later on as the $\tilde{p}_r$ will be calculated separately and iteratively with the alternating direction method.

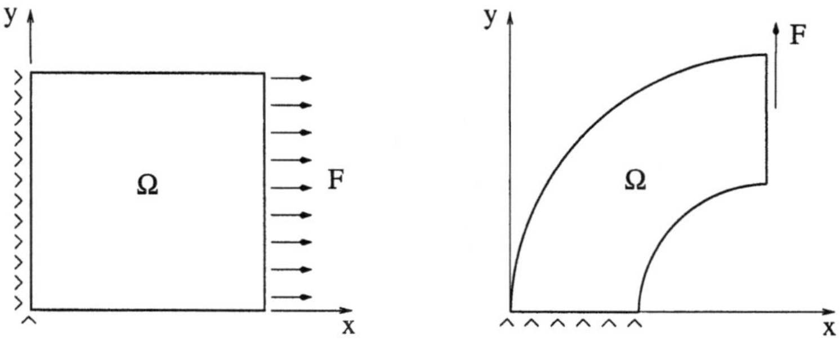

Fig. 2. Geometry of a beam (left) and the quarter of a ring (right) problems

The matrix in (14) is positive definite, thus the minimiser $(u, \tilde{p}_1, \tilde{p}_2)$ has to fulfill the necessary condition of the derivative being equal to zero:

$$\begin{pmatrix} B^T\mathbb{C}B & -B^T\mathbb{C} & -B^T\mathbb{C} \\ -\mathbb{C}B & \mathbb{C}+\mathcal{D}^1 & \mathbb{C} \\ -\mathbb{C}B & \mathbb{C} & \mathbb{C}+\mathcal{D}^2 \end{pmatrix} \begin{pmatrix} u \\ \tilde{p}_1 \\ \tilde{p}_2 \end{pmatrix} + \begin{pmatrix} -f - B^T\mathbb{C}(p_1^0 + p_2^0) \\ \mathbb{C}(p_1^0 + p_2^0) + \mathbb{Q}\alpha_1^0 \\ \mathbb{C}(p_1^0 + p_2^0) + \mathbb{Q}\alpha_2^0 \end{pmatrix} = 0. \tag{15}$$

Extracting the vector $(\tilde{p}_1, \tilde{p}_2)^T$ from the two lower lines in (15) and inserting it into the first one yields the Schur-Complement system in u:

$$\begin{aligned} &B^T(\mathbb{C} - \begin{pmatrix} \mathbb{C} \\ \mathbb{C} \end{pmatrix}^T \begin{pmatrix} \mathbb{C}+\mathcal{D}^1 & \mathbb{C} \\ \mathbb{C} & \mathbb{C}+\mathcal{D}^2 \end{pmatrix}^{-1} \begin{pmatrix} \mathbb{C} \\ \mathbb{C} \end{pmatrix})Bu \\ &= f + B^T\mathbb{C}((p_1^0 + p_2^0) - \begin{pmatrix} I \\ I \end{pmatrix}^T \begin{pmatrix} \mathbb{C}+\mathcal{D}^1 & \mathbb{C} \\ \mathbb{C} & \mathbb{C}+\mathcal{D}^2 \end{pmatrix}^{-1} \begin{pmatrix} \mathbb{C}(p_1^0 + p_2^0) + \mathbb{Q}\alpha_1^0 \\ \mathbb{C}(p_1^0 + p_2^0) + \mathbb{Q}\alpha_2^0 \end{pmatrix}). \end{aligned} \tag{16}$$

This linear system is solved by a multigrid preconditioned conjugate gradient method, see e.g. BRAMBLE [Bra95]. From the numerical tests we have seen that it is not necessary to use the multigrid preconditioner arising from the plasticity problem, the preconditioner for the related problem of elasticity is sufficient and much faster.

For the multigrid method, we use one Gauss-Seidel pre- and post-smoothing step in a V-Cycle, the system on the coarse grid is solved exactly. Furthermore, the nested iteration approach was used, which means that the starting values for the coarse grid correction are the restrictions of the fine grid functions.

5 Numerical Experiments

The algorithm was implemented in NGSolve - the finite element solver extension package of the mesh generator tool NETGEN developed in our group. Finite element basis functions were chosen as piecewise linear for the displacement u and piecewise constant for the plastic strains p_1 and p_2. Furthermore,

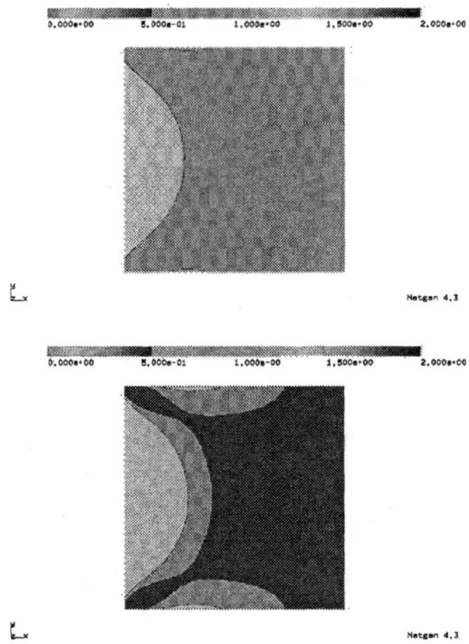

Fig. 3. Plasticity domains in the single-yield (left) and two-yield case (right) of the beam

the full multigrid method was used, i.e., we started with a coarse grid, solved the problem, refined the grid, solved the problem on the finer grid et cetera.

The algorithm was tested on two- as well as on three-dimensional domains, for both the single-yield and multi-yield case, see Figure 2 for the geometries.

The first testing geometry is the 2D beam of Figure 2 with the left edge fixed and the right edge charged with a force acting in the direction of the external normal vector. The second geometry tested is the 3D quarter of a ring from Figure 2 with constant thickness in the z-axis which is the same as the thickness of the ring in the 2D sketch. The quarter ring is fixed on the lower face and a force is acting upwards on the right face. The finest uniform mesh consists of 131 072 triangles (which corresponds to 658 428 degrees of freedom DOF in the calculation of u) for the 2D examples and 25 088 tetrahedra (122 334 DOF) for the 3D example. Figures 3 and 4 show the plasticity domains in the single-yield and in the multi-yield case. The elastic zones are colored light grey, the first plastic zones are middle-grey, and the second plastic zone is dark-grey.

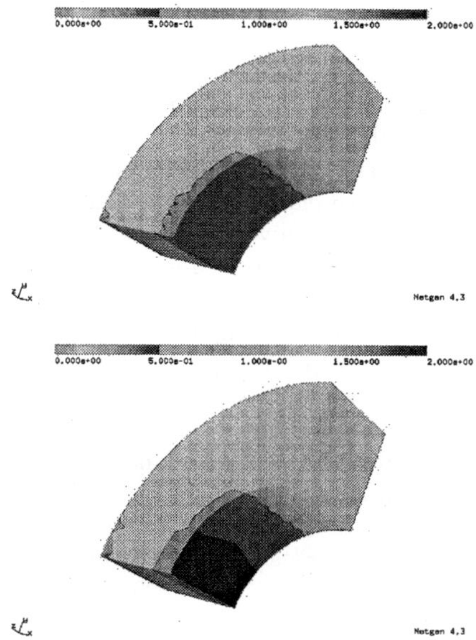

Fig. 4. Plasticity domains in the single-yield (left) and two-yield case (right) of the quarter of the ring

6 Conclusions and Future Work

In this paper a multi-yield plasticity model and its numerical computations were shown. The nonlinear iterative algorithm that uses a multigrid preconditioned solver was presented, its performance in 2D and 3D was demonstrated.

In the future we will extend the solution idea to a quasi-Newton algorithm, i.e., the Schur-Complement matrix will have some Hessian-type entries in order to improve the computational performance. This idea is already implemented for the single-yield case, where the numerical results demonstrate the faster algorithm performance with linear complexity. We expect the same result for the multi-yield case.

Another long-term aim is to identify the interfaces between the elastic and plastic zones and to refine the mesh adaptively in such a way, that the interface is approximated by the mesh. Then we expect an even faster performance of the algorithm.

7 Acknowledgment

The authors are pleased to acknowledge support by the Austrian Science Fund 'Fonds zur Förderung der wissenschaftlichen Forschung (FWF)' under grant SFB F013/F1306 in Linz, Austria.

References

[ACZ99] J. Alberty, C. Carstensen, and D. Zarrabi. Adaptive numerical analysis in primal elastoplasticity with hardening. *Comput. Methods Appl. Mech. Eng.*, 171(3-4):175–204, 1999.

[BCV03] M. Brokate, C. Carstensen, and J. Valdman. A quasi-static boundary value problem in multi-surface elastoplasticity: Part 1 - analysis. SFB Report 2003-16, Johannes Kepler University Linz, SFB "Numerical and Symbolic Scientific Computing", 2003. (and submitted).

[Bes58] J.F. Besseling. A theory of elastic, plastic and creep deformations of an initially isotropic material showing anisotropic strain-hardening, creep recovery and secondary creep. *J. Appl. Mech.*, 25:529–536, 1958.

[Bra95] J. H. Bramble. *Multigrid methods.* Pitman research notes in mathematical series. Longman, 1995.

[HR95] W. Han and B.D. Reddy. Computational plasticity: the variational basis and numerical analysis. *Computer methods in applied mechanics and engineering*, pages 283–400, 1995.

[HR99] W. Han and B. Reddy. *Plasticity: Mathematical Theory and Numerical Analysis.* Springer-Verlag New York, 1999.

[Ish54] A. Ju. Ishlinskii. The general theory of plasticity with linear hardening (in russian). *Ukrainian mathematical journal*, 6(3), 1954.

[Iwa66] W.D. Iwan. A distributed-element model for hysteresis and its steady state dynamic response. *J. Appl. Mech.*, 33:893–900, 1966.

[Kie03] J. Kienesberger. Multigrid preconditioned solvers for some elastoplastic problems. In *Lecture Notes in Computer Science.* Springer, 2003. (to appear).

[Kre96] P. Krejčí. *Hysteresis, Convexity and Dissipation in Hyperbolic Equations.* GAKUTO International Series, Mathematical Sciences and Applications, 1996.

[Mel38] E. Melan. Zur Plastizität des räumlichen Kontinuums. *Ingenieur-Archiv*, 9:116–126, 1938.

[Pra28] L. Prandtl. Ein Gedankenmodell zur kinetischen Theorie der festen Körper. *ZAMM*, 8:85–106, 1928.

[Pra49] W. Prager. Recent developments in the mathematical theory of plasticity. *J. Appl. Phys.*, 9:235–241, 1949.

[Val02] J. Valdman. *Mathematical and Numerical Analysis of Elastoplastic Material with Multi-Surface Stress-Strain Relation.* PhD thesis, Christian-Albrechts-Universität zu Kiel, 2002. published at www.dissertation.de in Berlin, Germany, 2002, ISBN 3-89825-501-8, download: http://www.sfb013.uni-linz.ac.at/~jan/plasticity.pdf.

[Vis94] A. Visintin. *Differential models of hysteresis.* Springer, 1994.

Celebrating Fifty Years of David M. Young's Successive Overrelaxation Method

David R. Kincaid

Department of Computer Sciences, University of Texas at Austin, Austin, Texas 78712 USA *kincaid@cs.utexas.edu*

Summary. It has been over fifty years since David M. Young's original work on the successive overrelaxation (SOR) methods. This fundamental method now appears in all textbooks containing an introductory discussion of iterative solution methods. (Most often the SOR method appears after a presentation of Jacobi iteration and Gauss-Seidel iteration and before the conjugate gradient iterative method.) We present a brief survey of some of the research of Professor David M. Young, together with his students and collaborators, on iterative methods for solving large sparse linear algebraic equations. This is not a complete survey but just a sampling of various papers with a focus on some of these publications.

Dr. David M. Young's doctoral thesis [27] was accepted in 1950 by his supervising Professor Garrett Birkhoff of Harvard University and his paper [28] based this work appeared in 1954. This is one of the landmark contributions in modern numerical analysis. The red-black ordering for matrices is of great importance in parallel computing. Gene Golub has said: "It's almost as if David could see into the future!"

David Young celebrated his 80th birthday on October 20, 2003 (`http://www.ma.utexas.edu/CNA/photos.html`).

1 Introduction

We present a brief survey of some of the work of Professor David M. Young, together with his students and collaborators, on iterative methods. Dr. David M. Young has been involved in research on iterative methods for solving large sparse linear algebraic equations for over forty years until his recent retirement. This is not a complete survey but just a sampling of various projects with a focus on some of his publications.

2 Successive Overrelaxation

From research first done at Harvard University, Young presented in his Ph.D. thesis [27], and in a subsequent paper [28], an analysis of the successive overrelaxation (SOR) method for the case where the coefficient matrix of the linear algebraic system $Au = b$ is consistently ordered [30]. An elliptic partial differential equations over a region with grid points numbered in the *natural ordering* (left-to-right and up) and using the standard five-point discretization stencil results in such a matrix system. In fact, any matrix system derived in this way has Young's *Property A* [30]. Moreover, a consistently ordered system can be obtained from one with Property A after a suitable permutation.

For a matrix with Property A, one can permute the rows and corresponding columns to obtain a *red-black system.* The *red-black ordering* corresponds to a red and black checkerboard ordering of the grid points. When A is a red-black matrix, it is consistently ordered and Young's equation [27] $(\lambda + \omega - 1)^2 = \omega^2 \mu^2 \lambda$ gives a relation between the eigenvalues λ of the iteration matrix $\mathcal{L}_\omega$ for the SOR method and the eigenvalues μ of the iteration matrix B for the Jacobi method. If A is symmetric positive definite, then the eigenvalues of B are real and less than 1 in absolute value and the optimum or *best* value of the acceleration factor ω is given by $\omega_b = 2/\left(1 + \sqrt{1 - S(B)^2}\right)$. Here $S(B)$ is the *spectral radius* of the Jacobi matrix B, which is the magnitude of the eigenvalue of largest absolute value of the matrix B. Moreover, the spectral radius of the SOR matrix with the optimum relaxation parameter $\omega = \omega_b$ is given by $S(\mathcal{L}_{\omega_b}) = \omega_b - 1 = r$.

For model problems involving the Poisson equation over a region with mesh points of grid size h, it can be shown that the number of iterations required for convergence of the SOR method is $n = \mathcal{O}(h^{-1})$ whereas the number of iterations is $n = \mathcal{O}(h^{-2})$ using either the Jacobi or Gauss-Seidel methods. In this situation, the SOR method is faster by an order of magnitude.

Work has been done on the choice of the optimum ω for the case where A is consistently ordered but not symmetric positive definite and where some of the eigenvalues of the Jacobi iteration matrix B are *complex eigenvalues.* Several programs are available for choosing the optimum ω if all of the eigenvalues of B are known or if one knows a convex region containing them. See Young and Eidson [35] and Young and Huang [39].

An extension of the SOR method is the *modified SOR (MSOR) method* for a linear system with a red-black coefficient matrix. The MSOR method involves the use of relaxation factors ω_1, ω_1', ω_2, ω_2', ..., where ω_i is used for the *red* components, and ω_i' is used for the *black* points, for each i. In Young, Wheeler, and Downing [34], it is shown that there are suitable values of ω_i and ω_i' that are as good, though not better than, the choice $\omega_i = \omega_i' = \omega_b$ for all i. On the other hand, other choices are more effective if one measures the effectiveness in terms of certain norms as shown in Young and Kincaid [42]. Chapters 8 and 10 of Young [30] cover the modified SOR method with fixed and variable parameters, respectively.

A number of other modifications and extensions have been made to the SOR theory. For instance in *group* or *block methods*, the unknowns are grouped into blocks and all values within a block are updated simultaneously. Usually, each inner iteration of a block method is done by a direct method since the matrices for the blocks are assumed to be easily solvable. For example, these matrices are tridiagonal in the case of the line SOR method when the five-point finite difference stencil is used. Also, faster convergence is obtained for line SOR methods than for point SOR methods, in general. Moreover, the general SOR theory has been applied to group iterative methods in Chapter 14 of Young [30].

Research on *norms* associated with the SOR method for the red-black system has resulted in new formulas. It has been shown that graph of the $D^{\frac{1}{2}}$-norm function for the SOR matrix $\mathcal{L}_{\omega_b}^m$ is not monotonically decreasing (it increases and then decreases), but the $A^{\frac{1}{2}}$-norm is indeed a monotonically decreasing function of m; however, it is still considerably larger than the spectral radius function $S(\mathcal{L}_{\omega_b}^m) = r^m$. See Young and Kincaid [42] and Chapter 7 in Young [30].

Corresponding to the SOR method is the *Symmetric Successive Overrelaxation (SSOR) method* in which an iteration consists of one iteration of the (forward) SOR method followed by one iteration of the (backward) SOR method. In the *Unsymmetric Successive Overrelaxation (USSOR) method*, different parameters may be used in the red and black equations, respectively. Young [29] presents convergence properties of the symmetric and unsymmetric successive overrelaxation methods and related methods.

3 Chebyshev Acceleration

The SOR method can be regarded as a way to accelerate the convergence of the Jacobi method in a certain sense. Another way of speeding up the convergence of the Jacobi method is to use an extrapolation method or a Chebyshev acceleration method, which is based on Chebyshev polynomials. These are general procedures and they can be applied to methods other than just the Jacobi method as shown in Hageman and Young [7].

Suppose the basic iterative method to be used in the acceleration procedure has the form $u^{(n+1)} = Gu^{(n)} + k$ where the eigenvalues μ of G are bounded such that $m(G) \leq \mu \leq M(G)$. Here $m(G)$ and $M(G)$ are the smallest and largest eigenvalues of G, respectively. Using the three-term relation for Chebyshev polynomials, the optimal *Chebyshev acceleration method* can be written as $u^{(n+1)} = \rho_{n+1}\{\gamma(Gu^{(n)} + k) + (1-\gamma)u^{(n)}\} + (1-\rho_{n+1})u^{(n-1)}$ where $\rho_{n+1} = (1-(\sigma/2)^2\rho_n)$, (with $\rho_1 = 1$ and $\rho_2 = (1-\sigma^2/2)$), $\sigma = [M(G) - m(G)]/[2 - M(G) - m(G)]$, and $\gamma = 2/[2 - M(G) - m(G)]$.

Varga [26] refers to this procedure as the *Chebyshev semi-iterative method.* One needs to choose estimates for $M(G)$ and $m(G)$, which may cause difficulties in some cases. In fact, the behavior of the acceleration procedure is often sensitive to these estimates and especially the one for $M(G)$. It can be shown that the optimum Chebyshev acceleration procedure is an order of magnitude faster than the optimum extrapolated procedure for σ close to one [7].

If one applies the Chebyshev acceleration procedure to the Jacobi method as the basic method for solving a linear system with a red-black coefficient matrix, then the computation can be simplified. This is done by rewriting the procedure in terms of only the red points or only the black points for the Jacobi method. Golub and Varga [6] refer to this as the *cyclic Chebyshev semi-iterative method.* The cyclic acceleration method is the original Chebyshev acceleration method with half of the calculations bypassed. The resulting method is equivalent to a special case of the modified SOR method with $\omega_i = \omega_i' = \rho_{n+1}$.

With an *adaptive Chebyshev acceleration procedure*, one continuously revises the iteration parameters as the iterative method proceeds. The algorithm fixes the smallest eigenvalue estimate $m_E \le m(G)$ and adaptively modifies the largest eigenvalue estimate M_E but keep $M_E \le M(G)$. The iterative procedure continues using these values of m_E and M_E until the observed convergence is much slower than expected in a certain sense. By solving a *Chebyshev equation*, the algorithm increases M_E but keeps $M_E \le M(G)$. This adaptive Chebyshev acceleration procedure is repeated until convergence is achieved according to the stopping test being utilized. Chebyshev polynomials are used in the algorithm for choosing these maximum and minimum eigenvalue estimates. Such a procedure was developed by Hageman and Young [7] and it was incorporated into the algorithms used in the ITPACK software packages [18].

It has been shown that, in some cases, one can obtain almost as good a convergence as with Chebyshev acceleration by the use of a *stationary second degree method* given by $u^{(n+1)} = \rho\{\gamma(Gu^{(n)}+k)+(1-\gamma)u^{(n)}\}+(1-\rho)u^{(n-1)}$. Here let $\rho = 1$ when $n = 0$. See some of the papers by Young and/or Kincaid on second-degree methods [14, 15, 20, 31].

4 Conjugate Gradient Acceleration

Conjugate gradient acceleration $u^{(n+1)} = \rho_{n+1}\{\gamma_{n+1}(r^{(n)} + u^{(n)}\} + (1 - \rho_{n+1})u^{(n-1)}$ is similar to Chebyshev acceleration except that the parameters used involve inner products:

$$\rho_{n+1} = [1 - (\gamma_{n+1}/\gamma_n\rho_n)\langle r^{(n)}, r^{(n)}\rangle / \langle r^{(n-1)}, r^{(n-1)}\rangle]^{-1},$$

(with $\rho_1 = 1$) and $\gamma_{n+1} = \langle r^{(n)}, r^{(n)}\rangle / \langle r^{(n)}, Ar^{(n)}\rangle$. Here $r^{(n)} = b - Au^{(n)}$ is the residual vector. As with Chebyshev acceleration method, conjugate gradient acceleration method can speed-up the Jacobi method and other methods. Conjugate gradient acceleration has some advantages over Chebyshev acceleration [7]. It can be shown that the convergence of conjugate gradient acceleration, measured in a certain norm, is at least as fast as that of Chebyshev acceleration. With conjugate gradient acceleration there are no parameter estimates; however, the basic iterative method may involve a parameter as in the case when SSOR is used as the basic method. Since the conjugate gradient acceleration requires the computation of inner products for each iteration, the work required per iteration may be somewhat greater than for Chebyshev acceleration. For basic methods that are not symmetrizable, the generalized conjugate gradient methods can be used to accelerate their convergence [7].

We can introduce a nonsingular matrix W as follows $[W(I-G)W^{-1}][Wu] = Wk$ which in terms of the original linear system is $[WQ^{-1}AW^{-1}][Wu] = WQ^{-1}b$. The *generalized conjugate gradient acceleration method* [7] corresponding to the basic iterative method is given by

$$u^{(n+1)} = \rho_{n+1}\{\gamma_{n+1}\delta^{(n)} + u^{(n)}\} + (1-\rho_{n+1})u^{(n-1)}$$

where

$$\rho_{n+1} = [1 - (\gamma_{n+1}/\gamma_n\rho_n)\langle W\delta^{(n)}, W\delta^{(n)}\rangle / \langle W\delta^{(n-1)}, W\delta^{(n-1)}\rangle]^{-1},$$

(with $\rho_1 = 1$),

$$\gamma_{n+1} = [1 - \langle W\delta^{(n)}, WG\delta^{(n)}\rangle / \langle W\delta^{(n)}, W\delta^{(n)}\rangle]^{-1},$$

and the *pseudo-residual vector* is $\delta^{(n)} = Gu^{(n)} + k - u^{(n)}$. The conjugate gradient acceleration method minimizes the $[W^TW(I-G)]^{\frac{1}{2}}$-matrix-norm of the error as compared with any polynomial acceleration procedure based. If A and Q are symmetric positive definite matrices and if $W^TW = Q$, then we minimize the $A^{\frac{1}{2}}$-matrix-norm of the error as in the conjugate gradient method. It can be shown that the average rate of convergence for the conjugate gradient method, when measured in the $[W^TW(I-G)]^{\frac{1}{2}}$-matrix-norm, is at least as large as that for the corresponding Chebyshev acceleration procedure. (See Hageman and Young [7].) As in Young [30], we denote the L-matrix norm by $||Q||_L = ||LQL^{-1}||_2$.

5 Nonsymmetric Systems

A difficult problem is solving the linear system when the coefficient matrix A is not necessarily symmetric positive definite or even symmetric. Three *generalized conjugate gradient acceleration methods* called ORTHODIR, ORTHOMIN, and ORTHORES were considered by Young and Jea [40]. It was shown that under fairly general conditions these methods converge, in exact arithmetic, in at most N iterations, where N is the order of the matrix. Also in Jea and Young [9], the *biconjugate gradient (BCG) method* as well as other forms of *Lanczos methods* were considered as generalized conjugate gradient acceleration methods corresponding to certain double linear systems involving A and A^T.

The *generalized minimum residual (GMRES) method* [25] is a widely used method for solving nonsymmetric linear systems. The method is generally very reliable although stagnation may occur in some cases. Moreover, for nonsymmetric systems, the amount of work required per iteration usually increases as the number of iterations increases. 'Recently, Young working with Chen and Kincaid developed various generalizations of the GMRES method and combined them with the Lanczos procedure. New iterative methods called

GGMRES, MGMRES, and LAN/MGMRES have been established [2, 3, 4, 5]. The GGMRES method is a slight generalization of the GMRES method. The GMRES method minimizes a norm of the residual and GGMRES minimizes a more general norm—both involve a minimization condition. Alternatively, the MGMRES method is a modification of the GMRES method applied to a symmetric indefinite linear system using a Galerkin condition. This latter method is related to the BCG method and to other variants of the Lanczos method. The LAN/GMRES method aims at combining the reliability of the GMRES method with the reduced work of a Lanczos-type method. When conducting initial numerical experiments on nonsymmetric linear systems arising from convection-diffusion problems, it was found that LAN/MGMRES was comparable with a number of other methods that are extensively used.

6 Software for Iterative Methods

Under the direction of Kincaid and Young, several research-oriented software packages were written as part of the ITPACK Project at the Center for Numerical Analysis. Beginning in the mid-1970s, there was an increased effort to develop iterative algorithms and portable public domain software. Software packages, such as the ITPACK 2C package, were developed, which included automatic procedures for handling choices that were causing difficulties for users of iterative methods. Automatic procedures included were developed for determining all necessary iteration parameters and for accurate and realistic stopping tests for iterative algorithms. Algorithms based on these procedures were described in the book by Hageman and Young [7]. Also, software from the ITPACK 2C package was modified and incorporated into the ELLPACK package at Purdue University for solving elliptic partial differential equations. (See Rice and Boisvert [24].)

While the ITPACK 2C package was intended primarily for linear systems where the coefficient matrix is symmetric positive definite or nearly so, other packages such as NSPCG [23] and PCG [10] were developed with the capability of handling nonsymmetric systems. Other ITPACK Project software include ITPACKV 2D [17], ITPACK 3A [45] and ITAPCK 3B [22], for example. See Kincaid and Young [19] for a review of the ITPACK software packages.

7 Alternating-Type Iterative Methods

To construct an alternating-type method for solving $Au = b$, we choose matrices H, V, and Σ such that $A = H + V + \Sigma$, where Σ is a diagonal matrix with positive diagonal elements. For any linear system of the form $(H + \rho\Sigma)v = w$ or $(V + \rho\Sigma)v = w$, we assume that $H + \rho\Sigma$ and $V + \rho\Sigma$ are nonsingular matrices for any positive real number ρ and so that for any vector w one can easily solve for v. To define an *alternating-type iterative method*, we choose positive numbers ρ and ρ' and, for a given $u^{(n)}$, we determine $u^{(n+\frac{1}{2})}$ and $u^{(n+1)}$ by $(H+\rho\Sigma)u^{(n+\frac{1}{2})} = b-(V-\rho\Sigma)u^{(n)}$ and $(V+\rho'\Sigma)u^{(n+1)} = b-(H-\rho'\Sigma)u^{(n+\frac{1}{2})}$. Thus, we have $u^{(n+1)} = T_{\rho,\rho'}u^{(n)} + k_{\rho,\rho'}$ where $T_{\rho,\rho'} = (V + \rho'\Sigma)^{-1}(H - \rho'\Sigma)(H + \rho\Sigma)^{-1}(V - \rho\Sigma) = I - (\rho + \rho')(V + \rho'\Sigma)^{-1}\Sigma(H + \rho\Sigma)^{-1}(H + V)$ and $k_{\rho,\rho'} = (\rho+\rho')(V+\rho'\Sigma)^{-1}\Sigma(H+\rho\Sigma)^{-1}b = (I-T_{\rho,\rho'})A^{-1}b$. Examples of alternating-type methods are the *alternating-direction implicit (ADI) method*, the *symmetric successive overrelaxation (SSOR) method*, and the *unsymmetric successive overrelaxation (USSOR) method*. With the ADI method, H and V are either tridiagonal or are permutationally similar to tridiagonal matrices and $\Sigma = I$. With the SSOR and USSOR methods, H and V are lower triangular and upper triangular matrices, respectively, and Σ is a diagonal matrix with positive diagonal elements.

In certain cases, the ADI method converges rapidly. For example, with problems involving Poisson's equation in the rectangle with a grid of mesh size h, the ADI method convergences in $n = \mathcal{O}(\log h^{-1})$ iterations using the optimum number of parameters and in $n = \mathcal{O}(h^{-1/m})$ iterations using the best m parameters. Recall that $n = \mathcal{O}(h^{-1})$ for the SOR method. The *commutative case* is when $HV = VH$, $H\Sigma = \Sigma H$, $V\Sigma = \Sigma V$. It holds for certain separable self-adjoint elliptic partial differential equations defined over rectangles. Given the commutativity condition and also bounds on the eigenvalues of H and V, necessary and sufficient convergence conditions related to choosing ADI parameters can be found in Birkhoff, Varga, and Young [1] and in Chapter 17 of Young [30]. Also see Young and Wheeler [48].

With a *nonstationary alternating-type iterative method*, the parameters ρ and ρ' may vary from iteration to iteration. We seek to determine the parameters $\{\rho_i\}$ and $\{\rho_i'\}$ so that $u^{(m)}$ is as close to the true solution $\overline{u} = A^{-1}b$ as possible. In practice, we seek to make the spectral radius $S\left(\prod_{i=1}^{m} T_{\rho_i,\rho_i'}\right)$ as small as possible. As an alternative to the (sequential) non-stationary method, we consider the *parallel alternating-type iterative method*. See papers by Young and Kincaid [43, 44].

8 Books

The classical research monograph *Iterative Solution of Large Linear Systems* [30] by Professor Young was published in 1971 and reprinted in 2003. Also, the book *Applied Iterative Methods* [7] by David M. Young in collaboration with Louis A. Hageman appeared in 1981 and was reprinted in 2004. Also, we should mention *A Survey of Numerical Mathematics*, in two volumes (1972 and 1973), written by David M. Young and Robert T. Gregory [36, 37].

References

1. Birkhoff, G., Varga, R. S., Young, D. M.: Alternating direction implicit methods, in F. Alt, M. Rubinoff, editors, *Advances in Computers*, Academic Press, New York, 189–273, 1962.
2. Chen, J-Y. Iterative solutions of large sparse nonsymmetric linear systems, Report CNA–285, Center for Numerical Analysis, University of Texas at Austin, January 1997.
3. Chen, J-Y., Kincaid, D. R., Young, D. M.: MGMRES iterative method, in J. Wang, M. B. Allen III, B. M. Chen, T. Mathew, editors, *Iterative Methods in Scientific Computation*, IMACS, New Brunswick, 15–20, 1998.
4. Chen, J-Y., Kincaid, D. R., Young, D. M.: GGMRES iterative method, in J. Wang, M. B. Allen III, B. M. Chen, T. Mathew, editors, *Iterative Methods in Scientific Computation*, IMACS, New Brunswick, 21–26, 1998.
5. Chen, J-Y., Kincaid, D. R., Young, D. M.: Generalization and modifications of the GMRES iterative method, *Numerical Algorithms*, **21** (1999) 119–146.
6. Golub, G. H., Varga, R. S.: Chebyshev semi-iterative methods, successive overrelaxation iterative methods, and second-degree Richardson iterative methods, Parts I & II, *Numer. Math.*, **3** 147–168, 1961.
7. Hageman, L. A., Young, D. M.: *Applied Iterative Methods*, Academic Press, New York, 1981. (Reprinted by Dover, New York, 2004.)
8. Jea, K. C.: Generalized conjugate gradient acceleration of iterative methods, Report CNA–176, Center for Numerical Analysis, University of Texas at Austin, February 1982.
9. Jea, K. C., Young, D. M.: On the simplification of generalized conjugate-gradient methods for nonsymmetrizable linear systems, *Linear Algebra and Its Applications*, 52/53:399–417, 1983.
10. Joubert, W. D.: PCG examples manual: A package for the iterative solution of large sparse linear systems on parallel computers, Report CNA–284, Center for Numerical Analysis, University of Texas at Austin, July 1996.
11. Kincaid, D. R.: An analysis of a class of norms of iterative methods for systems of linear equations, Ph.D. thesis, University of Texas at Austin, 1971. Also, Report CNA–18, Center for Numerical Analysis, University of Texas at Austin, May 1971.
12. Kincaid, D. R.: Norms of the successive overrelaxation method, *Math. Comp.*, 26(118):345–357, 1972.
13. Kincaid, D. R.: A class of norms of iterative methods for solving systems of linear equations, *Numer. Math.*, 20:392–408, 1973.
14. Kincaid, D. R.: On complex second-degree iterative methods, *SIAM J. Numer. Anal.*, 11(2):211–218, 1974.
15. Kincaid, D. R.: Stationary second-degree iterative methods, *Applied Numerical Mathematics*, 16:227–237, 1994.
16. Kincaid, D. R., Hayes, L. J.: editors, *Iterative Methods for Large Linear Systems*, Academic Press, New York, 1990.
17. Kincaid, D. R., Oppe, T. C., Young, D. M.: ITPACKV 2D User's Guide, Report CNA–232, Center for Numerical Analysis, University of Texas at Austin, May 1986.
18. Kincaid, D. R., Respess, J. R., Young, D. M., Grimes, R. G.: *ITPACK 2C: A FORTRAN Package for Solving Large Sparse Linear Systems by Adaptive Accelerated Iterative Methods*, *ACM Trans. Math. Software* **8**, 1982.

19. Kincaid, D. R., Young, D. M.: A brief review of the ITPACK project, *Journal Computer & Applied Mathematic*, 24:121–127, 1988.
20. Kincaid, D. R., Young, D. M.: Linear stationary second-degree methods for solution of large linear systems, in Th. M. Rassias, H. M. Srivasiava, A. Yanushauska, editors, *Topics in Polynomials of One and Several Variables and Their Applications*, World Scientific Publishing Co., River Edge, NJ, 609–629, 1993.
21. Kincaid, D. R., Young, D. M.: Note on parallel alternating-type iterative methods, in S. D. Margenov P. S. Vassilevski, editors, *Iterative Methods in Linear Algebra II*, IMACS, New Brunswick, 131–139, 1996.
22. Mai, T-Z., Young, D. M.: ITPACK 3B user's guide (preliminary version), Report CNA–201, Center for Numerical Analysis, University of Texas at Austin, January 1986.
23. Oppe, T. C. Joubert, W. D., Kincaid, D. R.: NSPCG user's guide, version 1.0: A package for solving large sparse linear systems by various iterative methods, Report CNA–216, University of Texas at Austin, Center for Numerical Analysis, April 1988.
24. Rice, J. R., Boisvert, R.: *Solving Elliptic Problems Using ELLPACK*, Spring-Verlag, New York, 1985.
25. Saad, Y., Schultz, M. H.: GMRES: A generalized minimal residual algorithm for solving nonsymmetric linear systems, *SIAM J. Scientific and Statistical Computing*, 7:856–869, 1986.
26. Varga, R. S.: *Matrix Iterative Analysis*, Prentice-Hall, Englewood Cliff, New Jersey, 1962.
27. Young, D. M.: *Iterative Methods for Solving Partial Difference Equations of Elliptic Type*, Ph. D. thesis, Harvard University, Mathematics Department, Cambridge, MA, May 1950.
28. Young, D. M.: Iterative methods for solving partial difference equations of elliptic type, *Trans. Amer. Math. Soc.*, 76:92–111, 1954.
29. Young, D. M.: Convergence properties of the symmetric and unsymmetric successive overrelaxation methods and related methods, *Math. Comp.*, 24(112):793–807, 1970.
30. Young, D. M.: *Iterative Solution of Large Linear Systems*, Academic Press, New York, 1971. (Reprinted by Dover, New York, 2003.)
31. Young, D. M.: Second-degree iterative methods for the solution of large linear systems, *J. Approximation Theory*, 5:137–148, 1972.
32. Young, D. M.: On the accelerated SSOR method for solving large linear systems, *Advances in Mathematics*, 23(3):215–271, 1977.
33. Young, D. M.: A historical review of iterative methods, in *A History of Scientific Computation* S. G. Nash, editor, Addison-Wesley, Reading, MA, 180–194, 1989. Also in Report CNA–206, Center for Numerical Analysis, University of Texas at Austin, February 1987.
34. Young, D. M., Downing, J. A., Wheeler, M. F.: On the use of the modified successive overrelaxation method with several relaxation factors, in W. A. Kalenich, editor, *Proceedings of IFIP 65*, Spartan Books, Washington, D.C., 1965.
35. Young, D. M., Eidson, H. D.: On the determination of the optimum relaxation factor for the SOR method when the eigenvalues of the Jacobi method are complex, Report CNA–1, Center for Numerical Analysis, University of Texas at Austin, September 1970.
36. Young, D. M., Gregory, R. T.: *A Survey of Numerical Mathematics*, Volume 1, Addison-Wesley, New York, 1972. (Reprinted by Dover, New York, 1988.)

37. Young, D. M., Gregory, R. T.: *A Survey of Numerical Mathematics*, Volume 2, Addison-Wesley, New York, 1973. (Reprinted by Dover, New York, 1988.)
38. Young, D. M., Hayes, L. J.: The accelerated SSOR method for solving large linear systems. Report CNA–123, Center for Numerical Analysis, University of Texas at Austin, May 1977.
39. Young, D. M., Huang, R.: Some notes on complex successive overrelaxation, Report CNA–185, Center for Numerical Analysis, University of Texas at Austin, July 1983.
40. Young, D. M., Jea, K. C.: Generalized conjugate gradient acceleration of non-symmetrizable iterative methods, *Linear Algebra and Its Applications*, 34:159–194, 1980.
41. Young, D. M., Jea, K. C., Mai, T-Z.: Preconditioned conjugate gradient algorithms and software for solving large sparse linear systems, Report CNA–207, Center for Numerical Analysis, University of Texas at Austin, March 1987.
42. Young, D. M., Kincaid, D. R.: Norms of the successive overrelaxation method and related methods, Report TNN–94, Computation Center, University of Texas at Austin, September 1969.
43. Young, D. M., Kincaid, D. R.: Parallel implementation of a class of nonstationary alternating-type methods, in D. Bainov, V. Covachev, editors, *Proceedings of the Third International Colloquium on Numerical Analysis*, VSP, Utrecht, The Netherlands, 219–222, 1995.
44. Young, D. M., Kincaid, D. R.: A new class of parallel alternating-type iterative methods, *Journal of Computational and Applied Mathematics*, 74:331–344, 1996.
45. Young, D. M., Mai, T-Z.: ITPACK 3A User's Guide (Preliminary Version), Report CNA–197, Center for Numerical Analysis, University of Texas at Austin, 1984.
46. Young, D. M., Mai, T-Z.: Iterative algorithms and software for solving large sparse linear systems, *Communications in Applied Numerical Methods*, 4(3):435–456, 1987. (Also, Report CNA–215, Center for Numerical Analysis, University of Texas at Austin, 1987.)
47. Young, D. M., Mai, T-Z.: The search for omega, in D. Kincaid, L. Hayes, editors, [16], 293–311.
48. Young, D. M., Wheeler, M. F.: Alternating direction methods for solving partial difference equations, in W. F. Ames, editor, *Nonlinear Problems of Engineering*, Academic Press, New York,1964.

On the Relational Database Style Parallel Numerical Programming *

Béla Kiss[1] and Anna Krebsz[2]

[1] Department of Mathematics, Széchenyi István University, H–9026 Győr, Egyetem tér 1., Hungary, E-mail: *bkiss@sze.hu*
[2] Department of Numerical Analysis, Eötvös Loránd University, H–1117 Budapest, Pázmány Péter Sétány 1.C, Hungary, E-mail: *krebsz@ludens.elte.hu*

Summary. A simplified relational database model implementation for parallel numerical computation is presented. The efficiency of this data model is demonstrated through the example of an optimally parallelized H–matrix (hierarchical–matrix) multiplier and its application as 3-dimensional Laplace preconditioning matrix.

1 Introduction

The numerical algorithms have become quite complex and require dynamic data structures. The general practice is to use a mixed data structure for this purpose [5]. In this paper a simplified version of the well–known relational database model is presented, which is developed for parallel numerical computation. This database model was introduced by E. F. Codd [3, 6]. The great advantage of this model is the flexible, simple and homogenous data structure. Moreover, the application of special geometric search trees (oct–tree, ADT, etc.) [4, 6, 13] can be avoided in such way. The simplified database model is implemented in object oriented manner by using the programming language $C++$.

The non–parallel version of this simplified relational database model for non–structural mesh generation is investigated in [11]. This paper demonstrates the efficiency of the parallel version in the case of an optimally parallelized H–matrix (hierarchical–matrix) multiplier. The applied H–matrix technique is originated from W. Hackbush and B. N. Koromskij [8, 9, 2]. A great advantage of the presented parallel H–matrix multiplier is that its implementation is independent of the geometrical complexity of the domains.

Our paper is organized as follows. Section 2 contains the description of the simplified relational database model. The parallelized H–matrix multiplier is discussed in Section 3. Section 4 is devoted to its application as Laplace preconditioning matrix on a very complex 3-dimensional domain.

* This work was partly supported by the Grants T043258 and T042826 of the Hungarian National Research Found.

2 Simplified Relational Database Model

In the case of the classical relational database model the **data files** are considered as **tables**, where the number of **columns** is fixed and the number of **rows** may vary. Each column contains elementary data from the same data type and can be referred to by a symbolic name. These symbolic names are called field identifiers. Each row (record) is identified by its row (record) number and a part of a row (record) by its field identifier.

The numerical problems usually require only one ordering on one data table. Hence it is worth to use combined data and index tables here to store the records in the adequate key order. This idea is implemented as a **RelDataBase** class with the following member functions.

Table 1. Member Functions of the Class **RelDataBase**

Member Function	Description
`Create(int TableId, int RecSize, KeyDesc* pKeyDesc);`	Creates a data table.
`Erase(int TableId);`	Erases a data table.
`Insert(int TableId, char* pRecord);`	Inserts a record.
`Update(int TableId, char* pRecord);`	Updates a record.
`Delete(int TableId, char* pRecord);`	Deletes a record.
`DeleteAll(int TableId, char* pRecord);`	Deletes all record.
`GetFirst(int TableId, char* pRecord);`	Gets the first.
`GetNext(int TableId, char* pRecord);`	Gets the next.
`GetLast(int TableId, char* pRecord);`	Gets the last.
`GetPrior(int TableId, char* pRecord);`	Gets the previous.
`GetCurrent(int TableId, char* pRecord);`	Gets the current.
`SeekLessEq(int TableId, char* pRecord);`	Seeks for less equal.
`SeekGreatEq(int TableId, char* pRecord);`	Seeks for greater equal.

For the sake of flexibility the records are called by reference, and the record pointers are casted to character type. The third parameter **KeyDesc** of the member function **Create** is a key descriptor structure for key segment identification (data type, beginning position).

```
struct KeyDesc
{
    char KeyFieldType[KEY_SEGMENT_MAX];
    char KeyFieldBegPos[KEY_SEGMENT_MAX];
};
```

The **integrated data and index tables** are realized by the well–known **Red–Black** and **B** balanced tree structures [4, 6]. Both tree structures are asymptotically optimal and widely used. The cost of the tree operations **Insert**, **Delete**, **Update** and **Seek** require less than or equal to $O(\log(n))$ arithmetic operations, where n denotes the number of tree–nodes. However, the cost of the sequential read in index order is only $O(1)$ in the practice.

We have applied the distributed database technique, using local simplified relational database on each process together with the message–passing standard **MPI** [7].

3 Parallel H–Matrix Multiplier

Let $\Omega \subset R^3$ be a given domain and let it be endowed with a regular, quasi–uniform tetrahedral mesh. Furthermore let $Q = \times_{i=1}^{3} [a_i, b_i]$ be a suitable chosen rectangular box which contains Ω. The coordinates of the minimal and maximal nodes of Q are denoted by **dRectMin** and **dRectMax**, respectively. We consider the H–matrix partitioning of this fictitious domain Q. Two main levels of the partitioning are introduced. The first one, **nLevelSubDom** is the level of subdomains. This level determines the size of greatest H–matrix blocks. The second one, **nLevelMesh**, is the level of the quasi–uniform mesh on Ω.

3.1 H–Matrix Partitioning

We have applied the next H–matrix partitioning algorithm.

```
//--------------------------------------------------------------------
void HMtxGen (int nLevelSubDom, int nLevelMesh,
          double dRectMin[3], double dRectMax[3])
{ int nLV, nLevel, nRowBlock[3], nColBlock[3];
  nLevel = 0;

  for(nLV = 0; nLV < 3; nLV++) { nRowBlock[nLV] = 0; nColBlock[nLV] = 0;}

  HSubMtxGen(nLevelSubDom,nLevelMesh,dRectMin,dRectMax,
    nLevel,nRowBlock,nColBlock);
}

//--------------------------------------------------------------------
void HSubMtxGen(int nLevelSubDom, int nLevelMesh,
        double dRectMin[3], double dRectMax[3],
        int nLevel, int nRowBlock[3], int nColBlock[3])
{int nLVi,nLVj,nLVk,nNextLevel,nRowSubBlocks[8][3],nColSubBlocks[8][3];

  if( nLevel == nLevelMesh)
      HMtxBlockWrite( nLevelSubDom,nLevelMesh,dRectMin,dRectMax,
      nLevel,nRowBlock,nColBlock);
  else
  { nNextLevel = nLevel+1;
      for(nLVi = 0; nLVi < 2; nLVi++)
        for(nLVj = 0; nLVj < 2; nLVj++)
          for(nLVk = 0; nLVk < 2; nLVk++)
          { nRowSubBlocks[4*nLVi+2*nLVj+nLVk][0]=2*nRowBlock[0]+nLVi;
            nRowSubBlocks[4*nLVi+2*nLVj+nLVk][1]=2*nRowBlock[1]+nLVj;
            nRowSubBlocks[4*nLVi+2*nLVj+nLVk][2]=2*nRowBlock[2]+nLVk;

            nColSubBlocks[4*nLVi+2*nLVj+nLVk][0]=2*nColBlock[0]+nLVi;
            nColSubBlocks[4*nLVi+2*nLVj+nLVk][1]=2*nColBlock[1]+nLVj;
```

```
            nColSubBlocks[4*nLVi+2*nLVj+nLVk][2]=2*nColBlock[2]+nLVk; }

      for(nLVi = 0; nLVi < 8; nLVi++)
      { for(nLVj = 0; nLVj < 8; nLVj++)
        { if((nLevel < nLeveSubDom) ||
            (dBlockCentrDistance(nRowSubBlocks[nLVi],
              nColSubBlocks[nLVj]) <= (sqrt(3.01)/2.0)))
            HSubMtxGen(nLevelSubDom,nLevelMesh,dRectMin,dRectMax,
              nNextLevel,nRowSubBlocks[nLVi],nColSubBlocks[nLVj]);
          else
            HMtxBlockWrite(nLevelSubDom,nLevelMesh,dRectMin,dRectMax,
              nNextLevel,nRowSubBlocks[nLVi],nColSubBlocks[nLVj]); } }
  }
}
```

3.2 H–Matrix–Vector Multiplication

Two tables are introduced for the matrix–vector computation by H–submatrices. The table **THMtx** contains the descriptions of the H–submatrix blocks. If the row and column blocks of a submatrix block are not too close to each other then this table is used for the submatrix–vector product computation as well. In this implicit case the appropriate variables are determined by the row and column blocks by using the table **TSimp** of tetrahedra. A tetrahedron belongs to an H–submatrix row or column block if its center of gravity belongs to the given block.

If a block is small and the row and column blocks are close to each other then the submatrix block is given explicitly. In this case the submatrix–vector product is computed by the application of the table **TExplHMtxBlocks**.

The record structures of tables **THMtx**, **TExplHMtxBlocks** and **TSimp** are the next ones.

```
//---------------------------------------------------------------------
//  Table : THMtx
//  Content:Geometric search table for H-submatrix blocks.
//  Key :   nCSubDomId+nRSubDomId+nSubMtxType+nSubMtxId
//          +nLevel+nColBlockCoords+nRowBlockCoords
struct RHMtx
{
    int     nRowBlockSubDomId; // Subdomain identifiers of the row
    int     nColBlockSubDomId; // and column blocks.
    int     nNeighbType;// Neighb. type of the subdomains (NBT_IDENT, NBT_NEAR, NBT_FAR).
    int     nSubMtxType;       // Submatrix type (BLOCK_EXPL, BLOCK_IMPL).
    int     nExplHMtxBlockId;  // Submatrix block id. in the explicit case.
    int     nLevel;            // Division level.
    int     nRowBlockCoords[3];// The block coordinates of the
    int     nColBlockCoords[3];// row and column blocks.
};
//---------------------------------------------------------------------
//  Table : TExplHMtxBlocks
//  Content:Explicitly given H-submatrix blocks.
//  Key :   nExplHMtxBlockId+nRVarId+nCVarId
struct RExplHSubMtx
{
    int     nExplHMtxBlockId;  // Explicit submatrix block identifier.
    int     nRVarId;           // Row variable identifier.
    int     nCVarId;           // Column variable identifier.
    double  dValue;            // Value.
```

```
};
//------------------------------------------------------------------------
// Table : TSimp
// Content:Geometric search table for terahedra.
// Key :    nSubDomId+nLevel+nBlockCoords+nSimpId
struct RSimp
{
    int     nSimpId;             // Tetrahedron identifier.
    int     nSubDomId;           // Subdomain identifier.
    int     nLevel;              // Fine mesh level.
    int     nBlockCoords[3];     // Search block coordinates.
    int     nLocation;           // Location (LOC_INSIDE, LOC_BOUNDARY).
    int     nNodeIds[4];         // Node identifiers.
};
```

The global matrix–vector product $\underline{y} = H \cdot \underline{x}$ is computed as a sum of local matrix-vector products of the form $\underline{y}_{L(i),R(j)} = H_{L(i),R(j),C(k)} \cdot \underline{x}_{L(i),C(k)}$. Here $\underline{x}_{L(i),C(k)}$ stand for the subvector of $\underline{x}$ determined by the cluster with parameters level $L(i)$ and column block coordinate $C(k)$. The vector $\underline{y}_{L(i),R(j)}$ is defined similarly. Only uniform H-matrices will be considered. In this case the order of the approximation is the same on each cluster determined by H–submatrix blocks. Then the implicitly given H–submatrix blocks can be expressed as $H_{L(i),R(j),C(k)} = \sum_{l=1}^{M} \sum_{m=1}^{M} \underline{a}_l \cdot \underline{b}_{l,m}^T$ and the matrix–vector products $\underline{y}_{L(i),R(j)} = H_{L(i),R(j),C(k)} \cdot \underline{x}_{L(i),C(k)}$ are computed as

$$\underline{y}_{L(i),R(j)} = \sum_{l=1}^{M} c_l \cdot \underline{a}_l = \sum_{l=1}^{M} \underline{a}_l \cdot c_l = \sum_{l=1}^{M} \underline{a}_l \cdot \left(\sum_{m=1}^{M} \underline{b}_{l,m}^T \cdot \underline{x}_{L(i),C(k)} \right),$$

where the constant M depends only on the order of the used approximation.

The distribution of the vectors among the processes is determined by the subdomain level H–matrix partitioning. The values $c_l = \sum_{m=1}^{M} \underline{b}_{l,m}^T \cdot \underline{x}_{L(i),C(k)}$ are computed on that subdomain which is identified by the H–submatrix column block. Three types of $\underline{y}_{L(i),R(j)} = H_{L(i),R(j),C(k)} \cdot \underline{x}_{L(i),C(k)}$, i.e. of local matrix-vector product computations by H-submatrix blocks are introduced. The field **nNeighbType** of the table **THMtx** serves this purpose. The possible values of this field are the following.

The **compact data exchange** denotes the exchange of single and grouped data together with the values c_l. In the grouped case we send only the parameters level $L(i)$ and block coordinate $R(j)$ of the row block cluster. The costs of the local and global variants of the compact data exchange are $O(\log(\frac{n}{p}))$ and $O(\log(p) \cdot p)$, respectively. Here n is the number of unknowns and p is the number of processes.

The computation of the global $\underline{y} = H \cdot \underline{x}$ product needs an additional step to correct the local $\underline{y}_{L(i),R(j)}$ values at the boundaries of neighbouring subdomains. For this purpose a simple local data exchange is applied by using the **MPI** functions **MPI_Send** and **MPI_Recv** too. The cost of this correction step is $O((\frac{n}{p})^{\frac{2}{3}})$.

Table 2. Possible Values of the Field **nNeighbType**

Value	Description
NBT_IDENT	Then the subvectors $\underline{x}_{L(i),C(k)}$ and $\underline{y}_{L(i),R(j)}$ belong to the same subdomain.
NBT_NEAR	Then the subvectors $\underline{x}_{L(i),C(k)}$ and $\underline{y}_{L(i),R(j)}$ belong to neighbouring subdomains in geometric sence. Here compact local data exchange is applied by using the table **THCommNear** and the **MPI** functions **MPI_Send** and **MPI_Recv**.
NBT_FAR	Then the subvectors $\underline{x}_{L(i),C(k)}$ and $\underline{y}_{L(i),R(j)}$ belong to non–neighbouring subdomains. Here compact global data exchange is applied by using the array **ArrHCommFar** and the **MPI** function **MPI_Allreduce**.

Since the computational cost of the values c_l and $\sum_{l=1}^{M} c_l \cdot \underline{a}_l$ is $O(\log(\frac{n}{p}) \cdot \frac{n}{p})$ on each subdomain, the total computational cost of the global matrix–vector product $\underline{y} = H \cdot \underline{x}$ is $O\left(\log\left(\frac{n}{p}\right) \cdot \frac{n}{p} + p\right)$ per subdomain.

The record structures of the table **THCommNear** and the array **ArrHCommFar** are the next ones.

```
//----------------------------------------------------------------------
//  Table : THCommNear
//  Content:Data exchange among neighbouring subdomains.
//  Key :   nSubDomId+nDataType+nVarId+nLevel+nBlockCoords
struct RHCommNear
{
    int     nSubDomId;          // Subdomain identifier.
    int     nDataType;          // Data type (DT_SINGLE,DT_GROUPED).
    int     nVarId;             // Variable id. (single case).
    int     nLevel;             // Division level(grouped case).
    int     nBlockCoords[3];    // Block coordinates(grouped case).
    double  dValues[M];         // Values.
};
//----------------------------------------------------------------------
//  Array : ArrHCommFar
//  Content:The array for data exchange among non-neighbouring
//          subdomains.
pArrCommFarArr = new double[nProcNum*M];
            // The number of processes multiplied by M.
```

3.3 *H*-Matrix Multiplier Object

Our *H*-matrix multiplier object is implemented in the following $C++$ class structure form.

```
//-------------------------------------------------------------------
class HMTX {
public:
    HMTX();
    ~ HMTX();
    int SetIOPuff(char* pIOPuff, int nIOPuffSize);
    int SetCommParam(int nProcId, int nProcNum);
```

```
    int SetDataBase(RelDataBase *pRDBase);  // Rel.database pointer.
    int SetGlobalTables(int tGSubDom,       // Subdomain descriptions.
                        int tGSubDomVar,    // Subdomain-varibale
                        int tGVarSubDom,    // connections.
                        int tGNode,         // Nodes.
                        int tGSimp,         // Simplices
                        int tGHMtx,         // H-matrices.
                        int tGExplHMtxBlocks,// Explicit H-matrix blocks.
                        int tGVectSo,       // Source vector.
                        int tGVectTa );     // Target vector.
    int SetLocalTables( int tLSubDom,       // Subdomain descriptions.
                        int tLSubDomVar,    // Subdomain-variable connections.
                        int tLHCommNear,    // Near-communication.
                        int tLNode,         // Nodes.
                        int tLSimp,         // Simplices.
                        int tLHMtx,         // H-matrices.
                        int tLExplHMtxBlocks,// Explicit H-matrix blocks.
                        int tLVectS,        // Auxiliary vector.
                        int tLVectSo,       // Source vector.
                        int tLVectTa, );    // Target vector.

    int ScatSubDomInfo( void );             // Scatters tGSubdom.
    int ScatSubDomVar( void );              // Scatters tGSubDomVar and
                                            // tGVarSubDom.
    int ScatNode( void );                   // Scatters tGNode.
    int ScatSimp( void );                   // Scatters tSimp.

    int ScatHMtx( void );                   // Scatters tHMtx.
    int ScatExplHMtxBlocks( void );         // Scatters tExplHMtxBlocks.

    int ScatVect(int tGVect, int tLVect);   // Scatters tGVect.
    int GathVect(int tGVect, int tLVect);   // Gathers tLVect.

    int CompMtxMult(int rLVectTa, int rLVectSo);
private:
    int CorrVectBlocks(int nProcId1, int nProcId2);       // The member functions
    int SendVectBlock(int nProcId1, int nProcId2);        // of the compact local
    int RecvAndAddVectBlock(int nProcId1, int nProcId2); // data exchange.

    int CorrVects(int nProcId1, int nProcId2);            // The member functions
    int SendVect(int nProcId1, int nProcId2);             // of the simple local
    int RecvAndAddVect(int nProcId1, int nProcId2);       // data exchange.
... };
```

4 Numerical Results

Our parallel H-matrix multiplier implementation has been tested as Laplace preconditioning matrix. The enclosed test results regard to quasi-uniform regular tetrahedral triangulations of the very complex water domain Ω of the RÁBA Euro 2 diesel engine. The shape of this domain can be seen in Figure 1.

We have tested the parallel solution of the three–dimensional Poisson equation

$$-\Delta u = f \quad \text{in} \quad \Omega, \qquad u = 0 \quad \text{on} \quad \partial\Omega$$

by using the preconditioned conjugate gradient (PCG) [5, 10, 12] algorithm.

The Poisson equation has been discretized by the finite element method, applying piecewise linear finite elements. The optimal preconditioning matrix

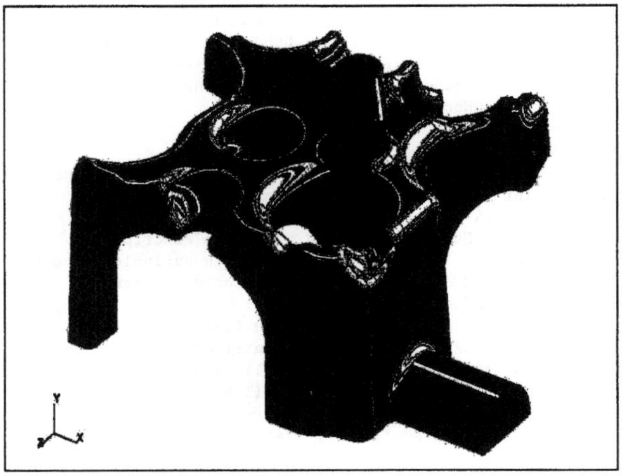

Fig. 1. The Test Domain

P was given in the form $P = V^2$, where the matrix V is the H–matrix approximation of the bilinear form

$$V(u,v) = \int_\Omega \int_\Omega \frac{u(\underline{x}) \cdot v(\underline{y})}{|\underline{x} - \underline{y}|^2} \, d\underline{y} \, d\underline{x},$$

which is an $H^{-1/2}$ norm representation. It is discretized by piecewise linear finite elements on the level of the given tetrahedral mesh, and by piecewise bilinear elements on the clusters, which are determined by the H-submatrix blocks.

Table 3. Test results

Nodes	Tetrahedra	Processes	Computers	Prec. matrix	Iterations	CPU time in sec.
20455	100986	1	1	D	17	92
		1	1	P	5	87
		8	8	D	17	54
		8	8	P	5	18
		64	16	D	17	23
		64	16	P	5	6
121441	403944	1	1	D	26	246
		1	1	P	7	224
		8	8	D	26	117
		8	8	P	7	59
		64	16	D	26	64
		64	16	P	7	12

Our computer cluster was a 16–machines cluster of IBM PC compatible machines with 900MHz Intel processor using the message–passing standard MPI under the operation system LINUX.

The computational results given in Table 3 correspond to the case when the right hand side of the Poisson equation is

$$f(x, y, z) = \sin(xyz)$$

and the error bound is $\varepsilon = 10^{-4}$. The preconditioning matrix D denotes the inverse of the main diagonal of the matrix of the discretized Laplace operator.

From our results we can conclude that the matrix P is an efficient preconditioning matrix, but its computation is fast enough in parallel environment only.

Remark 1. The H-matrix approximation of the inverse Laplace-matrix [1] would probably give a better convergence results, however, we found that its implementation is much more complicated.

References

1. Bebendorf, M., Hackbusch, W. (2002): Existence of H-Matrix Approximation to the Inverse FE-Matrix of Elliptic Operators with L^∞–Coeffitients, Max-Planck-Institute, Leipzig, Preprint no. 21.
2. Börm, S., Grasedyck, L., Hackbusch, W. (2002): Introduction to Hierarchical Matrices with Applications, Max-Planck-Institute, Leipzig, Preprint no. 18.
3. Codd, E. F. (1970): A Relational Model of Data for Large Shared Data Banks, Communication of the ACM, **13**(6), 377–387.
4. Cormen, T. H., Leiserson, C. E., Rivest, R. L. (1990): Introduction to Algorithms, MIT Press.
5. Douglas, C., Haase, G., Langer, U. (2003): A Tutorial on Elliptic PDE Solvers and Their Parallelisation, SIAM.
6. Garcia-Molina, H., Ullmann, J.D., Widom, J. (2000): Database System Implementation, Prentice Hall.
7. Gropp, W., E. Lusk, E. (2001): Installation and Users Guide to MPICH, a Portable Implementation of MPI, Technical Report ANL-01/x, Argonne National Laboratory.
8. Hackbush, W. (1999): A Sparse Matrix Arithmetic Based on H-matrices, Part I: Introduction to H-matrices, Computing **62**, 89–108.
9. Hackbush, W., Koromskij, B. N. (2000): A Sparse Matrix Arithmetic Based on H-matrices, Part II: Application to Multidimensional Problems, Computing **64**, 21–47.
10. Heinrich, V. (1993): Iterative Methods for Linear Systems of Equations, Hamburger Beiträge zur Angewandted Mathematics, Hamburg.
11. Kiss, B., Krebsz, A. (2002): On the Relational Database Type Numerical Programming, In: Proc. of the 3rd Int. Conf. on Engineering Comp. Technology,(eds.: Topping, B. H. V., Bittnar, Z.), Civil-Comp Press, 127–128 (Book + CD-ROM).

12. Samarskij, A. A., Nikolajev, E. S. (1989): Numerical Methods for Grid Equations, vol. 1 and 2, Basel, Bikhäuser.
13. Samet, H. (1990): Application of Spatial Data Structures, Addison-Wesley, Reading, MA.

A Dynamical System Describing Evolution of the Implicit Surfaces in Incompressible Viscous Liquids

Petr Klouček[1], Michel V. Romerio[2] and Jennifer L. Wightman[1]

[1] Department of Computational and Applied Mathematics, Rice University, 6100 Main Street, Houston, TX 77005, USA
kloucek@rice.edu and *wightman@rice.edu*

[2] Institut d'Analyse et Calcul Scientifique, École Polytechnique Fédérale de Lausanne, 1015 Lausanne, Switzerland
romerio@epfl.ch

Summary. We present an equation capable of describing the evolution of implicit surfaces (boundaries of gas bubbles) in liquids. The equation does not contain any convective term and thus it is independent of the velocity and pressure fields, given by e.g. Navier-Stokes equations. The equation itself is obtained by a series of variational problems. First, by interpreting the Euler implicit time-stepping scheme for the update of Lagrangian flow maps as a Monge-Kantorovich transference problem. This approach provides then a variational principle for the updates of an Eulerian phase Indicatrix in terms of the Wasserstein metric. The Euler-Lagrange equations corresponding to the later variational principle provide the thought after evolution equation for the implicit surfaces.

1 Classical Approach to Tracking of the Implicit Surfaces

Let us consider a mixture of liquid and gas. We assume that the gas is formed in bubbles, and we further require that both phases are immiscible. Let us denote by $\mathbf{X} = \mathbf{X}(\mathbf{x}, t)$ the Lagrangian flow map. Let the Eulerian Indicatrix function $\chi \in BV(\Omega, \{0,1\})$ represent the characteristic function of the domain Ω_G occupied by the gas. We set $\chi(\mathbf{x}, t) = 1$ for $\mathbf{x} \in \Omega_G$ and $\chi(\mathbf{x}, t) = 0$ otherwise. The connection between the Lagrangian and Eulerian formalism is provided by the relation

$$\partial_t \mathbf{X}(\mathbf{x}, t) = \mathbf{v}\left(\mathbf{X}(\mathbf{x}, t), t\right), \qquad \text{for all } \mathbf{x} \in \Omega, \quad t > 0, \tag{1}$$

where $\mathbf{v}(\mathbf{x}, t)$ is the velocity field in the Eulerian representation. This equation can be solved for a smooth $\mathbf{v}$ by the Cauchy-Lipschitz theorem. The required immiscibility has the form

$$\chi\left(\mathbf{X}(\mathbf{x}, t), t\right) = \chi(\mathbf{x}, 0) = \chi^0(\mathbf{x}), \qquad \text{for all } \mathbf{x} \in \Omega, \quad t > 0. \tag{2}$$

The transport equation for the Indicatrix follows immediately from the immiscibility requirement (2) and from the equation (1). Namely,

$$
\begin{aligned}
0 &= \frac{d}{dt}\chi(\mathbf{X}(\mathbf{x},t),t) \\
&= \partial_t\chi(\mathbf{X}(\mathbf{x},t),t) + \mathbf{v}(\mathbf{X}(\mathbf{x},t),t)\cdot\nabla\chi(\mathbf{X}(\mathbf{x},t),t), \quad \mathbf{x}\in\Omega,\ t>0.
\end{aligned} \tag{3}
$$

The symbol ∇ denotes gradients computed with respect to the configuration at $t = 0$.

The free surfaces are characterized by the set $S(\nabla\chi(\cdot,t))$, denoting the points of discontinuities in the gradient of χ. The transport equation (3) can thus be written formally (note that $\nabla\chi$ lives in the dual to L^∞!) as

$$
\partial_t\chi(\mathbf{x},t) + \mathbf{v}(\mathbf{x},t)\cdot\nabla\chi(\mathbf{x},t) = 0 \quad \text{on the free surface } S(\nabla\chi(\cdot,t)). \tag{4}
$$

The explicit dependence on the unknown, implicit, surface $S(\nabla\chi(\cdot,t))$ can be circumvented by introduction of the so-called *level set function* G, [11]. Namely, let $G(\mathbf{x},t)$ be a smooth function measuring the distance to the interface, being positive inside of a gas domain, and let θ denote the Heaviside function. Then $\chi = \theta\circ G$. Substituting this representation of the Indicatrix into the transport equation (4) yields

$$
\delta_0(G)\,(\partial_t G + \mathbf{v}\cdot\nabla G) = 0. \tag{5}
$$

Assuming that the velocity field is known for any $\mathbf{x}\in\Omega$, we obtain from (5) the following *level set formulation*, [11],

$$
\partial_t G(\mathbf{x},t) + \mathbf{v}(\mathbf{x},t)\cdot\nabla G(\mathbf{x},t) = 0, \qquad \mathbf{x}\in\Omega,\ t>0. \tag{6}
$$

Our goal is to derive a variational principle for χ. In other words we strive to derive a Helmholtz free energy depending on χ. A similar approach effort within the field models is reported in [?]. The selfcontained, i.e., independent of the subsequent velocity and pressure fields, evolution equation for the Indicatrix is then obtained as the Euler-Lagrange equation of the Lagrangian of the system. The key difference between the approach leading to (6) and our theory is that we look at the evolution of the implicit (free) surfaces as the Monge-Ampère transport problem.

2 Generalized Least Action Principle

The Lagrangian of our system includes the kinetic, potential, and surface energies. We assume that the density of the liquid and the density of the gas are constant. Thus

$$
\rho_0(\mathbf{x}) \overset{\text{def}}{=} \rho(\mathbf{x},0) = \rho(\mathbf{X}(\mathbf{x},t),t). \tag{7}
$$

We assume that the density ρ is given by

$$
\rho(\mathbf{X}(\mathbf{x},t),t) = \rho_G\chi(\mathbf{X}(\mathbf{x},t),t) + \rho_L(1-\chi(\mathbf{X}(\mathbf{x},t),t)), \quad \mathbf{x}\in\Omega,\ t>0, \tag{8}
$$

where ρ_G, ρ_L are the density of the gas and liquid, respectively. We consider a gas-liquid system with total energy of a Mumford-Shah type, [9], given by

$$\begin{aligned}
&\mathcal{E}(\mathbf{X},t) \stackrel{\text{def}}{=} \tfrac{1}{2} \int_{\Omega} \rho_0(\mathbf{x}) \left(\frac{d}{dt}\mathbf{X}(\mathbf{x},t)\right)^2 dx + E(\mathbf{X},t), \\
&\text{where} \\
&E(\mathbf{X},t) \stackrel{\text{def}}{=} \int_{\Omega} \rho_0(\mathbf{x})\, \mathbf{g}\cdot\mathbf{X}(\mathbf{x},t)\, dx + \alpha \left\|\nabla\chi(\cdot,t)\right\|(\Omega), \qquad \text{and} \\
&\chi\left(\mathbf{X}(\mathbf{x},t),t\right) = \chi(\mathbf{x},0).
\end{aligned} \tag{9}$$

The surface energy, $\alpha \left\|\nabla\chi(\cdot,t)\right\|(\Omega)$, i.e., the total variation of $\nabla\chi$, represents the perimeters of the subdomains occupied by the gas multiplied by a surface tension coefficient α which is a positive measured quantity. The vector $\mathbf{g}$ represents the gravitation force field.

We apply a generalization of the *Principle of Stationary Action*, [8], to the Lagrangian (9). The equlibrium dynamics obtained from this principle are given by the Euler equations with an added dissipative term. Moreover, the pressure drop across the gas-liquid interfaces satisfies the Laplace-Young equation. To account for the dissipative effects, we define the action as

$$A(\mathbf{X},t_1,t_2) \stackrel{\text{def}}{=} \int_{t_1}^{t_2} e^{\lambda(t-t_0)}\mathcal{E}(\mathbf{X},t)\, dt, \qquad t_2 \geq t_1 \geq t_0, \tag{10}$$

where $\lambda > 0$ represents the Rayleigh's friction dissipation coefficient, and $t_0 \geq 0$ is a given initial time. Then

$$\mathbf{X} \mapsto A(\mathbf{X},t_1,t_2) \tag{11}$$

represents a functional on the *configuration space* $\mathcal{M}$[1], given by

$$\begin{aligned}
\mathcal{M} \stackrel{\text{def}}{=} \{\mathbf{X} : \Omega \mapsto \Omega \,|\, \mathbf{X} &\text{ is one-to-one and onto, volume preserving,} \\
&\text{separately } C^1\text{-maps in both the gas and the liquid}\}.
\end{aligned} \tag{12}$$

We require the action to be stable on the flow maps $\mathbf{X}(.,t) \in \mathcal{M}$ with respect to variations which are compatible with the incompressibility and the immiscibility constraints. Therefore we consider a family of maps $\mathbf{X}_\tau$ preserving the Lebesgue measure such that $\mathbf{X}_0 = \mathbf{X}$. We set

$$\mathbf{Y} = \frac{d}{d\tau}\mathbf{X}_\tau\Big|_{\tau=0}.$$

[1] The configuration space is not a linear function space. Consequently, the differentiation has to be considered in the tangent space to $\mathcal{M}$ where summation is allowed.

Hence, $\mathbf{Y} \in T_{\mathbf{X}(.,t)}\mathcal{M}$. The tangent space to $\mathcal{M}$ at the point $\mathbf{X}(.,t)$ is given by

$$T_{\mathbf{X}(.,t)}\mathcal{M} = \{\mathbf{Z} \circ \mathbf{X}(.,t) \,|\, \mathbf{Z} \in T_{\mathbf{Id}}\mathcal{M},\ \mathbf{X}(.,t) \in \mathcal{M}\}, \tag{13}$$

where $\mathbf{X}(.,t) : \mathbf{x} \mapsto \mathbf{X}(\mathbf{x},t)$, and

$$\begin{aligned} T_{\mathbf{Id}}\mathcal{M} = \{\mathbf{Z} : \Omega \mapsto \mathbb{R}^3 \,|\, & \operatorname{div} \mathbf{Z}(\mathbf{x}) = 0 \text{ for } \mathbf{x} \in \Omega_G, \Omega_L, \\ & \mathbf{Z} \cdot \mathbf{n} = 0 \text{ on } \partial\Omega, \\ & [\![\mathbf{Z} \cdot \mathbf{n}]\!] = 0 \text{ on } \partial\Omega_L \cap \partial\Omega_G\}. \end{aligned} \tag{14}$$

The jump in $\mathbf{Z} \cdot \mathbf{n}$ is the usual difference of the values from the different sides of the gas-liquid interfaces. The structure of the linear space $T_{\mathbf{Id}}\mathcal{M}$, i.e., the divergence-free condition, is, by the Liouville theorem, equivalent to the preservation of the Lebesgue measure. Moreover the variations must satisfy

$$\mathbf{Y}(.,t_1) = \mathbf{Y}(.,t_2) = 0. \tag{15}$$

We implement the Principle of Stationary Action by requiring the Gâteaux derivative of the action to vanish on variations $\mathbf{Y} \in T_{\mathbf{X}(.,t)}\mathcal{M}$ satisfying (15), i.e.,

$$dA(\mathbf{X}, t_1, t_2, \mathbf{Y}) = 0, \qquad \text{for all } \mathbf{Y}(.,t) \in T_{\mathbf{X}(.,t)}\mathcal{M} \text{ satisfying (15).} \tag{16}$$

We find that (16) translates into the following problem:
Find $\mathbf{X}(.,t) \in \mathcal{M}$, $\frac{d}{dt}\mathbf{X}(.,t) \in L^2\left([0,T], \left(T_{\mathbf{X}(.,\mathbf{t})}\mathcal{M}\right)^*\right)$, *such that*

$$\begin{aligned} &\int_0^T \int_\Omega \rho(\mathbf{X}(\mathbf{x},t),t) \left(\frac{d}{dt}\mathbf{X}(\mathbf{x},t) \cdot \frac{d}{dt}\mathbf{Y}(\mathbf{x},t) - \lambda \frac{d}{dt}\mathbf{X}(\mathbf{x},t) \cdot \mathbf{Y}(\mathbf{x},t)\right) dx\, dt \\ &+ \int_0^T \int_\Omega \rho(\mathbf{X}(\mathbf{x},t),t)\mathbf{g} \cdot \mathbf{Y}(\mathbf{x},t)\, dx\, dt \\ &+ \alpha \int_0^T \int_{\partial\Omega_G(0)} H(\mathbf{X}(\mathbf{s},t))\mathbf{Y}(\mathbf{s},t) \cdot \mathbf{n}(\mathbf{X}(\mathbf{s},t))\, dS\, dt = 0, \\ &\qquad \textit{for all } \mathbf{Y}, \frac{d}{dt}\mathbf{Y} \in L^2\left([0,T], T_{\mathbf{X}(.,\mathbf{t})}\mathcal{M}\right), \quad \mathbf{Y}(.,0) = \mathbf{Y}(.,T) = 0. \end{aligned} \tag{17}$$

Here, H denotes the mean curvature. The space $\left(T_{\mathbf{X}(.,\mathbf{t})}\mathcal{M}\right)^*$ denotes the dual to $T_{\mathbf{X}(.,\mathbf{t})}\mathcal{M}$. The duality is given by the Riemannian metric induced by the kinetic energy. Consequently, we can identify the dual space with itself.

3 Evolution of free surfaces and Monge-Ampère Transport Problem

As the first step towards the sought after variational principle, we discretize in time the Lagrangian $\mathcal{E}$ by the implicit Euler scheme to proceed from a given state $\mathbf{X}^k$ to $\mathbf{X}^{k+1}$. The dissipative effects are included in the definition of the generalized action, but they are not included into the Lagrangian itself. To overcome this difficulty, we rescale the time coordinate. Namely, let $\tilde{\mathbf{X}}$ be given by

$$\tilde{\mathbf{X}}(\cdot,\tau) \stackrel{\text{def}}{=} \mathbf{X}(\cdot,t), \qquad \text{where} \qquad \tau \stackrel{\text{def}}{=} \frac{e^{\lambda(t-t_0)}-1}{\lambda}. \tag{18}$$

Then, using the substitution $t \to \tau$,

$$\begin{aligned} &\int_{t_1}^{t_2} \tfrac{1}{2} e^{\lambda(t-t_0)} \int_{\Omega} \rho_0(\mathbf{x}) \left(\frac{d}{dt}\mathbf{X}(\mathbf{x},t)\right)^2 dx\, dt \\ &= \int_{\tau_1}^{\tau_2} \tfrac{1}{2} \int_{\Omega} \rho_0(\mathbf{x}) \left(\frac{d}{d\tau}\tilde{\mathbf{X}}(\mathbf{x},\tau)\right)^2 dx\, (\lambda\tau+1)^2\, d\tau. \end{aligned} \tag{19}$$

Similarly,

$$\int_{t_1}^{t_2} e^{\lambda(t-t_0)} E(\mathbf{X})\, dt = \int_{\tau_1}^{\tau_2} E\left(\tilde{\mathbf{X}}(\mathbf{x},\tau)\right) d\tau. \tag{20}$$

Let $\tau_k = k\,\Delta\,\tau$, and let $\tilde{\mathbf{X}}^k(.) \stackrel{\text{def}}{=} \tilde{\mathbf{X}}(.,k\,\Delta\,\tau)$. In accordance with the PSA, we assume that $\tilde{\mathbf{X}}^{k+1}$, $\tilde{\mathbf{X}}^{k-1}$ are given states. We are looking for $\tilde{\mathbf{X}}^k$ which is a solution of the minimization problem

$$\begin{aligned} &\inf\left\{ (\lambda\tau_k+1)^2 \tfrac{1}{4} \int_{\Omega} \rho_0(\mathbf{x}) F\left(\tilde{\mathbf{X}}\right) dx + (\Delta\,\tau)^2 E\left(\tilde{\mathbf{X}}\right) \mid \tilde{\mathbf{X}} \in \mathcal{M} \right\}, \quad \text{where,} \\ &F\left(\tilde{\mathbf{X}}\right) \stackrel{\text{def}}{=} \left|\tilde{\mathbf{X}}(\mathbf{x}) - \tilde{\mathbf{X}}^{k+1}(\mathbf{x})\right|^2 + \left|\tilde{\mathbf{X}}(\mathbf{x}) - \tilde{\mathbf{X}}^{k-1}(\mathbf{x})\right|^2. \end{aligned} \tag{21}$$

Since

$$\Delta\,\tau = \frac{e^{\lambda(t_{k+1}-t_0)}-1}{\lambda} - \frac{e^{\lambda(t_k-t_0)}-1}{\lambda} = e^{\lambda(t_k-t_0)} \frac{e^{\lambda\,\Delta\,t_k}-1}{\lambda},$$

we have for $\lambda \ll 1$

$$e^{-\lambda(t_k-t_0)}\,\Delta\,\tau \approx \Delta\,t_k. \tag{22}$$

Hence, going back to the original time coordinate and using the above approximation, we obtain

$$\mathbf{X}^k \stackrel{\text{def}}{=} \operatorname*{Arginf}_{\mathbf{X}\in\mathcal{M}} \tfrac{1}{4} \int_{\Omega} \rho_0(\mathbf{x}) F(\mathbf{X})\, dx + (\Delta t_k)^2 E(\mathbf{X}). \tag{23}$$

Remark 3.1 (Weak form (17) and the variational principle (23)). *The variational formulation (23) is obtained from the PSA applied to the time interval (t_{k-1}, t_{k+1}) by using the trapezoidal rule to provide an integration rule in time. Computing the Gateaux derivative of the functional appearing in (23) and using the discrete integration by parts in time, we recover the implicitly in time discretized weak formulation (17).* □

Remark 3.2 (Time step $\triangle t_k$). *It follows from the above calculations that if $\lambda \ll 1$ then, from (18) and (22),*

$$\triangle t_k \approx \frac{\Delta \tau}{\lambda k \, \Delta \tau + 1}.$$

Thus, $\lim_{k \to +\infty} \triangle t_k = 0$. □

We first state the following result before we proceed to make the link between the minimizer of the variational problem (24) below and the flow maps.

Theorem 3.3. *Let us assume that Ω is bounded with piece-wise smooth boundary, and that $\chi^{k+1}, \chi^k \in \mathcal{K}$ are given. Then there exists a unique minimizer χ^k of the following variational problem:*

$$\inf \left\{ \mathcal{D}_W(\chi, \chi^{k+1}, \chi^{k-1}) + (\triangle t_k)^2 \, E_E(\chi) \,|\, \chi \in \mathcal{K} \right\}, \tag{24}$$

We use the following Definitions

$$\begin{aligned}
\mathcal{D}_W(\chi, \chi^{k+1}, \chi^{k-1}) &\stackrel{def}{=} \tfrac{1}{2}\rho_G \operatorname{dist}_W \left(\chi, \chi^{k+1}\right)^2 + \tfrac{1}{2}\rho_L \operatorname{dist}_W \left(1-\chi, 1-\chi^{k+1}\right)^2 \\
&\quad + \tfrac{1}{2}\rho_G \operatorname{dist}_W \left(\chi, \chi^{k-1}\right)^2 + \tfrac{1}{2}\rho_L \operatorname{dist}_W \left(1-\chi, 1-\chi^{k-1}\right)^2,
\end{aligned}$$

$$E_E(\chi) \stackrel{def}{=} \int_\Omega \rho_0(\mathbf{x})\mathbf{g} \cdot \mathbf{x} \, dx + \alpha \left\| \nabla \chi \right\| (\Omega),$$

$$\rho(\mathbf{x}) \stackrel{def}{=} \rho_G \chi(\mathbf{x}) + \rho_L(1 - \chi(\mathbf{x})),$$

$$\begin{aligned}
\mathcal{K} \stackrel{def}{=} \{\chi \in BV(\Omega, \{0,1\}) \,\big|\, &\exists \mathbf{X} \in \mathcal{N}\left(\chi^0, \chi\right) : \\
&\int_\omega \chi(\mathbf{x}) \, dx = \int_{\mathbf{X}^{-1}(\omega)} \chi^0(\mathbf{x}) \, dx, \ \forall \text{ open } \omega \subset \Omega \}.
\end{aligned}$$

Proof. The proof follows from the strict convexity of the Wasserstein distance on the relaxed space $\mathcal{K}_{\text{relaxed}}$ which is defined as $\mathcal{K}$ but with $BV(\Omega, [0,1])$ as the base space. The existence of the unique minimizer of (24) is then obtained by the density argument, which follows from the compatibility of the Wasserstein metric with a convergence in the L^p–spaces. The proof can be found in [7]. □

Remark 3.4. *We recall that the Monge-Ampère problem is formulated as follows*

$$\inf\left\{\int_{\Omega} \chi^k(\mathbf{x})\left|\nabla\phi(\mathbf{x})-\mathbf{x}\right|^2 dx \,\middle|\, \nabla\phi \in \mathcal{N}\left(\chi^k,\chi^{k+1}\right)\right\}, \tag{25}$$

where

$$\begin{aligned}
&\mathcal{N}\left(\chi^0,\chi^1\right) \stackrel{def}{=} \\
&\{\nabla\phi \in \mathcal{L}\left(\chi^0,\chi^1\right) \mid \phi \textit{ Lip. cont. and convex}, \nabla\phi \textit{ invertible for a. a. } \mathbf{x}\in\Omega\}, \\
&\mathcal{L}\left(\chi^0,\chi^1\right) \stackrel{def}{=} \Big\{M \in \mathcal{I}\left(\chi^0,\chi^1\right) \,\Big| \\
&\qquad \int_{M^{-1}(\omega)} \chi^0(\mathbf{x})\,dx = \int_{\omega} \chi^1(\mathbf{x})\,dx, \textit{ for all Borel subsets } \omega \textit{ of } \Omega\Big\}, \\
&\mathcal{I}\left(\chi^0,\chi^1\right) \stackrel{def}{=} \\
&\{M : \{\chi^0>0\} \subseteq \Omega \mapsto \{\chi^1>0\} \subseteq \Omega \mid M \textit{ measurable}\}.
\end{aligned} \tag{26}$$

□

The existence result of Theorem 3.3 serves as a basis for showing the solvability of the variational problem (23). In particular, it follows from [7] and [12] that the minimizer has the form

$$\mathbf{X}^k = \nabla\phi^{k-1,k} \circ \mathbf{X}^{\mathbf{k-1}}, \tag{27}$$

where $\phi^{k-1,k}$ is the unique solution of the Monge-Ampère problem (25). Moreover, the link mentioned above is provided by

$$\chi^k\left(\mathbf{X}^k(\mathbf{x})\right) = \chi^0(\mathbf{x}), \qquad \mathbf{x}\in\Omega. \tag{28}$$

4 Dynamical System

In order to obtain a dynamical system for the evolution of the Indicatrix, we perform two steps. First, we compute the Gâteaux derivative of (24) and then we, formally, take a limit as $\Delta t \to 0_+$. Obtaining the Gâteaux derivative is a delicate task, see [7] for complete details.

Let $\mathbf{X}^{k,k+1} \in \mathcal{M}$ be the optimal flow map of χ^k to χ^{k+1}. Similarly, let $\mathbf{X}_\tau \in \mathcal{M}$ be a family of optimal maps bringing χ^k onto χ_τ such that $\mathbf{X}_\tau\big|_{\tau=0} = \mathbf{X}^{k,k+1}$. We expand $\mathbf{X}_\tau$ around the point $\tau=0$. Up to the first order in τ, we have

$$\mathbf{X}_\tau(\mathbf{x}) = \mathbf{X}^{k,k+1}(\mathbf{x}) + \tau\frac{d}{d\tau}\mathbf{X}_\tau\left(\mathbf{X}^{k,k+1}(\mathbf{x})\right)\Big|_{\tau=0}. \tag{29}$$

Since $\mathbf{X}_\tau \in \mathcal{M}$ we have

$$\frac{d}{d\tau}\mathbf{X}_\tau\left(\mathbf{X}^{k,k+1}(\mathbf{x})\right)\big|_{\tau=0} \in T_{\mathbf{X}^{k,k+1}}\mathcal{M}. \tag{30}$$

Consequently, in view of (13),

$$\operatorname{div}\frac{d}{d\tau}\mathbf{X}_\tau(\mathbf{X}^{k,k+1}(\mathbf{x}))\big|_{\tau=0} = 0 \qquad \text{for a. a. } \mathbf{x} \in \Omega. \tag{31}$$

Hence, we consider all possible variations of the form

$$\begin{aligned} &\frac{d}{d\tau}\mathbf{X}_\tau\left(\mathbf{X}^{k,k+1}(\mathbf{x})\right)\big|_{\tau=0} = (\nabla f \circ \mathbf{h})(\mathbf{x}), \qquad \text{where} \\ &\mathbf{h} \in \mathcal{M}_k, \quad f \in W^{1,2}\left(\Omega, \mathbb{R}^1\right), \quad \text{and} \quad \operatorname{div}(\nabla f(\mathbf{h}(\mathbf{x}))) = 0 \quad \text{for a.a. } \mathbf{x} \in \Omega. \end{aligned} \tag{32}$$

It follows from [7] that for any $f \in W^{1,2}\left(\Omega, \mathbb{R}^1\right)$ there exists $\mathbf{h} \in \mathcal{M}_k$ such that the divergence free requirement in (32) is satisfied. Thus, we require χ^k to solve

$$\begin{aligned} \frac{d}{d\tau}\mathcal{D}_W(\chi_\tau, \chi^{k+1}, \chi^{k-1})\big|_{\tau=0} + (\triangle t_k)^2\frac{d}{d\tau}E_\epsilon(\chi_\tau)\big|_{\tau=0} = 0, \\ \text{for all } f \in W^{1,2}(\Omega, \mathbb{R}^1). \end{aligned} \tag{33}$$

The fundamental step is to show that within $\mathcal{O}(\triangle t)$ we have

$$\begin{aligned} &\frac{d}{d\tau}\frac{1}{2}\operatorname{dist}_W(\chi_\tau, \chi^{k+1})^2\big|_{\tau=0} + \frac{d}{d\tau}\frac{1}{2}\operatorname{dist}_W(\chi_\tau, \chi^{k-1})^2\big|_{\tau=0} \\ &\approx \int_\Omega \left(\chi^{k+1}(\mathbf{x}) - 2\chi^k(\mathbf{x}) + \chi^{k-1}(\mathbf{x})\right) f(\mathbf{x})\,dx \\ &\quad - \int_\Omega (\mathbf{x} - h(\mathbf{x}))\cdot\nabla f(\mathbf{h}(\mathbf{x}))\chi^k(\mathbf{x})\,dx - \int_\Omega \left(f(\mathbf{h}(\mathbf{x})) - f(\mathbf{x})\right)\chi^k(\mathbf{x})\,dx \\ &\quad \text{for all } f \in W^{1,2}\left(\Omega, \mathbb{R}^1\right), \quad h \in \mathcal{M}_k. \end{aligned} \tag{34}$$

The last two integrals vanish due to the transport equation. After some calculations, suitable approximations, and the limit procedures $\epsilon \to 0_+$, $\triangle t \to 0_+$, (again see [7] for details), we find that the dynamical system is given by a degenerate equation

$$\begin{aligned} &\rho_G\left(\partial_{tt}\chi(\mathbf{x},t) - \lambda\partial_t\chi(\mathbf{x},t)\right) = \operatorname{div}\left(\chi(\mathbf{x},t)\left((\rho_G - \rho_L)\,\mathbf{g} + \alpha\nabla H(\mathbf{x},t)\right)\right), \\ &\mathbf{x} \in S(\nabla\chi), \quad t > 0. \end{aligned} \tag{35}$$

Remark 4.1. *The sequence of solutions obtained by solving the discretized version of* (35) *is a sequence of approximate minimizers of* (24). *Hence, this sequence is written with respect to a moving frame given, implicitly, by the optimal flow maps. The connection between the successive solutions is given by*

$$\chi^{k+1}\left(\nabla\phi^{k,k+1}(\mathbf{x})\right) = \chi^k(\mathbf{x}), \qquad \mathbf{x} \in \Omega, \tag{36}$$

where $\nabla\phi^{k,k+1}$ is the solution of the Monge-Ampère transport problem. This relation has three implications. First, it shows that χ^{k+1} is the Eulerian representer of χ^k. The previous time level solution, χ^k, is written with respect to the Lagrangian frame of reference. Secondly, it follows from (35) *that we do not need to convert the two systems. Instead, we can use χ^{k+1} directly in the fixed, Eulerian frame of reference.*

The last implication of (36) *concerns the use of the solutions of* (35) *with other differential equations, say incompressible Euler equations, semidiscretized in time. The purpose of computing s^k is both the tracking of the phase boundaries and the computation of the evolving density. We have in the Lagrangian frame of reference*

$$\rho_{Lagrange}(\mathbf{x}, k\,\triangle\, t) = \rho_G\chi^k(\mathbf{x}) + \rho_L\left(1 - \chi^k(\mathbf{x})\right). \tag{37}$$

Since $\rho_{Lagrange}(\mathbf{x}, k\triangle t) = \rho_{Euler}(\nabla\phi^{k,k+1}(\mathbf{x}), k\triangle t)$, it follows from (36) *that the appropriate form of the density passed to the Euler equations written in the fixed frame of reference, in the time-semidiscrete form, is the following*

$$\rho_{Euler}(\mathbf{x}, k\,\triangle\, t) = \rho_G\chi^{k+1}(\mathbf{x}) + \rho_L\left(1 - \chi^{k+1}(\mathbf{x})\right). \tag{38}$$

In other words, the tracking has to be computed one $\triangle t$ ahead of the solution of the Euler equations. □

5 Appendix: Wasserstein Distance

The Wasserstein distance, denoted $\mathrm{dist}_W(.,.)$, is defined by, [13], [2],

$$\mathrm{dist}_W(s_0, s_1)^2 \stackrel{\mathrm{def}}{=} \inf_{\substack{\mu\circ\pi_{\mathbf{x}}^{-1}=s_0,\\ \mu\circ\pi_{\mathbf{y}}^{-1}=s_1}} \iint_{\Omega\times\Omega} |\mathbf{x}-\mathbf{y}|^2\, d\mu(\mathbf{x},\mathbf{y}). \tag{39}$$

In words, the infimum in (39) is taken over bistochastic measures with marginals s_0 and s_1, respectively. This means

$$\begin{aligned} \iint_{\Omega\times\Omega} \varphi(\mathbf{x})\, d\mu(\mathbf{x},\mathbf{y}) &= \int_{\Omega} \varphi(\mathbf{x}) s_0(\mathbf{x})\, dx, \qquad \text{for all } \varphi \in C(\overline{\Omega}),\\ \iint_{\Omega\times\Omega} \varphi(\mathbf{y})\, d\mu(\mathbf{x},\mathbf{y}) &= \int_{\Omega} \varphi(\mathbf{y}) s_1(\mathbf{y})\, dy, \qquad \text{for all } \varphi \in C(\overline{\Omega}). \end{aligned} \tag{40}$$

We assume that the functions s_0 and s_1 are bounded, non-negative, measurable functions with compact support. It follows from [Section 3, [2]] that the infimum in (39) is attainable and, c. f. [Proposition 3.1, [2]], that the unique optimal measure μ^* has the form

$$\mu^*(\mathbf{x},\mathbf{y}) = \delta\,(M^*(\mathbf{y}) - \mathbf{x})\, s_0(\mathbf{x})\, dx. \tag{41}$$

It can be shown, [Proposition 3.1, [2]], that the representating map is almost everywhere equal to the gradient of a convex Lipschitz continuous scalar function which has an, almost everywhere, invertible gradient. Substituting the measure μ^* back into (39), we obtain an alternative formulation of the Wasserstein distance more closely related to the Monge-Kantorovich problem, ([10], [4], [1], [5], [3], [6], [14]). Namely, if the marginals have the same mass, i. e. , $\int_\Omega s_0(\mathbf{x})\, dx = \int_\Omega s_1(\mathbf{x})\, dx$, it follows from the above discussion, that there exist ϕ^*, convex and Lipschitz continuous, such that $\nabla\phi^* \in \mathcal{L}(s_0, s_1)$ and invertible, solving

$$\inf\left\{ \int_\Omega s_0(\mathbf{x})\, |M(\mathbf{x}) - \mathbf{x}|^2\, dx \,\middle|\, M \in \mathcal{L}(s_0, s_1), M \text{ invertible} \right\}, \qquad \text{where}$$

$$\begin{aligned} \mathcal{I}(s_0, s_1) &\stackrel{\text{def}}{=} \{M : \{s_0 > 0\} \subseteq \Omega \mapsto \{s_1 > 0\} \subseteq \Omega \,|\, M \text{ measurable}\},\\ \mathcal{L}(s_0, s_1) &\stackrel{\text{def}}{=} \Big\{M \in \mathcal{I}(s_0, s_1) \,\Big|\, \int_{M^{-1}(\omega)} s_0(\mathbf{x})\, dx \\ &\qquad = \int_\omega s_1(\mathbf{x})\, dx, \text{ for all Borel subsets } \omega \text{ of } \Omega\Big\}. \end{aligned} \tag{42}$$

Thus

$$\mathrm{dist}_W(s_0, s_1)^2 = \int_\Omega \chi^0(\mathbf{x})\, |\nabla\phi^*(\mathbf{x}) - \mathbf{x}|^2\, dx. \tag{43}$$

We note that the requirement of the transport of s_0 onto s_1 by a volume preserving map can be written in a form

$$\int_\Omega s_1(\mathbf{x})\varphi(\mathbf{x})\, dx = \int_\Omega s_0(\mathbf{x})\varphi(M(\mathbf{x}))\, dx, \qquad \text{for all } \varphi \in C^0(\Omega), \tag{44}$$

which is to say that,

$$s_1(M(\mathbf{x})) = s_0(\mathbf{x}), \qquad \mathbf{x} \in \Omega, \tag{45}$$

for $M \in \mathcal{M}$. □

6 Acknowledgement

The authors would like to thank Professor Olivier Besson for his hospitality and support while we worked on the presented material in the Institut de

Mathématique, Université de Neuchâtel. We are also pleased to acknowledge the generous support for this work provided by the Alcan Technology Center. Petr Klouček was supported in part by the grant NSF DMS-0107539, by the grant from TRW Foundation and by the grant NASA SECTP. Michel Romerio was supported in part by Contract No. 74837-001-0349 from the Regents of University of California (Los Alamos National Laboratory) to William Marsh Rice University. Jennifer Wightman was supported in part by the grant NASA SECTP.

References

1. J.-D. Benamou and Y. Brenier. *A numerical method for the optimal time-continuous mass transport problem and related problems*, pages 1–11. Contemporary Mathematics: Monge-Ampére Equation:Application to Geometry and Optimization. American Mathematical Society, first edition, 1999. L. A. Caffarelli and M. Milman, eds.
2. Y. Brenier. Polar factorization and monotone rearrangement of vector-valued functions. *Comm. Pure Appl. Math.*, 64:375–417, 1991.
3. L. A. Caffarelli. Boundary regularity of maps with convex potentials. *Ann. of Math.*, 2(3):453–496, 1996.
4. B. Dacorogna and J. Moser. On a partial differential equation involving the jacobian determinant. *Ann. Inst. H. Poincaré Anal. Non Linéare*, 7(1):1–26, 1990.
5. L. C. Evans. Partial differential equations and monge-kantorovich mass transfer. *Lecture Notes*, pages 286–294, 1998.
6. W. Gangbo and R. J. McCann. The geometry of optimal transport. *Acta Math.*, (2):113–161, 1996.
7. P. Klouček, M. V. Romerio, and J. L. Wightman. The variational formulation of tracking free surfaces in liquids. *SIAM, J. Appl. Math.*, 2003. submitted.
8. C. Marchioro and M. Pulvirenti. *Mathematical Theory of Incompressible Nonviscous Fluids.* Applied Mathematical Sciences. Springer-Verlag, 1994.
9. J.-M. Morel and S. Solomini. *Variational Methods in Image Segmentation*, volume 14 of *Progress in Nonlinear Differential Equations and Their Applications.* Birkhäuser, Boston, 1st edition, 1995.
10. J. Moser. On the volume elements on a manifold. *Trans. Amer. Math. Soc.*, 120:286–294, 1965.
11. S. Osher and R. Fedkiw. *Level Set Methods and Dynamic Implicit Surfaces*, volume 153 of *Applied mathematical Sciences.* Springer Verlag, New York, 1 edition, 2003.
12. F. Otto. Evolution of microstructure in unstable porous media flow: a relaxation approach. *Comm. Pure Appl. Math.*, 2002. to appear.
13. S. T. Rachev. *Probability metrics and the stability of stochastical models.* Wiley, New York, 1 edition, 1991.
14. S. T. Rachev and L. Rüschendorf. *Mass transportation problem*, volume I, II of *Probability and its Application.* Springer-Verlag, New York, 1998.

Discrete Maximum Principles in Finite Element Modelling

Sergey Korotov[1] and Michal Křížek[2]

[1] Department of Mathematical Information Technology, P. O. Box 35, University of Jyväskylä, FIN–400 14, Finland, *korotov@mit.jyu.fi*

[2] Mathematical Institute, Academy of Sciences, Žitná 25, CZ–115 67 Prague 1, Czech Republic, *krizek@math.cas.cz*

Dedicated to Prof. Miloslav Feistauer on his 60th birthday.

Summary. Nonobtuse tetrahedral partitions and linear finite elements guarantee the validity of a discrete analogue of the maximum principle for a wide class of parabolic and elliptic problems in the three-dimensional space. In this paper we propose global and local refinement techniques which produce nonobtuse face-to-face tetrahedral partitions of a polyhedral domain.

1 Introduction

The maximum principle represents one of the most characteristic features of solutions of second order elliptic or parabolic problems. In this paper we survey some of our recent results concerning the discrete maximum principle for three-dimensional nonlinear elliptic boundary-value problems solved by linear tetrahedral finite elements. In particular, we introduce sufficient geometric conditions on tetrahedral meshes that guarantee the validity of the discrete maximum principle, and present algorithms for global and local refinements of meshes preserving these geometrical properties.

Linear tetrahedral finite elements are commonly used for solving second order boundary value problems, since they do not require a high regularity of the solution. The structure and properties of the associated stiffness matrix essentially depend on the dihedral angles between faces of tetrahedral elements. To see this fact, let us consider an arbitrary tetrahedron $ABCD$. Let p and q be two linear functions such that

$$p(A) = 1,\ p(B) = p(C) = p(D) = 0,$$

$$q(B) = 1,\ q(A) = q(C) = q(D) = 0.$$

Then a straightforward calculation leads to the following formula (see [11, p. 63] for the proof)

$$\nabla p \cdot \nabla q = -\frac{\mathrm{meas}_2 ACD\,\mathrm{meas}_2 BCD}{9\,(\mathrm{meas}_3 ABCD)^2}\cos\alpha, \tag{1.1}$$

where α is the angle between the faces ACD and BCD (see Fig. 1) and the symbol meas_d stands for d-dimensional measure. Scalar product (1.1) is, therefore, independent of all the other 5 dihedral angles of the tetrahedron $ABCD$. If $\alpha > \pi/2$ then the scalar product (1.1) is obviously positive. Hence, each obtuse dihedral angle of the tetrahedron $ABCD$ gives a positive contribution to the corresponding off-diagonal entry of the element stiffness matrix, when solving a boundary value problem with the Laplace operator by the finite element method.

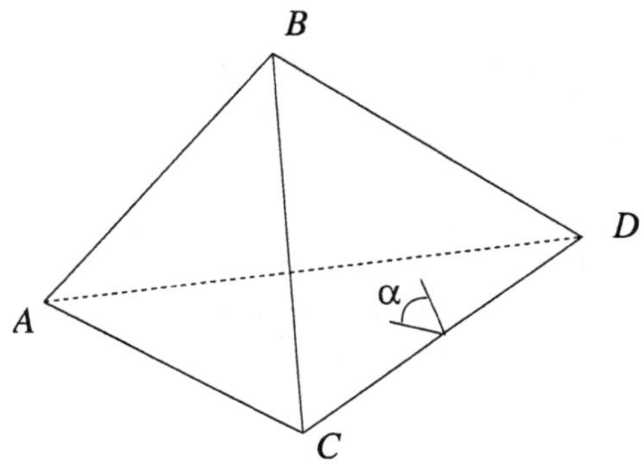

Fig. 1. A tetrahedron whose two faces include the angle α

Note that the same is also true (see [11]) for a wider class of nonlinear elliptic problems of the form

$$-\nabla \cdot (\lambda(x, u, \nabla u)\nabla u) = f(x) \quad \text{in} \quad \Omega, \tag{1.2}$$

$$u = 0 \qquad \text{on} \quad \partial\Omega, \tag{1.3}$$

where λ is a positive smooth function and Ω is a bounded polyhedral domain with Lipschitz boundary $\partial\Omega$. Equation (1.2) describes, for instance, a stationary nonlinear heat conduction or magnetic potential in ferromagnetic media (cf. [13]).

According to [5, p. 206], problem (1.2)–(1.3) satisfies the *maximum principle*, i.e., if $f \leq 0$ then the maximum of u over $\overline{\Omega}$ is attained on the boundary $\partial\Omega$ (see also [14]). It is thus natural to look for a class of finite elements such that the same implication is satisfied. Note that, e.g., for bilinear rectangular elements the discrete maximum principle can be violated (see [1, p. 254]). The same is true also for trilinear block elements (a simple example is given in [12, p. 562]). However, if we decompose these elements into nonobtuse triangles or tetrahedra keeping the number of degrees of freedom, the discrete maximum principle will hold.

2 Nonobtuse tetrahedralizations

A tetrahedron is said to be *nonobtuse* if all six dihedral angles between its faces are less than or equal to $\pi/2$. By a *tetrahedralization* we shall mean a face-to-face partition of $\overline{\Omega}$ into tetrahedra in the standard sense (see [3]). A tetrahedralization is said to be *nonobtuse* if it contains only nonobtuse tetrahedra.

By [3, p. 150], linear triangular nonobtuse elements guarantee the validity of the discrete maximum principle in the plane. This result is generalized into three-dimensional space in [11], namely, linear tetrahedral elements applied to problem (1.2)–(1.3) on nonobtuse partitions yield irreducibly diagonally dominant stiffness matrices (whose all off-diagonal entries are nonpositive). It is well known (see [15, p. 85]) that such matrices are monotone. Nonobtuse tetrahedral partitions thus guarantee the validity of the discrete maximum principle for problem (1.2)–(1.3) solved by linear elements, i.e., we have $u_h \leq 0$ provided $f \leq 0$, where u_h is the continuous and piecewise linear Galerkin approximation of the solution of (1.2)–(1.3). In other words, u_h attains its maximum on the boundary $\partial\Omega$, if the homogeneous Dirichlet boundary conditions are considered. Note that nonobtuse tetrahedralizations enable us to prove the L^∞-convergence of finite element approximations to the weak solution (see, e.g., [4]).

According to [7], an arbitrary polyhedron can be decomposed into nonobtuse tetrahedra. In order to improve the discretization error a given partition is refined locally or globally. That is why, the issue of preserving the nonobtusity appears during the refinement process.

In [6] we give a global refinement procedure yielding nonobtuse tetrahedra over the whole domain. This procedure is briefly described in Section 3. However, such a technique requires a large amount of computer memory to store the associated stiffness matrix. Therefore, in Section 4 we introduce a local refinement procedure yielding nonobtuse partitions that refine only near a given vertex, where a singularity of the exact solution may appear (see, e.g., [2], [8], [9]).

3 Global refinement techniques

Definition 3.1. A tetrahedron is said to be a *path tetrahedron* if it has three mutually perpendicular edges which do not pass through the same vertex.

The reason for the name of the above tetrahedron is that its three perpendicular edges form a "path" (see Fig. 2).

Proposition 3.2. *Any path tetrahedron is nonobtuse.*

For the proof see [6, p. 728–729].

In Fig. 2 we observe a typical shape of a path tetrahedron (all its right angles, solid and dihedral, are indicated there).

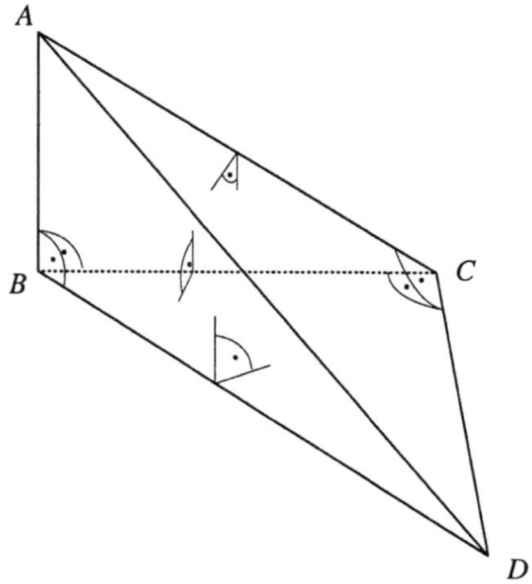

Fig. 2. A path tetrahedron

Theorem 3.3. *Let T be an arbitrary tetrahedron such that its circumcentre belongs to T, and let all faces of T be nonobtuse triangles. Then there exists a family of tetrahedral partitions of T containing only path tetrahedra.*

This theorem is proved in [6]. Note that each path tetrahedron satisfies the assumptions of Theorem 3.3, since its faces are right triangles and its circumcentre is the midpoint of the longest edge. In Fig. 3 we see a partition of a tetrahedron T, that satisfies the assumptions of Theorem 3.3, into path tetrahedra. They are defined in the following way. First we divide each face F of T into 6 or 4 right subtriangles by connecting the circumcentre of F with 3 vertices and 3 midpoints of sides of F. The common vertex of these subtriangles is the circumcentre of F. This kind of plane refinement we call *2d yellow* (see [6]). Denoting the circumcentre of T by G, we can define the path subtetrahedra as the convex hull of G and particular right subtriangles on the surface of T (compare with Fig. 3). We call such a kind of three-dimensional refinement *3d yellow.*

The advantage of the above approach is that a common face F of any two adjacent tetrahedra (satisfying the assumptions of Theorem 3.3) in a given tetrahedralization of a polyhedral domain is divided in a unique way. This enables us to develop global refinement techniques yielding only nonobtuse subtetrahedra (see [6] for details).

4 Local refinement techniques

The main idea of generating local refinement technique producing only nonobtuse partitions is exposed in the following theorem.

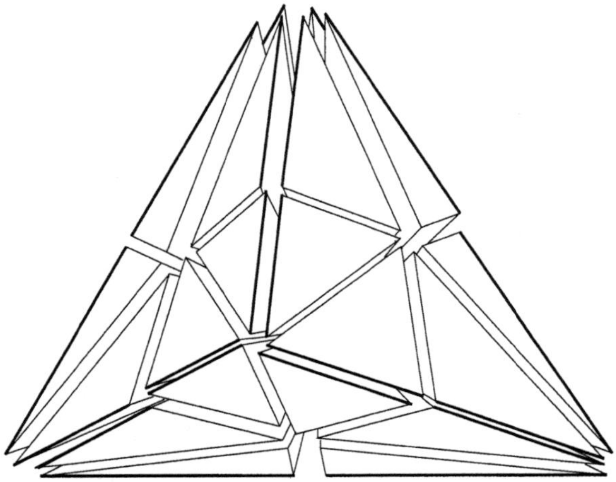

Fig. 3. Nonobtuse partition of a tetrahedron into path subtetrahedra

Theorem 4.1. *Let ABCD be a path tetrahedron whose edges AB, BC, and CD are mutually perpendicular. Then there exists an infinite family of nonobtuse tetrahedralizations of ABCD into path tetrahedra that locally refine ABCD in a neighbourhood of the vertex A.*

For a detailed constructive proof see [2] or [9]. Its main idea is sketched on Fig. 4. Using several sophisticated orthogonal projections, we first subdivide the tetrahedron $ABCD$ into five nonobtuse tetrahedra. Then we show that the path tetrahedron $ATSQ$ from Fig. 4 is similar to the original tetrahedron $ABCD$. The subtetrahedron $ATSQ$ can be now decomposed into 5 subtetrahedra in a similar way as $ABCD$, and thus we can get further refinement near the point A. In this manner, we obtain recursively the required infinite family of nonobtuse tetrahedralizations.

An algorithm for a local refinement of a cube producing only nonobtuse tetrahedra is presented in [8]. If several cubes meet at one point, then we can apply this algorithm to each of them so that the whole partition remains face-to-face. For instance, in Fig. 5 we see such a local refinement a polyhedral domain $\Omega = (-1,1)^3 \setminus [0,1)^3$ which presents the union of 7 cubes. Each cube is first divided in a standard way into 6 path tetrahedra having a common vertex in the reentrant corner of Ω. For each of the $7 \times 6 = 42$ tetrahedra we apply the algorithm given by Theorem 4.1 such that the partition of each path tetrahedron is just the mirror image of the partition of any adjacent tetrahedron having a common face. Note that the concave (reentrant) corner in Fig. 5 is called the *Fichera corner* or the *Fichera vertex*.

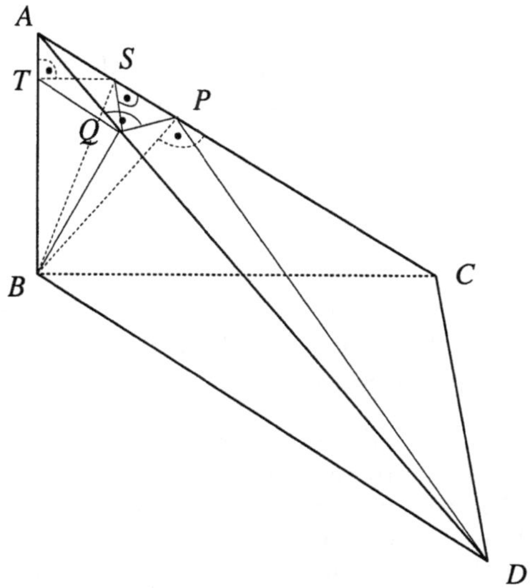

Fig. 4. Partition of a path tetrahedron into 5 path subtetrahedra

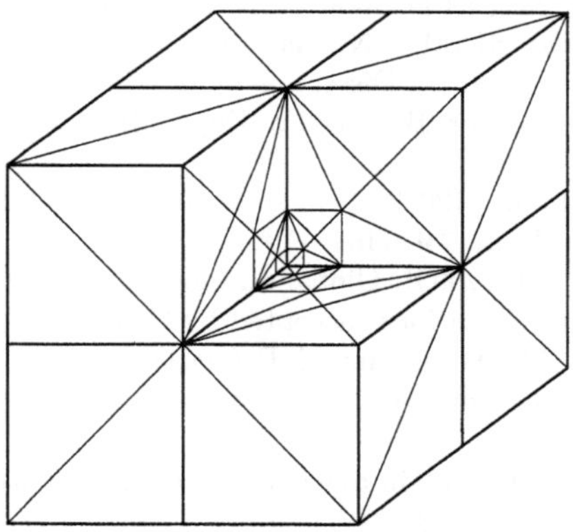

Fig. 5. Local refinement technique producing nonobtuse tetrahedra near the Fichera corner

5 Concluding remarks

Nonobtuseness of all dihedral angles of all tetrahedra in the partition represents only a sufficient condition to guarantee the discrete maximum principle for linear elements. In [10] we present a weakened condition for the shape of tetrahedra, which enables us to use tetrahedra, some of whose dihedral angles are slightly bigger than $\pi/2$. This condition is also only sufficient to get a monotone stiffness matrix and thus also the validity of the discrete maximum principle.

Acknowledgement
This paper was supported by Grant InBCT of TEKES, Finland and Grant No. A 1019201 of the Academy of Sciences of the Czech Republic.

References

1. Axelsson, O., Barker, V.A. (1984): Finite Element Solution of Boundary Value Problems. Theory and Computation. Academic Press, New York
2. Beilina, L., Korotov, S., Křížek, M. (2003): Application of the local nonobtuse tetrahedral refinement techniques near Fichera-like corners. Preprint 2003-02, Chalmers Finite Element Center, Göteborg, 1–16
3. Ciarlet, P.G. (1991): Basic error estimates for elliptic problems. In: Ciarlet, P.G., Lions, J.L. (eds) Handbook of Numer. Anal., vol. II. North-Holland, Amsterdam
4. Feistauer, M., Felcman, J., Rokyta, M., Vlášek, Z. (1992): Finite-element solution of flow problems with trailing conditions. J. Comput. Appl. Math., **44**, 131–165
5. Gilbarg, D., Trudinger, N.S. (1977): Elliptic Partial Differential Equations of Second Order. Springer-Verlag, Berlin
6. Korotov, S., Křížek, M. (2001): Acute type refinements of tetrahedral partitions of polyhedral domains. SIAM J. Numer. Anal. **39**, 724–733
7. Korotov, S., Křížek, M. (2002): Dissection of an arbitrary polyhedron into nonobtuse tetrahedra. Preprint nr. B 3/2002 of Dept. Math. Inform. Technol., Univ. of Jyväskylä, 1–6
8. Korotov, S., Křížek, M. (2003): Local nonobtuse tetrahedral refinements of a cube. Appl. Math. Lett., **16**, 1101–1104
9. Korotov, S., Křížek, M. (2003): Global and local refinement techniques yielding nonobtuse tetrahedral partitions (accepted by Comput. Math. Appl.), 1–10
10. Korotov, S., Křížek, M., Neittaanmäki, P. (2001): Weakened acute type condition for tetrahedral triangulations and the discrete maximum principle. Math. Comp. **70**, 107–119
11. Křížek, M., Lin Qun (1995): On diagonal dominance of stiffness matrices in 3D. East-West J. Numer. Math. **3**, 59–69
12. Křížek, M., Liu, L. (2003): On the maximum and comparison principles for a steady-state nonlinear heat conduction problem. Z. Angew. Math. Mech. **83**, 559–563
13. Křížek, M., Neittaanmäki, P. (1990): Finite Element Approximation of Variational Problems and Applications. Pitman Monographs and Surveys in Pure and Applied Mathematics vol. 50, Longman Scientific & Technical
14. Protter, M.H., Weinberger, H.F. (1967): Maximum Principles in Differential Equations. Prentice-Hall, New Jersey
15. Varga, R.S. (1962): Matrix Iterative Analysis. Prentice-Hall, New Jersey

A Posteriori Error Estimation in Terms of Linear Functionals for Boundary Value Problems of Elliptic Type

Sergey Korotov[1], Pekka Neittaanmäki[1] and Sergey Repin[2]

[1] Department of Mathematical Information Technology
University of Jyväskylä
P.O. Box 35, FIN–40014 Jyväskylä, Finland
korotov, pn@mit.jyu.fi

[2] V.A. Steklov Institute of Mathematics in St.-Petersburg
St.-Petersburg, 191011, Fontanka 27, Russia
repin@pdmi.ras.ru

Summary. The paper deals with a posteriori error estimation in terms of special problem-oriented quantities, represented as a linear functionals that control the behavior of a solution in certain subdomains, along some lines, or at especially interesting points. The method of estimating such quantities is based on the analysis of the adjoint boundary-value problems, whose right-hand sides are formed by the considered linear functionals. On this way, we propose a new effective approach based on two principles: (a) the original and adjoint problems are solved on non-coinciding meshes, and (b) the term presenting the product of gradients of errors of the primal and adjoint problems is estimated by using the "gradient averaging" technique. The model problem of elliptic type is analysed and the results of numerical tests are presented.

1 Introduction

A posteriori error estimates play an important role in modern numerical analysis. For finite element methods, they are usually obtained either by estimating a weak norm of the residual (see, e.g., [1], [2], [3], [4], [5], [12]) or by using special post-processing procedures (see, e.g., [12], [13]). For Galerkin approximations of linear elliptic problems, they estimate the error in the global (energy) norm and also provide an error indicators that are further used in various mesh adaptive procedures. Global error estimates give a general presentation on the quality of an approximate solution and a stopping criteria. However, for the engineering purposes, such an information is often not sufficient. In many cases, analysts are mainly interested not in the value of the total error, but in errors over certain subdomains, lines, or at special points. A possible way of estimating such errors is to introduce a linear functional ℓ associated with a "quantity of interest" and to obtain an estimate for the value of $< \ell, u - v >$, where u is the exact solution and v is the approximate one.

Known methods (see, e.g., [1], [6], [11]) find estimates of $< \ell, u - u_h >$ for a Galerkin approximation u_h by employing an additional (adjoint) problem, whose right-hand side is formed by the functional ℓ . If the Galerkin approximation of the adjoint problem is computed on the same mesh as u_h, then the functional $< \ell, u - u_h >$ is expressed via a certain integral functional, which can be estimated by using, e.g., "equilibrated residual method" (see, e.g., [1], [11]).

In the present work, we propose a new way of estimation of "quantities of interest". It is based on two principles: (a) the original and adjoint problems are solved on non-coinciding meshes, and (b) the term presenting the product of errors arising in the primal and adjoint problems is estimated by the gradient recovery technique widely used in various applied problems (see [7], [10], [12], [13]). This differs our approach from others, where it is usually assumed that the Galerkin approximations of the primal and adjoint problems are computed in the same finite dimensional subspaces.

The effectivity of the method suggested in this paper, strongly increases when one is interested not in a single solution of the primal problem for a concrete data, but analyzes a series of approximate solutions for a certain set of boundary conditions and various right-hand sides (such a situation is typical in the engineering design when it is necessary to model the behavior of a construction for various working regimes). In this case, the adjoint problem must be solved *only once* for each "quantity of interest", and its solution can be further used in testing the accuracy of approximate solutions of various primal problems.

2 General scheme

Introduce Hilbert spaces H and Y with scalar products and norms

$$\|\xi\|_H = (\xi, \xi)_H^{1/2}, \quad \|y\|_Y = (y, y)_Y^{1/2},$$

and the Banach space V, which is continuously embedded into H, with the norms denoted by $\| \cdot \|_V$. Let $\Lambda \in \mathcal{L}(V, Y)$, $\mathcal{A} \in \mathcal{L}(Y, Y)$, and

$$c_1 \|y\|_Y^2 \leq (\mathcal{A} y, y)_Y \leq c_2 \|y\|_Y^2 \quad \forall y \in Y, \qquad c_3 \|w\|_V \leq \|\Lambda w\|_Y \quad \forall w \in V_0, \tag{2.1}$$

where V_0 is a subspace of V and c_1, c_2, c_3 are positive constants. Given $f \in V_0^*$, consider the following problem:

<u>**Primal Problem ($\mathcal{P}$)**</u>: Find $u \in V_0$, such that

$$(\mathcal{A} \Lambda u, \Lambda w)_Y \; = < f, w > \quad \forall w \in V_0, \tag{2.2}$$

where $< \cdot, \cdot >$ denotes the duality pairing of the spaces V_0 and V_0^*.

Let ℓ be another element of V_0^*, which forms the "quantity of interest" $< \ell, u - \bar{u} >$, for an arbitrary element $\bar{u} \in V_0$ viewed as an approximation of u. In order to estimate the above defined quantity, we introduce another problem.

Adjoint Problem ($\mathcal{P}_a$): Find $v \in V_0$, such that

$$(\mathcal{A}^* \Lambda v, \Lambda w)_Y \;= <\ell, w> \quad \forall w \in V_0, \tag{2.3}$$

where $\mathcal{A}^*$ is the operator adjoint to $\mathcal{A}$, i.e., $(\mathcal{A}y, z)_Y = (y, \mathcal{A}^* z)_Y \quad \forall y, z \in Y$.

In the friequently encountered cases of the selfadjoint operators, the left–hand sides of (2.2) and (2.3) coincide and both the problems are associated with a functional of the type

$$J(w) \;=\; \frac{1}{2}(\mathcal{A}\Lambda w, \Lambda w)_Y + \; <\mu, w>,$$

which is known to have a unique minimizer on V_0 for any $\mu \in V_0^*$.

Proposition 1. *Let u and v be solutions of problems ($\mathcal{P}$) and ($\mathcal{P}_a$), respectively. Then for any $\bar{u}, \bar{v} \in V_0$*

$$<\ell, u - \bar{u} >=: E(\bar{u}, \bar{v}) = E_0(\bar{u}, \bar{v}) + E_1(\bar{u}, \bar{v}), \tag{2.4}$$

where

$$E_0(\bar{u}, \bar{v}) \;= <f, \bar{v}> - (\mathcal{A}\Lambda\bar{u}, \Lambda\bar{v})_Y, \tag{2.5}$$

and

$$E_1(\bar{u}, \bar{v}) \;=\; (\mathcal{A}\Lambda(u - \bar{u}), \Lambda(v - \bar{v}))_Y. \tag{2.6}$$

The proof can be found in [8].

The term $E_0(\bar{u}, \bar{v})$ is explicitly computable, whereas the term $E_1(\bar{u}, \bar{v})$ contains unknown solutions of Problems ($\mathcal{P}$) and ($\mathcal{P}_a$). Evidently, the term $E_0(\bar{u}, \bar{v})$ dominates if $\bar{v}$ is "sufficiently close" to the exact solution v of Problem ($\mathcal{P}_a$). Really, if $\bar{v} \to v$ in V, then $\bar{v} \to v$ in H, and $\Lambda(v - \bar{v}) \to 0$ in Y, so that

$$E_1(\bar{u}, \bar{v}) \to 0, \quad \text{and} \quad \mathrm{E_0(\bar{u}, \bar{v})} \;\to <\ell, \mathrm{u} - \bar{\mathrm{u}}>,$$

i.e., $E_0(\bar{u}, \bar{v})$ contains the major part of the quantity of interest.

Let V_h and V_τ be two finite-dimensional subspaces of V_0, and let $\bar{u} = u_h$, $\bar{v} = v_\tau$, where u_h and v_τ are solutions of the problems

$$(\mathcal{A}\Lambda u_h, \Lambda w_h)_Y \;= <f, w_h> \quad \forall w_h \in V_h, \tag{2.7}$$

$$(\mathcal{A}^* \Lambda v_\tau, \Lambda w_\tau)_Y \;= <\ell, w_\tau> \quad \forall w_\tau \in V_\tau. \tag{2.8}$$

In a particular case of $V_h \equiv V_\tau$, the term $E_0(u_h, v_\tau) = 0$ due to the orthogonality condition in problem (2.7). Thus, if meshes in problems (2.7) and (2.8) coincide, then the estimate has only one term containing the product of the (unknown) energy errors. On the contrary, an usage of non-coinciding meshes leads to another estimate that has two terms.

Finding a sharp approximation of the adjoint problem may require high computational costs. A more economical way is to use an approximate solution of the adjoint problem having approximately the same quality as the approximate solution of the primal one and recover unknown functions Λu and Λv by some post-processing techniques. Let G_h and G_τ be certain averaging operators defined on V_h and V_τ, respectively. We replace $E(u_h, v_\tau)$ by the directly computable functional

$$\widetilde{E}(u_h, v_\tau) := E_0(u_h, v_\tau) + \widetilde{E}_1(u_h, v_\tau), \tag{2.9}$$

where

$$\widetilde{E}_1(u_h, v_\tau) = (\mathcal{A}(\Lambda u_h - G_h(\Lambda u_h)), (\Lambda v_\tau - G_\tau(\Lambda v_\tau))). \tag{2.10}$$

If the operators G_h and G_τ provide proper recovery of Λu and Λv, then it is natural to await that the difference between $E_1(u_h, v_\tau)$ and $\widetilde{E}_1(u_h, v_\tau)$ is presented by the higher order terms and, thus, the latter quantity can be succesfully used instead of E_1. This observation is justified theoretically and numerically in what follows for a model problem of elliptic type.

3 Model elliptic type problem

3.1 Formulation of the problem

We define the operator Λ as $\nabla := (\frac{\partial}{\partial x_1}, \frac{\partial}{\partial x_2})$ and set

$$Y = L_2(\Omega, \mathbf{R}^2),\ H = L_2(\Omega),\ V = W_2^1(\Omega),\ V_0^* = H^{-1}(\Omega).$$

Consider the following problem: find u satisfying the system

$$-div\,(A\,\nabla u) = f \quad in \quad \Omega, \tag{3.1}$$

$$u = 0 \quad on \quad \partial\Omega. \tag{3.2}$$

In the above, Ω is a bounded and connected domain in $\mathbf{R}^2$ with a Lipschitz continuous boundary $\partial\Omega$, the symbol ν denotes the unit outward normal to the boundary, the matrix of coefficients $A = \{a_{ij}(x)\}_{i,j=1}^2$ is symmetric and meets the conditions

$$a_{ij}(x) \in L_\infty(\Omega), \quad A\,(x)\xi \cdot \xi \geq C\,\|\xi\|^2 \quad \forall \xi \in \mathbf{R}^2 \quad \forall x \in \overline{\Omega}, \tag{3.3}$$

where the dot denotes the scalar product in $\mathbf{R}^2$. Also, it is assumed that

$$f \in L_2(\Omega), \tag{3.4}$$

Hereafter, various constants, independent of h and τ, are denoted by one letter C.

Now, Problem $(\mathcal{P})$ consists of finding $u \in V_0$ such that

$$\int_\Omega A\,\nabla u \cdot \nabla w\; dx = \int_\Omega f w\; dx \quad \forall w \in V_0, \tag{3.5}$$

where $V_0 = \overset{\circ}{W}{}^1_2(\Omega)$. Let $\bar{u} \in V_0$ be an approximation of u. Assume that we are interested in the value of

$$< \ell, u - \bar{u} >= \int_\Omega \varphi(u - \bar{u})\, dx, \tag{3.6}$$

where supp $\varphi \subset \omega \subseteq \Omega$. Estimates of such quantities for one (or several) a priori given functions φ give an information about the behaviour of $u - \bar{u}$ in ω.

In this case, Adjoint Problem $(\mathcal{P}_a)$ consists of finding $v \in V_0$ such that

$$\int_\Omega A\nabla v \cdot \nabla w\, dx =< \ell, w > \quad \forall w \in V_0. \tag{3.7}$$

Suppose that $\bar{v} \in V_0$ is an approximation of v. Then, for problems (3.5) and (3.7), the relation (2.4) reads as follows:

$$< \ell, u - \bar{u} > = E_0(\bar{u}, \bar{v}) + E_1(\bar{u}, \bar{v}), \tag{3.8}$$

where

$$E_0(\bar{u}, \bar{v}) = \int_\Omega f\bar{v} dx - \int_\Omega A\nabla\bar{v} \cdot \nabla\bar{u}\, dx, \tag{3.9}$$

$$E_1(\bar{u}, \bar{v}) = \int_\Omega A\nabla(v - \nabla v) \cdot \nabla(u - \nabla u)\, dx. \tag{3.10}$$

Let V_h and V_τ be two finite-dimensional subspaces in V_0, constructed by the Courant type finite element discretization. The respective Galerkin approximations u_h and v_τ are formed by piecewise-affine continuous functions and defined on the domains

$$\Omega_h = \bigcup_{T_h^i \in \mathcal{T}_h} T_h^i, \qquad \Omega_\tau = \bigcup_{T_\tau^j \in \mathcal{T}_\tau} T_\tau^j \tag{3.11}$$

where the respective triangulations are denoted by $\mathcal{T}_h$ and $\mathcal{T}_\tau$, and their elements are denoted by T_h^i and T_τ^j, respectively. Consider the corresponding finite dimensional problems (we assume for a moment that $\Omega_h = \Omega_\tau = \Omega$):

Problem $(\mathcal{P}^h)$: Find $u_h \in V_h$ such that

$$\int_\Omega A\nabla u_h \cdot \nabla w_h\, dx = \int_\Omega f w_h\, dx \quad \forall w_h \in V_h, \tag{3.12}$$

Problem $(\mathcal{P}_a^\tau)$: Find $v_\tau \in V_\tau$ such that

$$\int_\Omega A\nabla v_\tau \cdot \nabla w_\tau\, dx =< \ell, w_\tau > \quad \forall w_\tau \in V_\tau. \tag{3.13}$$

Set $\bar{u} = u_h$ and $\bar{v} = v_\tau$. Then

$$< \ell, u - u_h >=: E(u_h, v_\tau) = E_0(u_h, v_\tau) + E_1(u_h, v_\tau).$$

On $\mathcal{T}_h$, we define the gradient averaging operator $G_h : L_\infty(\Omega_h, \mathbf{R}^2) \to W_2^1(\Omega_h, \mathbf{R}^2)$ as the operator that defines a vector-valued piecewise linear function by setting each nodal values as the mean values of ∇u_h on all elements incident with the corresponding nodal point. The averaging operator $G_\tau : L_\infty(\Omega_\tau, \mathbf{R}^2) \to W_2^1(\Omega_\tau, \mathbf{R}^2)$ is defined in the same manner. Nowadays, averaging operators of this type are widely used in the mathematical modelling (see, e.g. [1], [4], [7], [10], [12]).

Averaging near the boundary ($\Omega_h \neq \Omega_\tau \neq \Omega$) requires a more complicated analysis. However, in the literature this question was investigated (see, e.g. [7], where concrete forms of the averaging operators near the boundary are presented). In our further considerations, we use the results of the above paper and, therefore, impose the same conditions on the problem data and the structure of meshes. Namely, we assume that $\partial\Omega$ belongs to the class C^3, the coefficients a_{ij} are smooth and the sets Ω_h and Ω_τ, defined in (3.11), are such that

$$\Omega_h \subset \Omega_\tau \subset \Omega, \quad \mathrm{dist}\,\{\Omega_h, \partial\Omega\} \leq \mathrm{c}h^2, \quad \mathrm{dist}\,\{\Omega_\tau, \partial\Omega\} \leq \mathrm{c}\tau^2. \tag{3.14}$$

Further, we assume that functions u_h and v_τ as well as the averaged gradients $G_h(\nabla u_h)$ and $G_\tau(\nabla v_\tau)$ are extended by zero on $\Omega_{0h} := \Omega \setminus \Omega_h$ and $\Omega_{0\tau} := \Omega \setminus \Omega_\tau$.

For any pair $(w_h, w_\tau) \in V_h \times V_\tau$, we define the following functional

$$\widetilde{E}(w_h, w_\tau) := E_0(w_h, w_\tau) + \widetilde{E}_1(w_h, w_\tau), \tag{3.15}$$

where

$$\widetilde{E}_1(w_h, w_\tau) = \int_{\Omega_h} A(\nabla w_h - G_h(\nabla w_h)) \cdot (\nabla w_\tau - G_\tau(\nabla w_\tau))dx. \tag{3.16}$$

The functional $\widetilde{E}(u_h, v_\tau)$ is directly computable once the approximations u_h and v_τ are defined. Our aim is to show that $\widetilde{E}(u_h, v_\tau)$ is a simple and effective estimator of the quality $< \ell, u - u_h >$.

We can easily show that

$$|E_0(u_h, v_\tau)| \leq C\, h|u|_{2,2,\Omega}, \tag{3.17}$$

where $|u|_{2,2,\Omega}$ denotes the L_2-norm of the second derivatives of u, which exist due to the conditions imposed on Problem ($\mathcal{P}$), see below.

The estimate (3.17) is sharp, in the sense that it gives a correct asymptotical order of the term E_0, whereas the terms $E_1(u_h, v_\tau)$ and $\widetilde{E}_1(u_h, v_\tau)$ are asymptotically smaller ones. Really,

$$|E_1(u_h, v_\tau)| \leq C\, h\tau|u|_{2,2,\Omega}|v|_{2,2,\Omega}, \tag{3.18}$$

and

$$|\widetilde{E}_1(u_h, v_\tau)| = | \int\limits_{\Omega_h} A(\nabla u_h - G_h(\nabla u_h)) \cdot (\nabla v_\tau - G_\tau(\nabla v_\tau))\, dx| \leq$$

$$\leq C(\| \nabla u_h - \nabla u \|_{2,\Omega} + \| \nabla u - G_h(\nabla u_h) \|_{2,\Omega_h}) \times$$
$$\times (\| \nabla v_\tau - \nabla v \|_{2,\Omega} + \| \nabla v - G_\tau(\nabla v_\tau) \|_{2,\Omega_\tau}).$$

Therefore, if the superconvergence of the averaged gradients takes place, then the first terms in the round brackets dominate and, therefore, $\widetilde{E}_1(u_h, v_\tau)$ has the same asymptotic order $h\tau$.

From the above, we see that asymptotically $E_0(u_h, v_\tau)$ contains the main part of the quantity $< \ell, u - u_h >$, except the case $V_h \equiv V_\tau$, when this term vanishes.

It remains to compare $E_1(u_h, v_\tau)$ and $\widetilde{E}_1(u_h, v_\tau)$. We do this under additional assumptions used in [7]. Namely, it is assumed that f and ℓ belong to $W_2^2(\Omega)$, so that $u, v \in W_2^3(\Omega)$ and that triangulations $\mathcal{T}_h$ and $\mathcal{T}_\tau$ are composed of uniform elements, for which the superconvergence estimates

$$\| \nabla u - G_h(\nabla u_h) \|_{2,\Omega_h} \leq C\, h^{3/2}(\| u \|_{3,2,\Omega} + \| f \|_{2,2,\Omega}), \tag{3.19}$$

$$\| \nabla v - G_\tau(\nabla v_\tau) \|_{2,\Omega_\tau} \leq C\, \tau^{3/2}(\| v \|_{3,2,\Omega} + \| \ell \|_{2,2,\Omega}) \tag{3.20}$$

hold. Also, we use the superconvergence of $\Pi_h u$ and $\Pi_\tau v$, where Π_h and Π_τ are the nodal interpolation operators, that have been established (see [7]) under the same additional assumptions. In our case, they have the form

$$\| \nabla u_h - \nabla(\Pi_h u) \|_{2,\Omega_h} \leq C\, h^{3/2}(\| u \|_{3,2,\Omega} + \| f \|_{2,2,\Omega}), \tag{3.21}$$

$$\| \nabla v_\tau - \nabla(\Pi_\tau v) \|_{2,\Omega_\tau} \leq C\, \tau^{3/2}(\| v \|_{3,2,\Omega} + \| \ell \|_{2,2,\Omega}). \tag{3.22}$$

Proposition 2. *Let $\partial\Omega \in C^3$, $\mathcal{T}_h$ and $\mathcal{T}_\tau$ be two families of triangulations with above described properties, and $u, v \in W_2^3(\Omega)$. Then, for sufficiently small h and τ $(\tau < h)$, we have*

$$|E_1(u_h, v_\tau) - \widetilde{E}_1(u_h, v_\tau)| \leq C\, (h\tau^{\frac{3}{2}} + \tau h^{\frac{3}{2}} + h\tau(h+\tau)^{\frac{1}{2}-\frac{1}{m}} + \tau^2) + \mu(h, \tau), \tag{3.23}$$

where m is any positive integer greater than 2, $\mu(h,\tau)$ contains higher order terms, and the constant C does not depend on h and τ.

The proof can be found in [8].

Let us summarize the above analysis of the behavior of $E_0(u_h, v_\tau)$, $E_1(u_h, v_\tau)$, and $\widetilde{E}_1(u_h, v_\tau)$ with respect to sufficiently small h and τ $(h > \tau)$:

- the explicitly computable term $E_0(u_h, v_\tau)$ is the major one;
- the terms $E_1(u_h, v_\tau)$ and $\widetilde{E}_1(u_h, v_\tau)$ are subsidiary ones;
- the difference between $E_1(u_h, v_\tau)$ and $\widetilde{E}_1(u_h, v_\tau)$ has higher asymptotic order provided that the solutions of the problems $(\mathcal{P})$ and $(\mathcal{P}_a)$ are regular enough to guarantee the superconvergence of the Galerkin approximations.

The above observations suggest the following numerical strategy:

a) Define V_τ taking into an account the nature of the functional ℓ (e.g., by putting extra trial functions in a subdomain associated with it), and calculate v_τ.
b) Define V_h and calculate u_h.
c) Calculate $E_0(u_h, v_\tau)$ directly and use post–processed values of ∇u_h and ∇v_τ to estimate $E_1(u_h, v_\tau)$ replacing gradients by the averaged gradients.

3.2 Numerical experiments

In this subsection, we present numerical results for the model problem (more numerical tests on the subject can be found in [8]).

Test Problem: Find u such that

$$-\frac{\partial^2 u}{\partial x^2} - \frac{\partial^2 u}{\partial y^2} = f(x,y) \quad \text{in} \quad \Omega = (0,1) \times (0,1),$$

$$u = 0 \quad \text{on} \quad \partial\Omega,$$

where the exact solution is the infinitely smooth "hat-function":

$$u(x,y) = \begin{cases} \exp\left(16 - \frac{16}{1-(4x-3)^2}\right) \cdot \exp\left(16 - \frac{16}{1-(4y-3)^2}\right), & \\ \qquad\qquad\qquad \text{if } (x,y) \in (0.5, 1.0) \times (0.5, 1.0), & \\ 0, & \text{otherwise,} \end{cases}$$

As a quantity of interest, we take

$$< \ell, u - u_h >= \int_\omega (u - u_h)\, dx, \text{ where } \omega = (0.125, 0.250) \times (0.125, 0.250).$$

and define the effectivity index $I_{\text{eff}} = \frac{|\tilde{E}(u_h, v_\tau)|}{|< \ell, u - u_h >|}$.

The results of computations are presented in Table 1 below, where the symbol $N \times N$ ($N = 16, 32, 64, 128$) mean that the corresponding test is performed on a triangular mesh of thr square $(0,1) \times (0,1)$ consisting of $2 \times N \times N$ right triangles. The results demonstrate a good performance of the estimator.

Table 1. The results of performance of the estimator $\tilde{E}$.

Primal	Adjoint	E_0	$\tilde{E}_1$	$\tilde{E}$	$< \ell, u - u_h >$	I_{eff}
16×16	32×32	-0.00000501	-0.00000062	-0.00000439	-0.00000495	0.89
16×16	64×64	-0.00000496	-0.00000017	-0.00000479	-0.00000495	0.97
16×16	128×128	-0.00000495	-0.00000005	-0.00000490	-0.00000495	0.99

4 Final comments

1) The computations made in the item a) can be further used for estimation of errors of an approximate solution obtained on another mesh and for different righ-hand side functions f.
2) The approach is valid for another boundary conditions [8], is suitable for estimating local integral norms [8], and can be applied to problems in linear elasticity theory [9].

References

1. Ainsworth, M, Oden, J. T. (2000): A posteriori error estimation in finite element analysis. John Wiley & Sons, Inc.
2. Babuška, I., Rheinbold, W. C. (1978): Error estimates for adaptive finite element computations. SIAM J. Numer. Anal. **15**, 736–754.
3. Babuška, I., Rheinbold, W. C. (1978): A posteriori error estimates for the finite element method. Internat. J. Numer. Methods Engrg. **12**, 1597–1615.
4. Babuška, I., Strouboulis, T. (2001): The Finite Element Method and its Reliability. Oxford University Press Inc., New York.
5. Bangerth, W., Rannacher, R. (2003): Adaptive finite element methods for differential equations. Lectures in Mathematics ETH Zürich. Birkhäuser Verlag, Basel.
6. Becker, R., Rannacher, R. (1996): A feed-back approach to error control in finite element methods: Basic approach and examples. East-West J. Numer. Math. **4**, 237–264.
7. Hlaváček, I., Křížek, M. (1987): On a superconvergent finite element scheme for elliptic systems. Apl. Mat. **32**, 131–154.
8. Korotov, S., Neittaanmäki, P., Repin, S. (2003): A posteriori error estimation of goal-oriented quantities by the superconvergence patch recovery. J. Numer. Math. **11**, 33–59.
9. Korotov, S., Neittaanmäki, P., Repin, S. (2004): A posteriori error estimation in terms of linear functionals for problems in the elasticity theory. Russian J. Numer. Anal. Math. Modelling (to appear).
10. Křížek, M., Neittaanmäki, P. (1984): Superconvergence phenomenon in the finite element method arising from averaging gradients. Numer. Math. **45**, 105–116.
11. Oden, J. T., Prudhomme, S. (2001): Goal-oriented error estimation and adaptivity for the finite element method. Comput. Math. Appl. **41**, 735–756.
12. Verfürth, R. (1996): A review of a posteriori error estimation and adaptive mesh-refinement techniques. Wiley-Teubner.
13. Zienkeiewicz, O. C., Zhu, J. Z. (1987): A simple error estimator and adaptive procedure for practical engineering analysis. Internat. J. Numer. Methods Engrg. **24**, 337–357.

Numerical Solution of Flow in Backward Facing Step

Karel Kozel[1], Petr Louda[2] and Petr Sváček[1]

[1] Department of Technical Mathematics, Karlovo nám. 13, Faculty of Mechanical Engineering, CTU, Prague *kozelk@fsik.cvut.cz*
[2] Institute of Thermomechanics AS CR, Czech Academy of Science, Dolejškova 5, Prague

Summary. The work deals with numerical testing of two different numerical methods based on finite volumes (FV) and finite elements (FE) for different Reynolds numbers. The finite volume method is based on upwind scheme (third order) for convective terms and central second order for dissipative terms. Finite element method consists of stabilization of weak formulation for higher Reynolds numbers with the help of streamline-upwind (Petrov-Galerkin) modification.

Authors compare both numerical results with experiment for laminar $Re \in (100, 700)$, where steady solution exists, using the length of separation domain on lower wall as well as on upper wall for $Re \geq 400$.

1 Introduction

The work deals with the numerical solution of the 2D and 3D flow through backward facing step. The mathematical model is the system of Navier-Stokes or Reynolds averaged Navier-Stokes (RANS) equations with two-equation turbulence models. Concerning BFS flow in 2D, we focus ourselves on the transitional regimes from laminar to turbulent flow. The start of transition is investigated using the model of laminar flow, which allows us to make even the small differences between different numerical schemes visible. The end of transition is investigated using the RANS model. A comparison with measurement data is made. The laminar model is extended to 3D.

We are also interested in comparison of results obtained with finite volumes and finite elements schemes. We discuss the stabilization of finite elements with the help of advanced SUPG method in order to obtain robust solver for unsteady incompressible laminar flows.

2 Mathematical model

The equations used to solve laminar flow of an incompressible viscous fluid are the Navier-Stokes ones, which in 2D and Cartesian coordinates (x, y) have the form

$$RW_t + F_x + G_y = \tilde{R}\frac{1}{Re}(W_{xx} + W_{yy}), \tag{1}$$

where

$$W = col\,\|p, u, v\|\,, \quad R = \tilde{R} = diag\,\|0, 1, 1\|\,,$$
$$F = uf + col\,\|0, p, 0\|\,, \quad G = vf + col\,\|0, 0, p\|\,, \quad f = col\,\|1, u, v\| \tag{2}$$

Re is Reynolds number, u, v components of velocity vector, p static pressure divided by density, and $col\|\cdot\|$ denotes column vector.

In the case of turbulent flow, we solve the Reynolds averaged Navier-Stokes (RANS) equations to obtaind mean flow-field. The RANS equations have the form of (1) with (u, v), p being mean values of velocity and pressure and right hand side replaced by

$$\frac{\tilde{R}}{Re}(W_{xx} + W_{yy}) + S_x + T_y, \tag{3}$$

with

$$S = col\,\|0, \tau_{11}, \tau_{21}\|\,, \quad T = col\,\|0, \tau_{12}, \tau_{22}\|\,, \tag{4}$$

where τ_{ij} is Reynolds stress tensor. It is approximated using low-Re two-equation turbulence models mentioned in Section 5.

The formulation of laminar and turbulent 2D backward-facing step flow was completed by the boundary conditions in the inlet: fully developed channel flow, on the walls: zero velocity, and in the outlet: zero streamwise derivative of velocity.

3 Finite volume method

The system of Navier-Stokes or RANS equations is solved by means of artificial compressibility method, which replaces matrix R in (1) by a regular matrix. We have used

$$R = diag\,\|1/U_{max}, 1, 1\|\,, \tag{5}$$

where U_{max} is maximum velocity in the domain. The method is thus applicable for solution of steady flows only.

The Eq. (1) with (5) is solved using a cell-centered finite volume mehod. The mean value of W for the i,j finite volume (cell) $D_{i,j}$, $W_{i,j}$, will satisfy a numerical approximation of

$$R\frac{\partial W_{i,j}}{\partial t}|D_{i,j}| + \oint_{\partial D_{i,j}} F dy - G dx = \tilde{R}\frac{1}{Re}\oint_{\partial D_{i,j}} W_x dy - W_y dx, \tag{6}$$

where $|D_{i,j}|$ is area of the cell. The finite volumes compose structured (multi-block) grid consisting of quadrilaterals. The grid is orthogonal and for turbulent cases refined along walls.

3.1 Spatial discretization

The integrals in (6) are approximated using mid-point rule. Cell face values of F, G, W_x,
W_y thus need to be defined. The face between cells $D_{i,j}$ and $D_{i+1,j}$ is denoted by $i+1/2, j$, analogically the face between cells $D_{i,j}$ and $D_{i,j+1}$ will be denoted by $i, j+1/2$.

The discretization of convective and pressure terms consists of defining cell face velocities $u_{i+1/2,j}$, $v_{i,j+1/2}$ and pressure in the central way,

$$u_{i+1/2,j} = \frac{1}{2}(u_{i,j} + u_{i+1,j}), \quad p_{i+1/2,j} = \frac{1}{2}(p_{i,j} + p_{i+1,j}),$$
$$v_{i,j+1/2} = \frac{1}{2}(v_{i,j} + v_{i,j+1}), \quad p_{i,j+1/2} = \frac{1}{2}(p_{i,j} + p_{i,j+1}),$$

and an upwind biased "Monotone Upstream-centered Schemes for Conservation Laws" (MUSCL) [12] interpolation in the direction of grid lines, here the line $j = const$, for cell face momentum in (2):

$$f_{i+1/2,j} = f_i + \frac{1}{4}(1+\kappa)(f_{i+1} - f_i) + \frac{1}{4}(1-\kappa)(f_i - f_{i-1}), \quad u_{i+1/2} > 0,$$
$$f_{i+1/2,j} = f_{i+1} + \frac{1}{4}(1+\kappa)(f_{i+1} - f_i) + \frac{1}{4}(1-\kappa)(f_{i+2} - f_{i+1}), u_{i+1/2} \leq 0, \quad (7)$$

where the constant index j has been ommited in the r.h.s. and $f_{i,j+1/2}$ is obtained similarly in j-direction depending on $v_{i,j+1/2}$. We have used $\kappa = 1/3$, i.e. up to third order accurate upwind. The same form is used for turbulent flows as well, although the formal accuracy is degraded by non-regularity of the grid. On the other hand, the grid refinement is needed inside of shear layers, where the diffusive term is dominant. There was no need to use a limiter in Eq. (7) for momentum equations. In case of turbulence model equations a limiter is sometimes necessary to achieve better convergence in the non-turbulent regions. We have used a minmod one.

The approximation of cell face velocity derivatives $\tilde{R}W_x$, $\tilde{R}W_y$ needed in diffusive terms is central. It uses quadrilateral dual finite volumes constructed over each face of primary volume – the vertices are located at end of primary face and in centres of adjacent primary volumes. The mid-point rule quadrature formula is again used, with face value of velocity defined as average of values in vertices of dual cell [11].

3.2 Pressure stabilization

The finite volume is common for all unknowns. In order to avoid pressure-velocity decoupling in this collocated arrangement, a pressure diffusion term in the form of Laplacian of pressure is added to the r.h.s. of continuity equation. The magnitude of pressure diffusion is adjusted by the local resolution of physical diffusion. The term is also essential for robustness of the method [11].

The effect resembles the one in the pressure stabilized Petrov-Galerkin method, however in the present formulation the pressure diffusion does not vanish completely in steady state.

3.3 Discretization in time

The system (1) with R given in (5) is integrated in time by means of explicit methods – multi-stage Runge-Kutta methods and MacCormack method – or by implicit backward Euler method. In the implicit method, the system of algebraic equations is linearized by Newton method, which results in an block nine-diagonal system. The system is solved iteratively using a line Gauss-Seidel method, applying direct block tridiagonal system inversion on the lines. For turbulent flows, solely implicit method was used.

4 Finite Element Method

In order to clear up the finite element method approach, we re-write the system of the incompressible Navier-Stokes equations in nondimensional form

$$\begin{aligned} \frac{\partial \mathbf{u}}{\partial t} - \nu \Delta \mathbf{u} + (\mathbf{u} \cdot \nabla)\mathbf{u} + \nabla p &= 0, \qquad \text{in } \Omega \times (0,T), \\ \nabla \cdot \mathbf{u} &= 0, \qquad \text{in } \Omega \times (0,T). \end{aligned} \tag{8}$$

This system is equipped with suitable boundary conditions of either Dirichlet type (inlet and walls) and do-nothing boundary condition (outlet). The initial condition for velocity components has to be added. The time derivative of velocity components vanishes in the case of relevant stationary solution and thus can be ommited. On the other hand we use the time stepping scheme for finding the stationary solution. Nevertheless, the precise time stepping scheme is needed in the case of nonstationary fluid flow. We follow now with time and space discretization.

For time discretization we use second order implicit schemes, where

$$\frac{\partial \mathbf{u}}{\partial t} \approx \frac{3\mathbf{u}^{n+1} - 4\mathbf{u}^{n} + \mathbf{u}^{n-1}}{2\tau}$$

This leads to a solution of one nonlinear system in each time step. We use the second order linearization of the nonlinear convective term, which leads to the semiimplicit scheme

$$(\mathbf{u}\cdot\nabla)\mathbf{u} \approx \left[(2\mathbf{u}^n - \mathbf{u}^{n-1})\cdot\nabla\right]\mathbf{u}^{n+1}.$$

This leads to the following scheme

$$\frac{3\mathbf{u}^{n+1} - 4\mathbf{u}^n + \mathbf{u}^{n-1}}{2\tau} - \nu\Delta\mathbf{u}^{n+1} + \left((2\mathbf{u}^n - \mathbf{u}^{n-1})\cdot\nabla\right)\mathbf{u}^{n+1} + \nabla p^{n+1} = 0, \quad (9)$$

$$\nabla\cdot\mathbf{u}^{n+1} = 0,$$

equipped with appropriate boundary and initial conditions.

Next, the problem (9) is reformulated in a weak sense, which is suitable for the solution with the aid of the finite element method. Defining the velocity space $X = \left(H^1(\Omega)\right)^2$ and the pressure space $M = L_0^2(\Omega)$ it is easy to see that the solution $U = (\mathbf{u}, p)$ of problem (9) satisfies

$$a(U,V) = f(V), \qquad \forall\, V = (\mathbf{v}, q) \in (X, M) \quad (10)$$

where

$$a(U,V) = \frac{3}{2\tau}(\mathbf{u},\mathbf{v}) + \nu(\nabla\mathbf{u},\nabla\mathbf{v}) + \left(\left((2\mathbf{u}^n - \mathbf{u}^{n-1})\cdot\nabla\right)\mathbf{u},\mathbf{v}\right) - $$
$$- (p, \nabla\cdot\mathbf{v}) + (\nabla\cdot\mathbf{u}, q),$$
$$f(V) = \frac{1}{2\tau}\left(4\mathbf{u}^n - \mathbf{u}^{n-1}, \mathbf{v}\right) \quad (11)$$

and by $(\cdot,\cdot)$ we denote the scalar product in the space $L^2(\Omega)$. Moreover, we require that $\mathbf{u}$ satisfies the Dirichlet boundary conditions. The couple $(\mathbf{u}, p)$ represents a solution on time level $n+1$, i.e. $\mathbf{u}^{n+1} := \mathbf{u}$ and $p^{n+1} := p$.

Further, the use of the Galerkin FEM restrict the weak formulation from couple of spaces (X, M) to approximate spaces (X_h, M_h), i.e. find $U_h \in (X_h, M_h)$ such that

$$a(U_h, V_h) = f(V_h), \qquad \forall\, V_h = (\mathbf{v}_h, q_h) \in (X_h, M_h) \quad (12)$$

The couple (X_h, M_h) of finite element spaces should satisfy the BB condition, which guarantees the stability of a scheme: there exists a constant $c > 0$ such that

$$\sup_{\mathbf{w}\in X_h} \frac{(p, \nabla\cdot\mathbf{w})}{\|\mathbf{w}\|_{1,2,\Omega}} \geq c\|p\|, \qquad \forall p \in M_h. \quad (13)$$

We use Taylor-Hood family of finite elements P^{k+1}/P^k-approximation.

The discretization (10) of the convection part may lead to 2nd order accuracy, but the approximate solution may suffer from spurious oscillations for high Reynolds numbers. In order to avoid this drawback, we apply the stabilization via streamline-diffusion/Petrov-Galerkin technique (see, e.g., [5], [2]).

In the stabilized problem we consider some additional terms defined by

$$\mathcal{L}_{h,n}(U,V) = \sum_K \delta_K \left(\frac{3}{2\tau}\mathbf{u} - \nu\Delta\mathbf{u} + (\mathbf{w}\cdot\nabla)\,\mathbf{u} + \nabla p, (\mathbf{w}\cdot\nabla)\mathbf{v} \right)_K ,$$

$$\mathcal{F}_{h,n}(V) = \sum_K \delta_K \left(\frac{1}{2\tau}\left(4\mathbf{u}^n - \mathbf{u}^{n-1}\right), (\mathbf{w}\cdot\nabla)\mathbf{v} \right)_K , \tag{14}$$

where the function $\mathbf{w}$ stands for $\mathbf{w} = 2\mathbf{u}^n - \mathbf{u}^{n-1}$ and by $(\cdot,\cdot)_K$ we denote the scalar product in the space $L^2(K)$. The parameter δ_K is a function of local (element) Reynolds number $Re^{\mathbf{w}}$ based on the transport velocity $\mathbf{w}$:

$$\delta_K = \delta^* \frac{h}{2\|\mathbf{w}\|}\xi(Re^{\mathbf{w}}), \qquad \text{where} \qquad Re^{\mathbf{w}} = \frac{h\|\mathbf{w}\|_K}{2\nu} \tag{15}$$

and we set $\xi(Re^{\mathbf{w}}) = \frac{Re^{\mathbf{w}}}{Re^{\mathbf{w}}+1}$. The parameter δ^* is an additional free parameter. The resulting stabilized system then reads

$$a(U_h,V_h) + \mathcal{L}_{h,n}(U_h,V_h) + \sum_{K\in\tau_h} \tau_k\Big(\nabla\cdot\mathbf{u}_h, \nabla\cdot\mathbf{v}_h\Big) = f(V_h) + \mathcal{F}_{h,n}(V_h). \tag{16}$$

where we used the additional *grad* – *div*-stabilization, for details see, e.g., [2].

The space-time discretization leads to the solution of the following system of equations repeatedly for each time step

$$S\underline{\mathbf{u}} + 2\tau B\underline{p} = \underline{f}, \qquad B^T\underline{\mathbf{u}} = 0. \tag{17}$$

where $\underline{u} \in R^{n_h}$ and $\underline{p} \in R^{m_h}$ are vectors whose components represent degrees of freedom defining the velocity $\mathbf{u}$ and the pressure p, respectively, S is a non-singular $n_h \times n_h$ matrix and B is an $n_h \times m_h$ matrix. In practical computations we solve this system with the help of iterative methods or - for a smaller system - with the help of advanced direct methods, see, e.g., [6]. UMFPACK is a set of routines for solving unsymmetric sparse linear systems, $Ax = b$, using the Unsymmetric MultiFrontal method written in ANSI/ISO C.

5 Results

The Reynolds number is defined using 2/3 of maximum velocity in the inlet (i.e. bulk velocity in the laminar case), and step height.

The figure 1 shows the distance of reattachment point on the lower wall from the step. The grid has regular spacing in both directions equal to 0.05. The result of finite volume method on coarse grid with double spacing is also shown.

At higher Reynolds numbers, the laminar flow separates on the upper wall as well. The results are shown in Fig. 2. This separation is for higher Reynolds numbers predicted in better agreement with measurement than the primary one in Fig. 1.

For $Re > 6600$, the flow is fully turbulent and two dimensional in the mean. The primary separation length becomes independent of Re and according to [7] equal 8 step heights. Our numerical results give well higher values 8.83 using SST (Shear Stress Transport) turbulence model [9] and 9.05 using sst mC k-ϵ (modified Chien k-ϵ with shear stress transport) one [10]. Both turbulence models are low-Re two-equation ones, first one uses k-ω formulation, second one k-ϵ. Several profiles of streamwise component of velocity for these models are compared in the upper part of Fig. 3. The lower part of same figure shows the k-variable of turbulence models, which could be interpreted as kinetic energy of velocity fluctuations. Also the structure of primary separation highly depends on turbulence model, as shown in Fig. 4. However, high dependence on numerical method (and grid) should be mentioned as well – e.g. the small corner vortex for SST model disappears if the limiter in ω-equation is used.

Acknowledgements

The work has been partially supported by the Research Plan of MSMT No. 210000003 and by grants No. 101/03/0018 and No. 101/02/0684 of GA CR.

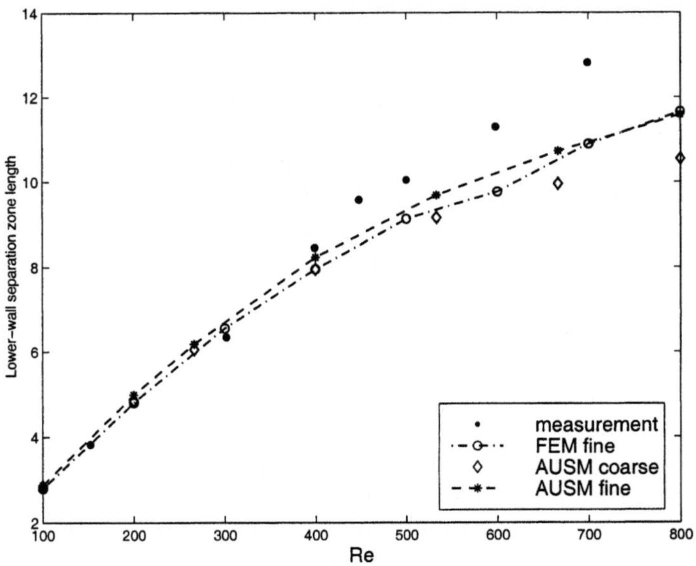

Fig. 1. Prediction of primary separation zone

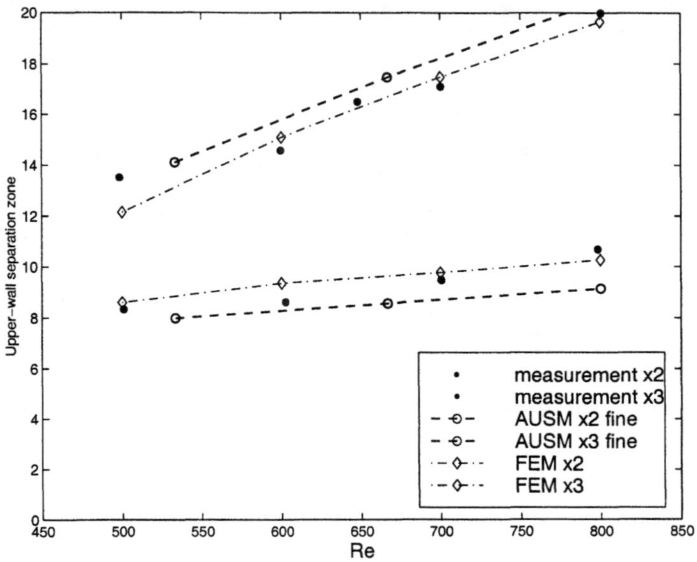

Fig. 2. Prediction of secondary separation zone

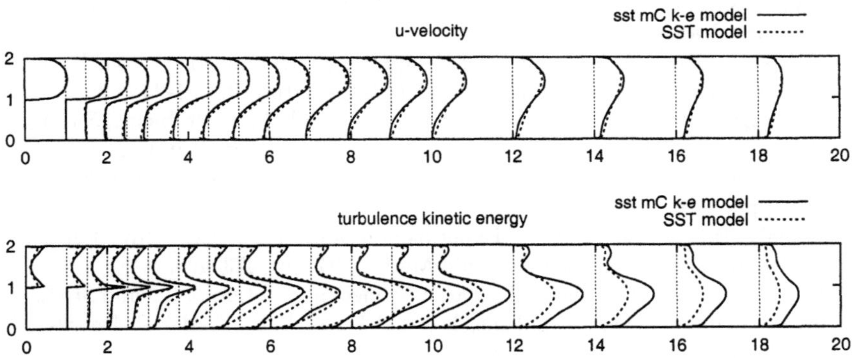

Fig. 3. BFS flow with SST and sst mC k-ϵ model, $Re = 6667$

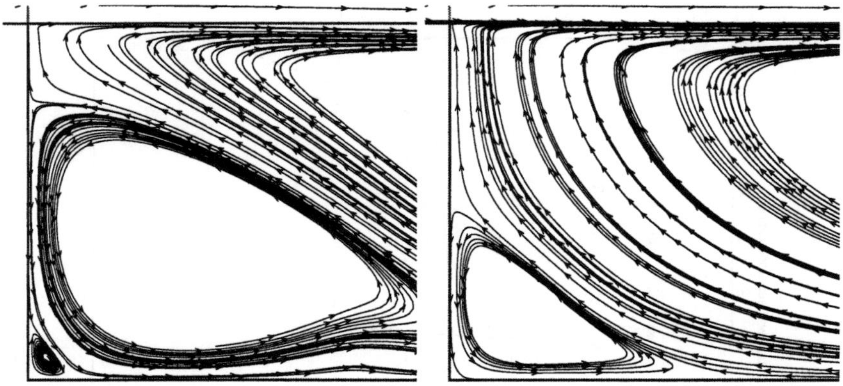

Fig. 4. Detail of streamlines for BFS flow, SST (left) and sst mC k-ϵ model (right), $Re = 6667$

References

1. Feistauer M. (1993): Mathematical Methods in Fluid Dynamics. Longman Scientific & Technical, Harlow.
2. Gelhard T., Lube G., Olshanskii M. A.: Stabilized Finite Element Schemes with LBB-stable elements for incompressible flows (preprint)
3. Sváček P., Feistauer M.(2003): Numerical Simulations of Laminar Viscous Incompressible Flow Over a Profile, Proc. Topical Problems of Fluid Mechanics, IT CAS, Prague.
4. Dolejsi V. (2001): Anisotropic Mesh Adaptation Technique for Viscous Flow Simulation, East-West Journal of Numerical Mathematics, **9(1)**:1-24.
5. Turek S. (1998): Efficient Solvers for Incompressible Flow Problems, Springer.
6. Davis T. A.,Duff I. S. (1999): A combined unifrontal/multifrontal method for unsymmetric sparse matrices, ACM Transactions on Mathematical Software, **25**, no. 1, pp. 1-19.
7. Armaly B.F. et al. (1983): Experimental and Theoretical Investigation of Backward-facing Step Flow, J. Fluid Mech., **127**, pp. 473–496.
8. Kozel K., Louda P., Příhoda J. (2001): Numerical Solution of 2D and 3D Incompressible Viscous Flow Problems, Internal Flows Vol. 2, (eds) P. Doerffer, 5th ISAIF Symposium, pp. 847–854.
9. Menter F. R. (1994): Two-Equations Eddy-Viscosity Turbulence Models for Engineering Applications, AIAA Journal, **32**, No.8, pp. 1598-1605.
10. Kozel K., Louda P., Příhoda J.(2003): Numerical Solution of Turbulent Backward Facing Step and Impinging Jet Flows, Modelling Fluid Flow '03, vol. I, (eds) T. Lajos, J. Vad, Budapest, pp. 694–701.
11. Louda P., (2002): Numerical Solution of 2D and 3D Turbulent Impinging Jet Flow, Ph.D. thesis, CTU Prague (in Czech).
12. Yee H. C. (1989): A Class of High-Resolution Explicit and Implicit Shock-capturing Methods, NASA Technical Memorandum 10 1088.

Periodicity Properties of Solutions to a Hysteresis Model in Micromagnetics

Martin Kružík

Institute of Information Theory and Automation, Academy of Sciences of the Czech Republic, Pod vodárenskou věží 4, CZ-182 08 Praha 8, Czech Republic.
kruzik@utia.cas.cz

Summary. We show the existence of a solution with a periodic integral average of the magnetization. This property is generic for both uniaxial as well as cubic ferromagnets. Our proof mainly uses properties of time-discrete approximations and a fixed point theorem.

1 Introduction

In this paper we show the existence of a solution to a hysteresis model of bulk ferromagnets established in [12] having time-periodic spatial averages of the magnetization. This holds under general assumptions for uniaxial as well as for cubic ferromagnets. The model is based on the Brown's theory of micromagnetics [1] which is here enriched by a suitable rate-independent dissipation mechanism. The basic assumption is that the transformation of the magnetization from one pole to another one requires a certain amount of energy. This energy is related to the coercive force H_c. The rate-independence allows for the description of pure hysteresis losses [5] and is well accepted for a fairly wide range of frequencies of external magnetic fields. There are also other attempts in the literature to build phenomenological rate-independent dissipation mechanisms into the models; see e.g. [13]. We assume sufficiently slow processes so that the released heat can be put off (hence the process is isothermal). The model fully relies on energy principles and it is based on the two main requirements, namely, *stability* (1) and the *energy inequality* (2). Roughly speaking, we say that $q = q(t)$ is a solution process if

$$\forall\, \tilde{q}:\ I(t, q(t)) \le I(t, \tilde{q}) + \mathcal{D}(q(t), \tilde{q}) \ \text{ and} \tag{1}$$

$$I(t, q(t)) + \mathrm{Var}(\mathcal{D}, q; s, t) \le I(s, q(s)) + \int_s^t \partial_t I(\theta, q(\theta))\, \mathrm{d}\theta\ , \tag{2}$$

where I is Gibbs' stored energy of the system, "Var" stands for the total variation, $\mathcal{D}$ is a dissipation functional ensuring a rate-independent response and $s \le t \in [0, T]$, where $[0, T]$ is the process time interval. The formulation of rate-independent evolutionary processes in continuum mechanics by means of (1) and (2) appeared in [10]. Its application to problems of rate-independent hysteresis in micromagnetics appeared in [12]. In particular, they proved the existence of a solution to a model capturing a virgin magnetization process.

Not much is known about properties of solutions to (1) and (2). Mielke and Theil [9, Th. 7.1] showed the uniqueness of the solution if I is smooth and uniformly convex in q. However, these assumptions do not hold in our application. As our functional I comes from the convexification of a function with multiple minima it is not strictly convex and it is affine along the easy axis. Nevertheless, analyzing time-discrete problems corresponding to (1) and (2) we can show periodic behavior of the average magnetization. The main tool is the Tychonoff fixed point theorem and uniqueness of the average magnetization in the time-discrete case.

2 Model

2.1 Stored energy and its relaxation

The theory of rigid ferromagnetic bodies [1] assumes that a *magnetization* $m : \Omega \to \mathbb{R}^n$, describing the state of a body $\Omega \subset \mathbb{R}^n$, $n = 2, 3$, is subjected to the *Heisenberg-Weiss constraint*, i.e. has a given (in general, temperature dependent) magnitude

$$|\, m(x)\,| = M_{\rm s} \quad \text{for almost all } x \in \Omega\ ,$$

where $M_{\rm s} > 0$ is the *saturation magnetization*, considered here as a constant (since temperature is considered constant, too).

In the no-exchange formulation, which is valid for large bodies [4], the Helmholtz free energy of a rigid ferromagnetic body $\Omega \subset \mathbb{R}^n$ consists of two parts. The first part is the *anisotropy energy* $\int_\Omega \varphi(m(x))\,{\rm d}x$ related crystallographic properties of the ferromagnet. A typical $\varphi : S := \{s \in \mathbb{R}^n;\ |s| = M_{\rm s}\} \to \mathbb{R}$ is a nonnegative function vanishing only at a few isolated points on S determining *directions of easy magnetization*, e.g. at two points for uniaxial materials or at six (or eight) for cubic ones. The second part of the Helmholtz energy, $\frac{1}{2}\int_{\mathbb{R}^n} |\nabla u_m(x)|^2\,{\rm d}x$, is the *energy of the demagnetizing field* ∇u_m self-induced by the magnetization m; its potential u_m is governed (after neglect of many terms in the full Maxwell system) by

$$\operatorname{div}\big(-\nabla u_m + m\chi_\Omega\big) = 0 \quad \text{in } \mathbb{R}^n\,, \tag{3}$$

where $\chi_\Omega : \mathbb{R}^n \to \{0, 1\}$ is the characteristic function of Ω. The demagnetizing-field energy thus penalizes non-divergence free magnetization vectors. Standardly, we will understand (3) in the weak sense, i.e. $u_m \in H^1(\mathbb{R}^n)$ will be called a weak solution to (3) if the integral identity $\int_{\mathbb{R}^n} \big(m\chi_\Omega - \nabla u_m(x)\big) \cdot \nabla v(x)\,{\rm d}x = 0$ holds for all $v \in H^1(\mathbb{R}^n)$, where $H^1(\mathbb{R}^n) \equiv W^{1,2}(\mathbb{R}^n)$ denotes the Sobolev space of functions from $L^2(\mathbb{R}^n)$ with all first derivatives (in the distributional sense) also in $L^2(\mathbb{R}^n)$. Altogether, the Helmholtz energy $E(m)$, has the form

$$E(m) = \int_\Omega \varphi(m(x))\,\mathrm{d}x + \frac{1}{2}\int_{\mathbb{R}^n} |\nabla u_m(x)|^2\,\mathrm{d}x\ . \tag{4}$$

If the ferromagnetic specimen is exposed to some external magnetic field $h = h(x)$, the so-called *Zeeman's energy* of interactions between this field and magnetization vectors equals to $H(m) := -\int_\Omega h(x)\cdot m(x)\,\mathrm{d}x$. Finally, the following variational principle governs equilibrium configurations:

$$\begin{cases} \text{minimize} \quad G(m) := E(m) - H(m) \\ \qquad\qquad\qquad = \displaystyle\int_\Omega \varphi(m(x)) - h(x)\cdot m(x)\,\mathrm{d}x + \frac{1}{2}\int_{\mathbb{R}^n} |\nabla u_m(x)|^2\,\mathrm{d}x\ , \\ \text{subject to (3)}, \ (m, u_m) \in \mathcal{A}\times H^1(\mathbb{R}^n)\ , \end{cases} \tag{5}$$

where the introduced notation G stands for *Gibbs' energy* and $\mathcal{A}$ is the set of admissible magnetizations

$$\mathcal{A} := \{m \in L^\infty(\Omega;\mathbb{R}^n);\ |\,m(x)\,| = M_\mathrm{s} \text{ for almost all } x\in\Omega\}\ .$$

As $\mathcal{A}$ is not convex we cannot rely on direct methods in proving the existence of a solution. In fact, the solution to (5) need not exist in $\mathcal{A}\times H^1(\mathbb{R}^n)$ Due to nonconvexity of $\mathcal{A}$ weak limits of minimizing sequences of (5) do not necessarily live in $\mathcal{A}\times H^1(\mathbb{R}^n)$.

It is, therefore, natural to look for an extension (=relaxation) of our problem in which we would properly describe the behavior of (5) along minimizing sequences. It is well-known [4] that such relaxation can be achieved by extending the Helmholtz energy by continuity on the convex set of the so-called *Young measures* [14]

$$\bar{E}(\nu) = \int_\Omega \varphi\bullet\nu\,\mathrm{d}x + \frac{1}{2}\int_{\mathbb{R}^n} |\nabla u_{(\mathrm{id}\,\bullet\,\nu)}(x)|^2\,\mathrm{d}x\ , \tag{6}$$

where $[v\bullet\nu](x) := \int_{\mathbb{R}^n} v(s)\nu_x(\mathrm{d}s)$ and $\mathrm{id} : \mathbb{R}^n \to \mathbb{R}^n$ is the identity. The set of Young measures $\mathcal{Y}(\Omega;S) \subset L^\infty_\mathrm{w}(\Omega;\mathrm{rca}(S)) \cong L^1(\Omega;C(S))^*$ is the set of all weakly measurable essentially bounded mappings $x\mapsto\nu_x : \Omega\to \mathrm{rca}(S)\cong C(S)^*$ such that ν_x is a probability Radon measure supported on the sphere S for a.a. $x\in\Omega$; the adjective "weakly measurable" means that $v\bullet\nu$ is Lebesgue measurable for any $v\in C(S)$. A natural embedding of a magnetization $m\in L^\infty(\Omega;\mathbb{R}^n)$, $|m(x)| = M_\mathrm{s}$, to $\mathcal{Y}(\Omega;S)$ is $\nu = i(m)$ defined by $\nu_x = \delta_{m(x)}$ with δ_s denoting the Dirac measure at $s\in S$. We say that a sequence $\{\nu^k\}_{k\in\mathbb{N}}\subset\mathcal{Y}(\Omega;S)$ converges weakly* to ν if $\lim_{k\to\infty}\langle\nu^k, f\rangle = \langle\nu, f\rangle$ for any $f\in L^1(\Omega;C(S))$ or, equally, for any $f = g\otimes v$ with $g\in L^1(\Omega)$ and $v\in C(S)$, where the tensorial notation means naturally $[g\otimes v](x,s) = g(x)v(s)$. From the last fact, we can also say that $\nu^k \to \nu$ weakly* if and only if w*- $\lim_{k\to\infty} v\bullet\nu^k = v\bullet\nu$ for all $v \in C(S)$. where the weak*-limit is understood in $L^\infty(\Omega)$. Considering the weak* topology on $L^\infty_{\rm w}(\Omega;{\rm rca}(S))$ makes $\mathcal{Y}(\Omega;S)$ a convex, metrizable compact set containing densely the set of admissible magnetizations $\mathcal{A}$ if embedded via i.

As shown in [4, 11], a correct relaxation (=natural extension) of (5) is

$$\begin{cases} \text{minimize} \;\; \bar{G}(\nu) := \bar{E}(\nu) - H(\mathrm{id}\bullet\nu)\;, \\ \text{subject to (3) with } m = \mathrm{id}\bullet\nu, \quad (\nu,u) \in \mathcal{Y}(\Omega;S)\times H^1(\mathbb{R}^n)\;, \end{cases} \tag{7}$$

The model (7) represents a so-called *mesoscopic level* model, because a minimizing Young measure ν records some, but not full information about spatial oscillations of a minimizing sequence of (5) around each "macroscopic" point x through volume fractions described as the probability distribution ν_x. This information makes possible to describe the effective magnetic properties by means of the first moment, the "macroscopic" magnetization $m = \mathrm{id}\bullet\nu$, and moreover seems sufficient for designing a dissipative mechanism in a good agreement with experiments, which will be just exploited further.

2.2 Rate-independent dissipation

For usual loading regimes and magnetically hard materials, one must consider certain dissipation. Our simplified standpoint is that the amount of dissipated energy within the phase transformation from one pole to the other can be described by a single, phenomenologically given number (of the dimension $\mathrm{J/m^3{=}Pa}$) depending on the coercive force $H_{\rm c}$. Hence, we need to identify the particular poles according to the magnetization vector. Inspired by [8, 10] and considering L poles ($L = 2$ for uniaxial magnets or 6 or 8 for cubic magnets), we define a continuous mapping $\mathfrak{L} : S \to \triangle_L$ where $\triangle_L := \{\xi \in \mathbb{R}^L;\ \xi_i \geq 0,\ i = 1,...,L,\ \sum_{i=1}^L \xi_i = 1\}$. In other words, $\{\mathfrak{L}_1,...,\mathfrak{L}_L\}$ forms a *partition of unity* on S such that $\mathfrak{L}_i(s)$ is equal 1 if s is in i-th pole, i.e. $s \in S$ is in a neighborhood of i-th easy-magnetization direction. Of course, $\mathfrak{L}(m)$ in the (relative) interior of $\triangle_L$ indicates m in the region where no definite pole is specified. Hence $\mathfrak{L}$ plays the role of what is often called an *order parameter*.

In terms of the mesoscopic microstructure described by the Young measure ν, the "mesoscopic" order parameter is naturally defined as

$$\lambda = \Lambda\nu := \mathfrak{L}\bullet\nu \tag{8}$$

where $[\mathfrak{L}\bullet\nu](x) := \int_S \mathfrak{L}(s)\nu_x(\mathrm{d}s)$. Thus Λ is just a continuous extension of the mapping $m \mapsto \mathfrak{L}(m)$, i.e. if $\{m_k\}$ converges to ν weakly* in $L^\infty_{\rm w}(\Omega;{\rm rca}(S))$, then $\mathfrak{L}(m_k) \rightharpoonup \Lambda\nu$ weakly* in $L^\infty(\Omega;\mathbb{R}^L)$.

To described phenomenologically the dissipative energetics, one must prescribe a (pseudo)*potential of dissipative forces* ϱ as a function of the rate of λ. For rate-independent processes, this potential must be convex and homogeneous of degree-1. Considering a (not necessarily Euclidean) norm $|\cdot|_L$

on $\mathbb{R}^L$, one can postulate $\varrho(\dot{\lambda}) = H_c|\dot{\lambda}|_L$, a constant $H_c > 0$ is the so-called coercive force. The energy needed to transform i-th pole to j-pole is then $H_c|e_i - e_j|_L$ with e_i the unit vector with 1 at the i-th position. The state of the specimen Ω (at a given time t) will be described by the couple $q = q(t) \equiv (\nu, \lambda) = (\{\nu_{x,t}\}_{x\in\Omega}, \lambda(\cdot, t))$. Let us denote by $\mathcal{Q}$ the convex set of admissible configurations:

$$\mathcal{Q} := \Big\{q = (\nu, \lambda) \in \mathcal{Y}(\Omega; S) \times L^\infty(\Omega; \mathbb{R}^L) \\ \lambda(x) \in \triangle_L, \;\; \Lambda\nu = \lambda \text{ for a.a. } x \in \Omega\Big\} \tag{9}$$

The total dissipation of the process between states $q_1, q_2 \in \mathcal{Q}$ is defined as [6, 12]

$$\mathcal{D}(q_1, q_2) := \int_\Omega H_c|\lambda_1 - \lambda_2|\,\mathrm{d}x, \qquad q_i = (\nu_i, \lambda_i) \in \mathcal{Q}\ , \ i = 1, 2. \tag{10}$$

For the analysis below, we will need to consider rather a certain *regularization* of the stored energy $\mathcal{E}$ which would control spatial smoothness of λ. For this, we will augment $\mathcal{E}$ by a higher-order term

$$\mathcal{E}_\rho(\nu, \lambda) := \bar{E}(\nu) + \begin{cases} \rho\|\lambda\|^2_{H^\alpha(\Omega;\mathbb{R}^L)} & \text{if } \lambda \in H^\alpha(\Omega; \mathbb{R}^L), \\ +\infty & \text{otherwise,} \end{cases} \tag{11}$$

where $H^\alpha(\Omega) \equiv W^{\alpha,2}(\Omega)$ denotes the usual Sobolev-Slobodetskiĭ space and where we assume

$$\alpha, \rho > 0, \text{ fixed.} \tag{12}$$

From now on, we will work with this regularized relaxed stored energy $\mathcal{E}_\rho$ rather than $\mathcal{E}$.

Let us abbreviate the Gibbs energy by

$$\mathcal{G}(t, q) := \mathcal{E}_\rho(q) - \langle \mathcal{H}(t), q\rangle\ , \tag{13}$$

where

$$\langle \mathcal{H}(t), q\rangle = \big[H(t)\big](\mathrm{id} \bullet \nu) = \langle \nu, h(\cdot, \cdot) \otimes \mathrm{id}\ \rangle. \tag{14}$$

Let us agree to identify quite naturally the mapping $t \mapsto \nu(t) = \{[\nu(t)]_x\}_{x\in\Omega}$ with a Young measure $(x, t) \mapsto \nu_{x,t}$.

Definition 1. *We say that a process $q = q(t)$ is stable if*

$$\forall \tilde{q} \in \mathcal{Q}: \qquad \mathcal{G}(t, q) \le \mathcal{G}(t, \tilde{q}) + \mathcal{D}(q(t), \tilde{q}) \tag{15}$$

for all $t \in [0, T]$.

Definition 2. *We say that the process $q = q(t)$ satisfies the energy inequality if for all $s, t \in [0, T]$, $s \leq t$,*

$$\underbrace{\mathcal{G}(t, q(t))}_{\substack{\text{effective Gibbs'}\\ \text{energy at time } t}} + \underbrace{\mathrm{Var}(\mathcal{D}, q; s, t)}_{\substack{\text{dissipated}\\ \text{energy}}} \leq \underbrace{\mathcal{G}(s, q(s))}_{\substack{\text{Gibbs' ener-}\\ \text{gy at time } 0}} - \underbrace{\int_s^t \left\langle \frac{\mathrm{d}\mathcal{H}}{\mathrm{d}t}, q(\theta) \right\rangle \mathrm{d}\theta}_{\substack{\text{reduced work of}\\ \text{external field}}} \ , \quad (16)$$

where the total variation over the time interval $[s, t]$ is defined standardly, without using explicitly any time derivative, as

$$\begin{aligned} \mathrm{Var}(\mathcal{D}, q; s, t) &:= \sup \sum_{i=1}^{I} \mathcal{D}(q(t_{i-1}), q(t_i)) \\ &\equiv \sup \sum_{i=1}^{J} \int_\Omega d(\lambda(t_{i-1}), \lambda(t_i)) \mathrm{d}x, \end{aligned} \quad (17)$$

where the supremum is taken over all $J \in \mathbb{N}$ and over all partitions of $[s, t]$ in the form $s = t_0 < t_1 < \ldots < t_{J-1} < t_J = t$.

In what follows "BV" stands for a space of maps with bounded variations.

Definition 3. *The process $q = q(t)$, $q \equiv (\nu, \lambda)$, will be considered as a solution if $\nu \in \mathcal{Y}(\Omega \times [0, T]; S)$, $\lambda \in \mathrm{BV}([0, T]; L^1(\Omega; \mathbb{R}^L))$ and $q(t) \in \mathcal{Q}$ for all $t \in [0, T]$, and it is stable in the sense* (15) *for all $t \in [0, T]$ and satisfies the energy inequality* (16) *for a.a. $s, t \in [0, T]$, $s \leq t$.*

The existence of a response q with the above mentioned properties was shown even in a more general case in [12] by a semi-discretization in time, using the implicit Euler scheme. For simplicity, let us consider an equidistant partition of the time interval $[0, T]$ with a time step $\tau > 0$, assuming T/τ an integer. Even more, we consider a sequence of τ's converging to zero and such that, τ_i/τ_{i+1} is integer, i.e. each next partition is a refinement of the preceding one.

Then we put $q_\tau^0 = q_0$, a given initial condition, and, for $k = 1, \ldots, T/\tau$ we define q_τ^k recursively as a solution of the minimization problem

$$\begin{cases} \text{Minimize} \quad I(q) := \mathcal{G}(k\tau, q) + \mathcal{D}(q_\tau^{k-1}, q) \\ \text{subject to} \quad q \equiv (\nu, \lambda) \in \mathcal{Q} \ , \end{cases} \quad (18)$$

where $\mathcal{Q}$ is from (9), $\mathcal{G}$ is from (13), and $\mathcal{D}$ from (10). If a solution (i.e. a *global* minimizer) to (18) is not unique, we just take an arbitrary one for q_τ^k. Then we define the piecewise constant interpolation $q_\tau \in L^\infty(0, T; L^\infty_{\mathrm{w}}(\Omega; \mathrm{rca}(S)) \times L^\infty(\Omega; \mathbb{R}^L))$ so that $q_\tau|_{((k-1)\tau, k\tau]} = q_\tau^k$ for $k = 1, \ldots, T/\tau$ while for $t = 0$ we put $q_\tau(0) = q_0$. Besides, assuming

$$h \in W^{1,1}(0, iT; L^1(\Omega; \mathbb{R}^n)),\ h(\cdot, t + jT) = h(\cdot, t)\ , \qquad i \in \mathbb{N},\ 1 \le j \le i-1,\ t \in [0, T] \tag{19}$$

we have certainly $\mathcal{H} \in C(0,T; L^1(\Omega; C(S))$ and we can define the piecewise constant approximation of $\mathcal{H}$, denoted by $\mathcal{H}_\tau$, by $\mathcal{H}_\tau(t) = \mathcal{H}(k\tau)$ for $t \in ((k-1)\tau, k\tau]$ and by $\mathcal{H}_\tau(t) = \mathcal{H}(0)$ for $t = 0$. Besides, we will still need the piecewise affine interpolation, denoted by $\mathcal{H}_\tau^{\text{aff}}$, i.e. $\mathcal{H}_\tau^{\text{aff}}$ is affine in time if restricted on the interval $[(k-1)\tau, k\tau]$ for $k = 1, ..., T/\tau$ and $\mathcal{H}_\tau^{\text{aff}}(k\tau) = \mathcal{H}(k\tau)$ for $k = 0, ..., T/\tau$. Also, we will naturally assume that the initial condition q_0 is admissible and even stable:

$$q_0 \in \mathcal{Q} \quad \text{and} \quad \mathcal{E}_\rho(q_0) \le \mathcal{E}_\rho(\tilde{q}) + D(q_0, \tilde{q}) + \langle \mathcal{H}(0), q_0 - \tilde{q} \rangle \qquad \forall \tilde{q} \in \mathcal{Q}; \tag{20}$$

note that it implies, in particular, that $\mathcal{E}_\rho(q_0) < +\infty$. The scheme (18) together with a suitable spatial discretization is also suitable for a numerical solution to our problem.

Let us define a sufficiently large set $\mathcal{P}$ where the values of all the processes $q_\tau(\cdot)$ will certainly live; here it is natural to put

$$\mathcal{P} := \Big\{ (\nu, \lambda) \in \mathcal{Q};\ \ ||\lambda||_{H^\alpha(\Omega; \mathbb{R}^L)} \le C_1 \Big\}; \tag{21}$$

the constant C_1 can be now considered arbitrary but sufficiently large, and will be fixed later, see (22). We will endow $\mathcal{P}$ by the (weak*×weak)-topology of $L^\infty_{\text{w}}(\Omega; \text{rca}(S)) \times H^\alpha(\Omega; \mathbb{R}^L)$. Clearly, the set $\mathcal{P}$ is compact.

The following can be found in [6] or in a slightly different form in [12].

Proposition 1. *Let* (12), (19) *and* (20) *hold. Let* $q_\tau = (\nu_\tau, \lambda_\tau)$ *be a solution constructed recursively from solutions to* (18) *at the prescribed time increments. The following a-priori estimates hold:*

$$||\lambda_\tau||_{L^\infty(0,T; H^\alpha(\Omega;\mathbb{R}^L) \cap L^\infty(\Omega;\mathbb{R}^L)) \cap \text{BV}([0,T]; L^1(\Omega;\mathbb{R}^L))} \le C_1, \tag{22}$$

$$||\nu_\tau||_{L^\infty(0,T; L^\infty_{\text{w}}(\Omega; \text{rca}(S)))} \le C_2, \tag{23}$$

$$||\mathfrak{G}_\tau||_{\text{BV}([0,T])} \le C_3, \tag{24}$$

where $\mathfrak{G}_\tau(t) := \mathcal{E}_\rho(q_\tau(t)) - \langle \mathcal{H}_\tau(t), q_\tau(t) \rangle$ *denote Gibbs' energy of the approximate trajectory.*

Proposition 1 shows that solutions to (18) live in $\mathcal{P}$ if C_1 is large enough. Let us denote

$$\hat{P} := \big\{ (\beta, \lambda) \in \mathbb{R}^n \times H^\alpha(\Omega; \mathbb{R}^L);\ \exists (\nu, \lambda) \in \mathcal{P} \text{ such that } \beta = \hat{\nu} \big\}\ , \tag{25}$$

where

$$\hat{\nu} := \frac{1}{\text{meas}\ \Omega} \int_\Omega \int_S A\nu_x(\mathrm{d}A)\,\mathrm{d}x \ = \frac{1}{\text{meas}\ \Omega} \int_\Omega m(x)\,\mathrm{d}x\ , \tag{26}$$

i.e., $\hat{\nu}$ is the average of the macroscopic magnetization over the specimen Ω, in particular $|\hat{\nu}| \le M_s$. We call $\hat{\nu}$ the $\Omega \times S$ average of ν. Endowing $\hat{P}$ with the Euclidean topology of $\mathbb{R}^n$ $\times$ the weak topology of $H^\alpha(\Omega;\mathbb{R}^L)$ we see that $\hat{\mathcal{P}}$ is convex, closed and bounded in the Banach space $\mathbb{R}^n \times H^\alpha(\Omega;\mathbb{R}^L)$ with the norm $\|(\beta,\lambda)\|_{\mathbb{R}^n\times H^\alpha(\Omega;\mathbb{R}^L)} = |\beta| + \|\lambda\|_{H^\alpha(\Omega;\mathbb{R}^L)}$. We define

$$\Theta : \mathcal{P} \to \hat{P}:\ \Theta(\nu,\lambda) = (\hat{\nu},\lambda)\ . \tag{27}$$

Lemma 1. *Let* (12) *hold. Let* $q_1 = (\nu_1,\lambda_1)$, $q_2 = (\nu_2,\lambda_2)$ *be two solutions of* (18) *in* $\mathcal{Q}$ *for some fixed* $1 \le k \le T/\tau$. *Then* $\Theta(q_1) = \Theta(q_2)$.

Proof. The assertion $\lambda_1 = \lambda_2$ follows from the convexity of I from (18) in q and the strict convexity of I in λ. Note also that I is strictly convex in $\nabla u_{(\mathrm{id}\,\bullet\,\nu)}$, i.e., in the magnetostatic field. Denoting m_1 and m_2 the macroscopic magnetization vectors corresponding to q_1 and q_2, respectively, the weak formulation of (3) gives $\int_{\mathbb{R}^n}(m_1(x) - m_2(x))\chi_\Omega(x)\cdot\nabla v(x)\,\mathrm{d}x = 0$ for any $v \in H^1(\mathbb{R}^n)$. As for any $w \in \mathbb{R}^n$ one can find $v \in H^1(\mathbb{R}^n)$ such that $\nabla v = w$ in Ω we have $\int_\Omega m_1(x)\,\mathrm{d}x = \int_\Omega m_2(x)\,\mathrm{d}x$ and the result follows. □

Let $Z : \hat{P} \to \hat{P}$ be defined as follows. For any $(\beta_0,\lambda_0) \in \hat{P}$ we take $q_0 = (\nu_0,\lambda_0) \in \mathcal{P}$ such that $\beta = \hat{\nu}_0$ and solve (18) for all $k = 1,\ldots,T/\tau$. Then we calculate $\beta_\tau^{T/\tau} = \hat{\nu}_\tau^{T/\tau}$ and we set $Z(\beta_0,\lambda_0) := (\beta_\tau^{T/\tau},\lambda_\tau^{T/\tau})$, where $q_\tau^{T/\tau} = (\nu_\tau^{T/\tau},\lambda_\tau^{T/\tau})$. Note that the image of (β_0,λ_0) does not depend on particular $(\nu_0,\lambda_0) \in \mathcal{P}$ having $\hat{\nu}_0 = \beta_0$ because solutions in (18) depend only on λ_0 in the initial condition.

Lemma 2. *The mapping* $Z : \hat{P} \to \mathbb{R}^n \times H^\alpha(\Omega;\mathbb{R}^L)$ *is weakly sequentially continuous in* $\mathbb{R}^n \times H^\alpha(\Omega;\mathbb{R}^L)$.

Sketch of proof. We can suppose that $T/\tau = 1$. Otherwise we write Z as a composition of analogously defined mappings from the $(k-1)$th step to the kth step. Having $(\beta_j,\lambda_j) \to (\beta_0,\lambda_0)$ in $\hat{P}$ for $j \to \infty$ we denote $F_j(q) := \mathcal{G}(\tau,q)+\mathcal{D}(q,q_j)$, and $F_0(q) := \mathcal{G}(\tau,q)+\mathcal{D}(q,q_0)$, $q_j = (\nu_j,\lambda_j)$, $j \ge 0$. We have $\lim_{j\to\infty} F_j = F$ uniformly in $\mathcal{P}$ and as F_j are sequentially lower semicontinuous on $\mathcal{P}$ by [2] any sequence of minimizers of $\{F_j\}_{j\ge1}$ contains a subsequence converging to a minimizer of F_0. As $\Omega\times S$ average of ν-components of minimizers of F_0 is given uniquely the whole sequence of $\Omega \times S$ averages of ν-components of minimizers of F_j converges to the $\Omega \times S$ average of the ν-component of minimizers of F_0. Altogether $\lim_{j\to\infty} Z(\beta_j,\lambda_j) = Z(\beta_0,\lambda_0)$. □

Lemma 3. *There is* $(\beta,\lambda) \in \hat{P}$ *such that* $Z(\beta,\lambda) = (\beta,\lambda)$.

Proof. $\hat{P}$ is a closed, convex and bounded subset of the reflexive Banach space $\mathbb{R}^n \times H^\alpha(\Omega;\mathbb{R}^L)$. As Z is weakly sequentially continuous in $\hat{P}$ the assertion follows from the Tychonoff fixed point theorem; cf. [3]. □

The following theorem establishes the existence of a solution to our problem, which has the same average magnetization at the time $t = 0$ and $t = T$.

Theorem 1. *Let* (12) *and* (19) *be valid. Then there are a process* $q = (\nu, \lambda) \in \mathcal{Y}(\Omega \times [0,T]; S) \times \mathrm{BV}([0,T]; L^1(\Omega;\mathbb{R}^L))$, $\Theta(q(0)) = \Theta(q(T))$ *and a net* $\{q_{\tau_\xi}\}_{\xi\in\Xi}$, $\Theta(q_{\tau_\xi}(0)) = \Theta(q_{\tau_\xi}(T))$ *such that:*
(i) $\lim_{\xi\in\Xi} \lambda_{\tau_\xi}(t) = \lambda(t)$ *strongly in* $L^2(\Omega;\mathbb{R}^L)$ *for all* $t \in [0,T]$,
(ii) $\lim_{\xi\in\Xi} \nu_{\tau_\xi}(t) = \nu(t)$ *weakly* in* $L^\infty_{\mathrm{w}}(\Omega;\mathrm{rca}(S))$ *for all* $t \in [0,T]$,
(iii) $\lim_{\xi\in\Xi} \mathcal{G}_{\tau_\xi}(t, q_{\tau_\xi}(t)) = \mathcal{G}(t,q)$ *for all* $t \in [0,T]$.
Moreover, $\lambda = \Lambda\nu$ *a.e. on* Ω *for every* $t \in [0,T]$ *and* q *thus obtained is a solution process according to Definition 3 and* $\Theta(q(0)) = \Theta(q(T))$.

Sketch of proof. The proof is similar to the one of [7, Th. 3.4] or [6, Prop. 3.13]. The point (i) relies on the weak* compactness of the set $L^\infty_{\mathrm{w}}(\Omega;\mathrm{rca}(S))^{[0,T]}$ due to the Tychonoff's theorem on compactness of product topologies. □

Theorem 2. *Let* (12) *and* (19) *hold. Then there is a solution process* $q = (\nu, \lambda) \in \mathcal{Y}(\Omega \times [0,iT]; S) \times \mathrm{BV}([0,iT]; L^1(\Omega;\mathbb{R}^L))$ *such that* $\Theta(q(t + jT)) = \Theta(q(t))$ *for any* $t \in [0,T]$ *and any* $1 \le j \le i-1$. *In particular,* $\int_\Omega m(x,t)\,\mathrm{d}x = \int_\Omega m(x, t + jT)\,\mathrm{d}x$.

Proof. The process q can be constructed by the T-periodic extension of the process whose existence was established in Theorem 1. Clearly, the extended definitions 1 and 2 hold for such process. □

Acknowledgment. This work was partly supported by the GA AV ČR grant No. A 107 5005.

References

1. Brown, W.F. Jr. (1966): Magnetostatic principles in ferromagnetism. Springer, New York
2. Dal Maso, G. (1993): An introduction to Γ-convergence. Birkhäuser, Boston
3. Deimling, K. (1985): Nonlinear Functional Analysis. Springer, Berlin
4. DeSimone, A. (1993): Energy minimizers for large ferromagnetic bodies. Arch. Rat. Mech. Anal., **125**, 99–143
5. Hubert, A., Schäfer, R. (1998): Magnetic Domains. Springer, Berlin
6. Kružík, M. (2003): Periodic solutions to a hysteresis model in micromagnetics. In preparation
7. Mielke, A., Roubíček, T. (2003): A rate-independent model for inelastic behavior of shape-memory alloys. Multiscale modeling and simulations, to appear. +

8. Mielke, A., Theil, F. (1999): Mathematical model for rate-independent phase transformations. In: Alber, H.-D., Balean, R., Farwig, R. (eds) Models of Cont. Mechanics in Analysis and Engineering. Shaker-Verlag Aachen
9. Mielke, A., Theil, F. (2003): On rate-independent hysteresis models. Nonlin. Diff. Eq. Appl., to appear
10. Mielke, A., Theil, F., Levitas, V. (2002): A variational formulation of rate-independent phase transformations using extremum principle. Arch. Rat. Mech. Anal., **162**, 137–177
11. Pedregal, P. (1997): Parametrized Measures and Variational Principles. Birkhäuser, Basel
12. Roubíček, T., Kružík, M. (2002): Mesoscopic model for ferromagnets with isotropic hardening. Zeitschrift f. Angew. Math. Phys., to appear
13. Visintin, A. (1997): Modified Landau-Lifshitz equation for ferromagnetism. *Physica B*, **233**, 365–369
14. Young, L.C. (1937): Generalized curves and existence of an attained absolute minimum in the calculus of variations. Comptes Rendus de la Société et des Lettres de Varsovie, Classe III, **30**, 212–234

Mixed Finite Element Method on Polygonal and Polyhedral Meshes

Yuri Kuznetsov[1] and Sergey Repin[2]

[1] Department of Mathematics, University of Houston, Houston, TX 77204
kuz@math.uh.edu

[2] St. Petersburg Department of V.A. Steklov Institute of Mathematics of the Russian Academy of Sciences, St. Petersburg 191011, Russia *repin@pdmi.ras.ru*

Summary. A new mixed finite element method for the diffusion equations on general polygonal and polyhedral meshes is presented. The basis vector functions in macrocells are designed by solving the local mixed finite element problems with the lowest order Raviart-Thomas elements. Numerical results for the Poisson equation on distorted prismatic meshes are given.

1 Introduction

This paper is a natural generalization of the authors' recent results [3] to the general diffusion equation with mixed type boundary conditions. The paper is organized as follows. In Section 2 we give the formulation of the problem and describe the meshes to be used. Section 3 is the most important in the paper. We describe an approach to designing macroelements in the space of fluxes by solving mixed finite element problems with the lowest order Raviart-Thomas elements [1], [4] on macrocells. A convergence result is given in Section 4. Finally, in Section 5 we present and discuss numerical results for a particular $3D$ test problem.

2 Problem formulation

We consider the diffusion problem in the form of the system of the first order differential equations

$$\begin{aligned} K^{-1}\,\bar{u} + \operatorname{grad} p &= 0 \\ \operatorname{div} \bar{u} + \qquad cp &= f \end{aligned} \tag{1}$$

in a bounded connected polygonal (polyhedral) domain Ω in $\mathbb{R}^d$, $d = 2$ $(d = 3)$, with homogeneous boundary conditions

$$\begin{aligned} p = 0 \quad &\text{on} \quad \Gamma_D, \\ \bar{u} \cdot \bar{n} = 0 \quad &\text{on} \quad \Gamma_N. \end{aligned} \tag{2}$$

Here Γ_D and Γ_N are the Dirichlet and the Neumann parts of the boundary $\partial\Omega$, $\bar{n}$ is the outward unit normal to $\partial\Omega$, $K = K(x)$ is the diffusion tensor, $K = K^T > 0$, $c = c(x)$ is a nonnegative function, and $f \in L_2(\Omega)$. We assume that Γ_D is a closed subset of $\partial\Omega$ consisting of a finite number of segments (polygons) in the case $d = 2$ $(d = 3)$.

The weak formulation of (1), (2) reads as follows: find
$\bar{u} \in V \equiv \{\bar{v} : \bar{v} \in H_{\text{div}}(\Omega),\ \bar{v} \cdot \bar{n} = 0 \text{ on } \Gamma_N\}$, $p \in Q \equiv L_2(\Omega)$
such that

$$\begin{aligned} \int_\Omega (K^{-1}\bar{u}) \cdot \bar{v}\, \mathrm{dx} - \int_\Omega p(\nabla \cdot \bar{v})\, \mathrm{dx} &= 0 \\ \int_\Omega (\nabla \cdot \bar{u}) q\, \mathrm{dx} + \int_\Omega cpq\, \mathrm{dx} &= \int_\Omega fq\, \mathrm{dx} \end{aligned} \tag{3}$$

for all $(\bar{v},\ q) \in V \times Q$.

Let Ω_h be a partitioning of Ω into m nonoverlapping polygonal (polyhedral) macrocells E_k:

$$\Omega_h = \bigcup_{k=1}^{m} E_k, \tag{4}$$

and V_h and Q_h be finite element subspaces of V and Q, respectively. In this paper we assume that the interface $\Gamma_{kl} = \partial E_k \cap \partial E_l$ between macrocells E_k and E_l is either a point, or a segment, or a simply connected polygon in the case $d = 3$ (see, for instance, Fig. 1). We also assume that for any function $p_h \in Q_h$ its trace on E_k is a constant, and for any vector-function $\bar{v} \in V_h$ its trace on E_k is a piecewise affine vector-function. Moreover, we assume that for any vector-function $\bar{v}_h \in V_h$ its normal component $\bar{v} \cdot \bar{n}_{kl}$ on Γ_{kl} is a constant, where $\bar{n}_{kl}$ is the unit normal vector to Γ_{kl} directed from E_k to E_l, $k > l$. An example of a polygonal mesh Ω_h is given in Fig. 1. The arrows on the interfaces show the normal component of the fluxes.

In the case of a triangular $(d = 2)$ or a tetrahedral $(d = 3)$ partitioning of Ω, V_h is the lowest order Raviart-Thomas space $\mathrm{RT}_0(\Omega_h)$ subject to the boundary condition on Γ_N.

The mixed finite element approximation to (1), (2) reads as follows: find $(\bar{u}_h,\ p_h) \in V_h \times Q_h$ such that

$$\begin{aligned} \int_\Omega (K^{-1}\bar{u}_h) \cdot \bar{v}\, \mathrm{dx} - \int_\Omega p_h(\nabla \cdot \bar{v})\, \mathrm{dx} &= 0 \\ \int_\Omega (\nabla \cdot \bar{u}_h) q\, \mathrm{dx} + \int_\Omega cp_h q\, \mathrm{dx} &= \int_\Omega fq\, \mathrm{dx} \end{aligned} \tag{5}$$

for all $(\bar{v},\ q) \in V_h \times Q_h$. The finite element problem results in the system of linear algebraic equations

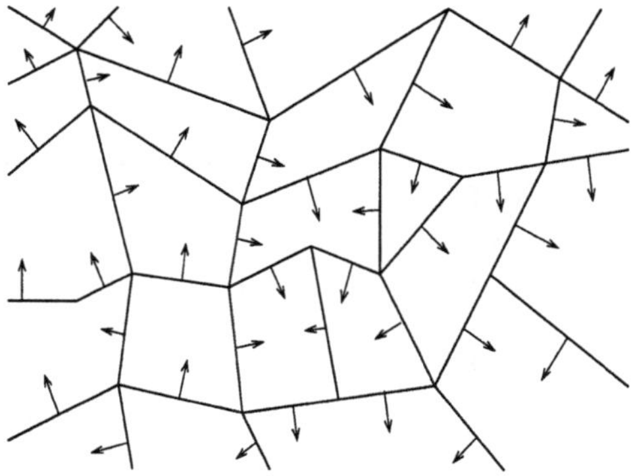

Fig. 1. Example of a polygonal mesh

$$\mathcal{A}\begin{pmatrix}\bar{u}\\ \bar{p}\end{pmatrix} = \begin{pmatrix}0\\ \bar{F}\end{pmatrix} \tag{6}$$

with a saddle point matrix

$$\mathcal{A} = \begin{pmatrix}M & B^T\\ B & -\Sigma\end{pmatrix} \tag{7}$$

where M is a symmetric positive definite matrix and Σ is a symmetric positive definite (or semidefinite) matrix.

3 New finite element space V_h

Let E be a macrocell in Ω_h and ∂E be a union of s segments, $d=2$ (polygons, $d=3$), $\Gamma_1, \ldots, \Gamma_s$. For the moment, we assume that any two adjacent segments (polygons) Γ_i and Γ_j, $i \neq j$, do not belong to the same line (plane). We recall that adjacent interfaces of a macrocell E_k with neighboring macrocells E_i and E_j may belong to the same line, $d=2$ (plane, $d=3$). A possible situation is shown in Fig. 2.

Let $E_h = \bigcup\limits_{i=1}^{t} e_i$ be a conformal partitioning of E into triangles e_i, $d=2$ (tetrahedra, $d=3$), and $\mathrm{RT}_0(E_h)$ be the lowest order Raviart-Thomas finite element space. Here t is the total number of cells. We consider the mixed finite element approximation to system (1) with right hand side $f \equiv \mathrm{const}_E$ in E, $c \equiv 0$ in E, and with boundary conditions $\bar{u} \cdot \bar{n}_E = \delta_{1,j}$ on Γ_j, $j = \overline{1,\ s}$, where $\bar{n}_E$ is the outward unit normal to ∂E: find $\bar{u}_{E,h} \in \mathrm{RT}_0(E_h)$, $\bar{u}_{E,h} \cdot \bar{n}_E = \delta_{1,j}$ on Γ_j, $j = \overline{1,\ s}$,
$p_{E,h} \in Q_{E,h} \equiv \{q : q \equiv \mathrm{const\ on}\ e_i,\ i = \overline{1,\ t}\}$
such that

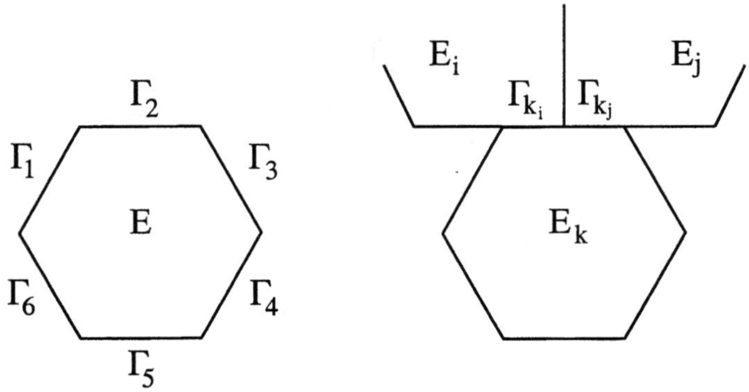

Fig. 2. A macrocell E and adjacent macrocells E_k, E_i, and E_j

$$\begin{aligned} \int_E (K^{-1}\bar{u}_{E,h}) \cdot \bar{v}\, \mathrm{dx} - \int_E p_{E,h}(\nabla \cdot \bar{v})\, \mathrm{dx} &= 0 \\ \int_E (\nabla \cdot \bar{u}_{E,h}) q\, \mathrm{dx} &= \mathrm{const}_E \int_E q\, \mathrm{dx} \end{aligned} \tag{8}$$

for all $\bar{v} \in \mathrm{RT}_0(E_h)$, $\bar{v} \cdot \bar{n}_E = 0$ on ∂E, $q \in Q_{E,h}$. Here $\delta_{i,j}$ denotes the Kronecker delta. The value of const_E in (8) is subject to the equation

$$\int_E \nabla \cdot \bar{u}_{E,h}\, \mathrm{dx} \equiv \int_{\partial E} (\bar{u}_{E,h} \cdot \bar{n}_E)\, \mathrm{dl} = \mathrm{const}_E \cdot |E|$$

which gives

$$\mathrm{const}_E = \frac{|\Gamma_1|}{|E|} \tag{9}$$

where

$$|\Gamma_1| = \int_{\Gamma_1} \mathrm{dl} \quad \text{and} \quad |E| = \int_E \mathrm{dx}.$$

Under condition (9) problem (8) has the unique solution $\bar{w}_1 = \bar{u}_{E,h}$.

Repeating the same procedure on the same mesh E_h with the boundary conditions $\bar{u}_{E,h} \cdot \bar{n}_E = \delta_{i,j}$ on Γ_j, $1 \le j \le s$, $i = 2, \ldots, s$, we get the vector-functions $\bar{w}_1$, $\bar{w}_2$, …, $\bar{w}_s$ which satisfy the conditions

$$\bar{w}_i \cdot \bar{n}_E = \delta_{i,j} \quad \text{on} \quad \Gamma_j, \quad 1 \le i,j \le s.$$

To this end, a vector function

$$\bar{w} = \sum_{i=1}^{s} \alpha_i \bar{w}_i \tag{10}$$

satisfies the boundary conditions

$$\bar{w} \cdot \bar{n}_E = \alpha_i \quad \text{on} \quad \Gamma_i, \quad i = \overline{1,\ s}. \tag{11}$$

The extension of the procedure to the case when adjacent Γ_i and Γ_j may belong to the same line, $d = 2$, or the same plane, $d = 3$, is straightforward.

To complete the description of the procedure for the designing of vector-functions $\bar{w}_i$, $i = \overline{1,\ s}$, we consider a situation where Γ_1 is partitioned into two segments, $d = 2$, (simply connected polygons, $d = 3$) $\Gamma_{1,D}$ and $\Gamma_{1,N}$. We assume that any cell e_i from E_h adjacent to the boundary ∂E satisfies the condition that any of its edges, $d = 2$ (faces, $d = 3$) belong either to the interior of E, or to Γ_D, or to Γ_N. Then, the requested vector function $\bar{w}_i$ is the solution of the problem:
find $\bar{w}_i \in \mathrm{RT}_0(E_h)$, $\bar{w}_i \cdot \bar{n}_E = 1$ on $\widehat{\Gamma}_i \equiv \Gamma_i \setminus \Gamma_N$, $\bar{w}_i \cdot \bar{n}_E = 0$ on $\partial E \setminus \widehat{\Gamma}_i$, $p_{E,h} \in Q_{E,h}$ such that the equations (8) are satisfied for any $\bar{v} \in \mathrm{RT}_0(E_h)$, $\bar{v} \cdot \bar{n}_E = 0$ on ∂E, $q \in Q_{E,h}$. The constant on the right hand side of (8) is defined by the formula

$$\mathrm{const}_E = \frac{|\widehat{\Gamma}_i|}{|E|} \tag{12}$$

where

$$|\widehat{\Gamma}_i| = \int_{\widehat{\Gamma}_i} \mathrm{dl}.$$

The extension of the procedure to other Γ_j, $2 \le j \le s$ is again straightforward.

The vector-functions $\bar{w}$ in (10) with new vector-functions $\bar{w}_i$ satisfies the conditions

$$\begin{aligned} \bar{w} \cdot \bar{n}_E &= \alpha_i \ \text{on} \ \widehat{\Gamma}_i, \quad i = \overline{1,\ s}, \\ \bar{w} \cdot \bar{n}_E &= 0 \quad \text{on} \ \partial E \cap \Gamma_N. \end{aligned} \tag{13}$$

We denote the designed space of finite element vector-functions $\bar{w}(x)$ for a macroelement E_k in Ω_h by $V_{k,h}$. The global space $V_h \subset H_{\mathrm{div}}(\Omega)$ for the mesh is defined by

$$V_h = \left\{ \bar{v} : \bar{v} \in \mathrm{RT}_0(\Omega_h),\ \bar{v}|_{E_k} \in V_{k,h},\ k = \overline{1,\ m} \right\} \tag{14}$$

4 Convergence

To prove the convergence result for the mixed finite element problem (5) with the designed space V_h we assume that

– all triangles (tetrahedra) $e_i^{(k)}$, $i = \overline{1,\ t_k}$, in the partitionings of E_k, $k = \overline{1,\ m}$, are regular shaped [1],

– the number of triangles (tetrahedra) t_k in the partitionings of E_k is bounded by a constant independent of k, i.e.,

$$\max_{1\leq k\leq m} t_k \leq \text{const.} \tag{15}$$

Then, the convergence result follows from [3] after some minor modifications.

Theorem 1. *Under the assumptions made, the finite element solution* $(\bar{u}_h,\ p_h)$ *of problem* (5) *converges to the solution* $(\bar{u},\ p)$ *of problem* (3) *in* $H_{\text{div}}(\Omega) \times L_2(\Omega)$, *i.e.*

$$\begin{aligned} ||\bar{u}_h - \bar{u}||_{H_{\text{div}}(\Omega)} &\longrightarrow 0 \\ ||p_h - p||_{L_2(\Omega)} &\longrightarrow 0 \quad \textit{as} \quad h \to 0 \end{aligned} \tag{16}$$

where

$$h = \max_{1\leq k\leq m} \operatorname{diam} E_k.$$

5 Numerical results

To illustrate the proposed method by numerical results we consider the Dirichlet boundary value problem for the Laplace equation:

$$\begin{aligned} \Delta p &= 0 \ \text{ in } \Omega, \\ p &= g \ \text{ on } \partial\Omega \end{aligned} \tag{17}$$

where Ω is a prism in $\mathbb{R}^3$ with the rectangular faces orthogonal to the $(x,\ y)$-plane and the triangular faces parallel to the $(x,\ y)$-plane, and a function g is the trace on $\partial\Omega$ of the test harmonic function $p = (x^2 - z^2) + x(y^2 - z^2)$.

Let Ω_h be a distorted prismatic mesh which is obtained from the reference rectangular prismatic mesh by perturbation of vertices along the vertical mesh lines. An example of such a distorted macrocell is given in Fig. 3. The choice of this mesh is relevant to geophysical applications (basin modeling and reservoir simulation) where distorted prismatic meshes are used for discretizations of diffusion type equations in layered formations.

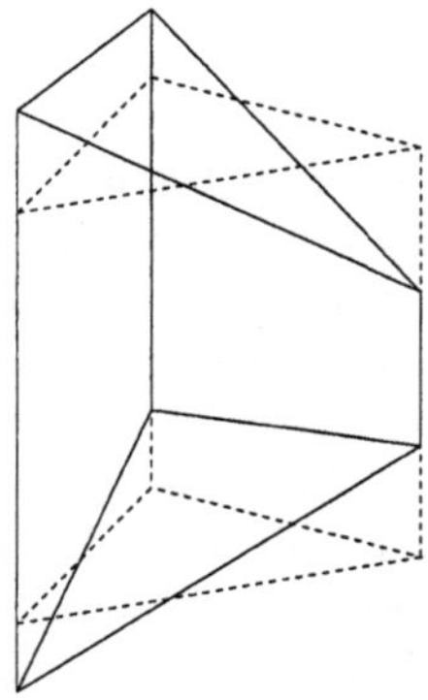

Fig. 3. Reference (dashed lines) and distorted (solid lines) prismatic macrocell

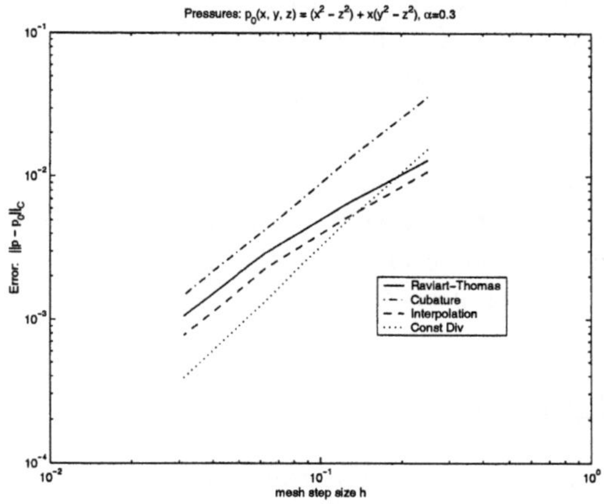

Fig. 4. Error for pressures in C-norm

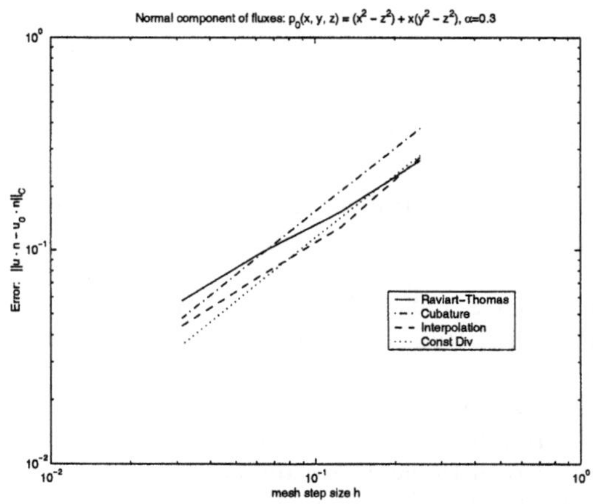

Fig. 5. Error for normal component of fluxes in C-norm

Four discretization methods have been utilized. The numerical results are shown in Fig. 4 and Fig. 5. The first one is the nonconforming mixed finite element method with the lowest order Raviart-Thomas elements. In fact, the finite elements for the space of fluxes are nonconforming only on triangular interfaces between distorted prisms where the continuity conditions for the normal component of fluxes are imposed in the center of mass of the triangles. On the quadrilateral interfaces which are orthogonal to (x, y)-coordinate plane, the normal component of the fluxes are constant.

The second method which is called "Cubature" in the figures is a modification of the mimetic discretization [5] adopted to the prismatic meshes. The third method is called "Interpolation". This is a conforming finite element method based on a modification of the Piola transformation adopted to prismatic meshes.

Finally, the method proposed in this paper is called "Const Div". The numerical results are given for the error functions in the discrete maximum norms. For the solution function p (called "pressure") we measure the errors for the mean values over the macrocells, and for the fluxes we measure the mean values of the normal components over the interfaces.

The numerical results show that all the methods provide the same order of accuracy of approximation to the solution of the differential problem (17). We recall that the method proposed in this paper can be used on arbitrary polygonal and polyhedral meshes.

6 Acknowledgments

The work was partly supported by a grant from Los Alamos Computational Sciences Institute (LACSI). The authors are grateful to O. Boyarkin for providing us with the results of numerical experiments.

References

1. Brezzi, F., Fortin, M. (1991): Mixed and hybrid finite element methods. Springer, Berlin
2. Kuznetsov, Yu. (2003): Spectrally equivalent preconditioners for mixed hybrid discretizations of diffusion equations on distorted meshes. J. Numer. Math., v. **11**, No. **1**, 61–74
3. Kuznetsov, Yu., Repin, S. (2003): New mixed finite element method on polygonal and polyhedral meshes. Russian J. Numer. Anal. and Math. Modelling, v. **18**, No. **3**, 261–278
4. Roberts, J.E., Thomas, J.-M. (1991): Mixed and hybrid methods. In: Ciarlet, P., Lions, J.-L. (eds) Handbook of Numerical Analysis II. Finite element methods, 523–639. Elsevier/North Holland, Amsterdam
5. Shashkov, M., Steinberg, S. (1996): Solving diffusion equations with rough coefficients in rough grids. J. Comp. Phys., **129**, 383–405

Semi-discrete Schemes for Hamilton-Jacobi Equations on Unstructured Grids

Doron Levy[1] and Suhas Nayak[2]

[1] Department of Mathematics, Stanford University, Stanford, CA 94305-2125, USA
dlevy@math.stanford.edu

[2] Department of Mathematics, Stanford University, Stanford, CA 94305-2125, USA
snayak@stanford.edu

Summary. We present a new semi-discrete central scheme for approximating solutions of Hamilton-Jacobi equations on unstructured meshes. This scheme extends the numerical Hamiltonians of Kurganov *et al.* to unstructured grids. Similarly to the previous works on structured grids, a semi-discrete formulation of central schemes is made possible due to estimates of the local speeds of propagation. The consistency of the method is obtained following Abgrall's calculations for the consistency of an upwind Lax-Friedrichs scheme on unstructured grids. We conclude with comments on high-order reconstructions.

1 Introduction

We present a new central-upwind scheme for approximating solutions of Hamilton-Jacobi (HJ) equations on unstructured grids. These equations are of the form

$$\phi_t + H(\nabla\phi) = 0, \qquad \mathbf{x} = (x_1, \dots x_d) \in \mathbb{R}^d, \tag{1}$$

where $\phi = \phi(\mathbf{x}, t)$, and with a Hamiltonian, H, that depends on $\nabla\phi$ and possibly on x and t. Since solutions of (1) may develop discontinuous derivatives even when the initial data is smooth, it is generally required to interpret the solution of (1) in a suitable weak sense. The corresponding theory, in the form of "viscosity solutions" has been significantly developed over the past two decades and we refer to [5, 6, 14, 15].

The increased understanding of the nature of solutions to HJ equations have turned the area of numerical methods for the HJ equations into an active research area. Converging first-order methods were introduced by Souganidis [19]. The order of accuracy of the methods was increased through an essentially non-oscillatory (ENO) reconstruction in the upwind schemes of Osher, Sethian and Shu [17, 18]. Weighted essentially non-oscillatory (WENO) reconstructions, which were first introduced for hyperbolic conservation laws [8, 16], were then used by Jiang and Peng [7] to even further increase the accuracy of the numerical approximations using a compact reconstruction. Extensions of the first-order and ENO upwind methods to unstructured grids were done by

Abgrall [1]. The numerical fluxes of Abgrall were combined with WENO reconstructions on triangular meshes by Hu and Shu in [20]. Another finite-volume scheme on unstructured grids was proposed by Kossioris *et al.* in [9].

While upwind schemes require solving Riemann problems (or at least approximating their solutions) on the interfaces between two cells, Godunov-type central schemes utilize evolution points that are located away from the discontinuous interfaces in order to avoid Riemann solvers altogether. Fully-discrete central schemes for HJ equations were first introduced by Lin and Tadmor in [12, 13], and further improved by Bryson and Levy in [2]. Fully-discrete schemes of order greater than two were derived in [3] using central-WENO reconstructions. Semi-discrete formulations of central schemes enjoy reduced numerical dissipation while keeping track over the local speeds of propagation of information that is propagating from the discontinuous interfaces. Second-order semi-discrete central schemes for HJ equations were derived by Kurganov and Tadmor in [11]. A more accurate estimate of the local speed of propagation was then utilized to reduce the numerical dissipation in [10]. The numerical flux of Kurganov, Noelle and Petrova, was combined with the WENO reconstructions of Jiang and Peng to obtain fifth-order, semi-discrete central schemes for multi-dimensional HJ equations in [4].

In this paper we extend our previous works on semi-discrete central schemes for multi-dimensional HJ equations to unstructured grids. Our derivation results with a new numerical flux that takes into account information regarding the local speeds of propagation of information from the discontinuous interfaces. Similarly to any other central scheme, the evolution points are taken away from the interfaces in order to avoid Riemann solvers. This numerical Hamiltonian should be viewed as a central version of the numerical Hamiltonian of Abgrall [1]. It can also be viewed as a generalization of the numerical Hamiltonians of Kurganov *et al.* [10, 11] that were derived for Cartesian grids. When we assume that the "unstructured" grid is Cartesian (which still is an admissible grid in our formulation) the scheme we obtain is different of the schemes obtained in [11, 10]. This difference results from a different averaging procedure we use with the values obtained at the different evolution points, and is reflected by a different coefficient in front of the dissipative term.

2 A Semi-discrete scheme for HJ equations

We consider two-dimensional Hamilton-Jacobi equations of the form

$$\phi_t + H(\nabla\phi) = 0, \qquad \mathbf{x} = (x_1, x_2) \in \Omega \subset \mathbb{R}^2, \tag{1}$$

augmented with initial values $\phi(x, t = 0) = \phi_0(x)$. We assume that $\mathcal{T}$ is a given triangulation of Ω. The grid points are denoted by x_α. For every grid point there are m_α angular sectors T_l^α that are ordered counterclockwise. For simplicity we will drop the α index from the triangles, and use the notation

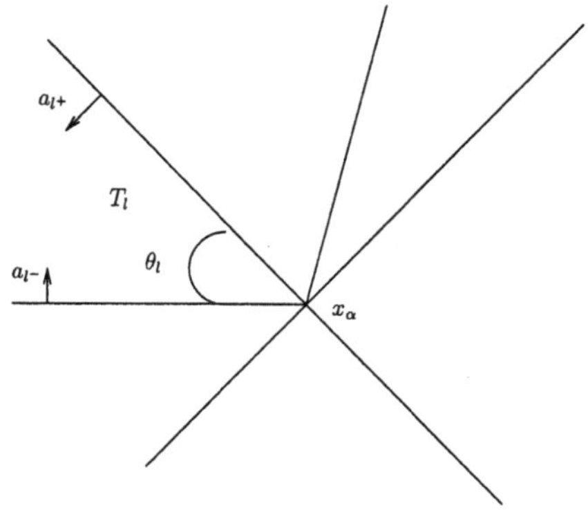

Fig. 1. Grid point x_α and its angular sectors. Each angular sector l has an associated angle θ_l

$T_l = T_l^\alpha$ whenever possible. Each node of our triangulation may be visualized as in Figure 1.

The semi-discrete scheme will be constructed as a limit of a fully-discrete scheme as the time-step tends to zero. We therefore assume a time step, Δt, and use the standard notation $t^n = n\Delta t$. Let φ_α^n denote the approximate value of $\phi(x_\alpha, t^n)$. We assume that for each time t^n, φ_α^n is given at each node x_α of the unstructured grid. We also assume that the values φ_α^n can be used to reconstruct a continuous piecewise-polynomial interpolant, i.e., a polynomial in each sector. We will comment below on a method for obtaining such a reconstruction. In either case, the numerical flux we develop is independent of the reconstruction step. We denote this reconstruction by $\tilde{\varphi}_\alpha$ and denote the approximation of $\nabla\varphi$ in each sector $\nabla\tilde{\varphi}_{\alpha,l}$.

The second ingredient we need is an estimation of the maximal speeds of propagation at the interfaces of each angular sector (in a direction that is perpendicular to the interface). For any given angular sector, T_l, the counter-clockwise speed of propagation is denoted by a_l^+ and the speed of propagation on the other interface is a_l^-. These speeds can be estimated by:

$$\begin{aligned} a_l^+ &= \max\left\{\max_{T_l}\left\{\nabla H(\nabla\varphi_{\alpha,l})\cdot \mathbf{n}_{l-1,l}\right\}, \max_{T_{l-1}}\left\{\nabla H(\nabla\varphi_{\alpha,l-1})\cdot \mathbf{n}_{l-1,l}\right\}\right\}, \\ a_l^- &= \max\left\{\max_{T_l}\left\{\nabla H(\nabla\varphi_{\alpha,l})\cdot \mathbf{n}_{l+1,l}\right\}, \max_{T_{l+1}}\left\{\nabla H(\nabla\varphi_{\alpha,l+1})\cdot \mathbf{n}_{l+1,l}\right\}\right\} \end{aligned} \tag{2}$$

where $\mathbf{n}_{j,l}$ is the normal vector on the interface between sectors T_j and T_l pointing into T_l.

We can now determine in every sector T_l around x_α an evolution point x_α^l that is located away from the interfaces (see Figure 2). This will be done using the distances that are defined through the local speeds of propagation $a_l^\pm$.

The distance of the evolution point x_α^l from x_α is denoted by d_l. Clearly, d_l depends on the local speeds of propagation $a_l^\pm$ and on the angle θ_l and is given by

$$d_l^2 = \frac{(a_l^-\Delta t)^2 + 2a_l^- a_l^+ \Delta t^2 \cos\theta_l + (a_l^+\Delta t)^2}{\sin^2\theta_l}. \tag{3}$$

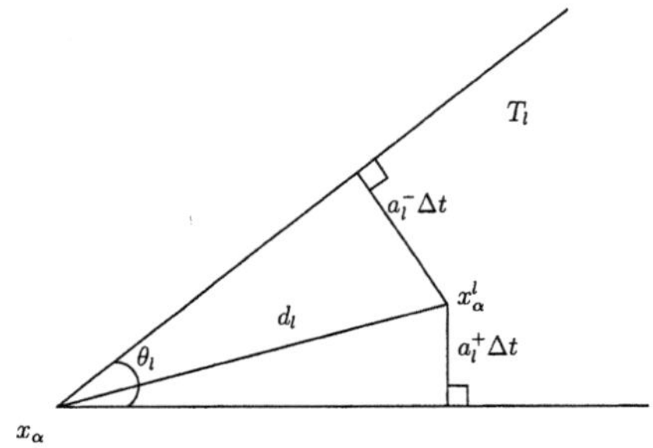

Fig. 2. Evolution point, x_α^l, derived from the maximal local speeds of propagation into T_l, a_l^+ and a_l^-

We then define $\hat{d}_l$ to be $d_l/\Delta t$, a quantity that does not explicitly depend on Δt.

The interpolant $\tilde{\varphi}(\mathbf{x}, t^n)$ is evolved to the next time step t^{n+1} at the points x_α^l, which are located away from the propagating discontinuities (assuming that the time step Δt is sufficiently small). From (1), this is given to first order in time by the Taylor expansion

$$\varphi(x_\alpha^l, t^{n+1}) = \tilde{\varphi}(x_\alpha^l, t^n) - \Delta t H(\nabla\tilde{\varphi}(x_\alpha^l, t^n)) + O(\Delta t^2), \tag{4}$$

where the approximation of the gradient $\nabla\varphi$ at x_α^l, $\nabla\tilde{\varphi}(x_\alpha^l, t^n)$, is obtained from the reconstruction.

The next step is to combine the values of φ at the different evolution points around x_α, x_α^l, into one value φ_α^{n+1}. This is done by writing a convex combination with weights $s_l \geq 0$ that are yet to be determined:

$$\varphi_\alpha^{n+1} = \frac{\sum_{l=1}^{m_\alpha} s_l \varphi(x_\alpha^l, t^{n+1})}{\sum_{l=1}^{m_\alpha} s_l}. \tag{5}$$

Using (4), we may express (5) as

$$\varphi_\alpha^{n+1} = \frac{\sum_{l=1}^{m_\alpha} s_l \left[\tilde{\varphi}(x_\alpha^l, t^n) - \Delta t H(\nabla\tilde{\varphi}(x_\alpha^l, t^n)\right]}{\sum_{l=1}^{m_\alpha} s_l}. \tag{6}$$

If we now define ρ_l to be the unit vector in the direction of x_α^l from x_α, we can use a Taylor expansion in space

$$\tilde{\varphi}(x_\alpha^l, t^n) = \tilde{\varphi}(x_\alpha, t^n) + d_l \rho_l \cdot \nabla\tilde{\varphi}(x_\alpha^l, t^n) + O(\Delta t^2).$$

Here by $\nabla\tilde{\varphi}(x_\alpha^l, t^n)$ we refer to the value of the gradient at x_α that is associated with the reconstruction in sector T_l at x_α^l. We may therefore rewrite (6) as the fully discrete scheme

$$\varphi_\alpha^{n+1} = \tilde{\varphi}_\alpha^n + \frac{\Delta t}{\sum_{l=1}^{m_\alpha} s_l} \sum_{l=1}^{m_\alpha} s_l \left[\hat{d}_l \rho_l \cdot \nabla\tilde{\varphi}(x_\alpha^l, t^n) - H(\nabla\tilde{\varphi}(x_\alpha^l, t^n))\right]. \tag{7}$$

In the limit $\Delta t \to 0$, (7) becomes a semi-discrete scheme:

$$\frac{d}{dt}\varphi_\alpha(t) = \frac{1}{\sum_{l=1}^{m_\alpha} s_l} \sum_{l=1}^{m_\alpha} s_l \left[\hat{d}_l \rho_l \cdot \nabla\tilde{\varphi}_\alpha^l(t) - H(\nabla\tilde{\varphi}_\alpha^l(t))\right], \tag{8}$$

where for each l, $\nabla\tilde{\varphi}_\alpha^l(t)$ denotes $\lim_{\Delta t\to 0} \nabla\tilde{\varphi}(x_\alpha^l, t^n)$. All that remains is to determine the coefficients s_l in (8). These coefficients will be obtained through a consistency condition. The consistency of the scheme implies that if the value of the gradient is identical in every sector that surrounds x_α then the numerical Hamiltonian should reduce to the differential Hamiltonian. Hence we are seeking for coefficients s_l such that

$$\sum_{l=1}^{m_\alpha} s_l \hat{d}_l \rho_l = 0. \tag{9}$$

Such coefficients indeed exist, and we can use, e.g., the results of Abgrall in [1] to find them. The observation that was made there was that if $\mu_{l+1/2}$ denoted a unit vector in a direction that is aligned with the interface between the sectors T_l and T_{l+1}, and if $\theta_l < \pi$ (which is the case with a triangulation, e.g.), then

$$\sum_{l=1}^{m_\alpha} \gamma_{l+\frac{1}{2}} \mu_{l+\frac{1}{2}} = 0, \tag{10}$$

provided that

$$\gamma_{l+\frac{1}{2}} = \epsilon \left[\tan\left(\frac{\theta_l}{2}\right) + \tan\left(\frac{\theta_{l+1}}{2}\right)\right], \tag{11}$$

for any $\epsilon > 0$. In order to incorporate (10)–(11) into our framework, we split each angle θ_l into two parts $\theta_l^\pm$ that are defined as

$$\theta_l^\pm = \arcsin \frac{a_l^\pm}{\hat{d}_l}, \tag{12}$$

(see figure 3).

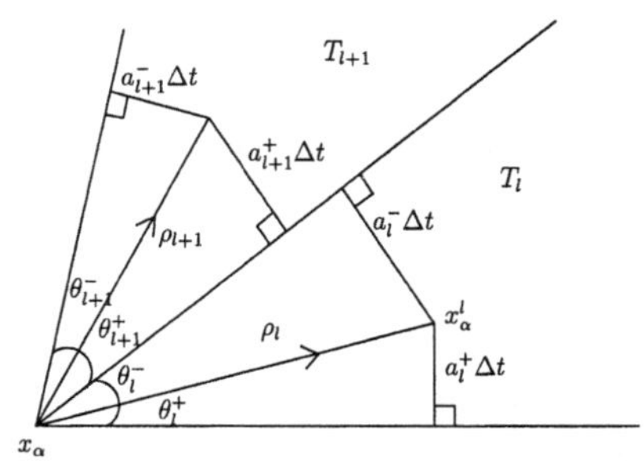

Fig. 3. The angles around x_α

The consistency condition (9) is then satisfied if the weights s_l are defined as

$$s_l = \frac{\epsilon}{\hat{d}_l}\left[\tan\left(\frac{\theta_l^+ + \theta_{l-1}^-}{2}\right) + \tan\left(\frac{\theta_l^- + \theta_{l+1}^+}{2}\right)\right], \tag{13}$$

where $\theta_l^{\pm}$ are given by (12). In our case, the coefficient ϵ will anyhow cancel out in the semi-discrete formulation (8), so it can be omitted from (13).

To summarize, if we define

$$\beta_l = \left[\tan\left(\frac{\theta_l^+ + \theta_{l-1}^-}{2}\right) + \tan\left(\frac{\theta_l^- + \theta_{l+1}^+}{2}\right)\right],$$

then the semi-discrete scheme is given by

$$\frac{d}{dt}\varphi_\alpha(t) = \frac{1}{\sum_{l=1}^{m_\alpha}\frac{\beta_l}{\hat{d}_l}}\sum_{l=1}^{m_\alpha}\beta_l\left[\rho_l\cdot\nabla\tilde{\varphi}_\alpha^l(t) - \frac{H(\nabla\tilde{\varphi}_\alpha^l(t))}{\hat{d}_l}\right]. \tag{14}$$

Remarks.

1. A simple version of the scheme can be obtained by assuming that all the speeds of propagation are identical. In this case, the local speeds are replaced by their maximum, i.e., $a = \max_l\{a_l^+, a_l^-\}$. In this case

$$\hat{d}_l = \frac{a}{\sin(\theta_l/2)},$$

and the semi-discrete scheme (14) can be written in the simpler form

$$\frac{d}{dt}\varphi_\alpha(t) = \frac{a}{\sum_{l=1}^{m_\alpha}\beta_l\sin\frac{\theta_l}{2}}\sum_{l=1}^{m_\alpha}\beta_l\left[\rho_l\cdot\nabla\tilde{\varphi}_\alpha^l(t) - \frac{\sin\frac{\theta_l}{2}}{a}H(\nabla\tilde{\varphi}_\alpha^l(t))\right]. \tag{15}$$

If, in addition to the velocities being identical, the angles are also assumed to be identical, i.e. $\theta = \theta_l$, $\forall l$, then (15) can be further simplified into

$$\frac{d}{dt}\varphi_\alpha(t) = \frac{1}{m_\alpha}\sum_{l=1}^{m_\alpha}\left[\frac{a}{\sin\frac{\theta}{2}}\rho_l\cdot\nabla\tilde{\varphi}_\alpha^l(t) - H(\nabla\tilde{\varphi}_\alpha^l(t))\right]. \tag{16}$$

2. In [1] Abgrall derived a Lax-Friedrichs-type scheme on triangular meshes. In our notation, his scheme is of the form

$$\frac{d}{dt}\varphi_\alpha(t) = \frac{a}{\pi}\sum_{l=1}^{m_\alpha}\beta_{l+\frac{1}{2}}\rho_{l+\frac{1}{2}}\cdot\left(\frac{\nabla\tilde{\varphi}_\alpha^l(t) + \nabla\tilde{\varphi}_\alpha^{l+1}(t)}{2}\right) - H\left(\frac{\sum_{l=1}^{m_\alpha}\theta_l\nabla\tilde{\varphi}_\alpha^l}{2\pi}\right). \tag{17}$$

Here $\rho_{l+1/2}$ is the unit vector in the direction of the interface between the sectors T_l and T_{l+1}, and $\beta_{l+1/2} = \tan(\theta_l/2)+\tan(\theta_{l+1}/2)$. The derivation of (17) involved evolution points that were located on the interfaces between the sectors. This resulted with the form of the dissipative term in (17) that contains averages of gradients in adjacent sectors. Also, the scheme (17) involves a Hamiltonian that is evaluated at the average of the derivatives that are computed in different sectors (with weights that are proportional to the angles). This term was postulated to be in this form, and could have taken different forms. In our case (14), this term is replaced by an average over the Hamiltonian that is evaluated in different sectors. In our case, the form of this term is dictated by the derivation of the scheme.

3. We would like to emphasize that the scheme (14) does not require Riemann solvers. This is due to the scheme's derivation, which places the evolution points away from the boundaries of the angular sectors around every point x_α.
4. If the grid is a Cartesian grid with equal spacing in the x- and y-directions, the number of angular sectors at each point is $m_\alpha = 4$, and $\sin(\theta/2) = \sin(\pi/4) = \sqrt{2}/2$. In the simple case where all velocities are taken to be identical in both direction, the scheme (16) becomes

$$\begin{aligned}\frac{d}{dt}\varphi_\alpha(t) &= \frac{a}{2}\left(\varphi_x^+ - \varphi_x^- + \varphi_y^+ - \varphi_y^-\right) \\ &-\frac{1}{4}\left[H\left(\varphi_x^+,\varphi_y^+\right) + H\left(\varphi_x^-,\varphi_y^+\right) + H\left(\varphi_x^+,\varphi_y^-\right) + H\left(\varphi_x^-,\varphi_y^-\right)\right],\end{aligned} \tag{18}$$

with the obvious notations, e.g. $H(\varphi_x^+,\varphi_y^+)$ is the Hamiltonian evaluated at the gradient at x_α that is taken from the first quadrant. The scheme (18) is identical to the consistent and monotone semi-discrete scheme that was derived for Cartesian grids in [4, 11].
5. The order of accuracy of the scheme (14) is determined by the order of accuracy of the reconstruction and the order of the ODE solver. Other than that, the formulation (14) is independent of the order of accuracy of the scheme. The scheme is a Godunov-type scheme with a global reconstruction that is evolved in time in evolution points that are located away from the interfaces. In practice, the final semi-discrete scheme (14) requires only the values of the gradient at the grid points x_α that are computed in the different angular sectors around x_α. Hence, all that one needs from the reconstruction is the values of these gradients. It is therefore possible to use the same reconstructions that were developed for upwind schemes for HJ equations on triangular meshes. For examples, a high-order (third- or fourth-order) weighted essentially non-oscillatory (WENO) reconstruction for HJ equations on triangular grids was derived by Zhang and Shu [20]. It can be incorporated as it is into the present framework.

3 Conclusion

We have derived a new semi-discrete central scheme for HJ equations on unstructured grids. This scheme is a generalization of the semi-discrete central schemes on Cartesian grids [4, 10, 11]. It is a Godunov-type scheme where a global reconstruction is evolved in time and then projected back to the grid points. Since the evolution is performed away from the interfaces of the angular sectors, there is no need to use exact or approximate Riemann solvers. The formal accuracy of the scheme depends on the accuracy of the reconstruction and the order of the ODE solver being used.

References

1. Abgrall, R. (1996): Numerical discretization of the first-order Hamilton-Jacobi equation on triangular meshes. Commun. Pure Appl. Math., **49**, 1339–1373.
2. Bryson, S., Levy, D.: Central schemes for multi-dimensional Hamilton-Jacobi equations. SIAM J. Sci. Comput. (to appear).
3. Bryson, S., Levy, D. (2003): High-Order Central WENO Schemes for Multi-dimensional Hamilton-Jacobi Equations. SIAM J. Numer. Anal., **41**, 1339–1369.
4. Bryson, S., Levy, D. (2003): High-order semi-discrete central-upwind schemes for multi-dimensional Hamilton-Jacobi equations. J. Comp. Phys., **189**, 63–87.
5. Crandall, M.G., Ishii, H., Lions, P.-L. (1992): User's guide to viscosity solutions of second order partial differential equations. Bull. Amer. Math. Soc., **27**, 1–67.
6. Crandall, M.G., Lions, P.-L. (1983): Viscosity solutions of Hamilton-Jacobi equations. Trans. Amer. Math. Soc., **277**, 1–42.
7. Jiang, G.-S., Peng, D. (2000): Weighted ENO schemes for Hamilton-Jacobi equations. SIAM J. Sci. Comp., **21**, 2126–2143.
8. Jiang G.-S., Shu C.-W. (1996): Efficient implementation of weighted ENO schemes. J. Comp. Phys., **126**, 202–228.
9. Kossioris G., Makridakis Ch., Souganidis P.E. (1999): Finite volume schemes for Hamilton-Jacobi equations. Numer. Math., **83**, 427–442.
10. Kurganov, A., Noelle S., Petrova G. (2001): Semi-discrete central-upwind schemes for hyperbolic conservation laws and Hamilton-Jacobi equations. SIAM J. Sci. Comp., **23**, 707–740.
11. Kurganov, A., Tadmor, E. (2000): New high-resolution semi-discrete central schemes for Hamilton-Jacobi equations. J. Comp. Phys., **160**, 720–742.
12. Lin, C.-T., Tadmor, E. (2000): L^1-stability and error estimates for approximate Hamilton-Jacobi solutions. Numer. Math., **87**, 701–735.
13. Lin, C.-T., Tadmor, E. (2000): High-resolution non-oscillatory central schemes for approximate Hamilton-Jacobi equations. SIAM J. Sci. Comp., **21**, 2163–2186.
14. Lions, P.L. (1982): Generalized solutions of Hamilton-Jacobi equations. Pitman, London
15. Lions, P.L., Souganidis, P.E. (1995): Convergence of MUSCL and filtered schemes for scalar conservation laws and Hamilton-Jacobi equations. Numer. Math., **69**, 441–470.
16. Liu, X.-D., Osher, S., Chan, T. (1994): Weighted essentially non-oscillatory schemes. J. Comp. Phys., **115**, 200–212.
17. Osher, S., Sethian, J. (1988): Fronts propagating with curvature dependent speed: algorithms based on Hamilton-Jacobi formulations. J. Comp. Phys., **79**, 12–49.
18. Osher, S., Shu, C.-W. (1991): High-order essentially nonoscillatory schemes for Hamilton-Jacobi equations. SIAM J. Numer. Anal., **28**, 907–922.
19. Souganidis, P.E. (1985): Approximation schemes for viscosity solutions of Hamilton-Jacobi equations. J. Diff. Equations, **59**, 1–43.
20. Zhang, Y.-T., Shu, C.-W. (2002): High-order WENO schemes for Hamilton-Jacobi equations on triangular meshes. Siam J. Sci. Comput., **24**, 1005–1030.

Numerical Simulation of Dislocation Dynamics

Vojtěch Minárik[1], Jan Kratochvíl[2], Karol Mikula[3] and Michal Beneš[1]

[1] Dept. of Mathematics, Faculty of Nuclear Sciences and Physical Engineering, Czech Technical University, Trojanova 13, 120 00 Praha 2, Czech Republic *minarikv@tjn.fjfi.cvut.cz*, *benes@tjn.fjfi.cvut.cz*

[2] Dept. of Physics, Faculty of Civil Engineering, Czech Technical University, Thákurova 7, 166 29 Praha 6, Czech Republic *kratochvil@fsv.cvut.cz*

[3] Dept. of Mathematics, Slovak University of Technology, Radlinskeho 11, 813 68 Bratislava, Slovakia *mikula@vox.svf.stuba.sk*

Summary. The aim of this contribution is to present the current state of our research in the field of numerical simulation of dislocations moving in crystalline materials. The simulation is based on recent theory treating dislocation curves and dipolar loops interacting by means of forces of elastic nature and hindered by the lattice friction. The motion and interaction of one dislocation curve and one dipolar loop placed in 3D space is considered. Equations of motion for a parametrically described dislocation curve are discretized by the flowing finite volume method in space. The interaction force is computed for each dipolar loop and along the discretized curve. The resulting system of ordinary differential equations is solved by a higher order time solver.

Physical background Plastic deformation of crystalline solids is a result of the motion of dislocations. The theory of dislocations is described in number of text books, e.g. [8]. Here we recall only the basic mobile properties of dislocations and the nature of their mutual interactions.

A dislocation is a line defect of the crystal lattice. Along the dislocation line the regular crystallographic arrangement of atoms is disturbed. The dislocation line is represented by a closed curve or a curve ending at the surface of the crystal. At low homologous temperatures the dislocations can move only along crystallographic planes (the slip planes) with the highest density of atoms. The motion results in mutual slipping of the neighboring parts of the crystal along the slip planes. The slip displacement carried by a single dislocation, called Burgers vector, is equal to one of the vectors connecting the neighboring atoms.

The displacement field of atoms from their regular crystallographic positions around a dislocation line can be treated (except the close vicinity of the line) as elastic stress and strain fields. On the other hand, a stress field exerts a force on a dislocation. The combination of these two effects causes the elastic interaction among dislocations.

One of the most distinguished features of plastic deformation at the microscale is a great overproduction of dislocations during a deformation process. Only a small fraction of generated dislocations is needed to carry plastic deformation, the rest is stored in the crystal. The deformed crystals supersaturated with dislocations tend to decrease the internal energy by mutual screening of their elastic fields. If dislocations possess a sufficient maneuverability provided by easy cross-slip (solids with wavy slip) the leading mechanism is an individual screening. The dislocations are stored in the form of dipoles which are transformed to prismatic dislocation dipolar loops of the prevailing edge character or such loops are directly formed (the experimental evidence is summarized in [9]).

The glide dislocations and the dislocation loops have much different characteristic length scales and mobile properties:

- While the segments of glide dislocations extend over distances of micrometers, the size of the prismatic dipolar loops is of the order of 10 nm.
- The glide dislocations are moved by the shear stress resolved in the slip plane, while the loops are drifted by stress gradients and/or swept by the glide dislocations. The loops being prismatic they can move along the direction parallel to the direction of their Burgers vector only.
- During deformation, the glide dislocations become curved. The local curvature of the glide dislocations seems to be one of the leading factors controlling the pattering [11, 12]. The loops can be approximately treated as rigid objects.

Due to the above mentioned complexity the formation of dislocation structures as a consequence of the interactions among dislocations is still an open problem. In this paper we will be concerned with a particular case: a dislocation curve interacting with a dipolar loop.

Dislocation Curve and Dipolar Loop We consider a plane dislocation segment with fixed ends; the segment represented by a plane curve can bow in a slip plane which is identified with the xz-plane of the coordinate system, i.e. $y = 0$. If the dislocation segment approaches a loop, the curve can pass by or the curve and the loop start to move together or the curve is stopped by the loop [13].

As the loop is allowed to move along the direction parallel to its Burgers vector **b** only (see Fig. 1a) just the force component in that direction causes the loop motion. Additionally, the lattice friction acts against the motion. The detailed condition for the dislocation curve and the loop is specified in Sect. .

The position of the loop is represented by 3 coordinates of its center. There are two types of dipolar loops: a *vacancy dipolar loop* and an *interstitial dipolar loop*; each type in two stable configurations [10] (see Fig. 2).

In vacancy loops a strip of atoms in regular crystallographic positions is missing. On the other hand; in interstitial loops an extra strip of atoms is added. For that reason the vacancy and interstitial loops produce different stress fields. The Burgers vectors of vacancy and interstitial loops have opposite directions. This fact we incorporate into our model by using negative value for the x-axis component of the Burgers vector for interstitial loop.

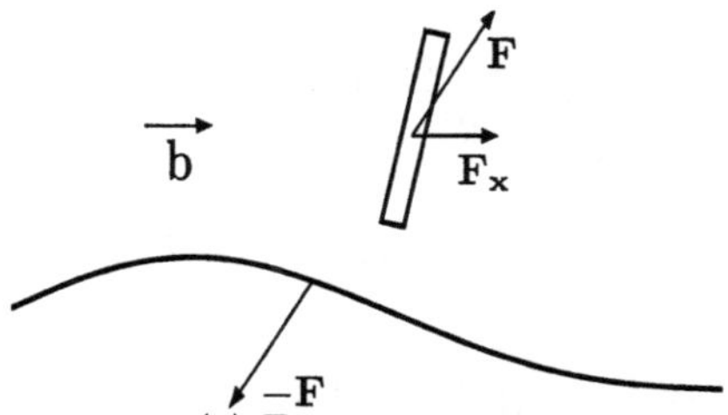

(a) Force interaction

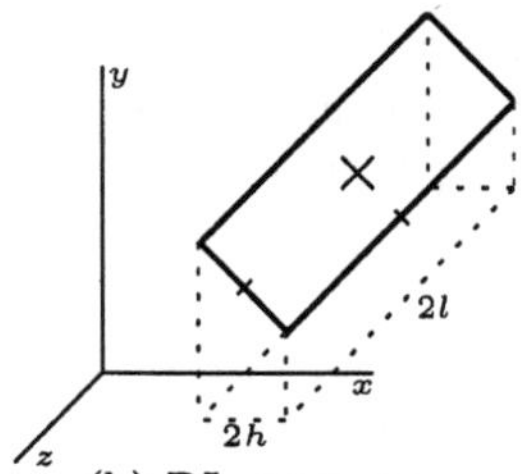

(b) DL geometry

Fig. 1. Dislocation dynamics problem geometry: (**a**) Dislocation curve and dipolar loop interaction; (**b**) Dipolar loop geometry

We denote the dipolar loop types and stable configurations by V_1, V_2 for the vacancy loop, and I_1, I_2 for the interstitial loop, respectively (see Fig. 2). In the mathematical model we represent the dipolar loop as a small rectangle

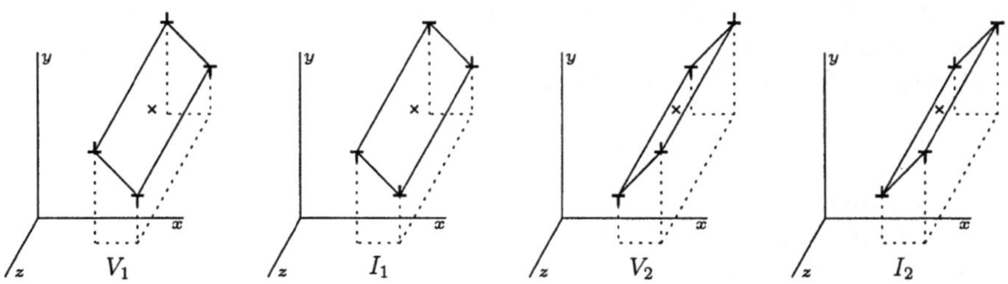

Fig. 2. Types and stable configurations of a dipolar loop. Longer sides of dipolar loop are parallel to the z-axis and lie in different layers of the atomic lattice.

with two longer sides parallel to the z-axis and two shorter sides parallel either to $[1,1,0]$ or $[1,-1,0]$ depending on the type of loop. The dimensions of dipolar loop are $2l$ and $2\sqrt{2}h$, respectively, see Fig. 1b.

Stress Field of Dipolar Loop Each type of dipolar loop produces a stress field. The formula for this field will be needed in the numerical simulation of the dislocation dynamics. In this work we use the stress field σ_{ij} presented by Kroupa et al [6, 7] (using Einstein's symbolic rule for sums):

$$\sigma_{ij} = -\frac{\mu}{4\pi(1-\nu)} \iint_A \frac{1}{\varrho^3} \left\{ \left[\frac{3(1-2\nu)}{\varrho^2} b_k \varrho_k \nu_n \varrho_n + (4\nu - 1) b_k \nu_k \right] \delta_{ij} \right. \tag{1}$$

$$+(1-2\nu)(b_i\nu_j + b_j\nu_i) + \frac{3\nu}{\varrho^2}\left[b_k\varrho_k(\nu_i\varrho_j + \nu_j\varrho_i) + \nu_k\varrho_k(b_i\varrho_j + b_j\varrho_i)\right]$$

$$\left. + \frac{3(1-2\nu)}{\varrho^2} b_k\nu_k\varrho_i\varrho_j - \frac{15}{\varrho^4} b_k\varrho_k\nu_n\varrho_n\varrho_i\varrho_j \right\} \mathrm{d}A \, .$$

In (1) we introduce following symbols ($i, j, k, n \in \{1, 2, 3\}$):

σ_{ij} $\sigma_{ij} = \sigma_{ij}(x, y, z)$ — components of the stress field tensor, depending on the position in space
μ shear modulus
ν Poisson's ratio
A area of the dipolar loop, with $\mathrm{d}A = 2h\sqrt{2}\mathrm{d}t$
b_i, b_j, b_k components of the Burgers vector
$\varrho_i, \varrho_j, \varrho_k, \varrho_n$ components of the relative position vector, $\varrho_1 = x, \varrho_2 = y, \varrho_3 = z$
ϱ relative distance from the dipolar loop, $\varrho = \sqrt{\varrho_1^2 + \varrho_2^2 + \varrho_3^2}$
$\nu_i, \nu_j, \nu_k, \nu_n$ components of the dipolar loop normal vector
δ_{ij} Kronecker symbol

The normal unit vector ν is chosen to be $\frac{1}{\sqrt{2}}[1, 1, 0]$ for dipolar loops of type V_1 and I_1, and $\nu = \frac{1}{\sqrt{2}}[1, -1, 0]$ for dipolar loops of type V_2 and I_2.

Dipolar Loop and Dislocation Curve Interaction The interaction force per unit length of dislocation line is given by the Peach-Koehler equation, which written for the i-th component reads:

$$f_i = \varepsilon_{ijk}\sigma_{jm}b_m s_k \, , \tag{2}$$

where we denote:

f_i i-th component of the interaction force per the unit length of the dislocation line
ε_{ijk} Levi-Civita symbol
σ_{jm} components of the stress field tensor at the dislocation position
b_m components of the Burgers vector
s_k components of unit vector **s** which has the direction of the dislocation line

Mathematical Model The dynamics of the system of a dislocation curve and a dipolar loop is governed by a system of two equations describing their motion. The motion law for the dislocation curve is represented by the well-known mean curvature flow equation (see e.g. [4, 1, 3])

$$B\nu = \kappa + F \tag{3}$$

where ν is the normal velocity of evolving curve, κ its curvature, B (drag coefficient) is a constant given by material and F represents external driving force.

The moving dislocation curve Γ_t can be parameterized by a smooth time dependent vector function $\mathbf{X} : I \times S \to R^2$, i.e., at any time t it is given as the Image($\mathbf{X}(.,t)$) $= \{\mathbf{X}(u,t), u \in S\}$ where S is a fixed parametrization interval and I is a time interval. For a smooth curve $|\mathbf{X}_u| > 0$ and for unit arc-length parametrization s, $ds = |\mathbf{X}_u| du$. Then $\mathbf{X}_s$ and $\mathbf{X}_s^{\perp}$ represent unit tangent and normal vectors, respectively. Using Frenet's formulae, the evolution equation (3) can be rewritten to the form of intrinsic diffusion equation [4, 1, 3, 5]

$$B\mathbf{X}_t = \mathbf{X}_{ss} + F\mathbf{X}_s^{\perp} \tag{4}$$

for the position vector $\mathbf{X}$. If we denote by $X^x(t,s)$ and $X^z(t,s)$ the components of the dislocation curve position vector in the xz-plane, then $\mathbf{X}_s^{\perp} = [X_s^z, -X_s^x]$. The equation (4) will be solved numerically to model a complicated dislocation curve dynamics.

Now we must explain exactly what is covered by the term F in (4). We know that F incorporates the interaction between the dislocation curve and the dipolar loop. To get detailed knowledge of F, we must go back to the Peach-Koehler equation (2). Assuming the dislocation curve can move only in the xz-plane and the dipolar loop can glide along the x-axis, we need to put f_x and f_z from the Peach-Koehler equation into the governing equations. Denoting the Burgers vectors of the dislocation curve and the dipolar loop $\mathbf{b}_{curve} = [b_{curve}, 0, 0]$ and $\mathbf{b} = [b, 0, 0]$, respectively, we get

$$f_x = \sigma_{xy} b_{curve} s_z \ , \quad f_z = -\sigma_{xy} b_{curve} s_x \ , \tag{5}$$

where $\sigma_{xy} = \sigma_{12}$, and s_x, s_z are the components of the dislocation curve's tangential vector, which can be also written as

$$[s_x, s_y, s_z] = \mathbf{X}_s^{\perp} = [X_s^z, 0, -X_s^x] \ . \tag{6}$$

Thus, the term $F = \sigma_{ext} b_{curve} + \sigma_{xy} b_{curve}$ covers the stress of the dipolar loop exerted on the dislocation curve, as well as any external stresses which may the material be exposed to. To be more precise, in order to obtain force vector acting at given position of the dislocation curve, one needs to multiply F and the dislocation curve's normal vector $\mathbf{X}_s^{\perp}$ at that position. We also explicitly write the dependence of F on the dislocation curve Γ_t because the curve position is required for the evaluation of its relative position to the dipolar loop. The obtained relative position is then used in the evaluation of σ_{xy}.

The stress σ_{xy} is given by (1), but we can simplify this formula for our specific situation – Burgers vector has only one non-zero component and the dipolar loop is a rectangle which has one of the two possible configurations. Under the assumption that the dipolar loop parameter h is very small, we can use Taylor expansion in (1) and integrate it to obtain an algebraic formula for the stress:

$$\sigma_{xy}(x,y,z) = -\frac{\mu hb}{2\pi(1-\nu)}\left\{\left[\frac{l-z}{\varrho_-}+\frac{l+z}{\varrho_+}\right]\left[\frac{x\pm y}{(x^2+y^2)^2}\left(\pm x+y-8\frac{x^2y}{x^2+y^2}\right)\right]\right.$$
$$+\left[\frac{l-z}{{\varrho_-}^3}+\frac{l+z}{{\varrho_+}^3}\right]\left[\pm\nu+\frac{xy}{(x^2+y^2)^2}\left(y^2-3x^2\mp 4xy\right)\right] \tag{7}$$
$$\left.+\left[\frac{l-z}{{\varrho_-}^5}+\frac{l+z}{{\varrho_+}^5}\right]\left[-\frac{3x^2y(x\pm y)}{x^2+y^2}\right]\right\},$$

where

$$\varrho_- = \sqrt{x^2+y^2+(l-z)^2}\,, \quad \varrho_+ = \sqrt{x^2+y^2+(l+z)^2}\,. \tag{8}$$

With the upper sign in (7) we get the stress formula for the dipolar loops V_1 and I_1, while with the lower sign we get the formula for V_2 and I_2 dipolar loops.

In order to obtain the equation governing the motion of the dipolar loop we have to sum all the stress contributions of the dislocation curve. It is enough to consider only the contributions in the x-axis direction because it is the only direction the dipolar loop is allowed to glide in.

The stress contribution of the dislocation curve can be obtained according to the action-reaction principle by simple reversing the sign of f_x in (5) and integrating along the dislocation curve:

$$F_x^c = \int_{\Gamma_t} \sigma_{xy} b_{curve} n_x \mathrm{d}s\,, \tag{9}$$

where n_x is the x-axis component of the dislocation curve element normal vector. Note it can be replaced using the derivatives of $\mathbf{X}$ with respect to the parametrization s since it holds $n_x = X_s^z$. Besides F_x^c there is one other kind of force — the friction force F_0 which is a constant given by the material and which acts against the gliding of the dipolar loop. Giving all the above information together, we come to the equation governing the gliding of the dipolar loop:

$$\frac{\mathrm{d}x}{\mathrm{d}t} = \frac{1}{BP}F_{x,total}(\Gamma_t, x(t))\,, \tag{10}$$

where $x(t)$ is the x-axis position of the dipolar loop, $P = 4(l+\sqrt{2}h)$ is the perimeter of the dipolar loop, and

$$F_{x,total}(\Gamma_t, x(t)) = \begin{cases} F_x^c - F_0 & \text{if } F_x^c > F_0 \\ 0 & \text{if } |F_x^c| < F_0 \\ F_x^c + F_0 & \text{if } F_x^c < -F_0\,. \end{cases} \tag{11}$$

The complete dislocation dynamics problem for one dislocation curve and one dipolar loop then follows when we put (4) and (10) together with initial and boundary conditions.

Numerical Scheme For discretization of the problem described earlier, we use the flowing finite volume method [5] in space and the method of lines [2] in time. Discrete solution is represented by a moving polygon given, at any time t, by plane points $\mathbf{X}_i, i = 0, ..., M$. The values $\mathbf{X}_0$ and $\mathbf{X}_M$ are prescribed in case of fixed ends of the curve. The segments $[\mathbf{X}_{i-1}, \mathbf{X}_i]$ are called flowing finite volumes. We construct also dual volumes $\mathcal{V}_i = \left[\mathbf{X}_{i-\frac{1}{2}}, \mathbf{X}_i\right] \cup \left[\mathbf{X}_i, \mathbf{X}_{i+\frac{1}{2}}\right]$, $i = 1, .., M-1$, where $\mathbf{X}_{i-\frac{1}{2}} = \frac{\mathbf{X}_{i-1}+\mathbf{X}_i}{2}$ (see Fig. 3).

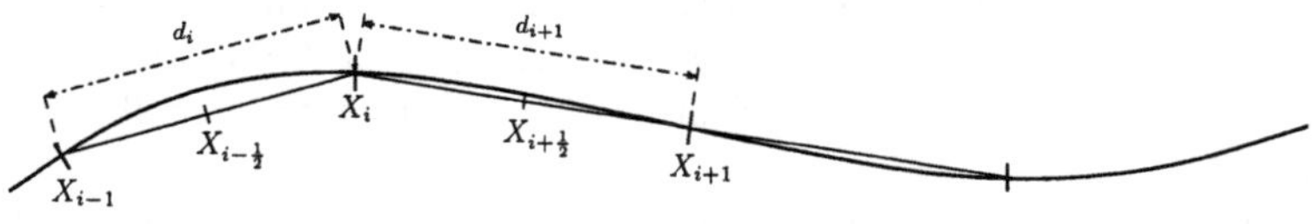

Fig. 3. Piecewise linear approximation of the dislocation curve

Integrating evolution equation (4) in dual volume $\mathcal{V}_i$ we obtain

$$\int_{\mathcal{V}_i} B\frac{\partial \mathbf{X}}{\partial t}\mathrm{d}s = \int_{\mathcal{V}_i} \frac{\partial^2 \mathbf{X}}{\partial s^2}\mathrm{d}s + \int_{\mathcal{V}_i} F\left(\frac{\partial \mathbf{X}}{\partial s}\right)^{\perp} \mathrm{d}s. \tag{12}$$

Then we simply get

$$B\frac{d_i + d_{i+1}}{2}\frac{\mathrm{d}\mathbf{X}_i}{\mathrm{d}t} = \left[\frac{\partial \mathbf{X}}{\partial s}\right]_{\mathbf{X}_{i-\frac{1}{2}}}^{\mathbf{X}_{i+\frac{1}{2}}} + F_i\left[\mathbf{X}^{\perp}\right]_{\mathbf{X}_{i-\frac{1}{2}}}^{\mathbf{X}_{i+\frac{1}{2}}} \tag{13}$$

where

$$d_i = |\mathbf{X}_i - \mathbf{X}_{i-1}| = \sqrt{(X_i^x - X_{i-1}^x)^2 + (X_i^z - X_{i-1}^z)^2} \tag{14}$$

and F_i is a constant approximation of F in dual volume $\mathcal{V}_i$, $F_i = \sigma_{xy}(\mathbf{R_i})b_{curve}$, where $\mathbf{R_i} = \mathbf{X}_i - [x(t), y, z]$ is the relative positional vector of $\mathbf{X}_i$ and the dipolar loop center. If we replace the terms on the right-hand side by finite differences and averaged values, respectively, we end up with the system of ordinary differential equations

$$B\frac{\mathrm{d}\mathbf{X}_i}{\mathrm{d}t} = \frac{2}{d_i+d_{i+1}}\left(\frac{\mathbf{X}_{i+1}-\mathbf{X}_i}{d_{i+1}} - \frac{\mathbf{X}_i-\mathbf{X}_{i-1}}{d_i}\right) + \frac{2}{d_i+d_{i+1}}F_i\frac{\mathbf{X}_{i+1}^{\perp}-\mathbf{X}_{i-1}^{\perp}}{2}, \quad i = 1, \ldots, M-1\,. \tag{15}$$

In discretization of the governing equation (9) for the dipolar loop we sum contributions of every curve segment to obtain

$$F_x^c = \sum_{i=0}^{M-1} \sigma_{xy}\left(X_{i+\frac{1}{2}}^x - x(t), -y, X_{i+\frac{1}{2}}^z - z\right) b_{curve}\left(X_{i+1}^z - X_i^z\right)\,, \tag{16}$$

where $[x(t), y, z]$ is the center of the dipolar loop at time t. Next, we use formula (11) applied to discrete F_x^c defined above and get

$$\frac{\mathrm{d}x}{\mathrm{d}t} = \frac{1}{BP} F_{x,total}(\mathbf{X}_0, \ldots, \mathbf{X}_M, x). \tag{17}$$

The complete discrete problem consists of (15) and (17) with accompanying initial and boundary conditions.

Results of Numerical Experiments We made several numerical simulations in which we used different settings. For the basic physical parameters we used values which were experimentally measured for nickel crystals at room temperature [10]: average length and width of the dipolar loop $l = 60$ nm, $h = 4$ nm, Burgers vector $b = 0.26$ nm, shear modulus $\mu = 80$ GPa, and Poisson's ratio $\nu = 0.33$. When not specified otherwise, we used drag coefficient $B = 10^{-5}$ Pa s.

In the simulations we were changing not only the type and initial position of the dipolar loop, but also the initial shape of the dislocation curve and the value of friction force. We observed following facts (not all of them can be demostrated here):

- For the dislocation curve with fixed ends the curvature acts against the external stress. Therefore, there exists some equilibrium shape the dislocation curve tends to.
- When no external stress is applied, the dislocation curve of any initial shape tends to the straight line (potential energy minimization). Adding dipolar loop, an oscillating motion (Fig. 4) of the dipolar loop as well as the dislocation curve can occur.
- The direction in which the dipolar loop leaves the dislocation curve's interaction region depends on the type of the dipolar loop. Simply, V_2 shifts to the left where V_1 shifts to the right.

Acknowledgements The first author was partly supported by the NSF grant NSF00-138 Award 0113555, the second author was partly supported by the project of the Grant Agency of the Czech Republic No. 201/01/0676, the third author was partly supported by the project KONTAKT No.ME654 and Grant Agency of the Czech Republic No.106/03/0826 and the fourth author was partly supported by the grant VEGA 1/0313/03.

References

1. Angenent, S.B., Gurtin, M.E. (1989): Multiphase Thermomechanics with an Interfacial Structure 2. Evolution of an Isothermal Interface. Arch. Rat. Mech. Anal., **108**, 323–391
2. Beneš, M., Mikula, K. (1998): Simulation of Anisotropic Motion by Mean Curvature - Comparison of Phase Field and Sharp Interface Approaches. Acta Math. Univ. Comenianae, Vol. **LXVII**, **1**, 17–42
3. Dziuk, G. (1994): Convergence of a Semi Discrete Scheme for the Curve Shortening Flow. Mathematical Models and Methods in Applied Sciences, **4**, 589–606
4. Gage, M., Hamilton, R.S. (1986): The Heat Equation Shrinking Convex Plane Curves. J. Diff. Geom., **23**, 69–96

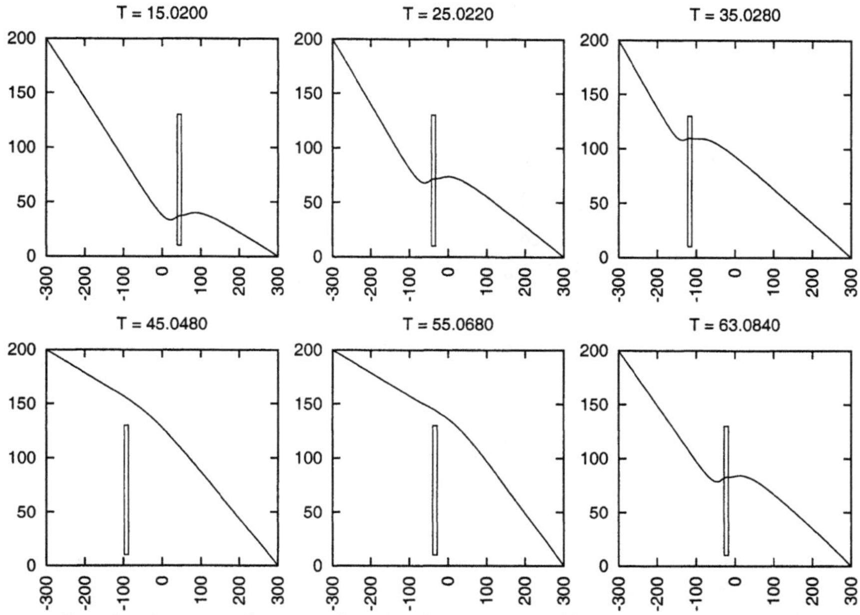

Fig. 4. Dipolar loop oscillations. Dipolar loop of type V_1 starts to glide to the left of the dislocation curve (timelevels $T = 15.02$, $T = 25.022$, $T = 35.028$). Then it reverses as the attractive force of the dislocation curve gains the control over the system for some time ($T = 45.048$, $T = 55.068$). Second reversing occurs before $T = 63.084$.

5. Mikula, K., Ševčovič, D. (2001): Evolution of Plane Curves Driven by a Nonlinear Function of Curvature and Anisotropy. SIAM J. Appl. Math., Vol. **61**, No. **5**, 1473–1501
6. Kroupa, F. (1965): Long-Range Elastic Field of Semi-Infinite Dislocation Dipole and of Dislocation Jog. phys.stat.sol., Vol. **9**, 27–32
7. Verecký, Š., Kratochvíl, J., Kroupa, F. (2002): The Stress Field of Rectangular Prismatic Dislocation Loops. phys.stat.sol. (a), Vol. **191**, 418–426
8. Hirth, J.P., Lothe, J. (1982): Theory of Dislocations. John Willey
9. Kratochvíl, J., Saxlová, M. (1993): Dislocation Pattern Formation and Strain Hardening in Solids. Physica Scripta, **T49**, 399–404
10. Tippelt, B., Bretschneider, J., Hähner, P. (1997): The Dislocation Microstructure of Cyclically Deformed Nickel Single Crystals at Different Temperatures. phys. stat. sol. (a), **163**, 11–26
11. Saxlová, M., Kratochvíl, J., Zatloukal, J. (1997): The Model of Formation and Disintegration of Vein Dislocation Structure. Materials Science and Engineering, **A234-236**, 205–208
12. Kratochvíl, J. (2001): Self-organization Model of Localization of Cyclic Strain into PSBs and Formation of Dislocation Wall Structure. Materials Sci. Eng. A, **A309-310**, 331–335
13. Kratochvíl, J., Kroupa, F., Kubin, L.P. (1999): The Sweeping of Dipolar Loops in Cyclic Deformation: Kinetic Diagrams. In: Proceedings of the 20th Risoe Int.

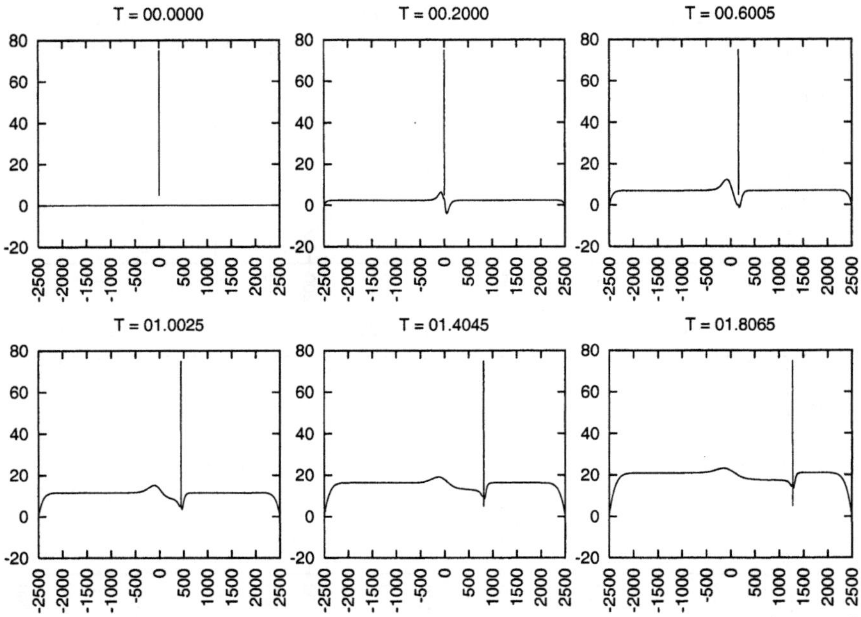

Fig. 5. Dipolar loop swept by the curve; on the other hand, the curve is distorted by the stress field of the loop. In this test there were used: $\mu = 80$ GPa, $\nu = 0.33$, $B = 10^{-4}$ Pa s, $b = 0.707$ nm, $l = 35$ nm, $h = \sqrt{2}$ nm, $F_0 = 4$ MPa m, applied stress $\sigma_a = -1.155$ MPa. Initial position of dipolar loop was $[0, 40, -30]$. The subsequent stages are shown for increasing time T.

Symposium on Material Science: Deformation-Induced Microstructures: Analysis and Relation to Properties. Risoe National Laboratory, Roskilde, Denmark, 387–392

Implicit FEM-FCT algorithm for compressible flows

Matthias Möller[1], Dmitri Kuzmin and Stefan Turek

Institute of Applied Mathematics (LSIII), University of Dortmund, Vogelpothsweg 87, D-44227 Dortmund, Germany, [1]*matthias.moeller@math.uni-dortmund.de*

Summary. The flux-corrected transport (FCT) methodology is generalized to implicit finite element schemes and applied to the Euler equations of gas dynamics. The underlying low-order scheme is constructed by applying scalar artificial viscosity proportional to the spectral radius of the cumulative Roe matrix. All conservative matrix manipulations are performed edge-by-edge which leads to an efficient algorithm for the matrix assembly. The outer defect correction loop is equipped with a block-diagonal preconditioner so as to decouple the discretized Euler equations and solve all equations individually. As an alternative, a strongly coupled solution strategy is investigated in the context of stationary problems which call for large time steps.

1 Introduction

The concepts of flux-corrected transport can be traced back to the celebrated SHASTA scheme proposed by Boris and Book [1] in the early 1970s. Later, their algorithm was superseded by Zalesak's multidimensional limiter [11] and carried over to finite elements by Löhner and his coworkers [7].

In recent publications [3], [4], [5] we presented a generalization of this approach to implicit finite element discretizations. A notable benefit of the new FEM-FCT formulation was the representation of anti-/diffusive terms as sums of skew-symmetric internodal fluxes. Moreover, an iterative limiting strategy was introduced which prevents the limiter from getting overly diffusive for large time steps. In this paper, we concentrate on flux correction for the Euler equations of gas dynamics and discuss various algorithmic aspects pertinent to the treatment of nonlinear hyperbolic systems.

2 FEM-FCT for scalar equations

As a model problem, consider the generic conservation law $\frac{\partial u}{\partial t} + \nabla \cdot (\mathbf{v}u) = 0$ where $\mathbf{v} = \mathbf{v}(\mathbf{x}, t)$ is a nonuniform velocity field. Let us employ standard Galerkin FEM for the discretization in space and interpolate the convective fluxes in much the same way as the sought solution [2]. After mass lumping, we get an ODE system given by

$$M_L \frac{\mathrm{d}u}{\mathrm{d}t} = Ku \qquad \Leftrightarrow \qquad m_i \frac{\mathrm{d}u_i}{\mathrm{d}t} = \sum_{j \neq i} k_{ij}(u_j - u_i) + \delta_i u_i, \quad \delta_i = \sum_j k_{ij} \quad (1)$$

where $M_L = \mathrm{diag}\{m_i\}$ denotes the 'lumped' mass matrix and $K = \{k_{ij}\}$ stands for the discrete transport operator. The second expression in (1) corresponds to a single node i where the first term in the right-hand side is engendered by the incompressible part of the transport operator. Provided that $k_{ij} \geq 0$, $\forall j \neq i$, a system of such form is called *local extremum diminishing* in the absence of the term $\delta_i u_i$ which vanishes for divergence-free velocity fields and is responsible for the physical growth of local extrema otherwise.

One major ingredient of any FCT algorithm is the nonoscillatory positivity-preserving low-order scheme which can be constructed by applying artificial diffusion $D = \{d_{ij}\}$ so as to render all off-diagonal entries of the linear operator $L = K + D$ nonnegative [3]. Hence, the optimal diffusion coefficients are given by

$$d_{ij} = d_{ji} = \max\{0, -k_{ij}, -k_{ji}\}, \qquad d_{ii} = -\sum_{j \neq i} d_{ij}. \qquad (2)$$

Since discrete diffusion operators have zero row/column sums [3] the diffusive term can be decomposed into skew-symmetric fluxes $(Du)_i = \sum_{j \neq i} f_{ij}$, where $f_{ij} = d_{ij}(u_j - u_i)$. Thus, the modifications in (2) are conservative. Starting with the Galerkin operator $L := K$, the low-order operator can be constructed by applying artificial diffusion edge-by-edge for each pair of nodes i and j whose basis functions have overlapping support

$$\begin{aligned} l_{ii} &:= l_{ii} - d_{ij}, \qquad & l_{ij} &:= l_{ij} + d_{ij}, \\ l_{ji} &:= l_{ji} + d_{ij}, \qquad & l_{jj} &:= l_{jj} - d_{ij}. \end{aligned} \qquad (3)$$

The *discrete upwinding* technique presented above carries over to multidimensions and yields the least diffusive linear LED scheme. However, **linear** monotonicity-preserving methods can be at most first-order accurate. As a consequence, compensating antidiffusion which constitutes the difference between the discretizations of high and low order

$$P_i = \sum_{j \neq i} f_{ij}, \qquad f_{ij} = -\left(m_{ij} \frac{\mathrm{d}}{\mathrm{d}t} + d_{ij}\right)(u_j - u_i), \quad f_{ji} = -f_{ij} \qquad (4)$$

is to be constructed in a **nonlinear** way so as to remove the excessive artificial diffusion.

After an implicit time discretization we obtain a nonlinear algebraic system

$$M_L \frac{u^{n+1} - u^n}{\Delta t} = \theta L u^{n+1} + (1 - \theta) L u^n + P(u^{n+1}, u^n), \qquad 0 < \theta \leq 1 \quad (5)$$

which can be solved via the fixed-point defect correction scheme

$$u^{(m+1)} = u^{(m)} + A^{-1} r^{(m)}, \qquad u^{(0)} = u^n, \qquad m = 0, 1, 2, \ldots. \qquad (6)$$

Here, A is a 'suitable' preconditioner and $r^{(m)} = b^{(m+1)} - Au^{(m)}$ is the defect vector for the m-th iteration cycle. The latter incorporates the constant right-hand side stemming from the low-order scheme plus compensating antidiffusion

$$b^{(m+1)} = b^n + \sum_{j \neq i} \alpha_{ij} f_{ij}^{(m)}, \qquad \text{where} \quad b^n = [M_L + (1-\theta)\Delta t L]u^n. \tag{7}$$

Varying the correction factors α_{ij} between zero and unity, one can blend the high-order method with the concomitant low-order one. The latter should be used in the vicinity of steep gradients where spurious oscillations are likely to arise. The construction of the solution-dependent correction factors and of the fully discretized antidiffusive fluxes $f_{ij}^{(m)}$ is elucidated in [5]. As an alternative to (7), an iterative limiting strategy was proposed for implicit FEM-FCT schemes operated at large time steps. Roughly speaking, the amount of previously accepted antidiffusion is taken into account so that only the rejected portion of the antidiffusive flux needs to be limited at subsequent defect correction steps.

3 Euler equations

Compressible flows are governed by the Euler equations which represent a system of conservation laws for the mass, momentum and energy of an inviscid fluid. These hyperbolic PDEs are typically written in divergence form

$$\frac{\partial U}{\partial t} + \nabla \cdot \mathbf{F} = 0, \qquad \text{where} \quad \nabla \cdot \mathbf{F} = \sum_{d=1}^{3} \frac{\partial F^d}{\partial x_d}. \tag{8}$$

The vector of conservative variables U and the triple of fluxes $\mathbf{F} = (F^1, F^2, F^3)$ for each direction of the Cartesian coordinate system are defined as follows

$$U = \begin{bmatrix} \rho \\ \rho \mathbf{v} \\ \rho E \end{bmatrix}, \quad \mathbf{F} = \begin{bmatrix} \rho \mathbf{v} \\ \rho \mathbf{v} \otimes \mathbf{v} + p\mathcal{I} \\ \rho H \mathbf{v} \end{bmatrix}. \tag{9}$$

Here, ρ, $\mathbf{v}$, p, E and $H = E + p/\rho$ stand for the density, velocity, pressure, total energy per unit mass and stagnation enthalpy, respectively. This system is completed by an equation of state $p = (\gamma - 1)\rho(E - |\mathbf{v}|^2/2)$, where $\gamma = c_p/c_v$ denotes the ratio of specific heats for a polytropic gas ($\gamma = 1.4$ for air).

By application of the chain rule, the Euler equations can be written in an equivalent quasi-linear formulation in terms of the Jacobian matrices $\mathbf{A} = (A^1, A^2, A^3)$

$$\frac{\partial U}{\partial t} + \mathbf{A} \cdot \nabla U = 0, \qquad \text{where} \quad \mathbf{A} \cdot \nabla U = \sum_{d=1}^{3} A^d \frac{\partial U}{\partial x_d}, \quad A^d = \frac{\partial F^d}{\partial U}. \tag{10}$$

4 Galerkin matrix assembly

In what follows, an efficient edge-based assembly technique for the standard Galerkin discretization of the Euler equations is presented. Let us start with the divergence form (8) and interpolate the fluxes using the group finite element formulation [2] which yields an ODE system similar to (1)

$$M_C \frac{\mathrm{d}\mathrm{U}}{\mathrm{d}t} = K\mathrm{U}. \tag{11}$$

Here, M_C denotes the block-diagonal consistent mass matrix for the coupled system and K is a discrete counterpart of the operator $-\mathbf{A} \cdot \nabla$ for the quasi-linear formulation (10). Let the entries of the mass matrix and the vector of coefficients coming from the discretization of space derivatives be defined as follows

$$m_{ij} = \int_\Omega \varphi_i \varphi_j \,\mathrm{d}\mathbf{x}, \qquad \mathbf{c}_{ij} = \int_\Omega \varphi_i \nabla \varphi_j \,\mathrm{d}\mathbf{x}. \tag{12}$$

As long as the mesh is fixed, the coefficients $\mathbf{c}_{ij}$ remain constant and thus the operator K can be assembled efficiently without resorting to a costly numerical integration.

Recall that basis functions sum to unity, so that the sum of their derivatives vanishes. Hence, the coefficients $\mathbf{c}_{ij}$ satisfy $\mathbf{c}_{ii} = -\sum_{j \neq i} \mathbf{c}_{ij}$ and the right-hand side of the five coupled equations for node i is given by

$$(K\mathrm{U})_i = -\sum_j \mathbf{c}_{ij} \cdot \mathbf{F}_j = -\sum_{j \neq i} \mathbf{c}_{ij} \cdot (\mathbf{F}_j - \mathbf{F}_i). \tag{13}$$

In his pioneering work on approximate Riemann solvers [9], Roe showed that the differences between the components of $\mathbf{F}$ and U are related by $\mathbf{F}_j - \mathbf{F}_i = \hat{\mathbf{A}}_{ij}(\mathrm{U}_j - \mathrm{U}_i)$, where the triple of matrices $\hat{\mathbf{A}}_{ij} = (\hat{A}^1_{ij}, \hat{A}^2_{ij}, \hat{A}^3_{ij})$ corresponds to the Jacobian tensor $\mathbf{A}$ evaluated for the special set of density-averaged variables

$$\hat{\rho}_{ij} = \sqrt{\rho_i \rho_j}, \qquad \hat{\mathbf{v}}_{ij} = \frac{\sqrt{\rho_i}\mathbf{v}_i + \sqrt{\rho_j}\mathbf{v}_j}{\sqrt{\rho_i} + \sqrt{\rho_j}}, \qquad \hat{H}_{ij} = \frac{\sqrt{\rho_i}H_i + \sqrt{\rho_j}H_j}{\sqrt{\rho_i} + \sqrt{\rho_j}}. \tag{14}$$

This enables us to express the nodal value $(K\mathrm{U})_i$ in terms of the conservative variables

$$(K\mathrm{U})_i = -\sum_{j \neq i} \mathbf{c}_{ij} \cdot \hat{\mathbf{A}}_{ij}(\mathrm{U}_j - \mathrm{U}_i), \qquad \text{where} \quad \mathbf{c}_{ij} \cdot \hat{\mathbf{A}}_{ij} = \sum_{d=1}^{3} c^d_{ij} \hat{A}^d_{ij}. \tag{15}$$

The dot product can be interpreted as a 'projection' of the triple $\hat{\mathbf{A}}_{ij}$ onto the numerical edge $\mathbf{ij}$. For our purposes, it is expedient to introduce the splitting $\mathbf{c}_{ij} \cdot \hat{\mathbf{A}}_{ij} = -(\mathrm{A}_{ij} + \mathrm{B}_{ij})$, where the two components of the *cumulative Roe matrix* are defined by [5]

$$\mathrm{A}_{ij} = \mathbf{a}_{ij} \cdot \hat{\mathbf{A}}_{ij}, \qquad \mathbf{a}_{ij} = -\frac{\mathbf{c}_{ij} - \mathbf{c}_{ji}}{2}, \tag{16}$$

$$\mathrm{B}_{ij} = \mathbf{b}_{ij} \cdot \hat{\mathbf{A}}_{ij}, \qquad \mathbf{b}_{ij} = -\frac{\mathbf{c}_{ij} + \mathbf{c}_{ji}}{2}. \tag{17}$$

A similar decomposition can be performed for the contribution of the edge **ij** to $(K\mathrm{U})_j$

$$\mathbf{c}_{ji} \cdot (\mathbf{F}_j - \mathbf{F}_i) = \mathbf{c}_{ji} \cdot \hat{\mathbf{A}}_{ij}(\mathrm{U}_j - \mathrm{U}_i), \qquad \text{where} \quad \mathbf{c}_{ji} \cdot \hat{\mathbf{A}}_{ij} = \mathrm{A}_{ij} - \mathrm{B}_{ij}. \tag{18}$$

Integration by parts reveals that in the interior of the domain $\mathbf{a}_{ij} = -\mathbf{c}_{ij}$ while $\mathbf{b}_{ij} = 0$ [5], so that only the skew-symmetric part A_{ij} needs to be evaluated for interior edges. The symmetric part B_{ij} only applies to the cumulative Roe matrices for boundary edges. According to (15)–(18), the contribution of the edge **ij** to the term $K\mathrm{U}$ reads

$$(\mathrm{A}_{ij} + \mathrm{B}_{ij})(\mathrm{U}_j - \mathrm{U}_i) \quad \longrightarrow \quad (K\mathrm{U})_i, \tag{19}$$

$$(\mathrm{A}_{ij} - \mathrm{B}_{ij})(\mathrm{U}_j - \mathrm{U}_i) \quad \longrightarrow \quad (K\mathrm{U})_j. \tag{20}$$

Together with the fact that the coefficients $\mathbf{c}_{ij}$ remain constant and thus can be assembled and stored once and for all during the initialization process, (19)–(20) suggest an efficient edge-based algorithm for the matrix assembly. The underlying data structure can be generated from the sparsity pattern of the finite element matrix and contains entries for all pairs of nodes whose basis functions have overlapping supports [5]. In contrast to the scalar case this connectivity exists not only between basis functions for different nodes but also between those for different variables. Hence, each coefficient of the discrete operator is given by a square matrix of dimension equal to the number of variables.

It can be readily inferred from (19)–(20) that the contribution of the numerical edge **ij** to the global matrix $K \in \mathbb{R}^{5N\times 5N}$ is given by

$$\begin{aligned} \mathrm{K}_{ii} &= -\mathrm{A}_{ij} - \mathrm{B}_{ij}, \qquad & \mathrm{K}_{ij} &= \mathrm{A}_{ij} + \mathrm{B}_{ij}, \\ \mathrm{K}_{ji} &= -\mathrm{A}_{ij} + \mathrm{B}_{ij}, \qquad & \mathrm{K}_{jj} &= \mathrm{A}_{ij} - \mathrm{B}_{ij}. \end{aligned} \tag{21}$$

These local Jacobians are evaluated edge-by-edge and their entries K_{ij}^{kl}, $k, l = 1, \ldots, 5$ are scattered to the corresponding positions in the 25 blocks $K_{kl} \in \mathbb{R}^{N\times N}$ [5].

5 Artificial viscosity

To a large extent, the ability of a FEM-FCT algorithm to withstand the formation of wiggles depends on the quality of the underlying low-order method. For scalar transport equations we derived the least diffusive positivity-preserving scheme by elimination of negative off-diagonal entries from the discrete transport operator.

In [5] the LED-principle was generalized to hyperbolic systems by rendering all off-diagonal matrix blocks positive semi-definite. If we perform mass lumping and replace the high-order operator K in (11) by the low-order one we obtain the ODE system

$$M_L \frac{\mathrm{d}\mathrm{U}}{\mathrm{d}t} = L\mathrm{U}, \qquad \text{where} \qquad \begin{array}{ll} \mathrm{L}_{ii} = -\mathrm{A}_{ij} - \mathrm{D}_{ij}, & \mathrm{L}_{ij} = \mathrm{A}_{ij} + \mathrm{D}_{ij}, \\ \mathrm{L}_{ji} = -\mathrm{A}_{ij} + \mathrm{D}_{ij}, & \mathrm{L}_{jj} = \mathrm{A}_{ij} - \mathrm{D}_{ij}. \end{array} \tag{22}$$

The low-order operator L is constructed in much the same way as (21) by applying tensorial artificial viscosity $\mathrm{D}_{ij} \in \mathbb{R}^{5\times 5}$ to the Roe matrices. The global matrix assembly can be adopted from the previous section. The missing symmetric boundary part B_{ij} of the cumulative Roe matrix is incorporated into the raw antidiffusive fluxes

$$\mathrm{F}_{ij} = -\left(\mathrm{M}_{ij}\frac{\mathrm{d}}{\mathrm{d}t} + \mathrm{D}_{ij} - \mathrm{B}_{ij}\right)(\mathrm{U}_j - \mathrm{U}_i), \qquad \mathrm{F}_{ji} = -\mathrm{F}_{ij}, \tag{23}$$

where $\mathrm{M}_{ij} = m_{ij}\mathrm{I}$ denotes the local diagonal mass matrix.

The hyperbolicity of the Euler equations implies that any linear combination of the three Jacobian matrices is diagonalizable with real eigenvalues, such that the cumulative Jacobian matrix admits the following factorization

$$\mathrm{A}_{ij} = R(\mathbf{a}_{ij})\Lambda(\mathbf{a}_{ij})R(\mathbf{a}_{ij})^{-1}. \tag{24}$$

Let us 'project' the density-averaged velocity $\hat{\mathbf{v}}_{ij}$ onto the numerical edge $\mathbf{ij}$ and define the local speed of sound as follows

$$\hat{v}_{ij} = \frac{\mathbf{a}_{ij}\cdot\hat{\mathbf{v}}_{ij}}{|\mathbf{a}_{ij}|}, \qquad \hat{c}_{ij} = \sqrt{(\gamma-1)\left(\hat{H}_{ij} - \frac{|\hat{\mathbf{v}}_{ij}|^2}{2}\right)}. \tag{25}$$

Here, $|\mathbf{a}_{ij}|$ denotes the Euclidean norm of the coefficient vector $\mathbf{a}_{ij}$. As a consequence, the diagonal matrix of eigenvalues can be readily computed as

$$\Lambda(\mathbf{a}_{ij}) = |\mathbf{a}_{ij}|\mathrm{diag}\left\{\hat{v}_{ij} - \hat{c}_{ij}, \hat{v}_{ij}, \hat{v}_{ij}, \hat{v}_{ij}, \hat{v}_{ij} + \hat{c}_{ij}\right\}. \tag{26}$$

In [5] we gave a detailed description of how to derive a generalization of Roe's approximate Riemann solver from (24) by elimination of negative eigenvalues.

A much cheaper alternative is to add scalar dissipation proportional to the spectral radius of the Roe matrix $d_{ij} = |\mathbf{a}_{ij}|(|\hat{v}_{ij}| + \hat{c}_{ij})$ [5]. The resulting artificial viscosity operator $\mathrm{D}_{ij} = d_{ij}\mathrm{I}$, which in fact is the same for all components, needs to be applied only to the five diagonal blocks of the finite element matrix. Numerical examples demonstrate, that in the framework of flux correction the final solution even benefits from this slightly overdiffusive low-order scheme because of an improvement in the phase accuracy [4], so that the application of a costly Riemann solver does not pay off.

6 Defect correction

After an implicit time discretization we obtain a nonlinear algebraic system similar to (5) which can also be solved by the defect correction scheme

$$\mathrm{U}^{(m+1)} = \mathrm{U}^{(m)} + [A^{(m)}]^{-1}\mathrm{R}^{(m)}, \qquad \mathrm{R}^{(m)} = \mathrm{B}^{(m+1)} - A^{(m)}\mathrm{U}^{(m)}. \tag{27}$$

In a practical implementation the 'inversion' of A is performed by applying some inner iteration to solve the linear subproblem for the solution increment (an improvement of the residual by 1–2 digits suffices) and update the last iterate thereafter

$$A^{(m)}\Delta\mathrm{U}^{(m)} = \mathrm{R}^{(m)}, \qquad \mathrm{U}^{(m+1)} = \mathrm{U}^{(m)} + \Delta\mathrm{U}^{(m)}, \qquad \mathrm{U}^{(0)} = \mathrm{U}^{n}. \tag{28}$$

The matrix in the left-hand side of this linear system can be replaced by a (block-diagonal) preconditioner $C^{(m)}$, so as to decouple the discretized Euler equations [5]. As an alternative we can apply the (preconditioned) BiCGSTAB method directly to the coupled system (28).

Let us split the low-order operator into its diagonal, subdiagonal and superdiagonal parts $A^{(m)} = W^{(m)} + J^{(m)} + E^{(m)}$. In [5], a block-Jacobi preconditioner $C^{(m)} = J^{(m)}$ was suggested for the defect correction scheme, so that only the five diagonal blocks need to be assembled and stored

$$C_{kk}^{(m)} = J_k^{(m)} = M_L - \theta\Delta t L_{kk}^{(m)}, \ \forall k \qquad \text{and} \qquad C_{kl}^{(m)} = 0, \ \forall l \neq k. \tag{29}$$

As a consequence, the linear system (28) resolves into a sequence of scalar subproblems which can be solved separately or at best in parallel. For this purpose an iterative method, e.g. (preconditioned) BiCGSTAB or geometric multigrid, is applied to

$$\begin{aligned} J_k^{(m)}\Delta u_k^{(m)} &= r_k^{(m)}, \qquad k = 1,\dots,5 \\ u_k^{(m+1)} &= u_k^{(m)} + \Delta u_k^{(m)}, \ u_k^{(0)} = u_k^n. \end{aligned} \tag{30}$$

However, this segregated solution approach disqualifies for larger time steps since severe convergence problems of the outer iteration can be observed.

Longing for the full potential of the iterative limiter and the unconditional positivity of fully implicit time-stepping we have been testing some coupled solution strategies for (28) by means of a preconditioned BiCGSTAB method. In this case, additional blocks of the low-order operator may need to be assembled and stored or there should be another (direct) way to carry out the matrix-vector multiplications for updating the residual.

Let the preconditioner for the BiCGSTAB solver be given by $W^{(m)} + J^{(m)}$, where

$$W_{kl}^{(m)} = M_L - \theta\Delta t L_{kl}^{(m)}, \ \forall l < k, \qquad \text{and} \qquad W_{kl}^{(m)} = 0, \ \forall l \geq k, \tag{31}$$

$$E_{kl}^{(m)} = M_L - \theta\Delta t L_{kl}^{(m)}, \ \forall l > k, \qquad \text{and} \qquad E_{kl}^{(m)} = 0, \ \forall l \leq k. \tag{32}$$

This corresponds to the block-Gauss-Seidel scheme

$$J^{(m)}\Delta\mathrm{U}^{(m+1)} = \mathrm{R}^{(m)} - W^{(m)}\Delta\mathrm{U}^{(m+1)} - E^{(m)}\Delta\mathrm{U}^{(m)}. \tag{33}$$

Swapping the sub- and superdiagonal block-matrices in the equation above gives another variation of this algorithm. The alternating application of both subdiagonal and superdiagonal block-matrices results in a symmetric block-Gauss-Seidel approach. Recall that $J^{(m)}$ in equation (33) is a block-diagonal matrix for which each block corresponds to a scalar problem similar to (30). They can be solved in much the same way as for the segregated solution approach. The design of 'optimal' preconditioners is a nontrivial task which constitutes an important field for future research.

7 Numerical examples

Let us illustrate the potential of the implicit FEM-FCT algorithm by considering a steady two-dimensional supersonic flow over a wedge. Here, the free-stream Mach number is $M = 2.5$ and the deflection angle is $\theta = 15°$. The results presented at the top of Figure 1 are computed on a mesh of 128×128 bilinear elements by the low-order scheme (left) and the implicit FEM-FCT algorithm (right), respectively. It can be readily seen that the shock wave is unacceptably smeared by the low-order method. Nevertheless, both the upstream and downstream Mach numbers are predicted correctly. The iterative flux limiter resolves the shock very precisely within as few as 3–4 elements.

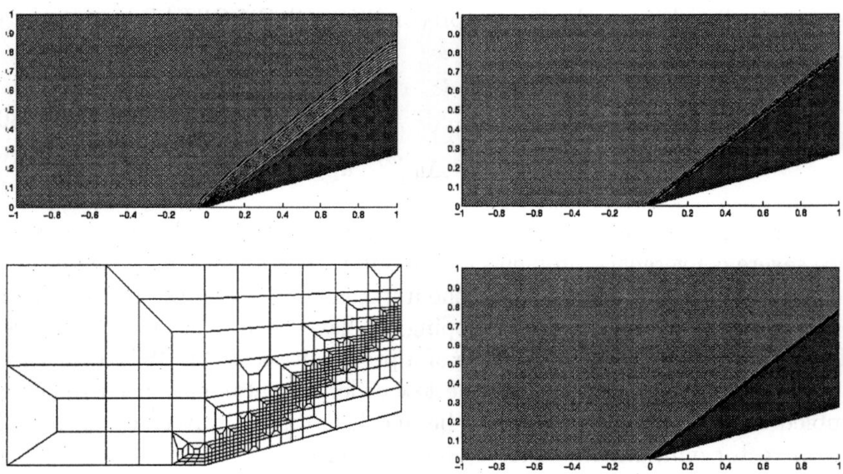

Fig. 1. Compression corner. Oblique shock at $M_1 = 2.5$, $\theta = 15°$.

To demonstrate the potential ability of our discrete FEM-FCT approach to deal with unstructured meshes, we constructed an adaptive coarse grid with the grid points clustered in the vicinity of the shock wave, see Figure 1 (bottom left). After two steps of global refinement this gives the computational mesh of approximately 10,000 vertices. The resulting numerical solution (bottom right) exhibits superb accuracy and remains absolutely free of oscillations.

Numerical results for a variety of standard gas dynamic test cases encompassing both transient and stationary flows are presented in [5]. Moreover, an in-depth investigation of scalar problems can be found in the same publication.

8 Conclusions

To our knowledge, most of the finite element schemes for solving the Euler equations on unstructured grids are explicit and, consequently, subject to an restrictive CFL condition. In this paper, an *implicit* high-resolution finite element scheme for hyperbolic systems was presented making use of the flux-corrected transport paradigm. The underlying low-order operator was constructed by applying scalar artificial viscosity proportional to the spectral radius of the cumulative Roe matrix for each edge of the sparsity graph. An efficient edge-based approach to matrix assembly was proposed. The design of suitable preconditioners for both segregated and coupled solution procedures was addressed. The performance of the new FEM-FCT algorithm was illustrated for a steady supersonic flow without and with adaptive mesh refinement. The development of robust and efficient iterative solvers for implicit FEM-FCT schemes including FAS-FMG multigrid [6],[8] and an analog of the *local MPSC* smoother for the incompressible Navier-Stokes equations [10] will be addressed in forthcoming publications.

References

1. J. P. Boris and D. L. Book, Flux-corrected transport. I. SHASTA, A fluid transport algorithm that works. *J. Comput. Phys.* **11** (1973) 38–69.
2. C. A. J. Fletcher, The group finite element formulation. *Comput. Methods Appl. Mech. Engrg.* **37** (1983) 225-243.
3. D. Kuzmin and S. Turek, Flux correction tools for finite elements. *J. Comput. Phys.* **175** (2002) 525-558.
4. D. Kuzmin, M. Möller and S. Turek, Multidimensional FEM-FCT schemes for arbitrary time-stepping. *Int. J. Numer. Meth. Fluids* **42** (2003) 265–295.
5. D. Kuzmin, M. Möller and S. Turek, High-resolution FEM-FCT schemes for multidimensional conservation laws. Technical report No. **231**, University of Dortmund, 2003, Submitted to: *Comput. Methods Appl. Mech. Engrg.*
6. P. W. Hemker and B. Koren, Defect correction and nonlinear multigrid for steady Euler equations. In: W.G. Habashi and M.M. Hafez (ed.). *Computational fluid dynamics techniques.* London: Gordon and Breach Publishers, 1995, 699–718.
7. R. Löhner, K. Morgan, J. Peraire and M. Vahdati, Finite element flux-corrected transport (FEM-FCT) for the Euler and Navier-Stokes equations. *Int. J. Numer. Meth. Fluids* **7** (1987) 1093–1109.

8. J.F. Lynn, *Multigrid Solution of the Euler Equations with Local Preconditioning.* PhD thesis, University of Michigan, 1995.
9. P. L. Roe, Approximate Riemann solvers, parameter vectors and difference schemes. *J. Comput. Phys.* **43** (1981) 357–372.
10. S. Turek, *Efficient Solvers for Incompressible Flow Problems: An Algorithmic and Comp utational Approach*, LNCSE 6, Springer, 1999.
11. S. T. Zalesak, Fully multidimensional flux-corrected transport algorithms for fluids. *J. Comput. Phys.* **31** (1979) 335–362.

A Singular Limit Method for the Stefan Problems

Hideki Murakawa[1] and Tatsuyuki Nakaki[2]

[1] Graduate School of Mathematics, Kyushu University, Japan *ma202051@math.kyushu-u.ac.jp*
[2] Faculty of Mathematics, Kyushu University, Japan *nakaki@math.kyushu-u.ac.jp*

Summary. This paper proposes an approximation scheme to a classical one-phase Stefan problem. The scheme is constructed to represent a singular limit of certain reaction-diffusion system approximating to the Stefan problem. To this end, a time-discrete operator-splitting methodology is used. Numerical experiments demonstrate that the scheme would be useful in practical computations.

1 Introduction

A classical one-phase Stefan problem is one of mathematical models which describe the melting of a body of ice maintained at temperature 0℃. One of attractive and important subjects of research on this problem is the analysis of interface between ice and water. The interface is hypersurface and the topological structure on ice and water regions may change. This fact induces numerical difficulties, and many numerical schemes are proposed to track the interfaces [1, 5, 8, 12, 13].

The aim of this paper is to propose an approximation scheme to the following one-phase Stefan problem, which is formulate by Eymard et al. [2, 4, 6]:

$$(SP)\begin{cases} w_t = \Delta(w^+) & \text{in } Q_T, \\ w^+ = A & \text{on } \partial\Omega \times (0,T), \\ w(x,0) = w_0(x) & \text{for } x \in \Omega, \end{cases}$$

where Ω is a bounded domain in $\mathbf{R}^d$ ($d \in \mathbf{N}$) with smooth boundary $\partial\Omega$, T is a positive number, $Q_T = \Omega \times (0,T]$, A and w_0 are given functions and $a^\pm = \max\{\pm a, 0\}$. The temperature of water is indicated by $u := w^+$, and $v := w^-$ is the function such that the support of v coincides with the mushy region, which have been studied by Bertsch et al. [2]. The ice-water interface is given by $\Gamma(t) := \partial\Omega^+(t) \cap \Omega$, where $\Omega^+(t) = \{x \in \Omega \mid u(x,t) > 0\}$.

The idea of our scheme is a time-discrete operator-splitting methodology to a reaction-diffusion equation which is given by Eymard et al. We also introduce a singular limit solution to construct an approximation scheme to Problem (SP). The advantages of our numerical method to our approximation scheme are

1. It has low computational cost,
2. It may be used on arbitrary geometries on Ω,
3. There is no artificial parameters,
4. We can track the interface even in high-dimensional problems,
5. Topological changes and complicated interfacial shapes can be handled easily.

In this paper, the convergence of the scheme is shown by numerical experiments in one-dimensional case. Quite recently, we propose a similar approximation scheme to moving boundary problems (including classical Stefan problems), and prove the convergence of the scheme [10]. We remark that our approximation scheme or numerical method is similar to the diffusion-generated approach for the mean curvature flow [3, 9] or the threshold competition dynamics method [7].

At the end of this section, we explain the detail of initial function w_0. The classical one-phase Stefan problem is often described by

$$\begin{cases} \hat{u}_t = \Delta \hat{u} & \text{in } \cup_{0<t\leq T} \hat{\Omega}(t) \times \{t\}, \\ \lambda V_\nu = -\dfrac{\partial \hat{u}}{\partial \nu} & \text{on } \cup_{0<t\leq T} \hat{\Gamma}(t) \times \{t\}, \\ \hat{u} = 0 & \text{on } \cup_{0<t\leq T} \hat{\Gamma}(t) \times \{t\}, \\ \hat{u} = A & \text{on } \partial\Omega \times (0,T], \\ \hat{u}(x,0) = u_0(x) & \text{for } x \in \hat{\Omega}_0, \\ \hat{\Omega}(0) = \hat{\Omega}_0, \end{cases} \tag{1}$$

where $\hat{\Omega}(t) \subset \Omega$ is an unknown domain which describes the water region, $\hat{\Gamma}(t) = \partial\hat{\Omega}(t) \cap \Omega$, λ is the latent heat, V_ν is the normal speed of the interface $\hat{\Gamma}(t)$, ν stands for the exterior unit normal vector to $\hat{\Gamma}(t)$, u_0 is a positive function representing the initial heat distribution, $\hat{\Omega}_0$ is an initial water region. We note that Problem (SP) is equivalent to Problem (1) if the interface $\Gamma(t)$ is smooth surface such that $\overline{\Gamma(t)} \subset \Omega$ for all $t \in (0,T]$, and vary smoothly with t, the function $w^+(t)$ smooth up to $\Gamma(t)$ for $t \in (0,T)$, and

$$w_0(x) = \begin{cases} u_0(x), & \text{if } x \in \hat{\Omega}_0, \\ -\lambda, & \text{otherwise} \end{cases} \tag{2}$$

(see [4]). Our numerical simulations are made by using (2).

2 Our scheme

First of all, we introduce our approximation scheme to the solution of Problem (SP).

Approximation Scheme 1. *Let N be a positive integer and $0 = t_0 < t_1 < \cdots < t_m < \cdots < t_N = T$. An approximation w^m to $w(t_m)$ is defined by*

$$w^m := \mathcal{K}(t_m) \circ \mathcal{K}(t_{m-1}) \circ \cdots \circ \mathcal{K}(t_1) w_0,$$

where $\mathcal{K}(t)$ is an operator with domain $L^\infty(\Omega)$ defined by

$$\mathcal{K}(t)z := \mathcal{H}_m(t)z^+ - z^- \qquad \text{for } t_m < t \leq t_{m+1} \quad (m = 0, 1, 2, \ldots, N-1),$$

and $\mathcal{H}_m(t)z$ denotes the solution of the following heat equation

$$\begin{cases} \bar{u}_t = \Delta \bar{u} & \text{in } \Omega \times (t_m, t_{m+1}], \\ \bar{u} = A & \text{on } \partial\Omega \times (t_m, t_{m+1}], \\ \bar{u}(x, t_m) = z(x) & \text{for } x \in \Omega. \end{cases} \tag{3}$$

Most of computational cost of Approximation Scheme 1 is spent on the heat equation (3), and the first advantage '1. Our numerical method has low computational cost' shown in Section 1 follows. If we use a finite element method or finite volume method to solve (3), one can say '2. Our numerical method may be used on arbitrary geometries on Ω'. Furthermore, we have '3. There is no artificial parameters' as shown the above.

In the following we explain the basic idea of our scheme (see also [11]). Eymard et al. [4] introduce the following reaction-diffusion system.

$$(RD_k) \begin{cases} U_t = \Delta U - kUV & \text{in } Q_T, \\ V_t = -kUV & \text{in } Q_T, \\ U = A & \text{on } \partial\Omega \times (0, T], \\ U(x, 0) = u_0(x) := w_0^+(x) & \text{for } x \in \Omega, \\ V(x, 0) = v_0(x) := w_0^-(x) & \text{for } x \in \Omega, \end{cases}$$

where k is a positive parameter. They show the relationship between Problems (RD_k) and (SP) in the following

Proposition 1 (Eymard, Hilhorst, van der Hout and Peletier [4]). *Let us assume*

$$\begin{cases} A \in W_2^{2,1}(Q_T) \cap C(\overline{Q_T}), \\ u_0(x) = A(x, 0) & \text{for } x \in \Omega, \\ 0 \leq v_0 \leq K & \text{in } \Omega, \end{cases}$$

for some positive constant K. Then (RD_k) has a unique weak solution $(U^{(k)}, V^{(k)}) \in W_2^{2,1}(Q_T) \times C^{0,1}([0, T]; L^\infty(\Omega))$ for every $k > 0$, and

$$U^{(k)} \to w^+, \quad V^{(k)} \to w^- \quad \text{as} \quad k \to \infty,$$

strongly in $L^2(Q_T)$, where w is a weak solution of Problem (SP).

To construct an approximate solution of Problem (RD_k) for $k = \infty$ (singular limit solution), we apply a standard operator-splitting methodology to Problem (RD_k), that is, we split Problem (RD_k) into diffusion and reaction parts as follows:

We construct families of functions $\{U^m\}_{m=0}^N$ and $\{V^m\}_{m=0}^N$ by defining
Step 1: Let $U^0(x) = u_0(x)$ and $V^0(x) = v_0(x)$.

Step 2: For $m = 0, 1, \dots, N-1$,
Step 2-1 (diffusion part): Let $\bar{U}^m(\cdot, t_{m+1}) = \mathcal{H}_m(t_{m+1})U^m$.
Step 2-2 (reaction part): Solve the following initial value problem:

$$\begin{cases} \hat{U}^m_t = -k\hat{U}^m\hat{V}^m & \text{in } \Omega \times (t_m, t_{m+1}], \\ \hat{V}^m_t = -k\hat{U}^m\hat{V}^m & \text{in } \Omega \times (t_m, t_{m+1}], \\ \hat{U}^m(x, t_m) = \bar{U}^m(x, t_{m+1}) & \text{for } x \in \Omega, \\ \hat{V}^m(x, t_m) = V^m(x) & \text{for } x \in \Omega. \end{cases} \tag{4}$$

Step 2-3: Put

$$\begin{cases} U^{m+1}(x) = \hat{U}^m(x, t_{m+1}), \\ V^{m+1}(x) = \hat{V}^m(x, t_{m+1}). \end{cases}$$

We are now in the position to construct an approximation scheme to Problem (SP). Let $\theta = k(t - t_m)$. Then (4) becomes

$$\begin{cases} \hat{U}^m_\theta = -\hat{U}^m\hat{V}^m & \text{in } \Omega \times (0, k\tau], \\ \hat{V}^m_\theta = -\hat{U}^m\hat{V}^m & \text{in } \Omega \times (0, k\tau]. \end{cases}$$

Passing to the limit as $k \to \infty$, that is $\theta \to \infty$, we obtain

$$\begin{cases} \lim_{\theta\to\infty} \hat{U}^m(x, \theta) = (\bar{U}^m(x, t_{m+1}) - V^m(x))^+, \\ \lim_{\theta\to\infty} \hat{V}^m(x, \theta) = (\bar{U}^m(x, t_{m+1}) - V^m(x))^-. \end{cases}$$

Here we use the facts that $(\hat{U}^m - \hat{V}^m)_\theta = 0$, $\bar{U}^m(x, t_{m+1}) \geq 0$ and $V^m(x) \geq 0$. From the above formal calculations, we obtain Approximation Scheme 1.

3 Numerical results

In this section, some numerical results obtained by our scheme are shown. We deal with a one-dimensional case with a known exact solution and a three-dimensional case. In the former, we show accuracy of our numerical interfaces by comparing with the exact ones. In the latter, it is demonstrated that the moving boundaries can be track even in high-dimensional problems.

3.1 One-dimensional results

Let $s(t)$ be the position of the interface at time t. We deal with a simple one-dimensional problem with $s(0) = 0$ and $u(0, t) = C$ for $t \geq 0$, where C is a positive constant (see [14]). In this case, it is already known that

$$s(t) = 2\alpha\sqrt{t}, \tag{5}$$

where α is a constant satisfying

$$2\alpha e^{\alpha^2}\int_0^\alpha e^{-z^2}dz = \frac{C}{\lambda}. \tag{6}$$

Let us compare our numerical interfaces with (5) when $\alpha = \lambda = 1$, $\Omega = (0,2)$ and $T = 1$. We compute a numerical approximation to the integral (6) using Simpson's rule, then we employ 4.0601569 as an approximation to C. Fixed uniform grids in space and time are adopted, that is, the spatial mesh size is $\delta x = 2/M$ and the time mesh size is $\delta t = 1/N$, where M is a positive integer. Let u_i^n and v_i^n $(0 \le i \le M,$ and $0 \le n \le N)$ be the approximations to $u(i\delta x, n\delta t)$ and $v(i\delta x, n\delta t)$, respectively, which implies that $w_i^n := u_i^n - v_i^n$ is the approximation to the weak solution $w(i\delta x, n\delta t)$ of Problem (SP). The initial data are given by

$$u_0^0 = C,\ v_0^0 = 0,$$
$$u_i^0 = 0,\ v_i^0 = 1 \quad (0 < i \le M).$$

We employ the implicit finite difference technique to obtain the numerical solution of heat equation. Then our numerical method deduces iterating the following two steps.

Step 1: Define $\tilde{u}_i^n$ by

$$\frac{\tilde{u}_i^n - u_i^n}{\delta t} = \frac{\tilde{u}_{i+1}^n - 2\tilde{u}_i^n + \tilde{u}_{i-1}^n}{\delta x^2} \qquad (0 < i \le M),$$
$$\tilde{u}_0^n = C,$$
$$\tilde{u}_{M+1}^n = \tilde{u}_{M-1}^n.$$

Step 2: Compute u_i^{n+1} and v_i^{n+1} $(0 \le i \le M)$ by

$$u_i^{n+1} = (\tilde{u}_i^n - v_i^n)^+,$$
$$v_i^{n+1} = (\tilde{u}_i^n - v_i^n)^-.$$

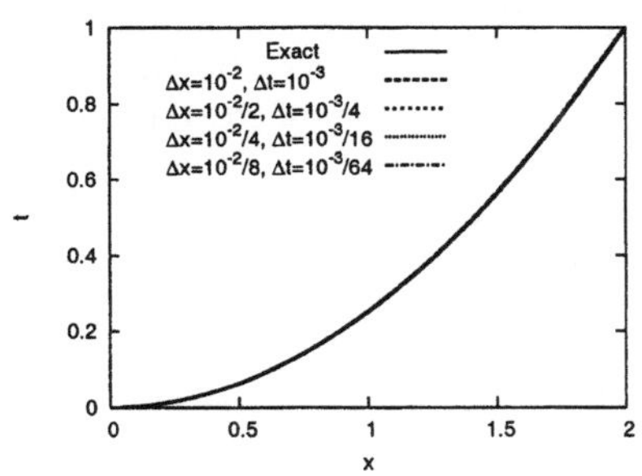

Fig. 1. Numerical interfaces for some meshes

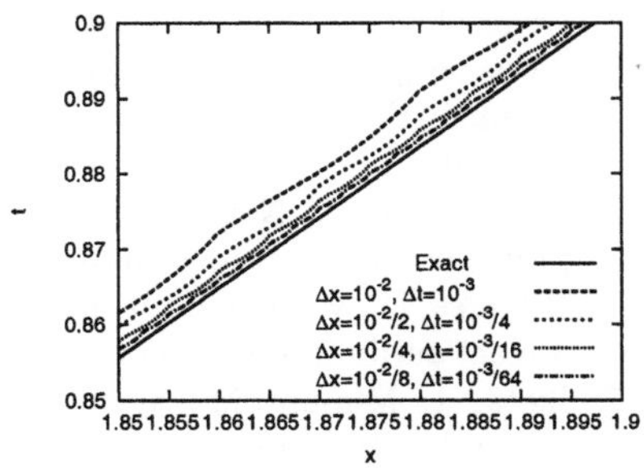

Fig. 2. Close-up of the numerical interfaces in Fig. 1

We show the exact and numerical interfaces in Fig. 1, where we employ the isosurface of level $-\lambda/2$ of $\{w_i^n\}_{0 \le i \le M,\ 0 \le n \le N}$ as the numerical interface.

Fig. 2 is a close-up of Fig. 1. One can say that the numerical interface converges to the exact one.

3.2 Three-dimensional simulation

Our numerical method can be easily adopted multi-dimensional problem. In this subsection we compute three-dimensional case with the computational domain $\Omega = (0,1)^3$. Fig. 3 shows our numerical simulation, which demonstrates how heat fluxes melt ice. In this simulation, we employ the explicit finite difference technique to obtain the numerical solution of heat equation because of our computational environment. The spatial mesh sizes are $\delta x = \delta y = \delta z = 1/100$ and the time step is $\delta t = 3 \times 10^{-6}$. The boundary condition is $u|_{\partial\Omega} = 1$. The initial data are shown at the top left of Fig. 3, where $(u_0(x,y,z), v_0(x,y,z)) = (1,0)$ holds in the liquid phase (black regions) and $(u_0(x,y,z), v_0(x,y,z)) = (0,1)$ does in the solid one (white ones).

We may observe that the interface with complex behavior can be computed, that is, one can say '4. We can track the interface even in high-dimensional problems' and '5. Topological changes and complicated interfacial shapes can be handled easily'.

References

1. Beckett, G., Mackenzie, J. A., Robertson, M. L. (2001): A moving mesh finite element method for the solution of two-dimensional Stefan Problems. J. Comp. Phys., **168**, 500–518
2. Bertsch, M., de Mottoni, P., Peletier, L.A. (1984): Degenerate diffusion and the Stefan problem. Nonlinear Analysis TMA, **8**, No.11, 1311–1336
3. Evans, L.C. (1993): Convergence of an algorithm for mean curvature motion. Indiana Univ. Math. J., **42**, No.2, 533–550
4. Eymard, R., Hilhorst, D., van der Hout, R., Peletier, L.A. (2000): A reaction-diffusion system approximation of a one-phase Stefan problem. In: Menaldi, J.L., Rofman, E., Sulem, A. (ed) Optimal control and partial differential equations. 156-170
5. Grossmann, C., Noack, A. (2002): Smoothing and Rothe's method for Stefan problems in enthalpy form. J. Comp. Appl. Math., **138**, 347–366
6. Hilhorst, D., van der Hout, R., Peletier, L.A. (1996): The fast reaction limit for a reaction-diffusion system. J. Math. Anal. Appl., **199**, 349–373
7. Ikota, R., Mimura, M., Nakaki, T.: A methodology for numerical simulations to a singular limit. preprint
8. Kawarada, H. (1989): Free boundary problem: Theory and numerical method (in Japanese). University of Tokyo Press
9. Merriman, B., Bence, J.K., Osher, S.J. (1994): Motion of multiple junctions: A level set approach. J. Comp. Phys., **112**, 334–363
10. Murakawa, H. (2004): A numerical method for Stefan type moving boundary problems (in Japanese). MA Thesis, Kyushu University, Japan

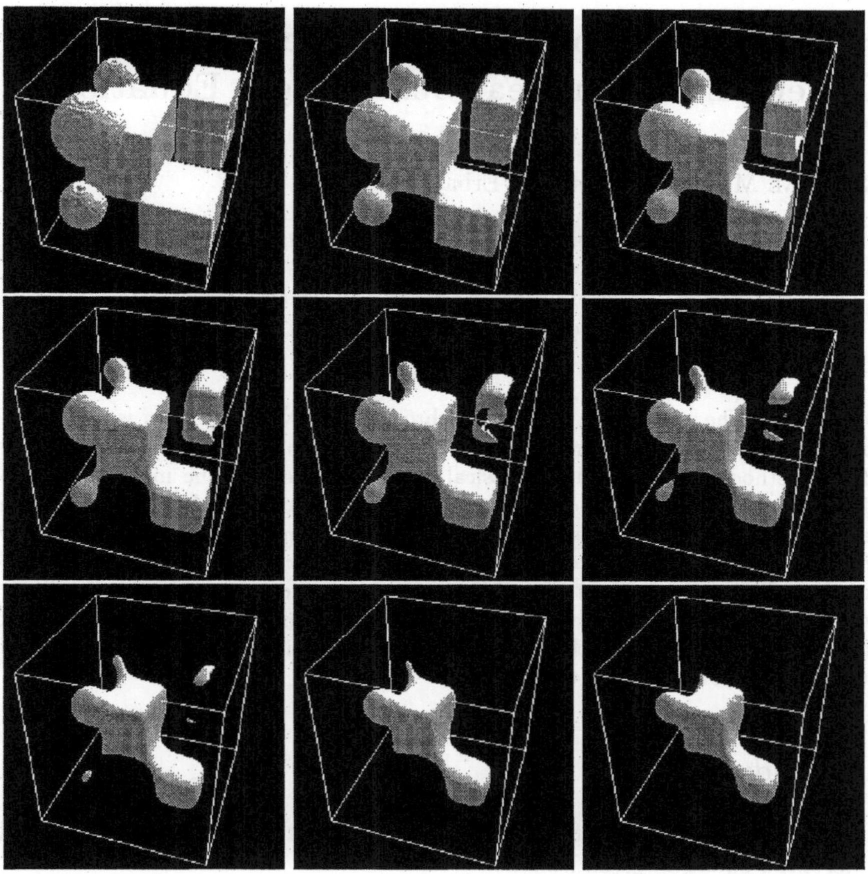

Fig. 3. Initial data and numerical solutions in a three-dimensional case. Time runs left to right and up to down.

11. Murakawa, H., Nakaki, T. (2003): A singular limit approach to moving boundary problems and its applications. Theoretical and Applied Mechanics Japan, **52**, 255–260
12. Nogi, T. (1974): A difference scheme for solving the Stefan problem. Publ. Res. Inst. Math. Sci. Kyoto Univ. series A, **9**, 543–505
13. Verdi, C., (1994): Numerical aspects of parabolic free boundary and hysteresis problems. Lecture Notes in Mathematics, **1584**, 213–284
14. Yamaguchi, M., Nogi, T. (1977): Stefan problem (in Japanese). Sangyotosho

Higher-Order Split-Step Schemes for the Generalized Nonlinear Schrödinger Equation

Gulcin M. Muslu[1] and Husnu A. Erbay[2]

[1] Istanbul Technical University, Department of Mathematics, Maslak 34469, Istanbul, Turkey *gulcin@itu.edu.tr*
[2] Istanbul Technical University, Department of Mathematics, Maslak 34469, Istanbul, Turkey *erbay@itu.edu.tr*

Summary. The generalized nonlinear Schrödinger (GNLS) equation is solved numerically by a split-step Fourier method. The first, second and fourth-order versions of the method are presented. A classical problem concerning the motion of a single solitary wave is used to compare the first, second and fourth-order schemes in terms of the accuracy and the computational cost. This numerical experiment shows that the split-step Fourier method provides highly accurate solutions for the GNLS equation. Furthermore, two test problems concerning the interaction of two solitary waves and an exact solution which blows up in finite time are investigated by using the fourth-order split-step scheme and particular attention is paid to the conserved quantities as an indicator of the accuracy.

1 Introduction

The generalized nonlinear Schrödinger (GNLS) equation is a nonlinear partial differential equation given by

$$iw_t + w_{xx} + q_1 \left|w\right|^2 w + q_2 \left|w\right|^4 w + iq_3(\left|w\right|^2)_x w + iq_4 \left|w\right|^2 w_x = 0, \qquad (1)$$

where $i = \sqrt{-1}$, w is a complex valued function of the spatial coordinate x and the time t, the parameters q_1, q_2, q_3 and q_4 are real constants and the subscripts t and x denote differentiation with respect to time and space, respectively. Compared to the usual nonlinear Schrödinger equation with a cubic nonlinearity, the GNLS equation possesses both cubic and quintic nonlinearities and nonlinear terms that contain derivatives. It has been derived as a model equation governing the modulation of a quasi-monochromatic wave train in a weakly nonlinear, dispersive medium.

Some properties of the GNLS equation are summarized here. Assume that w and all its derivatives converge to zero sufficiently rapidly as $x \to \pm\infty$. Solutions of the GNLS equation subjected to these boundary conditions are known to satisfy some conservation laws [1, 2]. According to these conservation laws, the conserved quantities

$$I_1 = \int_{-\infty}^{\infty} \left|w\right|^2 \, \mathrm{d}x \qquad (2)$$

$$I_2 = \int_{-\infty}^{\infty} \{|w_x|^2 - \frac{1}{2}(2q_3 + q_4)\, |w|^2\ \mathrm{Im}(ww_x^*) - \frac{1}{2} q_1\, |w|^4 + \frac{1}{6}[q_3(2q_3 + q_4) - 2q_2]\, |w|^6\}\ \mathrm{d}x \tag{3}$$

$$I_3 = \int_{-\infty}^{\infty} [2\ \mathrm{Im}(ww_x^*) - q_3\, |w|^4]\ \mathrm{d}x, \tag{4}$$

where the symbol * denotes complex conjugation, remain constant in time. Note that I_1 represents the theoretical L_2 norm of the system. Although the GNLS equation is generally known as a nonintegrable equation in the sense of the inverse scattering method, certain cases of the GNLS equation are completely integrable and such equations possess soliton solutions and an infinite number of conservation laws. For certain values of the coefficients and certain initial conditions, solutions of the general equation (1) experience finite-time blow up [1]. Only a few analytical solutions corresponding to some special cases of the GNLS equation are available [1, 2]. Therefore, numerical studies are essential to develop an understanding of the phenomena related to the GNLS equation.

One of the numerical methods employed for nonlinear dispersive wave equations is the split-step method proposed by Tappert [3]. The basic idea in the split-step method is to decompose the original problem into subproblems which are simpler than the original problem and then to compose the approximate solution of the original problem by using the exact or approximate solutions of the subproblems in a given sequential order. For nonlinear dispersive wave equations which are derived by balancing the effects of dispersion and nonlinearity, such as the GNLS equation that we will be solving, an appropriate approach is to split the original problem into linear and nonlinear subproblems which take into account purely dispersive and purely nonlinear effects, respectively. While various numerical methods have been employed for the numerical solutions of the cubic NLS equation in which the split-step method profits from the existence of a simple analytical solution for the nonlinear subproblem, less attention has been paid to the numerical solution of the GNLS equation of which the cubic NLS equation is a special case. A first-order split-step method was suggested by Pathria and Morris for the GNLS equation in [2].

The main purpose of this study is to introduce higher-order split-step Fourier schemes for the GNLS equation and is to compare these schemes from a computational efficiency viewpoint. To this end, the initial and boundary-value problem is decomposed into linear and nonlinear subproblems. A Fourier method is employed for the spatial discretizations of both linear and nonlinear subproblems. While the linear subproblem is treated exactly, a fourth-order Runge-Kutta scheme is used for the time integration of the nonlinear subproblem. Three different numerical schemes which are basically the first-order, second-order and fourth-order versions of the present split-step Fourier method are proposed. For an application of the present split-step Fourier method to the complex modified Korteweg-de Vries equation, we refer the reader to [4].

2 The Numerical Method

2.1 *Review of the split-step method*

It is best to present the split-step method as applied to a general evolution equation in the form

$$w_t = (\mathcal{L} + \mathcal{N})w, \tag{5}$$

where $\mathcal{L}$ and $\mathcal{N}$ are linear and nonlinear operators, respectively, and $\mathcal{L}$ and $\mathcal{N}$ do not commute with each other. For instance, we have

$$\mathcal{L} = i\,\frac{\partial^2}{\partial x^2}, \qquad \mathcal{N} = iq_1\,|w|^2 + iq_2\,|w|^4 - q_3(|w|^2)_x - q_4\,|w|^2\,\frac{\partial}{\partial x}$$

for the GNLS equation. If, for the moment, $\mathcal{L}$ and $\mathcal{N}$ are assumed to be t independent, a formally exact solution of equation (5) is given by

$$w(x, t + \Delta t) = \exp[\Delta t(\mathcal{L} + \mathcal{N})]w(x, t) \tag{6}$$

where Δt is the time step between the initial and final times. The linear equation $w_t = \mathcal{L}w$ and the nonlinear equation $w_t = \mathcal{N}w$ have known exact solutions

$$w(x, t + \Delta t) = \exp(\Delta t\mathcal{L})w(x, t) \tag{7}$$

and

$$w(x, t + \Delta t) = \exp(\Delta t\mathcal{N})w(x, t), \tag{8}$$

respectively. The main idea in the split-step method is to approximate the exact solution of equation (5) by solving the purely linear and purely nonlinear equations in a given sequential order, in which the solution of one subproblem is employed as an initial condition for the next subproblem. This may be realized by replacing the exponential operator $\exp[\Delta t(\mathcal{L} + \mathcal{N})]$ in equation (6) by a solution operator $\varphi_n(\Delta t)$ which includes an appropriate combination of products of the exponential operators $\exp(\Delta t\mathcal{L})$ and $\exp(\Delta t\mathcal{N})$. This produces a splitting error due to the noncommutativity of $\mathcal{L}$ and $\mathcal{N}$, and at this stage the celebrated Baker-Campbell-Hausdorf (BCH) formula is very useful to reduce noticeably the splitting error. In what follows, we study the first-, second- and fourth-order versions of the method. According to the BCH formula, the first-order approximation of the exponential operator in equation (6) is given by

$$\varphi_1(\Delta t) = \exp(\Delta t\mathcal{L})\exp(\Delta t\mathcal{N})\ . \tag{9}$$

Therefore, for the first-order version of the split-step method, the advancement in time is carried out in two steps. In the first step, a so-called intermediate solution is computed by advancing the solution according to the purely nonlinear equation. In the second step, the solution is advanced according to the linear dispersive equation in which the intermediate solution is used as an initial condition. In the second-order version of the method, the exponential operator in equation (6) is approximated by

$$\varphi_2(\Delta t) = \exp(\frac{1}{2}\Delta t\mathcal{N})\exp(\Delta t\mathcal{L})\exp(\frac{1}{2}\Delta t\mathcal{N}) \tag{10}$$

which is symmetric, that is, $\varphi_2(\Delta t)\varphi_2(-\Delta t) = 1$. A fourth-order splitting is given in the form

$$\varphi_4(\Delta t) = \varphi_2(\omega\Delta t)\varphi_2[(1-2\omega)\Delta t]\varphi_2(\omega\Delta t) \tag{11}$$

where $\omega = (2 + 2^{1/3} + 2^{-1/3})/3$. Note that the number of products of exponential operators increases with the order of decay of splitting error.

2.2 *Space discretization*

Application of the numerical method requires truncation of the infinite interval to a finite interval $[a, b]$. We assume that $w(x,t)$ satisfies the periodic boundary condition $w(a,t) = w(b,t)$ for $t \in [0,T]$. If the spatial period is, for convenience, normalized to $[0, 2\pi]$ using the transformation $X = 2\pi(x-a)/(b-a)$, the GNLS equation becomes

$$iw_t + \overline{p}w_{XX} + q_1 |w|^2 w + q_2 |w|^4 w + i\overline{q}_3(|w|^2)_X w + i\overline{q}_4 |w|^2 w_X = 0 \tag{12}$$

where

$$\overline{p} = (\frac{2\pi}{b-a})^2, \quad \overline{q}_3 = (\frac{2\pi}{b-a})q_3, \quad \overline{q}_4 = (\frac{2\pi}{b-a})q_4 \,. \tag{13}$$

The interval $[0, 2\pi]$ is divided into N equal subintervals with grid spacing $\Delta X = 2\pi/N$ where the integer N is even. The spatial grid points are given by $X_j = 2\pi j/N$, $j = 0, 1, 2, ..., N$. The approximate solution to $w(X_j, t)$ is denoted by $W_j(t)$. The discrete Fourier transform of the sequence $\{W_j\}$ is defined as

$$\hat{W}_k = \mathcal{F}_k[W_j] = \frac{1}{N}\sum_{j=0}^{N-1} W_j \exp(-ikX_j), \qquad -\frac{N}{2} \le k \le \frac{N}{2} - 1 \,. \tag{14}$$

The inversion formula for the discrete Fourier transform (14) is

$$W_j = \mathcal{F}_j^{-1}[\hat{W}_k] = \sum_{k=-\frac{N}{2}}^{\frac{N}{2}-1} \hat{W}_k \exp(ikX_j), \qquad j = 0, 1, 2, ..., N-1 \,. \tag{15}$$

Here $\mathcal{F}$ denotes the discrete Fourier transform and $\mathcal{F}^{-1}$ its inverse. These transforms can be realized efficiently via a fast Fourier transform (FFT) algorithm. For the FFT algorithm used here, the integer N must have only prime factors 2 and 3. In both linear and nonlinear subproblems we approximate spatial derivatives in both linear and nonlinear subproblems using discrete Fourier transforms.

2.3 *Time integration*

We consider a split-step method for the GNLS equation, in which the linear equation

$$w_t - i\overline{p}w_{XX} = 0 \tag{16}$$

and the nonlinear equation

$$w_t - iq_1 |w|^2 w - iq_2 |w|^4 w + \overline{q}_3(|w|^2)_X w + \overline{q}_4 |w|^2 w_X = 0 \tag{17}$$

are solved in a given sequential order corresponding to one of the splitting formulas (9)-(11). The linear equation (16) can be solved by means of the discrete Fourier transform and the advancements in time are performed according to

$$W_j^{m+1} = \mathcal{F}_j^{-1}[\exp(-i\overline{p}k^2\Delta t)\mathcal{F}_k[W_j^m]] \tag{18}$$

Here Δt is time step and W_j^m denotes the approximation to $w(X_j, m\Delta t)$. The spatial discretization of the nonlinear equation (17) by a Fourier pseudospectral method can be written as

$$\begin{aligned}\frac{dW_j}{dt} &= i\ (q_1 |W_j|^2 W_j + q_2 |W_j|^4 W_j) - \mathcal{F}_j^{-1}[ik\overline{q}_3\mathcal{F}_k[|W_j|^2]]W_j \\ &\quad -\mathcal{F}_j^{-1}[ik\overline{q}_4\mathcal{F}_k[W_j]]|W_j|^2, \qquad j = 0, 1, 2, ..., N-1.\end{aligned} \tag{19}$$

For the time integration of this equation, instead of using an approximate analytical technique [2] we adopt rather a different approach and employ a fourth-order Runge-Kutta method. Now the total error involved in integrating from time t to time $t + \Delta t$ will be the sum of the splitting error and the temporal discretization error of the nonlinear equation (17).

The first-order split-step Fourier method for the GNLS equation can be summarized as follows: Given the data W_j at any time step $t = t_m$, first advance the solution according to the nonlinear part, namely solve equation (19) using the fourth-order Runge-Kutta method for time integration. This becomes the initial data for the linear problem which is solved by the discrete Fourier transform as indicated by equation (18). The extension of the first-order split-step scheme based on equation (9) to the second-order and fourth-order split-step schemes based on equations (10) and (11), respectively, is straightforward.

3 Numerical Experiments

To gain insight into the performance of the suggested split-step schemes we perform the following three numerical experiments. The conservation properties of the split-step schemes are examined by calculating discrete analogues of the conserved quantities I_1, I_2 and I_3. The relative errors in discrete approximations to the conservation integrals (2), (3) and (4) are denoted by δ_1, δ_2 and δ_3, respectively, and they are defined by $\delta_1 = |\overline{I}_1 - \overline{I}_{10}|/|\overline{I}_{10}|$, $\delta_2 = |\overline{I}_2 - \overline{I}_{20}|/|\overline{I}_{20}|$ and $\delta_3 = |\overline{I}_3 - \overline{I}_{30}|/|\overline{I}_{30}|$ where $\overline{I}_1$,$\overline{I}_2$, $\overline{I}_3$ and $\overline{I}_{10}$,$\overline{I}_{20}$, $\overline{I}_{30}$ represent the calculated values of the conserved quantities I_1,I_2, I_3 at times t and $t = 0$, respectively.

3.1 *Solitary wave solution*

The purpose of the present numerical experiment is to verify numerically that the proposed split-step schemes exhibit the expected first-order, second-order and fourth-order convergence in time. The GNLS equation has a travelling solitary wave solution [1, 2, 5], which has the form

$$w(x,t) = [\frac{4}{4 + 3\sinh^2(x - 2t - 15)}]^{1/2} \exp[i\phi(x,t)] \tag{20}$$

$$\phi(x,t) = 2\tanh^{-1}[\frac{1}{2}\tanh(x - 2t - 15)] + x - 15 \tag{21}$$

for the choice of coefficients $q_1 = 1/2$, $q_2 = -7/4$, $q_3 = -1$, $q_4 = -2$. This solution represents a solitary wave initially at $x = 15$ moving to the right with velocity 2. The problem is first solved on the space interval $5 \le x \le 35$, as in [1, 2], for times up to $t = 3$. We present in Figure 1(a) the L_∞-errors of the first-order, second-order and fourth-order split-step schemes as a function of N for the final time $t = 3$ on a $\log_{10} - \log_{10}$ scale. We use the relation $\Delta t = \nu(\Delta x)^2$ to determine the value of Δt for a given Δx ($\equiv (b-a)/N$), where the value of ν is fixed at $\nu = 0.1$. We observe that the L_∞-errors decrease with increasing N until the boundaries start exerting their influence. The nondecreasing error behavior for the second-order and fourth-order schemes after the value of $N = 96$ is due to the limited space interval $5 \le x \le 35$. To show that this behavior can be eliminated by balancing the error due to boundary effects with the error due to internal resolution, we repeat the experiment of Figure 1(a) for the space interval $-20 \le x \le 60$. The results are presented in Figure 1(b) on a $\log_{10} - \log_{10}$ scale again. But this time the effect of the boundaries disappear and the L_∞-errors continue to decrease with increasing N.

To test whether the split-step schemes exhibit the expected convergence rates in time we perform some numerical experiments for various values of time step Δt and a fixed value of N. In these experiments we take $N = 512$ to keep spatial accuracy high. The results are shown in Table 1. We present the L_∞-errors for the terminating time $t = 3$. The convergence rates agree well with the expected rates for the first-order, second-order and fourth-order split-step schemes. The orders of decay of the L_∞-errors are the ones of the splitting formulae employed for the temporal integration.

To compare the proposed split-step Fourier schemes in terms of computational efficiency, we fix Δt and Δx and measure the computing times, the L_2-error, the L_∞-error and the conservation errors δ_1,δ_2 and δ_3 at the terminating time $t = 3$. Trapezoidal rule is used for the numerical quadrature of

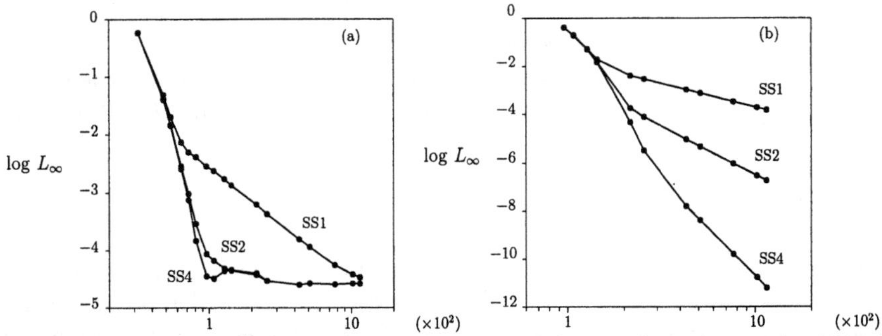

Fig. 1. The L_∞-errors at $t = 3$ as a function of the number of spatial grid points for the first-order (SS1), second-order (SS2) and fourth-order (SS4) split-step Fourier schemes. (a) The space interval: $5 \le x \le 35$, (b) The space interval: $-20 \le x \le 60$

.

Table 1. Comparison of the convergence rates in time for the first-order, second-order and fourth-order split-step Fourier schemes in the case of a single solitary wave $-20 \le x \le 60$.

	First-order		Second-order		Fourth-order	
Δt	L_∞	Order	L_∞	Order	L_∞	Order
0.0500	1.604E-2	-	2.079E-3	-	4.425E-4	-
0.0100	3.109E-3	1.019	7.975E-5	2.026	1.098E-6	3.727
0.0050	1.552E-3	1.002	1.991E-5	2.002	7.179E-8	3.935
0.0030	9.306E-4	1.001	7.167E-6	2.000	9.436E-9	3.972
0.0010	3.100E-4	1.001	7.962E-7	2.000	1.176E-10	3.991
0.0005	1.549E-4	1.001	1.990E-7	2.000	7.637E-12	3.945

the integrals. The results are represented in Table 2. Note that the computing times in Table 2 are normalized so that the computing time of the first-order split-step scheme is one unit. The results show that each of the conserved quantities is very well preserved by the split-step schemes. Furthermore, we observe that the computing time increases with the increasing order of the split-step method. We conclude that the fourth-order split-step scheme is computationally more efficient than the first-order and second-order schemes.

Table 2. Comparison of the L_∞-error, the L_2-error, the conservation errors δ_1, δ_2 and δ_3 and the computing times for the first-order, second- and fourth-order split-step Fourier schemes ($N = 512$, $-20 \le x \le 60$, $\Delta t = 0.61038 \times 10^{-3}$).

Method	L_∞	L_2	δ_1	δ_2	δ_3	Normalized cpu
First-order	1.892E-04	1.234E-04	1.691E-13	2.243E-07	9.984E-09	1.0
Second-order	2.966E-07	1.134E-07	2.198E-13	1.865E-13	1.381E-13	1.7
Fourth-order	1.716E-11	5.156E-12	5.913E-13	9.309E-14	3.985E-13	4.1

3.2 *Interacting Solitons*

In the second numerical experiment we study the interaction of two solitons for the integrable case of GNLS equation, in which the coefficients are $q_1 = 1$, $q_2 = 1$, $q_3 = -2$, $q_4 = 0$. The initial condition is given by

$$w(x,0) = \frac{1}{\sqrt{2}}\mathrm{sech}[\frac{1}{2}(x-15)]\exp i\{\frac{1}{4}(x-15)+\tanh[\frac{1}{2}(x-15)]\}$$
$$+\frac{1}{2\sqrt{2}}\mathrm{sech}[\frac{1}{4}(x-35)]\exp i\{-\frac{1}{2}(x-35)+\frac{1}{2}\tanh[\frac{1}{4}(x-35)]\} .$$

This equation corresponds to two solitons, the one initially located at $x = 15$ and moving to the right with speed 1/2 and the one initially located at $x = 35$ and moving to the left with speed 1. The exact values of the conserved quantities for this problem are $I_1 = 3$, $I_2 = 3/16$, and $I_3 = 0$. The problem is solved on the interval $-60 \leq x \leq 110$ for times up to $t = 20$ using the fourth-order split-step scheme. The numerical results show that the solitary waves are stable under the collision. Also each of the conserved quantities is very well preserved up to 10^{-12} for I_1 and I_2 and up to 10^{-10} for I_3 by the fourth-order split-step Fourier scheme. This behavior provides a valuable check on the numerical results.

3.3 *Blow-up*

For certain values of the coefficients and certain initial conditions, solutions to the GNLS equation experience finite time blow-up [1]. We now apply the fourth-order split-step scheme to a case of the GNLS equation in which the exact solution blows up in finite time. The initial condition is the Gaussian function $w(x,0) = \exp(-x^2)$ and the coefficients are $q_1 = -2$, $q_2 = 20$, $q_3 = 0$, $q_4 = 0$. The exact values of the conserved quantities I_1, I_2 and I_3 are $I_1 = \sqrt{\pi/2}$, $I_2 = \sqrt{\pi}(9\sqrt{2}+9-20\sqrt{6})/18$ and $I_3 = 0$ for this problem. In [1], it has been shown analytically that the exact solution $w(x,t)$ for this problem will blow up in finite time and furthermore, an upper bound on the blow-up time is $t \approx 1.7$.

In the present study, the above problem is solved on the interval $-7.5 \leq x \leq 7.5$ for times up to $t = 0.08$. We present the numerical results obtained using the fourth-order scheme on Table 3. Although a formal proof of the existence of the blow-up is not presented here, the numerical results strongly indicate that a blow-up is well underway by time $t = 0.08$. This is consistent with the numerical results presented in [1] and [5]. The fact that the three results about the predicted time of blow-up, which were obtained by totally different methods, are in complete agreement makes one believe in their correctness. As in [1] we conclude that the upper bound given in [1] is not sharp.

Table 3. Variation of discrete approximations of the conserved quantities I_1, I_2 and I_3 and $\mid W(0,t) \mid$ with time for the fourth-order split-step Fourier scheme ($N = 432$, $\Delta t = 0.005$).

t	$\overline{I}_1$	$\overline{I}_2$	$\overline{I}_3$	$\mid W(0,t) \mid$
0.00	1.253314	-2.684467	-9.793286E-17	1.000000
0.01	1.253314	-2.684467	-6.651917E-13	1.007348
0.06	1.253314	-2.684467	-4.673614E-12	1.526254
0.07	1.253314	-2.684448	-4.395637E-12	2.376429
0.08	1.253352	-2.829258	-6.085108E-10	3.430374

4 Conclusions

In this study we have applied the well-known split-step Fourier method to the GNLS equation. We have presented three split-step schemes in which the main difference among the three schemes is in the order of the splitting approximation used. The method is easy to implement on a computer and one can easily introduce higher-order splitting formulae to increase greatly the accuracy of split-step method. The numerical experiments reported here show that the fourth-order split-step Fourier scheme is advisable in situations where accuracy rather than the computational cost is of prime importance.

The numerical solutions obtained by using the present numerical schemes for the case of one solitary wave are compared with the exact solutions in order to assess the accuracy of these schemes. In addition, the performance of the numerical schemes has been monitored by computing both the conserved quantities and the computational costs. We have found that the numerical results are in a good agreement with the exact solutions and the results reported in the literature and that the schemes have remarkable conservation properties for global invariants. Moreover, the collision of two solitons and a finite time blow-up problem are investigated numerically and particular attention is paid to the behavior of the conserved quantities as an indicator of numerical difficulties.

The approaches presented in previously published papers related to numerical solutions of the GNLS equation have been mostly limited to the absence of the nonlinear derivative terms [6, 7, 8]. The numerical results presented above show that the nonlinear derivative terms do not create any special difficulties in the split-step Fourier method.

References

1. Pathria, D., Morris, J. Ll. (1989): Exact solutions for a generalized nonlinear Schrodinger equation, *Physica Scripta* **39**, 673-679.
2. Pathria, D., Morris, J. Ll. (1990): Pseudo-spectral solution of nonlinear Schrödinger equations, *J. Comput. Phys.* **87**, 108-125.

3. Tappert, F. (1974): Numerical solutions of the Korteweg-de Vries equation and its generalizations by the split-step Fourier method, *Lect. Appl. Math. Amer. Math. Soc.* **15**, 215.
4. Muslu, G. M., Erbay, H. A. (2003): A split-step Fourier method for the complex modified Korteweg-de Vries equation, *Computers Math. Applic.* **45**, 503-514
5. Robinson, M. P. (1997): The solution of nonlinear Schrödinger equations using orthogonal spline collocation, *Computers Math. Applic.* **33**, 39-57.
6. Cloot, A., Herbst, B. M., Weideman, J. A. (1990): A numerical study of the nonlinear Schrödinger equation involving quintic terms, *J. Comput. Phys.* **86**, 127-146.
7. Chang, Q., Jia, E., Sun, W. (1999): Difference schemes for solving the generalized nonlinear Schrödinger equation, *J. Comput. Phys.* **148**, 397-415.
8. Sheng, Q., Khaliq, A. Q. M., Al-Said, E. A. (2001): Solving the generalized nonlinear Schrödinger equation via quartic spline approximation, *J. Comput. Phys.* **166**, 400-417.

Numerical Methods and Simulation Techniques for Flow with Shear and Pressure Dependent Viscosity

Abderrahim Ouazzi[1] and Stefan Turek[2]

[1] Institute of Applied Mathematics, University of Dortmund, 44227 Dortmund, Germany. *Abderrahim.Ouazzi@math.uni-dortmund.de*
[2] Institute of Applied Mathematics, University of Dortmund, 44227 Dortmund, Germany. *Stefan.Turek@math.uni-dortmund.de*

Summary. In this note we present some of our recent results concerning flows with pressure and shear dependent viscosity. From the numerical point of view several problems arise, first from the difficulty of approximating incompressible velocity fields and, second, from poor conditioning and possible lack of differentiability of the involved nonlinear functions due to the material laws. The lack of differentiability can be treated by regularisation. Then, Newton-like methods as linearization technique can be applied; however the presence of the pressure in the viscosity function leads to an additional term introducing a new non-classical linear saddle point problem. The difficulty related to the approximation of incompressible velocity fields is treated by applying the nonconforming Rannacher-Turek Stokes element. However, then we are facing another problem related to the nonconforming approximation for problems involving the symmetric part of gradient: the classical discrete 'Korn's Inequality' is not satisfied. A new and more general approach which involves the jump across the inter-element boundaries should be used, which requires a small modification of the discrete bilinear form by adding an interface term, penalizing the jump of the velocity over edges. This is achieved via a modified procedure in the derivation of a Discontinuous Galerkin formulation. As a solver for the discrete nonlinear systems, a Newton variant is discussed while a 'Vanka-like' smoother as defect correction inside of a direct multigrid approach is presented. The results of some computational experiments for realistic flow configurations are provided, which contain a pressure dependent viscosity, too.

1 Introduction

The flowing of powders brings a new challenging and interesting problem to the CFD community: at very high concentrations and low rate-of-strain, grains are in permanent contact, rolling on each other. Therefore a frictional stress model must be taken into account. This can be done using plasticity and similar theories in which the material behavior is assumed to be independent of the velocity gradient or the rate-of-strain. This is in contrast to viscous Newtonian flow where stress specifically depends on a rate-of-strain. Furthermore, flowing powders do not exhibit viscosity and, again, this shows that a Newtonian rheology cannot describe granular flow accurately. It is assumed that the material is incompressible, dry, cohesionless, and perfectly rigid-plastic. Such properties are relevant for modelling the granular flows via special models for continuum mechanics, as for instance the Schaeffer model [9].

1.1 Equations of motion

The general equations of describing the motion of incompressible powders read:
Conservation of mass: $\frac{D\rho}{Dt} = \frac{\partial\rho}{\partial t} + \nabla\cdot(\rho\boldsymbol{u}) = 0$, $\frac{D*}{Dt}$ is the material derivative and $\boldsymbol{u}$ is the velocity vector.
Incompressible material: The bulk density, ρ, is a constant, so that $\nabla\cdot\boldsymbol{u} = 0$.
Equation of motion: $\rho\frac{Du}{Dt} = -\nabla\cdot\mathbf{T} + \rho g$ with $\mathbf{T} = \mathbf{S} + p\mathbf{I}$.

1.2 Constitutive equations

The constitutive equation is devoted to correlate between the deviatoric tensor, $\mathbf{S}$, and the velocity, through the second invariant of the rate deformation $D_{\mathrm{II}} = \frac{1}{2}\mathbf{D}:\mathbf{D}$, where the rate of deformation is given by $\mathbf{D} = \frac{1}{2}(\nabla\boldsymbol{u} + \nabla^T\boldsymbol{u})$.
Newtonian law: $\mathbf{S} = 2\nu_o\mathbf{D}$
Power law: $\mathbf{S} = 2\nu(D_{\mathrm{II}})\mathbf{D}, \quad \nu(z) = z^{\frac{r}{2}-1}, \quad r \geq 1$
Schaeffer's law: For a powder a constitutive equation which was first introduced by Schaeffer [9], has to obey a

- yield condition; $\|\mathbf{S}\| = \sqrt{2}p\sin\phi$,
- flow rule; $\mathbf{S} = \lambda\mathbf{D}$.

We use this correlation to obtain the constitutive equation

$$\mathbf{T} = \sqrt{2}p\sin\phi\frac{\mathbf{D}}{\|\mathbf{D}\|} + pI\,.$$

1.3 Generalized Navier-Stokes equations

The problem can be stated in the framework of the generalized incompressible Navier-Stokes equations:

$$\rho\frac{Du}{Dt} = -\nabla p + \nabla\cdot(\nu(p, D_{\mathrm{II}})\mathbf{D}) + \rho g, \quad \nabla\cdot\boldsymbol{u} = 0$$

If we define the nonlinear pseudo viscosity $\nu(\cdot,\cdot)$ as a function of the second invariant of the rate deformation D_{II} and the 'pressure' p, we can show that different materials can be ranged within different viscosity laws including powder;

- Power law defined for $\nu(z,p) = \nu_o z^{\frac{r}{2}-1}$
- Bingham law defined for $\nu(z,p) = \nu_o z^{-\frac{1}{2}}$
- Schaeffer's law (including the 'pressure') defined for $\nu(z,p) = \sqrt{2}\sin\phi\, pz^{-\frac{1}{2}}$

2 Problem formulation

Let us consider the flow of the stationary (!) generalized Navier-Stokes problem in (1.3) in a bounded domain $\Omega \subset \mathbb{R}^2$. If we restrict the set V of test functions to be divergence-free and if we take the constitutive laws into account, the above equations from (1.3) lead to:

$$\int_\Omega 2\nu(D_{\mathrm{II}}(\boldsymbol{u}),p)\mathbf{D}(\boldsymbol{u}) : \mathbf{D}(\boldsymbol{v})\,dx + \int_\Omega (\boldsymbol{u}\cdot\nabla\boldsymbol{u})\boldsymbol{v}\,dx = \int_\Omega \boldsymbol{f}\boldsymbol{v}\,dx, \qquad \forall \boldsymbol{v} \in V \tag{1}$$

It is straightforward to penalize the constraint $\operatorname{div}\boldsymbol{v} = 0$ to derive the equivalent mixed formulations of (1):

Find $(\boldsymbol{u},p) \in X \times M$ (with the spaces $X = H_0^1(\Omega)$ and $M = L^2(\Omega)$) such that:

$$\begin{aligned} \int_\Omega 2\nu(D_{\mathrm{II}}(\boldsymbol{u}),p)\mathbf{D}(\boldsymbol{u}) : \mathbf{D}(\boldsymbol{v})\,dx + \int_\Omega (\boldsymbol{u}\cdot\nabla\boldsymbol{u})v\,dx \quad + \int_\Omega p \operatorname{div}\boldsymbol{v}\,dx & \\ = \int_\Omega \boldsymbol{f}\boldsymbol{v}\,dx, \qquad & \forall \boldsymbol{v} \in X, \qquad (2) \\ \int_\Omega q \operatorname{div}\boldsymbol{u}\,dx = 0, \qquad & \forall q \in M, \end{aligned}$$

2.1 Nonlinear solver: Newton iteration

In this approach, the nonlinearity is first handled on the continuous level. Let $\boldsymbol{u}^l$ being the initial state, the (continuous) Newton method consists of finding $\boldsymbol{u} \in V$ such that

$$\begin{aligned} &\int_\Omega 2\nu(D_{\mathrm{II}}(\boldsymbol{u}^l),p^l)\mathbf{D}(\boldsymbol{u}) : \mathbf{D}(\boldsymbol{v})dx \\ &\quad + \int_\Omega 2\partial_1\nu(D_{\mathrm{II}}(\boldsymbol{u}^l),p^l)[\mathbf{D}(\boldsymbol{u}^l) : \mathbf{D}(\boldsymbol{u})][\mathbf{D}(\boldsymbol{u}^l) : \mathbf{D}(\boldsymbol{v})]dx \\ &\qquad + \boxed{\int_\Omega 2\partial_2\nu(D_{\mathrm{II}}(\boldsymbol{u}^l),p^l)[\mathbf{D}(\boldsymbol{u}^l) : \mathbf{D}(\boldsymbol{v})]pdx} \\ &\qquad = \int_\Omega \boldsymbol{f}\boldsymbol{v} - \int_\Omega 2\nu(D_{\mathrm{II}}(\boldsymbol{u}^l),p^l)\mathbf{D}(\boldsymbol{u}^l) : \mathbf{D}(\boldsymbol{v})dx, \quad \forall \boldsymbol{v} \in V\,, \end{aligned} \tag{3}$$

where $\partial_i\nu(\cdot,\cdot); i = 1,2$ is the partial derivative of ν related to the first and second variables, respectively. To see this, set $\boldsymbol{X} = \mathbf{D}(\boldsymbol{u}^l), \boldsymbol{x} = \mathbf{D}(\boldsymbol{u}), Y = p^l, y = p, F(\boldsymbol{x},y) = \nu(\frac{1}{2}|\boldsymbol{x}|^2, y)\boldsymbol{x}$ and $f(t) = F(\boldsymbol{X}+t\boldsymbol{x}, Y+ty)$, so that

$$\begin{aligned} \partial_{x_j}F_i(\boldsymbol{x},y) &= \partial_{x_j}\nu(\tfrac{1}{2}|\boldsymbol{x}|^2,y)x_jx_i + \nu(\tfrac{1}{2}|\boldsymbol{x}|^2,y)\delta_{ij} \\ \partial_y F_i(\boldsymbol{x},y) &= \partial_y\nu(\tfrac{1}{2}|\boldsymbol{x}|^2,y)x_i \end{aligned} \tag{4}$$

where δ_{ij} stands for the standard Kronecker symbol. Having

$$\begin{aligned} f_i'(t) &= \textstyle\sum_j \partial_{x_j} F_i(\boldsymbol{X}+t\boldsymbol{x}, Y+ty)x_j + \partial_y F_i(\boldsymbol{X}+t\boldsymbol{x}, Y+ty)y \\ &= \nu(\tfrac{1}{2}|\boldsymbol{X}+t\boldsymbol{x}|^2, Y+ty)\boldsymbol{x}_i \\ &+ \partial_1\nu(\tfrac{1}{2}|\boldsymbol{X}+t\boldsymbol{x}|^2, Y+ty)\langle \boldsymbol{X}+t\boldsymbol{x}, \boldsymbol{x}\rangle(\boldsymbol{X}_i+t\boldsymbol{x}_i) \\ &+ \partial_2\nu(\tfrac{1}{2}|\boldsymbol{X}+t\boldsymbol{x}|^2, Y+ty)y(\boldsymbol{X}_i+t\boldsymbol{x}_i) \end{aligned} \tag{5}$$

we decrease t towards zero, such that we obtain the Frechet derivative:

$$\begin{aligned} \nabla\cdot[\ & 2\nu(D_{\mathrm{II}}(\boldsymbol{u}^l), p^l)\mathbf{D}(\boldsymbol{u}) \\ &+ 2\partial_1\nu(D_{\mathrm{II}}(\boldsymbol{u}^l), p^l)(\mathbf{D}(\boldsymbol{u}^l):\mathbf{D}(\boldsymbol{u}))\mathbf{D}(\boldsymbol{u}^l) \\ &+ 2\partial_2\nu(D_{\mathrm{II}}(\boldsymbol{u}^l), p^l)p\mathbf{D}(\boldsymbol{u}^l)] \end{aligned} \tag{6}$$

2.2 New linear auxiliary problem

The resulting auxiliary subproblems in each Newton step consist of finding $(\boldsymbol{u}, p) \in X \times M$ as solutions of the linear (discretetized) systems

$$\begin{cases} A(\boldsymbol{u}^l, p^l)\boldsymbol{u} + \delta_d A^*(\boldsymbol{u}^l, p^l)\boldsymbol{u} + Bp + \delta_p B^*(\boldsymbol{u}^l, p^l)p = R_u(\boldsymbol{u}^l, p^l), \\ B^T\boldsymbol{u} \qquad\qquad\qquad\qquad\qquad\qquad\qquad\quad = R_p(\boldsymbol{u}^l, p^l), \end{cases} \tag{7}$$

where $R_u(\cdot,\cdot)$ and $R_p(\cdot,\cdot)$ denote the corresponding nonlinear residual terms for the momentum and continuity equations, and the operators $A(\boldsymbol{u}^l, p^l)$, B, $A^*(\boldsymbol{u}^l, p^l)$ and $B^*(\boldsymbol{u}^l, p^l)$ are defined as follows:

$$\langle A(\boldsymbol{u}^l, p^l)\boldsymbol{u}, \boldsymbol{v}\rangle = \int_\Omega 2\nu(D_{\mathrm{II}}(\boldsymbol{u}), p)\mathbf{D}(\boldsymbol{u}) : \mathbf{D}(\boldsymbol{v})\,dx \tag{8}$$

$$\langle Bp, \boldsymbol{v}\rangle = \int_\Omega p\nabla\cdot\boldsymbol{v}\,dx \tag{9}$$

$$\langle A^*(\boldsymbol{u}^l, p^l)\boldsymbol{u}, \boldsymbol{v}\rangle = \int_\Omega 2\partial_1\nu(D_{\mathrm{II}}(\boldsymbol{u}^l), p^l)[D(\boldsymbol{u}^l):D(\boldsymbol{u})][D(\boldsymbol{u}^l):D(\boldsymbol{v})]\,dx \tag{10}$$

$$\langle B^*(\boldsymbol{u}^l, p^l)\boldsymbol{v}, p\rangle = \int_\Omega 2\partial_2\nu(D_{\mathrm{II}}(\boldsymbol{u}^l), p^l)[D(\boldsymbol{u}^l):D(\boldsymbol{v})]p\,dx \tag{11}$$

3 Discretization

We consider a subdivision $T \in \mathcal{T}_h$ consisting of quadrilaterals in the domain $\Omega_h \in \mathbb{R}^2$, and we employ the rotated bilinear *Rannacher-Turek* element [5]. For any quadrilateral T, let (ξ, η) denote a local coordinate system obtained by joining the midpoints of the opposing faces of T. Then, in the *nonparametric* case, we set on each element T

$$\tilde{Q}_1(T) := span\left\{1, \xi, \eta, \xi^2 - \eta^2\right\}. \tag{12}$$

The degrees of freedom are determined by the nodal functionals $\{F_{\Gamma}^{(a,b)}(\cdot), \Gamma \subset \partial\mathcal{T}_h\}$,

$$F_{\Gamma}^{a} := |\Gamma|^{-1}\int_{\Gamma} v d\gamma \quad \text{or} \quad F_{\Gamma}^{b} := v(m_{\Gamma}) \quad (m_{\Gamma} \text{ midpoint of edge } \Gamma) \tag{13}$$

such that the finite element space can be written as

$$\begin{aligned} W_h^{a,b} := \{ & v \in L_2(\Omega_h), v \in \tilde{Q}_1(T), \forall T \in \mathcal{T}_h, v \text{ continuous w.r.t. all} \\ & \text{nodal functionals } F_{\Gamma_{i,j}}^{a,b}(\cdot), \text{ and } F_{\Gamma_{i0}}^{a,b}(v) = 0, \forall \Gamma_{i0}\}. \end{aligned} \tag{14}$$

Here, $\Gamma_{i,j}$ denote all inner edges sharing the two elements i and j, while Γ_{i0} denote the boundary edges of $\partial\Omega_h$. In this paper, we always employ version 'a' with the integral mean values as degrees of freedom. Then, the corresponding discrete functions will be approximated in the spaces

$$V_h := W_h^{a,b} \times W_h^{a,b}\,, L_h := \left\{ q_h \in L^2(\Omega), q_{h|T} = \text{const.}\,, \forall T \in \mathcal{T}_h \right\}. \tag{15}$$

Due to the nonconformity of the discrete velocities, the classical discrete 'Korn's Inequality' is not satisfied which is important for problems involving the symmetric part of the gradient [4]. Therefore, appropriate edge-oriented stabilization techniques (see [1, 2, 8], have to be included which directly treat the jump across the inter-elementary boundaries via adding the following bilinear form

$$\sum_{edges E} \frac{1}{|E|}\int_E [\phi_i][\phi_j] d\sigma \tag{16}$$

for all basis functions ϕ_i and ϕ_j of $W_h^{a,b}$. Taking into account an additional relaxation parameter $s = s(\nu)$, the corresponding stiffness matrices are defined via:

$$\langle \mathbf{S}\boldsymbol{u}, \boldsymbol{v}\rangle = s \sum_{E \in E_I \cup E_D} \frac{1}{|E|}\int_E [\boldsymbol{u}][\boldsymbol{v}] d\sigma \tag{17}$$

Here, the jump of a function $\boldsymbol{u}$ on an edge E is given by

$$[\boldsymbol{u}] = \begin{cases} \boldsymbol{u}^+ \cdot \boldsymbol{n}^+ + \boldsymbol{u}^- \cdot \boldsymbol{n}^- & \text{on internal edges } E_I\,, \\ \boldsymbol{u} \cdot \boldsymbol{n} & \text{on Dirichlet boundary edges } E_D, \\ 0 & \text{on Neumann boundary edges } E_N, \end{cases} \tag{18}$$

where $\boldsymbol{n}$ is the outward normal to the edge and $(\cdot)^+$ and $(\cdot)^-$ indicate the value of the generic quantity $(\cdot)$ on the two elements sharing the same edge.

4 Linear solver

This section is devoted to give a brief description of the involved solution techniques for the resulting linear systems. For the nonconforming Stokes element

$\tilde{Q}_1/Q_0$, a 'local pressure Schur complement' preconditioner (see [7]) as generalization of so-called 'Vanka smoothers' is constructed on patches Ω_i which are ensembles of one single or several mesh cells, and this local preconditioner is embedded as global smoother into an outer block Jacobi/Gauss-Seidel iteration which acts directly on the coupled systems of generalized Stokes, resp., Oseen type as described in [8]. If we denote by $\tilde{R}_u$ and $\tilde{R}_p$ the discrete residuals for the momentum and continuity equation which include the complete stabilisation term due to the modified bilinear form $\mathbf{S}$ as described in (17), one smoothing step in defect-correction notation can be described as

$$\begin{bmatrix} \boldsymbol{u}^{l+1} \\ p^{l+1} \end{bmatrix} = \begin{bmatrix} \boldsymbol{u}^{l} \\ p^{l} \end{bmatrix} + \omega^l \sum_i \begin{pmatrix} F + S^*_{|\Omega_i} & \tilde{B} + \delta_p \tilde{B}^*_{|\Omega_i} \\ \tilde{B}^T_{|\Omega_i} & 0 \end{pmatrix}^{-1} \begin{bmatrix} \tilde{R}_u(\boldsymbol{u}^l, p^l) \\ \tilde{R}_p(\boldsymbol{u}^l, p^l) \end{bmatrix} \tag{19}$$

with matrix $F = \tilde{A} + \delta_d \tilde{A}^*$ and $\tilde{A}, \tilde{B}, \tilde{A}^*$ and $\tilde{B}^*$ are the discrete matrices corresponding to the operators in (8), (9), (10) and (11). For the preconditioning step only a part of the matrix, i.e. $F + S^*$, is involved. All other components in the multigrid approach, that means intergrid transfer, coarse grid correction and coarse grid solver, are the standard ones and are based on the underlying hierarchical mesh hierarchy and the properties of the nonconforming finite elements (see [7] and [8] for the details).

5 Numerical tests

5.1 Newtonian case

In this case, the gradient and tensor formulations are equivalent; the accuracy and efficiency of the stabilized tensor discretization is checked by comparisons with the gradient formulation (see Table 1); the tests have been performed for the 'flow around cylinder' benchmark configuration [10]. For all three formulations the lift and drag forces are very similar.

Table 1. Efficiency of the stabilized nonconforming FEM: Lift and Drag forces

Level 5				
$1/\nu$		grad	tensor	stab. tensor
1	Drag	31252×10^{-1}	31221×10^{-1}	31231×10^{-1}
	Lift	30898×10^{-3}	30924×10^{-3}	30936×10^{-3}
	NL/MG	3/3	7/200	3/3
1000	Drag	55657×10^{-4}	55531×10^{-4}	55535×10^{-4}
($Re = 20$)	Lift	10180×10^{-6}	10259×10^{-6}	10277×10^{-6}
	NL/MG	11/4	11/12	11/3

5.2 Effect of convection

The average number of inner multigrid sweeps (MG) per outer nonlinear sweep (NL) increases with mesh refinement (see Table 2), due to the more dominant influence of the kernel function in the second order differential operator. Since,

Table 2. Nonlinear iteration (NL)/Averaged multigrid sweeps (MG) per nonlinear iteration for different viscosity parameter (*Re* numbers) and various formulations (gradient, tensor and stabilized tensor) and for different mesh levels

	$1/\nu$	1	10	1000
Level	Formulation	NL/MG	NL/MG	NL/MG
4	grad	3/3	4/3	11/4
	tensor	3/**15**	5/**17**	11/**4**
	stab. tensor	3/3	5/3	11/4
5	grad	3/3	4/3	11/3
	tensor	4/**140**	5/**35**	11/**10**
	stab. tensor	4/3	5/3	11/3
6	grad	3/3	4/3	11/3
	tensor	7/**200**	4/**161**	11/**12**
	stab. tensor	3/3	4/3	11/3

in contrast, the convection dominates with the increase of the Reynolds number, the average number of multigrid sweeps per nonlinear sweep decreases, as the influence of the kernel function is getting irrelevant. This may explain why many people from the CFD community did not pay much attention to this problem before.

5.3 Power law case

In this case the nonlinear viscosity has the form $\nu(z) = \nu_o z^{\frac{r}{2}-1}, z = D_{\mathrm{II}}$, and the gradient and tensor formulation are not equivalent any more. The quality of the solution is checked by comparisons with the well-known and stable conforming Q_2/P_1 approximation; the extended description can be seen in [3]. The accuracy of the nonconforming FEM is saved with the stabilized tensor discretization, see Table 3.

5.4 Pressure dependent viscosity

Finally, the nonlinear (pseudo) viscosity has the form $\nu(p, z) = \exp(\beta p)$, and we list the number of resulting nonlinear iterations and the averaged number of multigrid sweeps per nonlinear iteration for both Newton and Fixpoint methods as outer nonlinear solver. Table 4 shows that the presence of the new linear operator B^* cannot be ignored; otherwise, we destroy the efficiency of the Newton method which is necessary for the robust treatment of the significant nonlinearity.

Table 3. Comparison of the aproximation results for lift, drag and pressure difference for two FEM approaches, the stabilized nonconforming $\tilde{Q}_1/Q_0$ and the classical conforming Q_2/P_1 (see [3]).

Level	Elements	Drag	Lift	$\triangle p$	NN/NL	Drag	Lift	$\triangle p$	NN/NL
Power		$r = 1.5$				$r = 1.1$			
4	$\tilde{Q}_1/Q_0$	1594.20	14.25	24.56	9/2	916.02	3.7381	15.74	12/2
	Q_2/P_1	1635.80	14.39	25.09	8/140	953.94	3.9217	15.82	19/294
5	$\tilde{Q}_1/Q_0$	1615.60	14.43	24.81	8/2	935.13	3.9954	15.82	15/3
	Q_2/P_1	1637.60	14.44	25.07	9/723	957.64	4.0587	15.87	18/1162
6	$\tilde{Q}_1/Q_0$	1626.20	14.46	24.94	8/2	946.22	4.0592	15.85	13/5

Table 4. Corresponding results for the number of nonlinear iterations and the averaged number of linear sweeps per nonlinear cycle

$\nu(z,p) = \exp(\beta p)$		Fixpoint			Newton		
Level	β	0.1	0.3	0.5	0.1	0.3	0.5
5	stab. tensor	6/2	12/2	33/2	3/3	4/2	4/3
	gradient	6/2	11/2	34/2	3/3	4/2	4/3
6	stab. tensor	5/3	11/3	65/2	3/3	3/3	3/3
	gradient	5/3	9/3	76/2	3/3	3/3	5/3

6 Conclusion and outlook

We can conclude our present numerical analysis as follows:

- The proposed stabilization technique is stable and accurate for the used FEM spaces.
- The full (!) Newton method seems to be necessary for this type of nonlinear problem.
- The multigrid convergence behaviour for this new class of auxiliary linear subproblems is
  - (almost) identical for both gradient and deformation tensor formulations: ***The stabilization for nonconforming FEM works fine!***
  - depending on the involved pressure terms for both fixed point and Newton methods: ***More investigation should focus on the linear algebraic problem, beside the nonlinear solution procedure!***

In future, we will cover a wider range of granular materials (see [6] for a discussion):

- **General equation of motion for a powder**
$$\rho\frac{D\boldsymbol{u}}{Dt} = -\nabla p + \nabla\cdot\left[\frac{q(p,\rho)}{\|\mathbf{D}-\frac{1}{n}\nabla\cdot\boldsymbol{u}I\|}\left(\mathbf{D}-\frac{1}{n}\nabla\cdot\boldsymbol{u}I\right)\right] + \rho g, \text{ with}$$
- **Continuity equation**
$$\frac{\partial\rho}{\partial t} + \nabla\cdot(\rho\boldsymbol{u}) = 0, \text{ and}$$
- **Normality condition**

$$\nabla \cdot \boldsymbol{u} = \frac{\partial q(p,\rho)}{\partial p} \left\| \mathbf{D} - \frac{1}{n} \nabla \cdot \boldsymbol{u} I \right\|$$

– **The yield condition $q(p, \rho)$ is given by:**

Powder properties	Non-cohesive	Cohesive
Incompressible	$p \sin \phi$	$p \sin \phi + c \cos \phi$
Compressible	$p \sin \phi \left[2 - \frac{p}{\rho^{\frac{1}{\beta}}} \right]$	$p \sin \phi \rho^{\frac{1}{\beta}} - C \frac{(p-\rho^{\frac{1}{\beta}})^2}{\rho^{\frac{1}{\beta}}}$

References

1. Brenner, C. S. (2002), Korn's inequalities for piecewise H^1 vector fields, IMI Research Reports, **5**, 1–21
2. Hansbo, P. and Larson, M. G. (2002), Discontinuous Galerkin methods for incompressible and nearly incompressible elasticity by Nitsche's method, Computer Methods in Applied Mechanics an Engineering, **191**(17-18), 1895–1908
3. Hron, J., Ouazzi, A. and Turek, S. (2002), A computational comparison of two FEM solvers for nonlinear incompressible flow, proceedings of CISC2002 (to appear)
4. Knobloch, P. (2000), On Korn's inequality for nonconforming finite elements, Technische Mechanik, 205–214
5. Rannacher, R. and Turek, S. (1992), A simple nonconforming quadrilateral Stokes element, Numer. Meth. Par. Diff. Eq., **8**, 97–111
6. Tardos, G.I., McNamara, S. and Talu, I. (2003) Slow and intermediate flow of a frictional bulk powder in the couette geometry, in press, Powder Technology
7. Turek, S. (1998), Efficient solvers for incompressible flow problems: An algorithmic and computational approach, Springer,**6**, LNCSE
8. Turek, S., Ouazzi, A. and Schmachtel, R. (2002) Multigrid method for stabilized nonconforming finite elements for incompressible flow involving the deformation tensor formulation, JNM, **10**, 235–248
9. Schaeffer, D. G. (1987), Instability in the evolution equation describing incompressible granular flow, J. of Differential Equations, **66**, 19–50
10. Schäfer, M. and Turek, S. (1996) Benchmark computations of laminar flow around cylinder. In E.H. Hirschel, editor, Flow Simulation with High Performance Computers II, Note of Numerical Fluid Mechanics, **52**, 547–566

Piecewise Polynomial Approximations for Linear Volterra Integro-Differential Equations with Nonsmooth Kernels*

Arvet Pedas

Institute of Applied Mathematics, University of Tartu, Liivi 2, 50409 Tartu, Estonia *arvet.pedas@ut.ee*

Summary. The piecewise polynomial collocation method is discussed to solve linear Volterra-Basset integro-differential equations with weakly singular or other nonsmooth kernels. Using special graded grids, global convergence estimates are derived. The error analysis is based on certain regularity properties of the solution of the initial value problem.

1 Introduction

Volterra integral equations and integro- differential equations arise naturally in many mathematical models of various physical and biological phenomena. The study of their numerical methods has received considerable attention in the past. The survey articles [1, 2] and the monograph [3] convey a good picture of these developments and contain an extensive bibliography. The present paper is most closely related to the works [3, 4, 5, 6, 7, 8, 9] where a discussion about the convergence of collocation methods for the numerical solution of linear Volterra integro-differential equations with weakly singular kernels is given. In the present paper we extend these investigations to a wider class of equations. First we study the regularity properties of the solution (Section 2). Then we use these results in the construction and analysis of a piecewise polynomial collocation method for solving such equations numerically (Section 4). Using graded grids and an equivalent integral equation reformulation, we derive global convergence estimates for the numerical solutions. Our aim is to construct approximations which possess maximal convergence order on the whole interval of integration. The main results of the paper extend the corresponding results of [7, 8, 9] and are formulated in Theorems 1, 2 and 3.

* This work was supported by the Estonian Science Foundation (Research Grant No. 5859).

2 Integro-differential equation and smoothness of the solution.

Let $b \in \mathbf{R} = (-\infty, \infty)$, $b > 0$ and set $\Delta_b = \{(t,s) \in \mathbf{R}^2 : 0 \le t \le b, 0 \le s < t\}$, $\overline{\Delta}_b = \{(t,s) \in \mathbf{R}^2 : 0 \le s \le t \le b\}$. We consider an initial-value problem for a linear integro-differential equation of the form

$$y'(t) = p(t)y(t) + q(t) + \int_0^t K_1(t,s)y(s)ds + \int_0^t K_2(t,s)y'(s)ds\,, \quad 0 \le t \le b\,, \quad (1)$$

with given initial condition

$$y(0) = y_0\,, \quad y_0 \in \mathbf{R}\,. \quad (2)$$

Observe that, in contrast to "standard" Volterra integro-diffrential equations, the integrand $K_2(t,s)y'(s)$ in (1) depends on the derivative y' instead of the solution y itself. We assume that $K_1, K_2 \in W^{m,\nu}(\Delta_b)$, $p, q \in C^{m,\nu}[0,b]$, $m \in \mathbf{N} = \{1, 2, \ldots\}$, $\nu \in \mathbf{R}$, $\nu < 1$. Here $W^{m,\nu}(\Delta_b)$, $m \in \mathbf{N}$, $\nu < 1$, is defined as the set of all m times continuously differentiable functions $K : \Delta_b \to \mathbf{R}$ satisfying

$$\left| \left(\frac{\partial}{\partial t}\right)^i \left(\frac{\partial}{\partial t} + \frac{\partial}{\partial s}\right)^j K(t,s) \right| \le c \begin{cases} 1 & \text{if } \nu + i < 0\,, \\ 1 + |\log(t-s)| & \text{if } \nu + i = 0\,, \\ (t-s)^{-\nu-i} & \text{if } \nu + i > 0\,, \end{cases} \quad (3)$$

with a constant $c = c(K)$ for all $(t,s) \in \Delta_b$ and all non-negative integers i and j such that $i + j \le m$.

It follows from (3) (with $i = j = 0$, $0 \le \nu < 1$) that the kernels $K_1(t,s)$ and $K_2(t,s)$ of (1) may possess a weak singularity as $s \to t$. In case $\nu < 0$ the kernels K_1 and K_2 are bounded on Δ_b but their derivatives may be singular as $s \to t$. In particular, K_1 and K_2 may have the form

$$K_{\alpha,\beta}(t,s) = \kappa(t,s)(t-s)^{-\alpha}|\log(t-s)|^{\beta}, \quad 0 \le \alpha < 1, \beta \ge 0\,,$$

where $\kappa : \overline{\Delta}_b \to \mathbf{R}$ is a m times continuously differentiable function. Clearly, $K_{\alpha,0} \in W^{m,\alpha}(\Delta_b)$, $0 < \alpha < 1$,. $K_{0,1} \in W^{m,0}(\Delta_b)$ and $K_{\alpha,\beta} \in W^{m,\alpha+\varepsilon}(\Delta_b)$ for $0 \le \alpha < 1$, $\beta > 0$, with a small $\varepsilon > 0$ ($\varepsilon < 1 - \alpha$). Especially, if $K_1 = 0$ and $K_2 = K_{\alpha,0}$, $0 < \alpha < 1$, then equation (1) is of type which is often referred to as the Basset equation; the last one is playing important role in the mathematical modelling of the diffusion of discrete particle in a turbulent fluid (see, for example, [10, 11]).

The set $C^{m,\nu}[0,b]$, $m \in \mathbf{N}$, $\nu < 1$, consists of functions[1] $y \in C[0,b]$ which are m times continuously differentiable in $(0,b]$ and such that $\sum_{j=1}^{m} \sup_{0<t\le b} (w_{j-(1-\nu)}(t)|y^{(j)}(t)|) \le c$. Here

[1] By $C[a,b]$ we denote the Banach space of continuous functions $x : [a,b] \to \mathbf{R}$ with the norm $\|x\| = \max\{|x(t)| : a \le t \le b\}$. By $c, c_1, c_2, \ldots$ we denote positive constants, which may be different in different inequalities.

$$w_\lambda(t) = \begin{cases} 1 & \text{if } \lambda < 0\,, \\ (1+|\log t|)^{-1} & \text{if } \lambda = 0\,, \\ t^\lambda & \text{if } \lambda > 0\,, \end{cases} \qquad 0 < t \le b\,.$$

Equipped with the norm $\|y\|_{m,\nu} = \max\limits_{0\le t\le b} |y(t)| + \sum\limits_{j=1}^{m} \sup\limits_{0<t\le b} \left(w_{j-(1-\nu)}(t) \left| y^{(j)}(t) \right| \right)$, $C^{m,\nu}[0,b]$ is a Banach space. Thus, if a function y belongs to $C^{m,\nu}[0,b]$, $m \in \mathbf{N}$, $\nu < 1$, then its derivatives can be estimated by

$$\left| y^{(j)}(t) \right| \le c \begin{cases} 1 & \text{if } j < 1-\nu\,, \\ 1+|\log t| & \text{if } j = 1-\nu\,, \\ t^{1-\nu-j} & \text{if } j > 1-\nu\,, \end{cases} \tag{4}$$

where $0 < t \le b$ and $j = 0, 1, \ldots, m$. Note that $C^m[0,b]$, the set of m times continuously differentiable functions $y : [a,b] \to \mathbf{R}$, belongs to $C^{m,\nu}[0,b]$ for arbitrary $\nu < 1$.

Introducing a new unknown function

$$z = y'\,, \tag{5}$$

and using (2), equation (1) may be rewritten as a linear Volterra integral equation of the second kind with respect to z,

$$z(t) = \int_0^t K_1(t,s) \int_0^s z(\tau)d\tau ds + \int_0^t \left[p(t) + K_2(t,s) \right] z(s) ds + f(t)\,, \quad t \in [0,b]\,, \tag{6}$$

which may also expressed in the form

$$z(t) = \int_0^t K(t,s) z(s) ds + f(t)\,, \quad t \in [0,b]\,, \tag{7}$$

where

$$f(t) = q(t) + y_0 p(t) + y_0 \int_0^t K_1(t,s) ds\,, \quad t \in [0,b]\,, \tag{8}$$

and

$$K(t,s) = p(t) + \int_s^t K_1(t,\tau) d\tau + K_2(t,s)\,, \quad (t,s) \in \Delta_b\,, \tag{9}$$

We will employ (6) in the construction of numerical solutions for problem $\{(1),(2)\}$ (see Section 4). For the smoothness analysis of the solution of $\{(1),(2)\}$ is more convenient to use (7) which we write in the form $(I-T)z = f$, where I is the identity transformation and

$$(Tz)(t) = \int_0^t K(t,s) z(s) ds\,, \quad t \in [0,b]\,. \tag{10}$$

In the sequel, for given Banach spaces E and F we denote by $\mathcal{L}(E,F)$ the Banach space of linear bounded operators $A : E \to F$ with the norm $\|A\| = \sup\{\|Az\| : z \in E, \|z\| \le 1\}$.

Lemma 1. *Let $K_1, K_2 \in W^{m,\nu}(\Delta_b)$, $p \in C^{m,\nu}[0,b]$, $m \in \mathbf{N}$, $\nu \in \mathbf{R}$, $\nu < 1$. Then T is linear and compact as an operator from $L^\infty(0,b)$ to $C[0,b]$. Moreover, T is compact as an operator from $C^{m,\nu}[0,b]$ to $C^{m,\nu}[0,b]$.*

Proof. We present T (see (9) and (10)) in the form $T = T_{01}T_{02} + T_1 + T_2$, where the linear operators T_{01}, T_{02}, T_1 and T_2 are defined by settings

$$(T_{01}z)(t) = p(t)z(t)\,, \quad (T_{02}z)(t) = \int_0^t z(s)ds\,,$$

$$(T_1 z)(t) = \int_0^t L_1(t,s)z(s)ds \quad \text{with} \quad L_1(t,s) = \int_s^t K_1(t,\tau)d\tau\,,$$

$$(T_2 z)(t) = \int_0^t K_2(t,s)z(s)ds\,.$$

It follows from $K_1, K_2 \in W^{m,\nu}(\Delta_b)$ that $L_1(t,s)$ is bounded for $(t,s) \in \Delta_b$ and $K_2(t,s)$ is at most weakly singular: $|K_2(t,s)| \le c(t-s)^{-\alpha}$, $(t,s) \in \Delta_b$, $0 < \alpha < 1$. Therefore $T_1, T_2 : L^\infty(0,b) \to C[0,b]$ are compact. Clearly, $T_{01} \in \mathcal{L}(C[0,b], C[0,b])$ and $T_{02} : L^\infty(0,b) \to C[0,b]$ is compact. This implies $T_{01}T_{02} \in \mathcal{L}(L^\infty(0,b), C[0,b])$ is compact. In summary, $T_{01}T_{02} + T_1 + T_2 = T \in \mathcal{L}(L^\infty(0,b), C[0,b])$ is compact.

Further, it follows from $K_1 \in W^{m,\nu}(\Delta_b)$ that

$$\left(\frac{\partial}{\partial t} + \frac{\partial}{\partial s}\right)^j \int_s^t K_1(t,\tau)d\tau = \int_s^t \left(\frac{\partial}{\partial t} + \frac{\partial}{\partial \tau}\right)^j K_1(t,\tau)d\tau\,, \; j = 0,1,\ldots,m\,,$$

and hence $L_1 \in W^{m,\nu-1}(\Delta_b) \subset W^{m,\nu}(\Delta_b)$. Since $L_1, K_2 \in W^{m,\nu}(\Delta_b)$, $T_1, T_2 \in \mathcal{L}(C^{m,\nu}[0,b], C^{m,\nu}[0,b])$ are compact (see [7] for details). Since $1 \in W^{m,\nu}(\Delta_b)$, we also deduce that $T_{02} : C^{m,\nu}[0,b] \to C^{m,\nu}[0,b]$ is compact. If $y_1, y_2 \in C^{m,\nu}[0,b]$, $m \in \mathbf{N}$, $\nu < 1$ then, by (4), $y_1 y_2 \in C^{m,\nu}[0,b]$ and $\|y_1 y_2\|_{m,\nu} \le c\,\|y_1\|_{m,\nu}\|y_2\|_{m,\nu}$, with a constant c which is independent of y_1 and y_2. This implies $T_{01} \in \mathcal{L}(C^{m,\nu}[0,b], C^{m,\nu}[0,b])$. In summary $T_{01}T_{02} + T_1 + T_2 = T \in \mathcal{L}(C^{m,\nu}[0,b], C^{m,\nu}[0,b])$ is compact. Lemma 1 is proved.

The regularity of the solution of equation (1) is described in the following

Theorem 1. *Let $K_1, K_2 \in W^{m,\nu}(\Delta_b)$, $p, q \in C^{m,\nu}[0,b]$, $m \in \mathbf{N}$, $\nu \in \mathbf{R}$, $\nu < 1$. Then equation (7) has a unique solution $z \in C^{m,\nu}[0,b]$ implying that problem $\{(1), (2)\}$ has a unique solution $y \in C^{m+1,\nu-1}[0,b]$ for every $y_0 \in \mathbf{R}$.*

Proof. It follows from $p, q \in C^{m,\nu}[0,b]$ and $K_1 \in W^{m,\nu}(\Delta_b)$ that $f \in C^{m,\nu}[0,b]$. Indeed, $f = f_1 + f_2$, where (see (8)) $f_1(t) = q(t) + y_0 p(t)$, $t \in [0,b]$, and $f_2(t) = y_0 \int_0^t K_1(t,s)ds$, $t \in [0,b]$. Clearly, $f_1 \in C^{m,\nu}[0,b]$ and $f_2 = y_0 T1$, with T defined by (10). Since $1 \in C^{m,\nu}[0,b]$ and T is bounded as an operator from $C^{m,\nu}[0,b]$ to $C^{m,\nu}[0,b]$ (see Lemma 1), $f_2 \in C^{m,\nu}[0,b]$. As the homogenous equation $z = Tz$ has only the trivial solution $z = 0$, it follows from $f \in C^{m,\nu}[0,b]$ and Lemma 1 that $I - T$ has a bounded inverse $(I-T)^{-1} \in \mathcal{L}(C^{m,\nu}[0,b], C^{m,\nu}[0,b])$, and equation $(I-T)z = f$ has a unique solution $z = (I-T)^{-1}f \in C^{m,\nu}[0,b]$. In other words, $y' \in C^{m,\nu}[0,b]$, implying $y \in C^{m+1,\nu-1}[0,b]$. Theorem 1 is proved.

3 Piecewise polynomial interpolation

For given $N \in \mathbf{N}$, $r \in \mathbf{R}$, $r \geq 1$, let $\Pi_N^r = \{t_0, \ldots, t_N : 0 = t_0 < \ldots < t_N = b\}$ be a partition (a grid) of the interval $[0,b]$ given by the grid points

$$t_j = b(j/N)^r, \quad j = 0, \ldots, N. \tag{11}$$

Here r (also called the grading exponent) characterizes the non-uniformity of the grid Π_N^r: if $r > 1$ then the gridpoints (11) are more densely clustered near the left endpoint of the interval $[0,b]$. Let

$$\sigma_j = [t_{j-1}, t_j], \quad h_j = t_j - t_{j-1}, \quad j = 1, \ldots, N. \tag{12}$$

For given integers $m \geq 0$ and $-1 \leq d \leq m-1$, let $S_m^{(d)}(\Pi_N^r)$ be the spline space of piecewise polynomial functions on the grid Π_N^r:

$$S_m^{(d)}(\Pi_N^r) = \Big\{u : u\big|_{\sigma_j} \in \pi_m,\ j = 1, \ldots, N;$$
$$\big(u\big|_{\sigma_j}\big)^{(k)}(t_j) = \big(u\big|_{\sigma_{j+1}}\big)^{(k)}(t_j), 0 \leq k \leq d; j = 1, \ldots, N-1\Big\},$$

where π_m denotes the set of polynomials of degree not exceeding m and $u|_{\sigma_j}$ is the restriction of u to the subinterval σ_j, $j = 1, \ldots, N$. Note that elements of $S_m^{(-1)}(\Pi_N^r) = \{u : u|_{\sigma_j} \in \pi_m, j = 1, \ldots, N\}$ may have jump discontinuities at the interior points $t_1, \ldots, t_{N-1}$ of the grid Π_N^r.

In every subinterval σ_j $(j = 1, \ldots, N)$ we introduce $m \in \mathbf{N}$ interpolation points

$$t_{jl} = t_{j-1} + \eta_l h_j, \quad l = 1, \ldots, m \quad (j = 1, \ldots, N) \tag{13}$$

where $\eta_1, \ldots, \eta_m$ do not depend on j and N and satisfy

$$0 \leq \eta_1 < \ldots < \eta_m \leq 1. \tag{14}$$

To a given continuous function $z : [0,b] \to \mathbf{R}$ we assign a piecewise polynomial interpolation function $P_N z \in S_{m-1}^{(-1)}(\Pi_N^r)$ which interpolates z at the points (13): $(P_N z)(t_{jl}) = z(t_{jl})$, $l = 1, \ldots, m$; $j = 1, \ldots, N$. Thus, $P_N z$ is independently defined in every subinterval σ_j $(j = 1, \ldots, N)$ and $(P_N z)(t)$

may be discontinuous at $t = t_j$, $j = 1, \dots, N-1$; we may treat $P_N z$ as a two valued function at these points. Note that in case $\eta_1 = 0$, $\eta_m = 1$ (see (14)), $P_N z \in [0, b]$. We introduce also an interpolation operator P_N which assigns for every function $z \in C[0, b]$ its piecewise polynomial interpolation function $P_N z$.

Lemma 2. *[7, 8] Let $z \in C^{m,\nu}[0, b]$, $m \in \mathbf{N}$, $\nu \in \mathbf{R}$, $\nu < 1$. Then $\sup\limits_{t\in[0,b]} \big|z(t) - (P_N z)(t)\big| \le c\,\varepsilon_N^{(m,\nu,r)}$, where c a constant not depending on N and*

$$\varepsilon_N^{(m,\nu,r)} = \begin{cases} N^{-m} & \text{for } m < 1-\nu, r \ge 1; \\ N^{-m}(1+\log N) & \text{for } m = 1-\nu, r = 1; \\ N^{-m} & \text{for } m = 1-\nu, r > 1; \\ N^{-r(1-\nu)} & \text{for } m > 1-\nu, 1 \le r < m/(1-\nu); \\ N^{-m} & \text{for } m > 1-\nu, r \ge m/(1-\nu). \end{cases} \tag{15}$$

Lemma 3. *Let $T : L^\infty(0, b) \to C[0, b]$ be a linear compact operator. Then $\|T - P_N T\|_{\mathcal{L}(L^\infty(0,b), L^\infty(0,b))} \to 0$ as $N \to \infty$.*

Proof. An easy observation shows that

$$\big\|P_N\big\|_{\mathcal{L}(C[0,b], L^\infty(0,b))} \le c, \quad N \in \mathbf{N}, \tag{16}$$

where c is a constant not depending on N. On the base of (16) and Lemma 2 we obtain that $\|z - P_N z\|_{L^\infty(0,b)} \to 0$ as $N \to \infty$ for every $z \in C[0, b]$. Together with the compactness of $T : L^\infty(0, b) \to C[0, b]$ this yields the assertion of Lemma.

4 Collocation method

We look for an approximation v to the solution z of equation (6) in the space $S_{m-1}^{(-1)}(\Pi_N^r)$ determing $v = v^{(N,m,r)} \in S_{m-1}^{(-1)}(\Pi_N^r)$, $m \ge 1$, from the following conditions:

$$\begin{aligned} v(t_{jl}) = f(t_{jl}) + \int_0^{t_{jl}} K_1(t_{jl}, s) \int_0^s v(\tau) d\tau ds + \\ + \int_0^{t_{jl}} \big[p(t_{jl}) + K_2(t_{jl}, s)\big] v(s) ds, \quad l = 1, \dots, m; j = 1, \dots, N, \end{aligned} \tag{17}$$

with $\{t_{jl}\}$, given by (13). Having determined the approximation v for z, we can also determine the approximation u for y, the solution of initial value problem $\{(1),(2)\}$, setting (see (5))

$$u(t) = y_0 + \int_0^t v(s) ds, \quad t \in [0, b]. \tag{18}$$

Note that the choice of the collocation points (13) with $\eta_1 = 0$, $\eta_m = 1$ in (14) actually implies that the resulting collocation approximation v belongs to the smoother polynomial spline space $S_{m-1}^{(0)}(\Pi_n^r)$.

Theorem 2. *Let $y_0 \in \mathbf{R}$, $K_1, K_2 \in W^{m,\nu}(\Delta_b)$, $p, q \in C^{m,\nu}[0,b]$, $m \in \mathbf{N}$, $\nu \in \mathbf{R}$, $\nu < 1$, and assume that the collocation points* (13), *with the grid points* (11) *and parameters* (14), *are used.*

Then, for all sufficiently large N, say $N \geq N_0$, and for every choice of parameters (14) *with $\eta_1 > 0$ or $\eta_m < 1$, the equalities* (18) *and* (17) *determine unique approximations $u \in S_m^{(0)}(\Pi_N^r)$ and $v \in S_{m-1}^{(-1)}(\Pi_N^r)$ (with $v|_{\sigma_j} = (u|_{\sigma_j})'$, $j = 1, \dots, N$) to the solution y of problem* {(1), (2)} *and its derivative y', respectively. If $\eta_1 = 0$, $\eta_m = 1$, then $u \in S_m^{(1)}(\Pi_N^r)$ and $v = u' \in S_{m-1}^{(0)}(\Pi_N^r)$. For all $N \geq N_0$ the following error estimates holds:*

$$\left\| e^{(i)} \right\|_\infty \leq c\,\varepsilon_N^{(m,\nu,r)}\,, \quad i \in \{0, 1\}\,. \tag{19}$$

Here c is a constant not depending on N, $\varepsilon_N^{(m,\nu,r)}$ is given by (15) *and*

$$\left\| e^{(i)} \right\|_\infty = \max_{j=1,\dots,N} \left(\max_{t \in \sigma_j} \left| u_j^{(i)}(t) - y^{(i)}(t) \right| \right), \quad u_j = u\big|_{\sigma_j}\,, \quad i \in \{0, 1\}\,. \tag{20}$$

Proof. As we know from Section 2, problem {(1),(2)} is equivalent to the integral equation (7) where $z = y'$ and the forcing function f and the kernel K are given by (8) and (9), respectively. We rewrite (7) in the form $z = Tz + f$, with T defined by (10). We find that $f \in C^{m,\nu}[0,b] \subset L^\infty(0,T)$. It follows from Lemma 1 that T is compact as an operator from $L^\infty(0,b)$ to $L^\infty(0,b)$. Therefore, $z = Tz + f$ has a unique solution $z \in L^\infty(0,b)$. Moreover, on the base of Theorem 1 we obtain that $z \in C^{m,\nu}[0,b]$.

Further, conditions (17) are equivalent to the operator equation representation $v = P_N T v + P_N f$, with P_N defined in Section 3. From Lemma 3 and from the boundedness of $(I - T)^{-1}$ in $L^\infty(0,b)$ we obtain that $I - P_N T$ is invertible in $L^\infty(0,b)$ for sufficiently large N, say $N \geq N_0$. Moreover, the norms of $(I - P_N T)^{-1}$ are uniformly bounded in N:

$$\left\| (I - P_N T)^{-1} \right\|_{\mathcal{L}(L^\infty(0,b), L^\infty(0,b))} \leq c\,, \quad N \geq N_0\,. \tag{21}$$

Thus, for $N \geq N_0$, equation $v = P_N T v + P_N f$ provides a unique solution $v \in S_{m-1}^{(-1)}(\Pi_N^r)$ ($v \in S_{m-1}^{(0)}(\Pi_N^r)$ if $\eta_1 = 0$, $\eta_m = 1$). For v and z, the solutions of equations $v = P_N T v + P_N f$ and $z = Tz + f$ respectively, we have

$$v - z = (I - P_N T)^{-1}(P_N z - z)\,, \quad N \geq N_0\,. \tag{22}$$

Now (21) yields $\|v - z\|_{L^\infty(0,b)} \leq c\|P_N z - z\|_{L^\infty(0,b)}$, $N \geq N_0$, with a constant c which is independent of N. Applying Lemma 2 we obtain the estimate (19) with $i = 1$.

Further, due to (18) and (2), $y(t) - u(t) = \int_0^t [y'(s) - v(s)]ds$, $t \in [0,b]$. Applying (19) with $i = 1$ we obtain the estimate (19) with $i = 0$. Theorem 2 is proved.

Thus, according to Theorem 2, in case $m > 1-\nu$, the approximation order $\|y-u\|_\infty \le cN^{-m}$ is guaranteed for $r \ge m/(1-\nu)$. For ν close to 1, $\nu < 1$, this condition on r may be too restrictive. To obtain the order $\|y-u\|_\infty \le cN^{-m}$, the condition on r can be considerable relaxed, as shown in the following

Theorem 3. *Let* $y_0 \in \mathbf{R}$, $K_1 \in W^{m,\nu}(\Delta_b)$, $K_2 \in W^{m,\nu-1}(\Delta_b)$, $p, q \in C^{m,\nu}[0,b]$, $m \in \mathbf{N}$, $\nu \in \mathbf{R}$, $\nu < 1$, $m > 1-\nu$, *and assume that the collocation points* (13) *with the gridpoints* (11) *and parameters* (14) *are used. Then, with the notation of Theorem 2, we have the following estimates for the error* $y-u$*:*

1) if $1-\nu < m < 2-\nu$ *then* $\|y-u\|_\infty \le cN^{-m}$ *for* $r \ge 1$*;*
2) if $m = 2-\nu$ *then*

$$\|y-u\|_\infty \le c\begin{cases} N^{-m}(1+\log N) & \text{for } r = 1\,; \\ N^{-m} & \text{for } r > 1\,; \end{cases}$$

3) if $m > 2-\nu$ *then*

$$\|y-u\|_\infty \le c\begin{cases} N^{-r(2-\nu)} & \text{for } 1 \le r < m/(2-\nu)\,, \\ N^{-m}(1+\log N) & \text{for } \quad r = m/(2-\nu)\,, \\ N^{-m} & \text{for } \quad r > m/(2-\nu)\,. \end{cases}$$

Proof. Using the equality $(I-P_NT)^{-1} = I + (I-P_NT)^{-1}P_NT$, we rewrite the error (22) in the form $v - z = P_Nz - z + (I-P_NT)^{-1}P_NT(P_Nz-z)$, $N \ge N_0$. Due to continuity and boundedness of $K(t,s)$ on Δ_b, T is bounded, as an operator from $L^1(0,b)$ to $C[0,b]$ (see (9) and (10)). Together with (5), (16) and (21) we obtain that

$$|y(t)-u(t)| = \left|\int_0^t [z(s)-v(s)]\,ds\right| \le c\int_0^b \big|(P_Nz)(s)-z(s)\big|ds\,,$$

where $0 \le t \le b$. Since $z \in C^{m,\nu}[0,b]$, $m > 1-\nu$, then (see [12], p.116,[7, 8])

$$\max_{t\in\sigma_j}\big|(P_Nz)(t)-z(t)\big| \le c(t_j-t_{j-1})^{m+1}t_j^{1-\nu-m}\,, \quad j = 1,\ldots,N\,,$$

with $\{t_j\}$, given by (11). It follows from (11) that

$$(t_j-t_{j-1})^{m+1}t_j^{1-\nu-m} \le cN^{-r(2-\nu)}j^{r(2-\nu)-m-1}\,, \quad j = 1,\ldots,N\,.$$

Therefore, for $t \in [0,b]$, we have

$$\big|y(t)-u(t)\big| \le c\sum_{j=1}^N \int_{t_{j-1}}^{t_j} \big|(P_Nz)(s)-z(s)\big|ds \le c_1N^{-r(2-\nu)}\sum_{j=1}^N j^{r(2-\nu)-m-1}, \tag{23}$$

with c and c_1 not depending on N. Furthermore, for a number $\alpha \in \mathbf{R}$ we have

$$\sum_{j=1}^N j^\alpha \le c\begin{cases} N^{\alpha+1} & \text{if } \alpha > -1\,, \\ 1+|\log N| & \text{if } \alpha = -1\,, \\ 1 & \text{if } \alpha < -1\,, \end{cases} \tag{24}$$

where c is a constant which does not depend on N. Applying (24) with $\alpha = r(2-\nu) - m - 1$ to (23) it is easy to see that the statements of Theorem 3 hold.

Remarks. 1) The equalities (17) form a system of algebraic equations whose exact form is determined by the choice of a basis in $S_{m-1}^{(-1)}(\Pi_N^r)$ (or in $S_{m-1}^{(0)}(\Pi_N^r)$ if $\eta_1 = 0$, $\eta_m = 1$). For instance, in each subinterval $[t_{j-1}, t_j]$ ($j = 1, \dots, N$) we may use the representation $v(t_{j-1} + \tau h_j) = \sum_{l=1}^{m} c_{jl} L_l^{(m-1)}(\tau)$, $\tau \in [0,1]$, where $L_l^{(m-1)}(\tau)$ denotes the lth Lagrange fundamental polynomial of degree $m-1$ associated with the parameters (14), that is $L_l^{(m-1)}(\tau) = \Pi_{i \neq l}^{m}(\tau - \eta_i)/(\eta_l - \eta_i)$, $\tau \in [0,1]$. The conditions (17) then lead to a linear system of equations for the coefficients $c_{jl} = c_{jl}^{(N)}$, $l = 1, \dots, m$; $j = 1, \dots, N$.

2) Method {(17),(18)} where we have discretized the integral equation (6) is equivalent to the collocation method applied directly to problem {(1),(2)}. In the latter form the collocation method in more particular case ($K_2 = 0$, $K_1 \in W^{m,\nu}(\Delta_b)$, $0 < \nu < 1$, $p, q \in C^m[0,b]$, $m \in \mathbf{N}$) has been examined in [3, 4, 5, 6].

3) The convergence results established by Theorems 2 and 3 are derived under the assumptions that all needed integrals in (17) can be evaluated analytically. Since this is rarely possible in concrete applications, there arises the question how to approximate these integrals so that the resulting fully discretized collocation method converges under the same conditions and with the same rate as it is proved for the "exact" collocation method in Theorems 2 and 3. This question will be discussed elsewhere.

References

1. Brunner H. (1982): A survey of recent advances in the numerical treatment of Volterra integral and integro-differential equations. J. Comput. Appl. Math., 213–229.
2. Baker C. T. H. (2000): A perspective on the numerical treatment of Volterra equations. J. Comput. Appl. Math., **125**, 217–249.
3. Brunner H., van der Houwen P. J. (1986): The Numerical Solution of Volterra Equations, CWI Monographs 3, North-Holland, Amsterdam.
4. Makroglou A. (1981): A block-by-block method for Volterra integro-differential equations with weakly-sin gular kernels. Math. Comp., **37**, 95–99.
5. Brunner H. (1986): Polynomial spline collocation methods for Volterra integro-differential equations with weakly singular kernelss. IMA J. Numer. Anal., **6**, 221–239.
6. Tang T. (1993): A note on collocation methods for Volterra integro-differential equations with weakly singular kernels. IMA J. Numer. Anal., **13**, 93–99.
7. Brunner H., Pedas A., Vainikko G. (2001): Piecewise polynomial collocation methods for linear Volterra integro-differential equations with weakly singular kernels. SIAM J. Numer. Anal., **39**, 957–982.

8. Brunner H., Pedas A., Vainikko G. (2001): Spline collocation method for linear Volterra integro-differential equations with weakly singular kernels. BIT, **41**, 5, 891–900.
9. Parts I., Pedas A. (2003): Spline collocation methods for weakly singular Volterra integro-differential equations. In: Brezzi F., Buffa A., Corsaro S., Murli A. (eds) Numerical Mathematics and Advanced Applications, Enumath 2001, Springer-Verlag, Italia, Milano, 919–928.
10. McKee S., Stokes A. (1983): Product integration methods for the nonlinear Basset equation. SIAM J. Numer. Anal., **20**, 1, 143–160.
11. Brunner H., Tang T. (1989): Polynomial spline collocation methods for the nonlinear Basset equation. Computers Math. Applic., **18**, 5, 449–457.
12. Vainikko G. (1993): Multidimensional Weakly Singular Integral Equations. Lecture Notes in Mathematics, 1549, Spronger-Verlag, Berlin, Heidelberg, New-York.

On a Discontinuous Galerkin Method for Radiation-Diffusion Problems

Ilaria Perugia[1], Dominik Schötzau[2] and James Warsa[3]

[1] Dipartimento di Matematica, Università di Pavia, Via Ferrata 1, 27100 Pavia, Italy, email: *perugia@dimat.unipv.it.*

[2] Mathematics Department, University of British Columbia, Vancouver, BC V6T 1Z2, Canada, email: *schoetzau@math.ubc.ca.*

[3] Transport Methods Group, Los Alamos National Laboratory, Los Alamos, NM 87545, USA, email: *warsa@lanl.gov.* Supported by the U. S. Department of Energy.

Summary. In this paper, we show that the discontinuous finite element method recently developed by Warsa, Wareing and Morel for radiation-diffusion problems belongs to a class of generalized local discontinuous Galerkin methods. We then derive a priori error bounds for this method and numerically confirm them to be sharp.

1 Introduction

In the recent work [7] and [6], Warsa, Wareing and Morel introduced a discontinuous finite element method for the discretization of a radiation–diffusion problem that is represented by a system of two coupled first order equations for the zeroth and first angular moments of the particle distribution (the scalar flux and current, respectively). These so-called P_1 equations arise out of an angular Galerkin approximation to the Boltzmann transport equation based upon a spherical–harmonic trial space of first order; see [3] or [4] for more details. In the absence of time dependence (the case considered here) the resulting problem finds the current $\mathbf{J} = \mathbf{J}(\mathbf{x})$ and the scalar flux $\Phi = \Phi(\mathbf{x})$ satisfying

$$\nabla \Phi + 3\,\sigma_t(\mathbf{x})\mathbf{J} = 3\,\mathbf{Q}_1, \qquad \nabla \cdot \mathbf{J} + \sigma_a(\mathbf{x})\Phi = Q_0, \qquad \text{in } \Omega \subset \mathbb{R}^d, \tag{1}$$

subject to so-called vacuum and reflecting boundary conditions, respectively,

$$\frac{1}{4}\Phi - \frac{1}{2}\mathbf{J}\cdot\mathbf{n} = 0 \quad \text{on } \Gamma_V, \qquad \mathbf{J}\cdot\mathbf{n} = 0 \quad \text{on } \Gamma_R. \tag{2}$$

Here, Ω is a bounded polygonal ($d = 2$) or polyhedral ($d = 3$) domain with outward normal unit vector $\mathbf{n}$ on the boundary $\Gamma = \partial\Omega$, which is partitioned into two parts $\Gamma = \Gamma_V \cup \Gamma_R$ with disjoint interiors. The right-hand sides $Q_0 \in L^2(\Omega)$ and $\mathbf{Q}_1 \in L^2(\Omega)^d$ are the zeroth and first angular moments of an inhomogeneous source. We assume that the material coefficients σ_t and σ_a belong to

$L^\infty(\Omega)$ and satisfy $\sigma_t(\mathbf{x}) \geq \sigma_\star > 0$ and $\sigma_a(\mathbf{x}) \geq 0$ in Ω ($\sigma_a = 0$ in purely scattering subregions). For simplicity, we further assume that $\int_{\Gamma_V} ds > 0$.

The method of Warsa, Wareing and Morel approximates both the unknowns $\mathbf{J}$ and Φ by piecewise linear functions. It is designed in such a way that the radiation energy as well as the radiation momentum are conserved over each cell, as in standard, upwind Godunov schemes. In combination with efficient preconditioning techniques, the results in [5–7] indicate that the method of Warsa, Wareing and Morel can be applied to a wide range of problems.

In this note, we show that the discrete formulation of the P_1 equations of Warsa, Wareing and Morel belongs to the general class of mixed discontinuous Galerkin (DG) methods analyzed by Castillo, Cockburn, Perugia and Schötzau in [1]. This class extends and generalizes the local discontinuous Galerkin (LDG) method proposed by Cockburn and Shu [2]. In the original LDG approach the vector unknown can be eliminated from the equations in a local and element-wise manner. In contrast, the method of Warsa, Wareing and Morel belongs to the "truly" mixed variants of the LDG method described in [1] for which such a local elimination is no longer possible. While this can be seen as a shortcoming of truly mixed DG methods it in fact leads to better and nearly optimal convergence rates for the approximation of the vector variable; see [1]. Furthermore, as described in [6], the P_1 equations are used to accelerate the iterative convergence of the Boltzmann transport equation solution. The transport equation is discretized with a spatial DG method and in many cases effective acceleration strongly depends on how well the vector unknown of the acceleration equations is approximated. This application is what makes the use of the truly mixed LDG discretization for the P_1 equations necessary and motivates the study presented in this paper.

We apply the theoretical results in [1] and conclude that the method of Warsa, Wareing and Morel is well-posed. Moreover, the following a priori error bounds hold: for an approximation order $k \geq 0$, the method exhibits convergence rates in the mesh size of order $k + \frac{1}{2}$ in a suitable energy norm, and of order $k + 1$ in the L^2–norm of the scalar flux. We present a set of numerical convergence tests for a three dimensional model problem on tetrahedral meshes that verify the theoretical predictions. We must point out that these tests complete the tests in [1] where no numerical results were shown for truly mixed DG methods.

2 Discontinuous Galerkin discretization

In this section, we detail the mixed discontinuous Galerkin discretization proposed by Warsa, Wareing and Morel [6, 7] and cast the method in the setting of [1].

We consider shape regular meshes $\mathcal{T}_h$ of mesh size h that partition the domain Ω into triangle and/or parallelograms. We allow for irregular nodes, in general, but assume that the local mesh sizes are of bounded variation. Using the same notation as in [1], we let $\mathcal{E}_{\mathcal{I}}$ be the union of all interior faces of $\mathcal{T}_h$, $\mathcal{E}_V$ and $\mathcal{E}_R$ the union of all boundary faces of $\mathcal{T}_h$ on Γ_V and Γ_R, respectively, and set $\mathcal{E} = \mathcal{E}_{\mathcal{I}} \cup \mathcal{E}_V \cup \mathcal{E}_R$. For piecewise smooth vector– and scalar–valued functions $\mathbf{w}$ and u, we introduce the following trace operators. Let $e \subset \mathcal{E}_{\mathcal{I}}$ be an interior face shared by two elements K^+ and K^-, and write $\mathbf{n}^\pm$ for the outward normal unit vectors to the boundaries $\partial K^\pm$, respectively. Denoting by $\mathbf{w}^\pm$ and $u^\pm$ the traces on $\partial K^\pm$ taken from $K^\pm$, respectively, we define the jumps across e by $[\![\mathbf{w}]\!] = \mathbf{w}^+ \cdot \mathbf{n}^+ + \mathbf{w}^- \cdot \mathbf{n}^-$ and $[\![u]\!] = u^+\mathbf{n}^+ + u^-\mathbf{n}^-$, and the averages $\{\!\{\mathbf{w}\}\!\} = (\mathbf{w}^+ + \mathbf{w}^-)/2$ and $\{\!\{u\}\!\} = (u^+ + u^-)/2$. On a boundary face $e \subset \mathcal{E}_V \cup \mathcal{E}_R$, we set $[\![\mathbf{w}]\!] = \mathbf{w} \cdot \mathbf{n}$, $[\![u]\!] = u\,\mathbf{n}$, $\{\!\{\mathbf{w}\}\!\} = \mathbf{w}$ and $\{\!\{u\}\!\} = u$.

Let $\mathcal{T}_h$ be a triangulation $\mathcal{T}_h$ of Ω and $k \geq 0$ an approximation order. We wish to approximate $(\mathbf{J}, \Phi)$ by a piecewise polynomial function $(\mathbf{J}_h, \Phi_h)$; that is, $(\mathbf{J}_h, \Phi_h)|_K \in \mathcal{P}_k(K)^d \times \mathcal{P}_k(K)$ for all $K \in \mathcal{T}_h$. Here, $\mathcal{P}_k(K)$ denotes the set of polynomials of degree at most k on K. This approximation is defined by imposing that, for all elements $K \in \mathcal{T}_h$ and all test functions $(\mathbf{w}, u) \in \mathcal{P}_k(K)^d \times \mathcal{P}_k(K)$,

$$\begin{aligned} 3\int_K \sigma_t\, \mathbf{J}_h \cdot \mathbf{w}\, d\mathbf{x} - \int_K \Phi_h\, \nabla \cdot \mathbf{w}\, d\mathbf{x} + \int_{\partial K} \widehat{\Phi}_h\, \mathbf{w} \cdot \mathbf{n}_K\, ds &= 3\int_K \mathbf{Q}_1 \cdot \mathbf{w}\, d\mathbf{x}, \\ -\int_K \mathbf{J}_h \cdot \nabla u\, d\mathbf{x} + \int_{\partial K} u\, \widehat{J}_{h,K}\, ds + \int_K \sigma_a\, \Phi_h\, u\, d\mathbf{x} &= \int_K Q_0\, u\, d\mathbf{x}. \end{aligned} \tag{3}$$

Here, $\widehat{J}_{h,K}$ and $\widehat{\Phi}_h$ are the so–called numerical fluxes, that are approximations to the traces of $\mathbf{J} \cdot \mathbf{n}_K$ and Φ on the element interfaces and are chosen as follows (see [6]).

First, for an element K^+ and an interior face e shared by K^+ and a neighboring element K^-, we define the following inwardly and outwardly directed discrete flows (partial currents) by

$$J^{\text{in}}_{e,K^+} = \frac{1}{4}\Phi_h^- - \frac{1}{2}\mathbf{J}_h^- \cdot \mathbf{n}_{K^+}, \qquad J^{\text{out}}_{e,K^+} = \frac{1}{4}\Phi_h^+ + \frac{1}{2}\mathbf{J}_h^+ \cdot \mathbf{n}_{K^+}.$$

If the face e of K^+ is contained in $\mathcal{E}_V \cup \mathcal{E}_R$, the outwardly directed flow J^{out}_{e,K^+} is defined as $J^{\text{out}}_{e,K^+} = \frac{1}{4}\Phi_h^+ + \frac{1}{2}\mathbf{J}_h^+ \cdot \mathbf{n}_{K^+}$. Further, if the face e of K^+ belongs to $\mathcal{E}_V$, we set $J^{\text{in}}_{e,K^+} = 0$, whereas J^{in}_{e,K^+} is not needed for $e \subset \mathcal{E}_R$.

The numerical fluxes are then taken as

$$\widehat{J}_{h,K}|_e = (1-\xi)\,(J^{\text{out}}_{e,K} - J^{\text{in}}_{e,K}), \qquad \widehat{\Phi}_h|_e = 2\left[(1+\xi)\, J^{\text{out}}_{e,K} + (1-\xi)\, J^{\text{in}}_{e,K}\right], \tag{4}$$

with $\xi = 0$, if $e \subset \mathcal{E}_{\mathcal{I}} \cup \mathcal{E}_V$, and $\xi = 1$, if $e \subset \mathcal{E}_R$.

This completes the definition of the DG method proposed in [6, 7] for problem (1)–(2) (where only the case $k = 1$ was considered). Notice that, for e shared by K^+ and K^-, $\widehat{J}_{h,K^+}|_e = -\widehat{J}_{h,K^-}|_e$, whereas the definition $\widehat{\Phi}_h|_e$ does not depend on which side of e it is taken from. This is the reason for the subscript K in $\widehat{J}_{h,K}$. The choice in (4) of the numerical fluxes can be motived by physical arguments corresponding to the so–called Marshak approximation;

obviously the choice is not unique, but consistency with the intended application in which solution of the P_1 equations is embedded requires that we make this choice; see [6] for more details. Note that the fluxes can further be easily adapted to take into account inhomogeneous boundary data.

We now cast the fluxes in (4) in the setting of [1]. Denote by $\widehat{\mathbf{J}}_h$ a vector field such that $\widehat{\mathbf{J}}_h|_e \cdot \mathbf{n}_K = \widehat{J}_{h,K}$, for all $e \subset \partial K$ (the definition of $\widehat{\mathbf{J}}_h|_e$ no longer depends on which side it is taken from). Then we have

$$\begin{aligned}
&\widehat{\mathbf{J}}_h|_e = \{\!\{\mathbf{J}_h\}\!\} + \frac{1}{4}[\![\Phi_h]\!], && \widehat{\Phi}_h|_e = \{\!\{\Phi_h\}\!\} + [\![\mathbf{J}_h]\!], && \text{if } e \subset \mathcal{E}_{\mathcal{I}},\\
&\widehat{\mathbf{J}}_h|_e = \frac{1}{2}\mathbf{J}_h + \frac{1}{4}\Phi_h\,\mathbf{n}, && \widehat{\Phi}_h|_e = \frac{1}{2}\Phi_h + \mathbf{J}_h\cdot\mathbf{n}, && \text{if } e \subset \mathcal{E}_V,\\
&\widehat{\mathbf{J}}_h|_e = \mathbf{0}, && \widehat{\Phi}_h|_e = \Phi_h + 2\,\mathbf{J}_h\cdot\mathbf{n}, && \text{if } e \subset \mathcal{E}_R.
\end{aligned}$$

The form of these fluxes shows that the method in (3) belongs to the class of mixed DG methods investigated in [1] and the theoretical results there can be used to analyze it. In particular the formulation (3) is consistent and uniquely solvable; see [1, Proposition 2.1].

3 Error analysis

In this section, we discuss the a priori error bounds that are obtained for the formulation (3). For a piecewise smooth function $(\mathbf{w}, u)$, we define the seminorm

$$|(\mathbf{w},u)|_h^2 = 3\|\sigma_t^{\frac{1}{2}}\,\mathbf{w}\|_{0,\Omega}^2 + \|[\![\mathbf{w}]\!]\|_{0,\mathcal{E}_{\mathcal{I}}\cup\mathcal{E}_V}^2 + 2\|[\![\mathbf{w}]\!]\|_{0,\mathcal{E}_R}^2 + \|\sigma_a^{\frac{1}{2}}\,u\|_{0,\Omega}^2 + \frac{1}{4}\|[\![u]\!]\|_{\mathcal{E}_{\mathcal{I}}\cup\mathcal{E}_V}^2.$$

Whenever $\sigma_a > 0$, $|(\cdot,\cdot)|_h$ actually defines a norm. Our main result is given in Theorem 1, which derives directly from the analysis in [1, Theorem 2.2]. Strictly speaking, that analysis was carried out for $\sigma_a \equiv 0$ and simpler boundary conditions but extension to the current situation poses no difficulties.

Theorem 1. *Assume the exact solution* $(\mathbf{J}, \Phi)$ *of* (1)–(2) *to belong to* $H^{s+1}(\Omega)^d \times H^{s+2}(\Omega)$, *with* $s \geq 0$. *Let* $(\mathbf{J}_h, \Phi_h)$ *be the DG approximation obtained by* (3), *for an approximation degree* $k \geq 0$. *Then we have the error bound*

$$|(\mathbf{J}-\mathbf{J}_h, \Phi-\Phi_h)|_h \leq C\,h^{\min\{s,k\}+\frac{1}{2}}\,(\|\mathbf{J}\|_{s+1,\Omega} + \|\Phi\|_{s+2,\Omega}),$$

with a constant $C > 0$ *independent of the mesh size* h.
Furthermore, if the domain and the coefficients σ_t *and* σ_a *are sufficiently regular, we also have the* L^2*–bound*

$$\|\Phi - \Phi_h\|_{0,\Omega} \leq C\,h^{\min\{s,k\}+1}\,(\|\mathbf{J}\|_{s+1,\Omega} + \|\Phi\|_{s+2,\Omega}),$$

with a constant $C > 0$ *independent of the mesh size* h.

The order of convergence in the approximation of the seminorm $|(\cdot,\cdot)|_h$ is half a power of h better than for the standard LDG method, due to the truly mixed nature of the numerical fluxes; cf. [1]. Moreover, the above result also holds true for $k = 0$, i.e., for piecewise constant approximations. For the LDG method, no convergence has been observed either theoretically or numerically in this case.

4 Numerical results

In this section, we present the results of a series of numerical experiments that demonstrate the theoretical error estimates of Theorem 1.

4.1 Smooth solution

We start by testing the performance of the method for a smooth solution. We consider the radiation–diffusion system (1) on $\Omega = (0,1)^3$, with reflecting boundary conditions on the faces $\{x = 0\}$ and $\{x = 1\}$, and vacuum boundary conditions on the remaining boundary faces. The material coefficients are $\sigma_t = 1$ and $\sigma_a = 10^{-4}$. The right–hand sides Q_0 and $\mathbf{Q}_1$ are chosen so that the exact solution $(\mathbf{J}, \Phi)$ is a polynomial of order 7 and thus arbitrarily smooth. The corresponding numerical solutions for $k = 0$ and $k = 1$ are computed on a sequence of tetrahedral meshes $\{\mathcal{T}_i\}_{i=1}^{5}$ constructed by uniformly dividing the domain into Cartesian grids with 2^i, $i = 1, \dots, 5$, equal intervals in each dimension. Each cube in the grid is then subdivided into six tetrahedra of equal volume. The mesh size h_i on mesh $\mathcal{T}_i$ is therefore proportional to 2^{-i}.

In Table 1 we report the errors and the numerical convergence rates r_i in the $|\cdot|_h$–seminorm, in the L^2–norm of Φ, as well as in the L^2–norm of $\mathbf{J}$. Clearly, we obtain convergence order $k + \frac{1}{2}$ in the $|\cdot|_h$–seminorm, as well as order $k + 1$ in the L^2–norm for Φ. This confirms the theoretical results in Theorem 1. Also, the $\|\mathbf{J} - \mathbf{J}_h\|_{0,\Omega}$ part of the $|\cdot|_h$–seminorm actually converges more rapidly than the whole $|\cdot|_h$–seminorm, namely with the optimal rate $k + 1$. The convergence behavior of the different jump contributions to the $|\cdot|_h$–seminorm is shown in Figure 1. Here, we plot the different errors against $n = 2^i$, which is proportional to h_i^{-1}. We use the abbreviations I, R and V for $\mathcal{E}_{\mathcal{I}}$, $\mathcal{E}_R$ and $\mathcal{E}_V$, respectively. Comparison with the line $y = n^{-k-1/2}$ clearly shows that the jumps over interior faces converge with order $k + \frac{1}{2}$. On the other hand, comparison with the line $y = n^{-k-1}$ shows that all the boundary jumps exhibit a better convergence of the order $k + 1$. This indicates that the errors in the interior jump contributions are the ones that render the first estimate in Theorem 1 sharp. Finally, we note the similar behavior is observed for solutions that are piecewise in H^2-regular; for brevity, these numerics have been omitted.

Table 1. Smooth solution: errors and convergence rates.

		$\|(\mathbf{J}-\mathbf{J}_h, \Phi-\Phi_h)\|_h$		$\|\Phi-\Phi_h\|_{0,\Omega}$		$\|\mathbf{J}-\mathbf{J}_h\|_{0,\Omega}$	
	i	error	r_i	error	r_i	error	r_i
$k=0$	1	7.11 e–2	–	1.45 e–2	–	3.13 e–2	–
	2	4.91 e–2	5.34 e–1	8.90 e–3	7.00 e–1	1.90 e–2	7.18 e–1
	3	3.21 e–2	6.14 e–1	4.71 e–3	9.17 e–1	1.02 e–2	9.04 e–1
	4	2.14 e–2	5.82 e–1	2.40 e–3	9.73 e–1	5.21 e–3	9.67 e–1
	5	1.47 e–2	5.47 e–1	1.21 e–3	9.90 e–1	2.63 e–3	9.87 e–1
$k=1$	1	4.40 e–2	–	7.22 e–3	–	1.21 e–2	–
	2	1.91 e–2	1.20 e+0	2.36 e–3	1.61 e+0	3.71 e–3	1.71 e+0
	3	7.41 e–3	1.36 e+0	6.33 e–4	1.90 e+0	9.78 e–4	1.92 e+0
	4	2.74 e–3	1.43 e+0	1.61 e–4	1.97 e+0	2.47 e–4	1.99 e+0
	5	9.91 e–4	1.47 e+0	4.06 e–5	1.99 e+0	6.16 e–5	2.00 e+0

4.2 A two–material problem

In practice the coefficient σ_t may have strong discontinuities at the interfaces between different materials. In these cases, where the vector-valued quantities $\frac{1}{3\sigma_t}\nabla\Phi$ and $\mathbf{J}$ are typically smoother than the scalar flux Φ, the use of a truly mixed method, such as the one considered in this paper, is of particular importance. In this series of numerical experiments, we consider a problem with material discontinuities. The radiation–diffusion equations (1) are solved with $\sigma_a = 0$ and a discontinuous coefficient σ_t. We again set $\Omega = (0,1)^3$, and specify vacuum boundary conditions on the faces $\{x=0\}$ and $\{x=1\}$, with reflecting boundary conditions on the remaining boundary faces. We define $\sigma_t = a$, for $0 \le x \le 0.5$, $\sigma_t = b$, for $0.5 < x \le 1$, with positive parameters a and b, modeling a material discontinuity at $x = 0.5$. The right–hand sides $\mathbf{Q}_1$ and Q_0 are chosen so that the solution $(\mathbf{J}, \Phi)$ is given as follows. Denoting by $\varphi(y,z)$ the polynomial $\varphi(y,z) = y^2z^2$, we take $\mathbf{J}(x,y,z) = \left(\frac{3}{4}(a + x(b-a)) \cdot \varphi(y,z),\, 0,\, 0\right)$ and

$$\Phi(x,y,z) = \begin{cases} 3a(x-\frac{1}{2})\varphi(y,z) & \text{in } (0,0.5)\times(0,1)\times(0,1), \\ 3b(x-\frac{1}{2})\varphi(y,z) & \text{in } (0.5,1)\times(0,1)\times(0,1). \end{cases}$$

Notice that Φ is piecewise smooth, but only belongs to $H^{\frac{3}{2}-\varepsilon}(\Omega)$, $\varepsilon > 0$, whereas $\frac{1}{3\sigma_t}\nabla\Phi = \left(\varphi(y,z), (x-\frac{1}{2})\partial_y\varphi(y,z), (x-\frac{1}{2})\partial_z\varphi(y,z)\right)$ is a smooth function. We use the same sequence of tetrahedral meshes as the numerical experiments presented in Section 4.1.

We consider the two cases $a = 1$, $b = 0.01$ and $a = 100$, $b = 1$. Notice that the jump in the normal derivative of Φ at the surface $x = 0.5$ is equal to $3(a-b)y^2z^2$. This jump is almost two orders of magnitude larger in the case $a = 100$, $b = 1$ than in the case $a = 1$, $b = 0.01$. For this reason deterioration of the convergence rates due to the lack of smoothness of the exact solution might be more serious in the second case than in the first one.

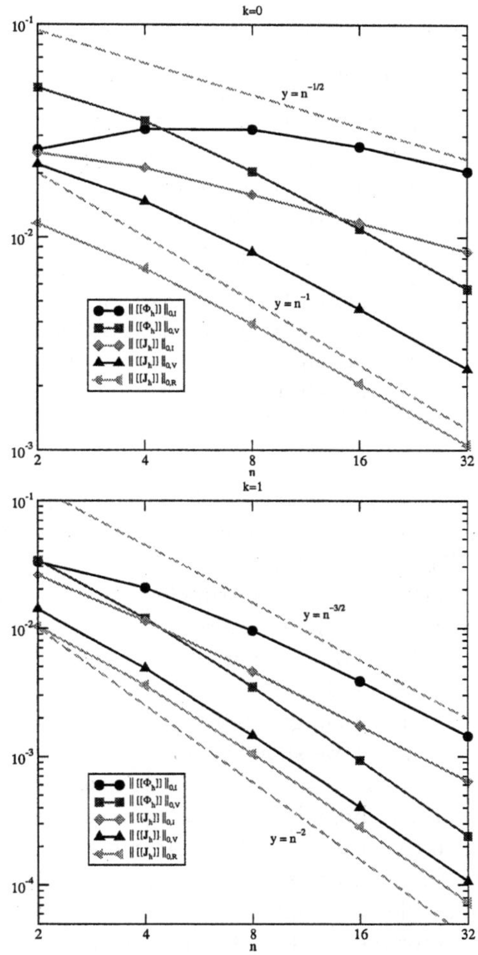

Fig. 1. Smooth solution: L^2-errors of the jumps in $\mathbf{J}$ and Φ

Table 2 shows the errors and the numerical convergence rates r_i in the $|\cdot|_h$–seminorm and in the L^2–norm of Φ, as well as in the L^2–norm of $\mathbf{J}$, for the case $a = 1$, $b = 0.01$, Order $k + \frac{1}{2}$ in the $|\cdot|_h$–seminorm, and order $k+1$ convergence in the L^2–norm for Φ are obtained. These results agree with the theoretical estimates in Theorem 1, even though the elliptic regularity assumptions of Theorem 1 for the estimate of the L^2–error in Φ are not satisfied. Furthermore, the L^2–error in $\mathbf{J}$ also converges with the optimal order $k + 1$. In Figure 2, the errors in the jumps of Φ and $\mathbf{J}$ are shown. Again, we see that the interior jumps converge at the rates $\mathcal{O}(h^{k+\frac{1}{2}})$ while the boundary jumps show the full convergence rate of $\mathcal{O}(h^k)$.

Table 2. Two–material problem, $a = 1$, $b = 0.01$: errors and convergence rates.

		$\|(\mathbf{J}-\mathbf{J}_h,\Phi-\Phi_h)\|_h$		$\|\Phi-\Phi_h\|_{0,\Omega}$		$\|\mathbf{J}-\mathbf{J}_h\|_{0,\Omega}$	
	i	error	r_i	error	r_i	error	r_i
$k=0$	1	1.76 e–1	–	6.15 e–2	–	5.14 e–2	–
	2	1.27 e–1	4.67 e–1	3.18 e–2	9.50 e–1	2.92 e–2	8.17 e–1
	3	8.95 e–2	5.04 e–1	1.61 e–2	9.85 e–1	1.54 e–2	9.21 e–1
	4	6.30 e–2	5.08 e–1	8.07 e–3	9.95 e–1	7.91 e–3	9.61 e–1
	5	4.44 e–2	5.05 e–1	4.04 e–3	9.98 e–1	4.01 e–3	9.80 e–1
$k=1$	1	9.42 e–2	–	1.59 e–2	–	1.39 e–2	–
	2	3.66 e–2	1.36 e+0	4.06 e–3	1.97 e+0	3.56 e–3	1.97 e+0
	3	1.33 e–2	1.46 e+0	1.02 e–3	1.99 e+0	8.73 e–4	2.03 e+0
	4	4.77 e–3	1.49 e+0	2.55 e–4	2.00 e+0	2.13 e–4	2.04 e+0
	5	1.69 e–3	1.49 e+0	6.38 e–5	2.00 e+0	5.26 e–5	2.02 e+0

Now consider the case $a = 100$, $b = 1$. The numerical rates in Table 3 indeed paint a different picture than the previous case shown in Table 2 for $a = 1$ and $b = 0.01$. For both $k = 0$ and $k = 1$ the rates in the $|\cdot|_h$–seminorm have deteriorated slightly, but they are still close to $k + \frac{1}{2}$. Similar behavior is observed in the L^2–norm of $\mathbf{J}$. These results are in agreement with Theorem 1. The differences in the L^2–errors in Φ are more remarkable: the convergence rates are smaller than 0.5 for $k = 0$ and better than 2 for $k = 1$. However, the oscillations in the numbers might indicate that the asymptotic regime has not been reached yet. Finally, we show in Figures 3 the behavior of the errors in the jumps of Φ and $\mathbf{J}$ for $k = 0$ and $k = 1$, respectively. For $k = 0$ the jumps of Φ on the reflective boundary in this case converge more slowly – and suboptimally – than in the case $a = 1$, $b = 0.01$, while the other jumps exhibit rates of convergence similar to those for the case $a = 1$, $b = 0.01$. For $k = 1$ the jumps of Φ on the reflective boundary exhibit convergence similar to those for $k = 0$. However, convergence is at the optimal rate of $\mathcal{O}(h^{\frac{3}{2}})$ that is, half an order less than the L^2–norm. All the other jumps show a convergence rates as for the $k = 0$ case.

5 Conclusions

The discontinuous Galerkin method of Warsa, Wareing and Morel introduced in [7] and [6] is a (truly) mixed discontinuous Galerkin method that belongs to the general class of methods analyzed in [1]. The analysis there ensures well-posedness and a priori error estimates, which have been verified in a series of numerical experiments. It is important for the specific applications in which solutions of the P_1 equations are needed that convergence has been demonstrated for the vector unknown, even in the case $k = 0$.

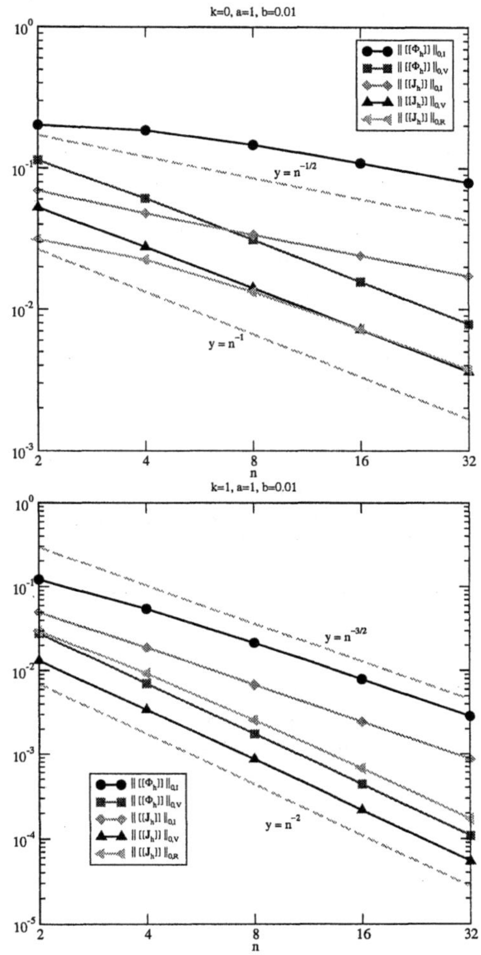

Fig. 2. Two–material problem, $a = 1$, $b = 0.01$: L^2-errors of the jumps in $\mathbf{J}$ and Φ

References

1. P. Castillo, B. Cockburn, I. Perugia, and D. Schötzau. An a priori error analysis of the local discontinuous Galerkin method for elliptic problems. *SIAM J. Numer. Anal.*, 38:1676–1706, 2000.
2. B. Cockburn and C.-W. Shu. The local discontinuous Galerkin method for time–dependent convection–diffusion systems. *SIAM J. Numer. Anal.*, 35:2440–2463, 1998.
3. B. Davison. *Neutron Transport Theory.* Clarendon Press, 1957.
4. E. E. Lewis and Jr. W. F. Miller. *Computational Methods of Neutron Transport.* American Nuclear Society, 1993.

Table 3. Two—material problem, $a = 100$, $b = 1$: errors and convergence rates.

	i	$\|(\mathbf{J}-\mathbf{J}_h, \Phi-\Phi_h)\|_h$ error	r_i	$\|\Phi-\Phi_h\|_{0,\Omega}$ error	r_i	$\|\mathbf{J}-\mathbf{J}_h\|_{0,\Omega}$ error	r_i
$k=0$	1	1.47 e+1	–	7.76 e+0	–	3.83 e+0	–
	2	1.05 e+1	4.84 e–1	5.05 e+0	6.21 e–1	2.01 e+0	9.29 e–1
	3	7.49 e+0	4.90 e–1	3.49 e+0	5.34 e–1	1.04 e+0	9.56 e–1
	4	5.32 e+0	4.95 e–1	2.59 e+0	4.29 e–1	5.33 e–1	9.57 e–1
	5	3.79 e+0	4.87 e–1	1.94 e+0	4.17 e–1	2.78 e–1	9.42 e–1
$k=1$	1	5.36 e+0	–	2.68 e+0	–	8.23 e–1	–
	2	2.09 e+0	1.36 e+0	6.91 e–1	1.95 e+0	2.17 e–1	1.92 e+0
	3	8.44 e–1	1.31 e+0	1.56 e–1	2.14 e+0	5.72 e–2	1.93 e+0
	4	3.43 e–1	1.30 e+0	3.35 e–2	2.22 e+0	1.53 e–2	1.90 e+0
	5	1.36 e–1	1.33 e+0	7.37 e–3	2.19 e+0	4.15 e–3	1.88 e+0

5. J. S. Warsa, M. Benzi, T. A. Wareing, and J. E. Morel. Preconditioning a mixed discontinuous finite element method for radiation diffusion. In F. Brezzi, A. Buffa, S. Corsaro, and A. Murli, editors, *Numerical Mathematics and Advanced Applications, Proceedings of ENUMATH 2001*, pages 967–978. Springer–Verlag, 2003.
6. J.S. Warsa, T.A. Wareing, and J.E. Morel. Fully consistent diffusion synthetic acceleration of linear discontinuous transport discretizations on three-dimensional unstructured meshes. *Nucl. Sci. Engrg.*, 141:236–251, 2002.
7. J.S. Warsa, T.A. Wareing, and J.E. Morel. Solution of the discontinuous P_1 equations in two-dimensional Cartesian geometry with two-level preconditioning. *SIAM J. Sci. Comput.*, 24:2093–2124, 2003.

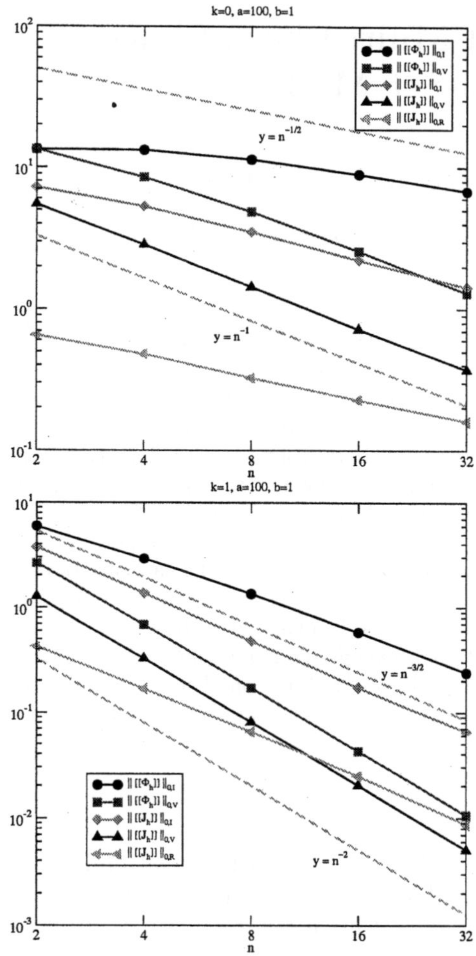

Fig. 3. Two–material problem, $a = 100$, $b = 1$: L^2-errors of the jumps in $\mathbf{J}$ and Φ

Modeling of Multi-Phase Flows with a Level-Set Method

Sander P. van der Pijl*, A. Segal and C. Vuik

Delft University of Technology, Delft, The Netherlands

Summary. Multi-phase flows are frequently modeled in engineering fluid mechanics. In this work incompressible two-phase flows are considered. The present research aims to model high density-ratio flows with complex interface topologies, typically air/water flows. Applications are mixtures of bubbles and droplets. Aspects which are taken into account are: a sharp front (density changes rapidly), arbitrary shaped interfaces, surface tension, buoyancy and coalescence of drops/bubbles. Attention is paid to mass-conservation and integrity of the interface.

A survey of available computational methods is performed in [1]. The computational method used in this paper is the Mass Conserving Level-Set method (MCLS, [2]). The MCLS method is based on the Level-Set methodology, using a VOF-function to conserve mass. This function is advected without the necessity to reconstruct the interface. The ease of MCLS is based on an explicit relationship between the Volume-of-Fluid function and the Level-Set function. The method is straightforward to apply to arbitrarily shaped interfaces, which may collide and break up.

1 Introduction

Various methods have been put forward to treat multi-phase flows. A classification is given in [1]. The two methods that are most suitable for the current research are the *Volume-of-Fluid* (VOF) method and the *Level-Set* method. For both methods a marker function is used to define the interface. In case of the Volume-of-Fluid method, a marker function, say Ψ, indicates the fractional volume of a certain fluid, say fluid '1', in a computational cell. It can be seen as the concentration of the marker particles of the MAC-method, when the number of particles goes to infinity.

An alternative to the Volume-of-Fluid method is the Level-Set method ([3, 4]). The interface is now defined by the zero level-set of a marker function, say Φ: $\Phi = 0$ at the interface, $\Phi > 0$ inside fluid '1' and $\Phi < 0$ elsewhere. The function Φ is chosen such that it is smooth near the interface. This eases the computation of interface derivatives. Also, methods available from hyperbolic conservation laws can be used to advect the interface. The interface is (implicitly) advected by advecting Φ as if it was a material constant:

$$\frac{\partial \Phi}{\partial t} + \mathbf{u} \cdot \nabla \Phi = 0. \qquad (1)$$

* Research granted by NWO.

The Level-Set method has some advantages over the Volume-of-Fluid method. Especially when solving the flow-field is concerned, since interface normals, curvature and distance towards the interface can be expressed easily in terms of Φ or derivatives of Φ. Also, advecting the interface is possible by the application of 'of-the-shelf' techniques available from hyperbolic conservation laws. For these reasons, the Level-Set method has been chosen as the basis of our work. However, mass-conservation is not an intrinsic property and is considered the major drawback of the Level-Set method. Our work focuses on a mass-conserving way to advect the interface by means of the Mass-Conserving Level-Set method (MCLS, [2]).

The MCLS method has a shared foundation with the CLSVOF method ([5, 6]) and to a lesser extend with the combined Level-Set/particle method ([7]) in the sense that it is based on Level-Set and additional effort is made to conserve mass. The difference with CLSVOF is that here there is no combination of two existing methods. The method takes full advantage from all additional information provided by the Level-Set function Φ, rather than coupling Level-Set with Volume-of-Fluid/PLIC. In fact we use the Volume-of-Fluid function Ψ as a help variable to conserve mass, without applying the difficult convection (namely interface reconstruction) which makes the VOF so elaborate. The key issue of our method is that we define a simple relationship between the Level-Set function Φ and Volume-of-Fluid function Ψ. This relation is obtained by assuming piecewise linear interfaces within a computational cell:

$$\Psi = f(\Phi, \nabla\Phi). \tag{2}$$

It makes the advection of the Volume-of-Fluid function Ψ easy (i.e. without interface reconstruction) and finding Φ from Ψ a straightforward task. This is carried out by well-known numerical tools, like Picard and Newton iterations. The PLIC method is not adopted (unlike CLSVOF), yet mass is conserved in the same manner. Note that the CLSVOF method might not be easily extendible to 3D space. Yet the extension of MCLS to three-dimensional space can be done in a straightforward way. Note also that with this approach, it is not necessary to smooth (or regularize) Ψ, which is common for other methods.

2 Governing Equations

Consider two fluids '0', and '1' in domain $\Omega \in \mathbb{R}^2$ which are separated by an interface S. Both fluids are assumed to be incompressible, i.e.:

$$\nabla \cdot \mathbf{u} = 0, \tag{3}$$

where $\mathbf{u} = (u, v)^t$ is the velocity vector and u and v are the velocities in x- and y-direction respectively. The flow is governed by the incompressible Navier-Stokes equations:

$$\frac{\partial \mathbf{u}}{\partial t} + \mathbf{u} \cdot \nabla \mathbf{u} = -\frac{1}{\rho}\nabla p + \frac{1}{\rho}\nabla \cdot \mu \left(\nabla \mathbf{u} + \nabla \mathbf{u}^t\right) + \mathbf{g}, \tag{4}$$

where ρ, p, μ and $\mathbf{g}$ are the density, pressure, viscosity and gravity vector respectively. The density and viscosity are constant within each fluid. Using the Level-Set function Φ these can be expressed as

$$\mu = \mu_0 + (\mu_1 - \mu_0)H(\Phi) \tag{5}$$

and similar for ρ, where the subscript indicates the corresponding fluid and H is the Heaviside step function.

2.1 Interface conditions

The interface conditions express continuity of mass and momentum at the interface:

$$\begin{aligned} [\mathbf{u}] &= 0 \\ [p\mathbf{n} - \mathbf{n} \cdot \mu \left(\nabla \mathbf{u} + \nabla \mathbf{u}^t\right)] &= \sigma\kappa\mathbf{n}, \end{aligned} \tag{6}$$

where the brackets denote jumps across the interface, $\mathbf{n}$ is a normal vector at the interface, σ is the surface tension coefficient and κ is the curvature of the interface. Clearly, the velocity $\mathbf{u}$ is continuous at the interface. If the viscosity μ is continuous at the interface, it can be shown that the derivatives of the velocity components are continuous too ([8, 9]). In that case Eqn. (6) reduces to $[\mathbf{u}] = \mathbf{0}$ and $[p] = \sigma\kappa$. To achieve that, the viscosity is forced to be continuous by smoothing Expression (5):

$$\mu = \mu_0 + (\mu_1 - \mu_0)H_\alpha(\Phi), \tag{7}$$

where H_α is the smoothed (or regularized) Heaviside step function

$$H_\alpha(x) = \begin{cases} 0 & x < -\alpha \\ \frac{1}{2}\left(1 + \sin(\frac{x}{\alpha}\frac{1}{2}\pi)\right) & |x| \leq \alpha \\ 1 & x > \alpha \end{cases} \tag{8}$$

and α is a parameter proportional to the mesh width. Here α is chosen as (following [10]) $\alpha = \frac{3}{2}h$, where h is the mesh width. According to [11], the viscosity is then smoothed over three mesh widths, provided $|\nabla\Phi| = 1$. Note that only the viscosity is smoothed, not the density ρ. Note also that when the density is not regularized, mass is conserved when the *volume* of a certain fluid or phase is conserved. In fact, the MCLS method conserves volumes by construction. Due to the non-regularized density-field, mass is conserved too. Instead of taking into account the pressure-jump at the interface due to the surface tension forces, the continuous surface force/stress (CSF, [12]) methodology is adopted.

3 Computational Approach

The Navier-Stokes equations are solved on a Cartesian grid in a rectangular domain by the pressure-correction method ([13]). The unknowns are stored in a Marker-and-Cell (staggered) layout ([14]). For the interface representation the Level-Set methodology is adopted. The interface conditions are satisfied by means of the continuous surface force (CSF) methodology. The discontinuous density field is dealt with similarly to the GhostFluid method for incompressible flow ([8]). Further information about the flow-field computations can be found in [2].

3.1 Interface advection

The strategy of modeling two-phase flows is to compute the flow with a given interface position and subsequently evolve the interface in the given flow field. In the foregoing, it has been described how the flow is computed with a given interface position. Next we consider the evolution of the interface.

Level-Set The interface is implicitly defined by a Level-Set function Φ. More precisely, the interface, say S, is the zero level-set of Φ:

$$S(t) = \left\{ \mathbf{x} \in \mathbb{R}^2 | \Phi(\mathbf{x}, t) = 0 \right\}. \tag{9}$$

The interface is evolved by advecting the Level-Set function in the flow field as if it were a material constant (Eqn. (1)):

$$\frac{\partial \Phi}{\partial t} + \mathbf{u} \cdot \nabla \Phi = 0. \tag{10}$$

A homogeneous Neumann boundary condition for Φ is imposed at the boundaries. It will be clear that accuracy of the approximation of Eqn. (10) determines the accuracy of the interface representation. The accuracy will also determine the mass errors. For this purpose, the discretization of the gradient of Φ can be either first order upwind, or second or third order ENO ([10, 11, 15]). In case of the first-order spatial discretization, a forward Euler temporal discretization is sufficient. In case of the higher order spatial discretization, a Runge-Kutta scheme is applied (e.g. [16]).

MCLS The difficulty with the Level-Set method is, that although Φ might be conserved, this does not imply that mass is conserved. On the other hand, with the Volume-of-Fluid method, mass is conserved when Ψ is conserved. In order to conserve mass with the Level-Set method, corrections to the Level-Set function are made by considering the fractional volume Ψ of a certain fluid within a computational cell. First the usual Level-Set advection is performed: first-order advection and unmodified re-initialization. Low order advection and re-initialization will ensure numerical smoothness of Φ. Furthermore, when the flow-field is computed, higher order accuracy might not be expected when the CSF method is applied and viscosity is regularized. In that respect, higher order discretization of Eqn. (10) will only lead to improved mass conservation for the pure Level-Set methods. Since the obtained Level-Set function $\Phi^{n+1,*}$ will certainly not conserve mass, corrections to $\Phi^{n+1,*}$ are made such that mass is conserved. This requires three steps:

1. the relative volume of a certain fluid in a computational cell (called 'volume-of-fluid' function Ψ) is to be computed from the Level-Set function Φ^n: $\Psi = f(\Phi, \nabla\Phi)$;
2. the volume-of-fluid function has to be advected conservatively during a time step towards Ψ^{n+1};
3. with this new volume-of-fluid function Ψ^{n+1}, corrections to $\Phi^{n+1,*}$ are sought such that $f(\Phi^{n+1}, \nabla\Phi^{n+1}) = \Psi^{n+1}$ holds.

These three steps will be explained subsequently.

Step 1: Volume-of-Fluid function A relationship between the Level-Set function Φ and the so-called volume-of-fluid function Ψ is found by considering the fractional volume of a certain fluid in a computational cell Ω_k. In this paper, a Cartesian mesh is employed consisting of computational cells $\Omega_k, k = 1, 2, \ldots$. By $\mathbf{x}_k = (x_k, y_k)^t$ the center node of Ω_k is meant and Δx and Δy are the mesh sizes in x and y direction respectively. The volume-of-fluid function Ψ_k is defined in terms of Level-Set function Φ by

$$\Psi_k = \frac{1}{\mathrm{vol}(\Omega_k)} \int_{\Omega_k} H(\Phi)\,\mathrm{d}\Omega. \tag{11}$$

where H is the Heaviside step function. The Level-Set function Φ is *linearized* around Φ_k, which leads to

$$\Psi_k = f(\Phi_k, \nabla\Phi_k). \tag{12}$$

Note that in contrast with other approaches, the Heaviside step function is not regularized. After some mathematical manipulations, the function f is evaluated as

$$f(\Phi_k, \nabla\Phi_k) = \begin{cases} 0 & \Phi_k \leq -\Phi_{max\,k} \\ \frac{1}{2}\dfrac{(\Phi_{max\,k} + \Phi_k)^2}{\Phi^2_{max\,k} - \Phi^2_{mid\,k}} & -\Phi_{max\,k} < \Phi_k < -\Phi_{mid\,k} \\ \frac{1}{2} + \dfrac{\Phi_k}{\Phi_{max\,k} + \Phi_{mid\,k}} & -\Phi_{mid\,k} \leq \Phi_k \leq \Phi_{mid\,k} \\ 1 - \frac{1}{2}\dfrac{(\Phi_{max\,k} - \Phi_k)^2}{\Phi^2_{max\,k} - \Phi^2_{mid\,k}} & \Phi_{max\,k} < \Phi_k < \Phi_{max\,k} \\ 1 & \Phi_k \geq \Phi_{max\,k}, \end{cases} \tag{13}$$

where

$$\begin{array}{ll} \Phi_{maxk} = \frac{1}{2}(|D_{xk}| + |D_{y_k}|) & \Phi_{midk} = \frac{1}{2}\Big| \, |D_{y_k}| - |D_{xk}| \, \Big| \\ D_{xk} = \Delta x \, \frac{\partial \Phi}{\partial x}\Big|_k & D_{y_k} = \Delta y \, \frac{\partial \Phi}{\partial y}\Big|_k , \end{array} \tag{14}$$

which are approximated by central differencing.

Step 2: Volume-of-Fluid advection At a certain time instant the volume-of-fluid function can be computed from Φ by means of Eqn. (12). The volume-of-fluid function after a time step is found by considering the flux of fluid F that flows through a boundary Γ of a computational cell during time-step Δt:

$$\Psi^{n+1}_{i+\frac{1}{2},j+\frac{1}{2}} = \Psi^{n}_{i+\frac{1}{2},j+\frac{1}{2}} - \frac{1}{\Delta x \Delta y}\big(F_{x i+1,j+\frac{1}{2}} - F_{x i,j+\frac{1}{2}} + F_{y_{i+\frac{1}{2},j+1}} - F_{y_{i+\frac{1}{2},j}}\big), \tag{15}$$

The fluxes are again computed by linearizing Φ (just like Eqn. (13)). In fact, the fluxes are computed by the straightforward application of f.

It is possible that fluid is fluxed more than once through different faces, which would cause unphysical values of Ψ. As reported in e.g. [1], this can be solved by employing either a multidimensional scheme or flux-splitting. For reasons of simplicity we have chosen for the second approach. The order of fluxing is: first in x-direction, then in y-direction. Currently the flux-splitting of [5] is adopted. As reported in [5], undershoots and/or overshoots can still occur, which leads to unphysical values of Ψ, namely < 0 and > 1. If these values are replaced by 0 and 1 respectively, mass errors arise which are of order 10^{-4}. This is also experienced in the current research. Mass errors are completely avoided by redistributing Ψ ([2]).

Step 3: Inverse function Having found a new Volume-of-Fluid function Ψ^{n+1}, the initial guess of the Level-Set function $\Phi^{n+1,*}$ (after Level-Set advection) is modified, such that mass is conserved within each computational cell. In other words, find $(\Phi_1, \Phi_2, \dots)$, such that

$$|f(\Phi_k^{n+1}, \nabla\Phi_k^{n+1}) - \Psi_k^{n+1}| < \epsilon \qquad \forall k = 1, 2, \dots , \tag{16}$$

where ϵ is some tolerance. It will be clear that due to the behavior of Ψ no unique solution Φ exists. However, a (small) correction to Φ^* is searched, where Φ^* comes from Level-Set advection. A solution Φ is found by the following iteration (until convergence): leave Φ unmodified in a grid point when the Volume-of-Fluid constraint is satisfied and make corrections locally when this constraint is not satisfied. This is achieved by using the inverse function g of f as given in Eqn. (13) with respect to argument Φ_k:

$$f(g(\Psi, \nabla\Phi), \nabla\Phi) = \Psi. \tag{17}$$

4 Applications

The behavior of the MCLS approach is shown by a couple of standard advection tests, with a prescribed velocity field. Thereafter, the method is applied to the complete set of equations by considering a falling drop and a rising bubble respectively.

4.1 Advection tests

Linear advection The first advection test is a circle, which is advected by an uniform velocity field. The velocity field is prescribed by $(u, v) = (0, -1)$. The dimensions of the computational domain are: $L_x = 10$ and $L_y = 100$. which is discretized by a 10x100-mesh. Initially a circle of radius R_0 is placed at $x = L_x/2$ and $y = L_y - 2R_0$. For the case of $R_0 = 4$ (a circle with a diameter of 8 mesh sizes), the relative mass is plotted in Fig. 1 as function of the traversed distance of the circle. First-order, second-order and third-order pure Level-Set

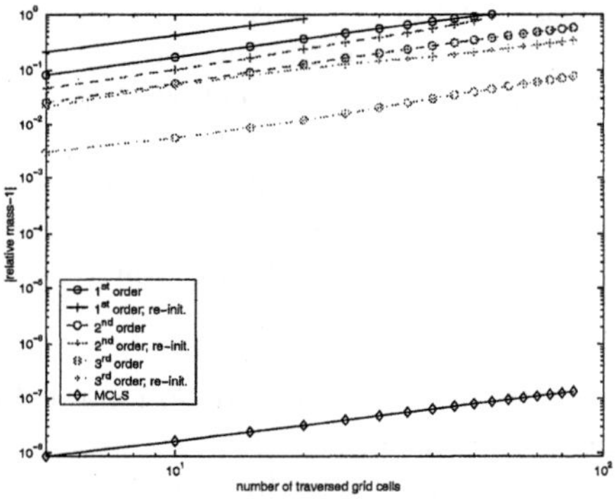

Fig. 1. Relative mass for the linear advection test; $\epsilon = 10^{-8}$ (every 10th iteration marked)

simulations (with and without re-initialization) are compared with the MCLS method. ENO discretization is adopted for the pure Level-Set method (see aforementioned references). The order of re-initialization is in agreement with the order of advection. The tolerance in the VOF advection is taken to be: $\epsilon = 10^{-8}$. Globally speaking it can be said that mass is always lost for the pure Level-Set advection. Mass losses are smaller for higher accuracy and re-initialization causes much higher mass losses. The MCLS method conserves mass up to the specified tolerance.

4.2 Air/water flow

In [8] a two-dimensional rising air bubble in water is considered. The dimensions and sizes are: $L_x = 0.02\ m$, $L_y = 1\frac{1}{2}L_x$, $R_0 = \frac{1}{6}L_x$, $x_0 = y_0 = \frac{1}{2}L_x$. The gravity and material constants are: $g = 9.8\ \frac{m}{s^2}$, $\sigma = 0.0728\ \frac{kg}{s^2}$, $\rho_w = 10^3\ \frac{kg}{m^3}$, $\rho_a = 1.226\ \frac{kg}{m^3}$, $\mu_w = 1.137\ 10^{-3}\ \frac{kg}{ms}$ and $\mu_a = 1.78\ 10^{-5}\ \frac{kg}{ms}$. where subscripts $_w$ and $_a$ indicate water and air respectively.

Results are shown in Fig. 2(a) for three different mesh sizes. We take $\epsilon = 10^{-8}$. Relative mass losses are of the same order and in agreement with the advection tests. Note that the number of grid cells is much smaller than in [8]. The results are the same for $t \leq 0.025$ for all mesh sizes. Thereafter small differences occur. The results compare well with [8]. The MCLS method seems to result in a more coherent structure at the highly curved part of the interface at $t = 0.05$. This is thought to be caused by the low resolution of the grids used here.

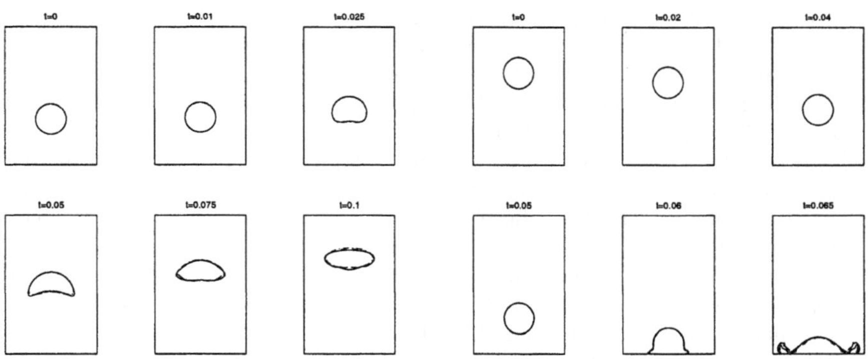

(a) Rising bubble (b) Falling droplet

Fig. 2. Air/water flows; $-\cdot-$: 30×45; $--$: 40×60; — : 60×90 mesh

In Fig. 2(b) results are shown for a falling droplet. The conditions are the same as for the rising bubble, except for the sign of Φ at $t = 0$ and $y_0 = L_x$. Mass conservation properties are the same as before. The result are the same until the droplet hits the bottom. Thereafter differences occur. This is thought to be due to limited number of grid cells available to capture the flow-phenomena near the wall. The results compare well with [8]. Note that the results in [8] span $t \leq 0.05$; no results after collision are presented.

5 Conclusion

The mass Conserving Level-Set (MCLS) has been presented. The method is based on the Level-Set methodology, where mass is conserved by considering the fractional volume of a certain fluid within a computational cell. Advection tests were used to compare the method with the Level-Set method. Mass is conserved up to a specified (vanishing) tolerance. The MCLS method combines the attractiveness of the Level-Set method with the mass-conserving properties of the Volume-of-Fluid methods, without adopting the latter. This makes the implementation much easier than for a Volume-of-Fluid (based) method, especially in three-dimensional space. The applicability of the MCLS method was illustrated by the application to air-water flows. It is possible to capture bubbles or droplets within a limited number of grid cells without mass losses up to the prescribed tolerance. This is an important feature, since future work will concern three-dimensional problems/geometries, where the amount of grid cells available to an individual entity will decrease considerably.

References

1. S.P. van der Pijl. Free-boundary methods for multi-phase flows. AMA report 02-13, Delft University of Technology, 2002. http://ta.twi.tudelft.nl.
2. S.P. van der Pijl, A. Segal, and C. Vuik. A mass-conserving level-set method for modeling of multi-phase flows. AMA report 03-03, Delft University of Technology, 2003. http://ta.twi.tudelft.nl.
3. W. Mulder, S. Osher, and J.A. Sethian. Computing interface motion in compressible gas dynamics. *Journal of Computational Physics*, 100:209–228, 1992.
4. S. Osher and R.P. Fedkiw. Level set methods: An overview and some recent results. *Journal of Computational Physics*, 169:463–502, 2001.
5. M. Sussman and E.G. Puckett. A coupled level set and volume-of-fluid method for computing 3D and axisymmetric incompressible two-phase flows. *Journal of Computational Physics*, 162:301–337, 2000.
6. M. Sussman. A second order coupled level set and volume-of-fluid method for computing growth and collapse of vapor bubbles. *Journal of Computational Physics*, 187:110–136, 2003.
7. D. Enright, R. Fedkiw, J. Ferziger, and I. Mitchell. A hybrid particle level set method for improved interface capturing. *Journal of Computational Physics*, 183:83–116, 2002.
8. M. Kang, R.P. Fedkiw, and X.-D. Liu. A boundary condition capturing method for multiphase incompressible flow. *Journal of Scientific Computing*, pages 323–360, 2000.
9. Z. Li and M.-C. Lai. The immersed interface method for the navier-stokes equations with singular forces. *Journal of Computational Physics*, 171:822–842, 2001.
10. M. Sussman, P. Smereka, and S. Osher. A level set approach for computing solutions to incompressible two-phase flow. *Journal of Computational Physics*, 114:146–159, 1994.
11. Y.C. Chang, T.Y. Hou, B. Merriman, and S. Osher. A level set formulation of eulerian interface capturing methods for incompressible fluid flows. *Journal of Computational Physics*, 124:449–464, 1996.
12. J.U. Brackbill, D.B. Kothe, and C. Zemach. A continuum method for modeling surface tension. *Journal of Computational Physics*, 100:335–354, 1992.

13. J.J.I.M. van Kan. A second-order accurate pressure correction method for viscous incompressible flow. *SIAM J. Sci. Stat. Comp.*, 7:870–891, 1986.
14. F.H. Harlow and J.E. Welch. Numerical calculation of time-dependent viscous incompressible flow of fluid with free surfaces. *Physics of Fluids*, 8:2182–2189, 1965.
15. M. Sussman, E. Fatemi, P. Smereka, and S. Osher. An improved level set method for incompressible two-phase flows. *Computers and Fluids*, 27:663–680, 1998.
16. M. Sussman and E. Fatemi. An efficient, interface-preserving level set redistancing algorithm and its application to interfacial incompressible fluid flow. *SIAM Journal on Scientific Computing*, 20(4):1165–1191, 1999.

Numerical Modeling of Bypass Flow

Vladimir Prokop[1] and Karel Kozel[2]

[1] Czech Technical University, Faculty of Mechanical Engineering
prokop@marian.fsik.cvut.cz
[2] Czech Technical University, Faculty of Mechanical Engineering

Summary. This paper deals with problem of numerical solution of laminar viscous incompressible stationary and unstationary flows through vessel with bypass. One could describe these problems using model of Navier-Stokes equations and find steady solution of unsteady system by using multistage Runge-Kutta method together with time dependent artificial compressibility method. Non-stationary solution is achieved from initial stationary solution by prescribing of nonstationary outlet conditions. Some results of numerical solution of cardiovascular problems are presented: stationary and unstationary 2D flows in a vessel and a bypass.

1 Mathematical model

In the cardiovascular system we could find many different types of vessels like large arteries, vessels of medium size and capillaries. They differ in diameter and in thickness and composition of the wall. In larger vessels the blood flow can be assumed to behave as an incompressible continuum. One can describe this type of flow using system of momentum and continuity equation written in conservation form:

$$\rho \frac{\mathrm{D}\mathbf{w}}{\mathrm{D}t} - \nabla \cdot \tau = \rho \mathbf{f} \tag{1}$$

$$\nabla \cdot \mathbf{w} = 0 \tag{2}$$

where τ is stress tensor of the fluid, $\mathbf{w}$ is velocity vector and $\mathbf{f}$ is vector of external forces, which is later not taken into account. Density of the fluid ρ is supposed to be constant in physiological conditions, although it depends on the red cells concentration. The functional dependence of the stress tensor τ on velocity vector $\mathbf{w}$ and the blood pressure p is descented by the following relations:

$$\tau_{ij} = -p\delta_{ij} + \sigma_{ij} \tag{3}$$

$$\sigma_{ij} = 2\mu\left(\frac{\partial w_i}{\partial x_j} + \frac{\partial w_j}{\partial x_i}\right) \tag{4}$$

Equations (3) and (4) describe Newtonian fluid. Important feature of blood flow is pulsatility caused by the periodic motion of the heart. It is also known [2] that there is scarcely any turbulence in vessels except some special cases.

The walls of a tube which is the model of a vessel are supposed to be riggid and the velocity vector $\mathbf{w}$ is null on them. Blood flow can be assumed to be laminar [2]. Indeed, in physiological conditions, the values of speed involved are low enough. Morover, generally, the periodicity of the flow, together with short length of vascular districts, do not give rise to fully developed turbulence. Reynolds number $Re = \frac{dw^*}{\nu}$ is important feature of the flow behaviour. Quantity w^* is characteristic velocity, $\nu = \mu/\rho$ is kinematic viscosity and d is a lenght scale. In large and medium human vessels, the Reynolds number ranges from 400 up to 10000. In stationary case pulsating nature of blood flow is not considered. Elasticity of vessel tubes is not considered in both cases. The flow could be then described as viscous, incompressible, laminar and stationary (unstationary) in 2D by the system of Navier-Stokes equations without influence of exterior forces and heat exchange. The system is written in conservative non-dimensional vector form rewritten from (1), (2)

$$\tilde{R}W_t + F_x + G_y = \tilde{R}\frac{1}{Re}\left(W_{xx} + W_{yy}\right), \tag{5}$$

where $W = (p, u, v)^T$ is the vector of solution, $\tilde{R} = \text{diag}||0, 1, 1||$ and $F = (u, u^2 + p, uv)^T, G = (v, uv, v^2 + p)^T$ denote inviscid fluxes, (u, v) is velocity vector, p denotes pressure, $Re = \frac{U^* L^* \rho^*}{\eta^*}$ is Reynolds number, where U^* is speed of upstream flows, L^* denotes width of a channel, η^* is a reference kinematic viskosity (* means reference dimensional quantity). For upstream boundary conditions we use velocity vector (u, v), along the walls vector of velocity is equal zero because of viscosity of fluid and impenetrability of wall, downstream boundary condition is $p = p_2$, which should ensure pressure gradient.

2 Numerical model

Solution of the system (5) is obtained using method of artificial compressibility, then equation of continuity is completed with term $\frac{1}{a^2}p_t$, where $a^2 \in R^+$. Rewritten in vector form the improved system (5) is following

$$W_t + F_x + G_y = \tilde{R}\frac{1}{Re}\left(W_{xx} + W_{yy}\right), \tag{6}$$

where $W = (\frac{p}{a^2}, u, v)^T$. Finding steady solution one could solve unsteady system (6) by finite volume method together with time dependent method. System of equations (6) could be solved using three stage Runge-Kutta method using given steady boundary conditions. At the inlet extrapolation of pressure is used. At the outlet the value of pressure is set constant for stationary case and for unstationary case the pressure is prescribed by sinus function in form:

$$p_2 = p_{20}(1 + \alpha \sin 2\pi\omega t), \tag{7}$$

where ω is a frequency and α is an amplitude. Multistage Runge-Kutha method is stabilized by artificial viscosity term (Jameson's type):

$$W^n_{i,j} = W^{(0)}_{i,j} \tag{8}$$

$$W^{(r)}_{i,j} = W^{(0)}_{i,j} - \alpha_r \Delta t \overline{R}W^{(r-1)}_{i,j}, (r = 1, \dots, m) \tag{9}$$

$$W^{n+1}_{i,j} = W^{(m)}_{i,j}, m = 3, \tag{10}$$

where

$$\overline{R}W^{(r-1)}_{i,j} = RW^{(r-1)}_{i,j} - DW^{(r-1)}_{i,j} \tag{11}$$

and coefficients $\alpha_1 = 0.5, \alpha_2 = 0.5, \alpha_3 = 1.0$, so the numerical method is second order in time and space. The form of steady residual $RW^n_{i,j}$ depends on the method used for solving space derivatives:

$$RW_{i,j} = \frac{1}{\mu_{ij}} \sum_{k=1}^{4} [(F^i_k - \frac{1}{Re} F^v_k) \Delta y_k - (G^i_k - \frac{1}{Re} G^v_k) \Delta x_k], \tag{12}$$

where $F^i = F, G^i = G$ and $F^v = (0, u_x, v_x)^T, G^v = (0, u_y, v_y)^T$. Artificial viscosity term $DW^n_{i,j}$ in this case depends on the second derivatives of pressure and is used for improving stability of the solution:

$$DW_{i,j} = E\big[\gamma_i (W_{i+1,j} - 2W_{i,j} + W_{i-1,j}) + \gamma_j (W_{i,j+1} - 2W_{i,j} + W_{i,j-1})\big] \tag{13}$$

$$E = \mathrm{diag}\|0, \epsilon_1, \epsilon_2\|, \epsilon_1, \epsilon_2 \in R \tag{14}$$

$$\gamma_i = \max(\gamma_{i1}, \gamma_{i2}), \quad \gamma_j = \max(\gamma_{j1}, \gamma_{j2}) \tag{15}$$

$$\gamma_{i1} = \frac{|p_{i+1,j} - 2p_{i,j} + p_{i-1,j}|}{|p_{i+1,j} + 2p_{i,j} + p_{i-1,j}|}, \gamma_{i2} = \frac{|p_{i,j} - 2p_{i-1,j} + p_{i-2,j}|}{|p_{i,j} + 2p_{i-1,j} + p_{i-2,j}|}, \tag{16}$$

$$\gamma_{j1} = \frac{|p_{i,j+1} - 2p_{i,j} + p_{i,j-1}|}{|p_{i,j+1} + 2p_{i,j} + p_{i,j-1}|}, \gamma_{j2} = \frac{|p_{i,j} - 2p_{i,j-1} + p_{i,j-2}|}{|p_{i,j} + 2p_{i,j-1} + p_{i,j-2}|}. \tag{17}$$

Time step is obtained from the formula, which is needed for the stability limitation of the RK method, where $CFL = 2$:

$$\Delta t \leq \min_{i,j,k} \frac{CFL \mu_{ij}}{\rho_A \Delta y_k + \rho_B \Delta x_k + \frac{2}{Re}\left(\frac{\Delta x^2 + \Delta y^2}{\mu_{ij}}\right)}, \tag{18}$$

where $\mu_{ij} = \int\int_{D_{ij}} dxdy$. The convergence of iterative process is followed using the behavior of residual in space L_2

$$\|RW^n_{i,j}\|_2 = \left\{ \frac{1}{MN} \sum_{i,j} \left(\frac{W^{n+1}_{i,j} - W^n_{i,j}}{\Delta t} \right)^2 \right\}^{1/2} \tag{19}$$

$$\text{resp. } \log \|RW^n_{i,j}\|_2,$$

where MN is the number of finite volumes.

3 Some numerical results

In this section we present numerical results achieved using above described numerical methods. Firstly we show results of stationary cases, where the outlet condition is stationary. For lower Reynolds numbers like 500 or 1000 we could see good convergence of scheme and smooth results, for higher Reynolds numbers we have good convergence when using artificial viscosity. We must take into acount that the edges between bypass and vessel are sharp. This problem should be eliminated in the future. We could see zones of separation in bypass after bifurcation and also in the domain after contraction of vessel. In the second part of our results we present our first results of unstationary flow. We start from steady solution for specific Reynolds number and using the same method as before only changing the outlet condition from stationary to unstationary we obtain results of unstationary flow. Results presented here are for 20 percent contraction of vessel and for bypasses which are about 40 or 30 percent of diameter of vessel. Also the distance between the vessel and bypass is in the presented cases constant. For unsteady solution is necessary to use $a \to \infty$ or computation in dual (artifitial) time.

Acknowledgement: This wok was partly supported by grant GAČR No. 201/00/0684 and Research Plan MSM 210000010.

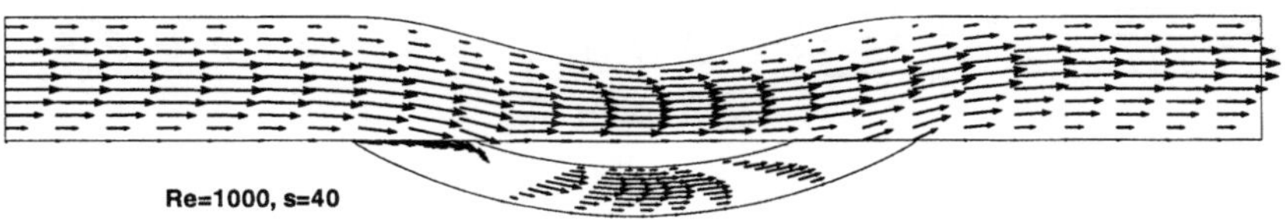

Fig. 1. Bypass flows no.1a, Re=1000, vector field of velocity

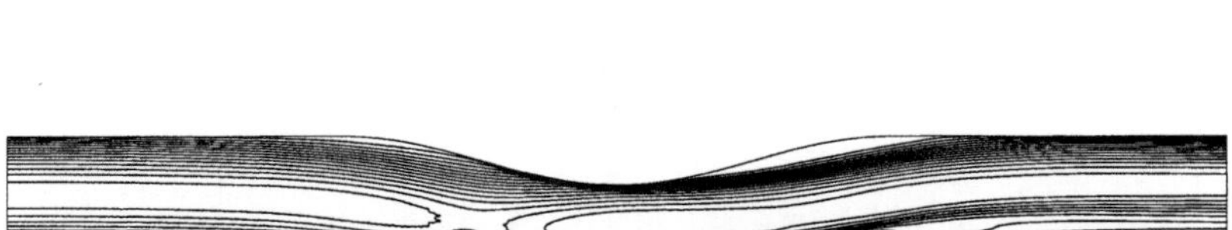

Fig. 2. Bypass flows no.1b, Re=1000, izolines of velocity

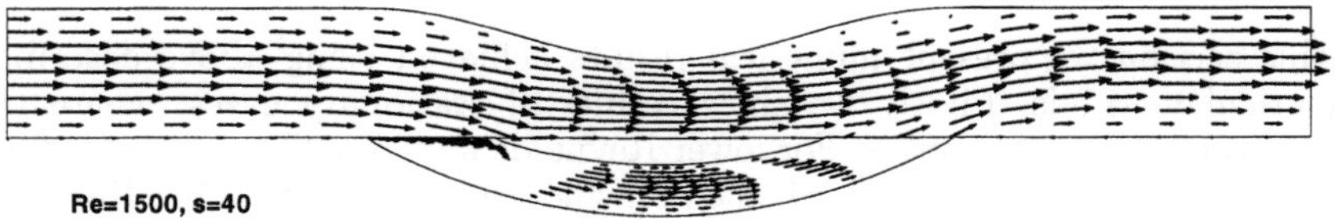

Fig. 3. Bypass flows no.2a, Re=1500, vector field of velocity

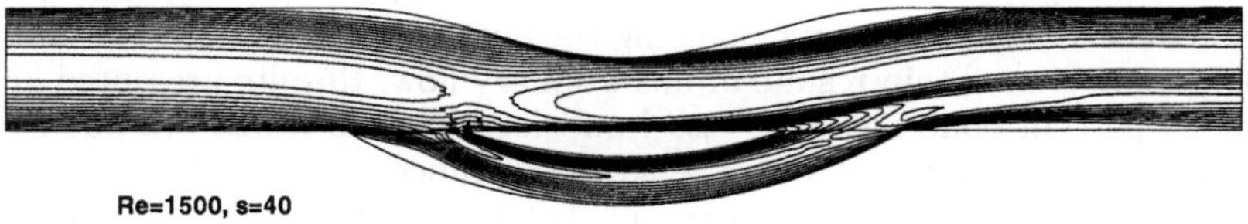

Fig. 4. Bypass flows no.2b, Re=1500, izolines of velocity

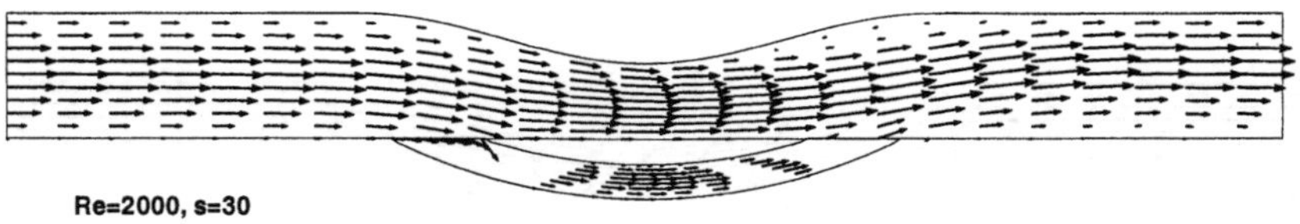

Fig. 5. Bypass flows no.3a ,Re=2000, vector field of velocity

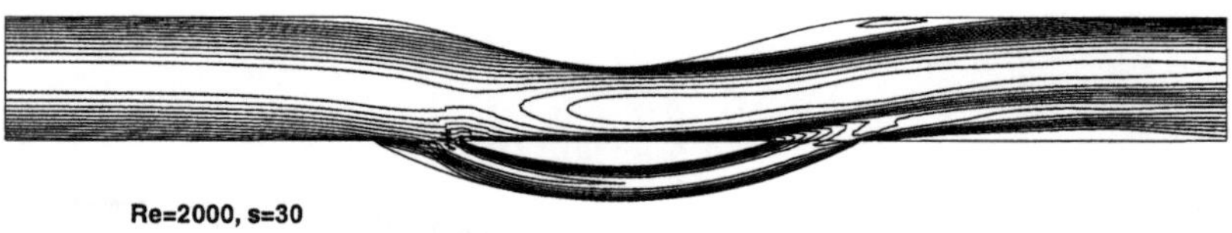

Fig. 6. Bypass flows no.3a ,Re=2000, izolines of velocity

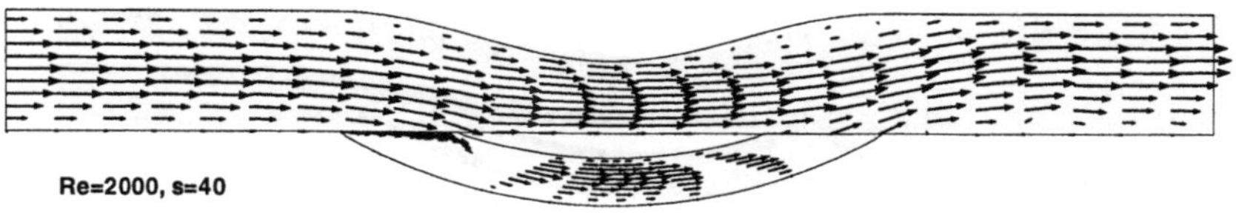

Fig. 7. Bypass flows no.4, Re=2000, vector field of velocity

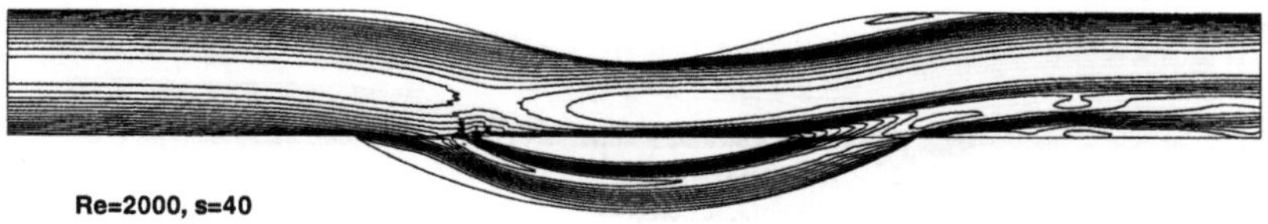

Fig. 8. Bypass flows no.4, Re=2000, izolines of velocity

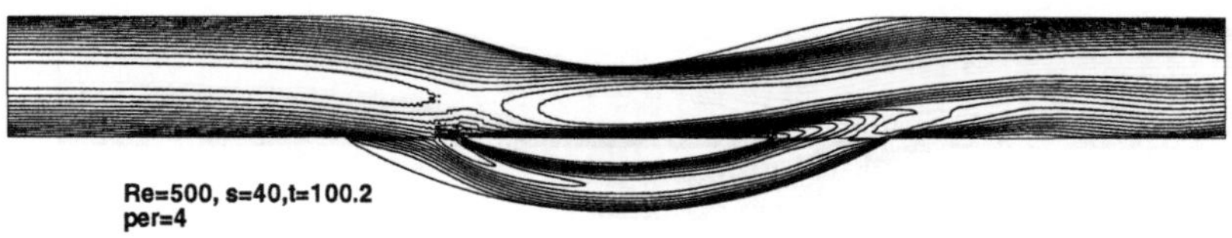

Fig. 9. Bypass unstationary flows no.1, Re=500, izolines of velocity

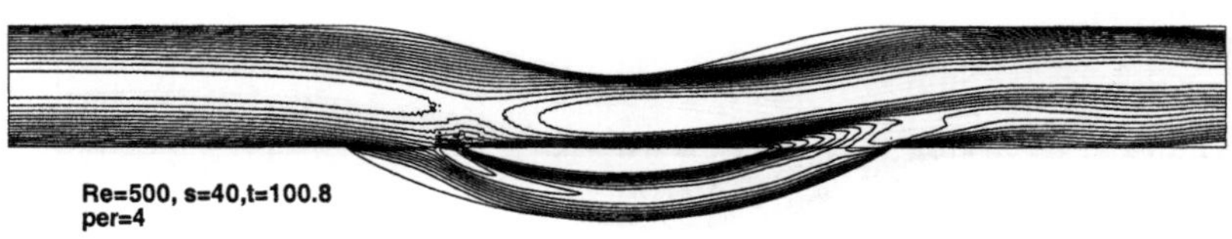

Fig. 10. Bypass unst. flows no.2, Re=500, izolines of velocity

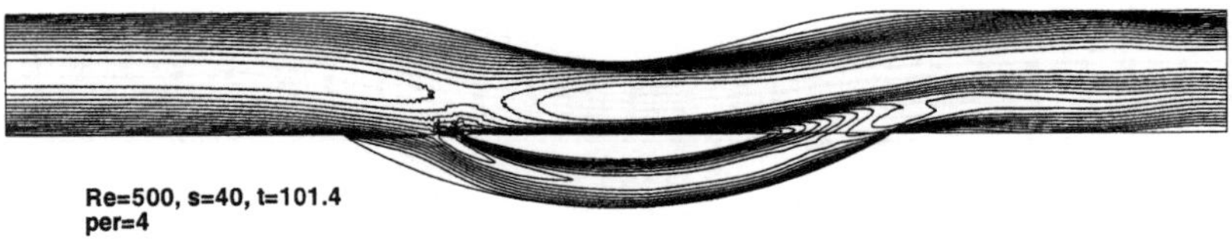

Fig. 11. Bypass unst. flows no.3, Re=500, izolines of velocity

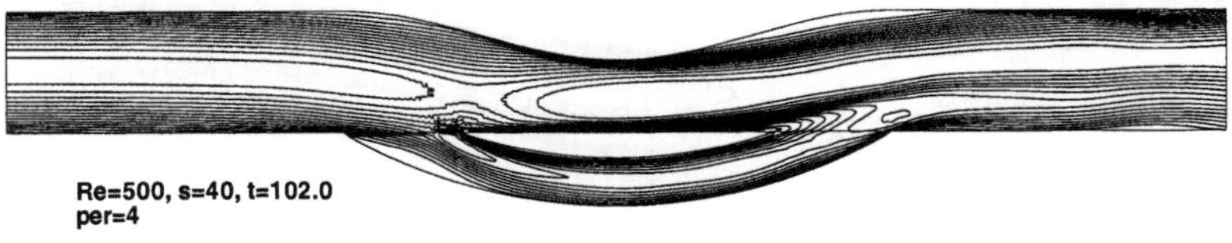

Fig. 12. Bypass unst. flows no.4, Re=500, izolines of velocity

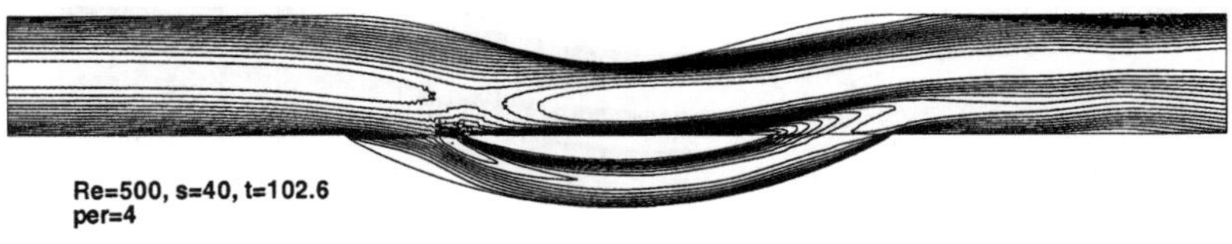

Fig. 13. Bypass unst. flows no.5 ,Re=500, izolines of velocity

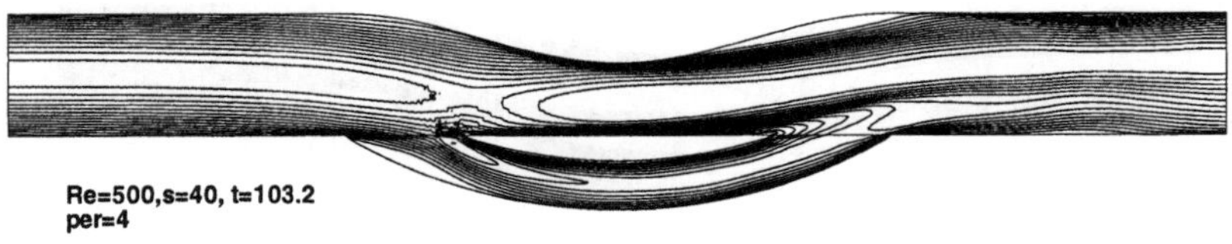

Fig. 14. Bypass unst. flows no.6 ,Re=500, izolines of velocity

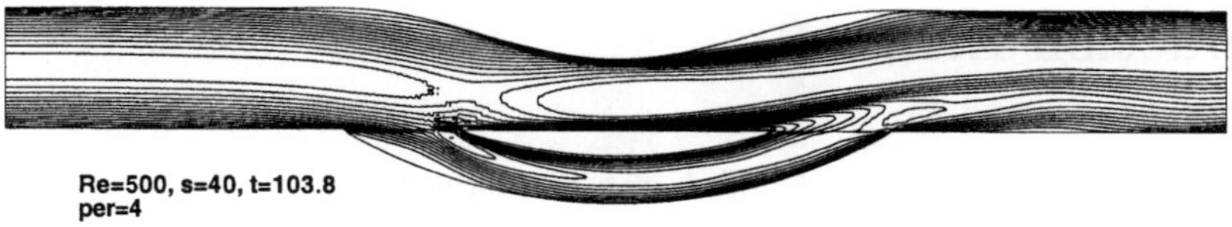

Fig. 15. Bypass unst. flows no.7, Re=500, vector field of velocity

References

1. R. J. LeVeque: *Numerical Methods for Conservation Laws*, Birkhauser Verlag, 1990
2. A. Quarteroni, M. Tuveri, A. Veneziani: *Computational Vascular Fluid Dynamics: Problems, Models and Methods*
3. R. Dvorak, K. Kozel: *Mathematical modeling in aerodynamics*,CVUT, 1996 (in Czech)
4. S. Canic, D. Mirkovic: *A Hyperbolic System of Conservation Laws in Modeling Endovascular Treatment of Abdominal Aortic Aneurysm*, International Series of Numerical Mathematics Vol.140, 2001
5. K. Kozel, V. Prokop: *Numerical solution of incompresible viscous flows through a channel with cavitas and a channel with bypass*, Proceeding of "The 2nd International Conference Of Applied Mathematics And Informatics At Universities 2002", Trnava, September 2002, p. 59 - 63

A Posteriori Estimation of Dimension Reduction Errors

Sergey Repin[1], Stefan Sauter[2] and Anton Smolianski[2]

[1] V.A. Steklov Institute of Mathematics, Fontanka 27, 191 011 St. Petersburg, Russia *repin@pdmi.ras.ru*

[2] Institute of Mathematics, Zurich University, CH-8057, Zurich, Switzerland *stas, antsmol@amath.unizh.ch*

Summary. A new a-posteriori error estimator is presented for the verification of the dimensionally reduced models stemming from the elliptic problems on thin domains. The original problem is considered in a general setting, without any specific assumptions on the domain geometry, coefficients and the right-hand sides. The estimator provides a guaranteed upper bound for the modelling error in the energy norm, exhibits the optimal convergence rate as the domain thickness tends to zero and accurately indicates the local error distribution.

1 Introduction

The method of dimension reduction is a popular approach frequently used by engineers for the approximate solution of the problems posed in *thin* domains. The term "thin" means that the size of the original physical domain along one coordinate direction is much smaller than along the others; this allows to make some simplifying assumptions on the behaviour of the exact solution and to replace the original, for instance, three-dimensional problem by a two-dimensional one. It is, however, clear that the solution of the new, "reduced" problem will, in general, differ from the solution to the original high-dimensional problem. Thus, the dimension reduction method unavoidably produces the error that can be referred to as the dimension reduction or the *modelling error*. The essential part of the model verification is, hence, a reliable *a posteriori* control of the dimension reduction error.

Despite the practical importance of the topic, only a few a posteriori estimators for the dimension reduction error have been introduced so far. In [10] and [2] (see also [1]) residual-type estimators were proposed and proved reliable and efficient under the assumptions that the right-hand side of the given equation is zero and the original domain is a plate with plane parallel faces. In [3] and [8] implicit estimators based on the solution of local Neumann problems were developed; the estimators were intended for hierarchical modelling and involved the solution of local three-dimensional problems.

In this work we propose a reliable and efficient a posteriori estimator for the dimension reduction error in the energy norm, having no specific assumptions on the right-hand side of the given equation and considering a general geometry of the given domain. We show that, for the zero-order dimension reduction method considered here, the estimator of Babuška and Schwab (see [1], [2]) can be obtained as a particular case of our estimator when the right-hand side of the equation is zero and the original domain is a plate with plane parallel faces. We demonstrate the optimal convergence of the estimator as the plate thickness tends to zero (although, it is worth noting that the proposed estimator preserves its reliability for any positive thickness). Finally, we observe how accurately the estimator indicates the local error distribution, thus, allowing for a local improvement of the model.

2 Problem setting

We consider a three-dimensional Lipschitz domain

$$\Omega := \{x \in \mathbb{R}^3 \mid (x_1, x_2) \in \widehat{\Omega}\,,\ d_\ominus(x_1, x_2) < x_3 < d_\oplus(x_1, x_2)\}\,,$$

where $\widehat{\Omega} \subset \mathbb{R}^2$ is its projection on the (x_1, x_2)-plane ($\widehat{\Omega}$ has the Lipschitz boundary $\widehat{\Gamma}$) and $d_\ominus$ and $d_\oplus$ are Lipschitz continuous functions: $\widehat{\Omega} \to \mathbb{R}$. The lower and upper faces of Ω are denoted by

$$\Gamma_\ominus := \{x \in \mathbb{R}^3 \mid (x_1, x_2) \in \widehat{\Omega}\,,\ x_3 = d_\ominus(x_1, x_2)\}$$

and

$$\Gamma_\oplus := \{x \in \mathbb{R}^3 \mid (x_1, x_2) \in \widehat{\Omega}\,,\ x_3 = d_\oplus(x_1, x_2)\}\,,$$

the lateral boundary by

$$\Gamma_0 := \{x \in \mathbb{R}^3 \mid (x_1, x_2) \in \widehat{\Gamma}\,,\ d_\ominus(x_1, x_2) < x_3 < d_\oplus(x_1, x_2)\}$$

(see Figure 1).

Remark. We consider $d_\ominus$ and $d_\oplus$ as explicit functions of (x_1, x_2)-coordinates only for the sake of simplicity. The generalization of the theory to the case of an arbitrary Lipschitzian domain Ω presents no difficulty from the conceptional point of view.

The assumption that the given domain Ω is "thin" can now be written as

$$\operatorname{diam} \widehat{\Omega} \gg \max_{\overline{\widehat{\Omega}}} d\,(x_1, x_2)\,, \tag{1}$$

where $d = d_\oplus - d_\ominus$ is the domain thickness, $d\,(x_1, x_2) \geq d_* > 0\ \ \forall (x_1, x_2) \in \overline{\widehat{\Omega}}$. Although the assumption is of purely qualitative nature, it serves as a basis for the derivation of the corresponding two-dimensional reduced model. We also have to notice that Figure 1 depicts a simplified case; in the geometrical definitions we do not assume the domain thickness $d\,(x_1, x_2)$ to be a constant.

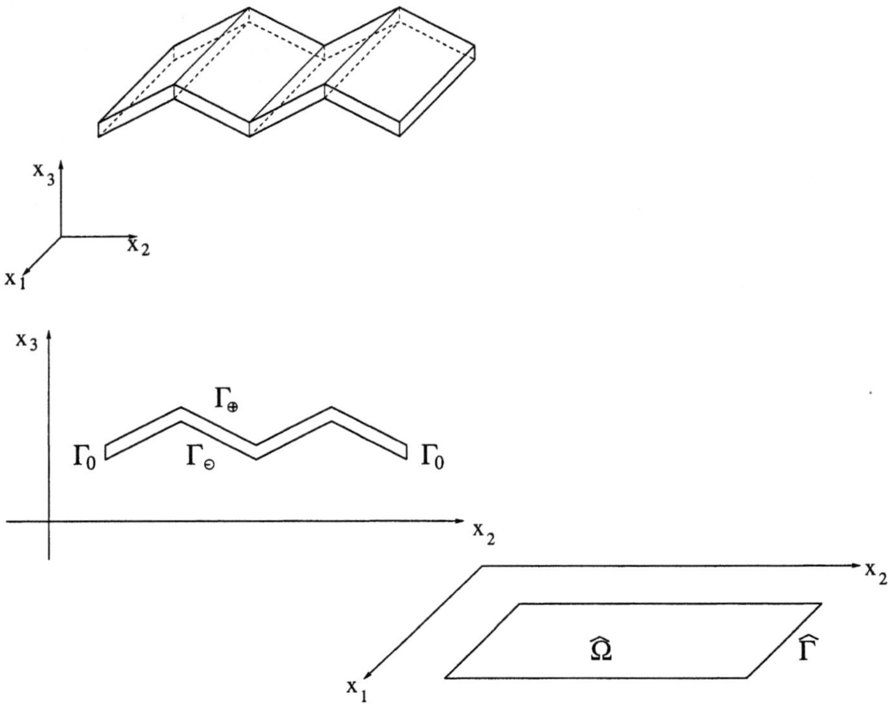

Fig. 1. Sketch of the domain geometry

In the domain Ω we consider a model elliptic problem

$$-\mathrm{Div}\,(\mathbf{A}\nabla u) = f \quad \text{in } \Omega\,, \tag{2}$$

$$u = 0 \quad \text{on } \Gamma_0\,, \tag{3}$$

$$\mathbf{A}\nabla u \cdot \boldsymbol{\nu}_\ominus = \widehat{F}_\ominus \quad \text{on } \Gamma_\ominus\,, \tag{4}$$

$$\mathbf{A}\nabla u \cdot \boldsymbol{\nu}_\oplus = \widehat{F}_\oplus \quad \text{on } \Gamma_\oplus\,, \tag{5}$$

where $f \in L_2(\Omega)$, $\widehat{F}_\ominus$, $\widehat{F}_\oplus \in L_2(\widehat{\Omega})$, $\boldsymbol{\nu}_\ominus$ and $\boldsymbol{\nu}_\oplus$ are outward normal vectors at $\Gamma_\ominus$ and $\Gamma_\oplus$ respectively. The matrix $\mathbf{A} = (a_{ij}(x))_{i,j=\overline{1,3}}$ with the components from $L_\infty(\Omega)$ is symmetric and uniformly positive definite, i.e. there exist constants $0 < c < C < \infty$ such that

$$c|\xi|^2 \leq \mathbf{A}(x)\xi \cdot \xi \leq C|\xi|^2 \quad \forall \xi \in \mathbb{R}^3\,, \text{ a. e. in } \Omega\,.$$

From now on we will frequently use the notation $\widehat{x} = (x_1, x_2)$, $x = (\widehat{x}, x_3)$, and all functions depending only on (x_1, x_2) will be marked by $\widehat{\ }$; in addition, we will distinguish between 3- and 2-dimensional divergence operator:

$$\mathrm{Div}\,\boldsymbol{\tau} = \boldsymbol{\tau}_{1,1} + \boldsymbol{\tau}_{2,2} + \boldsymbol{\tau}_{3,3}\,,\ \mathrm{div}\,\widehat{\boldsymbol{\tau}} = \widehat{\boldsymbol{\tau}}_{1,1} + \widehat{\boldsymbol{\tau}}_{2,2}\,.$$

The weak form of the problem (2)–(5) reads
<u>*Problem (P)*:</u> Find $u \in V_0 := \{v \in H^1(\Omega) \mid v = 0 \text{ on } \Gamma_0\}$ such that

$$\int_\Omega \mathbf{A}\nabla u \cdot \nabla w \, dx = \int_\Omega f \, w \, dx + \int_{\Gamma_\ominus} \widehat{F}_\ominus \, w \, ds + \int_{\Gamma_\oplus} \widehat{F}_\oplus \, w \, ds \qquad \forall w \in V_0 \,. \quad (6)$$

3 The reduced problem

The assumption (1) allows one to suppose that

$$\text{the exact solution } u \approx const \text{ in the } x_3\text{-direction.} \quad (7)$$

This gives rise to the so-called *zero-order reduced model* for the original problem (6). The model is very popular due to its simplicity and purely two-dimensional formulation. The discussion on the hierarchy of the reduced models of different orders can be found in, e.g., [9], [2].

Then, introducing the subspace

$$\widehat{V}_0 := \{v \in V_0 \mid \exists \widehat{v} \in H_0^1(\widehat{\Omega}) \text{ such that } v(x) = \widehat{v}(\widehat{x}) \text{ for a.e. } x = (\widehat{x}, x_3) \in \Omega\}$$

and the operation $(\widetilde{\ })$ of averaging in the x_3-direction

$$\forall g \in L_1(\Omega) \ : \quad \widetilde{g}(\widehat{x}) := \frac{1}{d\,(\widehat{x})} \int\limits_{d_\ominus(\widehat{x})}^{d_\oplus(\widehat{x})} g(\widehat{x}, x_3) \, dx_3 \ \text{ for a.e. } \widehat{x} \in \widehat{\Omega} \,,$$

we can deduce from (6) the *reduced problem* (see [7]) that reads
<u>*Problem* $(\widehat{P})$:</u> Find $\widehat{u} \in \widehat{V}_0$ such that

$$\int_{\widehat{\Omega}} d\,(\widehat{x}) \widetilde{\mathbf{A}}_p(\widehat{x}) \nabla \widehat{u} \cdot \nabla \widehat{w} \, d\widehat{x} = \int_{\widehat{\Omega}} d\,(\widehat{x}) \widehat{f}(\widehat{x}) \widehat{w} \, d\widehat{x} \qquad \forall \widehat{w} \in \widehat{V}_0 \,, \quad (8)$$

where $\widehat{f} = \widetilde{f} + \frac{\widehat{F}_\ominus \sqrt{1+|\nabla d_\ominus|^2} + \widehat{F}_\oplus \sqrt{1+|\nabla d_\oplus|^2}}{d}$ and $\widetilde{\mathbf{A}}_p(\widehat{x}) = (\widetilde{a}_{ij}(\widehat{x}))_{i,j=\overline{1,2}}$ is the averaged "plane" part $(\mathbf{A}_p(x) = (a_{ij}(x))_{i,j=\overline{1,2}})$ of the matrix $\mathbf{A}$.

It is clear that problem (8) is a two-dimensional elliptic problem with the homogeneous Dirichlet boundary condition:

$$-\text{div}\,(d(\widehat{x})\, \widetilde{\mathbf{A}}_p(\widehat{x}) \nabla \widehat{u}) = d(\widehat{x})\, \widehat{f}(\widehat{x}) \quad \text{in } \widehat{\Omega} \quad (9)$$

$$\widehat{u} = 0 \quad \text{on } \widehat{\Gamma} \,. \quad (10)$$

4 A posteriori estimation of the modelling error

In order to control the dimension reduction error $e := u - \widehat{u}$, we apply the functional-type a posteriori error estimate derived in [6] (see also [4] and [5])

to the original three-dimensional problem (6):
For all $\gamma > 0$, $\delta > 0$ and $y^* \in H_*(\Omega, \mathrm{Div})$ there holds

$$|||u - \widehat{u}|||^2 \leq (1+\gamma)\, M_1^2 + \left(1 + \frac{1}{\gamma}\right)(1+\delta) C_\Omega^2\, M_2^2 + \left(1+\frac{1}{\gamma}\right)\left(1+\frac{1}{\delta}\right) C_\Gamma^2 (1 + C_\Omega^2)\, M_3^2\,, \tag{11}$$

where $|||\cdot|||$ is the energy norm, $|||v||| := \left(\int_\Omega \mathbf{A}(x)\nabla v \cdot \nabla v\, dx\right)^{1/2}$ $\forall v \in V_0$, C_Ω is the constant from Friedrichs' inequality ($C_\Omega^{-2} = \inf\limits_{w\in V_0\setminus\{0\}} \frac{|||w|||^2}{\|w\|^2_{L_2(\Omega)}}$), C_Γ is the constant from the trace inequality ($C_\Gamma^2 = \sup\limits_{w\in V_0\setminus\{0\}} \frac{\|w\|^2_{L_2(\Gamma_\oplus)}+\|w\|^2_{L_2(\Gamma_\ominus)}}{|||w|||^2+\|w\|^2_{L_2(\Omega)}}$) and the functionals M_1^2, M_2^2, M_3^2 are defined as follows:

$$M_1^2 := \int_\Omega (\nabla\widehat{u} - \mathbf{A}^{-1}y^*)\cdot(\mathbf{A}\nabla\widehat{u} - y^*)\, dx\,,$$
$$M_2^2 := \|\mathrm{Div}\, y^* + f\|^2_{L_2(\Omega)}\,,$$
$$M_3^2 := \|\widehat{F}_\ominus - y^*\boldsymbol{\nu}_\ominus\|^2_{L_2(\Gamma_\ominus)} + \|\widehat{F}_\oplus - y^*\boldsymbol{\nu}_\oplus\|^2_{L_2(\Gamma_\oplus)}\,.$$

We emphasize that the estimate is valid for any positive numbers γ and δ and for any vector-function y^* from the space $H_*(\Omega, \mathrm{Div})$ defined as

$$H_*(\Omega, \mathrm{Div}) := \{y^* \in L_2(\Omega, \mathbb{R}^3) \mid \mathrm{Div}\, y^* \in L_2(\Omega)\,,\ y^*\cdot\boldsymbol{\nu}_\ominus \in L_2(\Gamma_\ominus)\,,\ y^*\cdot\boldsymbol{\nu}_\oplus \in L_2(\Gamma_\oplus)\}\,.$$

While the best possible option would be to take as y^* the exact flux $\mathbf{A}\nabla u$ (then M_2 and M_3 would vanish and M_1 would give us the energy norm of the exact error e), we have to restrict ourselves to choosing some computable quantity, i.e. not containing the unknown exact solution u. We approximate the flux by

$$y^* = \widetilde{\mathbf{A}}_p \nabla\widehat{u} + \boldsymbol{\tau}^*\,, \tag{12}$$

where $\boldsymbol{\tau}^* = \{0\,,\ 0\,,\ \psi(x)\}^T$, ψ is the auxiliary function from $L_2(\Omega)$ such that $\psi_{,3} \in L_2(\Omega)\,,\ \psi \in L_2(\Gamma_\ominus)$ and $\psi \in L_2(\Gamma_\oplus)$. Using (9), it is easy to verify that y^* from (12) belongs to $H_*(\Omega, \mathrm{Div})$. A discussion about other choices of y^* can be found in [7].

Substituting (12) into the functionals M_1^2, M_2^2, M_3^2, we obtain (see the details in [7])

$$M_1^2 = \tag{13}$$
$$\int_\Omega (b_{33}\psi^2 + 2(\mathbf{b}_3 \cdot \widetilde{\mathbf{A}}_p \nabla\widehat{u})\psi)\, dx + \int_{\widehat{\Omega}} d(\widehat{x})\, (\widetilde{\mathbf{B}}_p \widetilde{\mathbf{A}}_p - \mathbf{I})\nabla\widehat{u} \cdot \widetilde{\mathbf{A}}_p \nabla\widehat{u}\, d\widehat{x}\,,$$
$$M_2^2 = \tag{14}$$
$$\|f - \widetilde{f} - \frac{\widehat{F}_\ominus \sqrt{1+|\nabla d_\ominus|^2} + \widehat{F}_\oplus \sqrt{1+|\nabla d_\oplus|^2}}{d} + \psi_{,3} - \frac{\nabla d}{d} \cdot \widetilde{\mathbf{A}}_p \nabla\widehat{u}\|^2_{L_2(\Omega)}\,,$$
$$M_3^2 = \tag{15}$$
$$\|\widehat{F}_\ominus - \widetilde{\mathbf{A}}_p \nabla\widehat{u} \cdot \boldsymbol{\nu}_\ominus - \psi \nu_{\ominus 3}\|^2_{L_2(\Gamma_\ominus)} + \|\widehat{F}_\oplus - \widetilde{\mathbf{A}}_p \nabla\widehat{u} \cdot \boldsymbol{\nu}_\oplus - \psi \nu_{\oplus 3}\|^2_{L_2(\Gamma_\oplus)}\,,$$

where $\widetilde{\mathbf{B}}_p$ is the averaged "plane" part of the matrix $\mathbf{B} := \mathbf{A}^{-1}$ (i.e., if $\mathbf{B}(x) = (b_{ij}(x))_{i,j=\overline{1,3}}$, then $\mathbf{B}_p(x) = (b_{ij}(x))_{i,j=\overline{1,2}}$), the vector $\mathbf{b}_3 := \{b_{31}\,,\ b_{32}\}^T$ and $\mathbf{I}$ is the 2×2 identity-matrix.

Now we still have the freedom of choosing the auxiliary function ψ that in the case of the Poisson equation should, obviously, approximate the derivative $u_{,3}$ of the exact solution in the x_3-direction. The simplest choice is to take such a ψ that the term M_3 (i.e. the residual on the Neumann boundary condition) would be identically zero. This can be immediately achieved by letting $\psi(x) = \widehat{\alpha}(\widehat{x})\, x_3 + \widehat{\beta}(\widehat{x})$ with the coefficient functions $\widehat{\alpha}\,, \widehat{\beta} \in L_2(\widehat{\Omega})$ uniquely determined by the requirement $M_3 = 0$. Other options for the function ψ are considered in [7]. Then, minimizing the right-hand side of (11) with respect to the scalar parameters $\gamma > 0$ and $\delta > 0$, we arrive at the estimate

$$|||u - \widehat{u}||| \le M := M_1 + C_\Omega\, M_2\,, \tag{16}$$

where M_1 and M_2 are defined by (13) and (14). The error majorant M has been derived for quite general geometry of Ω and coefficient matrix $\mathbf{A}(x)$. However, to make the estimate more transparent, we consider two particular cases.

4.1 Plate of constant thickness

We assume that

$$d_\oplus = d_\ominus + d_0 \quad (d_0 = const > 0) \tag{17}$$

and, in addition, that

$$\mathbf{A} = \mathbf{A}(\widehat{x}) \ \ \text{(this immediately implies } \mathbf{B} = \mathbf{B}(\widehat{x})\text{)}\,, \tag{18}$$
$$a_{31} = a_{32} = 0 \ \ \text{(this yields } \mathbf{B}_p = \mathbf{A}_p^{-1}\,,\ b_{33} = a_{33}^{-1}\,,\ b_{31} = b_{32} = 0\text{)}\,. \tag{19}$$

With these assumptions the terms M_1 and M_2 in estimate (16) become simpler:

$$M_1 = \left(\int_\Omega a_{33}^{-1} \psi^2\, dx\right)^{1/2}, \qquad M_2 = \|f - \widetilde{f}\|_{L_2(\Omega)}\,. \tag{20}$$

One may notice that the integral in the first term M_1 of the error majorant M can be rewritten as

$$\int_{\Omega} a_{33}^{-1} \psi^2 \, dx = d_0 \cdot \int_{\widehat{\Omega}} a_{33}^{-1} (\widehat{\alpha}^2 \frac{d_{\oplus}^2 + d_{\oplus} d_{\ominus} + d_{\ominus}^2}{3} + \widehat{\alpha}\widehat{\beta}(d_{\oplus} + d_{\ominus}) + \widehat{\beta}^2) \, d\widehat{x} \,,$$

which means that the term M_1 is of order $\mathcal{O}(d_0^{1/2})$ when the plate thickness d_0 tends to zero. If $f \in L_\infty(\Omega)$, the second term M_2 is obviously of the same order $\mathcal{O}(d_0^{1/2})$, i.e. the whole estimator M converges to zero with the rate $\mathcal{O}(d_0^{1/2})$ as $d_0 \to 0$. This is the optimal convergence rate for the modelling error e in the energy norm, as was shown in [9] for the simpler case of a plate with plane parallel faces and $f = 0$. It is worth noting that, if $f \in C^1(\Omega)$, the second term in M is of higher order $\mathcal{O}(d_0^{3/2})$ as compared to the first term.

4.2 Plate with plane parallel faces

If, in addition to (18), (19), we strengthen the assumption (17) replacing it by

$$d_{\oplus} = \frac{d_0}{2} \,, \; d_{\ominus} = -\frac{d_0}{2} \quad (d_0 = const > 0) \,, \tag{21}$$

the auxiliary function ψ will take the simple form $\psi = \frac{\widehat{F}_{\oplus} + \widehat{F}_{\ominus}}{d_0} x_3 + \frac{\widehat{F}_{\oplus} - \widehat{F}_{\ominus}}{2}$ and the error estimate (16) will read

$$|||u - \widehat{u}||| \le \sqrt{\frac{d_0}{3}} \left(\int_{\widehat{\Omega}} a_{33}^{-1} (\widehat{F}_{\oplus}^2 + \widehat{F}_{\ominus}^2 - \widehat{F}_{\oplus} \widehat{F}_{\ominus}) \, d\widehat{x} \right)^{1/2} + C_{\Omega} \, \|f - \widetilde{f}\|_{L_2(\Omega)} \,. \tag{22}$$

If we set here $f = 0$, $a_{33} = 1$ and $\widehat{F}_{\oplus} = \widehat{F}_{\ominus} = \widehat{F}$, we obtain

$$|||u - \widehat{u}||| \le \sqrt{\frac{d_0}{3}} \, \|\widehat{F}\|_{L_2(\widehat{\Omega})} \,, \tag{23}$$

which is exactly the estimator of Babuška and Schwab (see [1]) for the zero-order reduced model. Thus, the latter estimator can be obtained as a particular case of the error majorant (16) if one makes the assumptions (18), (19), (21) and sets $f = 0$. This is a particularly interesting fact, since we advocate the estimation approach that is completely different from the one utilized in [1] (see the details in [7] and [6]).

5 Numerical example

In order to analyse the performance of the proposed error estimator, we consider a simple two-dimensional test problem in the "sine-shape" domain (see Figure 2 (left)) whose upper and lower faces are given by

$$d_{\oplus,\ominus}(x) = \sin(k\pi x) \pm \frac{d_0}{2}, \quad k = 1, 2, \dots,$$

where $d_0 > 0$ is the domain thickness. In this example, $\widehat{\Omega} = (0, 1)$ and $\Omega = \{(x, y) \in \mathbb{R}^2 \mid x \in \widehat{\Omega},\ d_\ominus(x) < y < d_\oplus(x)\}$. The considered problem is

$$\begin{aligned} -\Delta u &= f && \text{in } \Omega, \\ u &= 0 && \text{at } x = 0 \text{ and } x = 1, \\ \nabla u \cdot \boldsymbol{\nu}_{\oplus,\ominus} &= \widehat{F}_{\oplus,\ominus} && \text{at } y = d_{\oplus,\ominus}, \end{aligned}$$

and the right-hand sides of the equation and of the boundary condition are computed using the exact solution

$$u(x, y) = \sin(\pi x) \cdot y^m \quad (m = 1, 2, \dots).$$

The reduced problem (8) is, in this case, a one-dimensional Dirichlet problem that, of course, can be solved very accurately (in the present work, we address the estimation of the modelling error only, assuming that the discretization error stemming from the solution of the reduced problem is negligible).

Figure 2 (right) shows the convergence rates of the exact modelling-error in the energy norm ($|||e|||$) and of the error majorant M as the domain thickness d_0 tends to zero. It is clear that both the exact error and the majorant converge to zero with the theoretically predicted, optimal rate $\mathcal{O}(d_0^{1/2})$, and, moreover, the *effectivity index* $\frac{M}{|||e|||}$ demonstrates the asymptotics $\frac{M}{|||e|||} = 1 + \mathcal{O}(d_0)$. It is also important to note that the presented error estimator provides a reliable upper bound for the exact error at any positive values of the domain thickness d_0, i.e. also in the cases when the domain is not "thin" at all.

Finally, the local error distribution provided by the exact error and by the first, M_1-term of the majorant are depicted in Figure 3. The figure shows that already for rather large value of the domain thickness $d_0 = 0.1$ the majorant delivers a sufficiently accurate information on the location of the regions of the biggest modelling error, while for $d_0 = 0.05$ the exact and the estimated error distributions are practically coincident.

References

1. Babuška, I., Lee, I., Schwab, C. (1994): On the a posteriori estimation of the modeling error for the heat conduction in a plate and its use for adaptive hierarchical modeling. Appl. Numer. Math., **14**, 5–21
2. Babuška, I., Schwab, C. (1996): A posteriori error estimation for hierarchic models of elliptic boundary value problems on thin domains. SIAM J. Numer. Anal., **33**, 221–246
3. Oden, J.T., Cho, J.R. (1996): Adaptive *hpq*-finite element methods of hierarchical models for plate- and shell-like structures. Comput. Meth. Appl. Mech. Engrg., **136**, 317–345

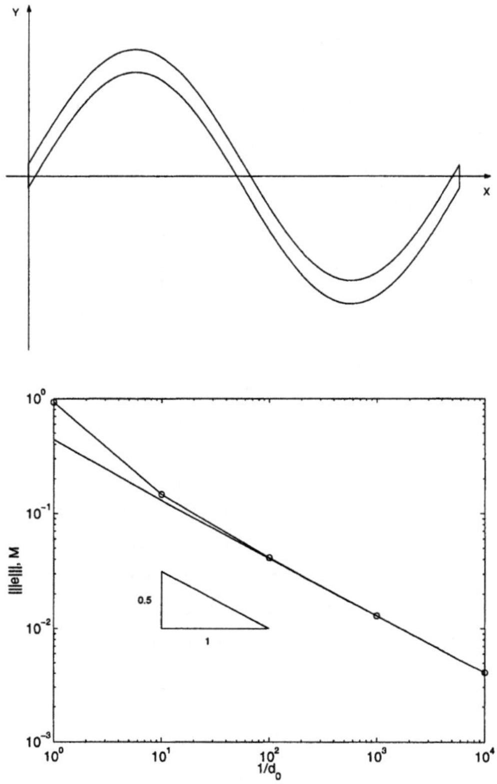

Fig. 2. (left) The domain geometry, (right) Convergence rate of the exact error and of the error majorant, $k = 2$, $m = 2$, the majorant is indicated by "o"

4. Repin, S.I. (2000): A posteriori error estimation for variational problems with uniformly convex functionals. Math. Comp., **69**, 481–600
5. Repin, S.I., Sauter, S.A., Smolianski, A.A. (2003): A posteriori error estimation for the Dirichlet problem with account of the error in the approximation of boundary conditions. Computing, **70**, 205–233
6. Repin, S.I., Sauter, S.A., Smolianski, A.A. (2003): A posteriori error estimation for the Poisson equation with mixed Dirichlet/Neumann boundary conditions. Preprint 02–2003, Institut of Mathematics, Zurich University, http://www.math.unizh.ch/index.php?preprint
7. Repin, S.I., Sauter, S.A., Smolianski, A.A. (2003): A posteriori estimation of dimension reduction errors for elliptic problems on thin domains. Preprint 18–2003, Institut of Mathematics, Zurich University, http://www.math.unizh.ch/index.php?preprint
8. Stein, E., Ohnimus, S. (1997): Coupled model- and solution-adaptivity in the finite-element method. Comput. Meth. Appl. Mech. Engrg., **150**, 327–350
9. Vogelius, M., Babuška, I. (1981): On a dimensional reduction method I. The optimal selection of basis functions. Math. Comp., **37**, 31–46

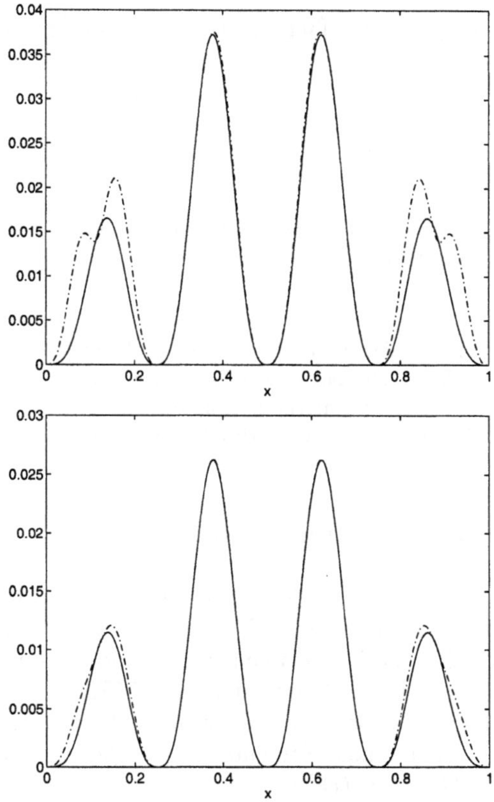

Fig. 3. Local error distribution provided by the exact error (solid line) and by the M_1-term of the majorant (dash-dot line), $k = 4$, $m = 4$: (left) $d_0 = 0.1$, (right) $d_0 = 0.05$

10. Vogelius, M., Babuška, I. (1981): On a dimensional reduction method III. A posteriori error estimation and an adaptive approach. Math. Comp., **37**, 361–384

Analysis of a Multi-Numerics/Multi-Physics Problem

Béatrice Rivière[1]

Department of Mathematics, University of Pittsburgh, 301 Thackeray, Pittsburgh PA 15260 *riviere@math.pitt.edu*

Summary. The coupled Stokes and Darcy flows problem is solved by locally conservative numerical methods. Discontinuous Galerkin methods are used in the Stokes region and discontinuous Galerkin methods coupled with mixed finite element methods are employed in the Darcy region. Optimal a priori error estimates are derived.

1 Introduction and Model Problem

In this work, a numerical method for solving the coupled problem of Stokes and Darcy equations is formulated and analyzed. The coupled system arises from the study of the interaction between surface and subsurface flow. Discontinuous finite elements and mixed finite elements (MFE) are used in subregions of the subsurface domain while only discontinuous finite elements are used in the surface domain. While mixed elements are popular and efficient on regular grids, discontinuous Galerkin (DG) methods are accurate and easily implementable on highly unstructured meshes. The proposed method enables to take advantage of one of these locally conservative methods in a particular subregion of the subsurface. This work is an extension of the coupling of DG for Stokes and MFE for Darcy [8], and DG for Stokes and Darcy [7]. Similar couplings are studied in the work of Layton, Schieweck and Yotov [5] and Discacciati and Quarteroni [2, 3]. Let Ω be a domain in $\mathbb{R}^2$, subdivided into three

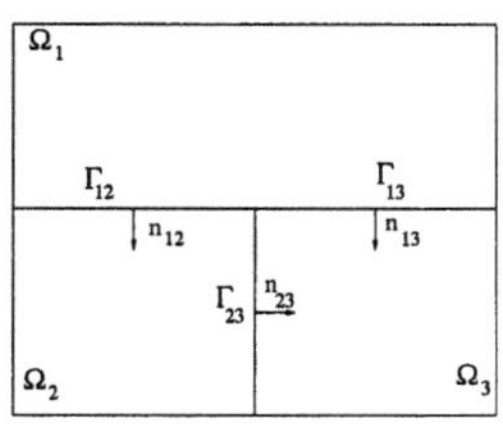

Fig. 1. Computational domain

subdomains Ω_1, Ω_2, and Ω_3. Denote by Γ_{ij} the interface between Ω_i and Ω_j for $i < j$ (see figure 1). Define also the boundary $\Gamma_i = \partial\Omega_i \cap \partial\Omega$, for $i = 1, 2, 3$. The velocity $\boldsymbol{u}_i$ (resp. pressure p_i) denotes the restriction of the fluid velocity

$\boldsymbol{u}$ (resp. pressure p) to the subdomain Ω_i. We assume that the fluid satisfies the Stokes equations in Ω_1.

$$-\nabla\cdot(2\mu\boldsymbol{D}(\boldsymbol{u}_1)-p_1\boldsymbol{I})=\boldsymbol{f}_1,\quad \text{in}\quad \Omega_1, \tag{1}$$

$$\nabla\cdot\boldsymbol{u}_1=0,\quad \text{in}\quad \Omega_1. \tag{2}$$

Here, $\boldsymbol{f}_1$ is an external force acting on the fluid, $\mu>0$ is the constant fluid viscosity and $\boldsymbol{D}(\boldsymbol{u})=\frac{1}{2}(\nabla\boldsymbol{u}+\nabla\boldsymbol{u}^T)$ is the strain tensor. Let $\boldsymbol{n}$ denote the unit outward normal vector to the boundary $\partial\Omega$. The single phase flow problem is solved on the region $\Omega_2\cup\Omega_3$.

$$\nabla\cdot\boldsymbol{u}_i=f_i,\quad \boldsymbol{u}_i=-\boldsymbol{K}\nabla p_i,\quad \text{in}\quad \Omega_i,\ i=2,3. \tag{3}$$

The permeability tensor $\boldsymbol{K}$ is symmetric, positive definite, bounded below and above uniformly. As boundary conditions, we consider $\boldsymbol{u}=0$ on the Stokes boundary Γ_1, and $\boldsymbol{K}\nabla p\cdot\boldsymbol{n}=0$ on the Darcy boundary $\Gamma_2\cup\Gamma_3$. At the interface, the transmissibility conditions arise from the mass conservation, the balance of forces across each interface, and the Beaver-Joseph-Saffman law [1, 9]. Denoting by $\boldsymbol{n}_{ij}$ (resp. $\boldsymbol{\tau}_{ij}$) the normal (resp. tangential) unit vector to the interface Γ_{ij}, for $i<j$, the conditions are written as:

$$\boldsymbol{u}_1\cdot\boldsymbol{n}_{12}=-\boldsymbol{K}\nabla p_2\cdot\boldsymbol{n}_{12},\quad \text{on}\quad \Gamma_{12}, \tag{4}$$

$$\boldsymbol{u}_1\cdot\boldsymbol{n}_{13}=\boldsymbol{u}_3\cdot\boldsymbol{n}_{13},\quad \text{on}\quad \Gamma_{13}, \tag{5}$$

$$\boldsymbol{u}_3\cdot\boldsymbol{n}_{23}=-\boldsymbol{K}\nabla p_2\cdot\boldsymbol{n}_{23},\quad p_2=p_3,\quad \text{on}\quad \Gamma_{23}, \tag{6}$$

$$p_1-2\mu(\boldsymbol{D}(\boldsymbol{u}_1)\boldsymbol{n}_{1i})\cdot\boldsymbol{n}_{1i})=p_i,\quad \text{on}\quad \Gamma_{1i},\quad i=2,3, \tag{7}$$

$$\boldsymbol{u}_1\cdot\boldsymbol{\tau}_{1i}=-2G(\boldsymbol{D}(\boldsymbol{u}_1)\boldsymbol{n}_{1i})\cdot\boldsymbol{\tau}_{1i},\quad \text{on}\quad \Gamma_{1i},\quad i=2,3. \tag{8}$$

As the pressure is unique up to an additive constant, we assume that $\int_\Omega p=0$. A weak solution $(\boldsymbol{u},p)$ of the coupled Stokes-Darcy equations exists (see proof in [5]). We assume that it is also a strong solution with enough regularity. Section 2 defines the discrete spaces and the numerical method. Section 3 contains the error analysis.

2 Numerical Method

The discontinuous Galerkin method is used in the region $\Omega_1\cup\Omega_2$, while the mixed finite element method is used in the region Ω_3. For $i=1,2,3$, let $\mathcal{E}_h^i$ be a non-degenerate quasi-uniform subdivision of Ω_i, let Γ_h^i be the set of interior edges and let h denote the maximum diameter of elements. We assume that the meshes match at the interface Γ_{13} match, but they may not match at the interfaces Γ_{12} and Γ_{23}. Given a fixed normal vector $\boldsymbol{n}_e$, on each interior edge, pointing from E_e^1 to E_e^2, the average $\{w\}$ and jump $[w]$ of function w is defined:

$$\{w\}=\frac{1}{2}(w|_{E_e^1})+\frac{1}{2}(w|_{E_e^2}),\quad [w]=(w|_{E_e^1})-(w|_{E_e^2}),\quad \forall e=\partial E_e^1\cap\partial E_e^2.$$

On a boundary edge, the jump and average coincide with the trace of the function. For any integer $k \geq 0$, the Sobolev space on a domain $\mathcal{O}$ is denoted by $H^k(\mathcal{O}) = \{v \in L^2(\mathcal{O}) : D^m v \in L^2(\mathcal{O}), \quad \forall |m| \leq k\}$, with norm $\|\cdot\|_{k,\mathcal{O}}$. We denote by $L^2_0(\mathcal{O})$ the space of square-integrable functions with zero average, with L^2 inner-product $(\cdot,\cdot)_{\mathcal{O}}$.

Let k_1, k_2 and k_3 be positive integers. The DG discrete spaces are

$$\begin{aligned}
\boldsymbol{X}_h^1 &= \{\boldsymbol{v}_1 \in (L^2(\Omega_1))^2 : \quad \forall E \in \mathcal{E}_h^1, \quad \boldsymbol{v}_1 \in (I\!P_{k_1}(E))^2\},\\
M_h^1 &= \{q_1 \in L^2(\Omega_2) : \quad \forall E \in \mathcal{E}_h^1, \quad q_1 \in I\!P_{k_1-1}(E)\},\\
M_h^2 &= \{q_2 \in L^2(\Omega_2) : \quad \forall E \in \mathcal{E}_h^2, \quad q_2 \in I\!P_{k_2}(E)\}.
\end{aligned}$$

The MFE discrete spaces are the classical ones, such as the Raviart-Thomas spaces [6]. We assume that the mixed velocity spaces $\boldsymbol{X}_h^3 \subset \boldsymbol{H}(div;\Omega_3)$ contains polynomials of degree k_3 and the pressure spaces $M_h^3 \subset L^2(\Omega_3)$ polynomials of degree $k_3 - 1$. We associate to these spaces the following norms:

$$\|\boldsymbol{v}_1\|^2_{X_h^1} = \sum_{E\in\mathcal{E}_h^1} \|\nabla \boldsymbol{v}_1\|^2_{0,E} + \sum_{e\in\Gamma_h^1\cup\Gamma_1} \frac{\sigma_{1,e}}{|e|}\|[\boldsymbol{v}_1]\|^2_{0,e} + \frac{1}{G}\sum_{e\in\Gamma_{12}} \|\boldsymbol{v}_1\cdot\boldsymbol{\tau}_{12}\|^2_{0,e},$$

$$\|q_2\|^2_{M_h^2} = \sum_{E\in\mathcal{E}_h^2} \|\nabla q_2\|^2_{0,E} + \sum_{e\in\Gamma_h^2} \frac{\sigma_{2,e}}{|e|}\|[q_2]\|^2_{0,e},$$

$$\|\boldsymbol{v}_3\|^2_{X_h^3} = \|\boldsymbol{v}_3\|^2_{0,\Omega_3}, \quad \|q_1\|^2_{M_h^1} = \|q_1\|^2_{0,\Omega_1}, \quad \|q_3\|^2_{M_h^3} = \|q_3\|^2_{0,\Omega_3}.$$

Here, $|e|$ denotes the measure of each edge e, the parameters $\sigma_{1,e}$ and $\sigma_{2,e}$ are positive constants defined later. Throughout the paper, C denotes a generic positive constant whose is independent of the mesh size h. We now define the global finite element spaces.

$$\boldsymbol{X}_h = \{\boldsymbol{v} = (\boldsymbol{v}_1, \boldsymbol{v}_3) : \boldsymbol{v}_i \in \boldsymbol{X}_h^i, \ \forall \eta \in \boldsymbol{X}_h^3\cdot\boldsymbol{n}_{13} \ \sum_{e\in\Gamma_{13}} \int_e \eta(\boldsymbol{v}_1 - \boldsymbol{v}_3)\cdot\boldsymbol{n}_{13} = 0\}, \tag{9}$$

$$M_h = \{q = (q_1, q_2, q_3) : q_i \in M_h^i, \quad q \in L^2_0(\Omega)\}. \tag{10}$$

We recall a result proved in [4] that generalize a Sobolev imbedding. There exists a constant C independent of h such that

$$\forall \boldsymbol{v}_1 \in \boldsymbol{X}_h^1, \quad \forall q_2 \in M_h^2, \quad \|\boldsymbol{v}_1\|_{0,\Omega_1} \leq C\|\boldsymbol{v}_1\|_{X_h^1}, \quad \|q_2\|_{0,\Omega_2} \leq C\|q_2\|_{M_h^2}. \tag{11}$$

We introduce the following bilinear forms $a_1 : \boldsymbol{X}_h^1 \times \boldsymbol{X}_h^1 \to I\!R$, $b_1 : \boldsymbol{X}_h^1 \times M_h^1 \to I\!R$, $a_2 : M_h^2 \times M_h^2 \to I\!R$, $a_3 : \boldsymbol{X}_h^3 \times \boldsymbol{X}_h^3 \to I\!R$ and $b_3 : \boldsymbol{X}_h^3 \times M_h^3 \to I\!R$:

$$a_1(\boldsymbol{u}_1,\boldsymbol{v}_1) = 2\mu \sum_{E\in\mathcal{E}_h^1}\int_E \boldsymbol{D}(\boldsymbol{u}_1):\boldsymbol{D}(\boldsymbol{v}_1) + \mu \sum_{e\in\Gamma_h^1\cup\Gamma_1} \frac{\sigma_{1,e}}{|e|}\int_e [\boldsymbol{u}_1]\cdot[\boldsymbol{v}_1]$$

$$-2\mu \sum_{e\in\Gamma_h^1\cup\Gamma_1}\int_e \{\nabla \boldsymbol{u}_1\boldsymbol{n}_e\}\cdot[\boldsymbol{v}_1] + 2\mu\epsilon_1 \sum_{e\in\Gamma_h^1\cup\Gamma_1}\int_e \{\nabla \boldsymbol{v}_1\boldsymbol{n}_e\}\cdot[\boldsymbol{u}_1]$$

$$+\frac{\mu}{G}\sum_{e\in\Gamma_{12}}\int_e (\boldsymbol{u}_1\cdot\boldsymbol{\tau}_{12})(\boldsymbol{v}_1\cdot\boldsymbol{\tau}_{12}) + \frac{\mu}{G}\sum_{e\in\Gamma_{13}}\int_e (\boldsymbol{u}_1\cdot\boldsymbol{\tau}_{13})(\boldsymbol{v}_1\cdot\boldsymbol{\tau}_{13}),$$

$$b_1(\boldsymbol{v}_1,p_1) = -\sum_{E\in\mathcal{E}_h}\int_E p_1\nabla\cdot\boldsymbol{v}_1 + \sum_{e\in\Gamma_h^1\cup\Gamma_1}\int_e \{p_1\}[\boldsymbol{v}_1]\cdot\boldsymbol{n}_e,$$

$$a_2(p_2,q_2) = \sum_{E\in\mathcal{E}_h^2}\int_E \boldsymbol{K}\nabla p_2\cdot\nabla q_2 + \sum_{e\in\Gamma_h^2}\frac{\sigma_{2,e}}{|e|}\int_e [p_2][q_2]$$

$$-\sum_{e\in\Gamma_h^2}\int_e \{\boldsymbol{K}\nabla p_2\cdot\boldsymbol{n}_e\}[q_2] + \epsilon_2\sum_{e\in\Gamma_h^2}\int_e \{\boldsymbol{K}\nabla q_2\cdot\boldsymbol{n}_e\}[p_2].$$

$$a_3(\boldsymbol{u}_3,\boldsymbol{v}_3) = \int_{\Omega_3}\boldsymbol{K}^{-1}\boldsymbol{u}_3\cdot\boldsymbol{v}_3, \quad b_3(\boldsymbol{v}_2,q_2) = \int_{\Omega_3} q_3\nabla\cdot\boldsymbol{v}_3.$$

By introducing the parameters ϵ_1, ϵ_2 that take the value 1 or -1, we allow for non-symmetric or symmetric bilinear forms a_1 and a_2. We assume that in the non-symmetric case ($\epsilon_1 = 1$), the parameter $\sigma_{1,e}$ is equal to 1 and in the symmetric case ($\epsilon_1 = -1$), the parameter $\sigma_{1,e}$ is bounded below by a sufficiently large positive value. The same assumptions hold true for ϵ_2 and $\sigma_{2,e}$. Combining the subdomains bilinear forms, we define

$$A(\boldsymbol{u},p;\boldsymbol{v},q) = a_1(\boldsymbol{u}_1,\boldsymbol{v}_1) + a_2(p_2,q_2) + a_3(\boldsymbol{u}_3,\boldsymbol{v}_3), \ \forall \boldsymbol{u},\boldsymbol{v}\in\boldsymbol{X}_h, \ \forall p,q\in M_h, \tag{12}$$

$$B(\boldsymbol{v},q) = b_1(\boldsymbol{v}_1,q_1) - b_3(\boldsymbol{v}_3,q_3), \quad \forall \boldsymbol{v}\in\boldsymbol{X}_h, \quad \forall q\in M_h. \tag{13}$$

It is easy to show that there is a coercivity constant $\kappa > 0$ such that

$$A(\boldsymbol{v},q;\boldsymbol{v},q) \geq \kappa(\|\boldsymbol{v}_1\|^2_{X_h^1} + \|\boldsymbol{v}_3\|^2_{X_h^3} + \|q_2\|^2_{M_h^2}), \quad \forall \boldsymbol{v}\in\boldsymbol{X}_h, \quad \forall q\in M_h. \tag{14}$$

Finally, we define a bilinear form $\Lambda : (\boldsymbol{X}_h\times M_h)\times(\boldsymbol{X}_h\times M_h)\to \mathbb{R}$ acting on the interfaces Γ_{12} and Γ_{23}.

$$\Lambda(\boldsymbol{u},p;\boldsymbol{v},q) = \int_{\Gamma_{12}}(p_2\boldsymbol{v}_1\cdot\boldsymbol{n}_{12} - q_2\boldsymbol{u}_1\cdot\boldsymbol{n}_{12}) + \int_{\Gamma_{23}}(q_2\boldsymbol{u}_3\cdot\boldsymbol{n}_{23} - p_2\boldsymbol{v}_3\cdot\boldsymbol{n}_{23}).$$

With these forms, the numerical method is: find $(\boldsymbol{U},P)\in\boldsymbol{X}_h\times M_h$ such that

$$\forall(\boldsymbol{v},q)\in\boldsymbol{X}_h\times M_h, \ A(\boldsymbol{U},P;\boldsymbol{v},q) + B(\boldsymbol{v},P) + \Lambda(\boldsymbol{U},P;\boldsymbol{v},q) = \tag{15}$$
$$= (\boldsymbol{f}_1,\boldsymbol{v}_1)_{\Omega_1} + (f_2,q_2)_{\Omega_2},$$

$$\forall q\in M_h, \quad B(\boldsymbol{U},q) = -(f_3,q_3)_{\Omega_3}. \tag{16}$$

Lemma 1. *If* $(\boldsymbol{u}, p)$ *is the strong solution of the coupled Stokes-Darcy flow problem* (1)-(8), *then* $(\boldsymbol{u}, p)$ *satisfy the variational equations: for all* $(\boldsymbol{v}, q) \in \boldsymbol{X}_h \times M_h$

$$A(\boldsymbol{u}, p; \boldsymbol{v}, q) + B(\boldsymbol{v}, p) + \Lambda(\boldsymbol{u}, p; \boldsymbol{v}, q) = \tag{17}$$

$$= (\boldsymbol{f}_1, \boldsymbol{v}_1)_{\Omega_1} + (f_2, q_2)_{\Omega_2} - \int_{\Gamma_{13}} p_3(\boldsymbol{v}_1 - \boldsymbol{v}_3) \cdot \boldsymbol{n}_{13},$$

$$B(\boldsymbol{u}, q) = -(f_3, q_3)_{\Omega_3}. \tag{18}$$

Proof. Multiplying the Stokes equation (1) by $\boldsymbol{v}_1 \in \boldsymbol{X}_h^1$, integrating by parts over one element E, summing over all elements in $\mathcal{E}_h^1$ and using the regularity of the strong solution yields:

$$\sum_E \int_E (2\mu \boldsymbol{D}(\boldsymbol{u}_1) : \boldsymbol{D}(\boldsymbol{v}_1) - p_1 \nabla \cdot \boldsymbol{v}_1)$$

$$- \sum_{e \in \Gamma_h^1} \int_e \{-p_1 \boldsymbol{I} + 2\mu \boldsymbol{D}(\boldsymbol{u}_1)\} \boldsymbol{n}_e \cdot [\boldsymbol{v}_1] + \epsilon \sum_{e \in \Gamma_h^1} \int_e \{2\mu \boldsymbol{D}(\boldsymbol{v}_1)\} \boldsymbol{n}_e \cdot [\boldsymbol{u}_1]$$

$$- \sum_{e \in \Gamma_{12} \cup \Gamma_{13}} \int_e (-p_1 \boldsymbol{I} + 2\mu \boldsymbol{D}(\boldsymbol{u}_1)) \boldsymbol{n}_{12} \cdot \boldsymbol{v}_1$$

$$- \sum_{e \in \Gamma_1} \int_e (-p_1 \boldsymbol{I} + 2\mu \boldsymbol{D}(\boldsymbol{u}_1)) \boldsymbol{n} \cdot \boldsymbol{v}_1 + \epsilon \sum_{e \in \Gamma_1} \int_e 2\mu \boldsymbol{D}(\boldsymbol{v}_1) \boldsymbol{n} \cdot \boldsymbol{u}_1 = \int_{\Omega_1} \boldsymbol{f}_1 \cdot \boldsymbol{v}_1.$$

Using the interface conditions (7), (8), we obtain

$$a_1(\boldsymbol{u}_1, \boldsymbol{v}_1) + b_1(\boldsymbol{v}_1, p_1) + \sum_{e \in \Gamma_{12}} \int_e p_2 \boldsymbol{v}_1 \cdot \boldsymbol{n}_{12} + \sum_{e \in \Gamma_{13}} \int_e p_3 \boldsymbol{v}_1 \cdot \boldsymbol{n}_{13} = (\boldsymbol{f}_1, \boldsymbol{v}_1)_{\Omega_1}.$$

Similarly, multiplying (3) by a test function $q_2 \in M_h^2$, integrating by parts on one element, summing over all elements in $\mathcal{E}_h^2$ and using the boundary condition yields:

$$a_2(p_2, q_2) - \int_{\Gamma_{23}} \boldsymbol{K} \nabla p_2 \cdot \boldsymbol{n}_{23} q_2 + \int_{\Gamma_{12}} \boldsymbol{K} \nabla p_2 \cdot \boldsymbol{n}_{12} q_2 = (f_2, q_2)_{\Omega_2}.$$

With the conditions (4) and (6), we obtain

$$a_2(p_2, q_2) - \sum_{e \in \Gamma_{12}} \int_e \boldsymbol{u}_1 \cdot \boldsymbol{n}_{12} q_2 + \sum_{e \in \Gamma_{23}} \int_e \boldsymbol{u}_3 \cdot \boldsymbol{n}_{23} q_2 = (f_2, q_2)_{\Omega_2}.$$

Rewriting (3) as $\boldsymbol{K}^{-1} \boldsymbol{u}_3 = -\nabla p_3$, we easily obtain by multiplying by $\boldsymbol{v}_3 \in \boldsymbol{X}_h^3$, integrating by parts and using (6):

$$a_3(\boldsymbol{u}_3, \boldsymbol{v}_3) - b_3(\boldsymbol{v}_3, p_3) - \sum_{e \in \Gamma_{13}} \int_e p_3 \boldsymbol{v}_3 \cdot \boldsymbol{n}_{13} - \sum_{e \in \Gamma_{23}} \int_e p_2 \boldsymbol{v}_3 \cdot \boldsymbol{n}_{23} = 0.$$

Finally, the regularity of $\boldsymbol{u}$ and a simple integration of (3) give:

$$b_1(\boldsymbol{u}_1, q_1) = 0, \quad b_3(\boldsymbol{u}_3, q_3) = (f_3, q_3)_{\Omega_3}.$$

The final result is obtained by adding the previous variational equations.

We now recall some approximation properties satisfied by the spaces $\boldsymbol{X}_h$ and M_h [4, 8]. Given $\boldsymbol{v} \in (H^1(\Omega_1 \cup \Omega_3))^2$, there exists $\tilde{\boldsymbol{v}} \in \boldsymbol{X}_h$ such that

$$\forall q \in M_h, \quad B(\boldsymbol{v} - \tilde{\boldsymbol{v}}, q) = 0, \tag{19}$$

$$\forall e \in \Gamma_h^1 \cup \Gamma_1, \quad \forall \boldsymbol{q} \in (\mathbb{P}_{k_1-1}(e))^2, \quad \int_e [\tilde{\boldsymbol{v}}_1] \cdot \boldsymbol{q} = 0, \tag{20}$$

$$\forall e \in \Gamma_{12} \cup \Gamma_{13}, \quad \forall \boldsymbol{q} \in (\mathbb{P}_{k_1-1}(e))^2, \quad \int_e (\boldsymbol{v}_1 - \tilde{\boldsymbol{v}}_1) \cdot \boldsymbol{q} = 0, \tag{21}$$

$$\forall e \in \Gamma_{23}, \quad \forall \eta \in \boldsymbol{X}_h^3 \cdot \boldsymbol{n}_{23}, \quad \int_e (\boldsymbol{v}_3 - \tilde{\boldsymbol{v}}_3) \cdot \boldsymbol{n}_{23} \eta = 0. \tag{22}$$

If in addition, $\boldsymbol{v}_1 \in (H^{k_1+1}(\Omega_1))^2$, $\boldsymbol{v}_3 \in (H^{k_3+1}(\Omega_3))^2$, the approximation $\tilde{\boldsymbol{v}}$ satisfies

$$\begin{aligned} \|\boldsymbol{v} - \tilde{\boldsymbol{v}}\|_{0,\Omega_1} \le C h^{k_1+1} |\boldsymbol{v}|_{k_1+1,\Omega_1}, \quad \|\boldsymbol{v} - \tilde{\boldsymbol{v}}\|_{X_h^1} \le C h^{k_1} |\boldsymbol{v}|_{k_1+1,\Omega_1}, \\ \|\boldsymbol{v} - \tilde{\boldsymbol{v}}\|_{X_h^3} \le C h^{k_3+1} |\boldsymbol{v}|_{k_3+1,\Omega_3}. \end{aligned} \tag{23}$$

Define also the L^2 projection $\tilde{p}$ of the pressure p. If $p_i \in H^{s_i}(\Omega_i)$, we have

$$\forall q \in M_h^i, \quad \forall E \in \mathcal{E}_h^i, \quad i = 1, 2, 3, \quad \int_E q(p - \tilde{p}) = 0, \tag{24}$$

$$m = 0, 1, \quad i = 1, 2, 3, \quad \|p_i - \tilde{p}_i\|_{m,\Omega_i} \le C h^{s_i - m} |p_i|_{s_i,\Omega_i}. \tag{25}$$

Lemma 2. *The discrete solution to* (15), (16) *exists and is unique.*

Proof. In a finite-dimensional setting, it suffices to show that the solution is unique. Set $f_i = 0$ and $(\boldsymbol{v}, q) = (\boldsymbol{U}, P)$ in (15), (16). This yields $A(\boldsymbol{U}, P; \boldsymbol{U}, P) + \Lambda(\boldsymbol{U}, P; \boldsymbol{U}, P) = 0$. Since $\Lambda(\boldsymbol{v}, q; \boldsymbol{v}, q) = 0$, we are left with $a_1(\boldsymbol{U}_1, \boldsymbol{U}_1) + a_2(P_2, P_2) + a_3(\boldsymbol{U}_3, \boldsymbol{U}_3) = 0$. This clearly implies that $\boldsymbol{U}_1 = \boldsymbol{U}_3 = 0$, and that P_2 is a global constant over Ω_2.

We now define

$$\bar{P} = \frac{1}{|\Omega_1| + |\Omega_3|} \left(\int_{\Omega_1} P + \int_{\Omega_3} P \right),$$

and consider $\hat{P} \in L^2(\Omega_1 \cup \Omega_3)$ such that $P = \hat{P} + \bar{P}$. Then the equation (15) becomes

$$\forall \boldsymbol{v} \in \boldsymbol{X}_h, \quad B(\boldsymbol{v}, \hat{P}) + (P_2 - \bar{P}) \left(\int_{\Gamma_{12}} \boldsymbol{v}_1 \cdot \boldsymbol{n}_{12} - \int_{\Gamma_{23}} \boldsymbol{v}_3 \cdot \boldsymbol{n}_{23} \right) = 0. \tag{26}$$

Since $\hat{P} \in L^2_0(\Omega_1 \cup \Omega_3)$, there is a function $\hat{\boldsymbol{z}} \in (H^1_0(\Omega_1 \cup \Omega_3))^2$ such that $-\nabla \cdot \hat{\boldsymbol{z}} = \hat{P}$. Denote by $\tilde{\boldsymbol{z}}$ the approximation of $\hat{\boldsymbol{z}}$ satisfying (19)-(22), and choose $\boldsymbol{v} = \tilde{\boldsymbol{z}}$ in (26). With property (19) and the regularity of $\hat{\boldsymbol{z}}$, we obtain

$$\|\hat{P}\|^2_{0,\Omega_1\cup\Omega_3} = B(\hat{\boldsymbol{z}}, \hat{P}) = B(\tilde{\boldsymbol{z}}, \hat{P}) = 0.$$

The equation (26), with $\hat{P} = 0$ becomes:

$$(P_2 - \bar{P})(\int_{\Gamma_{12}} \boldsymbol{v}_1 \cdot \boldsymbol{n}_{12} - \int_{\Gamma_{23}} \boldsymbol{v}_3 \cdot \boldsymbol{n}_{23}) = 0, \quad \forall \boldsymbol{v} \in \boldsymbol{X}_h.$$

Since P belongs to $L^2_0(\Omega)$, the constants P_2 and $\bar{P}$ are related by

$$|\Omega_2| P_2 + (|\Omega_1| + |\Omega_3|)\bar{P} = 0.$$

These two equations imply that $P_2 = \bar{P} = 0$, which concludes the proof.

We now finish this section with some trace and inverse inequalities needed for the analysis. Let E be a mesh element with diameter h_E. Let k be a positive integer. Then, there exists a constant C independent of h_E such that

$$\forall \phi \in I\!P_k(E), \quad \forall e \subset \partial E, \quad \|\phi\|_{0,e} \leq C h_E^{-1/2} \|\phi\|_{0,E}. \tag{27}$$

$$\forall \phi \in I\!P_k(E), \quad \forall e \subset \partial E, \quad \|\nabla\phi \cdot \boldsymbol{n}_e\|_{0,e} \leq C h_E^{-1/2} \|\nabla\phi\|_{0,E}. \tag{28}$$

3 Error estimates

Theorem 1. *Let $(\boldsymbol{u}, p)$ be the solution of the coupled problem (1)-(8) such that $\boldsymbol{u}|_{\Omega_i} \in (H^{k_i+1}(\Omega_i))^2$ for $i = 1, 3$, $p|_{\Omega_i} \in H^{k_i}(\Omega_i)$ for $i = 1, 3$, and $p|_{\Omega_2} \in H^{k_2+1}(\Omega_2)$. Then the discrete solution $(\boldsymbol{U}, P)$ of (15), (16) satisfies the following estimate*

$$\|\boldsymbol{U}_1 - \boldsymbol{u}_1\|_{X_h^1} + \|P_2 - p_2\|_{M_h^2} + \|\boldsymbol{U}_3 - \boldsymbol{u}_3\|_{X_h^3} \leq C h^{k_1}(|\boldsymbol{u}|_{k_1+1,\Omega_1} + |p|_{k_1,\Omega_1})$$
$$+ C h^{k_2} |p|_{k_2+1,\Omega_2} + C h^{k_3}(|\boldsymbol{u}|_{k_3+1,\Omega_3} + |p|_{k_3,\Omega_3}).$$

Proof. Define $\tilde{\boldsymbol{u}}$ the approximation of $\boldsymbol{u}$ satisfying the properties (19)-(23) and $\tilde{p}$ the approximation of p satisfying (24)-(25) Define $\boldsymbol{\chi} = \boldsymbol{U} - \tilde{\boldsymbol{u}}$ and $\xi = P - \tilde{p}$ and $\boldsymbol{\chi}_i, \xi_i$ their restrictions to the subdomains Ω_i. Subtracting (17) and (18) from (15) and (16), choosing $\boldsymbol{v}_1 = \boldsymbol{\chi}_1, q_2 = \xi_2$ and $\boldsymbol{v}_3 = \boldsymbol{\chi}_3$ and using the coercivity result (14), we obtain

$$\kappa(\|\boldsymbol{\chi}_1\|^2_{X_h^1} + \|\xi^2\|^2_{M_h^2} + \|\boldsymbol{\chi}_3\|^2_{X_h^3}) \leq a_1(\boldsymbol{u}_1 - \tilde{\boldsymbol{u}}_1, \boldsymbol{\chi}_1) + a_2(p_2 - \tilde{p}_2, \xi_2)$$
$$+ a_3(\boldsymbol{u}_3 - \tilde{\boldsymbol{u}}_3, \boldsymbol{\chi}_3) + b_1(\boldsymbol{\chi}_1, p_1 - \tilde{p}_1) - b_3(\boldsymbol{\chi}_3, p_3 - \tilde{p}_3) - b_1(\boldsymbol{u}_1 - \tilde{\boldsymbol{u}}_1, \xi_1)$$
$$+ b_3(\boldsymbol{u}_3 - \tilde{\boldsymbol{u}}_3, \xi_3) + \int_{\Gamma_{12}} (p_2 - \tilde{p}_2)\boldsymbol{\chi}_1 \cdot \boldsymbol{n}_{12} - \int_{\Gamma_{12}} \xi_2(\boldsymbol{u}_1 - \tilde{\boldsymbol{u}}_1) \cdot \boldsymbol{n}_{12}$$
$$+ \int_{\Gamma_{23}} \xi_2(\boldsymbol{u}_3 - \tilde{\boldsymbol{u}}_3) \cdot \boldsymbol{n}_{23} - (p_2 - \tilde{p}_2)\boldsymbol{\chi}_3 \cdot \boldsymbol{n}_{23} + \int_{\Gamma_{13}} p_3(\boldsymbol{\chi}_1 - \boldsymbol{\chi}_3) \cdot \boldsymbol{n}_{13}.$$

The pressure term $b_3(\chi_3, p_3 - \tilde{p}_3)$ vanishes because of (24). The terms $b_1(\boldsymbol{u}_1 - \tilde{\boldsymbol{u}}_1, \xi_1)$ and $b_3(\boldsymbol{u}_3 - \tilde{\boldsymbol{u}}_3, \xi_3)$ vanish because of property (19) of the approximation $\tilde{\boldsymbol{u}}$. We now bound the remaining terms. By using Cauchy-Schwarz, trace inequalities, and the approximation results (23) and (25), we can bound the first three terms

$$a_1(\boldsymbol{u}_1 - \tilde{\boldsymbol{u}}_1, \chi_1) + a_2(p_2 - \tilde{p}_2, \xi_2) + a_3(\boldsymbol{u}_3 - \tilde{\boldsymbol{u}}_3, \chi_3) \leq \frac{1}{8}\|\chi_1\|^2_{X_h^1}$$
$$+Ch^{2k_1}|\boldsymbol{u}|^2_{k_1+1,\Omega_1} + \frac{1}{8}\|\xi_2\|^2_{M_h^2} + Ch^{2k_2}|p|^2_{k_2+1,\Omega_2}$$
$$+\frac{1}{8}\|\chi_3\|^2_{X_h^3} + Ch^{2k_3+2}|\boldsymbol{u}|^2_{k_3+1,\Omega_3}.$$

Because of the L^2 projection, the first pressure term is reduced and bounded as follows:

$$b_1(\chi_1, p_1 - \tilde{p}_1) = \sum_{e\in\Gamma_h^1\cup\Gamma_1} \int_e \{p_1 - \tilde{p}_1\}[\chi_1]\cdot \boldsymbol{n}_e$$
$$\leq \frac{1}{8} \sum_{e\in\Gamma_h^1\cup\Gamma_1} \frac{\sigma_e}{|e|}\|[\chi_1]\|^2_{0,e} + Ch^{2k_1}|p|^2_{k_1,\Omega_1}.$$

The remaining terms are the interface terms. Using the approximation result (25), the trace inequality (27) and Cauchy-Schwarz's inequality, we obtain

$$\int_{\Gamma_{12}} (p_2 - \tilde{p}_2)\chi_1 \cdot \boldsymbol{n}_{12} \leq C\|\chi_1\|_{0,\Omega_1} h^{k_2}|p|_{k_2+1,\Omega_2} \leq \frac{1}{8}\|\chi_1\|^2_{X_h^1} + Ch^{2k_2}|p|^2_{k_2+1,\Omega_2}.$$

Similarly,

$$\int_{\Gamma_{23}} (p_2 - \tilde{p}_2)\chi_3 \cdot \boldsymbol{n}_{23} \leq \frac{1}{8}\|\chi_3\|^2_{M_h^3} + Ch^{2k_2}|p|^2_{k_2+1,\Omega_2}.$$

With the bound (11) and the approximation result (23), we have

$$\int_{\Gamma_{12}} \xi_2(\boldsymbol{u}_1 - \tilde{\boldsymbol{u}}_1)\cdot \boldsymbol{n}_{12} \leq \frac{1}{8}\|\xi_2\|^2_{M_h^2} + Ch^{2k_1}|\boldsymbol{u}|^2_{k_1+1,\Omega_1}.$$

The approximation result (23), the inverse inequality (27) and the bound (11) give

$$\int_{\Gamma_{23}} \xi_2(\boldsymbol{u}_3 - \tilde{\boldsymbol{u}}_3)\cdot \boldsymbol{n}_{23}$$
$$\leq h^{-1/2}\|\xi_2\|_{0,\Omega_2} h^{k_3-1/2}|\boldsymbol{u}_3|_{k_3+1,\Omega_3} \leq \frac{1}{8}\|\xi_2\|^2_{M_h^2} + Ch^{2k_3}|\boldsymbol{u}_3|_{k_3+1,\Omega_3}.$$

Define $\tilde{p}_3^e$ the L^2 projection of p_3 with respect to the L^2 inner product on the edge e. Since χ belongs to $\boldsymbol{X}_h$ and by definition of the projection, we have

$$\int_{\Gamma_{13}} p_3(\boldsymbol{\chi}_1 - \boldsymbol{\chi}_3)\cdot \boldsymbol{n}_{13}$$
$$= \sum_{e\in\Gamma_{13}} \int_e (p_3 - \tilde{p}_3^e)(\boldsymbol{\chi}_1 - \boldsymbol{\chi}_3)\cdot \boldsymbol{n}_{13} = \sum_{e\in\Gamma_{13}} \int_e (p_3 - \tilde{p}_3^e)\boldsymbol{\chi}_1 \cdot \boldsymbol{n}_{13}.$$

Assume that each edge e of Γ_{13} is shared by the elements $E_e^1 \in \mathcal{E}_h^1$ and $E_e^3 \in \mathcal{E}_h^3$. Define the constant $c_e = |e|^{-1}\int_e \boldsymbol{\chi}_3 \cdot \boldsymbol{n}_{13}$.

$$\sum_{e\in\Gamma_{13}} \int_e (p_3 - \tilde{p}_3^e)\boldsymbol{\chi}_1 \cdot \boldsymbol{n}_{13} = \sum_{e\in\Gamma_{13}} \int_e (p_3 - \tilde{p}_3^e)(\boldsymbol{\chi}_1 \cdot \boldsymbol{n}_{13} - c_e)$$
$$\leq Ch^{k_3}\sum_{e\in\Gamma_{13}} |p_3|_{k_3,E_e^3}\|\nabla\boldsymbol{\chi}_1\|_{0,E_e^1} \leq \frac{1}{8}\|\!|\nabla\boldsymbol{\chi}\|\!|_{0,\Omega_1}^2 + Ch^{2k_3}|p_3|_{k_3,\Omega_3}^2.$$

The theorem is obtained by combining all bounds.

Theorem 2. *Under the assumptions of Theorem 1 and the additional assumption that for any element $E \in \mathcal{E}_h^3$, the edge $\partial E \cap \Gamma_{23}$ belongs to only one element of $\mathcal{E}_h^2$, the approximation P satisfies the error estimate*

$$\|P - p\|_{0,\Omega} \leq Ch^{k_1}(|\boldsymbol{u}|_{k_1+1,\Omega_1} + |p|_{k_1,\Omega_1})$$
$$+Ch^{k_2}|p|_{k_2+1,\Omega_2} + Ch^{k_3}(|\boldsymbol{u}|_{k_3+1,\Omega_3} + |p|_{k_3,\Omega_3}).$$

Proof. Subtracting (17) from (15), the error equation is

$$\forall \boldsymbol{v} \in \boldsymbol{X}_h, \quad B(\boldsymbol{v},\xi) = a_1(\boldsymbol{u}_1 - \boldsymbol{U}_1, \boldsymbol{v}_1) + a_3(\boldsymbol{u}_3 - \boldsymbol{U}_3, \boldsymbol{v}_3)$$
$$+\int_{\Gamma_{12}} (p_2 - P_2)\boldsymbol{v}_1 \cdot \boldsymbol{n}_{12} - \int_{\Gamma_{23}} (p_2 - P_2)\boldsymbol{v}_3 \cdot \boldsymbol{n}_{23} + \int_{\Gamma_{12}} p_3(\boldsymbol{v}_1 - \boldsymbol{v}_3)\cdot \boldsymbol{n}_{13}. \tag{29}$$

Define $\bar{\xi} = (|\Omega_1| + |\Omega_3|)^{-1}\int_{\Omega_1\cup\Omega_3} \xi$ and the function $\hat{\xi} \in L_0^2(\Omega_1 \cup \Omega_3)$ by $\hat{\xi} = \xi - \bar{\xi}$.

$$B(\boldsymbol{v},\xi) = B(\boldsymbol{v},\hat{\xi}) + B(\boldsymbol{v},\bar{\xi}) = B(\boldsymbol{v},\hat{\xi}) - \bar{\xi}(\int_{\Gamma_{12}} \boldsymbol{v}_1 \cdot \boldsymbol{n}_{12} - \int_{\Gamma_{23}} \boldsymbol{v}_3 \cdot \boldsymbol{n}_{23}).$$

Let $\hat{\boldsymbol{v}} \in (H_0^1(\Omega_1 \cup \Omega_3))^2$ such that $-\nabla\cdot\hat{\boldsymbol{v}} = \hat{\xi}$ and $\|\hat{\boldsymbol{v}}\|_{X_h^1} + \|\hat{\boldsymbol{v}}\|_{X_h^3} \leq C\|\hat{\xi}\|_{0,\Omega_1\cup\Omega_3}$. Choose in (29) $\boldsymbol{v} = \tilde{\boldsymbol{v}}^h$ the approximation of $\hat{\boldsymbol{v}}$, defined by (19)-(23):

$$\|\hat{\xi}\|_{0,\Omega_1\cup\Omega_3}^2 = a_1(\boldsymbol{u}_1 - \boldsymbol{U}_1, \tilde{\boldsymbol{v}}_1^h) + a_3(\boldsymbol{u}_3 - \boldsymbol{U}_3, \tilde{\boldsymbol{v}}_3^h)$$
$$+\int_{\Gamma_{12}} (p_2 - P_2)\tilde{\boldsymbol{v}}_1^h \cdot \boldsymbol{n}_{12} - \int_{\Gamma_{23}} (p_2 - P_2)\tilde{\boldsymbol{v}}_3^h \cdot \boldsymbol{n}_{23} + \int_{\Gamma_{12}} p_3(\tilde{\boldsymbol{v}}_1^h - \tilde{\boldsymbol{v}}_3^h)\cdot \boldsymbol{n}_{13}.$$

The additional assumption for the meshes at Γ_{23} is needed to bound $\int_{\Gamma_{23}}(p_2 - P_2)\tilde{\boldsymbol{v}}_3^h \cdot \boldsymbol{n}_{23}$. Using the fact that $\int_e \tilde{\boldsymbol{v}}_3^h \cdot \boldsymbol{n}_{23} = 0$ for $e = \partial E \cap \Gamma_{23}$ and $E \in \mathcal{E}_h^3$, and the inequality $\|\tilde{\boldsymbol{v}}^h\|_{X_h^1} + \|\tilde{\boldsymbol{v}}^h\|_{X_h^3} \leq C\|\hat{\xi}\|_{0,\Omega_1\cup\Omega_3}$, we obtain:

$$\|\hat{\xi}\|_{0,\Omega_1\cup\Omega_3} \leq C\|\boldsymbol{u}_1 - \boldsymbol{U}_1\|_{X_h^1} + C\|\boldsymbol{u}_3 - \boldsymbol{U}_3\|_{X_h^3} + C\|p_2 - P_2\|_{M_h^2} + Ch^{2k_3}|p_3|^2_{k_3,\Omega_3}.$$

We now bound $\|\bar{\xi}\|_{0,\Omega_2\cup\Omega_3} = \bar{\xi}(|\Omega_1| + |\Omega_3|)^{1/2}$. Since $P \in L_0^2(\Omega)$, we have

$$\bar{\xi} = \frac{-1}{|\Omega_1| + |\Omega_3|}\int_{\Omega_2}(P_2 - \tilde{p}_2) \leq \frac{|\Omega_2|}{|\Omega_1| + |\Omega_3|}\|P_2 - \tilde{p}_2\|_{0,\Omega_2}$$
$$\leq C\|P_2 - p_2\|_{X_h^2} + Ch^{k_2}|p_2|_{k_2+1,\Omega_2}.$$

Combining the results of Theorem 1 with the bounds above, give the optimal error bound.

As a concluding remark, one can introduce Lagrange multipliers on the interfaces, and obtain an equivalent method. This allows the decoupling of each subdomain, which may be advantageous for a parallel implementation.

References

1. G.S. Beavers and D.D. Joseph. Boundary conditions at a naturally impermeable wall. *J. Fluid. Mech*, 30:197–207, 1967.
2. M. Discacciati and A. Quarteroni. Analysis of a domain decomposition method for the coupling of stokes and darcy equations. In Brezzi et al., editor, *Numerical Analysis and Advanced Applications - Enumath 2001*, pages 3–20. Springer, Milan, 2003.
3. M. Discacciati and A. Quarteroni. Convergence analysis of a subdomain iterative method for the finite element approximation of the coupling of stokes and darcy equations. *Computing and Visualization in Science*, 2004. to appear.
4. V. Girault, B. Rivière, and M. Wheeler. A discontinuous Galerkin method with non-overlapping domain decomposition for the Stokes and Navier-Stokes problems. *Mathematics of Computation*, 2003, to appear.
5. W.J. Layton, F. Schieweck, and I. Yotov. Coupling fluid flow with porous media flow. *SIAM J. Numer. Anal.*, 40(6):2195–2218, 2003.
6. R. A. Raviart and J. M. Thomas. A mixed finite element method for 2nd order elliptic problems. In *Mathematical Aspects of the Finite Element Method, Lecture Notes in Mathematics*, volume 606, pages 292–315. Springer-Verlag, New York, 1977.
7. B. Rivière. Analysis of a discontinuous finite element method for the coupled Stokes and Darcy problems. Technical Report TR-MATH 03-11, University of Pittsburgh, 2003.
8. B. Rivière and I. Yotov. Locally conservative coupling of Stokes and Darcy flows. *SIAM J. Numer. Anal.* revised.
9. P. Saffman. On the boundary condition at the surface of a porous media. *Stud. Appl. Math.*, 50:292–315, 1971.

The Discontinuous Galerkin Method for Singularly Perturbed Problems

Hans-Görg Roos[1] and Helena Zarin[2]

[1] Institute of Numerical Mathematics, Dresden University of Technology, Germany *roos@math.tu-dresden.de*

[2] Department of Mathematics and Informatics, University of Novi Sad, Serbia and Montenegro *helena@im.ns.ac.yu*

Summary. A nonsymmetric discontinuous Galerkin method with interior penalties is considered for convection–diffusion problems with parabolic layers. On an anisotropic mesh with bilinear elements we prove error estimates (uniformly in the perturbation parameter) in an integral norm associated with this method. On different types of interelement edges we derive the values of discontinuity–penalization parameters. Numerical experiments support the theoretical results.

1 Introduction

Consider the following boundary value problem with homogeneous Dirichlet boundary conditions

$$\begin{cases} -\varepsilon\triangle u + b_1 u_x + c u = f & \text{in} \quad \Omega = (0,1)^2\,, \\ u = 0 & \text{on} \quad \Gamma = \partial\Omega\,. \end{cases} \tag{1}$$

Here $0 < \varepsilon \ll 1$ represents a perturbation parameter and b_1, c and f are real–valued functions defined on $\overline{\Omega}$. Assuming

$$b_1(x) \ge \beta_1 > 0\,, \qquad c(x) \ge 0\,, \qquad c(x) - \frac{1}{2}\frac{\partial b_1}{\partial x}(x) \ge \gamma > 0\,, \qquad x \in \overline{\Omega}\,, \tag{2}$$

the problem (1) belongs to the class of convection–diffusion problems whose solutions beside an exponential layer exhibit also parabolic layers.

In contrast to convection–diffusion problems with only exponential layers, not much is known concerning sharp estimates for derivatives of u. However, for the problem (1) with $b_1(x) = 1$ and $c(x) = 0$, in [10] the following result was proved using elliptic decompositions (in the corresponding result from [13] more compatibility is required):

Theorem 1. *If $b_1(x) = 1$, $c(x) = 0$ and $f \in C^{3,\alpha}(\overline{\Omega})$, for some $\alpha \in (0,1)$, satisfies the compatibility condition $f(0,0) = f(0,1) = f(1,0) = f(1,1) = 0$, then the solution u of the boundary value problem* (1) *can be decomposed as $u = S + E_1 + E_2 + E_3$, where for all $(x,y) \in \overline{\Omega}$ and $0 \le i+j \le 3$*

$$\left|\frac{\partial^{i+j} S}{\partial x^i \partial y^j}(x,y)\right| \le C\,, \qquad \left|\frac{\partial^{i+j} E_1}{\partial x^i \partial y^j}(x,y)\right| \le C\varepsilon^{-i} e^{-(1-x)/\varepsilon}\,,$$

$$\left|\frac{\partial^{i+j} E_2}{\partial x^i \partial y^j}(x,y)\right| \le C\varepsilon^{-j/2}\left(e^{-\tilde{\gamma}y/\sqrt{\varepsilon}} + e^{-\tilde{\gamma}(1-y)/\sqrt{\varepsilon}}\right),$$

$$\left|\frac{\partial^{i+j} E_3}{\partial x^i \partial y^j}(x,y)\right| \le C\varepsilon^{-(i+j/2)} e^{-(1-x)/\varepsilon}\left(e^{-\tilde{\gamma}y/\sqrt{\varepsilon}} + e^{-\tilde{\gamma}(1-y)/\sqrt{\varepsilon}}\right),$$

with a constant $\tilde{\gamma} > 0$.

A diverse nature of the vector field $\mathbf{b} = (b_1, 0)$ produces a complete change in the behaviour of the boundary layers. According to Theorem 1, beside the regular boundary layer of exponential type at the outflow $x = 1$, there are also parabolic layers placed near the sides $y = 0$ and $y = 1$. In the sequel we assume that the solution decomposition from the previous theorem exists also for the more general problem (1)–(2).

The only result known so far for finite element methods (FEMs) on layer–adapted meshes for the problem (1) is the estimate

$$\varepsilon^{1/2}|u - u_h|_{H^1(\Omega)} \le CN^{-1}\ln N\,,$$

from [10] for the Galerkin FEM with linear or bilinear elements on a Shishkin mesh (see also the survey paper [6]). Surprisingly, up to now there is no result in the literature for the streamline–diffusion finite element method (SDFEM). Furthermore, for the SDFEM the optimal choice of the stabilization parameter near parabolic layers is an open task, see discussion in [4] and [5]. This was also a reason for us to investigate some alternative discretization technique. Here, for numerical solving of (1)–(2) we use the h–version of a nonsymmetric discontinuous Galerkin finite element method with interior penalties (the NIPG method), [2], [7], [8], [9]. The technique from [2] is applied, but with a bilinear interpolant in error splitting instead of an L^2–projection onto a finite element space. This allows us to use the well–known interpolation error estimates for the problem (1)–(2). We finally show that on a specially chosen shape–irregular mesh (Shishkin mesh), this method yields error bound that is uniform in the perturbation parameter. Since our discretization involves the layer–adapted mesh whose construction directly uses information from the solution decomposition, the technique for proving ε–uniform error estimates from this paper cannot be applied on more general (nonrectangular) domains Ω, when such a decomposition is not available.

2 The nonsymmetric discontinuous Galerkin method

Following [2] and the notation therein, let $\mathcal{T}$ be a general partitioning of the domain $\Omega = (0,1)^2$ consisting of disjoint open axiparallel rectangles κ such that $\overline{\Omega} = \bigcup_{\kappa\in\mathcal{T}} \kappa$. In contrast to [2] we allow anisotropic (shape–irregular) meshes, but assume that there are no hanging nodes.

The broken Sobolev space of composite order $\mathbf{s} = \{s_\kappa : \kappa \in \mathcal{T}\}$ is defined with $H^{\mathbf{s}}(\Omega,\mathcal{T}) = \left\{v \in L^2(\Omega) : v|_\kappa \in H^{s_\kappa}(\kappa), \forall\kappa \in \mathcal{T}\right\}$. Let us assume that each $\kappa \in \mathcal{T}$ is an affine image of a fixed reference element $\hat{\kappa} = (-1,1)^2$, i.e. $\kappa = F_\kappa(\hat{\kappa})$. Then the finite element space is

$$S(\Omega,\mathcal{T},\mathbf{F}) = \left\{v \in L^2(\Omega) : v|_\kappa \circ F_\kappa \in \mathcal{Q}_1(\hat{\kappa})\right\} ,$$

where $\mathbf{F} = \{F_\kappa : \kappa \in \mathcal{T}\}$ and $\mathcal{Q}_1(\hat{\kappa})$ is the space of bilinear functions defined on $\hat{\kappa}$.

Let $\mathcal{E}$ be the set of all open one–dimensional element interfaces associated with $\mathcal{T}$, and $\mathcal{E}_{\text{int}} \subset \mathcal{E}$ be the set of all edges $e \in \mathcal{E}$ contained in Ω. Also, let $\Gamma_{\text{int}} = \{x \in \Omega : x \in e \text{ for some } e \in \mathcal{E}_{\text{int}}\}$. Then for each $e \in \mathcal{E}_{\text{int}}$ we define the jump and the mean value of a function $v \in H^1(\Omega,\mathcal{T})$ across e by $[v]_e = v|_{\partial\kappa\cap e} - v|_{\partial\kappa'\cap e}$ and $\langle v\rangle_e = (v|_{\partial\kappa\cap e} + v|_{\partial\kappa'\cap e})/2$, respectively. Here e is a common edge for elements κ and κ', and $\partial\kappa$ denotes the union of all open edges of κ. With each $e \in \mathcal{E}_{\text{int}}$ we associate the unit normal vector ν pointing from κ to κ'; if $e \subset \Gamma$, we take ν to be the unit outward normal vector μ. We also define the inflow and outflow parts of $\partial\kappa$ by $\partial_-\kappa = \{x \in \partial\kappa : \mathbf{b}(x)\cdot\mu_\kappa(x) < 0\}$, $\partial_+\kappa = \{x \in \partial\kappa : \mathbf{b}(x)\cdot\mu_\kappa(x) \geq 0\}$, respectively, where $\mu_\kappa(x)$ represents the unit outward normal vector to $\partial\kappa$ at the point $x \in \partial\kappa$.

For any element $\kappa \in \mathcal{T}$ and $v \in H^1(\kappa)$, we denote by v_κ^+ the interior trace of $v|_\kappa$ on $\partial\kappa$. In the case $\partial_-\kappa\backslash\Gamma \neq \emptyset$, for some $\kappa \in \mathcal{T}$, for each $x \in \partial_-\kappa\backslash\Gamma$ there exists a unique $\kappa' \in \mathcal{T}$ such that $x \in \partial_+\kappa'$. Now for a function $v \in H^1(\Omega,\mathcal{T})$ and for some $\kappa \in \mathcal{T}$ with the property $\partial_-\kappa \setminus \Gamma \neq \emptyset$, we define the outer trace v_κ^- of v on $\partial_-\kappa \setminus \Gamma$ as the inner trace $v_{\kappa'}^+$ such that $\partial_+\kappa' \cap (\partial_-\kappa \setminus \Gamma) \neq \emptyset$. The jump of v across $\partial_-\kappa \setminus \Gamma$ is defined by $\lfloor v \rfloor_\kappa = v_\kappa^+ - v_\kappa^-$. In order to simplify the notation, in the sequel we omit indices in the terms $[v]_e$, $\langle v\rangle_e$ and $\lfloor v\rfloor_\kappa$.

Now, the weak formulation of (1) that corresponds to the NIPG method reads

$$\begin{cases} \text{find } \; u_h \in S(\Omega,\mathcal{T},\mathbf{F}) \text{ such that} \\ B(u_h,v_h) = \displaystyle\sum_{\kappa\in\mathcal{T}} \int_\kappa f v_h \, \mathrm{d}x\,, \quad \text{for all } v_h \in S(\Omega,\mathcal{T},\mathbf{F})\,, \end{cases} \tag{3}$$

where the bilinear form B is given by

$$\begin{aligned} B(v,w) = &\sum_{\kappa\in\mathcal{T}} \left(\varepsilon \int_\kappa \nabla v\cdot\nabla w \,\mathrm{d}x + \int_\kappa (\mathbf{b}\cdot\nabla v + cv)w \,\mathrm{d}x\right) \\ &+ \int_\Gamma \sigma v w \,\mathrm{d}s + \int_{\Gamma_{\text{int}}} \sigma[v][w] \,\mathrm{d}s \\ &+ \sum_{\kappa\in\mathcal{T}} \left(-\int_{\partial_-\kappa\cap\Gamma} (\mathbf{b}\cdot\mu)v^+w^+ \,\mathrm{d}s - \int_{\partial_-\kappa\backslash\Gamma} (\mathbf{b}\cdot\mu_\kappa)\lfloor v\rfloor w^+ \,\mathrm{d}s\right) \\ +\varepsilon &\int_\Gamma (v(\nabla w\cdot\mu) - (\nabla v\cdot\mu)w) \,\mathrm{d}s + \varepsilon \int_{\Gamma_{\text{int}}} ([v]\langle\nabla w\cdot\nu\rangle - \langle\nabla v\cdot\nu\rangle[w]) \,\mathrm{d}s\,, \end{aligned}$$

for $v, w \in H^1(\Omega, \mathcal{T})$. Here σ is called the *discontinuity–penalization* parameter, and is defined by $\sigma|_e = \sigma_e$, $e \in \mathcal{E}$, where σ_e is a nonnegative constant. In the sequel we shall present the exact choices of σ_e, for all edges $e \in \mathcal{E}$.

As in [2], assuming that $u \in H^2(\Omega, \mathcal{T})$ and that u and $\nabla u \cdot \nu$ are continuous across each interior edge e, we obtain that B satisfies the Galerkin orthogonality property. Also, the bilinear form allows one to introduce the so–called DG–norm

$$\|v\|_{\mathrm{DG}}^2 = B(v,v)\,, \qquad v \in H^2(\Omega, \mathcal{T})\,.$$

In [3] the authors proved the existence and uniqueness of the solution u_h of the discrete problem (3). All error estimates in that paper and in [2] are derived in the DG–norm.

Specifying the result from [2] to the problem (1) on a *shape–regular* mesh of maximum element diameter h, we obtain

$$\|u - u_h\|_{\mathrm{DG}}^2 \le C\left(\varepsilon h^2 + h^3 + h^4\right)\|u\|_{H^2(\Omega)}^2\,, \tag{4}$$

with the choice $\sigma_e = \varepsilon h_e^{-1}$, where h_e represents the length of an edge $e \in \mathcal{E}$. In general, the estimate (4) is useless when $\varepsilon \to 0$, see Theorem 1. Therefore, we use an a priori constructed layer–adapted mesh (Shishkin mesh) and show that on such a partitioning robust convergence is guaranteed. For simplicity, we use the standard conforming Shishkin mesh and avoid hanging nodes.

3 The discretization mesh and the interpolation error

Discretization mesh. For the discretization of the boundary value problem (1), here we use an anisotropic tensor–product Shishkin with $(N+1)\times(N+1)$ mesh nodes, that is adapted to the layers at $x = 1$, $y = 0$ and $y = 1$. Let N be an integer divisible by 4 and let λ_x and λ_y be mesh transition parameters defined by

$$\lambda_x = \min\left\{\frac{1}{2}, 2\frac{\varepsilon}{\beta_1}\ln N\right\}, \qquad \lambda_y = \min\left\{\frac{1}{4}, 2\sqrt{\frac{\varepsilon}{\tilde{\gamma}}}\ln N\right\},$$

where β_1 is the lower bound for the function b_1 and $\tilde{\gamma}$ is a constant from the solution decomposition (Theorem 1). The domain Ω is split into $\Omega = \Omega_{11} \cup \Omega_{12} \cup \Omega_{21} \cup \Omega_{22}$, with

$$\Omega_{11} = [1-\lambda_x, 1] \times ([0,\lambda_y] \cup [1-\lambda_y, 1])\,, \qquad \Omega_{12} = [1-\lambda_x, 1] \times [\lambda_y, 1-\lambda_y]\,,$$
$$\Omega_{21} = [0, 1-\lambda_x] \times ([0,\lambda_y] \cup [1-\lambda_y, 1])\,, \qquad \Omega_{22} = [0, 1-\lambda_x] \times [\lambda_y, 1-\lambda_y]\,.$$

Then the intervals $[0, 1-\lambda_x]$ and $[1-\lambda_x, 1]$ are uniformly dissected into $N/2$ subintervals to give the mesh Ω_x^N in x-direction, while for Ω_y^N we dissect $[0, \lambda_y]$ and $[1-\lambda_y, 1]$ into $N/4$ subintervals and $[\lambda_y, 1-\lambda_y]$ into $N/2$ subintervals.

Taking the tensor of Ω_x^N and Ω_y^N, we obtain our final rectangular Shishkin mesh.

On such a constructed partitioning we need to introduce several types of edges (depending on the type, we later determine the discontinuity–penalization parameter σ_e for each edge $e \in \mathcal{E}$). The edges of type I belong to the set $(1-\lambda_x, 1] \times ([0, \lambda_y) \cup (1-\lambda_y, 1])$. They are part of the layer region (more precisely, they lie in the corner layers) and their length is either $h_x = 2\lambda_x N^{-1}$ or $h_y = 4\lambda_y N^{-1}$. Assuming $\lambda_x = 2\varepsilon/\beta_1 \ln N$, $\lambda_y = 2\sqrt{\varepsilon/\tilde{\gamma}} \ln N$ and $\varepsilon \leq CN^{-1}$, we have $h_x \ll H_x := 2(1-\lambda_x)N^{-1}$ and $h_y \ll H_y := 2(1-2\lambda_y)N^{-1}$. Therefore we describe these edges as short; other edges are called long. Edges of type II are also short and belong to the rest of the layer region $\Omega_{12} \cup \Omega_{21}$. Since this region contains also long edges, we refer to them as being of the third type. Type III also contains long edges near the layers from the set $\Omega_{22}^{\dagger} = \overline{\Omega_{22} \setminus \Omega_{22}^{*}}$. Finally, edges of type IV are long and lie in $\Omega_{22}^{*} = [0, 1-\lambda_x - H_x] \times [\lambda_y + H_y, 1-\lambda_y - H_y]$.

Interpolation error. As it was already stated, in the error analysis instead of an L^2–projection onto a finite element space, here we shall use a bilinear interpolant u^I of u that vanishes on Γ. In order to obtain an ε–uniform estimate for $\|u-u_h\|_{\mathrm{DG}}$, we shall use various bounds on interpolation error $\eta = u - u^I$. For example, in [11] one can find L^2– and L^∞–estimates of η

$$\|\eta\|_{L^2(\Omega)} \leq CN^{-2}, \qquad \|\eta\|_{L^\infty(\overline{\Omega \setminus \Omega_{22}})} \leq CN^{-2}\ln^2 N, \qquad \|\eta\|_{L^\infty(\Omega_{22})} \leq CN^{-2},$$

while [14] and [15] contain results for $\nabla\eta$. All these results hold under the assumption $\sqrt{\varepsilon}\ln^2 N \leq C$ and they are proved using the decomposition from Theorem 1 and the technique from [1].

In the following section we present some of the key points of the error analysis for the NIPG method on the Shishkin mesh when applied to the problem (1). More details can be found in [14] and in the forthcoming paper [15].

4 Error analysis

We start the error analysis by first introducing the error decomposition $u - u_h = \eta + \xi$, $\eta = u - u^I$, $\xi = u^I - u_h$. The final estimate for $\|u-u_h\|_{\mathrm{DG}}$ is further obtained from the triangle inequality.

First notice that from the Galerkin orthogonality property one has $\|\xi\|_{\mathrm{DG}}^2 = -B(\eta, \xi)$. Analyzing each term in $B(\eta, \xi)$ with the technique from [2], we conclude that $\|\xi\|_{\mathrm{DG}}$ can be estimated by terms that depend on the interpolation error η, data functions, ε and the mesh. Collecting these expressions and the expressions from $\|\eta\|_{\mathrm{DG}}$, we conclude that in the final error bound $\|u-u_h\|_{\mathrm{DG}}$ the following terms are to be estimated:

$$I_1 \equiv \left(\sum_{\kappa\in\mathcal{T}} \varepsilon \|\nabla\eta\|^2_{L^2(\kappa)}\right)^{1/2}, \qquad I_2 \equiv \left(\sum_{\kappa\in\mathcal{T}} \|\sqrt{c}\eta\|^2_{L^2(\kappa)}\right)^{1/2},$$

$$I_3 \equiv \left(\sum_{e\subset\Gamma^{\mathrm{H}}} \int_e \frac{\varepsilon^2}{\sigma_e} (\partial_y\eta)^2 \,\mathrm{d}s + \sum_{e\subset\Gamma^{\mathrm{V}}} \int_e \frac{\varepsilon^2}{\sigma_e} (\partial_x\eta)^2 \,\mathrm{d}s\right)^{1/2},$$

$$I_4 \equiv \left(\sum_{e\subset\Gamma^{\mathrm{H}}_{\mathrm{int}}} \int_e \left(\frac{\varepsilon}{k_\kappa} + \frac{\varepsilon}{k_{\kappa'}}\right) (\eta^2|_{\partial\kappa\cap e} + \eta^2|_{\partial\kappa'\cap e}) \,\mathrm{d}s \right.$$

$$\left. + \sum_{e\subset\Gamma^{\mathrm{V}}_{\mathrm{int}}} \int_e \left(\frac{\varepsilon}{h_\kappa} + \frac{\varepsilon}{h_{\kappa'}}\right) (\eta^2|_{\partial\kappa\cap e} + \eta^2|_{\partial\kappa'\cap e}) \,\mathrm{d}s\right)^{1/2},$$

$$I_5 \equiv \left(\sum_{e\subset\Gamma^{\mathrm{H}}_{\mathrm{int}}} \int_e \frac{\varepsilon^2}{\sigma_e} ((\partial_y\eta)^2|_{\partial\kappa\cap e} + (\partial_y\eta)^2|_{\partial\kappa'\cap e}) \,\mathrm{d}s \right.$$

$$\left. + \sum_{e\subset\Gamma^{\mathrm{V}}_{\mathrm{int}}} \int_e \frac{\varepsilon^2}{\sigma_e} ((\partial_x\eta)^2|_{\partial\kappa\cap e} + (\partial_x\eta)^2|_{\partial\kappa'\cap e}) \,\mathrm{d}s\right)^{1/2},$$

$$I_6 \equiv \left(\int_\Gamma \sigma\eta^2 \,\mathrm{d}s\right)^{1/2}, \qquad I_7 \equiv \left(\int_{\Gamma_{\mathrm{int}}} \sigma[\eta]^2 \,\mathrm{d}s\right)^{1/2},$$

$$I_8 \equiv \left|\sum_{\kappa\in\mathcal{T}} \int_\kappa (\mathbf{b}\cdot\nabla\xi)\eta \,\mathrm{d}x\right|,$$

$$I_9 \equiv \left(\sum_{\kappa\in\mathcal{T}} \|c_0^{-1} c\eta\|^2_{L^2(\kappa)}\right)^{1/2}, \qquad I_{10} \equiv \left(\sum_{\kappa\in\mathcal{T}} \|c_0\eta\|^2_{L^2(\kappa)}\right)^{1/2},$$

$$I_{11} \equiv \left(\sum_{\kappa\in\mathcal{T}} \|\eta^+\|^2_{\partial_+\kappa\cap\Gamma}\right)^{1/2}, \qquad I_{12} \equiv \left(\sum_{\kappa\in\mathcal{T}} \|\eta^-\|^2_{\partial_-\kappa\backslash\Gamma}\right)^{1/2},$$

$$I_{13} \equiv \left(\sum_{\kappa\in\mathcal{T}} \|\eta^+\|^2_{\partial_-\kappa\cap\Gamma}\right)^{1/2}, \qquad I_{14} \equiv \left(\sum_{\kappa\in\mathcal{T}} \|\eta^+ - \eta^-\|^2_{\partial_-\kappa\backslash\Gamma}\right)^{1/2}.$$

First it can be proved that, [14],

$$I_1 \le CN^{-1}\ln N, \qquad I_2 \le CN^{-2}.$$

The treatment of the terms $I_3, I_4, \ldots, I_7$ depends on the type of edge; more precisely, we look for the contribution from each edge to these terms. Let $I_{j,\mathrm{I}}$,

Table 1. Different types of edges and corresponding parameters σ_e

edge type		parameter σ_e
I, II	horizontal	$\sqrt{\varepsilon}N/\ln N$
I, II	vertical	$N/\ln N$
III	horizontal	$N\ (e\subset\Omega_{22}^{\dagger}),\ \sqrt{\varepsilon}N/\ln N$ (otherwise)
III	vertical	$N\ (e\subset\Omega_{22}^{\dagger}),\ N/\ln N$ (otherwise)
IV	horizontal	εN
IV	vertical	εN

$I_{j,\mathrm{II}}$, $I_{j,\mathrm{III}}$ and $I_{j,\mathrm{IV}}$ denote the contributions of edges of the type I, II, III and IV to the terms I_j, $j=3,4,\ldots,7$, respectively. Assuming that the values of the discontinuity–penalization parameters are the same for all edges of the same type, then from L^∞–interpolation error estimates for η and $\nabla\eta$ we have

$$I_{3,\mathrm{I}}+I_{6,\mathrm{I}}\le C\varepsilon^{\frac14}N^{-\frac32}\ln^2 N\,,\qquad I_{4,\mathrm{I}}+I_{5,\mathrm{I}}+I_{7,\mathrm{I}}\le C\varepsilon^{\frac14}N^{-1}\ln^2 N\,,$$
$$I_{3,\mathrm{II}}+I_{6,\mathrm{II}}\le C\varepsilon^{\frac14}N^{-\frac32}\ln^2 N\,,\qquad I_{4,\mathrm{II}}+I_{5,\mathrm{II}}+I_{7,\mathrm{II}}\le C\varepsilon^{\frac14}N^{-1}\ln^2 N\,,$$
$$I_{3,\mathrm{III}}+I_{6,\mathrm{III}}\le CN^{-\frac32}\ln^{\frac32} N\,,\qquad I_{4,\mathrm{III}}+I_{5,\mathrm{III}}+I_{7,\mathrm{III}}\le CN^{-1}\ln^{\frac32} N\,,$$
$$I_{3,\mathrm{IV}}+I_{6,\mathrm{IV}}\le C\varepsilon^{\frac12}N^{-\frac32}\,,\qquad I_{4,\mathrm{IV}}+I_{5,\mathrm{IV}}+I_{7,\mathrm{IV}}\le C\varepsilon^{\frac12}N^{-1}\,.$$

The choices of the discontinuity–penalization parameter σ_e are summarized in Table 1.

We proceed with the term I_8 which reduces to

$$I_8=\left|\sum_{\kappa\in\mathcal{T}}\int_\kappa b_1(\partial_x\xi)\eta\ \mathrm{d}x\right|\le\left|\int_{\Omega_{11}\cup\Omega_{12}} b_1(\partial_x\xi)\eta\ \mathrm{d}x\right|+\left|\int_{\Omega_{21}\cup\Omega_{22}} b_1(\partial_x\xi)\eta\ \mathrm{d}x\right|\,.$$

The first term in I_8 can be estimated with

$$\left|\int_{\Omega_{11}\cup\Omega_{12}} b_1(\partial_x\xi)\eta\ \mathrm{d}x\right|\le C\|\partial_x\xi\|_{L^2(\Omega_{11}\cup\Omega_{12})}\|\eta\|_{L^2(\Omega_{11}\cup\Omega_{12})}$$
$$\le C\sqrt{\varepsilon}N^{-2}\ln^2 N\|\nabla\xi\|_{L^2(\Omega_{11}\cup\Omega_{12})}\le CN^{-2}\ln^2 N\|\xi\|_{\mathrm{DG}}\,,$$

while for the second term we have

$$\left|\int_{\Omega_{21}\cup\Omega_{22}} b_1(\partial_x\xi)\eta\ \mathrm{d}x\right|\le C\|\partial_x\xi\|_{L^2(\Omega_{21}\cup\Omega_{22})}\|\eta\|_{L^2(\Omega_{21}\cup\Omega_{22})}$$
$$\le CH_x^{-1}\|\xi\|_{L^2(\Omega_{21}\cup\Omega_{22})}N^{-2}\le CN^{-1}\|\xi\|_{\mathrm{DG}}\,.$$

Thus,

$$I_8 \le CN^{-1}\|\xi\|_{\mathrm{DG}}\,.$$

At the end, it can be proved, [14],

$$\begin{aligned} I_9 &\le CN^{-2}\,, & I_{10} &\le CN^{-2}\,,\\ I_{11} &\le CN^{-2}\ln^2 N\,, & I_{12} &\le CN^{-3/2}\ln^2 N\,,\\ I_{13} &\le CN^{-2}\ln^2 N\,, & I_{14} &\le CN^{-3/2}\ln^2 N\,. \end{aligned}$$

Collecting previously given estimates for $I_1, I_2, \dots, I_{14}$, we observe that the long edges from the layer region (type III edges) produce an error of the highest order. Therefore, the main result for the NIPG method on anisotropic Shishkin mesh for the convection–diffusion problem (1) with regular and parabolic layers reads

Theorem 2. *Let u be a solution of the convection–diffusion problem (1) and let u_h be a solution of the discrete problem (3) on the Shishkin mesh. Assuming $\sqrt{\varepsilon}\ln^2 N \le C$ and (2), and choosing the penalty parameter as in Table 1, we have*

$$\|u - u_h\|_{\mathrm{DG}} \le CN^{-1}\ln^{3/2} N\,.$$

5 Numerical experiments

In the sequel we experimentally verify the theoretical result from Theorem 2. We test the discontinuous Galerkin finite element method (3) on the anisotropic Shishkin mesh when applied to the problem (1) with $b_1(x) = c(x) = 1$. The right–hand side f is chosen such that the function

$$u(x,y) = x\left(1 - e^{-(1-x)/\varepsilon}\right)\left(1 - e^{-y/\sqrt{\varepsilon}}\right)\left(1 - e^{-(1-y)/\sqrt{\varepsilon}}\right)$$

is the exact solution.

Table 2 presents the maximum values for $\varepsilon = 10^{-3}, \dots, 10^{-8}$ of different norms of the error $e_h \equiv u - u_h$. The first column corresponds to the DG–norm $\|e_h\|_{\mathrm{DG}}$, while the second corresponds to an estimate of the maximum norm of the error, denoted by $\|e_h\|_{d,\infty}$. We compute this estimate using an auxiliary mesh that contains 25 uniformly distributed mesh points per element. We also compute the values of the L^2–norm of the error $\|e_h\|_{L^2(\Omega)}$, as well as the L^∞–norm of the jumps $\|[u_h]\|_\infty$ along the edges. We observe that the numerical results for the DG–norm of the error e_h are better than those predicted in Theorem 2. These results also indicate second–order accuracy of the jumps along interior edges and $3/2$ as the order of convergence of the L^2–norm of the error.

For this test problem, the NIPG method on anisotropic Shishkin mesh is inferior in $\|e_h\|_{d,\infty}$ compared to the bilinear Galekin FEM and bilinear SDFEM (with the streamline–diffusion parameters as in [12], or in [5]). Nevertheless,

Table 2. The NIPG method on the Shishkin mesh for the test problem

	$\|e_h\|_{DG}$		$\|e_h\|_{d,\infty}$		$\|e_h\|_{L^2(\Omega)}$		$\|[u_h]\|_\infty$	
N	*error*	*rate*	*error*	*rate*	*error*	*rate*	*error*	*rate*
8	2.200(−1)	0.850	3.465(−1)	0.572	5.099(−3)	0.762	2.113(−1)	1.301
16	1.220(−1)	1.135	2.332(−1)	0.759	3.007(−3)	1.016	8.576(−2)	1.725
32	5.555(−2)	1.346	1.378(−1)	0.954	1.487(−3)	1.299	2.594(−2)	1.996
64	2.185(−2)	1.488	7.115(−2)	1.200	6.044(−4)	1.485	6.504(−3)	2.133
128	7.788(−3)	–	3.097(−2)	–	2.159(−4)	–	1.483(−3)	–

it can be expected that this method, or a more general DGFEM, will exceed the streamline–diffusion FEM for the problems with more complicated layer structures where the flexibility of the DGFEM with respect to the mesh is useful.

References

1. Dobrowolski, M., Roos, H.–G. (1997): A priori estimates for the solution of convection–diffusion problems and interpolation on Shishkin meshes. J. Anal. Appl., **16**, 1001–1012
2. Houston, P., Schwab, C., Süli, E. (2002): Discontinuous *hp*–Finite Element Methods for Advection–Diffusion–Reaction Problems. SIAM J. Numer. Anal., **39**, 2133–2163
3. Houston, P., Schwab, C., Süli, E. (2000): Discontinuous *hp*–finite element methods for advection–diffusion problems. Technical Report NA–00/15, Oxford University Computing Laboratory, Oxford, UK
4. Kopteva, N.V. (2001): How accurate is the streamline–diffusion FEM inside parabolic layers? Lecture presented at the 19th Biennial Dundee Conference on Numerical Analysis
5. Linß, T. (2002): Anisotropic meshes and streamline–diffusion stabilization for convection–diffusion problem. Preprint MATH-NM-11-2002, Institut für Numerische Mathematik, Technische Universität Dresden, Germany
6. Linß, T. (2003): Layer–adapted meshes for convection–diffusion problems. Comp. Meth. Appl. Mech. Eng., **192**, 1061–1105
7. Oden, J.T., Babuška, I., Baumann, C.E. (1998): A discontinuous *hp* finite element method for diffusion problems. J. Comp. Phys., **146**, 491–519
8. Prudhomme, S., Pascal, F., Oden, J.T., Romkes, A. (2000): Review of *a priori* error estimation for discontinuous Galerkin methods. TICAM Report 00–27, Texas Institute for Computational and Applied Mathematics, Austin, USA
9. Rivière, B., Wheeler, M.F., Girault, V. (2001): A priori error estimates for finite element methods based on discontinuous approximation spaces for elliptic problems. SIAM J. Numer. Anal., **39**, 902–931
10. Roos, H.-G. (2002): Optimal convergence of basic schemes for elliptic boundary value problems with strong parabolic layers. J. Math. Anal. Appl., **267**, 194–208
11. Roos, H.-G., Skalický, T. (1997): A comparison of the finite element method on Shishkin and Gartland–type meshes for convection–diffusion problems. CWI Quarterly, **10**, 277–300

12. Roos, H.-G., Stynes, M., Tobiska, L. (1996): Numerical Methods for Singularly Perturbed Differential Equations. Springer-Verlag, Berlin
13. Shishkin, G.I. (1992): Discrete Approximation of Singularly Perturbed elliptic and Parabolic Equations. Russian Academy of Sciences, Ural Section, Ekaterinburg, (in Russian)
14. Zarin, H. (2003): Finite element methods for singularly perturbed problems with special emphasis on discontinuities. PhD Thesis, University of Novi Sad, Serbia and Montenegro
15. Zarin, H., Roos, H.–G. (2003): Interior penalty discontinuous approximations of convection–diffusion problems with parabolic layers. submitted to Numer. Math.

A Finite-Volume Mass- and Vorticity-Conserving Shallow-Water Model using Penta-/Hexagonal Grids

William Sawyer and Rolf Jeltsch

Swiss Federal Institute of Technology (ETH Zurich) *sawyer@sam.math.ethz.ch*

Summary. A finite-volume scheme using penta-/hexagonal (PH) grids is presented for the shallow water model on the sphere. The irregular structure of the PH grid presents new challenges, e.g. for the calculation of energy gradients. Radial basis functions (RBFs) are employed for the accurate and efficient approximation of these values. The resulting algorithm is shown to be mass- and vorticity-conserving, and initial numerical results are presented.

1 Introduction

We present a mass- and vorticity-conserving finite-volume algorithm to solve the shallow-water equations on the sphere (SWES), where the sphere is spanned by a grid of pentagonal and hexagonal cells (hereafter referred to as a PH grid).

The vector-invariant formulation of the SWES is

$$\frac{\partial h}{\partial t} + \nabla \cdot (\mathbf{v}h) = 0 \tag{1}$$

$$\frac{\partial \mathbf{v}}{\partial t} + (\zeta + f)\mathbf{k} \times \mathbf{v} + \nabla(\kappa + \Phi) = 0, \tag{2}$$

where $\mathbf{v}$ is the velocity field, $\zeta = \mathbf{k} \cdot (\nabla \times \mathbf{v})$ is the relative vorticity, $f = 2\omega \sin\theta$ the Coriolis parameter (ω is the angular velocity of the earth, θ the latitude), $\mathbf{k}$ the vertical unit vector, h the depth of the fluid (directly proportional to the fluid mass in the cell), κ the kinetic energy, and $\Phi = gh$ the potential. $\eta = \zeta + f$ is also referred to as the absolute vorticity.

This paper is an extension of work by Lin and Rood [10] who proposed a mass- and vorticity-conserving algorithm for a standard orthogonal latitude-longitude (lat-lon) grid. While [10] has proved successful in atmospheric modeling, it makes assumptions about the orthogonality of the underlying grid. Such grids have a singularity at the poles, which must be dealt with specially. On PH grids, which have no pole problem, the algorithm is not immediately applicable.

A short overview of PH grids is given in Section 2. In Section 3 it is then shown that the inherent limitation in [10] to orthogonal grids lies only in its advection algorithm. Assuming the availability of an appropriate advection algorithm for PH grids, a related formulation exists for general non-orthogonal grids. One problem with such grids, however, is the additional difficulty calculating the gradients in equation (2). A technique for calculating these to higher order is described. First numerical results of tests proposed in [14] are presented in Section 4.

Although only the two-dimensional SWES are discussed here, the ultimate goal of our research is to create a three-dimensional solver for atmospheric dynamics. A technique for extending the 2D SWES to the 3D problem is treated in [8] and will not be treated further here.

2 Overview of Penta-/Hexagonal Grids

The use of penta-/hexagonal grids – which cover the sphere with twelve spherical pentagons and an arbitrary number of hexagons – for atmospheric modeling is not new. A summary of finite-difference methods on such grids was given by Williamson in [13]. A milestone in the use of PH grids was achieved by Heikes and Randall [4] who built much of the foundation for a genuine General Circulation Model (GCM). Baumgartner, Majewski, et al. [6] built a production model based on these grids which now provides numerical weather forecasts at the German Weather Service (DWD).

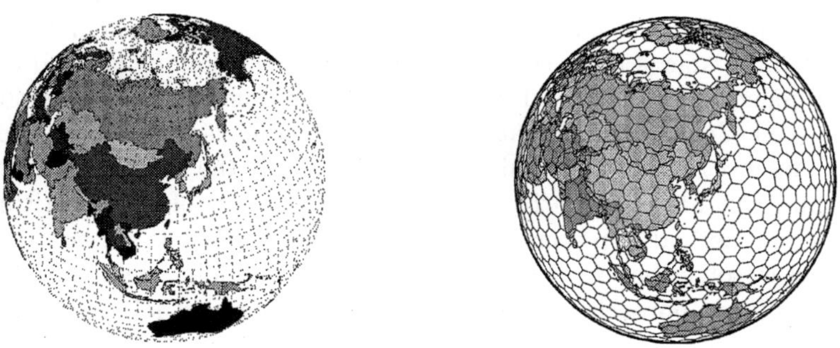

Fig. 1. The standard lat-lon grid (left) illustrates the Mercator projection of surface coordinates to the rectangular domain $[-\pi/2, pi/2] \times [-\pi, \pi]$. The PH grid (right) instead decomposes the sphere into 12 spherical pentagonal cells and the rest into hexagonal cells.

A PH grid can be obtained by first distributing n points over the globe in a relatively even manner. A common method, used in [4, 6] among others, is to start with an icosahedron (consisting of 20 equilateral triangles). Each triangle is then subdivided recursively into 4 equilateral triangles until the desired resolution is obtained. This procedure gives rise to an *icosahedral grid.* Heikes et al. [4] use roughly the same technique although the grid is *twisted* to maintain a northern/southern hemisphere symmetry. From this grid a PH grid can be constructed, e.g. by considering the perpendicular bisectors of each triangle's edge. In this case, the cells form the *Voronoi* complement grid of the icosahedral grid. The Voronoi cell C_k is the set of points on the sphere equidistant or closer to triangle vertex k than any other.

Icosahedral grids are not the only basis for the class of PH grids. Given a non-degenerate set of points on the sphere, the spherical convex hull can be constructed. The facets of this hull are triangles. Care must be taken with the choice of the triangle vertices so that the intersection (subsequently referred to as the *cell vertex*) of the three perpendicular bisectors will be inside the triangle. Around any given triangle vertex the cell vertices will form either pentagons or hexagons (not necessarily regular, but by their construction convex). Finally, the surface is "inflated" to form a sphere (see Figure 1). Several issues then arise due to the spherical geometry which are described in [11] but will be passed over here for sake of brevity.

A non-degenerate distribution of four or more points yields a convex hull with triangular faces. With twelve or more points, the Voronoi dual diagram will be a PH grid. The points can be evenly distributed by minimizing a potential function, in which each point on the sphere can be considered as a point charge. An exhaustive study of such distributions has been made by Sloane, et al. [3]. Finally the cell centers can be defined as the cell's barycenter; there are numerical advantages for this choice. The barycenter does not necessarily coincide with the corresponding triangle vertex.

Heikes and Randall [5] point out that even if the perpendicular bisector of the triangle edge is used, the midpoint of the chord joining the cell centers will not generally coincide with the midpoint of the chord connecting the cell vertices (cell edge midpoint or *flux point*). This fact leads to problems if simple finite differences are used to calculate, for example, the gradient in the flux point, and can have a negative influence on the order of the algorithm. In [5] a revised method of placement for the triangle vertices is suggested which minimizes a (somewhat artificial) norm describing midpoint alignment. In [12], a more founded approach is taken: the vertices are conceptually connected by springs and allowed to adjust their positions by spring dynamics. The resulting configuration still has the advantages of the icosahedral grid, while improving the algorithm's order.

Several authors, e.g. [6, 13], have pointed out the advantages of PH grids for atmospheric modeling. In the first place, they avoid the "pole" problem of lat-lon grids, namely the convergence of the meridians at the poles (small grid cells can violate a CFL condition). Moreover, PH grid cells only have edge neighbors. That is, there is no pair of cells which share only a cell vertex. This allows a straightforward application of finite volumes, as will be seen subsequently.

3 Shallow Water Model on the Sphere with PH Grids

By taking the curl and divergence of equation (2), respectively, we arrive at the *vorticity/divergence* form of the SWEs:

$$\frac{\partial \eta}{\partial t} + \nabla \cdot (\mathbf{v}\eta) = 0 \tag{3}$$

$$\frac{\partial \delta}{\partial t} + \nabla \cdot [\eta \mathbf{k} \times \mathbf{v}] + \nabla^2 (\kappa + \Phi) = 0. \tag{4}$$

An expression for the velocity can be found by solving elliptic equations to determine the *stream function* and *velocity potential*:

$$\begin{aligned} \text{Vorticity:} &\quad \eta = \triangle\psi \quad \text{with stream-function} \quad \psi \\ \text{Divergence:} &\quad \delta = \triangle\chi \quad \text{with velocity potential} \quad \chi \\ \text{Velocity:} &\quad \mathbf{v} = \mathbf{k} \times \nabla\psi + \nabla\chi. \end{aligned}$$

This approach is taken in [6], for example, at considerable computational expense, and it seems worthwhile to look for a more computationally expedient method.

For lat-lon grids, an explicit mass- and vorticity-conserving shallow-water model was proposed in [10]. To derive this algorithm, equation (2) was formulated in lat-lon coordinates, thus inherently limiting it to orthogonal grids. The elegance of the method lies in the numerical treatment of the vector-invariant formulation under consideration of the vorticity equation (3), which implies that vorticity η is merely advected. Thus a local constraint is imposed: local changes in the vorticity can only effect the region within the propagation zone of the advection.

The explicit time stepping method for the grid cell $\Omega_{i,j}$ at the i,jth location, presented in [10], is:

$$h^{n+1} = h^n + F(u^*, \Delta t; h^\theta) + G(v^*, \Delta t; h^\lambda) \tag{5}$$

$$u^{n+1} = u^n + \Delta t \left\{ \mathcal{Y}(v^*, \Delta t; \eta^\lambda) - \frac{1}{A\Delta\lambda\cos\theta}\delta_\lambda \left[\kappa^* + \overline{\overline{\Phi^{n+1}}^{\theta}}^{\lambda} \right] \right\} \tag{6}$$

$$v^{n+1} = v^n - \Delta t \left\{ \mathcal{X}(u^*, \Delta t; \eta^\theta) + \frac{1}{A\Delta\theta}\delta_\theta \left[\kappa^* + \overline{\overline{\Phi^{n+1}}^{\lambda}}^{\theta} \right] \right\}, \tag{7}$$

where F and G are flux-form operators in longitude and latitude, specifically the difference of incoming and outgoing fluxes $F = \mathcal{F}_{i+1/2} - \mathcal{F}_{i-1/2}$ and $G = \mathcal{G}_{j+1/2} - \mathcal{G}_{j-1/2}$; $\delta_\lambda(.)$ and $\delta_\theta(.)$ are grid differences, while the $\overline{(.)}^\lambda$ and $\overline{(.)}^\theta$ are grid means, defined in the edge midpoints in λ and θ respectively.

The source terms $\mathcal{X}$ and $\mathcal{Y}$ in equations (6) and (7) are the time-averaged fluxes for the finite-volume discretization of equation (2). Ignoring higher order terms, these can be approximated for some scalar quantity q^n at time step n as:

$$\mathcal{X}(u^*, \Delta; q^n) = \frac{1}{\Delta t} \int_t^{t+\Delta t} u^* \, q \, dt \quad \mathcal{Y}(v^*, \Delta; q^n) = \frac{1}{\Delta t} \int_t^{t+\Delta t} v^* \, q \, dt. \tag{8}$$

The discrete, cell-averaged relative vorticity is defined as:

$$\zeta_i = \frac{1}{A \Delta\lambda \cos\theta} \delta_\lambda [v^n] - \frac{1}{A \Delta\theta \cos\theta} \delta_\theta [u^n \cos\theta]. \tag{9}$$

Crucial to the algorithm is the use of staggered grids (C- and D-*grids*, described in [7]), on which the velocities are defined in the same point as the flux. Thus there is a one-to-one dependency between fluxes and velocities. If the fluxes depended on more than one velocity value, it would be necessary to solve for velocity from a system of equations. But in this case it is straightforward to calculate *predictor* values, u^* and v^* (on a C-grid) which are then inserted into the corrector step, on a D-grid, in (5), (6) and (7). As this technique is discussed at length in [10], we will pass over this topic.

There is some additional complexity at the pole, where a *pole cap* cell (a regular polygon with $m = 2\pi/\Delta\lambda$ vertices) needs to be considered separately. The poles are also treated in a way which conserves mass and vorticity.

After some analysis of the algorithm in [10], several key features present themselves. The global conservation of discrete mass follows directly from equation (1). With some calculation (see [11]), it can be shown that the global discrete vorticity is conserved as well. Furthermore, orthogonality is inherently assumed in the algorithm, particularly in its use of a flux-form Semi-Lagrangian (FFSL) advection scheme presented in [9]. FFSL is a finite-volume scheme which splits the time-averaged fluxes of the generic scalar quantity q into an east-west $\mathcal{X}$ and north-south $\mathcal{Y}$. This splitting can only realistically be applied on an orthogonal grid.

Our goal is to obtain a similar algorithm to solve the SWES which is not limited to orthogonal grids, nor even to grids which are orthogonal to their dual grid (i.e. possess the Voronoi-Delaunay property). To this end it is necessary to introduce nomenclature which is sufficiently general for PH grids.

Ω_i	cell with index i
f_i	Coriolis parameter in Ω_i
$\mathcal{N}_i$	index set of neighboring cells to Ω_i
h_i^n	fluid depth ("mass") in cell Ω_i at time step n
η_i^n	discrete absolute vorticity in cell Ω_i at time step n
$l_{i,k}$	length of edge $\{i,k\}$
$\mathbf{n}_{i,k}$	outward vector perpendicular to edge $\{i,k\}$
$\mathbf{t}_{i,k}$	counter-clockwise unit vector along edge $\{i,k\}$
$w^n_{\perp_{i,k}}$	"C-grid" wind speed $\perp$ to edge $\{i,k\}$
$w^n_{\parallel_{i,k}}$	"D-grid" wind speed $\parallel$ to edge $\{i,k\}$
$\mathcal{F}^n_{i,k}(q,\mathbf{w},\Delta t)$	flux of quantity q with wind $\mathbf{w}$ for Δt through edge $\{i,k\}$
$\mathcal{G}^n_{i,k}(q)$	discrete approx. to gradient of q at flux-point $\{i,k\}$

Using this notation it is possible to formulate expressions for the discrete vorticity and divergence, which are simply generalizations of the formulas in [10]:

$$\eta_i = \frac{\sum_{k\in\mathcal{N}_i} w_{\parallel_{i,k}} l_{i,k}}{|\Omega_i|} + f_i \quad \delta_i = \frac{\sum_{k\in\mathcal{N}_i} w_{\perp_{i,k}} l_{i,k}}{|\Omega_i|}. \tag{10}$$

These are second order if the cell center is the barycenter.

At this point the one-to-one correspondence with [10] is lost. In [10], the C- and D-grid winds (corresponding to $w^n_{\perp_{i,k}}$ and $w^n_{\parallel_{i,k}}$) are interchanged by two averagings. This only works if the grid is orthogonal, thus for non-orthogonal grids another approach must be found. In [1] a linear reconstruction (a Raviart-Thomas element of order 0) for the wind inside the cell is made, thus retaining only the edge-parallel winds as unknowns. However there is a limitation: the coefficients of the linearization can only be determined in the case of a triangular grid. Unfortunately, for PH grids the linear system is over-defined, and higher order approximation would have to be employed.

We propose instead to work with both $w^n_{\perp_{i,k}}$ and $w^n_{\parallel_{i,k}}$, treating each equally. In order to do this, we need the help of equation (4) to determine $w^n_{\perp_{i,k}}$. The predictor-corrector character of algorithm in [10] is retained to provide higher order in time. This gives rise to the following half-time updates:

$$h_i^{n+1} = h_i^n - \frac{\Delta t}{|\Omega_i|}\mathcal{F}^n_{i,k}(h_i, w_\perp^{n+1/2}\mathbf{n}_{i,k}, \Delta t) \tag{11}$$

$$w^{n+1}_{\parallel_{i,k}} = w^n_{\parallel_{i,k}} - \Delta t\left\{\mathcal{F}^n_{i,k}(\eta_i, w_\perp^{n+1/2}\mathbf{n}_{i,k}, \Delta t) + \mathcal{G}_{\parallel_{i,k}}(\kappa+\Phi)\right\} \tag{12}$$

$$w^{n+1}_{\perp_{i,k}} = w^n_{\perp_{i,k}} - \Delta t\left\{\mathcal{F}^n_{i,k}(\eta_i, -w_\parallel^{n+1/2}\mathbf{n}_{i,k}, \Delta t) + \mathcal{G}_{\perp_{i,k}}(\kappa+\Phi)\right\}. \tag{13}$$

Inserting the above updates into the definitions for discrete vorticity, we find:

$$\eta_i^{n+1} = \frac{1}{|\Omega_i|} \sum_{k\in N_i} \left(w_{\|_{i,k}}^n - \Delta t \left\{ \mathcal{F}_{i,k}^n(\eta_i, w_{\perp_{i,k}}^n \mathbf{n}_{i,k}, \Delta t) + \mathcal{G}_{\|_{i,k}}(\kappa+\Phi) \right\} \right) l_{i,k} + f_i$$

$$= \eta_i^n - \frac{\Delta t}{|\Omega_i|} \left\{ \sum_{k\in N_i} \mathcal{F}_{i,k}^n(\eta_i, w_{\perp_{i,k}}^n \mathbf{n}_{i,k}, \Delta t) l_{i,k} - \sum_{k\in N_i} \mathcal{G}_{\|_{i,k}}(\kappa+\Phi) l_{i,k} \right\}$$

$$= \eta_i^n - \frac{\Delta t}{|\Omega_i|} \sum_{k\in N_i} \mathcal{F}_{i,k}^n(\eta_i, w_{\perp_{i,k}}^n \mathbf{n}_{i,k}, \Delta t) l_{i,k} - \Delta t \, CG_i$$

$$\sum_{\forall i} \eta_i^{n+1} |\Omega_i| = \sum_{\forall i} \eta_i^n |\Omega_i|$$

$$-\Delta t \underbrace{\sum_{\forall i} \sum_{k\in N_i} (\mathcal{F}_{i,k}^n(\eta_i, w_{\perp_{i,k}}^n \mathbf{n}_{i,k}, \Delta t) + \mathcal{G}_{\|_{i,k}}(\kappa+\Phi)) l_{i,k}}_{0} .$$

Similarly, we arrive at the anticipated form for the updated divergence:

$$\delta_i^{n+1} = \delta_i^n - \frac{\Delta t}{|\Omega_i|} \sum_{k\in N_i} \mathcal{F}_{i,k}^n(\eta_i, -w_\| \mathbf{n}_{i,k}, \Delta t) l_{i,k} - \frac{\Delta t}{|\Omega_i|} \sum_{k\in N_i} \mathcal{G}_{\perp_{i,k}}(\kappa+\Phi) l_{i,k}.$$

A clear hurdle in the above PH grid updates are the calculations of the gradients $\mathcal{G}$. Not only should these be as accurate as possible, they should also satisfy the constraint that the discrete curl of G_i, namely CG_i, is identically 0. Given a generic scalar cell mean quantities q_i, it is possible to find a higher order expression for the edge gradients [11] at the flux point. For example, one can define a *spherical radial basis function* (SRBF, see [2]) defined by a local set of q_i clustered around the flux point:

$$q(\mathbf{x}) = \sum_{j=1}^{N} \lambda_k \Upsilon(r_j) + \lambda_{N+1} + (\lambda_{N+2}, \ldots, \lambda_{N+d+1})^T \mathbf{x} \quad \text{with} \quad r_j = \|\mathbf{x} - \mathbf{x}_j\|, \tag{14}$$

where $\Upsilon(r)$ is a radial basis function, e.g. $\Upsilon(r) = r^2 \log r / 16\pi$ for *thin-plate splines*, and the λ are found such that $q(\mathbf{x}_j) = q_j$. If one considers the subset q_j from cells centered on (and including) cell Ω_i and uses the resulting SRBF to approximate all the edge-parallel gradients of the central cell, the discrete curl-gradient CG_i disappears:

$$CG_i(q) = \sum_{k\in N_i} (\nabla q \cdot \mathbf{t})_{i,k} l_{i,k}$$

$$= \sum_{j=1}^{N} \lambda_j \frac{\Upsilon'(r_j)}{r_j} \mathbf{r}_j \cdot \underbrace{\sum_{k\in N_i} \mathbf{t}_{i,k} l_{i,k}}_{0} + (\lambda_{N+2}, \lambda_{N+3}) \cdot \underbrace{\sum_{k\in N_i} \mathbf{t}_{i,k} l_{i,k}}_{0} = 0.$$

But the use of this gradient approximation will not produce consistent gradients between neighboring cells, which is a requirement. On the other hand, it is possible to use *different* SRBFs $q^{(k)}(x)$ — each built from a subset $q_j^{(k)}$ clustered around a flux point $\{i,k\}$ — to approximate the gradients. In this case there is only one consistent gradient value per flux point. However, CG_i does not necessarily disappear.

$$CG_i(q) = \sum_{k\in\mathcal{N}_i} (\nabla q^{(k)} \cdot \mathbf{t})_{i,k} l_{i,k}$$

$$= \sum_{k\in\mathcal{N}_i} \left(\sum_{j=1}^{N} \lambda_j^{(k)} \frac{\Upsilon'(r_j^{(k)})}{r_j^{(k)}} \mathbf{r^{(k)}}_j + (\lambda_{N+2}^{(k)}, \lambda_{N+3}^{(k)}) \right) \cdot \mathbf{t}_{i,k} l_{i,k} \quad \neq 0 \text{ in general.}$$

SRBFs offer a flexible interpolation method with scattered data and do not require specific grid qualities, such as the orthogonality of the grid with its dual (the Voronoi-Delaunay property). If a higher order approximation, e.g. with SRBFs, for q in each cell vertices is used, then a second order gradient approximation can be found which fulfills $CG_i \equiv 0$ and is consistent between cells.

$$V_{i,j,k} = \text{Some approximation of vertex value shared by cells } i,j \text{ and } k \tag{15}$$

$$\mathcal{G}_{\|_{i,k}(q)} = \frac{V_{i,k,s(i,k)} - V_{i,k,p(i,k)}}{l_{i,k}} \tag{16}$$

$$CG_i(q) = \sum_{k\in N_i} V_{i,k,s(i,k)} - V_{i,k,p(i,k)} = 0, \tag{17}$$

where $p(i,k)$ and $s(i,k)$ are the indices of the preceding and successive neighbor cells of i and k around i in counter-clockwise fashion. The final formulation of the predictor is

$$h_i^{n+1/2} = h_i^n - \frac{\Delta t}{2|\Omega_i|} \sum_{k\in N_i} \mathcal{F}_{i,k}^n(h_i, w_\perp^n \mathbf{n}_{i,k}, \frac{\Delta t}{2}) \tag{18}$$

$$w_{\|_{i,k}}^{n+1/2} = w_{\|_{i,k}}^n - \frac{\Delta t}{2} \left\{ \mathcal{F}_{i,k}^n(\eta_i, w_\perp^n \mathbf{n}_{i,k}, \frac{\Delta t}{2}) + \mathcal{G}_{\|_{i,k}}(\kappa + \Phi) \right\} \tag{19}$$

$$w_{\perp_{i,k}}^{n+1/2} = w_{\perp_{i,k}}^n - \frac{\Delta t}{2} \left\{ \mathcal{F}_{i,k}^n(\eta_i, -w_\|^n \mathbf{n}_{i,k}, \frac{\Delta t}{2}) + \mathcal{G}_{\perp_{i,k}}(\kappa + \Phi) \right\}, \tag{20}$$

with the time-averaged kinetic and potential energies,

$$\kappa^* = \frac{1}{2}(\mathbf{v}^{n+1/2} \cdot \mathbf{v}^{n+1/2}) \quad \Phi^* = g h^{n+1/2}. \tag{21}$$

The corrector step is

$$h_{i,k}^{n+1} = h_{i,k}^{n} - \frac{\Delta t}{|\Omega_i|} \sum_{k \in N_i} \mathcal{F}_{i,k}^{n}(h_i, w_{\perp}^{n+1/2} \mathbf{n}_{i,k}, \Delta t) \tag{22}$$

$$w_{\|_{i,k}}^{n+1} = w_{\|_{i,k}}^{n} - \Delta t \left\{ \mathcal{F}_{i,k}^{n}(\eta_i, w_{\perp}^{n+1/2} \mathbf{n}_{i,k}, \Delta t) + \mathcal{G}_{\|_{i,k}}(\kappa^* + \Phi^*) \right\} \tag{23}$$

$$w_{\perp_{i,k}}^{n+1} = w_{\perp_{i,k}}^{n} - \Delta t \left\{ \mathcal{F}_{i,k}^{n}(\eta_i, -w_{\|}^{n+1/2} \mathbf{n}_{i,k}, \Delta t) + \mathcal{G}_{\perp_{i,k}}(\kappa^* + \Phi^*) \right\} . \tag{24}$$

Thus for PH grids, the SWES problem, as in [10], can be completely determined from gradient approximations and advective fluxes $\mathcal{F}_{i,k}$ across cell boundaries. Fluxes between vertex neighbors are avoided since these do not exist on PH grids. In this paper the flux determination is considered a "black box": the advection problem on PH grids is far from simple, and finding a second order monotone method for solving it is a matter of our current research.

4 Numerical Tests and Discussion

A prototype of the suggested algorithm was programmed in Matlab; later it will be programmed in Fortran 90. We employ the tests suggested in [14], in particular the Cosine Bell and Rossby-Haurwitz tests. For the initial tests, a simple first-order, hence rather diffusive, advection algorithm was implemented. The gradient approximation is performed using SRBF approximations and vertex differences from equation (16). The results for the cosine bell rotation are illustrated in Figures 2 and 3.

References

1. L. Bonaventura. Development of the ICON dynamical core modelling strategies and preliminary results. Unpublished manuscript, 2003.
2. G. Fasshauer and L. Schumaker. *Scattered Data Fitting on the Sphere*, pages 117–166. Vanderbilt Univ. Press, 1998.
3. R. H. Hardin, N. J. A. Sloane, and W. D. Smith. Spherical Codes. Book in preparation; see `http://www.research.att.com/~njas/electrons/`, 2003.
4. R. Heikes and D. A. Randall. Numerical Integration of the Shallow-water Equations of a Twisted Icosahedral Grid. Part I: Basic Design and Results of Tests. *Monthly Weather Review*, 123:1862–1880, 1995.
5. R. Heikes and D. A. Randall. Numerical Integration of the Shallow-water Equations of a Twisted Icosahedral Grid. Part I: Basic Design and Results of Tests. *Monthly Weather Review*, 123:1881–1887, 1995.
6. D. Majewski, D. Liermann, P. Prohl, B. Ritter, M. Buchhold, T. Hanisch, G. Paul, and W. Wergen. The Operational Global Icosahedral-Hexagonal Gridpoint Model GME: Description and High-Resolution Tests. *Mon. Wea. Rev.*, 130:319–338, Feb. 2002.

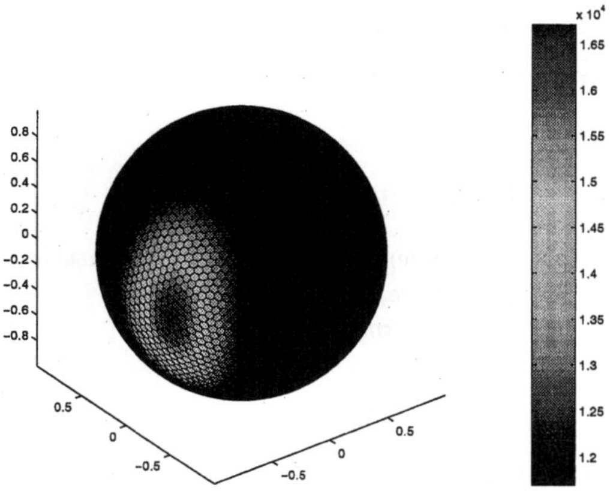

Fig. 2. The Cosine-Bell test from [14] is applied to a grid containing 4482 cells. After one full rotation in approx. 600 hrs., the peak has diffused considerable, but retains its basic form.

Hours	Ratio of peak	Ratio of min
0	1.00	1.00
60	0.75	0.96
120	0.63	0.90
180	0.55	0.88
240	0.51	0.85
300	0.47	0.92
360	0.45	0.90
420	0.44	0.89
480	0.41	0.95
540	0.39	0.96
600	0.38	0.99

Fig. 3. For the test in Fig. 2, the table contains the ratios, as a function of time, of the maxima (i.e. cone peak) and minima to the starting values

7. F. Mesinger and A. Arakawa. *Numerical Methods used in Atmospheric Models*, volume 1, chapter 1. GARP, 1976. Publication Series No. 17.
8. S.-J. Lin. A finite-volume integration method for computing pressure gradient force in general vertical coordinates. *Q. J. R. Met. Soc.*, 123:1749–1762, 1997.
9. S.-J. Lin and R. B. Rood. Multidimensional Flux Form Semi-Lagrangian Transport Schemes. *Mon. Wea. Rev.*, 124:2046–2070, 1996.
10. S.-J. Lin and R. B. Rood. An explicit flux-form semi-Lagrangian shallow water on the sphere. *Q. J. R. Met. Soc.*, 123:2477–2498, 1997.
11. W. Sawyer. A Shallow Water Model on the Sphere using Penta-/Hexagonal Grids. Research report, SAM ETHZ, 2004. In preparation.
12. H. Tomita, M. Tsugawa, M. Satoh, and K. Goto. Shallow Water Model on a Modified Icosahedral Geodesic Grid by Using Spring Dynamics. *J. Comp. Phys.*, 174:579–613, 2001.
13. D. Williamson. Difference Approximations for Fluid Flow on a Sphere. *GARP Publication Series*, 2(17):51–120, Sept. 1978.
14. D. Williamson, J. Drake, J. Hack, R. Jacob, and P. Swarztrauber. A Standard Test Set for Numerical Approximations to the Shallow Water Equations in Spherical Geometry. *J. Comp. Phys.*, 102:211–224, 1992.

Application of Parallel Computing Techniques for Problems of Degenerated Diffusion

Milan Šenkýř, Jiří Mikyška and Michal Beneš

Department of Mathematics, Faculty of Nuclear Sciences and Physical Engineering, Czech Technical University, Trojanova 13, 120 00 Prague, Czech Republic
Contact e-mails: *kiwi@kmlinux.fjfi.cvut.cz, mikyska@kmlinux.fjfi.cvut.cz, Michal.Benes@fjfi.cvut.cz*

Summary. In this contribution, we discuss parallelization of the problem of curve dynamics in plane. Related PDEs are based on the levelset method introduced in [5], and on the phase-field method described in [1]. Numerical schemes use a finite-difference discretization in space and explicit time solvers. Parallel algorithms are designed for systems with distributed memory, and are based on the domain splitting. The achieved results indicate strength and efficiency of the described approach in case of such highly nonlinear problems.

1 Mean-curvature flow

We study the following motion law for closed planar curves denoted as Γ:

$$v_\Gamma = -g(\theta)\kappa_\Gamma + F, \tag{1}$$

in the direction of the Euclidean normal vector to Γ. Here, $\mathbf{n}_\Gamma$ denotes the normal vector to Γ, v_Γ the normal velocity, κ_Γ the mean curvature, F a forcing term, and g is a suitable positive 2π-periodic function of curve anisotropy, θ is the angle between $\mathbf{n}_\Gamma$ and a prescribed direction. We take $g(\theta) = \psi(\theta)+\psi''(\theta)$, where $\psi(\theta) = 1+\zeta\cos(N_{fold}\,\theta)$, ζ is the anisotropy strength and N_{fold} a type of symmetry. The equation (1) in the form of the Gibbs-Thompson law is contained in the modified Stefan problem. For details, we refer the reader to [1]. In [4], we may find an application in noise filtering, edge detection and morphing of computer-processed image data.

Hamilton-Jacobi equation. Assume that the curve $\Gamma(t)$ is represented by a levelset of a function $P = P(t,x)$, i.e., $\Gamma(t) = \{x \in \mathbb{R}^2 \mid P(t,x) = const.\}$. We can express the quantities appearing in (1) by means of P:

$$\mathbf{n}_\Gamma = -\frac{\nabla P}{|\nabla P|}, \quad v_\Gamma = \frac{\partial_t P}{|\nabla P|}, \quad \kappa_\Gamma = div(\mathbf{n}_\Gamma).$$

Then, we can introduce the Hamilton-Jacobi equation (see [5, 3])

$$\frac{\partial P}{\partial t} = g(\theta)|\nabla P|\nabla\cdot(\frac{\nabla P}{|\nabla P|}) + |\nabla P|F. \tag{2}$$

Allen-Cahn equation. An extensive experience with non-linear reaction-diffusion equations led to the development of a phase-field approximation of (1) by the Allen-Cahn equation [2], or by a modified Allen-Cahn equation [1]. The evolution of the levelset $\frac{1}{2}$ of its solution approximates the evolution of the manifold $\Gamma(t)$, as discussed in [1].

First, we denote a rectangular domain $\Omega = (0, L_1) \times (0, L_2) \subset \mathbb{R}^2$, $[x, y] \in \Omega$, the time variable $t \in (0, T)$. The problem for an unknown function $p = p(t, x, y)$ reads as follows

$$\xi \frac{\partial p}{\partial t} = g(\theta) \left(\xi \Delta p + \frac{1}{\xi} f_0(p) \right) + F(u) \xi |\nabla p|, \text{ in } (0, T) \times \Omega,$$
$$p|_{\partial\Omega} = 0 \text{ on } (0, T) \times \partial\Omega, \quad p|_{t=0} = p_{ini}(x) \text{ in } \Omega.$$

Here, $\xi > 0$ is a parameter related to the thickness of the interface layer (it is usually set to a value $<< 1$). The polynomial $f_0(p) = ap(1-p)(p - \frac{1}{2})$ with $a > 0$ is derived from the double-well potential w_0 as $w_0' = -f_0$. The function $F = F(x, y)$ is bounded. The function p_{ini} is an initial condition. We refer the reader to [1], for details concerning the equation and physical background of it.

Numerical schemes. We treat the PDE problems (2) and (3), both closely related to (1), by several numerical schemes implemented by means of parallelization tools for the systems with distributed memory. The problems are solved in a spatial domain $\Omega = (0, L_1) \times (0, L_2)$, which is discretized by a rectangular uniform grid with mesh sizes h_1, h_2 in directions x and y.

We introduce the following notations for a given function u:

$$h_1 = \frac{L_1}{N_1},\ h_2 = \frac{L_2}{N_2},\ u_{ij} = u(ih_1, jh_2),$$
$$\omega_{\mathbf{h}} = \{[ih_1, jh_2] \mid i = 1, \ldots, N_1 - 1;\ j = 1, \ldots, N_2 - 1\},$$
$$\bar{\omega}_{\mathbf{h}} = \{[ih_1, jh_2] \mid i = 0, \ldots, N_1;\ j = 0, \ldots, N_2\},\ \gamma_{\mathbf{h}} = \bar{\omega}_{\mathbf{h}} - \omega_{\mathbf{h}},$$
$$u_{\bar{x},ij} = \frac{u_{ij} - u_{i-1,j}}{h_1}, \quad u_{x,ij} = \frac{u_{i+1,j} - u_{ij}}{h_1}, \quad u_{\mathring{x},ij} = \frac{u_{i+1,j} - u_{i-1,j}}{2h_1},$$
$$u_{\bar{y},ij} = \frac{u_{ij} - u_{i,j-1}}{h_2}, \quad u_{y,ij} = \frac{u_{i,j+1} - u_{ij}}{h_2}, \quad u_{\mathring{y},ij} = \frac{u_{i,j+1} - u_{i,j-1}}{2h_2},$$
$$u_{\bar{x}x,ij} = \frac{1}{h_1^2}\left(u_{i+1,j} - 2u_{ij} + u_{i-1,j}\right), \quad u_{\bar{y}y,ij} = \frac{1}{h_2^2}\left(u_{i,j+1} - 2u_{ij} + u_{i,j-1}\right),$$
$$\bar{\nabla}_h u = [u_{\bar{x}}, u_{\bar{y}}], \quad \mathring{\nabla}_h\, u = [u_{\mathring{x}}, u_{\mathring{y}}], \quad \Delta_h u = u_{\bar{x}x} + u_{\bar{y}y}.$$

Direct discretization of the levelset equation. The curvature expressed in terms of second-order derivatives

$$\kappa_\Gamma = -\frac{\partial^2_{xx} P\, (\partial_y P)^2 - 2\, \partial^2_{xy} P\, \partial_x P\, \partial_y P + \partial^2_{yy} P\, (\partial_x P)^2}{((\partial_x P)^2 + (\partial_y P)^2)^{3/2}},$$

allows us to use central differences to approximate both first- and second-order derivatives. We then propose an explicit scheme in the following form (n is the time-level index, τ is the time step):

$$P_{ij}^{n+1} = P_{ij}^{n} + \tau\, g(\theta) \frac{P_{\bar{x}x}\,(P_{\mathring{y}})^2 - 2\,P_{\mathring{x}\mathring{y}}\,P_{\mathring{x}}\,P_{\mathring{y}} + P_{\bar{y}y}\,(P_{\mathring{x}})^2}{(P_{\mathring{x}})^2 + (P_{\mathring{y}})^2} + \tau\sqrt{(P_{\mathring{x}})^2 + (P_{\mathring{y}})^2}F,$$

which is subject of a regularization when $(P_{\mathring{x}})^2 + (P_{\mathring{y}})^2 = 0$. The relationship of τ and h is given by a stability condition.

The equation (1) defines the motion law on $\Gamma(t)$ only. On the other hand, the function P is obtained from the equation (2) valid in Ω. In our work, we extend the forcing term F from (1) to (2) as it is. Other extensions (construction of extension velocities) are discussed, e.g. in [6].

Discretization of the regularized levelset equation. Let $\varepsilon > 0$ be a small regularization parameter. Instead of (2), we solve the following problem:

$$\frac{\partial P}{\partial t} = \sqrt{\varepsilon^2 + |\nabla P|^2}\ \mathrm{div}\left(\frac{\nabla P}{\sqrt{\varepsilon^2 + |\nabla P|^2}}\right) + F\sqrt{\varepsilon^2 + |\nabla P|^2},\ \text{in } (0,T)\times\Omega,$$

$$\left.\frac{\partial P}{\partial n}\right|_{\partial\Omega} = 0 \text{ on } (0,T)\times\partial\Omega, \quad P|_{t=0} = P_{ini}(x) \text{ in } \Omega.$$

It can be approximated using the following explicit nine-point-stencil finite-difference scheme:

$$P_{i,j}^{k+1} = P_{i,j}^{k} + \tau Q(\mathring{\nabla}_h\, P_{i,j}^k)\,\Bigg(F_{i,j}^k + \left(\frac{P^k_{\bar{x},i+1,j}}{h_1 Q(\nabla P^k_{i+\frac{1}{2},j})} - \frac{P^k_{\bar{x},ij}}{h_1 Q(\nabla P^k_{i-\frac{1}{2},j})} + \frac{P^k_{\bar{y},i,j+1}}{h_2 Q(\nabla P^k_{i,j+\frac{1}{2}})} - \frac{P^k_{\bar{y},ij}}{h_2 Q(\nabla P^k_{i,j-\frac{1}{2}})}\right)\Bigg),$$

where $Q(u,v) = \sqrt{\varepsilon^2 + u^2 + v^2}$, $\nabla P^k_{i+\frac{1}{2},j} = [P^k_{x,i,j}, \frac{1}{2}(P^k_{\mathring{y},i+1,j} + P^k_{\mathring{y},i,j})]$, $\nabla P^k_{i-\frac{1}{2},j} = [P^k_{\bar{x},i,j}, \frac{1}{2}(P^k_{\mathring{y},i,j} + P^k_{\mathring{y},i-1,j})]$, $\nabla P^k_{i,j+\frac{1}{2}}$ and $\nabla P^k_{i,j-\frac{1}{2}}$ evaluated analogically. The scheme is only conditionally stable.

Discretization for the Allen-Cahn equation is derived by spatial finite differences. Nodal values then remain functions of time, for which we obtain a system of ODEs (the semi-discrete scheme) in the following form:

$$\xi^2 \frac{\mathrm{d}p^h}{\mathrm{d}t} = \xi^2 g(\theta)\left(\Delta_h p^h + f_0(p^h)\right) + \xi^2 |\nabla_h p^h| F \text{ on } \omega_{\mathbf{h}},$$

$$p^h\,|_{\gamma_{\mathbf{h}}} = 0, \quad p^h(0) = p_{ini}.$$

The equations are numerically solved by the Runge-Kutta-Mersn 4-th order method with adaptive time step. The scheme has been analyzed in [1] from the convergence viewpoint.

2 Parallelization techniques

The above described algorithms are parallelized by means of the Message Passing Library MPI both using Fortran 77/90 and C programming languages. Computations using MPI, version 1.1 were performed on the supercomputing systems IBM SP3 and Cray T3E at CINECA[1], IBM SP, IBM SP2 at the Czech Technical University in Prague, and computations using LAM MPI library[2] were performed on a local network of Linux PC workstations at the Czech Technical University in Prague. In both approaches described below, the computational task is performed by one or more processes, each of them running either on a separate processor (a hardware unit, or virtual unit in the emulated mode).

Cartesian domain splitting is an approach where a rectangular domain Ω is decomposed into rectangular subdomains, each of them treated by one process. Boundaries of subdomains overlap by one grid line, on which they exchange data. The amount of communication between processes depends on the blocking strategy. We tested the *row-wise blocking strategy*, where the domain is decomposed row-wise. Each block interacts with neighbouring blocks during a timestep. The other tested strategy was the *chequerboard blocking.* In this case, each block communicates with maximum eight neighbours during a timestep.

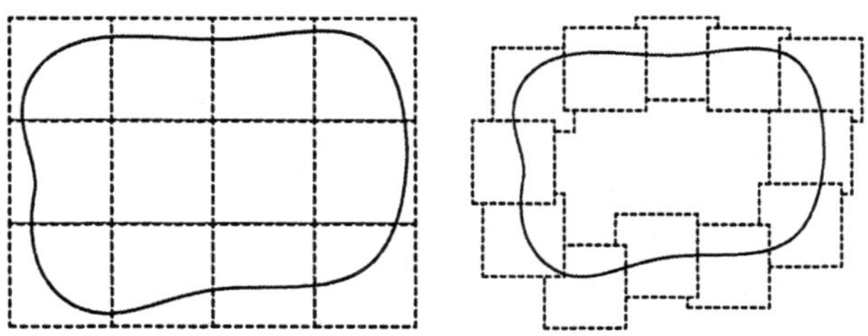

Fig. 1. Cartesian domain splitting (left), and narrow-band splitting (right)

Narrow-band technique introduced in [6] explores the fact that we are interested only in the evolution of the curve $\Gamma(t)$. It is therefore enough to follow the evolution of $P = P(t, x)$ in the vicinity of the levelset $\Gamma(t)$. The presented approach provides a significant speedup. On the other hand, it is less accurate and more difficult to implement, because it requires a reconstruction of the narrow band when Γ approaches its edge (the operation is called *reinitialization*). In our implementation, we cover the curve by overlapping squares of a constant width which are assigned to processes in an intuitive way (Fig. 1).

[1] Supercomputing Center of Italian Universities, Bologna

[2] LAM MPI, Local Area Multicomputer is an open source implementation of MPI standard, `http://www.lam-mpi.org`

For example, in case of 64 covering squares and 16 processes, the first four squares are computed by the first process, the second four squares by the second process etc. Consequently, the narrow band created by such squares is not of constant width. The processes exchange data for all nodes, where the squares overlap. The approach is easy to implement including processing of the grid by parts small enough to fit them in the fast cache memory of processors. For the purpose of algorithm evaluation, we define the following quantities:

$$\text{Speedup} = \frac{\text{run time in a single process}}{\text{run time in } n \text{ processes}}, \quad \text{Eff.} = \frac{\text{Speedup}}{\text{number of processes}}.$$

Speedup and efficiency of parallelization for the direct algorithm of the levelset equation - Study 1 (IBM SP). In this study, we consider a circle of the initial radius $R_0 = 1.35$ placed in a domain $(0,4) \times (0,4)$, which shrinks according to the law $v_\Gamma = -\kappa_\Gamma$ (see Figure 2a). The domain is discretized with the mesh size 0.02 in both directions, the time step is $\tau = 4 \cdot 10^{-5}$. Number of time steps is 22500, the computation stops right before the shrinking time $T = 0.9$ (see [1],). The code is parallelized by means of the domain splitting. The results achieved on the IBM SP system are shown in Table 1.

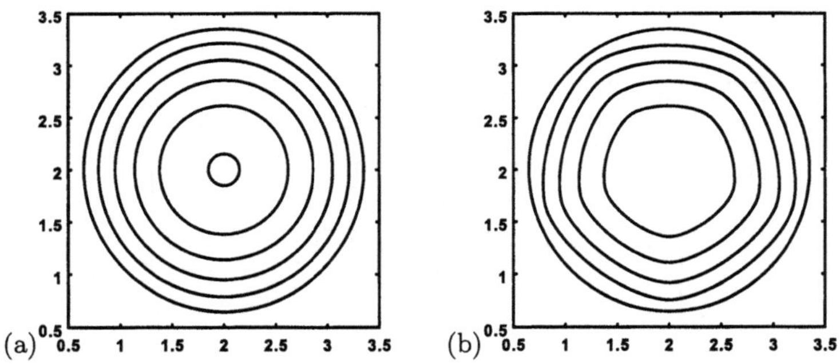

Fig. 2. (a) A circle in $(0,4) \times (0,4)$ shrinking from the initial radius $R_0 = 1.35$ to the radius $R_T = 0.15$ according to the isotropy law $v_\Gamma = -\kappa_\Gamma$. (b) An initial circle of the radius $R_0 = 3.0$ deforming itself according to the 5-folded anisotropy law described by Eq.(3), where $N_{fold} = 5$ and $\zeta = 0.025$.

Study 2 (IBM SP). The above given problem (see Figure 2a) was recomputed using several choices of the mesh size and the time step. As it can be seen from Table 2, efficiency of parallelization depends on the size of data exchanged between processes (e.g., it is faster to send 200kB of data than twice 100kB, due to an initiation).

Study 3 (IBM SP and Linux network). In this case, the initial condition (a circle with the initial radius $R_0 = 1.35$) evolves according to (1) with $F = 0$, $N_{fold} = 5$ and $\zeta = 0.025$ as indicated in Figure 2b). With numerical parameters

Table 1. The results of parallelization efficiency on IBM SP.

Number of processes	Mesh nodes per process	CPU time per process	Mesh nodes total	Communication mesh nodes	Eff.
1	40000	908	40000	0	-
4	10000	258	40000	400 (1.0%)	88%
8	5000	149	40000	800 (2.0%)	76%
12	3333	113	40000	1000 (2.5%)	67%
16	2500	94	40000	1200 (3.0%)	60%

Table 2. Efficiency depends on the mesh size. Computation performed on the IBM SP system (CPU time per process and efficiency).

Mesh size	200 x 200	267 x 267	400 x 400	667 x 667
Time step	$4.0 \cdot 10^{-5}$	$2.3 \cdot 10^{-5}$	$1.0 \cdot 10^{-5}$	$3.6 \cdot 10^{-6}$
Iterations	22500	40000	90000	250000

Mesh size \ Processes	1	4	8	12	16
200 x 200	908	258 (88%)	149 (76%)	113 (67%)	94 (67%)
267 x 267	2585	697 (93%)	392 (83%)	277 (78%)	231 (70%)
400 x 400	14171	3657 (97%)	1915 (93%)	1343 (88%)	1058 (84%)
667 x 667	98904	25574 (97%)	12889 (96%)	8740 (94%)	6775 (91%)

$h_1 = h_2 = 0.01$, $\tau = 1.1 \cdot 10^{-5}$ and 64286 time levels, it terminates at $t = 0.72$. The curve is covered by squares 35 points wide. Due to the curve shrinking, number of active nodes in the narrow band decreases from $\sim$ 33000 to 16000 as shown in Table 3. Compared to the domain splitting, efficiency is lower. This is caused by the fact, that the overlapping areas between processes are larger, and even the number of active nodes increases with the number of processes. On the other hand, the computation is faster, as only a part of the grid is active, and the absolute amount of exchanged data is smaller.

Table 3. Narrow-band approach on IBM SP and Linux network applied to an anisotropic circle shrinking.

Number of processes	Min. no. of active nodes	Avg. no. of active nodes	Max. no. of active nodes	Communication mesh nodes (% of Avg)	CPU time per process (IBM SP)	CPU time per process (linux cluster)
1	16226	26390	33268	0	15219	2608
4	16366	26407	33252	560 (2.1206%)	5009 (76%)	884 (74%)
8	16383	26497	33368	1120 (4.2269%)	3260 (58%)	576 (57%)
12	16435	26517	33409	1680 (6.3356%)	2613 (49%)	456 (48%)
16	16581	26620	33536	2240 (8.4147%)	2365 (40%)	-

Speedup and efficiency of parallelization for the regularized levelset equation - Study 4 (IBM SP3). We considered the test problem shown in Figure 3. For a given number of processes, it is possible to split the domain either into $NPROC$ rows, or into $NPROCX$ columns and $NPROCY$ rows in the checquerboard blocking ($NPROCX \times NPROCY = NPROC$). Unless $NPROC$ is a prime, there are several possibilities for selecting $NPROCX$ and $NPROCY$. Tables 4 and 5 present the runtimes, numbers of communication nodes, and efficiencies for both of the above mentioned blocking strategies attained on an IBM SP3 machine. It is clear from Table 5 that the checquerboard blocking is superior to the row-wise blocking for higher numbers of processes, which is due to a lower number of communicating nodes.

Table 4. Results of parallelization for the row-wise blocking.

Number of processes	Mesh nodes per process	CPU time per process (s)	Mesh nodes total	Communication mesh nodes	Eff.
1	360000	2656	360000	0	-
4	90000	778	360000	1800(0.5000%)	85%
9	40000	307	360000	4800(1.3333%)	96%
16	22500	180	360000	9000(2.5000%)	92%
25	14400	145	360000	14400(4.0000%)	73%
36	10000	125	360000	21000(5.8333%)	57%

Table 5. Results of parallelization for the chequerboard blocking.

Number of processes	Mesh nodes per process	CPU time per process (s)	Mesh nodes total	Communication mesh nodes	Eff.
1	360000	2656	360000	0	-
4	90000	684	360000	1201(0.3336%)	97%
9	40000	320	360000	2404(0.6677%)	92%
16	22500	171	360000	3609(1.0025%)	94%
25	14400	122	360000	4816(1.3377%)	87%
36	10000	85	360000	6025(1.6736%)	87%

Speedup and efficiency of parallelization for the Allen-Cahn equation - Study 5 (IBM SP3). In this computation, we studied the isotropic curve evolution starting at a four-folded pattern in a spatial domain $(0,2)\times(0,2)$. The curve approaches the circle of radius $R = 0.6$ according to the law $v_\Gamma = -\kappa_\Gamma + F(x)$ where the forcing F is a suitable radially symmetric and linear function. Other parameters are $\xi = 0.01$, $h_1 = h_2 = 0.00995$. The curve evolution is in Figure 4(a). The domain was divided into 1, 4, and 16 rectangular subdomains,

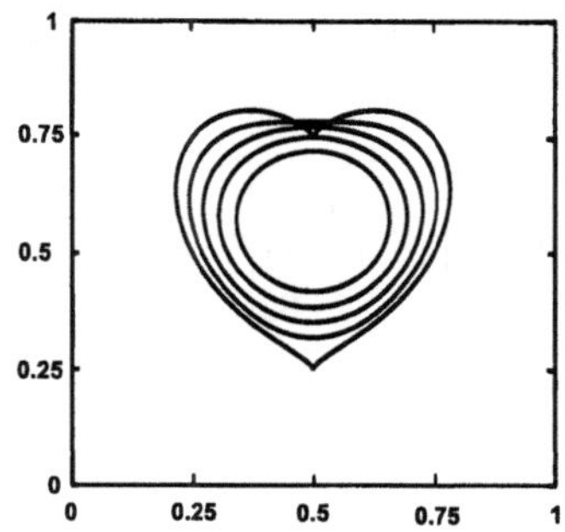

Fig. 3. Evolution of the cardioida curve is driven by the regularized levelset equation, solved in the unit square with $\varepsilon = 10^{-8}$, grid 600×600, $\tau = 10^{-6}$ and 20000 time levels

and the computation was repeated with corresponding number of processes. The mesh size and the total number of mesh points remained the same, the number of mesh points per process decreased, the number of communication mesh points increased, both with increasing number of processes. Measurement results are in Table 6.

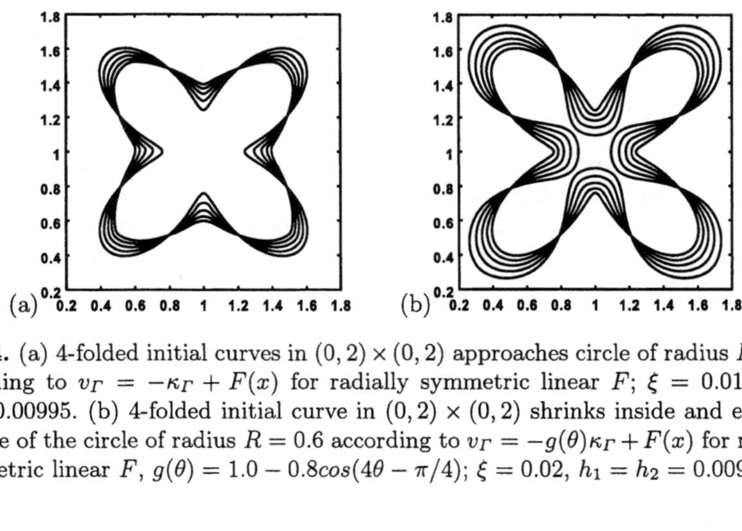

Fig. 4. (a) 4-folded initial curves in $(0,2) \times (0,2)$ approaches circle of radius $R = 0.6$ according to $v_\Gamma = -\kappa_\Gamma + F(x)$ for radially symmetric linear F; $\xi = 0.01$, $h_1 = h_2 = 0.00995$. (b) 4-folded initial curve in $(0,2) \times (0,2)$ shrinks inside and expands outside of the circle of radius $R = 0.6$ according to $v_\Gamma = -g(\theta)\kappa_\Gamma + F(x)$ for radially symmetric linear F, $g(\theta) = 1.0 - 0.8cos(4\theta - \pi/4)$; $\xi = 0.02$, $h_1 = h_2 = 0.00995$.

Study 6 (CRAY T3E). The computation, performed on CRAY T3E, studies the anisotropic curve evolution starting at a four-folded leaf-like curve placed in a spatial domain $(0, 0.4) \times (0, 0.4)$. The curve shrinks inside a circle of radius $R_0 = 0.1$, and expands outside of it thanks to a spatially dependent choice of F in the law $v_\Gamma = -g(\theta)\kappa_\Gamma + F\ g(\theta) = 1.0 - 0.8cos(4\theta - \pi/4)$; $\xi = 0.02$, $h_1 = h_2 = 0.00995$. The curve evolution is in Figure 4(b). The domain was divided into 1, 4, 16, 25 and 64 rectangular subdomains, and the

Table 6. Table of parameters for the use of IBM SP3 - Study 5.

Number of processes	Mesh elements per process	CPU time per process	Mesh elements total	Communication mesh elements	Eff.
1	40401	118.11	40401	0	-
4	10201	29.89	40401	401(0.9925%)	99%
16	2601	10.01	40401	1197(2.9628%)	74%

computation was repeated with corresponding number of processes. Mesh size and total number of mesh points remained the same, number of mesh points per processes decreased, number of communication mesh points increased, both with increasing number of processes. Measurement results are in Table 7.

Table 7. Table of parameters for the use of CRAY T3E - Study 6.

Number of processes	Mesh elements per process	CPU time per process	Mesh elements total	Communication mesh elements	Eff.
1	40401	37.55	40401	0	-
4	10201	9.65	40401	401(0.99%)	97%
16	2601	3.23	40401	1197(2.96%)	73%
25	1681	2.48	40401	1592(3.94%)	61%
64	625	2.20	40401	2765(6.84%)	27%

Acknowledgment. The first author was partly supported by the project No. 159 of the Czech-Slovak Science Programme, the second author was partly supported by the project CTU 0309914 of the Czech Technical University in Prague, and the third author was partly supported by the project No. 201/01/0676 of the Grant Agency of Czech Republic. The authors gratefully acknowledge technical support of the Center for High Performance Computing at the Czech Technical University in Prague (Project No. 145), and technical support of the CINECA - High Performance Computing Centre, Bologna (Programme Minos).

References

1. M. Beneš, *Mathematical and computational aspects of solidification of pure substances*, Acta Mathematica Universitatis Comenianae **70, No. 1** (2001), 123–152.
2. C.M. Elliott, M. Paolini, and R. Schätzle, *Interface estimates for the fully anisotropic Allen-Cahn equation and anisotropic mean curvature flow*, Math. Models Methods Appl. Sci. **6** (1996), 1103–1118.
3. L.C. Evans and J. Spruck, *Motion of level sets by mean curvature I*, J. Diff. Geom. **33** (1991), 635–681.

4. V. Minárik and M. Beneš, *Numerical solution of degenerate parabolic equations of Hamilton-Jacobi type within the context of computer image processing*, ALGORITMY 2002, Proceedings of contributed papers and posters (Bratislava), 2002, pp. 162–170.
5. J.A. Sethian, *Level set methods*, Cambridge University Press, New York, 1996.
6. ———, *Evolution, implementation, and application of level set and fast marching methods for advancing fronts*, Journal of Computational Physics **169** (2001), 503–555.

The Finite Element Analysis of an Elliptic Problem with a Nonlinear Newton Boundary Condition

Veronika Sobotíková

Department of Mathematics, Faculty of Electrical Engineering,
Czech Technical University, Technická 2, 166 27 Praha 6
Czech Republic
sobotik@math.feld.cvut.cz

Summary. Results of the study of an elliptic 2D problem with a nonlinear Newton boundary condition are presented. The problem is discretized with the use of the FEM and the integrals are evaluated by numerical quadratures. In the case of a nonpolygonal domain the main attention is paid to the effect of a piecewise linear approximation of the boundary. The error estimate for the solution of the discrete FE problem is derived.

1 Introduction

A number of problems in science and technology can be described by partial differential equations with a nonlinear Newton boundary condition, see e.g. [1], [8], [5] and [7]. In this contribution we deal with a finite element approximation of one of such problems which can be met in the modelling of electrolysis of aluminium with the aid of the stream function. In this case the nonlinear term in the boundary condition has a "polynomial" behaviour and describes turbulent flow in a boundary layer.

Let us introduce some notations which we will use later. Let $G \subset \mathbf{R}^2$ be a bounded domain with a Lipschitz-continuous boundary ∂G. By $\overline{G}$ we denote the closure of G and by n the unit outward normal to ∂G. We use the well-known Lebesgue and Sobolev spaces $L^p(G), L^p(\partial G), W^{k,p}(G)$, $H^k(G) = W^{k,2}(G), W^{k,p}(\partial G)$ for $k \in \{0,1,2,\dots\}$ and $p \in [1,\infty]$ (see e.g. [6]). By $\|.\|_{k,p,G}$ and $\|.\|_{k,p,\partial G}$ we denote the standard norms in $W^{k,p}(G)$ and $W^{k,p}(\partial G)$, respectively. Then $\|.\|_{0,p,G}$ and $\|.\|_{0,p,\partial G}$ mean, of course, the norms in $L^p(G)$ and $L^p(\partial G)$. The symbols $|.|_{k,p,G}$ and $|.|_{k,p,\partial G}$ stand for the seminorms in $W^{k,p}(G)$ and $W^{k,p}(\partial G)$. The space $(H^1(G))^*$ is the dual space to $H^1(G)$ and $\langle .,. \rangle$ is the duality pairing between $(H^1(G))^*$ and $H^1(G)$. If $K \subset \mathbf{R}^n$, we denote by $P_k(K)$ the space of all polynomials on K of degree less than or equal to k.

2 Continuous Problem

Let $\Omega \subset \mathbf{R}^2$ be a bounded domain. We suppose the boundary $\partial\Omega$ of Ω to be Lipschitz-continuous, and moreover, let the boundary of Ω be piecewise of the class C^3.

We consider the following boundary value problem:

Find $u : \overline{\Omega} \to \mathbf{R}$ such that

$$-\triangle u = f \qquad \text{in } \Omega, \tag{1}$$

$$\frac{\partial u}{\partial n} + \kappa |u|^{\alpha} u = \varphi \qquad \text{on } \partial\Omega, \tag{2}$$

where $f : \Omega \to \mathbf{R}$ and $\varphi : \partial\Omega \to \mathbf{R}$ are given continuous functions, and $\kappa > 0,\ \alpha \geq 0$ are given constants.

In a standard way we can introduce a weak formulation of the above problem:

A function $u : \Omega \to \mathbf{R}$ is said to be a *weak solution of problem* (1) – (2), if

$$\begin{aligned} &\text{a)}\ \ u \in H^1(\Omega), \\ &\text{b)}\ \ a(u,v) = L(v) \quad \forall v \in H^1(\Omega). \end{aligned} \tag{3}$$

The forms a and L from (3,b) are defined for $u, v \in H^1(\Omega)$ by

$$\begin{aligned} a(u,v) &= b(u,v) + d(u,v), \\ L(v) &= L^{\Omega}(v) + L^{\Gamma}(v), \end{aligned}$$

where

$$\begin{aligned} b(u,v) &= \int_{\Omega} \nabla u \cdot \nabla v \, dx, \\ d(u,v) &= \kappa \int_{\partial\Omega} |u|^{\alpha} u \, v \, dS, \\ L^{\Omega}(v) &= \int_{\Omega} f v \, dx, \\ L^{\Gamma}(v) &= \int_{\partial\Omega} \varphi v \, dS. \end{aligned}$$

It was shown in [2] that the operator $A : H^1(\Omega) \to (H^1(\Omega))^*$ corresponding to the nonlinear form a by

$$\langle A(u), v \rangle = a(u,v) \qquad \forall u, v \in H^1(\Omega)$$

is uniformly monotone, Lipschitz-continuous on every bounded subset of $H^1(\Omega)$ and coercive; the linear form L defining the right-hand side of (3,b) is continuous. Hence, by the monotone operator theory, problem (3) has exactly one solution.

3 Finite Element Discretization

We discretize the problem with the use of the finite element method. Let us consider a system $\left\{\Omega_h\right\}_{h\in(0,h_0)}$, $0 < h_0 < 1$, of polygonal approximations of Ω. As Ω need not be convex and thus the inclusion $\Omega_h \subset \Omega$ need not be valid, we suppose that Ω^* is a bounded domain with Lipschitz-continuous boundary such that $\overline{\Omega} \subset \overline{\Omega}^*$ and $\overline{\Omega}_h \subset \overline{\Omega}^*$ for every $h \in (0, h_0)$. Let the right-hand side f of the equation (1) be defined on the whole domain Ω^*.

On the domains Ω_h we consider triangulations $\mathcal{T}_h$ formed by a finite number of closed triangles T. We say that $T \in \mathcal{T}_h$ is a *boundary triangle*, if T has a side $S \subset \partial\Omega_h$. By s_h we denote the set of all sides $S \subset \partial\Omega_h$ of all boundary triangles $T \in \mathcal{T}_h$. We suppose that all vertices of $\mathcal{T}_h$ are in $\overline{\Omega}$, that all vertices lying on $\partial\Omega_h$ belong to $\partial\Omega$ too, and that every boundary triangle has exactly two vertices lying on $\partial\Omega_h$. Moreover, let all points from $\partial\Omega$ where the condition of C^3-smootheness of $\partial\Omega$ is not satisfied be vertices of $\mathcal{T}_h$. Finally, we suppose the intersection of $\partial\Omega$ and $\partial\Omega_h$ to be formed only by sides and vertices of triangles from $\mathcal{T}_h$.

We denote by h_T the length of the maximum side of $T \in \mathcal{T}_h$. Let us set

$$\tilde{h} = \max_{T\in\mathcal{T}_h} h_T.$$

We assume the index h to be chosen in such a way that $h = \tilde{h}$.

In what follows we denote by $|T|$ the area of a triangle $T \in \mathcal{T}_h$ and by $|S|$ the length of a side $S \in s_h$.

In order to be able to prove the solvability of the discrete problem and the convergence of the method, we assume that:

a) The system $\left\{\mathcal{T}_h\right\}_{h\in(0,h_0)}$ of triangulations is *regular*, which means that the magnitudes of inner angles of all triangles $T \in \mathcal{T}_h$ are bounded from zero by a positive constant ϑ_0 independent of $h \in (0, h_0)$.

b) The triangulations $\mathcal{T}_h, h \in (0, h_0)$, *locally* satisfy the *inverse assumption* at $\partial\Omega$:
There exists $\nu > 0$ such that for every $h \in (0, h_0), S \in s_h$, we have

$$|S| > \nu h.$$

Due to these two assumptions there exists a constant $\lambda > 0$ such that

$$h^2 > |T| > \lambda h^2$$

for every boundary triangle $T \in \mathcal{T}_h$ and every $h \in (0, h_0)$.

An approximate solution of problem (3) will be sought in the space of linear triangular conforming elements $H_h \subset H^1(\Omega_h)$:

$$H_h = \{v_h \in C(\overline{\Omega}_h);\ v_h|_T \in P_1(T)\ \forall T \in \mathcal{T}_h\}.$$

The forms a and L defining the weak solution are discretized in two steps. First we integrate in all integrals over Ω_h and $\partial\Omega_h$ instead of over Ω and $\partial\Omega$, respectively. In the second step we apply quadrature formulae to evaluate these integrals. We suppose the formula used for the integration over triangles to be exact for all constant functions and the formula used for the integration over sides to be exact for all linear functions and to be monotone, i.e., all its coefficients are positive. In such a way we come to the following approximations of the forms defining the weak solution ($v_h, w_h \in H_h$):

$$\begin{aligned} a_h(v_h, w_h) &= b_h(v_h, w_h) + d_h(v_h, w_h), \\ L_h(w_h) &= L_h^{\Omega}(w_h) + L_h^{\Gamma}(w_h), \end{aligned}$$

where

$$\begin{aligned} b_h(v_h, w_h) &= \int_{\Omega_h} \nabla v_h \cdot \nabla w_h \, dx, \\ d_h(v_h, w_h) &= \kappa \sum_{S \in \mathcal{S}_h} |S| \sum_{\mu=1}^{m} \beta_\mu (|v_h|^\alpha v_h w_h)(x_{S,\mu}), \\ L_h^{\Gamma}(w_h) &= \sum_{S \in \mathcal{S}_h} |S| \sum_{\mu=1}^{m} \beta_\mu (\varphi_h w_h)(x_{S,\mu}), \\ L_h^{\Omega}(w_h) &= \sum_{T \in \mathcal{T}_h} |T| \sum_{\mu=1}^{M} \omega_\mu (f w_h)(x_{T,\mu}). \end{aligned}$$

Here φ_h is an approximation of the function φ from the boundary condition, which will be introduced later.

Let us note that in the approximation of the form b we do not use numerical integration, as we integrate constant functions.

Now we can define an *approximate solution of problem* (3) as a function $u_h : \Omega_h \to \mathbf{R}$ such that

$$\begin{aligned} &\text{a)}\ u_h \in H_h, \\ &\text{b)}\ a_h(u_h, v_h) = L_h(v_h) \qquad \forall v_h \in H_h. \end{aligned} \tag{4}$$

4 Ideal Triangulation

Since the problem is nonlinear in the boundary condition, we meet some difficulties in the analysis of the FEM. These are caused especially by the fact that in general the boundaries of Ω_h and Ω may not be identical, the union of all triangles of $\mathcal{T}_h$ may not form $\overline{\Omega}$, and that we seek the approximate solution in a space which may be different from the space in which we look for the weak solution. We handle these difficulties with the aid of Zlámal's ideal elements (see [12]).

First we introduce the concept of the ideal triangle. Let us have a triangulation $\mathcal{T}_h$ of the domain Ω_h and let $T \in \mathcal{T}_h$. If T is not a boundary triangle, we set $T^{id} = T$ and we denote its vertices by $P_1,\ P_2,\ P_3$ in any way. If T is a boundary triangle, then we numerate its vertices in such a way that the vertices P_1 and P_3 lie on the boundary. Then we replace in the triangle T the straight side $S = \overline{P_1 P_3} \subset \partial\Omega_h$ by the arc $\Sigma_S = \widehat{P_1 P_3} \subset \partial\Omega$ to get the curved ideal triangle T^{id}. We denote the set of all ideal triangles by $\mathcal{T}_h^{id}$ and call it *the ideal triangulation associated with* $\mathcal{T}_h$. It is obvious that the union of all ideal triangles T^{id} from $\mathcal{T}_h^{id}$ forms $\overline{\Omega}$.

As the functions from the space H_h are not defined on the whole domain Ω, we need to modify somehow these functions. We proceed in the following way:

Let us consider the reference triangle K in the (ξ_1, ξ_2) - plane with the vertices $R_1 = (0,0),\ R_2 = (1,0)$ and $R_3 = (0,1)$. We denote by X^0 the affine mapping which maps the triangle K one-to-one on the triangle T in such away that $X^0(R_i) = P_i$ for $i = 1,2,3$, and let Ξ^0 be its inverse.

By Zlámal [12], there exists such a mapping X, which maps one-to-one the reference triangle K on the ideal triangle T^{id}, that it as well as its inverse Ξ are of the class C^2. With the aid of these mappings we can now define a *function* $\hat{w}$ *associated with a function* $w \in H_h$ by

$$\hat{w}(x) = w(X^0(\Xi(x))) \qquad \forall x \in T^{id}.$$

It is obvious that we can construct this function $\hat{w}$ for any function w defined on $\overline{\Omega}_h$ and not only for $w \in H_h$.

In a similar way we can introduce a function $\hat{\gamma} : \partial\Omega \to \mathbf{R}$ *associated* with a function γ defined on $\partial\Omega_h$. We put

$$\hat{\gamma}(x) = \gamma(X^0(\Xi(x))) \qquad \forall S \in \mathcal{S}_h,\ \forall x \in \Sigma_S.$$

Moreover, we can define for every function $\gamma : \partial\Omega \to \mathbf{R}$ its *approximation* γ_h given on $\partial\Omega_h$ by

$$\gamma_h(x^0) = \gamma(X(\Xi^0(x^0))) \qquad \forall S \in \mathcal{S}_h,\ \forall x^0 \in S.$$

In such a way we approximate the function φ from the boundary condition.

Remark: It is obvious that $\hat{w}|_T = w|_T$ for every $T \in \mathcal{T}_h \cap \mathcal{T}_h^{id}$.

It was shown in [4] that

$$\left| \int_{\partial\Omega} \hat{v}(x)dS - \int_{\partial\Omega_h} v(x)dS \right| \le c_5 h \int_{\partial\Omega_h} |v(x)|dS$$

for every $v \in L^1(\partial\Omega_h)$. Moreover, the following relation between the seminorms of w and $\hat{w}$ in the Sobolev spaces $W^{1,p}(S)$ and $W^{1,p}(\Sigma)$ is valid:

$$|w|^p_{1,p,S} = \left(\frac{|S|}{|\Sigma|}\right)^{1-p} |\hat{w}|^p_{1,p,\Sigma},$$

where

$$1 \le \frac{|\Sigma|}{|S|} < 1 + ch.$$

This estimates allows us to derive relations (5, a, b) between norms and seminorms of w and $\hat{w}$ from the following Lemma 1. The proof of (5, d) can be found in the same paper [4], for the proof of (5, c) see [10].

Lemma 1. *Let $p \in [1, \infty)$. Then there exist positive constants $C_1 = C_1(p)$, $C_2 = C_2(p)$, $C_3 = C_3(p)$, C_4 and $h_1 \in (0, h_0]$ such that for every $h \in (0, h_1)$*

$$\begin{array}{llll} \text{a)}\ C_1^{-1}\|\hat{w}\|_{0,p,\partial\Omega} & \le \|w\|_{0,p,\partial\Omega_h} & \le C_1\|\hat{w}\|_{0,p,\partial\Omega} & \forall w \in L^p(\partial\Omega_h), \\ \text{b)}\ C_2^{-1}|\hat{w}|_{1,p,\partial\Omega} & \le |w|_{1,p,\partial\Omega_h} & \le C_2|\hat{w}|_{1,p,\partial\Omega} & \forall w \in W^{1,p}(\partial\Omega_h), \\ \text{c)}\ C_3^{-1}|\,\hat{w}\,|_{1,p,\Omega} & \le |\,w\,|_{1,p,\Omega_h} & \le C_3|\,\hat{w}\,|_{1,p,\Omega} & \forall w \in H_h, \\ \text{d)}\ C_4^{-1}\|\hat{w}\|_{1,2,\Omega} & \le \|w\|_{1,2,\Omega_h} & \le C_4\|\hat{w}\|_{1,2,\Omega} & \forall w \in H_h. \end{array} \quad (5)$$

Remark: In what follows, whenever we will refer to any result from [2] or [3], where the domain Ω is polygonal, the polygonality of Ω was not used in the associated proof.

5 Existence of Approximate Solutions.

Let $f \in W^{1,q}(\Omega^*)$ for some $q > 2$ and $\varphi \in W^{1,r}(\partial\Omega)$ for some $r > 1$. By [2] the forms a_h are under these assumptions continuous and strictly monotone. Moreover, it was shown in [4] that the forms L_h are continuous and that there exists a positive constant C_5 such that

$$|L(\hat{v}_h) - L_h(v_h)| \;\le\; C_5 h\|v_h\|_{1,2,\Omega_h}$$

for every $h \in (0, h_0)$, $v_h \in H_h$. Concerning the coercivity of the forms a_h, we have the following lemma:

Lemma 2. *Let h_1 and C_4 be as in Lemma 1. Then there exist $h_2 \in (0, h_1]$ and $C_6 > 0$ such that*

$$a_h(v_h, v_h) \ge C_6\|v_h\|^2_{1,2,\Omega_h}$$

for every $h \in (0, h_2)$ and $v_h \in H_h$ with $\|v_h\|_{1,2,\Omega_h} \ge C_4$.

Remark: In what follows, we suppose h_0 to be so small that $h_2 = h_1 = h_0$ in Lemmas 1 and 2.

The coercivity of a_h was obtained by the aid of the following estimate of the error in the approximation of the form d:

$$|d(\hat{v}_h, \hat{w}_h) - d_h(v_h, w_h)| \le C_7 h^{1/2-\alpha/r}\|v_h\|^{\alpha+1}_{1,2,\Omega_h}\|w_h\|_{1,2,\Omega_h}$$

$$\forall h \in (0, h_0),\ \forall v_h, w_h \in H_h,$$

where $C_7 = C_7(r)$ is a positive constant, $r \in [1, \infty)$.

On the basis of the above mentioned properties of the forms L_h and a_h, similar to those of the forms L and a, the monotone operator theory gives:

Theorem 1. *Discrete problem (4) has for every $h \in (0, h_0)$ exactly one solution $u_h \in H_h$.*

(See [4], Theorem 4.1.)

6 Convergence of the Finite Element Method

The monotonicity of the forms a_h is sufficient in the proof of the existence and uniqueness of the approximate solutions. However, for the derivation of the error estimate of the method we need somewhat stronger result. By [11], the forms a_h are "almost uniformly monotone" in the following sense:

$$a_h(v_h, v_h - w_h) - a_h(w_h, v_h - w_h) \geq \sigma(\| \hat{v}_h - \hat{w}_h \|_{1,2,\Omega})$$

$$\forall v_h, w_h \in H_h,$$

where the function

$$\sigma(t) = \begin{cases} \bar{c}\, c^* t^{\alpha+2} & \text{for } 0 \leq t \leq (c^*)^{-\frac{1}{\alpha}} \\ \bar{c}\, t^2 & \text{for } t \geq (c^*)^{-\frac{1}{\alpha}} \end{cases}$$

is nonnegative and increasing in $[0, \infty)$.

In [11] also another estimate of the error in the approximation of the form d was derived:

We have ($1 < p \leq \infty$, $C_8, C_9 = C_9(p)$ are positive constants):

$$|d(\hat{v}_h, \hat{w}_h) - d_h(v_h, w_h)| \leq$$
$$\leq \left(C_8 h \|v_h\|_{1,2,\Omega_h}^{\alpha+1} + C_9 h^{1-\frac{1}{p}} |v_h|_{1,p,\Omega_h} \|v_h\|_{0,\infty,\Omega_h}^{\alpha} \right) \|w_h\|_{1,2,\Omega_h}$$

for all $h \in (0, h_0)$ and $v_h, w_h \in H_h$. (We set $\frac{1}{\infty} = 0$.)

On the basis of the above inequalities an abstract error estimate was established. In what follows, we denote by u the weak solution and by u_h, $h \in (0, h_0)$, solutions of the discrete problem.

Lemma 3. *Let $Q(t) = \frac{\sigma(t)}{t}$ for $t > 0$, $Q(0) = 0$, Q_{-1} be the inverse of Q and let $1 < p \leq \infty$. Then there exist positive constants C_{10}, C_{11}, C_{12} and $C_{13} = C_{13}(p)$ such that*

$$\|u - \hat{u}_h\|_{1,2,\Omega} \leq \|u - \hat{v}_h\|_{1,2,\Omega} +$$
$$+ Q_{-1} \left\{ C_{10} h + C_{11}(1 + \|u\|_{1,2,\Omega}^{\alpha} + \|\hat{v}_h\|_{1,2,\Omega}^{\alpha}) \|u - \hat{v}_h\|_{1,2,\Omega} + \right.$$
$$\left. + C_{12} h (\|v_h\|_{1,2,\Omega_h} + \|v_h\|_{1,2,\Omega_h}^{\alpha+1}) + C_{13} h^{1-\frac{1}{p}} |v_h|_{1,p,\Omega_h} \|v_h\|_{0,\infty,\Omega_h}^{\alpha} \right\}$$

for all $h \in (0, h_0)$ and $v_h \in H_h$.

Finally, we can formulate the main result, the error estimate for the solution of the discrete problem (see [11], Theorem 4.1):

Theorem 2. *Let the solution u of the continuous problem satisfy $u \in H^2(\Omega)$ and let $u_c \in H^2(\Omega^*)$ be an extension of the function u on $\mathbf{R}^2$. Then for every $p \in (2, \infty)$ there exist $\tilde{h} \in (0, h_0]$ and a positive constant $C = C(p, \|u_c\|_{2,2,\Omega^*})$ such that*

$$\|u - \hat{u}_h\|_{1,2,\Omega} \leq C h^{\frac{1-1/p}{\alpha+1}}$$

for all $h \in (0, \tilde{h})$.

(For the existence of u_c see, e.g., [9], Theorem 3.10.)

Remark: If there exists such a function $u_c \in H^2(\Omega^*) \cap W^{1,\infty}(\Omega^*)$ that $u_c|_\Omega = u$, it is possible to show that the rate of convergence of the method is $O(h^{\frac{1}{\alpha+1}})$. It is an open question, whether this stronger assumption is necessary for improving the error estimate from Theorem 2.

Acknowledgment

The support of the Research Project J04/98/210000010 of Ministry of Education of the Czech Republic is gratefully acknowledged.

References

1. Bialecki, R., Nowak, A.J. (1981): Boundary value problems in heat conduction with nonlinear material and nonlinear boundary conditions. Appl. Math. Modelling, **5**, 417–421
2. Feistauer, M., Najzar, K. (1988): Finite element approximation of a problem with a nonlinear Newton boundary condition. Numer. Math., **78**, 403–425
3. Feistauer, M., Najzar, K., Sobotíková, V. (1999): Error estimates for the finite element solution of elliptic problems with nonlinear Newton boundary condition. Numer. Funct. Anal. Optim., **20**, 835–851
4. Feistauer, M., Najzar, K., Sobotíková, V. (2001): On the finite element analysis of problems with nonlinear Newton boundary condition in nonpolygonal domains. Appl. Math., **46**, 353–382
5. Ganesh, M., Graham, I.G., Sivaloganathan, J. (1994): A pseudospectral three-dimensional boundary integral method applied to a nonlinear model problem from finite elasticity. SIAM J. Numer. Anal., **31**, 1378–1414.
6. Kufner, A., John, O., Fučík, S. (1977): Function spaces. Academia, Prague
7. Křížek, M., Liu, L., Neittaanmäki, P. (1999): Finite element analysis of a nonlinear elliptic problem with a pure radiation condition. In: Sequeira, A., de Veiga, H.B., Videman, J.H. (eds) Applied Nonlinear Analysis. Kluwer, Amsterdam, 271–280
8. Moreau, R., Ewans, J.W. (1984): An analysis of the hydrodynamics of alluminium reduction cells. J. Electrochem. Soc., **31**, 2251–2259

9. Nečas, J. (1967): Les méthodes directes en théorie des équations elliptiques. Academia, Prague
10. Sobotíková, V. (1996): Finite elements on curved domains. East-West J. Numer. Math., **4**, 137–149
11. Sobotíková, V. (2003): An error estimate for the finite element solution of an elliptic problem with a nonlinear Newton boundary condition in nonpolygonal domains. Numer. Funct. Anal. Optim., **24**, 621–635
12. Zlámal, M. (1973): Curved elements in the finite element method. I. SIAM J. Numer. Anal., **10**, 229–240

Automatic Goal-Oriented *hp*-Adaptivity Without Error Estimates

Pavel Šolín[1] and Leszek Demkowicz[2]

[1] CAAM, Rice University *solin@rice.edu*
[2] ICES, The University of Texas at Austin *leszek@ices.utexas.edu*

Summary. We propose and test a fully automatic, goal-oriented *hp*-adaptive strategy for elliptic problems. The method combines two approaches: the standard goal-oriented adaptivity based on a simultaneous solution of the primal and dual problem, and a recently proposed automatic *hp*-adaptive strategy based on minimizing the projection-based interpolation error of a reference solution.

1 Introduction

Nowadays, the theory of the *hp*-version of the finite element method is well-established and founded on solid results mostly due to the efforts of Babuška and coworkers. However, the practical realization of fully automatic and robust 3D *hp*-adaptive algorithms still presents many serious difficulties mainly due to excessive programming complexity.

We would like to introduce a novel fully automatic algorithmic approach to goal-oriented *hp*-adaptivity for elliptic problems. The methodology does not rely on estimates of error or its higher derivatives, and it is capable of achieving exponential convergence not only in the asymptotic but also in *preasymptotic* range of error level. Due to the limited length of this paper, only basic ideas of the approach can be presented, but many details on both theory and computer implementation can be found, e.g., in [3, 4, 5, 7, 8].

2 Different roles of error estimation in *h*-, *p*- and *hp*-adaptivity

Error estimation forms an essential part of most *h*- and *p*-adaptive finite element algorithms. Recall that *h*-adaptivity is based on spatial refinement of elements with largest contributions to the error and that *p*-adaptivity achieves the reduction of the error by increasing the polynomial order in elements. In both cases, thanks to the low number of options an element can be refined (in most cases only one), the estimate of *magnitude* of error in elements is sufficient to guide the adaptive process.

The situation, however, changes with *hp*-adaptivity that allows both for pure *p*-refinements and spatial refinements with suitable redistribution of the polynomial order to element sons. Typically one has several options to choose from, and the number of possibilities increases dramatically as the polynomial order of elements in the mesh gets higher. The situation is illustrated in Fig. 1.

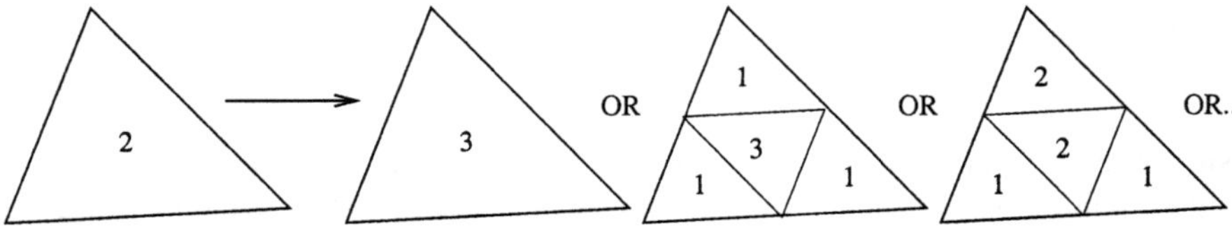

Fig. 1. Several options for *hp*-refinement of a quadratic triangular element. The numbers in the elements indicate their polynomial orders.

It is clear that information about the magnitude of the error in elements is not enough to drive an automatic *hp*-adaptive algorithm – in order to decide between a pure *p*-refinement and a (let us call it) genuine *hp*-refinement, and for the selection of optimal polynomial orders in the element sons, one needs information about the actual *shape* of the error.

There are several possible approaches to do this, all of them in some sense utilizing information about higher-order spatial derivatives of the error function. In [2] the authors apply standard error estimates to the element sons, [6] uses duality-based estimates of higher derivatives of the error in a goal-oriented algorithm. We will follow the idea [4] and calculate an approximation to the error function by means of *reference solutions*.

3 Reference solutions and approximate error function

Consider a bounded domain $\Omega \subset \mathbb{R}^2$ with a Lipschitz continuous boundary and a standard boundary value problem

$$b(u, v) = f(v) \quad \text{for all } v \in V, \tag{1}$$

where $V = V(\Omega)$ is a Hilbert space, b a symmetric bilinear positive-definite elliptic form over $V \times V$ and $f \in V'$ a linear form.

By $\tau_{h,p}$ and $u_{h,p}$ we denote the coarse mesh and the coarse mesh approximation to the exact solution u, respectively. For simplicity let us say that the mesh $\tau_{h,p}$ covers the domain Ω exactly.

Assume that one can use the values of the coarse mesh approximation $u_{h,p}$ to calculate a function u_{ref} that approximates the exact solution u essentially better than $u_{h,p}$ itself. Then the difference

$$err_{h,p} = u_{ref} - u_{h,p} \tag{2}$$

gives a meaningful approximation of the true error

$$e_{h,p} = u - u_{h,p}, \tag{3}$$

and the function u_{ref} is called *reference solution.*

There are various ways to calculate reference solutions. For example, highly accurate approximations based on *Babuška's extraction formulae* (postprocessing formulae applicable mainly to lower-order elements) were used to guide *hp*-adaptivity in [5]. A robust way to obtain reference solutions for elliptic problems without any limitation on the polynomial order was proposed by Demkowicz [4]. The function u_{ref} is defined as approximate solution on a uniformly *hp*-refined mesh, i.e. on a mesh where all elements are refined so that $h \to h/2$ and $p \to p+1$. Since $u_{h,p}$ already contains useful information about lower frequencies in the solution, the higher frequencies identifying $u_{h/2,p+1}$ can be obtained with a reasonable amount of work using a two-grid solver.

4 Projection-based interpolation

This elementwise local technique, that plays an essential role in the presented automatic adaptive algorithm, generalizes the standard Lagrange (vertex) *interpolation* to higher-order finite elements by combining it with *projection* on spaces generated by hierarchic higher-order shape functions. We can confine ourselves to a reference domain, since this is where almost all operations in an *hp*-adaptive code are performed. Choose, for example, a reference triangle $\hat{T}$. Let p^b denote the polynomial order in the interior of $\hat{T}$ and p^1, p^2 and p^3 the polynomial orders related to its edges e_1, e_2 and e_3, respectively. By v_1, v_2, v_3 denote the vertices of $\hat{T}$. Consider a sufficiently regular function w defined in $\hat{T}$.

The projection-based interpolant $w_{h,p}$ is constructed in three steps: First one calculates the *vertex interpolant* $w^v_{h,p}$ as a linear combination of vertex shape functions $\varphi^{v_1}, \varphi^{v_2}$ and φ^{v_3}, such that

$$w^v_{h,p}(v_j) = w(v_j), \quad j = 1, \ldots, 3. \tag{4}$$

In the next step one subtracts the vertex interpolant from the original function w and defines a new function

$$w^{(1)} := w - w^v_{h,p} \tag{5}$$

that vanishes at all vertices. The *edge interpolant* $w^e_{h,p}$ is computed in the form of a sum of contributions over all edges,

$$w^e_{h,p} = \sum_{k=1}^{3} w^{e_k}_{h,p}. \tag{6}$$

Each function $w^{e_k}_{h,p}$, $k = 1, \ldots, 3$, is a linear combination of edge shape functions $\varphi_2^{e_k}, \ldots, \varphi_{p^k}^{e_k}$, such that

$$\|w^{(1)} - w^{e_k}_{h,p}\|_{H^{1/2}_{00}(e_k)} \tag{7}$$

is minimal. The minimum is achieved if and only if the trace of the difference $(w^{(1)} - w^{e_k}_{h,p})|_{e_k}$ is normal to the traces of all edge functions $\varphi_2^{e_k}, \ldots, \varphi_{p^k}^{e_k}$ on the edge e_k in the norm $H^{1/2}_{00}(e_k)$. Hence the discrete minimization problem (7) translates for each edge e_k, $k = 1, \ldots, 3$, into a system of $p^k - 1$ linear algebraic equations. The norm $H^{1/2}_{00}(e_k)$ is defined using harmonic extensions of functions defined on edges to the element interior, and therefore is difficult to evaluate exactly. For practical computations one can replace it by a weighted H^1_0-norm [4, 8].

In the last step one defines a new function

$$w^{(2)} := w^{(1)} - w^{e}_{h,p}. \tag{8}$$

Notice that this function generally does not vanish on edges. The bubble interpolant $w^b_{h,p}$ is obtained by projecting $w^{(2)}$ on a space generated by the bubble functions of orders $p \le p^b$ in the H^1-seminorm. Minimization of the difference

$$|w^{(2)} - w^b_{h,p}|_{H^1} \tag{9}$$

in the H^1-seminorm translates, analogously as in the previous case, into $(p^b - 1)(p^b - 2)/2$ linear algebraic equations.

Finally the projection-based interpolant $w_{h,p}$ is defined as the sum of the vertex, edge and bubble interpolants,

$$w_{h,p} = w^v_{h,p} + w^e_{h,p} + w^b_{h,p}. \tag{10}$$

5 Adaptivity as elementwise minimization of approximate error

At the beginning of each mesh adaptation step one has at his disposal the following information: the coarse mesh $\mathcal{T}_{h,p}$, the coarse mesh solution $u_{h,p}$, the uniformly refined mesh $\mathcal{T}_{h/2,p+1}$, the reference solution $u_{ref} \equiv u_{h/2,p+1}$ and the approximate error function $err_{h,p} = u_{ref} - u_{h,p}$. The question is how to use the function $err_{h,p}$ to adapt the mesh $\mathcal{T}_{h,p}$ in an optimal way.

Recall from Fig. 1 that there always are several possibilities an element can be *hp*-refined. With no further information on the exact solution u it is not known a-priori which *hp*-refinement is the optimal one. Therefore a possible strategy could be to parse through all element refinement options, always creating a new mesh $\mathcal{T}^*_{h,p}$, computing a new approximate solution $u^*_{h,p}$

and selecting the element refinement option that maximizes the value of error drop $\triangle err_{h,p}$,

$$\begin{aligned}\triangle err_{h,p} &= \|err_{h,p}\|_{e,\Omega} - \|err^{*}_{h,p}\|_{e,\Omega} \\ &= \|u_{ref} - u_{h,p}\|_{e,\Omega} - \|u_{ref} - u^{*}_{h,p}\|_{e,\Omega} \\ &= \sum_{i=1}^{N} \|u_{ref} - u_{h,p}\|_{e,K_i} - \|u_{ref} - u^{*}_{h,p}\|_{e,K_i}.\end{aligned} \tag{11}$$

Here $\|.\|_{e,\Omega}$ means the standard energy norm

$$\|w\|^2_{e,\Omega} = b(w,w) \tag{12}$$

that is defined for every form b satisfying assumptions listed at the beginning of Section 3.

Unfortunately, $\triangle err_{h,p}$ is a global quantity and (11) cannot be maximized elementwise. Obviously one cannot afford to solve the global discrete problem for each element and all its *hp*-refinement options.

However, at the cost of introducing an asymptotically negligible error one can replace the coarse mesh solution $u_{h,p}$ in (11) with the coarse mesh interpolant of the reference solution $\Pi_{h,p}u_{ref}$ and at the same time the function $u^{*}_{h,p}$ by the interpolant $\Pi^{*}_{h,p}u_{ref}$ of the reference solution to the mesh $\tau^{*}_{h,p}$. This transforms (11) to

$$\triangle ERR_{h,p} = \sum_{i=1}^{N} \|u_{ref} - \Pi_{h,p}u_{ref}\|_{e,K_i} - \|u_{ref} - \Pi^{*}_{h,p}u_{ref}\|_{e,K_i}. \tag{13}$$

This is a step of crucial importance. Thanks to locality of the projection-based interpolation operators described in Section 4, the approximate error $ERR_{h,p}$ *can be minimized elementwise.* In other words, one does not have to solve global discrete problems for all investigated element refinement options. For each element K_i, $i = 1, 2, \ldots, N$, the global problem related to maximization of $\triangle err_{h,p}(\Omega)$ is replaced with a local problem of maximizing $\triangle ERR_{h,p}(K_i)$, where

$$\triangle ERR_{h,p}(K_i) = \|u_{ref} - \Pi_{h,p}u_{ref}\|_{e,K_i} - \|u_{ref} - \Pi^{*}_{h,p}u_{ref}\|_{e,K_i}. \tag{14}$$

In each mesh optimization step the maximum of interpolation error decrease rates over all elements is calculated and only elements with rates exceeding, e.g., 1/3 of the maximum, are selected for refinement. The implementation, however, involves many important details that exceed the scope of this paper – we refer, e.g., to [4, 7].

6 One-dimensional illustration

The algorithm can be best illustrated in one spatial dimension. Competitive refinements have a very natural structure here: Adding one DOF to a linear element ($p = 1$) can be done either as a $p \to p+1$ refinement or as a $h \to h/2$ refinement with $p_L = 1, p_R = 1$. Adding one DOF to a quadratic element can be done either as a $p \to p+1$ refinement or as a $h \to h/2$ refinement with $p_L = 1, p_R = 2$ or $p_L = 1, p_R = 2$. Adding one DOF to a cubic element results into four analogous options and so on.

Consider, e.g., the Poisson problem $-u'' = f$ in the interval $\Omega = (0, \pi)$ with homogeneous Dirichlet boundary conditions. The function f is chosen in such a way that the function $u(x) = k(1-x)^m \sin(nx)$, $k = 2$, $m = 2$, $n = 5$, depicted in Fig. 2, is the exact solution.

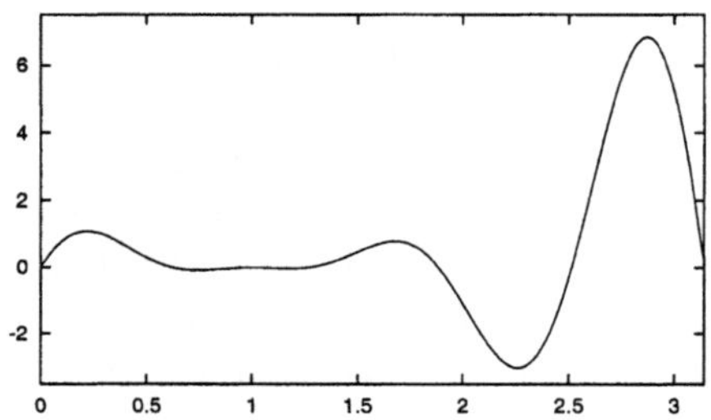

Fig. 2. The exact solution u

The following results were obtained by means of a one-dimensional automatic goal-oriented *hp*-adaptive C++ code MESHOPT[1]. One starts from an equidistant mesh consisting of three quadratic elements. The convergence curve in H^1-seminorm is shown in Fig. 3.
Figs. 4 – 7 visualize a few first steps of the automatic *hp*-adaptive algorithm.

7 Incorporation of goal-oriented adaptivity

In comparison with the adaptivity in energy norm which attempts to minimize the energy of the residual of the approximate solution, the goal-oriented approach attempts to control concrete features of the solved problem (*quantities of interest*). Very often quantities of interest can be represented as bounded linear functionals of the solution, see basic literature on goal-oriented adaptivity, or, for a review, [7, 8].

[1] MESHOPT can be downloaded free of charge at the web page of the second author `http://www.caam.rice.edu/~solin`.

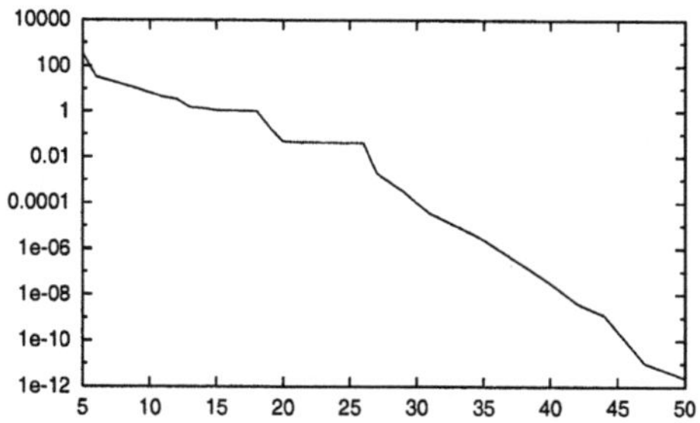

Fig. 3. Convergence curve. x-axis: number of DOF, y-axis: $|u - u_{h,p}|^2_{H^1(0,\pi)}$ in decimal logarithmic scale

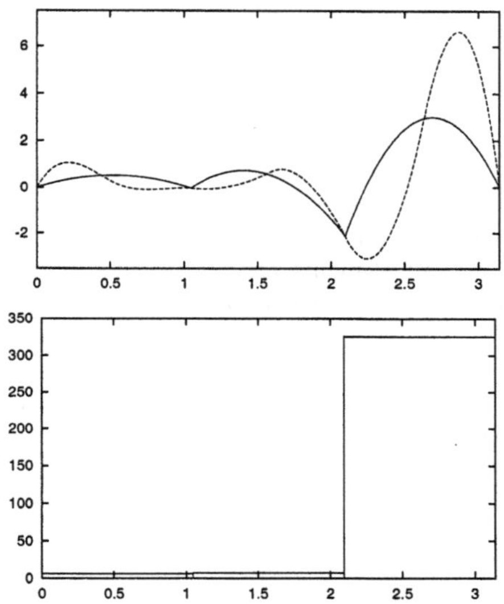

Fig. 4. Left: $u_{h,p}$ (solid line) and $u_{ref} \equiv u_{h/2,p+1}$ (dashed line). Right: projection error decrease rates. Only element K_3 exceeds 1/3 of maximum and is selected for refinement. Largest decrease of interpolation error on K_3 is achieved by means of its p-refinement.

7.1 Dual problem and estimate of error in goal

Let us recall the basic ideas leading to the formulation of the dual problem, since they will be used to incorporate the goal-oriented adaptivity into the energy-driven *hp*-adaptive strategy discussed in Section 5.

Consider problem (1) and its discrete version $b(u_{h,p}, v_{h,p}) = f(v_{h,p})$ for all $v_{h,p} \in V_{h,p}$ where $V_{h,p} \subset V$ is a polynomial finite element approximation of space V. Define the error $e_{h,p} = u - u_{h,p}$ and consider the residual $r_{h,p}(v_{h,p}) = f(v_{h,p}) - b(u_{h,p}, v_{h,p})$. Relate the residual $r_{h,p}$ to the error in the quantity of interest, i.e. find $G \in V''$ such that $G(r_{h,p}) = L(e_{h,p})$. By reflexivity, G can be related to an element v in the original space (*influence function*),

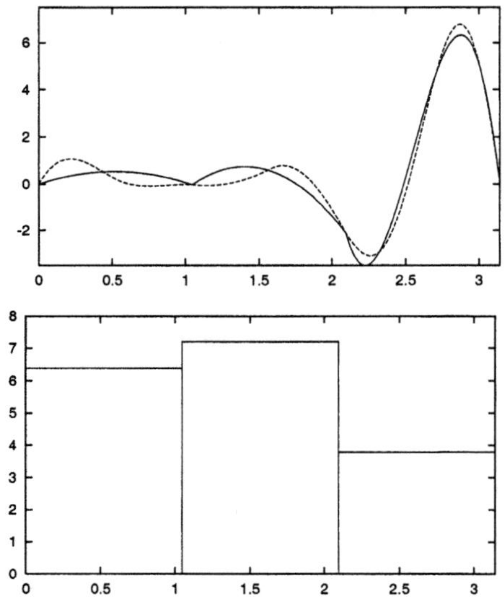

Fig. 5. Left: $u_{h,p}$ (solid line) and $u_{h/2,p+1}$ (dashed line). Right: projection error decrease rates. This time, all elements are selected for refinement. Automatically selected best combination is p-refinement of K_1 and K_2 and hp-refinement with $p_L = p_R = 2$ for K_3.

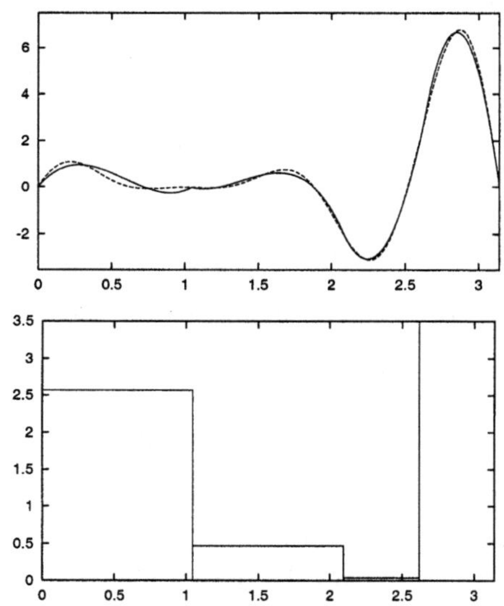

Fig. 6. Left: $u_{h,p}$ (solid line) and $u_{h/2,p+1}$ (dashed line). Right: projection error decrease rates. Elements K_1 and K_4 will be p-refined.

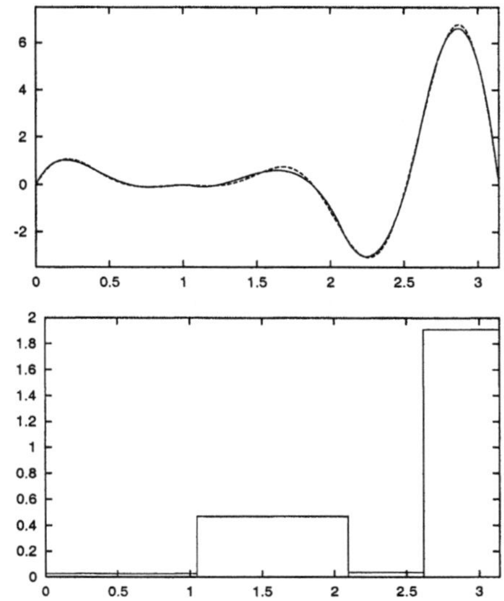

Fig. 7. Left: $u_{h,p}$ (solid line) and $u_{h/2,p+1}$ (dashed line). Right: projection error decrease rates. Element K_4 will be p-refined. In the following step (not depicted), element K_2 will be selected for hp-refinement with $p_L = p_R = 2$. And so on...

$$G(r_{h,p}) = r_{h,p}(v) = f(v) - b(u_{h,p}, v) = b(u,v) - b(u_{h,p}, v) \\ = \underbrace{b(e_{h,p}, v) = L(e_{h,p})} \tag{15}$$

where v is the solution to the *dual problem*: Find $v \in V$ such that

$$b(u,v) = L(u) \tag{16}$$

for all $u \in V$. Consider the discrete dual problem $b(u, v_{h,p}) = L(u)$ for all $u \in V_{h,p}$. Estimate the error in the quantity of interest by means of the errors *in energy norms* for both the primal and dual problem:

$$|L(u) - L(u_{h,p})| = |L(u - u_{h,p})| = |b(u - u_{h,p}, v)| \tag{17}$$

$$= |b(u - u_{h,p}, v - v_{h,p})| \le C \sum_{K \in \tau_{h,p}} ||u - u_{h,p}||_{e,K} ||v - v_{h,p}||_{e,K}.$$

Standard orthogonality property for the error in the solution was used.

7.2 Elementwise minimization of approximate error in goal

Recall that the energy-driven hp-adaptive algorithm from Section 5 minimizes the global error in energy norm by elementwise maximizing the drop of projection-based interpolation error (13) of the reference solution u_{ref} from the coarse mesh $\tau_{h,p}$ to the next optimal mesh $\tau^*_{h,p}$.

The estimate (17) shows that the error $|L(u) - L(u_{h,p})|$ in goal is controlled by errors of both the primal and dual solutions in the energy norm. Therefore, an *hp*-adaptive algorithm will minimize the error in goal if instead of elementwise maximizing (13) it will elementwise maximize the product

$$\triangle ERR_{h,p}^{goal}(K_i) = \triangle ERR_{h,p}^{primal}(K_i) \triangle ERR_{h,p}^{dual}(K_i) \tag{18}$$

where

$$\triangle ERR_{h,p}^{primal}(K_i) = \|u_{ref} - \Pi_{h,p} u_{ref}\|_{e,K_i} - \|u_{ref} - \Pi_{h,p}^{*} u_{ref}\|_{e,K_i} \tag{19}$$

and

$$\triangle ERR_{h,p}^{dual}(K_i) = \|v_{ref} - \Pi_{h,p} v_{ref}\|_{e,K_i} - \|v_{ref} - \Pi_{h,p}^{*} v_{ref}\|_{e,K_i}. \tag{20}$$

Here $v_{ref} \equiv v_{h/2,p+1}$ is the reference solution to the dual problem, calculated on the uniformly *hp*-refined mesh $\tau_{h/2,p+1}$.

8 Reference to numerical examples

Performance of the presented goal-oriented *hp*-adaptive strategy is quite impressive in comparison with the same algorithm applied to standard goal-oriented adaptivity or to standard (energy-driven) *hp*-adaptivity. However, there is little space here for a sufficient presentation of a realistic model problem and meaningful discussion of numerical results. We refer the reader to a recent book [8] together with the paper [7] that uses the presented automatic goal-oriented *hp*-adaptive strategy to resolve a challenging industrial application related to axisymmetric Maxwell's equations.

Acknowledgment

The work of the first author was supported by the Grant Agency of the Czech Republic under Grant No. GP102/01/D114. The second author acknowledges the financial support of Air Force under Contract F49620-98-1-0255.

References

1. Ainsworth M., Senior, B. (1997): Aspects of an *hp*-adaptive Finite Element Method: Adaptive Strategy, Conforming Approximation and Efficient Solvers. Comput. Methods Appl. Math. Engrg., **150**, 65-87.
2. Ainsworth M., Senior, B. (1997): Aspects of an *hp*-adaptive Finite Element Method: Adaptive Strategy, Conforming Approximation and Efficient Solvers. Comput. Methods Appl. Math. Engrg., **150**, 65-87.

3. Demkowicz, L., Oden, J. T., Rachowicz, W., Hardy, O. (1989): Toward a Universal *hp*-Adaptive Finite Element Strategy. Part 1: Constrained Approximation and Data Structure. Comput. Methods Appl. Math. Engrg. **77**, 79-112.
4. Demkowicz, L., Rachowicz, W., Devloo, Ph. (2002): A Fully Automatic *hp*-Adaptivity. J. Sci. Comput. **17**, Nos.1-3, 127-155.
5. Rachowicz, W., Oden, J. T., Demkowicz, L. (1989): Toward a Universal *hp*-Adaptive Finite Element Strategy. Part 3: Design of *hp* Meshes", Comput. Methods Appl. Math. Engrg. **77**, No. 2, 181-212.
6. Heuveline, V., Rannacher, R. (2003): Duality-based adaptivity and the $h-p$ finite element method. In: M. Feistauer (ed) Proceedings of ENUMATH (August 18 - 22, 2003, Prague, Czech Republic), Springer Berlin Heidelberg
7. Šolín, P., Demkowicz, L. (2003): Goal-oriented *hp*-adaptivity for elliptic problems. Comput. Methods Appl. Math. Engrg., accepted.
8. Šolín, P., Segeth, K., Doležel, I. (2003): Higher-Order Finite Element Methods. Chapman & Hall/CRC Press, Boca Raton, London, New York, Washington, D.C.

A Compression Method for the Helmholtz Equation

Mirjam Stolper[1] and Sergej Rjasanow[2]

[1] Saarland University, FR. 6.1 Mathematik, 66041 Saarbrücken
stolper@num.uni-sb.de
[2] Saarland University, FR. 6.1 Mathematik, 66041 Saarbrücken
rjasanow@num.uni-sb.de

Summary. The collocation boundary element method for the Dirichlet boundary value problem is considered. In order to solve efficiently the resulting linear systems for several wave numbers, the adaptive cross approximation (ACA) method is applied to the matrices. In particular, the algorithm is reformulated for complex problems and the so-called Fourier method is used to compute the necessary entries. Finally, some numerical examples for the solution are presented.

1 Introduction

The Helmholtz equation

$$\left(\Delta + \kappa^2\right) u = 0\,, \quad \kappa = \frac{\omega}{c} \in \mathbb{R}^+\,,$$

arises in many physical problems related to wave propagation. In acoustic applications, ω and c are the frequency and the speed of the sound, and u corresponds to the pressure field. We are interested in the solutions of the associated exterior Dirichlet boundary value problem (BVP)

$$\begin{aligned} \Delta\, u(y) + \kappa^2 u(y) &= 0\,, \qquad x \in \mathbb{R}^3 \setminus \Omega, \\ u(y) &= g(y)\,, \qquad x \in \Gamma\,, \\ \left(\frac{\partial}{\partial r} - i\kappa\right) u(y) &= o\left(\,|y|^{-1}\right) \text{ for large } |y| = r\,. \end{aligned} \tag{1}$$

for a spectrum of real wave numbers $0 \le \kappa \le \kappa_{max}$, where κ_{max} is corresponding to the highest frequency. In (1), $\Gamma = \partial\Omega$ denotes the smooth boundary of the bounded, simply connected domain Ω, and g is a given function. Using Boundary element methods (BEM) to treat these problems, we need to solve a large linear system for each wave number. The memory requirement for each problem is $Mem = O(N^2)$ and a naive procedure for the matrix-vector multiplications (using an iterative solver) is given by $Op = O(M\,N^2)$, where M denotes the number of frequencies, and N is the number of degrees of freedom by BEM discretisation. Typical values are $N = 10^3 - 10^4$ for the dimension of the problem and $M = 10 - 10^2$ for the wave numbers of interest.

Whereas we introduced in [10] and [11] a numerical method, for computing the associated matrices, which is based on the Fourier transform with respect to the wave number κ, we now discuss an efficient method for solving the resulting linear systems. In particular, we apply the Adaptive Cross Approximation (ACA) method (see e.g. [1], [2]) to the transformed matrices and discuss the behaviour of the compression factors in dependence on the wave number.
The paper is organised as follows.
In Sect. 2, we consider the boundary integral formulation for the problem and its discrete form. A review of the Fourier method for computing the matrices is presented in Sect. 3. In Sect. 4, we reformulate the ACA algorithm for the complex problem. Finally, we give some numerical results (Sect. 5).

2 Boundary Integral Formulation and Collocation Method

We define the combined single- and double-layer potential

$$(\mathcal{B} - \mathrm{i}\eta\mathcal{A})\,[f] \equiv \int_\Gamma \frac{\partial G(x,y,\kappa)}{\partial n_x} f(x) - \mathrm{i}\eta\, G(x,y,\kappa) f(x)\, \mathrm{d}F_x\,, \quad y \in \mathbb{R}^3 \setminus \Gamma\,,$$

where $\eta \in \mathbb{R}^+$, and $G(x,y,\kappa)$ is the fundamental solution of the Helmholtz equation defined by

$$G(x,y,\kappa) = \frac{1}{4\pi} \frac{\mathrm{e}^{\mathrm{i}\kappa|x-y|}}{|x-y|}\,, \quad x,y \in \mathbb{R}^3.$$

Note that the fundamental solution and thus the above potential satisfy the Sommerfeld radiation condition.
Using this potential to treat the Dirichlet BVP (1), we need to solve the boundary integral equation (BIE)

$$\left(\frac{1}{2}\mathcal{I} + \mathcal{B} - \mathrm{i}\eta\mathcal{A}\right)[f](y) = g(y)\,, \quad y \in \Gamma\,. \tag{2}$$

For all wave numbers κ, the following uniqueness theorem holds.

Theorem 1. *The exterior Dirichlet BVP (1) has a unique solution for all $g \in H^s(\partial\Omega)$, $s \in \mathbb{R}$,*

$$u(y) = \int_\Gamma \left[\frac{\partial G(x,y,\kappa)}{\partial n_x} - \mathrm{i}\eta\, G(x,y,\kappa)\right] f(x)\, \mathrm{d}F_x \quad \in H^{s+\frac{1}{2}}_{loc}(\Omega^c)\,. \tag{3}$$

In (3), $f \in H^s(\partial\Omega)$ denotes the unique solution of the BIE (2).

For more details on the analysis of the Helmholtz equation, we refer the reader to [3], [5].
In order to solve the equation (2) numerically, the surface Γ is discretised using a system of N plane, triangle panels $\Gamma \approx \Gamma_h = \bigcup_{j=1}^{N} \Gamma_j$. It is natural to use the approximate function f_h in the following form $f \approx f_h(x) = \sum_{j=1}^{N} \mathrm{v}_j \varphi_j(x)$, where the associated ansatz functions φ_j, $j = 1, \ldots, N$, are piecewise constant on Γ_j.
Therefore, the BIE above leads to

$$\left(\frac{1}{2} I + B(\kappa) - \mathrm{i}\eta A(\kappa)\right) \mathrm{v} = \varrho \,, \tag{4}$$

where the elements of the matrices $A, B \in \mathbb{C}^{N\times N}$ are defined by

$$a_{ij}(\kappa) = \frac{1}{4\pi} \int\limits_{\Gamma_j} \frac{\mathrm{e}^{\mathrm{i}\kappa r}}{r} \mathrm{d}F_x \text{ and } b_{ij}(\kappa) = \frac{1}{4\pi} \int\limits_{\Gamma_j} \frac{\mathrm{e}^{\mathrm{i}\kappa r}}{r^3} (\mathrm{i}\kappa r - 1) < n_x, x - y_i > \mathrm{d}F_x .$$

where $r := |x - y_i|$. The vector $\mathrm{v} \in \mathbb{C}^N$ and the right-hand side of the systems are given by $(\mathrm{v})_j = \mathrm{v}_j$ and $(\varrho)_i = g(y_i)$, where y_i, $i = 1, \ldots, N$, denote the corresponding collocation points. It should be remarked that the matrices in (4) explicitly depend on the wave number κ.

3 Fourier-Method

In order to compute the matrices in Eq.(4) for several wave numbers, we first apply the inverse Fourier transformation ($\kappa \to \xi$) to the matrices.
The elements of the inverse Fourier transformed Matrix $\breve{A}(\xi) = \mathcal{F}^{-1}_{\kappa,\xi}[A](\xi) \in \mathbb{R}^{N\times N}$ are given by

$$\breve{a}_{ij}(\xi) = \frac{1}{4\pi} \frac{1}{\xi} \int\limits_{\Gamma_j} \delta(\xi - r) \mathrm{d}F_x \,, \quad i, j = 1, \ldots, N, \tag{5}$$

as a consequence of $\mathcal{F}^{-1}_{\kappa,\xi}[\mathrm{e}^{i\kappa r}](\xi) = \mathcal{F}^{-1}_{\kappa,\xi}[1](\xi - r) = \delta(\xi - r)$. In the case of the double-layer potential matrix, we remark

$$\mathcal{F}^{-1}_{\kappa,\xi}[\mathrm{e}^{i\kappa r}(i\kappa r - 1)](\xi) = \left(r\frac{d}{dz} - 1\right) \delta(z)\Big|_{z=\xi-r}$$

and obtain $\breve{B}(\xi) \in \mathbb{R}^{N\times N}$ with

$$\breve{b}_{ij}(\xi) = \frac{1}{4\pi} \int\limits_{\Gamma_j} \frac{1}{r^3} \left(r\frac{d}{dz} - 1\right) \delta(z)\Big|_{z=r-\xi} (n_x, x - y_i) \mathrm{d}F_x \,, \quad i, j = 1, \ldots, N. \tag{6}$$

Since we consider plane triangle elements Γ_j, the analytical computation of the entries (5) and (6) is based on an appropriate transform of the coordinates. For more details on the analytical computation, we refer the reader to [11]. According to this paper, we restrict our further discussion to the special case printed in **Fig.** 1

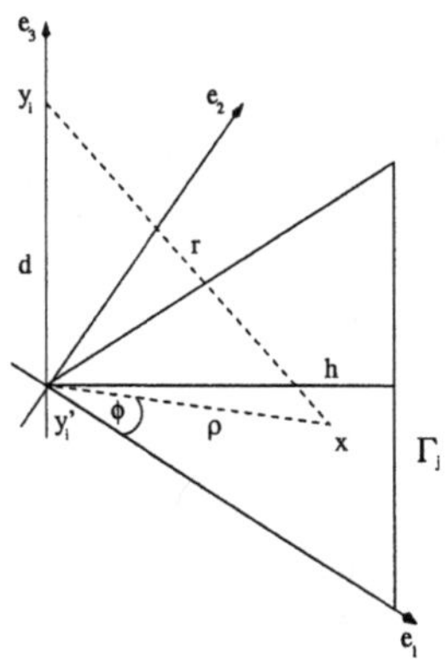

Fig. 1. The computation of the entries of the matrices

and obtain the following analytical expressions:

$$\breve{a}_{ij}(\xi) = \frac{1}{4\pi}\alpha_{ij}(\xi)\mathbb{1}_{[\xi_{min},\xi_{max}]}(\xi)\,, \text{ where } \alpha_{ij}(\xi) := \Phi - \arccos\frac{h}{\sqrt{\xi^2-d^2}}\,, \quad (7)$$

and

$$\breve{b}_{ij}(\xi) = \frac{1}{4\pi}\Big(\Phi\frac{d}{\xi}\,\delta(\xi-|d|) - \underbrace{\frac{d}{\xi^2-d^2}\,\frac{h}{\sqrt{\xi^2-d^2-h^2}}}_{\beta_{ij}(\xi)}\mathbb{1}_{[\xi_{min},\xi_{max}]}(\xi)\Big)\,,\; i \neq j\,. \quad (8)$$

Notice that the diagonal elements of $\breve{B}(\xi)$ are zero.
In (7) and (8), ξ_{max} denotes the maximal distance of the vertex of the triangle to the origin, and ξ_{min} is defined by the minimum of $\xi_0 := \sqrt{d^2+h^2}$ and the minimal distance of the vertex of the triangle to the origin. Further, Φ is the maximal angle between the triangle and the e_1-axis.
The expression above implies that each element has a local support. Therefore the matrices $\breve{A}(\xi)$ and $\breve{B}(\xi)$ have sparse structures for a fixed $\xi \in [\,0\,, \operatorname{diam}(\Gamma)\,]$. It should be remarked that $\alpha_{ij}(\xi)$ and $\beta_{ij}(\xi)$ are integrable functions with anti-derivatives $\sigma_{ij}(\xi)$ and $\gamma_{ij}(\xi)$. Furthermore, $\beta_{ij}(\xi)$ becomes singular at the point $\xi = \xi_0$.

We return to matrices which depend on the wave number by applying the Fourier transformation ($\xi \to \kappa$), i.e. $C(\kappa) = \mathcal{F}_{\xi,\kappa}\,[\breve{C}](\xi)](\kappa)$ with $C = A$ and

$C = B$ respectively. Using the expressions (7) and (8), we obtain new matrices $\tilde{A}(\kappa), \tilde{B}(\kappa) \in \mathbb{C}^{N\times N}$, where its elements are defined by

$$\tilde{a}_{ij}(\kappa) = \frac{\mathrm{e}^{\mathrm{i}\kappa\xi_0}}{4\pi}\left(\sigma_{ij}(\xi)\Big|_{\xi_{min}}^{\xi_{max}} + \sum_{l=1}^{m_p} w_l \bar{\alpha}_{ij}(\xi_l)\mathrm{sinc}(\kappa\xi_l)\mathrm{e}^{\mathrm{i}\kappa\xi_l}\right) \tag{9}$$

and $\tilde{b}_{ii}(\kappa) = 0$,

$$\tilde{b}_{ij}(\kappa) = \frac{\Phi}{4\pi}\,\mathrm{sgn}(d)\,\mathrm{e}^{\mathrm{i}\kappa|d|} - \frac{\mathrm{e}^{\mathrm{i}\kappa\xi_0}}{4\pi}\left(\gamma_{ij}(\xi)\Big|_{\xi_{min}}^{\xi_{max}} + \sum_{l=1}^{m_p} w_l \bar{\beta}_{ij}(\xi_l)\mathrm{sinc}(\kappa\xi_l)\mathrm{e}^{\mathrm{i}\kappa\xi_l}\right). \tag{10}$$

In particular, we need to solve the linear system

$$\left(\frac{1}{2}I + \tilde{B}(\kappa) - \mathrm{i}\eta\tilde{A}(\kappa)\right)\nu = \varrho\,. \tag{11}$$

Due to the independence of the wave number κ, the antiderivative $\sigma_{ij}(\xi)$ and the approximation function $\bar{\alpha}_{ij}(\xi)$ in (9) and, similarly, $\gamma_{ij}(\xi)$ and $\bar{\beta}_{ij}(\xi)$ in (10) need to be calculated only once. Thus we will treat the respective linear systems (11) for several $\kappa \leq \kappa_{max}$ using always these computed data.

4 Application of the ACA

In order to efficiently solve the above linear system, we consider the combination of the described method with an approximation method. In particular, such numerical methods provide an approximation of the solution in almost linear complexity by solving a perturbed linear system, in which the matrix is compressed. Schemes such as fast multipole [4], panel clustering [9], and $\mathcal{H}$-matrices [7], [8] are based on explicitly given kernel approximations by degenerate kernels, i.e. a finite sum of separable functions. In contrast, the ACA algorithm [1], which uses the $\mathcal{H}$-matrix format, is purely algebraic and relies on a small part of the system matrix for its blockwise approximation by low-rank matrices.
Theoretical and numerical aspects for the ACA of collocation matrices that contain asymptotically smooth kernels are discussed in detail in [2]. It should be pointed out that the devised algorithms can formally be applied to matrices for which the kernel is not asymptotically smooth but can be approximated by a degenerated function, e.g. the kernel of the Helmholtz operator. The following estimate is shown in [6].

Lemma 1. *Let $x, y \in \mathbb{R}^3$ and $\nu := |y|/|x| < \nu_0 < 1$. It holds*

$$\left|\frac{\mathrm{e}^{\mathrm{i}\kappa|x-y|}}{|x-y|} - L_p(x,y)\right| \leq c\frac{e^{\kappa|x|}}{|x|}p^2\nu^p\,,$$

where

$$L_p(x,y) = i\kappa \sum_{n=0}^{p-1} \frac{2n+1}{4\pi} h_n^{(1)}(\kappa|x|) j_n(\kappa|y|) P_n(\frac{x \cdot y}{|x||y|})$$

and c only depends on ν_0. In (12), j_n denote the spherical Bessel functions, $h_n^{(1)}$ the spherical Hankel functions and P_n correspond to the Legendre polynomials.

According to the paper [2], we outline the method for the Helmholtz equation, i.e. we reformulate the so-called partially pivoted ACA for the collocation matrices of the Helmholtz equation.
Let $C \in \mathbb{C}^{m' \times n'}$ be a given block. The method produces vectors $u_l \in \mathbb{C}^{m'}$, $v_l \in \mathbb{C}^{n'}$, $l = 1, \ldots, k$, from which the approximant S_k can be formed

$$S_k = \sum_{l=1}^{k} u_l v_l^*$$

and it holds, $C = S_k + R_k$. In particular, we obtain the following algorithm.

Algorithm (partially pivoted ACA)
Let $S_0 = 0$ and $R_0 = C$.
Z denotes the index-set of the computed rows of R_k, $k \geq 0$. Let $Z := \emptyset$ and $i_1 := 1$.

For $k = 0, 1, \ldots$ compute

1. $k := k + 1$
2. $\star$ If $i_k > m'$, then exit.
   Else
3. $Z := Z \cup \{i_k\}$ and $\tilde{v}_k = C^* e_{i_k} - \sum_{l=1}^{k-1} \overline{(u_l)_{i_k}} v_l$
4. Find j_k, so that $(\tilde{v}_k)_{j_k} = \max_j |(v_k)_j|$
5. If $(\tilde{v}_k)_{j_k} = 0$, then $i_k := i_k + 1$ and go to $\star$
6. $v_k = (\tilde{v}_k)_{j_k}^{-1} \tilde{v}_k$ and $u_k = C e_{j_k} - \sum_{l=1}^{k-1} \overline{(v_l)_{j_k}} u_l$
7. $S_k = S_{k-1} + u_k v_k^*$
8. Find i_{k+1}, so that $(u_k)_{i_{k+1}} = \max_{i \notin Z} |(u_k)_i|$

until the stopping criterion is fulfilled.

Since the block C will not be generated completely, the Frobenius norm of the approximant S_k will be used to obtain the stopping criterion. An appropriate stopping criterion is to terminate the iteration, if for a given $\varepsilon > 0$ at step k it holds that

$$||u_k||_F||v_k||_F \le \varepsilon||S_k||_F \,, \tag{12}$$

where the value of $||S_k||_F$ can be recursively computed as follows:

$$||S_k||_F^2 = ||S_{k-1}||_F^2 + 2\sum_{l=1}^{k-1} u_k^* u_l v_l^* v_k + ||u_k||_F^2||v_k||_F^2 \,.$$

Note, that we need $O((m'+n')k^2)$ operations to generate the approximant S_k and its memory requirement is given by $O((m'+n')k)$.

5 Numerical Experiments

(I) (II) (with $\mathrm{diam}(\Gamma) = 0.8m$)

Fig. 2. The surfaces

We first consider the time-harmonic acoustic scattering of a given incoming plane wave

$$u^I = \mathrm{e}^{\mathrm{i}(x\cdot d)}\,, \quad d^T = (0,0,1)$$

by a soft-sound domain Ω. Then the total acoustic wave takes the form $u = u^I + u^S$, where u^S denotes the scattered wave satisfying the Sommerfeld radiation conditions. Further, the homogeneous Dirichlet boundary condition for u holds, $g = -u^I$. In particular, we present numerical experiments for the BIE (4). Since we chose the surface of the unit sphere, see **Fig.** 2 (I), the solution $f(y)$ is known,

$$f(y) = \sum_{n=0}^{\infty} \mathrm{i}^n (2n+1) \frac{j_n(\kappa)}{\mathrm{i}\kappa(\kappa j_n'(\kappa) - \mathrm{i}\eta j_n(\kappa))h_n^{(1)}(\kappa)} P_n((y\cdot d))\,.$$

The solution, which arises from the Fourier method, is denoted by f_{FT}.
We apply the algorithm to a family of surfaces converging to the unit sphere. The sequence is generated by recursive refinement of the meshes dividing each of the surface triangle in four and projecting the new knots to the unit sphere, cf. [2].

For the approximation of the blocks we used the above algorithm and ε in (12) is chosen 10^{-4}, while the relative accuracy of un-preconditioned GMRES was 10^{-6}.
We always consider the compression factors (CF), i.e. the ratios of the amount of storage needed when using the approximant and the amount of storage for the original matrix.
In the tables below, the factors are printed for different N, the first one for a fixed wave number $\kappa = \pi/2$ (Table 1) and the second one for various wave numbers with $\kappa \cdot h \approx 0.36$ (Table 2).

Table 1. Compression factors for $\kappa = \pi/2$

N	h	CF (%)	$\|\|f - f_{FT}\|\|_{\mathbb{L}_2}$	$\|\|f - f_{FT}\|\|_{\mathbb{L}_2} / \|\|f\|\|_{\mathbb{L}_2}$	#$It.$
80	0.97	100	3.91E-01	6.14E-02	5
320	0.51	93	1.07E-01	1.62E-02	5
1280	0.25	46	2.75E-02	4.09E-03	5
5120	0.13	18	6.96E-03	1.03E-03	5
20480	0.06	6	1.93E-03	2.86E-04	4

Table 2. Compression factors for various κ with $\kappa \cdot h \approx 0.36$

N	κ	CF (%)	$\|\|f - f_{FT}\|\|_{\mathbb{L}_2}$	$\|\|f - f_{FT}\|\|_{\mathbb{L}_2} / \|\|f\|\|_{\mathbb{L}_2}$	#$It.$
80	0.62	100	2.7E-01	3.74E-02	4
320	1.20	92	1.05E-01	1.54E-02	5
1280	2.28	49	2.65E-02	4.25E-03	5
5120	4.60	22	1.16E-02	2.24E-03	6
20480	9.24	8	1.93E-02	3.87E-03	9

Figure 3 shows the behaviour of the compression factors in dependence on the wave number for different N.
In the graph, the curves correspond top down to the cases $N = 80$, $N = 320$, $N = 1280$ and $N = 5120$. We decide that the factors increase with a larger wave number.
Similarly, for the surface printed in **Fig.** 2 (II), we consider the problem

$$\mathcal{A}\,[f](y) = \left(-\frac{1}{2}\,\mathcal{I} + \mathcal{B}\right)[w](y) \tag{13}$$

for the exterior Dirichlet BVP for the Helmholtz equation using collocation with piecewise constant ansatz functions. Since we chose $w = G(x, y_0, \kappa)$ with

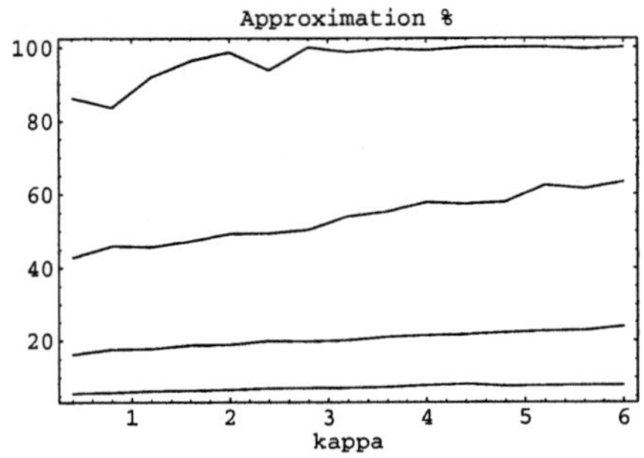

Fig. 3. Compression factors in dependence on the wave number

$x \in \Gamma$ and $y_0 \in \Omega$, the solution of the equation (13) is known to be $f = \partial_{n_x} G(x, y_0, \kappa)|_{x \in \Gamma}$. Table 3 shows the behaviour of the compression factors of the single- and double-layer potential matrices for various wave numbers with $\kappa \cdot h \approx 0.5$ for different N.

Table 3. Compression factors for various κ with $\kappa \cdot h \approx 0.5$

N	κ	$CF(SL)$ (%)	$CF(DL)$ (%)	$\|\|f - f_{FT}\|\|_{\mathbb{L}_2}$	$\|\|f - f_{FT}\|\|_{\mathbb{L}_2}/\|\|f\|\|_{\mathbb{L}_2}$	$\#It.$
1356	11	50	64	6.52E-02	1.08E-01	38
5424	23	23	28	9.34E-02	1.15E-01	66
21696	46	9	10	3.18E-01	2.27E-01	150

References

1. Bebendorf, M. (2000): Approximation of boundary element matrices. Numer. Math., **86**, 565–589
2. Bebendorf, M., Rjasanow, S. (2003): Adaptive low-rank approximation of collocation matrices. Computing, **70**, 1–24
3. Chen, G., Zhou, J. (1992): Boundary element methods. Academic Press Ltd., London
4. Cheng, H., Greengard, L., Rokhlin, V. (1999): A fast adaptive multipole algorithm in three dimensions. J. Comput. Phys., **155**, 468–498
5. Colton, D. L., Kress, R.(1992): Integral equation methods in scattering theory. Krieger Publishing Company, Malabar
6. Goreinov, S. A. (1999): Mosaic-skeleton approximation of matrices generated by asymptotically smooth and oscillatory kernels. In: Tyrtyshnikov, E. (ed.) Matrix Methods and Algorithms, INM RAS, Moscow (in Russian)
7. Hackbusch W. (1999): A sparse matrix arithmetic based on $\mathcal{H}$-matrices. I. Introduction to $\mathcal{H}$-matrices. Computing, **62**, 89–108

8. Hackbusch, W., Khoromskij, B. N. (2000): A sparse $\mathcal{H}$-matrix arithmetic. II. Application to multi-dimensional problems. Computing, **64**, 21–47
9. Hackbusch, W., Nowak, Z.P. (1989): On the fast matrix multiplication in the boundary element method by panel clustering. Numer. Math., **54**, 463–491
10. Köhl, M., Rjasanow, S. (2003): Multifrequency analysis for the Helmholtz equation. Comput. Mech., **32**, 234–239
11. Stolper, M. (2004): Computing and compression of the boundary element matrices for the Helmholtz equation. J. Numer. Math., **12**, 55–75, in Press.

Application of a Stabilized FEM to Problems of Aeroelasticity

Petr Sváček[1] and Miloslav Feistauer[2]

[1] Department of Technical Mathematics, Faculty of Mechanical Engineering, Czech Technical University Prague, Karlovo n. 13, 121 35 Praha 2, Czech Republic *svacek@karlin.mff.cuni.cz*

[2] Faculty of Mathematics and Physics, Charles University Prague, Sokolovská 83, 186 75 Praha 8, Czech Republic *feist@karlin.mff.cuni.cz*

Summary. This paper is concerned with modelling of fluid-structure interaction. We consider two-dimensional viscous incompressible flow past a moving airfoil, which is considered as a solid body with two degrees of freedom, allowing vertical and torsional oscillations of the airfoil. The fluid flow is simulated by the Navier-Stokes equations in the Arbitrary Lagrangian-Eulerian formulation, discretized by the finite element method. We describe the SUPG stabilization of the FEM, time discretization, equations describing the motion of the airfoil and the solution of the discrete problem. The solution of a test problem is presented.

1 Formulation of a flow problem in a moving domain

We analyze numerically two-dimensional viscous incompressible flow past a moving airfoil, which is considered as a solid body with two degrees of freedom, allowing vertical and torsional oscillations of the airfoil. The study of this problem plays an important role in the design of aerospace vehicles. The aero-elastic stability of aerospace vehicles and the aero-elastic responses represented by dynamic load prediction and vibration levels in wings, tails and other aerodynamic surfaces have a great impact on the design as well as in the cost and operational safety.

We assume that $(0,T)$ with $T > 0$ is a time interval and by Ω_t we denote a computational domain occupied by the fluid at time t. By $\mathbf{u} = \mathbf{u}(x,t)$ and $p = p(x,t)$, $x \in \Omega_t$, $t \in (0,T)$, we denote the velocity and the kinematic pressure (i.e, dynamic pressure divided by the density of the fluid), respectively, and ν will denote the kinematic viscosity.

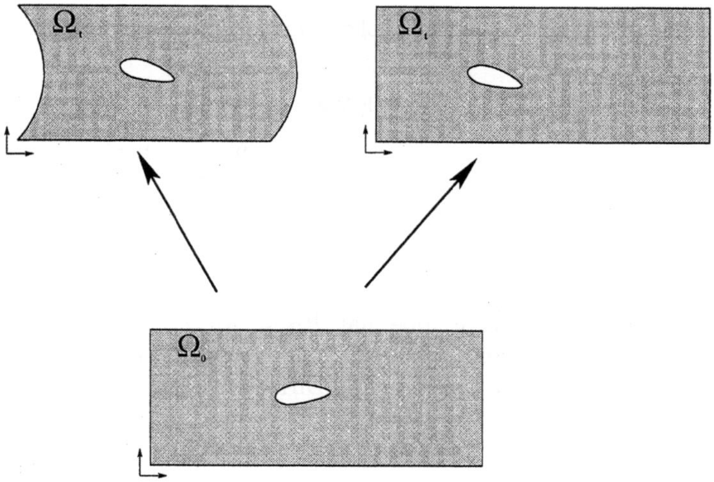

Fig. 1. The Lagrangian(left) and Arbitrary-Lagrangian-Eulerian mapping(right)

In order to simulate flow in a moving domain, we employ Arbitrary Eulerian-Lagrangian (ALE) method. Let us denote by Ω_{ref} the computational domain at chosen fixed time – *reference* or *original configuration.* (We can set, e.g. $\Omega_{\text{ref}} = \Omega_0$.) A one-to-one mapping of the reference configuration onto the computational domain Ω_t at time t – *current configuration* – is denoted by $\mathcal{A}_t$, i.e.

$$\begin{aligned} \mathcal{A}_t : \Omega_{\text{ref}} &\longrightarrow \Omega_t, \\ X &\mapsto x(X,t) = \mathcal{A}_t(X). \end{aligned} \tag{1}$$

Based on this mapping we can compute the *domain velocity* $\tilde{\mathbf{w}}$ at all points X of the reference configuration Ω_{ref} for each time level:

$$\tilde{\mathbf{w}}(X,t) = \frac{\partial}{\partial t} x(X,t), \tag{2}$$

which can be transformed to the space coordinates x by the relation

$$\mathbf{w} = \tilde{\mathbf{w}} \circ \mathcal{A}_t^{-1}, \quad \text{i.e.} \quad \mathbf{w}(x,t) = \tilde{\mathbf{w}}(\mathcal{A}_t^{-1}(x),t) \tag{3}$$

With the aid of ALE mapping we compute the so-called *ALE derivative* $\frac{D^{\mathcal{A}}}{Dt}$, which is anologous to the *material derivative* in the Lagrangian approach. For a function $f : \Omega_t \times (0,T) \rightarrow R$, we set

$$\frac{D^{\mathcal{A}}}{Dt} f(x,t) = \frac{\partial \tilde{f}}{\partial t}(X,t), \qquad X = \mathcal{A}_t^{-1}(x), \tag{4}$$

where $\tilde{f} = f \circ \mathcal{A}_t$. We find that

$$\frac{D^{\mathcal{A}}}{Dt} f = \frac{\partial f}{\partial t} + (\mathbf{w} \cdot \nabla) f. \tag{5}$$

Now we reformulate the Navier-Stokes equations in the ALE form

$$\frac{D^{\mathcal{A}}}{Dt}\mathbf{u} + [(\mathbf{u}-\mathbf{w})\cdot\nabla]\,\mathbf{u} + \nabla p - \nu\triangle\mathbf{u} = 0 \quad \text{in } \Omega_t, \tag{6}$$

$$\nabla\cdot\mathbf{u} = 0 \quad \text{in } \Omega_t. \tag{7}$$

This system is equipped with the initial condition

$$\mathbf{u}(x,0) = \mathbf{u}_0, \quad x \in \Omega_0, \tag{8}$$

and boundary conditions. We assume that $\partial\Omega_t = \Gamma_D \cup \Gamma_O \cup \Gamma_{Wt}$, where Γ_D, Γ_O and Γ_{Wt} are mutually disjoint. On Γ_D, representing the inlet and, possibly, impermeable fixed walls, we prescribe the Dirichlet boundary condition

$$\mathbf{u}|_{\Gamma_D} = \mathbf{u}_D. \tag{9}$$

We denote by Γ_{Wt} the boundary of the airfoil at time t. (See Fig. 1 showing schematically the difference between the Lagrangian and Arbitrary Lagrangian-Eulerian mapping.) On Γ_{Wt} we assume that the fluid velocity $\mathbf{u}$ equals the velocity $\tilde{\mathbf{u}}_\Gamma$ of the profile:

$$\mathbf{u}|_{\Gamma_{Wt}} = \tilde{\mathbf{u}}_\Gamma = \mathbf{w}|_{\Gamma_{Wt}}. \tag{10}$$

The part Γ_O of the boundary represents the outlet, where we prescribe the "do-nothing" boundary condition

$$-(p - p_{ref})\mathbf{n} + \nu\frac{\partial\mathbf{u}}{\partial\mathbf{n}} = 0 \quad \text{on } \Gamma_O, \tag{11}$$

where $\mathbf{n}$ is the unit outer normal to $\partial\Omega_t$ and p_{ref} is a prescribed reference outlet pressure.

2 Discretization

There is a number of possibilities how to carry out the space-time discretization ([5], [10]). In order to develop a stable, accurate scheme, which can easily treat complicated boundaries, we apply the finite element method (FEM). By $Re = UL/\nu$ we define the Reynolds number. Here U is a reference velocity (usually the far field velocity) and L is the length of the airfoil. The relevent Reynolds numbers in our applications are quite large, namely between 10^5 and 10^6. (For such regimes the flow is usually turbulent, but we simulate the flow with the aid of the classical Navier-Stokes equations without any turbulence model.) In order to obtain a physically acceptable numerical solution, it is not possible to use a standard Galerkin FEM, but we have to introduce a suitable stabilization. Here we apply the *streamline diffusion method* (also called SUPG method) together with *grad-div* stabilization of pressure, following [6], [9].

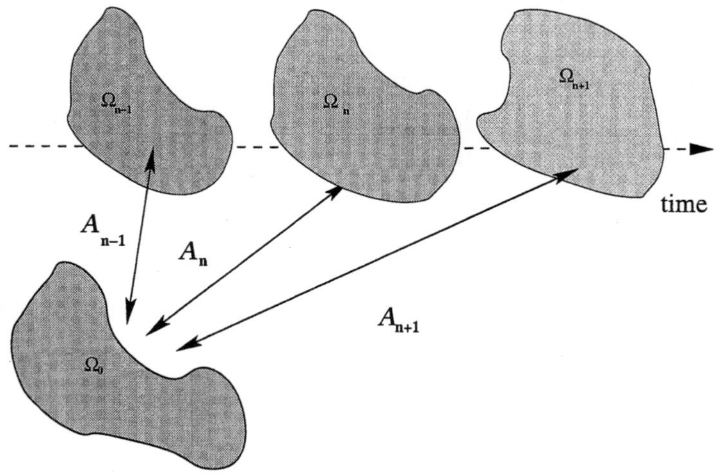

Fig. 2. Time discretization on a moving domain.

2.1 Time discretization

First let us describe the time discretization of the problem. We consider a partition $0 = t_0 < t_1 < \cdots < T$, $t_k = k\tau$, with a time step $\tau > 0$, of the time interval $[0, T]$ and approximate the solution $\mathbf{u}(t_n)$ (defined in Ω_{t_n}) at the time instant t_n by $\mathbf{u}^n$. For the time discretization we use a second-order two-step scheme using the computed approximate solution $\mathbf{u}^{n-1}$ in $\Omega_{t_{n-1}}$ and $\mathbf{u}^n$ in Ω_{t_n} for the calculation of $\mathbf{u}^{n+1}$ in the domain $\Omega_{t_{n+1}}$. With a given ALE mapping $\mathcal{A}_t$ we have

$$\mathcal{A}_{t_{n-1}}(X) = x^{n-1}, \quad \mathcal{A}_{t_n}(X) = x^n, \quad \mathcal{A}_{t_{n+1}}(X) = x^{n+1}, \tag{12}$$

where $X \in \Omega_{\text{ref}}$ is a given point from the reference configuration, e.g. a node of the triangulation. (See Fig. 2.)

Now we define the approximation of the ALE derivative at time t_{n+1} and point x^{n+1} by

$$\begin{aligned}\frac{D^{\mathcal{A}}\mathbf{u}}{Dt}(x^{n+1}, t_{n+1}) &\approx \frac{3\tilde{\mathbf{u}}^{n+1}(X) - 4\tilde{\mathbf{u}}^n(X) + \tilde{\mathbf{u}}^{n-1}(X)}{2\tau} \\ &= \frac{3\mathbf{u}^{n+1}(x^{n+1}) - 4\mathbf{u}^n(x^n) + \mathbf{u}^{n-1}(x^{n-1})}{2\tau}.\end{aligned} \tag{13}$$

and obtain the problem for the unknown functions $\mathbf{u}^{n+1} : \Omega_{t_{n+1}} \to R^2$ and $p^{n+1} : \Omega_{t_{n+1}} \to R$:

$$\begin{aligned}&\frac{3\mathbf{u}^{n+1}(x^{n+1}) - 4\mathbf{u}^n(x^n) + \mathbf{u}^{n-1}(x^{n-1})}{2\tau} \\ &+\Big(\big(\mathbf{u}^{n+1}(x^{n+1}) - \mathbf{w}^{n+1}(x^{n+1})\big) \cdot \nabla\Big)\mathbf{u}^{n+1}(x^{n+1}) \\ &-\nu\Delta\mathbf{u}^{n+1}(x^{n+1}) + \nabla p^{n+1}(x^{n+1}) = 0, \\ &\operatorname{div}\mathbf{u}^{n+1}(x^{n+1}) = 0,\end{aligned} \tag{14}$$

where $\mathbf{w}^{n+1} \approx \mathbf{w}(t_{n+1})$. This problem is equipped with the boundary conditions (9) – (11) on $\partial\Omega_{t_{n+1}}$. Taking into account that $\mathcal{A}_{t_{n+1}}(\mathcal{A}_{t_i}^{-1}(x^i)) \in \Omega_{t_{n+1}}$, we can transform equations (14) completely to the domain $\Omega_{t_{n+1}}$:

$$\begin{aligned}
&\frac{3\mathbf{u}^{n+1} - 4\hat{\mathbf{u}}^n + \hat{\mathbf{u}}^{n-1}}{2\tau} \\
&+ \left((\mathbf{u}^{n+1} - \mathbf{w}^{n+1}) \cdot \nabla\right) \mathbf{u}^{n+1} - \nu \Delta \mathbf{u}^{n+1} + \nabla p^{n+1} = 0 \quad \text{in } \Omega_{t_{n+1}}, \\
&\operatorname{div} \mathbf{u}^{n+1} = 0 \quad \text{in } \Omega_{t_{n+1}},
\end{aligned} \tag{15}$$

where $\hat{\mathbf{u}}^i = \mathbf{u}^i \circ \mathcal{A}_{t_{n+1}} \circ \mathcal{A}_{t_i}^{-1}$. This system is again equipped with the boundary conditions (9) – (11).

2.2 Space discretization

In what follows, we shall carry out the space discretization of the problem to find approximations of functions $\mathbf{u} := \mathbf{u}^{n+1}$ and $p = p^{n+1}$ defined in the domain $\Omega_{t_{n+1}}$, satisfying system (15) and boundary conditions (9) – (11). To this end, we reformulate this problem in a weak sense. Let us set $\Omega = \Omega_{t_{n+1}}$ and define the velocity spaces $W = (H^1(\Omega))^2, X = \{\mathbf{v} \in W; \mathbf{v}|_{\Gamma_D \cap \Gamma_{Wt}} = 0\}$ and the pressure space $M = L_0^2(\Omega) = \{q \in L^2(\Omega); \int_\Omega q\, dx = 0\}$. Then it is easy to find that the solution $U = (\mathbf{u}, p)$ of problem (15) satisfies

$$a(U, U, V) = f(V), \qquad \forall\, V = (\mathbf{v}, q) \in (X, M). \tag{16}$$

Here

$$\begin{aligned}
a(U^*, U, V) &= \frac{3}{2\tau}(\mathbf{u}, \mathbf{v}) + \nu(\nabla\mathbf{u}, \nabla\mathbf{v}) + \left(\left((\mathbf{u}^* - \mathbf{w}^{n+1}) \cdot \nabla\right)\mathbf{u}, \mathbf{v}\right) \\
&\quad - (p, \nabla \cdot \mathbf{v}) + (\nabla \cdot \mathbf{u}, q), \\
f(V) &= \frac{1}{2\tau}\left(4\hat{\mathbf{u}}^n - \hat{\mathbf{u}}^{n-1}, \mathbf{v}\right) - \int_{\Gamma_O} p_{ref}\mathbf{v} \cdot \mathbf{n}\, dS, \\
U &= (\mathbf{u}, p),\ V = (\mathbf{v}, q),\ U^* = (\mathbf{u}^*, p),
\end{aligned} \tag{17}$$

where $(\cdot,\cdot)$ denotes the scalar product in $L^2(\Omega)$. Moreover, we require that $\mathbf{u}$ satisfies the Dirichlet boundary conditions (9), (10). The couple $(\mathbf{u}, p)$ represents the solution on the time level t_{n+1}, i.e. $\mathbf{u}^{n+1} := \mathbf{u}$ and $p^{n+1} := p$.

In order to apply the Galerkin FEM, we shall restrict the weak formulation from the spaces W, X, M to approximate spaces W_h, X_h, M_h, $h \in (0, h_0)$, $h_0 > 0, X_h = \{\mathbf{v}_h \in W_h; \mathbf{v}_h|_{\Gamma_D \cap \Gamma_{Wt}} = 0\}$. Hence, we want to find $U_h = (\mathbf{u}_h, p_h) \in W_h \times M_h$ such that $\mathbf{u}_h$ satisfies approximately conditions (9), (10) and

$$a(U_h, U_h, V_h) = f(V_h), \qquad \forall\, V_h = (\mathbf{v}_h, q_h) \in X_h \times M_h. \tag{18}$$

The couple (X_h, M_h) of the finite element spaces should satisfy the *Babuška–Brezzi (BB) condition*, which guarantees the stability of the scheme: there exists a constant $c > 0$ such that

$$\sup_{\mathbf{w}\in X_h} \frac{(p, \nabla\cdot\mathbf{w})}{|\mathbf{w}|_{H^1(\Omega)}} \geq c\|p\|_{L^2(\Omega)}, \qquad \forall p \in M_h, \quad h \in (0, h_0). \tag{19}$$

We proceed in the following way. Assuming that Ω is polygonal, by $\mathcal{T}_h$ we denote a triangulation of Ω with standard properties from the FEM. The pressure space M is then approximated by the space of piecewise polynomial functions of degree $\leq k$:

$$p \approx p_h \in M_h = \{q \in M \cap C(\overline{\Omega}); q|_K \in P^k(K), \forall K \in \mathcal{T}_h\} \tag{20}$$

and the velocity space W and X are approximated by the spaces of piecewise polynomial functions of degree $\leq k+1$:

$$\begin{aligned} \mathbf{u} \approx \mathbf{u}_h \in W_h &= \{\mathbf{v} \in W \cap \left(C(\overline{\Omega})\right)^2; \mathbf{v}|_K \in \left(P^{k+1}(K)\right)^2, \forall K \in \mathcal{T}_h\} \\ X_h &= W_h \cap W. \end{aligned} \tag{21}$$

This couple (X_h, M_h) satisfies the BB condition (see, [13]).

In practical computations we use the Taylor-Hood P^2/P^1 elements.

3 Stabilization of the FEM

The standard Galerkin discretization (18) may produce approximate solutions suffering from spurious oscillations for high Reynolds numbers. In order to avoid this drawback, we apply the stabilization via streamline-diffusion/Petrov-Galerkin technique (see, e.g., [10], [9], [6]). We define the stabilization terms

$$\begin{aligned} \mathcal{L}_h(U^*, U, V) &= \sum_{K\in\mathcal{T}_h} \delta_K \Big(\frac{3}{2\tau}\mathbf{u} - \nu\Delta\mathbf{u} + (\overline{\mathbf{w}}\cdot\nabla)\,\mathbf{u} + \nabla p, (\overline{\mathbf{w}}\cdot\nabla)\mathbf{v}\Big)_K, \\ \mathcal{F}_h(V) &= \sum_{K\in\mathcal{T}_h} \delta_K \Big(\frac{1}{2\tau}\left(4\hat{\mathbf{u}}^n - \hat{\mathbf{u}}^{n-1}\right), (\overline{\mathbf{w}}\cdot\nabla)\mathbf{v}\Big)_K, \\ U &= (\mathbf{u}, p), \quad V = (\mathbf{v}, q), \quad U^* = (\mathbf{u}^*, p), \end{aligned} \tag{22}$$

where the function $\overline{\mathbf{w}}$ stands for the transport velocity $\overline{\mathbf{w}} = \mathbf{u}^* - \mathbf{w}^{n+1}$, $(\cdot,\cdot)_K$ denotes the scalar product in $L^2(K)$ and $\delta_K \geq 0$ are suitable parameters. Moreover, we introduce the pressure stabilization terms

$$\mathcal{P}_h(U, V) = \sum_{K\in\mathcal{T}_h} \tau_K(\nabla\cdot\mathbf{u}, \nabla\cdot\mathbf{v})_K, \quad U = (\mathbf{u}, p), \quad V = (\mathbf{v}, q), \tag{23}$$

with suitable parameters $\tau_K \geq 0$.

The *stabilized discrete problem* reads: Find $U_h = (\mathbf{u}_h, p_h) \in W_h \times M_h$ such that $\mathbf{u}_h$ satisfies approximately conditions (9), (10) and

$$\begin{aligned} &a(U_h, U_h, V_h) + \mathcal{L}_h(U_h, U_h, V_h) + \mathcal{P}_h(U_h, V_h) = f(V_h) + \mathcal{F}_h(V_h), \\ &\forall V_h \in X_h \times M_h. \end{aligned} \tag{24}$$

The parameter δ_K is defined on the basis of the transport velocity $\overline{\mathbf{w}}$ as

$$\delta_K = \delta^* \frac{h_K}{2\|\overline{\mathbf{w}}\|_{L^\infty(K)}} \xi(Re^{\overline{\mathbf{w}}}), \tag{25}$$

where

$$Re^{\overline{\mathbf{w}}} = \frac{h_K \|\overline{\mathbf{w}}\|_{L^\infty(K)}}{2\nu} \tag{26}$$

is the local Reynolds number and h_K is the size of the element K measured in the direction of $\overline{\mathbf{w}}$. The factor $\xi(\cdot)$ is a monotonically increasing function of $Re^{\overline{\mathbf{w}}}$ such that for local advection dominance ($Re^{\overline{\mathbf{w}}} > 1$) $\xi \to 1$ and for local diffusion dominance ($Re^{\overline{\mathbf{w}}} < 1$) $\xi \to 0$. The parameter $\delta^* \in (0, 1]$ is an additional free parameter. We set, e.g.

$$\xi(Re^{\overline{\mathbf{w}}}) = \min\Big(\frac{Re^{\overline{\mathbf{w}}}}{6}, 1\Big). \tag{27}$$

The choice of the parameters τ_K is again different in diffusion and convection dominated regions. We put

$$\tau_K = \tau^* h_K \|\overline{\mathbf{w}}\|_{L^\infty(K)} \quad \text{and} \quad \tau_K = 0 \tag{28}$$

for local advection dominance and local diffusion dominance, respectively, where $\tau^* \in (0, 1]$. For theoretical analysis of such a choice we refer to [6], [9].

The nonlinear problem (24) is (on each time level) solved iteratively. Starting from an initial approximation $U_h^{(0)}$ and assuming that already iterate $U_h^{(k)}$ has been computed, we define $U_h^{(k+1)} \in W_h \times M_h$ by

$$\begin{aligned} & a(U_h^{(k)}, U_h^{(k+1)}, V_h) + \mathcal{L}_h(U_h^{(k)}, U_h^{(k+1)}, V_h) \\ & + \mathcal{P}_h(U_h^{(k+1)}, V_h) = f(V_h) + \mathcal{F}_h(V_h), \\ & \quad \forall V_h \in X_h \times M_h. \end{aligned} \tag{29}$$

For each time level t_{n+1} we set

$$U_h^{(0)} := (2\hat{\mathbf{u}}^n - \hat{\mathbf{u}}^{n-1}, \hat{p}^n). \tag{30}$$

As numerical experiments show, only a few iterations (29) have to be computed on each time level.

Obviously, problem (29) is linear. It is equivalent to the linear algebraic system

$$S\underline{u} + 2\tau B\underline{p} = f, \qquad B^T\underline{u} = 0, \tag{31}$$

where $\underline{u} \in R^{n_h}$ and $\underline{p} \in R^{m_h}$ are vectors whose components represent degrees of freedom defining the velocity $\mathbf{u}$ and the pressure p, respectively, S is a nonsingular $n_h \times n_h$ matrix and B is an $n_h \times m_h$ matrix. The solution of this system was realized by the direct solver UMFPACK ([1], [2]), which works sufficiently fast for systems with up to 10^5 equations. For larger systems the domain decomposition approach or algebraic multigrid ([11], [12]) will be used.

4 Description of the airfoil motion

The airfoil can oscillate in the vertical direction and in the angular direction around the so-called elastic axis. This vertical and torsional motion is described by the linearized system of ordinary differential equations (see [8], [3])

$$\begin{aligned} m\ddot{H} + k_H H + S_\alpha \ddot{\alpha} &= -F, \\ S_\alpha \ddot{H} + I_\alpha \ddot{\alpha} + k_\alpha \alpha &= M, \end{aligned} \tag{32}$$

where the following notation is used: H – vertical displacement (oriented downward), α – angle of rotation around the elastic axis, k_H – displacement stiffness, S_α – static moment round the elastic axis, I_α – inertia moment, k_α – torsional stiffness.

The force F acting in the vertical direction and the torsional moment M are defined by

$$\begin{aligned} F &= -\int_{\Gamma_{Wt}} \sum_{j=1}^{2} \tau_{2j} n_j dS, \\ M &= -\int_{\Gamma_{Wt}} \sum_{i,j=1}^{2} \tau_{ij} n_j r_i^{\text{ort}} dS, \end{aligned} \tag{33}$$

where

$$\begin{aligned} \tau_{ij} &= \rho \left(-p\delta_{ij} + \nu \left(\frac{\partial u_i}{\partial x_j} + \frac{\partial u_j}{\partial x_i} \right) \right), \\ r_1^{\text{ort}} &= -(x_2 - x_{T2}), \quad r_2^{\text{ort}} = x_1 - x_{T1}, \end{aligned} \tag{34}$$

$\mathbf{n} = (n_1, n_2)$ is the unit outer normal to $\partial\Omega_t$ on Γ_{Wt} (pointing into the airfoil) and $x_T = (x_{T1}, x_{T2})$ is the given position of the elastic axis (lying in the interior of the airfoil) and ρ is the fluid density.

System (32) is tranformed to a first-order ODE system and then solved by the second-order Runge-Kutta method. We proceed in such a way that the computed approximate solution U_h of (24) on time levels t_n and t_{n-1} and the corresponding force F and moment M are extrapolated and used for obtaining H and α at t_{n+1}. This allows us to determine the mapping $\mathcal{A}_{t_{n+1}}$, the domain $\Omega_{t_{n+1}}$ and approximate the domain velocity $\mathbf{w}^{n+1}$. Then we pass to (24) on the next time level t_{n+1}.

5 Numerical results

We have performed a number of numerical simulations. In this paper we present the results obtained for the profile NACA $63_2 - 415$. The length of the profile is 0.3 m, far field velocity is 20 $m\,s^{-1}$ and the air kinematic viscosity give

Reynolds number equal to 400000. Fig. 5 shows the position of the moving profile and velocity isolines at four time instants. We can see conspicuously von Karmán vortices leaving the airfoil. Moreover, the oscillations of vertical position h of the airfoil and the angle α of rotation around the elastic axis are shown and values corresponding to the above airfoil positions are marked.

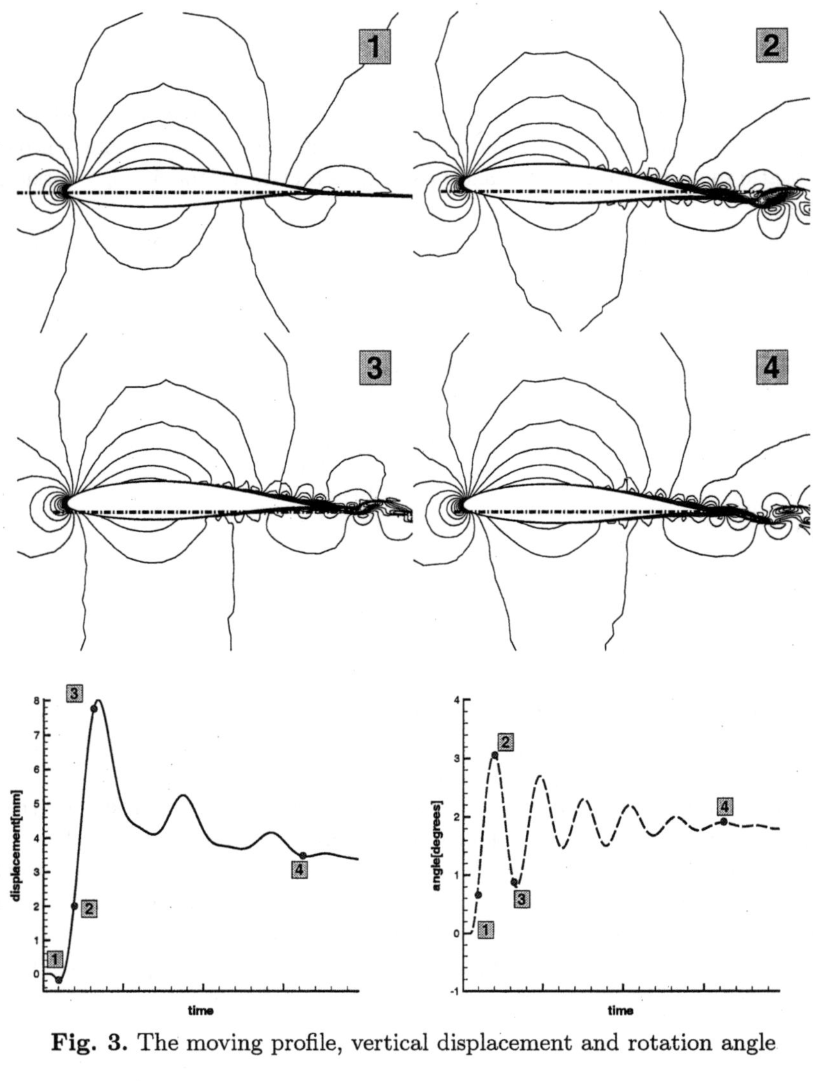

Fig. 3. The moving profile, vertical displacement and rotation angle

Acknowledgments The authors acknowledge the financial support of the Grant Agency of the Czech Republic by the project No. 101/02/0391 "Numerical simulation and experimental research of aeroelasticity of aircrafts considering large displacements". This research was also supported under Research Plans MSM 21000003 and MSM 113200007 of the Ministry of Education of the Czech Republic.

References

1. *Davis T. A.*: A column pre-ordering strategy for the unsymmetric-pattern multifrontal method, Technical report TR-03-006. Submitted to ACM Trans. Math. Software.
2. *Davis T. A., Duff I. S.*: A combined unifrontal/multifrontal method for unsymmetric sparse matrices, ACM Transactions on Mathematical Software, vol. 25, no. 1, pp 1-19, 1999.
3. *Dowell E. H.*: A Modern Course in Aeroelasticity. Kluver Academic Publishers, Dodrecht, 1995.
4. *Elman H., Silvester D.*: Fast nonsymmetric iterations and preconditioning for Navier-Stokes equations. SIAM J. Sci. Comput. 17(1), pp. 33-46, 1996.
5. *Feistauer M.*: Mathematical Methods in Fluid Dynamics. Logman Scientific & Technical, Harlow, 1993.
6. *Gelhard T., Lube G., Olshanskii M. A.*: Stabilized finite element schemes with LBB-stable elements for incompressible flows (preprint)
7. *Heywood J. G., Rannacher R., Turek S.*: Artificial boundaries and flux and pressure conditions for the incompressible Navier-Stokes equations. Stochastische mathematische modelle, Universität Heidelberg, (681), July 1992. (preprint)
8. *Horáček J.*: Nonlinear formulation of oscillations of a profile for aero-hydroelastic computations, Proceedings of the Colloquium Dynamics of Machines 2003, I. Dobiáš, editor. Institute of Thermomechanics, Prague, 2003.
9. *Lube G.*: Stabilized Galerkin finite element methods for convection dominated and incompressible flow problems, Num. Anal. and Math. Model., Banach Center publications (29), Warszawa, 1994.
10. *Turek S.*: Efficient Solvers for Incompressible Flow Problems, Springer Berlin 1998.
11. *Wathen A., Silvester D.*: Fast iterative solution of stabilised Stokes systems, part I: Using simple diagonal preconditioners. SIAM J. Numer. Anal., 30(5), pp. 630-649, 1993.
12. *Wathen A., Silvester D.*: Fast iterative solution of stabilised Stokes systems, part II: Using general block preconditioners. SIAM J. Numer. Anal., 31, pp. 1352-1367, 1994.
13. *Verfürth R.*: Error estimates for mixed finite element approximation of the Stokes equations. R.A.I.R.O. Analyse numérique/Numerical analysis, 18(2), pp. 175-182, 1984.

A Numerical Approach to the Dynamical Behavior of Initiated Pulses in Some Nonlinear Diffusion Equations

Kenji Tomoeda

Department of Applied Mathematics and Information,
Osaka Institute of Technology
Asahi-ku, Osaka, 535-8585, Japan *tomoeda@ge.oit.ac.jp*

Summary. The dynamical behavior of the pulses, which is governed by the interaction between diffusion and absorption, shows the several phenomena. The remarkable ones are the *pulse splitting phenomena* which accompany *pulse connecting phenomena*. In this paper such phenomena are investigated from numerical points of view, and the mathematical justification is stated.

1 Introduction

We consider the propagation of thermal waves in an one-dimensional absorbing medium in which there is an interaction between diffusion and absorption. To describe such a propagation we may use the nonlinear diffusion equation with absorption, which is well-known as the description of the flow of the liquids through the homogeneous porous medium and is represented in the form of the initial value problem:

$$v_t = (v^m)_{xx} - cv^p, \qquad x \in \mathbf{R}^1, \quad t > 0, \tag{1.1}$$

$$v(0,x) = v^0(x), \qquad x \in \mathbf{R}^1. \tag{1.2}$$

Here we have the following assumptions:

(i) $m(>1)$, $p(>0)$, and $c(\geq 0)$ are constants and $m+p \geq 2$;
(ii) $v^0(x) \in C^0(\mathbf{R}^1)$ is nonnegative and has compact support.

In a heated plasma v denotes the temperature and $-cv^p$ describes the losses caused by radiation. We may take $p = 0.5$ for bremsstrahlung radiation and $0.5 \leq p \leq 2$ for synchrotron radiation [14]. The diffusion rate of (1.1) vanishes at points where $v = 0$. This degeneracy causes the occurrence of the finite propagation of the support.

From analytical points of view, Aronson [1], Oleinik, Kalashnikov and Chzou Yui-Lin [13], Kalashnikov [7],[8], and Herrero and Vázquez [6] proved the existence and uniqueness of a weak solution and the property of the finite propagation of the support under the assumptions stated above. Moreover, $v(t,x)$ is smooth in the open set $\mathcal{P}(v) = \{(t,x)|v(t,x) > 0 \text{ and } t > 0\}$, and has the following properties:

(P-1) For $c = 0$, or $c > 0$ and $p \geq 1$ the diffusion is active and supp $v(t, \cdot)$ monotonously expands as t increases;

(P-2) For $c > 0$ and $0 < p < 1$ the absorption is active and the solution vanishes identically at some finite time $T^* > 0$.

In Case (P-1) supp $v(t, \cdot)$ never splits into any multiple connected components for $t > 0$, when supp $v^0(x)$ is connected. Thus the *pulse splitting phenomena* never appear. In Case (P-2) there is a possibility of the *pulse splitting phenomena* caused by absorption, when $v^0(x)$ has two local maxima. Rosenau and Kamin [14] suggested this possibility by numerical computation. Chen, Matano and Mimura [3] constructed the pulse splits into multiple connected components in a finite time. This motivates us to investigate the more detail of the behavior of pulses. For this end we continue numerical computation and find the following phenomena, where the initial pulse $v^0(x)$ has two local maxima and a connected compact support:

(NS-1) The *pulse splitting phenomena* appear, and thereafter these two pulses evolve separately until one of them vanishes (see the left hand side in Fig. 1);

(NS-2) The *pulse splitting phenomena* never appear for $t > 0$ (see the right hand side in Fig. 1);

(NS-3) After the *pulse splitting phenomena* appear, these pulses become connected, and thereafter the *pulse splitting phenomena* appear again (see Fig. 2).

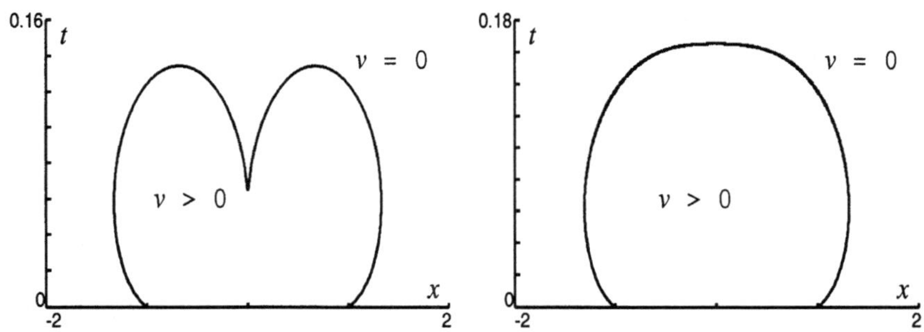

Fig. 1. Numerical support splitting and non-splitting phenomena

When $m + p = 2$ and $0 < p < 1$, we obtain some sufficient conditions under which the phenomena (NS-1) and (NS-2) appear ([12],[15]). However, we are unable to answer the question mathematically whether the phenomenon (NS-3) is true or not. In this paper, we try to justify a part of it, which is as follows.

(NS-4) The *pulse splitting phenomena* appear, and thereafter these pulses become connected.

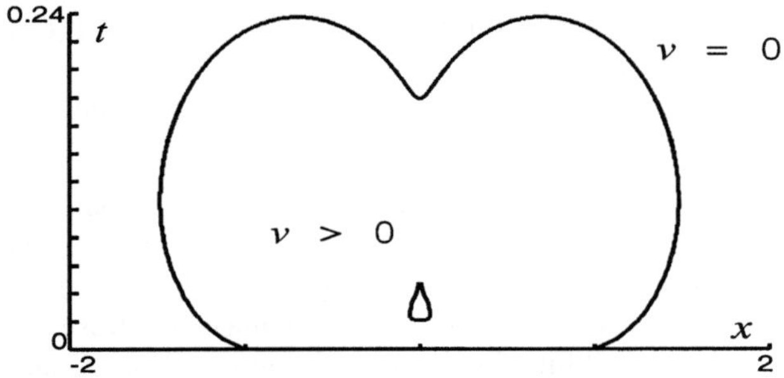

Fig. 2. Numerical support splitting phenomena with connecting property

We call such phenomena *pulse splitting phenomena with connecting property*, and assume the following conditions.

CONDITION A. $c > 0$, $m + p = 2$ and $0 < p < 1$ hold;

CONDITION B. i) $v^0(x) \in C^0(\mathbf{R}^1)$ is a nonnegative function with compact support and $\left((v^0(x))^{m-1}\right)_x \in L^\infty(\mathbf{R}^1) \cap BV(\mathbf{R}^1)$.

ii) $\left((v^0(x))^{m-1}\right)_x$ is absolutely continuous on $\mathbf{I} = \{x | v^0(x) > 0\}$ and ess.inf$_{\mathbf{I}} \left((v^0(x))^{m-1}\right)_{xx}$ is finite.

Our proof is based on the finite difference scheme([10],[11],[12]), the comparison theorem [2] and Kersner's exact solution [9]. Unfortunately, in the case where $m + p \neq 2$, $m > 1$ and $0 < p < 1$, we are unable to find any exact solution and to succeed in constructing the finite difference scheme with convergence. This is the reason why we are concerned with the specific case stated in Condition A.

2 Finite difference schemes

We put $u = v^{m-1}$ and rewrite (1.1)–(1.2) as follows:

$$u_t = muu_{xx} + a(u_x)^2 - c', \tag{4.1}$$

$$u(0, x) = u^0(x) \equiv (v^0(x))^{m-1}, \tag{4.2}$$

where $a = \dfrac{m}{m-1}$, $c' = (m-1)c$ and the term of absorption is written as the constant $-c'$ by the assumption $m+p = 2$. Our difference scheme approximates the problem (4.1)–(4.2) instead of (1.1)–(1.2), and is described as follows:

Find the sequence $\{u_h^n\}_{n=1,2,\cdots} \subset V_h$ for each $u_h^0 \in V_h$ such that

$$u_h^{n+1} = S_{h,k} u_h^n \quad \text{for } n = 0, 1, 2, \cdots, \tag{4.3}$$

where $\ell(u_h^0) = \ell(u^0)$, $r(u_h^0) = r(u^0)$ and $u_h^0(x_i) = u^0(x_i)$ for all $i \in \mathbf{Z}$, h is a space mesh width and V_h is the set of the nonnegative continuous functions $u_h = u_h(x)$ with the following properties:

(i) u_h has compact support;
(ii) u_h is linear on each interval $[x_i, x_{i+1}]$ $(i \in \mathbf{Z})$, where

$$x_i = \begin{cases} ih & \text{for } i \in \mathbf{Z} \setminus \{L-1, R+1\}, \\ \ell & \text{for } i = L-1, \\ r & \text{for } i = R+1. \end{cases} \tag{4.4}$$

$$L = L(\ell) \equiv \min\{i \in \mathbf{Z} \mid ih > \ell\}, \qquad \ell = \ell(u_h), \tag{4.5}$$

$$R = R(r) \equiv \max\{i \in \mathbf{Z} \mid ih < r\}, \qquad r = r(u_h). \tag{4.6}$$

$S_{h,k}$ is somewhat complicated form and its detail is stated in [10],[11] and [12]. We omit the description of $S_{h,k}$. The variable time step $k = k_{n+1} \equiv t_{n+1} - t_n$ $(t_0 = 0)$ determined by

$$k = \frac{1}{c'}\max(u_L, u_{L+1}) \text{ for the approximation to the left interface,} \tag{4.7}$$

or

$$k = \frac{1}{c'}\max(u_R, u_{R-1}) \text{ for the approximation to the right interface.} \tag{4.8}$$

The left and right numerical interfaces are defined by

$$\ell_n = \ell(u_h^n) \quad \text{and} \quad r_n = r(u_h^n) \quad \text{for} \quad n = 0, 1, 2, \cdots, \tag{4.9}$$

respectively. When $S_{h,k}u_h^{n^*} \equiv 0$ holds for some integer $n^* > 0$, we put the numerical extinction time $T_h^* = t_{n^*+1} \equiv t_{n^*} + k_{n^*+1}$, and stop the numerical computation. We define the left(resp. right) numerical interface curves $\ell_h(t)$(resp. $r_h(t)$) by piecewise-linearly interpolating (t_n, ℓ_n)(resp.(t_n, r_n))$(0 \le n \le n^*)$.

We obtain the basic estimates which enable the proofs of the convergence of the numerical solutions and the convergence of the numerical interface curves([10], [11], [12]). The former can be proved by Graveleau and Jamet's argument used in the proofs of Lemma 6.1 and Theorem 7.1([5]). The latter can be proved by applying the idea of DiBenedetto and Hoff [4] to our difference scheme. Moreover, we obtained the interface equation(see Main Theorem in [11]). We state the basic estimates and the convergence of numerical solutions without proof.

Theorem 1 (Basic estimates). *Under Condition A assume $u_h^0 \in V_h$. Then u_h^n either becomes extinct or belongs to V_h for each $n \ge 0$, and the following estimates hold for all $n \ge 0$:*

$$T_h^* \le t_n + \frac{\|u_h^n\|_\infty}{c'}, \tag{4.10}$$

$$\ell_0 - a\|(u_h^0)_x\|_\infty t_n \le \ell_n \le r_n \le r_0 + a\|(u_h^0)_x\|_\infty t_n, \quad \textit{if } u_h^{n+1} \not\equiv 0, \tag{4.11}$$

$$0 \le u_h^n(x) \le \max(\|u_h^0\|_\infty - c't_n, 0) \quad \textit{on } \mathbf{R}^1, \tag{4.12}$$

$$\|(u_h^n)_x\|_\infty \le \|(u_h^0)_x\|_\infty, \tag{4.13}$$

$$TV((u_h^n)_x) \le TV((u_h^0)_x), \tag{4.14}$$

$$\|(u_h^{n+1} - u_h^n)/k_{n+1}\|_{L^1(\mathbf{R}^1)} \le (m+a)\|u_h^0\|_\infty TV((u_h^0)_x)$$
$$+c'(r_0 - \ell_0 + 2a\|(u_h^0)_x\|_\infty t_n), \tag{4.15}$$
$$\inf_{i\in\mathbf{Z}} \delta^2 u_i^0 \le \inf_{i\in\mathbf{Z}} \delta^2 u_i^n, \tag{4.16}$$

where $\|\cdot\|_\infty$ *denotes* $\|\cdot\|_{L^\infty(\mathbf{R}^1)}$.

Theorem 2 (Convergence of Numerical Solutions). *Under Conditions A and B let* $\{h\}$ *be an arbitrary sequence which tends to zero. Then, there exists the unique weak solution* v *of* (1.1)–(1.2), *and*

$$\|v_h - v\|_{L^\infty(\mathcal{H})} \longrightarrow 0 \quad \text{and} \quad |T_h^* - T^*| \longrightarrow 0 \quad \text{as} \quad h \to 0, \tag{4.17}$$

where $\mathcal{H} = [0,\infty) \times \mathbf{R}^1$, $v_h = (u_h)^{1/(m-1)}$, $u_h(t,x) = u_h^n(x)$ *on* $[t_n, t_{n+1}) \times \mathbf{R}^1$ *for all* t_n *and* h, *and* T^* *is the extinction time.*

3 Kersner's exact solution

When $m + p = 2$ and $m > 1$, there is Kersner's exact solution [9] given by

$$K(t,x,\rho,\sigma) = (b_1 t + b_2(\sigma))^{-\frac{1}{m-1}} \tag{2.1}$$
$$\times \left[a_1(\rho,\sigma)(b_1 t + b_2(\sigma))^{\frac{2}{m+1}} - a_2(b_1 t + b_2(\sigma))^2 - x^2\right]_+^{\frac{1}{m-1}},$$

where

$$a_1(\rho,\sigma) = \frac{c(m-1)^4\sigma^2 + 4m^2\rho^2}{4m^2\{(m-1)\sigma\}^{\frac{2}{m+1}}}, \quad a_2 = \frac{c(m-1)^2}{4m^2}, \tag{2.2}$$
$$b_1 = \frac{2m(m+1)}{m-1}, \quad b_2(\sigma) = (m-1)\sigma, \tag{2.3}$$

$\rho > 0$ and $\sigma > 0$ are arbitrary numbers and $[g]_+ = \max\{g, 0\}$. This solution satisfies (1.1)–(1.2) with $v^0(x) = K(0,x,\rho,\sigma)$ in the weak sense and becomes a classical solution in the open set wherein $K(t,x,\rho,\sigma) > 0$. It is easily seen that supp $K(0,x,\rho,\sigma) = [-\rho,\ \rho]$ and the right and left interface curves ζ_+ and ζ_- are written as follows, respectively:

$$\zeta_\pm(t) = \pm\left\{a_1(\rho,\sigma)(b_1 t + b_2(\sigma))^{\frac{2}{m+1}} - a_2(b_1 t + b_2(\sigma))^2\right\}^{\frac{1}{2}}. \tag{2.4}$$

When

$$\sigma < \frac{2\rho}{(m-1)^2}\sqrt{\frac{m}{c}} \tag{2.5}$$

holds, supp $K(t,x,\rho,\sigma)$ expands for $t \in [0,\ \tilde{T}(\rho,\sigma)]$, where

$$\tilde{T}(\rho,\sigma) = \frac{1}{b_1}\left[\left\{\frac{a_1(\rho,\sigma)}{(m+1)a_2}\right\}^{\frac{m+1}{2m}} - b_2(\sigma)\right]. \tag{2.6}$$

For $t > \tilde{T}(\rho,\sigma)$ supp $K(t,x,\rho,\sigma)$ shrinks and $K(t,x,\rho,\sigma)$ identically vanishes at the extinction time $t = T^*(\rho,\sigma)$ given by

$$T^*(\rho,\sigma) = \frac{1}{b_1}\left[\left\{\frac{a_1(\rho,\sigma)}{a_2}\right\}^{\frac{m+1}{2m}} - b_2(\sigma)\right]. \tag{2.7}$$

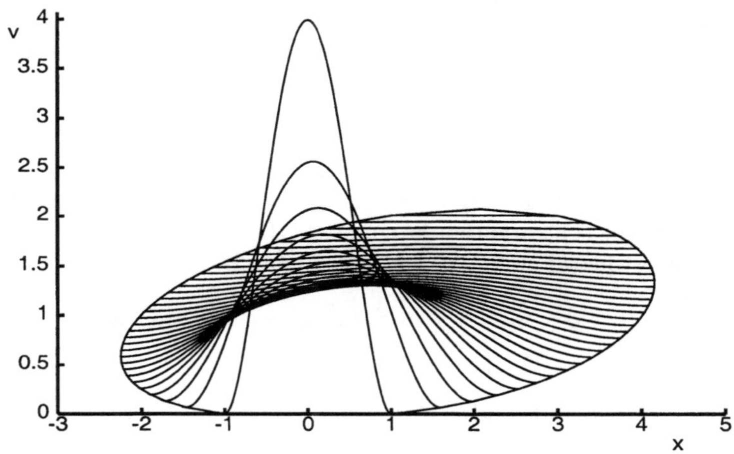

Fig. 3. Kersner's Exact Solution and Free Boundary with $m = 1.5$, $p = 0.5$, $c = 1$

4 Pulse splitting phenomena with connecting property

For a positive number γ we define an even function $v^0(x)$ by

$$v^0(x) = \begin{cases} K(t, x+\rho+\gamma, \rho, \sigma) & \text{on } (-\infty, 0], \\ v^0(-x) & \text{on } (0,\infty). \end{cases} \tag{3.1}$$

Let $\eta \in (0,\rho)(0 < \eta < \rho)$ be an arbitrary fixed constant and ε be an arbitrary positive number such that $\varepsilon < K(0, \rho-\eta, \rho, \sigma)$. For $v^0(x)$ we introduce an even function $v^0_\varepsilon(x)$ satisfying the following

CONDITION C. i) $v^0_\varepsilon(x) = v^0_\varepsilon(-x)$ and $v^0(x) \le v^0_\varepsilon(x)$ hold on $\mathbf{R}^1$;

ii)

$$v^0_\varepsilon(x) = \begin{cases} K(0, x+\rho+\gamma, \rho, \sigma) & \text{on } (-\infty, -\gamma-\eta], \\ \varepsilon & \text{on } [-\gamma, 0], \end{cases} \tag{3.2}$$

and $v_\varepsilon^0(x)$ is a decreasing function on $[-\gamma-\eta,-\gamma]$;

iii) $v_{\varepsilon'}^0(x) \le v_\varepsilon^0(x)$ holds for $\varepsilon' \le \varepsilon$;

iv) $v_\varepsilon^0(x)$ is sufficiently smooth in its support $[-2\rho-\gamma,\ \gamma+2\rho]$ and

$$\|u_{\varepsilon x}^0\|_\infty \le \|u_x^0\|_\infty, \quad TV(u_{\varepsilon x}^0) \le (TV(u_x^0)), \quad \text{ess.inf}\, u_{\varepsilon xx}^0 \ge \text{ess.inf}\, u_{xx}^0, \tag{3.3}$$

where $u^0(x) = (v^0(x))^{m-1}$ and $u_\varepsilon^0(x) = (v_\varepsilon^0(x))^{m-1}$.

Let $v(t,x)$ and $v_\varepsilon(t,x)$ be the solutions of (1.1)–(1.2) with $v(0,x)=v^0(x)$ and $v_\varepsilon(0,x)=v_\varepsilon^0(x)$, respectively. Then we obtain the theorem, which justifies the appearance of the phenomena (NS-4).

Theorem 3. *Under Conditions A and C let γ, ρ and σ be constants such that* supp $v(t,x)$ *becomes connected at $t=\tilde{T}(\rho,\sigma)$. Then, for some $\tilde{\varepsilon}$, there exist $\tilde{t}$ $(0<\tilde{t}<\tilde{T}(\rho,\sigma))$ and $\tilde{x}$ $(-\gamma\le\tilde{x}\le\gamma)$ such that $v_{\tilde{\varepsilon}}(\tilde{t},\tilde{x})=0$ holds, and* supp $v_{\tilde{\varepsilon}}(\tilde{T}(\rho,\sigma),\cdot\,)$ *is connected.*

Proof. Putting $\mathbf{S}=[0,\ \tilde{T}(\rho,\sigma)]\times[-\gamma,\gamma\ ]$, we show that $\mathbf{S}$ contains at least one point $(\tilde{t},\tilde{x})$ such that $v_{\tilde{\varepsilon}}(\tilde{t},\tilde{x})=0$ for some positive constant $\tilde{\varepsilon}$. For this end we assume the contrary; that is, suppose $v_\varepsilon(t,x)>0$ on $\mathbf{S}$ for $\varepsilon>0$. Then, the following estimates hold by Theorems 1 and 2.

$$0 \le u_\varepsilon(t,\cdot) \le \max(\|u_\varepsilon^0\|_\infty - c't,\ 0) \quad \text{on } \mathbf{R}^1, \tag{3.4}$$

$$\|u_{\varepsilon x}(t,\cdot)\|_\infty \le \|u_{\varepsilon x}^0\|_\infty, \tag{3.5}$$

$$\int_{-\gamma}^{\gamma} |u_{\varepsilon xx}(t,x)|dx \le TV(u_{\varepsilon x}(t,\cdot)) \le TV(u_{\varepsilon x}^0)), \tag{3.6}$$

where $u_\varepsilon=(v_\varepsilon)^{m-1}$. By using these inequalities and Condition C we obtain

$$\begin{aligned}
\int_{-\gamma}^{\gamma} u_\varepsilon(t,x)dx &= \int_{-\gamma}^{\gamma} u_\varepsilon(0,x)dx \\
&\quad + \int_0^t\int_{-\gamma}^{\gamma} \left\{m u_\varepsilon(t,x)u_{\varepsilon xx}(t,x) + a(u_{\varepsilon x}(t,x))^2 - c'\right\} dxdt \\
&= 2\gamma\varepsilon^{m-1} \\
&\quad - \int_0^t \left\{2\gamma c' - (m-2)a\int_{-\gamma}^{\gamma} u_\varepsilon(t,x)u_{\varepsilon xx}(t,x)dx - a\Big[u_\varepsilon(t,x)u_{\varepsilon x}(t,x)\Big]_{-\gamma}^{\gamma}\right\} dt \\
&\le 2\gamma\varepsilon^{m-1} - \left\{2\gamma c' - a \max_{[0,t]\times[-\gamma,\gamma]} u_\varepsilon(t,x)\Big((2-m)TV(u_x^0) + 2\|u_x^0\|_\infty\Big)\right\} t \\
&\qquad\qquad \text{for } t\in[0,\tilde{T}(\rho,\sigma)].
\end{aligned} \tag{3.7}$$

Let d_1 be an arbitrary fixed positive constant such that

$$d_1 < \frac{2\gamma c'}{a\Big((2-m)TV(u_x^0) + 2\|u_x^0\|_\infty\Big)}. \tag{3.8}$$

Then, by the continuity of the solution $v_\varepsilon(t,x)$ and the comparison theorem on the initial data([2]) there exist positive constants ε_1 and $\tilde{T}_1 < \tilde{T}(\rho,\sigma)$ such that

$$\max_{[0,t]\times[-\gamma,\gamma]} u_\varepsilon(t,x) < d_1 \quad \text{for } t < \tilde{T}_1 \quad \text{and} \quad \varepsilon < \varepsilon_1 < \min\left(d_1, K(0,\rho-\eta,\rho,\sigma)\right) \tag{3.9}$$

We put

$$\tilde{T}_2(\varepsilon) = \frac{2\gamma\varepsilon^{m-1}}{\left\{2\gamma c' - ad_1\Big((2-m)TV(u_x^0) + 2\|u_x^0\|_\infty\Big)\right\}}, \tag{3.10}$$

and choose $\tilde{\varepsilon} < \varepsilon_1$ such that $\tilde{T}_2(\tilde{\varepsilon}) < \tilde{T}_1$. Hence, it follows from (3.7) that

$$\int_{-\gamma}^{\gamma} u_{\tilde{\varepsilon}}(t,x)dx < 0 \quad \text{for } t \in [T_2(\tilde{\varepsilon}), \tilde{T}_1], \tag{3.11}$$

which is a contradiction. Thus, $v_{\tilde{\varepsilon}}(\tilde{t},\tilde{x}) = 0$ holds for some $(\tilde{t},\tilde{x}) \in \mathbf{S}$. It is clear by the comparison theorem that supp $v_{\tilde{\varepsilon}}(\tilde{T}(\rho,\sigma),\cdot)$ becomes connected, which completes the proof.

5 Acknowledgments

This work was supported by Japan Society for the Promotion of Science through Grant-in-Aid (No. 13440038) for Scientific Research (B).

References

1. D.G. Aronson, The porous medium equation, In some Problems in Nonlinear Diffusion(eds. A. Fasano and M. Primicerio), Lecture Notes in Mathematics **1224**, Springer-Verlag,1986.
2. M. Bertsch, A class of degenerate diffusion equations with a singular nonlinear term, *Non-linear Anal.*, **7**(1983),117-127.
3. X.-Y. Chen, H. Matano and M. Mimura, Finite-point extinction and continuity of interfaces in a nonlinear diffusion equation with strong absorption, *J. reine angew. Math.*, **459**(1995),1–36.
4. E. DiBenedetto and D. Hoff, An interface tracking algorithm for the porous medium equation, *Trans. Amer. Math. Soc.*, **249**(1984),463–500.
5. J.L. Graveleau and P. Jamet, A finite difference approach to some degenerate nonlinear parabolic equations, *SIAM J. Appl. Math.*, **20** (1971), 199–223.
6. M.A. Herrero and Vázquez, The one-dimensional nonlinear heat equation with absorption: Regularity of solutions and interfaces, *SIAM J. Math. Anal.*, **18** (1987), 149–167.
7. A.S. Kalashnikov, The propagation of disturbances in problems of non-linear heat conduction with absorption, *Zh. Vychisl. Mat. i Mat. Fiz.*, **14** (1974), 891–905.

8. A.S. Kalashnikov, Some problems of the qualitative theory of non-linear degenerate second-order parabolic equations, *Russian Math. Surveys*, **42** (1987), 169–222.
9. R. Kersner, The behavior of temperature fronts in media with nonlinear thermal conductivity under absorption, *Vestnik. Mosk. Univ. Mat.*, **33** (1978), 44–51.
10. M. Mimura, T. Nakaki and K. Tomoeda, A numerical approach to interface curves for some nonlinear diffusion equations, *Japan J. Appl. Math.*, **1** (1984), 93–139.
11. T. Nakaki and K. Tomoeda, A finite difference approach to the interface equation for some nonlinear diffusion equations with absorption, *Proc. Japan Acad.*, **77**, Ser. A(2001), 32–37.
12. T.Nakaki and K.Tomoeda, A finite difference scheme for some nonlinear diffusion equations in absorbing medium: support splitting phenomena, SIAM J. Numer. Anal., **40**(2002),945-964.
13. O.A. Oleinik, A.S.Kalashnikov and Chzou Yui-Lin, The Cauchy problem and boundary value problems for equations of the type of nonstationary filtration, *Izv. Acad. Nauk SSSR Ser. Mat.*, **22** (1958), 667–704.
14. P. Rosenau and S. Kamin, Thermal waves in an absorbing and convecting medium, *Physica*, **8D** (1983), 273–283.
15. K. Tomoeda, The behavior of impulsively initiated thermal waves in an absorbing medium, Dyn. Contin. Discrete Impuls. Syst. Ser. B, **10**(2003),151–164.

Fully Two-dimensional HLLEC Riemann Solver and Associated Difference Schemes

Pavel Váchal[1], Richard Liska[2] and Burton Wendroff[3]

[1] Czech Technical University in Prague, Brehova 7, 115 19 Prague 1, Czech Republic *vachal@galileo.fjfi.cvut.cz*
[2] Czech Technical University in Prague, Brehova 7, 115 19 Prague 1, Czech Republic *liska@siduri.fjfi.cvut.cz*
[3] Los Alamos National Laboratory, Group T-7, MS B284, Los Alamos, NM 87545, USA *bbw@lanl.gov*

Summary. Fully two-dimensional, sixteen state, approximate Riemann solver of HLLEC type for Euler equations with sixteen states and central waves which approximate contact discontinuities has been developed. The Riemann solver is used in Godunov and WAF finite difference schemes. Numerical example of their perfomance is presented.

1 Introduction

The aim of this paper is to describe the class of HLLE Riemann solvers and propose a fully two-dimensional extension of the one-dimensional HLLC Riemann solver from [8].

In 1983, Harten, Lax and van Leer suggested [3] to approximate the solution of a Riemann problem by three constant states, separated by two waves, propagating with constant speeds. A particular algorithm for computation of these wave speeds were presented five years later by Einfeldt [2]. Since the assumption of two waves is correct only for hyperbolic systems of two equations, Toro, Spruce and Speares [8] added one more wave, creating the 4-state one-dimensional HLLC solver. In fluid dynamics, this new wave corresponds to a contact discontinuity. Later, Wendroff presented a series of 9-state solvers, extending the 3-state HLLE approach to two dimensions [10], [11]. In this paper, we construct a contact-corrected version of this approach, adding six new waves.

The outline is as follows: First, we introduce the basic principles of HLLE and HLLC Riemann solvers in the simple one-dimensional case. Then we proceed to two dimensions, describe the 9-state solver from [10], [11] and introduce its new, contact-corrected version. Finally, we demonstrate the application of our new solver in two particular numerical methods, namely the Godunov-type difference scheme and the WAF (Weighted Average Flux) approach, and present results of selected numerical tests.

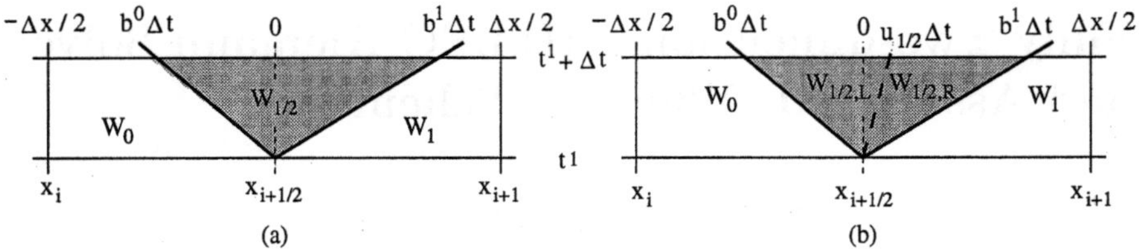

Fig. 1. 1D solvers: (a) 3-state HLLE; (b) 4-state HLLC

2 One Dimension

Methods presented in this paper have been derived for the Euler equations. In one dimension we have

$$\begin{pmatrix} \rho \\ \rho u \\ E \end{pmatrix}_t + \begin{pmatrix} \rho u \\ \rho u^2 + p \\ (E+p)u \end{pmatrix}_x = \mathbf{0}, \tag{1}$$

where ρ is the density, u fluid velocity, p pressure and E the density of the total energy. Completing the system (1) by the equation of state for an ideal polytropic gas $p = (\gamma-1)\left(E - \frac{1}{2}\rho u^2\right)$, we can write it in the general differential form $\mathbf{w}_t + \mathbf{f}(\mathbf{w})_x = \mathbf{0}$.

Consider the initial Riemann problem

$$\mathbf{w}(x,t_1) = \begin{cases} W_0 = (\rho_0, \rho_0 u_0, E_0)^T & \text{for} \quad x_i \le x < x_{i+\frac{1}{2}} \\ W_1 = (\rho_1, \rho_1 u_1, E_1)^T & \text{for} \quad x_{i+\frac{1}{2}} < x \le x_{i+1} \end{cases}. \tag{2}$$

Following [11], we will approximate the solution at $t_1 < t < t_1 + \Delta t$ with three constant states W_0, $W_{\frac{1}{2}}$ and W_1, divided by two waves, propagating with constant speeds b^0 and b^1. The situation is shown in Fig. 1(a). With this layout and notation, the integral form of the conservation law over this space-time domain becomes

$$\begin{aligned} \frac{\Delta x}{2}(W_0 + W_1) = {} & \left(\frac{\Delta x}{2} + b^0 \Delta t\right) W_0 + \left(b^1 \Delta t - b^0 \Delta t\right) W_{\frac{1}{2}} + \\ & + \left(\frac{\Delta x}{2} - b^1 \Delta t\right) W_1 + \left[f(W_1) - f(W_0)\right] \Delta t \,. \end{aligned} \tag{3}$$

Note that data below the plots in Fig. 1 show the absolute position, while data above the plots are relative distances to the center of the staggered cell. Such simplified notation will be especially advantageous later in 2D.

Solving (3) for $W_{\frac{1}{2}}$ gives

$$W_{\frac{1}{2}} = \frac{b^1 W_1 - b^0 W_0 - f(W_1) + f(W_0)}{b^1 - b^0}. \tag{4}$$

To have the scheme completely determined, we must decide, how to choose wave speeds b^0 and b^1. We will again follow [11] and use the Einfeldt speeds, based on Roe averages, as summarized in [7]. Now we know all the states and wave speeds of the approximate Riemann solver. Such a solver can be used in several difference schemes, as we will see in Sect. 4.

As mentioned above, Toro et al. [8] extended this solver to a 4-state version, called HLLC, which resolves stationary contact discontinuities exactly. The main idea is to split the intermediate state by a third wave, representing a contact. Since velocity and pressure stay unchanged across contact discontinuities, the two central states will differ only in density (and, of course, total energy, but this can be computed from pressure and other known state variables). So, consider the initial Riemann problem (2). Then the 4-state solver can be summarized by the following algorithm:

1. Compute Einfeldt wave speeds b^0, b^1 from Roe averages as in [7].
2. Compute the intermediate state from (4) the same way as in the 3-state solver (Fig. 1(a)). Denote this state $W_{\frac{1}{2}} = (\rho_{\frac{1}{2}}, \rho_{\frac{1}{2}} u_{\frac{1}{2}}, E_{\frac{1}{2}})^T$.
3. Compute $u_{\frac{1}{2}}$, $\rho_{\frac{1}{2}}$ and $p_{\frac{1}{2}}$ from $W_{\frac{1}{2}}$.
4. Assume $u_{\frac{1}{2}}$ and $p_{\frac{1}{2}}$ to be preserved across the contact discontinuity. Then the contact has to move also with speed $u_{\frac{1}{2}}$ and densities in the two new intermediate regions can be computed from the scalar Rankine-Hugoniot condition for the mass conservation, applied to the left, resp. right wave:

$$\rho_{\frac{1}{2},L} = \rho_0 \frac{u_0 - b^0}{u_{\frac{1}{2}} - b^0} \quad , \quad \rho_{\frac{1}{2},R} = \rho_1 \frac{u_1 - b^1}{u_{\frac{1}{2}} - b^1}.$$

5. Compute both intermediate states (Fig. 1(b)):

$$W_{\frac{1}{2},L} = \begin{pmatrix} \rho_{\frac{1}{2},L} \\ \rho_{\frac{1}{2},L} u_{\frac{1}{2}} \\ \frac{1}{2}\rho_{\frac{1}{2},L}\left(u_{\frac{1}{2}}\right)^2 + \frac{p_{\frac{1}{2}}}{\gamma-1} \end{pmatrix} \quad , \quad W_{\frac{1}{2},R} = \begin{pmatrix} \rho_{\frac{1}{2},R} \\ \rho_{\frac{1}{2},R} u_{\frac{1}{2}} \\ \frac{1}{2}\rho_{\frac{1}{2},R}\left(u_{\frac{1}{2}}\right)^2 + \frac{p_{\frac{1}{2}}}{\gamma-1} \end{pmatrix}. \tag{5}$$

Note that there is also another way to evaluate $p_{\frac{1}{2}}, E_{\frac{1}{2},L}$ and $E_{\frac{1}{2},R}$ only from Rankine-Hugoniot conditions [6, 7, 1].

3 Two-dimensional Riemann Solvers

In two dimensions, the Euler Equations have the form

$$\begin{pmatrix} \rho \\ \rho u \\ \rho v \\ E \end{pmatrix}_t + \begin{pmatrix} \rho u \\ \rho u^2 + p \\ \rho u v \\ (E+p)u \end{pmatrix}_x + \begin{pmatrix} \rho v \\ \rho u v \\ \rho v^2 + p \\ (E+p)v \end{pmatrix}_y = \mathbf{0} \tag{6}$$

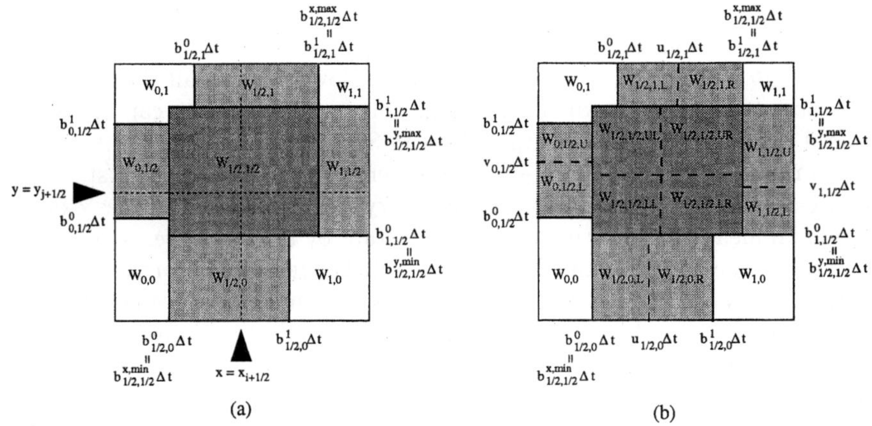

Fig. 2. 2D Riemann solvers: (a) 9-state HLLE with highlighted initial condition; (b) 16-state HLLEC

with the equation of state for ideal polytropic gas $p = (\gamma - 1)(E - \frac{1}{2}\rho(u^2 + v^2))$, where v is the fluid velocity in the y-direction.

First, let us summarize the fully two-dimensional 9-state HLLE Riemann solver, originally presented in [11]. We want to solve the 2D Riemann problem, formed by two partitions at $x = x_{i+\frac{1}{2}}$ and $y = y_{j+\frac{1}{2}}$, subdividing the domain $[x_i, x_{i+1}] \times [y_j, y_{j+1}]$ into four regions with constant states. As the system evolves in time, we approximate the solution as shown in Fig. 2(a). Waves, propagating from the center with constant speeds, split the domain into 9 regions. For Δt sufficiently small, none of the waves in the y-direction will reach the edge $y = y_j$, so that the state $W_{\frac{1}{2},0}$ is affected only by the one-dimensional Riemann problem in the x-direction, given by $W_{0,0}$ and $W_{1,0}$. Analogous considerations along the other three edges lead us to following method: The corner states ($W_{0,0}$, $W_{1,0}$, $W_{0,1}$, $W_{1,1}$) stay undisturbed. We take these to compute the edge states ($W_{\frac{1}{2},0}$, $W_{\frac{1}{2},1}$, $W_{0,\frac{1}{2}}$, $W_{1,\frac{1}{2}}$), using a 3-state solver based on the one-dimensional HLLE solver from section 2. Then we do not need to solve the real two-dimensional problem in the central area, since $W_{\frac{1}{2},\frac{1}{2}}$ is given by a two-dimensional conservation law, applied to the whole staggered cell.

Such 9-state Riemann solver works well for problems containing shocks and rarefaction waves, but its weakness is poor resolution of contact discontinuities. Our goal was to modify it to a solver which resolves contact surfaces better (stationary contacts even exactly), and which degenerates to the 1D 4-state solver from Sect. 2 for one-dimensional initial problems in coordinate directions (i.e. if there is only one interface in the 2D initial condition).

The most straightforward way is to follow Toro's one-dimensional approach and split each of the intermediate edge regions (i.e. states $W_{\frac{1}{2},0}$, $W_{\frac{1}{2},1}$, $W_{0,\frac{1}{2}}$, $W_{1,\frac{1}{2}}$) by a wave corresponding to a contact discontinuity. It is obvious that to fulfill the degeneracy condition mentioned above we must subdivide also the central region (i.e. state $W_{\frac{1}{2},\frac{1}{2}}$) by two additional waves. This gives us the setup for a 16-state solver as shown in Fig. 2(b).

It is not difficult to compute speeds of splitting waves and values of new intermediate states along the edges. Basically, we use the same process as in one dimension (5), taking also transversal velocity into account. In [9] it is proven that we can simply copy these from the nearest corner states, that is along the edge $y = y_j$ we simply take $v_{\frac{1}{2},0,L} = v_{0,0}$ and $v_{\frac{1}{2},0,R} = v_{1,0}$, along edge $x = x_i$ we take $u_{0,\frac{1}{2},L} = u_{0,0}$ and $u_{0,\frac{1}{2},U} = u_{0,1}$, etc.

What we have to do is to set speeds of waves splitting the central region. One possibility is to take the velocities from the conservation laws. We compute the central state as if it would not be further subdivided, then divide the momentum in each direction by density and use these velocities as speeds of the splitting waves. Besides this one, there are several other approaches to set these partitions.

The only remaining issue is to compute new states in the four central regions. First, we estimate the velocities. To fulfill the condition of degeneracy, we take for each direction the average of velocity in edge states and corner states with interface parallel to this direction, weigthed by length of these interfaces. That means, for $u_{\frac{1}{2},\frac{1}{2},UR}$ we take the velocities $u_{0,1}$, $u_{\frac{1}{2},1,L}$, $u_{\frac{1}{2},1,R}$ and $u_{1,1}$, weighted by the lengths of the interfaces of their regions with upper right central region, while for $v_{\frac{1}{2},\frac{1}{2},UR}$, we taked weighted average of $v_{1,0}$, $v_{1,\frac{1}{2},L}$, $v_{1,\frac{1}{2},U}$ and $v_{1,1}$. Then we estimate central densities. This time, for each central state, we compute the density from 1D Rankine-Hugoniot conditions across each interface with surrounding regions (i.e. with corner and edge regions), and weigh the results again by lengths of interfaces. Since we know the total mass in the central states, which must be conserved, we multiply now all the four density estimates by a suitable factor to satisfy the mass conservation law. Similarly, we correct the momentum estimates (corrected density multiplied by estimated velocity) in both directions by adding a suitable amount to each state, in order to keep momentum conserved. Finally, we suppose pressure to be the same in all four central regions and compute its value from the conservation law for total energy.

Let us summarize the complete algorithm of our new 2D 16-state HLLEC Riemann solver (consult Fig. 2(b)):

1. Compute wave speeds $b^0_{\frac{1}{2},0}$, $b^1_{\frac{1}{2},0}$, $b^0_{\frac{1}{2},1}$, $b^1_{\frac{1}{2},1}$, $b^0_{0,\frac{1}{2}}$, $b^1_{0,\frac{1}{2}}$, $b^0_{1,\frac{1}{2}}$ and $b^1_{1,\frac{1}{2}}$, using Einfeldt formulas from [7].
2. Compute the intermediate (light shaded) states for each of four 1D problems along edges.
   a) Using the 1D 4-state HLLC solver with appropriate 2D fluxes, compute the longitudinal velocities (i.e. $u_{\frac{1}{2},0}$, $u_{\frac{1}{2},1}$, $v_{0,\frac{1}{2}}$, $v_{1,\frac{1}{2}}$,) and both densities (i.e. $\rho_{\frac{1}{2},0,L}$, $\rho_{\frac{1}{2},0,R}$, $\rho_{\frac{1}{2},1,L}$, $\rho_{\frac{1}{2},1,R}$, $\rho_{0,\frac{1}{2},L}$, $\rho_{0,\frac{1}{2},U}$, $\rho_{1,\frac{1}{2},L}$ and $\rho_{1,\frac{1}{2},U}$).

b) Copy the transversal velocities always from the nearest corner state (i.e. $v_{\frac{1}{2},0,L} = v_{0,0}$, $u_{0,\frac{1}{2},L} = u_{0,0}$, etc.
c) Compute the intermediate pressures (i.e. $p_{\frac{1}{2},0}, p_{\frac{1}{2},1}, p_{0,\frac{1}{2}}, p_{1,\frac{1}{2}},$) so, that the total energy in "1D-stripe" along each edge is conserved.

3. Compute the central state $W_{\frac{1}{2},\frac{1}{2}}$ from the 2D integral conservation law as if it was not further subdivided. These values will be used to assure conservation and split the central region.
4. Split the central region into four pieces (UL, UR, LL, LR) divided by waves with speeds based on velocities from the previous step.
5. Estimate eight central velocities (i.e. $u_{\frac{1}{2},\frac{1}{2},\alpha}$, $v_{\frac{1}{2},\frac{1}{2},\alpha}$, $\alpha \in \{UL,LL,UR,LR\}$) from adjacent edge and corner regions, so that the solver degenerates to 1D solver for 1D initial conditions in x-, resp. y-direction.
6. Estimate four central densities (i.e. $\rho_{\frac{1}{2},\frac{1}{2},\alpha}$, $\alpha \in \{UL,LL,UR,LR\}$), using Rankine-Hugoniot conditions across the interfaces with surrounding (edge and corner) states, weighted by lengths of particular interfaces.
7. Correct central densities from step 6 so, that the integral mass conservation law over the whole staggered cell is satisfied.
8. Correct central velocities from step 5 so, that integral momentum conservation laws in both coordinate directions over the whole staggered cell are satisfied.
9. Finally, compute central pressure $p_{\frac{1}{2},\frac{1}{2}}$ (assuming it is equal for all four central states) so, that the total energy is conserved.

4 Application in Difference Schemes

In this section, we demonstarte how our new Riemann solver can be practically used in difference schemes. In particular, we first present the simplest, first order Godunov-type method and then a more accurate WAF scheme.

4.1 Godunov-type Scheme

The scenario of the simplest Godunov scheme is as follows:

1. Construct the uniform rectangular mesh with spatial steps Δx and Δy, with cells centered at (x_i, y_j). Dual (staggered) mesh will be formed by cells centered at $(x_{i+\frac{1}{2}}, y_{j+\frac{1}{2}})$.
2. Discretize the initial condition, so that the values are constant inside each original cell. The interfaces form 2D Riemann problems in staggered cells.
3. Choose the time interval Δt so, that these Riemann problems stay separated in their respective staggered cells during the whole time step.
4. Let the Riemann problems evolve and compute their approximate solutions after Δt with the HLLEC Riemann solver from Sect. 3.
5. Create new constant states by integrating the solutions over the original cells. (For us, this means only weighted averaging.)
6. Repeat from step 3 until the solution at desired time is reached.

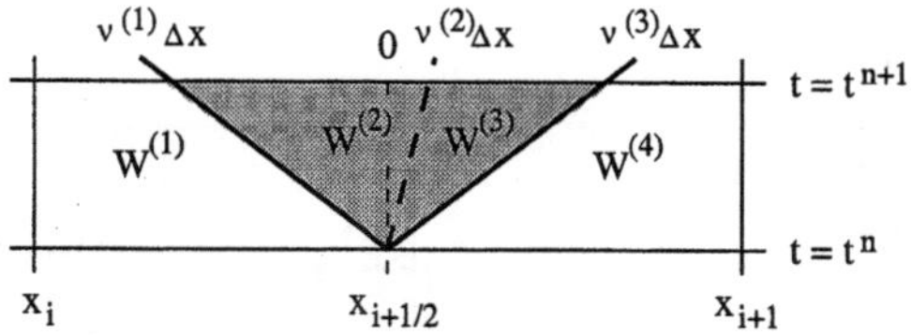

Fig. 3. 1D WAF scheme with a 4-state approximate Riemann solver

4.2 WAF Scheme

Here we apply our 2D Riwmann solver in Weighted Average Flux (WAF) methods [6, 7]. We want to solve the system of Euler equations in one dimension. We start as with the Godunov approach: we create a uniform mesh and replace the initial condition by a piecewise constant function, so that it forms a set of Riemann problems at cell interfaces. Then we let the system evolve in time and apply the approximate 4-state HLLC solver on each Riemann problem as in the Fig. 3. To obtain the value in cell center x_i at new time level $(t^{n+1} = t^n + \Delta\mathrm{t})$, we use conservative formula

$$W_i^{n+1} = W_i^n - \frac{\Delta\mathrm{t}}{\Delta\mathrm{x}}\left(F_{i+\frac{1}{2}}^n - F_{i-\frac{1}{2}}^n\right), \tag{7}$$

where $F_{i-\frac{1}{2}}^n$ and $F_{i+\frac{1}{2}}^n$ are weighted average fluxes in the staggered cell at left, resp. at right. These can be computed at each time level from real physical fluxes in four states of the solver as

$$F_{i+\frac{1}{2}} = \sum_{k=1}^{4} \alpha^k f\left(W_{i+\frac{1}{2}}^{(k)}\right). \tag{8}$$

To avoid excessive oscillations, weights α^k are taken as

$$\alpha^k = \frac{1}{2}\left(\phi_{i+\frac{1}{2}}^{(k)} - \phi_{i+\frac{1}{2}}^{(k-1)}\right) \quad , \quad k = 1, \ldots, 4 \tag{9}$$

with $\phi_{i+\frac{1}{2}}^{(0)} = -1$, $\phi_{i+\frac{1}{2}}^{(4)} = 1$, and the limiter functions

$$\phi_{i+\frac{1}{2}}^{(k)} = \mathrm{sgn}\left(\nu_{i+\frac{1}{2}}^{(k)}\right)\bar{\phi}_{i+\frac{1}{2}}^{(k)} \quad , \quad k = 1, \ldots, 3. \tag{10}$$

One can use the *van Leer* limiter, *minmod*, or many others listed for example in [7].

In two dimensions, the conservative formula is

$$W_{i,j}^{n+1} = W_{i,j}^n - \frac{\Delta\mathrm{t}}{\Delta\mathrm{x}}\left(F_{i+\frac{1}{2},j}^n - F_{i-\frac{1}{2},j}^n\right) - \frac{\Delta\mathrm{t}}{\Delta\mathrm{y}}\left(G_{i,j+\frac{1}{2}}^n - G_{i,j-\frac{1}{2}}^n\right), \tag{11}$$

and we compute each weighted average flux from cell-sized region centered at the point given by its indices (i.e. flux $F_{i+\frac{1}{2},j}$ from rectangle $[x_i, x_{i+1}] \times$

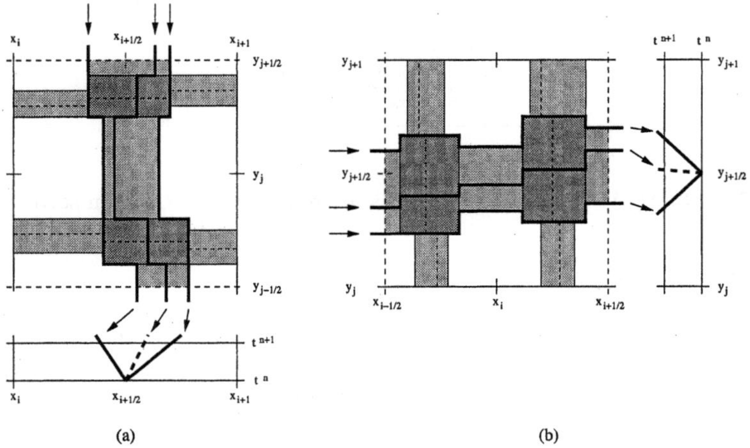

Fig. 4. 2D WAF scheme: (a) computation of $F_{i+\frac{1}{2},j}$; (b) computation of $G_{i,j+\frac{1}{2}}$

$[y_{j-\frac{1}{2}}, y_{j+\frac{1}{2}}]$, etc.). Rather than to introduce 2D limiters and derive the WAF method in two dimensions, we use the 1D WAF approach in each appropriate direction, with each of the four fluxes taken as an average from a suitable set of states. This is shown schematically in Fig. 4 and described more precisely in [9].

5 Numerical Results

In Fig. 5, we can see results for the density, as computed by the Godunov scheme with 9-state solver, and by both Godunov and WAF schemes with the new HLLEC solver. The test is a slightly modified version of Test 16 from [4], originally presented and described in [5]. The domain $[0,1] \times [0,1]$ is split by two partitions, located at $x = 0.5$ and $y = 0.5$, into four regions with constant states, forming a 2D Riemann problem. Initial values

$\rho_{UL} = 1.0222$	$p_{UL} = 1.0$	$\rho_{UR} = 0.5313$	$p_{UR} = 0.4$
$u_{UL} = -0.7179$	$v_{UL} = 0.1$	$u_{UR} = 0$	$v_{UR} = 0.1$
$\rho_{LL} = 0.8$	$p_{LL} = 1.0$	$\rho_{LR} = 1.0$	$p_{LR} = 1.0$
$u_{LL} = 0$	$v_{LL} = 0.1$	$u_{LR} = 0$	$v_{LR} = 0.8276$

are chosen so that the analytical solution of one-dimensional problem along each edge of the domain is a simple wave: from the top in clockwise direction: left-propagating rarefaction, upward moving shock, stationary contact, upward moving contact.

As expected, our new two-dimensional HLLEC Riemann solver gives better results than the 9-state HLLE. The most visible improvement is a sharper resolution of contact discontinuities, the stationary contact is resolved even

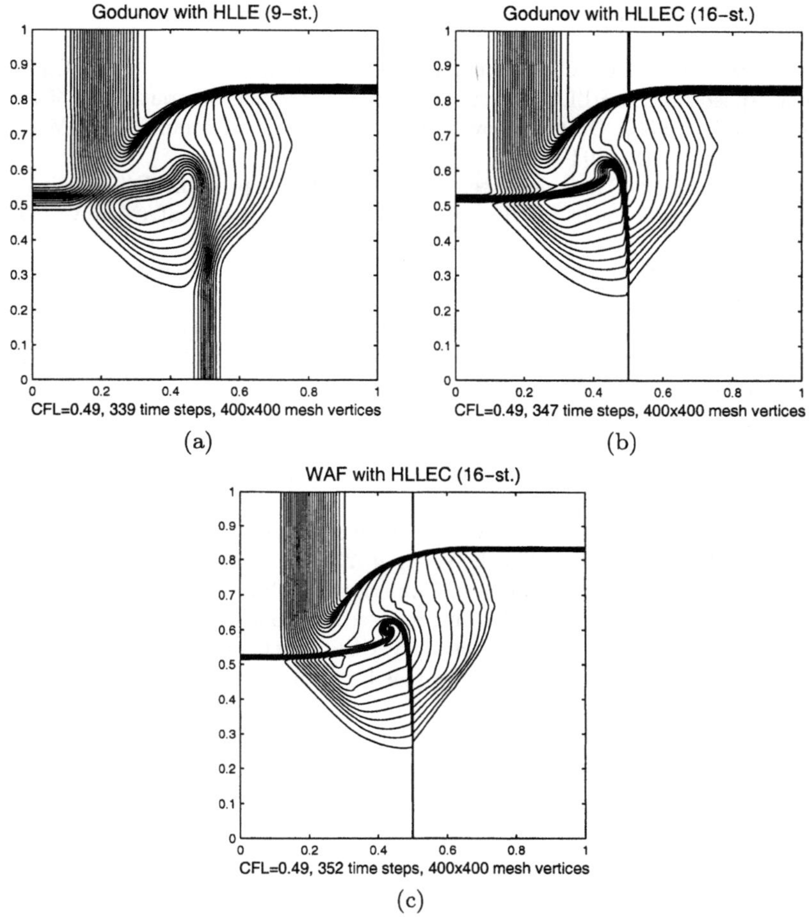

Fig. 5. Contour plots of density at time $t = 0.2$ as computed by particular difference schemes

exactly. However, also shock and rarefaction wave are also treated better by the new solver. The price we have to pay for this improved resolution are slight but visible artifacts at initial locations of discontinuities, in the middle of the upper edge in Fig. 5 (b),(c). Comparing both schemes which use the HLLEC solver, we see that WAF performs better than Godunov, which is not surprising, since it is a second order accurate scheme.

Acknowledgements. First two authors were supported in part by the Czech Grant Agency grant 201/00/0586.

References

1. P. Batten, N. Clarke, C. Lambert and D. M. Causon (1997): On the Choice of Wavespeeds for the HLLC Riemann Solver, SIAM J. Sci. Comput., **18**, 1553–1570
2. Einfeldt B. (1988): On Godunov-type Methods in Gas Dynamics. SIAM J. Numer. Anal., **25** (2), 294–318
3. Harten A., Lax P.D., van Leer B. (1983): On Upstream Differencing and Godunov-type Schemes for Hyperbolic Conservation Laws. SIAM Rev., **25** (1), 35–61
4. Lax P.D., Liu X.-D. (1998): Solution of Two-Dimensional Riemann Problems of Gas Dynamics by Positive Schemes. SIAM J. Sci. Comp., **19** (2), 319–340
5. Schulz-Rinne C.W., Collins J.P., Glaz H.M. (1993): Numerical Solution of the Riemann Problem for Two-dimensional Gas Dynamics. SIAM J. Sci. Comput., **14**, 1394–1414
6. Toro E.F. (1989): A Weighted Average Flux Method for Hyperbolic Conservation Laws. Proc. R. Soc. London A, **423**, 401–418
7. Toro E.F. (1997): Riemann Solvers and Numerical Methods for Fluid Dynamics. Springer Verlag, Berlin, Heidelberg
8. Toro E.F., Spruce M., Speares W. (1994): Restoration of the Contact Surface in the HLL-Riemann Solver. Shock Waves, **4**, 25–34
9. Váchal, P. (2003): Some Aspects of Numerical Resolution of Contact Discontinuities in Conservation Laws. MS Thesis, Czech Technical University, Prague
10. Wendroff B. (1999): A Two-dimensional HLLE Riemann Solver and Associated Godunov-type Difference Scheme for Gas Dynamics. Computers and Math. with Applications, **38**, 175–185
11. Wendroff B. (1999): Approximate Riemann Solvers, Godunov Schemes, and Contact Discontinuities. In: Godunov Methods. Theory and Applications. Oxford, UK, Oct. 18-22, 1999

Deflation Accelerated Parallel Preconditioned Conjugate Gradient Method in Finite Element Problems

Fred J. Vermolen, Kees Vuik and Guus Segal[1]

Delft University of Technology, Department of Applied Mathematical Analysis, Mekelweg 4, 2628 CD, Delft, The Netherlands, *F.J.Vermolen@math.tudelft.nl*

Summary. We describe the algorithm to implement a deflation acceleration in a preconditioned Conjugate Gradient method to solve the system of linear equations from a Finite Element discretization. We focus on a parallel implementation in this paper. Subsequently we describe the data-structure. This is followed by some numerical experiments. The experiments indicate that our method is scalable.

1 Introduction

Large linear systems occur in many scientific and engineering applications. Often these systems result from a discretization of model equations. The systems tend to become very large for three-dimensional problems. Some models involve time and space as independent parameters and therefore it is necessary to solve such a linear system efficiently at each time-step.

In this paper we only consider symmetric positive definite (SPD) discretization matrices. Since the matrices are sparse in our applications, we use an iterative method to solve the linear system. In order to get a fast convergence of the method we use a preconditioned Conjugate Gradient method, where incomplete Choleski factorization is used as a preconditioner. This method is very suitable for parallellization.

The present study involves a parallellization of the Conjugate Gradient method in which the inner products, the matrix-vector multiplication and preconditioning are parallellized. This parallellization is done by the use of domain decomposition, where the domain of computation is divided into subdomains and the overall discretization matrix is divided over the subdomains. To each subdomain we allocate a processor. A well-known problem is that the parallellized method is not scalable: the number of CG-iterations and wall-clock time increase as the number of subdomains increases. To make the method scalable one uses a coarse grid correction (see for an overview and introduction Smith et al [6]) or the deflation method. In [3] it is shown that deflation gives a larger acceleration to the parallel preconditioned CG-method. The idea to use deflation for large linear systems of equations is not new. Among others, Nicholaides [4] and Vuik et al [8, 1, 10, 9] apply this method to solve large ill-conditioned linear systems. The result of deflation is that the components of the solution in the direction of the eigenvectors corresponding to the extremely small eigenvalues are projected to zero. The effective condition number of the resulting singular system becomes more favourable. In the present paper we deal with "algebraic" deflation vectors. For more details on the various types of deflation vectors we refer to [8] and [7].

We assume that the domain of computation Ω consists of a number of disjoint subdomains Ω_j, $i \in \{1, \ldots, m\}$, such that $\cup_{j=1}^{m} \overline{\Omega}_j = \overline{\Omega}$. To each subdomain we allocate a processor and a deflation vector, z_j, for which we define

$$z_j = \begin{cases} 1, & \text{for } (x,y) \in \Omega_j, \\ 0, & \text{for } (x,y) \in \Omega \setminus \Omega_j. \end{cases} \tag{1}$$

In case of Finite Volume methods we have to distinguish between cell-centered and vertex-centered discretization. In the cell-centered the deflation vector is not defined on the interfaces between consecutive subdomains. In the vertex-centered case, however, we have an overlap at the interface points. In this paper we use a Finite Element discretization, which is always vertex-centered by its construction. The subdomains may be considered as "super"-elements consisting of a set of finite elements. The global stiffness matrix is never constructed, only the "super"-element matrix is constructed. Matrix-vector multiplication is carried out per "super"-element and only after adding of the contributions of each "super"-element the global vector is obtained. In this way parallellization of the Finite Element method can be done in a natural way. For the interface points we use the concept of "average overlap", which is explained as follows: Given a deflation vector z_j on an interfacial node that is shared by Ω_j and p neighbours of Ω_j, then we set at this point:

$$z_j = \frac{1}{p+1}. \tag{2}$$

The deflation method is applied successfully to problems from transport in porous media where coefficients abruptly change several orders of magnitude [9]. In the present paper we consider a Galerkin Finite Element discretization of the Laplace equation with a Dirichlet and a Neumann boundary condition at Γ_D and Γ_N respectively (note that $\Gamma_N \cup \Gamma_D = \partial\Omega$):

$$\begin{cases} -\Delta u = f, & (x,y) \in \Omega \\ u = \tilde{u}(x,y), \text{ for } (x,y) \in \Gamma_D, & \dfrac{\partial u}{\partial n} = 0, \text{ for } (x,y) \in \Gamma_N \end{cases} \tag{3}$$

where u denotes the solution and $\tilde{u}$ represents a given function. The resulting discretization matrix is symmetric positive definite. The domain is divided into subdomains and the resulting system of linear equations is solved by the use of a parallellized Deflated ICCG. In the text the algorithm is given and the issues of data-structure for the parallellization of the solution method are described. Subsequently, we describe some numerical experiments. For more mathematical background we refer to [1, 10, 7].

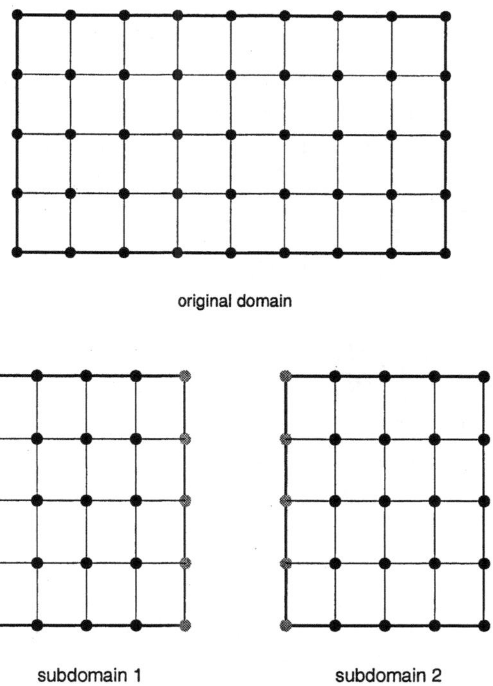

Fig. 1. Domain decomposition for a vertex centered discretization

2 Deflated Incomplete Choleski preconditioned Conjugate Gradient Method

In this section we describe the deflated method for the symmetric positive definite discretization matrix A. Let $Z = (z_1 \dots z_m)$ represent the matrix whose columns consist of the deflation vectors z_j, as defined in equations (1) and (2). The matrix Z is chosen such that its column space approximates the eigenspace of these eigenvectors that correspond to the smallest eigenvalues. We, then, define the projection $P := I - AZ(Z^T AZ)^{-1} Z^T := I - AZE^{-1}Z^T$. It is shown in [7] that the matrix PA is positive semi-definite (and hence singular). Kaasschieter [2] showed convergence of the Conjugate Gradient method for cases in which the matrix is singular. Let b be the right-hand side vector and x be the solution vector, then we solve

$$Ax = b. \tag{4}$$

After application of deflation by left multiplication of the above equation by P, we obtain

$$PAx = Pb. \tag{5}$$

Since PA is singular the solution is not unique. We denote the solution that is obtained by use of the ICCG method on equation (5) by $\tilde{x}$. To get the solution x we use

$$x = (I - P^T)x + P^T x. \tag{6}$$

It is shown in [7] that $P^T\tilde{x} = P^T x$, hence the solution of equation (5) can be used. The second part $(I - P^T)x = Z(Z^T AZ)^{-1} Z^T b$ is relatively cheap to compute. Hence the solution x is obtained by addition of the two contributions, i.e. $x = P^T\tilde{x} + Z(Z^T AZ)^{-1} Z^T b$. For completeness we give the algorithm of the Deflated ICCG:

Algorithm 1 (DICCG [9]):

$k = 0, \underline{\tilde{r}}_0 = P\underline{r}_0, \underline{p}_1 = \underline{z}_0 = L^{-T}L^{-1}\underline{\tilde{r}}_0$

while $||\underline{\tilde{r}}_k||_2 > \varepsilon$

$k = k+1, \quad \alpha_k = \dfrac{\underline{\tilde{r}}_{k-1}^T \underline{z}_{k-1}}{\underline{p}_k^T PA\underline{p}_k}$

$\underline{x}_k = \underline{x}_{k-1} + \alpha\underline{p}_k, \quad \underline{\tilde{r}}_k = \underline{\tilde{r}}_{k-1} - \alpha_k PA\underline{p}_k$

$\underline{z}_k = L^{-T}L^{-1}\underline{\tilde{r}}_k, \quad \beta_k = \dfrac{\underline{\tilde{r}}_k^T \underline{z}_k}{\underline{\tilde{r}}_{k-1}^T \underline{z}_{k-1}}$

$\underline{p}_{k-1} = \underline{z}_k + \beta_k\underline{p}_k$

end while

The inner products, matrix vector multiplication and vector updates in the above algorithm are easy to parallellize. Parallellization of the incomplete Choleski preconditioned Conjugate Gradient method has been done before by Perchat et al [5]. We use a restriction and a prolongation operator and block preconditioners for the preconditioning step in the above algorithm. Note that it is necessary to have a symmetric preconditioner. This is obtained by choosing the restriction and prolongation matrices as transposes of each other. Let $\underline{r}_k$ be the residual after k CG-iterations and N be the total number of subdomains, then overall preconditioning is expressed in matrix form by:

$$\underline{z}_k = \left(\sum_{i=1}^{N} R_i^T M_i^{-1} R_i\right) \underline{r}_k. \tag{7}$$

Here $\underline{z}_k$ represents the updated residual after preconditioning. Further, R_i and M_i respectively denote the restriction operator and block preconditioner. We will limit ourselves to the issues of the implementation of the data-structure needed for the parallel implementation of deflation. The above algorithm is a standard ICCG except for the lines that contain the matrix P.

3 Data-structure of the deflation vectors for parallellization

To create P and Pv we need to make z_j and to compute Az_j, $z_i^T Az_j$ and Pv. To do this efficiently we make use of the sparsity pattern of the deflation vectors z_j.

We create the vectors z_j in the subdomain Ω_j only and send essential parts to its direct neighbours. We explain the data-structure and communication issues for a rectangular example. In the explanation we use the global numbering from the left part of Figure 2. Note that in the implementation the local numbering is used in the communication and calculation part. The global numbering is used for post-processing purposes only. The example can be generalized easily to other configurations. The situation is displayed in Figure 2.

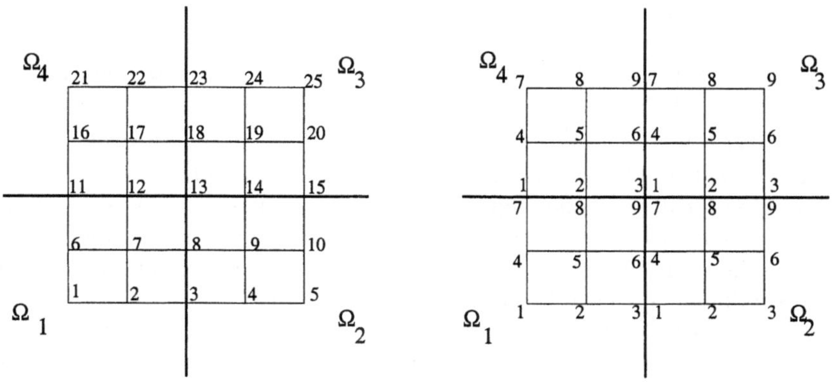

Fig. 2. A sketch of division of Ω into subdomains $\Omega_1, \ldots, \Omega_4$. Left figure represents the global numbering of the unknowns, right figure represents the local numbering.

In Figure 2 on each subdomain Ω_i a deflation vector z_i is created. For all the interface nodes (say numbers 3, 8, 13, 12, 11 of Ω_1) the number of neighbours is determined, then equation (2) is applied to determine the value of the corresponding entry of z_1. This implies that it is necessary to have a list of interface nodes for each subdomain and to have a list with the number of neighbouring subdomains on which a particular interface node is located. However, this information is not sufficient. For example the vector Az_1 must be computed and also be multiplied with vectors z_j. The vector Az_1 has non-zero entries not only inside the domain Ω_1 and on its interfaces, but also in its direct neighbours, i.e. the points 4, 9, 14, 16, 17, 18, 19. Therefore we also need to extend the vectors z_1 with these points to have a well defined matrix-vector multiplication. In the same way the vectors Az_j, $j \in \{2,3,4\}$ have non-zero entries in Ω_1. For the other deflation vectors we proceed analogously.

We further explain the computational part which is relevant to processor 1, i.e. subdomain ω_1 only, the other subdomains are dealt with similarly. For example Az_2 will have a non-zero contribution in all interface points of Ω_1 and Ω_2 but also in the points 2, 7 and 12. All vectors in common points of any subdomain and Ω_1 are given the global value, i.e. the value that is the result of addition. This requires communication between Ω_1 and this particular subdomain. This is in contrast to the matrix A, which is stored only locally per subdomain without the addition at common interfaces. So to compute Az_2 on Ω_1 we need an extra list of neighbouring points of Ω_2 in domain Ω_1 that are not on the common interface. This, however, is not sufficient. For example the value of Az_2 in nodal point 12 also has a contribution of Ω_4. So in Ω_1 we need an extra list of points on common sides of Ω_1 and Ω_j that are direct neighbours of Ω_k $(j \neq k)$ but are not on the interface of Ω_1 and Ω_k. For example node 12 is a common point of Ω_1 and Ω_4 and a neighbour of Ω_2, further Ω_4 is a neighbour of Ω_2. After communication and addition of the values of Az at these particular nodes, the matrix E, consisting of the inner products, is calculated and sent to processor 1.

Then, for a given vector v we compute at each processor its inner product with z_i $(z_i^T v)$. Then all these inner products are sent to processor 1 (Ω_1) where $y = E^{-1}\left(z_1^T v \; z_2^T v \; z_3^T v \; z_4^T v\right)^T$ is computed by Choleski and subsequently the results are sent to all the other neighbouring processors. Then, $Pv = v - ZAy$ is computed locally. All the steps are displayed schematically in algorithm 2, where we explain the situation for a case with two processors. Note that E has a profile structure where the profile is defined by the numbering pattern of the subdomains. Hence for a block structure in two dimensions we obtain a similar sparsity pattern for E as for a two-dimensional discretization. If, however, a layered structure is used, then E gets the same pattern as a one-dimensional discretization matrix.

Algorithm 2 (Parallellization of Deflation) $P = I - AZ(Z^TAZ)^{-1}Z^T$.

Processor 1		Processor 2
Make z_1		Make z_2
	Communication	
Make Az_1, Az_2^Γ		Make Az_2, Az_1^Γ
	Communication	
$Az_1 = Az_1 + Az_1^\Gamma$ $E_{11} = z_1^T Az_1$ $E_{12} = z_1^T Az_2^\Gamma$		$Az_2 = Az_2 + Az_2^\Gamma$ $E_{22} = z_2^T Az_2$ $E_{22} = z_2^T Az_1^\Gamma$
	Send E to proc 1	
Choleski decomp E		
	$Pv =$ $v - AZE^{-1}Z^Tv$	
Compute $z_1^T v$		Compute $z_2^T v$
	Send $z_2^T v$	
$y = E^{-1}\begin{pmatrix} z_1^T v \\ z_2 v \end{pmatrix}$		
	Send y to proc 2	
$v - y_1 Az_1 - y_2 Az_2^\Gamma$		$v - y_2 Az_2 - y_1 Az_1^\Gamma$

4 Numerical experiments

To illustrate the advantage of the deflation method we present the number of CG-iterations and wall-clock time as a function of the number of layers (left and right graphs respectively in Figure 3). We start with one layer and extend the domain of computation with one horizontal layer, which is placed on top. This is done consecutively up to 7 layers. In the examples we choose the number of elements the same in each layer. The problem size increases as the number of layers increases. It can be seen that if deflation is not used then the convergence will take more time since the number of CG iterations increases. The use of deflation yields that the number of iterations and wall-clock time for the parallel case does not depend on the number of layers. This is also observed for the number of CG-iterations for the sequential computations. This makes the method scalable.

Further, we present the number of iterations as a function of the number of layers as in the preceeding example for three methods: no projection, coarse grid correction and deflation. The results are shown in Figure 4 (left graph). It can be seen that both the coarse grid correction and the deflation methods are scalable, however deflation gives the best results. This is in agreement with the analysis as presented in [3] where it is proven that the deflated method converges faster than the coarse grid correction. The same behaviour is observed if the domain is extended in a blockwise distribution of added subdomains (see the right graph in Figure 4).

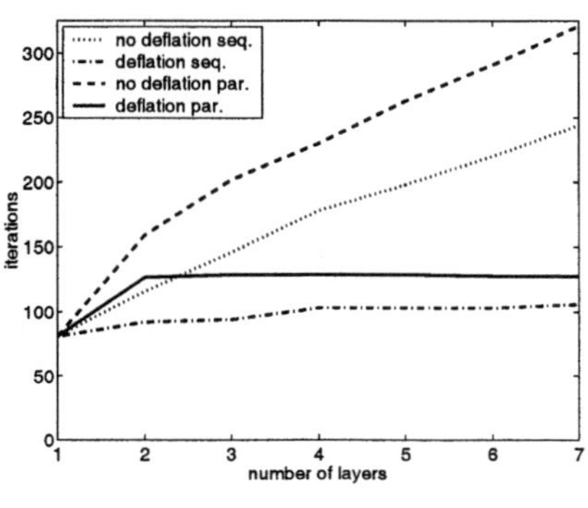

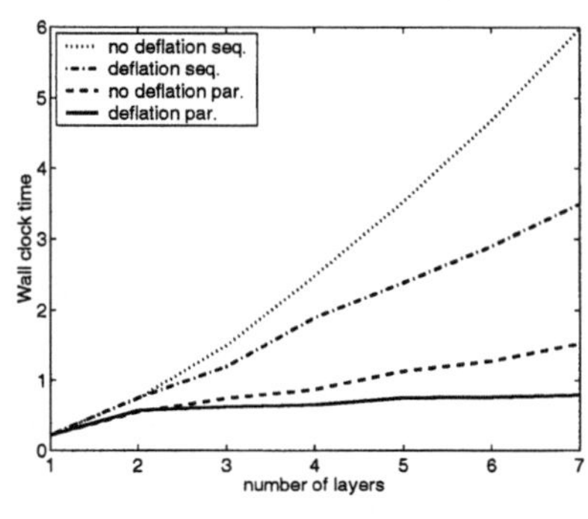

Fig. 3. Left figure: The number of iterations as a function of the number of layers for deflated and non-deflated parallellized and sequential ICCG method. Right figure: The wall clock-time as a function of the number of layers for deflated and non-deflated parallellized and sequential ICCG method.

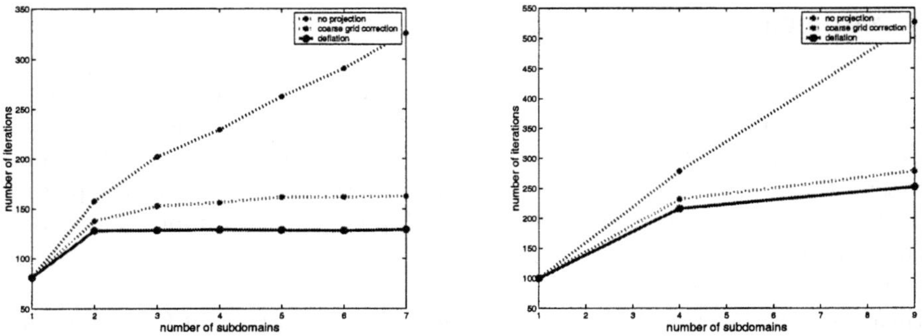

Fig. 4. The number of iterations as a function of the number of layers for the parallellized ICCG method for three methods: no projection, coarse grid correction and deflation. Left graph: layered extension of the domain of computation, Right graph: blockwise extension of the domain of computation

5 Conclusions

The deflation technique has been implemented successfully in a parallellized and sequential ICCG method to solve an elliptic problem by the use of finite elements. The domain decomposition can be chosen blockwise and layerwise. Some numerical experiments are shown in the present paper. Further, the number of iterations and wall-clock time become independent of the number of added layers if deflation is applied in a parallel ICCG method. Hence deflation is favourable in both sequential and parallel computing environments.

References

1. J. Frank and C. Vuik. On the construction of deflation-based preconditioners. *SIAM J. Sci. Comput.*, pages 442–462, 2001.
2. E. F. Kaasschieter. Preconditioned Conjugate Gradients for solving singular systems. *Journal of Computational and Applied Mathematics*, 24:265–275, 1988.
3. R. Nabben and C. Vuik. A comparison of deflation and coarse grid correction applied to porous media flow. Technical report 03-10, Delft University of Technology, Delft University of Technology, Delft, The Netherlands, 2003.
4. R.A. Nicholaides. Deflation of Conjugate Gradients with applications to boundary value problems. *SIAM J. Numer. Anal.*, 24:355–365, 1987.
5. E. Perchat, L. Fourment, and T. Coupez. Parallel incomplete factorisations for generalised Stokes problems: application to hot metal forging simulation. Report, EPFL, Lausanne, 2001.
6. B. Smith, P. Bjørstad, and W. Gropp. *Domain Decomposition.* Cambridge University Press, Cambridge, 1996.
7. F.J. Vermolen, C. Vuik, and A. Segal. Deflation in preconditioned Conjugate Gradient methods for finite element problems. *J. Comput. Meth. in Sc. and Engng.*, to appear, 2003.

8. C. Vuik, A. Segal, L. el Yaakoubli, and E. Dufour. A comparison of various deflation vectors applied to elliptic problems with discontinuous coefficients. *Appl. Numer. Math.*, 41:219–233, 2002.
9. C. Vuik, A. Segal, and J. A. Meijerink. An efficient preconditioned CG method for the solution of a class of layered problems with extreme contrasts in the coefficients. *J. Comput. Phys.*, 152:385–403, 1999.
10. C. Vuik, A. Segal, J. A. Meijerink, and G. T. Wijma. The construction of projection vectors for a Deflated ICCG method applied to problems with extreme contrasts in the coefficients. *J. Comput. Phys.*, 172:426–450, 2001.

Advantages of Binomial Checkpointing for Memory-reduced Adjoint Calculations

Andrea Walther[1] and Andreas Griewank[2]

[1] Institute of Scientific Computing, Technische Universität Dresden, Germany
awalther@math.tu-dresden.de
[2] Institute of Mathematics, Humboldt University Berlin, Germany
griewank@mathematik.hu-berlin.de

Summary. Checkpointing techniques become more and necessary for the computation of adjoints. This paper presents the more common multi-level checkpointing as well as the less known binomial checkpointing. The checkpointing approaches are compared with respect to the number of time steps the adjoint of which can be calculated, the run-time needed for the adjoint calculation and the memory requirement. Some examples illustrate the shown results

1 Introduction

For many time-dependent applications, the corresponding simulations can be performed using ordinary or partial differential equations. Furthermore, quite often there are quantities that influence the result of the simulation. Throughout, we assume that these quantities are control functions, for example heating in and/or at the boundary of a domain. To compute an approximation of the simulated process for a time interval $[0, T]$, one applies an appropriate integration scheme given by

$$y_0 = y^0 \,, \quad y_i \;=\; F_i(y_{i-1}, u_{i-1}, t_{i-1}) \qquad i = 1, \ldots, N \,,$$

where $y_i \in \mathbf{R}^n$ denotes the state and $u_i \in \mathbf{R}^m$ the control at time t_i for a given time grid $t_0, \ldots, t_N$ with $t_0 = 0$ and $t_N = T$. The operator $F_i : \mathbf{R}^n \times \mathbf{R}^m \times \mathbf{R} \mapsto \mathbf{R}^n$ defines the time step to compute the state at time t_i. Note that we do not assume a uniform grid. To optimize a specific criterion or to obtain a desired state, the cost functional

$$\hat{J}(u) = J(y(u), u)$$

measures the quality of $y(u)$ and $u = (u_1, \ldots, u_N)$.
Here, $y(u) = (y_1(u), \ldots, y_N(u))$ describes the dependence of the state y on the control u. For applying a calculus-based optimization method, one may use an adjoint integration

$$\bar{y}_N = 0 \,, \quad \bar{y}_{i-1} \;=\; \bar{F}_i(\bar{y}_i, y_{i-1}, u_{i-1}, t_{i-1}) \qquad i = N, \ldots, 1 \,, \tag{1}$$

motivated by the adjoint differential equation that belongs to the differential equation describing the state. Subsequently or concurrently, the desired derivative information $\hat{J}_u(u)$ can be reconstructed from $\bar{y} = (\bar{y}_0, \ldots, \bar{y}_N)$. The specific choice of the adjoint steps $\bar{F}_i$ depends on the forward integration and whether one prefers the continuous adjoint or the discrete adjoint formulation, see e.g. [7, 3, 5]. For the purpose of this paper, it is only important to note that the adjoint integration has to be performed backwards in time and that the complete forward trajectory $y = (y_0, \ldots, y_{N-1})$ is required. Hence, storing all states $(y_0, \ldots, y_{N-1})$ during the forward integration and reading them in reverse order during the adjoint integration forms one simple possibility to overcome this difficulty. Then the computing time for the adjoint calculation consists of the evaluation of N time steps F_i storing the state y_{i-1} and the evaluation of N adjoint steps $\bar{F}_i$.

The storage requirement of the basic approach to calculate adjoints is proportional to the number N of time steps. If we want to calculate the adjoint of a real-world problem with thousands of time steps this memory requirement of the basic approach may become a serious problem. For example, for computing 3D flows with unstructured grids one may need easily 10 to 100 MBytes to store only one state vector y_i [10]. Therefore, it is reasonable to assume that due to their size, only a very limited number of intermediate states can be kept in memory. They may serve as checkpoints, such that the required information for the backward integration is generated piecewise during the adjoint calculation. Sections 2 and 3 present two different checkpointing techniques. The resulting run-times and memory requirements are compared in Section 4. Finally, some conclusions and an outlook are given in Section 5.

2 Uniform Checkpoint Distribution

To distribute the checkpoints equidistantly over the given number of time steps forms one obvious solution to the storage requirement problem. Subsequently the adjoints are computed for each of the resulting groups of time steps separately. Denoting the number of checkpoints used by c, the corresponding calculation of the adjoint values can be performed using the following algorithm where the counter i is identified with the state y_i:

Two-level Checkpointing

Initialization: Reserve space for c_1 checkpoints, store the initial state y_0 in the first one and set

$$c_2 = \begin{cases} \lceil N/(c_1+1) \rceil & \text{if} \quad c_1 \lceil N/(c_1+1) \rceil < N \\ \lfloor N/(c_1+1) \rfloor & \text{else} \end{cases}$$

Advance: Starting from the initial state, advance to state $c_1 \cdot c_2$ by performing the time steps F_i, $1 \le i \le c_1 \cdot c_2$. While integrating forward, store the states $(j-1)\, c_2$ in the checkpoints j for $j = 2, \ldots, c_1$.

Reverse:
do $p = c_1$, 0, -1
Evaluate the time steps F_i, $p \cdot c_2 < i < N$ storing the states i, $p \cdot c_2 \leq i < N-1$,
perform the adjoint steps $\bar{F}_i$, $N \geq i > p \cdot c_2$ to calculate the adjoints,
set $N = p \cdot c_2$, if $p > 0$ read the contents of checkpoint p.
end do

Fig. 1 sketches the two-level checkpointing for $N = 16$ time steps and $c = c_1 + c_2 = 6$. Throughout, the time steps are plotted along the vertical axis and the computing time required for the adjoint calculation is represented by the horizontal axis. Each solid horizontal line including the horizontal axis itself represents a checkpoint. The time, when a state is stored in a checkpoint, is marked with a black circle for the first level and with a black square for the second level. The slanted black lines represent the evaluation of time steps. The adjoint steps are drawn as dashed slanted lines. Finally, black arrows depict the usage of a state y_i for an adjoint step F_{i+1} without performing the corresponding time step F_i. This adjoint calculation is possible due to the assumed structure (1) of the adjoint steps. Note, that it may be required to evaluate F_N once to initialize the adjoints. This evaluation can be introduced right after the evaluation of F_{N-1} for $p = c_1$. For illustration purposes, we suppose throughout that all time steps and all adjoint steps have the same temporal complexity normalized to 1. However, to apply the presented optimal checkpointing techniques, only the identical temporal complexity of all time steps is required. In this example, 24 time steps are performed. Hence, the number

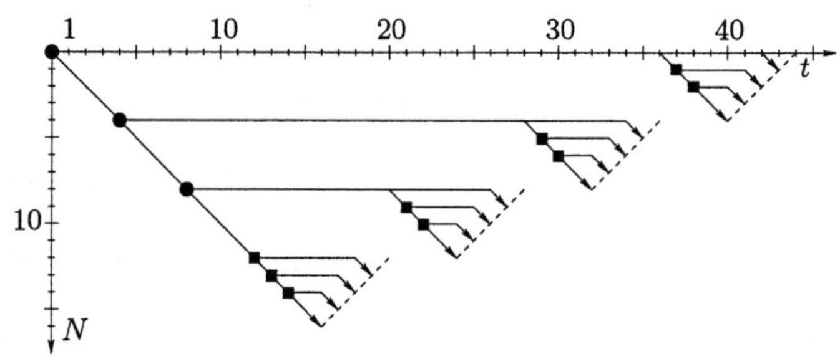

Fig. 1. Two-level checkpointing for $N = 16$ time steps and $c = c_1 + c_2 = 6$ checkpoints

of additional time step evaluations caused by the two-level checkpointing compared to the basic approach equals 9. Furthermore, at most 6 states have to be kept in memory.

The two-level checkpointing has been proposed several times in the literature, e.g. [11, 1], and is easy to implement. Naturally, one can apply two-level checkpointing repeatedly for the groups of time steps that are separated by

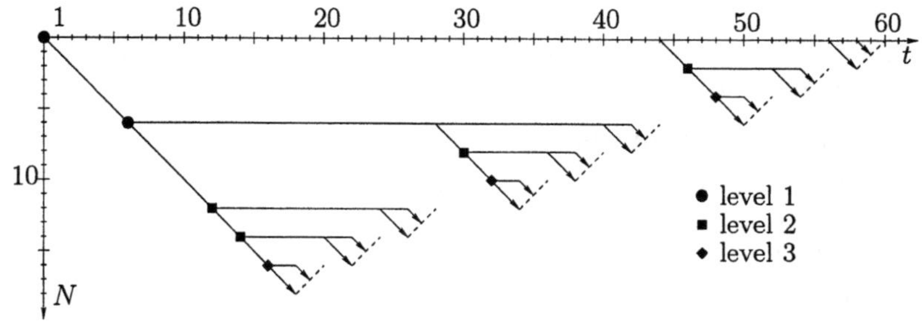

Fig. 2. Three-level checkpointing for $N = 18$ time steps and $c = 5$ checkpoints

equidistant checkpoints. This approach is called *multi-level checkpointing* [6] and sketched by Fig. 2 for the three-level case. The multi-level checkpointing is defined by the number of levels r, the number of checkpoints c_i that are uniformly distributed at level i, $i = 1, \ldots, r-1$, and the number of states c_r that have to be stored at the highest level r. Hence, the parameters of the adjoint calculation shown in Fig. 2 are $c_1 = 2$, $c_2 = 2$, and $c_3 = 1$. For a given r-level checkpointing, one easily derives the following identities

$$N_r = \prod_{i=1}^{r}(c_i+1), \quad M_r = \sum_{i=1}^{r} c_i, \quad T_r = \sum_{i=1}^{r} \frac{c_i N_r}{c_i+1} = r\,N - \sum_{i=1}^{r} \prod_{\substack{j=1 \\ j\neq i}}^{r} (c_j+1)\,,$$

where N_r denotes the number of time steps for which the adjoint can be calculated using the specific r-level checkpointing. The corresponding memory requirement equals M_r. The number of time step evaluations required for the adjoint calculation is given by T_r, since at the first level $c_1 N_r/(c_1{+}1)$ time steps have to be evaluated to reach the second level. At the second level, one group of time steps is divided into c_2+1 groups. Hence, $c_2(N_r/c_1+1)/(c_2+1)$ time steps have to be evaluated in each group to reach the third level. Therefore, we obtain $(c_1+1)c_2(N_r/c_1+1)/(c_2+1) = c_2 N_r/(c_2+1)$ at the second level and so on. It follows that each time step F_i is evaluated at most r times. Hence, if we apply two-level checkpointing, each time step is evaluated no more than two times.

The two- as well as the multi-level checkpointing technique have the drawback that at each level the checkpoints are not reused. Each checkpoint stores at each level only one state and becomes idle as soon as the data that is stored in the checkpoint has been used for the adjoint calculation. A method that reuses the checkpoints as soon as possible is proposed in the next section.

3 Binomial Checkpoint Distribution

When one applies the checkpointing technique proposed in [5], the adjoint values are again generated piece by piece but only one state is employed for the adjoint calculation at any time. Therefore, the checkpointing procedure has to be adapted as follows:

Binomial Checkpointing

Initialization: Reserve space for c checkpoints and store the initial state y_0 in the first one.
do $p = N$, 1, -1
  Advance: Starting from the last checkpoint assigned, advance to state $p - 1$.
  If checkpoints are free, set as many of them as possible to states i along the way.
  Reverse: Perform the adjoint step $\bar{F}_p$ to calculate the adjoint.
  If state $p - 1$ is stored in a checkpoint, free the checkpoint up.
end do

The memory requirement of this checkpointing procedure equals $M_b = c$. Naturally, the question arises where one should place the checkpoints in the action "Advance" of the algorithm to minimize the number of time step evaluations. The application of the routine revolve ensures that the initiated checkpointing process is provably optimal with respect to the run-time increase for a given number of checkpoints [5]. More specifically, for the structure (1) of the adjoint steps considered here, the following complexity result holds:

Theorem 1. *Let N be the total number of time steps for which the adjoint has to be calculated. Suppose, up to c checkpoints are available at any time. Then the minimal number of time step evaluations needed for the adjoint calculation equals*

$$T_b = N\,r - \binom{c+r}{r-1},$$

where r the unique integer satisfying

$$\binom{c+r-1}{r-1} < N \leq \binom{c+r}{r}. \tag{2}$$

The proof of Theorem 1 (see [5]) constructs recursively checkpointing schedules that attain the minimal number T_b. For the optimal checkpointing procedures the positions of the checkpoints are given by binomial coefficients. This fact explains the name *binomial checkpointing.* Furthermore, the proof of Theorem 1 shows that each time step F_i is evaluated at most r times. Hence, r has the same meaning as in the previous section. It was proved earlier that a logarithmic growth of memory and run-time can be achieved using binomial checkpointing by providing an appropriate number of checkpoints [4].

The routine revolve implements the optimal binomial checkpointing and can be incorporated easily in an existing adjoint calculation [2, 8]. Moreover, one can build a heuristic based on revolve such that the adjoint calculation using binomial checkpointing becomes applicable also if the number of time steps is not known a-priori, e.g. due to adaptive time stepping, and/or if the temporal complexity of the time steps is not constant, e.g. due to implicit methods [9].

One optimal checkpointing schedule computed with revolve for $N = 16$ time steps and $c = 3$ checkpoints is shown in Fig. 3. Once more, it might be

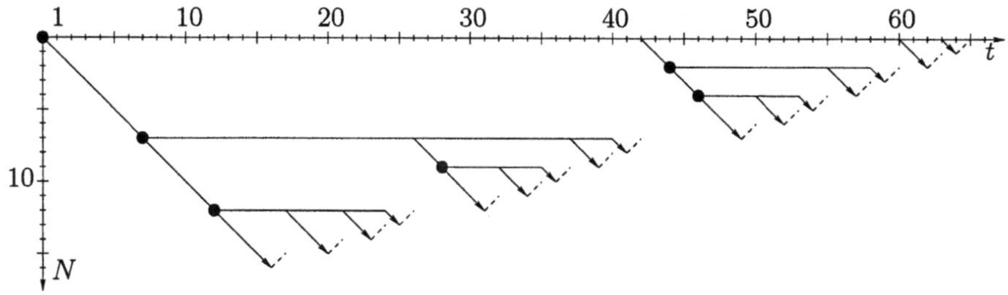

Fig. 3. Binomial Checkpointing for $N = 16$ time steps and $c = 3$ checkpoints

necessary to evaluate F_N once to initialize the adjoints. Since the situation is the same for the multi-level checkpointing and does not influence the results in the sequel, we ignore throughout the evaluation of F_N. For the example shown above, the number of time step evaluations equals $T_r = 33$. Compared to the two-level checkpointing, the computing time for the adjoint calculation increases by less than 50 %. Furthermore, only 3 states have to be kept in memory. Hence, the storage requirement is reduced by 50 %. The relation between the two checkpointing approaches will be discussed in more detail in the next section.

4 Comparison of Both Checkpoint Distributions

The integer r has the same meaning for both checkpointing approaches, namely the maximal number of times any particular time step F_i is evaluated during the adjoint calculation. Hence for comparing both approaches, assume at the beginning that r has the same value and that the same amount of memory is used, i.e. $M_r = M_b = c$.

Now, we examine the maximal number of time steps N^* for which an adjoint calculation can be performed using the two approaches. Assuming that r is a divisor of c and $M_r = c$, one obtains the identity

$$N_r^* = \left(\frac{c}{r} + 1\right)^r = \left(\frac{c+r}{r}\right)^r \quad \text{with} \quad c_i = \frac{c}{r}, \quad i = 1, \ldots, r ,$$

for the uniform checkpoint distribution because of the structure of N_r. Theorem 1 yields

$$N_b^* = \binom{c+r}{r} = \prod_{i=0}^{r-1}\left(\frac{c}{r-i}+1\right)$$

for the binomial checkpoint distribution. Obviously, one has

$$\frac{c}{r}+1 < \frac{c}{r-i}+1 \qquad \text{for } 0 < i \leq r-1 \,.$$

These inequalities yield $N_r^* < N_b^*$ if $r \geq 2$. Hence for all $r \geq 2$ and and a given c, binomial checkpointing allows the adjoint calculation for a larger number of time steps compared to uniform checkpointing. In more detail, using Stirling's formula we obtain

$$\frac{N_b^*}{N_r^*} \approx \binom{c+r}{r}\left(\frac{c}{r}+1\right)^{-r} = \frac{1}{\sqrt{2\pi r}}\left(\frac{c}{r}+1\right)^{c} \approx \frac{1}{\sqrt{2\pi r}}\exp(r)\,.$$

Hence, the ratio of N_b^* and N_r^* grows exponentionally in r without any dependence on the number of available checkpoints. Fig. 4 shows N_r^* and N_b^* for the most important values $2 \leq r \leq 5$. Since r denotes the maximal number

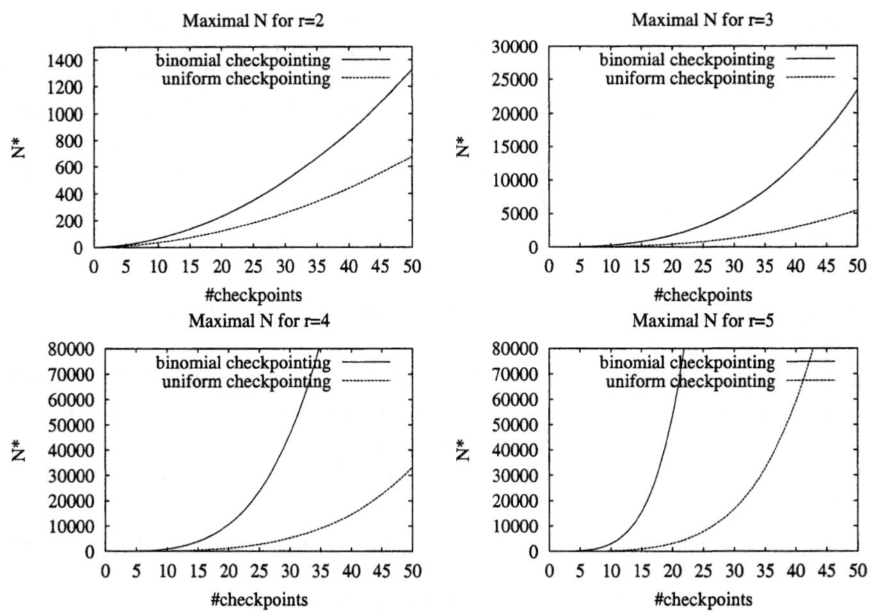

Fig. 4. N_r^* and N_b^* for $r = 2, 3, 4, 5$

of times each time step is evaluated, we have the following upper bounds for the number of time steps evaluated during the adjoint calculation using r-level checkpointing and binomial checkpointing, respectively:

$$T_r = c\left(\frac{c}{r}+1\right)^{r-1} < r\,N_r^* \qquad \text{and} \qquad T_b = rN_b^* - \binom{c+r}{r-1} < r\,N_b^* \,.$$

For example, it is possible to compute the adjoint for $N = 23000$ time steps with only 50 checkpoints, less than $3N$ time step evaluations, and N adjoint steps using binary checkpointing instead of three-level checkpointing, where $N_3^* \leq 5515$. If we allow $4N$ time step evaluations then 35 checkpoints suffice to compute the adjoint for 80000 time steps using binomial checkpointing, where $N_4^* \leq 9040$. These number are only two possible combinations taken from Fig. 4 to illustrate the really drastic decrease in memory requirement that can be achieved if binomial checkpointing is applied.

However, usually the situation is the other way round, i.e. one knows N and/or c and wants to compute the adjoint as cheap as possible in terms of computing time. Here, the first observation is that r-level checkpointing introduces an upper bound on the number of time steps the adjoint of which can be computed, because the inequality $N \leq (c/r+1)^r$ must hold. Furthermore, binomial checkpointing allows for numerous cases also a decrease in run-time compared to the uniform checkpointing. For a given r-level checkpointing and $M_r = c$, one has to compare T_r and T_b. Let r_b be the unique integer satisfying (2). Since at least one checkpoint has to be stored at each level, one obtains the bound $r \leq c$. I.e., one must have $c >= \log_2(N)$ to apply uniform checkpointing. Therefore, the following combinations of r and r_b are possible for the most important, moderate values of r:

$$r = 3 \Rightarrow r_b \in \{2,3\}, \qquad r = 4 \;\Rightarrow\; r_b \in \{3,4\}, \qquad r = 5 \;\Rightarrow\; r_b \in \{3,4,5\}\,.$$

For $3 \leq r \leq 5$, one easily checks that $T_r > T_b$ holds if $r_b < r$. For $r = r_b$, one can prove the following, more general result:

Theorem 2. *Suppose for a given N and a r-level checkpointing with $M_r = c$ that the corresponding r_b satisfying (2) coincide with r. Then, one has*

$$\begin{aligned} T_2 = 2N - c - 2 \;&=\; T_b \qquad && \textit{if } r = r_b = 2 \\ T_r > T_b & && \textit{if } r = r_b > 2. \end{aligned}$$

Proof: For $r_b = r = 2$ the identity $T_2 = T_b$ is clear. For $r = r_b > 2$, the inequality

$$\begin{aligned} \sum_{i=1}^{r}\prod_{\substack{j=1\\ j\neq i}}^{r}(c_j+1) &= \frac{(r-1)!}{(r-1)!}\left(\prod_{j=1}^{r-1}(c_j+1) + (c_r+1)\sum_{i=1}^{r-1}\prod_{\substack{j=1\\ j\neq i}}^{r-1}(c_j+1)\right) \\ &< \frac{1}{(r-1)!}\prod_{i=2}^{r}\left(\sum_{j=1}^{r}c_j + i\right) \;=\; \binom{c+r}{r-1} \end{aligned}$$

holds. Using the definitions of T_r and T_b, this relation yields immediately $T_r > T_b$. ∎

Hence, except for the case $r = r_b = 2$, where T_r and T_b coincide, the run-time caused by binomial checkpointing is less than the one caused by multi-level checkpointing if $r = r_b$.

5 Conclusions

This article discusses several checkpointing techniques, namely multi-level checkpointing and binomial checkpointing. A detailed analysis of the number of time steps the adjoint of which can be calculated, the run-time needed for the adjoint calculation and the memory requirement is given.

One can conclude that binomial checkpointing allows adjoint calculations with a surprisingly small fraction of the memory needed by the basic approach. This storage reduction causes only a very moderate increase in run-time. On the other hand, we see that r-level checkpointing induces for a given number of checkpoints an upper bound on the number of time steps the adjoint of which can be computed. This upper bound can only be increased by introducing a next level of checkpointing. In addition it is shown that the run-time required for the adjoint calculation with r-level checkpointing exceeds the run-time needed for binomial checkpointing for the most important values of $r > 2$, whereas for $r = 2$ both methods yield the same run-time. However, for $r = 2$ and a given amount of memory, binomial checkpointing allows the adjoint computation for a larger number of time steps. Hence, even for $r = 2$ binomial checkpointing is preferable.

Moreover, it is quite often the case that the number N of time steps is not known a-priori, for example due to an adaptive time stepping method. Then, it becomes difficult to distribute the checkpoints for the two- or multi-level checkpointing such that the minimal run-time is attained. For binomial checkpointing the extension a-revolve deals with the unknown number of time steps by using a heuristic for the checkpoint placements. In addition, a-revolve can also handle time steps with varying temporal complexity. For time steps the cost of which do not change drastically, the heuristics implemented in a-revolve work well such that the corresponding adjoint calculation is only a few percentages slower than the one based on revolve [9]. Hence, binomial checkpointing provides memory-reduced adjoint calculation also in more general situations.

References

1. Charpentier, I. (2001): Checkpointing schemes for adjoint codes: Application to the meteorological model Meso-NH. SIAM J. Sci. Comput., **22**, 2135-2151
2. Gockenbach, M., Reynolds, D., Symes, W. (2002): Efficient and Automatic Implementation of the Adjoint State Method. Trans. Math. Soft., **28**, 22-44
3. Griesse, R. (2003): Parametric Sensitivity Analysis in Optimal Control of a Reaction-Diffusion System–Part II: Practical Methods and Examples. To appear in Opt. Meth. Soft.

4. Griewank, A. (1992): Achieving logarithmic growth of temporal and spatial complexity in reverse automatic differentiation. Opt. Meth. Soft., **1**, 35-54
5. Griewank, A., Walther, A. (2000): Revolve: An implementation of checkpointing for the reverse or adjoint mode of computational differentiation. Trans. Math. Soft., **26**, 19-45
6. Heimbach, P., Hill, C., Giering, R. (2003): An efficient exact adjoint of the parallel MIT general circulation model, generated via automatic differentiation. To appear in Future Generation Computer Systems.
7. Hinze, M. (1999): Optimal and instantaneous control of the instationary Navier-Stokes equations. Habilitationsschrift, Fachbereich Mathematik, Technische Universität Berlin
8. Hinze, M., Walther, A. (2002): Discrete approximation schemes for reduced gradients and reduced Hessians in Navier-Stokes control utilizing an optimal memory-reduced procedure for calculating adjoints. Preprint MATH-NM-06-2002, Tech. Uni. Dresden. Submitted
9. Hinze, M., Sternberg, J. (2003): A-Revolve: An adaptive memory- and run-time-reduced procedure for calculating adjoints; with an application to the instationary Navier-Stokes system. To appear in Opt. Math. Soft.
10. Kaltenbach, H.-J., Jürgens, W., Spille, A. (2001): Numerische Simulation, Beeinflussung und Eigenmoden-Analyse einer abgelösten Strömung mit Querkomponente. In: Ergebnisberichte SFB 557 TP A6
11. Kubota, K. (1998): A Fortran77 preprocessor for reverse mode automatic differentiation with recursive checkpointing. Opt. Meth. Soft., **10**, 319-335

An Efficient Multigrid FEM Solution Technique for Incompressible Flow with Moving Rigid Bodies

Decheng Wan, Stefan Turek and Liudmila S. Rivkind

Institute of Applied Mathematics LS III, University of Dortmund,
Vogelpothsweg 87, 44227 Dortmund, Germany

Summary. This paper uses the fictitious boundary method described in [1] for the solution of incompressible flow with moving rigid bodies in complex geometries. The method is based on a special treatment of Dirichlet boundary conditions inside of a FEM approach in the context of a hierarchical multigrid scheme such that the flow can be efficiently computed on a fixed computational mesh while the solid boundaries are allowed to move freely through the given mesh. In this paper, we focus on the calculations of the drag and lift forces acting on the moving solid bodies which are not captured by the mesh. The comparison between the present and benchmark results for the flow around a circular cylinder with different Reynolds numbers is first presented, and then the result for a circular cylinder oscillating in a channel is given. The simulation results compared with corresponding reference results are found to be very reasonable and satisfactory.

1 Introduction

Incompressible flow problems with moving rigid bodies in complex geometries have drawn attention of numerous investigators. Their studies have been motivated by the desire to understand the fundamental physics of such flows as well as their practical importance in various areas. The phenomena of such flow problems are visible everywhere around our living environments such as: flow around high-rise building, the drag force induced by driving car accelerating in the wind, ocean current interaction with the offshore structures, sedimentation flow in estuary and sand flow in desert, etc.

From the numerical point of view, incompressible flow with moving rigid bodies in complex geometries is quite hard to simulate, since it can require a huge amount of time for the generation or deformation of the computational grid when the corresponding boundaries are complex or changing. Such problems have motivated the development of numerous algorithms, which can be broadly classified into two families. One of them is a 'body-conformal approach' which always keeps the computational mesh in accordance to the geometrical details [2, 3]. Another one is a 'fixed grid approach' in which case the mesh is (arbitrarily) fixed and internal objects are allowed to move freely through the mesh [4, 5]. One big advantage of such 'fixed grid approaches' over the conventional 'body-conformal approaches' is that the computational mesh remains unchanged such that the CPU cost per time step can be significantly decreased - less computational effort due to saving the expensive mesh generation - and that such techniques can be easily incorporated into standard CFD codes which mostly allow fixed computational grids without local adaptivity only; however, the resulting accuracy is not clear. Therefore, the overall aim is to deal successfully with the moving boundaries such that the accuracy of the numerical approximation is sufficiently high while at the same time also the computational cost is decreased.

In the spirit of the 'fixed grid approaches', a simple and efficient 'fictitious boundary method' for the detailed simulation of incompressible flow with complex geometries and/or moving interfaces was developed in the paper [1]. The method is based on a fixed unstructured FEM background grid. It starts with a coarse mesh which contains already many of geometrical fine-scale details, and employs a (rough) boundary parametrization which sufficiently describes all large-scale structures with regard to the boundary conditions. Then, treat all fine-scale features as interior objects such that the corresponding components in all matrices and vectors are unknown degrees of freedom which are implicitely incorporated into all iterative solution steps (see [1]).

In this paper, we used the fictitious boundary method for the simulation of incompressible flow with moving rigid bodies in complex geometries. In many cases, the calculation of forces acting on the moving rigid bodies is very important for the further study of the interaction between fluid and body, like in particulate flow, sedimentation flow, and fluid-structure flow, etc. However, in the fictitious boundary method, it is not so easy and straightforward to compute these interesting forces, because the drag coefficient C_d and lift coefficient C_l acting on the moving solid bodies are a very delicate quantity: they include the results directly on the wall surface of the moving rigid bodies which is represented implicitly in the fictitious boundary method due to the use of a fixed grid rather than a body-conformal grid. Therefore, the integral of forces only over the wall surface of rigid bodies cannot be implemented directly in the fictitious boundary method. For overcoming this difficulty, a volume integral instead of the conventional surface integral for the calculation of the C_d and C_l by introducing an auxiliary function [7] or two additional functions [8] is suggested. Obviously, in such volume integral calculations, the reconstruction of the wall surface of the moving rigid bodies can be avoided. In this paper, we use the Duchanoy's idea of the volume integral [7], and expand his implementation in a finite volume method into the finite element method and the fictitious boundary method.

2 The fictitious boundary method

The details of the fictitious boundary method have been described in [1]. For the following considerations, let Ω be a bounded domain with a piecewise smooth boundary Γ. The equations to be solved are the incompressible Navier-Stokes equations

$$\rho\frac{\partial \mathbf{u}}{\partial t} + \rho\mathbf{u}\cdot\nabla\mathbf{u} = \mathbf{f} - \nabla p + \mu\nabla^2\mathbf{u}\,, \qquad \nabla\cdot\mathbf{u} = 0\,, \tag{1}$$

where $\mathbf{u}$ is the velocity, p the pressure, μ the dynamic viscosity coefficient, ρ the density, $\mathbf{f}$ the source term which may include the gravitational force. The above equations are to be solved with $\mathbf{u}(\mathbf{x},t) = \mathbf{u}_\partial(\mathbf{x},t)$ on parts of the boundaries of the flow domain where $\mathbf{u}_\partial(\mathbf{x},t)$ is the prescribed boundary velocity, including time-dependent moving boundaries. The details of solving such incompressible flow problems can be found in the FeatFlow software [10, 11] which is based on (nonconforming) FEM discretizations, adaptive implicit time-stepping, nonlinear Newton-like methods, (geometrical) multigrid solvers (for velocity and pressure separately) on quite arbitrary domains.

In the following part, we give the description of a volume integral approach for the calculation of the drag coefficient C_d and lift coefficient C_l acting on the moving solid bodies. Let S be the wall surface of the rigid bodies, $\mathbf{n}_S$ be the inward pointed unit normal with respect to Ω and tangential vector $\tau = (n_y, -n_x)$. The drag and lift forces are usually calculated by a surface integral as follows

$$F_D = \int_S \left(\mu\frac{\partial \mathbf{u}_\tau}{\partial \mathbf{n}_S}n_y - p\,n_x\right)ds\,, \qquad F_L = -\int_S \left(\mu\frac{\partial \mathbf{u}_\tau}{\partial \mathbf{n}_S}n_x + p\,n_y\right)ds\,, \tag{2}$$

while the drag and lift coefficient are calculated via

$$C_d = \frac{2F_D}{\rho\bar{U}^2 D}\,, \qquad C_l = \frac{2F_L}{\rho\bar{U}^2 D}\,, \tag{3}$$

where $\bar{U}$ is the characteristic velocity, and D the characteristic length.

From Eq.(3) and Eq.(2), we can see that the surface integral around the wall surface of the rigid bodies should be conducted for the calculation of the C_d and C_l. However, in the present fictitious boundary method, the shapes of the wall surface of the moving rigid bodies is implicitly imposed in the fluid field. If we reconstruct the shapes of the wall surface of the moving rigid bodies, it is not only a time consuming work, but also the accuracy is only first order due to a piecewise constant interpolation. For overcoming this problem, we use the following method to calculate the C_d and C_l in which the surface integral is replaced by a volume integral. We define a parameter α as

$$\alpha(\mathbf{x}) = \begin{cases} 1 & \text{for} \quad \mathbf{x}\in\Omega_c\,, \\ 0 & \text{for} \quad \mathbf{x}\in\Omega\,, \end{cases} \tag{4}$$

where $\mathbf{x}$ denotes the coordinates of the edge midpoints of cells, Ω_c is the domain occupied by the rigid bodies, Ω is the fluid domain, the whole domain is $\Omega_T = \Omega \cup \Omega_c$. The importance of such a definition of the parameter can be seen from the fact that the gradient of α is zero everywhere except at the wall surface of the rigid bodies, and equal to the normal vector $\mathbf{n}$ defined on the grid [7, 12], i.e.

$$\mathbf{n} = -\nabla\alpha\,. \tag{5}$$

The total stress tensor σ of the fluid flow is

$$\sigma = -p\mathbf{I} + \mu\left[\nabla\mathbf{u} + (\nabla\mathbf{u})^T\right]\,. \tag{6}$$

Hence the forces acting over the wall surface of the rigid bodies can be computed by

$$\mathbf{F}_T = \int_{\Omega_T} \sigma\,\mathbf{n}\,d\Omega = -\int_{\Omega_T} \sigma\,\nabla\alpha\,d\Omega\,. \tag{7}$$

The drag force and lift force can be obtained from the Eq.(7),

$$F_D = -\int_{\Omega_T}\left[\mu\left(\frac{\partial u}{\partial x}\frac{\partial \alpha}{\partial x} + \frac{\partial u}{\partial y}\frac{\partial \alpha}{\partial y}\right) - p\,\frac{\partial \alpha}{\partial x}\right]d\Omega\,, \tag{8}$$

$$F_L = -\int_{\Omega_T}\left[\mu\left(\frac{\partial v}{\partial x}\frac{\partial \alpha}{\partial x} + \frac{\partial v}{\partial y}\frac{\partial \alpha}{\partial y}\right) - p\,\frac{\partial \alpha}{\partial y}\right]d\Omega\,. \tag{9}$$

Therefore through Eq.(8), Eq.(9) and Eq.(3) we can calculate the new drag and lift coefficients (C_d and C_l) via the volume integral over the whole domain Ω_T instead of the surface integral over the wall surface of the rigid bodies in Eq.(2). The integral over each element covering the whole domain Ω_T is evaluated with the standard 3×3 point Gaussian quadrature. Since the gradient $\nabla\alpha$ is non-zero only at the wall surface of the rigid bodies, thus the volume integrals need to be computed only in one layer of mesh cells around the rigid bodies. It is convenient for the present fictitious boundary method to calculate the C_d and C_l.

3 Numerical tests

This section consists of two parts. The first part presents a quantitative examination for the benchmark case of flow around a circular cylinder with $Re = 20$ and 100 solved by the present fictitious boundary method. The second part gives the computing results for a circular cylinder oscillating in a channel. For comparison, corresponding reference results are also presented.

3.1 Flow around a circular cylinder

We consider the benchmark case of flow around a circular cylinder described in the paper [9]. The body-conformal mesh of Fig. 1 (a) is used to provide reference results, while the channel meshes of Fig. 1 (b) and (c) are employed by the present fictitious boundary method. The channel height is $H = 0.41$ m, the cylinder diameter $D = 0.1$ m. The Reynolds number is defined by $Re = \bar{U}D/\nu$ with the mean velocity $\bar{U} = 2U(0, H/2, t)/3$. The kinematic viscosity of the fluid is given by $\nu = \mu/\rho = 10^{-3}\, m^2/s$ and its density by $\rho = 1\, kg/m^3$. The inflow profiles are parabolic with different U such that the resulting Reynolds numbers are $Re = 20$ (steady case) and $Re = 100$ (nonsteady case).

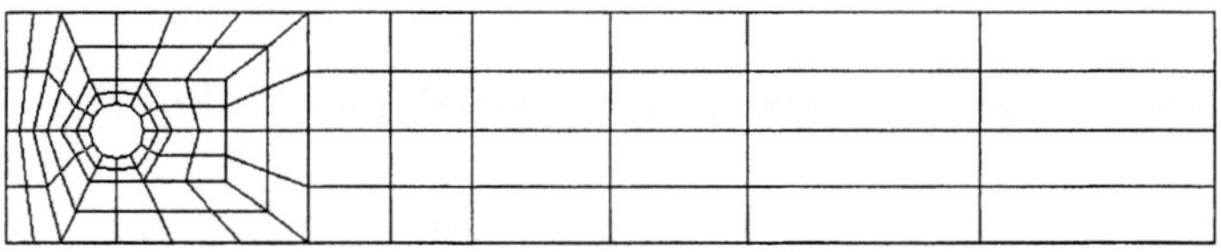

(a) body-conformal mesh (LEVEL = 2)

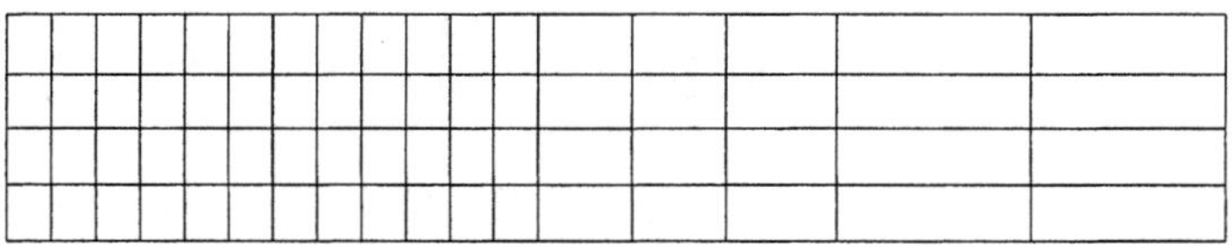

(b) channel mesh I (LEVEL = 1)

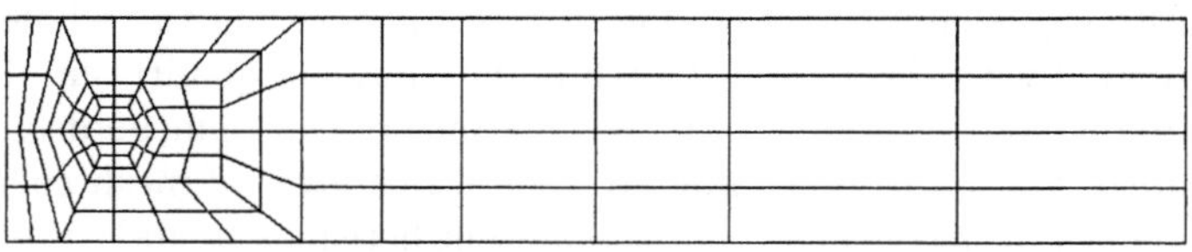

(c) channel mesh II (LEVEL = 2)

Fig. 1. Different coarse meshes

Table 1. The number of elements for different refined meshes

LEVEL	3	4	5	6	7
body-conformal mesh	384	1536	6144	24576	98304
channel mesh I	1088	4352	17408	69632	278528
channel mesh II	416	1664	6656	26624	106496

We first perform a stationary simulation ($Re = 20$), based on the body-conformal mesh, the channel mesh I and the channel mesh II, respectively.

The shown coarse meshes will be successively refined by connecting opposite midpoints. Table 1 gives the number of elements for these meshes after such global refinements. Here LEVEL corresponds to the number of refinements. The following Table 2 shows the comparison of the drag coefficient C_d and the lift coefficient C_l based on the body-conformal mesh, the channel mesh I and the channel mesh II, respectively. The calculation of C_d and C_l based on the body-conformal mesh uses the surface integral formula in Eq.(2) which is referred as reference results, while for the cases of using the channel mesh I and the channel mesh II, the volume integral formula in Eq.(8) and Eq.(9) are employed. In this table, the results calculated from LEVEL = 3 to LEVEL = 7 are all shown together. The corresponding benchmark values in [9] are also listed in the table. From the comparisons, it can be seen that the results calculated by the present fictitious boundary method agree sufficiently well ($\sim 1\%$) with both the reference results. The results for the channel mesh I are found to be not completely satisfying, while the results for channel mesh II are improved since there is local refinement of the mesh near the wall surface of the cylinder. The results for such low Reynolds number simulations show that an appropriate global grid refinement as well as adequate local mesh adaptation are necessary. The present fictitious boundary method proves to be competitive with the standard approaches for such typical CFD applications.

Table 2. Comparison of C_d and C_l for $Re = 20$

	C_d			C_l		
LEVEL	Ref.	ch. mesh I	ch. mesh II	Ref.	ch. mesh I	ch. mesh II
3	0.53450D+01	0.55296D+01	0.54196D+01	0.56128D-02	0.12165D-01	0.24435D-02
4	0.55066D+01	0.53537D+01	0.54207D+01	0.84683D-02	0.10742D-01	0.67612D-02
5	0.55404D+01	0.54278D+01	0.55161D+01	0.98915D-02	0.61455D-02	0.89128D-02
6	0.55581D+01	0.55012D+01	0.55571D+01	0.10384D-01	0.99024D-02	0.94709D-02
7	0.55683D+01	0.55421D+01	0.55640D+01	0.10554D-01	0.97706D-02	0.10192D-01
Ref. [9]	0.55795D+01			0.10618D-01		

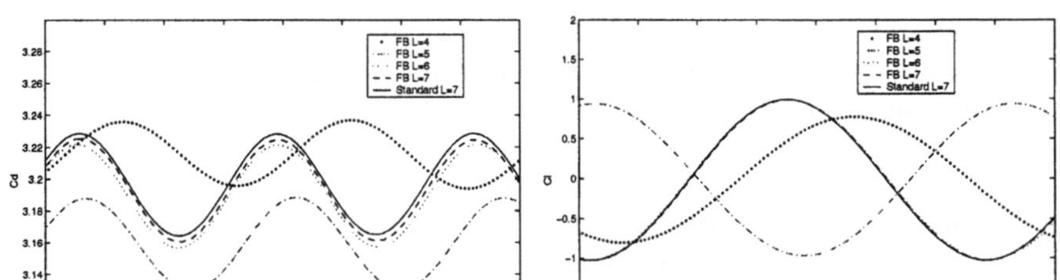

Fig. 2. Periodical results of C_d and C_l for $Re = 100$

Next, we further examine the resulting accuracy for a medium range Reynolds number $Re = 100$ which leads to periodical time-dependent vortex shedding behind the cylinder. Since we are mainly interested in the spatial accuracy of the fictitious boundary method, in particular capturing the important effects near the cylinder, we try to eliminate the temporal discretization error by choosing very small time steps. Then, we proceed the nonstationary simulations until a fully periodical flow behaviour of all quantities has been observed. Finally, we compare the results for one period. Figure 2 shows the results of the C_d and C_l. The "Standard $L = 7$" means that the body-conformal mesh of Fig. 1 (a) with 7 level refinement was used to provide the reference result, while "FB $L = 3 \sim 7$" represents the results obtained by the present fictitious boundary method using the channel mesh I in Fig. 1 (b) with different refinement levels. From these figures, we can see that the various results are identical with regard to the reference results. The results also show that the present fictitious boundary method leads to comparative results like the standard approaches with 'body-fitted' meshes. It can also be claimed that results with higher accuracy can be reached via local mesh adaptivity (see for example the concept of aligned adaptive computational meshes in [1]).

3.2 An oscillating cylinder in a channel

To demonstrate the ability of the present fictitious boundary method to handle flows with complex moving boundaries, we have chosen a flow configuration with a cylinder undergoing sinusoidal transverse oscillation in a channel with specified amplitudes and frequencies. The channel mesh of Fig. 3 (a) is employed by the present fictitious boundary method. The computational domain size is (2.2×0.41). The mean location of the cylinder center (X_0, Y_0) is $(1.1, 0.2)$ relative to the left bottom corner of the domain. The cylinder diameter D is equal to 0.1. No-slip is prescribed on the left, right, top and bottom boundaries. The cylinder is oscillating sinusoidally such that the location of its center (X_c, Y_c) is given by $(X_c(t) = X_0 + A\sin(2\pi f\, t),\ Y_c(t) = Y_0)$, where t is the time, and $A = 0.25$ and $f = 0.25$ are amplitude and frequency of the oscillation, respectively. The kinematic viscosity of the fluid is given by $\nu = \mu/\rho = 10^{-3}\, m^2/s$ and its density by $\rho = 1\, kg/m^3$. The fluid in channel is initially at rest. Since there is no benchmark result available for comparison, we carried out a reference calculation to provide comparing data. In the reference calculation, the body-conformal mesh of Fig. 3 (b) is used, we fix the cylinder but set the coordinate system moving with the same motion but with opposite moving direction of the moving cylinder in the calculation of the fictitious boundary method. Table 3 gives the number of elements for the channel mesh and body-conformal mesh in Fig. 3 with different numbers of refined levels.

Fig. 4 gives contour plots for the vorticity distribution obtained by the fictitious boundary method based on the channel mesh. These pictures show that the flow in the channel is disturbed by the oscillating cylinder, and the

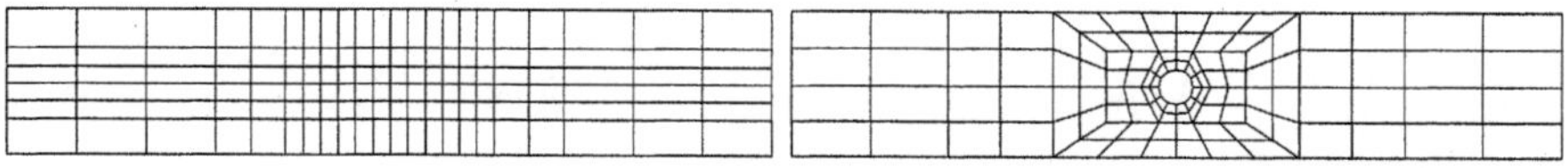

(a) channel mesh (LEVEL = 1) (b) body-conformal mesh (LEVEL = 2)

Fig. 3. Coarse meshes used for the oscillating cylinder in a channel

vortex is generated periodically in the wake of the cylinder. The range of wakes becomes longest when the cylinder is at the end of the moving direction ($t = t_0 + \frac{1}{4}T$, $t_0 + \frac{3}{4}T$, T is time period), while when the cylinder is in the middle position of its oscillation, the flow is seriously perturbed and becomes more complex ($t = t_0$, $t_0 + \frac{2}{4}T$). Fig. 5 illustrates the comparison of the drag coefficient C_d and lift coefficient C_l between the results of the fictitious boundary method based on the channel mesh and the reference calculation based on the body-conformal mesh. The results calculated from LEVEL = 4 to LEVEL = 7 are all shown together. The corresponding coefficients C_d and C_l for one period between $t = 19.79$ to 23.79 are shown in Fig. 5 (c) and (f), the solid line represents the results of the reference calculation based on the body-conformal mesh at LEVEL = 7, while the dash line denotes the results obtained by the fictitious boundary method based on the channel mesh at LEVEL = 7. From the comparisons, we can see that both FB and Ref. results are identical with the increase of the mesh refinements. The FB results calculated by the present fictitious boundary method are agreeable very well with the reference results, although the FB results exhibit small oscillations due to the non-aligned cylinder movement through the grid lines.

Table 3. The number of elements for different refined meshes

LEVEL	4	5	6	7
channel mesh	8448	33792	135168	540672
body-conformal mesh	1792	7168	28672	114688

(a) $t = t_0$ (b) $t = t_0 + \frac{1}{4}T$

(c) $t = t_0 + \frac{2}{4}T$ (d) $t = t_0 + \frac{3}{4}T$

Fig. 4. Vorticity contour plot for an oscillating cylinder in a channel

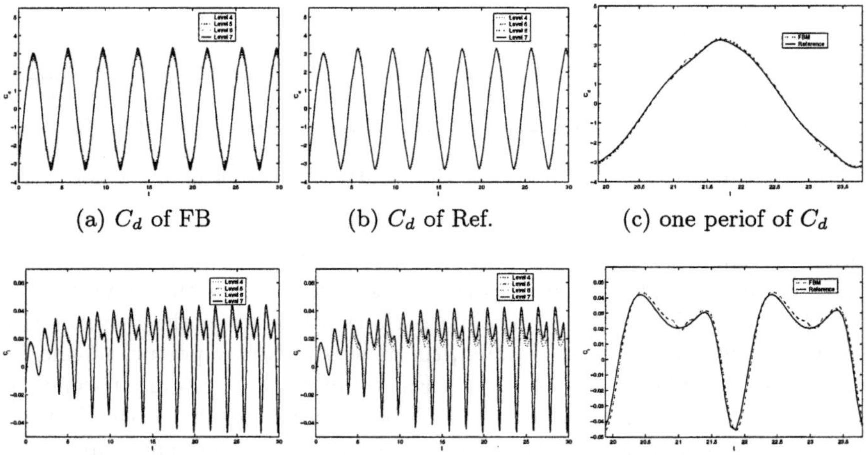

(a) C_d of FB (b) C_d of Ref. (c) one periof of C_d

(d) C_l of FB (e) C_l of Ref. (f) one periof of C_l

Fig. 5. The comparison of C_d and C_l between fictitious boundary (FB) and reference (Ref.)

4 Conclusions

The presented fictitious boundary method for simulating incompressible flows with moving rigid bodies in complex geometries has been validated in a two-dimensional configuration. We showed that the use of an arbitrarily fixed (unstructured) FEM background mesh is accurate enough to calculate those sensitive quantities (drag coefficient and lift coefficient) on the wall surface of the cylinder . Comparisons of the results using a body-fitted mesh and a fixed structured mesh show good agreement. The advantage of the present method is that since the body motion is independent of the mesh, problems associated with mesh reconfiguration and motion are avoided, computations on a fixed grid are cheaper than on a body-fitted one, and finally, the extension of the method to 3D is straightforward. It is also worthy to note that the availability of the present method to accurately compute the forces acting on the moving rigid bodies provides a good and solid base for further study of particulate flow as well as the interaction between fluid and structure as proposed by Glowinski in the paper [4].

References

1. Turek, S., Wan, D.C. and Rivkind, L.S.: The Fictitious Boundary Method for the implicit treatment of Dirichlet boundary conditions with applications to incompressible flow simulations. Lecture Notes in Computational Science and Engineering, Volume 35, Springer Verlag, (2003)

2. Feng, J., Hu, H.H and Joseph, D.D.: Direct simulation of initial value problems for the motion of solid bodies in a Newtonian fluid; part 2, Couette and Poiseuille flows. J. Fluid Mesh. **277** (1994) 271
3. Hu, H.H.: Direct simulation of flows of solid-liquid mixtures. Int. J. Multiphase Flow. **22** (2) (1996) 335
4. Glowinski, R., Pan, T.W., Hesla, T.I. and Periaux, J.: A fictitious domain approach to the direct numerical simulation of incompressible viscous flow past moving rigid bodies: application to particulate flow. J. Comput. Phys. **169** (2001) 363
5. Glowinski, R.: Handbook of numerical analysis (Volume IX): Numerical methods for fluids (Part 3). Ciarlet, P.G and Lions, J.L. editors, North-Holland, (2003)
6. Heywood, J., Rannacher, R. and Turek, S.: Artificial boundaries and flux and pressure conditions for the incompressible Navier–Stokes equations. Int. J. Numer. Meth. Fluids **22** (1996) 325
7. Duchanoy, C. and Jongen, T.R.G.: Efficient simulation of liquid-solid flows with high solids fraction in complex geometries. Compters and Fluids **32** (2003) 1453
8. John, V.: Higher order finite element methods and multigrid solvers in a benchmark problem for the 3D Navier-Stokes equations. Int. J. Num. Meth. Fluids **40** (2002) 775
9. Schäfer, M. and Turek, S.: Benchmark computations of laminar flow around cylinder. in E.H. Hirschel (editor) Flow Simulation with High-Performance Computers II. Volume 52 of Notes on Numerical Fluid Mechanics, Vieweg (1996) 547
10. Turek, S.: Efficient Solvers for Incompressible Flow Problems. Springer Verlag, Berlin-Heidelberg-New York (1999)
11. Turek, S. et al.: **FEATFLOW** . Finite element software for the incompressible Navier–Stokes equations: User Manual, Release 1.2 (1999) (www.featflow.de)
12. Brackbill, J.U., Kothe, D.B. and Zemach, C.: A continuum method for modelling surface tension. J. Comput. Phys. **100** (1992) 335

Higher-Order FEM for a System of Nonlinear Parabolic PDE's in 2D with A-Posteriori Error Estimates

Martin Zítka[1], Karel Segeth[2] and Pavel Šolín[3]

[1] Academy of Sciences, Prague *zitka@math.cas.cz*
[2] Academy of Sciences, Prague *segeth@math.cas.cz*
[3] Rice University, Houston, Texas *solin@rice.edu*

Summary. Initial-boundary value problems for systems of nonlinear parabolic partial differential equations arise in many important practical applications in electromagnetics, chemistry, modelling of diffusion and heat transfer processes and other fields. We are concerned with their solution by means of the method of lines with higher-order finite element spatial discretization on unstructured triangular meshes. Obviously, development of realistic a-posteriori error estimates plays an essential role in the application of a strategy of this type.

1 Introduction

Initial-boundary value problems for systems of nonlinear parabolic partial differential equations (PDE's) arise in many important practical applications in electromagnetics, chemistry, modelling of diffusion and heat transfer processes and many other fields[1].

We are concerned with the numerical solution of such problems by the method of lines (MOL) combined with fully automatic *hp*-adaptive finite element (FE) discretization on unstructured triangular meshes in space. This approach has the potential of reducing the size of discrete problems significantly while preserving the accuracy of results.

Until now, automatic *hp*-adaptivity has been applied almost exclusively to stationary problems (see, e.g., [1, 4, 2] and references therein). It is our aim to extend the promising automatic *hp*-adaptive strategies for elliptic problems [7, 8] to parabolic equations and in this paper we present two basic steps towards this goal:

- Efficient implicit time-adaptive higher-order FE solver PARSYS_2D for systems of nonlinear parabolic PDE's with general boundary conditions for all solution components.

[1] This work was supported by the Grant Agency of the Czech Republic under projects No. GP102/01/D114 and 201/01/1200.

– A-posteriori error estimates appropriate for the class of evolutionary problems studied.

2 Definition of the problem

Let $\Omega \subset \mathbb{R}^2$ be a bounded domain with piecewise-polynomial Lipschitz-continuous boundary and $J = (0,T]$ a finite time interval. We consider a system of N_{eq} nonlinear parabolic equations

$$\begin{aligned} \dot{\boldsymbol{u}}(\boldsymbol{x},t) - \nabla \cdot (\boldsymbol{a}(\boldsymbol{u}, \nabla \boldsymbol{u}, \boldsymbol{x}, t)\nabla \boldsymbol{u}(\boldsymbol{x},t)) = \boldsymbol{f}(\boldsymbol{u}, \nabla \boldsymbol{u}, \boldsymbol{x}, t) \quad & \text{in } \Omega, \quad t \in J, \\ \boldsymbol{u}(\boldsymbol{x},0) = \boldsymbol{v}(\boldsymbol{x}) \quad & \text{in } \Omega, \\ u_i(\boldsymbol{x},t) = u_i^D(\boldsymbol{x},t) \quad & \text{on } \Gamma_i^D, \\ a_i(\boldsymbol{u}, \nabla \boldsymbol{u}, \boldsymbol{x}, t)\partial u_i/\partial n = g_i(\boldsymbol{x},t) \quad & \text{on } \Gamma_i^N, \end{aligned} \tag{1}$$

($\dot{u}$ stands for $\partial u/\partial t$) where $\boldsymbol{u} = (u_1, \ldots, u_{N_{\text{eq}}})$ is the solution, $\boldsymbol{a}$ and $\boldsymbol{f}$ are smooth vector-valued functions, n is the unit outward normal to the boundary and $\partial\Omega = \overline{\Gamma_i^D \cup \Gamma_i^N}$ for all $i = 1, \ldots, N_{\text{eq}}$. Further we denote $\boldsymbol{g} = (g_1, \ldots, g_{N_{\text{eq}}})$. The ∇ (nabla) operator is defined as usual, $\nabla = (\partial/\partial x_1, \partial/\partial x_2)$. The vector-valued coefficient $\boldsymbol{a} = (a_1, \ldots, a_{N_{\text{eq}}})$ is bounded,

$$0 < \mu \leq a_i(.) \leq M, \qquad 1 \leq i \leq N_{\text{eq}}, \tag{2}$$

and both $\boldsymbol{a}$ and $\boldsymbol{f}$ are Lipschitz-continuous,

$$||\boldsymbol{a}(\boldsymbol{r}) - \boldsymbol{a}(\boldsymbol{s})|| \leq L||\boldsymbol{r} - \boldsymbol{s}||, \tag{3}$$

$$||\boldsymbol{f}(\boldsymbol{r}) - \boldsymbol{f}(\boldsymbol{s})|| \leq L||\boldsymbol{r} - \boldsymbol{s}|| \quad \forall \boldsymbol{r}, \boldsymbol{s} \in \mathbb{R}^{3N_{\text{eq}}+3}. \tag{4}$$

Without loss of generality we restrict ourselves to homogeneous Dirichlet boundary conditions for the formulation of the variational problem. In the nonhomogeneous case, an appropriate vector-valued lift function is chosen that yields an additional contribution to the right-hand side (see, e.g., [8] for details). The variational form of (1) reads

$$\begin{aligned} (\dot{\boldsymbol{u}}(\boldsymbol{x},t), \varphi) + & (\boldsymbol{a}(\boldsymbol{u}, \nabla \boldsymbol{u}, \boldsymbol{x}, t)\nabla \boldsymbol{u}(\boldsymbol{x},t), \nabla\varphi) \\ & = (\boldsymbol{f}(\boldsymbol{u}, \nabla \boldsymbol{u}, \boldsymbol{x}, t), \varphi) + (\boldsymbol{g}(\boldsymbol{x},t), \varphi)_{\Gamma^N}, \quad \forall \varphi \in V \\ \boldsymbol{u}(\boldsymbol{x},0) & = \boldsymbol{v}(\boldsymbol{x}), \end{aligned} \tag{5}$$

where $t \in (0,T)$ and the form of the space $V \subset (H^1)^{N_{\text{eq}}}$ is dictated by Γ_i^D, $1 \leq i \leq N_{\text{eq}}$. The symbol $(.,.)$ stands for the $L^2(\Omega)$ scalar product.

Following the concept of MOL, we discretize the spatial variable first and leave the temporal variable continuous. Consider a finite element mesh $\tau_{h,p}$ covering $\bar{\Omega}$, where a polynomial order $p(K_i) \geq 1$ is associated with each element $K_i \in \tau_{h,p}$, $1 \leq i \leq N_{\text{elem}}$. Let $V_{h,p}(\Omega)$ be an appropriate piecewise-polynomial subspace of V. We pose the semidiscrete problem to find $\boldsymbol{u}_{h,p} \in V_{h,p}(\Omega)$ for all $t \in J$, such that

$$(\dot{\boldsymbol{u}}_{h,p}(t), \chi(\boldsymbol{x})) + (\boldsymbol{a}(\boldsymbol{u}_{h,p}, \nabla \boldsymbol{u}_{h,p}, \boldsymbol{x}, t)\nabla \boldsymbol{u}_{h,p}(\boldsymbol{x}, t), \nabla \chi(\boldsymbol{x}))$$
$$= (\boldsymbol{f}(\boldsymbol{u}_{h,p}, \nabla \boldsymbol{u}_{h,p}, \boldsymbol{x}, t), \chi(\boldsymbol{x})) + (\boldsymbol{g}(\boldsymbol{x}, t), \chi(\boldsymbol{x}))_{\Gamma^N}, \qquad (6)$$
$$\boldsymbol{u}_{h,p}(\boldsymbol{x}, 0) = \boldsymbol{v}_{h,p}(\boldsymbol{x}),$$

for all $\chi \in V_{h,p}(\Omega)$ and $t \in J$. Here, $\boldsymbol{v}_{h,p} \in V_{h,p}(\Omega)$ is the H^1-projection of $\boldsymbol{v} \in V(\Omega)$.

3 Finite-dimensional spaces

Let $\tau_{h,p}$ be a mesh on Ω consisting of N_{elem} disjoint triangles K_i, $i = 1, \ldots, N_{\text{elem}}$. The polynomial order $p(K_i) \geq 1$ is assigned to each element K_i. Local polynomial orders on edges are determined using the *minimum rule* (minimum of polynomial orders on adjacent elements). Polynomial orders on all elements are collected into a vector $\boldsymbol{p} = \{p(K_1), p(K_2), \ldots, p(K_{N_{\text{elem}}})\}$. We further define

$$V^{\boldsymbol{p}}_{h,p,0}(\Omega) = \{\varphi \in [H^1_0(\Omega)]^{N_{\text{eq}}},\ \varphi_k|_{K_i} \in \mathcal{P}_{p(K_i)}(K_i),\ i = 1, \ldots, N_{\text{elem}},$$
$$\varphi_k|_{e_j} \in \mathcal{P}_{p(e_j)}(e_j);\ j = 1, \ldots, N_{\text{e}};\ k = 1, \ldots, N_{\text{eq}}\}. \qquad (7)$$

By $V^q_{h,p,0}(\Omega)$ we denote the space $V^{\boldsymbol{p}}_{h,p,0}(\Omega)$ corresponding to a uniform distribution of the polynomial order $\boldsymbol{p} = \{q, q, \ldots, q\}$.

The design of hierarchic basis functions of $V^{\boldsymbol{p}}_{h,p,0}(\Omega)$ is well-known (see, e.g., [8]). The basis consists of *vertex functions* (associated with mesh vertices), *edge functions* (associated with edges where $p(e) \geq 2$) and *bubble functions* (associated with element interiors where $p(K) \geq 3$). To simplify the explanation, we introduce the following notation:

- $B^v_{\tau_{h,p},1}$ ... the set of all vertex functions in the basis,
- $B^e_{\tau_{h,p},q}$ ... the set of all edge functions of the polynomial order q,
- $B^b_{\tau_{h,p},q}$ ... the set of all bubble functions of the polynomial order q,
- B_{e_i} ... the set of all edge functions associated with the edge e_i,
- B_{K_i} ... the set of all bubble functions associated with the triangle K_i.

4 A-posteriori error estimation

We extend a technique for a-posteriori estimation that was first proposed in [3] and further developed in [5] (in both cases for 1D problems). For the sake of clarity we restrict the explanation to the scalar case ($N_{\text{eq}} = 1$) with homogenous Dirichlet conditions on the whole boundary $\partial\Omega$, with a and f only depending on the exact solution u.

The error of the solution to the semidiscrete problem is defined as usual,

$$e(\boldsymbol{x}, t) = u(\boldsymbol{x}, t) - u_{h,p}(\boldsymbol{x}, t). \qquad (8)$$

By $\tilde{u}_{h,p}$ we denote the elliptic projection of the exact solution to the space $V^p_{h,p,0}(\Omega)$ (i.e. $V^q_{h,p,0}(\Omega)$ with $q = p$). The main idea of the error estimation approach can be outlined as follows: The problem is solved with $u_{h,p} \in V^p_{h,p,0}(\Omega)$. Then the problem is solved once again with $\overline{u}_{h,p} \in V^{p+1}_{h,p,0}(\Omega)$ (i.e. $V^q_{h,p,0}(\Omega)$ with $q = p+1$). The estimate is based on the difference $\overline{u}_{h,p} - u_{h,p}$.

Let us begin with the identity

$$(\dot{e}, \chi) + (a(u_{h,p}+e)\nabla e, \nabla\chi) = (f(u_{h,p}+e), \chi) - (\dot{u}_{h,p}, \chi) - (a(u_{h,p}+e)\nabla u_{h,p}, \nabla\chi) \tag{9}$$

for almost every $t \in J$ and all functions $\chi \in H^1_0$, that follows directly from (8), (5) and (6). Now our aim is to introduce an easily computable error estimate $\bar{E}(\boldsymbol{x}, t)$ that should be close to the quantity e.

Definition 1. *Let us define the space*

$$\widehat{V}^q_{h,p,0}(\Omega) = \operatorname{span}\{B^e_{\tau_{h,p,q}} \cup B^b_{\tau_{h,p,q}}\}.$$

We define three estimates $\bar{E}_{\mathrm{PN}}$, $\bar{E}_{\mathrm{PL}}$ and $\bar{E}_{\mathrm{EL}}$ as functions associated with $\widehat{V}^{p+1}_{h,p,0}(\Omega)$ by the means of the identity (9).

Definition 2. *We say that a function $\bar{E} = \bar{E}_{\mathrm{PN}}$ is a nonlinear parabolic error estimate if the identity*

$$\begin{aligned}(\dot{\bar{E}}, \widehat{\chi}) &+ (a(u_{h,p} + \bar{E})\nabla\bar{E}, \nabla\widehat{\chi}) \\ &= (f(u_{h,p} + \bar{E}), \widehat{\chi}) - (\dot{u}_{h,p}, \widehat{\chi}) - (a(u_{h,p} + \bar{E})\nabla u_{h,p}, \nabla\widehat{\chi})\end{aligned} \tag{10}$$

holds for almost every $t \in J$ and all functions $\widehat{\chi} \in \widehat{V}^{p+1}_{h,p,0}(\Omega)$, and if the identity

$$(a(v)\nabla\bar{E}, \nabla\widehat{\chi}) = (a(v)\nabla(v - u_{h,p}), \nabla\widehat{\chi}) \tag{11}$$

holds for $t = 0$ and all functions $\widehat{\chi} \in \widehat{V}^{p+1}_{h,p,0}(\Omega)$.

Definition 3. *We say that a function $\bar{E} = \bar{E}_{\mathrm{PL}}$ (or $\bar{E} = \bar{E}_{\mathrm{EL}}$) is a linear parabolic (elliptic) error estimate if the identity*

$$(\dot{\bar{E}}, \widehat{\chi}) + (a(u_{h,p})\nabla\bar{E}, \nabla\widehat{\chi}) = (f(u_{h,p}), \widehat{\chi}) - (\dot{u}_{h,p}, \widehat{\chi}) - (a(u_{h,p})\nabla u_{h,p}, \nabla\widehat{\chi}) \tag{12}$$

or

$$(a(u_{h,p})\nabla\bar{E}, \nabla\widehat{\chi}) = (f(u_{h,p}), \widehat{\chi}) - (\dot{u}_{h,p}, \widehat{\chi}) - (a(u_{h,p})\nabla u_{h,p}, \nabla\widehat{\chi}) \tag{13}$$

holds for almost every $t \in J$ and all functions $\widehat{\chi} \in \widehat{V}^{p+1}_{h,p,0}(\Omega)$, and if identity (11) holds for $t = 0$ and the same functions $\widehat{\chi}$.

Further, let us introduce a function $\tilde{e}_{h,p} \in \widehat{V}^{p+1}_{h,p,0}(\Omega)$ such that $\tilde{u}_{h,p} + \tilde{e}_{h,p}$ is the elliptic projection of u to $V^{p+1}_{h,p,0}(\Omega)$.

Definition 4. *By* $\tilde{e}_{h,p}(\boldsymbol{x},t)$ *we denote a function in* $\widehat{V}^{p+1}_{h,p,0}(\Omega)$ *such that*

$$(a(u)\nabla(\tilde{u}_{h,p}+\tilde{e}_{h,p}),\nabla\widehat{\chi}) = (a(u)\nabla u,\nabla\widehat{\chi}) \tag{14}$$

holds for almost every $t\in J$ *and all functions* $\widehat{\chi}\in\widehat{V}^{p+1}_{h,p,0}(\Omega)$.

Definition 5. *By* θ *we denote the difference*

$$\theta(\boldsymbol{x},t) = \tilde{u}_{h,p}(\boldsymbol{x},t) - u_{h,p}(\boldsymbol{x},t). \tag{15}$$

Further we define functions $\eta = \eta_{\mathrm{PN}},\eta_{\mathrm{PL}},\eta_{\mathrm{EL}}$ *for* $\bar{E} = \bar{E}_{\mathrm{PN}},\bar{E}_{\mathrm{PL}},\bar{E}_{\mathrm{EL}}$ *such that*

$$\eta(\boldsymbol{x},t) = \tilde{e}_{h,p}(\boldsymbol{x},t) - \bar{E}(\boldsymbol{x},t). \tag{16}$$

The following proposition compares our estimate with the quantity $\tilde{e}_{h,p}$. By $\|.\|_r$ we denote the $H^r(\Omega)$ norm.

Lemma 1. *Let* $\bar{E}_{\mathrm{PN}}\in\widehat{V}^{p+1}_{h,p,0}(\Omega)$ *be the error estimate given by* (10) *and* (11), *and* $\tilde{e}_{h,p}\in\widehat{V}^{p+1}_{h,p,0}(\Omega)$ *be the function defined by* (14). *Let* $\|\theta(.,t)\|$ *and* $\|\eta_{\mathrm{PN}}(.,t)\|$ *be nondecreasing functions of the variable* t *and let*

$$\|\eta_{\mathrm{PN}}(.,0)\| \le Ch^{p+1}. \tag{17}$$

Let $\tilde{e}_{h,p}$ *and* $\bar{E}_{\mathrm{PN}}$ *depend on* t *in a sufficiently smooth way. Then there exists a constant* $C>0$ *such that*

$$\|\eta_{\mathrm{PN}}\|_1 \le Ch^{p+1}. \tag{18}$$

Remark 1. Analogous propositions can be shown also for $\|\eta_{\mathrm{PL}}\|_1$ and $\|\eta_{\mathrm{EL}}\|_1$ with minor differences in the proof.

Proof. We start with the formula

$$\begin{aligned}(\dot{\bar{E}},\widehat{\chi}) &+ (a(u_{h,p}+\bar{E})\nabla\bar{E},\nabla\widehat{\chi}) + (\dot{u}_{h,p},\widehat{\chi}) + (a(u_{h,p}+\bar{E})\nabla u_{h,p},\nabla\widehat{\chi})\\ &- (\dot{\tilde{u}}_{h,p},\widehat{\chi}) - (\dot{\tilde{e}}_{h,p},\widehat{\chi}) - (\dot{\widehat{\rho}},\widehat{\chi}) - (a(u)\nabla(\tilde{u}_{h,p}+\tilde{e}_{h,p}),\nabla\widehat{\chi})\\ &= (f(u_{h,p}+\bar{E}),\widehat{\chi}) - (f(u),\widehat{\chi}),\end{aligned} \tag{19}$$

that follows from (5), (14) and (10), and use

$$\widehat{\rho}(\boldsymbol{x},t) = u(\boldsymbol{x},t) - \tilde{u}_{h,p}(\boldsymbol{x},t) - \tilde{e}_{h,p}(\boldsymbol{x},t). \tag{20}$$

By rearranging terms, substituting $\eta\in\widehat{V}^{p+1}_{h,p,0}(\Omega)$ for $\widehat{\chi}$, applying the Schwarz inequality and using the fact that

$$\|I_{h,p}w-w\| + h\|\nabla(I_{h,p}w-w)\| \le Ch^{p+1}\|w\|_{p+1} \quad \text{when } w\in H^{p+1}\cap H^1_0, \tag{21}$$

where $I_{h,p}$ is the interpolation operator from H^1_0 into $V^p_{h,p,0}(\Omega)$, we arrive to

$$\frac{1}{2}\frac{\mathrm{d}}{\mathrm{d}t}\|\eta\|^2 + S\|\nabla\eta\|^2 \le C_1(t)\|\eta\|^2 + C_2(t)h^{2p+2} \tag{22}$$

where $S > 0$. We omit the term with $\|\nabla\eta\|^2$ on the left-hand side. Integrating on both parts of (22), we find

$$\|\eta(t)\|^2 \leq \|\eta(0)\|^2 + C_3(t)h^{2p+2} + 2\int_0^t C_1(s)\|\eta(s)\|^2\,\mathrm{d}s \tag{23}$$

if the corresponding functions are Lebesgue integrable on the interval $[0,T]$. Employing now (17), assuming that the corresponding functions are again Lebesgue integrable on $[0,T]$, and applying Gronwall's lemma, we obtain the bound

$$\|\eta(t)\|^2 \leq C_4(t)h^{2p+2} + C_5(t)h^{2p+2} \leq Ch^{2p+2}. \tag{24}$$

Turning now back to the inequality (22) and assuming that $\|\eta\|$ is nondecreasing, we can write

$$\frac{\mathrm{d}}{\mathrm{d}t}\|\eta\|^2 \geq 0. \tag{25}$$

Using (24) and (25), we finally obtain from (22) that

$$\|\nabla\eta\|^2 \leq Ch^{2p+2}. \tag{26}$$

The statement of the lemma follows directly from (24) and (26). □

For each of the three above error estimates let us introduce an *effectivity index* Θ_{PN}, Θ_{PL}, and Θ_{EL}, respectively, defined as

$$\Theta = \frac{\|\bar{E}\|_1}{\|e\|_1}. \tag{27}$$

The following theorem contains a principal result related to $\lim_{h\to 0}\Theta$:

Theorem 1. *Let $u(t) \in H^{p+1} \cap H_0^1$ and $u_{h,p}(t) \in V_{h,p,0}^p(\Omega)$ be the solutions of* (5) *and* (6) *for all $t \in J$. Let $\bar{E} \in \widehat{V}_{h,p,0}^{p+1}(\Omega)$ be the solution of* (10), (11) *(for $\bar{E}_{\mathrm{PN}}$),* (11), (12) *(for $\bar{E}_{\mathrm{PL}}$), or* (13) *(for $\bar{E}_{\mathrm{EL}}$). Then, under appropriate regularity assumptions,*

$$\lim_{h\to 0}\Theta = 1 \tag{28}$$

for almost every $t \in J$, where Θ is Θ_{PN}, Θ_{PL} or Θ_{EL}.

Proof. The proof is rather technical (see [10]). Its main idea is the same as in the 1D case presented in [5]. □

5 Numerical results

5.1 Model problem

To verify the theoretical results, we set up a model problem (1), where we chose the analytical solution as

$$u(\boldsymbol{x}, t) = \cos(50(x_1 + x_2) - t)x_1 x_2(1 - x_1)(1 - x_2) \tag{29}$$

and the coefficient $a = a(s) = 1 + s$, $s \in \mathbb{R}$. The right-hand side function f was calculated to yield the exact solution u.

The values of Θ_{EL} and Θ_{PL} for $t = 0.4$, $h = \sqrt{13/36}$ and for different p's and mesh parameters are summarized in Tables 1 and 2. The implicit Euler formula with the step $\Delta t = 0.02$ has been used for the time integration. According to Theorem 1, the values should tend to 1 as $h \to 0$.

	$p = 1$	$p = 2$	$p = 3$	$p = 4$
h		4.09651988491	2.68370145897	2.07009170501
$h/2$	1.01140677283	1.00652321679	0.98450775723	0.96472314783
$h/4$	1.00251576073	0.99884836859	0.99764998322	0.99629420488
$h/8$	0.99790790301	0.99587534487	0.99996695628	1.01800491423

Table 1. Values of Θ_{EL} for different h's and p's.

	$p = 1$	$p = 2$	$p = 3$	$p = 4$
h		4.08160620473	2.67782363272	2.06631766115
$h/2$	1.00953184295	1.00589904628	0.98419486064	0.96447454547
$h/4$	1.00221228302	0.99868374826	0.99752249029	0.99620694415
$h/8$	0.99774286135	0.99582769340	0.99993253125	1.01798017133

Table 2. Values of Θ_{PL} for different h's and p's.

6 Brief description of PARSYS_2D

During the first stage of the project we implemented a robust higher-order FE solver PARSYS_2D for systems of nonlinear parabolic PDE's in 2D analogously as we did in 1D (see the PARSYS package [6]). Implicit time-integration methods allow the solver also to solve systems of elliptic PDE's after leaving out the time-derivative. We use our software XGEN (see [9]) to generate unstructured triangular meshes of very good quality on domains with arbitrarily complicated geometries. All the software is available in Internet[2].

[2] The mentioned C++ software packages PARSYS, PARSYS_2D and XGEN can be downloaded free of charge from the web page of Pavel Šolín, `http://www.caam.rice.edu/~solin`.

6.1 Definition of the solved system of PDE's

It is not trivial to find an efficient and user-friendly way to define a system of non-linear PDE's. We approached this problem by allowing the user to specify a large number of matrix- and vector-valued coefficients. The solved equation has a general form

$$
\begin{aligned}
C\frac{\partial U(\boldsymbol{x},t)}{\partial t} &= \frac{\partial}{\partial x_1}\left(P_1\frac{\partial U}{\partial x_1}(\boldsymbol{x},t)\right) + \frac{\partial}{\partial x_1}\left(P_2\frac{\partial U}{\partial x_2}(\boldsymbol{x},t)\right) + \frac{\partial}{\partial x_2}\left(P_3\frac{\partial U}{\partial x_1}(\boldsymbol{x},t)\right) \\
&+ \frac{\partial}{\partial x_2}\left(P_4\frac{\partial U}{\partial x_2}(\boldsymbol{x},t)\right) + \frac{\partial}{\partial x_1}\left(P_5 U(\boldsymbol{x},t)\right) + \frac{\partial}{\partial x_2}\left(P_6 U(\boldsymbol{x},t)\right) \\
&+ P_7 U(\boldsymbol{x},t) + F,
\end{aligned}
\tag{30}
$$

where $U(\boldsymbol{x},t) = (U_1(\boldsymbol{x},t),\ldots,U_{N_{\text{eq}}}(\boldsymbol{x},t))$ is the solution vector and $U_i \in H^1(\Omega)$ for $t \in J$, $i = 1,\ldots N_{\text{eq}}$. The user is requested to prescribe seven $N_{\text{eq}} \times N_{\text{eq}}$ matrices

$$P_i = P_i\left(U, \frac{\partial U}{\partial x_1}, \frac{\partial U}{\partial x_2}, \boldsymbol{x}, t\right), \quad i = 1,\ldots,7,$$

together with a diagonal matrix

$$C = \operatorname{diag}\left(C_1(U, \frac{\partial U}{\partial x_1}, \frac{\partial U}{\partial x_2}, \boldsymbol{x}, t), \ldots, C_{N_{\text{eq}}}(U, \frac{\partial U}{\partial x_1}, \frac{\partial U}{\partial x_2}, \boldsymbol{x}, t)\right)$$

and a source term vector

$$F = \left(F_1(U, \frac{\partial U}{\partial x_1}, \frac{\partial U}{\partial x_2}, \boldsymbol{x}, t), \ldots F_{N_{\text{eq}}}(U, \frac{\partial U}{\partial x_1}, \frac{\partial U}{\partial x_2}, \boldsymbol{x}, t)\right)^T.$$

Remark 2. In (30), we apply $\partial/\partial x_1$ and $\partial/\partial x_2$ to vectors. By this notation we mean that these operators are applied to each component of the vector.

The vector-valued initial condition has the standard form $U_i(\boldsymbol{x},0) = U_i^0(\boldsymbol{x})$, $i = 1,\ldots,N_{\text{eq}}$. To each solution component U_i, $i = 1,\ldots,N_{\text{eq}}$, we can prescribe either Dirichlet or Neumann boundary conditions. The Dirichlet boundary conditions have the form

$$U_i(\boldsymbol{x},t) = U_i^D(\boldsymbol{x},t), \quad \boldsymbol{x} \in \partial\Omega, \quad i = 1,\ldots N_{\text{eq}}. \tag{31}$$

The Neumann boundary conditions are prescribed in the form

$$n_1(P_1\frac{\partial U}{\partial x_1} + P_2\frac{\partial U}{\partial x_2} + P_5 U) + n_2(P_3\frac{\partial U}{\partial x_1} + P_4\frac{\partial U}{\partial x_2} + P_6 U) = g(\boldsymbol{x},t), \tag{32}$$

where $\boldsymbol{n} = (n_1, n_2)$ is the outer unit normal to the boundary. One can prescribe various types of boundary conditions on different parts of the boundary.

6.2 Spatial and temporal discretization

The solver distributes the polynomial orders $p(K_i)$ in the mesh $\tau_{h,p}$ accordingly to user's input and uses the above mentioned minimum rule to decide about local polynomial orders associated with edges. Partially novel Lobatto-based hierarchic shape functions (see [8]) up to the 10th order were used in order to obtain finite elements with excellent conditioning properties. For the sake of efficiency, economical Gauss quadrature points and weights from the CD-ROM [8] up to the 20th polynomial order were utilized.

The spatial semidiscretization yields a system of nonlinear ordinary differential equations (ODE's). The user can choose either to use the built-in implicit Euler scheme which is only first-order accurate, or to utilize absolutely stable implicit higher-order adaptive ODEPACK subroutines. These subroutines are very sophisticated and one of their significant advantages is that they are based on explicit evaluation of the right-hand side of the ODE system – thus, no linearization whatsoever is needed by the user. The solvers are capable of numerically obtaining information about the Jacobi matrix of the right-hand side that is needed for the backward differentiation formula (BDF) on which they are based.

6.3 Outlook

The combination of *hp*-adaptive higher-order FE discretization in space with the adaptive higher-order time-integration methods of ODEPACK offers exciting perspectives for the numerical solution of large evolutionary problems. We hope to report on progress on the implementation of automatic *hp*-adaptive strategies in our 2D solver soon.

References

1. Demkowicz, L., Oden, J. T., Rachowicz, W., Hardy, O. (1989): Toward a universal *hp*-adaptive finite element strategy. Part 1: Constrained approximation and data structure, Comput. Methods Appl. Math. Engrg., **77**, 79–112
2. Demkowicz, L., Rachowicz, W., Devloo, Ph. (2002): A fully automatic hp-adaptivity, J. Sci. Comput., **17**, 127–155.
3. Moore, P.K. (1994): A posteriori error estimation with finite element semi- and fully discrete methods for nonlinear parabolic equations in one space dimension. SIAM J. Numer. Anal., **31**, 149–169
4. Rachowicz, W., Oden, J. T., Demkowicz, L. (1989): Toward a universal *hp*-adaptive finite element strategy. Part 3: Design of *hp* meshes, Comput. Methods Appl. Mech. Engrg., **77**, 181–212
5. Segeth, K. (1999): A posteriori error estimation with the finite element method of lines for a nonlinear parabolic equation in one space dimension. Numer. Math., **83**, 455–475

6. Segeth, K., Šolín, P., Kočiřík, M. (2002): Some algorithmic aspects of higher-order finite element schemes in multidimensions. In: Software and Algorithms of Numerical Mathematics **14**. (Proceedings of Summer School, Kvilda 2001), Plzeň, University of West Bohemia, 199–221
7. Šolín, P., Demkowicz, L. (2003): Goal-Oriented *hp*-Adaptivity for Elliptic Problems, Comput. Methods Appl. Mech. Engrg., accepted
8. Šolín, P., Segeth, K., Doležel, I. (2003): Higher-Order Finite Element Methods. Chapman & Hall / CRC Press, Boca Raton, FL
9. Šolín P. (2000): On a mesh generation technique based on a special smoothing procedure for uniform inner point distribution, Acta Technica CSAV **45**, 397–417
10. Zítka, M. (2002): Higher-Order FEM for a System of Nonlinear Parabolic PDE's, Diploma Thesis, Charles University, Prague

MIX
Papier aus verantwortungsvollen Quellen
Paper from responsible sources
FSC® C105338

If you have any concerns about our products,
you can contact us on
ProductSafety@springernature.com

In case Publisher is established outside the EU,
the EU authorized representative is:
Springer Nature Customer Service Center GmbH
Europaplatz 3, 69115 Heidelberg, Germany

Printed by Libri Plureos GmbH
in Hamburg, Germany